47

125.00

GERMANY

CZECH REPUBLIC

SLOVAKIA

Rhine

Danube

AUSTRIA

HUNGARY

SWITZ.

SLOVENIA

Drava

CROATIA

SERBIA

Bergamo

Milan

Vicenza

Verona

Padua

BOSNIA -

Sava

Turin

Pavia

Mantua

Venice

HERCEGOVINA

YUGOSLAV.

Po

Bobbio

Modena

Ferrara

Pula

Sarajevo

Genoa

Bologna

Ravenna

MONACO

Lucca

Rimini

SAN MARINO

Zadar

Split

Florence

Urbino

Pisa

Arezzo

Gubbio

Hvar

Siena

Tiber

Assisi

MONTENEGRO

Elba

Orvieto

ADRIATIC

Todi

SEA

Dubrovnik

CORSICA

Chieti

Kotor

Tarquinia

Tivoli

ALB.

Rome

Palestrina

ITALY

Ostia Antica/Portus

Cassino

Durres

Benevento

Capua

Caserta

Cumae

Herculaneum

Naples

Pompeii

Capri

Salerno

Metapontum

Taranto

Paestum

Velia

SARDINIA

TYRRHENIAN SEA

IONIAN
SEA

Cagliari

Nora

Palermo

Monreale

SICILY

Taormina

Selinus

Agrigento

Gela

Syracuse

MALTA

TUNISIA

MEDITERRANEAN SEA

D0221120

INTERNATIONAL DICTIONARY OF
HISTORIC PLACES

INTERNATIONAL DICTIONARY OF HISTORIC PLACES

INTERNATIONAL DICTIONARY OF
HISTORIC PLACES

VOLUME 3
SOUTHERN EUROPE

Editor
TRUDY RING

Associate Editor
ROBERT M. SALKIN

Photo Editor
SHARON LA BODA

FITZROY DEARBORN PUBLISHERS
CHICAGO AND LONDON

Copyright © 1995 by
FITZROY DEARBORN PUBLISHERS

All rights reserved including the right of reproduction in whole or in part in any form.
For information, write to:
FITZROY DEARBORN PUBLISHERS
70 East Walton Street
Chicago, Illinois 60611
U.S.A.
or
11 Rathbone Place
London W1P 1DE
England

Library of Congress Cataloging-in-Publication Data:

International dictionary of historic places / editor, Trudy Ring;
   associate editor, Robert M. Salkin.
       p. cm.
     Essays on the history of 1,000 historic places.
     Includes bibliographical references and index.
     Contents: v. 1. Americas — v. 2. Northern Europe — v.
3. Southern Europe — v. 4. Middle East and Africa — v. 5 Asia and
Oceania.
      ISBN 1-884964-05-2 (set): $125 (per vol.) — ISBN 1-884964-00-1
(v. 1). — ISBN 1-884964-01-X (v. 2). — ISBN 1-884964-02-8 (v. 3).
 — ISBN 1-884964-03-6 (v. 4). — ISBN 1-884964-04-4 (v. 5).
     1. Historic sites. I. Ring, Trudy, 1955— II. Salkin, Robert M., 1965—
CC135.I585   1995
973—dc20   94-32327
                                                 CIP

British Library Cataloguing-in-Publication Data

International Dictionary of Historic Places, Volume 3—Southern Europe
I. Ring, Trudy   II. Salkin, Robert M.
970

ISBN 1-884964-02-8

First published in the U.S.A. and U.K. 1995
Typeset by Acme Art, Inc.
Printed by Braun-Brumfield, Inc.

Cover photograph: Palazzo Vecchio
Courtesy of A.P.T., Florence
Cover designed by Peter Aristedes, Chicago Advertising and Design

Frontispiece and endpaper maps by Tom Willcockson, Mapcraft

CONTENTS

HISTORIC PLACES BY COUNTRY

ALBANIA
Durrës

ANDORRA

BOSNIA-HERCEGOVINA
Sarajevo

BULGARIA
Sofia

CROATIA
Dubrovnik
Hvar
Pula
Split
Zadar

CYPRUS
Citium
Famagusta
Salamis

GIBRALTAR

GREECE
Aegina
Argos
Athens: Acropolis
Athens: Agora
Cephalonia
Corinth
Crete
Delos
Delphi
Eleusis
Epidaurus
Euboea
Gortyn
Kastoria
Kavala/Philippi
Keos
Kos
Messenia
Metéora
Mistra
Mount Athos
Mount Helicon
Mycenae
Nauplia
Naxos
Olympia/Elis

Phigalia/Bassae
Rhodes
Samothrace
Sparta
Thermopylae
Thessaloníki
Thíra
Tiryns
Vergina

ITALY
Agrigento
Arezzo
Assisi
Benevento
Bergamo
Bobbio
Bologna
Cagliari
Capri
Capua
Caserta
Cassino
Chieti
Cumae
Elba
Ferrara
Florence: Fiesole
Florence: Piazza del Duomo
Florence: Piazza della Signoria
Florence: Ponte Vecchio
Gela
Genoa
Gubbio
Herculaneum
Lucca
Mantua
Metapontum
Milan
Modena
Monreale
Naples
Nora
Orvieto
Ostia Antica/Portus
Padua
Paestum
Palermo
Palestrina
Pavia
Pisa
Pompeii

EDITOR'S NOTE

Fitzroy Dearborn Publishers' *International Dictionary of Historic Places* is designed to provide detailed and accurate information on places that have been the site of important events in human history and that have been preserved for the benefit of future generations. These places are not merely reconstructions of history, but are sites where history was made—be it political, military, architectural, artistic, religious, or social history.

The dictionary includes five volumes, which combined will cover nearly 1,000 sites worldwide. Volume 3 is devoted to Southern Europe, from the Iberian Peninsula to Turkey. Some of the sites date to prehistoric times, while others were marked by events in the classical period, the Middle Ages, the Renaissance, or the modern era. Some of them were abandoned centuries ago and have been rediscovered by modern archaeologists; others have been inhabited continuously and have witnessed historic events up to the present time. Volume 3 follows Volume 1, Americas, and Volume 2, Northern Europe. Future volumes in the series will be Volume 4, Middle East and Africa, and Volume 5, Asia and Oceania.

The entry on each site includes a detailed essay explaining the events that occurred there and their historical significance, as well as providing information on what the site offers to contemporary visitors. Headnotes for each essay provide the site's geographic location, a concise description, and the address of an information office at the site, or, when no such office is available, the address of a central contact for the area. (Because of ongoing conflict at a few sites, no site office or contact was available for them.) We expect this information to assist persons who are traveling to any of the places, as well as those who wish to write or call the sites to request material. Each entry also includes a section on further reading. This section is a selective listing of relevant published works, as recommended by the author of the entry.

The entries were compiled from publicly available sources, including books, magazine and newspaper articles, and, in some cases, material supplied by the site offices. We thank the site offices for supplying this material, as well as illustrations. The site offices' assistance in no way constitutes their endorsement of the facts presented. Our contributors and editors, however, have made every effort to ensure the accuracy of each entry.

We wish to thank a few people who have been particularly helpful, either by answering question after question, by providing large quantities of photographs, or by obtaining especially hard-to-find photos. We hereby acknowledge the extraordinary contributions of Fay Ishtar of the Embassy of Turkey, Washington, D.C.; Mary Kay Hartley of the Italian Government Tourist Office, New York; Olga Kefalogiannis of the Greek National Tourist Office, New York; Maria Theodorou of the Consulate General of Greece, Chicago; Hilary Ilijas of the Art and Culture Council of America, Toronto; Tanja Posavec of the Embassy of Croatia, Washington, D.C.; Alessandra Cicala and Giuseppina Valente of the United States Information Service; Giovanna Mango of the Municipio di Capua; and Vincenzo De Gennaro of the Museo Provinciale Campano, Capua.

Finally, for their valuable editorial assistance, we thank Mary F. McNulty, Marijke Rijsberman, Jennifer Schellinger, and Paul E. Schellinger.

—Trudy Ring

CONTRIBUTORS

Philippe Barbour

Richard Bastin

Bernard A. Block

Jessica M. Bowen

Elizabeth Brice

Elizabeth E. Broadrup

Holly E. Bruns

Monica Cable

Dellzell Chenoweth

Maria Chiara

Olive Classe

Christopher P. Collier

Sina Dubovoy

Amira Dzirlo

Jeffrey Felshman

John A. Flink

Lawrence F. Goodman

Sarah M. Hall

Mark D. Hanafee

William Harms

Patrick Heenan

Pam Hollister

Jeff W. Huebner

Tony Jaros

Rion Klawinski

Lisa Klobuchar

Manon Lamontagne

Monique Lamontagne

Cynthia L. Langston

Gregory J. Ledger

Clarissa Levi

Claudia Levi

Nicolette Loizou

Jean L. Lotus

Mary F. McNulty

Caterina Mercone Maxwell

Julie A. Miller

Hyunkee Min

Laurence Minsky

Paul Mooney

L. R. Naslund

Michael D. Phillips

Marijke Rijsberman

Trudy Ring

Robert M. Salkin

June Skinner Sawyers

Paul E. Schellinger

Kenneth R. Shepherd

Jill I. Shtulman

James Sullivan

Hilary Collier Sy-Quia

Jeffrey M. Tegge

Patricia Trimnell

Randall J. Van Vynckt

Aruna Vasudevan

Patricia Wharton

Thomas Wiloch

Beth F. Wood

Peter C. Xantheas

INTERNATIONAL DICTIONARY OF
HISTORIC PLACES

Aegina (Attica, Greece)

Location: A small triangular-shaped island in the center of the Saronic Gulf, twenty miles southwest of Athens, overlooking the Isthmus of Corinth.

Description: Aegina, which became a first-rate naval power in the seventh century B.C., is credited with being one of the first Greek states to introduce coinage and develop a system of weights and measures. After a military defeat by Athens in the fifth century B.C., Aegina lost its prominence. Over the ensuing centuries, Aegina passed through Macedonian, Aetolian, Pergamene, Roman, Venetian, and Ottoman Turk control. During the Greek War of Independence, Aegina's main town, also named Aegina, briefly became the capital of Greece. Today, the island is a holiday resort destination for the residents of Athens.

Site Office: Archaeological Museum of Aegina
Aegina, Attica
Greece
(297) 22248

Archaeological evidence suggests that Aegina has been inhabited since the late Neolithic times, about 4000 B.C. Remains reveal striking cultural unity between the Aeginetans and the peoples of the Peloponnese, the Cyclades Islands, and southern mainland Greece. The first settlers on Aegina are believed to have migrated to the island from the Peloponnese. Later, additional inhabitants arrived from the Near East. The island was initially referred to as Oenone, but its name was later changed to Aegina in honor of a nymph of that name who supposedly was carried to the island by Zeus.

Aegina prospered during three periods in history: the Middle and Late Bronze Age, from 2000 to 1100 B.C.; the classical Greek age, from 600 to 400 B.C.; and sporadically under the Venetians, from the thirteenth to the eighteenth century A.D. During its first period of prosperity, Aegina came into contact with the Minoan and Mycenaean civilizations. Minoan civilization was centered on the Greek island of Crete. Mycenaean civilization later usurped the Minoans and was centered in southern Greece on the Peloponnese. Its main city was Mycenae. Later, during the classical Greek age, this period of history would be dimly remembered as the Greek heroic age. Little is known about this period of Greek history, and even less is known about Minoan and Mycenaean influence on the island of Aegina.

Settlers arrived on Aegina from Argos, Crete, and Thessaly around the thirteenth century B.C. Some 200 years later, Aegina fell to the invading Dorian Greeks led by Deiphontes of Argos. The Dorians were a major division of the ancient Greeks distinguished by their well-marked dia-

lect. It appears they came from northern and northwestern Greece. Mycenaean and Minoan civilizations, already in decline, fell to the conquering Dorians.

Aegina was subsequently abandoned for a brief period, but colonizers once again came to the island from the Peloponnese, perhaps Epidaurus, around 950 to 900 B.C. Aegina broke away from Epidaurian control and forged close ties with Argos. The Aeginetans had been taking to the seas for centuries and now developed one of the premiere centers of trade in the ancient world, having commercial relations with the Cyclades Islands, Anatolia, and Egypt. The Aeginetans shipped items produced on the island, primarily bronze ware and pottery, and transported merchandise from other parts of the ancient world that had less developed naval fleets.

Aegina benefited from its geographic position in the Aegean Sea. Until the advent of steamships centuries later, the Aegean was a focal point of exchange between Europe and Asia. When trade later expanded into western Europe and the Black Sea, the Isthmus of Corinth became the bridge that handled much of the trade between eastern Europe, western Europe, the Black Sea, and Asia. Its proximity to the isthmus allowed the island to remain active in trade and reap financial rewards.

In the seventh century B.C., the Aeginetans shook off Argos's control and remained independent until 455 B.C. Aeginetan independence was occasionally disrupted by an ongoing feud between the island state and Athens. During Aegina's period of independence it produced two firsts in the ancient world. Between 656 and 650 B.C., the Aeginetans developed a system of weights and measures that was adopted throughout Greece. It was the oldest system of weights and measures in the classical world. Approximately twenty-five years later, around 625 B.C., Aegina struck its first coins. The coins were called "silver tortoises." They are believed to be among the first coins ever minted in the Greek world. They circulated widely and have been found in virtually every major port in the eastern Mediterranean. They remained the primary currency for all of the Peloponnese, except Olympia, for more than two centuries.

Aegina differed from most other Greek states in that land ownership was not concentrated in the hands of the aristocracy. Aegina instead had a stable mercantile class and, unlike the rest of ancient Greece, suffered little unrest over control of land.

The governmental structure on Aegina, however, was an oligarchy. Regarding the life of Aeginetan citizens below the social stratum of officeholders, little is known. Some evidence suggests that even long-term residents of the island may have been denied political rights, had limited political rights, or completely lacked citizenship.

Between the mid-sixth and mid-fifth centuries B.C., the individual Greek states matured in power, prosperity, and culture. They became increasingly conscious of their com-

Street scene on Aegina
Photo courtesy of Greek National Tourist Organization

mon heritage and institutions. They worshiped not only their own local gods but also the various gods of the Greek culture. They no longer faced ill-organized opposition from foreign invaders. Instead, formidable opposition from the Persian, Carthaginian, and Etruscan empires threatened their way of life. Various Greek states withstood well-organized foreign attacks by forming coalitions. These coalitions were headed by Sparta, Gela, Hieron, and Athens. The various coalitions also helped bring increased stability to the Peloponnese, the Aegean area, and Sicily. The latter part of this period saw a balance of power form between three relatively equal coalitions: Athens and its Ionian allies, Sparta and its Peloponnesian allies, and Corinth with Syracuse and Akragas.

The island of Aegina inspired Pindar, considered to be the greatest ancient Greek lyrical poet, to write at least seven of his finest odes, including his eighth Pythian ode. Most of Pindar's commissions came from the Greek aristocracy, especially the Aeacids of Aegina. In all, one-fourth of his choral odes were done for Aeginetans. Pindar's odes celebrated victories achieved at four prominent athletic festivals: the Pythian, Olympic, Isthmian, and Nemean games. Many of his writings have been lost or are in fragments with only four completed books in existence. These four surviving books constitute approximately one-fourth of his total work and are believed to contain his true masterpieces.

Aegina developed into a major naval power in the seventh century B.C., and Athens began actively opposing it in the sixth century B.C. The two rival Greek states clashed often over the next two centuries. The Athenian leader Solon passed laws attempting to restrict Aeginetan trade. Later, troubles worsened when Boetia and Thebes appealed to Aegina for support in their difficulties with Athens. The Aeginetans hesitated, but when the Thebans suffered heavy losses to Athens they decided to act. Aegina led a series of raids along the Attic coast hoping to distract Athens and give Thebes and Boetia time to recuperate. Instead, the move led to intermittent warfare between Athens and Aegina that lasted until 481 B.C.

War was compounded by commercial jealousies. Athens was angered over the establishment of an Aeginetan station at the Egyptian port of Naucratis. Naucratis, the center of relations between various Greek states and Egypt, was a Greek settlement and trading station on the western branch of the Nile River. It gave the various Greek states involved at the site exclusive rights in trade between Greece and Egypt.

The Aeginetans in turn were angered over Athenian domination in the export of Greek vases. The trade was quite lucrative—and made the Athenians quite wealthy—because the vases often contained such items as unguents, oils, perfumes, and wine.

Information on the war between Aegina and Athens is scarce. Much of what is known comes from a chronology by Herodotus. Herodotus reported that Athens was very much helped in its troubles with Aegina by the discovery of an unusually rich vein of silver at Laureum. Themistocles, the archon of Athens, convinced the Athenians in 483–82 B.C. to use the money from the silver to fortify the port of Piraeus and build additional ships. Themistocles also set forth a comprehensive naval policy for the Athenian state. His actions helped strengthen Athens and prepare the state for continuing troubles with Aegina and for later troubles with Persia.

The threat of Persian invasion finally ended the war between Aegina and Athens. They allied themselves with other Greek states in 481 B.C. to resist the Persians. The Persian leader Xerxes I invaded Greece in 480, scoring a victory at Thermopylae and burning Athens. The Athenians organized an evacuation of their women, children, and other non-combatants, sending many of them to refuge in Aegina. Aegina supplied ships that were instrumental in the defeat of Xerxes at Salamis in 480. The Aeginetans also were present at the victory over the Persians at Plataea in 479. Although the Persian Wars dragged on for another ten years, these two battles marked the beginning of the end of the Persian threat to Greece.

Between the years of 469 to 464 B.C., Aegina aided its fellow Peloponnesian League member Sparta in putting down a revolt in Messenia. In 459 B.C., Aegina sided with Sparta and Corinth in a war against Athens and its allies. In the opening operations of this conflict, Corinth, angered over an Athenian alliance with its neighbor Megara, spearheaded the most vigorous resistance to Athens. The Athenians began naval operations in the Saronic Gulf in 458 B.C. and defeated a Peloponnesian fleet off Cecryphaleia, a small island between Aegina and the Peloponnesian shore. The loss was a serious blow to both Corinth and Aegina, since the two furnished the most ships in the naval contingent.

The Athenians won a second decisive victory over the Aeginetans and their allies off the coast of Aegina. The Athenians captured Aegina and seized seventy Aeginetan ships. Frantic to help the besieged Aeginetans, a number of Peloponnesian League members engaged Athens in a series of diversionary battles. Corinth and Megarid also sent 300 hoplites (infantrymen) to aid the Aeginetans. It was not enough. After a protracted siege, the Aeginetans finally surrendered.

Terms of the surrender were quite harsh. The Aeginetans were forced to dismantle protective walls encircling the island, thus leaving them defenseless; hand over all remaining ships; pay a yearly tribute to Athens; and join the Athenian-led Delian League. The Thirty-Year Treaty signed between the Delian and Peloponnesian League members brought an end to the warfare by 445 B.C. Under the agreement, Aegina continued as a tribute-paying member of the Athenian alliance, but was given a guarantee by Athens of autonomy.

Hostilities flared up once again between the Athenian alliance and the Peloponnesian League in 431 B.C., inaugurating the Peloponnesian War. The Aeginetans had continued to resent Athens, and they helped persuade Sparta to enter the war against their enemy. In retaliation, the Athenians once again invaded Aegina and defeated the Aeginetans. Then they expelled the Aeginetans from the island and colonized it. It has been speculated that not all the Aeginetans were expelled; some who had collaborated with the invading Athenian forces may have been exempted from this fate and granted Athenian citizenship. In addition to retaliation, Athens had a second purpose in expelling the Aeginetans. The island gave Athens a convenient base from which to colonize the Peloponnese region.

Pericles, then leader of Athens, also may have used the expulsion of the Aeginetans and the island's subsequent colonization by Athenians to appease the war-weary populace at home with promises of money or land in Aegina. The settlement of Aegina was done by allotment and involved the colonization of both its main town and surrounding countryside.

One of the more famous members of the Athenian colony, according to some, was Plato. It has been alleged that he was born on the island shortly after his family arrived on Aegina. He later went to Athens with the rest of the Athenian colony after the island was recaptured and resettled by its native inhabitants in 405–404 B.C.

Upon their forcible removal from Aegina, Sparta gave the Aeginetans a new home in the town of Thyrea. The Aeginetans' stay at Thyrea did not last long. In 424 B.C., it fell to the Athenians, and its inhabitants were moved from there to Athens. In 405–404 B.C., acting on behalf of the Aeginetans, Lysander of Sparta captured Aegina, evicted the Athenian settlers, and repopulated it with the Aeginetans.

Aegina never fully recovered from its devastating loss to Athens, however. From 322 to 229 B.C., Aegina was under Macedonian control; then it joined the Achaean League, a military federation centered at Achaea, a rural region of Greece in the Peloponnese. It represented the Greek city-states' response to the superior might of the Hellenistic kingdoms and Rome, but it was formed too late to be effective in opposing Roman conquest. In 221 B.C., Aegina was taken over by the Romans and made a part of the Aetolian League, which enslaved the island's residents. One year later, Rome sold the island to Attalus I, king of Pergamum. Pergamene rule brought about a brief period of prosperity on Aegina until the island, along with the rest of the Pergamene kingdom, was annexed by the Romans in 133 B.C.

The ensuing centuries saw only minor activity on the island. In the 30s B.C., Roman triumvir Mark Antony briefly placed Aegina under Athenian control. Aegina later suffered a destructive siege by Germanic invaders in A.D. 267. Venetians took over the island in the Middle Ages, and once again Aegina prospered somewhat. Later, in 1718, the island was ceded to Turkey in the Treaty of Passarowitz. It remained in Turkish hands until 1826.

In 1811, archaeologists working atop Mount Ayios Ilias on Aegina excavated the temple of Athena Aphaia, a Doric-styled structure first erected in 570 B.C. and destroyed sixty years later. The remains seen today are from a reconstruction in the fifth century B.C. The structure is an excellent example of early classical Greek architecture. Excavations also uncovered some of ancient Greece's most important statues. The statues were found on the east and west pediments of the temple and represented the first modern European contact with archaic Greek art. Stylistic differences among the seventeen statues have caused them to be given varying construction dates by many scholars. Ten of the seventeen statues are from the western pediment and are well preserved. Five more are from the eastern pediment and are in less good condition. The remaining two are in fragments, and the exact positions they occupied in the temple are unknown. If one accepts the different dates attributed to the statues, then the western pedimental decorations were completed prior to the Persian Wars and the eastern ones following the Battle of Marathon. The seventeen statues were removed from the temple to the Munich Glyptothek in Germany. Fragments of another group of statues from the site are displayed in the National Archaeological Museum in Athens.

The identity of the divinity for whom the temple of Athena Aphaia was constructed has been much debated. Sculptures in the front of the temple refer to the goddess Athena, but an important dedicatory inscription honors a local deity Aphaia. Aphaia was honored with additional inscriptions cut elsewhere into the temple's marble. Most likely, locals assimilated Athena with their own goddess Aphaia and continued to refer to the temple as belonging to Aphaia even after it had been officially dedicated to Athena.

Additional archaeological sites have been uncovered and studied on Mount Ayios Ilias, including the remains of a single column from a temple dating from 520 to 500 B.C. The temple may have been dedicated to Apollo or Poseidon. Other remains excavated on Mount Ayios Ilias and in the surrounding area include a Thessalian settlement from approximately the thirteenth century B.C., a propylon from the sixth century B.C., and a Pergamene temple. At the foot of Mount Ilias and to the east are a theatre and a stadium. Other historic buildings on Aegina include a medieval castle and a seventeenth-century monastery.

During the Greek War of Independence the main town of Aegina became the capital of Greece from 1826 to 1828. The island has stayed largely out of the historic spotlight since, and today Aegina's primary industries are fishing and tourism. The ancient tradition of pottery making continues, and some farming is conducted on the island. Aegina is a popular destination for both foreign tourists and nearby Athenians.

Further Reading: Thomas J. Figueira's *Athens and Aegina in the Age of Imperial Colonization* (Baltimore, Maryland, and London: Johns Hopkins University Press, 1991) chronicles much of the hostilities between Athens and Aegina, with a very thorough account of the expulsion of the Aeginetans and the colonization of the island. N. G. L. Hammond's *A History of Greece to 322 B.C.* (Oxford: Clarendon, and New York: Oxford University Press, 1959; third edition, 1986) is a complete history of ancient Greece, often touching upon Aegina, the island's troubles with Athens, and the Peloponnesian War. Raphael Sealey's *A History of the Greek City States from 700–338 B.C.* (Berkeley and Los Angeles, and London: University of California Press, 1976) contains a small, yet valuable amount of Aeginetan history. Reynold Higgins's *The Aegina Treasure: An Archaeological Mystery* (London: British Museum Publications, 1979) provides an interesting account on the artwork of ancient Aegina and the strange paths some of the artwork took to get where it is today.

—Peter C. Xantheas

Agrigento (Agrigento, Italy)

Location: Two miles from the central southern coast of Sicily, about sixty miles southeast of Palermo.

Description: Site of one of the most important Greek Sicilian cities and unusually rich in Greek ruins, Agrigento has been of modest importance since Roman times, owing its continued existence to sulfur mining and exports of agricultural products.

Site Office: A.A.S.T. (Azienda Autonoma Soggiorno e Turismo)
Via Atenea 123
92100 Agrigento, Agrigento
Italy
(922) 20454

The ancient Greek city of Akragas was officially founded in 582 or 581 B.C. by the citizens of Gela, another Greek colony about forty miles east on Sicily's southern coast. For decades before the official founding, Gela had apparently been trading in the area with native tribes, whose history is shrouded in legend but to whose presence the archaeological record bears ample witness in the form of tombs and pottery. By the time the Greeks colonized Sicily, the native peoples were concentrated in the interior, and one of the centers of their civilization lay at present-day Sant'Angelo Muxaro, almost twenty miles north of Agrigento. Some archaeologists have identified this site with the legendary Kamikos, capital and stronghold of King Kokalos, a figure in Cretan legends.

As one legend has it, Kokalos received Daedalus after his famous airborne escape from Crete and employed the inventor as his chief engineer. Hearing of Daedalus's services to Kokalos, King Minos of Crete traveled to Sicily to reclaim Daedalus, but he was tricked, killed by Kokalos's daughters, and buried in a mausoleum at what later became Heraclea Minoa, a little farther up the coast from Akragas. The legend was sufficiently alive in the minds of the later Greek settlers that, in the early fifth century B.C., Theron of Akragas claimed to have found King Minos's remains and shipped them back to Crete for a more honorable burial.

The Gelans who founded Akragas hailed from Rhodes and Crete, and they brought new immigrants from these Greek islands with them to establish their new settlement. They built an acropolis on a rocky ridge between two rivers, the Hypsas (now Drago) to the west and the Akragas (now San Biagio) to the east. The residential center was built on the land gently sloping down toward the sea. The southern boundary of the city was formed by a second ridge from which the land drops down precipitously to the extended beach below. The earliest sanctuary at Akragas, possibly predating the founding of the city and devoted to the Chthonic deities, was located on the southern ridge, which later became the site of the agora and the major temples. Parts of the acropolis, also known as the Rupe Atenea, or Rock of Athens, were walled very early on. Two temples, dedicated to Zeus Atabyrios and to Athena Lindia, were built on the acropolis. Caves below the east end of the Rupe Atenea may have been devoted to the worship of Demeter and Persephone. The residential section, apparently laid out on a grid plan right from the start, was also fortified with walls and gates at regular intervals.

Much of the fortification of the city was probably built under the tyrant Phalaris, whose reign lasted from about 570 to 555, and was originally meant as protection against the native Sicans and the Carthaginians, who held trading posts in western Sicily. Little is known with certainty about Phalaris, whose cruelty has become the subject of many legends. The poet Pindar, for instance, mentions the brazen bull in which Phalaris was said to roast alive all those who incurred his wrath. Ancient historians allege that Phalaris managed to take power by misappropriating a large sum of money meant for the construction of a temple, which he used to hire a mercenary army. At Phalaris's bidding, these soldiers massacred the important men of the city while taking women and children hostage and leaving Phalaris without serious opposition. Whether these tales of the tyrant's perfidy are based on truth cannot be ascertained, but it is known that Phalaris considerably enlarged the territory controlled by Akragas through his campaigns against the Sicans.

The city flourished from the moment of its founding, in large part because of its fertile hinterland. Already in the late sixth century B.C., Akragas came to rival Syracuse, always the most successful of the Greek cities on Sicily. Much of the wealth of Akragas resulted from the export of fruit and other agricultural products, particularly to North Africa. As evidence of its prosperity, Akragas began minting silver coins that bore the image of a crab, the city's emblem, in the sixth century. By the end of the same century, the Akragantines started construction of two very large temples, meant as an outward measure of the city's wealth as much as a token of devotion to the gods. One of these, the Temple of Hercules, was partially restored in 1924, when eight of its columns were set upright once again. Even less of the other temple, the Olympieion or Temple of Zeus, has been preserved, although what remains is especially interesting. Contemporary architectural principles of temple construction were not suited to a temple this large, so that several special devices were necessary to make the building structurally sound. One of these devices involved the use of giant male figures, or *telamones*, in addition to columns to bear the weight of the huge entablature. One of these *telamones*, now

Temple of the Dioscuri at Agrigento
Photo courtesy of A.A.S.T., Agrigento

lying prone among the scattered stones of the former temple, has been restored.

Early in the fifth century, the fortifications of Akragas took on a new aspect and a new importance as they came to function as protection first against Greek neighbors and, later, foreign invaders. Akragas's first internecine conflict was with Selinus, a Greek settlement farther west along the coast, over Selinus's satellite community of Heraclea Minoa. Akragas won the dispute—with the military aid of Syracuse, it seems—and extended its territories as far west as the Halycus River (now Platani). The next territorial expansion was achieved under the leadership of the tyrant Theron, who took Himera, a Greek settlement on the north coast of the island, in 483. As a result of this conquest, Akragas controlled a slice of central Sicily from the south to the north of the island and became a power second only to Syracuse, which controlled all the territories to the east. Attempting to break the hegemony of Syracuse and Akragas, Carthage sent a huge armada to Sicily at the request of the former tyrant of Himera, but the Greek allies inflicted a resounding defeat on the Carthaginian forces at Himera in 480, so confirming their power over their respective territories. The victors also brought home great numbers of slaves—enemy soldiers captured on the battlefield—and the resulting influx of cheap labor sparked another building boom in the city. Another victory, this time over Motya, a Phoenician settlement on an island off the west coast, enhanced the city's wealth and splendor even more.

Theron was succeeded as tyrant of Akragas by his son Thrasydaeus, a particularly rash, lawless, and violent man, who hastened the end of tyranny in the city by his ill-judged attack on Syracuse. Thrasydaeus's army was badly beaten, and Thrasydaeus himself fled to Megara in Greece, where, nevertheless, he was tried and executed. The philosopher Empedocles, a native of Akragas, is reputed to have participated in the establishment of democratic government in the city, but his role is not well verified. Democracy brought some decision-making power to those Akragantines who held citizenship but did not extinguish the hunger for power and territorial expansion. In 445 Akragas again declared war on Syracuse, with the same result as before.

Despite the turmoil, the fifth century B.C. was kind to Akragas, whose reputation in the Greek world continued to grow. The city was praised not only for its wealth and exuberant lifestyle, but also for its vineyards, its olives, its horses, and, finally, its sculpture. Following the defeats by Syracuse, Akragas maintained a well-advised neutrality in the squabbles between other Greek cities and during the Greek invasions of 427 and 415. The last decade of the fifth century, however, brought disaster to much of southern Sicily, and this time Akragas did not escape unharmed. After a Carthaginian expedition destroyed Selinus in 409, the Akragantines took in the Selinuntine refugees, feeding and sheltering 26,000, according to the ancient historian Diodorus Siculus, but undoubtedly they did not imagine that a similar fate was in store for themselves. In 407 the Carthaginian army returned, and it laid siege to Akragas in 406.

Syracuse came to the city's aid, but after seven months of fighting—among themselves as well as with the enemy—the allies abandoned the city to be sacked by the Carthaginians while the people fled to Leontini (modern-day Lentini), on the east coast. The invading army meanwhile marched on to Syracuse, taking other cities on its way, and finally negotiating a treaty that established a Carthaginian *epikrateia*, or province, in a large stretch along the southern coast. Akragas was part of this province.

Under the terms of the treaty, the people who returned home had to pay tribute and were not permitted to rebuild the city's fortifications. In the wake of the Carthaginian invasion, Akragas survived for some seventy years in severely reduced circumstances. Although it did regain its independence, the city never recovered its former glory, wealth, or population. Even the most conservative of the ancient estimates of the population of Akragas before 406 puts the inhabitants at 100,000. Even today, Agrigento has fewer than 60,000 inhabitants, and the modern city is situated only on the ancient acropolis hill, leaving the extended residential area of ancient Akragas no more than a tourist attraction.

However, the year 340 B.C. brought some promise of a renaissance, when Timoleon of Syracuse brought in new immigrants as part of a program to revitalize the stretch of Greek Sicily that had been laid waste by the Carthaginian invasions. Some of the new colonists were to bring the hinterland under cultivation once again, while the rest were to help rebuild the city. Many improvements in the residential area and in public buildings date from this period. The walls and gates were also restored, and before long they served to defend the city from Syracusan attack. In 311—by which time Akragas had allied itself with Carthage against the depredations of Syracuse—the new tyrant of Syracuse, Agathocles, laid siege to the city. He was unable to breach the defenses, however, in part because of the intervention of the Carthaginian fleet.

By some strange logic, Agathocles then made war on all of Carthaginian Sicily, resolving to drive the Africans out of the island altogether. While the Carthaginians laid siege to Syracuse, Agathocles headed a successful military expedition to North Africa and soon beleaguered Tunis. After some years of difficulties with supplies and reinforcements, both invading armies found themselves in an untenable position, however. Carthage ceded all of Sicily to Syracuse, which in return agreed to withdraw from North Africa. By the most surprising tactics, Agathocles had in fact managed to take possession of Akragas, and, fortunately for the Akragantines, he had done so by wreaking havoc elsewhere.

Nevertheless, Akragas was not satisfied with the results of Agathocles' unusual military adventure, and in 308 the city rebelled against Syracuse. Agathocles put down the rebellion the following year, but by then the Carthaginians were back in Sicily. A treaty in 306 B.C. once again recognized a Carthaginian province, although this time it was much less extensive and was concentrated in the eastern part of the island. Akragas was left a possession of Syracuse, and so it remained until about 276, in spite of another fruitless rebellion following Agathocles' assassination in 289.

Although Agathocles' reign had not by any means been quiet and peaceful, his death was the beginning of a century of far more devastating upheaval in Sicily. Carthage saw another chance to expand its influence on the island during the chaos that followed Agathocles' death. The Mamertines, mercenary soldiers operating mostly from a base in Messina, became another destabilizing force as they started raiding Greek cities all over the island. They partially destroyed Gela in one raid, prompting the Gelans to appeal to Phintias of Akragas for help. Phintias razed Gela and moved its inhabitants to a new settlement that bore his name (now Licata). In the midst of the chaos, the Sicilian Greeks brought in another foreign invader—Pyrrhus, he of the Pyrrhic victories, which had taken place in Italy just previously—to restore order and aid them against Carthage. Landing at Tauromenion (now Taormina) in 278, Pyrrhus swept through the island in two years until he was halted at Lilybaeum (now Marsala), the last Carthaginian stronghold on the far western tip of the island. Backed by the superior Carthaginian fleet, Lilybaeum was too much for Pyrrhus, who was fast becoming unpopular with the Greek cities because he proved much less amenable to Greek autonomy than they had foreseen. When Pyrrhus withdrew from Sicily in 276, he left Syracuse in solid control of the eastern end of the island, while Carthage was poised to recoup its losses in western Sicily.

Akragas had profited by the Mamertine raiding and had not been seriously affected by the Pyrrhic invasion, but the First Punic War that followed was to be the end of its brief renaissance. When the Romans invaded Sicily in 264, Carthage established a base at Akragas, so making it one of the prime military targets for the Roman army. After a six-month siege, Rome took the city, selling 25,000 of its inhabitants into slavery to finance its continued advances against Carthage. Unfortunately, Carthage was back in Akragas by 254 and this time took care to burn the city.

Carthage again established itself in the rebuilt city in the Second Punic War, and this time Akragas was betrayed by mercenaries in the Carthaginian camp in 211. Again the Romans sold the inhabitants together with all the booty they could lay hands on. Following the end of the Punic War, the city, its name now changed to Agrigentum, became more or less autonomous. The Roman government replenished the population in 197 and a slow Latinization began. Although by late Roman times most of the inhabitants still spoke Greek, new houses generally followed Roman building conventions. Apart from a moment of unrest during a slave revolt in the second century B.C., all was quiet in the now modest town.

Christianity arrived in Agrigentum in approximately A.D. 390 at the latest. One of the Greek temples, the misattributed Temple of Concord on the southern ridge, was converted to a church, and catacombs were built at the western end of the city. By this time, Agrigentum survived only because it had begun mining the rich sulfur deposits in its hinterland, which provided some stimulus to its faltering economy. Not long after, the Roman Empire crumbled, the Vandals briefly controlled Sicily, and then the island passed into Byzantine hands. None of these developments left much of a mark on Agrigentum, which still had a distinctly Greek flavor. Byzantine domination, of course, enhanced the Hellenic elements of Agrigentine culture, since the Christian churches all adopted the Byzantine liturgy, in spite of attempts by the western papacy to retain a firm footing in this island so close to the Vatican.

Real changes came in the ninth century with the Arab conquest of Sicily, which brought some fifty years of fighting, plague, and famine. By 828, the Arabs controlled much of the western part of Sicily and recolonized Agrigentum. A group of Berbers settled in and around the city, now called Girgenti. Although the established population was free to practice its own faith, mosques were built in the city, and much business began to be conducted in Arabic. The Girgenti Berbers had come to cultivate the land, and soon they developed an animosity toward the Arab rulers, who never made Sicily their home. Commerce and agriculture flourished in this period, as the Arabs brought superior farming techniques and improvements in irrigation. By the eleventh century, as Arab rule began to disintegrate, an Arab family was trying to make the area around Girgenti into an independent emirate. They might have been successful had it not been for the arrival of Roger I (the Norman), who landed at Messina in 1060 and quickly took control of northeastern Sicily.

In 1087, Roger captured Girgenti. Although the Moslem majority of Girgenti enjoyed religious freedom under the Normans, the Norman conquest slowly Latinized the city once again. A bishopric was established immediately after the conquest, and most of the bishops came from Normandy and other parts of western Europe. These bishops performed most local administrative functions and were primarily responsible for the administration of justice. They also authorized the destruction of many of the surviving Greek temples, to be used as building material for the new churches. Archaeologists suspect that the Girgenti cathedral and the Church of Santa Maria dei Greci, both on the ancient acropolis hill, were built on the foundations of the Temple of Zeus Atabyrios and the Temple of Athena Lindia, respectively. The cathedral, originally built in the eleventh century in the Romanesque style but reconstructed many times in subsequent centuries, was never vaulted, and its painted wooden ceiling is one of its principal attractions. The much smaller Santa Maria dei Greci still bears the traces of the integration of different cultures that characterized Norman rule. The typically Norman wood ceiling is juxtaposed with Byzantine frescoes.

By the late twelfth century, Girgenti was still largely Moslem, and so the bishop of Girgenti found himself in an exceedingly uncomfortable position during the Moslem uprising of 1189 and 1190. By then the central government favored Christian traders over Moslem farmers, as anti-Moslem sentiment began to make itself more clearly felt. The Moslem uprising quickly spread through the west of the island, and Girgenti was taken by Moslem bands. Holding the bishop prisoner, they converted the cathedral to a barracks. The uprising was no more than a temporary dislocation, however, and could not prevent the end of the strong Arab influence in the city.

For the next 600 hundred years or so, Girgenti quietly continued its existence, while the crown (which passed into Spanish hands in the fourteenth century), the bishop, and the local barons fought for control over the city and its surrounding land. Girgenti was once again incorporated into the royal demesne in the 1390s, but most of the fifteenth and sixteenth centuries saw warring among the local aristocracy over land-holdings. Early in the seventeenth century, the bishop bought up much of Girgenti and neighboring Licata. But in 1647 a popular revolt, sparked by food shortages and originating in Palermo, spread across the island. All of Sicily had seriously declined since the passing of Arab rule, and it was the common people who suffered. In 1647, a local mob broke into the episcopal palace of Girgenti, where the bishop was hiding with his food supplies. The popular movement lacked cohesion, however, and passed without bringing permanent change. Early in the eighteenth century, the monks of Girgenti revolted against royal authority, but their short-lived rebellion did not, and was never meant to, improve the lot of the citizens.

Ultimately, it was sulfur that assured the survival of Girgenti, since it provided a steady export trade, despite the fact that the city did not have a real harbor. The bishop of Girgenti ordered a breakwater built from the remains of the Olympieion, but by 1840, some fifty years later, the project was still unfinished. The harbor at Porto Empedocle, some eight miles west along the coast, served as port of call to most of the sulfur traders. In 1808, when demand for sulfur rose tremendously, King Ferdinand I granted mining rights to the local aristocracy, giving a boost to the city's depressed economy. But Girgenti had to wait for an improved infrastructure until the 1920s, when Mussolini began a program of building roads and bridges. At the same time, Girgenti's name was changed to Agrigento in an effort to Latinize Sicily. By that time, however, the locally mined sulfur was far outpriced by sulfur mined elsewhere, and Agrigento became primarily a tourist town. Some expansion took place in the 1950s and 1960s, but that was due in large part to real estate speculation that brought a building boom on unsafe lands, with illegal building licenses. In 1965, a landslide caused major damage to the illegally constructed high rises, but damage to the architectural treasures of the city was limited.

Further Reading: Pierre Sébilleau's *Sicily,* translated by Oliver Coburn (New York: Oxford University Press, and London: Kaye and Ward, 1968; as *La Sicile,* Paris: Arthaud, 1966) contains a somewhat impressionistic but serviceable description of modern Agrigento and has many beautiful photographs of the ancient ruins. A brief sketch of Agrigento in antiquity and a good description of the archaeological remains are available in *Sicily: An Archaelogical Guide* by Margaret Guido (New York and Washington: Praeger, and London: Faber, 1967). The three volumes of *A History of Sicily* by M. I. Finley and Denis Mack Smith (London: Chatto and Windus, and New York: Viking, 1968) place the history of Agrigento in the context of Sicilian history in general.

—Marijke Rijsberman

Alalakh (Hatay, Turkey)

Location: On the Orontes River between Antakya (Antioch), Turkey, and Aleppo, Syria, in the Turkish province of Hatay, near the Turkish/Syrian border.

Description: Archaeological site with the ruins of seventeen ancient cities on successive levels. The oldest dates to before the early Bronze Age, the latest to 1194 B.C.

Site Office: Tourist Information Office
Vali Ürgen Alanı, No. 47
Antakya, Hatay
Turkey
(326) 2160610

Alalakh, also known as Tel Açana, is the site of a succession of ancient cities on the Orontes River in Amūq, the plain of Antioch. It is near the spot where the south-to-north-flowing Orontes hooks to the west and flows into the Mediterranean. This fertile plain is in the Hatay Province of Turkey, a narrow strip along the Mediterranean coast that protrudes southward into Syria.

Due to the abundant resources of the plain and the forests of the Amanus Mountains (Mountains of Cedar), the area was settled early. Alalakh was inhabited from before the early Bronze Age until 1194 B.C. Trade routes linking Aleppo with the Mediterranean, and the Middle East with Anatolia and the Aegean, crossed at Alalakh.

Excavations conducted from 1936 through 1949 by Sir Leonard Woolley for the British Museum revealed towns on seventeen levels, which Woolley labeled I through XVII with I at the top. Little has been learned about the earliest civilization at the site, as the underground water table has affected Level XVII. There is evidence, however, that the inhabitants had bronze tools, and two different types of pottery were found. The first was wheel-thrown pottery newly introduced in this period; the second was a very different and older type of pot that implies an earlier group living there who were driven off by the Level XVII dwellers. Place names in the area also suggest an earlier people as many are derived from an unidentified language revealed in the earliest layers at Alalakh.

Temples played an important role throughout Alalakh's history, with one in each layer. Palaces were also found at several levels, the first at Level XII (2700–2350 B.C.). It was probably copied from a Mesopotamian palace, as it had such features as a mud-brick colonnade, not previously seen in Amūq. At Level VII, the majestic palace of Yarim-Lim was found, its state rooms decorated with frescoes. The frescoes employed methods similar to those found in the palace at Knossos, which was built at least 100 years

later. Also found was Yarim-Lim's burial chamber, which consisted of a huge pit fifty feet deep, with a shaft in the center an additional thirty feet deep. The funeral urn was placed in the shaft and the shaft filled with stones. Then the pit was filled with successive layers of ceremonial buildings, each burned and then topped with another layer. All of this formed the foundation for the Royal Chapel, which contained a statue of Yarim-Lim.

It was with the cuneiform tablets, found in an archive room of the palace on Level VII, that written information on the history of this time in Syria first appeared. Deciphering the tablets revealed the name of the town as Alalakh in the kingdom of Yamhad. Nearby Aleppo was Yamhad's capital, and Alalakh its major port. Alalakh had become a semi-independent state when Yarim-Lim aided the king of Aleppo in putting down a rebellion. As a reward the king gave Yarim-Lim sovereignty over Alalakh, although as a vassal state to Aleppo. The nature of Alalakh's government prior to Yarim-Lim is unknown. It may have been ruled from Aleppo or have been in the hands of rebel relatives of the king.

A letter from 1770 B.C. provided a glimpse of the political environment at the time. It said that no king could be strong enough on his own and that Yarim-Lim, "the Man of Alalakh," had twenty vassal kings. Since ten or fifteen was the norm, this gave him enough power to provide military aid to King Hammurabi of Babylon. The extent of Yarim-Lim's domain is unknown; it may have encompassed only Amūq. Level VII covers the reigns of Yarim-Lim and his son, Ammitaquam, both of whom reigned a very long time. A revolt ended Yarim-Lim's dynasty in the seventeenth century B.C. The Chapel Royal was destroyed and the statue of Yarim-Lim damaged, although the head survived and was found in the twentieth century A.D. Ammitaquam was not content to be "the Man of Alalakh," a vassal of Aleppo, so he declared his kingdom independent.

The kings of Alalakh and Aleppo were Amorites (Western Semites) and the largest part of the population Semitic. Some time before 2000 B.C., a warring Indo-European people, the Hurrians, arrived from Iran. They settled between the upper Tigris and Euphrates Rivers and set up numerous minor principalities in a kingdom they called Mitanni. By 1900 B.C. they had conquered territory all the way south to Syria. During Level VII, Hurrians were still a minority group in Alalakh's population, but were gaining importance in government and religion. Terminology used on the tablets indicates the scribes may have been Hurrian, a Hurrian calendar was in use, and King Ammitaquam had a Hurrian wife.

Aleppo was an important trading city, along with its vassals encompassing a domain from the Mediterranean to the Euphrates, and presenting a very attractive target for the

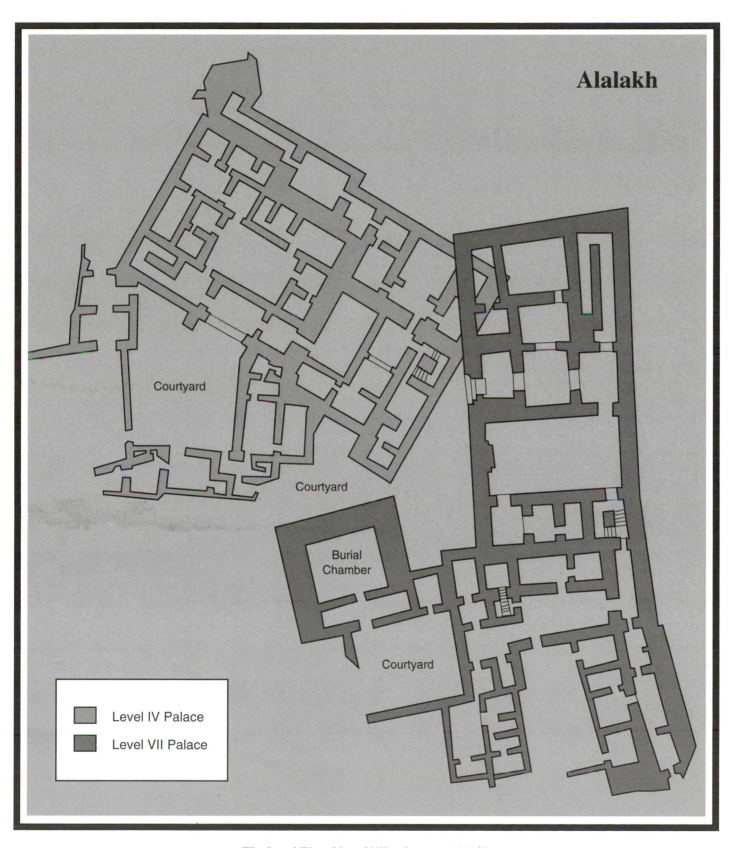

Alalakh

Courtyard

Courtyard

Burial
Chamber

Courtyard

Level IV Palace
Level VII Palace

The Level IV and Level VII palaces at Alalakh
Illustration by Tom Willcockson

Hittite ruler Hattusilis I in the seventeenth century B.C. Rather than attack Aleppo directly, he first attacked and destroyed Alalakh (Level VII), cutting Aleppo off from the sea. It does not appear that Aleppo came to Alalakh's aid against the Hittite attack. It is possible that Ammitaquam's proclamation of sovereignty is evidence of internal dissent in Yamhad, which Hattusilis exploited. Accounts of the conquest, written for Hattusilis, were found during excavations at Hattusas, the Hittite capital.

The Hittites formed the first significant kingdom in Asia Minor. Although they are thought to have come from the Caucasus with their chariots around 2000 B.C., their exact origin has not been confirmed. They settled in Anatolia and without resorting to violence dominated the native people, the Hatti, to whom they were probably related. Although they subsequently waged war to build, expand, and maintain their empire, they were known for their humanity.

Hittites eventually controlled territory stretching from the Euphrates on the east, and the Black Sea on the north to the Mediterranean. Since their empire enveloped far-flung towns in sometimes harsh terrain, they maintained a decentralized form of government, administered by a bureaucracy of scribes. Some towns were self-governing, while others belonged to vassal kingdoms. Many of the trading and diplomatic practices of the Middle East were derived from the Hittites. They had laws appropriate for their farming economy, with restitution rather than revenge the thrust of their criminal law. Although sometimes compared to the Code of Hammurabi of Babylon, Hittite laws were less cruel and more just than any in Mesopotamia at the time.

Texts were found at Alalakh written in both Hittite and Akkadian (a Semitic language also known as Assyro-Babylonian), covering subject matter from astrology, divination, and conjuration to word lists, and mundane business and administrative documents. Due to the number of tablets and inscriptions, Alalakh has proven particularly helpful in documenting life in the area.

The political situation changed again under Mursilis I, who succeeded Hattusilis. He conquered northern Syria, destroying Aleppo, and bringing the kingdom of Yamhad to an end. In 1595 B.C. he also conquered Babylon, ending the Hammurabi dynasty. His success did not last, however. As he returned with the wealth and goods of Babylon, he was killed by his usurper brother-in-law and a period of internal instability in the Hittite kingdom resulted. Weakened by dissension and continuing harassment by the Hurrians, the Hittite empire lost much of its territory in Syria and Babylonia, although it remained in power in the area around Hattusas. During this period Alalakh was alternately subject to Egypt and the Hurrians. Egyptian artifacts from the Egyptian Middle Kingdom (2160–1700 B.C.) found at Alalakh show a relationship with Egypt, the exact nature of which is unclear.

No records were left by Mursilis himself. With the end of the Amorite dynasties, including that of Hammurabi, records throughout the area became scarce. Difficulties in dating events in the area persist because of this undocumented

period. When record-keeping resumed, society had changed. From the destruction of Level VII, the Hurrian presence increased until Level IV in the middle of the millennium, when Hurrians dominated and Alalakh was a Hurrian society. The gods were Hurrian and only a tiny minority of the populace was West Semitic. The Hittites had returned to their homeland.

The leading citizens of Alalakh were *mariyanna*, aristocrats who owned the horses and light chariots that were so important for warfare. (They were also used by kings and nobles for hunts on the steppes.) Society was divided into castes, with the king and *mariyanna* at the top, knights, craftsmen and other professionals in the middle, and below them freeborn citizens who could be pressed into service on public works projects and drafted into the army. At the end of the period of obscurity, King Tuthmosis I of Egypt invaded the area. Dominance of Mesopotamia and Asia Minor was again split between the Hurrians and Egypt.

During Level IV (1550–1473) B.C., Alalakh was the capital of the kingdom of Mukish, a vassal of the kingdom of Mitanni, a confederation of Hurrian princes. Parattarna, and later his son Saustatar, were the kings of Mitanni during this period. Kings of Alalakh were Idrimi, his son Niqmepa, and his grandson Ilimilimma. A statue of King Idrimi with autobiographical carvings dates from this period, although it was found buried under the floor of the temple in Level I.

The inscriptions revealed that when Idrimi's father, the king of Aleppo, died, Idrimi and his brothers took refuge with his mother's family in Emar. Disappointed with his brothers, Idrimi took "his horse, his chariot and his groom" and went to Canaan, but again had to flee from his enemies. He wandered for seven years, often taking refuge with the semi-nomadic bedouins who traveled the steppes outside the cities. Although these *Sutu* sometimes attacked travelers, they sheltered Idrimi.

When the omens showed the time was right for his return, Idrimi raised an army and sailed with them to Mukish, where he was rejoined by his brothers and proclaimed king of Neya, Mukish, Ama'u, and Alalakh. Parattarna acknowledged Idrimi's kingship and accepted his homage as a vassal. Legal documents found in the Level IV palace show that the vassal kings had a certain degree of freedom. They could wage war, make treaties with neighbors, and even accept vassals of their own, as long as they remained subservient to the king of Mitanni. Idrimi's kingship had to be ratified by Parattarna before it took effect.

The inscriptions further revealed that Idrimi was pious and ruled wisely for thirty years. He was particularly kind to the *Sutu,* in gratitude for their earlier aid. Although he never won back his father's throne in Aleppo, he was a great king. In a war against the Hittites he took seven of their towns and returned to Alalakh with much booty.

Conflict between Mitanni and Egypt continued under Idrimi's son, Niqmepa, as Egypt persisted in its efforts to conquer Syria. Northern Syria remained under Mitanni control, while central Syria was under tenuous Egyptian control.

When Tuthmosis III of Egypt conquered Alalakh's neighbor, Nukhashshe, Alalakh sent ambassadors and appears to have become an Egyptian vassal.

The reign of Tuthmosis IV signaled an end of the hostilities. Egypt and Mitanni became allies, possibly due to the threat posed by the resurgent power of the Hittites and Assyrians. (A revitalized Hittite empire, the Middle Kingdom, arose in the fifteenth century B.C.) The relationship was secured by the marriage of the pharaoh to a Mitanni princess. Alalakh was formally ceded to Mitanni, probably as a result of the marriage. This peace lasted until 1400, when Suppiluliumas, the great Hittite warrior, reconquered the area for the Hittites, and Alalakh became a Hittite military stronghold.

Suppiluliumas did not establish firm Hittite rule; city-states were made vassals. An area stretching from the Mediterranean to the Euphrates had been controlled from Aleppo. To neutralize its power, Suppiluliumas gave Aleppo to one of his sons. To keep an eye on Alalakh and other Mediterranean cities, he set up his favorite son at Charchemish, Aleppo's great rival. Charchemish was surrounded by vassal states, including a reorganized and reduced kingdom of Mitanni. Suppiluliumas was in Alalakh to sign a treaty with Ugarit and there are other signs that the Hittites occupied Alalakh, which was probably also a vassal state. After Suppiluliumas's death, there was a revolt, and excavations at Level III reveal that all of the Hittite temples were burned. One of Suppiluliumas's successors, Tudhaliyas IV, who occupied the Hittite throne about 100 years later, was shown with his queen on a carved relief of very high quality found in the excavations at Alalakh.

A type of pottery unique to Alalakh appeared in the late fourteenth century B.C. in Level II. These elegant goblets were a variant of Nuzi Ware (Hurrian Ware), called Açana Ware, with intricate floral designs showing Cretan influence. Level II also shows Alalakh was importing Mycenean-style pottery from Cyprus, a practice that continued until the end of Level I.

Level I, on top of the mound created by the previous ruins, was a walled city, with houses crowded together. The northwest end was higher due to the more substantial ruins of centuries of palaces. On this level there was evidence that Alalakh again revolted and enjoyed independence for a short time until the Sea Peoples arrived in 1194 B.C. The twelfth to ninth centuries B.C. were a period of general turmoil in Anatolia, the Aegean, and the Mediterranean, called the Dark Ages by some historians. For various reasons, perhaps the effects of the Trojan War, overpopulation, and exhaustion of the land, there was a general movement of pillaging refugees, including the Sea Peoples. Troy, Hattusa, and other cities were destroyed when these groups ravaged the Aegean and the eastern Mediterranean. Their raids signaled the end of Mycenean rule in Greece and a reduction of the Hittite empire to a few small kingdoms. The second wave of Sea Peoples destroyed Alalakh around 1200 B.C. The invaders were finally routed in 1191 B.C. near Egypt.

The site of this final annihilation of Alalakh, Level I, yielded the inscribed statue of Idrimi, which provided a great deal of information about the earlier period. Idrimi must have been revered, as his statue was preserved and maintained in a place of honor in successive temples. In the last destruction of the town, the statue was thrown from its pedestal and broken, but someone lovingly buried it to protect the remains from further desecration. Alalakh was never rebuilt, and was all but forgotten until Sir Leonard Woolley began his excavations. The Hittite Empire also vanished from history, surviving only in Old Testament references to the Hittim. When they were rediscovered in the 1860s, the extent of their empire, and sophistication of their culture and government were totally unexpected. Another archaeological surprise came when the first written tablets were discovered in 1906, revealing the Hittite language to be of Indo-European origin.

After the demise of Alalakh, the region was controlled in succession by Neo-Hittite States, the Assyrians, Neo-Babylonians, Persians, Macedonians, Romans, Byzantines, and even Christian Crusaders. Under Selim I (Selim the Grim), the Ottoman sultan/caliph who pushed southward into Syria and Palestine, the territory was incorporated into the Ottoman Empire. The narrow strip of land around Alalakh is now the Hatay province of Turkey.

Further Reading: Richard Stoneman's *A Traveller's History of Turkey* (Moreton-in-Marsh, Gloucestershire: Windrush, and Brooklyn, New York: Interlink, 1993) is an interesting and very readable history of Turkey from prehistory through the 1990s. *The Cultural Atlas of Mesopotamia and the Ancient Near East* by Michael Roaf (Oxford: Equinox, and New York: Facts on File, 1990) provides a wealth of information about the history, technology, art, and culture of the Near East from 12,000 B.C. through the first century B.C. This book has hundreds of beautiful color photographs, maps, and illustrations. *Ancient Mesopotamia, Portrait of a Dead Civilization* by A. Leo Oppenheim (Chicago: University of Chicago Press, 1964; revised edition completed by Erica Reiner, 1977) is a fairly detailed account of life in ancient times, based on evidence provided by archaeology. There are some specific references to the Alalakh finds.

—Julie A. Miller

Alcobaça (Leiria, Portugal)

Location: West-central Portugal, near the Atlantic Ocean, eighteen miles south-southwest of Leiria and sixty miles north of Lisbon.

Description: Agricultural town that developed around a former Cistercian monastery; a full range of monastic buildings includes a large church significant in architectural history.

Site Office: Alcobaça Municipal Tourism Board
Praça 25 de Abril
2460 Alcobaça, Leiria
Portugal
(62) 42377

The early history of the former Cistercian monastery of Santa Maria a Velha at Alcobaça is cloudy. Facts, if they ever were known by the monks who wrote the histories of the institution, were either inadvertently or deliberately superseded by fictions based on myth, misunderstanding, or simply wishful thinking. Some of the legends of massacres or miracles are so improbable that modern historians can reasonably dismiss them, but sometimes the details needed to fill in the resulting gaps are lacking.

The importance of Alcobaça, however, remains undeniable: the monastery—agriculturally progressive, education-minded, politically astute—worked closely with the early government to provide the fledgling kingdom of Portugal with badly needed cultural and economic capital. Moreover, the huge abbey church, ranking among the most significant monuments in a country whose population is almost entirely Roman Catholic, has had a profound influence on the general scheme of European religious architectural history.

A necessary condition for the establishment of a Christian institution of such prominence was the removal of the Moslems from the area, a process that accelerated in 1095 when Alfonso VI of Castile granted the county of Porto to Henry of Burgundy, who was married to Alfonso's daughter Theresa. Under this arrangement, Henry was technically a vassal of the Castilian king, but as long as Henry contributed the required support to the realm, he was able to maintain Portugal's de facto independence.

Meanwhile, the Cistercian order was founded by St. Robert of Molesmes at Cîteaux in France in 1098. Under an austere Benedictine rule, Robert and his monks sought to reintroduce ascetic values to a monasticism that had become much too worldly to allow devotion to a life of piety and meditation. By 1115 a young monk from Cîteaux named Bernard had established a similar monastery at Clairvaux. Bernard of Clairvaux rose to prominence in both church and civil politics until his death in 1153, and it was he who adapted the proportions and structure of the budding Gothic church architecture of Burgundy to an austere, practical ideal for Cistercian structures throughout Europe. In the Bernardine schema, the church was considered not the "house of God," but rather an *oratorium*, a place for the soul's communion with God.

Upon Henry of Burgundy's death, his widow Theresa became regent of his territories. Fearful that his mother's ties to Castile would cost Portugal its relative independence, Henry and Theresa's son Afonso I Henriques wrested control of the area from Theresa to become count of Portugal in 1128. By 1139 he had assumed the title King Afonso I, a full-scale declaration of independence from Castile; the clergy supported Portugal's separation from Castile, and a vassalage to the Holy See was established in 1143. One of Afonso's own sons became a monk at Alcobaça, and was buried there in the 1220s.

The religious orders, specifically the English Crusaders who were on their way to the Holy Land, worked closely with Afonso during the 1140s to conquer the Moslem areas of Portugal. Afonso encouraged Christian resettlement of his new territories by financing monasteries; the first two Cistercian establishments were at Tarouca and Sever. After the retaking in 1147-48 of Santarém, Lisbon, and the coastal Estremadura region, Afonso provided a generous endowment to found a Cistercian center at the confluence of the Alcoa and Baça rivers.

Alcobaça's first settlement included at least twelve Cistercian monks sent in 1153 from Clairvaux under the leadership of Abbot Ranulphus. The monastic buildings at Alcobaça, begun on a grand scale in 1178 and dedicated in 1223, were essentially laid out according to the Cistercian ideal plan. This complex lay to the north of the church, which was consecrated in 1252. The church may be the most precise and best preserved image of the destroyed Clairvaux III, both in regard to the Burgundian structural system and to the proportions. French masters almost certainly were responsible for building the church or at least directing its construction, as the architectural methods demonstrated there were not indigenous to Portugal. After Pontigny in France, Alcobaça— at about 350 feet in length—is the largest surviving church of the Cistercian order.

The imposing interior of the church at Alcobaça features some of the best stone carving in Portugal. The church's construction notably demonstrates the so-called hall church design. The church's three groin vaults of about sixty-five feet are nearly identical in height, presenting a squarish exterior. The side aisle vaults are only slightly narrower and lower than the central nave vault, an unusual approach to Cistercian church architecture; in Burgundy, for

The abbey at Alcobaça
Photo courtesy of Portuguese National Tourist Office

example, the churches featured the more typical side aisles as obviously secondary to the main space, carrying the vault loads down to the ground. The Alcobaça church required additional buttresses on the southern and eastern sides to provide the necessary stability. On the north, the cloister adjoins the church.

In a sense Alcobaça is a hybrid of Cistercian architectural developments, with the plan following the Burgundian model, the choir elevation showing influences from Pontigny in southern France, and the exterior having a more robust, Iberian character, like the Cistercian abbey at Santes Creus in Catalonia. Although the church at Alcobaça was the first Portuguese building completely in the Gothic style, the Cistercian approach to Gothic design proved less influential throughout the country than the designs of the later mendicant orders such as the Dominicans.

As with other monasteries, Alcobaça's initial charge was to repopulate its vicinity, providing religious, social, and economic direction for the country. The monasteries generally devoted themselves to a life of well-organized rural work, one of their major tasks being the development of unproductive land. The Cistercians practiced a methodological approach to agriculture, providing an admirable example as cultivators and as landlords to the local farmers. The monks assiduously collected relevant manuscripts and botanical specimens to provide a scientific basis for bringing fertility to otherwise barren lands.

Just as the major Cistercian centers in France had sent monks out to found new monasteries throughout Europe, so too Alcobaça began to expand its own influence throughout Portugal with the establishment of a daughter-house at Bouro in 1174. This was the first of twelve such affiliates subject to Alcobaça's authority. The prior of Alcobaça attended to the business propositions of these affiliations with due attention. Alcobaça stipulated in charters to daughter-houses, for example, that orchards and vines be planted, and, where appropriate, that salt pans be constructed for the extraction of salt from seawater (dried and salted fish were a mainstay of medieval Portugal's economy). In exchange for each charter, Alcobaça received an annual tribute comprising a percentage of agricultural produce and salt. In its 1422 charter to the monks at a daughter house in Alfeizerão, the annual tribute to Alcobaça was stated to be one-fifth of the fruit and one-quarter of the salt.

In addition to its wide-ranging agricultural pursuits, the monastery participated actively in Portugal's seafaring economy. Alcobaça's ports included Pederneira, nearest to the monastery, a center for fishing and shipbuilding. Other ports the monastery owned were Paredes, San Martinho, Alfeizerão, and Selir. By 1254 Alcobaça was collecting profits from whale oil at its ports.

The actual Christian reconquest of the region of present-day Portugal was not completed until the mid-thirteenth century, and Alcobaça's geographic location near the Tagus River frontier between Portugal and the western sections of the Moslem al-Andalus empire made the monastery vulnera-

ble when Moslems attempted to regain their lost territory in 1195. They occupied Alcobaça, but despite stories claiming the massacre of the entire monastic community, the evidence indicates that the monastery was affected only marginally by the invasion. In that same year, monks from Alcobaça founded a second daughter-house at Seiça, and just five years later they settled a community at Maceira Dão.

Alcobaça developed slowly during the remainder of the twelfth century. In addition to awarding generous land grants, however, Afonso and his successors allowed the abbey numerous privileges, and by the mid-thirteenth century the abbey was well on its way to becoming Portugal's wealthy and powerful religious center. The monastic buildings were completed in 1223, and the immense church was dedicated on November 20, 1252, with the prominent bishops of Lisbon and Coimbra presiding at the ceremony. Expansion continued with the addition of the cloister and annexes in the early fourteenth century.

Scholarly pursuits were a luxury that could not widely be afforded, yet Alcobaça's library was founded with the monastery itself. As well, the scriptorium for copyists was in operation early on, by the end of the twelfth century. Alcobaça rapidly ascended to dominance in Portugal's intellectual culture during the thirteenth century. The monks under Abbot Estevam Martins established Portugal's first public school in 1269. The curriculum included grammar, logic, and theology. Under the auspices of King Dinis (ruled 1279–1325), Abbot Martinho II of Alcobaça and the abbots of other monasteries promoted the foundation of the University of Lisbon in 1290, which moved to Coimbra north of Alcobaça in 1308.

Alcobaça's scriptorium not only provided copies of extant works, but contributed substantially to the nation's literary tradition. Although the birth of formal Portuguese prose is traced to a series of fifteenth-century royal historiographers, it is possible that these writers based their own styles on the many translations from the Latin, Italian, and French that had been made at Alcobaça between 1350 and 1490. Alcobaça's royal connections would have made such manuscripts readily available to these later writers affiliated with the Aviz court.

Alcobaça's political ties served it well. The abbot was a prominent person in Portugal, possessing such titles as "Lord of Winds and Waters" (Senhor de águas e ventos) due to the number of windmills and watermills on Alcobaça's estates, and "Lord of the Coverts" (Senhor dos coutos), which were wooded preserves for royal hunters. Additionally, the abbot held the right to dispense criminal and civil justice in the villages and seaports of which he was lord. Alcobaça's abbot sat on the *Cortés,* was grand chaplain of the court, and had spiritual jurisdiction over the Order of Christ (a military order created in 1319 that inherited the possessions of the Templar knights) and the Aviz dynasty, which ruled Portugal from 1385 to 1580.

Members of the royal family were buried in the abbey, and their tombs are outstanding examples of Portuguese

Gothic sculpture. Alcobaça's ties to royalty enhanced the legend-making that went on there; two of the most outstanding royal tombs of Portugal perpetuate the ill-fated romance of King Peter I and Inés de Castro.

The intimate association of the abbot at Alcobaça with the ruling class no doubt helped him contend with the rivalries of church and state. A century of conflicts between Portugal's rulers and the church began in earnest not long after the founding of Alcobaça, under the reign of Sancho I (ruled 1185–1211); after Afonso Henriques's generous endowment, notable royal patronage of Alcobaça recommenced only after Sancho's death. Most of these contentions, however, pitted the kings against powerful bishops in the towns, which were owned by the crown. (Alcobaça's charter, on the other hand, stipulated that the lands would revert to the crown only if the monks vacated them.) Abbeys such as Alcobaça were able to continue developing the country's agricultural and educational potential. The Cistercians, in particular, managed to prosper despite the church's dwindling authority vis-à-vis the civil realm; other monastic orders initially were not as successful economically, although they later superseded the Cistercians in influence.

As overseas discoveries of the fourteenth and fifteenth centuries enriched the country, the decline of the old monastic orders such as the Cistercians began to spread, accelerating in the mid–fifteenth century; the rule of Manuel I (from 1495 to 1521) marked a period of political eminence and ecclesiastical decline. Ironically, it was during Manuel's reign that the church facade at Alcobaça received its flamboyant Gothic remodeling.

Despite its political decline, Alcobaça continued to exert its influence in both temporal and spiritual affairs. From the sixteenth to the eighteenth centuries, the abbey was expanded to the south and to the east. It maintained an agricultural school and model farms, and also opened mines and pioneered in metallurgy, ceramics, glasswork, weapons manufacture, and other industries. Portugal's first pharmacy was established at Alcobaça during the fifteenth century; the conspicuous absence of herbal texts in the abbey's otherwise substantial legacy of manuscripts suggests that the pharmacy may have been a non-monastic enterprise. The abbey also set up one of the first printing presses in Portugal. The port of Alfeizerão, probably the most important of Alcobaça's holdings, had an active shipbuilding economy that during the reign of Manuel I in the early sixteenth century owned eighty ships.

In a significant rift with the Cistercian leadership in France, in 1567 Alcobaça—under the direction of the king's brother, Cardinal Henry—became head of the Portuguese Congregation of St. Bernard with its own abbot general and triennial chapters. As part of the commendatory system that the government had introduced to the religious orders a century before, this meant that the abbatial title to Alcobaça would be granted in commendation to Portuguese nobility, an arrangement that was anathema to the Cistercian ideal of apolitical spiritual pursuits. When the aged and feeble Cardinal Henry became king in 1578 upon the untimely death of his nephew, the adventurer King Sebastian, he was unable to secure dispensation from Rome to allow him to break the vow of celibacy in order to produce an heir, and Portugal's annexation by Philip II of Spain was inevitable.

With the diminishing political influence of the church in general and the incursion of other ambitious religious orders, especially the Jesuits, into Portugal, Alcobaça lost its moral leadership in the seventeenth and eighteenth centuries. Yet as an enterprise the monastery was still successful enough to embellish its church in 1725 with a baroque facade (incorporating parts of the Early Gothic portal) by the Italian Giovanni Turriano.

It appears that the monks enjoyed a comfortable enough life, even sumptuous perhaps in the culinary realm. The activity emanating from the immense kitchen at Alcobaça certainly impressed William Beckford, an English patron of the arts and architectural dilettante who published a diary of visits he had made in 1795 to Alcobaça and the nearby monastery of Batalha. Although Beckford was given to hyperbole and flights of fancy, the outsized kitchen that can still be seen at Alcobaça indeed suggests a certain food-related fervor.

During the French incursion into Portugal in 1810, General André Masséna housed a divisional headquarters at Alcobaça in his campaign against the Portuguese and British forces under Arthur Wellesley (the future Duke of Wellington who defeated Napoléon at Waterloo). The French pilfered any valuables that were readily at hand, including copies of the works of the first-century Jewish historian Flavius Josephus.

The French military presence itself was not unusually detrimental to the abbey; the battles between Napoléon's forces and the English took place elsewhere. But political ideas from north of the peninsula slowly but surely had filtered down to Portugal. By the post-Napoleonic period, the Portuguese were ready for their own revolution, a belated one thanks to the ongoing English involvement in the country's affairs that followed the departure of the French. The new liberalism that began with the revolution of 1820—which eventually drove the conservative, church-backed monarchy to its overseas domain in Brazil—did not bode well for Alcobaça.

A new monasticism was evolving from the downtrodden religious orders' instinct for survival. Monasteries appeared obviously anachronistic—certainly reactionary and possibly even parasitic—following the Enlightenment and the industrial revolution, and stringent measures were required to combat the public vitriol toward the religious targets. The religious orders throughout Portugal in the nineteenth century—indeed, throughout Europe in the wake of the liberal revolutions—began to devote themselves to a more ascetic type of life not so offensive to the populace; the Roman Catholic orders needed to reestablish their place in the world, and learned to offer a viable, apparently humble alternative to the rampant materialism, urban ills, and spiritual impoverishment associated with the industrial juggernaut.

In Portugal, however, the church had naturally backed the conservatives as the struggle for power played out between liberal and conservative elements within the royal family. The conservatives lost, and for the Cistercian monastery of Santa Maria a Velha at Alcobaça, it was too late. During the summer of 1833, the monks uncertainly continued to occupy the abbey, but finally were forced to leave Alcobaça. On October 16, 1833, the townspeople sacked the monastery for one final time.

The government intervened to salvage most of the library's manuscripts. The loss of any irreplaceable manuscripts dealt a blow to the integrity of the collection, of course, but relative to the devastation of monastic life throughout Portugal, the most valuable part of the library at Alcobaça remained fairly unscathed. The destruction of printed works, on the other hand, was much more substantial during the forays into the abbey by first the French and, later, the local revolutionaries.

The government officially dissolved the Cistercian order in Portugal in 1834. While some of the other orders managed to return to their properties, the monks of Alcobaça had been driven out for good. The government used the monastery as a military barracks in subsequent years.

Only in 1930, a century after the abbey's ignominious denouement, was the national treasure at Alcobaça restored, mostly as a museum. It is arguable how valid the various restorations have been. The desire to recapture the spirit, the presumed purity of the original Cistercian conditions in medieval Portugal, has led to the removal from the complex of many of the important embellishments from later, more prosperous centuries; the intricate rococo woodwork by William Elsden is gone, for example. The massive abbey church now presents a picture of austerity that may convey the intentions of the ascetic Bernard of Clairvaux but perhaps has less to do with the powerful abbots who readily consorted with royalty (when they were not royalty themselves).

Nonetheless, the church's essential architectural character remains intact, providing an important source for research into Cistercian architecture. Even though the Alcobaça church's height did not relate to the church known as Clairvaux III, for example, the plan faithfully followed that model; this enables an analysis of a plan that would have been lost when the French church was destroyed.

The medieval cloister is a tourist attraction, while other parts of the complex serve to house the aged; the church itself continues to serve a parish. Most of the library's magnificent manuscripts reside at the National Library in Lisbon.

Further Reading: *The Fundo Alcobaça of the Biblioteca Nacional, Lisbon* (Collegeville, Minnesota: Hill Monastic Manuscript Library, 1988–90) by Thomas L. Amos is a three-volume compendium detailing the contents of the abbey library's surviving manuscripts, which are in the National Library at Lisbon. This impressive work includes a well-balanced history of the monastery; bibliophiles will enjoy the separate history devoted to the library and scriptorium. *The Cistercians: Ideals and Reality* by Louis J. Lekai (Kent, Ohio: Kent State University Press, 1977) provides a comprehensive look by an authority on the religious order that settled Alcobaça in the context of greater Roman Catholic issues. *History of Portugal* by A. H. de Oliveira Marques (New York and London: Columbia University Press, 1972) is a two-volume standard history that presents the national context. *Excursion à Alcobaça et Batalha* by William Beckford (Lisbon: Livraria Bertrand, 1956), in English and French, is a curious travel memoir dominated by the romantic impressions of an eccentric English patron of the arts in the late eighteenth and early nineteenth centuries.

—Randall J. Van Vynckt

Altamira (Santander, Spain)

Location: On the northern coast of Spain, in the province of Santander eighteen miles west of the city of Santander and two miles from the village of Santillana del Mar, in the Cantabrian region.

Description: The caves of Altamira (Cuevas de Altamira) contain some of the finest prehistoric artwork in Europe; the area includes a number of such caves, as well as museums, historic villages, and spectacular scenery.

Site Office: Centro de Investigacion de Altamira
39330 Santillana del Mar, Santander
Spain
(942) 81 80 05

In the summer of 1879 Marcelino San de Sautuola, a historian, was exploring a cave in the region of his summer estate on the northern coast of Spain. He had found some small bones and flints in the cave floor in previous years and hoped to discover more. One day he was accompanied by his young daughter, Maria, who amused herself by exploring the dark passages while he performed his patient excavations. Suddenly he heard a joyful shout: "Look, Papa, look at the painted bulls!"

Thus was discovered one of the most remarkable and one of the most controversial finds in archaeology: the magnificent prehistoric paintings in the cave of Altamira, the first discovered in Europe.

The Cantabrian region of northern Spain stretches along the coast from the Basque country in the east to the Asturias in the west. Its capital is the coastal town of Santander, an important port since the Middle Ages. The region is dominated by the great mountains called the Cordillera Cantabrica, with altitudes between 5,000 and 7,000 feet, and which include the Picos de Europa, one of the most spectacular mountain chains in Europe. These mountains, essentially an extension of the Pyrenees, run roughly parallel to the coastline only fifteen to thirty miles from the Bay of Biscay, and are composed of Cretaceous limestones. It is a perfect setting for caves.

The Romans encountered an Iberian tribe living in the rugged mountains and named them the Cantabri. The Romans regarded them as the fiercest warriors on the Iberian Peninsula, and spent ten years trying to subdue them. By 19 B.C. the Cantabri were close to extinction.

The Cantabri were not the first people to make their home in the mountains overlooking the bay. Thirteen thousand years ago most of Europe still lay beneath the glaciers of the Ice Age, but a small strip of land between the Cordillera Cantabrica and the Bay of Biscay was clear. Here people lived as hunter-gatherers, retiring to the mouths of area caves for protection from the harsh weather and predatory animals. Here they also painted and engraved animals and other figures on the cave walls.

The main cave of Altamira, just southwest of Santillana del Mar in the hills above the town, was discovered in 1868 when a hunter in pursuit of a lost dog uncovered the mouth. Its treasures remained unknown until it was explored by Sautuola. An interest in natural history led to his fascination with the prehistoric antiquities beginning to be unearthed in the region. He carried out some modest excavations of the cave in 1875, discovering some flints and animal bones that served as evidence of human habitation. Back in Madrid he showed his findings to a prominent scholar of prehistory, Juan Vilanova, who confirmed Sautuola's suspicions and offered advice on further excavation. In 1879 Sautuola returned to Altamira with high hopes, but even he was not prepared for the find his daughter made on the cavern roof. It was a large fresco, about sixty by thirty feet, covered with animals, beautifully observed and gracefully painted. Most were animals no longer native to Spain, predominantly bison, realized in warm reddish browns accented with black.

To this point the only prehistoric artwork discovered had been small carvings. It is to Sautuola's credit that he was able to make the connection between the modest artifacts and remains so far discovered and the spectacular paintings, concluding that the paintings had been produced by the same people. But many questions remained. Returning to Madrid he again consulted Vilanova, who resolved to examine Sautuola's discoveries for himself.

Vilanova's first task upon arriving at Altamira was to prove that the animals pictured had indeed lived at the time the artifacts were created; to do this he began a methodical excavation of the cave in order to classify the fossilized skeletons Sautuola had reported. Not only did he identify the remains as belonging to all the animals pictured, he discovered that some of the bones themselves had been engraved with pictures of the same animals. The period of the fossils, and therefore of the paintings, was positively identified as prehistoric.

In 1880 Sautuola published his findings. The discovery caused an uproar among prehistorians. Today the concept of prehistoric cave paintings is taken for granted, and it is difficult to understand how impossible it seemed to nineteenth-century archaeologists. They did not believe that Stone Age people were intelligent enough, cultured enough, or technologically capable of producing such work. Sautuola's work was met with skepticism, suspicion, and accusations of outright fraud. Those scholars most renowned in the field were his most vocal opponents, and in spite of Vilanova's support, many believed the paintings were forgeries. Unanswered questions, such as how the painters could have pro-

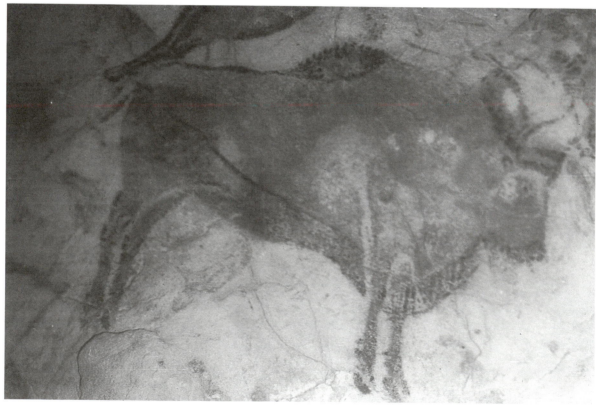

Cave paintings in Altamira
Photos courtesy of Tourist Office of Spain, Chicago

duced light without leaving traces of smoke on the ceiling, buttressed opposing arguments.

Most adamantly opposed was Emile Cartailhac, president of the Prehistoric Society of France. In 1881 he sent a representative who visited that cave in Altamira and pronounced the paintings of recent origin. Not until 1887 when Edouard Piette, a French authority, pronounced them prehistoric did the scholarly world begin to take Sautuola seriously. It was the discovery of other painted caves, both in the Santander region and, more importantly, in France, that finally convinced the French experts. In 1902 Cartailhac published a famous article, "'Mea culpa' d'un sceptique," in the journal L'Anthropologie, in which he freely admitted his error in judgment. The vindication came to late, however; Sautuola had died in 1888, and Vilanova in 1893.

During the early twentieth century, excavation and research continued at Altamira. Ironically, the project was headed by Cartailhac himself. Henri Breuil, an artist working as part of the team, published copies of the cave paintings in 1906. In 1908 the first exposition of European prehistoric art was held, generating much publicity for Altamira and its treasures. After interest was expressed by King Alfonso XIII, the Duke of Alba led an effort that in 1925 resulted in the construction of a road, the installation of electric lighting, and the grading of the path inside the cave. Research came to a halt in the 1930s as a result of the Spanish Civil War, a catastrophe from which Spanish archaeology did not fully recover for twenty years.

Meanwhile, the publicity and controversy surrounding the discoveries at Altamira led to the exploration of the many other caves in the area, with the result that dozens were found to contain prehistoric engravings and paintings, including a few of the same caliber as Altamira. Since the resumption of exploration in the 1970s, new examples of cave art have been discovered throughout the northern Spanish provinces. Along with similar finds in southwestern France, the province of Cantabria offers a virtual laboratory for the study of prehistoric art.

The main cave at Altamira is 886 feet long, shaped like a backwards Z with a broken and drooping tail. The entrance, or vestibule, is the base of this tail. Some time long ago the vestibule roof collapsed, obstructing the view of this great chamber. This was where Sautuola made his initial excavations, and where early people sought shelter from the Ice Age weather. At the rear of the vestibule is the opening to a side chamber known as the Great Hall of Paintings, the location of the ceiling fresco of bison. The scholar M. J. Dechelette termed this chamber the Sistine Chapel of Quaternary [Prehistoric] Art. Although most of the animals depicted on the ceiling are bison in various poses, the largest animal shown is a red deer, seven and one-third feet long. The ingenious artist made use of the natural curves and projections of the rock surface to give some of the animals a three-dimensional effect. A recent stylistic analysis has brought some experts to the conclusion that a single artist produced the entire ceiling, although other scholars believe

the ceiling displays a series of five layers of paintings produced over time.

For pigments, the prehistoric artist used naturally occurring mineral sources. For example, iron oxide produces ochres of various hues, from yellow to red. For black the source was manganese dioxide. White pigments, although rarer, were also used at Altamira, concocted from a mixture of mica and illite. Containers of pigment have been found in many caves, including Altamira. The most likely binding agent for these pigments was water. They were probably applied with animal hair brushes, similar to those used by artists today.

Returning to the main corridor of the cave and continuing back through its passages, one can see engravings, paintings, and carvings in the soft surface of the rock. Most of the figures represented are deer and bison, primary food sources for the cave's inhabitants. Although human figures are rare, they are not entirely absent. In all, approximately 150 painted figures line the cave walls.

A second cave at Altamira was discovered during construction of a roadway in 1928. Explorers found a human skeleton, old but not prehistoric, and a wealth of the natural formations common to limestone caverns, but no paintings.

The findings at Altamira are typical of Cantabrian cave art in that most of the paintings represent animals, especially deer, bison, and horses. Among the animals depicted, there is a roughly equal number of bison and red deer, but the artists seem to have lavished more effort on the bison than on the deer. The latter tend to be shown in smaller engravings (with the exception of the large deer on the bison ceiling). Some scholars speculate that the art in the caves was at least partially associated with hunting rituals. Red deer were a predominant food source, but bison, being so much larger and providing so much food per animal, would have been a highly sought-after target. As such they may have merited special attention from the artists.

In addition to this representational art in the cave, there is also strictly symbolic art that distinguishes Cantabrian from French cave art and whose meaning is not understood. It is extremely difficult for twentieth-century humans to grasp the purpose and meaning of art produced 10,000 to 15,000 years ago. Since the discovery of the paintings at Altamira, scholars have continuously debated this question, offering a score of explanations from art for art's sake, to hunting magic, to shamanism, to protowriting. For now, the intent behind the art remains a mystery.

Although perfectly preserved for thousands of years, the paintings of Altamira have been seriously damaged by the masses of viewers who have visited over the last century. The Spanish authorities closed the cave completely from 1977 to 1982 and were relieved to see some improvement in the internal conditions. Today written permission must be obtained from the Centro de Investigacion de Altamira for viewing the cave; a maximum of ten people per day are permitted inside. Fortunately, a wealth of caves in the area offer alternatives for the interested visitor, and there is also a museum with exhibits on prehistoric Cantabria. Twenty miles south of Santander are the caves of Puente Viesgo, including the Cueva de Castillo. It is a large cave with many paintings and carvings, including one of the two paintings of elephants known in Spain.

Santillana del Mar is the village closest to Altamira, and a national monument in its own right. The architecture of the village dates back to the ninth century, and examples of structures from the thirteenth through the eighteenth centuries are plentiful. Many houses bear heraldic shields proclaiming the ancestry of former owners. A twelfth-century church, the Colegiata de Santa Juliana, was built in the Romanesque style and has an interesting cloister. There is also a museum, the Museo Diocesano, with a display of religious art and artifacts, located in the Monasterio Regina Coeli.

Not far from Altamira is the provincial capital, Santander, a medieval port city located on an inlet of the Bay of Biscay. A tremendous fire swept through the city in 1941, virtually destroying the entire community. Santander rebuilt itself and is today a popular destination for Spanish tourists, as well as one of Spain's principal seaports. The Universidad Internacional Menendez Pelayo is also located here. The Museo de Prehistoria y Arqueologia includes artifacts and photographs of the paintings from Altamira, as well as other sites throughout the province. The Museo Maritimo includes an aquarium, and the Museo de Bella Artes includes a portrait of Fernando VII by Goya.

Although much of the cathedral was damaged by the fire, the structure below ground remains, including a chapel with unusual Romanesque vaulting. For those who prefer modern architecture, the twentieth-century neo-Gothic *palacio* built for Alfonso XIII is now a part of the university.

Further Reading: *The Cave of Altamira* by J. Carballo, translated by Madelein D. Lorch (Santander: Patronato de las Cuevas Prehistoricas, 1963) is outdated but still a useful introduction by the man who founded the Museo de Prehistorica y Arqueologia. For a more scholarly perspective, *Iberia Before the Iberians: The Stone Age Prehistory of Cantabrian Spain* by Lawrence Guy Straus (Albuquerque: University of New Mexico Press, 1992) examines the results of recent research in the Cantabrian caves, including the geological, biological, and chronological context of the people who painted them. A spectacular photograph of the Altamira ceiling appears in Paul G. Bahn's and Jean Vertut's *Images of the Ice Age* (New York: Facts on File, 1988), which also contains Vertut's photographs of other prehistoric art from around the world. Bahn supplies a useful interpretive text. Noel W. Smith examines the message and the meaning of the artwork in *An Analysis of Ice Age Art: Its Psychology and Belief System,* American University Studies, series 20, volume 15 (New York: Peter Lang, 1992).

—Elizabeth Brice

Ampurias (Gerona, Spain)

Location: On the coast of the Gulf of Roses in Catalonia, twenty-two miles northwest of Gerona.

Description: Also known as Empúries; archaeological site (with museum), displaying the remains of a harbor town founded by Greeks and conquered by the Romans and Visigoths; together with two nearby resort towns named for it.

Contact: Tourist Information Centre
Placa de les Escoles 1
L'Escala, Gerona
Spain
(972) 77 06 03

Ampurias is the official name, in standard Castilian Spanish, for the place known as Empúries in Catalan, the language of the modern inhabitants of the region in which it stands. Both names derive from the ancient Greek "Emporion" and the Latin "Emporiae" or "Emporium," which means a place of trade. This designation is found also in the names of two nearby towns—San Martín de Ampurias and Castelló de Ampurias—as well as Ampurdàn, the plain stretching inland from the Costa Brava, the "wild coast" of the Catalan province of Gerona.

The ancient town uncovered by archaeologists in the twentieth century appears to have been founded by ancient Greeks around the middle of the sixth century B.C., along with the town of Roses nearby. It has been suggested, from the ambiguous references to the area in the writings of the Greek geographer Strabo, that the native Iberians had already settled there by the time the Greeks arrived; this settlement may have been called Indica. The Greek colonists may have come from what is now Marseilles, then the Greek colony of Massalia, which itself had been founded only a short time before by settlers from Phocaea in Asia Minor. It is not clear why they established this new outpost, since the area is removed from the routes they used for transporting tin. They may instead have been involved in developing trade in other metals or in agricultural products with the Iberian peoples inland.

For whatever reasons, the settlement seems to have prospered. Most scholars agree that the first location the Greeks chose was an island near the point where the Río Fluviá then met the Mediterranean, probably known to them as Kypsela, and that this may well have proved too small to accommodate the growing population within fifty years of their arrival. Some of the Greeks—perhaps even all of them—moved to a new settlement, Emporion, on the mainland, across the harbor to the south of Kypsela, near where they had already established their cemetery. Archaeologists now refer to Kypsela and the original Emporion as, respectively, Palaeopolis (Old City) and Neapolis (New City).

Emporion, which eventually grew to encompass both Kypsela and whatever remained of Indica, may have had around 1,300 people living in its six-acre walled center and several hundred more living around it. Archaeological excavations of its remains have so far uncovered a section of its harbor, its town gate, and parts of the city wall, perhaps from the third century B.C. This was built of stones without mortar and is of the type known as "Cyclopean," in reference to the Greek legend that such walls had been built by the Cyclopes, the mythical one-eyed giants. Inside the ruins of these walls can be seen the remains of a temple dedicated to Asclepius, the god of medicine, among which can now be seen a copy of the original statue of the god; another temple, which was probably dedicated to Hygeia, his wife; and, beyond the square on which both stood, a temple of Zeus Serapis, a deity who combined features of Zeus, the king of the Olympian gods, and of Amon-Ra, his Egyptian counterpart. From here a colonnaded road leads to the agora, the public square in which markets and meetings were held, characteristic of ancient Greek cities. Artifacts found at the site include mosaic designs from private houses, as well as fragments of pots and vases from Corinth and Athens. These findings may indicate the mixed origins of the settlers or may be evidence of extensive trade.

While Greek traders were settling on the Mediterranean coast, at Emporion and elsewhere, colonies were being established inland in southern Spain, from the sixth century B.C. onward, by settlers from Carthage, the Phoenician city on the north coast of Africa. Over the course of about 300 years, the Carthaginians came to dominate trade between the Iberian Peninsula and other countries on the shores of the Mediterranean, and it is likely that Emporion was already economically dependent on them by the time of the First Punic War (264–241 B.C.) between Carthage and the rising power of Rome. Between 239 and 228 B.C., the Carthaginians extended their military control of the peninsula into the north, conquering Emporion but probably leaving it in the hands of its Greek inhabitants. Only two years after this conquest was completed, however, the Romans, now turning their attention to the western Mediterranean, came to an agreement with the Carthaginians, who relinquished to them the province of Septimania, in which Emporion stood.

The agreement did not last long. In 218 B.C., at the beginning of the Second Punic War, which lasted until 201 B.C., Emporium, as the Romans called it, was the place where the Roman general Publius Scipio Africanus landed troops in an attempt to force the Carthaginian general Hannibal to fight in Spain rather than Italy. In 210 Hannibal passed through Septimania on his way to invade Italy, and some accounts suggest that the arrival of the Romans took place in this year. By 201 Rome had control of much of the peninsula, having

Roman ruins at Ampurias
Photo courtesy of Instituto Nacional de Promocion del Turismo

pushed the Carthaginians back to the islands of Ibiza and Minorca. Since excavations have revealed that the Greek inhabitants of Emporion had a giant catapult inside the town, it is possible that they resisted these invasions.

By 195 B.C., however, the Roman conquest of Septimania was complete, and in that year Emporium became the base for military operations led by General Cato the Elder against those inland groups of native Iberians who still resisted. The town now expanded inland, to the west, especially after veterans of Julius Caesar's military campaigns were given homes and land around Emporium in the middle of the first century B.C. While Greeks may have continued living and trading around the harbor, the Romans built a new center on higher ground. Excavations here, which still continue, have uncovered a forum (the Roman equivalent of an agora) and a main avenue passing through it, the main gate (its threshold marked by the wheels of chariots and carts), and parts of the Roman city wall, made of stone and concrete and

as high as sixteen feet, around the forty-eight-acre site. Within this area there can also be seen the remains of the town's amphitheatre, which was approximately half the size of the Colosseum in Rome; a temple for the worship of Augustus and of Rome itself; and a basilica, a building housing law courts and a market. But the outstanding remains of the Roman city are two houses near the gate, probably built shortly before the end of the Roman republic and the accession of the first emperor, Augustus. One of them contains mosaics of an actor's mask and a partridge, and a portrait of a woman who may have been its owner; whoever they were, its residents were influential enough to arrange for a breach in the city wall to give the house a view over the Mediterranean. The other house, built in a style deeply influenced by Greek architecture, contains an altar decorated with paintings of a cock and serpents and a mosaic design of black and white stars. Both have been restored in recent decades; cypresses, laurels, and potted plants have been placed in their gardens.

The Roman city of Emporium, which may have had up to 10,000 inhabitants inside its walls and more along the coast, declined in status after Augustus established Tarraco, now Tarragona, early in the first century A.D. It was attacked by barbarians as early as A.D. 300, and it may have been abandoned around that date or soon afterward, leaving a much smaller group to establish what is now San Martín de Ampurias, between the site of Kypsela (Paleopolis) and the silted remnants of the harbor. From around 410, further invasions of Spain by Vandals, Cimbri, Teutons, and Ambrones probably made life extremely difficult for these people, but the district was given some stability once the Visigoths, unlike the other invaders, established permanent settlements in the fifth century, after having swept down from Toulouse and established their headquarters successively at Narbonne, Barcelona, and, in 510, Toledo. A basilica, this time meaning a church, was built near the former agora in the lower town of Neapolis, probably during the fifth century. Excavations of tombs inside the remains of this building suggest that those buried here were Christian Visigoths, not Greeks or Romans. This basilica became the seat of a Christian bishop soon after the Visigoths accepted Christianity.

After all these invasions and changes of location, Ampurias was probably reduced to a small settlement on the edge of the area of Greek and Roman remains. In 717 the Moslem Arab and Berber armies, which had entered Spain six years before, reached what is now Catalonia and destroyed Ampurias, probably as punishment for its inhabitants' resistance to this latest invasion. Having crossed the Pyrenees into France in 720, the Moslems were defeated at Poitiers in 732 and withdrew from most of northeastern Spain. By 785, when the Franks seized the province of Gerona, Ampurias (or, in Catalan, Empúries) had become the name of a county within it, centered on the town of Castelló de Ampurias, on the Muga River to the north of the ancient city. Gerona and the provinces around it formed the *limes Hispanicus* or "Spanish march," the embattled edge of the empire of Charlemagne that was to become the nucleus of modern Catalonia, as settlers were encouraged to return from 801 onward, when the Franks seized Barcelona from the Moors. As for the town of Ampurias itself, its end is almost as mysterious as its beginning. Although it was revived after the Moslem invasions, references to it cease during the ninth century, when it appears to have been decisively ruined and finally abandoned. It is likely that it fell victim to a series of raids by the pirates of Norman and Moslem origins who harassed much of the western Mediterranean at that time from their bases on what Europeans called the Barbary Coast, the Maghreb region of northwestern Africa.

For more than 1,000 years afterward, Ampurias survived only as a group of ruins, partially hidden under the soil; as the name of a small and isolated Servite monastery established in the Middle Ages between the Greek and Roman sections of the Neapolis site; and as an element in two other place names and two titles, one secular and the other religious. While San Martín de Ampurias continued as a small

fishing village near the monastery, protected by a wall that still stands today, Castelló de Ampurias became the residence of the bishops and the counts of Ampurias, many of whom played prominent parts in the politics of medieval Catalonia. In 878 Wilfred Guifré, count of Cerdanya-Urgell, became ruler of Barcelona. As the Franks, divided during the conflicts among Charlemagne's heirs, lost control of the region, the dynasty Count Wilfred founded gradually extended its domain, retaining the Frankish feudal system unknown in the rest of what is now Spain. In 1054 the count of Ampurias submitted to Raymond Bérenger I, count of Barcelona from 1035 to 1076, although the nominal rule of the kings of France over the whole region continued until Louis IX gave up his claims to Catalonia in 1258. During this formative period of Catalan culture, Castelló de Ampurias acquired such notable monuments as the Church of Santa María, built in the Gothic style during the thirteenth and fourteenth centuries and incorporating an ancient Romanesque tower; its town hall, also in the Gothic style; the Palace of the Counts; the Church of Santa Magdalena; the Monastery of La Merced; and a bridge built in 1354, all of which can still be seen today.

Modern Spain was created between 1479 and 1492 through the marriage of Ferdinand II of Aragon, whose realms included Catalonia, and Isabella of Castile-León, and their conquest of the last Moslem stronghold at Granada. In relation to the state they founded, Catalonia became a neglected and suspect backwater; both San Martín de Ampurias and Castelló de Ampurias declined in status and population. During the seventeenth century, the site of Ampurias itself was used as a source of building materials for L'Escala, a fishing village specializing in anchovies that stands on a promontory one mile to the south of the site and is now a holiday resort.

The Greek and Roman remains of Ampurias were first excavated by the archaeologist Emilio Gandia y Ortega in 1907. Many of the more important artifacts uncovered by him and his successors are displayed in museums of Gerona and Barcelona, perhaps the most outstanding of which are two Greek statues, one of Aphrodite, the goddess of love (although only her torso remains) and the other of Asclepius, from the temple dedicated to him. The Museo Monográfico, housed inside the former church of the Servite monastery, contains a number of mosaics, such as a depiction of the sacrifice of Iphigenia by her father King Agamemnon before the Trojan War, as well as reproductions of some of the artifacts displayed in Barcelona.

In recent years, some twenty-five centuries after the Greek port of Emporion was founded, the coast on which it stands, then a sparsely populated region remote from the centers of Mediterranean civilization, has become the busy and prosperous Costa Brava, popular with vacationers from the rest of Spain itself and from northern Europe. The Greek and Roman ruins of Ampurias form a historic "island" within the region, a quiet and mysterious space separate from the resorts that stand near it and carry its name. This special memorial status was recognized symbolically in 1992, when the Olympic flame, carried from Greece on its way to the Olympic Games at Barcelona, was brought to land at the site. Yet there was also some unintended irony in the gesture: the town must have been too small and far away ever to have had any role in the ancient games at Olympia, and it is unlikely that Roman legionaries or Visigothic warriors would have appreciated the Olympic ideals, ancient or modern. Like so many other archaeological sites, Ampurias can thus be used to represent a kind of continuity—in this case between the classical Mediterranean cultures and modern Europe—that can obscure the enormous differences between the present and the distant past, inhabited as it was by a succession of peoples whose names and lives are mostly unknown and whose religious beliefs and social attitudes are extremely difficult to reconstruct.

Further Reading: *Atlas of Classical Archaeology,* edited by M. I. Finley (London: Chatto and Windus, and New York: McGraw-Hill, 1977), contains an admirably clear and detailed description of the ruins. Among the many travel guides to the area, *Blue Guide: Spain* by Ian Robertson (London: Black, and New York: Norton, 1989) contains a useful section on the site.

—Patrick Heenan

Anazarbus (Adana, Turkey)

Location: The ancient city of Anazarbus is located atop a limestone escarpment above the modern village of Anavarza, forty-four miles northeast of the Turkish city of Adana, in the province of Adana.

Description: Since its founding in the early first century B.C., Anazarbus has been a fortified holding of the Romans, Byzantines, Arabs, and Armenians. After the Armenian Kingdom of Cilician Armenia was defeated in the late fourteenth century, the site was abandoned. The site today contains extensive ruins from all periods of its history.

Contact: Tourist Information Office
Atatürk Cad., No. 13
Adana, Adana
Turkey
(322) 3591994 or 3524886

The village of Anavarza lies approximately forty-four miles northeast of Adana on the plains of Cilicia Campestris, or Smooth Cilicia. Although the town is of little importance today, the nearby site of Anazarbus was once a strong and powerful fortress dating to pre-Christian times.

The date of Anazarbus's founding is difficult to ascertain. Some historians and archaeologists date the site to the seventh century B.C. to the former Kundu, a city that rebelled against the powerful ruling Assyrians. Little evidence can be found to uphold that theory, however, and experts generally date Anazarbus only to the first century B.C., citing as evidence a number of coins found in the vicinity.

Even the origin of the site's name is shrouded in uncertainty. The modern Turkish name for the area is Anavarza, but variants include Anazarbe, Anazarba, Anarzab, and Ayn Zarba. Archaeologist Michael Gough has hypothesized that the name comes from the Persian word *nabarza,* or unconquered. This term was commonly associated with Mithras, the Indo-Persian god of light and creation. The cult of Mithras, made up entirely of males, was popular among soldiers who spread the sect's precepts throughout the Roman Empire. Similarities in doctrine between the cult of Mithras and Christianity made the cult a powerful rival for several centuries. The name Anazarbus was probably given first to the steeply sloping 2.7-mile limestone escarpment that dominates the site, and then eventually encompassed the settlement at its foot as well.

In the first century B.C. Anazarbus was probably part of the empire of Tarcondimotus, who at that time ruled a large portion of the Cilician plains of Asia Minor from his capital at the ancient city of Hieropolis Castabala. When King Tarcondimotus was killed in battle against the Romans in 31 B.C., the Roman Octavian (later Emperor Augustus) proclaimed Tarcondimotus II ruler of the principality. Grateful for Augustus's allowing him to rule, Tarcondimotus II renamed Anazarbus "Caesarea" or "Caesarea by Anazarbus" as a tribute to the emperor. After Tarcondimotus's dynasty died out in the early first century A.D., Anazarbus came under direct Roman rule.

Anazarbus prospered under the Romans. The town grew, and by the early third century it was considered equal to the city of Tarsus. A coin commemorating Anazarbus was issued in A.D. 221 when the notorious Emperor Elagabalus, known for his involvement in sexual scandals, was one of the city's rulers.

What once was a prosperous Roman town today lies in ruins. But among the broken walls and rubble can be found archaeological treasures from the once-powerful Anazarbus. At the base of the escarpment lie the city walls, once protecting the fortress thought to be unconquerable. High above the city on the side of a large hill is a cave containing an inscription dating to A.D. 153 by the priestess Regina, invoking the blessing of the gods Zeus, Hera Gamelia, and Ares.

Below the cave and about 500 meters north lies an ancient tomb dating to the first century A.D., commonly called the Tomb of the Furies. Inside the tomb, according to an inscription over it, lie two eunuchs, servants of Queen Julia the Younger, who herself died in A.D. 17. Near the Tomb of the Furies on the north side of the town runs the main aqueduct believed to have been built during the reign of Domitian in the first century A.D. This aqueduct was the primary water supply for Anazarbus, bringing fresh water from the upper reaches of the Sumbas River. Inside the city's walls just south of the aqueduct lies a baths complex built during Byzantine rule.

Just inside the walls of Anazarbus stand the impressive remains of a Roman triumphal arch. Thought to date to the reign of Septimius Severus in the late second and early third centuries A.D., the original arch measured about thirty yards by ten yards and consisted of a major central arch supported by smaller side arches. Elaborately constructed and decorated with six Corinthian columns and pilasters, the original design included a concrete structure faced with fine ashlar and tufa.

Some of the most impressive remnants of Roman rule are the ruins of the stadium. Lying outside the city walls, the stadium at Anazarbus was massive, measuring more than 400 yards long and more than 70 yards wide. Spectators at the stadium sat in seats hewn out of the rock face and were most likely protected from the elements by a canopy supported by beams placed directly in the rock; visitors to Anazarbus may still see holes for the canopy supports cut into the rock above the top row of seats. The inscription above the seats has been dated to the third or fourth century A.D.

City gate at Anazarbus
Photo courtesy of Embassy of Turkey, Washington, D.C.

Near the stadium are a number of sarcophagi and rock-tombs, as well as traces of an ancient Roman road connecting Anazarbus with other important cities such as Flaviopolis and Hieropolis Castabala. The road was elevated onto a causeway in some spots, probably to prevent seasonal flooding. A Roman bridge spanned a river just outside of the escarpment. Ruins just off of the road suggest that Anazarbus also possessed an amphitheatre backed by the rock face. An underground passageway cut into the rock can still be seen, although it is filled with rubble.

A rock cleft dividing Anazarbus, thought to have been created in late Roman or early Byzantine times, holds a number of sarcophagi and tombs, as well as an inscription from the Forty-Sixth Psalm, "God is our refuge and stronghold." The words of the psalmist are believed to be in reference to earthquakes. In later years, Anazarbus suffered greatly from recurring earthquakes. The psalm itself continues, urging people not to be afraid "though earth should tumble about us, and the hills be carried away into the depths of the sea." Aside from indicating that Anazarbus was an early victim to natural disasters, the inscription also indicates that Christianity had taken root in the city by the third or fourth century A.D.

Just beyond the cleft lie the fragmented remains of a theatre. Unfortunately, the destruction of the theatre is great enough to prevent accurate dating or estimates as to its size and importance. Nearby lie more sarcophagi and two reliefs cut into the rock face. The first relief details several athletes doing gymnastics while their instructor stands holding a wreath and palm frond. The other relief shows a winged deity on a dolphin's back; a funerary inscription is written above the deity's head.

Aside from the remaining ruins, little is known about Anazarbus during the late Roman and Byzantine eras. It apparently was an important city in the early fifth century, during the reign of Theodosius II. Early in the sixth century, Anazarbus became known as Justinopolis. Then, in 526, after a particularly strong earthquake destroyed the town, it was rebuilt and renamed Justinianopolis, a tribute to the Emperor Justinian I, who assisted with the city's reconstruction.

Ruins from this era include the large baths complex located near the first-century aqueduct and a nearby church. This church, built during the early Byzantine era, was a basilica dedicated to the Apostles. Another church, Kaya Kilises or Church on the Rocks, was cut into the rock face surrounding Anazarbus and decorated with Byzantine mosaics. The defensive wall that still stands outside the city—albeit in a crumbling state—was built and fortified during the Byzantine period.

Anazarbus's location on the outskirts of the Byzantine Empire made it difficult to protect. When skirmishes between Byzantine and Arab forces began, Anazarbus was particularly vulnerable, and it fell under Arab control in the eighth century. The new overlords once again renamed the city, this time calling it Ayn Zarba. The city's defenses were strengthened by Hārūn ar-Rashīd in 796, and the fortifications can still be seen on the escarpment above the lower walled city.

The reinforced area consists of three sections. The first division is the first enceinte, which was constructed under ar-Rashīd and housed the town's garrison. When the city was under siege, citizens from the lower city sought refuge in the protective walls of this enceinte. The second enceinte and a three-story tower were added under the Armenians.

Arabic rule of Anazarbus was dissolved in the eleventh century with the establishment of Cilician Armenia. Beginning in 1080, Armenian control of the Cilician plains lasted for several centuries. Again, Anazarbus prospered, reaching its zenith in 1100 when Toros I made the town capital of his empire.

During Armenian rule, changes were again made at Anazarbus. In addition to expanding the fortifications, the Armenians built a small church within the first enceinte. The Church of the Armenian Kings is in poor condition, but portions of an inscription along the top of the church's walls can still be read. The inscription mentions prayers for the health of one's children and invokes the memory of one's parents. Farther along, a memorial mentioning Rupen III is dated 1188. When the church was visited by traveler Victor Langlois and recorded in his memoirs in the mid-nineteenth century, remarkable frescoes still decorated the inside of the church. Since then, however, natural forces and vandalism have ruined a large number of these frescoes.

Anazarbus reigned as the first city of the Armenian Kingdom for nearly 100 years, after which imperial control and power shifted first to Tarsus, then to Sis. No longer the seat of government, the city of Anazarbus began to deteriorate. Cilician Armenia itself fell to the Mamlūks of Egypt in 1375. With the Armenian kingdom gone, the centuries-old city of Anazarbus was abandoned and never again inhabited. The ruins of Anazarbus still stand near the village of Anavarza, a tribute to the great powers—Roman, Byzantine, Arab, and Armenian—that once controlled the area.

Further Reading: Charles Burney and David Marshall Lang's *The Peoples of the Hills* (New York: Praeger, and London: Weidenfeld and Nicolson, 1971) offers a detailed look at the ancient history surrounding Ararat and the Caucasus and includes a look at the Armenian Kingdom that once encompassed Anazarbus. Readers wishing to learn more about Armenian Cilicia will find Avedis K. Sanjian's *Colophons of Armenian Manuscripts, 1301–1480* (Cambridge, Massachusetts: Harvard University Press, 1969) an invaluable resource. *Blue Guide: Turkey, The Aegean and Mediterranean Coasts* by Bernard McDonagh (London: Black, and New York: Norton, 1989) includes a comprehensive history of Anazarbus, as does John Freely's *The Companion Guide to Turkey* (London: Collins, 1979; Englewood Cliffs, New Jersey: 1984).

—Monica Cable

Andorra

Location: In the eastern Pyrenees Mountains on the border between France and Spain, Andorra is located midway between Barcelona and Toulouse, in the Catalan region.

Description: The Principality of Andorra is an autonomous country. With a total surface area of just over 180 square miles, Andorra ranks among the larger of Europe's mini-states (Andorra, Vatican City, San Marino, Liechtenstein, Luxembourg, Monaco, and Gibraltar), with only Luxembourg, at 998 square miles, taking up more space. Andorra has been essentially self-governing since the turn of the ninth century, with the bishop of Urgell and president of France serving as co-princes but usually refraining from exercising direct control over the small state. A written constitution ratified in May 1993 has made these roles almost entirely honorary, and in July of the same year Andorra became the 184th member of the United Nations, finally independent after nearly 1,200 years of pseudo, if de facto, sovereignty.

Site Office: Syndicat d'Initiativa, Oficina de Turisme
Carrer Dr. Villanova
Andorra la Vella
Andorra
(33) 20214

The mountainous co-principality of Andorra has maintained its territory and identity since its founding around the ninth century A.D. by virtue of inaccessibility, indifference to the outside world, and, of course, the unwavering stewardship and protection of at least one, if not always both, of the powerful neighbors on whose frontier it sits. Situated between France and Spain, 8,000 feet above sea level, the tiny alpine state has not only survived, but flourished and even profited by the revolutions, upheavals, and struggles for power that characterize the regional history of Western Europe.

Yet despite their integral location between two of the world's premiere powers, Andorrans have over the centuries consistently proven themselves to be content to enjoy their ringside seat on the course of history and leave its making to others. Andorrans have never had their own military because they have never chosen to go to war. There are no universities in Andorra. Although both France and Spain built paved roads to the northern and southern borders of the co-principality in the 1920s, the Andorrans did not get around to paving their portions until decades later—they had no desire to leave, nor did they see much point in making it any easier for outsiders to get in. The calculated advent of tourism in the 1950s changed that attitude, but only a little, and very, very slowly.

As of the mid-1990s Andorra still lacked a railroad station or an airport.

Andorrans have taken advantage of history even as they have avoided it. Historically neutral, Andorra has maintained its open borders with France and Spain even when the rest of the border has been closed. It has always maintained a defense agreement with at least one of the two countries. Through shrewd negotiations with its overwhelmingly powerful neighbors, Andorra became a tax haven and duty-free district long before the terms were even coined. Andorran-run Catalan-speaking schools are relatively new (in 1982, Catalan was introduced as the official language of instruction in Andorra); most schools in the principality have been administered by either the Spanish, the French, or the Catholic Church. While the allegation that Andorrans have historically played one patron against the other in order to get their way is probably true, the hardy mountain folk have always succeeded and France and Spain have always succumbed. Usually this arrangement has been to the benefit of all involved, but since Andorra has always needed its patrons more than they have needed it, it is Andorra that has reaped the lion's share of that benefit.

Local legend, which remains unsubstantiated, says that Andorra was created around the beginning of the ninth century (suggested dates include 784, 801, 803 and 806) by a charter of the Emperor Charlemagne as a gift to the people of the valleys of Andorra who helped guide his armies through the treacherous topography during his successful effort to push the invading Arabs further in the direction of Africa, from whence they had come in the century before. This violent tide was neither the first nor the last to sweep through the Pyrenees; the Celts, Vandals, Visigoths, and Romans had all preceded the Arabs, and such events as the Spanish Inquisition and World War II were still to come.

In many ways, Andorra was already an entity unto itself long before Charlemagne formalized its borders. Part of the "grey area" between France and Spain, Andorra was inhabited at the time by somewhere between a few hundred and a few thousand ethnic Catalans. Catalans, like their neighbors to the west, the Basques, were neither Spanish nor French, spoke their own language, maintained their own traditions, and had carved their own civilization out of the interior of the Pyrenees. Fairly small as mountain ranges go, the Pyrenees still provided a significant buffer between the Catalans and the larger, potentially overwhelming civilizations to the north and south. To a large extent this division remains true today, with a strong Basque separatist movement active in both France and Spain. The Catalan population, by comparison, gained its own territory, however small, when Andorra was created in 806.

St. Serni, one of Andorra's historic churches
Photo courtesy of Syndicat d'Initiativa, Oficina de Turisme, Andorra

The co-principality of Andorra is a collection of valleys surrounded by peaks reaching 10,000 feet in height. Only about 4 percent of the land is arable, and the people tend to congregate in the low, flat valleys between the peaks, leaving much of the country untouched. Four major rivers traverse the country: the North Valira, the East Valira, the Valira, and the Madriu. These rivers converge at the village of Escaldes (named for its thermal springs), where they become the Great Valira, which then runs through the capital city of Andorra la Vella.

Sometime around 843, it has been suggested, Charlemagne's grandson Charles II granted authority over Andorra to the counts of the Spanish territory of Urgell. In 1007, Ermengol I, then count of Urgell, granted half of Andorra to the monastery of St. Cerni, making the bishops of the see of Urgell the territory's primary rulers. In 1133, Ermengol IV gave the remainder of Andorra's territory to the bishops as well, putting the Catholic Church firmly in control of the tiny city-state.

But all was not well in Urgell. Various neighboring territories had made threatening advances toward the Catholic See, forcing its officials to seek protection from the Caboet family, with whom the bishops jointly ruled the territory of Andorra. In 1159 a treaty was signed whereby Andorra would remain under the jurisdiction of the bishop of Urgell, but would be granted in fief to the House of Caboet, thereby ensuring its proper administration and, it was hoped, long-term security. The bishops of Urgell have always played a role in the administration of Andorra, and still do to this day.

In 1185, Arnalda de Caboet, heiress to the Caboet estates, married Viscount Arnau de Castellbo, bringing with her the fiefdom of Andorra. Forces in the see of Urgell resisted this transfer of power, but in 1186 they acquiesced, granting Castellbo the valleys of St. Joan, Caboet, and Andorra on a perpetual lease at a fixed rent, thereby affirming the shift in Andorra's secular administration.

In 1208 this relationship evolved yet again when Ermessenda de Castellbo, the daughter of Arnalda de Caboet and Viscount Castellbo, married Count Roger Bernard II of Foix, thus transferring the fiefdom of Andorra to the House of Foix, based on the French side of the Pyrenees. For the next seventy years the nature of Andorra's administration remained in dispute. In 1226 the people of Andorra swore their allegiance to the bishops of Urgell. By 1239 the situation had deteriorated to such a degree that the count of Urgell had to come to the aid of the bishop, under attack by Roger Bernard II. In 1246 a truce was signed between the see and Roger IV of Foix, but the dispute continued, forcing Pope Innocent IV in 1251 to send papal emissaries to the territory to act as mediators. In 1258 Roger IV made a formal appeal to Pope Alexander IV regarding the less-than-diplomatic conduct of the see.

In 1269 the situation grew even more complicated when Roger Bernard III signed a treaty of alliance with Prince Peter of Aragon, turning the Andorran dispute into a full-fledged medieval morass. Later that same year Roger Bernard III was granted absolute rights over the Viscounty of Castellbo by James I of Aragon after Arnau de Castellbo and his daughter were accused of heresy. Not that this meant much, for in 1272 James I took formal exception to the behavior of the count of Foix. By 1277, Peter III, then king of Aragon, requested a cessation of hostilities between Foix and the see of Urgell.

In 1278 the agreement was finally reached that would peacefully govern Andorra until 1993. After 400 years of bucolic tranquility and seventy years of aristocratic bickering, Roger Bernard III and the bishop of Urgell signed the first *pareatge* outlining the rights and responsibilities of the co-rulers of Andorra, finally making the tiny state the co-principality it has been ever since. In 1288 a new *pareatge* was signed to clear up remaining differences between the co-princes, but the agreement of 1278 is today regarded as the founding date of the modern Principality of Andorra.

The *pareatge* gave equal rights over the administration of Andorra to the bishop of Urgell and the count of Foix. Initially, the residents of the principality had no official say in the government of their territory, which, given the tenor of the times, was not unusual. They did, however, have to pay an annual tribute to each of the princes consisting of 460 pesetas, 6 hams, 6 cheeses, and 12 hens to the bishop and 960 francs to the count. The numbers were derived in part from the fact that Andorra at the time was composed of six parishes, a number that stood firm until the parishes were redrawn and a seventh added in 1978.

Like the borders of the parishes, virtually everything about Andorra was to stay fixed for the next 700 years. Wars raged on both sides of the border, countries and even civilizations fell, rose, grew, and shrank as the centuries passed, but the only event to have a truly significant impact upon the tiny state was the incorporation of Spain into the European Economic Community in 1986.

Yet the Andorrans did not stand idly by as history marched past them. Their principality proved to be a useful tool both to themselves and to their co-princes. For example, beginning a trend that has continued ever since, in 1391 the Deputation of Catalonia formally allowed the people of Andorra to import and export goods from Catalonia without paying duties. In 1419 the bishop of Urgell passed a law stating that Andorrans were not required to pay legal costs incurred during any trial at which they were found innocent. Already, Andorrans had begun to reap the privileges of their inconspicuousness.

The year 1419 also saw a major step forward in the process of creating a democratic tradition in the principality when the landed men of Andorra petitioned the bishop for permission to form an administrative council composed of men of the territory. The bishop, Francesc Tovia, acquiesced, and the proposal was later ratified by the count of Foix. Thus was formed the Land Council, a democratically elected council vested with the necessary authority to manage the day-to-day internal affairs of the principality. This legislature would eventually evolve into the General Council composed of two members from each of the six—or seven, after 1978—parishes and an equal number of members elected by national vote. The General Council is led by a Syndic General and Subsyndic General, elected by the council from within its own ranks.

Andorra was granted numerous additional tax and trade privileges in succeeding years. In 1468 Andorrans were exempted from taxes paid on goods brought to Andorran markets from outside the principality. In 1516 they were granted the right to hold a tax-exempt fair in Andorra la Vella every two weeks. In 1534 the House of Foix exempted Andorrans from paying taxes on imported goods brought into the territory on roads owned by the House of Foix. In 1538, Charles I, king of Spain (and Holy Roman Emperor as Charles V), even granted Andorrans permission to continue to trade with both France and Spain during a war between the two countries. Andorra was to be granted many more such privileges in the future.

In 1589, Henry II of Foix ascended to the throne of France, becoming King Henry IV and placing French control over Andorra in the hands of the crown. Generally, this shift in authority from a count to the supreme authority of France meant little to the Andorrans. In 1622 they were asked to provide men and money for the French military, and in 1640 King Louis XIII abolished the rule of the bishop of Urgell in the principality. By 1650 the bishops had been reinstated.

The new arrangement proceeded with relatively few glitches until the French Revolution in 1789. In 1793 the Revolutionary government refused to accept the Andorrans' annual dues, effectively annulling French control. During the ensuing war between France and Spain from 1793 to 1814, Andorra was effectively under Spanish control, with the see of Urgell urging Andorrans to shy away from French influences and the king of Spain agreeing to defend Andorra in case of French attack. Oddly, Napoléon restored the French feudal link with Andorra in 1806 with the consent of then-enemy Spain, thus shifting French authority once again, this time into the hands of the president of the French Republic. "Andorra is too fantastic," Napoléon is alleged to have said about the principality he could easily have crushed. "Let it remain as a museum piece."

During the nineteenth century Andorra's always-present smuggling industry blossomed into a full-fledged mainstay of the economy. Traditionally blessed with enviable tax privileges, Andorrans had long since viewed it as their right to trade in whatever stood to produce a profit. Their numerous tax exemptions coupled with differing laws on either side of the border made it easy for the enterprising mountain people to ignore statutes completely. During the Carlist Wars in Spain from 1832 to 1843, for example, Andorra played a pivotal role as a conduit for weapons, refugees, and supplies between Spain and France. The Andorrans did not feel compelled to play this role for the duration of the war, however. In 1835 they declared themselves neutral and asked for protection from France.

Smuggling has always been practiced openly in Andorra. The principality itself is open territory, and a long-established code of silence ensures safety while on home turf. But by far the greatest advantage Andorrans have held is their dual citizenship rights in both France and Spain. Laws preventing the trade of a particular commodity between the two countries rarely applied to Andorra, allowing enterprising Andorrans to act as intermediaries for parties on either side. Sometimes, such transactions were not even in violation of the letter of the law, but they were almost always in violation of its spirit. Allegedly, local authorities actually sold the rights to trade in certain contraband articles to individual Andorrans, although, for obvious reasons, documentation of such practices does not exist.

Historically, individual families jealously guarded their secret trails and mountain paths from Andorra to France and Spain. In the eighteenth century Andorrans turned tobacco smuggling into a huge business, much to the chagrin of their neighbors. This practice continued well into the twentieth century. When the border between France and Spain was closed in the years following World War II, only Andorrans could easily move back and forth between the two countries. The rise of the Franco regime in Spain created great demand for French products, from automobiles to cosmetics. Strictly speaking, such trade was illegal, but neither France nor Spain could stop Andorrans from traveling to France to buy goods and then traveling to Spain to sell them at enormous profits.

The entire smuggling industry is so deeply ingrained in Andorran culture that an unwritten code of conduct has evolved between the Andorran smugglers and the French and Spanish border guards charged with stopping them. If a border guard catches sight of smugglers, the code says that he is to fire his gun once in the air as fair warning. At this signal the smugglers are expected to drop their contraband and run. If they do so, they will not be chased. The contraband will be confiscated and the smugglers will be free to try again another day. Because of this pragmatic approach, actual violence between smugglers and border guards has been very rare.

Smuggling is only one part of the distinctive body of Andorran national tradition. For centuries, the backbone of the principality's political system rested with the "cap de casa," or "head of the house." The cap de casa was obliged to keep a weapon at the ready in case the council voted to mobilize the citizen militia, and only the cap de casa was obliged to serve. The cap de casa could also name his heir, who, although usually the eldest son, could be anybody. This heir would then, by law, inherit all but a quarter of the cap de casa's estate in addition to his rights and responsibilities. Usually, other family members agreed to give up their quarter share of the estate in the name of keeping the family together.

The sale of land has also been highly regulated. One law held that anyone who sells land retains the right to repurchase the parcel at the original sale price should the new owner choose to sell. If the new owner sells to someone else, he too retains the right. Sometimes, several former landowners, or their heirs, would sue to invoke this right simultaneously, leading the council to modify the law to cover only one-fourth of the original sale, thereby reducing the number of lawsuits created by the tradition.

Andorran citizenship is another complex issue. For centuries, nobody could apply for Andorran citizenship unless that person's family had lived in Andorra for at least three generations. Marrying an Andorran heiress would also suffice. Even today, Andorra's strict citizenship laws mean that only about one-quarter of the principality's population of 55,000 is considered to be officially Andorran.

In 1842 an entrepreneur named Louis Langlois applied for permission to build a casino in Andorra, causing deep political division among the General Council. Internal dissent and opposition from the French prefect of Ariège led to the plan's demise. In 1917 a similar scheme was floated by a consortium of Austrian promoters who wanted to establish a worldwide lottery in Andorra. This time, the General Council voted in favor of the plan, but the co-princes turned it down.

In 1866 the New Reform gave the vote to all heads of households and redefined the duties of the General Council and the smaller Communal Councils in charge of administering the parishes. Anti-reformist forces staged an armed rebellion, but it was quickly put down. On April 5, 1933, a group of young Andorran men stormed the General Council and demanded universal male suffrage. After a series of flareups during which the council was temporarily suspended, universal male suffrage was granted in August with support from France and in spite of vigorous opposition by the see of Urgell. The law was rescinded in 1941 when the representative of Marshal Pétain changed the French policy. In 1947 universal male suffrage was permanently reinstated.

Andorra stayed officially neutral during both world wars, giving asylum to refugees from both sides and prospering in the smuggling business. Allied agents, Nazi Gestapo officers, Spanish Republicans, leaders of the French Underground, and Allied aviators shot down over France all passed through Andorra. Only Hitler's Gestapo ever violated the principality's neutrality, and then usually only in clandestine ways, such as kidnapping Allied aviators in the night.

The postwar world brought change to Andorra as surely as it brought change to the major players on the world stage. In 1952 Andorra signed the Geneva Convention on intellectual property. This small act may not seem significant in light of other changes taking place throughout the world, but the intellectual property convention was the first international agreement ever independently signed by Andorra.

In the later 1950s the Andorrans decided that the fast-growing tourism business could provide economic stability, largely replacing smuggling or gambling. With its high mountain peaks, some of which are covered by snow for more than eight months of the year, development of a skiing industry seemed the logical choice. Year-round recreational facilities were also developed, and the principality's largely duty-free status was heavily promoted abroad. The success of this venture, certainly in terms of jobs created, is borne out in the

population explosion of the latter half of the twentieth century. In 1949, the principality was home to approximately 5,900 people, a number that had remained constant, give or take a thousand, for centuries. According to the 1960 census, there were 8,392 residents. By 1974 the number had grown to 25,000, and twenty years later the figure had more than doubled to 55,000. Today, tourism and associated retailing are by far Andorra's most important industries, followed by manufacturing and agriculture. Tobacco products, leather, lumber, textiles, furniture, building materials, iron, lead, and stone are among Andorra's most important sources of revenue.

The late twentieth century has been as important to Andorra's development as was the late thirteenth century. In 1970 universal female suffrage was introduced. The voting age was also lowered from twenty-five to twenty-one and the third-generation rule was relaxed to extend the vote to second-generation Andorrans. In 1977 this was extended to include first-generation Andorrans at least twenty-eight years of age. In 1985 the voting age was reduced to eighteen.

In 1981 the co-princes requested the formation of an executive council under the auspices of the General Council to prepare plans for institutional reform of the principality. In January 1982, following the General Council election of December 1981, the new president of the council, Oscar Ribas Reig, appointed a council of six ministers charged with providing Andorra with a written constitution, protecting local industry and encouraging private investment. The formation of the Executive Council alone represented a major step toward true checks-and-balances democracy in Andorra.

The new government met with several obstacles almost immediately. Severe storm damage suffered in November 1982 coupled with a worldwide recession led the council to enact a personal income tax in August 1983, despite great opposition. Later proposals for additional taxes on bank deposits, hotel rooms, and sales of property forced the government to resign in April of 1984.

The new government elected in the general election of 1985 had to relax Andorra's traditional laws against trade unions when Spain entered the European Economic Community (now the European Union) in January 1986, making Andorra, as a co-principality ruled by two member nations, a de facto member itself. In June 1988 the first indigenous Andorran trade union was founded by a French and Spanish consortium, but it was officially abolished in July by the General Council, which still had not passed defining legislation regarding the legalization of trade unions. In 1988 Andorra also formally enacted the Universal Declaration of Human Rights, adopted by the United Nations in 1948.

Andorra signed a comprehensive trade agreement with the EEC in June 1990 that gave the principality the freedom to buy and sell industrial goods within the community under a uniform customs tariff, meaning that Andorrans retained their historical tax breaks by being able to trade with the community as a member nation without actually being one. Although Andorra is treated as a third-party state for the purposes of trade in agricultural goods, its status as a duty-free zone for goods imported via EEC countries has served as a catalyst for new investment.

In 1990 the General Council, again led by Reig, voted unanimously to work again toward the creation and implementation of a written constitution. The co-princes agreed, as long as the document was enacted only through popular referendum. Once again, in January 1992, the government collapsed before completing its mission. Two rounds of voting on April 5 and 12 re-elected Reig's coalition, who again elected him president.

A draft constitution was prepared and presented to the electorate in March 1993 and was overwhelmingly supported by the voters. In the referendum, 75.7 percent of the 9,123 eligible voters cast votes, resulting in approval by 74.2 percent of the citizenry.

The constitution retained the bishop of Urgell and president of France as heads of state, but with mostly ceremonial powers. The co-princes retained veto power only over treaties with Spain or France that affect Andorran borders or security. All native Andorrans and foreigners with at least twenty years' residence in the principality were entitled to sovereign Andorran citizenship and the right to vote. Political parties and trade unions were also legalized. An independent judiciary was formed, the General Council was given the right to craft independent foreign policy, and Andorra was free to join international organizations in its own right. After signing a treaty of cooperation with the newly independent state, Spain and France in late 1993 established embassies in Andorra.

Notable historic sites in Andorra include the House of the Valleys (the headquarters of Andorran government) in the capital city of Andorra la Vella; the Sanctuary of Our Lady of Merixtell in Cantillo; the ruins of castles in Santa Coloma and St. Julià de Lòria; and many ancient houses and historic churches, especially from the Romanesque period, throughout the small country. Andorra hosts numerous cultural festivals, particularly during the summer, and draws many visitors seeking to take advantage of its excellent skiing, hunting, and fishing.

Further Reading: Thorough sources of information on Andorra are notoriously difficult to find in places other than France or Spain, but the *World Bibliographical Series* (Oxford: Clio, and Santa Barbara, California: ABC-CLIO, 1993) edition on Andorra, compiled by Barry Taylor, contains an exhaustive listing of books and articles in print, and is a good place to start. Additionally, the *Europa World Year Book*, thirty-second edition, two volumes (London: Europa Publications, and Detroit: Gale, 1994), is an excellent source of up-to-date information on politics, economic data, and other vital statistics. Modern Andorra is covered in many travel guides to France and Spain, and is a favorite topic among travel writers looking for a new place to explore.

—John A. Flink

Antalya (Antalya, Turkey)

Location: Located on the northwestern coast of the Gulf of Antalya, below the Taurus Mountains in southwest Turkey.

Description: A major port and trading center since ancient times, with remains of Hellenistic, Roman, Byzantine, and earlier Turkish settlements.

Site Office: Tourist Information Office
Cumhuriyet Caddesi Özel İdare İş Hanı Altı
Antalya, Antalya
Turkey
(242) 2411747

According to legend, notably endorsed by the ancient Greek historian Herodotus, Pamphylia, the region that includes Antalya, was settled by warriors from the Greek city-state of Mycenae, who are said to have explored the coast on their way back from taking part in the Trojan War. There were certainly Greek towns and villages in the region, Mycenaean or not, as is indicated by the statues, pottery, mosaics, and other objects discovered at a number of sites in Pamphylia and now displayed in the city's Museum of Archaeology and Ethnography. These towns are thought to have been founded in the twelfth century B.C. and reorganized by Ionian Greek settlers in the seventh century B.C. These settlements were not strong enough to resist occupation by Lydians and then by Persians, but survived until the time of Alexander the Great, who seized Pamphylia in 334 B.C.

Antalya itself was originally Attaleia, the city of Attalus II, a king of Pergamum who founded it in 159 or 158 B.C. as a Mediterranean port for his kingdom. This foundation may be attested by the partly ruined towers, still to be seen either side of the Roman Gate of Hadrian in Antalya, which are believed to have been constructed as part of the original city wall. Attalus's kingdom did not last for long after his death in 139, for his successor, Attalus III, who died in 133, left most of his lands to the Roman Empire, which was already advancing into Asia Minor (as the Romans, and many later Europeans, called Anatolia, now the Asian portion of the Republic of Turkey). Pamphylia was not inherited by the Romans, but instead fell into anarchy. The Romans moved in and took over the region anyway, gradually eliminating the pirates who were interfering with the city's trade and extending their rule into the country-side, a process more or less completed when the military commander Pompeius (Pompey) led a campaign against the remaining pirates in 67 B.C.

The Roman presence lasted approximately 500 years and has left many memorials, including parts of the harbor wall and some ruins of military buildings on Kalekapisi Meydani, now the city's main square. In A.D. 130 Emperor Hadrian visited the city, an event still commemorated by the Gate of Hadrian, a triple-arched marble and granite structure on the ancient road now called Atatürk Caddesi. Part of the Greek inscription originally placed on the gate can now be seen many miles away, in the Ashmolean Museum in the British city of Oxford. On the same ancient road as the gate, there also stands a round tower on a square base, the Hidirlik Kulesi, which was built around the same time. It is now approximately forty-three feet high but was probably much taller originally. It may once have been a lighthouse or even a tomb, just as the Castel Sant'Angelo in Rome, which it resembles, was originally the tomb of the same Emperor Hadrian.

It is possible that Christianity had come to Antalya as early as the first century, when Sts. Paul, Barnabas, and Mark are said to have visited the region, perhaps even landing in Attaleia itself. In any case, the oldest former church remaining in the city was dedicated to the Virgin Mary, perhaps on the site of a Roman temple, and completed during the fifth century. It was converted into a mosque around 800 years later, and since it was damaged by fire in 1851 it has been known as the Kesik Minare Camii, the mosque with the truncated minaret (tower).

From the early fourth century, western Anatolia was part of the Eastern Roman Empire, controlled by emperors who resided in Byzantium (later called Constantinople and now known as İstanbul) and whose rule survived long after the collapse of the Western Roman Empire in 476. During the eighth and ninth centuries they ordered the reinforcement of the city walls, first built around Attaleia by the Romans, to protect its residents against Arab expeditions into Pamphylia. In the eleventh century, around the time that the Christian Western European monarchies began sending crusading armies to reconquer Palestine and other parts of the Middle East, Attaleia became the seat of a bishop of the Eastern Orthodox Church and the center of Christian activity in the region. Because of its relative proximity to Italy, the city became one of the conduits for transporting crusaders and their supplies eastward; in 1149 it served as a place of refuge for the crusaders who were being driven out of Palestine by Moslem forces.

Meanwhile, as the power of the Byzantine Empire declined, more and more Turkish immigrants and invaders entered Anatolia from the east. Among these the best organized and most powerful were those led by the Seljuk dynasty, descended from military commanders in the service of the caliphs of Baghdad. From around 1048 numerous Turkish settlers came to Anatolia to escape the Seljuk empire established in what are now Iran and Iraq. Then, from around 1065 onward, the Seljuk forces themselves entered Anatolia, driving the earlier arrivals farther and farther west and defeating

Yivli Minare Mosque, Antalya
Photo courtesy of Embassy of Turkey, Washington, D.C.

the Byzantine armies sent against them. Attaleia was first occupied by the Seljuk Turks early in the twelfth century, when it acquired its first Turkish name, Adalia, later changed to Antalya. After a confused period during which the Byzantine Empire revived and drove the Seljuks out of western Anatolia, only to be attacked by western European Crusaders who occupied Byzantium itself in 1204, the Seljuks seized the opportunity to sweep back into western Anatolia, under their Sultan Giyaseddin Keyhusrev (or Kai-Khusraw). When they reached Antalya again, in 1207, they found the city under the control of an Italian, Aldobrandini of Tuscany, who had been appointed governor by Byzantium but had then made himself independent; he departed, and Keyhusrev's warriors moved in. Two major buildings in Antalya date from this second period of Seljuk rule: the Yivli Minare Mosque, established in approximately 1230, and the Karatay Medrese, a theological college established by Karatay, a vizier (minister) of the Seljuk sultan, in approximately 1250. The mosque has a minaret made of brick, set out in a fluted pattern and decorated with blue tiles; it is around 125 feet high and has become a symbol of Antalya.

Seljuk domination of the Turkish and other peoples of Anatolia was relatively brief; in 1243 their empire fell under the control of the Mongols as part of their enormous but fragile empire spread across western Asia and eastern Europe. When the Mongol empire in turn fractured into a number of rival states, the Seljuks found themselves unable to reassert their control over the Turks. Anatolia came to be governed by a number of independent *beys* (military commanders) who carved out smaller states for themselves. Thus, by the late fourteenth century, Antalya was normally subject to the principality of Hamit, although it was often under threat from the neighboring, and more powerful state of Qaraman to the east. Finally, both Qaraman and Hamit came under threat from the increasingly powerful Osmanli or Ottoman Turks, who had conquered much of the Balkans from their bases in northwestern Anatolia and who now turned to extend their power over the other Turkish states. In 1390 Hamit was conquered by the forces of the Ottoman ruler Bayezid I (Yildirim), who was declared sultan four years later.

From 1402 to 1413, following the destruction of the Ottoman Empire by yet another invader, Timur (Tamerlane), Antalya passed into the hands of the rulers of Qaraman; then, once the Ottomans under their sultan Mehmed I had reasserted their authority, it became an Ottoman city once again in 1416, when Mehmed defeated the last remaining independent Turkish states. The Qaramans continued to resist Ottoman supremacy, attacking Antalya one final time in 1422, but their leader Mehmet Bey died during the attack and his family surrendered to the sultan. Antalya was now part of an empire that, unlike its immediate predecessors, was to last for 500 years. It had retained its role as a major port and point of contact with the Black Sea, the Middle East, and Africa throughout the ups and downs of Byzantine, Seljuk, and Hamit rule. Under the Seljuks it had also begun to develop commercial relations with Cyprus, then a dependency of

Venice, and with other city-republics of northern Italy, whose rulers and merchants agreed with their Turkish counterparts to set aside religious differences for the sake of mutual profit. As a result, Antalya had long been a relatively cosmopolitan settlement. The Arab traveler Ibn Baṭṭūṭa, who visited the city in the middle of the fourteenth century, noted that the port was surrounded by a wall and dominated by a multinational body of Christian merchants, while the rest of the city, although mainly inhabited by Moslem Turks, also had Greek and Jewish districts.

The prosperity of Ottoman Antalya was necessarily dependent on the fortunes of the surrounding countryside and on the empire's relations with other countries. Accordingly, the 200 years following the Ottoman conquest of Constantinople in 1453 and its transformation into the new Ottoman capital, İstanbul, were a kind of golden age for Antalya. During this period, the empire defeated the rival Safavid Empire of Iran, challenged Portuguese incursions into the Indian Ocean, and expanded to take in most of Arab North Africa, Cyprus, and other Mediterranean islands, as well as a large part of Hungary. Under a government that tolerated religious minorities—in striking contrast with almost all of the Christian states of the period—and maintained commercial relations with France, many Italian cities, and those other states with which it was not actively at war, Antalya's merchants were able to build on past successes, helped by the fact that, as residents of an Ottoman city, they and their employees were not subjected either to military conscription or to most of the taxes that burdened the peasants in the countryside. The magnificent Murat Pasha Mosque, constructed in the sixteenth century, and many of the houses in the old harbor district date from this period.

Antalya was not entirely at peace, however. In 1472 what proved to be the last expeditionary force of European Crusaders arrived in the city, breaking the defensive chain that had been laid across the harbor but failing to defeat the soldiers sent to hold the city's fortress against them. When they sailed away they took the chain with them; it is now kept in the Vatican City. In 1511 Antalya was briefly the headquarters of a rising against Sultan Bayezid II (Adlî) led by Sah Kulu, a member of the militant Ṣafavid sect of Islam that had originated in Iran in the early fourteenth century and then spread into Anatolia. Bayezid's son Korkut, the governor of the city, was engaged in his own private war with his brother Ahmet and was unable to prevent Sah Kulu from sending followers to other Turkish cities. Later the same year, however, Bayezid's forces re-took Antalya, and after Sah Kulu's death in battle the movement was dispersed and defeated.

By the time the city's Tekeli Mehmet Mosque was built in the middle of the eighteenth century, Antalya was undergoing relative decline along with the Ottoman Empire itself. The city's historic links with the Black Sea and western Asia were broken as the Russian Empire grew to challenge the Ottomans; the eastern Mediterranean trade in which its merchants had once been prominent became less and less important to those European countries that now had colonial empires far away; both the production and the consumption of the rural populations of Anatolia and North Africa, on whom its trade depended, were severely damaged by ever-increasing tax burdens, epidemic disease, and, in many areas, the exhaustion of the soil by over-intensive farming. Much of its trade had steadily drained away to İstanbul, which had gradually expanded its direct trade with Alexandria from the early sixteenth century. By the middle of the nineteenth century both the friends and the enemies of the Ottoman Empire had become accustomed to calling it the "Sick Man of Europe," and Antalya, still governed by conservative guilds of merchants unwilling and unable to take new risks, shared in the sickness.

The Iskele Camii, the Harbor Mosque built on the seafront in Antalya in the late nineteenth century, was the last great building to be completed in the city under the Ottomans. In the city's hinterland, agriculture began to revive, particularly after 1897, when the empire's loss of Crete to Greece was followed by an influx of Turkish refugees from that island, many of whom settled around Antalya. Yet, no longer exempt from taxes or military service, the people of Antalya shared in suffering the impact of repeated government bankruptcies, the disintegration of the empire as many of its European subjects gained independence during the nineteenth century, and the military defeats inflicted in the Second Balkan War (1912–13) and then in World War I.

By the end of 1918 effective Ottoman rule was confined to the area of modern Turkey, and the Arab lands were being divided between the forces of Britain and France under the Sykes-Picot Agreement of 1916. In 1917, in return for endorsing this agreement, their ally Italy had been promised a share of Anatolia. Accordingly, Italian troops occupied Antalya itself, along with the rest of southwestern Anatolia, in 1919, while British, French, and Greek forces poured into the rest of Turkey.

In July 1921 nationalist forces led by Mustafa Kemal, who later renamed himself Kemal Atatürk, expelled the Italians from Antalya. Their expulsion was followed by the removal of most of the Greek-speaking population of Anatolia, including Antalya's, in an exchange of populations between Greece and the new Republic of Turkey established in 1922. Their centuries-long presence in the city is now commemorated by a display of nineteenth- and twentieth-century Greek Orthodox icons in the Museum of Archaeology and Ethnography.

Since 1922 Antalya has been the capital of an *il* (province) named for it, from which come the sesame, sunflower seeds, and cotton that are processed in the city and the citrus fruit, olives, sugar beets, and other agricultural products that are exported from it. In recent years the city, which now has a population of around 260,000, has also become internationally famous as one of the centers of the tourist industry on the "Turkish Riviera." The restoration of the old harbor as a tourist attraction, with improved access and a number of hotels, stores, and restaurants, was completed in 1985. The city that the tourists now visit has grown far beyond its historic center and yet has succeeded in preserving, both in the old harbor and in the streets around it, much of ancient Attaleia and medieval Adalia.

Further Reading: There are no book-length treatments in English of the city of Antalya. For a general overview of the Ottoman rulers of the city, see Stanford Shaw's *History of the Ottoman Empire and Modern Turkey,* volume 1, *Empire of the Gazis: The Rise and Decline of the Ottoman Empire 1280–1808* (Cambridge and New York: Cambridge University Press, 1976). One of the better tourist guides to the region is John Freely's *Classical Turkey* (London: Viking, and San Francisco: Chronicle Books, 1990), part of the Architectural Guides for Travellers series.

—Patrick Heenan

Antioch (İsparta, Turkey)

Location: In central-western Turkey, approximately 225 miles southeast of İstanbul and 100 miles north of the Mediterranean coast. Approximately one-half mile north of the modern town of Yalvaç, Pisidian Antioch was built on the southern slopes of the Sultan Dalğari Mountains, north of Lake Egirdir.

Description: Ruins are all that remain of Pisidian Antioch, which was deliberately destroyed in the early eighth century and completely abandoned in the thirteenth century. The ruins, discovered by Europeans in 1833, were intermittently studied until the mid-1920s. Surveys were conducted in 1962 and 1982; however, excavation of important Roman landmarks did not begin until 1987. These had not been completed in the 1990s, but ruins of many structures are now visible. Inhabited by numerous groups before being colonized by the Romans, Antioch was the site of preaching by Sts. Paul and Barnabas in the first century A.D. and thus played an important role in the early spread of Christianity.

Site Office: Yalvaç Museum
Yalvaç, İsparta
Turkey

The lake district of Pisidia, a fertile valley protected on the north by the harsh crags of the Sultan Dalğari Mountains, was inhabited by various communities of people before the city of Antioch was founded, as it has been in the nearly 1,300 years since the city was leveled. In 2000 B.C., the Hittites occupied the region; 200 years later, they were replaced there by Luvians. A thousand years later, in 800 B.C., the area was in the center of the Phrygian kingdom. Homer mentions the Phrygians in the *Iliad* as being allies of Troy. Medes, Lydians, and Persians had all laid claim to Pisidia before the years 336 to 323 B.C., when Alexander the Great conquered most of the known world.

Alexander was succeeded by five of his generals, each of whom took power over a section of the empire. The oldest of them, Antigonus, who ruled part of Asia Minor, attempted to unite the disparate parts of the empire, but was killed by three of the other generals in 301 B.C. at the battle of Ipsus. Seleucus I, who had "inherited" the east, was one of the three newly made kings who had allied against Antigonus; in 280 B.C. he battled another of the kings, Lysimachus, and won. With his major rivals out of the way, Seleucus was on the verge of conquering land on a near-Alexandrian scale when he stepped off a boat in Europe and was immediately assassinated. His empire stayed in the family, however, and his son, Antiochus I (the first of several Seleucid kings named Anti-

ochus), succeeded to the throne. The Seleucid dynastic line lasted until 64 B.C.

In all, sixteen cities in the Seleucid Empire were named after Antiochus I. One of those was Syrian Antioch, founded by Seleucus I in 301. It gained renown as the third most important city in the Roman Empire, and later became the city where Christianity was first named. Smaller and less famous than Syrian Antioch, Pisidian Antioch also played an important role in the spread of Christianity—perhaps the most important of any Roman city.

Pisidian Antioch may have been established by Seleucus I or Antiochus I; the date of its establishment is not known, nor is much of its history during Seleucid rule. Gauls moved into the area in such numbers that the province next door to Pisidia was named Galatia. Some of them must have wound up in Antioch. Greek traders in Antioch made deals with descendants of the Phrygians and Medes, with people of mixed or unknowable origin, and with a substantial community of Jews, who established a large synagogue. Phrygians worshiped at the nearby Temple of Mên, dedicated to the moon god. Tent-making, a solid middle-class profession (akin somewhat to modern house-building), was a notable industry there. The city was a natural trading post: fortified by mountains and nourished by the lakes, it hosted traders heading west from the Mediterranean coast and was a way station on the eastern frontier of civilization.

A stadium and a theatre were built in the city sometime before the birth of Christ. Granted independence from the Seleucids by the Treaty of Apameia in about 188 B.C., Pisidian Antioch was marked by tolerance and diversity. Its outlook, like its language, was Greek.

Sometime between 39 and 36 B.C., with Rome stretching to the borders of the former Alexandrian Empire (and beyond), Pisidian Antioch became part of the kingdom of Galatia, ruled by King Amyntas. At his death in 25 B.C., Galatia became a Roman province under Augustus, and Antioch was renamed Colonia Caesarea. A squad of veteran Roman legionnaires was dispatched to the city, which lay square on the great highway, Via Sebasta. Latin became the official language of government, although the residents still used Greek, and the city was made over by the colonizers in the image of Rome. Seven hills were designated districts, miniature *Vici,* as in Rome; construction of walls, baths, and colonnaded streets was begun; and statues of notable Romans lined the new avenues. At the death of Augustus in A.D. 14, a large temple to his memory was built on the highest hill in the city.

In A.D. 46, two men passed through the city gates. One had recently changed his name and his religion. Formerly known as Saul of Tarsus, he claimed to have been converted by the spirit of Jesus Christ, who had spoken to him person-

Roman aqueducts at Antioch

The main street of ancient Antioch
Photos courtesy of İsparta Tourist Office

ally. Calling himself the thirteenth apostle, he arrived from Cyprus, where he had begun using his Roman name, Paul, by way of Perga, where he had been incapacitated by a mysterious illness—perhaps malaria, perhaps epilepsy (Paul later said that he had been "sick" during his first mission in Galatia). Paul and his companion Barnabas had come to Antioch to spread the word about this new Jewish sect, only recently given a name, and to seek converts among both Jews and Gentiles. By then, the Christians had been promoting Christ as the promised Messiah to the Jews for years, but none had made a concentrated effort to convert Gentiles. The mission to Pisidian Antioch would be the first such attempt.

Paul had special credentials and talents that fit particularly well for this mission, in this city. Unlike Barnabas, and most of the residents of Pisidian Antioch, Paul was a Roman citizen, with all of the privileges that came with it. (The empire counted approximately 80 million subjects, but only 6 million of these were granted Roman citizenship.) The two men thus had no trouble with the sentries at the gate. Paul, the son of tent-makers in Tarsus, spoke fluent Greek, which made proselytizing among the people in Pisidian Antioch more likely to meet with success. As a former member of the Pharisaic sect of Judaism—stricter and more orthodox than the other main branch of Judaism, the Sadducees—Paul had been trained from childhood in the myriad arguments of Mosaic law. His teacher, Gamaliel I, was the grandson of the great Rabbi Hillel, and Paul himself, as an important Pharisee, had led the way in persecuting Christians in Jerusalem, personally witnessing the stoning of Stephen. It was unlikely that any provincial rabbi would be able to best him in a religious debate.

His timing was right, too. Though the death of the emperor Tiberius in 37 had been commemorated in Antioch with the construction of a new public square, most of the empire had rejoiced at his passing. His successor, Germanicus, was emperor for only a few months before he was succeeded by Caligula, who exceeded Tiberius in cruelty and insanity. Thought to be epileptic (curiously, some modern scholars posit that Paul was epileptic as well), Caligula made his horse a consul, proclaimed himself God, and commanded the empire to worship him. He was murdered by guards in 41. The next emperor, Claudius, decreed an end to Caligula worship, and a new era of religious tolerance began. Claudius was on the throne when Paul and Barnabas entered Antioch.

Alone among the apostles, Paul had not known Jesus before the crucifixion, although he affirmed that meeting him in spirit was as valid a criterion for apostleship as knowing him in the world. He was also alone among the apostles in his training, in his citizenship, and in his zeal to spread the new faith to everyone in the world. Soon after arriving in Antioch from Perga (where a third member of the missionary party, John Mark, had inexplicably deserted), and just after completing a difficult journey over the wild and barren Taurus Mountains, Barnabas and Paul went to Sabbath services at the synagogue.

Attached to the ancient west wall of the city, the synagogue was just a few hundred yards south of a site where a Roman bath complex was being built. After services, Paul and Barnabas were asked to speak to the worshipers, most of whom had been born Jewish. The congregation also contained some converts to Judaism from among the Gentiles (to the Jews, Gentiles were anyone who was not Jewish; to the Romans, anyone who was not Roman).

Judaism was a difficult religion to gain entrance to: converts had to study and know the law, and be circumcised to mark them as Jews. Paul began (as told in Acts 13:16 of the King James Bible): "Men of Israel, and ye that fear God, give audience," and went on to tell the assemblage that the Messiah had arrived, that neither knowledge of the Mosaic law nor undergoing circumcision were important to God, that all men were brothers, and that faith was enough to guarantee resurrection. He described the situation in Jerusalem, the persecution of the Christians by Romans and Jews alike. The congregation listened intently, and invited him to speak again the next week.

The following Friday evening, the Sabbath service at the synagogue was packed to overflowing, with more Gentiles than Jews cramming every available space to hear the new preacher. The crowds of Gentiles were enough to create anger and consternation among the Jews, who rejected Paul and Barnabas and would not allow them to speak in the synagogue again.

Acts 13:46 quotes Paul as saying to them: "It was necessary that the word of God should first have been spoken to you: but seeing ye put it from you, and judge yourselves unworthy of everlasting life, lo, we turn to the Gentiles." Though the decision that Paul be apostle to the Gentiles (as Peter would be apostle to the Jews) had already been made in Jerusalem before his arrival in Antioch, the reaction of Antioch's Jews set him firmly on the course.

Paul and Barnabas continued to preach in Antioch for the next few months, building the new religion in a city that was also in the process of being built from the ground up. Even the streets they walked, in sandals and rags, were still under construction. As the military and administrative center for that part of Galatia, Antioch was an important point of dissemination for the new religion. Acts 13:49 reports, "the word of the Lord was published throughout all the region." Eventually, the Jews prevailed upon the Romans to expel Paul and Barnabas from the city, and ban them from all of Pisidia. But the word that went out of Antioch was eventually spread to the rest of the world.

Though there is no record of Paul visiting the city again, the fact that he did return to the Christian communities created during his travels in Galatia suggests that he did. With the death of Claudius in 54, Nero became the Roman emperor, and his reign was as intolerant and cruel as Claudius's had been benign. Both Judaism and Christianity were outlawed. Paul was put to death in Rome sometime after the Great Fire in 64, but the nucleus of Christian converts continued to grow, as did Antioch.

In the first century A.D., an aqueduct nearly five miles long and approximately fifteen feet high was completed in Antioch by the Romans. It led to a monumental fountain, called the Nymphaeum. The Propylon, three great archways decorated with statues and friezes depicting Augustan victories and Pisidian slaves, connected the Augustus and Tiberius Squares. The baths were completed: the complex included a Palaestra, or exercising area; a changing room; cold, warm, and hot rooms; general service rooms; a boiler room; and a store.

Both the stadium and the theatre were renovated and enlarged by the Romans, the stadium sometime during the second century. The events presented there also changed during the Roman era, from wrestling and boxing matches to wild animal fights and gladiator battles. The theatre, built on a hill near the center of the city, seated 5,000 people in 26 rows. A tunnel running under the seats, unique in the Roman world, was part of Antioch's main street, called *Cardo Maximus*.

Around the beginning of the third century, the synagogue where Paul had extended his first invitation to the Gentiles was replaced by a small church, the Church of St. Paul. Sometime later it was succeeded by a larger one. In 295, Latin ceased to be the official language of the colony. Government was conducted in Greek once again. The third century was, perhaps, the most prosperous time in the history of Antioch. While the emperor, Diocletian, was halving other Roman provinces (and doubling the cost of their maintenance), the province of Pisidia was enlarged, with Antioch becoming its capital. At around the same time, the city ceased to be a colony.

In the early fourth century, under Emperor Constantine the Great, Christianity became the official Roman religion. By then, however, the empire was in serious trouble. Economic disorganization and high interest rates led to more land being held by fewer people, and by 395 the empire was formally split between east and west. In 410, Rome was sacked by the Visigoths; by 476, the Western Empire was dead. Antioch retained its importance in the Byzantine Empire: another church built in the center of the city in the fifth century demonstrated the importance of Christianity to both the city and Byzantium.

With the rise of Islam in the late sixth century and the beginning of the Arab conquests in the seventh, Antioch, while important to the Byzantine rulers in Constantinople, was not important enough to defend. The date of its demolition is variously given as 713 or 720; what is certain is that Byzantine defense efforts against the first *jihad* (holy war) were concentrated on defending Constantinople. Constantinople repulsed the Arab attack in 718, and the Arab armies avenged the defeat on the rest of Asia Minor.

Most likely, the Arab armies swept through Antioch en route to Constantinople, perhaps in 713, and created further destruction on their return in 720. While they took more than 700 prisoners at that time, they did not take the city. However, Antioch was not a city anymore: people still lived there (and remained until the fourteenth century when many of the inhabitants moved on to Yalvaç), but Pisidian Antioch practically disappears from Byzantine records in the 720s.

Investigations of the site began in the nineteenth century. An Englishman, F. V. Arundell, pinpointed the location of Antioch in 1833. Others studied Antioch sporadically during the remainder of the century. Another Englishman, Sir William Ramsay, and an American, D. M. Robinson, teamed up to excavate Antioch prior to World War I and again in the mid-1920s. They excavated much of the Roman colony, but the two men differed on the interpretations of their finds, and no satisfactory account of their work has been published. Many of the artifacts they found, however, are on display at the İstanbul Archaeological Museums. Extensive excavation resumed in the 1980s, with a team headed by Stephen Mitchell and assisted by the Yalvaç Museum. Excavation is still in progress in the 1990s, but many of the ruins of ancient Antioch are visible, including the city walls, the western gate, the Propylon, the Nymphaeum, the Roman baths, the temple of Augustus, the theatre, the stadium, a colonnaded street, the Temple of Mên, and the Church of St. Paul. Some smaller objects from the site are displayed in the Yalvaç Museum.

Further Reading: *Pisidian Antioch: The Journeys of St. Paul to Antioch* by Mehmet Taşlialan (Yalvaç: Yalvaç Museum, 1991) is a short but useful history of Antioch, with pictures and descriptions of the ruins. Expect many spelling and grammatical errors in the English translation. An entertaining and lively description of first-century Antioch can be found in *Paul the Traveller: St. Paul and his World* by Ernle Bradford (New York: Barnes and Noble, 1993). Bradford's retelling of the story of Paul's life is compelling and filled with interesting detail. *Paul the Convert: The Apostolate and Apostasy of Saul the Pharisee* by Alan F. Segal (New Haven, Connecticut, and London: Yale University Press, 1990) is an exhaustive study of the links between Paul's Pharisaic training and his mission to the Gentiles. It says little about Pisidian Antioch, however. Antioch's importance in the early spread of Christianity derives in the first place from the Acts of the Apostles, contained in the New Testament. This provides no description of Antioch, but contains information concerning the missions there of the Apostle Paul.

—Jeffrey Felshman

Aphrodisias (Aydın, Turkey)

Location: In southwestern Turkey, approximately 90 miles east of the Aegean Sea, 250 miles south of İstanbul, and 78 miles north of the Mediterranean Sea.

Description: Ancient city dedicated to Aphrodite; site of excellently preserved Greco-Roman architecture, epigraphy, sculpture, and religious artifacts; under archaeological excavation since 1961 by Dr. Kenan T. Erim and other scholars.

Site Office: Aphrodisias Museum
Geyre Koyu
Karacasu, Aydın
Turkey
(256) 448 8003

The site of Aphrodisias, an ancient city dedicated to the goddess of love, covers an area of approximately 200 acres and is essentially flat except for two conical mounds in the south-central area. Sophisticated radiocarbon analysis of artifacts from these areas, known as the Acropolis and the Pekmez Mounds, indicates that the area of Aphrodisias has been inhabited since approximately 5800 B.C., with evidence of continuous occupation from about 3000 B.C. Prehistoric deposits excavated from the base of the Acropolis between 1967 and 1973 reveal that it was not a true acropolis but an artificial habitation mound, or *hoyuk*, that resulted from a buildup of a series of prehistoric settlements. Excavation of the smaller Pekmez Mound, which lies east of the Acropolis, has unearthed small stone idols, evidence of worship of a goddess probably associated with mother nature or fertility divinities prevalent in Anatolia and other parts of western Asia dating as far back as Neolithic times.

Among the original recorders of the history of Aphrodisias was the first-century B.C. geographer Strabo, who included it among the cities of the neighboring province of Phrygia. In the sixth century A.D., a grammarian and encyclopedist named Stephanus of Byzantium mentioned the city by its ancient name Ninoë and asserted that the site had been founded by the Pelasgian Leleges, who called it Lelegonopolis. He claimed that later its name was changed to Megalè Polis (the Great City), then finally to Ninos or Ninoë. Although his theory about the Leleges is questionable, archaeological evidence has verified the connection with Ninos. In 1977 a team directed by the archaeologist Kenan T. Erim partially uncovered a large basilica off the southwest corner of the agora (marketplace) of Aphrodisias. Relief panels that decorated the balustrade of an upper level depicted figures identified with inscriptions, including the names of Ninos and Semiramis. Based on stylistic and archaeological comparisons, the reliefs were dated to the second half of the third century B.C., implying that there was a tradition of connection with Ninos within the cult during that time.

In the Hellenistic period, which extended from 323 to 31 B.C., "Ninoë" was changed to "Aphrodisias." One theory suggests that the change may have been a calculated political move by the city's priests or elders during the late Hellenistic period, when Romans moved to take over western Asia. Romans claimed ancestry from the Trojan prince Aeneas, the son of the goddess Venus, who was equated with the Greek goddess Aphrodite. Another reason for the name change may have been the increasing reputation of the Aphrodisian sanctuary, which was probably constructed in the first century B.C. Aphrodisias was, at that time, a sacred site featuring the sanctuary and secondary buildings that housed priests or religious attendants who tended to the needs of the cult, and enough workers to maintain fields and the requirements of other associated properties. Worship of the original temple goddess was most likely influenced by the Akkadian Nin and the Mesopotamian goddess of love and war, Ishtar or Astarte, eventually transforming to become Aphrodite. By the late Hellenistic period, the temple's fame had spread beyond Caria (the region surrounding Aphrodisias) and western Asia Minor into the Roman Empire. Appian, a second-century A.D. historian, reported that in 81 B.C. the Roman dictator Lucius Cornelius Sulla offered a golden crown and a double axe to Aphrodite at the Carian shrine after being advised to do so by the oracle of Apollo at Delphi.

Aphrodisias was drawn into the Roman internal conflicts following the assassination of Julius Caesar in 44 B.C. The assassins, Brutus and Cassius, escaped Italy and fled to Asia Minor, where they invaded various communities. Aphrodisias apparently remained loyal to Caesar's faction, and there is evidence that the sanctuary and private properties at the site were attacked and damaged by the armies and supporters of Brutus and Cassius. Much of the credit for keeping the city loyal to Caesar goes to an Aphrodisian by the name of Zoilos, who may have been inherited as a slave from Caesar's household. Zoilos was later freed by Octavian, who, along with Mark Antony and Lepidus, ruled Rome in the Second Triumvirate following Caesar's death, and who eventually became the emperor Augustus.

In 1969 a team excavating the northern wall of the theatre found numerous inscriptions detailing the historical and sociological background of the area during the late Hellenistic period (the structure is therefore known as the Archive Wall). A letter by Octavian inscribed in Greek on the wall referred to Zoilos as a close and "esteemed friend" and to Aphrodisias as "one city from all of Asia I have selected to be my own." Other epigraphic documents from the Archive

The stadium at Aphrodisias
Photo courtesy of Embassy of Turkey, Washington, D.C.

Wall record missives from Roman emperors dating from the first century B.C. to the third century A.D., attesting to special privileges issued via a triumviral decree that granted the city freedom, tax-free status, and an increase in protected asylum within the sanctuary.

With its tax-free status and its ever-increasing popularity as a destination for religious pilgrimage, Aphrodisias began a long period of prosperity in the late first century B.C. The city gained in reputation as a cultural and artistic center as well as a center for religious activity. Among its citizens in the first centuries A.D. were Xenocrates, who wrote discourses on medicine; Chariton, an early writer of ancient romances; and Alexander of Aphrodisias, one of the most respected commentators on Aristotle. Schools of philosophy and oratory also evolved there during this period.

One of the most significant events, however, was the increased attraction of sculptors to the site and the establishment of a school of sculpture there. Quality marble was abundant in the area, a factor that undoubtedly contributed to the growth of the artistic community. Miners working at Baba Dag and quarries east of the city arrived early in the day to complete the task of cutting raw marble for the sculptures. Because of the weight—one cubic meter of Baba Dag marble can weigh as much as three tons—the stone cutters would rough-cut sarcophagi, statues, or busts at the site to reduce heaviness, and then move the pieces to city workshops by oxcart. The marble was multicolored, some milky white, some bluish gray, and some veined, so that it was possible for an expert to cut a piece of various colors and create a composition in which one section contrasted sharply with another.

Although it is not clear where, exactly, the large number of artists came from, it is thought that, along with indigenous and Greek sculptors, many had moved from the aesthetically oriented Attalid Kingdom of Pergamum. Many artists and sculptors who were employed by the kings of Pergamum were left without work after the death of Attalus III, who willed his empire and treasures to the Romans near the end of the second century B.C.

In the congenial environment of Aphrodisias, these artists developed workshops and studios, advanced their level of technique and style, and developed a reputation beyond Anatolia and Rome. Reliefs, monuments, sarcophagi and marble works were commissioned not only by wealthy Romans but also by citizens and municipalities throughout the Mediterranean. Although there are few surviving literary texts regarding ancient art—and especially the art of sculpture—evidence of the particular style and popularity of the Aphrodisian school has been uncovered in numerous southern European and Mediterranean locations. Names of artists, accompanied by the ethnic adjective "Aphrodisieus," have been found on signed bases, statues, and sculpted fragments in Rome and various parts of Italy, the Aegean, Greece, and other regions. Remarkable figures of old and young centaurs were discovered in the emperor Hadrian's villa at Tivoli. They reveal signatures of the Aphrodisian sculptors Aristeas and Papias. These figures are now at the Capitoline Museum in Rome.

In the early 1940s, noted Italian scholar Maria F. Squarciapino challenged the accepted scholarly assumption that the sculptors of Aphrodisias were merely copyists of earlier masterworks. Subsequent archaeological analysis has confirmed that a true school of sculpture, unique in its methods, developed at this site and thrived for an exceptionally long period. The most noted achievement of the Aphrodisian sculptors lay in the area of portraiture, but they also showed such superb skill in architectural decorative sculpture that certain motifs have become closely identified with them. Among these are the "peopled-scroll" pilasters adorning the Hadrianic and Theatre Baths and the ever-present mask-and-garlands friezes. Much of the work reflects an earlier style, theme, or tradition associated with the classical sculpture of their Hellenistic antecedents, especially those of Pergamum, but nearly all pieces maintained an originality of approach and a particular Aphrodisian flavor.

During the second century A.D., Aphrodisias was one of the most famous and affluent cities in Asia Minor, hosting a steady current of religious pilgrims and curious travelers and housing a population of 50,000 to 60,000 residents. The situation began to deteriorate, however, in the late third century as political and administrative changes took place. Aphrodisias eventually lost its special tax-free and autonomous status. Recent epigraphic documents uncovered by Erim's team reveal that Caria and Phrygia were combined to form a single province in the 250s, and that Aphrodisias probably served as the core administrative center of that merged unit.

By the fourth century, with the fall of Rome in the West and the rise of Byzantium in the East, Christian rulers attempted to eradicate the pagan worship of Aphrodite. The city's administration promoted establishment of an archbishopric, and an Aphrodisian bishop, Ammonius, is known to have participated in the Council of Nicaea as early as 325 A.D. Nevertheless, references have been found revealing that pagan sacrifice still took place there as late as 484. In spite of the friction that must have been unavoidable between the religious cultures, benefactors from both sides were still making endowments to the city to help meet civic costs as late as 529 A.D. The city's school of sculpture had greatly declined by this time, owing at least in part to the Byzantine Empire's low regard for the plastic arts.

Sometime during the late sixth century or early seventh century, Christian rulers attempted to change the name of the city to Stavropolis, or "City of the Cross," in order to disassociate it from Aphrodite. The temple of Aphrodite was converted into a Christian basilica, and the original classical architectural style was altered. This transformation required the complete elimination of large blocks of the shrine, or *cella,* that housed the cult statue of the goddess Aphrodite, and the movement and re-erection of several Ionic columns surrounding this *cella* to create a basilica plan. In addition to these and various other changes, the temple suffered serious earthquake damage in the seventh century. Repairs and alterations were made in the eleventh or twelfth century, but raids by Seljuk Turks and Turkoman horsemen between the eleventh and thirteenth centuries ultimately destroyed the temple.

Byzantine commentators of the twelfth and thirteenth centuries, such as Nicetas Choniates and George Pachymeres, testify that the city of Aphrodisias, also known then as Caria, was raided and captured at least four times by the Turks. In 1453 the Byzantine capital of Constantinople fell to the Ottomans, and western Anatolia became a small part of the vast Ottoman Empire, which at one time extended as far west as Vienna.

The site was essentially abandoned after the thirteenth century, when remaining inhabitants voluntarily resettled elsewhere. The city and the area around it were absorbed by the principality, or *beylik,* of ancient Tralles, now known as Aydın. Other sources assert that it may also have become part of the *beylik* of Menteşe for a time. Eventually a Turkish village named Geyre (a corruption of "Caria") developed among the ruins of ancient Aphrodisias when settlers were attracted to the area because of fertile soil and plentiful water. Travelers to this area have made reference to the village of Geyre from the seventeenth century onward.

Aphrodisias has been of interest to scholars and archaeologists since at least the mid-eighteenth century, and the site was partially excavated by French and Italian teams in the first half of the twentieth century. These initial excavations were short-lived, however, and were abandoned entirely just before World War II. In 1956 the area southeast of Aphrodisias sustained a severe earthquake. Though neither the village of Geyre nor the ruins of the ancient city sustained much damage, authorities—who sensed the archaeological potential of the site—decided at that time to resettle the makeshift houses and buildings of the village approximately one and one-quarter miles to the west of the Byzantine fortification walls of Aphrodisias. Even after the initiation in 1961 of large-scale excavations under Dr. Erim, however, construction of the new village moved slowly. Only by 1979, when the on-site museum was opened, had most of the inhabitants of the older village of Geyre been transferred to their new community.

The Byzantine wall, which separates the city from the surrounding necropolises, is about two miles in circumference and still stands in several places. Also well-preserved is the huge stadium, located in the northern section of the site, which includes about thirty tiers of seats and could hold almost 30,000 spectators. The stadium was probably built in the first or second centuries A.D. and would have been used primarily for athletic events, festivals, and public meetings. During the Byzantine period the east end was modified into a smaller arena area with protective walls, suggesting that gladiatorial shows and animal exhibitions or circus performances were held here. One of the best preserved features at this site is the theatre. This building, located on the eastern slope of the Acropolis, held nearly 8,000 spectators and served numerous functions. It is also the site of the Archive Wall.

Restoration continues at Aphrodisias. The great Tetrapylon near the Temple of Aphrodite has been carefully reassembled, as has the Sebasteion, a complex dedicated to the cult of the immortalized emperor Augustus and his descendants. The museum inaugurated in 1979 has been filled. For the present, new finds are being kept in a warehouse at the project compound. The sculptures and reliefs discovered in the six large halls of the Baths of Hadrian are now in the İstanbul Archaeological Museum.

Many fields and orchards on and around the site are still privately owned and farmed. The project planners hope to purchase these lands whenever possible so that research can continue. They also plan to reconstruct, roof, and protect the buildings of Aphrodisias and to put most of the uncovered finds back into the site to create what Erim termed "an archaeological haven."

Further Reading: *Aphrodisias: City of Venus Aphrodite* by Kenan T. Erim (London: Muller, Blond and White, and New York: Facts on File, 1986) is a beautifully illustrated and in-depth description of the history of Aphrodisias based on archaeological finds. *Aphrodisias de Carie: Colloque Du Centre de Recherches Archéologiques de L'Université de Lille III,* edited by Juliette de la Geniere and Kenan T. Erim (Paris: Éditions Recherche sur les Civilisations, 1987) and *Aphrodisias Papers 2,* edited by R. R. R. Smith and Kenan T. Erim (Ann Arbor: University of Michigan Press, 1991) collect papers presented at colloquiums at the University of Lille in 1985 and at New York University in 1989, respectively. These articles tend to concentrate on interpretation rather than description of the finds, and some are written in French, Italian, or German.

—Holly E. Bruns

Arezzo (Arezzo, Italy)

Location: Arezzo, capital city of the province of the same name, is situated on a sloping hill in the Apennines, in the eastern Tuscany region of north-central Italy. It lies in a fertile plain near the junction of the Chiani and Arno Rivers, fifty-five miles southeast of Florence.

Description: Ancient Arretium, one of the twelve cities of the Etruscan League by the mid–seventh century B.C., and the center of a prosperous bronze industry, was an important Roman colony in the first century B.C., famous for its red-glazed clay Arretine pottery. Sacked by the Goths in the third century A.D., it became part of the Lombard Kingdom in the sixth century. Arezzo then became part of the Frankish "march" of Tuscany in the eighth century, later the duchy, or margraviate, of Tuscany. With the eleventh-century land conflicts between the papacy and the Holy Roman Empire, Arezzo became a free commune, or city-state, flourishing through the fourteenth century until finally falling to the Florentine Republic. It was part of the Duchy, and then Grand Duchy, of Florence, from the sixteenth to eighteenth centuries, dominated by the Florentine Medici dynasty, and then the Austrian house of Habsburg-Lorraine. After a short period of French Napoleonic rule in the early nineteenth century, when Tuscany was known as the Etruscan Republic, or Etruria, the Habsburg grand dukes were restored. Arezzo became part of the united Kingdom of Italy in 1861. The city has many medieval and Renaissance buildings, and is rich in art treasures, most notably the *Legend of the True Cross* frescoes by fifteenth-century painter Piero della Francesca.

Site Office: A.P.T. (Azienda di Promozione Turistica)
Piazza Risorgimento 116
CAP 52100 Arezzo, Arezzo
Italy
(575) 23952

Arezzo dates to ancient times. The modern city occupies the site of the original Etruscan settlement, though its Etruscan name is not known. Called Arretium by the Romans, it was an important Etruscan town by the mid–seventh century B.C., one of the twelve confederate cities of Etruria. It was the center of a large and flourishing bronze industry; the most famous of all Etruscan bronzes, the *Chimera,* was found here in the sixteenth century and has been dated to the mid-to-late fifth century B.C. It is now in the Museo Archeologico in Florence.

Archaeological evidence indicates that Arretium was probably a dependency of Clusium (the modern Chiusi), an Etruscan territory, until about 300 B.C., when it became independent. It was also about this time that Arretium developed from an agricultural into an industrial city. In 390, the Celtic Senone tribe invaded Etruria and besieged Clusium, warring with Rome for the next century. The Romans subdued the Etruscan cities and neighboring Umbria by 295, forming the Northern Confederates. In 284, the Senones defeated the Romans near Arretium. In 283, Roman forces, led by consul Publius Cornelius Dolabella, were sent to avenge the insult; the Senones were routed and expelled, and Rome annexed their lands. In Arretium, the Romans supported the local aristocracies against the lower classes.

Arretium was once the terminus of the strategic Roman highway Via Cassia. In the Second Punic War between Rome and Carthage in the late third century B.C., Arretium was a Roman military base, and contributed to the defeat and death of Hasdrubal, Hannibal's brother, in central Italy. The city also supplied thousands of spears, lances, and helmets, as well as bronze fittings for warships, for the military expedition of Scipio Africanus to North Africa in 205; his forces decisively defeated Hannibal in 202, thus ending the Second Punic War. Afterward, Arretium lost much of its commercial importance when the Via Cassia was extended through Florence to meet the Via Emilia at Bologna. In the power struggles between Roman political leaders at the start of the first century B.C., Arretium sided with Marius against Sulla, who seized the town in about 80; he razed the Etruscan-built brick fortifications and eventually established a thriving Roman colony there.

Few ancient relics remain in Arezzo save for traces of a Roman amphitheatre in the garden of the fifteenth-century Convent of San Bernardo, where the fine Archaeological Museum is housed. Archaeologically, the town is especially famous for its red-glazed clay Arretine pottery, produced here in factories between approximately 30 B.C. and A.D. 30. The highly prized vases and tableware, decorated with reliefs, were among the town's main sources of wealth, and became the standard ware throughout the Roman world; they were even exported to the east coast of India. The Archaeological Museum contains a collection of these Roman vases, as well as a number of Etruscan bronzes.

Arretium's fortunes rose and declined with Rome's Western Empire, which fell in A.D. 476. Meanwhile Goths, Lombards, and Franks were in turn fighting their way south across the Alps. The town was sacked by Goths in the mid–third century and became part of the Lombard Kingdom in northern Italy in approximately 570. Frankish king Pépin III invaded Italy under papal authority in 754; his son Charlemagne was crowned king of the Franks and Lombards in 774, thus begin-

Etruscan sculpture at Arezzo's Archaeological Museum
Photo courtesy of Soprintendenza Archeologica per la Toscana

ning the Carolingian dynasty. At this time, Tuscany—the region to which Arezzo belonged—was a Frankish "march," or frontier district, ruled by the counts of Lucca.

With the breakup and decline of the Carolingian dynasty through the ninth century, local Tuscan counts began to assume the power of dukes or margraves (the hereditary princes of border provinces). The tenth century saw the rise of the Attoni house of Canossa. A member of this house, Boniface, received the margraviate of Tuscany from German king Conrad II in 1027, becoming one of the major princes of Italy. He also acquired Upper Lorraine through his marriage to Beatrice. After Boniface was assassinated in 1052, Beatrice governed Tuscany until her death in 1076; their daughter, Matilda of Tuscany, known as the Great Countess, then became margravine, or heiress, of her father's dominions.

In the eleventh century, absent emperors and ambitious popes began dividing the country against itself; many towns, resisting foreign imperial rule, paid allegiance to their bishops; some powerful local landowners served the popes, others the emperors, often changing loyalties. The controversy between the empire and the papacy coincided in northern Italy with the rise of the communes, or independent city-states; a number of the more civilized, prosperous towns took advantage of the struggles to declare autonomy at the expense of their bishops or foreign lords. Arezzo established a free commune in 1098 and flourished, extending its territorial power over several neighboring towns.

The conflict worked further to the benefit of the communes in the early twelfth century, when Countess Matilda, a longtime papal supporter, changed allegiances and made the emperor, and not the pope, the heir to her extensive estates in 1111. Matilda's death in 1115 led to a long contest between the papacy and the empire over her inheritance, a struggle that allowed the Tuscan city-states to consolidate their independence until the unity of the march, or margraviate, was broken. By the beginning of the thirteenth century, Tuscany—indeed, most of medieval Italy—had become a battleground between the Ghibellines and the Guelphs. Arezzo had joined the Ghibelline faction, which was allied with the Holy Roman Emperor and the local aristocracy. Their rivals, the Guelphs, were allied with the Holy See and the emerging merchant classes.

The Guelph-allied Republic of Florence, then the leading city-state in Tuscany, was determined to annex the cities of Pisa and Arezzo, strongholds of Ghibellinism. In 1289, at the Battle of Campaldino, the Ghibelline cities, led by the Aretines, were defeated by Guelph forces, led by Florence; the young Florentine poet Dante Alighieri is said to have taken part in the battle. Peace was made in 1293, and the Arezzo commune prospered, reaching its zenith in the early fourteenth century. The commune declined, however, after Florence finally succeeded in buying the town in 1384.

With the Guelphs victorious, the Republic of Florence consolidated and expanded its power through the fifteenth century, an era that saw the rise of the rich and powerful Medici dynasty, which began with Giovanni de' Medici and his son Cosimo. Starting in 1434, the house of Medici—bankers, merchant oligarchs, patrons of the arts, and benevolent despots—would dominate Tuscan life for three centuries, first as rulers of Florence and the Florentine Republic, and then of the Duchy, and ultimately the Grand Duchy of Tuscany. Arezzo would remain within the Florentine sphere of influence until the French occupation of 1798–99.

The late Middle Ages and early Renaissance gave birth to Arezzo's many fine historical monuments, mainly churches, palaces, and a hilltop citadel; above all, Arezzo is a small city rich in hidden art treasures. The old town, built by the Medicis, is pentagonal in shape and surrounded by ancient and medieval walls, the oldest parts of which were built on a 1,053-foot hill; on its summit stands the Fortezza Medicea, built between 1541 and 1561 by Antonio Sangallo, a noted architect and military engineer. Streets fan down the hill from the remains of the Medici fortress. Palazzos and houses were built from the fourteenth to sixteenth centuries around the sloping Piazza Grande, the former city center. They include the open-balconied Palazzo delle Logge, now a museum, designed and built from 1540 to 1573 by native Giorgio Vasari, a noted painter, biographer, and architect; and the Palazzo della Fraternità dei Laici, which houses a museum and art gallery. The Palazzo del Tribunal, by R. Cerrotti, and the Palazzo Pretorio, decorated with many coats of arms and housing the public library, also date from the fourteenth century.

Arezzo was also the birthplace of Petrarch, the great lyric poet; Guido d'Arezzo, a monk who invented the musical notation system; and the ribald satirist and dramatist Pietro Aretino, said to have been expelled from town as a young man for writing blasphemous sonnets. Other distinguished citizens included noted botanist Andrea Cesalpino and Leonardo Bruni, scholar, historian, and chancellor of Florence. In Roman times, Maecenas, the patron of Horace and Virgil, was also born here.

A bishopric, Arezzo is home to several churches, including the Romanesque Church of San Domenico, begun in 1275, with a crucifix by Cimabue (Bencivieni di Pepo); the Romanesque-Gothic cathedral, begun in 1286, with a modern facade built from 1900 to 1914, and which contains the tombs of St. Donatus and Bishop Guido Tarlati; the Romanesque Church of Santa Maria della Pieve, begun about 1287, which has a notable campanile and facade, both completed in 1333 by Marchionne, and which also contains a polyptych by Pietro Lorenzetti; the Gothic Church of San Francesco, which was founded in 1322; and the Renaissance Church of Santa Maria delle Grazie, with an altar by Andrea della Robbia.

The choir, or Cappella Maggiore, of the Church of San Francesco, located on the Via Cavour in the center of town, contains what is perhaps the most preeminent of Italy's hidden art treasures, equal to anything in Florence: the magnificent fresco cycle *The Legend of the True Cross,* by Piero della Francesca. Astonishingly, Piero's art has only come to be appreciated within the last century, and he is now considered to be one of the great Italian painters of the fifteenth century, a master of measured perspective and spatial depth. His Arezzo fresco series, painted between 1452 and approximately 1459, is his greatest achievement. The frescoes were neglected in Piero's own time, however, primarily due to their lack of typically Florentine vigor; it was only in the late nineteenth century that their austerely geometric compositions and their purity of color and light began to be truly appreciated. And, though Aretine historian Vasari wrote the first biography of Piero in 1550, it was not until the early twentieth century that Piero's vision was shown to influence the direction of Renaissance painting throughout northern Italy.

Born to a family of artisans and merchants in Borgo San Sepolcro (now known as Sansepolcro), a small town near Arezzo, Piero apprenticed in the 1430s in Florence, then the liveliest center of the early Renaissance. Giovanni Bacci, a humanist and member of a wealthy merchant family that owned the San Francesco chapel in Arezzo, first commissioned the Florentine Gothic artist Bicci de Lorenzo to decorate the church. Bicci began working in 1447 but died in 1452, having completed only the vaulted ceiling and parts of the choir's entrance arch. Piero probably began work on the frescoes immediately afterward; within seven or eight years, before his journey to Rome in 1459, he had covered the walls of the church's choir with the most modern and technically sophisticated frescoes yet produced in Italy during the fifteenth century. *The Legend of the True Cross* mural cycle, whose subject matter was perhaps proposed to Bicci and Piero by the Franciscan order of Arezzo, is drawn from a medieval Christian legend as set forth by Genoese bishop Jacopo de Voraigne in *The Golden Legend,* a thirteenth-century text that recounts the miraculous story of the wood of Christ's Cross. The legend was popular in the Middle Ages, especially in Franciscan churches.

Piero also completed church and court frescoes, some now lost, in Florence, Borgo San Sepolcro, Monterchi, Rome, Ferraro, Rimini, and Urbino, as well as the Confraternity of the Annunciation of Arezzo. He died in Borgo in 1492.

In 1798, during the Napoleonic Wars, a French Revolutionary force invaded the Grand Duchy of Tuscany, then ruled by Austria. The grand duke fled, and, in March 1798, a provisional French government transformed the grand duchy into the Etruscan Republic. The rapid French conquest of Italy sparked national uprisings, often led by priests and bishops. Horrified at the new regime's brutal and irreligious character, a counterrevolution erupted in and around Arezzo; two peasants—a man and a woman, believed to be San Donato and the "madonna of Comfort"—organized an armed band that developed into the Aretine army. They drove the French and their sympathizers from the countryside, committing many atrocities, and even occupied Florence. A revivified French army reentered Florence and dispersed the bands by October 1800; after Aretine harassment, the troops were welcomed, even though they committed atrocities, too. After the fall of Napoléon, Tuscany was again ruled by Austria for a time, then became part of a unified Italy in 1861.

Arezzo suffered major damage during World War II and was captured by the Allies in July 1944. Today, the city has a basic agricultural economy and is the market center for the surrounding area, primarily for silkworm eggs, wool textiles, olive oil, and wine. It has rail repair shops, furniture factories, clothing and footwear factories; goldware and lace are exported. Arezzo is also a center of the antiques trade.

Further Reading: Though slightly outdated, *A Short History of Italy,* edited by H. Hearder and D. P. Waley (Cambridge and New York: Cambridge University Press, 1963), provides a compact, readable introduction to the country, from pre-Roman times to the present. The three-volume *Italy in the Making* by G. F. H. and J. Berkeley (Cambridge: Cambridge University Press, 1932–40) will be of use to historical scholars. While *Piero della Francesca* by Alessandro Angelini (Florence: Scala, 1985; translated by Lisa Pelletti, New York: Riverside, and London: Constable, 1990) serves as a general introduction to the life and work of the fifteenth-century artist, *The Enigma of Piero* by Carlo Ginzburg, translated by Martin Ryle and Kate Soper (London: Verso/New Left Books, and New York: Routledge, Chapman and Hall, 1981; as *Indagini su Piero,* Torino: Einaudi, 1981) treats Piero's Arezzo fresco cycle—the masterpiece for which the town is primarily known—in much greater depth.

—Jeff W. Huebner

Argos (Argolis, Greece)

Location: In the northeast Peloponnese region in southern Greece, about four miles north of the Gulf of Argos, in the department of Argolis.

Description: Argos is one of the oldest cities of Greece, and was one of the most important in ancient times. Ruins are scattered throughout the present-day city of Argos, which has largely been constructed on the ancient city site. An agora (marketplace), amphitheatre, and other buildings have been excavated on and near Mount Larissa, northwest of the central city. The Heraeum, a temple to the Greek goddess Hera, is located about eight miles north of the present-day city.

Site Office: Argos Museum
Olga Street
Argos, Argolis
Greece
(751) 28819

Perhaps the oldest of Greek cities, Argos was founded in approximately 2000 B.C., although the site was inhabited for at least a thousand years before that date. Argos came to prominence with the collapse of the Mycenean civilization at the close of the Late Helladic Age, and the subsequent fall of the nearby cities of Mycenae and Tiryns. With the conquest of the area by the Dorians from the north, Argos became a regional power in the Peloponnese, rivaling Sparta in influence over the Hellenic world. The city reached its peak of influence in the seventh century B.C. when it flourished under King Pheidon. Argos appears as a major Greek city in the works of Homer and other ancient writers.

The early history of Argos is lost in legend. Perhaps the most famous of these legends tells of Adrastus, a son of King Talaus of Argos. Leaving the city at a young age, Adrastus journeyed to his grandfather Polybus in Sicyon, where he married and, upon Polybus's death, became king of Sicyon. Reconciling with his family, Adrastus later returned to Argos to serve as king. Along with Polynices, Tydeus, Amphiaraus, and three others, Adrastus launched an attack on Thebes to restore Tydeus to his rightful throne in that city. On their journey, the "Seven against Thebes," as they came to be known, visited Nemea, where Polynices, son of the legendary King Oedipus, died from a serpent's bite. The Nemean games were established to honor Polynices. The attack on Thebes failed, with all being killed save for Adrastus. But Adrastus bided his time and, ten years later, he returned to Thebes with the sons of his slain companions. This time they were victorious, although Adrastus lost his son in the second battle. Adrastus died of grief in Megara on the way home from Thebes.

Only scattered references to Argos and its history are found in Greek written accounts, and these accounts were recorded several centuries after the events they recount. If legendary stories are to be believed, Argos seems to have been ruled by three dynasties—those of Inachos, Danaos, and Atreus—and a total of some twenty kings during a period of about 600 years before the Doric invasion of 1200 B.C. Legend claims that Phoroneus, the son of Inachos (said to be a river spirit as well as a political leader), was the city's founder. Phoroneus was head of the Pelasgians, an ancient Indo-European people who inhabited Greece and the eastern islands of the Mediterranean. Because the Pelasgians lived in small farming communities or as nomads and did not have a central gathering place, Phoroneus founded Argos, which legend claimed was the world's first city. He named his son Argos as well. The dynasty founded by Phoroneus ruled Argos until the arrival of a new people led by Danaos, possibly leader of a tribe expelled from Egypt. Danaos's family members ruled the Argolis, the northeast plain of the Peloponnese, as a group. An early legend tells of the Princess Danae—daughter of Akrisios, ruler of Argos—who was visited by the god Zeus in the form of a shower of gold. By Zeus, she gave birth to Perseus who went on to slay the dreaded Medusa and is said to have founded the city of Mycenae. The dynasty founded by Danaos came to an end when Atreus, leader of a people from Asia Minor, conquered Argos and established his rule over the whole of Argolis. When the Dorians invaded the region King Tisamenos, the last of the Atreus dynasty, fought vainly against them.

With the Doric conquest of the region and the subsequent collapse of the Mycenean culture, Argos became an important political and cultural center. The Dorians so dominated Argolis that by the ninth century B.C., a local legend claimed that Hercules was an ancestor of the Dorians. By the seventh century B.C. Argos was the center of Dorian political and military power. The Dorians made the city their primary settlement for two reasons: the secure fortress on nearby Mount Larissa provided protection for the inhabitants of Argos, and the city had a reliable supply of fresh drinking water.

Argos was also a leader in the manufacture of pottery and metalwork. An early Geometric pottery, similar to that found in other Greek cities, evolved by 800 B.C. to include rustic scenes and human figures. Fine silver jewelry, detailed bronze items, cauldrons, and eating utensils were also manufactured in Argos.

Two Argive sculptors achieved lasting fame for their work. Ageladas, who lived in the late sixth or fifth century B.C., was a sculptor who worked in bronze. His most famous work was a statue of Zeus for the temple on Mount Ithome; this statue was pictured on coins from Messene. Ageladas was

Ancient ruins on Argos
Photo courtesy of Greek National Tourist Organization

the teacher of the Greek sculptors Phidias, Myron, and Polyclitus. Polyclitus rose to have a wide influence over later sculptors in Greece and Rome. Often depicting Greek athletes or gods, Polyclitus's works realistically present the human form and were later used as models for beginning sculptors. Copies of his work have been identified in several Roman cities as well. Polyclitus wrote a book on the subject of rhythm and proportion in sculpture. His most famous works include a towering seated version of the goddess Hera housed at the Heraeum, a temple located eight miles north of Argos. This statue is known from its representations on ancient coins.

The first Dorian leader, Temenos, ruled Argolis from his capital in Argos. The first nine kings to follow him belong to the prehistoric period of Argos. The tenth king in the Temenid dynasty, Pheidon, became the most powerful leader in the Peloponnese. He introduced a system of weights and measures, minted coins to promote foreign trade, and, because of his claim to be a descendant of Hercules, demanded to be put in charge of the games established by the legendary hero. Pheidon also organized the games of the twenty-eighth Olympiad, probably in 668 B.C. More important, Pheidon forced a number of other Greek cities to cooperate with his political ambitions. Pheidon's political ambitions were shared by his son and grandson, who followed him on the throne of Argos. The city's expanding power frightened its rivals. Several neighboring cities formed a union with the intention of standing firm against Argos's growing influence.

But it was Sparta to the south, another ambitious city expanding its influence, that eventually initiated open conflict with Argos. Both cities desired control over the plain of Cynouria, which lies to the south of Argos. Conflicts began during the eighth century B.C. In 669 B.C., the Argives won a decisive battle for the region at Hysiai. During the Second Messenian War in the mid–seventh century B.C., Argos sided with the Arcadians, Pisatans, and Messenians against Sparta. Sporadic outbursts of violence eventually led to another battle between the Spartans and Argives near the city of Thyrea. This battle gave Sparta control of the region and a stronghold from which it could launch attacks. In 494 B.C. Sparta clashed with the Argives again in the battle of Sepeia, near the city of Tiryns, killing so many Argive soldiers that for a time, the slaves of the city were obliged to govern. Legend claims that following this battle a female poet, Telesilla, led a defense of Argos's city walls by old men, women, and children, thus saving the weakened city from conquest.

During the Persian Wars, in which most of the Greek cities fought against the powerful Persian Empire to the east, Argos was so weakened by its defeat at Sepeia that it remained neutral. Not until the Persian Wars were over did Argos again involve itself in Hellenic military affairs. The city formed a strategic alliance with Corinth, Mantinea, and Elis in a bid to strengthen its position against Sparta. During a battle at Mantinea in 418 B.C., the Argives and their allies lost to the Spartans. The following year, when Argos began construction of a long defensive wall from the city to the Gulf of Argos, the Spartans forcibly prevented its completion. Argos joined with the nearby city of Corinth during the Corinthian War from 395 B.C. to 386 B.C. against Sparta, a war that ultimately freed many Greek cities from Spartan domination.

Argos's many military ventures left it weakened, but it continued to involve itself in the various conflicts among Greek cities. When Macedonia conquered the Greek cities during the third century B.C., Argos joined the Achaean League formed by the victorious Macedonians. This membership led in 146 B.C. to the city being treated as an enemy by the conquering Romans, although it was spared the brunt of Roman vengefulness and eventually served as the headquarters for Roman administration of Greece. After severe attacks by the Goths in A.D. 395, the Romans rebuilt the city. Argos, which had lost much of its significance, became a part of the Byzantine Empire after the collapse of Rome, was captured by the Franks in 1210, was held in fief by Athens from 1246 to 1261, and was again a part of the Byzantine Empire from 1261 until 1460. The Ottoman Empire then ruled the city until 1830, except for a brief period of Venetian domination in the late seventeenth and early eighteenth centuries. Argos played a major role once more as a stronghold of the Greek independence movement in the 1820s. This aroused the wrath of the Ottoman general Ibrahim Pasha, who destroyed much of Argos in 1825. After the war Argos became part of an independent Greece and was rebuilt once again.

The distinguishing features of Argos today are the sites relating to its ancient history. Among King Pheidon's conquests during Argos's period of expansion were the nearby cities of Mycenae and Prosymna, which were destroyed. The Argives seized a temple to the goddess Hera, one of the oldest cult sites in the Argolis, located near Prosymna. A new temple to the deity—called the Heraeum—was designed by the architect Eupolemos and constructed by the Argives about 410 B.C. An eight-mile-long road was built from Argos to the Heraeum. The site soon became the leading Hellenic sanctuary to Hera, bringing the city renown throughout the Hellenic world. The original temple had been built about 680 B.C. on a small hill; the new temple was at the base of the hill and surrounded by several other buildings, including a bathhouse, a guest house for visitors to the site, and a building where symposia were held. A giant gold and ivory statue of Hera, the work of the sculptor Polyclitus, was the new temple's primary feature.

The original temple to Hera was one of the first Greek buildings to be surrounded by columns. It had a roof tiled with terra-cotta instead of the usual wood and thatch, and was perhaps the first Greek building to have columns of stone rather than the earlier wood. Its remains, primarily the building's foundations and scattered columns, are located at the highest terrace on the site. The temple to Hera built by the Argives is located on the terrace immediately beneath. Originally the terrace had been used for an annual festival honoring Hera. The dimensions of the new temple, some 130 feet by 66 feet, made it an impressive showcase of Argive wealth and power. Also found at the site are the foundations of several smaller buildings as well as the remains of stone staircases and stoas, or arcades.

The beliefs and activities of the Hera cult are not known for certain, but it is thought that they focused on the mysteries of the afterlife. Similar to the cult found at Eleusis, which worshipped the goddess Demeter, members of the Hera cult envisioned their goddess as a deity who knew much about death and the afterlife. One legend illustrates the belief: a woman who asked the goddess to bestow upon her two sons the best fate available found the next morning that the two young men had died in their sleep. Hera was, like Demeter, a goddess of fertility as well as the afterlife, and she ruled over the earth's plants and animals.

Remains of the Heraeum were first uncovered in 1854 by A. R. Rangavis and C. Bursian. In 1892, Charles Waldstein, head of the American School of Classical Studies in Athens, began work at the site with a team of some 200 workers for three years. His findings were published as *The Argive Heraeum*. The first volume, covering the buildings, appeared in 1902; the second volume, concerning the artifacts found at the site, appeared in 1905.

The ancient city of Argos today lies largely buried under more recent construction. Because of this, excavations of the city began only early in this century. The early findings were first published in the *Bulletin de Correspondence Hellenique*. Since 1952 the French School of Archaeology has conducted regular excavation work in the city and has published numerous monographs and reports detailing its findings.

Among the ruins that can be found throughout the present-day city and on the outskirts of town are those on Mount Larissa, a hill northwest of the central city, where there are fortified walls dating from the Mycenean period. At the hill's summit is a fort called the Kastro, first constructed by the Byzantines, who incorporated earlier stone walls built by the Greeks into the structure. The fort was later used and expanded by the Franks and Venetians as well. Two walls enclose the fort: an outer wall encircling the hilltop and an inner wall surrounding a central tower. At one point the fort held off Frankish attacks for seven years. The foundations of ancient temples to Zeus Larissaios and Athena Polias are located within the fort's walls.

Just southeast of Larissa lies an ancient Greek amphitheatre. Dating from the fourth century B.C., and displaying signs of later Roman reconstruction, the theatre was capable of holding some 20,000 spectators, ranking it among the largest theatres in all the ancient world. Its eighty-nine tiers of seats are cut into the rock on the slope of Mount Larissa.

Just northeast of the theatre lies the Kriterion, a building where events from an ancient legend are said to have taken place. According to Greek mythology, Danaos took refuge in the city of Argos with his fifty daughters to avoid the fifty sons of his brother Aegyptus, who wished to marry the girls. The family was found and the girls were forcibly married to their cousins. Danaos secretly ordered his daughters to murder their husbands and, except for Hypermnestra, they obeyed him. The daughters were punished in the underworld for their crimes by being forever condemned to fetch water in jars that leaked so badly that the water always spilled out. Danaos brought Hypermnestra to trial at the Kriterion for failing to kill her husband as he had instructed.

Southwest of Larissa along Tripoleos Street lie the remains of an ancient agora, or marketplace, built in the fifth century B.C. Nearby are found Roman baths dating to the second century A.D., a Roman villa from A.D. 400, and an Odeon, or concert hall, dating to the first century A.D. The area was destroyed during the Gothic invasion of Greece in A.D. 395 and rebuilt by the Romans. A fifth-century B.C. shrine to Aphrodite, enclosed by a stone wall, was apparently destroyed by the Goths and not rebuilt.

Northeast of Larissa is Aspis, a low hill resembling a warrior's shield lying on the ground. Some of the oldest remains are to be found here, including Mycenean tombs, sixth-century B.C. walls, and prehistoric houses dating to about 2000 B.C. North of Aspis lies a sanctuary to the Greek gods Athena and Apollo, known to the Argives as Apollo of the Ridge and Sharp-Eyed Athena. The ruins of the sanctuary consist of four terraces, although the buildings themselves are so ruined, and have been so mixed with the remains of later Christian and Byzantine churches built on the site, that the effect is chaotic. An altar to Apollo measuring some fifty by fifteen feet is located on the lowest terrace. A cistern can still be seen in the ruins of Athena's temple. Some scholars believe that a temple to Leto—mother of both Apollo and Athena—may also form part of this complex.

The museum at Argos, housed in a building donated by the French School of Archaeology, is on Olga Street. Pottery, armor, figurines, musical instruments, and other items found during excavations of the city's historic sites are displayed here. Outside the museum in the courtyard are mosaic floors taken from the Roman villa located near the amphitheatre. The mosaics depict the Greek god Dionysus and the seasons and months of the year.

Further Reading: *Mycenae-Epidaurus-Tiryns-Nauplion* (Athens: Clio, 1978) is a complete guide to the archaeological sites in the department of Argolis, including those found at Argos. Offering similar information is *The Peloponnese: A Traveller's Guide to the Sites, Monuments and History* (Athens: Ekdotike Athenon S.A., 1984) by E. Karpodini-Dimitradi. Michael Grant's *The Rise of the Greeks* (New York: Scribner's, and London: Weidenfeld and Nicolson, 1987) devotes a chapter to the history of Argos at its peak. *A History of Argos to 500 B.C.* (Minneapolis: University of Minnesota Press, 1976) by Thomas Kelly challenges long-held beliefs about the history of the city, including the belief that Argos and Sparta were adversaries.

—Thomas Wiloch

Aspendus (Antalya, Turkey)

Location: Between the cities of Perga and Side on the
Mediterranean coast of Turkey in the historic
Pamphylian region.

Description: An ancient city most famous for its Roman
theatre, perhaps the best-preserved structure of its
kind.

Contact: Tourist Information Office
Vali Ürgen Alani, No. 47
Antalya, Antalya
Turkey
(326) 2160610

Most of the ancient structures visible in Aspendus today date from its period of rule by the Romans, but the area was inhabited for several centuries before the Romans came. Tradition holds that Achaeans colonized the area after the Trojan War. Some believe, however, that it was settled originally by the Hittites. Later, in the seventh century B.C., Greeks from Argos colonized the site. For a time, the city was a member of the Athenian Maritime League, also known as the Delian League.

The Eurymedon River, which runs through Aspendus, was in the very early days navigable as far as the city. The river provided recreation and livelihood for the townspeople. The Mediterranean Sea, the river's source, is only some seven miles from the hilltops where the city sits. Due to the city's defensible location atop two hills, and its favorable climate, it was a thriving export center for the products of the region, which included salt, olive oil, cattle, silks, rugs, and small, valuable figurines carved of lemon wood and decorated with ivory. It also was a strategic naval base. It was the site of many battles, including two significant ones during the Persian Wars, in which Persia attempted to annex all of Greece. The Athenian Admiral Cimon led Greek forces in the two battles in 468 B.C. One was fought on the sea and the other on land. Cimon soundly defeated the Persians and captured many of their ships, but did not stop at that. He sent diversionary troops dressed as Persians up the Eurymedon while he and his men landed farther up the coast and succeeded in capturing portions of the Persian army.

Alexander the Great arrived in Pamphylia in 338 B.C. Alexander started his conquest of Asia and the Persians from the Aegean coast and traveled along the southern coast of Anatolia as far as Side. The terrain of the area was difficult; not until he came into the Pamphylian plain was navigation any easier and he was able to subdue four great cities of the region, Perge, Sillyum, Aspendus, and Side. After Side, Alexander turned north to continue his conquest, leaving one of his trusted men in charge of the Pamphylian area. The rulers of the city of Aspendus appeared to support Alexander, but when he left, they fortified the city and strengthened the citizenry within its walls. Alexander was too smart for them, however, and he returned and set up an encampment on the banks of the Eurymedon River and took the city again. This time, among other types of punishment, he levied higher taxes than those agreed upon previously to support his army. Aspendus apparently had poor relations with Side and other neighboring cities; Alexander had to set up a court to deal with complaints concerning land grabs by the Aspendians. Alexander left behind a vital Hellenistic legacy at Aspendus, although no substantial architectural remains of earlier civilizations have yet been uncovered. Some coins have been found that bear the name of the city and date to the fifth and fourth centuries B.C. Aspendus was one of the few cities of Turkey that was minting silver coins by the early fifth century B.C.

The Romans conquered Aspendus and began their period of supremacy there in the second century B.C. Aspendus continued to prosper under the Romans, who emphasized cultural development of the city, as evidenced by the magnificent structures still standing today. The well-preserved and famous Roman theatre of Aspendus was built in the second century A.D., during the reign of Marcus Aurelius. It stands as a reminder of the great architects of Roman times. The architects of Hellenistic times made many advances in the building of theatres, but the Romans, and the native architects they influenced, achieved an incredible harmony of structure. Greek theatres tended to blend into the countryside; and it was hard to tell where they stopped and the countryside began. The theatre at Aspendus is a distinct entity, however—an architectural frame. Its architect was named Zeno; not to be confused with philosophers of the same name, he was a native of Aspendus. It is dedicated (as inscribed over the entrance) "to the gods of the country and to the imperial house." The theatre is built almost entirely on barrel-vaulted arches. The seating area is well integrated with the stage building, giving the audience an impression of being isolated from the outside world and totally wrapped up in the theatrical experience.

As was true of many structures of this time, the theatre is a mix of architectural styles. The auditorium is in the Greek horseshoe shape and was built, in part, against a hillside, conforming with the Greek custom. But, in true Roman style, the entrances to the theatre are roofed and parallel to the auditorium. The stage building is imposing at 82 feet high and 369 feet wide. The entire cavea and stage of the theatre are still intact, as are some of the decorative niches that once held statues. The stage is huge, and behind it is a wall that is 120 yards long by 80 feet high. This wall has three doors and contains columns rising two stories with entablature and pediment, all of which remain astoundingly well preserved. Classi-

The Aspendus theatre
Photo courtesy of Embassy of Turkey, Washington, D.C.

cal theatre was performed on this stage, and philosophical, religious, and political speeches were also delivered there.

For a brief period in the thirteenth century, after the Seljuk Turks had taken over the region, the stage building of the theatre was used as a royal residence. During this time the tower-like entrance to the theatre was built. Part of the stage was covered with plaster and painted with red designs, and the niches were tiled with turquoise and black. Portions of the theatre were reinforced during the thirteenth century, but some interior columns, which had collapsed in an earthquake, were not rebuilt.

Mustafa Kemal Atatürk, first president of the modern Turkish republic, spearheaded a restoration of the theatre and advocated its continued use, not as a museum, but serving its original purpose as a performance space. The theatre, seating 15,000, continues to be used today for plays, concerts, and athletic competitions. Its acoustics are still a marvel; amplification devices are unnecessary.

Also among the Roman remains of Aspendus is an ancient aqueduct, in surprisingly good condition. The aqueduct brought water into the city from the mountains to the north. With its two massive water-control towers still standing, the aqueduct is considered the best-preserved structure of its type in Asia Minor.

Other notable Roman remains in Aspendus include a gymnasium, baths, a stadium, a marketplace with arcaded shops, and various government buildings. An ancient bridge over the Eurymedon, said to have been tall enough to allow ships to pass beneath it, eventually crumbled and was replaced with a Seljuk-designed structure in the Middle Ages.

Aspendus was abandoned at some point, but details of the abandonment remain unknown. The modern town of Belkiz grew up nearby. Still, Aspendus draws many visitors who come not only to admire the architecture, but to view performances in the ancient theatre.

Further Reading: History and architectural description of Aspendus can be found in such works as Freya Stark's *Gateways and Caravans: A Portrait of Turkey* (New York: Macmillan, 1971; as *Turkey,* London: Thames and Hudson, 1971); Gilbert Horodin's *Turkey: The Cradle of Europe* (Lincolnwood, Illinois: Passport, 1992); and John Freely's *Classical Turkey* (San Francisco: Chronicle Books, 1990; Harmondsworth, Middlesex: Penguin, 1991). An excellent piece on the theatre is contained in *International Dictionary of Architects and Architecture,* volume 2, edited by Randall J. Van Vynckt (Detroit and London: St. James Press, 1993).

—Patricia Wharton

Assisi (Perugia, Italy)

Location: In Perugia province in the Umbria region of central Italy, situated on the spur of Mount Subasio above the valleys of the Topino and Chiascio Rivers, east of the city of Perugia.

Description: Famous as the birthplace of St. Francis and St. Clare, Assisi is a major religious center that attracts millions of worshipers annually. Major landmarks include the St. Francis Basilica, a majestic double church, and its frescoes painted by some of Italy's most famous painters.

Site Office: A.P.T. (Azienda di Promozione Turistica di Assisi)
Piazza del Commune, 12
CAP 06081 Assisi, Perugia
Italy
(75) 812450

Built upon the slopes of Mount Subasio, Assisi's origins date back to the time when central Italy was first inhabited by an ancient people called Umbrians. According to legend, the nucleus of the town was originally a temple dedicated to the goddess Minerva, Assisi's protector. Legends also recount how a strange and mysterious people, the Etruscans, invaded the region, pushing the Umbrians south of the Tiber, and established their city of Perugia just west of Assisi.

From then until the coming of the Romans, a continuous state of war existed between the two cities. Assisi's Umbrian defenders developed an elaborate subterranean network of passages, extending for miles under a walled fortress town. Much of this earlier construction is still evident today. In 309 B.C., Etruscans and Umbrians suspended their conflict and united against the new Roman foe, but within a year Perugia and all of Umbria were conquered by the Romans.

Assisi's isolated position on the slope of Mount Subasio made it less attractive than its neighbor Perugia as a Roman imperial city since it held a strategic position on the road from Rome to Ravenna in northern Italy. Assisi, nevertheless, prospered under Roman occupation: numerous temples, a theatre, and a circus were all built there by the Roman occupiers. Incorporated as it was under the Pax Romanus for six centuries, much of modern-day Assisi lies upon the foundations of earlier Roman structures. Roman mosaic floors and frescoed walls as well as busts of the Caesars, Roman gods, and goddesses still adorn many Assisian homes to this day.

Assisi also contributed to the great flourishing of Roman classical culture, having sent one of its sons, Sextus Propertius (c. 50 B.C.) to Rome to become a well-known poet. Chroniclers make little or no mention of Assisi's history in this period, and it would appear that nothing remarkable occurred in the Italian village until the coming of Christianity.

Assisi probably became a diocese in the third century A.D. Although later chroniclers associated it with many martyred early Christian bishops, the only real claim that can be substantiated is that of St. Rufinus, whose cult is mentioned by St. Peter Damian and who became Assisi's first patron saint. Assisi's relative isolation also seems to have spared it much of the destruction of advancing invaders who descended upon Rome. Only in 545 did the citizens confront any real danger when Assisi was besieged by Totila and later capitulated when its defender Siegried, a Goth, was killed. Assisi received its first bishop, Aventius, a legate of the Ostrogoths of Justinian, and lived in relative peace while managing its own affairs for the next two centuries.

In the late eighth century, Assisi was once again besieged by a foreign army, this time the Gauls, led by Charlemagne in 773. Much of central Italy had followed the Lombard dukes in their opposition to the papacy. Charlemagne as champion of the pope laid waste to much of the Umbrian countryside. When Assisi fell, Charlemagne quickly put its inhabitants to the sword. Assisi was repopulated with Christians sympathetic to Rome. It was probably at this time that Assisi's castle, the Rocca d'Assisi, was built upon the hillside peak above the town, from where its keep and crenellated towers looked down upon the people.

Throughout the Middle Ages, the castle became more a symbol of oppression than of protection. In the twelfth century, Assisi was once again embroiled in a bitter conflict between church and state. Supporters of the church were known as Guelphs (from the German Welf), while a group called the Ghibellines supported Holy Roman Emperor Frederick I Barbarossa, who for most of the latter half of the twelfth century had subjugated much of central Italy to his rule. Many of the Italian towns resented rule by the German Crown and appealed to the papacy for protection. Assisi became an imperial possession in 1195 when Conrad of Swabia took possession of the castle where he resided with his charge, Frederick II, who later would be one of the most bitter opponents to the pope. Assisians turned to Pope Innocent III to rid themselves of their German despot. Throughout much of this period, moreover, Assisi was involved in continual war with its neighbor, Perugia. Yet within these years of conflict and despotism, Assisi was to become the center of a religious movement that would capture the hearts of most of Christendom.

Francesco di Pietro di Bernardone, later St. Francis (the "seraphic saint"), was born in Assisi in 1181 or 1182. The son of a cloth merchant, Francis's youth was marked by a love of life and a spirit of worldliness that made him a leader among the Assisian youth. In 1202, he took part in the warfare between Assisi and Perugia and was taken prisoner and held for almost a year, falling gravely ill upon his return. Once

St. Francis Basilica
Photo courtesy of A.P.T., Assisi

recovered, he tried to return to combat. In late 1205 at Spoleto, he abandoned this notion, following instead a dream or vision that bade him return to Assisi and await a new call to a new service. While awaiting this call, Francis had many experiences that would change his life. While praying in a grotto near Assisi, he had a vision of Christ. During a pilgrimage to Rome, he exchanged his own garment with that of a beggar and begged for alms in front of St. Peter's Basilica. Later, he overcame his own fear by kissing the hand of a leper.

Francis finally received what he believed to be his call to service. While praying outside the chapel of San Damiano (St. Damian) just outside the gate of Assisi, he heard the crucifix order him to repair the chapel, which was then falling into ruin. Taking this charge seriously, he went home and gathered much of his father's cloth from his shop. After selling the cloth, he tried to give the proceeds to the bewildered priest of San Damiano. When his father learned of Francis's actions, he confined his son to the family home.

Francis was later brought before an episcopal tribunal and in the presence of the bishop not only renounced his father but all material goods. In his zeal to serve God, he literally tore the clothes from his back and returned all that he had been given by his father.

After a brief sojourn in a monastery on Mount Subasio and a brief visit among lepers, Francis once again undertook the cause of rebuilding San Damiano in May of 1207. He begged for stone in the public squares and open places of Assisi. By the following spring, San Damiano had been restored. Encouraged by this achievement, Francis immediately began repairing other Assisian places of worship—most notably Santa Maria degli Angeli (St. Mary of the Angels), also called the Portiuncula, which was to become the center of the Franciscan movement.

Francis began his mission in February 1209. As the number of friars increased, he ventured to Rome to seek formal approval from Pope Innocent III to establish a new order. Innocent III had expressed some reservation about the rule of life, fearing it too strict a rule to follow. He nevertheless approved the founding of the Franciscan order April 16, 1206. The Franciscans continued their work, mostly in Umbria, establishing themselves at Rivo-Toro, about a one-hour walk from Assisi, and later at the Portiuncula.

Friars who joined the new order were called upon to follow the teaching of Jesus and to walk in his footsteps. They were expected to continue to work for their upkeep, accepting as payment only food for their daily intake. This love of poverty that marked the first generation of Franciscans was inspired by the example of St. Francis himself. Later Franciscans would find the standard set by their founder a greater challenge, but the initial success of the Franciscan ministry was due in large part to the strength of Francis himself and to the simplicity of his teaching.

In 1212, Francis began a second order for women. He had been encouraged by the devotion of a young noblewoman named Clare who, in defiance of her father, left her paternal home and joined Francis at the Portiuncula. Francis installed her and other women of Assisi in the church of San Damiano. The women were provided with a religious habit similar to that of the Franciscans and a similar set of rules that excluded the call to missionary life. In exchange for labor and alms for upkeep, the Poor Clares (as the women came to be called) were to perform various domestic services for the Franciscans. As the number of Franciscans increased, the order quickly expanded to include all of Italy and beyond. In 1221, another order was established, the Third Order of Brothers and Sisters of Penance, a lay fraternity intended for those unable to leave their families and homes for a monastic existence but who could still carry out the principles of Franciscan life.

Francis did not limit his endeavors to Christendom. In the spring of 1212, he set out for the Holy Land to continue his mission but was shipwrecked on the east coast of the Adriatic Sea, on the coast of Slavonia. He returned the following winter and continued his work in central Italy. The following year he attempted again to reach the Moors of Spain but was forced to abandon this venture due to illness. He remained in Italy, where his presence was needed, for the next few years. Francis had been persuaded against venturing to France by Cardinal Ugolino (later Pope Gregory IX). His dream of seeing the Holy Land was finally realized in 1219 when he joined the crusaders on their voyage to Egypt. He reportedly met the Saracens at Damietta, where it is said he preached to the sultan, who was so impressed he gave permission for Francis to visit the sacred places of the Holy Land.

In his absence, the Franciscan Order had reached a crisis. Its number of members had increased to more than 5,000, and the once simple guiding rule was now proving too demanding for many friars. Church officials as well as many Franciscans themselves now believed that the rule had to be relaxed and formalized. Rumors of Francis's death added greater urgency to the crisis. When word of the problems at home reached him in Syria, he immediately set sail for Italy.

Upon his arrival in Italy in 1220, Francis appointed Peter Cantanii and later Elias of Cortona as vicars in charge of administering the order. They were chiefly responsible for the restructuring of the Franciscan Order and creating a new rule. The Rule of 1223 was less strict than the first Rule of 1209. The Roman curia had played an active role in its reform, and it reflected the Vatican's wishes.

Francis quickly withdrew from the day-to-day operations of the order. In summer of 1224, he retreated to La Verna (Alverna) in the mountains near Assisi. While in a mystical trance, he received the stigmata, the signs of the five wounds of Jesus, on his hands, his feet, and his side. Francis took great care to hide the stigmata, and news of it only broke following his death two years later. For the remainder of his life, he lived in constant pain and in near total blindness from an eye disease that he had contracted in the Holy Land. After medical treatments in Rieta and Siena failed, he was brought back to the Portiuncula in Assisi where he died on October 3, 1226. He was buried temporarily in the church of St. George in Assisi. Two years later he was canonized as a saint by Gregory XI, who also laid the first stone of the new basilica, built in his honor by his successor and friend, Brother Elias. His body was moved to its permanent resting place in 1230.

The Basilica of San Francesco commemorates the life of St. Francis. A majestic double church, it was completed and consecrated in 1253. The basilica contains frescoes by some of Italy's greatest masters, including Cimabue, Pietro Lorenzetti, Simone Martini, and reportedly Giotto. Some of Francis's followers regarded the opulence of the basilica as a direct contradiction to his teachings.

For hundreds of years it was generally believed that Giotto was the master behind the twenty-eight frescoes of the cycle of St. Francis's life in the Upper Church. In 1912, Friederick Rintelen, an eminent German art scholar, argued against this attribution by contrasting the St. Francis cycle with the known Giotto frescoes in the Arena Chapel at Padua. His findings unleashed a controversy that has endured to this day. Among the numerous theories that have surfaced since

Rintelen's declarations, one contends that Giotto may have originally started the cycle, which was then completed by others. Milliard Meiss, after careful study of the cycle, has argued that at least three different styles can be found in the frescoes, one of which may indeed be that of Giotto. Despite the controversy, all art scholars agree that the St. Francis Basilica is indeed a treasure trove of Italian art.

In the fourteenth and fifteenth centuries, Assisi fell under the powerful noble families that dominated Italy—the Visconti, the Montefeltro, and the Sforza. In the bitter conflicts that marked this period, it was often sacked and divided before lapsing into relative peace as part of the States of the Church. Assisi regained its stature as one of the spiritual centers in Europe in the twentieth century when the crypt tomb of St. Francis was restored in preparation for the 700th anniversary of his death. More than 2 million pilgrims passed by the tomb and renewed Assisi as a major pilgrim site. In recognition, the Germans designated Assisi as a hospital town in 1944 in order to spare it any destruction. Today, despite the 2 million visitors who flock to the mountain city every year, Assisi has managed to retain the sense of tranquil refuge and isolation it has enjoyed throughout much of its history.

Further Reading: Lina Duff Gordon's *The Story of Assisi* (London: Dent, 1900; New York: Dutton, 1913) is a thorough, though somewhat dated, history of this Umbrian city. Of the many biographies of St. Francis, Paul Sabatier's *Life of St. Francis of Assisi*, translated by Louise Seymour Houghton (London: Hodder and Stoughton, and New York: Scribner's, 1894; as *Vie de S. Françoise d'Assise*, Paris: Fischbuchner, 1894) is perhaps one of the best. Those interested in the controversy concerning the frescoes in the St. Francis Basilica should consult Milliard Meiss, *Giotto and Assisi* (New York: New York University Press, 1960).

—Manon Lamontagne

Athens (Attica, Greece): Acropolis

Location: In the heart of Athens, the capital of Greece.

Description: A large, limestone rock used by inhabitants of the city-state of Athens for shelter, defense, and worship. Now an archaeological site; sections are open to the public for viewing.

Site Office: Acropolis Museum
Athens, Attica
Greece
(1) 323-6665

The Acropolis of Athens is a timepiece in and of itself, a source of history that spans the several thousand years that people have inhabited the city and its surrounding areas. Just by looking at the layers of natural and man-made rock, historians are able to gain a better understanding of the groups who built, rebuilt, and destroyed the structure during their periods of rule. From the number and scale of monuments erected on top of the rock throughout various stages of Athenian history, it is easy to see that the Acropolis has been the site most sacred to every group of people who inhabited the city of Athens over the past 7,000 years.

Both the size and the physical characteristics of the Acropolis have made it attractive since its earliest occupation. The rock, primarily made up of semi-crystalline limestone, measures about 886 feet by 512 feet by 200 feet high at its greatest dimensions. It has steep sides and a large, flat top. While the Aegean Sea is easily accessible, it was purposefully set at some distance from the Aegean to avoid enemy or pirate raids by sea. The natural springs that flow throughout the rock were also attractive to the first inhabitants during the Neolithic period, around 5000 B.C.

The Greek historian Thucydides detailed the way Greek cities were traditionally founded and planned out; Athens and its Acropolis fit his description perfectly. The word "Akropolis" means "upper city," the central site around which city-states were built. Most of the buildings in the city-state were located near the base of the Acropolis so it would be easy for the inhabitants to take refuge on the more heavily protected sides and top of the rock in the event of a war or a raid. Throughout its existence the Acropolis has also housed the most important buildings in the city-state, be it the palaces of rulers, the city's treasury, or, in later times, the religious monuments, thus making them easier to protect in times of conflict.

The earliest known structures to appear on the Acropolis were walls built by the Mycenaeans, toward the end of the Bronze Age. The principal building during this time was the Royal Palace, which stood near the Old Temple of Athena. The Mycenaean civilization was based on a monarchic system of government; the most important buildings were the palace, citadel, and private residences built to house the king and his advisers. Naturally, these buildings were constructed on top of the Acropolis. These first crude attempts at fortification of the palace are now known as cyclopeian walls, large stones unevenly stacked on top of each other that were filled in with smaller stones and a clay mixture. Additional fortifications guarded the Acropolis's main western approach. Evidence of substantial protection of the Acropolis at this time indicates that Athens during the Mycenaean period was an important political and economic force in the region.

Mycenaean civilization flourished until 1200 B.C., when catastrophe struck, but exactly what this catastrophe was is not fully known. According to Egyptian records, it was caused by an invasion by the Sea Peoples. For the next 150 years, sub-Mycenaean people roamed the region, but not much remains of their culture and little is known about them.

Following the Mycenaean era, in the years from 1050 B.C. to 750 B.C., no major construction was done; all that survives from the era is the pottery. A group known as the Dorians had moved onto the Greek peninsula, but did little in the way of further development of the Acropolis. A revolving-door parade of tyrants followed the Dorians during a period known as the Archaic period, once again little affecting the actual structure of the Acropolis.

By the end of the sixth century B.C., Persia had established itself as the major power in the East. The Persian king Darius I had already captured Greek territory in Asia Minor and Thrace, and now set his sights on taking the Greek peninsula. Following several victories on the northern part of the peninsula, his troops began to move toward Athens, hoping to engage the Greeks at Marathon, where they believed they would have the tactical advantage. Outnumbered three to one, troops from Athens moved to meet the advancing Persians and dealt them a staggering defeat; the year was 490 B.C. This event quickly became key to the Acropolis; following the defeat at Marathon, Darius left his dream of controlling the Greek peninsula to his son, Xerxes I, who set out to destroy the Greeks with a new zeal, using the full thrust of the Persian forces. One of Xerxes's particular goals was the destruction of Athens, which had provided the troops that had shocked and destroyed his father's forces at Marathon ten years earlier. Xerxes methodically planned his advance toward Athens, and executed it to perfection with the help of an Athenian traitor at Thermopylae. The door had been opened for an advance on Athens.

News had traveled to the Athenians that a huge Persian force was advancing toward their city. The city was soon deserted, as women and children were evacuated, and all the available men were away with the naval fleet. A few citizens remained in the city after misconstruing the advice of the Oracle at Delphi, which told them to flee immediately on ships made of wood. These few interpreted the message to

Overview of the Acropolis

The Erechtheion
Photos courtesy of The Chicago Public Library

mean that if they hid themselves on top of the sacred Acropolis behind wooden planks, they would be saved. The Persians advanced into the city unabated, destroyed the token resistance, and burned and ruined the Acropolis.

All that remained of Athens after Xerxes's victory was its naval fleet, which bided its time and planned retaliation against the Persians. The navy was able to lure and surprise the Persian fleet at Salamis; the victory won there and subsequent victories at Plataea and Mycale in 479 B.C. drove the Persians from the peninsula and set the stage for a prosperous time in Athens, which included a construction plan for the Acropolis that would change the face of the rock forever.

Following the battle of Plataea, the residents of Athens returned to find their Acropolis destroyed. Plans were laid out in the short term to patch up some of the damage, including repairing the gateway and converting a part of the Old Temple of Athena into a treasury. The old fortification walls were almost completely destroyed, while other walls were in need of reinforcement. This work was completed, but some of the damaged areas were left as they were to remind the Athenians of the dark time of the Persian invasion.

The Athenians and Persians finally made peace in 449 B.C. With the new peace, funds that had been reserved for defense became available, and under the guidance of Pericles, a new beautification and building phase began. Not only did Pericles want to fully reconstruct the buildings on the Acropolis, he also wanted to build new structures to honor the gods for protecting them from the Persians, and to symbolize the victory in the wars. One of Pericles' dreams was the construction of a new temple on the Acropolis for Athena, the patron goddess of the city.

Pericles assigned the tasks of building the monuments to Ictinus and Callicrates, and the decoration of the monuments to his good friend Phidias. The monument dedicated to Athena is known as the Parthenon, and work began on it immediately; the structure was finished and dedicated in 438 B.C. The Parthenon, meaning "the virgin's apartment," was the largest and most intricate structure of the building program; its completion was a symbol of the rebirth of the city of Athens after the Persian occupation.

With the exception of a wooden roof, the Parthenon was constructed completely of marble, and is a masterpiece of the Doric style of architecture. It is situated on the highest point of the rock, midway between the east and west ends, on the site of past monuments to Athena, which had been destroyed by either the Persians or other invading groups. Inside the structure a forty-foot-high statue of Athena was designed and constructed by Phidias in gold and ivory. All traces of this great statue have disappeared; the only evidence of its existence is in historical writings. During ancient times, not only did the Parthenon serve as a place of homage and worship to Athena, but also as the treasury for all of Athens.

The monument has experienced numerous alterations throughout its existence. Alexander the Great presented gilded shields of bronze, taken from the Persians, to the statue during his reign. They were later removed by the Athenian military commander Lachares during the siege by Macedonian king Demetrius I Poliorcetes in 298 B.C. The statue of Athena had disappeared by the fifth century A.D. In the sixth century A.D. the Parthenon was converted by Justinian into the Church of Holy Wisdom, and later became the metropolitan church of Athens. It was converted to a Roman Catholic church when Athens was occupied by the Franks from the thirteenth through the mid-fifteenth centuries, and then later became a mosque. In the seventeenth century, the Turks converted the Parthenon into an ammunition dump, which exploded when a mortar was fired by the besieging Venetians. Columns on both the north and south sides were destroyed, along with nearly all of the cella and its frieze. Further damage was sustained during subsequent wars, particularly during World War II.

Other important structures developed and erected during the time of Pericles include the Propylaea, the Temple of Athena Nike, and the Erechtheion. The Propylaea is the marble entranceway to the Acropolis, built in Doric and Ionic style by the architect Mnesicles under the rule of Pericles during the years 437 B.C. to 432 B.C. It runs for 150 feet across the western end of the Acropolis, still the only entrance to the summit. It was never completed, having been abandoned at the start of the Peloponnesian War and never resumed. Historians speculate that Mnesicles could not carry out his original plans for the structure's design because of the enormous cost of finishing the Parthenon, and the fact that a war was on the horizon. The remains of the structure indicate a lack of the precise adornment that is found on the Parthenon, along with stunted building patterns on the east and southwest ends of the wall.

The Propylaea was kept in nearly its original state until the thirteenth century, when it was used as a palace by the Byzantines. Additions to the structure were made by both the Franks and Turks, but these were removed in 1836 by the Greek ruler Pittakis. During their reign, the Turks also covered the vestibule and converted it into a powder magazine, but it was struck by lightning in the seventeenth century, causing substantial damage to the east portico and two of the columns. Reconstruction and refurbishment took place in the early twentieth century.

The Temple of Athena Nike (meaning "bringer of victory") stands on a platform reaching above the southern end of the Propylaea. On Pericles' order, Callicrates was assigned as the architect for the project, which began on a small scale in 449 B.C. to celebrate the new peace and victory over Persia. The temple, which is still in existence today, was constructed from 427 B.C. to 424 B.C. It is a small structure, standing only twenty-seven feet long by eighteen feet wide, surrounded by eight columns designed in the Ionic style. An eighty-six-foot-long frieze surrounds the temple, depicting scenes of mythology and famous Greek battles. This work was destroyed by the Turks because of its high position as an artillery site in 1686, but was perfectly reconstructed in the 1800s and then dismantled and reconstructed in hopes of strengthening its fragile structure during the years 1936 to 1940. It is no longer open for public viewing.

The Erechtheion was the last major part of Pericles' construction program. Its original construction was delayed by the start of the Peloponnesian War, but was nearly completed by the year 409 B.C. The site on which the monument is constructed has historical significance for Athens. It is said to be the spot where the gods Poseidon and Athena battled for supremacy of the city; where Kekrops, the founder of Athens, was buried; and also where Athena's ward Erechtheus (the monument's namesake) was buried. The structure, built in Ionic style, was damaged by a fire in 406 B.C., and not rebuilt until 395 B.C. It was converted into a church around the sixth century A.D., and later converted into a home for the harem of the Turkish commandant of the Acropolis in the 1400s. Wars and weather tore away at the structure, and in 1979, a massive restoration project was undertaken. It was successful in returning the Erechtheion nearly to its original form, and was finished in 1987.

In recent years acid rain and normal weathering have begun to wear away the marble of which the monuments are constructed. In addition, iron clamps used in an attempt to strengthen the structures during the nineteenth century actually caused further weakening, and the large amount of visitor traffic is beginning to wear away the floor surfaces.

The Greek government responded to the problems by toughening environmental standards, closing off most of the Acropolis to visitors, and undertaking an attempt at restoration. The only building that may now be entered by visitors is the Propylaea, and no automobile traffic is allowed on any of the approaching roads. Archaeologists have concluded that any attempts at protection of the marble will do more harm than good, leaving the Acropolis to continue to deteriorate. Limited attempts at gentle restoration do continue however, including one to the Parthenon, which was begun in 1983 and expected to conclude in 1995.

Because of the stringent rules for visitors, only a small portion of the Acropolis can now be viewed closely. The Acropolis Museum is on site, and is full of objects that have been removed from other Acropolis structures in order to prevent further deterioration. Several notable sculptures are on display inside the museum, including one of Athena leaning on her lance.

Further Reading: An excellent look at Greek culture, architecture, and history is *Greek People* by Robert B. Kebric (Mountain View, California: Mayfield, 1989). Additional books that focus on Athens and its important historical structures are *These Were the Greeks* by H. D. Amos and A. G. P. Lang (Chester Springs, Pennsylvania: Dufour, and Amersham, Buckinghamshire: Hulton, 1979) and *The Stones of Athens* by R. E. Wycherly (Princeton, New Jersey: Princeton University Press, 1979).

—Tony Jaros

Athens (Attica, Greece): Agora

Location: Situated in Athens, the capital of Greece, near the Acropolis, to the south of the Areopagus Hills.

Description: "Agora" literally means "Gathering Place" or "Assembly." The Agora was a central part of the ancient Greek city, used for religious, political, cultural, and social events. It is estimated that development of the site began in the sixth century B.C. Whenever Athens was invaded, the Agora was pillaged, most notably in 480 B.C. by the Persians, and in 267 B.C. by the Herulians. Socrates and St. Paul both held discussions in the Agora. A Roman Agora was built in the first century B.C.

Site Office: Agora Museum
148 Ermou Street
Athens, Attica
Greece
(1) 346-3552

The Agora is one of the most visited historic sites in Greece. It is, however, almost impossible to grasp the importance of the site from the archaeological ruins that exist today. The Agora was an integral part of the Greek city. It was the center of social, cultural, and political life. The importance of the Athenian Agora reportedly dates back to the time of the hero-king Theseus, who according to Greek mythology was the founder of Athens. From its beginnings, Athens was divided into two parts: the Acropolis, the higher city, and the Asty, the lower city, which included the Agora and living quarters.

Originally occupying open ground to the northwest of the Acropolis, the Agora became the official site of athletic and artistic events, political discourses, markets, festivals—any social event taking place in Athens. The arrangement of the Athenian roads, in particular the Panathenaic Way, which led from the main gate (the Diplyon) to the Acropolis, dictated the position of the Agora. As the reputation and influence of Athens grew both in artistic and commercial terms, the Agora increased in size and importance. The legislative and political reforms implemented by Solon the Law-Giver in the sixth century B.C. heralded a period of sustained growth. As the borders of Athens expanded, the Agora became the focal point of the city, and with its increasing importance came the construction of buildings and roads. Solon moved the site of the Agora to the Agoraios Colonos in the sixth century B.C.

Following the Persian invasion of Athens in 480 B.C., the Agora was virtually destroyed, and most of the newly constructed buildings lay in ruins. During the fifth century B.C., most of the west side of the Agora was restored and civic buildings were erected on the north and south sides of a square surrounded by stoae or porticoes on three sides. The Thesseion, the Temple of Hephaistos, was also built during this period.

During the fifth century B.C., Athens was fortified. The Themistocleon wall and the Long Walls were erected; the latter effectively joined two cities by linking Athens to the port of Piraeus. Consequently, the Agora became the administrative center of Athens.

The second half of the fifth century B.C. marked an increase in building activity that would alter the appearance of the city. New buildings appeared on the Acropolis and on the south side of Athens. As the political and cultural reputation of the city became known, Athens attracted many enemies. The enmity that existed between Athenians and Lacedaemonians culminated in the outbreak of the Peloponnesian War, which began in 431 B.C. and ended with the Spartan victory in 404 B.C. During the fourth century B.C., the city's fortifications were further strengthened by the addition of breastwork and a moat.

Over the next 200 years Athens suffered from factional rivalry. However, the Hellenistic period and the second century B.C. marked a period of great construction, during which the Agora assumed its classic size and shape, following the erection of huge colonnades to the north and south of the site.

In 86 B.C. Athens was invaded by the Romans under General Sulla. The fortified walls were completely destroyed, and Athens effectively lay open to invasion for the next 200 years. The southern side of the Agora was desecrated, and reconstruction was slow.

During the Roman occupation of the city, many important edifices were erected. In 15 B.C. the Roman politician Agrippa commissioned the building of the Oedion, a concert hall. The Romans were also responsible for dismantling several impressive rural temples and buildings and relocating them to more central sites. The Temple of Ares is thought to have been removed from Acharnai at the foot of Mount Parnas to the Agora.

One of the most significant buildings of this period was the Agora of Caesar and Augustus, more popularly known as the Roman Agora. Built to the east of the Greek Agora, it was erected during the reign of the emperor Augustus in the first century B.C. Conceived as the new commercial center, the Roman Agora was linked to the old Agora in A.D. 100 by a street lined with stoae, which began at the Library of Pantainos and ended at the West Propylon. Later, latrines (Agoranomeion) and the Horologion of Andronicus Cyrrhestes were added to the east of the Roman Agora.

During the second century A.D., Athens was further developed as Hadrian's new city evolved. Among the build-

The Agora
Photo courtesy of Greek National Tourist Organization

ings constructed were the Panhellenion and the Odeion of Herodes Atticus. In A.D. 267, Athens was raided by Herulians, northern barbarians. Despite the earlier reparation and fortification of the city's ramparts by Emperor Valerian, the city was once more nearly destroyed. The stone from the destroyed city was used to construct a new defense wall north of the Acropolis. Extending from the Acropolis to the Roman Agora and Hadrian's Library, this wall enclosed the main center of Athens until well into the nineteenth century.

The Agora ceased to be an important area when its function as administrative center ended in the fifth century A.D. Most of the construction associated with the Agora after this time was linked to Athens's function as a cultural and philosophical center. A large number of houses were built south of the Agora, and in A.D. 400 a large gymnasium was built among the rubble of previously glorious buildings. This burst of activity ended in A.D. 529 with the rise of Christianity and the closure of the philosophical schools by Justinian.

During the late sixth century, Athens was pillaged by the Slavic peoples. After this the area surrounding the Agora was largely deserted. Gradually this once popular site became covered with silt. The Agora suffered once again during the Greek war of Independence in the 1820s.

From the 1830s onward, politicians and scholars called for a serious excavation of the ancient city. In the nineteenth century a group of Greek archaeologists excavated part of the gymnasium. The most methodical and successful excavation was carried out under the auspices of John D. Rockefeller Jr., who funded the American School of Classical Studies to begin work in Athens. With the cooperation of the Greek government, the school began work on the area around the Agora in 1931, interrupted only by World War II. Today the visitor to the Agora benefits greatly from the efforts of both Greek and American archaeologists.

The Stoa of Attalos, originally built in 150 B.C., was restored by the American School between 1953 and 1956, and inaugurated as the Agora Museum in 1956. Originally erected by King Attalos II of Pergamum as a gift to the people of the city, the Stoa of Attalos served the dual purpose of shielding the east side of the main square, and providing a meeting place for wealthy Athenians. Two stories high, the stoa housed twenty-one shops on each level. It had forty-five Doric columns on its lower floor and forty-five Ionic columns on its upper level. Today the facade is plain Pentelic marble instead of the red and blue painted exterior that the original stoae would have had. The restored model of the stoa has been adapted to house the almost 200,000 objects retrieved from the Agora site, which include reliefs from the Temple of Ares, and a fifth-century B.C. torso of Athena.

On the north side of the western section of the Agora lies the Temple of Thesseion, one of the best-preserved buildings of the site. Built by the architect Ictinus (who is also thought to have designed the Parthenon) in approximately 449 B.C., the Thesseion has thirty-four Doric columns. It houses a famous frieze depicting the labors of Hercules. During the fourteenth century, the temple was converted into the Church of Agios Georgios. Originally dedicated to Hephaestus, the god of metallurgy, the temple was surrounded by forges and foundries. In the Roman period it was surrounded by a garden.

To the northeast of the Thesseion stands the Stoa of Zeus Eleutherios, where Socrates discussed his political and philosophical theories. North of this is the Stoa of Basileios (Royal Portico) where the Archon Basileios held court. It is also the site of the Lithos, the stone at which the Archons took their oaths of honor.

In the period from 1980 to 1982 the west end of the Stoa Poikile (Painted Stoa) was discovered. The stoa was decorated with murals by painters of the famous fifth-century B.C. school of artists including Polygnotus, Micon, and Panainos, the brother of Phidias. This building gave its name to the Stoics, the school of philosophy founded by Zeno of Citium, who taught there around 300 B.C.

To the southeast of the Temple of Thesseion lies the New Bouleuterion (Council House), which was created by Solon the Law-Giver to house the meetings of the council (Boule). It was on this platform, cut out of the hill of Kolonos, that the council discussed important legislative matters. The minutes from these meetings were archived in the Metroon, the sanctuary of the Mother of the Gods (Meter Theon).

On the west side of the southern portion of the Agora lies the circular Tholos. Built between 470 and 460 B.C., it was the lodging and canteen of the legislative body, the Council of Four Hundred. In the main square of the Agora, for the most part covered by a railway, lies the Altar of the twelve Gods. Created by Peisistratus the Younger around 522 B.C., it became a place of sanctuary. It was also the place from which distances to other towns and sites were measured.

To the southeast along the Panathenaic Way, the markings and arranged foundations of the Temple of Ares are visible. Its design is also attributed to Ictinus. Towering over this to the south are the Stoa of the Giants. Originally conceived as decorations for the facade of the Odeion, they were built around 15 B.C. by the Agrippa. The Giants were originally made up of six statues, a combination of Giants (human and snake) and Tritons (human and fish).

The ruins of the Odeion and the later gymnasium lie behind the stoa. The Odeion was originally surrounded on three sides by a two-story portico and was entered through the stoa that lay to the south. Only a few remnants remain of the original building, which seated approximately 1,000 people. The Giants were later reused in the facade of the gymnasium built in the fifth century.

Across the Panathenaic Way and in front of the Stoa of Attalos lies the area where the Library of Pantainos originally stood. Built around A.D. 102 by Titus Flavius Pantainos, it was dedicated to both Emperor Trajan and Athena. Outside the Agora lies the fifth-century prison where Socrates was held before his death by hemlock. Although little remains of the original Agora, the Agora Museum, with its detailed explanations and restorations by archaeologists who worked on the excavation of the Agora, help the visitor to gain an insight into the life and society of ancient Athens.

Further Reading: *Greece: A Travel Survival Guide* (Hawthorne, Australia, and Berkeley, California: Lonely Planet, 1994) is a good basic introduction to the country. *Athens* by Michael Anz (London: n.p., 1926) is a pictorial guide to the city. *The Athenian Agora: Excavations in the Heart of Classical Athens* by John M. Camp (London and New York: Thames and Hudson, 1986) is a detailed examination of the site.

—Aruna Vasudevan

Ávila (Ávila, Spain)

Location: On the Adaja River, a tributary of the Duero, in central Spain; 54 miles west of Madrid; at 3,715 feet above sea level, the highest city in Spain.

Description: The capital of the province of Ávila, surrounded by magnificent twelfth-century walls and built on a hill on the banks of a mountain river.

Site Office: Oficina de Turismo
Plaza de la Catedral, 4
05001 Ávila, Ávila
Spain
(18) 21 13 87

The city of Ávila has been described by Miguel de Unamuno, the Spanish philosopher and writer, as "a diamond of stone made golden by the suns of centuries and centuries of suns." Constructed from the light-brown granite common to many of the cities of old Castile, Ávila does indeed have the appearance of glowing in the sun on its perch atop a rocky hill. The city was founded by Raimundo de Borgoña in the eleventh century on the banks of the Adaja River. Situated at such an altitude and so far from the coast, the city is well known for the extremity of its climatic variation. However, despite the interminable, bitterly cold winters, and the scorching summers that frequently bring water shortages to the treeless, arid province, much of the city's economy is based on agriculture and the manufacture of agricultural products.

Ávila's traceable historical origins date from the Romanization of the Celtic tribes of central Spain. Although undoubtedly best known for its magnificently preserved town walls, the city's most marked characteristic is its visible sense of devotion to its patron saint, Teresa of Jesus. Apart from the number of monuments dedicated to the saint, the local delicacy, *yemas de santa Teresa,* made exclusively from egg yolks and sugar, and her famous marble statue, there are reminders of her works and energetic spirit all over the city. Born in Ávila, Teresa entered the Carmelite order and in 1554 underwent a strong religious experience, after which she was frequently subject to ecstatic and other mystical experiences. She went on to re-found the *discalced* (barefoot) branch of the Carmelite order according to its original spirit. Before her death, she founded seventeen female and fifteen male monasteries. She was canonized in 1622 and made a doctor of the church in 1970. For some time her spiritual director was St. John of the Cross, an important mystical poet and theologian in his own right and now the city's second patron saint.

Of the many religious festivals celebrated in Ávila, the most ostentatiously observed is that of St. Teresa, which begins annually on October 8 and culminates on her feast day, October 15. The festivities include open-air dancing and immensely popular street entertainments and banquets. All the festivals in Ávila have religious origins, with many, particularly those held in the summer, featuring the traditional bullfights and bull runs, and processions conducted in folk costume.

If Ávila's origins are remote, the town's better-documented history begins after the Moorish conquest and the Christian reconquest by Alfonso VI, who started the massive repopulation of the nearly deserted province, enticing peasants down from the mountains in the northeast in exchange for privileges. Around the year 1080, Ávila began to prosper. Wealth and population eventually diminished, in the eighteenth and nineteenth centuries, as the oppressive manorial and feudal systems came to hinder progress. At the beginning of the twentieth century, the mining industry had all but disappeared, and the population was less than 12,000.

On the road from Salamanca to Ávila there is a notice in the form of a greeting to the visitor proclaiming "Ávila, the greatest walled city in the world." The impressive crenellated walls surrounding the old city do indeed dominate the surrounding landscape and can be perceived in their splendor from a distance. It is quite easy to believe that no other city walls in existence are equal in magnificence and preservation. The huge blocks of stone used are of the same golden granite as many of the city's buildings. Construction began in 1090 under the direction of Raimundo de Borgoña, who chose the designers and the thousands of Christian workers and Moorish slaves for the great task. The cement was blessed by the bishop of Oviedo, and the walls followed a Romanesque design with traces of Moorish decoration.

Entirely intact, the perimeter of the walls measures more than one and one-half miles and is interrupted every fifteen feet or so by one of ninety fortified lookout towers or by one of nine splendid fortified doorways. The sentry path can be entered by way of a stair on the Alcázar castle doorway and provides a good view of the surrounding area. This doorway and that of San Vicente were the first to be constructed and prevented access to the city from the plain to the east, the only side of which the terrain does not provide a natural defense. The rectangular shape of the walls is reminiscent of a fortified Roman camp; indeed, it is widely held that Ávila was of some importance under the Romans and was surrounded with walls during that period. Those walls may, in part or in whole, coincide with the perimeter of the eleventh- and twelfth-century walls that remain today.

The north part of the wall has few openings, although it is here that the most primitive one is located: the Portillo del Mariscal. The most splendid doorways are those that mark the entrance to the Alcázar and to the cathedral. The elegant San Vicente doorway is also particularly beautiful. It opens onto the basilica of San Vicente, which is made of blocks cut from red and yellow sandstone. The enormous basilica took two centu-

City walls of Ávila
Photo courtesy of Instituto Nacional de Promocion del Turismo

ries to build, from the twelfth to the fourteenth, and is said to have been constructed over the site where St. Vincent of Saragossa and his two sisters, Sabina and Cristeta, were martyred in the third century during the persecution of Decius.

Inside is the magnificent martyrs' tomb, an interesting twelfth-century Gothic work with oriental touches such as a roof in the shape of a pagoda. The sides of the tomb are engraved with the story of the saint's life, his capture in Ávila with his sisters, and their stripping, torture, and martyrdom.

It is a work of immense technical skill and evocative power. On this sepulchre, knights rested their hands in homage and swore oaths, and for many years it was also on the martyrs' tomb that criminals were made to rest their hands when pleading innocent or guilty to their crime, in the belief that an untruth would be impossible here. The "Catholic monarchs," Ferdinand and Isabella, put an end to this ceremony, but a sense of the supernatural is said to have pervaded the basilica and to have surrounded the tomb for centuries.

Another basilica legend involves the underground chapel and the crypt of St. Vincent, which also houses the chapel of St. Peter of the Boat. St. Peter died in the twelfth century on the Rio Tormes and a mule carried him to the city in the exact direction of the basilica when it suddenly fell down dead, leaving a mark on the ground that can still be seen today under a protective guard. This was interpreted as a miracle and the basis on which the basilica was accepted as the burial place of litigants.

Ávila particularly flourished at a time when the Romanesque style was at the height of its popularity. The great number of Romanesque works is also explained by the fact that the period coincided with that of the repopulation of the city in the twelfth and the thirteenth centuries. Among the most important of the Romanesque works is the cathedral, built both as a place of worship and a fortress. It was known as *El templo y la fortaleza,* symbolizing the cross and the sword. The cathedral was constructed during Alfonso VII's reign, although it had been started in 1157. It was built of the same yellow granite as the walls, and enhances the austere appearance of the ramparts and double battlements that are incorporated into the east side of the city walls.

In direct contrast to the buttressed exterior, the inside of the cathedral is the sumptuous setting for a number of remarkable works of religious art. The entrances open onto a central nave and two aisles. The long thin nave gives on to the masterpiece of the sculptor Vasco de la Zarza: the alabaster tomb of Cardinal Alonso de Madrigal, the fifteenth-century theologian and bishop of Ávila popularly known as "el tostado" (the swarthy). Best known for his prolific literary output (the main body of which contains 34 works consisting of 6,000 pages), Madrigal is depicted in the act of writing. The sepulchre is dated 1520 and the cardinal's remains were transported there from their primitive burial place on February 10, 1530.

The cathedral also contains an altarpiece from 1500, made in gilded wood and carved in a variety of styles with Italianate pilasters. Surrounding the large altar are five Renaissance panels depicting saints, finely illuminated by Romanesque windows. Above the choir stalls, in which the polyphonist composer Tomás Luis de Vitoria once learned to sing, are elaborate carved wooden panels with a row of wooden statues. With the sensitive carving of the decorative *trascoro* running behind the choir stalls (which was begun in 1536 and took eleven years to complete), the impressive ensemble is regarded as one of the major achievements in sixteenth-century Spanish carving. The chapter museum, which contains important paintings, is lit by well-known stained-glass windows.

Outside the city walls is the Gothic Dominican monastery of St. Thomas. Construction began in 1483. The building was originally erected as a cool summer residence for Ferdinand and Isabella, who furnished it magnificently and embellished its interior not suspecting that the building would house the sepulchre of their only son, the *infante* John, who died at the age of nineteen in 1497. Queen Isabella commissioned the magnificent sepulchre from Dominico Fancelli, the Florentine who also constructed the mausoleum of the Catholic monarchs themselves in the Capilla Real at Granada.

The monastery also became the home of the University of Ávila, which was founded in 1482 and suppressed in 1807. The monastery also served as the seat of the Inquisition, which in Spain was not an ecclesiastical but a civil court. The sepulchre of the celebrated inquisitor Tomás de Torquemada, who died in 1498, is in the sacristy, and on the main facade is an enormous "H" standing for the concept of *Hispanidad,* the quality of being Spanish. From the central cloister of the famous trio, novices' cloister, silent cloister, and cloister of the Catholic monarchs, a staircase leads to the choir gallery. The fifteenth-century Gothic choir stalls are carved in walnut with canopies and arabesque. The monastery contains the celebrated retable of Pedro Berruguete, court painter to Ferdinand and Isabella. Ten of his panels from Ávila, now in the Prado, demonstrate the power of Flemish inspiration in Spain as early as 1500, sixteen years before the Flemish Charles I became king of Spain.

Further Reading: Apart from the numerous tourist guides, atlases, gazetteers, and encyclopedia articles, there is very little in English on the town of Ávila, although there are many lives of Teresa of Ávila and commentaries on her works and spiritualities. On Ávila itself, there is a classic early–seventeenth-century work, *Historia de las grandezas de Avila* by L. Ariz. Felix Hernández Martín has written a well-illustrated account of the town, *Ávila,* fourth edition (Leon: Editorial Everest, 1975). Of more general histories of Spain relevant to Ávila, important accounts include *The Kingdom of León-Castilla under King Alfonso VI, 1065–1109* by Bernard F. Reilly (Princeton, New Jersey: Princeton University Press, 1966); *Enrique IV and the Crisis of Fifteenth-Century Castile, 1425–80* by William D. Phillips Jr. (Cambridge, Massachusetts: Medieval Academy of America, 1978); and L. P. Harvey's *Islamic Spain, 1250–1500* (Chicago: University of Chicago Press, 1990).

—Clarissa Levi

Badajoz (Badajoz, Spain)

Location: Four miles from the border of Spain and Portugal, along the Guadiana River.

Description: Capital of the province of Badajoz, in the region of Estremadura; occupied since prehistoric times; capital of a Moorish kingdom; after Christian reconquest of Spain, area frequently contested between Spanish and Portuguese; site of battles during the War of the Spanish Succession, Peninsular War, and Spanish Civil War; retains much Moorish and medieval architecture; neighboring town of Mérida has extensive Roman and Visigothic remains.

Site Office: Tourist Information Centre
Plaza de Libertad 3
Badajoz, Badajoz
Spain
(924) 31 53 53

Badajoz and its environs offer a microcosm of Spanish history. The site has been occupied since prehistoric times. Excavations nearby on San Cristóbal Hill have uncovered human traces dating from the Lower Paleolithic period. Badajoz itself was not founded until the Roman period, when it was established as the town of Colonia Pacensis. However, Roman remains in the city are few. Much more can be seen in the nearby town of Mérida, which contains the best-preserved architectural remnants of the period. Mérida was founded around 25 B.C. as a colony for retired soldiers of the Fifth and Tenth Legions and was given the name Augusta Emerita. Soon it became a provincial capital, containing about 50,000 inhabitants, which made it the largest city on the Iberian Peninsula. It also served as a political and cultural center, and this function is reflected in Mérida's extensive Roman remains.

Mérida is dominated by the Roman theatre and amphitheatre near the center of the town. They currently form part of a public park. The theatre was constructed between 24 and 16 B.C. by General Agrippa, Augustus Caesar's son-in-law. It contains seating for about 6,000 spectators in the classic tradition: a vast stone semicircle with a pit for the orchestra or chorus. During the reign of Emperor Hadrian (between A.D. 117 and 138) the back wall of the stage, which had been damaged in a fire, was rebuilt and decorated with a colonnade with statues placed between the columns. Behind the stage was a promenade where spectators could stroll during intermissions between the acts. The theatre has been restored and serves the community as the site of a summer drama festival.

The amphitheatre was completed around 8 B.C. and was intended for staging gladiatorial combats, chariot races, and *naumachiae,* mock battles between ships for which the amphitheatre was filled with water. It originally had seats for around 14,000 spectators, but most of these have disappeared, with the stonework from them going to repair other local sites. Some of the seats have been restored, however, in the area around the *vomitoria,* the tunnel-like exits through which the large crowds passed. The National Museum of Roman Art (Museo Nacional de Arte Romano) stands in the same park area as the amphitheatre and the theatre. It was constructed in 1985 and houses archaeological finds from the city, including jewelry, coins, and pottery, statues from the theatre and bits of the frieze from the city's forum. Also included are mosaics from some of the local houses.

Some distance away to the northwest, toward the little Rio Albarregas, lie the ruins of the Roman hippodrome (Circo Romano), the only one known to exist in Spain. It is still unrestored and is largely overgrown. Across from the theatre and amphitheatre are the remains of the Roman baths (termas), and a Roman house dating from the first century A.D., which retains many of its original wall paintings and mosaics. Some of these are in excellent condition, including a famous one called *Autumn,* depicting the treading of grapes for wine. Others show scenes from the daily life of the villa or depict geometrical patterns.

Northeast of the town are the remains of the town's principal Roman aqueduct, the Aqueduct of the Miracles (Acueducto de los Milagros). Some thirty-seven piers still exist, supporting ten arches that still stand up to three stories high. It was built to provide Mérida with water from the Prosérpina reservoir located about six miles outside the town. Some distance to the south stand Trajan's Arch (Arco de Trajano), a triumphal arch which marked the site of the Roman town's northern gate, and the Temple of Diana (Templo de Diana). Unlike many other pagan temples, the Temple of Diana was not turned into a Christian church, but was instead incorporated into a private dwelling during the sixteenth century. Historians have discovered that the temple was not in fact dedicated to the Roman Diana, but to another, unknown, goddess. Still farther south lie the bullring (Plaza de Toros) and the remains of the Casa del Mithraeo, a Roman villa in which the secret rites of the god Mithras were practiced.

North of most of the Roman ruins are buildings that date from the time of the Visigothic occupation of Spain. The Visigoths were a Christianized Germanic tribe that originated in what is now the Ukraine. Pressures from tribes farther east, especially the Huns, during the fourth century A.D. forced them to migrate into Roman territory. In A.D. 376, the Visigoths requested permission from the emperor at Constantinople to enter imperial territory. Although permission was granted, the Visigoths soon began to attack their new neigh-

The Alcazaba at Badajoz

bors within the empire. Emperor Valens gathered a force and attacked the Visigoths at Adrianople in the Balkans in 378. He was defeated and killed. In 410, under their leader Alaric, the Visigothic tribes sacked and burned Rome. Eventually, under pressure from other Germanic tribes, the Visigoths left Italy and settled in what is now southern France and Spain. The Visigothic Museum (Museo Visigótico), housed in a former church, contains many fragments of stone carvings from the period between the decline of the Roman Empire and the Moslem conquest of the Iberian Peninsula. Also in the area, and dating from Roman times, is the Church of St. Eulalia the Martyr (Mártir Santa Eulalia), dedicated to a Christian child who was killed in 304 after insulting a local official. She is reputed to have been roasted alive over a slow fire.

The most imposing relics of the Roman period in Mérida, however, are the bridge (Puente Romano), and the nearby castle that protects it. The bridge is one of the longest Roman bridges surviving, measuring about 866 yards and consisting of sixty-four arches spanning both forks of the Guadiana. Although the fortress was built originally by the Romans, it is known nowadays by its Moorish name: the Alcazaba. The Moors conquered Mérida in 713 and began to construct the fort from remnants of the earlier Roman and Visigothic buildings on the site. The building was completed in 855. After the town was reoccupied by Christians in 1229, Alfonso IX of Léon granted the Alcazaba and Mérida itself to the knights of the Order of Santiago.

It was under the Moorish rulers—the Umayyad caliphs of Córdoba, and the Almoravids and Almohads from North Africa—that Badajoz, called Badaljóz by the Moors, began its rise in importance and Mérida began its decline. In the early eighth century, Moslem forces loyal to the Umayyad caliph in Damascus invaded Spain from North Africa. When the Umayyads were overthrown in the east around 750, their power continued in Spain. It was not until around 1021 that the Umayyad dynasty finally toppled. The Moors established Badajoz as the capital of a small kingdom, or *taifa*, after the collapse of the caliphate. Among the remnants of Moorish culture preserved in Badajoz is its own Alcazaba, which served as a fortress against the rival emir of Mérida. It was taken over during the Christian period by the dukes of La Roca, and presently houses the local archaeological museum.

For some time the caliphate of Córdoba survived only as a loose collection of about twenty independent states. In the eleventh century Christian forces, under Alfonso VI of Castile, began pushing into what had been Moorish territory. In 1086, however, the Almoravids, a radical Islamic sect based in northwest Africa, invaded Spain to assist the local Moorish rulers and drive back the Christian attacks. Originally a warrior brotherhood of Berber nomads, they united the Islamic states against the Christians and then brought southern Spain under the rule of their kingdom in what is now Morocco. In 1147 the Almohads, another reforming sect, overthrew the Almoravids and brought about a brief final flourishing of Islamic culture in Spain. The Almohad rulers erected a large eight-sided keep in Badajoz called the Torre de Espantaperros, which offers a commanding view of the surrounding countryside.

Despite the protection provided by the Espantaperros tower, the town of Badajoz was captured by Alfonso IX of León around 1229. The Gothic-style Cathedral of San Juan, just off the main square, or Plaza de España, was begun shortly after the Christian reconquest, and was completed in 1284. It underwent significant alterations in the sixteenth, seventeenth, and eighteenth centuries. It houses many works of art, including paintings by Luis de Morales, a Renaissance-era artist and native of the town. Works by Morales also are displayed in two other Badajoz churches, San Andrés and La Concepción, and in the city's Provincial Museum of Art. Other famous natives of Badajoz include two conquistadors of the New World, Pedro de Alvarado and Sebastián Garciliaso de la Vega y Vargas.

Badajoz, because of its position near the Portuguese border, remained a place of contention between Portugal and Spain for many centuries after the Christian reconquest. It was seized by the Portuguese in 1385, 1396, and 1542. Badajoz served as the headquarters of the Spanish Habsburg king Philip II during his conquest of Portugal in 1580. Philip also endowed Badajoz with a structure that has become its symbol, the fortified town gate known as the Puerta de Palmas, built during his reign. The gate, with its two crenellated towers, later served as a prison. Near the gate is the Puente de Palmas, a bridge built during the same era atop Roman foundations.

Despite the efforts made to fortify Badajoz, it continued to be in the middle of conflicts between the Spanish and Portuguese; the Portuguese besieged and captured it for a time in 1660. Then in 1705, during the War of the Spanish Succession, Badajoz was besieged again. The war, sparked when King Charles II died without an heir, involved the adherents of three rival claimants to the Spanish throne; it was marked by military campaigns all over Europe, but the battles fought within Spain generally were indecisive. The war ended with Philip V, grandson of the French king Louis XIV, on the throne of Spain.

Badajoz was the site of warfare again a century later, during that portion of the Napoleonic Wars known as the Peninsular War. Napoléon invaded Spain in 1808, putting his brother Joseph on the throne; Spain and Portugal then rebelled, and British forces came to assist them. The French commander Nicolas-Jean de Dieu Soult besieged Badajoz beginning in 1810 and finally took it the following year. British forces under Viscount William Carr Beresford tried to recover Badajoz later that year and failed. Then in March 1812, Arthur Wellesley, the future duke of Wellington, made a surprise attack with a force that greatly outnumbered the French defenders. Wellesley drove out the French and captured the town; he lost 4,760 of his 15,000 soldiers, the French 1,200 out of 5,000.

Badajoz saw conflict once more in 1936, during the Spanish Civil War. This was a particularly brutal episode: nationalist troops marched Badajoz's loyalist defenders, perhaps as many as 1,800, into a bullring and massacred them. Some loyalists tried to escape across the Portuguese border, but Portuguese troops returned them to the nationalists.

Modern Badajoz and the surrounding area are marked by extreme poverty. The soil is not favorable to agriculture, and there is little industry, although the latter has increased somewhat in recent years, thanks to efforts to exploit the commercial possibilities of the Guadiana. One advantage of the dearth of industry, however, is that Badajoz and its neighboring communities, such as Mérida, have retained much of their historic charm. The region also contains several national parks and nature reserves. Its stark beauty has been celebrated by the novelist Camilo José Cela in *The Family of Pascal Duarte*, the story of a criminal raised in the countryside near Badajoz, and by the seventeenth-century painter Francisco de Zurbarán.

Further Reading: Although relatively little has been published in English about early Spanish history, there are some general sources that may prove helpful. For the Roman period, see Peter Brown's *The World of Late Antiquity: A.D. 150–750* (London: Thames and Hudson, and New York: Harcourt Brace Jovanovich, 1971). For the Visigothic invasions, see *The Barbarian Invasions: Catalyst of a New Order*, edited by Katherine Fischer Drew (Gloucester, Massachusetts: Peter Smith, 1970). For the Moorish occupation, see Philip K. Hitti's *History of the Arabs*, (London: Macmillan, 1937; tenth edition, London: Macmillan, and New York: St. Martin's, 1970), especially Part 4: "The Arabs in Europe: Spain and Sicily." The history of Renaissance and modern Spain are portrayed in Henry Kamen's *Golden Age Spain* (London: Macmillan, and Atlantic Highlands, New Jersey: Humanities Press International, 1988) and Raymond Carr's *The Spanish Civil War* (New York: Norton, 1986; as *Images of the Spanish Civil War*, London: Allen and Unwin, 1986).

—Kenneth R. Shepherd

Barcelona (Barcelona, Spain)

Location: On the Mediterranean coast, near the French border, some 400 miles east of Madrid.

Description: Capital of the province of the same name, Barcelona is arguably the most dynamic city in Spain and perhaps in all of Europe. Recently host to the Summer Olympics, the city is renowned for its contemporary architecture, famous artists, and distinctive regional identity.

Site Office: Tourist Information Office
Estacion de Francia
08003 Barcelona, Barcelona
Spain
(93) 3195758

Like so many port towns in Europe and particularly those on the Mediterranean, Barcelona's long history is replete with conquest, reconquest, occupation, and unrest. The current site was originally a Carthaginian settlement that dates from the third century B.C. In 218 B.C. the ruling family of Carthage established Barcino, which became a Roman colony in 133 B.C. The settlement grew rapidly into an important commercial center. In turn the Vandals, Visigoths, Moors, and Franks invaded the area.

In the tenth century, under the jurisdiction of the king of Aquitaine, Raymond Bérenger I united the counties of the region of Catalonia and established Barcelona as the seat of power. In the twelfth century Raymond Bérenger III married into the royal house of Provence, and that part of France was incorporated into the Catalan state. His son married a princess of Aragon, establishing the Catalan dynasty of the crown of Aragon. In the thirteenth century under the rule of James I of Aragon, Catalonia became an independent state with its capital in the prosperous city of Barcelona. Deposit banking originated in the city, which also was a significant maritime center, rivaling the Italian ports.

In the fifteenth century the union of Ferdinand and Isabella, the Catholic monarchs, ended the Aragon succession to the throne, and the much stronger Castile became the dominant power in the emerging Spain. Catalonia suffered a loss of autonomy during this period, but maintained its importance as an economic center. It was to Barcelona that Columbus returned after his voyage of discovery to the New World, and there he was received by the Catholic monarchs. Relations between Barcelona and the Crown were relatively peaceful throughout the golden age of Spain. The region maintained sufficient autonomy to satisfy the people but did not demand too much to worry the Crown. Catalonia prospered both economically and culturally.

As Spain's power declined in the seventeenth century, there ensued a period of significant revolt for Catalonian independence, in the 1640s and 1650s, but this movement eventually was quashed by the national government. By the mid–eighteenth century, Catalonia in particular had lost much of its traditional personality. Catalan had almost been lost as a unique language, and the social fabric of Catalonia had been seriously weakened. During this same period, however, the population of the area mushroomed, and the expanding textile industry brought continued prosperity to the region.

In the late nineteenth and early twentieth centuries, Catalonia experienced a period of rebirth with a standardization and modernization of language as well as the establishment of many cultural projects. The city played host to an international exhibition in 1888 and again in the late 1920s, bringing it considerable worldwide attention. It was also during this same period, however, that Barcelona began to suffer from an ongoing series of fires and bombings, first at the hands of anarchists, later in its bid for independence, and finally as part of the Civil War. In 1909 the "Tragic Week" marked the people's opposition to the conscription of thousands of young Catalans into the army to fight in the war against Morocco. Protesters burned a school, a monastery, and other religious buildings in a rebellion that they hoped would spark a movement for independence throughout the region. The uprising was quickly put down by the military, however. The revolt garnered the city a new name—"The Rose of Fire"—a name that continued to seem appropriate for several decades to come in strife-torn Barcelona.

Throughout the 1920s and 1930s, Catalonia struggled to establish itself as an independent state. That struggle included plans to host a "People's Olympics" in 1936 to serve as an alternative to the Olympics hosted by Hitler in Berlin. The coup staged by Francisco Franco and the nationalists in that very year, however, put an end to these plans and sent the city into a state of siege as it attempted to resist Franco's takeover of Spain. By 1938 the city was suffering from severe shortages of food and supplies and increasingly frequent and intense bombings by the nationalists. Franco's forces finally entered the city in January 1939. As many expected, Barcelona and Catalonia suffered from severe reprisals at the hands of the Franco regime. Thousands were killed, thousands of others fled to France, and those who remained in Barcelona endured severe food rationing and massive outbreaks of typhus and tuberculosis. The Catalan language was banned.

The suffering in Barcelona continued through the years of World War II and beyond. The nearly forty years of rule by Franco is now referred to as the "Franquist Night"—a time of darkness and suffering for the once prosperous city.

In 1951, Barcelona became the scene of intense opposition to Franco and his regime as the result of an increase in the cost of tram tickets in the city. This led to widespread

Monument to Columbus in Barcelona
Photo courtesy of Instituto Nacional de Promocion del Turismo

strikes and rekindled a spirit of nationalism among the people of Barcelona and greater Catalonia. By this time the early injunctions against Catalan language and culture had been eased, and huge numbers of immigrants were arriving in Barcelona from other parts of Spain and Europe. The city began an intense period of growth and rebuilding, and by the 1970s it began to attract prominent citizens, especially from Latin America, among them the authors Gabriel García Marquez, Mario Vargas Llosa, and José Donoso. After Franco's death in 1975 the Catalan assembly once again pushed for independence but instead was granted limited autonomy.

The establishment of a constitutional monarchy and democratic government in Spain brought new life to Barcelona and the entire country. Plans and projects that had been abandoned in 1939 were revived, exiles began to return, and the city took on a new life, recovering much of its past in the process.

The economic boom of the 1980s was felt throughout Spain but perhaps nowhere more strongly than in Barcelona. The city was largely restored and massive public works projects were undertaken. All of this activity culminated in Barcelona's hosting of the 1992 Summer Olympics during the 500th anniversary of the first voyage of Columbus. During the Olympic celebrations, the city and its art and culture were featured prominently throughout the world.

Perhaps no other city with as long a history has been so fully shaped by conscientious city planning as has Barcelona. In the 1850s a master plan was prepared for Barcelona by Ildefons Cerda. This plan was particularly influential since it outlined the growth of the city outside the old walls. For several decades there had been an injunction against new construction outside these walls. When that restriction was lifted there was an explosion of growth and construction, guided by Cerda's plan. Later, Jaussely in 1905 and Le Corbusier (Charles-Édouard Jeanneret) in 1932 each proposed influential urban plans for the city. Elements of both of these plans were incorporated into the growth and construction of modern Barcelona. Even in the recent preparations to host the Summer Olympics of 1992, Barcelona undertook an ambitious project of revitalization and transformation under the direction of Oriol Bohgias that included the creation and renovation of more than 200 public spaces. The city is now filled with carefully planned public areas that are home to works from some of the world's most important artists.

One of the most significant revitalization efforts has taken place along the waterfront near the famous monument to Christopher Columbus. An area that had been rather rundown has been transformed into an esplanade lined with palm trees and sand beaches. Gone are the rusted railroad tracks and rocky shore. People are returning to the harbor area, and other private revitalization efforts are now occurring nearby.

Barcelona moves at a different pace and seems to be on a different clock than most places. People in the city take time to stroll and enjoy the sites. Because of the afternoon heat in the summers, siestas are common and many shops and businesses shut down in the afternoon. It is late in the evening that the city comes alive, the busiest dinner hours in restaurants starting around 10:00 P.M. It is common to find shops and clubs open until very late. Likewise, the city is unusual in that it is a world of two languages, Spanish and Catalan. Street signs, labels, radio, television, and newspapers are in both and many people speak both.

The two most important parts of the city both historically and culturally are the very compact core area known as the Barrio Gótico—this is essentially the old city and is a maze of narrow streets—and the Ramblas, the main boulevard.

The Ramblas is a wonderful stretch that covers the city from the rather seedy end at the port to the quite rich area at the other end. Somerset Maugham called the Ramblas the most beautiful street in the world and most visitors would agree. It is filled with people of every sort and is lined with shops and cafes shaded by giant trees. It is truly the hub of the city; one can find activity along the Ramblas at all hours.

The Barrio Gótico is made up of buildings from the fourteenth and fifteenth centuries and is bustling with shops, bars, and nightlife. Among its important structures are the Town Hall and the Cathedral. The Town Hall is a combination of structures. The main facade dates from 1847; a much older Gothic facade on the north side dates from the early fifteenth century. The interior of the building is richly appointed and contains both rooms for civic functions and a chapel.

The cathedral was begun in 1298 and was not completed until the middle of the fifteenth century. The structure dominates the Barrio, and within its walls lie the wooden sarcophagi of Raymond Bérenger I and his wife. Of particular interest are the coats of arms on the upper tier of stalls, representing the Knights of the Golden Fleece who assembled in the cathedral in 1519. Much of the decoration within the cathedral is in a poor state of repair, but there is important sculptural work by Bartolomé Ordóñez, Pedro Vilar, and Giovanni Pisano.

Two things that should not be missed in Barcelona are the buildings designed by Antonio Gaudi y Cornet and the Picasso Museum. Among the many buildings designed by Gaudi, the most important is the famous Sagrada Familia Cathedral. George Orwell called it "one of the most hideous buildings in the world." Gaudi began work on the cathedral in 1883 and dedicated the rest of his life to the completion of the structure. When he was killed in a trolley accident in 1926, the building was only one-quarter finished. It remains unfinished today, but there always seems to be a little work taking place on the building. This enormous neo-Gothic structure has a rather organic feel, almost like oozing mud. It has elicited responses like those of Orwell as well as great praise for its novelty and foreshadowing of the expressionist style. While work continues slowly, funded by private donations, many of Gaudi's plans have been lost, and there is ongoing debate as to whether or not additional work should be done. Whatever one thinks of the cathedral, and regardless of its incomplete state, it has become a symbol of Barcelona and is a dominant part of the city's rich architectural heritage.

The Picasso Museum is home to a large collection of his early works. Two former palaces house the works done by Picasso before the start of the Spanish Civil War. Among the holdings of the museum are more than 80 oil paintings and more than 600 drawings, as well as notebooks and sketch pads. None of the works in the museum is considered to be a masterpiece, but the sheer size of the collection is significant and it does provide insight into the artistic development of this important artist.

In addition to Picasso and Gaudi, other famous artists from Catalonia include Salvador Dalí, Joan Miró, opera stars Montserrat Caballé and José Carreras, and cellist Pablo Casals. Something of the region's independent spirit seems reflected in both the artists and their art.

Long considered the second city of Spain, Barcelona no longer plays second fiddle to Madrid. In recent years it has emerged from the shadow of the Spanish capital and has established an identity apart from the rest of the country.

Barcelona is a city with a long history but it is also a city that embraced the age of modernism, with its highly wrought ornamental style, thus giving the city its unique appearance. Its ongoing commitment to design, style, and urban planning has made Barcelona perhaps the most dynamic city in Europe.

Further Reading: Although much has been written on Barcelona, the great majority of books are in Spanish and a few are in Catalan. In English, however, two recent books, in combination, provide a fairly complete overview of the city. The first, Felipe Fernández-Armesto's *Barcelona: A Thousand Years of the City's Past* (Oxford and New York: Oxford University Press, 1992) is a good historical introduction to the city. The second, *Walks in Picasso's Barcelona* by Mary Ellen Jordan Haight and James J. Haight (Salt Lake City: Gibbs Smith Publishers, 1992) is an easy-to-use guide that discusses the major points of interest in the city based on seven possible walking tours a visitor might choose to take.

—Michael D. Phillips

Batalha (Leiria, Portugal)

Location: About seventy miles north of Lisbon, just inland from Portugal's Atlantic coast.

Description: Site of Portugal's most majestic monastery, built to commemorate Portugal's victory over Castile in 1385; a symbol of Portuguese independence.

Site Office: Tourist Information Office
Largo Paulo IV
Batalha, Leiria
Portugal
(44) 96180

The small town of Batalha, with fewer than 8,000 inhabitants, is appropriately named after the Portuguese word for "battle," as it lies near the site of Portugal's most decisive military victory, the Battle of Aljubarrota. This victory over the encroaching Castilian army of Spain was the foundation of Portugal's independence. After his victory, King John I of Portugal built the Mostairo de Santa Maria de Vitória (Monastery of Our Lady of Victory) to commemorate his country's seemingly miraculous triumph. Today, this architectural wonder, commonly called simply "Batalha," is on UNESCO's list of World Heritage Sites and is the symbol of the independent Portuguese state.

International intrigue and intermarriage brought Spain and Portugal into conflict in the fourteenth century. The stage was set for problems with the death of King Dinis in 1325. As king, Dinis was notable for his restoration of ancient castles, his interest in improving the agriculture of his country, and his founding the chivalric Order of Aviz.

After Dinis's death, his less capable son, Afonso IV, came to the Portuguese throne. Portugal and Spain had always been closely related by their proximity and their royal bloodlines, but Afonso strengthened the ties by arranging the marriage of his son Peter—Portugal's future King Peter I—to Constança, the daughter of a Castilian leader. By his Spanish wife, Peter had a son, Ferdinand; but by another woman, Peter fathered an illegitimate child, John.

As the rightful heir to the throne, Ferdinand inherited the position of king upon Peter's death. John, on the other hand, was made grand master of the Order of Aviz. Ferdinand married Leonor Teles de Meneses, and the couple had a daughter, Beatriz. Ferdinand was even less of a ruler than his father, preferring to leave the affairs of his realm in the hands of a courtier.

Since the end of the reign of Dinis, the people of Portugal grew more and more dissatisfied with their monarchs. The era marked by Ferdinand's rule was especially unsettled, as the intrigues left over from Europe's Hundred Years War continued to play themselves out in the form of secret alliances and military associations. The Portuguese people grew concerned as the Castilians of Spain began to play a role in ruling Portugal. As Ferdinand lay dying, the Princess Beatriz, still a child, was married to John I of Castile, part of the Trastámara line of Spain. Upon Ferdinand's death in 1383, his queen Leonor claimed the regency in their daughter's name and ruled Portugal accompanied by a council that was largely Castilian.

With Portugal's throne willed to a young girl, John I of Castile began his preparations to take control of the country. Portugal's regent, Leonor—possibly at the request of Juan Fernández Andeiro, Leonor's lover and a Castilian himself—sent a proclamation throughout the countryside that the child Beatriz and John I be recognized as Portugal's new rulers.

Andeiro and John I were highly unpopular among the Portuguese, and Leonor's decree giving John virtual control of the country sparked riots in Lisbon. The protesters chose John, grand master of the military Order of Aviz, as their leader, to be assisted by Nuno Álvares Pereira, a nobleman famed for his military prowess. The Portuguese nobles, however, still supported Leonor and Beatriz—thus supporting John of Castile—possibly because they saw the Trastámara line as one that would protect their age-old control over their lands.

Faced with resistance to his taking the Portuguese crown, John of Castile invaded the country, reaching Santarém in January 1384. Leonor welcomed John and resigned her regency in his favor. More than fifty Portuguese towns and castles followed Leonor's lead and accepted John's rule. Facing unfavorable odds, the Order of Aviz asked England for help and received the support of John of Gaunt and numerous British merchants who had profited greatly from Portuguese trade and were unwilling to see the country taken by Spain.

By spring of that same year, Portuguese nobles began to fear John of Castile. Concerned for their country's independence, a number of officials turned to John of Aviz and pledged their allegiance to his cause. Soon, even Leonor became disenchanted with the Castilian and was accused by him of plotting his murder and subsequently banished.

His popularity rapidly waning, John of Castile devoted his kingdom's treasury to a military invasion of Portugal and amassed an army to storm Lisbon. He set up camp outside the city; the army guarded the city gates while Castilian ships blockaded the river, allowing no supplies in or out of Lisbon. The city of Lisbon seemed doomed to fall to Castile when the army was stricken by an outbreak of the plague. With the death toll reaching 200 a day, the Castilians were forced to abandon Lisbon and retreat.

With the siege lifted and the peril temporarily gone, John of Aviz now had the time to convince the remaining

The monastery at Batalha
Photo courtesy of Portuguese National Tourist Office

Portuguese nobles of his right to the throne. Although the Parliament's vote was not unanimous, in April 1385 he was crowned King John I of Portugal, initiating the House of Aviz.

John of Castile invaded again that summer in yet another attempt to secure Portugal as part of his realm. John had, over the previous months, replenished the ranks of his plague-decimated army with new Castilian recruits and most of the Portuguese nobility. By most accounts, John's force numbered nearly 20,000 and included a light cavalry. The Portuguese force, on the other hand, consisted of only some 7,000 troops accompanied by 500 or so English bowmen sent by John of Gaunt.

Despite the great odds against them, the Portuguese prepared for battle, taking up positions in a valley near Aljubarrota, where they waited for the Castilians to attack. Confident in their military superiority, the Castilians marched steadily toward the battlefield under the hot midday sun. As the Castilians slowly advanced, the Portuguese shifted their positions and designed a new battle strategy. To fortify their stance, Nuno Álvares commanded his troops to dig a series of trenches. The hot sun took its toll on both sides: the Portuguese were exhausted from hours of digging while the Castilians had already marched nearly seven miles before the battle began.

The attack began in earnest at sunset on August 14. John of Castile's advisers had urged him not to fight the same day as the long march, but the Castilian had lost control over his ranks. Despite their overwhelming numbers, the Castilians were defeated in less than an hour by the ready and waiting Portuguese. By nightfall, the best Castilian knights and a large majority of the Portuguese barons lay dead on the field at Aljubarrota.

During the wait for the Castilian forces, John of Aviz and Nuno Álvares had made a number of oaths. Nuno Álvares, suffering under the sun from a raging thirst, had sworn that henceforth water should always be available at the site for passers-by. To this day, a pitcher of fresh water is placed daily at Aljubarrota. The king, praying for victory, had promised to restore an old church at Braga. After defeating the huge Castilian army, John of Aviz saw his victory as nothing less than a miracle. According to legend, he threw his lance into the air and vowed to build a fabulous church to the Virgin Mary wherever the lance landed. In truth, his throw itself would have been a miracle; the site of the Batalha Abbey is nearly three miles from the actual battlefield.

Nevertheless, John's determination to construct the most beautiful church imaginable was no myth. The abbey is today considered to be the most splendid monument in Portugal. Construction on the abbey began in 1388 and continued for several centuries under a number of different architects. Batalha's elaborate exterior of weathered limestone is resplendent with innumerable pinnacles and spires, and the interior is no less impressive.

The first building stage of the great monument was carried out by two architects—the Portuguese Afonso Domingues, and Houguet, a Norman—and resulted in the church, the Founder's Chapel, the Royal Cloister, and the Chapter House. Domingues was in charge of the original plans, and, selecting a fine limestone, drafted a Gothic design based on an abbey at Alcobaça. John's new queen, the English Philippa of Lancaster, however, urged Houguet to add a very English perpendicular Gothic style, which gives Batalha its English feel.

The church itself seems out of place in its plainness, especially after the flamboyant exterior of the abbey. There is little decoration here, save for the colored light scattered as the sun passes through stained glass. The high ceiling of the church proper soars above the sanctuary, apparently anchored against flight by a number of piers. Mateus Fernandes, the Manueline architect who worked on the Royal Cloister and Unfinished Chapels, lies buried in this part of Batalha.

The Capelo do Fundador, or Founder's Chapel, is a square room off of the church. King John I and his Queen Philippa are buried here, as are two of their sons: Dom Fernando, killed in an attack on Tangier; and Dom Henrique, commonly known as Prince Henry the Navigator. Atop the tomb of the king and queen is a statue of the two holding hands with the king clothed in his heraldic armor.

The Claustro Real, or Royal Cloister, is one of the marvels of Batalha. Constructed in the pure, simple Gothic style by Afonso Domingues and intended as a place of meditation and prayer, a later House of Aviz ruler Manuel I (reigned 1495–1521) found the cloister uninviting and cold and ordered it redone by his architect Mateus Fernandes. Fernandes took Manuel's order to heart and redid the Royal Cloister in an extravagantly beautiful Moorish-influenced style known as Manueline, after the king. In the cloister sprouted fantastic carved vegetation and vines that crept up the columns and arches and half covered the windows. Sunlight filters in through window grilles representing a tangle of lotus, clover, laurels, artichokes, and poppies, occasionally interrupted by a double cross. The cloister that once had a decidedly English accent is now flavored by the embellishments of Morocco. A three-tiered water basin in the corner of the cloister fills the room with the quiet, cool murmur of water.

Adjacent to the Claustro Real through an arched doorway stands the Chapter House, the site of Portugal's tomb of the unknown soldier. One of the last works of the original architect, Afonso Domingues, this part of Batalha was also one of his greatest achievements. The sixty-foot-high ceiling is constructed with unsupported vaulting, an unbelievable feat of architecture during his time.

According to several stories, Domingues was quite old and blind when the chapter house was constructed. Unconvinced that his computations were correct, a number of younger architects challenged Domingues regarding the ceiling and its lack of intermediate support. Domingues's contemporaries felt the design was so unstable that convicted criminals were brought in to do what seemed to be a dangerous construction project. When the ceiling was completed and the workers refused to remove the scaffolding, Domingues put both his reputation as a master builder and his life on the line: he spent a night alone in the chapter house, under this magnificent ceiling—and lived. His likeness can be seen in the corner of the chapter house, under the unsupported vaultings that still stand today.

The most splendid part of Batalha is the Capelas Imperfeitas—the Unfinished Chapels. From John I on down the line of Aviz, the Portuguese rulers were haunted by their determination to build the most spectacular abbey in honor of the Virgin and in remembrance of their country's victory over Castile. After John's death, his son King Edward began construction of the Unfinished Chapels. The original plan for the chapels was quite impressive: a central, domed octagon surrounded by seven large chapels and one large entrance portal. These seven chapels were, in turn, to have six pentagonal chapels between their walls.

Begun under Edward, the Capelas Imperfeitas received their embellishments under Manuel. Although records are unclear, experts believe that two architects worked under Manuel on the Chapels: Mateus Fernandes and the Frenchman Boytac. Fernandes is credited with the large portal leading to the chapels. The fabulous carvings of the portal appear as delicately carved as a piece of ivory, taking on a distinctively eastern flavor. Fernandes's work on the Unfinished Chapels has been compared with great Hindu temples in India and the temple of Angkor Wat.

Upon Fernandes's death in 1515, Manuel contacted Boytac and commissioned him to continue work on the chapels. Boytac is credited for finishing the portion of the chapels that exists today. The buttresses rising from the chapels into empty air were originally intended to support an octagonal dome covering the central area. Extending nearly ten feet above the unfinished octagon, the surfaces of the buttresses are covered with intricately carved and distinctively Manueline vegetation and ivy. Construction on the chapels was abandoned when Manuel died in 1521.

Despite never having been completed, the Unfinished Chapels remain one of the most intricately decorated and moving portions of all of Batalha. Originally intended as a resting place for the famous members of the House of Aviz, the Capelas Imperfeitas hold only a few tombs, including those of Edward and his wife Leonor.

What began as a tribute to the Virgin ended as a monument to Portuguese architecture and national independence. The Mostairo de Santa Maria de Vitória, better known simply as Batalha, despite discoloration and damage inflicted both by the passing centuries and motor traffic, remains the most beautiful and famous monument in Portugal, and a tribute to the victorious House of Aviz.

Further Reading: John Dos Passos's *The Portugal Story* (Garden City, New York: Doubleday, 1969; London: Hale, 1970) explores three centuries of Portuguese culture and history, including the rise of the House of Aviz. Other historical accounts are provided by *A Concise History of Portugal* by David Birmingham (Cambridge and New York: Cambridge University Press, 1993), Marion Kaplan's *The Portuguese* (London: Viking, 1991; New York: Cambridge University Press, 1993), and Charles Nowell's *Portugal* (Englewood Cliffs, New Jersey, and Hemel Hempstead, Hertfordshire: Prentice-Hall, 1973). Readers especially interested in the amazing architecture of Batalha will enjoy Sacheverell Sitwell's detailed account of the monastery in *Portugal and Madeira* (London: Batsford, 1954; New York: Hastings, 1955). Those readers planning to visit Batalha will find a number of guidebooks invaluable, especially the very detailed *Fielding's Portugal* by A. Hoyt Hobbs and Joy Adzigian (Redondo Beach, California: Fielding, 1994).

—Monica Cable

Benevento (Benevento, Italy)

Location: In the Campania region and Benevento province of southern Italy, thirty miles northeast of Naples.

Description: Originally a Samnite city called "Malies" or "Maloenton," Benevento became a Roman colony in 268 B.C. The city (renamed "Beneventum") prospered during the Roman Empire at the crossroads of the Via Appia and Via Appia Traiana, major military and trade routes between Rome and the Adriatic Sea. After the Lombard invasion of A.D. 568, Benevento gained importance as capital of an independent Lombard duchy. Later Benevento fell under Byzantine, papal, and Napoleonic rule before being annexed to unified Italy in 1860. The city suffered heavy damage from Allied bombing during World War II, but its most famous Roman monument, Trajan's Arch, survived intact.

Site Office: E.P.T. (Ente Provinciale per il Turismo)
Via Giustiniani, 36
CAP 82100 Benevento, Benevento
Italy
(824) 25424

Since its founding in ancient times, Benevento has played many civic roles: Samnite city, Roman colony, independent Lombard capital, Byzantine and papal possession, Napoleonic principality, and modern manufacturing city. Despite centuries of political turbulence, Benevento has retained its ancient character. Today, remnants of Samnite culture and Roman power define the city's historic appeal.

Benevento stands on a wide plateau between the Calore and Sabato Rivers in southern Italy's Campania region. Although excavations have dated early settlement at the site to the Bronze Age (eighth or seventh centuries B.C.), little is known of its ancient inhabitants. Benevento entered recorded history in the third century B.C. as the Samnite capital city called "Malies," or "Maloenton" in a Greek dialect. The Samnites, an agrarian people who spoke Oscan, lived in the rugged southern Apennine Mountains. The Samnites had a fierce reputation. Around 350 B.C., undaunted by growing Roman power, they became militarily aggressive. While many tribal nations (including Etruria and Latium) succumbed to Rome's territorial conquests, the Samnites attempted some expansion of their own.

Samnite seizures of cities on Campania's coast directly threatened Roman power, and the challenge did not go unanswered. By 341 B.C., Roman forces had driven the Samnites from coastal Campania. The Samnite capital Maloenton (interpreted by the Romans as "Maleventum") itself became the target of Roman revenge in 321 B.C. The Roman army, however, never reached the city. Several miles west of Maloenton, at the Caudine Forks, the Romans entered a narrow ravine without bothering to send for reconnaissance. Samnite soldiers stationed on the ravine's slopes immediately surrounded the Romans. Ancient historians disagree as to whether the armies clashed or the Romans surrendered without a fight; in any case, the Romans admitted defeat and were forced to publicly display their submission. Although the Romans finally won their war with the Samnites in 290 B.C., the "Battle of the Caudine Forks" was long remembered as one of the most humiliating defeats ever suffered by the proud Roman army.

Having subdued their chief rival in central Italy, the Romans embarked on a campaign of territorial expansion in the south. Greek colonies in southern Italy (known as "Magna Graecia"), alarmed at the Roman threat to their independence, appealed to their homeland for protection. In 281 B.C. King Pyrrhus of Epirus, an ambitious and skilled commander, brought his army to the colonists' aid. The Greek and Roman armies first met in battle at Heraclea in 280 B.C. Pyrrhus's forces comprised 20,000 infantry, 3,000 cavalry, and 20 elephants. Pyrrhus unleashed the elephants as the Romans were losing ground; frightened by the massive animals, which they had never seen before, the Roman soldiers fled in fear. Roman losses were great, but Pyrrhus's army suffered heavy casualties as well.

The following year, the Greek army fought the Romans at Asculum in Apulia. Again, Pyrrhus's elephants charged the Romans, who retreated in panic; again, losses were serious on both sides. Although the battle was deemed Pyrrhus's victory, it was a costly one. These battles gave rise to the term "Pyrrhic victory." Pyrrhus's final encounter with the Romans came at Maloenton (modern Benevento) in 275 B.C. By then the Romans had learned to retaliate against Pyrrhus's elephants by pelting them with "fire darts." The Greek army suffered tremendous losses. Resoundingly defeated, Pyrrhus returned to Epirus, leaving southern Italy to the Roman conquerors.

After surviving military challenges from the Samnites and other tribes, Rome ensured its dominance by founding Latin colonies at strategic locations across Italy. Maloenton, the Samnite capital, was reestablished as a Latin colony in 268 B.C. The Romans chose to colonize Maloenton (which they renamed "Maleventum") for two main reasons: to assert their presence in Samnite territory, thereby weakening the Samnite tribal confederation; and to take advantage of Maleventum's physical advantages. Surrounding rivers and ravines made Maleventum easy to defend, and a growing network of roads converged there, making the city a transportation and communication center.

Maleventum did possess one troublesome characteristic: its name. The Latin prefix "male" means "bad," and the

Trajan's Arch
Photo courtesy of E.P.T., Benevento

name Maleventum suggested "bad air" or "ill wind." Following their custom of giving auspicious names to new colonies, the Romans literally changed "bad" to "good," making "Maleventum" into "Beneventum." Given their good fortune in defeating King Pyrrhus and his tribal allies in battle at Beneventum in 275 B.C., the Romans may have considered the choice especially fitting.

As a Latin colony, Beneventum retained its independence and was not technically part of the Roman state. In reality, however, the influx of new colonists from Rome transformed the city from Samnite capital to Roman outpost. Latin became the first language for all residents, both colonists and Samnite inhabitants. The city government, although it ruled with almost complete autonomy, reflected Roman political structures. Beneventum's government, like Rome's, included a popular assembly, a senate, and two chief executives. These two officials, elected annually, were given the Roman title *praetor*. Even Beneventum's groundplan underwent significant reconstruction to reflect Roman civic standards. The streets were reorganized on a centuriation grid, a Roman system of standard rectangular landplots within a linear grid of streets. The entire town was divided into wards that were named after wards in Rome. New public spaces—the forum, the comitium, and Capitoline temple—mimicked those in Rome, and a Roman style of architecture and decoration was favored for new buildings. Not surprisingly, within two or three generations, Beneventum became a thoroughly Latinized community.

Beneventum prospered as a Latin colony, largely because of its location at the intersection of at least five major transportation routes. The most important road through Beneventum was the legendary Via Appia (Appian Way), which began in Rome and stretched 360 miles to Brundisium, a port city on the "heel" of the Italian "boot." From there, travelers could sail across the Adriatic to Greece. This road provided safe and efficient travel for politicians, soldiers, and merchants crossing the Italian peninsula.

While Beneventum prospered economically from traffic on the Via Appia and other roads, the Romans used the highways to broaden their realm of influence and facilitate travel to and from Rome. During the Samnite wars, Rome asserted its presence in Campania by extending the Via Appia from Capua, another Samnite stronghold, to Beneventum around 300 B.C. Once Rome had colonized Beneventum several decades later, it incorporated existing rural roads into its growing network of highways. One ancient road, stretching north from Beneventum into Pentri tribal lands, served as a drover's road for herding livestock. The Romans paved the road between 125 and 120 B.C., creating the Via Minucia, a vital link to the interior Italian peninsula. Two centuries later, in A.D. 114, a mule track leading east from Beneventum was upgraded by the Romans to become the Via Appia Traiana (Trajan Way). This route, built by decree of Emperor Trajan, provided an alternate route from Beneventum to Brundisium. While the Via Appia passed through rather mountainous terrain in southern Italy, the Via Appia Traiana crossed the mountains east of Beneventum, reached the Adriatic coast at Barium, and offered a flat, easy path southward to Brundisium. Though slightly longer in length, the Via Appia Traiana cut travel between Beneventum and Brundisium by one day.

The growing network of Roman roads facilitated travel throughout the empire, and Rome continued to pursue territorial expansion. Rome's acquisitions, however, sparked a conflict with similarly ambitious Carthage, a city-state on Africa's north coast. The Carthaginian general Hannibal led his infamous invasion from Spain through the Alps and into Italy, testing Roman political unity as his troops marched down the peninsula (218–216 B.C.). Although some Latin colonies supported Hannibal and his promise to end Roman rule, most remained staunchly faithful to Rome. Many colonies sent troops to fight the Carthaginian invaders. Beneventum itself became a stronghold for the Roman army and paid for its loyalty when Hannibal's troops destroyed its fields. In the end, Hannibal retreated from Italy and Rome was saved, due in part to the strong support of its Latin colonies.

Hannibal's invasion was only the first of many to vex the Italian peninsula. Later invasions spelled the end of Roman rule. The Huns, invaders from central Asia, arrived in Europe around A.D. 370, forcing the resettlement of the Germanic peoples they displaced. These Germanic peoples then invaded the Italian peninsula. The Visigoths captured Rome in 410. Another Germanic people, the Ostrogoths, descended on Italy after their own liberation from the Huns in 451. Beneventum fell victim to Totila, king of the Ostrogoths, who partially destroyed the city in 452. Beneventum, along with the rest of the Italian peninsula, was under Ostrogoth control by 493. Roman Emperor Justinian, who ruled from the Eastern Empire's capital at Constantinople, waged war to recover the Western Empire in the mid-sixth century. He succeeded in defeating the Ostrogoths, but twenty years of war left the Italian peninsula exhausted and ripe for invasion.

Another Germanic people, the Lombards, successfully took advantage of Italy's weakness. Originally from northwestern Germany, the Lombards had migrated to Pannonia (modern Hungary), from which they invaded northern Italy in 568. The conquerors met little opposition, and their tactics earned them a reputation as barbarians. The Lombards had too few men and too little administrative capacity to conquer and rule the entire peninsula, and their resulting realm incorporated two separate regions of Italy. They occupied the Po Valley in northern Italy by 569 and captured the city of Pavia, their new capital, after a three-year siege. In 571, the Lombards also gained control of a large region of southern Italy, including Beneventum. Between the Lombards' northern and southern territories, a large swath of land from Ravenna to Rome remained under Byzantine-Roman Catholic authority.

The Lombards soon divided their territory into "dukedoms," with the southern region splitting into the duchies of Spoleto (under Faroald) and Benevento (under Zotto). The

Duchy of Benevento included most of southern Italy, except for the "heel" (Calabria) and "toe" (Buttium) of the "boot" and the Roman duchy of Naples. Benevento the city became capital of the duchy of the same name. The Duchy of Benevento enjoyed a considerable degree of independence from the northern Lombard kingdom. When Lombard armies threatened the Roman Catholic territory in central Italy, the pope negotiated a peace not only with Lombard king Agilulf but with the dukes of Spoleto and Benevento as well. In the next century, an ambitious duke of Benevento named Grimoald actually invaded the northern kingdom and seized the Lombard crown. During this chaotic period, the Byzantine emperor Constans II attempted to overthrow the Lombards and regain Italy for the empire. His army attacked Benevento, which was ably defended by Grimoald's son. The Byzantine army failed and retreated.

Lombard kings soon increased their efforts to control the Roman Catholic territory that separated northern and southern Lombard lands. King Liutprand managed to gain control of much of the Exarchate of Ravenna. He also expanded his influence in the southern duchies by putting his own supporters in power there. A subsequent king, Aistulf, brought the Duchies of Spoleto and Benevento firmly under control of the northern kingdom in 752. His next objective was to conquer Rome. Pope Stephen II, afraid of a Lombard invasion, appealed to the Frankish Carolingian dynasty for help. The Frankish ruler Pépin III, whose army was the most powerful of its time, defeated the Lombards and reestablished the exarchate as a papal territory in 755. When the next Lombard king, Desiderius, retaliated by regaining Ravenna in 772, the Frankish king Charlemagne arrived to aid the papacy. His army returned the exarchate to Roman Catholic control and proceeded to vanquish the Lombards and conquer their northern kingdom in 774. Although he gained the Lombard crown, Charlemagne was not able to take the Duchy of Benevento.

With Charlemagne repelled and northern Lombard interference silenced, the Duchy of Benevento finally became an independent state. It maintained this independence for centuries. The duchy was not immune to internal power struggles, however. Civil war raged in the 840s, ending with the duchy's subdivision into three smaller territories: Salerno, Capua, and Benevento. Constant warfare between these smaller princedoms weakened their power to resist external attack. The Saracens, invaders from northern Africa, conducted a series of raids that devastated parts of southern Italy in the ninth and tenth centuries. During this time the Byzantine Empire reasserted its claim to southern Italy. By the early eleventh century, the empire's forces succeeded in driving the Saracens out of Apulia; most of southern Italy, including the city of Benevento, became the Byzantine "Theme of Lombardy."

Byzantine authority over the city of Benevento proved short-lived. In 1051 Pope Leo IX acquired the city, bringing it under papal sovereignty. Benevento's citizens apparently offered the city voluntarily as part of a territorial exchange between the pope and Byzantine emperor. The assertion of papal authority in Benevento infuriated its former Lombard prince, who summoned the Normans as his allies in an attack on his former holdings. The Normans, a Scandinavian people who had invaded part of northern France in the ninth century, had won respect in Italy as sporadic fighters against the Saracens. The Normans had gained control of territory in Apulia and begun building a military force.

In 1053, at Civita in the Capitanata, the former prince's Norman allies met a force of Byzantines, Italians, and Lombards led by the pope himself. Despite their much smaller numbers, the Normans readily defeated the pope's forces. The treaty they negotiated after the battle gave them great political advantage. They returned the pope in state to Benevento, asked his pardon for their victory, and allowed the city to remain a papal possession. In return, the Normans obtained the pope's blessing for their future conquests. Armed with papal approval, the Normans proceeded to conquer the entire southern half of the Italian peninsula and Sicily by 1130. The city of Benevento retained its status as a papal enclave not subject to outside rule. It remained under papal authority from the eleventh to the nineteenth centuries.

The Normans maintained control of southern Italy for less than a century. Germanic invasions between 1190 and 1194 ended Norman rule and established the German Hohenstaufen dynasty in the Kingdom of Sicily. The Hohenstaufen emperor Frederick II's attempt to acquire more territory on the Italian peninsula threatened the Papal States, now expanded to central and northeastern Italy. Adversarial relations between the papacy and Hohenstaufens came to a head at Benevento in 1266. The new pope, Urban IV, a Frenchman, turned to his home country for military support against the Hohenstaufens. Charles, count of Anjou, summoned an army of 30,000 and marched to Italy. There, on the plain of Grandella just north of Benevento, the French and German armies clashed. Manfred, the Hohenstaufen king, was killed, and his forces fled. Charles's victorious army viciously sacked Benevento, despite its status as papal enclave. This victory ended Hohenstaufen rule and established the French as rulers of southern Italy, which became the Angevin Kingdom of Naples.

The city of Benevento remained under papal control while the surrounding Kingdom of Naples was transferred from France to Aragon to the Habsburgs to Austria. Napoléon invaded Italy in 1796, but his army was repelled by the Austrians. He rallied in 1800, successfully conquering Italy within a year. Napoléon even annexed the Papal States to his "Kingdom of Italy" in 1808. Benevento's status as papal enclave was rescinded in 1806, and Napoléon named his atheistic foreign minister Talleyrand Prince of Benevento. Napoléon's reign lasted only until 1815, when Benevento was restored to the papacy.

Benevento's alliance with the papacy ended in 1860, when it was annexed to a newly unified Italy. The nation's role in World War II brought devastation to Benevento, which sat directly in the path of the Allied armies. As in Roman times, Benevento in the early twentieth century was a transportation hub, with rail lines from all directions converging in the city. Allied airplanes bombed Benevento's train station district repeatedly as the Allies advanced from the south.

Today, Benevento is a small, semi-industrial city trying to preserve its historical inheritance. Remnants of the Roman era are visible throughout the city, most notably Trajan's Arch and the Roman theatre. Trajan's Arch, located at the site where the Via Appia entered the city, is the best preserved triumphal arch built in the Roman Empire. Fifty feet high and carved from Greek Parian marble, the arch marks a crossroads where other important roads met the Via Appia. The arch was built in A.D. 114 as a monument to Emperor Trajan, who conquered Dacia and Mesopotamia for Rome. The boldly propagandistic relief carvings, still easily readable, depict scenes from Trajan's career: the emperor greeting a barbarian ambassador, being crowned by the goddess of victory, receiving a thunderbolt from Jove, and finally joining the gods after his death. The arch, also called the Porta Aurea (Golden Gate), is truly Benevento's archaeological crown jewel.

Under Trajan's successor, Emperor Hadrian (who ruled from A.D. 117 to 138), the Romans constructed an enormous public theatre in bustling Beneventum. This forum for classical drama is a semicircular theatre with a diameter of 270 feet and 19 rows of stone seats. After enlargement during Caracalla's reign, the theatre could accommodate 20,000 spectators. Although development has now surrounded the site, the first and part of the second of the theatre's three tiers survive, as do remnants of stage buildings and a corridor behind the auditorium.

Benevento maintains ties to its earlier Samnite past as well. Many local businesses use the Italianized name "Sannio" to proclaim their local identity. The Museo Sannio, located in Benevento's oldest church (Santa Sofia), has a strong collection of the city's archaeological treasures. These include Samnite relics, Greco-Roman sculpture, coins, and Egyptian statuary recovered from a Roman temple dedicated to the Cult of Isis. Santa Sofia, the cathedral that houses the museum, dates to the eighth century, when it was built for Lombard Duke Gisulfo.

Further Reading: E. T. Salmon's *The Making of Roman Italy* (London: Thames and Hudson, and Ithaca, New York: Cornell University Press, 1982) and *Roman Colonization under the Republic* (London: Thames and Hudson, 1969; Ithaca, New York: Cornell University Press, 1970) provide excellent, detailed, eminently readable accounts of the early Roman Empire and its expansion. Janet Penrose Trevelyan's *A Short History of the Italian People* (New York: Pitman, and London: Allen and Unwin, 1956) surveys Italian history from the barbarian invasions to the present in an informed, engaging manner. Representing a different approach, Dora Jane Hamblin and Mary Jane Grunsfeld recount their journey along the Appian Way by interweaving the history of the road and the lands through which it runs in *The Appian Way, A Journey* (New York: Random House, 1974).

—Elizabeth E. Broadrup

Bergamo (Bergamo, Italy)

Location: In northern Italy, approximately thirty miles northeast of Milan, on a hill at the foot of the Bergamasque Alps and extending out to the Po plain.

Description: The seat of the Bergamo province in the Lombardy region of Italy, Bergamo was, for the 350 years before the Napoleonic era, the westernmost point of the Republic of Venice and, before that, a member of the Lombard League. It sits at an altitude of 822 to 1205 feet.

Site Office: A.P.T. (Azienda di Promozione Turistica del Bergamasco)
Via Vittorio Emanuele, 20
CAP 24100 Bergamo, Bergamo
Italy
(35) 213185 or 210204

In many ways, modern-day Bergamo comes across as two separate towns: a modern, sprawling city below (Città Basa) and a medieval, walled city above (Città Alta). As travel writer Kate Simon says, "Bergamo . . . is distinctly two small cities, in character, in mood, in looks; and is linked only by steep roads, a funicular, a name and handsome pride."

But the two Bergamos also share a history of having been dominated, except for two very brief periods of independence, by outside powers. Bergamo has been ruled by Milan, Spain, Napoleonic France, Austria, and most importantly Venice, among many others. It has been invaded, rebuilt, and fortified—only to be invaded again. This pattern continued until the very costly construction of its greatest landmark—the walls surrounding Città Alta, which, ironically, were completed too late in the history of warfare to be of any defensive use. Bergamo did not become part of a unified Italy (at that point called the Kingdom of Sardinia) until June 8, 1859, when Giuseppe Garibaldi arrived at its gates.

Bergamo's history dates to approximately 1200 B.C., when a Ligurian tribe settled on one of the hills that today make up Città Alta. They named the settlement "Barra," which leads some to believe that this tribe lived in the section of Bergamo now called "Ferra." Around 600 B.C., the Etruscans took control of—and possibly fortified—the settlement, but it did not stay in their hands for very long. Approximately fifty years later, in 550 B.C., the Gauls invaded and took control. With this change of power, the town's name was changed to "Berghem," taken from the Gallic (Celtic) words for mount (*berg*) and house (*hem*). This name remained unchanged until 196 B.C., when the Romans defeated the Gauls in battle. After their victory, the Romans Latinized Berghem, spelling it "Bergomum." However, even to this

day, some residents still use Berghem as a moniker for Bergamo.

The Romans already played a role in Bergamo before their victory in 196 B.C. In fact, in approximately 425 B.C. the town was destroyed by a tribe of Sienna Gauls and subsequently rebuilt, and enlarged, with the help of the Romans. After the Roman victory over the Gauls, Bergamo's ties to Rome were further strengthened during the Social War of 91-87 B.C. when it sent aid to Rome and was in turn rewarded. In 49 B.C., Julius Caesar granted the citizens of Bergamo the rights of Roman citizenship. The Gallic legacy, however, was felt again in A.D. 286 when Alessandro—the patron saint of Bergamo—was put to death by Maximian for refusing to attack a colony of Gallic Christians.

Little of note happened in Bergamo for the next 100 years, until approximately the year 400. Possibly on his march toward Rome in 401, King Alaric and his band of Visigoths raided Bergamo, destroying much of it. Bergamo was again raided—twice—in 450, the first by Genseric and his Vandals (also possibly on their way toward Rome) and, separately, by the Alans.

Just twenty-six years later, in 476, Odacer—newly crowned as a "king"—invaded and took control of Bergamo. But this rule was short lived. In 489, Theodoric, an Ostrogoth sent to retake Italy from Odacer, scored several decisive victories—most notably in Milan—and was able to gain control of, and thus defend, Bergamo. Theodoric then set up his own kingdom in Italy. After Theodoric's death, Bergamo and other areas within Italy were controlled by his daughter, Amalasuntha, first as regent for her son, then after her son's death, as a joint ruler with her husband. The Goths revolted. This gave Byzantine emperor Justinian an excuse to invade. During this time, however, the citizens of Bergamo were able to rebuild the town. Bergamo was again attacked in 569 when the Lombard king Aloin arrived and found the town deserted. Over the next 200 years, the native population inter-married with the Lombards and life was relatively peaceful.

In 774 Bergamo became part of the Carolingian Empire, and by 800, as a feudal seat, it was the site of a great market. Peace lasted less than 100 years, however: in 894 the town was sacked by Arnulf of Carinthia, who had invaded Italy at the request of Pope Formosus.

Approximately 200 years later, Bergamo became a free commune. Citizens controlled the government, holding consuls as needed. Under the commune, which lasted until the thirteenth century, one of Bergamo's greatest landmarks was constructed: the Basilica di Santa Maria Maggiore. It was begun in 1137 as an offering of thanks by citizens who had survived the drought and famine that had struck in 1133. Built on the site of a former church dedicated to St. Mary, the

Bergamo's medieval walled section
Photo courtesy of A.P.T. di Bergamo e Provincia

Basilica di Santa Maria Maggiore has been called the first communal building in Italy. It was used as a barracks during infighting among townspeople, as a meeting hall for civil servants, and as a storehouse. Not counting the nearly constant additions and rebuilding—including the main altar, consecrated in 1499—the basilica was completed around 1200. Today it is considered a masterpiece of Romanesque architecture.

Although the citizens of Bergamo traditionally supported their emperors, they joined with Verona in 1165 to rise against Frederick Barbarossa (Frederick I). Then in 1167, Bergamo along with fifteen other communes in Lombardy formalized their alliance against Barbarossa with the Oath of Pontida. Called the Lombard League, this alliance was able to defeat Barbarossa in 1176. With the Peace of Constance in 1183, however, Bergamo returned to its pro-imperial stance and, in fact, the citizens of Bergamo warmly welcomed Barbarossa a year later.

Around 1200, two political factions—the Guelphs (papal supporters) and the Ghibellines (imperial supporters)—began fighting in Italy. The conflicts between two prominent families in Bergamo, the Colleoni (who supported the Guelphs) and the Suardi (who supported the Ghibellines), sparked much destruction and violence. In 1295, the Suardis—with help from Milan's Visconti family—forced the Colleonis out of Bergamo. But the fighting continued and even intensified. In 1331, therefore, King John of Bohemia, invited to restore peace in Bergamo, assumed control of the city and revoked its citizens' communal rights. But King John's dynasty was short-lived. The next year, Azzone Visconti took Bergamo and for the next seventy-five years imposed heavy taxes on the citizenry.

The Guelph Malatesta family of Rimini interrupted Visconti rule in 1407, but the Viscontis, as part of a Ghibelline revolt led by Filippo Maria Visconti, again gained control of Bergamo in 1419. Just seven years later, the people in the valleys around Bergamo—who were Guelph supporters—offered their allegiance to Venice (which was neither Guelph nor Ghibelline). Then in 1428, the people of Bergamo also offered themselves to Venice.

Filippo Maria Visconti attacked Bergamo again in 1437. Coming to Bergamo's aid was Bartolomeo Colleoni, then a great Venetian mercenary. As writer Paul Hofmann points out, "His famous equestrian statue by Andrea Verrocchio on the Campo Santi Giovanni e Paolo, a small square to the east of the Grand Canal in Venice, proves how high his repute was there." Born near Bergamo and spending much of his later life there (his fortress was only a few miles outside of the town), Colleoni was responsible for the construction of one of the town's greatest chapels, the Cappella Colleoni, constructed between 1470 and 1476. Connected to Santa Maria Maggiore (in fact, part of that church was removed to make room for the chapel), it was built by Giovanni Antonio Amado to house Colleoni's tomb, as well as the tomb of Colleoni's daughter, Meda. Amado, one of the most famous architects and sculptors of his day, also sculpted Colleoni's and Meda's tombs. Works by other great artists grace the Cappella Colleoni as well, including frescoes by Tiepolo.

Between 1509 and 1529, Bergamo again fell under the rule of a quick succession of different foreign powers—including the French and the Spanish—before being retaken by Venice. The answer to Bergamo's woes seemingly came in 1561, when the Republic of Venice started building walls to protect Città Alta, replacing its earlier fortifications. Designed by Pallavicino and boasting sixteen bastions, the new walls were massive; hundreds of houses and many monuments were torn down to make way for them. When the citizens of Bergamo were on the verge of rebelling at this destruction, Venice was forced to send in approximately 15,000 soldiers. Venice had good reason to protect its costly new project: it designed these walls to be aesthetically strik-

ing as well as militarily superior. However, by the time the walls were completed more than fifty years later, the advent of cannons had made them useless as a defense.

During the volatile sixteenth century, Bergamo also played host to something more upbeat: the then immensely popular commedia dell'arte. This improvisational form of comedy (often credited to Venice, actually born in Bergamo) featured an array of stock characters that are recognizable even today, including Harlequin, Pantaloon, Columbine, and Pulcinella. These characters were used in impromptu plays, where they behaved the same way from one production to the next (that was apparently part of the commedia dell'arte's attraction). In tracing its history, Vernon Bartlett notes that "some experts claim that this form of entertainment is copied from the Roman pantomimus." However, Bartlett also pointed out that it had a local flavor: "Harlequin is said by some to have been a real person from one of the mountain valleys to the north of Bergamo; Pantaloon came from Venice and his name came from St. Pantaleone, that city's patron saint; Pulcinella came from Naples." Evolving into the English Punch and Judy shows (Punch is the Pulcinella character), the commedia dell'arte continues to influence us today; its characters are still a mainstay of masked balls.

The eighteenth century brought many important changes to Bergamo. On March 13, 1797, Venetian rule ended and the Republic of Bergamo was born. Bergamo's freedom, however, did not last long—just six short months: the French gained control of the town on October 17th as part of the treaty of Campoformio.

Bergamo began to make its artistic mark in the early nineteenth century. Gaetano Donizetti, the great operatic composer, spent much of his life in Bergamo. Giacomo Carrara, a local count, founded an art school in the city and upon his death left his private art collection to the school for the start of a museum. With donations by other collectors, this museum now houses important works by Lorenzo Lotto, Titian, Anton van Dyck, Pisanello, Giovanni Bellini, and Sandro Botticelli, as well as Piccio, who was one of the school's most famous students.

After the defeat of Napoléon in 1814, Bergamo was controlled by Austria (which had also briefly held Bergamo shortly before the turn of the century). However, consistent with Bergamo's history of domination, Austria's rule did not last for long. In 1848, Bergamo helped Milan in its revolt against the Austrians; a decade later, on June 8, 1859, Garibaldi and his troops entered Bergamo, freeing the town from Austrian rule. In return, 178 people from Bergamo joined Garibaldi. These troops helped unite all of modern-day Italy. (It has been said that more people from Bergamo would have joined Garibaldi had it not been for the parish priests in the nearby valleys who prevented them.)

The modern, lower part of Bergamo, the Città Basa, comprises two large suburbs and a fairground. This area was largely developed during an urban renewal project in the 1920s. Overseeing the project was architect Marcello

Piacentini, who was liked by Mussolini. Thus as Paul Hofmann writes, the "pompously monumental neo-Roman style that critics would later call Mussolini Modern" was born in Bergamo.

One pope is associated with Bergamo. John XXIII (pope from 1958 to 1963) was born in a nearby valley in 1881 and studied for the priesthood in a seminary in Città Alta. According to Paul Johnson, John XXIII "looked to Bergamo—not Rome—as his capital."

Today Bergamo boasts a population of approximately 130,000. Its industries include textiles, cement, and printing. Its Venetian past is very much alive: although Milan is much closer in distance, Bergamo's dialect and traditions continue to be closer to those of Venice.

Further Reading: While most of the available books and magazine articles in English that treat Bergamo tend to be boosterish, several provide reliable information. These include *Baedeker's Italy* (Englewood Cliffs, New Jersey: Prentice-Hall, and Basingstoke, Hampshire: Automobile Association, 1991); "Kitchen Pope/Warrior Pope" by Paul Johnson, which appeared in *Time* (New York), December 26, 1994/January 2, 1995; *Italy: The Places in Between* by Kate Simon (New York: Harper and Row, 1984); *Birnbaum's Italy 1993,* edited by Alexandra Mayes Birnbaum (New York: Harper Perennial, 1993); *Northern Italy* by Vernon Bartlett (New York: Hastings, 1973); *Cento Città: A Guide to the "Hundred Cities and Towns" of Italy* by Paul Hofmann (New York: Holt, 1988); and *Bergamo in Its History and Its Art* by Renato Ravanelli, translated by Ulisse Belotti (Bergamo: Grafica e Arte Bergamo, 1986).

—Laurence Minsky

Bobbio (Piacenza, Italy)

Location: In the province of Piacenza, in the Emilia-Romagna region of northern Italy; at the southern end of a large valley of the Apennine mountain chain drained by the Trebbia, near the Trebbia's confluence with its tributary, the Bobbio River; 68 miles south of Milan, 347 miles northwest of Rome.

Description: Ancient city, known to Romans as Bobium; site of Hannibal's first victory against the Romans in the Second Punic War; notable for the monastery of St. Columban, a seat of learning and culture in the Middle Ages; today a small town with slightly fewer than 4,000 inhabitants, with narrow, winding streets, large squares, and medieval buildings.

Site Office: A.P.T. (Azienda di Promozione Turistica)
Piazza S. Francesco
CAP 29022 Bobbio, Piacenza
Italy
(523) 936178

Bobbio has been in existence since at least the times of the Roman Empire, when the city was known as Bobium. It was near Bobium in 218 B.C. that the Carthaginian general Hannibal, in spite of a loss of a number of his troops and elephants during the bitter winter, gained victory in his first direct conflict with the armies of Rome at the Trebbia River during the Second Punic War.

Hannibal arrived in the area of the Po River by mid-November, his original force of 46,000 foot soldiers and cavalry reduced to 28,000 by the hardships of crossing through the Alps. He continued south, engaging in a series of cavalry skirmishes with Roman troops under the command of Scipio Africanus, also known as Scipio the Elder. Scipio was wounded in the fighting, and the command of the Roman forces fell to Sempronius Longus. The Roman army had about 40,000 foot soldiers and cavalry. The two armies drew near Bobbio in late December, probably about the time of the winter solstice, and camped on opposite sides of the Trebbia, Hannibal's troops on the west, the Romans on the east.

Hannibal sent his Numidian cavalry across the Trebbia in an attempt to draw the Romans into battle. At first, Sempronius responded only with cavalry, and the Numidians retreated across the river. Sepronius then ordered his entire army across the Trebbia, which was bitterly cold and swollen chest-high by rain. Battle lines formed on the west bank of the Trebbia. Hannibal, making a rare decision to use his elephants in actual combat (the only other occasion was at Zama) distributed them equally on his left and right flanks, both to deter flank attacks against his forces and to help him make flank attacks against the Romans, should there be an opportunity to do so.

Battle ensued. The Carthaginian cavalry drove off the smaller Roman cavalry, thus exposing the flanks of the Roman infantry to attack. The two infantry lines then engaged in heated combat, but Hannibal had a surprise in store for the Romans. His younger brother, Mago, who had been hiding along with 2,000 hand-picked troops, attacked the Romans from the rear. The Romans were thrown into disarray and their flanks crumbled, with the soldiers retreating toward the river. Those in the Roman center put up a determined resistance for a time, even breaking through the center of Hannibal's line at one point, but when they realized their position was untenable they, too, retreated. Many of the Roman infantrymen were killed by Hannibal's elephants and cavalry along the river bank; the few infantrymen who could escape, and most of the cavalry, retreated to Placentia (modern Piacenza).

Hannibal's battle losses were primarily among the Celtic portion of his infantry. However, many of his men, horses, and elephants died from exposure to the rains and snow that followed the battle. Responsibility for the Roman defeat was laid largely at the feet of Sempronius, for sending his men into battle across the frigid river. Over the succeeding years of the war, the tables eventually turned; the Romans decisively defeated Hannibal at Zama in 202 B.C., and the war ended the following year, with Rome victorious and Carthage never to recover.

In the sixth century A.D., Bobbio came under control of the Lombards, a Germanic people that invaded Italy in the year 568 and established a kingdom in the Po valley. During this period Bobbio's place in history was further secured by the establishment of a monastery there by the Irish missionary St. Columban.

St. Columban was born into a noble family of Leinster, Ireland around the year 543 and studied at the monastery of Comgall in Bangor, Ireland. There he was educated by a monk named Sinell in many disciplines, including Latin. The writings of St. Columban are among the oldest surviving works of the Irish monks.

In the late sixth century St. Columban came ashore on the European continent near what is now the small village of St. Coulomb in France, between St. Malo and Mont St. Michel. One of a group of twelve monks, he brought with him a belief in the rigors of monastic life. His mission was to establish a series of monasteries on the continent, following the Celtic model, devoted to prayer and missionary work among the peasants of Europe. He and his band traveled eastward into Gaul. In the region of the Vosges, on lands donated to him by King Guntram of Burgundy, St. Columban spent the next several decades establishing a succession of

Interior of St. Columban's monastery
Photo courtesy of A.P.T., Bobbio

monasteries at, among other places, Annegray, Luxeuil, and Fontaines.

During his time in Gaul, St. Columban came into conflict with the local bishops, mostly over the reckoning of the date of the Easter celebration, and thereby the entire liturgical calendar, which is calculated from the date of Easter. St. Columban, typical of the independent behavior that would later mark the beginning of his work at Bobbio, refused to appear before the Council of Chalon-sur-Saône, called to discuss this matter.

In 610, St. Columban was expelled from Gaul by King Theuderic II, supposedly for not approving the legitimization of Theuderic's out-of-wedlock son. He attempted to return to the British Isles but a storm at sea forced his return to Gaul. There he rejoined a band of his monks at Neustria and, skirting the lands under King Theuderic's control, traveled east to Switzerland, then south.

Making his way to Milan, he arrived at the court of the Lombard king, Agilulf, where he remained for some months. The king's wife, Queen Theodolinda, was a Catholic Christian. Agilulf, under the influence of his wife and at the suggestion of a man named Jocundus, gave lands to St. Columban at the junction of the Trebbia and Bobbio rivers, where a partially ruined church was standing, to found another monastery.

St. Columban was influential in expounding the tenets of Irish Christianity on the European continent, and his monasteries, despite the rigorous and penitential life led by his monks, grew both in fame and in the number of their adherents. St. Columban is credited with founding more than forty monasteries in continental Europe, but the monastery at Bobbio proved to be one of the most significant.

The monastery of Bobbio functioned under the strict, some would say harsh, regulations prescribed by its founder. These regulations were known as the Rule of St. Columban and were followed at all his other monasteries as well. As the abbot of Bobbio, St. Columban also had the authority to develop his own list of penances, which were rather severe. The penance for gluttony was a bread-and-water diet for a week (for a layman) or forty days (for a monk). Certain sexual transgressions might bring a penance of up to seven years. St. Columban also utilized the device of public confession which, along with penance, the monks of Bobbio saw not only as punishment but also as tools to refine the character of the penitent and to ameliorate the desire for revenge on the part of those who believed that they had been wronged by the publicly punished sinner. St. Columban believed in the necessity of the rehabilitation and conversion of those who had not fully accepted the Irish version of Christianity. The monastery at Bobbio was instrumental in converting the Lombard kingdom to the Catholic form of Christianity, thereby helping to bring about the end of the religious duality that existed in Italy between the Catholic Church and the German form of Arian heresy. At the time Columban's Bobbio monastery was founded, Arianism held great sway over the Lombard kingdom, even though it had been denounced at the Council of Nicaea in 325. St. Columban was vigorous in his arguments against Arianism. Although its tenets varied from time to time, the followers of Arianism, named after Arius, a fourth-century presbyter from the church of Alexandria, held that the Father and Son were distinct beings in the Trinity and the Son, although divine, was not equal to the Father. Also, they believed that the Son had existed prior to His appearance on earth and that the Messiah was not a real man but a divine being enclosed in flesh. King Agilulf, himself a follower of Arianism, is believed to have converted to Catholic Christianity prior to his death.

In his final days at Bobbio, St. Columban assumed the life of a hermit. He took up residence in a cave near the monastery grounds, joining his brethren only on Sundays and saint's days during religious services. St. Columban died at Bobbio in 615 and was interred at the monastery.

In spite of his death, the monastery continued to grow and flourish. In the year 643, the community of monks numbered 150. The monks lived and worked in a cluster of ten multi-storied buildings, the center of which was the church. An additional thirty buildings of a single story housed the lay workers of the monastery. The monastery owned twenty-eight farms located nearby, in addition to other properties farther away. The geographic area under control of the monastery was sufficiently large to require, in addition to the church located at the monastery, the services of another seven parish churches. The agricultural production of the monastery was impressive. In one year, for instance, the monastery had a surplus available for sale of 2,000 bushels of corn, 1,600 cartloads of hay, 2,700 liters of oil, and 5,000 pigs and cattle from the monastery alone. Other farms under the monastery's control produced a surplus of an additional 3,600 bushels of corn, 800 amphoras of wine, and various other products.

In the course of time, the Rule of St. Benedict, which was seen as more humane and less harsh than the Rule of St. Columban, which had taken firm hold in southern Italy, spread northward. In Lombardy, Benedictine rule was at first permitted, then encouraged. Eventually, Pope Gregory the Great ordered that the Rule of St. Benedict be followed in all monasteries and adhered to by all those who entered monastic life. In 670, by order of the Council of Autun in Burgundy, the monastery of Bobbio came under Benedictine rule.

In addition to providing for more humane treatment of the monastic community, the Rule of St. Benedict, with its emphasis on study and scholarship, contained the seeds of a cultural awareness that made possible the preservation of essential works of classical literature and fostered the propagation of Christian texts. Although the scribes of Bobbio, at work in the monastery's scriptorium, considered the copying of texts to be manual labor, they played an essential role in this preservation and propagation. The scriptorium at Bobbio reached a very high level of activity during the seventh and eighth centuries and was especially famed for the codices it produced. A codex was a bound collection of parchment sheets, the very first type of book. Codices had appeared during the first century, as a replacement for the papyrus roll.

Although St. Columban had not intended that any of the monasteries he founded be a center of culture, Bobbio grew to become one of the great seats of learning and held one of the largest libraries to be amassed during the Middle Ages. The books in the library at Bobbio were housed separately from the scriptorium, under lock and key and supervised by the armarius, or librarian. The library at Bobbio housed both secular and religious books and preserved works of classical literature, the majority of which were in Latin. Among these was the *Mediceus* of Virgil, which subsequently went to the Laurentian Library in Florence.

When the text on a particular piece of parchment was no longer considered worth saving, the parchment was washed, scraped and reused. This reused parchment is referred to as a palimpsest and the library at Bobbio contained a number of them. When the parchment is treated by modern chemicals, the original writing reappears with the newer text superimposed over it. The most famous palimpsest from Bobbio was that on which was written St. Augustine's commentary on the Psalms. After chemical treatment, it was found that St. Augustine's commentary was superimposed over the treatise *De republica* by Cicero.

The library at Bobbio also held many writings relating to German Arianism. Apparently, St. Columban thought it wise to know the enemy he preached and wrote against. These writings include the Paris Codex, which contain the memoirs of Auxentius on the Arian teachings of Ulfilas, a fourth-century missionary to the Goths. Also preserved is a fragment of the Gothic Bible of Ulfilas as well as his commentary on the gospel of St. John.

The library at Bobbio also contained many liturgical books, hymnals, martyrologies, and other works of Christian literature. The books varied in format and size; one, the Bobbio Missel, was unusual for its time in that, with pages measuring only about seven inches by three and one-quarter inches, it was easily portable, much the way a paperback book is today. The Bobbio Missel is now in the Bibliothèque Nationale in Paris.

Although its light shone for nearly 1,000 years as a keeper of the cultural and intellectual flame, the monastery at Bobbio fell into decline in the fifteenth century. Its library was broken up and dispersed, among other places, to the Ambrosian library of Milan, to Turin, and to the Vatican. Finally, the monastery of Bobbio was suppressed by the forces of the French, under Napoléon. Today, however, the monastery is open to visitors, and the relics of St. Columban are displayed in its museum. The existing buildings of the monastery date from the eleventh and thirteenth centuries and were restored in 1600.

For most of the rest of its history, Bobbio was a pawn in the power politics of Italy. In 1176 the townspeople took part in the historic Battle of Legnano, in which the citizen militias of the Lombard cities thwarted an attempted invasion of northern Italy by the forces of Holy Roman Emperor Frederick I (Barbarossa). Subsequently, in 1212 and again in 1219, Bobbio was taken control of by the town of Piacenza. However, Bobbio did eventually come under the rule of the Holy Roman Empire.

As in all of Italy during the Middle Ages, the control of Bobbio alternated between the Guelphs, a papal party that opposed the authority of the German emperors of the Holy Roman Empire in Italy, and the Ghibellines, Italian aristocrats who supported the German emperors. The death in 1343 of King Robert of Naples, who had assisted the the Guelphs in Bobbio, weakened their hold and the town came under the control of the aristocratic Visconti family of Milan. In 1748, under the terms of the Treaty of Aix-la-Chapelle, Bobbio became a part of the region of Savoy, and in 1860 it became part of a unified Italy.

Today Bobbio exists in relative obscurity and is visited primarily for its thermal baths, which are open from May to October. Its notable historic structures, in addition to the monastery of St. Columban, include its cathedral, with a mix of Romanesque and baroque architectural features; the medieval Ponte Vecchio (old bridge) over the Trebbia River; and houses and other buildings dating from the Middle Ages.

Further Reading: *Monks and Civilization: From the Barbarian Invasions to the Reign of Charlemagne* by Jean Décarreaux, translated by Charlotte Haldane (Garden City, New York: Doubleday, and London: Allen and Unwin, 1964; as *Les Moines et la Civilisation,* Paris: Arthaud, 1962) is a thorough history of early monasticism. The work describes everyday life in a medieval monastery and emphasizes the role that the monks played in the preservation of classical literature and intellectual thought throughout the period of the Dark Ages. Several individual monks are profiled in depth, including St. Columban and St. Benedict. *Italy and Her Invaders* by Thomas Hodgkin (Oxford: Clarendon, 1870; fourth edition, London: Oxford University Press, 1931) is an eight-volume, comprehensive work on the history and the effects of the repeated invasions experienced by Italy from the end of the Roman Empire through the first millennium. One volume of the work is devoted to the Lombard Kingdom of northern Italy, of which Bobbio was part. *The Lombard Communes* by W. F. Butler (London: Unwin, and New York: Scribner's, 1906) is a political history of the cities of northern Italy during the Middle Ages. Hannibal's victory at the Trebbia is described in detail in *Hannibal's War: A Military History of the Second Punic War* by J. F. Lazenby (Warminster, Wiltshire: Aris and Phillips, 1976).

—Rion Klawinski

Bologna (Bologna, Italy)

Location: The province of Bologna lies on the southern edge of the Po Plain and on the lower slopes of the Apennine foothills, in the center of northern Italy. The city is fifty-one miles north of Florence.

Description: Called La Dotta (the Learned: Bologna was home to the first university in Europe), La Grassa (the Fat: with reputedly the best cuisine in Italy), and La Rossa (the Red: referring to both the color of its brick buildings and its politics); known to have been inhabited since the Bronze Age and having been through several peaks of power and cultures since then, Bologna is now a prosperous trade and industry center, with the best preserved historic center in Italy after Venice, distinguished by twenty-one miles of porticoes—the greatest number in the world.

Site Office: A.P.T. (Azienda di Promozione Turistica di Bologna e Provincia)
Via Marconi, 45
CAP40122 Bologna, Bologna
Italy
(51) 239660

The site of Bologna has been inhabited since at least the Bronze Age, as evidenced by the foundations of huts and other remains from that period, which have been found in the center of the present-day city. The Bronze Age Indo-Europeans who occupied the territory were followed by the Italic people, and in the Iron Age what has become known as the Villanovan Civilization thrived, taking its name from a village near the city of Bologna called Villanova di Castenaso. In the nineteenth century, chamber tombs with false domes and vaults were discovered there, along with other remains indicating that this was an important Villanovan metalworking, trading, and agricultural center.

The Etruscans arrived in Italy in approximately 900 B.C., and from the end of the sixth century to the mid–fourth century B.C., Bologna, then called Velzna (or Felsina in Latin) was an important Etruscan town, trading with Greece via Ravenna, and with Rimini and Spina. The layout of some of Bologna's streets is still based on the Etruscan design. In 350 B.C., the Boïan Gauls, a nomadic tribe skilled in art and metalworking, captured Velzna after a violent struggle.

In 190 B.C. the Romans took Bologna from the Gauls, colonizing the area as Rome's "Eighth Region" to provide the Empire with food, in particular, with grain. They renamed the region Bononia. As they did with Verona, Trieste, Adria, and Padua, the Romans maintained Bononia as a nominally independent state, and in 187 Marcus Aemilius Lepidus ordered construction of the Via Aemilia, a road connecting Bologna

with Piacenza and Rimini as part of a defensive military system, parts of which remain in the city today, some of them preserved in subways underground. The Romans also built an aqueduct eleven miles long, still capable of providing the city with water.

Ravenna became capital of the Western Empire after the fall of Rome, and for several centuries Ravenna dominated Bologna. During this period, Bishop Petronius, later the city's chief patron saint (Sts. Peter and Ambrose are others), did much to revive Bologna's fortunes with civic building in the fifth century A.D. In the sixth century, the Benedictines established four monasteries, followed by the Carmelites, Augustinians, Lateranenses, Dominicans, and Franciscans. Eventually, there were approximately 100 monasteries in the city, a triumph for the church.

In 756, Pépin III, king of the Franks, gave Bologna to the church, which maintained suzerainty until 1860, in theory if not always in practice. In 1024 Milan established the first commune (free city-state), and Bologna followed suit soon after, although still nominally under papal control. The city-state thrived as an important trade center for Byzantium, Rome, and Lombardy.

In approximately 1050 the first university in Europe was founded at Bologna, initially consisting of the *glossatori*, who glossed or annotated the law codes of Justinian. The *glossatori* attracted students, and soon several small schools were established, often based in the teacher's house. For a considerable time, the new academics managed to maintain their independence from the church and the Austrians, both of which called on the scholars to give legal judgments during their ongoing conflicts. Students from Bologna traveled all over Europe. One of them, Vacarius, founded the School of Law at Oxford in 1144. In the course of the eleventh and twelfth centuries, Bologna reigned supreme in legal theory throughout Europe, with Rome the greatest center of practice. The dramatic increase in population accompanying the establishment of the university led to the building of the city's famous and extensive porticoes. The structures allowed a whole new floor to be attached to existing buildings without narrowing the streets, and provided shelter for pedestrians below.

The twelfth century was a golden age for Bologna. The city was one of the principal members of the Lombard League, formed in 1167 to support Pope Alexander III against the marauding Holy Roman Emperor Frederick I (Barbarossa). The Santo Stefano Complex dates from this time, a unique Romanesque grouping of four churches (once there were seven), a Benedictine cloister, and two chapels. It had been started in the fifth century A.D. in the time of St. Petronius, who is buried in the most spectacular of the four churches, the Church of the Holy Sepulchre. Its Romanesque cloister contains a bath-sized tub in which Pontius Pilate is said to have washed his hands after condemning Christ. The

Arca di San Domenico, Bologna
Photo courtesy of A.P.T. di Bologna e Provincia

Church of Sts. Vitale and Agricola contains some Roman capitals and columns, and the Church of the Crucifix and Church of the Trinity complete the group.

Bologna's best-known structures, the two leaning towers, also date from the twelfth century, when noble families competed to build fortified towers as symbols of their wealth and prestige. The tower built by the Asinelli family measures 320 feet high and leans 7.5 feet from the perpendicular; that built by the Garisenda family now stands only 165 feet high (it was shortened in 1360 for safety), with a 10-foot tilt. In the rush to erect the latter tower, the Garisendas failed to prepare a proper foundation, resulting in the severe tilt and forcing the family to abandon the project.

By the thirteenth century the university had 10,000 students. Taddeo Alderotti, the greatest doctor of the century and one of the first translators of Aristotle, was a professor at Bologna, and students there included two of the greatest thirteenth-century poets, Guido Guinizelli and Cino da Pistoia. Frederick II's chief notary, Pier della Vigne, also trained here. Tombs of the *glossatori* were placed in and near two new churches, San Domenico and San Francesco. San Domenico was built in 1251 to house the remains of the Spanish saint, Dominic, founder of the order of Preaching Friars, who died in the town in 1221, after having built a convent on the site. Many artists worked on his tomb, known as the Arca di San Domenico, including Nicola Pisano and his school and Nicolo di Bari, who changed his name to Niccolò dell'Arca as a result; there are several examples of his terra-cotta sculptures in the city. Michelangelo sculpted an angel and Sts. Petronius and Proculus for the tomb, and the chapel contains a St. Catherine by Filippino Lippi and work by Guido Reni and Il Guercino (Giovanni Francesco Barbieri). The Church of San Francesco was built between 1254 and 1263 and is one of the oldest Gothic churches in the country. Another Gothic church from the thirteenth century is Santa Maria dei Servi, known as "the jewel," which contains a Madonna by Cimabue (Bencivieni di Pepo), the founder of the Florentine School.

The Commune of Bologna decided to build a new square, the Piazza Maggiore, to mark the center of the town instead of the old, adjoining Roman Forum, and over the next four centuries the square was surrounded by splendid palaces. The Palazzo Communale was begun in 1287, built around the Clock Tower, which had just been acquired from Accursio, a well-known jurisconsult. The second part of the Palazzo was built in 1425, and a huge bronze of the Bolognese pope Gregory XIII, sponsor of the Gregorian calendar, was placed there. There is also a sixteenth-century staircase by Donato Bramante, and a picture gallery with medieval paintings, and Bolognese and Emilian art. This palace has been the seat of the municipal government ever since it was built.

Adjoining the Piazza Maggiore is the Piazza del Nettuno, which contains the Fountain of Neptune, created in the sixteenth century by Giambologna (Jean Boulogne), a Flemish sculptor. On this square is the Podestà Palazzo (Governors' Palace), which consists of a portico with nine arches supporting a large, single hall. The Torre dell'Arengo

(Arengo Tower) rises above it, containing the Arengo bell, second largest and heaviest in Italy, which has rung for celebrations, wars, and deaths since 1483. The hall was frescoed from 1911 to 1928 with scenes showing the history of the city.

Another of the main palaces on the square is the Palazzo di Re Enzo (Palace of King Enzo), begun in approximately 1245 to be the seat of the city-state. However, on May 26, 1249, Bologna and its Guelph allies, which traditionally included Reggio and Ferrara, faced their Ghibelline foes, Modena, Parma, and Cremona, at the Battle of Fossalta. Bologna defeated the imperial army of Frederick II and captured the emperor's natural son, Enzio, titular king of Sardinia, who was to remain in this splendid palace-prison until his death twenty-three years later in 1272. An annual feast was held until the eighteenth century to mark the anniversary of his capture.

The Battle of Fossalta marked the peak of the commune's power. In 1257 it passed the famous law, Paradisus, emancipating 6,000 serfs, but in 1278 Bologna unwillingly became part of the Papal States. This led to a century of violent struggles between the Milanese Viscontis, rebel Viscontis, other powerful local families, and the papal legates, all vying for power. In 1337 Taddeo Pepoli took control, succeeded by the Viscontis, and the papacy prevailed at other times. That Bologna's cathedral is dedicated to St. Peter, patron saint of Rome, is an indication of papal power in the city. Said to have been founded in the ninth or tenth century, then destroyed by fire, rebuilt, then damaged by an earthquake in 1222, the cathedral was renovated between 1392 and 1406. The saint of the Bolognese people, however, was Petronius, who had done so much for the city in the fifth century, and his basilica was begun in 1390, by order of the city administration, who intended it to be larger than St. Peter's in Rome. The papacy put a stop to that, but it is still huge, 432 feet long and 185 feet wide, with a facade in red and white, the colors of Bologna's emblem.

At the beginning of the fifteenth century the Bentivoglio family succeeded in gaining control of the city, still nominally part of the Papal States. The Bentivoglios were never princes, though they aspired to the lifestyle of their allies and protectors, the Medicis of Florence and the Sforzas. They ruled Bologna as "first citizens" via the senate. Despite the usual catalogue of murders, splendor, and illegitimate children, their regime was generally a time of peace for the citizens, and the city prospered. The Bentivoglios extended the Canale del Reno, which completed the waterway from the Po River to the city. Bologna was well provided with corn, and at this time the flax, hemp, and silk industries grew to become the basis of the city's economy until the early eighteenth century. In the seventeenth century there were more than 100 silk mills in the town, and the Bolognese mill was invented for silk spinning.

In the last quarter of the fifteenth century much reconstruction work was done on the city buildings. New work was almost always carried out in brick, for its components clay, selenite, and sandstone were the materials available locally and had been used since at least Roman times. Architects and sculptors were employed from Lombardy, Venezia, and Tuscany, and painters from Ferrara, Modena, and all over Italy came to decorate the Bentivoglios' city. Francia (Francesco Raibolini) was the forerunner of the School of Bologna and acknowledged by Raphael as a master of devotional art. His altarpieces, commissioned by the Bentivoglios, are in several churches and galleries in the city.

The Church of San Giacomo Maggiore, begun in 1267 in the Romanesque style and extended in the late thirteenth century in Gothic style, was extended again between 1478 and 1498 by the Bentivoglios to make room for their burial chamber. There is an underground passage leading from the church to what is now the Teatro Communale, then the site of the Palazzo Sanuti-Bevilacqua, the Bentivoglio palace, which was torn apart by the population when the family was ousted by Pope Julius II in 1506. Both the palace and San Giacomo Maggiore were richly adorned with art, and the Bentivoglio Chapel in the church contains Francia's best altarpiece and frescoes by Lorenzo Costa.

After the fall of the Bentivoglios in 1506, Bologna became the capital of the Northern Papal States, second only to Rome, until the arrival of Napoléon in 1797. Julius commissioned a statue of himself from Michelangelo, which was placed in front of the people's church, San Petronius. The population pulled this down at the first opportunity, and the bronze was sold to the pope's great enemy, Alfonso I of Ferrara, who had it melted down and cast into an enormous cannon, which he called Julius. The city still retained its own senate, composed of representatives of the city nobles, but the unpopular popes maintained tight control. The papal court often met in Bologna and on February 24, 1530, Holy Roman Emperor Charles V had Pope Clement VII crown him emperor and king of Italy in the Basilica of San Petronius. He was the last emperor ever to be crowned by a pope, ending a tradition begun by Charlemagne 700 years previously.

Bologna was the birthplace of a new movement in art when, in the 1580s, Lodovico, Agostino, and Annibale Carracci, a Bolognese family of painters, opened their academy, whose mission was to end Mannerism and return to nature for inspiration. Lodovico remained in Bologna all his life while his cousins Agostino and Annibale went to Rome in the 1590s. Supported by the popes, they dominated art in Counter-Reformation Rome for some time. This was also the time when Giambologna, the greatest sculptor in Florence after Michelangelo left, was creating the Fountain of Neptune in Piazza Nettuno, the first of many he did, including a great series for the gardens of the Medici villas, developing fountain art into one of the great baroque arts.

Toward the end of the sixteenth century, the senate increased its power and made another attempt to enlarge San Petronio. The Vatican, however, ordered Bologna's St. Peter's Cathedral to be remodeled in baroque style, symbolizing its control over the city. The pope also insisted on the building of the Archiginnasio, in which all the different schools of the university could be united. This led to another great increase in the number of students in the city (and more porticoes) and further expansion of the university itself. Here, Francesco Bonaventura Cavaliere, a pupil of Galileo's, began research that led to the infinitesimal calculus of Gottried Leibniz and Isaac Newton, while Ulisse Aldrovandi was one of the leading figures of the age in zoology. In 1655 the astronomer Gian Domenico Cassini traced the meridian on the floor of San Petronius and designed its huge astronomical clock, which tells the time in days, months, and years by a shaft of light through an oculus in the roof.

Art flourished, with a second Bolognese School developing. Domenichino (Domenico Zampieri) had trained at the Carracci Academy. He went to Rome in 1602 and to Naples in 1631. Another academy pupil was Francesco Albani, who also went to Rome in 1602. Guercino, one of Lodovico's students, became a leading baroque religious painter, both in Bologna and Rome. Guido Reni, who had joined the academy at the age of 20, worked in his home town from 1603 to 1607, spent seven years in Rome, and then returned to reign supreme in Bologna from 1614 until his death in 1642. In the nineteenth century he was thought to be Italy's greatest artist, known as "the Divine," an anti-baroque, Raphaelesque painter.

The papacy retained control of the city until the end of the eighteenth century. This century was the age of Bologna's great theatre designers and *trompe l'oeil* interior decorators. Ferdinand Bibiena and his four sons, including Antonio, were the most famous of these, and in 1763 the Teatro Communale was inaugurated. The building, the scenery, and the costumes for its first performance were all designed by Antonio Bibiena. The city also has a collection of eighteenth-century puppet theatres in one of its many museums.

In 1797 Napoléon Bonaparte established his Cisalpine Republic with its capital at Milan. His sister, Elisa Baciocchi, is buried in San Petronio; another mark left on Bologna by Napoléon is the neoclassical Villa Aldini, built overlooking the city, marking the place where he stood to admire it. One effect of the new regime was the reform of the university, which was instituted as a National University in 1797 and moved to the Palazzo Poggi in 1803, joining the Institute of Science and creating a new University Zone. The botanic garden was moved there, too, and a new importance was given to science studies.

Another effect of Napoléon's republic was that Bologna moved from being the Papal States' second city to become the regional capital of the Emilia, the center of a rich agricultural province. Agricultural production was encouraged by the sale of ecclesiastical properties, creating a land market. While some of the monastic buildings were destroyed after Bonaparte's arrival, he started the movement that led, during the nineteenth and twentieth centuries, to some thirty-one monasteries and thirty-eight convents being converted to different uses, such as prisons, barracks, schools, hospitals, and art galleries.

As a result of the Congress of Vienna in 1815, the church, supported by the Austrians, regained Bologna and maintained some degree of control until 1859. However, the Italian Socialist movement was born here in the nineteenth century, and there were several rebellions, most notably on February 5, 1831, when the papal legate was forced to hand over power to a provisional government until the end of March. Another rebellion began on August 8, 1848. In 1859 the pope's troops were chased out of Bologna, and in 1860 the city was united with Piedmont and the Kingdom of Sardinia, leading in due course to the unification of Italy.

Theatre and music continued to thrive through the nineteenth century. Bologna was the birthplace of Ottorino Respighi, the composer whose music was popularized by Arturo Toscanini, also from the Emilia. Giacchino Antonio Rossini studied at Bologna's Conservatrio G. B. Martini from 1806 to 1810 and spent much of his life in the city. Another famous Bolognese citizen was Guglielmo Marconi, who carried out his first experiments with radio at the Villa Grifoni; the city's airport bears his name. In 1890 painter Giorgio Morandi was born in Bologna. He studied there and then taught in the city throughout his life, and in the 1940s won international recognition for his gentle still lifes, influenced by the Metaphysical painters Giorgio De Chirico and Carlo Carrà. There are seventy-nine of Morandi's paintings in the Gallery of Modern Art, together with work by Umberto Boccioni, Carrà, Max Ernst, De Chirico, and Jackson Pollock. The National Picture Gallery is one of the best in Italy, with work by Bolognese masters from the fourteenth century onward, as well as other European masters. There are several paintings of eighteenth-century Bolognese everyday life by Guiseppe Crespi.

Since World War II Bologna has been run very efficiently by the Communist Party, thriving as Italy's fifth most important industrial center, despite the August 1980 right-wing terrorist bombing of the railway station that killed 85 people and injured more than 200. In 1990 it hosted the Soccer World Cup, and each year the International Children's Book Fair is held there, along with numerous trade fairs. The university thrives, particularly in the sciences, and has a great number of museums, all based on research done by staff and students.

Further Reading: *The Bentivoglio of Bologna: A Study in Despotism* by Cecilia M. Ady (Oxford and London: Oxford University Press, 1937) details the politics of the region in the fifteenth century and the exploits of the ruling family. *Annibale Carracci: A Study in the Reform of Italian Painting around 1590* by Donald Posner (New York and London: Phaidon, 1971) is a two-volume work on the Bologna School of Art and its influence. *The Universities of Europe in the Middle Ages* by Hastings Rashdall (Oxford: Clarendon, 1895; revised by F. M. Powicke and A. B. Emden, 1936) provides a full account of Bologna's university. *Coexistence, Communism and its Practice in Bologna, 1945–1965* by Robert H. Evans (Notre Dame, Indiana: University of Notre Dame Press, 1967) examines the Communist Party in the city.

—Beth F. Wood

Burgos (Burgos, Spain)

Location: In northern Spain, on the Rio Arlanzon, about 130 miles north of Madrid by the main road to France.

Description: The capital of the old Spanish kingdom of Castile, now capital of the province of Burgos, one of nine provinces in what is known officially as the Autonomous Community of Castile and León. The home of legendary Spanish hero El Cid (Rodrigo Díaz de Vivar), it contains many examples of Spanish Gothic architecture; chief among them is the city's great cathedral. It was the site of significant fighting in the Peninsular War of the early nineteenth century.

Site Office: Tourist Information Centre
7, Alonzo Martínez Square
09003 Burgos, Burgos
Spain
(947) 20 18 46

The city of Burgos clothes itself in history, art, and myth with equal splendor, and justifiably so. It was a medieval Spanish capital and the center of Spanish politics, literature, and language, home of a larger-than-life Spanish hero better known through legend than history, and it remains the proud possessor of one of the finest Gothic cathedrals in Europe, amid other period architecture.

Founded as an outpost of the kingdom of Asturias in 884, Burgos gained prominence as Spanish Christian nobles of the region began to unite in their fight against the Islamic Spanish kingdom of Andalusia to the south. Burgos became the capital of the new countship of Castile, named for its abundance of castles, which declared itself an independent kingdom in 1035. From then until 1561, when Madrid was named the new court city, Burgos was at the center of Spanish politics and Castilian culture.

Castilian cultural life was to have a significant impact on modern Spanish culture. It was in Castile that the *romancero*, the popular literature of poetry and song, arose and flourished, and that a legal system that preferred local customs over old written codes was established. The Castilian language began to be studied and regularized, and soon assumed a prominence equal to Latin. By the fifteenth century Castilian had become the language of the neighboring kingdoms of León and Aragon and had begun to replace the traditional dialects of the south. This was the origin of the modern language we know as Spanish, and the inhabitants of Burgos are still said to speak the purest Castilian in Spain. During the Middle Ages Burgos also was a major stop on the pilgrimage route to Santiago de Compostela.

The most famous citizen of Burgos is undoubtedly Rodrigo Díaz de Vivar, known to history and legend as El Cid

(from the Arabic for "lord"). Also known as Campeador, or "winner of battles," his life has been mythologized to the point that it is difficult to disentangle the real from the romanticized. He was made famous to posterity in the first great Castilian literary epic, *The Poem of the Cid*. Born around 1043, Rodrigo was the son of Diego Lainez, a minor Castilian noble with an estate at the village of Vivar, outside Burgos. Burgos was then the court city of Ferdinand I, and after Diego Lainez's death, Rodrigo was raised at Ferdinand's court under the patronage of Sancho, Ferdinand's eldest son and heir. Burgos was thus the setting for Rodrigo's youth and for his first successes. When Sancho succeeded his father as Sancho II in 1065, Rodrigo was made the royal standard-bearer and proved himself an able military commander, contributing significantly to Sancho's seizure of the kingdom of León from his younger brother Alfonso.

After Sancho's death in 1072, the exiled Alfonso returned to claim the now joint throne of Castile and León as King Alfonso VI. Despite understandable suspicion on the part of his former enemy, Rodrigo remained at the court in Burgos for ten years, retaining the king's favor and wedding his niece, Jimena, daughter of the count of Oviedo. Eventually the relationship between king and conqueror began to deteriorate as Rodrigo's ambitions collided with his master's. The last straw was Rodrigo's unauthorized attack upon Toledo, a Moorish kingdom under Alfonso's protection. Rodrigo was banished from Castile and left his homeland for Saragossa, another Moorish kingdom. Although eulogized as a leader of the Christian revolt against the Moors, in fact Rodrigo spent the next several years serving them as military commander and political advisor. During this time his natural abilities for intrigue were aided by the knowledge he gained of Moslem law, politics, and traditions. In 1087 Alfonso was forced to seek a rapprochement with the Cid. Spain was being invaded from the south by the Moslem Almoravids, and he needed Rodrigo's military leadership. But the Cid was deeply involved in the affairs of eastern Spain and used his new-found favor to gain title to the kingdom of Valencia. In 1089 Alfonso agreed to grant him any lands he took from the Moors; when their agreement broke down Rodrigo continued his campaign in his own interest. After five years of political intrigue and armed conflict, Valencia surrendered in 1094. Rodrigo defended his hard-won prize against attacks by the Almoravids later that same year and again in 1097. He died in Valencia in 1099. Jimena returned to Burgos three years later, bringing the Cid's body to the monastery of San Pedro de Cardeña. It lay there until 1919, when the bodies of both were entombed in the cathedral of Burgos.

The cathedral was begun in the thirteenth century with donations from the Mesta, a group of wealthy wool producers. The cathedral's famous steeple and spires have a delicate

Casa del Cordón
Photo courtesy of Instituto Nacional del Promocion del Turismo

grace that is distinctively Burgos. Considered one of the finest examples of Gothic church architecture in Europe, it is on the UNESCO list of World Heritage Sites. It was the project of King Ferdinand III, who held his court and kept several palaces in Burgos, in conjunction with Maurice, the bishop of Burgos. The first stone was laid by king and prelate on June 20, 1221, and by 1230 the apse was sufficiently completed for services to be held there. It was likely another twenty years before work began on the nave, and the cathedral was not finally completed until the sixteenth century.

Throughout those centuries many architects, both known and unknown, contributed to the rich ornamentation, soaring lines, and devout ambiance.

The original structure was built of white limestone during the first full blossoming of the Gothic style, and while probably inspired by French Gothic it boasts a distinctive Spanish flavor. It was built upon a hill slope, so that the northern section is higher than the southern and western sections. The traditional plan of the building is that of the Latin cross, formed by nave and transept. Over the centuries the original layout has become obscured by the addition of fifteen chapels slipped into every conceivable niche. The nave, with its massive pillars and groin vaulting, is supported by exterior arched buttresses. The principal entrance is in the impressive western facade. Above the three doors is a rose window and a gallery of statues. On either side are towers topped by the spires with their famous tracery. The two other facades, on the north and south, are simpler, yet adorned with statuary and carved doorways.

The fourteenth century saw the addition of a second cloister and two large chapels, the Chapel of St. Catherine and the Corpus Christi Chapel. The original cloister off the chancel was converted into offices and small chapels. During the following century master architect Hans of Cologne introduced the so-called Florid Style Gothic. He built a chapel for his patron, Bishop Don Alonso de Cartagena, and was responsible for the beautiful spires on the west towers. The great lantern over the crossing was constructed in the sixteenth century by Juan de Vallejo after the collapse of its predecessor.

The cathedral's interior is made dazzling by the glass skylight in the dome at the crossing and by the wealth of metalwork and gilt adornment. It is also the repository for many impressive tombs designed by the same architects who built the cathedral itself, and for artifacts of both history and legend. A less sober feature of the interior is the Papamoscas, or Flycatcher Clock above the west door; a bird emerges to open its mouth as the hour is struck.

In addition to the Cid's tomb, the cathedral houses his marriage contract and a chest made famous in legend. Supposedly the Cid filled a chest with sand and gave it the Jewish moneylenders of Burgos, telling them it was filled with gold and using it as collateral for a loan. Eventually, the legend assures us, he repaid the loan.

Many other reminders of El Cid's connection to Burgos are found throughout the city. Next to the cathedral, the most famous landmark of Burgos is the statue of El Cid in the Plaza General Primo de Rivera. El Solar del Cid marks the site of the house where Rodrigo lived during his years in Burgos, and it was at the Church of Santa Agueda that Rodrigo demanded the new King Alfonso swear he had no complicity in his brother Sancho's death.

Toledo for many years shared with Burgos the distinction of being a Castilian capital; disputes as to which was more important ended only with the establishment of the royal court at Madrid in 1561. After the decline of its political importance Burgos continued for a time to play a prominent commercial role, due largely to the powerful Mesta. In 1494 this Burgos guild had gained jurisdiction over all Castile's foreign trade, including the crucial exports of Merino wool. When Spanish trade later shifted south to Seville, the city declined until the eighteenth century under Charles III, who carried out extensive programs to improve commerce.

Burgos was the site of two significant battles during the war known to the English as the Peninsular War and the Spanish as the War for Independence. Napoléon Bonaparte's desire to defeat Great Britain led him to attack British ally Portugal by incursions across northern Spain, an ally of France. The temptation to conquer Spain was too great, however, and Napoléon soon seized portions of Spain and threatened King Charles IV. Britain came to Portugal's rescue and allied itself with Spanish rebels determined to repel the French invaders. As part of the long and complicated campaign, Burgos was captured by the French in 1808.

In September 1812 Arthur Wellesley, the future duke of Wellington, laid siege to Burgos, but the siege was an exercise in miscalculation, mismanagement, and poor strategy. Having brought only a few cannon from Madrid, Wellesley sent unsupported allied troops against the castle of Burgos, whose defenders decimated the infantry ranks. An outlying bastion was captured almost by accident, however, and Wellesley took unwarranted confidence from this success, using the bastion to launch another unsupported infantry charge that was again repelled with high casualties. Wellesley then ordered his troops to dig a tunnel to the enemy walls and detonate a mine beneath them. This the troops laboriously did, only to discover, upon charging once again, that they had blown up the wrong wall; in digging the tunnel, they had discovered the buried remains of an ancient castle and mistook it for the current fortifications. Their mine left the French defenses quite unscathed, and the attackers were once again driven off the field by withering French fire.

Finally, with more guns and another subterranean mine, the allies managed to breach the castle wall. After intense fighting they gained a foothold in the outer defenses, but they were assaulted by French troops who poured out of the castle into the trenches. Again, the allies suffered heavy losses.

After another attempt to break through the French defenses on October 18, Wellesley was forced to lift the siege, leaving valuable guns behind for the French. The French had suffered about 300 casualties, the allies more than 2,000. It was not until 1813 that the French left Burgos as part of their general withdrawal toward the Pyrenees. Upon evacuating, Joseph Bonaparte, Napoléon's brother and commander of the French forces, had the castle blown to bits.

During the Spanish Civil War in the late 1930s, General Francisco Franco made Burgos his temporary capital and military base from which he launched campaigns against Madrid and the Basque country. Burgos has remained a conservative city; in 1970 it was the site of the "Burgos trials" of dissident Basques, who were convicted of crimes against the nation without receiving a fair hearing. A more benign manifestation of that conservatism is in Burgos's preservation of its many sites of both historical and architectural significance.

In addition to the cathedral, Gothic houses of worship include the churches of San Esteban, damaged in 1813; San Gil; San Lesmes; San Nicolas de Bari, with its magnificent sixteenth-century altarpiece by Simon de Colonia; and Nuestra Senora la Real Y Antigua de Gamonal. At the Casa del Cordón, a home built by a city official in the fifteenth century, Ferdinand and Isabella received Christopher Columbus upon his return from his second voyage to America. The remains of the city's castle, destroyed by the 1813 explosion, are visible, along with portions of its wall, as are three gates, Santa María, San Esteban, and San Martin. The Arco de Santa María dates from the early Renaissance and includes sculptured portraits of Holy Roman Emperor Charles V and El Cid.

Burgos boasts a number of museums. The Museo Arqueológico is housed in the Casa De Miranda, built in the sixteenth century, and exhibits prehistoric, Roman, and medieval artifacts. The Museo Marceliano de Santa María displays more than 150 paintings by the Burgos artist whose name it bears. It is located in the old San Juan Monastery.

On the western outskirts of the city is the Museo-Monasterio de las Huelgas Reales. Built as a summer palace, it was converted into a convent for Cistercian nuns from the aristocracy by Alfonso VIII in 1187. Several Castilian kings were knighted in its church, and Alfonso and his wife Eleanor are buried there. The site includes the Museo de Telas (textile museum), a remarkable collection of thirteenth-century clothing. On the eastern edge of Burgos is the Cartuja de Miraflores, founded as a charter house by Juan II in 1441 and later remodeled by Simon de Colonia. The church altarpiece, by the great medieval artist Gil de Siloe, was gilded with the first gold brought back from Spanish America. Five miles

from Miraflores is the Monastery of San Pedro de Cardeña. Founded in 899 by Trappist monks, it was rebuilt in the eleventh century. It was here that the Cid left his wife when he entered into exile, and it was here they were both buried until a few decades ago. Thirty-six miles south of Burgos lies one of the most famous monasteries in Spain, the eleventh-century Monastery of Santo Domingo de Silos, as famous for its Romanesque cloister encrusted with sculpture and for its early production of written Castilian as for its Gregorian chants. The complex also includes an eighteenth-century baroque church. Founded by Dominicans, it was reoccupied in 1880 by French Benedictine brothers.

Further Reading: Literary historian Richard Fletcher provides an account of his search for the historic El Cid in *The Quest for El Cid* (New York: Knopf, 1990; Oxford: Oxford University Press, 1991). He explores both the literary traditions and historical documentation in his attempt to find the real Rodrigo Díaz de Vivar. For a thorough discussion of the complex campaigns of the Peninsular War, see *The Spanish Ulcer: A History of the Peninsular War* by David Gates (New York and London: Norton, 1986).

—Elizabeth Brice

Bursa (Bursa, Turkey)

Location: Northwest Turkey, in the foothills of Mount Uludağ, south of the Sea of Marmara.

Description: Established around 200 B.C., Bursa began as a small provincial town in Bithynia. When the Ottoman Turks made Bursa their first capital in 1326, the city became a political and religious center. Early Ottoman rulers built grand mosques and mausoleums there, monuments to their Islamic faith and imperial aspirations. After the Ottoman capital moved to İstanbul, Bursa developed a thriving silk trade.

Site Office: Tourist Information Office
Fevzi Çakmak Caddesí, Fomara İshanı, Kat: 2
Bursa
Turkey
(224) 2542274 or 2505966

Bursa rose to prominence in the fourteenth century as the first capital of the Ottoman Empire. Although famous for its distinctly Ottoman character, the city's historic roots are ancient. Bursa stands on the fertile plains of northwestern Anatolia (Asia Minor) in a region known as ancient Bithynia. King Prusias I of Bithynia founded the town in 183 B.C., calling it Prusa in his own honor. Its name later evolved to Bursa. Prusa grew into a small settlement on the slopes of Mount Uludağ, an imposing peak also known as Bithynia's Mount Olympus (several mountains in ancient Greece bore this mythical name).

Bursa's only notoriety during Hellenistic times derived from its hot springs. Visitors came to sample the waters, to consult with the well-known physician Asclepiades, and to make offerings at the shrine of the healing god Asclepius. The kingdom of Bithynia became a province of the Roman Empire in 76 B.C. Little information survives from this period, although it is known that the geographer Pliny the Younger governed Bursa under Emperor Trajan. Bursa's hot springs again proved an attraction to the subsequent Byzantine rulers. Emperor Justinian I built a palace and bath complex near Bursa, making the town a fashionable resort destination.

The Byzantines' long struggle against Islamic invaders from the east eventually involved Bursa. In 1075, the Islamic Seljuk Turks captured Bursa, but they lost control during successive centuries of fighting for dominance in Anatolia. By the thirteenth century, the Byzantine Empire had been partitioned. Bursa's new rulers, the Nicaean emperors, refortified the city's defensive walls. These defense works were soon tested by a rising Islamic power, the Ottoman Turks.

The origins of the Ottoman Turks are unclear. Few accounts from contemporary sources have survived, and the Ottoman sultans' penchant for mythologizing their ancestors further obscures historical fact. The story embraced by the sultans traces their origins to a Turkoman tribe in northeastern Iran. The tribal ruler, Süleyman Sah, was said to have escaped Mongol invaders in the early thirteenth century by fleeing westward. After Süleyman died crossing the Euphrates into Syria, most of the tribe returned to their homeland. One of Süleyman's sons, however, continued on his father's route, leading about 400 followers into Anatolia.

This son, Ertugrul, is credited with founding the Ottoman dynasty in Anatolia. According to the sultans' account, Ertugrul and his followers entered military service for the Seljuks of Rum, to aid in their struggles against the Christian Byzantines as well as the Mongols threatening invasion in eastern Anatolia. As a reward for this support, the Seljuk sultan was said to have given Ertugrul land in northern Phrygia, on the border of Byzantine Bithynia. The Ottoman sultans promoted this account of alliance with the Seljuks because it validated their claims to power in Anatolia.

Modern scholars have developed another explanation of the Ottomans' historical origins. Newly discovered thirteenth-century sources trace the Ottoman ancestors' arrival in Anatolia to the eleventh century, as part of a Turkoman migration from the east. These accounts depict the Ottoman ancestors as roaming mercenaries who for 200 years sold their military support for the highest profit. This less noble account undermines the Ottomans' fabled connection to the Seljuks.

Both versions of Ottoman origins agree, however, that by 1280 the Ottomans had control of lands in northern Phrygia. Ertugrul died around that time, passing leadership of the territory to his son Osman, born in Sögüt around 1258. Osman had aspirations to increase the territory under his control, but he moved cautiously until about 1300, when the once powerful Seljuk Empire in Anatolia collapsed. Osman seized two former Seljuk forts, Eskisehir and Karacahisar, which guarded the routes leading from his barren Phrygian territory to the fertile lands of Bithynia. Capturing Bursa, the Byzantine capital of Bithynia, soon became Osman's goal.

Osman sought to isolate Bursa by controlling the lands around it. His first major campaign cut communications between Bursa and the neighboring city of Nicaea. With the land route to Constantinople severed, Bursa's only way of communicating with the Byzantine capital was by sea, through Mudanya and other ports on the Sea of Marmara. By 1308, Osman had expanded his domain to include lands from the Black Sea to the Sea of Marmara. Bursa, though surrounded by Ottoman-controlled territory, held out against Osman. The recently refortified defensive walls posed a significant challenge to Osman's forces, and Bursa's remaining sea route to Constantinople kept supplies coming to the embattled city. When Osman seized the port of Mudanya in

Green Mausoleum at Bursa
Photo courtesy of Embassy of Turkey, Washington, D.C.

1321, however, Bursa's vital link to Constantinople was severed. Yet the city fought so determinedly against the Ottomans that Bursa survived another five years of siege before falling at last to Osman's son Orhan on April 6, 1326.

By claiming Bursa as their new capital, the once nomadic Ottomans established an independent political state with defined borders, a settled population, and a formidable army. Orhan asserted his status as ruler in traditional Middle Eastern ways, by minting coins and ordering recitation of Friday public prayers in his name only.

The Ottomans began transforming Byzantine Bursa into an Islamic capital through a program of public building that continued until the mid-fifteenth century, when the capital moved to Constantinople (renamed İstanbul). Mosques and imperial mausoleums, built in styles that honored both Byzantine and Seljuk traditions, soon graced the thriving city. The tombs of Osman and Orhan, founders of the Ottoman dynasty, stand in the heart of Bursa. Osman's descendants built regal mosques, symbols not only of the sultans' power but also of their devotion to the Islamic faith. Two such mosques, Ulu Camii (Great Mosque) and Yesil Camii (Green Mosque), have become sacred landmarks in modern Bursa.

From their capital at Bursa, the Ottomans launched new campaigns of territorial expansion. Orhan's son Murad

I led an army across the Dardanelles into Europe, where in 1362 he captured Adrianople (Edirne), a city that served for a time as the Ottoman capital. Murad I later expanded Ottoman influence into the Balkans by defeating the Serbians at Kosova in 1389. Murad I's successor, Bayezid I, turned back a European counter-offensive in a victory over Hungarian King Sigismund's Crusader army at Nicopolis in 1396. Bayezid ordered construction of Bursa's magnificent Ulu Camii, the mosque with twenty domes, to commemorate this victory.

Bayezid I sought to bring all Anatolia under unified Ottoman control, a quest begun by his grandfather Orhan. He had almost succeeded when he faced a challenge by Tamerlane (also known as Timur), the last great Mongol conqueror. Tamerlane's army resoundingly defeated the Ottoman forces in a battle near Ankara around 1401. Bayezid was taken prisoner, and Tamerlane occupied Bursa for about a year.

A decade of civil war followed Tamerlane's occupation of Bursa, but in 1413 Sultan Mehmed I assumed power and reestablished order. The Ottomans again turned their attention to territorial gain, launching attacks on the city to which they had long aspired, Constantinople. Mehmed I's grandson, Mehmed II (the Conqueror), earned his nickname by capturing Constantinople in 1453. The Ottomans transferred their capital to the more cosmopolitan İstanbul, leaving in Bursa the tombs of their founders and the landmarks of their rise to power.

Bursa's prominence as a political center faded with the transfer of capital status to Edirne and subsequently to İstanbul. The importance that Bursa had acquired as a commercial center, however, sustained the city economically and fostered a silk industry that continues to prosper today.

Vital trade routes linking Europe to the Orient and the Middle East via Anatolia developed in the thirteenth century. When the Ottomans came to power and established their capital at Bursa, these trade routes shifted to converge at the newly prominent city. Sultan Bayezid I's campaign to bring all of Anatolia under Ottoman rule also boosted Bursa's commercial fortunes. By expanding his territory in the east, Bayezid gained control of the heavily traveled silk caravan route. Merchants bringing silk from Persia now brought their goods over land to Bursa and then to seaports beyond. Bayezid also commanded trade routes between Bursa and the Mediterranean coast after capturing the main port cities through which Arabian and Indian goods entered Anatolia.

Ottoman control of major trade routes ensured security for Moslem merchants traveling from Iran and Arabia. As eastern products came to market in Bursa, European merchants eager to trade with the Moslems flocked to the city. Persian silk, produced raw in northern Iran and woven in Bursa, was in great demand by Europeans. Italian merchants from Venice, Genoa, and Florence traded woolen cloth for woven silk, which they sold in Europe for tremendous profit. Moslem merchants also offered spices (particularly pepper), sugar, dyes, musk, and Chinese porcelain. They traded these products for gold and silver coins as well as fine velvets and brocades manufactured in Bursa.

Although Europeans and Moslems traded a variety of goods in Bursa, it was silk that made the city rich. As the foremost marketplace for fine Persian silk, Bursa profited from Europe's increasing demand for silk during the fifteenth century. Raw silk imported from Asterabad and Gilan in northern Iran was woven on looms in Bursa, then sold to European merchants. By the sixteenth century, more than 1,000 looms operated in the city, producing silk for the domestic market as well as international export. When Ottoman wars with Persia caused raw silk shortages, Bursa's merchants began to breed silkworms locally. The city thus became capable of producing raw silk as well as weaving it for export. Bursa remained one of the Ottoman Empire's major trade centers (and most populous city in Anatolia) until well into the sixteenth century. Its economy suffered, however, as Europe developed its own silk manufacturing industry. The growth of the Aegean port city İzmir (formerly Smyrna) in the seventeenth century further eclipsed trading rival Bursa.

Bursa remained quietly prosperous in the centuries following its glorious era as Ottoman capital. Political strife did touch the city after World War I, however, when the invading Greek army occupied Bursa and caused consider-

able physical damage. Atatürk and his forces drove the Greeks from Turkey in 1922, a year before the modern Turkish Republic was created. To stimulate economic renewal in Bursa, the new government reestablished the textile industry and promoted other enterprises. Today Bursa's industries produce goods as diverse as carbonated beverages, knives, cannon, bathtubs, and automobiles, as well as textiles.

Modern Bursa, one of the largest cities in Turkey, retains a strong Ottoman character. The town's many mosques and imperial tombs, built by the earliest sultans, reflect Bursa's heritage as the first Ottoman capital. Orhan I, Bursa's conqueror, ordered construction of the first imperial mosque within a decade of claiming the city. Built between 1335 and 1339, Orhan Camii (Mosque of Orhan) incorporates elements of both Byzantine and Seljuk architecture. This reliance on architectural cues from the Ottomans' predecessors indicates that a distinctive Ottoman style had not yet developed. Yet the small mosque is not wholly derivative; Orhan Camii is the first example of a cross-axial *eyvan* mosque. (*Eyvan* indicates a vaulted or domed space recessed from a main hall.) The main dome of Orhan Camii, which arches over the central prayer room, is supported by two *eyvans,* which flank the central hall on opposite sides. Orhan Camii has not survived to the modern age unchanged. A Karaman invasion in 1413 damaged the mosque, but it was soon rebuilt and further restored in the eighteenth century.

The founder of the Ottoman dynasty, Osman, did not live to see his son capture besieged Bursa in 1326. Osman died in Söğüt in 1324. His son Orhan I moved the body to Bursa soon after his forces took the city, burying it in a former church near the city's southern defensive walls. The converted mosque became Orhan's burial site as well after his death in 1359. Both tombs were leveled in the devastating earthquake of 1855. Modern reconstructions were built in 1863.

Orhan's grandson, Sultan Bayezid I, succeeded in expanding Ottoman territory in Anatolia and into Europe. Preparing to battle a Crusader army at Nicopolis in Macedon in 1396, Bayezid swore he would build twenty mosques if he defeated the Europeans. The Ottomans were victorious, and Bayezid commenced construction of an impressive new mosque with twenty domes (a compromise solution) in Bursa. The Ulu Camii (Great Mosque), completed in 1399, has a facade of honey-colored limestone quarried on Mount Uludağ. Inside, rows of columns support the twenty domes above the spacious prayer hall. An intricately carved walnut pulpit (*mimber*), one of the finest of its kind, graces the prayer hall. Artful Islamic calligraphy spelling out the names of the holy disciples of Islam decorates the mosque's walls.

Bayezid I also oversaw construction of one of the first imperial mosque complexes in the Ottoman Empire. The complex (*külliye*) on Bursa's eastern edge included a mosque (Yildirim Bayezid Camii), schools, hospital, royal palace, and Bayezid's tomb (*türbe*). Bayezid's mosque (built 1391–95) is constructed on the cross-axial *eyvan* plan originated in the Orhan Camii, and has a marble facade with arched portico and single minaret. Bayezid's small, sparse tomb reflects his status as a ruler; his humiliating defeat by Tamerlane at Ankara in 1402 left him a captive and the Ottoman Empire in chaos. The 1855 earthquake that ravaged much of Bursa's Ottoman architecture took its toll on Bayezid's complex. The mosque was minimally restored in 1878 and 1948. The only other surviving buildings are the royal tomb and one school.

Bayezid's son Mehmed I struggled for years to bring civil war and unrest to an end. He succeeded and became sultan in 1413, eleven years after his father's defeat by Tamerlane. The new ruler commissioned a mosque that many consider the finest in Turkey. Bursa's Yesil Camii (Green Mosque) has won admiration for its simplicity of form united with spectacular interior ornamentation, including stone carvings and the colored tiles for which it is named. Architect Haci Ivaz Pasa employed the established Ottoman cross-axial *eyvan* groundplan; here, a small entrance dome leads into a central court flanked by side *eyvans,* and a main prayer *eyvan* rises four steps above the central court. A *mihrab* or prayer niche rises almost fifty feet in the main *eyvan.*

Yesil Canii's exterior is austere. Intricate carvings in the marble entryway and around windows allude to Seljuk decorative patterning. The windows are lined subtly with green Iznik tiles, which hint at the explosion of rich color within the mosque. Inside, Yesil Camii is a sea of color. Blue and green tiles line the ceilings, the *mihrab,* upper-level galleries where the sultan and his family worshiped, and all walls to a height of six feet. Many tiles bear floral patterns, again reminiscent of Seljuk style. Ornately carved geometric designs enliven the walnut doors, shutters, and *mimber* (pulpit). The small entrance *eyvan* contains an oculus, which casts sunlight on the marble ablutions fountain (*sadirvan*) below. This magnificent mosque represented the Ottomans' cultural maturity as well as their devotion to the Islamic faith.

Just across the street from Yesil Camii stands Mehmed I's tomb, the Yesil Turbë (Green Mausoleum). Mehmed died in 1421, and his tomb was built by his son Sultan Murad II in 1424. This lavishly decorated mausoleum was originally faced with the striking green Iznik tiles used in the Yesil Camii's interior. The 1855 earthquake destroyed these tiles, and they were replaced with blue tiles from Kütahya. Many original Iznik tiles covering the mausoleum's interior, however, have survived intact. Their glorious green color washes across not only the walls but also Mehmed I's own sarcophagus, which is lined with İznik tiles. Religious inscriptions in golden calligraphy also adorn the sarcophagus.

The last imperial complex built in Bursa is the Muradiye, in honor of Sultan Murad II, whose father, Mehmed I, built the Yesil Camii in Bursa (Murad II's son, Mehmed II, conquered Constantinople in 1453). The Muradiye included a mosque, schools, and tombs. Although the mosque was built on the traditional cross-axial *eyvan* plan, it lacked the grandeur and ornamentation of earlier Ottoman mosques. Even while Ottoman attention was turning to the conquest of Constantinople, both Murad II and Mehmed II the Conqueror must still have regarded Bursa as their home, for both chose to be buried here in the Muradiye. Murad II's simple tomb has an oculus in the dome so that sun and rain may enter, as was his wish. Two less fortunate members of the Ottoman dynastic family are buried here as well: Cem Sultan, Mehmed II's younger son, who schemed unsuccessfully for years to take the crown from his brother, Sultan Bayezid II; and Mustafa, oldest son of Süleyman the Magnificent, who was strangled in his father's tent, accused of being a traitor. Soon after the Muradiye was built, the Ottomans captured Constantinople and made it their capital, leaving Bursa with an architectural legacy it has sought to preserve.

Today Bursa is a prosperous city in which historic monuments and modern industry coexist. The city is often called "Green Bursa," a nickname that reflects its ample parkland, its historic mosques and mausoleums, and its character as an Islamic city where the Ottoman Empire was born.

Further Reading: Scholarly accounts of the history of the Ottoman Empire include Stanford Shaw's detailed yet expansive *History of the Ottoman Empire and Modern Turkey, Volume 1: Empire of the Gazis: The Rise and Decline of the Ottoman Empire 1280–1808* (Cambridge and New York: Cambridge University Press, 1976). An equally detailed examination of the Ottomans' beginnings and rise to power is Halil Inalcik's *The Ottoman Empire: The Classical Age 1300–1600,* translated by Norman Itzkowitz and Colin Imber (New York: Praeger, and London: Weidenfeld and Nicolson, 1973). Visitors' guides that provide in-depth historical background and commentary on Turkish sites include John Freely's *Companion Guide to Turkey* (London: Collins, 1979; Englewood Cliffs, New Jersey: Prentice Hall, 1984), and Andrew Mango's *Discovering Turkey* (London: Batsford, and New York: Hastings House, 1971).

—Elizabeth E. Broadrup

Cádiz (Cádiz, Spain)

Location: Seventy-six miles southwest of Seville on Spain's Atlantic coast near Gibraltar.

Description: One of the oldest inhabited cities in the Western world, Cádiz is and has been an important port city for centuries. A cultural melting pot, Cádiz now consists of both a historic old quarter and a more modern industrial area.

Site Office: Tourist Information Centre
Calderón de la Barca, 1
11004 Cádiz, Cádiz
Spain
(956) 211313

Cádiz is a bustling port city in Spain and for many years was among the world's most important. Its long history and continuous habitation make it one of the oldest cities in the Western world. Cádiz is the capital of the province of the same name and a major city of the region of Andalusia. It is situated in an area known for its lush landscape and delightful climate. The old city sits on a small promontory extending into the ocean, and it is connected to the mainland by a narrow sandy strip. At the end of a peninsula, Cádiz separates the Bay of Cádiz from the Atlantic Ocean. The city is surrounded by very high walls, some of which are up to fifty feet tall.

History and mythology are intertwined in the origins of Cádiz. Among the twelve works of Hercules was the separation of Europe and Africa. That event, which allowed the people of the Mediterranean to sail into the Atlantic, is said to have brought about the first settlement at Cádiz, one of the southernmost points on the Iberian Peninsula. Recent archaeological work indicates that the port may have been used as early as 2000 B.C., as evidenced by the excavation of Egyptian pottery from this period. Cádiz, however, is generally believed to have been founded about 1100 B.C. by the Phoenicians, who called the settlement Gadir. The port was then dominated by the Carthaginians, and it eventually fell to the Romans who named the site Julia Augustana Gaditana.

This Roman settlement became famous for its monopoly in the salted fish trade and the skill of its dancing women. References to these slave women and their dancing appear in the works of many classical authors, including Juvenal, Martial, and Statius. The type of dance performed by these women appears to have been something of a cross between modern Middle-Eastern belly dancing and Andalusian flamenco. It is certain, however, that it was highly erotic and had a hypnotic effect on the audience. Martial writes of the Gaditana dancers "wantonly shaking without ceasing their lascivious loins in trained measure."

After the Romans, Cádiz was occupied by the Visigoths, Arabs, and Normans. The area was nearly deserted when King Alfonso X of Castile claimed it in the thirteenth century and began a program of repopulation. Cádiz did not reach its glory, however, until the time of the Spanish voyages to America. Columbus's second journey to America set sail from Cádiz, and shortly thereafter Cádiz became the home of the Spanish fleet. It soon became the richest port in Europe. As a result of its great wealth and the flow of silver from Mexico and Peru into Cádiz, it became a favorite target of pirates. It also became a target of Spain's enemies, and it was at Cádiz that the Spanish fleet was burned by Sir Francis Drake in 1587. The city was again sacked by the British in 1596, but it was rebuilt, fortified, and was able to repel subsequent attacks in 1626, 1656, and 1702.

In 1720, Cádiz acquired monopoly trading rights with the New World and once again became extremely wealthy. Near the end of the eighteenth century the port suffered from the British blockade of Napoléon. Finally the French and Spanish fleets that had set sail from Cádiz suffered a stunning defeat at the hands of Admiral Horatio Nelson at Trafalgar in 1805, ending Spain's dominance of the seas.

During the Peninsular War (1808–14), in which England fought Napoléon's efforts to control the Iberian Peninsula, Cádiz resisted the French advance, and from 1810 to 1813 the town became the capital of occupied Spain. It was there that the first Spanish constitution was signed in 1812. This constitution was rejected by Ferdinand VI when he was restored to the throne in 1814. In 1820, the town rose in revolt, reestablished the constitution, and maintained it for three years against the will of the monarchy. Ferdinand visited Cádiz in 1823 and was imprisoned there. The French came to the king's aid and freed him in the battle of Trocadero, after which he was returned to the throne and resumed his absolute rule. Cádiz, however, remained a center of liberal and even radical politics. In 1883 fourteen anarchists were executed in Cádiz, having been accused of plotting to massacre the upper classes of Andalusia.

Cádiz was never to regain the importance and wealth it had in the eighteenth century. The ende of the nineteenth century was marked by something of a retreat by Cádiz into the issues of the larger Andalusian region and by the start of the twentieth century a strong regionalist movement had begun to take hold in the area. As with other such movements in Spain, this was cut short by the outbreak of the Spanish Civil War. Interestingly, much of the distinctive culture of the region has been adopted by the nation as a whole and is now thought of as uniquely Spanish—the flamenco, for example. Although for many years Cádiz was an important and wealthy port city, the region has generally lagged behind the rest of

A view of Cádiz
Photo courtesy of Instituto Nacional de Promocion del Turismo

the country economically, but it has had a considerable cultural impact on Spain.

In 1936 Cádiz played an important role in the early days of the Spanish Civil War. The war actually started in northern Africa, with the uprising of Spanish military leaders who were stationed there. The first revolts to take place in Spain itself occurred in the Andalusian provinces of Cádiz, Sevilla, and Córdoba. With the revolt of key military figures stationed there, Cádiz and the surrounding area were thus among the first parts of the country to "fall" to the national-

ists, and the success of the armies throughout the region made it a nationalist stronghold. It appears that political repression and executions went on for years in Cádiz, although the records of this are not complete nor entirely reliable.

Opposition to the Franco regime was slow to coalesce in Cádiz in large part because of mass migration from the area to Catalonia during the 1940s and 1950s. Cádiz went through a period of considerable decline during those decades and it was not until the 1970s that it began to experience a revival. Spain's economic boom of the 1980s returned a level of prosperity to Cádiz that it had not seen for more than a century, and considerable growth took place in the new section of town. Still, Cádiz will never again be the vital link between the Old World and the New, nor will it hold a monopoly on the trade of salted fish or other commodities.

Modern-day Cádiz is a city of some 175,000 people. The city itself is divided between the old quarter and a modern industrial sector. It is considered something of the Spanish counterpart to Marseilles, France—a bustling melting pot, teeming with life and filled with sailors from many lands who are seeking adventure. That spirit is exemplified every February, when the end of winter is greeted in the city with one of the world's largest carnival celebrations. Long a home of liberalism and tolerance, Cádiz has been noted for its large red-light district, as well as its sizable gay community. Voters in Cádiz consistently support leftist parties, and it seems somehow appropriate that Cádiz should be home to some of the wildest nightlife in Spain.

Because of the many battles and sackings experienced by Cádiz, there are few structures that are very old within the city. Nearly all the sites of historic interest in Cádiz lie within the old quarter, which is a maze of streets and alleys. Fortunately for tourists, all roads eventually lead to the oceanfront.

The most interesting structure in the city is the Cathedral (New Cathedral), a monumental baroque building dating from the eighteenth century and designed by Vicente Acero. In the crypt is buried Cádiz's most famous son, the composer Manuel de Falla. The treasury contains a magnificent collection of Spanish silver and embroidery, as well as a chalice by Benvenuto Cellini and fine paintings by Francisco de Zurbarán and Bartolomé Esteban Murillo. One particular item, the Custodia del Millón, is reputed to be set with a million precious stones. The cathedral is located in the Plaza Pio XII with cobblestone streets radiating out from the square.

Next to the cathedral is a reconstructed church known as Santa Cruz or the Old Cathedral. A thirteenth-century parish church once stood on the site but was destroyed in 1596. During the seventeenth century it was rebuilt next to the New Cathedral.

Another smaller church, the Capilla Santa Catalina, has several paintings by Murillo, including the *Stigmata of St. Francis*. It was while working in this church that Murillo fell from a high scaffold; he died shortly thereafter in Seville. Another small church, La Santa Cueva, has three significant paintings by Francisco Goya. The Hospital de Nuestra Señora del Carmen is home to El Greco's *The Ecstasy of St. Francis*.

The Oratorio de San Felipe Neri is yet another interesting religious structure. This small oval building, which dates from 1671, was the site of the reading of the constitution of 1812. A plaque commemorates the occasion. In addition, it contains a painting entitled *Immaculate Conception* by Murillo.

The Museo Provincial de Bellas Artes is home to important works of art by Peter Paul Rubens, Murillo, Jan van Eyck, and Rogier van der Weyden as well as a significant collection of twenty-one paintings by Zurbarán. These paintings originally hung in the Carthusian monastery at Jerez; along with collections in Seville and Guadalupe, they are the only such complete series by this artist. Zurbarán's paintings are distinctly Spanish and are noted for their intensity. Even when painting well-known figures, Zurbarán used Spanish monks as his models. The museum also contains an important archaeological section filled with antiquities from the Phoenician, Carthaginian, and Roman periods of Cádiz's history. Among its interesting holdings are a marble sarcophagus in human form from the Phoenician period and a capital decorated with volutes that appears to have come from a temple of the ancient Gadir.

One other interesting historical structure in Cádiz is the Puerta de Tierra, the only remaining old city gate. One enters the old quarter by passing through this gate, which dates from 1755. Much of the surrounding city wall no longer stands, so the gate is somewhat out of context.

The seaside promenade runs along the north and east sides of Cádiz and provides a striking and panoramic view of the Atlantic. The nearby public gardens are a gathering spot for locals and visitors alike and are characterized by beautiful plantings and populated with wild monkeys. From the vantage of the promenade one can look over what has been referred to as Cádiz's "underwater museum."

In the relatively shallow waters of this port it has been estimated that there are at least 2,500 old shipwrecks as well as several submerged settlements. Because of the muddy ocean floor, many of these old wooden vessels appear to be preserved almost entirely intact. Some ships date as far back as the Phoenician period, and there is a particularly large number from the sixteenth century, when Cádiz was heavily involved in trade with the New World.

Constant dredging of the port as well as land reclamation around Cádiz are destroying these relics much more quickly than they can be examined and catalogued. In addition, many of the most accessible sites have been plundered by divers in the past several decades. It seems likely that much of Cádiz's history lies buried in the ocean floor; unfortunately, little has been done on a large scale to preserve it. In addition to the underwater archaeological work in Cádiz, some excavations have been carried out on land, revealing Phoenician and Roman tombs and a Roman aqueduct.

A unique local phenomenon with a richly recounted history is the Solano, a very dry east wind that blows over Cádiz. It has been reported that in the past this occurrence caused a type of temporary lunacy and loss of control. The antidote to the Solano was a trip to the shore for a dip in the sea, with men and women going separately.

Cádiz today blends new and old in a relaxed atmosphere. Once a great economic force, Cádiz appears now to be content as a provincial capital, bastion of tolerance, and bustling center of many cultures.

Further Reading: Books about Cádiz in English are few and far between, but there are numerous articles from both popular and academic sources that discuss some specific topics concerning the city. Because of its importance in Spanish trade, commerce, and particularly the administration of its American colonies, many texts on these subjects dedicate numerous pages to Cádiz and especially its economic importance to Spain. A good overview of this history is to be found in John Lynch's two-volume *Spain under the Hapsburgs* (Oxford: Blackwell, and New York: New York University Press, 1981) and *The Hispanic World in Crisis and Change, 1598–1700,* also by Lynch (Oxford: Blackwell, 1992; Cambridge, Massachusetts: Blackwell, 1994).

—Michael D. Phillips

Cagliari (Cagliari, Italy)

Location: Cagliari, the capital of the Cagliari province and of the island region of Sardinia in Italy, is located on the south-central coast of Sardinia at the northern end of the Gulf of Cagliari, about 270 miles west-southwest of Naples and 375 miles south of Genoa.

Description: The site of Cagliari was probably occupied in prehistoric times by the proto-Sardinian peoples known as the Nuraghics. Seafaring Phoenicians were the island's first recorded settlers, establishing a coastal colony at Cagliari in the ninth to eight century B.C. It became a prosperous Carthaginian stronghold in the sixth century B.C., and was colonized by the Romans in the third century B.C. Caralis, as the Romans called it, became the most important town in Sardinia. Despite heavy Allied bombing during World War II, Cagliari has remained Sardinia's principal port and commercial center.

Site Office: E.P.T. (Ente Provinciale per il Turismo)
Piazza Deffenu, 9
CAP 09125 Cagliari, Cagliari
Italy
(70) 654811 or 651698

Although cave dwellings and grottoes near Cagliari indicate that the site has been inhabited since Neolithic times, the first settlers to leave a significant mark in the area were the mysterious Bronze Age Nuraghic peoples, who had their roots in the island's first prehistoric populations. These peoples are so named for the thousands of *nuraghi*, or conical basalt stone towers, that dominate the island landscape and have been dated between 1500 and 400 B.C. The origins of the Nuraghics are uncertain as they left no written records. The first recorded settlers in southern Sardinia were Ibero-Balearics from Spain and seafaring Phoenicians from Africa. Phoenician shippers and traders, interested in metal ores from Sardinian mines, founded trading colonies along the coasts in the ninth or eighth century B.C., one of which became the town of Caralis (Cagliari).

The island was occupied by Carthage, which had assumed dominance over western Phoenicia, in the sixth century B.C. Caralis became the principal Carthaginian stronghold in Sardinia, presumably named for the god Sardus. From about 500 B.C., the Carthaginians' increasing military grip on the island drove the Nuraghics into the mountains. After the First Punic War with Carthage in the second century B.C., the Romans took advantage of a Carthaginian mercenary revolt and demanded surrender of Sardinia; they took Caralis in 238 B.C., populating it with Roman colonists. The native tribes were subdued after several bloody campaigns. The island became (with Sicily) one of Rome's first overseas subject *provinciae*, under a praetor or propraetor, and a major Roman granary. Roman occupation of the city was to last some 700 years.

The city of Caralis probably obtained full civic rights from Julius Caesar in the first century B.C. In imperial Roman times, after 27 B.C., Caralis became the most important town on the island, owing primarily to its superior sheltered harbor; a detachment of the fleet of Misenun (an ancient naval port near Naples) was stationed there. While Sardinia passed back and forth between the control of a Roman *praefectus* and the Senate in the first century A.D., Caralis was the only Sardinian city to retain Roman civic rights. In the fourth and fifth centuries, the city was probably the seat of the Roman governor, and it became the chief point of the Sardinian road system.

The older part of town, founded by the Phoenicians (and fortified by the Pisans in the early fourteenth century), is situated on the slopes of a steep hill and is now known as Castello. It was probably also the site of the Carthaginian acropolis. The Roman *municipium* was located in the lower town's western and Marina quarters. These areas contain a great Roman amphitheatre, the most notable Roman ruin in Sardinia, constructed out of a natural rock depression in the second century; the remains of Roman houses, including that of Tigellius, a famous Sardinian poet and singer who was much favored by the Roman emperors; and an extensive limestone-cliff necropolis, with vertical rock-hewn chambers, dating to Phoenician and Punic times. Cagliari's archaeological museum, housed in the hill-topping Pisan castle, contains many Nuraghic bronze statuettes and other artifacts from local Carthaginian and Roman sites, as well as Byzantine, Arab, and Spanish antiquities.

The Vandals, crossing from Spain into Africa, invaded Sardinia in about 456; an occupying Roman force failed to expel them in 468. The Vandals' eighty-year rule spurred a cultural revival, largely because Vandal king Thrasamund had banished up to 120 North African bishops to Cagliari. One of them, the bishop of Hippo, brought the relics of St. Augustine, while another, St. Fulgentius, founded a monastery, the remains of which are near the Church of San Saturnino. The oldest and most important early Christian monument in Sardinia, San Saturnino (also known as the Church of Santi Cosimo e Damiano) was transformed in 470 from a small pagan Roman temple. The basilica was built in the shape of a Greek cross with a dome to designate the site where Saturnus (Saturnino) was martyred during the reign of Diocletian. The church, rebuilt from 1089 to 1119, was damaged by Allied bombing in May 1943, but reconstructed with the original stones by 1952.

San Saturnino

Roman amphitheatre at Cagliari
Photos courtesy of E.P.T., Cagliari

The Byzantines, under General Belisarius, defeated the Vandals in 533–34 and claimed Sardinia for the duke Cyril; it became one of the seven provinces of Byzantine Africa. The Goths, under Totila, invaded the island in 550, but it was recovered for the Byzantine Empire three years later. Byzantine rule, largely ineffective, lasted until the early eighth century, when Sardinia suffered a series of attacks by the Arab Saracens. The raids for plunder and prisoners began in 711 and continued through the eighth and ninth centuries. At this time, Cagliari was reduced to a village, as those inhabitants not taken into slavery fled into the island's interior. Finally, between 1000 and 1015, Saracen chief Mujahid el-'Amiri succeeded in establishing himself at Cagliari; though not the last Arab attack, it was the most serious attempt at conquest. In 1016, however, Mujahid was defeated at sea by the combined fleets of Pisa and Genoa.

Sardinia was virtually independent from the eighth to twelfth centuries, and eventually divided itself into four self-governing regions, or *giudicati,* for defensive purposes. The *giudicati,* one of which was Cagliari, only became clearly defined political territories in the eleventh century. They gradually developed into hereditary principalities, nearly always at war with each other over the next seven centuries. Meanwhile, the mainland city-states of Pisa and Genoa had become the foremost of Italian maritime republics in the eleventh century; but after they had driven the Saracens from Sardinia, they struggled bitterly for domination over the island. Initially, the Genoese established political and commercial zones of influence in the north and west, while the Pisans controlled the south and east, including the city of Cagliari.

The papacy also claimed sovereignty over Sardinia. Pope Innocent II appointed the archbishop of Pisa as primate of the island in 1133; Holy Roman Emperor Frederick I made a Genoese-backed *giudice* king of Sardinia in 1164. His rule, however, was ineffective. Soon after the peace of 1169 brought temporary truce to the republics, the *giudicati* of Cagliari and Gallura (in the northeast) passed through marriage to Pisan families and eventually to the republic. Pisa added more territory through an arranged marriage in 1238, but after Genoa soundly defeated Pisa in the 1284 naval Battle of Meloria, the Pisans were again left with only Cagliari and Gallura.

The Pisans dominated Cagliari from the twelfth century to the early fourteenth century. They made it one of the strongest centers of the western Mediterranean, building most of the principal medieval monuments and churches. In the early fourteenth century, the Pisans built an imposing castle and fortifications on the heights of the medieval upper town, topped by two famous defensive towers at opposite ends of the ramparts: the San Pancrazio (1305) and the Elefante (1307). The latter tower, by Sard architect Giovanni Capula, is so named for a statue of an elephant set on a high ledge.

Nearby is the Romanesque Cathedral of Santa Cecilia, which contains some of Cagliari's most important works of art. It was originally built between 1257 and 1312,

but was rebuilt with a baroque interior from 1669 to 1702; neo-Pisan touches were added in 1933. The cathedral has two of its original transept doors, as well as a stone pulpit from the Pisa Cathedral that was given to Cagliari in 1312. This twelfth-century masterpiece of Pisan sculpture, by Maestro Guglielmo, is decorated with New Testament scenes. The cathedral also has tombs and a sanctuary crypt whose rock-hewn walls depict bas-relief figures of Sardinian saints carved by seventeenth-century Sicilian artists. The niches contained the ashes of early Christian martyrs. In *Sea and Sardinia* D. H. Lawrence wrote of the cathedral, "It must have been a fine old pagan stone fortress once. Now it has come, as it were, through the mincing machine of the ages, and oozed out baroque and sausagey."

In 1297, Pope Boniface VIII invested King James II of Aragon with Sardinia. It was not until 1323, however, when the king's son, Infante Don Alfonso (later Alfonso IV), invaded Sardinia with the pope's blessing. Alfonso's admiral sailed into the Cagliari harbor with a substantial fleet, disembarking more than 10,000 troops. The Pisans tried to kidnap Alfonso's wife, but failed; they capitulated, and the Aragonese held them as feudal subjects, demanding annual tributes.

By 1326, Alfonso had driven the Pisans from Cagliari; by 1421, the Aragonese had abolished the *giudicati* system. Many years of warfare ensued, but, after the Battle of Macomer in 1478, Sardinia was wholly subjugated under the Aragonese crown, governed by viceroys who resided at Cagliari. A year later, Aragon was united to the Spanish monarchy under Ferdinand and Isabella; for the next two centuries, Sardinia languished under Spanish despotism and neglect. The economy of the island and Cagliari declined, and the population decreased.

In 1708, during the War of the Spanish Succession, Cagliari was bombarded by a British fleet on behalf of Austria; the city surrendered. As a result of the war, the Austrian Habsburgs replaced the Spaniards as the dominant power in Italy. Austrian archduke Charles, who became Holy Roman Emperor Charles VI, gained possession of Sardinia by the Treaty of Utrecht in 1713. Though the Spanish briefly captured Sardinia four years later, the 1718 Treaty of London ceded the island to Victor Amadeus II, duke of Savoy, as compensation for his loss of Sicily. Victor formally took possession of the island in 1720; he and his successors were known as the kings of Sardinia until 1861, when their kingdom was united with the Italian state and they became the kings of Italy.

Under the House of Savoy, whose seat of power was in Piedmont (the nation was also called the Kingdom of Sardinia-Piedmont), the island experienced somewhat of an economic and social revival. Although Victor Amadeus II and his succeeding son Charles Emmanuel III (reigned 1730–73) ruled as despots, abolishing all free institutions, they substantially increased Sardinia's prosperity. In 1793, during the French Revolutionary Wars, a French naval fleet bombarded Cagliari, but was repelled by the islanders and withdrew. In 1799, after the French annexed Piedmont,

Charles Emmanuel IV, the young ruler of Savoy and Sardinia, sought refuge for a few months in Sardinia. In 1802, Charles abdicated in favor of his brother, Victor Emmanuel I, who lived in Cagliari from 1806 to 1814.

The sanctuary of the San Cecilia Cathedral includes two notable mausoleum chapels where members of the House of Savoy are entombed. One contains the tomb of Louise Marie Joséphine, wife of King Louis XVIII of France, who died in exile in England in 1810. Because she was the daughter of Victor Emmanuel, king of Sardinia, her body was brought to Cagliari a year after her death. The other tomb is of Charles Emmanuel of Savoy, infant son of Victor Emmanuel I. The cathedral stands in the Piazza Palazzo between two ornate, imposing buildings also from the Savoy era. The Governor's Palace, formerly the royal palace of the House of Savoy, was built by Davisto in 1769 and contains frescoes by Domenico Bruschi. The other building is the even larger Archbishop's Palace, also built by Davisto.

In the 1815 Congress of Vienna, following the collapse of the Napoleonic Empire, the Kingdom of Sardinia was reconstituted under Victor Emmanuel I to include its former mainland territories, including Piedmont; France ceded Savoy. The kingdom expanded to include almost all of Italy during the nineteenth century. The French Revolution inspired a popular demand for a Sardinian constitution, but Victor abdicated rather than grant it, and Charles Felix, his brother and successor, put an end to the rebellion. However, reform-minded King Charles Albert, from another branch of the family, led the movement to expel the Austrian Habsburgs from mainland Italy and to establish a united Italy. In 1847, Sardinia was joined to the other Piedmont provinces of the House of Savoy, enabling it to participate in the benefits of the new Piedmont-Sardinia constitution granted by Charles Albert in March 1848. This so-called Statuto Albertino be-

came the foundation for a free government, and the basis for the Constitution of United Italy, proclaimed by the first Italian parliament in March 1861. Charles Albert's son, Victor Emmanuel II (who had become Sardinian king in 1849) was proclaimed first king of Italy and Sardinia became part of the unified Italian state.

Cagliari's harbor was expanded in 1938. During World War II, the city served as the military headquarters of Sardinia, with important Italian naval and air bases used for Mediterranean attacks. The bases, however, were destroyed by heavy Allied bombing in 1943; the Allies occupied the city that year. A movement for autonomy revived after the war, and Sardinia became a self-governing *regione* of Italy in February 1948. Cagliari, the island's principal port and commercial center, has shipyards, and industries producing cement, ceramics, superphosphates, and wine, as well as flour mills and sugar refineries; agricultural produce and salt extraction are also important activities. The city's chief exports are wine, salt, and other minerals.

Further Reading: *A Short History of Italy,* edited by H. Hearder and D. P. Waley (Cambridge and New York: Cambridge University Press, 1966), provides a compact, readable introduction to the nation's history, from classical to modern times. The book is especially strong on the Kingdom of Sardinia's contribution to the nineteenth-century independence movement and the drive for national unification. *Sardinia* by Virginia Waite (London: Batsford, and North Pomfret, Vermont: David and Charles, 1977) is a general introduction to the island, a region-by-region guide to historical lore and present-day attractions. *The Western Mediterranean World* by James M. Houston (Harlow, Essex: Longman, 1964; New York and Washington: Praeger, 1967) and *The Fall and Rise of Modern Italy* by Serge Hughes (Westport, Connecticut, and London: Greenwood, 1983) are of more scholarly historical interest.

—Jeff W. Huebner

Calahorra (Logroño, Spain)

Location: Thirty miles southeast of the city of Logroño.

Description: An ancient market town, with Roman remains and medieval church buildings; birthplace of the authors Quintilian and Aurelius Prudentius.

Contact: Tourist Information Center
    Miguel Villanueva, 10
    26001 Logroño, Logroño
    Spain
    (941) 29 12 60

Calahorra is situated on the southern bank of the Rio Cidacos, where it meets with the Ebro River, in the province of Logroño. The early history of the entire region of the upper Ebro is largely unknown and subject to a variety of competing theories. According to the writings of the Greek geographer Strabo, the earliest remaining written source of information on the site, the region was inhabited in his time (first century B.C.) by an indigenous people known as the Vascones. While many later historians have regarded these people as the ancestors of the modern Basques of northeastern Spain, others have questioned whether the Basques have inherited anything other than the name (which, regardless, is not their own name for themselves). Again, while the term most commonly used by historians for the prehistoric peoples of Spain is "Iberians," and while it is often asserted that they probably entered the peninsula from north Africa, in fact the term is a convenient but questionable label for a wide variety of cultures that may well have had quite different origins

It is generally agreed, however, that between the ninth and sixth centuries B.C., the Iberians in northern Spain, presumably including the mysterious Vascones or their ancestors, intermingled with the Celts, an incoming Indo-European people, to create what historians call the Celtiberians. It was during this obscure period that the site now known as Calahorra acquired the name "Calagoricos." The existence of the place-name element "gorri" in the modern Basque language has been taken as evidence of continuity between the Celtiberians and the Basques, although it may well be an example of linguistic borrowing, with no ethnic or other implications. In any case, it is agreed that Celtiberians continued to live at Calagoricos and other sites in the isolated Ebro region while the Greeks established colonies on the coast. The Greeks were followed by the Carthaginians, who arrived during the sixth century B.C. and eventually dominated trade between the Iberian Peninsula and other Mediterranean areas, and finally by the Romans, who challenged Carthaginian power and fought them in the First Punic War between 264 and 241 B.C.

By then the inhabitants of the Ebro region, who it is probably safe to call Vascones, are likely to have been drawn into trading relations with Carthage. At some point between 239 and 228 B.C., they may well have come under direct threat from the Carthaginians' campaigns to bring northern Spain under military control. It is generally believed that the Vascones joined with other indigenous peoples in helping the Romans to oust the Carthaginians from the peninsula during the Second Punic War (218–201 B.C.). However, this too may be questioned, since the Romans themselves, whose intention was to conquer the whole of the peninsula for themselves, are the only source for this belief. Indeed, it took the Romans almost 200 years to overcome the sporadic but fierce resistance of the Celtiberians and unite what they called "Hispania" under Roman rule.

The Romans eventually divided their new territory into two provinces, Hispania Ulterior (Farther Spain) in the southwest and Hispania Citerior (Hither Spain), later called Tarraconensis, in the central and northern parts of the peninsula. In 195 B.C. Roman forces led by General Cato the Elder fought their way through Hispania Citerior, capturing Calagoricos in 186 and renaming it Calagurris Nassica. As usual in their subject territories, they signed with the local leaders requiring the latter to provide conscripts for the Roman army and pay Roman taxes. These alone may well have been the basis for the convenient claim that their new subjects had welcomed the Romans, or even allied with them during the Second Punic War, but the widespread Celtiberian revolts that erupted between 155 and 133 B.C. cast doubt on that version of events.

Sixty years later, opposition to Roman rule came from rebel forces within the occupying army itself. Between 80 and 71 B.C., Sertorius, a Roman general who gathered his own small army, was able to attract support among the indigenous people of Hispania, including, in 77 B.C., the Celtiberians of Calagurris. He managed to seize control of large parts of north central Hispania through what would now be called guerrilla warfare, but the Roman authorities sent their leading general, Gnaeus Pompeius Magnus (Pompey), to put down the rebellion. In 73 B.C. Sertorius was betrayed and murdered. The inhabitants of Calagurris held out against Pompey's forces until the following year. According to legend, they resorted to cannibalism during the long siege when supplies failed and a famine struck the town. When the Romans entered the city, they found only one survivor, a woman, who was eating a human arm. She had apparently kept the town's bonfires burning in an attempt to convince the Romans that there were many more survivors with her.

The conquest of Hispania was complete by the year 13 B.C., under the rule of Augustus, the first Roman emperor. By the end of the third century A.D., extensive colonization had

Altar in Calahorra's cathedral
Photo courtesy of Instituto Nacional de Promocion del Turismo

helped to make it one of the richest of the Roman territories, and Calagurris was one of the two most important cities in the Ebro Valley, along with Pamplona. The Romans undertook major changes to the infrastructure, building roads between the towns and increasingly bringing the region into the empire. The excavated remains of an aqueduct and a circus can be seen at Calahorra today. Education became an important means to assimilate the leading members of the local population to Roman culture, and Calagurris contributed its share of prominent writers and poets. One such writer was Marcus Fabius Quintilius (Quintilian), born in Calagurris around A.D. 35. He became famous throughout the empire as the holder of the Chair of Rhetoric in Rome and as tutor to Pliny the Younger and several relatives of the emperors.

Prosperity did not last, however, either in the wider empire or in Hispania. By the end of the third century, economic decline was taking hold in Tarraconensis.

Calagurris was also significant in the spread of Christianity in the region, and martyrs and saints have an important

place in its history and legends. Two famous martyrs from Calagurris were the brothers Emeterius and Celedonius, who were tortured and killed because of their rejection of the official Roman religion. They are believed to be buried in the Church of Capilla de Casa Santa, built on the site of the former Roman prison; pilgrimages are still made to the site on August 31 every year. The major source of information on these patron saints of Calahorra, and on other martyrs, is the writings of the Christian poet Aurelius Prudentius, born in Calagurris in 348. His twelve-part Latin poem *Peristephanon,* which commemorates the sacrifices made by the Spanish martyrs and saints in the cause of Christianity, may have been commissioned to mark the elevation of Calagurris to the status of bishopric and the consecration of its first cathedral, around 400.

During the fifth century Hispania became the battleground for various Germanic invaders, including the Vandals and the Suevi; by 456 the largest group of these incoming peoples, the Visigoths, had ousted the last of their rivals. By the time the last, powerless Western Roman emperor died in 476, the Visigoths had carved out their own kingdom in southern Gaul (France) and northern Spain. Calagurris remained under Visigothic control while this kingdom shrank as a result of battles with the Franks, even after 531, when the region now known as La Rioja came under severe attack. The Gothic Church of San Andrés was built during this period, although only the portal and a statue of the apostle Andrew remain from the original building today.

Challenged by the Franks and faced with internal power struggles, the Visigoths were ill prepared for the invasion by the Moslem armies of Arabs and Berbers that began in the early eighth century. It was the Moslem rulers who gave Calagurris its modern name of Calahorra. They remained in control of the region for most of the period up to 1045, in spite of the creation and expansion of a number of small Christian kingdoms in northeastern Spain, among which both Navarre and Léon made several attempts to capture Calahorra. In 920 the king of Navarre, Sancho I Garcés of Pamplona, conquered Calahorra and other nearby towns, but a Moslem force under 'Abd ar-Rahman III an-Nāṣir retaliated against Sancho, forced him out of the town, and burned it to the ground. The Moslems went on to destroy Sancho's capital city, Pamplona, in 924. There is some dispute about what happened next: it is possible that for some time Calahorra and Upper Rioja were administered by the Navarrese under the control of the Moslems.

By the beginning of the eleventh century, however, the Christian regions were united under the rule of Sancho III Garcés (called el Mayor), king of Navarre and Castile-Léon, who appears to have driven the Moslems out of most of the region of La Rioja in the years up to his death in 1035. In 1045 Calahorra once again became a bishopric, as part of Navarre, which had passed to Sancho's eldest son Garcia III, who also took over formerly Castilian areas of La Rioja. His brother Ferdinand I, who had inherited Castile-Léon, waged war against Garcia to regain these areas, and in 1054 Garcia was killed in battle. The whole of La Rioja, including Calahorra, reverted to Castile-Léon. In 1092 the town as well as the region were plundered by Rodrigo Díaz de Vivar, "El Cid," who had turned from fighting the Moslems to rebelling against the Christian king Alfonso VI and who was to become the subject of an epic poem written in the twelfth century.

Calahorra remained under Castilian rule throughout the conflicts that broke out repeatedly as the Christian kingdoms of Spain annexed or repelled one another. These affected the town most directly and destructively during the reign of Peter I, "the Cruel," king of Castile, from 1350 onward. This notorious tyrant was challenged both by his half-brother Henry II of Castile and by their cousin Ferdinand, and the disputes among their supporters led to civil wars and interventions by forces from Aragon, France, and England. In 1366, with help from King Peter IV of Aragon and a French army under Bertrand du Guesclin, Henry seized power and was crowned as King Henry II in Calahorra. Peter the Cruel responded by allying himself with Prince Edward of England, the so-called Black Prince, who was then in control of large parts of France. But the massacres he ordered to be carried out in Calahorra and elsewhere are said to have alienated the English, who withdrew and left him to defeat and death, at Henry's hands, in Toledo in 1369.

Calahorra, attacked and damaged by both Peter's and Henry's forces, recovered from this eruption of violence, and its city walls, which still stand today, were soon rebuilt. Work also continued on the partly Gothic, partly Renaissance cathedral, one of the glories of Calahorra. It now contains, among many notable treasures, an alabaster altar believed to have been made by the leading sculptor of the day, Master Guillén; forty-one mirrors, decorated with Dutch engravings on copper, in its sacristy; numerous paintings of Christ and the saints; and a monstrance, known as El Cipres, which was donated by Henry IV in 1467.

In 1479, soon after the cathedral was completed, Calahorra came under the joint rule of King Ferdinand II of Aragon and his wife Queen Isabella I of Castile-Léon, who gradually replaced the rivalries and insecurity of the medieval kingdoms with a newly centralized Spain, although many older royal and regional institutions were at least formally maintained. Spain's expansion as an imperial power, which saw the Castilian flag planted in the Americas and which accelerated under their grandson, the emperor Charles V, brought wealth and stability to the La Rioja region. Agriculture flourished there, including the production of the wine it continues to export today; so did the inquisition, which located one of its principal courts at Calahorra under Charles's son Philip II. One of Spain's best known historians of the inquisition, Antonio Llorente (1756–1823), was a native of Calahorra and began his researches there.

During the seventeenth century, Calahorra had a population of about 5,000. Like many rural Castilian towns, it was dependent on the agriculture of its hinterland for its prosperity. In 1665 the local farmers rebelled against the corruption of the royal officials, the *regidores,* who were mostly members of rich families able to buy their offices and find ways to avoid paying taxes. After driving the major and the leading families out of Calahorra during skirmishes in which nobody was killed, the farmers allied with some of the local priests and with the town's guilds of skilled craftsmen and merchants to appoint their own mayor. The rebels remained in control of the town and the surrounding countryside for three months, until forces led by Don Alonso de Llano y Valdes, the governor of the neighboring region of Navarre, took control of the town once more on behalf of the Castilian authorities.

Calahorra next became a site of political upheaval during the civil war that broke out in 1833 over the succession to the throne following the death of King Ferdinand VII; it continued until 1839. Ferdinand's brother Don Carlos and his supporters, the Carlists, did not accept the claims of Ferdinand's daughter Isabella and took control of Calahorra because of its importance as a bishopric. Initially there was strong support for the Carlists in La Rioja and across Old Castile generally, but this was not enough to withstand the military response of Isabella's supporters. In April 1834 the Carlists were able to recapture Calahorra briefly, but the overwhelming strength of the Royalist forces pushed the Carlists northward into the Basque country, where they continued to fight until their ultimate defeat.

In 1890 the bishopric of Calahorra was united with that of Logroño to form one diocese, an indication of a relative decline in Calahorra's status and population. By the beginning of the twentieth century, however, an industrial base was slowly beginning to develop around the town, with sugar refining factories in addition to the vegetable canneries built in Calahorra and Logroño. The economic dislocation caused by the Great Depression and then by the Spanish Civil War of 1936 to 1939 led to a decline in the canning industry in the region. Still, Calahorra endured these conditions as well as the subsequent Franco dictatorship, just as it had survived through the previous centuries. It became a center for trade and processing of agricultural products, in particular peppers, olive oil, and canned vegetables such as tomatoes and artichokes. Today, some two decades after Franco's death and the restoration of parliamentary democracy, Calahorra, with its population of around 18,000, continues to play a prominent role as the second-largest city in La Rioja, a region of around 225,000 people (of whom nearly half live in the capital, Logroño). The economic prosperity of the region and the modern development that has accelerated since the 1970s have led the Calahorra authorities to replace what was left of its ancient Roman walls with new property developments. Yet much of its long and varied history remains visible, especially through the Roman artifacts displayed in its museums and through its cathedral, the Churches of San Andres and Capilla de Casa Santa, and its other medieval buildings.

Further Reading: Leonard A. Curchin's *Roman Spain: Conquest and Assimilation* (London and New York: Routledge, 1991) is an outstanding and detailed study of Hispania. Roger Collins's *The Basques* (Oxford: Blackwell, 1986; Cambridge, Massachusetts: Blackwell, 1987) includes some discussion of the historical links between Calahorra and the Basque region nearby. Joseph F. O'Callaghan's *History of Medieval Spain* (Ithaca, New York, and London: Cornell University Press, 1975) clarifies much of the complexity and confusion in the development of Castile and its rival Christian kingdoms.

—Monique Lamontagne

Capri (Napoli, Italy)

Location: An island in the Bay of Naples about eight miles from the Italian mainland.

Description: Famous since ancient Roman times for its spectacular scenery and favored as a resort by emperors, writers, and artists.

Site Office: A.A.S.T. (Azienda Autonoma di Cura, Soggiorno e Turismo)
CAP 80073 Capri, Napoli
Italy
(81) 837 0424

The island of Capri, which measures about ten miles by five miles, was originally attached to the Italian mainland but was separated from it by a series of volcanic eruptions, probably during the most recent Ice Age. Human settlement began early: pottery and stone tools from the Paleolithic Era have been found there, predating the distinctive Neolithic pots, decorated with red stripes, which have been discovered in caves and grottoes and labeled as examples of the "Capri style." But the history of the culture that made these pots is unknown, while that of succeeding centuries has become inextricably entangled with myth and legend. It is said, for example, that the island was settled by Teleboans, Greek pirates led by King Telon, and by the Sirens, the monsters, half-women and half-birds, who attempted to lure Odysseus with their songs in Homer's *Odyssey*. Today the Scoglio delle Sirene (Sirens' Rock) overlooks the sea near the southern port village of Marina Picolla. Even the origin of the name Capri—with the emphasis on the first syllable—is unknown. One plausible suggestion is that it comes from the Etruscan prefix word *capr-*, which means "rocky"; another is that it comes from the Greek word *kapros,* "wild boar."

The earliest documentary references to Capri, by the Greek historian Strabo, who wrote in the first century B.C., do not solve these mysteries but do plausibly claim that the northern port area of what is now Marina Grande and the town of Capri, built on higher ground now reached by a funicular railway, were the first areas settled. It is also known that the Etruscans built a defensive wall around the town of Capri in the fourth century B.C., before the Greeks, who had already been settling along the shores of Southern Italy from around 1000 B.C., took possession later in the same century and made Capri part of their trade routes from the Sorrento Peninsula. In 29 B.C. the island, known to the Romans as Caprae, was purchased from the Greek city of Naples by Octavian, later Emperor Augustus, when he returned from his campaigns in Egypt.

Augustus was enchanted by its natural beauty and revitalizing quality and did more than any other Roman emperor to develop the island, building roads and aqueducts; several residences, including his Palazzo a Mare (sea-palace); the Torre del Faro, a lighthouse on the northeastern shore that was used to send messages to the mainland until it was destroyed in an earthquake in A.D. 37; and the Scala Fenicia, a rock staircase of 880 steps, which links the lower town of Capri and the higher town of Anacapri two miles to the west and which may have been a replacement for a Greek stairway. The island became so popular as a holiday resort for many Roman soldiers that Augustus referred to it as Apragopolis, "the land of layabouts." Although much of the architecture was Roman, Augustus allowed the culture and language of the island to remain distinctively Greek.

Augustus's stepson Tiberius inherited the island in A.D. 14. Twelve years later he made his permanent home on Capri to take advantage of its mild climate and its seclusion. In effect, Capri became the capital of the Roman Empire. Tiberius had several villas built, including the Villa Damecuta and the Villa Iovis on the hills of Il Capo, as well as what are now called the Bagni di Tiberio (Baths of Tiberius). According to the Roman writers Suetonius and Tacitus, during his years on Capri Tiberius engaged in sadomasochism and pedophilia and arranged for his victims to be thrown into the sea from a rock, known as Salto di Tiberio, which is now linked to the Villa Iovis by a pathway named Via Tiberio; the writers also claim that his death in A.D. 37 resulted from a smothering ordered by his rival and successor, Caligula. Whatever the truth of these tales—and many scholars would argue that the facts were exaggerated or falsified to destroy Tiberius's reputation after his death—the legend of Tiberius has been associated with yet another version of the origins of the island's name, the claim that it comes from the word *capra,* "goat," used to refer to the emperor's hairy body and sexual promiscuity.

The eruption of Mount Vesuvius that destroyed the cities of Pompeii and Herculaneum in A.D. 79 also destroyed much of the spectacular Roman architecture on Capri. Although the island remained part of the Western Roman Empire until the death in Naples of its last ruler, Romulus Augustulus, in 476, it never again received the attention that Augustus and Tiberius had devoted to it. During the sixth century Capri was under the authority of the Duchy of Naples, and the Catholic Church established Benedictine monasteries on the island. In the seventh century Bishop Costanzo (Constance) visited the island and died near Marina Grande; he later became its patron saint, his feast day being celebrated every year on May 14. The Byzantine Basilica of San Costanzo, between Marina Grande and Anacapri, was built on the site of his tomb in the twelfth century. Since 1231, however, the town of Anacapri has celebrated its own patron saint Sant'Antonio de Padova (St. Anthony of Padua) on June 13.

Villa Damecuta on Capri
Photo courtesy of A.A.S.T., Capri

In 868 Capri came under the authority of the Doges of Amalfi, then a major Italian port and city-state. In 994 Pope John XVI appointed the first bishop of Capri and a cathedral was built in the town of Capri on the present site of the Church of Santo Stefano (built in 1682 by Marziale Desiderio). This confirmed the town's long-standing claim to be the political and religious capital of the island and also provided a focus for resistance to the attacks of Saracen (Arab) and other pirates that continued through the eleventh century, although many of the islanders preferred to seek refuge in the island's many caves and grottoes.

In 1138 the Normans conquered the island, making it part of their Kingdom of Sicily, which was inherited by the Holy Roman Emperor Henry VI in 1194, during his long struggle with the papacy for supremacy over Italy. His descendants controlled southern Italy, including the island, until Charles of Anjou, a brother of the French king Louis IX and an ally of the papacy, became King Charles I of Naples and

Sicily in 1266. Following the separation of Sicily from this kingdom by the Aragonese in 1302, Capri remained within it, under the Angevin dynasty that Charles had founded. By this time the population of the island numbered about 600, of whom most lived in or near the capital. Fishing was their main source of food, since large parts of the island were not suitable for farming; even so, Capri depended on heavy subsidies from the monarchy in Naples and the islanders often experienced hardship and at times came close to famine. It was under Queen Joan I that the island began to prosper again. In 1371 she funded the building of the Certosa di San Giancomo, a Carthusian monastery located on the richest land on the island (it was restored in 1933). Its monks continued to be favored by the crown and thus controlled much of the economic development of the island for many years. During an outbreak of the conflict between the Anjous and Aragon in 1435, Capri joined the side of the Aragonese and by 1442 it was under Aragon's control as part of the reunited Kingdom of Naples and Sicily. The Aragonese kings continued to favor the island with the privileges issued under the previous authorities, such as exemption from royal dues and the right to control the fishing in its waters.

In 1517 Capri passed with the rest of southern Italy into the control of Charles, the Habsburg prince who was heir to the Aragonese and who was also to become Holy Roman Emperor Charles V, the most powerful monarch in sixteenth-century Europe. His zeal to protect Europe against the expanding Ottoman Turkish empire led to Capri forming part of the front line between the two great powers of the Mediterranean. In 1535 Khayir ad-din (nicknamed Barbarossa, meaning Redbeard, by Europeans), the admiral of the Turkish fleet, attacked all the islands within the Bay of Naples, and in 1544 the town of Capri and the castle in Anacapri, which dated to the eighth century, were both severely damaged in a second attack. The castle has since become known as the Castello di Barbarossa. The Turkish fleet attacked yet again in 1553, this time damaging the Carthusian monastery of Certosa, while yet again many islanders found refuge in the caves and grottoes. Subsequently a militia was formed on the island and its fortifications were improved.

By 1561 the population of the island was nearly 1,500, with fishing remaining the predominant industry. The economic problems that afflicted much of southern Italy, mainly because of the success of the English and Dutch in taking away most of Italy's vital trade during the sixteenth and seventeenth centuries, were accompanied on Capri by a long-running power struggle between the bishops and the Carthusian monastery. The monks, who were not under the authority of the bishops, were the wealthiest landowners on the island, extracted levies on fishing, and controlled most agricultural exports. Among these they monopolized the wine trade, becoming known for their liqueur Certosino, and the olive trade, which was becoming an increasingly important part of the island's economy. Meanwhile, the bishops were nicknamed the "Bishops of Quails" since they were left with just a small income from the trade in these birds, which nest on Capri in the spring and fall. They found themselves presiding over one of the poorest dioceses in the Catholic Church and unable to support the numerous parishes on the island. In 1656, when half the population of the island was wiped out during a five-month outbreak of the plague, the bishop and his priests could not manage to care for all the sick, but their requests for help from the monks were turned down. Many corpses were dumped over the monastery walls in protest.

In 1707 the Kingdom of Naples and Sicily was given to Charles VI, a Habsburg archduke from Austria, but in 1734 it fell to the French Bourbon prince Charles VII, who became King Charles IV of Naples and Sicily and later King Charles III of Spain. Much of the excavation of the Roman antiquities of Capri was carried out under Charles and his successors. During the later part of the eighteenth century the economy of the island improved, with greater exports of white wine from Anacapri and red wine from Capri. Fruits that had been grown for local consumption also began to be exported, in addition to the olive oil and quails that the Certosa monks did not control. The island also began to attract its first foreign residents, who were mostly British.

The liberal Parthenopean Republic set up at Naples in 1799 after French Revolutionary intervention attracted the support of many islanders, including Bishop Gamboni, who established a seminary and four vocational schools for both men and women on Capri. After the Bourbons regained their territory, with the aid of the Russians and of the British led by Admiral Horatio Nelson, those who had supported the republic were imprisoned or executed and Bishop Gamboni, the last bishop appointed to the island, left Capri after his brief imprisonment there and died in Venice in 1811. In February 1806 the French reconquered Naples and the island of Capri under the authority of Napoléon's brother Joseph, who was newly declared King of Naples. Plans were made to fortify the island, but by May 1806 the British fleet had recaptured it in the name of the Bourbon dynasty. The British and their Corsican laborers began to strengthen the fortifications with a view to making Capri the eastern-Mediterranean equivalent of Gibraltar, and the creation of a garrison at Anacapri brought the number of troops on the island to 1,800. The British perhaps became overconfident: they were not prepared to repel the 3,000 soldiers of France and Naples who attacked the island in 1808. The subsequent French victory over the British is one of the many inscribed on the Arc de Triomphe in Paris; a commemorative medallion was issued in 1811. Under the French occupation that followed the victory, strict controls were imposed over trade with the mainland, which reduced the illegal exports on which some of the islanders had come to depend. On the other hand, the French finally took away the land and power of the monks at Certosa, as well as from the nuns at the convents of Santa Theresa in Anacapri and Capri, which had been established in the seventeenth century and which many islanders believed to have been riddled with corruption.

In 1815 Capri returned once more to Bourbon control, after the defeat of Napoléon allowed King Ferdinand I (formerly Ferdinand IV) to return to Naples and proclaim his Kingdom of the Two Sicilies. Forty-five years later the Kingdom of Italy was established under the Savoy dynasty of Piedmont and the islanders became legally, as they had long been culturally, Italians. During the later part of the nineteenth century many institutions on the island underwent great changes. The old monastery of Certosa became a prison and then a home for disabled war veterans from the Sicilian Revolutions of 1848–49. The Convent of Santa Theresa in the town of Capri also became a hospital for disabled veterans. The Church of Santo Stefano, no longer a cathedral now that there were no bishops, became a parish church in the archdiocese of Sorrento. Although some of the property from the monks had been given to the islanders, the income derived from it went straight to the bishop of Ischia, while other possessions were sold to outsiders. The economy had begun a rapid decline as early as 1839, when many of the vineyards were destroyed by plant lice, and travel restrictions were enforced to such an extent that Capri became largely a penal colony.

Most islanders turned to begging from foreign tourists, whose numbers had begun to increase after 1826, when two German artists had painted and written about the beauty of the island, in particular the Grotto Azurra (Blue Grotto) located near Anacapri. Inns were established to cater to the many famous painters, poets, and writers who began to flock to the island. By 1841 the population of the town of Capri was nearly 2,000, while Anacapri had slightly less than 1,500 residents. By the end of the nineteenth century tourism was becoming the major industry. Capri was promoted in many tourist guides to southern Italy from 1855, and in particular Baedeker's guide (from 1867), which emphasized the many grottoes as the most spectacular attractions apart from the Roman ruins. A steamboat service from Naples was introduced, and in 1874 a road was built to link the two towns of Capri and Anacapri, which had previously been linked only by the rock staircase. With the spread of liberal ideas from the foreigners and integration with the administration in Italy, the people of Capri were slowly achieving better standards of living, improving both their literacy and their health, though not enough to prevent a cholera outbreak in 1873 and a typhus epidemic in 1881. By 1881 the population of the island was nearly 5,000, of which slightly more lived in Capri rather than in Anacapri. Electrification of the towns continued to expand from 1892, when an electrical funicular railway was built between the town of Capri and its major port Marina Grande.

By the mid-1890s the island had earned a reputation for its liberal attitudes toward sex, especially with the legalization of homosexuality in Italy beginning in 1891. Because of this it attracted those who felt that their sexual orientation was not acceptable in their home countries. Among many writers, Somerset Maugham and E. F. Benson were frequent visitors. Capri also attracted Fritz Krupp, the son of the German arms manufacturer, who had a villa built that overlooked the sea and is now inside the Gardens of Augustus near the village of Marina Piccola. He was noted for his wild parties and the entertaining of young boys. A scandal broke out in 1902, culminating in Krupp's death, reportedly a suicide.

By 1905 there were more than 30,000 tourists a year visiting Capri, especially from Germany, and an increasing number of new non-Italian residents, including such refugees from the absolutism of the Russian empire as the writer Maksim Gorky, who brought his family to the island in 1906 and even established a short-lived school for revolutionaries there. Lenin visited Gorky in 1908, nine years before he led the Bolsheviks to victory in the Russian Revolution, and later extolled the beauty of the island.

After the Fascists took over Italy following World War I, Capri was governed by appointees commissioned directly from Rome. Under Mussolini several further excavations of Roman treasures were carried out between 1932 and 1937, and the island became a holiday resort for many Fascist officials. Ordinary tourists continued to flock to the island, and the popular song of 1934, "The Isle of Capri," did much to promote its attractions. Accordingly, its hotels, restaurants, and other services flourished despite worldwide depression. During World War II a garrison was established on the island in 1941, although it was for the most part used by German soldiers as a resort. The island did provide refuge to many mainlanders who fled during the numerous Allied bombings. After the liberation of Rome and much of southern Italy in 1944, the American forces took over the administration of the island and it became a rest and recreation center for American soldiers until 1945.

In the postwar period the island has continued to attract tourists and many who wanted to have second homes there, including the British singer and film actress Gracie Fields and the novelists Albert Moravia and Graham Greene, who helped to make Capri a fashionable destination for the rich and famous. It also has attracted the Camorra, the organized crime syndicates based in Naples, which have used the island for their trade in contraband goods. Capri today, even with a minimum (winter) population of more than 12,000 and intensified commercial exploitation of its resources, retains its reputation as one of the loveliest islands in the Mediterranean.

Further Reading: James Money's *Capri: Island of Pleasure* (London: Hamish Hamilton, 1986) is a very interesting guide to the island's history and people. Norman Douglas's fictional book *South Wind* (London: Secker, 1917; New York: Dodd, 1918) did much to raise the profile of the island, especially since it was based on the notorious and often scandalous behavior of some of the island's residents; he was to write several more books on Capri.

—Monique Lamontagne

Capua (Caserta, Italy)

Location: In southern Italy's Caserta province and Campania region, sixteen miles northeast of Naples, near the Volturno River.

Description: Built by the Etruscans between 800 and 500 B.C., ancient Capua became the luxurious capital of the Campania region and the peninsula's second-greatest city. The magnificent amphitheatre where Spartacus fought as a gladiator and a cathedral that survived a ninth-century Saracen pillage are among the extant vestiges of ancient Capua's glorious past.

Contact: E.P.T. (Ente Provinciale per il Turismo)
Palazzo Reale
CAP 81100 Caserta, Caserta
Italy
(823) 322233

Ancient Capua was once the capital of southern Italy's Campania region, second only to Rome in importance. This beautiful and opulent city became a leading industrial and commercial center, renowned for its bronze work and perfumes. First settled by the Oscans, an indigenous tribal group, the site that would become Capua (now known, in full, as Santa Maria di Capua Vetere) was transformed into a coveted city by the Etruscans between 800 and 500 B.C. Livy wrote that the Etruscans founded twelve colonies in Italy to correspond with the twelve cities of Etruria, and that Capua, which they first named Volturnum, was the greatest of these.

It is likely that the Etruscans first came to Capua by land and, over time, by sea as well, as part of their expansion into lush, fertile Campania. The earliest settlers may have followed an easy and direct land route that ran from Capua northward along several river valleys that eventually led to the Tiber.

The Etruscan colonization process was not formalized, nor was it cooperative with respect to the expansionist efforts of the various city-states. Instead, the expansion into Campania reflected the multiplicity of agendas (and consequential lack of interdependence) of these city-states. Evidence exists, however, that the Etruscan city of Vulci played a part in the foundation of Capua. Even in pre-urban times, Vulcentine products were used in Capua's cemeteries. Capua itself was thought to have founded Nola, an important road-station seventeen miles to the southeast.

According to historian Velleius Paterculus, the Etruscans founded Capua in about 800 B.C., but Cato the Elder places the date at 500 B.C. It is highly probable that Cato's date refers to a reorganization or enlargement of the city. The Etruscanization of Capua most likely began with the amalgamation of several villages. Recent investigations have suggested that there had been an uninterrupted process of development at this site from approximately 800 B.C., as evidenced by the discovery of an eighth-century necropolis at nearby Sant'Angelo in Formis, near a sanctuary of Artumes. From at least 800 B.C. onwards, Capua's development paralleled that of other Etruscan cities.

Greek culture was inextricably woven into the industry and art of Etruscan Capua. Architectural terra-cottas, such as temple decorations, from the city (dating from as early as the sixth century B.C.), are similar in design to those of the coastal Greek colonies of Cumae and Pithacusae, which had been founded at about the same time as Capua. Furthermore, excavations have produced an inscribed seal and bronze figures from the eighth century, samples of Capua's renowned metalwork, and an Etruscan helmet from about 650 B.C. that demonstrate a strong Greek influence, although it is believed that the copper used was imported from Etruria.

One of Capua's most important roles was to protect Etruscan trade links with Sybaris, the great Greek city in the southeastern Gulf of Taras (Taranto), which maintained its own northern trade route. Through Sybaris, the Capuans had contact with Mietus in Ionia, and this contact probably survived in spite of the destruction of Sybaris in 510 B.C. This connection is suggested by the Capuan adoption of the grid-iron system associated with Hippodamus, the town planner who was a colonist and perhaps the designer of the new pan-Hellenic foundation at Thurii in 443 B.C., which rose beside the site where Sybaris once stood.

The rivalry between Capua and Cumae, the latter of which controlled the seas, was symptomatic of the tension between the Etruscans and Greeks in southern Italy and would eventually lead to the fall of Capua and the end of the short-lived Etruscan Golden Age in Campania. In 524 B.C., the Etruscans launched a full-scale attack on Cumae, with the domination of Campania at stake. Greek historian Dionysius of Halicarnassus wrote that an army of half a million Etruscans faced defeat at the hands of Aristodemus of Cumae, who led the city to triumph with only a fraction of the number of troops, killing the enemy's general in one-on-one combat.

Aristodemus was eventually overthrown, and in 474 B.C. the Etruscans tried to take advantage of the situation by again assaulting the rival city. Cumae appealed to Hiero I of Syracuse, who led the Greeks to victory in the naval battle. The Etruscans suffered great losses and proceeded to also abandon their land campaign. Hiero was richly rewarded with the island of Pithacusae (modern-day Ischia), on which his garrison settled. This event came as a severe blow to both the homeland and to Etruscan Campania, since the area was now cut off by both land and sea from the northern city-states.

The Capua Cathedral following bombardment in World War II
Photo courtesy of Municipio di Capua

This defeat was to seal the fate of Capua, which, along with its allied cities, became isolated from its mother country and was forced to hire mercenary troops from among the native mountaineers of Samnium, an infertile inland region now known as the Abruzzi. Before long, the quest for more food and land drove their countrymen down from the mountains. The Samnites demanded land for settlement, and neither the Etruscans nor the Greeks could stop the flow of invaders into their territories. Capua's rulers attempted to protect the city with huge earthworks, but these efforts were to prove useless. The mercenaries serving in the garrison had already betrayed Capua to their fellow Samnites.

On the night of a public festival in 428 B.C., when the city's inhabitants were sleeping off the effects of strong wine, the mercenaries opened the gates, admitting the band of marauders who then slaughtered the Capuans throughout the night. The rest of the region faced a similar fate. In 423 B.C., Capua's rival, the city of Cumae, was also lost to the same enemy.

Ironically, a century later, the descendants of that first wave of fierce mountain peasants, long since urbanized, faced the same threat of invasion from the inhabitants of their old Samnite tribal centers, who were again driven into the plain by the pressures of overpopulation. Frightened by the prospect of such an incursion, the Capuan Samnites turned to Rome for help in 343 B.C., offering control of their own city in return. The first of three Samnite Wars, lasting a total of fifty years, began that year.

A Roman army under Valerius Corvus entered Campania and inflicted heavy losses on the Samnites near Mons Gaurus, between Cumae and Naples, sixteen miles south of Capua. No longer desperate for help, the Capuans decided to turn against the force that had saved them and invited the Latins to join them in forming a league and waging war against their common protector.

Rome prevailed in 340 B.C and abolished the Latin League (also known as the Latin Confederacy. Capua then fell under Roman control as a municipium, or self-governing community, and its people were granted limited Roman citizenship without, however, the right to vote. Rome conceded various types of autonomy to the former members of the Latin League, presumably to keep them divided and to prevent any form of political connection among the cities.

Capua retained its importance, however, and became physically linked to the capital city by the Appian Way in 312 B.C. Constructed under the aegis of Appius Claudius Caecus, the Appian Way, Rome's first paved road, was 132 miles long and extended from Rome southeastward through Latium and Campania.

During the Second Punic War (218 to 201 B.C.), the Capuans once again turned against Rome. When the great Carthaginian general, Hannibal, entered Campania with the intention of attacking Naples (and securing a much-needed seaport), he was deterred by the size of the city's walls. He instead marched to wealthy Capua, which had a faction that was very disgruntled with Rome. Livy wrote that Hannibal had learned of the opportunity that might await him in Capua from three reputable Campanians, released allied prisoners, who assured him that he had a chance to win over the city's allegiance. At first, the majority of Capuans remained loyal to Rome, but not for long. Opposition to the Roman rule increased as the city received oppressive demands for money, grain, foot soldiers, and cavalry.

In spite of a faction that did not wish to betray Rome, when Hannibal arrived at the city's gates, the Capuans promptly expelled the Roman garrison and invited him to an elaborate banquet. Rome's most important ally then made a treaty with the empire's greatest nemesis. In return for the city's defection, Hannibal agreed that no Carthaginian official would have any authority in Campania and that no Campanian was to serve, except as a volunteer, in the Carthaginian army. Capua's magistrature and laws could not be changed, and Hannibal was to release 300 Roman prisoners, who could be exchanged for Capuans sent to serve in the Roman army. Following the city's defection, Hannibal cut the great consular road, the Via Latina, which ran through Campania into Lucania, thus restricting Roman movement along the Volturnus.

Although Capua's defection was Hannibal's most important political coup, causing a serious disruption of the Roman confederacy, this victory soon proved to be a military millstone. Entirely unable to stand up to the Romans on their own, the Capuans burdened Hannibal with constant appeals for help. This placed Hannibal in the uncharacteristic role of defender and weakened his role as an aggressor.

While Hannibal was occupied with capturing nearby Casilium, vital because it had the only bridge over the lower Volturno, the Capuans again attempted to gain Cumae, but were severely defeated by Sempronius Gracchus. When Casilium surrendered to Hannibal, he placed a Carthaginian garrison in the city and gave it to the Capuans, afterward retiring to his winter quarters in Capua. Livy reported that the city's soft luxury and sensuality weakened Hannibal's battle-hardened men during their winter rest, and that they would emerge robbed of their strength, discipline, and moral fiber by overindulgence in wine and prostitution.

In 214 B.C., while Hannibal was away at Arpi, Rome sent two of its consuls, Fabius and Marcellus, to attack Capua. First, Fabius ravaged the nearby countryside, gathering the corn for himself. Called back into Campania by the Capuans, Hannibal determined that the Roman preparations for the siege were not far advanced, and he created a diversion by marching on to Puteoli (Pozzuoli). He made a third attempt to take Nola, but was thwarted by Marcellus. However, his strategy did prevent the Romans from regaining Capua and made it possible for the harvest to be gathered.

Two years later, as Hannibal took the city of Tarentum, the Romans again had their opportunity to lay siege to Capua. With food supplies dangerously declining, the Capuans appealed to Hannibal for assistance. He instructed Hanno, his trusted lieutenant, to resupply the city. Hanno avoided the Roman-held positions and camped near Beneventum, where he assembled some 2,000 wagons filled with supplies. However, Fulvius Flaccus led a spoiling attack that forced Hanno to abandon the effort and retreat to Bruttium. The Romans assembled two consular armies at Beneventum, only a day's march from Capua, and closed in on the city.

Hannibal came to the rescue after having ensured the security of Tarentum, forcing the Romans to withdraw in two directions. However, when Hannibal went south again in an attempt to gain the Tarantine harbor citadel and the vital port of Brundisium (modern Brindisi), the Roman besiegers returned and gained a stronghold in traitorous Capua in 211 B.C. The Roman camp consisted of two lines: the first one encircled Capua at a distance of about a quarter of a mile from its walls; the second was concentric to the first. The enclosed space contained living quarters for some 60,000 troops and their equipment. Nonetheless, a Capuan deputation somehow managed to slip through the Roman lines and caught up with Hannibal at Brundisium with yet another desperate cry for help. Although the Capuan cavalry had repeatedly interrupted the construction of siege works, when Hannibal arrived, he was unable to crack the double ring of trenches and ramparts that encircled the doomed city.

Unable to raise the siege by assault or to supply his cavalry following the destruction of nearby pastures, Hannibal attempted to save Capua by marching on Rome. He hoped that the armies would leave Capua to rescue the capital city. However, it was obvious that his chances of penetrating Rome's fortifications were slim, and the Romans saw no reason to recall their troops from Capua. Legend says that Hannibal threw a spear at the city as a symbolic gesture. Hannibal's bluff failed and in 211 B.C. the Romans regained the city.

The Capuans began to prepare for their impending punishment. Twenty-seven senators dined at the home of Vibius Virrius, and ended the evening by poisoning themselves. The next day, the Jupiter gate, which faced the main Roman camp, was opened, and one of the legions with two squadrons of cavalry rode in to take the city. All weapons were collected, the gates were shut, and no one was permitted to leave. The surviving Capuan senators were tortured and beheaded. The population was sold into slavery, and all Capuan land and buildings became Roman property. Traders and artisans moved into the city where they lived under Roman rule with no political or civic rights. In 194 B.C. Rome founded the colonies of Volturnum and Liternum on Capuan territory.

In 73 B.C., Capua again became a launching point of rebellion against Rome, with the start of the Gladiatorial War, or the Third Servile War. To satisfy Roman lust for bloody entertainment, Lentulus Battiades ran a gladiators' school at Capua, attended by slaves who were trained to fight one another to the death or to face the jaws of famished, wild beasts for the sake of Roman amusement. One day 200 gladiator slaves attempted escape, 78 successfully. They elected Spartacus, a Thracian slave, as their leader. Spartacus attempted to rally the millions of slaves throughout Italy, and organized an army of 7,000. According to Plutarch, Spartacus hoped to lead the slaves to freedom across the Alps, but his followers insisted on staying behind to pillage. Spartacus was slain in battle, and 6,000 of his followers were crucified along the Appian Way.

Although Capua suffered during the Roman civil wars in the final decades of the republic, after 27 B.C. it began to thrive again under the empire, when Capua became Rome's most prosperous city. However, Capua would again be forced to endure the onslaught of marauders and face final destruction in the dark years ahead. In A.D. 456, the magnificent city was sacked by Vandals under Genseric. About 400 years later, Moslem invaders, the Saracens, razed Capua but left standing the church of Santa Maria. The medieval and modern town that developed on the site of the fallen city was named Santa Maria Capua Vetere, in honor of the church. In A.D. 856, some citizens of ancient Capua settled three miles southeast, and founded another town called Capua on the site of Casilinum, which had been deserted in the second century A.D.

Evidence of ancient Capua's glorious and bloody past haunts the modern-looking Santa Maria Capua Vetere. The Duomo (Santa Maria) houses fifty-one columns from ancient Capuan temples. The imposing amphitheatre in which Spartacus fought, built under Augustus and restored by Hadrian, Antoninus Pius, and Mussolini, still stands. Second in size only to the Colosseum in Rome, this structure was four stories high and was surrounded by eighty arches, two of which remain. Over the centuries, the amphitheatre was exploited for building materials. In the eleventh century A.D., a Norman palace was built with stones removed from the ancient structure. In 1923, a subterranean mithraeum (temple of Mithras), with well-preserved frescoes, was discovered beneath the amphitheatre. In 1976, explorations near the monument revealed a military camp from the second century B.C., thought to be that of Hannibal's forces.

Various architectural ornaments that were once part of the amphitheatre now grace other public places. A second-century mosaic pavement with Tritons and Nereids was removed and set in a park outside of the monument. Several statues were placed in the Museo Archeologico Nazionale of Naples, and seven of the busts of deities which adorned the keystone of the arches now appear in the facade of modern Capua's town hall. One of the most intriguing treasures from the ancient city's earliest days is a tile (now in Berlin) that contains an Etruscan inscription and more than 300 legible words.

The new town of Capua became, in the Middle Ages, an important fortified outpost of the kingdom of Naples. As such it was often the object of struggle, and in 1501 it fell to the French and their ally Cesare Borgia during wars over the kingdom. The kingdom changed hands several times over the centuries, but Capua remained a part of it until 1860, when the kingdom became part of a unified Italy. The unification was made possible in large part by the efforts of nationalist and military leader Giuseppe Garibaldi, who occupied Capua in October 1860 after winning the Battle of the Volturno against the Bourbon rulers of Naples. Another Battle of the Volturno took place during World War II, in 1943, after which British troops occupied the town. Some damage was done by World War II bombing to the Museo Campano, which houses many sculptures and artifacts from ancient Capua, but the museum has since been restored. The new town's cathedral also suffered serious damage in World War II, but it has been rebuilt as well. Just northeast of the town are the Cappella dei Morti, which memorializes the victims of Cesare Borgia's siege, and the eleventh-century Basilica of Sant'Angelo in Formis, which has many Byzantine-influenced frescoes.

Further Reading: *The Roman Republic* by Isaac Asimov (Boston: Houghton Mifflin, 1966) is a concise and delightfully written account of this era and some of its distinctive personalities. *The Punic Wars* by Nigel Bagnall (London: Hutchinson, 1990) and *Hannibal: Challenging Rome's Supremacy* by Sir Gavin De Beer (London: Thames and Hudson, and New York: Viking, 1969) offer in-depth information on the great general's impact on the events of the ancient world. *The Romans: 850 B.C.-A.D. 337* by Donald R. Dudley (New York: Knopf, and London: Hutchinson, 1970), *The Etruscans* by Michael Grant (London: Weidenfeld and Nicolson, and New York: Scribner's, 1980) and *The Etruscans* by Werner Keller, translated by Alexander and Elizabeth Henderson (New York: Knopf, 1974; London: Cape, 1975; originally published as *Denn sie entzündeten*, Munich and Zurich: Droemer Knaur, 1970) provide some of the more extensive information available on ancient Capua. *Greece and Rome: Builders of Our World* compiled and published by the National Geographic Society (Washington, D.C.: 1968) is marvelously illustrated with photographs and artwork recreating historical events.

—Caterina Mercone Maxwell

Caserta (Caserta, Italy)

Location: In southern Italy's Campania region, some fifteen miles north of Naples.

Description: Known as the "Versailles of Naples" for its Bourbon palace, Caserta is the capital of the province of the same name. The old town (Caserta Vecchia) was founded by the Lombards and is located three miles northeast of the modern town.

Site Office: E.P.T. (Ente Provinciale per il Turismo)
Palazzo Reale
CAP 81100 Caserta, Caserta
Italy
(823) 322233

Caserta Vecchia, the old city in the hills northeast of the modern town, was first settled by the Lombards in the eighth century. In the eleventh century, the Normans took much of southern Italy from the Lombards, and Caserta Vecchia also fell into their hands. Never a significant city, Caserta Vecchia still retains its medieval appearance, in large part because of centuries of economic stagnation in the region. The old city's sole claim to fame—the Cathedral of San Michele—is owed to the Norman conquerors, who built the original structure between 1123 and 1153 in the Norman Romanesque style. Details of the original decoration, particularly the exterior sculpture and some mosaics inside the church, bear witness to the eclectic style encouraged by Norman architectural patrons in southern Italy and Sicily.

In the thirteenth century, Caserta was incorporated into the kingdom of Naples, under Holy Roman Emperor Frederick II. Later, Naples and its territories passed into Spanish hands. From the mid-sixteenth to the mid-eighteenth century, the kingdom was administered through a viceregency, a state of affairs that did nothing to alleviate the general backwardness of the region. The year 1734, however, brought a turn of events that was to affect Caserta profoundly. Don Carlos de Borbón (later to become King Charles III of Spain), then eighteen years old, arrived in Naples to become King Charles IV of Naples and Sicily. As a scion of the great house of the Bourbons, stronghold of absolute monarchy, Charles had great plans for Naples, which he found in a highly unsatisfactory condition. As many as one in seven of his Neapolitan subjects actually lived in the city of Naples, and the rest of the kingdom, sparsely populated and mired in poverty, was unable to sustain the capital in appropriately magnificent fashion. The royal palaces at Naples were in a state of disrepair and surrounded by the most atrocious slums.

Charles began a program to decentralize his kingdom, to stimulate the economy, and to build and rebuild palaces and royal villas to give physical embodiment to the magnificence of the absolute monarch. Much was indeed accomplished during Charles's reign. New roads were built, for instance, and harbors were enlarged in many of the Neapolitan ports. And it is to Charles's program of revitalization that modern Caserta owes its very existence. The new town of Caserta began its history as La Torre, a village belonging to the Caetani family of Sermonetta. It was at the time even more modest than Caserta Vecchia, but it had the good fortune to find itself along the Appian Way, the old Roman road rich in history. Charles had the Via Appia restored from Naples north to La Torre, renamed the village Caserta Nuova, and chose it for the site of what was to become the new capital and administrative center of the kingdom of Naples. Plans were developed for a reconstruction of the village on a grand scale, and the jewel in this urban crown was to be a new royal residence. In the end, Charles was recalled to Spain in 1758 to assume the Spanish throne, and he was unable to realize his ambitions for Caserta Nuova. Only the palace, known as the Reggia, was built in accordance to his wishes.

Charles, an amateur architect, had been involved in the design of the building with the architects Mario Gioffredo and Luigi Vanvitelli. Gioffredo had been responsible for the initial designs, executed according to the strictest geometrical principles, which Giambattista Vico had determined to be the most perfect expression of absolute monarchy. Vanvitelli modified these designs, changing the groundplan from the original square to a rectangle with four internal courtyards. He also enlivened Gioffredo's severely regular facades, although symmetry still reigns supreme in Vanvitelli's design. Both designs played an important role in the development of European palace architecture in the eighteenth century, even though only Vanvitelli's plans were realized. The French court, for instance, seriously debated the possibility of building an "envelope" of geometrical facades around Versailles to emulate Gioffredo's design. Several architects were commissioned to draw up designs for such an envelope.

Equally significant were the Reggia gardens, laid out following a design analogous to that of the palace and filled with fountains, cascades, reflecting pools, formal gardens, and statuary that connected the Bourbons to deities of ancient myth. These gardens, ironically, were an attempt to embody an ideal urban environment.

Construction on the royal palace was begun in 1752, and the first stone was laid by the king on his thirty-sixth birthday, January 20. To mark this occasion, squadrons of cavalry and regiments of infantry were stationed along the perimeter of the future Reggia, and two cannon with artillerymen were placed at each corner. Vanvitelli acted as master of the works during the first twenty years of the construction. Convicts, galley slaves, and an army of free workmen carried

The palace at Caserta
Photo courtesy of The Chicago Public Library

the project forward until Charles's departure for Spain, when the work slowed down considerably. Construction temporarily ground to a complete halt in 1764, during a severe plague and famine, when poor and homeless people occupied the half-finished building.

Luigi Vanvitelli died in 1773, when construction was beginning to near completion. His son, Carlo, continued the construction but encountered various difficulties and was unable to complete the building in full accordance with his father's plan. A central dome and four corner towers had to be eliminated from the design. The gardens were laid out by Martin Biancour to Luigi Vanvitelli's designs. The palace was finally finished in 1774 under Ferdinand I of Naples. Caserta's five-story palace is rectangular, with four interconnecting courtyards, a sumptuous chapel, 1,200 rooms, a magnificent theater, and forty-three staircases. Even in its own day, the great building stirred controversy. Although many admired the palace, some contemporaries—those less impressed with the idea of absolute monarchy and more taken with the Enlightenment ideals of individual freedom—considered it an expression of royal megalomania.

Although Caserta Nuova never did become the new capital of the kingdom, the village benefited greatly from the massive construction project and grew considerably. During

the long reign of Ferdinand IV, the Reggia was enlivened by theatrical performances, hunting parties, and balls, but more importantly Ferdinand stimulated the local economy by building a model town at San Leucio, two miles north of Caserta Nuova. Intended as a social experiment, San Leucio was constructed around the silk worm culture, which Ferdinand first introduced to the area. Caserta soon became famous for its silks. Today, the production of natural and artificial silks—some still woven on the original looms—still form an important part of Caserta's economy.

During the Risorgimento, the struggle for Italian unification, Caserta's palace became the headquarters of the nationalist leader Giuseppe Maria Garibaldi. After capturing Naples, Garibaldi prepared at Caserta for his offensive against the army of King Francis II of Naples and for his march on Rome. Although the Reggia was the largest palace in Europe, Garibaldi lived simply and frugally in one of the building's smallest rooms. A platoon of redshirts guarded him against the hordes of office seekers, political intriguers, business speculators, and admirers, who attempted to force their way in to see him at all hours. Garibaldi spent as much time as he could away from the palace, riding along the front line and reconnoitering the enemy's position. In 1860, the two armies clashed around Caserta in what has become known as the Battle of the Volturno, the first time Garibaldi commanded a large army and fought a pitched defensive battle.

Under insistent orders from the king, Marshal Ritucci, the commander of Francis's royal troops, launched his offensive along the Volturno in the hope of defeating Garibaldi's army and recapturing Naples. Although the Bourbon troops in the area outnumbered the nationalists by more than two to one, Ritucci only deployed 28,000 of the 50,000 royal soldiers at his disposal. Garibaldi's most important task was to convince his 20,000 men to stand fast and not retreat. Garibaldi succeeded, despite the fact that he could not speak personally with every soldier, as he had done in most of his other battles.

Ritucci had drawn up his forces in a wide semicircle around Caserta. This arrangement gave Garibaldi the advantage of being able to fight on a shorter interior line and to switch his reserves from one sector of the battle to the other. On his left wing, Garibaldi had 7,000 soldiers near Capua, holding the village of Santa Maria and the advance post of Sant' Angelo. His right wing of 5,000 held Maddaloni, and 280 men held an outpost at Castel Morrone on the extreme right. A detached force of 5,000 Bourbon troops was able to occupy Caserta, but this was an isolated success that finally did not affect the progress or the outcome of the battle. The royal forces surrendered after two days of fighting.

The Battle of the Volturno saved Naples from the Bourbon troops, but the city's natives showed little gratitude to their liberators. Garibaldi's English friend and biographer, Jessie Mario, was in charge of the frontline hospital at Caserta. She and her helpers did their best to care for the more than 1,300 wounded, but they were unable to cope with the great numbers. Many of the injured men were taken to Naples, where they fared very badly at the hands of the city's inhabitants, who had been unwilling to rise in support of their liberators and now refused them even a drink of water. The filthy conditions and incompetent care at the Neapolitan hospitals were responsible for a great number of deaths among the injured.

After the unification of Italy, the royal palace at Caserta was occasionally visited by the Savoyard kings. Victor Emmanuel III presented the palace to the state in 1921.

During World War II, the Reggia served for a time as the headquarters of the Allied High Command. The unconditional surrender of the German forces was signed here on April 29, 1945, and was received by Field Marshal Harold Alexander. Parts of the palace are now used as a commercial school and an air force officers' training school.

Caserta has become the capital of the Caserta province and, with its approximately 75,000 inhabitants, is a city of middling size. It is an agricultural and trade center with glass and food-processing industries. However, the city and its surrounding villages share in the general poverty that has held most of southern Italy in its grip for centuries. What industrialization has taken place at Caserta is insufficient to provide full employment.

In an ironic footnote to western history, Caserta saw a few days of rioting in September 1968. While protests against the Vietnam War were taking place elsewhere, and the students of Paris mounted a socialist revolution, the Casertans abandoned law and order because the local soccer team had not been promoted to the second division of the Italian soccer league. Citizens had apparently expected their elected government representatives to deliver on promises to realize such a promotion. When word reached the city that the promotion was not forthcoming, the fans began a spree of looting and vandalism. Although the disappointment over the soccer team served as a pretext, the hardships of southern Italian life, even in the twentieth century, most likely were the real cause of the rioting.

Further Reading: *Architecture, Poetry, and Number in the Royal Palace at Caserta* by George L. Hersey (Cambridge, Massachusetts, and London: MIT Press, 1983), provides a wide-ranging account of the Caserta Reggia and the ideas that went into its making. *Risorgimento: The Making of Italy* by Edgar Holt (London: Macmillan, 1970; as *The Making of Italy: 1850–1870*, New York: Atheneum, 1971) is a useful guide to this period of Italian history, providing references to the role of Caserta in the unification. *Garibaldi* by Jasper Ridley (New York: Viking, and London: Constable, 1974) is an excellent biography of the great liberator.

—Caterina Mercone Maxwell and Marijke Rijsberman

Cassino (Frosinone, Italy)

Location: The modern town of Cassino lies on the old Via
Casilina, about eighty miles southeast of Rome
and about sixty-five miles northwest of Naples.
Situated on the Rapido River about eighteen miles
from the Tyrrhenian Sea, the town is to be found
at the foot of Monte Cairo in the Abruzzi
Mountains. Just to the west lies Monte Cassino,
site of the first abbey of the Benedictine order.

Description: Cassino began its long history as a Volsci
settlement in the fifth century B.C. and
subsequently gained importance as a Roman town.
In the early sixth century, it became prominent as
the site of the archabbey of the Benedictine order.
For this reason, Cassino is often accounted the
birthplace of monasticism in the West. The old
town was destroyed during World War II, and the
new town of Cassino was built just south of its
original location.

Site Office: A.A.S.T. (Azienda Autonoma di Soggiorno e
Turismo)
Via Condotti, 6
CAP 03043 Cassino, Frosinone
Italy
(766) 21292 or 25692

Cassino originated in the fifth century B.C. as a dwelling
place of the Volsci people, who were then numerous in central
and southern Italy. The Volsci built a citadel on the summit
of Monte Cassino, but they could not prevent a Roman
takeover in 312 B.C. The town prospered under the Romans,
who called the settlement Casinum and built a temple to
Apollo within the citadel. Nothing remains of the temple to
Apollo, but ruins of an amphitheatre, a theatre, and a mausoleum still bear witness to the Roman presence. With the
advent of Christianity in the Roman Empire, Cassino took on
added significance, becoming the seat of a bishopric in the
fifth century A.D. However, the area was poorly defended and
was subject to barbarian attack. After years of neglect, the
citadel was deserted, although the town managed to retain a
few straggling inhabitants for centuries.

It was in that state that Benedict of Norcia, better
known as St. Benedict, found Casinum around 529. Benedict,
the son of a wealthy landowner in the province of Perugia,
had retreated from the world to study and be nearer to God
when he was a young man. Living in a cave above Subiaco
near Rome for several years, Benedict had attracted a number
of disciples and formed with them a small monastic community. When the community became too large to be accommodated at Subiaco, Benedict moved his monks to Casinum,
building churches in the hilltop citadel on the remains of the

Roman temple to Apollo and on the site of the ancient Roman
altar. Traces of these first two churches were discovered in
the 1950s, during restorations of the abbey. After the move to
Monte Cassino, Benedict codified the regulations governing
the community in the so-called Benedictine Rule, which
remained the foundation for Benedictine monasticism for
many centuries. The rule required alternating periods of
prayer and work, which was focused primarily on evangelism, scholarship, and education. Reaching out to society at
large through its emphasis on evangelism and education, the
order was to become a focus of spiritual and cultural development in western Europe.

Although Benedict died at Monte Cassino in 547, the
community continued its peaceful existence for a few decades. Some time in the 580s, however, the Germanic tribe of
the Lombards invaded the area and destroyed the abbey, while
the monks fled to Rome. Enjoying the patronage of Pope
Gregory the Great, Benedict's disciples kept the order alive
at Rome but did not return to Monte Cassino. Then, in 717,
Petronax of Brescia rebuilt Benedict's monastery, which
quickly became an important center of scholarship and a
model community for Benedictine settlements across Western Europe. The monastery obtained papal privileges that
made it accountable only to the Vatican and began receiving
donations from wealthy secular patrons, so that soon the
abbey controlled huge stretches of land and became a formidable political force. The order's influence to the north was
solidified when Charlemagne, who visited Monte Cassino in
787, conferred imperial privileges on the Benedictines and so
gave them pride of place among religious houses in the
Frankish empire.

As the hilltop complex grew with the addition of a
three-nave basilica, the town down the mountain slopes from
the flourishing abbey still failed to regain its former prosperity. Around 866 it was abandoned altogether, and a new town
was built at the foot of Monte Cairo. Fortifications were built
on the Rocca Ianula, of which only traces remain. The new
settlement was first called Eulogomenopolis, but it received
the more Italianate name of San Germano later and was
finally named Cassino in 1871, following the unification of
Italy. Less than two decades after the reconstruction of the
town in the 860s, life was once again disrupted, this time by
Moslem invaders. Although the town apparently did not
suffer extensive damage, the raiders left the Monte Cassino
abbey in ruins. Repeating history, the monks fled to Capua,
not to return until the mid-tenth century, when they refounded
the Monte Cassino monastery and began rebuilding the
abbey. Disaster struck again with a Norman attack in 1030.
However, reconstruction continued steadily through the eleventh century, which saw Monte Cassino lead Europe in manuscript production and illumination. Ancient pagan and

The Abbey of Monte Cassino following Allied bombings of World War II

The Abbey of Monte Cassino during reconstruction in the 1950s
Photos courtesy of The Chicago Public Library

Christian texts, many of them unique, were preserved in the Monte Cassino library. In this period, the foundation was laid for the fabulous collection of manuscripts, art, and, later, books that was to play a role in the battles for Monte Cassino during World War II. Many Monte Cassino monks attained high ranks in the Catholic church during the same period. In fact, the abbey produced three popes in the eleventh century. Many distinguished visitors were guests at the abbey, which could also boast an ever-growing endowment of lands.

A decline in Monte Cassino's fortunes began early in the twelfth century, exacerbated by struggles for dominance in southern Italy involving Norman and Germanic invaders. In 1139, for instance, Pope Innocent II took refuge in the Cassino stronghold of Rocca Ianula, but the Norman Roger II of Sicily managed to breach the defenses, taking Innocent prisoner. As political and military turbulence shook the area, discipline among the Benedictines weakened. Many monks and even the abbots absented themselves from the abbey for long periods. Even though peace was partially restored in the Latium region in 1230, when Pope Gregory IX concluded a peace agreement with the Holy Roman Emperor Frederick II at the Rocca Ianula fortress, the Benedictines did not recover their interest in the birthplace of their order.

A final blow to the abbey's pre-eminence came on September 9, 1349, when an earthquake destroyed the complex. Although rebuilding soon began under the auspices of Pope Urban V, a Benedictine, the abbey did not regain its former glory until the sixteenth and seventeenth centuries. By that time scholars began to return to the abbey, bringing Monte Cassino back to its former significance as a center of culture and art. The reconstructed abbey complex was also of architectural and artistic importance. The church that was finished in the early eighteenth century, for instance, was one of the chief glories of the Neapolitan baroque style and was decorated with frescoes by Luca Giordano. In the late eighteenth century, the abbey narrowly escaped another destruction, surviving the Napoleonic invasion of the 1790s without incident. The town of Cassino was not so fortunate, however, being sacked by the invading French army in 1799.

The unification of Italy in the nineteenth century brought upheavals to monastic organization in the country, and the Monte Cassino abbey underwent a drastic change of status. The abbey itself became state property and was declared a national monument. Fortunately, its magnificant library and art collection, among the most important in the world, were left intact. The Benedictine monks were charged with the administration of the complex and its lands as well as the library and art collection. The abbey has continued since then as an important center of historical and theological scholarship, although the joint British and American invasion of Italy in 1943 disrupted its scholarly quiet and occasioned the removal of most of the library holdings to the Vatican.

Following Allied victories in North Africa and Sicily, the British landed an invasion army on the coast of Calabria at the southern tip of Italy in September of 1943. Five days later Italy surrendered, but German troops continued to put up resistance. American forces landed south of Naples on the day after the Italian surrender, taking Naples after heavy fighting. Both Allied armies continued north to Rome but were slowed by bitter winter weather and nearly impassible mountain terrain. By mid-November, Allied forces were nearing the German defense barrier known as the Gustav Line. Running west to east across Italy, the Gustav Line followed the course of the Garigliano, Rapido, and Sangro Rivers, running past Cassino. The valley of the Liri River, which lies immediately north of Cassino, quickly became the focus of Allied attack. Monte Cassino, on its heights above the southern end of the Liri Valley, attracted German attention as a key defensive position and a potential observation post.

Two German army officers, Maximilian Becker and Julius Schlegel, independently went to the Monte Cassino abbey in October 1943, both intending to persuade the otherworldly monks to move their treasures to safety before the fighting reached Cassino. Among the abbey's invaluable possessions were eleven paintings by Titian, two paintings by Goya, and a painting by El Greco, besides a wealth of lesser art works and rare manuscripts. Since both German officers were under the command of Hermann Göring—well known as the prime looter of art in Nazi-occupied Europe—Abbot Diamare of Monte Cassino must have been faced with a difficult decision. In the end Becker and Schlegel convinced the abbot to let the Germans take the literary and art treasures north to Rome to be given to the Vatican or stored in safe locations.

Within a week the Germans were loading the art collection and library holdings onto trucks, leaving space for the monks, most of whom were to be relocated. The evacuation was completed during the first week of November. More than 100 truckloads were taken to Rome and to the German supply depot at Spoleto. As may have been expected, fifteen cases of art objects from Monte Cassino were delivered to Hermann Göring as a present for his fifty-first birthday. These were recovered after the war from an abandoned salt mine in Austria. The remainder of the Monte Cassino holdings were sent to the Vatican and various museums for safekeeping, and for the most part they survived intact. They never were returned to Monte Cassino, however. The Göring Division claimed credit for rescuing the Monte Cassino treasures after the end of the war, and Becker and Schlegel quarreled for years in the media over which one of them had played the most important role.

Meanwhile the Germans were preparing for an all-out defense of Cassino and the Monte Cassino heights. The civilian population of Cassino was evacuated, and the entire area, including the heights, was fortified with shelters, bunkers, dug-in tanks, tank traps, mines, and machine gun nests. Observation posts were established on Monte Cassino to direct artillery fire and the infrequent German air attacks. Whether or not the Germans ensconced themselves in the abbey is not known. The German army was aided in its defenses by the winter weather: heavy snowfall in the mountains and mud pools in the valleys turned even routine supply operations behind the lines into a difficult and dangerous undertaking for the Allies. In the end, the Gustav Line was breached and the road to Rome opened only after four bloody battles over almost six months.

In January of 1944, a massive offensive was launched by the Allies at Cassino. Additional British and American troops were landed at Anzio Beach, about thirty-five miles south of Rome. The Anzio landing was meant to draw German troops away from the Cassino front and to outflank the strong defenses of the Gustav Line. However, the Anzio forces hesitated to move forward and were pinned down for months in a narrow beachhead. The Americans at Cassino failed to cross the Rapido River, suffering many casualties in the attempt. A second offensive in February, mainly by New Zealand and Indian units, also made little progress.

At this time, the Allies became convinced—whether rightly or wrongly is still unclear—that the Monte Cassino abbey itself had been converted into a German observation post. In one of the most controversial actions of the war, the American Air Force launched an air raid on the abbey in mid-February. Heavy bombers, commonly used to attack entire cities, rained bombs on the monastery and almost destroyed it. The Germans later claimed that they were not in the abbey before the bombing. They did occupy its ruins afterwards, and savage fighting was later necessary to dislodge them and capture the Monte Cassino heights.

Another massive Allied air raid annihilated the town of Cassino in March. The accompanying ground assault on the town resulted in especially intense fighting. German soldiers who had served in the east claimed that it was worse than anything they had encountered in Russia. In the end, this third Allied attack failed because of stubborn German resistance. The last and successful battle began on May 11, after local Allied forces had been built up to outnumber the Germans three to one. Parallel assaults by the British Eighth Army at Cassino and the American Fifth Army south of Monte Cassino heights were mounted, together with a Polish assault on the monastery and a French flanking attack. The plan succeeded, and on May 18 a Polish detachment occupied the abbey of Monte Cassino. By June 4, American units occupied Rome. Two days later, the invasion of Normandy drew attention away from the Italian front. The battles of Cassino and Monte Cassino were tragic episodes, examples of war in its harshest form, a part of what has been called by military experts the most tactically absurd and senseless campaign of World War II. The British and Polish military cemeteries at Cassino still bear witness to many of the lives lost in 1943 and 1944.

With funding from the Italian government, restoration of the Archabbey of Monte Cassino was begun soon after the war. In 1964 Pope Paul VI consecrated the new monastery, proclaiming St. Benedict the patron saint of Europe on the same occasion. The severe exterior of the present monastery, which consists of four communicating cloisters laid out in a square, gives no indication of the wealth of the decor within. The basilica is particularly noteworthy for its rich marble, stucco, and mosaic decorations. The only structure to have survived the American air raid is the basilica crypt, which holds the remains of Benedict and his sister, St. Scholastica. The town of Cassino was reconstructed on a site slightly south of its prewar location and features distinctive white stone buildings, as it has since Roman times.

Further Reading: Information relating to Monte Cassino is concentrated most heavily on events in 1943 and 1944. The most detailed source on the Monte Cassino bombing is *Monte Cassino* by David Hapgood and David Richardson (New York: Congdon and Weed, and London: Angus and Robertson, 1984). Charles Connell's *Monte Cassino: The Historic Battle* (London: Elek, 1963) relates the history of the Allied campaign with emphasis on the role of the Polish forces. *Cassino: Portrait of a Battle* by Fred Majdalany (London: White Lion, 1973; second edition, London: W. H. Allen, 1981) is a thorough, well-written history of the Cassino battles. Janus Piekalkiewicz's *Cassino: Anatomy of the Battle* (London: Orbis, and Indianapolis, Indiana: Bobbs Merril, 1980) is notable for its illustrations.

— Bernard A. Block and Marijke Rijsberman

Cephalonia (Cephalonia, Greece)

Location: One of the seven Ionian Islands in the Ionian Sea off the west coast of Greece; forms, with Ithaca, the department of Cephalonia.

Description: The largest of the Ionian Islands, Cephalonia covers 302 square miles and is separated from Ithaca by a mile-wide channel. Chief town of the island and capital of the department is Argostolion. Cephalonia was associated with Odysseus (of Homeric fame) and has ancient ruins that provide evidence of thousands of years of occupation; it also has many historic monuments erected by the various foreign powers that have colonized the island.

Site Office: Tourist Information Office
Argostolion, Cephalonia
Greece
(671) 22248 or 22466

Cephalonia is the largest of the seven Ionian Islands formed from the summits of a limestone ridge running parallel to the mainland of Greece. Cephalonia has a long history of human habitation and has been caught in conflicts involving the Greek city-states, the Romans, various medieval conquerors, adversaries of the Napoléon, and the participants in World War II.

Archaeological finds date human existence on the island to at least 50,000 B.C. Fiskardo (or Phiscardo) Man, as he has been labeled by archaeologists, indicates that early civilizations on Cephalonia had much in common with those in western Sicily and Epirus. An investigation of remains on the two hills of the acropolis of Kranea uncovered a prehistoric deposit containing handmade Bronze Age pottery. In the southern part of the island, at Pronnos, prehistoric shards have been found.

Probably the most important archaeological discovery on Cephalonia was that of Mycenaean chamber and beehive tombs. These graves, cut deep into the limestone, date from about 1400 to 1100 B.C. This was a period when Achaeans were expanding into western Greece, and it is probable that they colonized the Ionian islands of Ithaca, Zante, and Cephalonia as well. Numerous skulls found inside the tombs show signs of head trauma, indicating that war occurred on the island at least as early as the Bronze Age.

The great epic poet, Homer, is said to have called the island "Sami," and it is suspected that the "glittering Samos" of the *Odyssey* is a reference to the mountains of Cephalonia. The neighboring island of Ithaca was the home of the mythical Odysseus, and it is presumed that Cephalonia may have been included in the kingdom of this hero. Some have questioned whether or not Homer's epic poems have any valuable

historic information aside from linguistic and literary history. Obviously some parts of the *Iliad* and *Odyssey* are fictional, but many of the situations and portraits of civilizations in these narratives have proven to be based on fact. At Nea Scala in the southeastern point of the island there are two notable mosaics dated from the Mycenaean era or earlier, which local citizens claim may have been part of Odysseus's palace.

Little else is known about Cephalonian history until the fifth century B.C. In 456, Athens began a powerful offensive against the coasts of the Peloponnese region. Tolmides, commanding a large, handpicked force of marines, ravaged the island of Cerigo, numerous coastal towns of Peloponnesus, and captured the islands of Zante and Cephalonia from mercantile Corinth. In 431 Corinth's ally Sparta, which had been developing a formidable army, seized Cephalonia from the Athenians. This year marked the beginning of the Peloponnesian War, which lasted until 404 B.C. The conflict between the rival Greek city-states of Athens and Sparta was fought in two phases. In the first phase (431–421 B.C.) Athenian sea power was matched by Spartan land power and a stalemate resulted. During the second phase (418–404 B.C.) the Athenians suffered major losses. Sparta enlisted Persian aid to build an aggressive fleet to force the final surrender of Athens.

During the fifth century B.C. the historian Thucydides made reference to Cephalonia, discussing the island's four city-states of Sami, Pali, Kranea, and Pronnos. Archaeological remains of Cyclopean walls have been found at the sites of each of these cities. It was at Sami, the most powerful of the city-states, located on the middle eastern coast of the island facing Ithaca Channel, that the Romans met with resistance when they landed on Cephalonia in 189 B.C. Piracy in the Ionian Sea had interfered with Roman communications for some time prior to the Roman invasion of this island. By 189 B.C., when the empire had finally defeated Carthage and suppressed the Aetolian League, it was in a position to challenge the marauders and attempt to pacify Cephalonia and its surrounding waterways. Under the consul Marcus Fulvius, the Roman military invaded the island and received a quick and peaceful surrender of the four city-states. At the last moment the citizens of Sami revolted. The reason for the sudden resistance is unknown. Perhaps the residents feared being evicted from their homes and their city, as Sami was in a strategic position geographically. Whatever their rationale, in spite of pleas from the Romans, the Sameans would not change their minds. Fulvius responded by besieging the city for four months, after which Sami was captured and ravaged, and the remaining inhabitants were sold into slavery.

Strabo, the geographer and historian, recorded that Caius Antonius, who was the colleague of Cicero in the consulship, was charged with extortion and exiled to

136

A harbor on Cephalonia
Photo courtesy of Greek National Tourist Organization

Cephalonia in the early part of the first century B.C. While there, Antonius is said to have ruled over the inhabitants as if the island were his personal property. Cephalonia continued to be used as a place of exile after it passed to the Byzantine Empire, following the division (and, eventually, the fall) of the Roman realm.

In the Middle Ages Byzantine territory became attractive to a variety of invaders. In 1084, the Norman duke Robert Guiscard joined forces with his son, Bohemund, on the island of Corfu. Bohemund had seized Corfu in 1083, and Robert wintered his army there in 1084 while planning to resume his advance on the Byzantine lands. The army was overcome that same winter, not by Byzantine forces, but by an outbreak of the plague. Robert did not contract the disease himself, however, until 1085, when he had set forth with his remaining forces to join his advance troops who had already reached Cephalonia. He died of plague on the island; the town where he died was later named "Phiscardo," which derives from "Guiscard."

During the next 800 years Cephalonia, like its neighboring Ionian Islands, served as a trophy to be seized, traded, and governed by the Normans, Venetians, Romans, and an assortment of dukes and counts seeking riches. In the late twelfth century Cephalonia was occupied by the Roman noble Matteo Orsini, and the Orsini family ruled the island until the fourteenth century, when the Tocco family, who originated at Benevento near Naples, was assigned Leukas and the southern Ionian Islands in 1357 by the king of Naples. The last of the Tocco rulers, Leonardo III, reigned from 1448 until he was ousted by the Turks in 1479. Upon landing at Cephalonia the Turks executed all ducal officials, burned the fortress of St. George, and deported most of the peasants to Constantinople.

In 1499, Venetian and Spanish armies under "el Gran Capitan," Gonzalo Fernández de Córdoba, surrounded and captured the St. George fortress and slaughtered the Turkish soldiers posted there. Venetians took possession of Cephalonia at that time and found the island so depopulated that they made land grants to Greek families fleeing the Turks. From that point Venice held Cephalonia for nearly 300 years, until in 1797, when it was taken by the French. The British then captured it from the French in 1809. The British made Cephalonia, along with other Ionian Islands, a protectorate.

The British soon began building a network of roads and improving water supplies. The most influential British administrator ever to have served on Cephalonia was Colonel (later Sir) Charles Napier, who later conquered Sind. He governed Cephalonia from 1822 until 1833. He was responsible for numerous construction projects, including the open quay of Argostolion, a model prison, and the Ayios Theodori lighthouse. He also is credited with reforming the island's justice system and protecting peasants against oppression by the aristocracy.

Napier befriended Lord Byron in 1823 when the famous poet took residence at Argostolion and Metaxata from August to December of that year. He had come to Greece to aid in the fight for independence from the Ottoman Empire, and had already helped raise large sums of money for the cause. While waiting for a sign from the Greek mainland indicating that circumstances were favorable to commit himself further, he resided in Metaxata. At the end of 1823, Byron left for the mainland, where he was offered a military command; however, he died of a fever at Mesolóngion shortly thereafter.

Rebellion against the British erupted on the Ionian Islands in 1848 and was especially vigorous on Cephalonia. Technically the islands were politically independent while under British military protection, but in fact the British controlled the islands' political affairs, including public appointments, policies, monies, the press, and the police. Sir John Colborne, Lord Seaton, who served as Lord High Commissioner of the islands from 1843 to 1849, was a proponent of liberal reforms, some of which were approved by the British Crown. The Ionians were not satisfied with the pace of change, however, and revolted in 1848. Seaton's successor, Sir Henry Ward, was confronted by another rebellion in 1849, and the British decided to reverse the liberalization process.

A decade later, the islands had become unmanageable and a political embarrassment to Britain. W. E. Gladstone, sent to the islands in 1858 as High Commissioner Extraordinary, evaluated the situation and concluded that the reforms suggested by Seaton were indeed necessary, but he could not muster support for them. Finally the British gave up trying to govern the islands and ceded them to Greece in 1864.

Cephalonia's twentieth-century history is marked by two significant events. During World War II the island was occupied by the Italian army. Italy surrendered to the Allies in the fall of 1943, but in anticipation of this event the German army had begun to occupy Cephalonia and other Italian-held bases in and around Greece. General Gandin, the commander of the Italian Acqui Division and associated forces on Cephalonia, received orders from the Italian government to resist any further advances by the Germans. At the same time the Italian command in Athens, dominated by the Germans, directed Gandin to place himself and his division under German command. The general, uncertain about which command to obey, played for time, hoping the Allies, who were at that time in the Aegean, would come to his aid. No help arrived, however, and the Germans soon attacked the Italians. The Italians were at a disadvantage, having no air power at Cephalonia while the Germans had a strong air force, but they resisted for about a week before surrendering on September 24. After the battle was over, a German firing squad executed nearly 5,000 Italian soldiers. Many Italians still refer to the soldiers as the "martyrs of Cephalonia," and even German historians have called the event a war crime.

Ten years later, the island suffered another tragedy when it was devastated by an earthquake in August 1953. Only Phiscardo, at the extreme northern tip of Cephalonia, escaped damage. Today, with the exception of remains around Phiscardo, many of the island's ancient architectural monuments have been lost. There are, however, many noteworthy sites to interest travelers, and new construction of beach-side hotels and restaurants has continued to take place since the earthquake. At Argostolion, visitors will find the Archaeology Museum containing a room of Mycenaean artifacts and the Koryalenios Historical Folklore Museum containing icons and Venetian records of the islands.

On the little peninsula north of Argostolion there are interesting *katavothri,* or "swallow-holes," where the sea is sucked into two large tunnels under the ground. No one was ever certain where the water went after it entered the holes until Australian geologists dumped a large quantity of dye into the water at the entrance in 1963. Fifteen days later the dye reappeared in the lake of the Melissani Cave and, on the other side of the island, at Karavomylos, near Sami.

Across the peninsula from Argostolion stands the reconstructed lighthouse of Ayios Theodori, and just beyond the village of Mazarakata, southwest of Argostolion, there are a number of Mycenaean beehive and chamber tombs dating from about 1400 to 1100 B.C. Ruins of the medieval fortress of St. George and the town that once surrounded it are located south of Argostolion; this site suffered extensive earthquake damage in 1636 and was abandoned in the following century. At Metaxata, the house where Lord Byron lived was destroyed by the 1953 earthquake, but a plaque on the site memorializes the poet-adventurer.

Further Reading: *The Ionian Islands: Zakynthos to Corfu* by Arthur Foss (London: Faber, 1969; Levittown, New York: Transatlantic Arts, 1970) is an accessible history of the islands and includes interesting photographs and detailed essays on the author's visits to particular sites of interest. *Inside Hitler's Greece: The Experience of Occupation, 1941–44* by Mark Mazower (New Haven and London: Yale University Press, 1993) provides a detailed view of Greece under Italian and German occupation, including the events on Cephalonia in 1943. *The White Flag* by Marcello Venturi, translated from the Italian by William Clowes (London: Blond, and New York: Vanguard, 1966; as *Bandiera bianca a Cefalonia,* Milan: Garzanti, 1967) provides an account of the demise of the Italian Acqui Division on Cephalonia.

—Holly E. Bruns

Chieti (Chieti, Italy)

Location: The capital of the province of Chieti in the Abruzzi region of Italy; 140 miles east/northeast of Rome, 152 miles northeast of Naples and 9 miles west of the Adriatic Sea.

Description: Occupied since pre-Roman times; contains extensive Roman ruins; home of Gian Pietro Carafa, Catholic reformer who founded the religious Order of the Theatines and later became Pope Paul IV; today the site of numerous historically significant churches and other buildings and several cultural institutions, including the prestigious Museo Nazionale Archeologico di Antichità; the Pinacoteca Provinciale, a provincial art gallery; and the Museo di Arte Sacra, which is notable for examples of the local woodcarving art.

Site Office: E.P.T. (Ente Provinciale per il Turismo)
Via B. Spaventa, 29
CAP 66100 Chieti, Chieti
Italy
(871) 65231 or 65232

In ancient times Chieti was known by its Latin name, Teate. Prior to its alliance with Rome, Chieti was a center of the Marrucini, an ancient people who inhabited a small area of the Italian coast on the Adriatic Sea. Archaeological remnants of this culture have been excavated, dating to the fifth and fourth centuries B.C. Although the Marrucini were related to the Samnites, another ancient Italian tribe, they were not officially aligned with the Samnite league. However, the Marrucini did participate in the Second Samnite War against Rome. At the end of the conflict in 304 B.C., Chieti entered the Roman alliance, after which the town was referred to as Teate Marrucinorum. Chieti was loyal to Rome during the Second Punic War, but more than a century later, in 91 B.C., the town attempted to revolt against Rome. After the Roman victory, Chieti was returned to the Roman fold and its residents were enrolled into the Roman tribe Arensis. Subsequently, Chieti became a Roman municipium.

Roman rule left its mark on Chieti; it still has the remains of Roman temples, including one dedicated to the Dioscuri, baths, and a theatre. The remains of three Roman temples were unearthed in 1935 behind the local post office and the remnants of a large Roman cistern can found in the Strada Marrucini, south of the city. The cistern measures 180 feet by 43 feet and is comprised of subterranean compartments connected by a series of arcades. Cut into the slope of a hillside, the cistern provided water to the thermal baths below.

Like the rest of the Roman Empire, Chieti suffered in the barbarian invasions during the declining years of the empire. The town was rebuilt, however, under the auspices of another invader from the north, Theodoric the Great, king of the Ostrogoths, after he conquered Italy toward the end of the fifth century A.D. The region was Christianized by this time, and Chieti became the seat of a bishop.

Subsequently Chieti, like all of Italy, experienced a long succession of invaders. The town became part of the Lombard kingdom after the Lombard invasion of Italy in 568. Chieti was invaded again, in 802, by armies under the command of Pépin, the son of Charlemagne who had been crowned the Carolingian king of Italy in 781 at the age of four. The town eventually came under control of the Hohenstaufen family, the family to which Holy Roman Emperor Frederick I (Barbarossa) belonged. Thus Chieti, as almost all of Italy, became entangled in the long struggle between the Guelphs, a papal party that opposed the authority of the German emperors of the Holy Roman Empire in Italy, and the Ghibellines, aristocrats who supported the German emperors. Through the Middle Ages, the control of Chieti continued to pass through many hands, including those of the French Angevins, the Spanish kingdom of Aragon, and the Caracciolo family of Naples. Chieti played a major role during the Counter-Reformation (also known as the Catholic Reformation), the effort of the Catholic Church to reform itself in the face of the Protestant Reformation led by Martin Luther in the sixteenth century, due to the influence of one of its bishops, Gian Pietro Carafa. Born on June 28, 1476, into one of the oldest and most influential baronial families of Naples, Gian Pietro Carafa was an ascetic, an ardent zealot of Roman Catholic orthodoxy, and eager to participate in the activity of the Counter-Reformation.

At the age of fourteen, he had run away from home and attempted to join the Dominican order. Prevented from doing so by his father, Gian Pietro nevertheless gained permission from his family to study theology. Carafa rose swiftly in the church hierarchy under the guidance of his uncle, Olivero Carafa. By 1500, Carafa was a papal attendant, a rank just below bishop, and by 1503, he held the title of protonotary. In 1505, at the age of twenty-nine, Carafa was named bishop of the see at Chieti.

Although appointed bishop in 1505, Gian Pietro was not anointed until 1506, after which he was sent immediately to Naples by Pope Julius II to act as papal nuncio to Ferdinand II of Aragon, newly arrived in Italy. Carafa did not return to Chieti until 1507, at which time he began his first attempts to reform what he saw as a troubled diocese. For five years Carafa struggled to raise the standards of his diocese, beginning to initiate reforms and at the same time resisting the encroaching influence of the Spaniards. So single-minded were his efforts that he failed to attend the first four meetings of the fifth Lateran Council, called by Pope Julius II to

Chieti's cathedral and bell tower
Photo courtesy of Italian Government Tourist Board - E.N.I.T.

examine reform. In 1515, Caraffa was sent to Spain by the Pope Leo X on church business and did not return to Chieti until 1520.

Also at that time, a religious confraternity called the Roman Oratory of Divine Love was under the direction of Cajetan of Thiene, a priest and member of the curia. The members of the oratory, through prayer, preaching, partaking in the sacraments and good works, sought both inner renewal of themselves and external reform of the church. However, Cajetan of Thiene eventually came to the realization that a new religious order was needed, created especially to rein in the moral laxity that he saw permeating Roman Catholicism and to combat the rising tide of Lutheranism. Cajetan of Thiene actively pursued Gian Pietro Carafa's aid and advice in this matter and Carafa, the type of individual needed to put Cajetan's ideas into practice, wholeheartedly embraced the concept of a new religious order. However, Carafa could not participate in its creation as long as he continued to hold his bishop's seat at Chieti, along with the archbishop's seat he also held at Brindisi, to which he had been confirmed in 1518. Therefore, he sought release from them.

In 1524, Pope Clement VII allowed Carafa to resign both of his offices and authorized the founding of a new community of Clerks Regular. Carafa, Cajetan, and two other members of the Oratory of Divine Love took their first vows at St. Peter's in Rome on September 14, 1524. The Order of Clerks Regular came to be commonly known as the Order of Theatines, from the Latin name for Carafa's bishopric in Chieti, where the order's first chapter house was founded, with Gian Pietro Carafa as its superior.

The Theatines were not a cloistered group of monastically inclined priests. Perhaps taking a cue from active Protestant reformers, they went into the world and preached and engaged in pastoral work. The Theatine Rule required vows of chastity, obedience and, above all, poverty, in direct opposition to the covetousness that was seen as the chief evil infesting Roman Catholicism. The Theatine Rule specifically forbade the possession of personal property or the begging of alms for support. The Theatines lived in a community, sharing a common house under the leadership of a superior who was elected yearly. The admittance of a novitiate required the unanimous consent of the entire chapter. The order abstained from meat, adhered scrupulously to the fasts proclaimed by the church, and practiced self-mortification. Their rules also required them to wear the ordinary black cassocks of secular priests and to be tonsured. They attended to the sick, the poor, and the illiterate. A member was not allowed to leave the chapter house alone. A companion was required and only after offering a prayer and receiving the superior's blessing was travel outside the house allowed. Conversing with women was to be avoided by the Theatines, except when made unavoidable by some necessity or by the prescribed laws regarding charity and then only with the direct approval of the superior. Carafa fully realized that only through the power of the papacy and the church hierarchy through which he had risen so swiftly could an effective Counter-Reformation movement be mounted. Therefore the Theatines were closely aligned with the Holy See in Rome.

It is in some measure ironic that, due to the vow of poverty taken by the Theatines and the fact that they were not allowed to beg alms for their support, membership in the order was realistically restricted to the sons of noble or otherwise wealthy families. However, this circumstance served Carafa's vision of the Theatines well. Although Theatine chapter houses were also founded in Rome, Venice, and Naples, in 1527 there were only fourteen members of the order and by 1533 there was a total of only twenty-one members. Nevertheless, the Theatines were an influential order in Roman Catholicism far beyond their actual numbers. In fulfilling his vision, Carafa's Theatine order became, according to historian Ludwig Pastor. a "carefully chosen circle of men... a corps d'élite... not so much a seminary for priests as at first might have been supposed, as a seminary for bishops who rendered weighty service to the cause of Catholic reform."

It is estimated that more than 200 bishops of the Catholic Church have emerged from the ranks of the Theatines. Their reformist spirit was so great that any cleric favoring internal reform in the church was dubbed a Theatine. The Theatines became the model for a number of other reform-minded orders, including the Somaschi, the Barnabites, and the highly powerful Roman Catholic Society of Jesus, known as the Jesuits, founded in 1534 by St. Ignatius Loyola and dedicated to zealous missionary work and education.

After his ascension to cardinal in 1536, Gian Pietro Carafa continued to aggressively pursue internal church reform and was made chief of the newly reactivated inquisition. At the age of seventy-nine, he was elected to the papacy in 1555, taking the name Pope Paul IV. As pope, he remained vehemently anti-Protestant and rejected as heretical the Peace of Augsburg, which recognized the co-existence of Catholics and Lutherans in Germany. His intolerant views and harsh treatment of the prisoners of the inquisition so fired hatred for him that, upon his death in 1559, rioters destroyed the headquarters of the inquisition, released the prisoners held there, and toppled and mutilated a statue of Carafa that had been erected on the Capitol in Rome.

In addition to producing a pope, Chieti and its surroundings have produced a number of artists and writers. The painter Franceso-Paolo Michetti was born in a small village on the outskirts of Chieti in 1852. He is famous for his renderings of the native peasants of the Abruzzi region. The first work that brought him fame was *The Procession of the Corpus Domini at Chieti*, done in 1876. The writer Pasquale de' Virgilii was born at Chieti in 1812. Virgilii was a romantic in spirit and, taking his inspiration from Lord Byron, traveled extensively. Pursuing the Greek ideal, Virgilii championed the liberal literary and journalistic movement in Naples, where he wrote both poetry and prose. Although today his five volumes of poetry remain largely unread, no less than Victor Hugo believed Virgilii to be the equal and spiritual successor of Dante.

In the nineteenth century, Chieti became part of a unified Italy. During World War II, King Victor Emmanuel

III, fleeing from Rome with the entire royal household in the face of Allied advances, stayed in Chieti while waiting for the steamer ship in which he was to leave Italy. Italian troops of the Legnano Division were called to form a protective triangle around the towns of Chieti, Ortona, and Pescara. The Italian Resistance was active in the Maiella Mountains, which overlook Chieti and the Italian Corps of Liberation campaigned in the Abruzzi and helped the Eighth Army of the Allied forces liberate Chieti.

Present-day Chieti has developed into a city with a growing academic and cultural reputation. In addition to the Museo Nazionale Archeologico di Antichità, the Pinacoteca Provinciale, and the Museo di Arte Sacra, the town also houses a large provincial library and a theological school. Chieti also has numerous architectural treasures, including several churches. The church of Santa Maria Mater Dominii holds a wooden Madonna, carved by the Italian artist Gagliardelli. The medieval church of Santa Maria del Tricalle, built in 1317 in an unusual octagonal shape, overlays the site of the temple of Diana Trivia. The most important medieval structure, in the Piazza Vittorio Emanuele, is the town's Duomo, or cathedral. Built between 1335 and 1498, the cathedral contains a baroque pulpit and stalls, a silver statue of St. Justin in the Tresoro by Nicola da Guardiagrele, and frescoes from the fifteenth century. The cathedral's Gothic bell tower was damaged during an earthquake in 1706 and was not restored until 1935. The Palazzo Municipale, in the same piazza, was built in 1517 as the palace of the Valignani family. The Palazzo Municipale was rebuilt in a neoclassical style during the nineteenth century.

Artifacts from Chieti's antiquity are housed in the city's respected Museo Nazionale Archeologico di Antichità. The museum's holdings are extensive and only a small portion of them are on display at any given time. The museum is divided into four sections. The archeological section is devoted to burial cults and contains artifacts from patrician tombs of the sixth and seventh century B.C. The museum also houses artifacts, including masks and pearls, of Phoenician-Punic origin unearthed from a burial site at Teramo in 1973. The pearls are of major importance as similar objects have only been found at Carthage, in Spain, and in Sardinia. Also of importance from the Teramo site is a stone stela that contains the first known written evidence of the term "Sabine." Along with the famous Warrior of Capestrano sculpture found in nearby L'Aquila, there are also a number of additional sculptures and sculpture fragments from Chieti itself and dating from the fifth century B.C. The museum also contains sections devoted to anthropology and geology, and a collection of more than 15,000 coins from the Abruzzi region, ranging from the Roman period to the mid-nineteenth century. The coins, arranged in chronological order, indicate the area's prosperity, derived from lucrative trade in wool, silk, and spices.

Further Reading: *Central Italy: An Archeological Guide*, edited by R. F. Paget and Glyn David (London: Faber, and Park Ridge, New Jersey: Noyes, 1973) is a clearly written, comprehensive guide to the archaeological remains of the prehistoric, pre-Roman, and Roman periods in central Italy. *The Catholic Reformation* by Pierre Janelle (Milwaukee: Bruce, 1963; London: Herder, 1964) accounts, in detail, the Catholic Church's attempt to reform itself in the face of the Protestant Reformation, and how this reform effort manifested itself in art, in literature, and in the structure of the church itself. Other works that deal with this matter include *The Counter Reformation* by Arthur Geoffrey Dickens (London: Thames and Hudson, and New York: Harcourt, Brace, 1969) and *The Catholic Reformation: Savonarola to Ignatius Loyola* by John C. Olin (New York: Harper, 1969). The origins of the Theatine order are handled in much greater detail in volume ten of the forty-volume *History of the Popes from the Close of the Middle Ages (Geschichte der päpste dem ausgang des mittelalters)* by Ludwig Pastor (Freiburg im Breisgau and St. Louis: Herder, 1891–1924; volumes 1–2, London: Hodges; volumes 3–34, London: Paul, Trench, Trübner; volumes 35–40, London: Routledge and Kegan Paul).

—Rion Klawinski

Citium (Larnaca, Cyprus)

Location: The ancient city of Citium lies buried beneath the modern-day city of Larnaca, in the district of Larnaca on the southeastern coast of Cyprus.

Description: Originally Citium was thought to have been a Phoenician settlement dating back to the eighth century B.C.; recent excavations, however, show the site to have been occupied as early as 1800 B.C. Excavations at the site are difficult, but archaeological and historical evidence shows Citium to have been an important city for a number of successive civilizations, including the Greeks and the Phoenicians.

Site Office: Cyprus Tourism Information
Plateia Vasileos Paviou
Larnaca, Larnaca
Cyprus
(4) 654322

Under the present-day city of Larnaca on the island of Cyprus lies Citium, or Kition, a city of uncertain origin and ethnicity, rumored through legends to have been founded by Khittim, grandson of Noah. Citium's location under a modern city makes excavation difficult, if not impossible. Remains from the Bronze Age, the period during which Citium is now believed to have been populated, generally are found ten to thirteen feet below ground; the foundations dug at Larnaca rarely go deeper than about six feet.

The island of Cyprus is thought to have been inhabited as early as the Neolithic period (from 4000 to 3000 B.C.). The south coast was populated first, as it housed several natural harbors, was sheltered from strong winds, and lies nearer to the lands of Egypt and Palestine. Cyprus was also rich in minerals, contributing to the wealth of the island during the Bronze Age.

The site of Citium lies just inland of a large, protected bay. At the time of its founding, a large marsh separated Citium from the bay, providing the town with a prime harbor spot. Citium shared the southern coast of Cyprus with a number of other cities, among these one of the largest Late Bronze Age towns of the area, that of Hala Sultan Tekké, inhabited from the sixteenth to the twelfth centuries B.C.

Until the mid-twentieth century, archaeologists assumed that Citium was a town founded by the sea-going Phoenicians around the eighth century B.C. This assumption was difficult to disprove owing to the difficulties in excavating at the site previously mentioned. Further support of the Phoenician hypothesis came from the knowledge that Citium was the only ancient city of Cyprus lacking a Greek foundation legend; myth had it that Citium was established instead by Belos, the king of Sidon.

Despite the inherent difficulties of excavation, archaeologists and historians were intrigued by Citium. Digs carried out in the area of Citium in the 1950s have uncovered a number of Early Bronze Age materials. The remains of the city wall, public buildings, and a number of workshops have been found. The wall, built of mud bricks, ran along a natural ridge. Several tombs have been discovered within the soft rock of the city walls. Despite the common occurrence of bronze artifacts in similar tombs near Citium, the graves found on the actual city site contain only pottery. Evidence shows that the tombs were looted in antiquity—the bronze items were likely removed and reused, if for nothing other than scrap metal.

Except for these tombs, no other traces of Early Bronze Age civilization have been found in Citium. This is not surprising, however, when the building methods of the Late Bronze Age peoples are taken into account. Archaeological evidence concludes that the Late Bronze Age settlers in the area razed whatever was left standing of the previous town and then built their own city on the now-cleared site. Proof for this hypothesis has been found in the northern section of the old city, where Early Bronze Age pottery has been found between Late Bronze Age floorings and the bedrock. Thus, despite the absence of an obvious Early Bronze Age settlement at Citium, the excavated tombs indicate that the town was inhabited as early as 1800 B.C.

The remains of those buried in the tombs shed light on this era's Citians, too. All skulls found in the Citium tombs had been deformed in a manner commonly associated with the use of a cradle board—one side of the skull was flattened while the other side bulged out accordingly. The skulls of the females found in the tombs, however, had additional deformations of the cranium: the skulls were flattened at the top. These deformations were probably caused by the women having carried heavy loads on top of their heads. A number of flattened skulls were found alongside valuable jewelry and other artifacts, indicating that the women with these cranial deformations were not of a low class, but rather that this was a common practice among Cypriots of this era.

It is generally believed that the once-prosperous city of Citium was abandoned suddenly in the thirteenth century B.C. The absence of any indication of destruction suggests that the town came to no violent end, but that the citizens left due to other causes—most likely a drought similar to the one that afflicted Greece at the same time. A new population apparently arrived soon, however. Archaeological excavation has uncovered areas of Citium in which the earlier city walls and building foundations were covered with soil approximately ten inches deep. Upon this soil lay the foundations of new buildings, constructed in a completely different architectural fashion.

Ruins of ancient Citium
Photo courtesy of Cyprus Tourism Organization

Stories spun from the time of the Trojan War tell of towns founded on the island of Cyprus by Greek heroes. The fact that both the architectural styles and pottery fragments found atop the early thirteenth-century Citium were entirely new to Cyprus tends to support the hypothesis of Achaean migration to Citium. The pottery found at the site is particularly compelling, as it bears remarkable stylistic similarity to pieces used during the late thirteenth century B.C. in the Aegean, especially the Peloponnese.

The remains of a pair of twin temples dating to the thirteenth or twelfth century B.C. have been unearthed in Citium, along with several nearby workshops constructed for the smelting of copper. The proximity of the workshops to the holy temples initially surprised archaeologists, but further excavation indicated that not only were the workshops near the temple, they were also placed so that they had direct access to the temples through doorways and narrow passageways.

Discoveries at the nearby site of Enkomi have led experts to make a number of assumptions for Citian, and indeed for Cypriot culture of this era as a whole. A number of statues have been discovered, statues standing on bases made in the form of ingots. One statue, a male, stands bearing a shield and spear, and has been termed the "Ingot God," protector of the copper mines of Cyprus. Another statue, this one of a nude goddess, is believed to have symbolized the fertility of Cypriot copper mines. The discovery of these deities, so intimately entwined with copper production, has supported the hypothesis that Cyprus's copper mining and subsequent smelting may have all been carried out under the watchful eye of the temples. This relationship between metallurgy and religion enforced the worship of both a male and a female deity throughout the island of Cyprus, whose economic well-being rested primarily upon the production of copper.

The temples at Citium were destroyed in the early twelfth century B.C., most likely during an invasion by the Sea Peoples. These people wrought havoc throughout the Mediterranean, antagonizing Egypt, destroying the Hittites, and razing a number of cities on Cyprus. The fate of these conquerors is not precisely known, although some claim the Sea Peoples merged with the Phoenicians, and subsequently ruled the region for several centuries.

Despite destruction at the hands of the Sea Peoples, Citium revived as Achaeans returned to Cyprus that same century. Interestingly enough, the exact same pattern of life, destruction, and rebirth is echoed at Enkomi. The newly settled Achaeans rebuilt Citium following the foundations of previous constructions, maintaining the prominence of the temples and copper workshops, indicating their continued value to the settlement. Pottery shards found in the strata associated with this time period include pieces in the Aegean style that were clearly made on the island of Cyprus, supporting the hypothesis of Citium's recolonization by the Achaeans after the attack by the Sea Peoples.

Archaeologists believe that the strata associated with this Aegean pottery corresponds to the period of the so-called Dorian invasion in Greece—actually an era from the early twelfth to middle eleventh centuries B.C., during which Peloponnesians sought refuge and better lives throughout the eastern Mediterranean. Again, the discovery of bits of pottery corroborates the theory of Achaean immigration, as other shards found in this strata bear striking similarities to pottery found in the buried layers of the citadel of Mycenae.

After the rebuilding of Citium by the Achaeans, the city suffered destruction once again. This time, experts believe, an earthquake in approximately 1075 B.C. leveled both Citium and a number of other Late Bronze Age towns on Cyprus. Following the now centuries-old tradition, Citium was once again rebuilt, but this time on a much larger scale. Archaeological evidence indicates that although Citium was rebuilt very soon after the devastating earthquake, other Cypriot towns of the same era, such as Enkomi, were not. The damage caused by the earthquake must have been considerable, perhaps even to the extent that harbors crucial to the seaside towns began to silt up.

But despite the grand reconstruction of Citium and its temples, the city appears to have been abandoned shortly thereafter. No items are found in the strata of this rebuilt portion of Citium that date later than 1000 B.C. No evidence of metallurgical activities—whether industrial or religious in nature—has been found either. The reason for the abandonment of Citium is uncertain, although it too can probably be attributed to the damage of the earlier earthquake. As with Enkomi, Citium's inner harbor and channel linking the town to the sea probably silted up over the years, causing the inhabitants of the town to relocate.

After years of building and rebuilding by the Greeks, Citium at last lay vacant. However, in keeping with past tradition, this vacancy was of short duration. At some point, still uncertain in the minds of both historians and archaeologists, the expanding empire of the Phoenicians extended its control over the remains of Citium, building it again from rubble into a prosperous city.

The Phoenicians may have reached Cyprus, exerting their influence on the cities and towns, as early as the twelfth century B.C. The earliest Phoenician pottery found on the island, however, dates only to the ninth century B.C. The earliest settlement of the Phoenicians located near Citium at Bamboula Hill, however, appears to have begun around 1000 B.C. While it may be impossible to determine the exact period in which the Phoenicians arrived at Citium, it is quite certain that they were there, and that they built for themselves a mighty city.

As the Achaeans had done, the Phoenicians constructed their city on the foundations of past ruins. The most remarkable building of this era was the reconstruction of the temple at Citium. This time, the temple was not built to serve as a place for both worship and copper smelting, but as the holy temple of Astarte, a Phoenician goddess similar in role to the Aphrodite of the Greeks. Astarte's temple in Citium was undoubtedly an important one, thought to be as impressive as Aphrodite's temple at Paphos, also on Cyprus, and probably known throughout the region.

The temple is thought to have been destroyed by fire sometime after its construction in the ninth century B.C. Ash-laden strata taken from the temple floor would appear to enforce this hypothesis. Owing to its importance, Astarte's temple was quickly rebuilt and became even wealthier and more important than before. During this period, from 800 to 600 B.C., the king of Citium, at that time called Khardihadast, paid tribute to Tyre, while the king of Tyre, Hiram II, paid tribute to King Tiglath-pileser III of Assyria. By the late seventh century B.C., Cyprus was occupied by the Assyrians. Sargon II erected a stele at Citium, boasting of his domination of the "seven kings of Yatnan," the Assyrian name for Cyprus.

Despite occupation by a number of lords, trade at Citium flourished. A number of pottery fragments found at Citium are thought to have been brought to the island from Athens. Other articles found in Astarte's temple are Egyptian in design. In addition to being active commercially, Citium was quite possibly made the headquarters for the Assyrian navy, thus taking advantage of the Phoenician expertise of sailing and helping to keep all of Cyprus loyal to Assyria. Assyria eventually became part of the Persian Empire.

While Persian rule of Cyprus was accepted by Phoenician Citium, other predominantly Greek Cypriot towns were not so tolerant. Athens supported Cyprus's struggle for independence and in the mid-fourth century B.C. sent General Cimon to assist the Cypriot freedom fighters. Cimon was killed shortly thereafter while attacking Citium, and Greece made peace with Persia, signing what became known as the Peace of Kallias.

The fighting resumed half a century later, and Cyprus was finally captured by the Athenians in 388 B.C. Greek rule of Citium was brief, however, as the treaty known as the Peace of Antalcidas, signed in 386 B.C., returned Cyprus to Persian hands.

Macedonian expansion under Alexander the Great in the late fourth century B.C. ended Persian rule of Cyprus. In the chaos after Alexander's death, Cyprus became the battleground for regional control by Antigonus and Ptolemy of Egypt. Citium's King Pumiathon aligned himself with Antigonus, and the town was subsequently attacked and conquered by Ptolemy and his allies. Once again, this historic town was razed. During this chaotic time, Citium's most famous citizen was born. Zeno (335–263 B.C.), founder of the famous Stoic school of Greek philosophy, was born at Citium of Phoenician lineage.

Citium never fully recovered from the blow dealt it by Ptolemy, and although the city was again rebuilt, it never regained its former splendor and importance. The town continued to suffer the effects of earthquakes, and by medieval times the once strategic harbor had silted up. The citizens of Citium eventually built the new city of Larnaca.

Despite the cycle of destruction going back to the Bronze Age, the worst devastation to Citium came in 1879 at the hands of the British during their occupation of Cyprus. Eager to dry up marshy lands causing malaria epidemics, the Royal Engineers were ordered to fill the marshes using a "mound of rubbish" that lay nearby. That "mound," then known as Bamboula Hill, was the Acropolis of Citium, the single most important, prominent, and historic site of the entire city. The officer in charge of the carnage submitted a report on the proceedings. It is only from this report speaking of fine masonry and broken pottery that archaeologists may determine what has been lost.

Further Reading: Vassos Karageorghis's *Kition: Mycenaean and Phoenician Discoveries in Cyprus* (London: Thames and Hudson, 1976; as *View from the Bronze Age: Mycenaean Phoenician Discoveries at Kition*, New York: Dutton, 1976) is a detailed look at the archaeological findings at the site. Karageorghis, a native of Cyprus, uses the information gleaned from years of research to reconstruct life at Citium throughout the Bronze Age. Wilson E. Strand's *Voices of Stone: The History of Ancient Cyprus* (Nicosia: Zavallis, 1974) provides a similar look at Citium and the island of Cyprus as a whole. Readers looking to visit the ancient Citium will find useful information in both Ian Robertson's *Blue Guide: Cyprus* (London: Black, and New York: Norton, 1990) and Robert Bulmer's *Passport's Illustrated Travel Guide to Cyprus* (Lincolnwood, Illinois: Passport, 1993).

—Monica Cable

Cnidus (Muğla, Turkey)

Location: On Cape Krio at the tip of the Datça peninsula of southwestern Turkey, where the Aegean Sea meets the Mediterranean, about twenty-four miles west of the town of Datça.

Description: Ruins of ancient city noted for its commerce and scholarship; once the site of the famous statue of Aphrodite by the sculptor Praxiteles, one of the greatest examples of classical sculpture.

Contact: Tourist Information Office
Hükümet Binasi, Iskele Mah.
Datça, Muğla
Turkey
(252) 712546 or 7123163

Cnidus has been deserted since the seventh century A.D., its ruins have been plundered repeatedly, and its greatest treasure, Praxiteles' statue of Aphrodite, has long been lost. At one time, however, it was among the wealthiest and most prominent cities on the west coast of Anatolia and was a center of culture and learning.

Cnidus—alternately spelled Knidos—was founded by Greek settlers, perhaps as early as 900 B.C.; one account attributes the settlement to Dorians, another to colonists from Argos. The original site of Cnidus was near the present-day village of Datça, farther east on the peninsula than the later site. Only fragments of fortification walls remain of the original settlement. It was a prosperous community, however; the residents of Cnidus exploited the peninsula's sea access, trading with Egypt and establishing colonies on Sicily and the island of Lipari. As with many other wealthy cities, it had its own building at the shrine to Apollo at Delphi, containing a treasury of artworks and other gifts to the god. By the sixth century Cnidus had become one of the most prominent cities in the Greek world. It also was a center for worship of Apollo by Cnidus and its fellow members of the Dorian Hexapolis, a federation of Greek settlements in the region.

In the mid–sixth century Persia began its conquest of the Greek cities in the vicinity of Cnidus. The Cnidians attempted to cut off the Persian invasion by literally cutting off their peninsula from the mainland; they dug a deep channel at the peninsula's narrowest point. Many of those who worked on the channel were injured by flying shards of rock. Taking the injuries as a sign that perhaps the project was not a good idea, the Cnidians asked the Delphic oracle if they should continue. The oracle's pronouncement was that they should abandon it: "Zeus would have made an island if he wished." The Cnidians ceased digging and surrendered their city to the Persians.

Greek cities' rebellion against Persian rule, and Persia's attempt to put down the rebellion, resulted in the Persian Wars in the early fifth century. Cnidus, along with most other Greek settlements in Asia, joined the Athenian-led Delian League to fight the Persians. The Athenian admiral Cimon based his fleet at Cnidus and sailed from there to defeat the Persians at the battle of the Eurymedon River in 468 B.C. Cnidus remained allied with Athens for several years, but when Athens and another ally—Sparta—had a falling-out that resulted in the Peloponnesian War (431–404 B.C.), Cnidus was among the cities that supported Sparta. Sparta won the war and became the preeminent power in Greece. Within a decade, however, Corinth, Athens, and other leading cities rebelled against Spartan domination. This rebellion culminated in the Corinthian War, in which a key naval battle was fought off the coast of Cnidus in 394 B.C. The Athenian admiral Conon, commanding a fleet from Persia—which had come in on the side of the rebels—decisively defeated the Spartans in the battle, which effectively ended Spartan rule in Asia Minor. Under the terms of the King's Peace, the treaty ending the war, Cnidus was returned to Persian rule in 386 B.C.

Despite the disputes over its possession, Cnidus remained extremely prosperous. With their sea trade flourishing to an almost unmanageable degree, the Cnidians built a new city at Cape Krio on the tip of the Datça peninsula about 360 B.C. This new location offered the benefit of two natural harbors—one opening to the Aegean, one to the Mediterranean—so that the city could handle an increased amount of business. The new site had another advantage: when unfavorable winds kept ships from being able to round the cape, the sailors had to wait for days there—and while waiting, they would spend a good deal of money in the city. The city expanded onto a nearby island, which the Cnidians joined to the mainland by constructing a causeway.

Soon after moving to the new site, Cnidus obtained the work of art that brought it lasting fame. Around 360 the Athenian sculptor Praxiteles, regarded as the greatest of his time, produced two statues of Aphrodite, goddess of love—one showing her clothed, one portraying her nude. He sold the clothed one to the city of Kos, the nude one to Cnidus. Some accounts say the nude one was intended for Kos as well, but the people of Kos found it too shocking to accept. The nude, for which Praxiteles had used his mistress as the model, was remarkably realistic; it was one of the first to exploit the fleshlike appearance of marble—previously, bronze had been the preferred material for representations of the human form—and it became the standard against which portrayals of female beauty were measured. The goddess was shown in a demure yet alluring pose. "Aphrodite's expression," writes historian Michael Grant, "grave and calm, was said to convey

Foundations of the Aphrodite temple
Photo courtesy of Embassy of Turkey, Washington, D.C.

a hint of invitation, and her gesture concealing the genital area drew erotic attention to it all the same."

Cnidus made the statue one of the leading tourist attractions of the era. It was displayed in a temple on a steep hill above the two harbors; the statue was placed so that it could be viewed from any direction. Travelers came to Cnidus just to see the statue and present gifts to the goddess. Some tried to buy the statue, one admirer even offering to pay all the city's debts in exchange for it. The Cnidians, however, preferred to hold on to their treasure.

The statue, like most of the work of Praxiteles, has disappeared. (The only Praxiteles sculpture to have survived is one of the god Hermes holding the infant Dionysus, and there is even some disagreement as to whether this is an original or an early, particularly fine copy.) The Aphrodite of Cnidus was most likely destroyed by zealous early Christians who not only disapproved of paganism but feared the cult of Aphrodite encouraged immoral sexual behavior. Several copies of the Aphrodite statue were produced in the Roman era, however, and these are displayed in various museums—including, ironically, one at the Vatican, considered to be the best copy. Images of Aphrodite on Cnidian coins also provide an idea of the statue's appearance. Twentieth-century Amer-

ican archaelogists have found the remains of the Aphrodite temple and what they believe to be the plinth on which the statue stood.

Cnidus's other claims to fame at this time included its medical school and its wine. The medical school developed about the same time as that on the Greek island of Kos, just a short voyage from Cnidus, and produced many renowned doctors, including Ctesias, who practiced at the royal court in Persia. Cnidian wine was reputed to be useful in the healing arts as well; it was believed to improve digestion and enrich the blood. It was one of the major exports from Cnidus's harbors. Other exports included vinegar, onions, cabbage, medicinal oils, and reeds for pens.

One of the most accomplished natives of the city was the mathematician, astronomer, and philosopher Eudoxus of Cnidus. Born about 400 B.C., he studied in Athens (with Plato) and in Egypt, then traveled widely before settling in Athens. He developed methods of calculating areas and volumes, and set forth theories to explain the movement of the planets. His mathematical work influenced Euclid; his astronomical efforts laid the foundation for Ptolemy's work. Another Cnidus native was the architect Sostratus, who in approximately 279 B.C. built the Pharos, a huge lighthouse in

Alexandria, Egypt, one of the seven wonders of the ancient world.

Alexander the Great drove the Persians out of Cnidus in 334 B.C. The city's ownership was contested after his death in 323. It came for a time under the rule of Egypt's Ptolemaic dynasty (who commissioned the Pharos) and for a while was under Rhodes. It assisted Rome in fighting King Antiochus III of Syria; the Romans had warned Antiochus not to invade Europe, but he did so just the same. Cnidus's friendly relationship with Rome meant that it was peacefully incorporated into the empire in the second century B.C. It received the status of a free city within the empire, and the Romans carried out numerous construction projects in Cnidus. After Christianity became the religion of the empire, Cnidus became an episcopal see, and with the division of the empire it became part of the Byzantine realm. Ruins of several Byzantine-era churches are in the vicinity.

Cnidus's sea access, which had been the source of its greatest glory, also was the source of its downfall. It was extremely vulnerable to pirate raids, which caused the inhabitants to abandon their city in the seventh century A.D.

Cnidus was not entirely forgotten, however. Its remains were plundered for a variety of building projects, including a residence for the nineteenth-century Egyptian ruler Muḥammad ʿAli Pasha. In the eighteenth and nineteenth centuries, a growing interest in classical lands led European intellectuals to travel to Cnidus and write descriptions of it. The first formal excavations of the site were carried out by the Englishman Charles T. Newton in 1857. He took several treasures away from Cnidus, including a magnificent statue of Demeter, the earth goddess, who was venerated by Cnidians to the same degree as Aphrodite. The statue is now in the British Museum. In the twentieth century, the most significant work at the site has been done by Long Island University under the direction of Iris Cornelia Love.

Buildings that have been uncovered at Cnidus, in addition to the temple of Aphrodite, include a shrine dedicated to Demeter and her daughter, Kore, constructed about 330 B.C.; a theatre with portions dating from Roman times and some from an earlier period; a Roman tomb; a lighthouse; courtyards; private houses; and portions of the city walls. These can be viewed on site; some artifacts from Cnidus have been taken to museums at Bodrum and İzmir. Still, it is the lost treasure of Cnidus, the Aphrodite of Praxiteles, for which the city is best known. It has inspired writers and poets into the twentieth century. After viewing a copy at the Metropolitan Museum of Art in New York, the poet Robert Francis pondered the subject in "Aphrodite as History":

> Though the marble is ancient
> It is only an ancient
> Copy and though the lost
> Original was still more ancient
> Still it was not Praxiteles
> Only a follower of Praxiteles
> And Praxiteles was not the first.

Further Reading: Cnidus and other ancient Turkish sites are explored in *Ancient Civilizations and Ruins of Turkey from Prehistoric Times until the End of the Roman Empire* by Ekrem Akurgal, translated by John Whybrow and Mollie Emre (İstanbul: Mobil Oil Türk A.S., 1969; fourth edition, İstanbul: Haset Kitabevi, 1978). Michael Grant's *A Short History of Classical Civilization* (London: Weidenfeld and Nicolson, 1991) provides a context in which to view the development of Cnidus. Also of interest is a series of articles by Iris Cornelia Love, all entitled "Preliminary Reports of the Excavations at Knidos, 1969–1971," published in the *American Journal of Archaeology* (New York), volume 74, 1970, pages 149–55; and volume 76, 1972, pages 61–76 and 393–405.

—Trudy Ring

Coimbra (Coimbra, Portugal)

Location: Along the Mondego River in west-central Portugal, about 120 miles north-northeast of Lisbon. Coimbra is in the Beira historic province and is the administrative center of the Coimbra district.

Description: An ancient city known for its medieval university and picturesque setting on a hillside along the Mondego. With a population of about 75,000 people, it is Portugal's fifth largest city. Coimbra has few industries, and the university, which enrolls about 13,000 students, forms the mainstay of the local economy. The city has many historic structures. Especially noteworthy are some of the older university buildings, particularly the magnificent baroque library built in the eighteenth century.

Site Office: Região de Turismo do Centro
Largo da Portagem
3000 Coimbra, Coimbra
Portugal
(39) 33028 or 33019

Coimbra has been endowed with a rich historical legacy. The city is the birthplace of six kings, and its ancient monastery of Santa Cruz houses the tombs of Portugal's first two kings. The first of these, Afonso Henriques, established Coimbra as the capital of the nascent Portuguese kingdom in 1139. The capital remained at Coimbra until 1260, when it was transferred to Lisbon. Much of Coimbra's heritage derives from the history of its university, which was permanently established in the city in 1537. Since then, Portuguese rulers have nurtured the development of the university. The list of those who have studied at Coimbra includes many members of Portugal's political and intellectual elite, such as Portugal's preeminent poet, Luis de Camões (1525–1579); Antonio Egas Moniz, who won the Nobel Prize in medicine in 1949; and Antonio Salazar, premier of Portugal from 1932 to 1968. The University of Coimbra had a virtual monopoly on higher education in Portugal until the twentieth century, thus assuring the city's national prominence.

Archaeological evidence indicates that the Mondego River valley has been inhabited for at least 5,000 years. Artifacts dating to Europe's Neolithic age, about 5,000 to 7,000 years ago, have been unearthed near the modern city of Coimbra. Settlements arose in the region from the ninth to seventh centuries B.C., during Europe's Iron Age.

Nine miles south of Coimbra lies Conimbriga, site of a major Roman settlement. Conimbriga has a close historical link with Coimbra, for it is from this settlement that Coimbra derives its modern name. Before the Romans arrived in the second century B.C., Conimbriga was the site of an ancient Lusitanian settlement. Some scholars believe that it was settled by the Conii, a pre-Celtic Lusitanian tribe.

Roman legions under Decimus Junius Brutus Callaicus captured Conimbriga in 138 B.C. The city later became linked by a Roman road to Olissipo (Lisbon) in the south and Augusta Braccara (Braga) in the north. This road also passed through Aeminium, a Roman settlement on the site of the modern city of Coimbra. Little is known about Aeminium, largely because archaeological research at Coimbra, unlike that done at the Conimbriga site, has been hindered by the area's urban landscape and population density. Excavations at Conimbriga, however, have yielded enough evidence to indicate that the city was quite prosperous. Conimbriga had a forum, a temple, an aqueduct, an extensive bathing complex, and palatial Roman manor houses.

Conimbriga reached its zenith during the second and third centuries A.D. Germanic tribes began periodically attacking the city in the late third century. Roman administration wavered, and in 468 Suevian tribesmen under Remismund sacked Conimbriga. The devastation wrought by the Suevi was so great that many of the city's surviving inhabitants evacuated the city and sought refuge in Aeminium. The birth of modern Coimbra and the decline of the affluent Roman settlement to the south dates from this episode. In about 589 Aeminium changed its name to Coimbra. By the end of the sixth century, the Visigoths had displaced the Suevi throughout the entire Iberian Peninsula. Also by this period, Coimbra had been made a bishopric, evidence of its growing stature.

The Moslem invasion of the Iberian Peninsula began in 711. The Moors advanced northward and occupied Coimbra sometime between 712 and 713. During the Portuguese reconquest era, 850 to 1249, Coimbra was on the frontier between Moslem and Christian realms, and possession of the city shifted between the two. The Moors held the city until 878, when Christian forces under Alfonso III of Asturias retook it. Christian control ended in 987, when Almanzor sacked the town. The city was virtually destroyed, remaining unoccupied and in ruins for several years until it was rebuilt by the Moors. In 1064, after a six-month siege, Ferdinand of Castile recaptured Coimbra for good. The Moors, however, continued to pose a threat and nearly recaptured the city in 1117. Today, evidence of the Moorish occupation of Coimbra can be seen in the remnants of the walls surrounding the upper town and one of its gates, the Almedina.

By the end of the eleventh century, Henry of Burgundy, the son-in-law of Alfonso VI of León-Castile, had acquired a fiefdom extending from the Minho River in the north to the Moslem frontier south of Coimbra. Vassalage to

Coimbra's university

Ruins of Roman spa near Coimbra
Photos courtesy of Portuguese National Tourist Office

the Spanish sovereign ended under Henry's son, Afonso Henriques, who in 1139 established an independent Portuguese kingdom with its capital at Coimbra. The city prospered as a well-populated mercantile center during the twelfth century. Two of its oldest surviving structures were built during this century: the monastery of Santa Cruz (built from 1131 to 1154), and the Old Cathedral (completed in the 1170s).

The reconquest of Portugal was completed with the recapture of the Algarve in 1249. Administration of the kingdom was moved from Coimbra to Lisbon in 1260. Despite losing political importance, Coimbra remained a significant cultural and religious center, largely due to the prominence of the monastery of Santa Cruz. In 1288 the prior of Santa Cruz, along with other religious leaders in Portugal, petitioned Pope Nicholas IV to permit the establishment of a *studium generale* (university) in Portugal. In 1290 the pope issued a bull granting permission, and King Denis established a university in Lisbon. The university encountered several problems, including the hostility of the people of Lisbon toward the students, not an uncommon phenomenon for medieval universities. In 1308 the university was transferred to Coimbra, where it remained until 1338, when it was moved back to Lisbon. In 1354 Coimbra once again became the seat but lost it to Lisbon again in 1377, where it was to remain until the sixteenth century.

After the last transfer of the university to Lisbon, Coimbra's prominence faded somewhat. Like many European cities, it was ravaged by the plague during the late 1340s and 1350s. The only major institution established during this century besides the university was Santa Clara Convent, founded by Queen Isabel, wife of King Denis. It was at Santa Clara that in 1325, after her husband's death, Queen Isabel retired. After her death eleven years later, she was entombed there. Isabel, noted as a peacemaker in a time of dynastic strife, was later canonized as St. Elizabeth of Portugal. She maintains a special place in the heritage of Coimbra as the patron saint of the city. Because of repeated flooding in the old Convent of Santa Clara, St. Elizabeth's tomb was moved to the new Convent of Santa Clara in the seventeenth century.

Coimbra regained its greatness as a center of culture and vitality during the sixteenth century. This was largely due to King John III, who ruled from 1521 to 1557. John brought Portugal to new heights in the late Renaissance. In 1537 he moved the seat of Portugal's chief university back to Coimbra for good. In fact, the king donated the royal palace of Alcaçova for the purpose of housing classrooms. The university was staffed by Portuguese scholars as well as scholars from other European countries, who were encouraged to come to Coimbra to teach. The city soon became an important center of Humanism. Along with the university, a school of sculpture was established in Coimbra in the 1530s. Most of the sculptors came from Normandy. Especially noteworthy was Jean de Rouen, whose work included an altarpiece in the chapel of the Old Cathedral. The Coimbra school of sculpture was responsible for significant improvements made to the

monastery of Santa Cruz that included its portico, a pulpit, and a new edifice.

The sixteenth century also brought with it the Counter-Reformation and its chief instrument, the Holy Office of the Inquisition, established in Portugal in 1536 by King John III. During the following decades, three main centers of the inquisition emerged in Portugal: Lisbon, Coimbra, and Évora. One of the victims of the inquisition in Coimbra was the university. Specifically targeted was the College of Arts, established as an independent entity in 1548 but incorporated into the university shortly afterward. Teachers were brought in from Paris and Bordeaux, and the college became renowned as a center of the arts. Officials of the Portuguese inquisition soon became suspicious of this liberal center of learning, and in 1558 the college was handed over to the Jesuits, the guardians of orthodoxy.

Despite the inquisition, or perhaps because of it, scholarship in Aristotelian philosophy and Thomist theology flourished in the latter part of the sixteenth century in Coimbra. A group of scholars at the College of Arts known as the Conimbricenses became famous throughout the continent. Their comments on Aristotle, published in eight volumes between 1592 and 1606, had a significant influence among the learned circles of Europe.

During the sixteenth century, the population of Coimbra grew from between 5,000 and 7,000 in 1532 to between 16,000 and 20,000 by the 1620s. With growth came urban development and new construction. Streets were widened, and at the close of the sixteenth century an aqueduct was built to improve the city's supply of water. Also during this period construction of a new cathedral and an episcopal palace began. Building continued apace in the eighteenth century, financed largely by the treasure arriving from the Portuguese overseas empire in Brazil. King John V was a great patron of the university. Under his direction the opulent baroque library was constructed between 1717 and 1728. Also built during this period was the distinctive clock tower of the university, which would come to be one of the city's chief landmarks.

In contrast to the improvement of the university's physical stature, its prestige as an institution of higher learning gradually declined. Despite the advances brought by the scientific revolution in the eighteenth century, the university's curriculum remained unchanged, harkening back to the mid-sixteenth century. The backwardness of higher education in Portugal was noted by many affluent Portuguese who had traveled and studied throughout Europe. Major reforms of the university's curriculum occurred during the reign of Joseph I, 1750–77. The architect of the reforms was Joseph's chief minister, Carvalho de Mello, the Marquis of Pombal. In 1772 Pombal instituted a wide variety of changes, emphasizing programs in the natural sciences and mathematics. Laboratories were added for physics and chemistry. Also added were an astronomical observatory and botanic gardens.

The opening of the nineteenth century brought war to the university town astride the Mondego. In 1807 Napoléon's armies, with the cooperation of the Spanish, invaded Portugal. The following year Britain, long the traditional ally of the Portuguese, landed troops at the mouth of the Mondego under General Arthur Wellesley (later to become Duke of Wellington). Coimbra became a base for Wellesley's troops. In the following year Coimbra was captured by French forces under General André Masséna, but was retaken shortly afterward.

Coimbra, especially the university, was deeply involved in the political changes occurring in Portugal during the early twentieth century. In 1903 students at the university joined with workers in a strike protesting labor conditions. The students struck again in 1908, this time in a campaign against the monarchy. In 1910 the Portuguese monarchy fell and the country became a republic. The Portuguese Republic proved to be a complete failure. From 1910 until 1926, when it fell, the republic had forty-five different governments and only one president who had completed his four-year term.

Meanwhile, in 1910 Antonio Salazar enrolled at the University of Coimbra. He completed his doctoral studies in finance and economics at the university in 1918. He had begun teaching at the university as a doctoral student in 1916 and continued instructing until 1928. As the republican experiment foundered in Portugal, opposition to the regime emerged at the university. In 1926 the republic fell and Portugal came under a military dictatorship. By 1932 Salazar had seized the reins of power as premier.

During the 1940s, under the direction of Salazar's government, much of Coimbra's upper town was reconstructed to give it a more modern appearance. Among the casualties of this program were several of its ancient streets and many buildings of historical and artistic value, including some university buildings. New utilitarian university buildings, lacking the charm of the older structures, were constructed. Among these were buildings for the Faculty of Arts, Medicine, and Science. The value of the reconstruction program has remained questionable to this day.

Because much of the city's prestige was based upon its university and the monopoly it held, Coimbra's greatness began to decline slowly after the founding of new universities in Portugal. The city's historical importance, however, will never fade. From its time as a Roman settlement to its apogee in the late Renaissance to the growth and changes of the modern and contemporary periods, the city has figured prominently in Portuguese history.

Further Reading: Most books on Portuguese history will have some information on Coimbra. Among the outstanding histories that include detailed treatment of the city are A. H. De Oliveira Marques's two-volume *History of Portugal* (New York and London: Columbia University Press, 1972) and H. V. Livermore's *A New History of Portugal* (London and New York: Cambridge University Press, 1966). Both provide readers with a thorough index, and Marques's index is annotated. For a general overview of Portuguese history, the introduction found in Douglas L. Wheeler's *Historical Dictionary of Portugal* (Metuchen, New Jersey, and London: Scarecrow, 1993) is invaluable. Specific information about the University of Coimbra is best sought in books concerning the history of universities in Europe. Although rather dated, Hastings Rashdall's multivolume *The Universities of Europe in the Middle Ages* (Oxford: Clarendon, 1895; revised by F.M. Powice and A.B. Emden, Oxford: Clarendon, 1936; new edition, London: Oxford University Press, 1946) provides a good section on the University of Coimbra. A detailed description of the university as well as a comprehensive historical account can be found in *A Universidade De Coimbra* (published through the office of the rector of the University of Coimbra, 1988). This book, written by several professors at the university, includes many splendid color photographs that enhance the text, which is provided in Portuguese, French, English, and German. There are a great many travel books that have adequate information on Coimbra. Among the more interesting and informative are Nick Timmons' *Off the Beaten Track, Portugal* (Ashbourne, Derbyshire: Moorland, and Old Saybrook, Connecticut: Globe Pequot, 1994), and *Birnbaum's Portugal 1993* (New York: HarperCollins, 1992) edited by Alexandra Mayes Birnbaum.

—Jeffrey M. Tegge

Córdoba (Córdoba, Spain)

Location: Southwest Spain, on the Guadalquivir River, on the road from Madrid to Seville; about 130 miles northeast of the Atlantic and 100 miles north of the Mediterranean; in the province of Córdoba and region of Andalusia.

Description: An ancient city that has evidence of Roman occupation together with an exceptional town center of about a square mile that has changed little from the period of Moorish domination, during which Córdoba was for a time the capital of Spain. Attractions include a massive mosque, one of the finest in Islam, which received the addition of a full-scale Christian cathedral in the sixteenth century; La Plaza de la Corredera (Colt Square), which has survived virtually intact from the seventeenth century; a Roman bridge, and extensive Moslem and Christian medieval fortifications.

Site Office: Tourist Information Centre
Córdoba City Hall
Judá Leví Square
Córdoba, Córdoba
Spain
(957) 47 20 00

Córdoba is situated in southern Spain on the Guadalquivir River, some 130 miles from the Atlantic coast. The river is no longer suitable for navigation, but was an important source of income before the tenth century A.D. Ruined mills left useless by the fall in the water level are visible in the center of the river. To the south of the city lies a fertile plain, long an agricultural center. A few miles to the north the Sierra Morena mountain range begins, and Córdoba is often seen as being the link between the sierra and plain areas. Like most southern Spanish cities, Córdoba has been inhabited for well over 2,000 years. It was important in Roman times, but reached particular splendor during the period of Moslem rule, when it was the capital of Spain. Many of the buildings constructed at this time still exist, as do nearly all the substantial Renaissance and baroque contributions to the city's architectural heritage. Evidence of settlement in the fifth century B.C. has been found in nearby hills: the Turdetans probably occupied the area until the middle of the second century B.C. These early inhabitants may have given the city its name, which appears to have either a Turdetan or Iberian root.

The outcome of the Second Punic War (218 to 201 B.C.), in which Rome emerged victorious in its battle with Carthage for control of the Mediterranean region, resulted in Roman domination of Spain, and the rise of Córdoba as a city: in all probability it was founded by the general Claudius Marcelius sometime between 171 and 152 B.C. The city was used as a base in operations against Viriathus, the native leader who directed resistance against the Romans in what is now the Estremedura region. As a result the city became one of the wealthiest in southern Spain, rivaling cities such as Hispalis (Seville) and Gades (Cádiz).

Part of the prolonged civil war that destabilized Rome in the first century B.C. took place in southern Spain. The people of Córdoba backed Pompey's side, and later that of his sons, but with their defeat nearby in the battle of Munda in 45 B.C. the Córdobans suffered the wrath of Julius Caesar. Some sources speak, probably with exaggeration, of 20,000 executions and mass enslavement after the battle. With the end of the civil war and the triumph of Augustus, Roman influence spread north through the Iberian Peninsula. With the growth came administrative reforms. Córdoba became capital of the new southern province known as "Betica" (or "Baetica") after the Guadalquivir River, then called "Betis." The city was given the title of "Colonia Patricia."

Córdoba has ample evidence of the peace and prosperity that the Roman world enjoyed in the first and second centuries A.D. A large Roman temple has been partially reconstructed next to the Town Hall; the Roman bridge in Córdoba is still used; and three Roman cemeteries have been discovered on the edge of the old city. Córdoba had shrines to the gods Attis and Cybele; these were probably used in ceremonies involving bulls—these ceremonies being the predecessors of modern bullfighting. Córdoba also gave the Rome of the first century A.D. two of its greatest thinkers: the philosopher and writer Seneca the Younger, who was forced to commit suicide by the Emperor Nero in 65 A.D.; and the poet Lucan, nephew of Seneca, who also took his life in the same year. In the later Roman period, San Acisclo and Santa Victoria were victims of Diocletian's oppression of Christians in the early century: they later became the patron saints of Córdoba. Emperor Constantine the Great, whose Edict of Milan in 313 ended the persecution of Christians, was advised by Hosius, who was also Córdoban. But the decline of the empire was reflected in the fortune of Córdoba. The capital was transferred to Hispalis (Seville) in the fourth century. With the collapse of Rome, southern Spain fell under the domination of the Byzantine Empire and the Vandals before being taken by another Germanic tribe, the Visigoths, in 571. Shortly afterward Córdoba backed the Visigothic Catholic prince Hermenegild in the rebellion against his father, King Leovigild (who was an Arian Christian). The rebellion was quashed, and Hermenegild martyred.

In 711 the Moslems crossed the Straits of Gibraltar and invaded Spain. Later in the century Córdoba became the chief city in Moslem Spain with the arrival of 'Abd ar-Raḥmān I. He

The Mezquita
Photo courtesy of Instituto Nacional de Promocion del Turismo

had fled Damascus, where his Umayyad family had been overthrown by the 'Abbasids, in 750. Upon arriving in Spain he ousted the governor and in 756 proclaimed himself emir, with his capital at Córdoba. The foundations of the power of al-Andalus, the name given to Moslem Spain, were laid during 'Abd ar-Rahmān I's reign. The most lasting of his achievements was the building of a fabulous mosque, known by the Spanish word *mezquita*. Later additions to the mosque provide evidence of the power of the men who ruled in Córdoba: particularly 'Abd ar-Rahmān II (reigned 822 to 852), 'Abd ar-Rahmān III an-Nāṣir (reigned 912 to 961) and the dictator Abū 'Āmir al-Manṣūr (reigned 978 to 1002). The mosque grew to become a forest of columns and arches; the design may have been inspired by the Roman aqueduct at Mérida.

'Abd ar-Rahmān II had to fight wars on four fronts: against the Gascons of Narbonne; the Franks in what is now Catalonia; the Asturians on the northern, Atlantic coast; and against the Norsemen, who raided the coast and penetrated as far inland as Seville in 844. At home he was faced with the difficult problem posed by Christians who deliberately sought martyrdom by cursing Muhammad in public, thus challenging the Moslem tradition of tolerance toward Jews and Christians as long as they did not engage in public blasphemy. The Council of Córdoba, convened in 852 to deal with the problem, ruled that those who courted martyrdom were guilty of suicide. On the lighter side, 'Abd ar-Rahmān II was a patron of the arts: he brought to Córdoba the Persian musician Ziryab, who is credited with adding the fifth string to the lute.

'Abd ar-Rahmān III came to power in 912, at a time when Córdoban supremacy was challenged by a rebellious local warlord. He successfully countered this and other threats, such as that posed by the Fatimids in North Africa. In 929 he was strong enough to proclaim himself caliph, giving him spiritual authority over all believers in al-Andalus. After this date there was little interference in Spain from the Moslems in the Middle East. 'Abd ar-Rahmān III ruled from a palace above the city. Named Medina Azahara, it was nearly a mile long and half a mile wide, and was held up by 5,000 columns. Not much remains of the building, casting doubt upon the accuracy of the description.

Later in the tenth century al-Manṣūr—"the Conqueror"—usurped the power of the legitimate caliph, Hisham II. Al-Manṣūr was famous for his expeditions against the Christian kingdoms in the north. But al-Manṣūr's insistence that his son should succeed him—when many did not believe al-Manṣūr to be a legitimate ruler—intensified factionalism among the aristocracy of al-Andalus, and within thirty years of his death Moslem Spain had broken up into a group of independent kingdoms, known as taifas. The taifas were unable to withstand the pressure from the Christians of Castile. They turned to tribes from North Africa to help in their fight against the Christians: first the Almoravids (1085 to 1145) and then the Almohads (1160 to 1220s) came to their aid, only to stay on as rulers. Both the Almoravids and the Almohads chose Seville as their capital in Spain.

Nonetheless, Córdoba continued to flourish as a center of learning. The twelfth-century philosopher Ibn Rushd, also known as Averroës, was born and lived in the city, writing commentaries on Aristotle that were to have a profound effect on medieval Christian theology. The Jewish scholar, philosopher, and doctor Moses ben Maimon, known as Maimonides, was born in Córdoba in 1135, but after an outbreak of persecution he went into exile in Egypt, where he became physician to Saladin. The Judería or Jewish quarter, near the mosque in the center of town, has changed little from the time of Maimonides. A small fourteenth-century synagogue can still be visited.

Moslem Córdoba probably had a population in excess of 100,000 people, though some writers have greatly inflated that figure. Its total area may have approached twelve square miles. The defensive ditch built by al-Manṣūr was twelve miles long. In the Moslem era the city became renowned for its silversmithing and leatherwork; these remain important Córdoban products today.

In 1212 the Castilians broke through the Sierra Morena mountain range at the battle of Navas de Tolosa, leaving the Guadalquivir valley open to attack. On June 29, 1236, Córdoba surrendered to the Castilian king Ferdinand III. Residents were obliged to leave their homes, and the city and surrounding lands were repopulated by the Christian conquerors. As in Ferdinand's time, much of Andalusia is still divided into *latifundia,* huge farms owned by absent landlords and tilled by underpaid labor. The system dates from Roman times, was modified by the Arabs, and then was used by Ferdinand as a way of rewarding his generals: for centuries the Castilian nobility lived off lands in Andalusia given to them in the thirteenth century.

After the Christian reconquest, the city of Córdoba went into decline. In the spring and summer of 1349 it was ravaged by the Black Death, then the scourge of Europe. Córdoba was the site of an important agricultural market and was the home of powerful noble families, but was overshadowed politically by Seville and Granada—which remained Moslem until 1492.

The campaign to reconquer Granada did much to revive Córdoba's fortunes. Córdoba was used as a logistical base for the campaign, and for a time was the residence of the Catholic monarchs of Spain, Ferdinand and Isabella. This explains the frequent visits by Christopher Columbus, who was looking for funding for his trans-Atlantic expedition. Columbus's audiences with the king and queen are thought to have taken place in the Alcazar, a medieval fortress that still exists and contains a museum. Much of the medieval city wall can also still be seen.

The wars of reconquest and the subsequent campaigns in Italy brought the Córdoban general Gonzalo Fernández de Córdoba to prominence. Known as the Great Captain, Fernández de Córdoba developed the use of pikemen in military formations that were to revolutionize European warfare. A century later a descendant of the same name became governor of Milan.

Córdoba's mosque had undergone minor alterations and been used for Christian worship since the fall of the city to the Castilians in 1236. In the 1520s, however, the construction of a full-scale cathedral was begun in the middle of the mosque, breaking the effect of the forest of arches: its construction has been condemned by later commentators, and even Holy Roman Emperor Charles V is said to have criticized the work done in his name when he saw it. The cathedral was not finished until the seventeenth century.

In 1683, the Plaza de la Corredera (Colt Square) was constructed to rival the great plazas of Madrid and Salamanca. For more than two centuries the plaza, which has survived virtually intact, was the focus of city life, being used for religious and political celebrations, as well as bullfights and fireworks. Censuses of the time show that the city's population rose from 27,900 in 1530 to 52,247 in 1571; this figure had slumped to 26,330 in 1618. After that the figures for the seventeenth century tell a story of demographic stagnation, exacerbated by the prolonged plagues of 1649–52 and 1676–84: in the latter outbreak 6.5 percent of the population in the southern part of the province died. In 1800 Córdoba's population was estimated at 40,000. Another factor in the decline of Córdoba, and Spain in general, after the sixteenth century was the Córdoban aristocracy's strong support for the "hidalgo code," by which trade or commerce was forbidden to nobles.

Miguel de Cervantes, who describes a Córdoda inn (now housing the city's cultural department) in his novel *Don Quixote,* wrote of the miserable state of Córdoba and its inhabitants, many of whom were reduced to begging. Reports by eighteenth-century travelers convey the same impression of a decaying city. By the nineteenth century, however, the very decadence of southern Spain made it attractive to visitors from northern Europe. The romantics put Córdoba on the tourist map, and they revived the work of the poet Luis de Góngora y Argote (1561–1627), who spent all his life in Córdoba. The narrator's encounter with Carmen in Prosper Mérimée's novel of the same name takes place on the banks of the Guadalquivir in Córdoba. The plot of *Carmen* also points to another phenomenon in the area: the presence of highwaymen. The *bandoleros* flourished during the occupation by the French in 1810, which many of them opposed, and in the years of disruption and econonmic crisis after Napoléon's forces left.

The problem of roadside bandits disappeared with the arrival of the railroad in the area. The line to Seville was opened in 1859, and the route to Málaga was inaugurated six years later. The railroad consolidated Córdoba's position as capital of the surrounding province and assisted in the exploitation of nearby copper mines. But in the nineteenth century agriculture still represented the mainstay of the local economy. The system of large estates or *latifundia* brought with it labor problems, particularly after the sale of church and common lands in the 1830s. These, according to one historian, "rapidly took on the worst characteristics of the feudal *latifundia*." In these conditions it was not surprising that Córdoba should be the scene of the first congress of Mikhail Bakunin's Anarchist International, in 1872, and that the province should experience considerable rural violence and strikes.

Spain's difficulties culminated in the Civil War. The 1936 military uprising that began the war was successful in the city of Córdoba, and part of the province. On July 28, eleven days after the revolt began, a republican column of 5,000 men led by General Miaja advanced upon the city, but turned back when it was discovered that nationalist reinforcements were due to arrive. Miaja was swiftly removed from his post as a result of this decision. The nationalist repression in Córdoba from the beginning of the civil war until the end of 1938 was directed by Colonel Ciriaco Cascajo Ruiz. Between 30,000 and 32,000 people are believed to have been executed at this time, many of them without trial.

During the years that Spain was ruled by dictator Francisco Franco, the city of Córdoba experienced strong population growth: from 56,000 in 1900 it had grown to 135,000 in 1940 and then to 232,000 in 1970. However, the movement from the countryside to large cities is reflected in the dramatic fall in the population of the province from 613,000 to 498,000 in the 1960s alone.

In the twenty years following Franco's death in 1975, Córdoba was the only city in democratic Spain to elect communist mayors. The first such mayor, Julio Anguita, went on to head the Communist Party at a national level. Córdoba's most important contributions to culture in the twentieth century have come from the painter José Antonio Romero de Torres (1879–1930). Romero's nudes can be seen in the museum named after him, and he is even commemorated in a popular local song. Córdoba also produced two of the century's most important bullfighters—Manolete (Manuel Rodríguez Sánchez), who by the time of his death in the ring in 1947 was considered by many to be the greatest "torero" ever, and "El Córdobes," who became bullfighting's first television star in the 1960s. There is a large municipal bullfighting museum in the Jewish quarter of the old city.

Further Reading: José Manuel Cuenca Toribio's *Historia de Córdoba* (Córdoba: Lugue, 1993) provides an overview of the city's history. The Roman presence in the city is covered by Ibañez Castro's *Córdoba Hispano-Romana* (Córdoba: Lugue, 1983). Because of the importance of Moslem Córdoba, life in the city during the early Middle Ages is extensively chronicled by works on Moorish, or Moslem, Spain. Gabriel Jackson's *The Making of Medieval Spain* (London: Thames and Hudson, and New York: Harcourt Brace Jovanovich, 1972) and W. Montgomery Watt's *A History of Islamic Spain* (Edinburgh: Edinburgh University Press, 1965; Chicago: Aldine, 1966) are among the best works published in English. Recent history in Córdoba has not been well covered, although Gerald Brenan's *The Spanish Labyrinth* (Cambridge: Cambridge University Press, and New York: Macmillan, 1943; second edition, 1950) is recognized as the definitive work on the social and economic background to the Civil War, and is particularly strong on rural unrest in Andalusia.

—Richard Bastin

Corinth (Corinth, Greece)

Location: On the north shore of the Isthmus of Corinth, which joins the Greek mainland with the Peloponnese, a large peninsula in southern Greece.

Description: Much of the ruined city's agora, or central marketplace, has been excavated. Among the buildings uncovered are the Temple of Apollo, a theatre, and the fortifications atop nearby Acrocorinthus, an 1,800-foot peak.

Site Office: Corinth Museum
Corinth, Corinth
Greece
(741) 31207

Corinth boasts a rich and at times turbulent history stretching back to approximately 900 B.C. Among the wealthiest cities of ancient Greece, Corinth was an important cultural, trading, and manufacturing center, as well as a major naval power. The city was a participant in the Peloponnesian War, which pitted the allies of Athens against those of Sparta, and in the Corinthian War, which freed Greek cities from Spartan rule. A frequent target of invaders over the centuries, Corinth has been ruled by Romans, Turks, Franks, Venetians, and others. The ruins of ancient Corinth can still be seen near the present-day city that bears the same name.

Corinth's origins reach to the late Stone Age, about 4000 B.C., when the area was settled by a pre–Indo-European people. Its name has the characteristic "-nth" ending found in the language of the Mycenaeans, who lived in the region before the Greeks arrived from the north around 2000 B.C. In Greek myth, Corinth is associated with Sisyphus, the condemned thief who must roll a stone up a hill, only to have it roll down again and again. It is also associated with Bellerophon, the rider of the flying horse Pegasus. Corinthian coins displayed a picture of Pegasus and because of this were nicknamed "colts."

With the conquest of the area by the Dorians in approximately 900 B.C., Corinth's story enters the realm of documented history. Aletes, leader of the Dorians who conquered the city, established a dynasty that ruled Corinth for many years. These early leaders were instrumental in expanding Corinth by merging several neighboring villages to form the city. The Dorian leaders also reduced all non-Dorians to a lowly status in society; they were known as "Wearers of Dogskin Caps."

After several generations, the Corinthian dynasty founded by Aletes became, through intermarriage, a dynasty of the Bacchis family. This family eventually abolished the monarchy and set up its own aristocratic government about 747 B.C. The government was composed of a 200-member assembly, an 80-member council, a steering committee of 8, and a president who was elected each year. All of these branches of the new government were dominated by the Bacchis family and its allies. An early, flattering history of the family was written by the noted Greek poet Eumelus, himself a Bacchis. The Bacchiad rulers were ambitious to increase Corinthian wealth and power, and, through their efforts at colonization and trade, the city enjoyed much of its material success.

Around 658 B.C., Cypselus, a distant relative of the Bacchis family, seized power absolutely and killed a number of his family rivals. To win the goodwill of neighboring cities that might see him as a usurper, Cypselus donated generously to the Olympian games and to the temple at Delphi. He redistributed lands owned by the aristocracy to a number of supporters and, with the subsequent increase of full citizens this produced, organized them into eight tribes determined by territory. He also published a code of laws for the city. Cypselus's son Periander, who ruled Corinth after his death, was responsible for building the diolkos, or boat rampway, across the Isthmus of Corinth.

Because of its strategic position, Corinth played a major role in Greek trade and politics from its earliest days, becoming a commercial center by the sixth century B.C., and establishing several colonies throughout the eastern Mediterranean. These colonies included Syracuse, a city on the eastern coast of Sicily founded in 735 B.C.; Corcyra (Corfu), an island on the northwest coast of the Greek mainland where a colony was founded in 735 B.C.; and Potidaea, a city in northern Greece near Macedonia, established in 609 B.C. The colonies provided Corinth with a steady supply of cattle. They also helped to relieve the pressures of an excess population by drawing away some of Corinth's citizens who hoped for a better life in a new city. So prosperous did Corinth become that Homer refers to "wealthy Corinth" in his *Iliad*.

Although healthy trade relations accounted for a great deal of Corinth's wealth, it was not the only reason for the city's prosperity. Corinth's location played a major role as well. Situated on a central isthmus where much of the area's commercial traffic flowed, travelers from the Greek mainland to the Peloponnese were obliged to pass through Corinthian territory. Corinth's central location also made it a focal point for sea trade between such far-flung points as Syria, Sicily, Egypt, and the Balkans. The isthmus itself, between three and four miles in width, divides the Gulf of Corinth, which leads to the Adriatic Sea, and the Saronic Gulf, which opens onto the Aegean Sea. At one time the Corinthians planned to build a canal across the isthmus and connect the two seas, but this proved to be an impossible project. They built instead the diolkos, a paved stone

Temple of Apollo at Corinth
Photo courtesy of Greek National Tourist Organization

rampway over which boats could be pulled across the itshmus on rollers. By taking this route, Greek merchants could save much traveling time and avoid the ever-present danger of pirates. A toll was charged for this service and the cargoes taxed, practices that added handsomely to Corinth's finances. In times of war or conflict with a rival city, Corinth could close the isthmus to shipping traffic completely, a threat that served to remind the city's rivals of Corinth's power in the region.

Because of its dependence on sea trade, Corinth was obliged to build a powerful naval force to protect its commerce. The city waged a constant sea campaign against pirates preying on ships in the area. In 664 B.C., it led a naval force against the Corcyrians, the first recorded naval war. The three primary Greek warships—the penteconter, bireme, and trireme—were invented in Corinth. In addition to its sea power, the city's land defenses were formidable enough to ward off possible attack. On the western edge of the isthmus was the smaller city of Lechaion, connected to Corinth by common defensive walls, known as the Long Walls, a mile and a half long. These walls were considered an engineering marvel of the time. Atop Acrocorinthus, an 1,800-foot peak

next to the city proper, stood a fortress that was, due to the steep incline, virtually impregnable to military attack.

Corinth was a leader among Greek cities not only in trade but also in the manufacture of many goods, especially those of an artistic nature. Statuettes made of ivory, bronze, and stone were manufactured there, and Corinthian artists were the only ones in Greece to experiment with large terra-cotta sculptures. Pliny the Elder claimed that a Corinthian named Cleanthes invented line drawing, while other traditions say that painting was invented in the city. Corinthian pottery was among the first to be painted by hand. The city's wool, perfumes, vases, and such bronze items as weapons and mirrors were popular items throughout the Greek world. Because of the extensive clay beds in the region, Corinth was able to produce pottery at an early date. For some 200 years, Corinthian clay pottery—colorful, shiny, and decorated with curvilinear designs and the silhouettes of animals—was commonly found as far away as Syria. The peak of the city's manufacturing prominence was about 600 B.C.

Corinth also played a major role in the development of early Greek architecture. A type of ancient Greek column, decorated with an unfolding leaf pattern at the top, is still called Corinthian. At nearby Piraeus, a small city annexed by Corinth in about the eighth century B.C., has been found one of the earliest examples of Greek architecture. A temple dedicated to Hera, this building was constructed by Corinthian architects in the late eighth century B.C. A temple of Apollo, the largest of its kind in the Greek world, was constructed in Corinth about 700 B.C., while at nearby Isthmia was a shrine to Poseidon dating from the same period.

The city's great wealth brought it a dubious reputation as a center of luxurious pleasures and vices. Archaeologists have discovered a Corinthian-style wine jar with explicit love scenes depicted on it, and the Corinthian aryballoi, small oil bottles, were routinely decorated with an abstract design based on the female genitalia. Particularly notorious was the city's Temple of Aphrodite, which housed sacred prostitutes. As elsewhere in the ancient world, the activities of these prostitutes were believed to encourage the fertility of the land. Because of the many sailors and travelers passing through Corinth, its reputation as a center of sin and vice was spread far and wide, perhaps out of proportion to the reality of the situation.

Corinth's growth as a commercial and manufacturing center eventually brought it into a series of debilitating conflicts with its aggressive neighbor Athens, a city on the nearby Greek mainland, which had also become a commercial and naval power. The first of these conflicts occurred in 460 B.C. when Megara, a small state on the isthmus normally allied with Corinth, formed an alliance with Athens. This betrayal spurred Corinth to declare war. In the subsequent clash, Corinth's army and navy were defeated in major battles by superior Athenian forces.

During the 450s, Athens concluded diplomatic alliances, sometimes by threat of force, with a number of cities along the Gulf of Corinth. This Athenian League, as the allied cities were called, began to threaten Corinth's valuable trade with its western colonies. Athens was again openly at odds with Corinth when, in 435 B.C., Corinth and its colony Corcyra (Corfu) came into conflict over Epidamnus (Durrës), a colony that Corcyra had in its turn founded. In a major naval battle, Corcyra's formidable fleet bested the Corinthians, but this defeat was not a decisive end to the war. As Corinth rebuilt its fleet in anticipation of renewing the war and teaching its wayward colony a lesson, Corcyra formed a defensive alliance with Athens in 433 B.C. This alliance led Athens later that year to send warships when Corcyra met the Corinthian navy in battle near the island of Sybota. The Athenian presence saved the Corcyrians from certain defeat. Following this battle, Athens then intervened in a disagreement between Corinth and its colony Potidaea, drawing Corinth's further anger. When Athens cut trade with Megara, which had switched sides and become a Corinthian ally again, Corinth appealed to Sparta for assistance in the growing dispute, a move that precipitated the Peloponnesian War.

The Peloponnesian War lasted off and on for some thirty years, with Greek cities allied with either Athens or Sparta fighting for supremacy over the Hellenic world. The war extended from the gates of Athens, where Spartan armies were a frequent threat, to far-flung colonies in Sicily. Corinth, a member of the Peloponnesian League allied with Sparta, suffered the loss of allies along the Gulf of Corinth as the Athenians captured Olpae, Anactorium, and Oeniadae. Trade, vital to the city's finances, was devastated by the war. In 404 B.C. Sparta finally forced terms on a defeated Athens and the Peloponnesian War came to a close. Following the war, as the defeated cities of the Athenian League were forced into a common alliance, all of Greece came under the domination of the victorious Spartans.

This alliance soon began to rankle some of its members. In 395 B.C., Thebes began a revolt against Spartan dominance in northern Greece. The revolt soon spread to the cities of Argos, Athens, and Corinth. By 394 B.C., this revolt, called the Corinthian War, had freed northern Greece, and Thebes was planning an invasion of the Peloponnese. The battle between Sparta and the rebels centered on Corinth and the isthmus connecting the Peloponnese with the mainland. At one point Corinth's defensive walls and harbors were taken and the city blockaded. The turning point of the siege came when an Athenian captain named Iphicrates and a band of mercenaries attacked a battalion of 600 well-armed Spartans outside the city gates. Using hit-and-run tactics to tire the armored soldiers, the mercenaries then launched a strong attack on the Spartans and killed virtually the entire battalion. For the Spartans, with only six battalions in their army, this was a catastrophic loss. Together with a major loss soon after this to the Persians at Cnidus in Asia Minor, the Spartans became too weak to continue the war. In 387 B.C. cities across Greece—including Corinth—freed themselves from Spartan domination.

In 344 B.C. Corinth again faced a military challenge when its Sicilian colony of Syracuse came under attack by Carthage, an ambitious city in North Africa. Corinth sent an expedition led by Timoleon to save the city. Because he had killed his own brother when the latter had tried to seize power by force, Timoleon was held in suspicion by some Corinthians. But when the expedition to Syracuse routed the Carthaginians and saved the colony, Timoleon won the full esteem of Corinth and the general retired to Sicily to live out his final years as a respected war hero.

In 338 B.C. King Philip II of Macedonia, Alexander the Great's father, conquered Greece and established the Pan-Hellenic League with headquarters at Corinth. Following Philip's assassination, Alexander quickly visited Corinth to quell any thoughts of rebellion. While there, he is said to have met with Diogenes the Cynic, a philosopher who taught a frugal way of life and is famed for wandering Corinth with a lit lamp, searching for one honest man. When confronted by Alexander the Great, who asked the philosopher if there was anything the king could do on his behalf, Diogenes answered, "Only step out of my sunlight."

Following the death of Alexander, Corinth was a prominent member of the Achaean League, a confederation of Greek cities established in 234 B.C. As the Roman Empire spread its influence in the eastern Mediterranean, it came into competition with the Achaean League for dominance. In 198 B.C. the Romans conquered Macedonia, which left Corinth the leading member among the remaining cities of the league. When the league unwisely attacked the Romans with an ill-matched army, it was severely beaten. Worse, this inspired the Romans to take vengeance for the assault. In 146 B.C., the Roman general Lucius Mummius conquered Corinth and, determined to make it an example to the rest of the Achaean League, ruthlessly destroyed the city. Those inhabitants not killed were rounded up and sold into slavery, the city's buildings were leveled, and the numerous city treasures were stolen and taken to Rome. For a century after this destruction, Corinth lay in ruins. Only grave robbers, looting the cemeteries for antiquities to sell in Rome, visited the area.

Corinth was forgotten until 44 B.C., when Julius Caesar ordered that the city be rebuilt to serve as Rome's provincial capital in Greece. In this role, Corinth once again became a bustling center of activity. Its population soared from a previous high of about 90,000 to more than 700,000. The Romans rebuilt the Temple of Apollo and much of the old city area. These reconstructions were done in the Roman style, blending this style with the original Greek design of the city. The Romans constructed new buildings around the central marketplace and laid limestone roads leading to the market. They also constructed a number of shops, six small temples, four basilicas, and many bathhouses, and refurbished the Greek theatre originally built in the fourth century B.C. Next to the theatre, and joined to it by a colonnaded court, was constructed an Odeion or music hall. A new Roman-style amphitheatre—the only one in Greece—was also built at this time. Here they staged gladiator battles and animal shows similar to those held at Rome's Colosseum. Later Roman improvements included an aqueduct from Lake Stymphalis. Emperor Nero tried in vain to build a canal across the isthmus in A.D. 67, but such a canal would not be completed until 1893. The apostle St. Paul lived in Corinth for about eighteen months in A.D. 51–52. He worked as a tent maker and leather worker when not preaching at the rostrum in the marketplace.

Following a series of devastating earthquakes, Corinth began a slow decline as a center of commerce in the Greek world. It suffered from a number of invasions as the region changed hands after the fall of the Roman Empire. In the thirteenth century Corinth was captured by Geoffroi de Villehardouin and Otho de la Roche during the Fourth Crusade. The city was taken by the Turks in 1458 and again in 1712, by the Knights of Malta in 1612, and by the Venetians in 1687, finally becoming a part of a restored Greece in 1822.

Since 1896 archaeologists have been excavating the ruins of Corinth, most of which date back only to the Roman period. Most of the earlier buildings were thoroughly destroyed at the time of the Roman invasion of the city. Today, excavated ruins are found atop Acrocorinthus, on the nearby slopes, in the old Roman city, and between the city and the coast.

The fortifications on Acrocorinthus range from a fourth-century B.C. Greek tower to a medieval moat and include examples of buildings constructed by the Turks, Franks, Venetians, and others who have ruled the city over the centuries. On the slopes of Acrocorinthus can be found the Sanctuary of Demeter, a small temple, while in the old Roman city below can be found the Temple of Apollo, which still retains seven of its original thirty-eight columns. The agora, or central marketplace, of Corinth has been uncovered, revealing a central square lined with many shops and small temples; the South Stoa, a building believed to have been constructed by King Philip II of Macedonia to serve as headquarters for his Pan-Hellenic League; and the Fountain of Peirene, a spring-fed fountain still decorated with second-century B.C. frescoes of swimming fish. Among other sites excavated are the Odeion, the original Greek theatre, and the rostrum where St. Paul preached. The Long Walls leading from the top of Acrocorinthus to the harbor at Lechaion can still be traced as well. The museum at Corinth houses artifacts from the earliest prehistoric times up to the Roman era, including sculpture, coins, pottery, and Roman portrait statues.

Further Reading: *Corinth: A Brief History and a Guide to the Excavations* (Athens: American School of Classical Studies, 1969) is a summary of the work done by the American team of archaeologists at Corinth. *The Greek Stones Speak: The Story of Archaeology in Greek Lands* (London: Methuen, and New York: St. Martin's, 1962) by Paul MacKendrick provides an overview of all archaeological work in Greece. *A History of Greece* (London: Methuen, 1929; ninth edition, London: Methuen, and New York: Barnes and Noble, 1957) by Cyril E. Robinson is a standard work in the field. Michael Grant's *The Rise of the Greeks* (London: Weidenfeld and Nicolson, 1987; New York: Scribner's, 1988) looks at Greek history city by city.

—Thomas Wiloch

La Coruña (La Coruña, Spain)

Location: Northern coast of Spain on the Bay of La Coruña.

Description: Capital of the Spanish province of the same name, La Coruña is a commercial port of considerable importance. Possibly Phoenician in origin, the town saw both Roman and Moslem periods before becoming a part of Spain. The port harbored the Spanish Armada, was burned by Sir Francis Drake, and was the site of an important British battle against Napoléon's forces. It was once the home of Picasso.

Site Office: Tourist Information Office
Dársena de la Marina
15001 La Coruña, La Coruña
Spain
22 1822

Reputedly founded by Phoenician traders and later occupied by Roman and Moslem settlers, La Coruña (Corunna in English) is most famous in modern times for the strategic importance of its harbor, which has proven a gateway of conquest for forces invading and leaving the Iberian peninsula. In 1386, John of Gaunt landed an army of 7,000 men in La Coruña to stake his claim to the throne of Castile. In 1588, La Coruña's harbor sheltered the flagging armada on its journey to England, and in 1589 Francis Drake and John Norreys returned to capture and burn much of the town. In 1747 and 1805 the bay witnessed minor naval victories by the British over the French, and in 1809 a British army commanded by the mawkishly eulogized John Moore held off Marshal Soult's French forces long enough to evacuate the British troops from Spain. Though Moore himself died in the battle, he achieved immortality in a dramatic monument and a sentimental poem by the Irishman Charles Wolfe.

Capital of the Galician province of La Coruña and one of the principal ports of the northern coast of Spain, La Coruña consists of an upper and lower town built on a narrow neck of land separating the Bay of Orzan from the Bay of Coruña. While the commercial life of La Coruña is centered primarily in the lower part of town, known as the Pescaderia or Ciudad Nueva, the ancient upper part of town, known as the Ciudad Vieja, contains most sites of interest to the visitor. Located on the eastern side of the bay and completely sheltered, the harbor has been fortified since the fifteenth century. The most notable of these fortifications is the Castillo de San Antón, built in 1589. In the eighteenth century, San Antón functioned as a jail, housing, among others, the political dissident and economist Melchor de Macanaz, who wrote prolifically during his confinement, which lasted from 1748 until his death in 1760. Today, the fort contains an archaeo-

logical museum. On land, the upper part of town was once protected on three sides by sixteenth-century bastions, whose traces can still be found.

The town's Roman roots are evident in the Torre de Hercules, a lighthouse that dates to antiquity but was restored in the eighteenth century and heightened to nearly 400 feet. The tower, located to the north, was built during the reign of Emperor Trajan (A.D. 98–117), although local legend maintains it was built by the Phoenicians. Monuments from medieval times in the Ciudad Vieja include the Santa Maria del Campo (constructed 1215–1302), and the twelfth-century Church of Santiago. Late Gothic relief work can be seen above the gateway of the nearby Convent of Las Barbaras, which depicts St. Michael weighing souls. Among La Coruña's Galician baroque structures are the eighteenth-century Church of Santo Domingo and the Jardin de San Carlos, which houses the granite tomb of John Moore.

Architecturally, the old town is probably most noteworthy for its *miradores,* or glazed balconies, which give the streets the appearance of rows of greenhouses. The Pescaderia also contains a tangle of broad and picturesque streets where the visitor can find the eighteenth-century Church of San Nicolas, worked on by architects Simon Rodriguez and Clemente Sarela, the Church of San Jose, designed by Fernando de Casas and Sarela, and the modernist Town Hall, built in 1904. Among the figures who traveled these streets are novelist Emile Pardo Bazan (1851–1921), liberal statesman and exiled author Salvador de Madariaga (1886–1978), and Picasso, who lived in La Coruña from 1891 to 1895.

It was once generally agreed that La Coruña was first settled by Phoenicians, though new evidence suggests the possibility of more ancient Celtic origins. Perhaps the earliest extant mention of the city is in the first-century history of Pomponius Mela, who identifies a seaport called Ardobrica in the country of Artabri. The name Portus Artabrorum was assigned to the bay on which the city was located. Some later histories claim the port was captured by the Romans in 60 B.C. and named Ardobicum Corunium, while others trace the present name to the medieval moniker Caronium and others still to the Latin Columna. Whatever its earliest history, the town fell into Moorish hands in the eighth century and again in the tenth until the defeat of al-Manṣūr in 1002. For nearly 400 years thereafter, except for a brief period of Portuguese occupation, La Coruña remained under Spanish control.

During the latter half of the fourteenth century, while England and France jockeyed for alliances against one another, the kingdoms of Iberia found themselves increasingly entangled in the hostilities between the two northern foes. It was a state of affairs that would affect La Coruña's history for the next 500 years and would, in 1386, bring to La Coruña's shore 7,000 English troops under John of Gaunt, son of King

Torre de Hercules at La Coruña
Photo courtesy of Turgalicia

Edward III. In 1366, Edward, the Black Prince (Gaunt's brother and heir apparent to the English throne), resolved to intervene in Castile's civil war on behalf of Peter the Cruel against Peter's half-brother Henry of Trastámara, who was allied with Charles V of France. The Black Prince hoped that by securing a strong Castilian alliance, he could receive for his duchy of Aquitaine much needed sea and land support against the French threat.

Although Peter himself was captured and killed by Henry in 1369, Gaunt's marriage to Peter's daughter and heiress Constance in 1371 solidified the English claim over Castile. In 1372, Edward III recognized Gaunt as king of Castile, though the claim was used by the English mainly as a bargaining chip to win the house of Trastámara from its French alliance. Over the next ten years, Henry—and later his son, John I, who assumed the Castilian throne after Henry's death in 1379—refused to abandon his French pact in favor of an English one. As a result, war seemed imminent, and England began to forge other Iberian alliances, including one with Ferdinand I of Portugal. However, because of economic difficulties at home and Charles V's successes in the war of 1369–75—and despite Franco-Castilian raids on England's shore—the English invasion seemed perpetually on hold.

Soon enough, however, events in Iberia again grew ominous. While Ferdinand I was ailing, his wife, Queen Leonor, and her lover, Andeiro, made a treaty with John I, offering to him in marriage Ferdinand's young daughter and heiress Beatriz. After the marriage, John rejected the terms of regency and declared himself king of Portugal. Revolts ensued, and Ferdinand's brother John of Aviz was elected by the rebels as regent. In 1384 Lisbon repelled a Castilian attack, and the following year John of Aviz was proclaimed king of Portugal. He soon renewed the English alliance, upon which Gaunt persuaded Parliament—as well as Edward III's successor, Richard II—to finance an invasion. Though the amount of financial aid fell vastly short of what Gaunt needed to ensure John of Castile's overthrow, he believed a series of successes on the peninsula might make Parliament forget its parsimony, and on July 9, 1386—already well into the campaigning season—he set sail from Plymouth with a force of 7,000 young and mostly untested Englishmen. Though eager to prove themselves, they had in fact little chance of achieving the expedition's goals.

What were those goals? Ostensibly, they were to save the Castilian throne from illegitimacy and the Castilian church from the schism that had compelled rival popes Urban VI of Rome and Clement VII of Avignon to excommunicate one another in 1377. (Because of John of Castile's French alliance, Castilian bishops recognized the See of Avignon.) Despite these professed aims, however, historians have argued that, having scant hope of actually winning Castile, Richard and Gaunt intended to "sell" their dynastic claim as "dearly" as they were able, in Anthony Goodman's words. Gaunt's forces arrived in La Coruña on July 25, the feast of Santiago. Though Gaunt's strategists had discussed a landing

on the friendly shores of Portugal, it seemed to them that an immediate foothold on the enemy's coast would seem more intimidating to John of Castile and more honorable to the young English soldiers. Practical benefits were associated with the landing on Santiago's feast day: for one, the citizens of La Coruña were engaged in celebrations and unprepared to defend the port; secondly, Santiago's rescue from schism might earn the saint's favor, it was thought, and sway Spanish sentiment toward Gaunt. Initially, La Coruña showed no intention to admit Gaunt. However, after the English had disembarked and entered the undefended town of Santiago de Compostela—whose citizens immediately recognized Gaunt as king and Urban as pope—La Coruña itself submitted to the same terms. The rest of Galicia soon surrendered, and through the winter of 1386–87 the English ruled uncontested in the province.

The next year Gaunt suffered bad luck in his attempt to move against more heavily defended portions of John's kingdom and eventually ceded his claim. Nevertheless, the occupation of La Coruña and the region proved significant for two reasons. First, the English army's honorable conduct in Galicia and its strict enforcement of the ordinances of war earned the good will of Galician inhabitants. Such Alexandrian restraint was remembered by Gaunt's grandson Henry V decades later in his invasion of Normandy and became protocol for English troops throughout their history of conquest. Secondly, Gaunt's ability to sway Galician sentiment weakened John of Castile's position and helped bring about a favorable settlement for the English. In the end, Gaunt married his daughter Catherine to the young nobleman who became King Henry III of Castile and León.

Over the next centuries La Coruña continued to find itself at the crossroads of European political affairs. In 1520 Charles V embarked from the town on his first visit to the Low Countries. Habsburg heir to the throne of Austria, heir by marriage to the lands of the Low Countries, Charles had become king of Spain in 1516 and the Holy Roman Emperor in 1519. At its height, his combined empire was the most extensive in Europe since the rule of Charlemagne. It was his son, Philip II, who set sail from La Coruña to wed Mary Tudor and who later launched the so-called Invincible Armada that, in 1588, took shelter in the harbor of La Coruña before sailing for England.

In May of 1588 the armada, which consisted of 130 ships, 2,630 cannon, and 27,500 men, started from Lisbon for a rendezvous off the coast of England. Said to be the greatest fleet ever assembled, the armada lacked in mobility what it boasted in strength, and by July 9 the hulking ships still had not cleared the Spanish coast. By that time the food and water on the galleys and pinaces had gone bad, and Alonso Pérez de Guzmán, duque de Medina-Sidonia, the indecisive commander of the fleet, ordered all ships to put in at La Coruña to replenish the victuals. The order was received by the central body of the fleet but not by the vanguard, which continued to sail for the channel; and that evening, as most of the ships began to anchor in La Coruña, a violent squall ripped through the armada and scattered along the coast those ships that had not succeeded in anchoring or moving into deeper waters. The mishap greatly weakened the Spanish force, and Medina-Sidonia complained to Philip that, with the combined setbacks from the storm, illness, and lack of supplies, the armada was now at one-third its former strength. Seeing the fleet's current prospects as hopeless, he petitioned the king for a postponement of the expedition. Philip refused, instead ordering Medina-Sidonia to make the necessary repairs, replenish victuals, and set sail as soon as possible. The armada remained in La Coruña for a month, crippled and vulnerable to the English fleet.

Meanwhile in Plymouth, Sir Francis Drake pushed for movement against the armada, though as yet the English were unaware of the fleet's disaster at La Coruña. Though conventional naval strategy called for a defensive posture by the smaller English force, Drake felt that a standoff in the channel, where there was little room for movement and where the more massive and better-armed Spanish vessels were at an advantage, would be disastrous. He proposed sailing upwind of the armada—that is, to the south of the fleet—where, if the armada made its move, the quicker English could engage them in open seas. Drake's plan met at first with resistance, but later, after the displaced portion of the armada had been spotted near England and it was surmised that the Spanish fleet had somehow been broken up, Drake received clearance to proceed toward the peninsula.

John Ruskin once wrote there is no such thing as bad weather—only different kinds of good weather. Such was the case when, after several failed attempts, the English fleet on July 7 finally found favorable winds and raced southeastward where the armada, reorganized into only a semblance of its original strength, waited uneasily for those same winds to shift toward England. That the English on the July 9 came nearly within striking distance of the ships in La Coruña is certain. Had the winds held for another day or two, observers wrote, the English would have pinned the armada in its harbor, and La Coruña would forever have been a monument to the demise of Philip's ambition. However, the winds again shifted favorably for the Spanish, blew the English back to Plymouth, and guaranteed the armada a different—though equally ill-starred—fate off the coasts of England and Ireland.

The next year, after the virtual destruction of the armada, Drake and John Norreys returned to Iberia with 20,000 men and 16,000 guns to disrupt Spanish trade and, according to Drake's commission, "to distress the king of Spain's ships." Drake saw the latter order, which called for a series of raids against smaller ports that sheltered the scattered remains of an already-defeated Spanish fleet, as overly cautious and instead advocated an invasion of Lisbon, the heart of what was left of Spanish maritime might. Hindsight reveals that such an attack would have brought about the fall of the Portuguese capital. However, Drake and Norreys found themselves on April 24, 1589, within sight of La Coruña, defended by only a small garrison. No more than five significant Spanish warships were sheltered in the harbor.

Drake immediately issued plans for attack. Because the town lay on a promontory and isthmus, it was inaccessible to the English force except by sea—via the harbor—and by the narrow neck that connected the lower town to the mainland. Reconnaissance revealed two lines of defense: the upper town was guarded by bastions that separated it from the rest of the neck, and the lower town was cut off from the mainland by similar walls spanning the base of the isthmus. In preparation for a land attack, Drake's advance guards landed on the eastern side of the bay opposite the town and, despite minor resistance, made their way along the shore to the outskirts of La Coruña. There the guards were joined by reinforcements and camped for the night, their position guaranteeing that the garrison in La Coruña could not interfere with the rest of the landing operations. That night and the next morning, Norreys brought ashore a number of field batteries, which, positioned against the harbor, soon forced the Spanish ships to move to the adjoining bay.

Meanwhile, Drake and Norreys planned a three-pronged attack against the town. At low tide, two small groups were to attempt to wade around opposite ends of the isthmus walls while, simultaneously, a force of 1,200 was to assault the lower town by sea. At midnight guns were fired to signal the attack, and the seaborne troops came ashore with ease. Trying to defend the shore, the Spanish garrison allowed the wall to be flanked on one end without resistance. On the opposite end, however, the English were driven back three times with heavy losses before troops from the boats arrived to support them. Soon, the lower town was in English hands and the defenders fled to the upper town, leaving behind 500 prisoners, who were executed, as well as the commander of the garrison, Don Juan de Luna, who was held alive.

The next morning the Spaniards burned or abandoned their ships in the bay, and the English looted the lower town for supplies. According to Drake's orders, the mission had been completed, yet for days the winds kept the English ships in the harbor, where they were continually harassed by Spanish batteries in the upper town. For this reason, Norreys resolved to capture the upper town, and on the 28th he sent an ultimatum under white flag to the Spaniards. The garrison responded with cannon, killing a drummer boy. Fearful of invoking more English ire, the Spaniards hanged the gunner responsible for the murder, swinging his body before the walls, but the gesture left Norreys and his men nonplussed. For five days they battered and mined the bastions until two breaches had been opened. When they finally charged the upper town, however, a tower at one of the breaches collapsed on the English troops, killing many of then, while the remaining breach proved too difficult to climb. The standoff outside the upper town remained unchanged.

Just after the failed charge of the upper town, Norreys learned of a drastic change in the situation. A Spanish force of 8,000 had arrived to rescue the trapped garrison and was encamped across the Mero River just to the east of where the English had first landed. Leaving Drake five regiments to guard the upper town, Norreys the next morning set out with seven regiments of his own to check the threat at the English rear. By ten o'clock his troops had secured the near side of the river but still faced a perilous situation. The Spaniards on the far bank had built a formidable breastwork that commanded the one approach to them: a two-hundred-yard bridge, upon which only three men could pass abreast. Despite the danger, the English made their charge. Armed with pikes and led by Norreys' brother Edward, they assaulted the bridge under a hail of musket fire and for a time reached the far bank. Edward himself was the first across, received a saber to the head, and had to be rescued. The battle raged, and three times the English were driven back. Affronted by his brother's wounding, Norreys himself at last took a pike and, dressed only in doublet and hose, led the English in ranks of three across the bridge. The young nobles rallied behind him and swept the breastworks, driving the Spaniards from their camp and chasing them across the countryside. Afterward, the English claimed to have slain 1,000 of the fleeing enemy and to have destroyed the surrounding land.

Back in the harbor, the winds had grown fair and Drake had made plans to abandon La Coruña. Before the English left, they burned the lower town and the remaining shipping, leaving an unpleasant legacy that would not be erased for another 210 years. By that time, Spain and England would be allies in the Peninsular War against Napoléon, and a young English general, Sir John Moore, would save a British army from certain destruction, and pave the way for the peninsula's eventual liberation from Napoléon, paying for it with his own life.

In late 1808, Moore's already weary army of 25,000 made a bold advance at Burgos in an attempt to break Napoléon's line of communication and divert the Emperor from his thrust toward Lisbon and Andalusia. Though the advance was crucial to the survival of Spanish resistance in the south, it left the British exposed to French attack and severed from naval support. By late December, Moore had no choice but to withdraw his army to the coast of Galicia—a long, inclement march through the mountains with Marshal Nicolas-Jean de Dieu Soult's advance cavalry in close pursuit. The retreat was as punishing as Moore feared. Bad roads slowed troop movement, British morale had bottomed out, and the combination of poor supplies and harsh winter weather left Moore's army so depleted that on January 2 Napoléon, believing the British to be finished, returned to Paris, leaving the command to Soult. Still Moore pressed toward La Coruña, and on January 11 his troops emerged from the mountains, bloody-footed and starving, to see the masts of naval transports waiting to take them from Spain.

As Moore rode into La Coruña and set up his headquarters on the Canyon Grande, he quickly recognized a difficult battle would have to be fought before his army could embark. For one thing, the majority of ships had not yet arrived and those now in the bay were mostly hospital and store vessels. In addition, the bulk of Soult's 25,000 infantry and 6,000 cavalry—supported by Michel Ney's 16,000 troops—were a two-day march from the town. After the retreat, Moore retained only 18,000 battle-ready men, no cavalry, and, because most of the guns were quickly loaded onto ships to avoid possible capture, only minimal artillery. For two days the British prepared their position two miles south of La Coruña along a ridge protected on its left by the Mero estuary but open to attack from the right. Commanding the position was a chain of hills running from northwest of the town to Mount Mero—excellent ground, but too wide for the British troops to hold. By the 15th, Soult had assumed the high ground, and the next day, the French cannon opened fire. The battle had begun.

Moore had predicted the strength of the French attack would be on his right and had placed extra troops there under Henry William Paget and Alexander MacKenzie Fraser. He was correct. Though the initial thrust was toward the British center at a village called Elvina, eight regiments of cavalry quickly tried to flank Paget and Fraser but were repulsed. Moore himself, however, rode to Elvina, where the British line was about to break under heavy French fire. His arrival rallied his troops. One regiment, the 42nd, had begun to fall back. Moore placed himself at their head and called a charge. "My brave Highlanders," he cried. "Remember Egypt! Think on Scotland!" The Scotsmen raised bayonets and advanced, driving back the French. Those present wrote later that Moore on his caramel-and-white horse seemed to be everywhere that day, exhorting his men at the center and then dashing off to the right flank. On the right Paget and Fraser came under heavy attack but held the lines, and as darkness fell it seemed clear that neither side would incur strategic loss. That was when a round-shot toppled Moore from his horse, severing his left arm and tearing open his left breast.

Moore knew the wound was fatal. The ball had pierced so deeply that his collarbone lay splintered in half and his left lung was exposed. "My good man," he told his surgeon, "you can do me no good. It is too high up." The surgeon agreed and was ordered to attend to the other wounded. Moore was taken in a blanket back to headquarters, where he asked if the French were beaten and muttered to his servant: "You know, Anderson, I have always wanted to die this way." Later that evening, he did.

On the field, the troop positions remained as they had before the fighting. In the night, however, the British stole onto the ships, leaving bivouac fires burning to confuse the enemy. Moore's body was taken to the landward wall of La Coruña and buried in a soiled military cloak. Later, when Soult entered the town, he ordered guns fired in honor of his opponent. The British departed Spain entirely, but not for long. The lessons learned from Moore's expedition paved the way for future operations against Napoléon, and within the year Sir Arthur Wellesley—later the Duke of Wellington—began the peninsular campaign that would ultimately expel the French from Spain.

Within days of his death, now-forgotten bits of doggerel eulogizing Moore made their appearances. Eight years later, however, an unknown Irish curate named Charles Wolfe published the "The Burial of Sir John Moore," eight stanzas now carved on the general's monument in La Coruña. Grandiose and gushy, the poem nonetheless befits the legend it has long since surpassed in fame:

> Not a drum was heard, not a funeral note,
> As his corse to the rampart we hurried;
> Not a solider discharged his farewell shot
> O'er the grave where our hero we buried.

Further Reading: *Moore of Corunna* by Roger Parkinson (London: Hart Davis, MacGibbon, 1976) provides a vivid account of the campaigns and death of Moore, as does *Sir John Moore* by Carola Oman (London: Hodder and Stoughton, 1953) and—for a drier military analysis—*Sir John Moore's Peninsular Campaign, 1808–1809* by noted military historian D. W. Davies (The Hague: Martinus Nijhoff, 1974). For an account of Gaunt's Iberian involvement, see *John of Gaunt, The Exercise of Princely Power in Fourteenth-Century Europe* (Harlow, Essex: Longman, and New York: St. Martin's, 1992) by Anthony Goodman. A colorful and indispensable rendering of Drake's expeditions is to be found in *Drake and the Tudor Navy* by Julian Corbett (London: Longmans, Green, 1917).

—Paul Mooney

Covadonga (Oviedo, Spain)

Location: Below the Picos de Europa Mountains, two miles south of Cangas de Onís in the region of Asturias.

Description: A pilgrimage site, held by many to be the birthplace of modern, Christian Spain, and namesake of the nearby national park.

Site Office: Tourist Information Center
El Repelao
Covadonga, Oviedo
Spain
(985) 584 60 13

The tiny village of Covadonga stands on a rocky spur in a valley below the rugged Picos de Europa, the highest range of the Cantabrian Mountains (the Cordillera Cantábrica), which dominate both the landscape and the long, complex history of the region (or principality) of Asturias. The early history of the region, and of Covadonga within it, is largely unknown: this inhospitable and historically remote land was sparsely populated in ancient times, and those who did live there left few artifacts and had no written language. However, archaeological excavations during the twentieth century have uncovered such mysterious and evocative remains as the prehistoric paintings on the walls of the Tito Bustillo Cave, twenty-three miles south of Covadonga. Discovered in 1968, these paintings confirm that Asturias has been inhabited for at least 20,000 years. At some later date, the region became the ancestral home of an indigenous people now known by their Latin name as the Astures, one of a number of groups in the Iberian Peninsula who are commonly referred to by historians as "Iberians." Although it is often claimed that they entered the peninsula from north Africa, many Iberians, including the Astures, who were especially remote from Spain's southern shores, possibly migrated from elsewhere in Europe.

The harsh mountain terrain and the rocky coastline on which the Astures settled provided them with some security and autonomy. Greek colonists established trading posts on the southern and eastern coasts, until approximately the sixth century B.C., and the Carthaginians, who followed soon after from their home in north Africa, began to use settlements in southern Spain as bases for controlling trade between the peninsula and other Mediterranean areas. It appears likely that Asturias remained unaffected by these developments, even when, in the third century B.C., the Romans and the Carthaginians fought two Punic Wars, partly on Iberian territory. But at some point in this contest for control of the western Mediterranean both sides learned that gold was to be found in the Cantabrian Mountains: the isolation of the Astures could not last much longer.

Having driven out the Carthaginians, the Romans turned to the task of extending their power throughout the peninsula in a series of campaigns that were to last for nearly two centuries. The Romans met with the fiercest resistance in the northwest part of the peninsula. Indeed, although the modern regions of Cantabria, Asturias, and Galicia officially formed part of the Roman province of Hispania Citerior (Hither Spain), the Astures, among other indigenous peoples, made stable government impossible, conducting raids on Roman camps and settlements until approximately 29 B.C., according to the contemporary Greek writer Plutarch. Three years later, under orders from Augustus, the first Roman emperor, a major punitive expedition was launched against the Astures and their allies the Cantabri. Conditions were extremely difficult for the imperial legionaries; rough and unfamiliar terrain, disease, and vermin hampered the Romans, who nevertheless succeeded in defeating the Astures that year and during further conflicts that erupted between 24 and 16 B.C. By 13 B.C. the last remaining resistance had been overcome, and Augustus's conquest of Hispania was complete.

Roman rule continued, at least formally, for another 400 years or so, with Asturias incorporated into the province of Gallaetia around the end of the third century A.D. Although it is likely that the Romans passed through the valley in which Covadonga now stands, like the Astures before them they created no permanent settlements in what remained very much of a remote outpost. Once the Roman Empire had begun its steady decline toward its collapse in the fifth century, Hispania became the battleground for various Germanic invaders. The Suevi occupied Gallaetia until approximately 585, when the Visigoths ousted them, and although they probably did not conquer the Astures, the Visigoths may well have Christianized them following the Visigoths' own conversion during the sixth century. Throughout the succeeding period of Visigothic rule, however, the Astures continued to resist any other form of assimilation.

Covadonga, which may have been settled by the Astures centuries before but which, even if so, had remained unimportant, suddenly enters the historical record at the end of the Visigothic era. The Moslem armies of Arabs and Berbers that had entered the peninsula from north Africa in 711 found it easy to overthrow the divided and weakened Visigoths and advanced into Asturias around 714. They captured the region in a rapid onslaught, forcing the Asturian fighters to take refuge in the hills. One of their battles, however, was to become a Spanish legend, and led to the naming of the city. During the month of May, in either 718 or 722, around 300 Asturian Christians, led by one Don Pelayo, took refuge in a mountain cave, where they prayed to the Virgin Mary for aid in their fight against the Moslems. The

Shrine in the sacred cave of Covadonga
Photo courtesy of Instituto Nacional de Promocion del Turismo

cave appears to have been named Auseva; in later times, it was known as the Santa Cueva (Holy Cave) or the Covadominica (Cave of Our Lady), and, finally, Covadonga.

According to the legend, Pelayo and his men swiftly defeated the Moslem forces sent against them by the emir of Córdoba. Whatever the immediate significance of the battle, the story of the refuge was to inspire many future Asturians to continue the resistance against the Moslem invaders, and Covadonga became a favored destination of Christian pilgrims. The defeat of the Moslems at the Battle of Covadonga has been seen as the catalyst for the creation of a Christian Spain; Pelayo has long been seen as one of the heroes of the Reconquista (reconquest), the centuries-long process by which a number of Christian kingdoms gradually coalesced and drove the Moslems out of Spain. Pelayo is said to have been buried along with members of his family inside the Santa Cueva, although it is not clear whether he died during the battle or later.

From approximately 730, when some of the Berbers in the Moslem armies rebelled against their leaders, the people of the region were increasingly successful in pushing the Moslems out. They benefited also from the southward withdrawal of the main Moslem forces after their defeat at Poitiers, in southern France, in 732. Their retreat gave Alfonso I, Pelayo's son-in-law and his successor as leader of the local Christians, the opportunity to place himself at the head of an autonomous Christian principality of Asturias, around 739. He established his court at Cangas de Onís, near Covadonga, and built the memorial Chapel of Our Lady of Battles (Virgen de las Batallas) near the cave. This has been restored twice, following damage by fires, in 1777 and again in 1936. It is possible that he also ordered the building of the first set of steps leading up to the cave (since replaced by marble stairs) and the placing of a statue of the Virgin inside the cave, although the statue that stands there today dates from the eighteenth century. He was buried in the chapel following his death in 757, alongside his wife Ermesinda (Pelayo's daughter).

Covadonga appears to have remained a pilgrimage center, at least during times of relative peace and security, throughout the Middle Ages, although it was never as important, either in the religious or economic life of Spain, as Santiago de Compostela in Galicia, where the supposed tomb of the apostle James attracted visitors from all over Europe. The sacred cave and the memorial chapel were remote, especially after the capital of the Christians of northeastern Spain moved first from nearby Cangas de Onís to Oviedo, and then, in 914, to Léon. Alfonso's principality was absorbed into the new and steadily expanding kingdom also known as Léon (or Léon-Asturias) and remained within it as it was united with, divided from, and reunited once more with the larger kingdom of Castile (itself originally a Léonese province). The single kingdom of Castile-Léon created in 1230 was the largest of the Christian states that, from 1252, controlled all of the peninsula except for Granada. Within it Asturias remained, as ever, sparsely populated and uninvolved in major political events (although its founding role is still reflected in the title Prince of Asturias, held by the eldest son of the reigning monarch). Covadonga's special standing as the birthplace of the Reconquista remained high. It was reinforced after 1492, when King Ferdinand of Aragon and his wife Queen Isabella of Castile-Léon, having achieved a union of their realms by 1479, completed the Reconquista by driving the last Moslem forces out of Granada.

Ferdinand and Isabella eventually joined their distant and legendary ancestor Pelayo as national heroes of Spain and supporters of the Catholic faith. The state they founded continued virtually unaltered well into the nineteenth century, largely (though not wholly) resisting the secular ideas gradually spreading through Europe. Between 1877 and 1891 the government sponsored the building of the Basilica of Nuestra Señora de las Batallas (Our Lady of Battles), a new church designed in a mixture of neo-Romanesque and neo-Gothic styles, near the sacred cave of Covadonga and connected to it by a tunnel. Opposite this church stands a museum containing treasures donated to the chapel and the basilica over the centuries, most notably a crown (for the Virgin Mary) decorated with hundreds of diamonds.

Unlike so many other church buildings throughout Europe, neither this basilica nor its predecessors ever became the focus of trade or settlement, for the terrain remained inhospitable to large human populations. The undisturbed landscapes of the region received official recognition in 1918, when the Montaña de Covadonga National Park was established near the village. It covers an area of sixty-five square miles of wooded land with two lakes, Lago Enol and Lago de la Ercina. Inside the park can be found wild boar, wildcats, bears, and wild horses. But it is not entirely uninhabited by human beings: every year a Shepherds' Festival is held inside the park, at Lago de la Ercina.

During the 1920s Covadonga was used as a venue for rallies by a growing number of right-wing political groups. The symbolic significance of the Battle of Covadonga proved useful to such groups, most of which were to be absorbed into the Falangist movement later led by General Franco, as a means of attracting support, especially among the military, and of vilifying liberals, socialists, communists, and anarchists as equivalents of the medieval Moslems whose secularism and demands for toleration and reform threatened the Catholic Spain created by Pelayo and his successors. During and after the Spanish Civil War (1936–39), Franco was frequently and explicitly portrayed as a new Pelayo, a crusader leading a renewed Reconquista. In October 1937 his nationalist forces, keen to seize control of the whole of the Cantabrian coast from the pockets of loyalist resistance (supporters of the secular republic founded in 1931), marched into Asturias. Many local people burned their own villages in the hope of slowing the enemy's advance, but Covadonga was soon captured and the region conquered. Between the completion of his Reconquista in 1939 and his death in 1975, Franco frequently spent holidays at the Hotel Pelayo in Covadonga, enjoying fishing in the lakes nearby.

Today, with Franco dead, his Falangist movement dispersed, and democracy restored, Covadonga has lost its political significance for almost all Spaniards, but it remains a popular destination both for pilgrims to the cave and basilica and for visitors passing through on their way into the national park.

Further Reading: Leonard A. Curchin's *Roman Spain: Conquest and Assimilation* (London and New York: Routledge, 1991) and Joseph F. O'Callaghan's *History of Medieval Spain* (Ithaca, New York, and London: Cornell University Press, 1975) provide comprehensive coverage of their respective periods; O'Callaghan in particular usefully explains the historical and cultural context of the pilgrimage site at Covadonga.

—Monique Lamontagne

Crete (Greece)

Location: Equidistant from three continents—Europe, Africa, and Asia—in the center of the eastern Mediterranean Sea, the island's long, thin land mass (10 to 40 miles wide) stretches over 160 mountainous miles and is the southernmost point in Europe. Crete lies along the great earthquake belt that links Mount Vesuvius with Sicily, mainland Greece, and Anatolian Turkey.

Description: Crete was the home of the first civilized European society, the Minoans, who built the city of Knossos; the ruins of this city, lying near the modern city of Iráklion, have been extensively excavated. Crete has many other historic sites, representing periods of rule by a variety of occupying groups.

Site Office: Greek National Tourist Organization
1 Xanthoudidou Street
Iráklion, Hērákleion
Greece
(81) 228-203

Sometimes referred to as "the Sixth Continent," the island of Crete has had an eventful history including the earliest European civilization, the Minoans. Centuries after the fall of that ancient culture, the people of Crete were subjected to successive occupations by the Romans, Byzantines, Saracens, Venetians, and Ottoman Turks. Survivors of such a long history of hardships, Crete's islanders retain a spirit of independence that is duly reflected in their vibrant culture.

Many of the greatest tales of Greek mythology were set on the island of Crete, and in recent years scholars have revealed unexpected truths in those tales. The first civilized Cretans, the Minoans, are named for the king Minos, son of Zeus (who is said to have been spared the cannibalism of his father Kronos while hiding in a cave on Crete). The name Minos has recently been attributed to a possible line of real-life royalty. The Minoans mounted the world's first maritime empire, which flourished during the second millennium B.C.

As early as the seventh millennium B.C., Neolithic tribes populated the island, prospering from its agricultural fertility. Later, Bronze-Age Anatolians settled both Crete and mainland Greece, importing their skills at toolmaking. It appears that this self-contained civilization gradually befriended the Egyptians, a contact that greatly influenced the Cretan concepts of art and worship. In order to trade the products of their successful farming and metalworks, the islanders moved from their caves and hilltops to seaside harbors, launching a fleet of ships to establish commercial routes with surrounding destinations. This new culture has been dubbed the Minoans.

By the year 2000 B.C., these Cretans had developed a centralized government headed by a king, and they had instituted tree-cutting methods and the use of the potter's wheel. Their religious beliefs revolved around goddess cults, which worshiped the symbolic female figure and put faith in the magical and the sacrificial. On the island, three distinct regions were already blooming: the central, southern, and eastern territories, congregating in towns such as Palaikastro, Mokhlos, and Gournia. Eventually, the island came to be regarded as a single entity, with roadways connecting the growing palatial capitals at Knossos, Phaistos, and Mallia. Legend has it that the latter two cities were ruled by Minos's brothers Rhadamanthy and Sarpedon, respectively. Despite the setback of a natural disaster around 1700 B.C.—most probably an earthquake—the Minoan advancements of this period laid a sturdy foundation for the spectacular cultural development in the centuries to come.

Shortly after the disaster, Crete boasted a population of approximately 1 million, a figure that doubles that of the present day. The sprawling palace at Knossos alone provided homes for at least 100,000. The ruins of that compound's interwoven corridors bear a startling resemblance to a site described in a famous tale of Greek mythology.

According to legend, Zeus's brother Poseidon gave Minos a white bull to sacrifice. When the king chose to keep the fine animal for himself and sacrifice another, an enraged Poseidon induced Minos's wife Pasiphaë to fall in love with the bull. Their coupling produced the Minotaur, a grotesque creature with the head of a bull and the body of a man. In order to isolate the beast, Minos ordered the inventor Daedalus to design the Labyrinth. At war with Athens, Minos secured from that city a periodic tribute of fourteen youths to feed to the Minotaur. When Theseus, son of Athens's King Aegeus, volunteered himself for the tribute, Minos's lovestruck daughter Ariadne gave him a thread with which he could mark his trail into the Labyrinth. Inside, Theseus slew the Minotaur and emerged, triumphant, from the maze.

The palace at Knossos, spreading over as many as six acres and rising to four floors and higher, resembles a labyrinth; when the British explorer Arthur Evans excavated those ruins at the turn of the twentieth century, he believed he had discovered the labyrinth of the Minotaur. What he did find—advanced pottery and artwork that depicted leisurely scenes—portrayed a wealthy, comfortable, and well-organized society. The Minoans' seaward trade brought copper from Cyprus, gold from Egypt, and copper and tin from what is now the Czech Republic, as well as a certain worldliness resulting from contact with the Scythians, Hittites, Babylonians, and others.

Sometime during the fifteenth century B.C., a second terrible devastation visited the prosperous Minoans. Theories

The Minoan palace at Knossos
Photo courtesy of Greek National Tourist Organization

abound—was it an earthquake? A war? The eruption of the volcano on the nearby island of Thíra? Was it some combination of the three? Scholars supporting the theory of a volcanic eruption believe it to have been of sufficiently enormous proportions to provide factual basis for the legend of the long-lost continent called Atlantis. Knossos's burial under rubble indicates natural disaster, but scorch marks on some of the palace walls imply that the hand of man played a role in the ancient's city's demise as well.

Evidence of the timely migration of many Cretans to Tunisia, Palestine, and elsewhere suggests that the islanders were aware of an impending catastrophe. Meanwhile, King Agamemnon's Mycenaeans, warlike mainland Greeks who had demonstrated their prowess by sacking Troy, were themselves driven from power by the invasion of Dorian soldiers from the Balkan Peninsula. The Mycenaeans took refuge on the islands of the Mediterranean, including Crete. Mycenaean domination was followed by a period of Dorian influence on Crete, and for the next several centuries the island was of little political or cultural importance. It was divided into numerous small city-states, which often were in conflict with one another. Many of the Eteorcretans (the

ancestral Cretans) and other islanders retreated to remote outposts, where they preserved many aspects of their archaic cultures. The Cretans' support of Persia in its war with Alexander the Great did little to improve the island's standing with the mainland, which in the meantime had developed a highly sophisticated civilization.

Records of activities on Crete are somewhat sparse from the late fourth century B.C. (the Hellenistic period) until the Saracen invasion of the ninth century A.D. Roman influence over Greek factions, growing considerably in the years before Christ, finally reached Crete in the year 67 B.C., when Crete was conquered by Metellus Creticus. In A.D. 64, St. Titus arrived on Crete to perform missionary work, and the remains of his church in the island's Roman capital of Gortyn are the oldest remnant of Cretan Christianity. Until its end in A.D. 337, the Roman occupation of Crete saw vast improvements in the island's water systems, and a modest increase in numbers of theatres and sanctuaries.

Crete, like so many other Aegean lands, was directly influenced by the rise of the Byzantine Empire in Constantinople. In the seventh century, Arab Saracens used Crete as a stepping-stone in their northward advance on Constantinople; in 824 they forced their leadership on the island, creating a stronghold at Iráklion (which was at the time the port of nearby Knossos; today it is the island's capital). The Saracens greatly fortified the city's Castro district, renaming it Rabdh-al-Khandah ("Fortress with a Moat"). Centuries later, under Venetian occupation, the fort's name evolved to Candia, a name by which the outside world knew the city and the entire island for several centuries.

Systematically destroying the Greco-Roman capital city at Knossos, the Islamic Saracens retained power in Crete for 137 years, and noble Greek families were dispatched from the mainland to the island to reassert Christian ideals. In 961, the future Byzantine Emperor Nicephorus II Phocas won the island back, intimidating his enemy by vaulting the severed heads of Saracens over the walls of their defense. Although the Byzantines brought a welcome stability to the island, native Cretans staged an unsuccessful bid for autonomy during the years 1090 to 1092, adding another episode to the long Cretan history of resistance to invaders.

Byzantium's relatively inauspicious reign on Crete lasted until 1212, when, during the crusades, Genoese merchants, having been awarded the commercially strategic island by the Byzantine emperor, promptly sold it to Venetian economic powers. The Venetians brought a vibrant economy to Crete, but their dictatorship had to quell no less than fourteen uprisings over four centuries. The first occurred almost immediately, after the new Venetian rulers confiscated estates to give to settlers from Italy. Another revolt took place in 1363, when rebellious islanders declared their homeland the Republic of St. Titus.

After the fall of Constantinople to the Ottoman Turks in 1453, Crete welcomed many of that city's refugees, including scholarly Byzantines who renewed interest in the legacies of ancient Greece. Inspired by the Italian Renaissance, they forged a cultural revival, rooted in Minoan traditions. The Venetians permitted the Byzantines to found a school for the study of ancient Greek manuscripts and other artifacts, at the monastery of St. Catherine in Candia.

With the occupation of Crete by Turkish forces, however, the scholars fled once again. In 1645, the Ottoman Turks sent 50,000 troops to begin besieging Candia. Although the city's great fort withstood years of attack, the Turks resolved to vanquish the city in 1666. The Italian hero Francesco Morosini commanded an army of Venetians, Greeks, and Cretans against the Turks, killing more than 100,000 of the invaders in fighting that dragged into 1669. At that time, the pope sent out a call for Christian forces to protect Candia; 8,000 French soldiers arrived, but their aid was not enough to prevent the Ottomans from finally toppling the defense.

With the increasing oppression of the working classes, who had brought prosperity to Crete, the island's economy fell into decline. Religious balance on the island tilted in the direction of Islam, due in no small part to a tax exemption extended only to Islamic worshipers; it is estimated that Islamic believers on Crete outnumbered Christians nearly four to one by the beginning of the eighteenth century. With a diminution of contact overseas, Crete became to Christian Europe a forgotten destination off the coast of Greece.

One hundred years later, however, Hellenism began to enjoy a revival, providing the seeds of a renewed Greek nationalism, one that nonetheless endured years of disruption, as foreign influence in Greek provinces continued. In 1824–25, the viceroy of Egypt, Muḥammad ʿAlī Pasha, asserted his rule over Crete, which the Turks had given him in gratitude for his military ventures against the rebellious Greeks. When Greece finally won its independence at the end of the decade, Crete was not part of the new nation. The British took it from Egypt, then returned it to Turkish rule in 1841. Cretans continued to desire autonomy; to express their furor over the continuation of enforced dependency, hundreds of Cretans committed suicide on November 8, 1866. The date is now commemorated as National Day.

Three years later, attendees of the Paris Conference validated Crete's autonomy, but not until November 1898 did the Turkish presence on the island end completely. By the time, the Cretans had mounted at least ten serious insurrections in 100 years. After the turn of the century, the Cretan minister of justice and foreign affairs, Eleuthérios Venizélos, led the movement for the island's reunification with the Greek nation, a union made official in 1913. Venizélos rose to national prominence when he orchestrated an enormous exchange of native peoples with the Turks; more than 1 million Greeks residing in Asia Minor were reassimilated into their homeland, many of them onto the island of Crete. A revered Cretan statesman, Venizélos ruled as premier of Greece in the late 1920s and early 1930s, but was sent into exile after instigating an unsuccessful military revolt in 1935.

Meanwhile, Crete was rapidly becoming an important site for archaeologists. Well-preserved examples of Minoan pottery had been discovered near Iráklion in 1878, an event that sparked a great surge of interest in the Cretan culture that had died out more than 3,000 years earlier. The German Heinrich Schliemann, who had previously proven the historical existence of the mythic Trojan War and other Homeric legends, was the first to dig at Knossos, a few miles south of Iráklion. Schliemann hoped to excavate the so-called Kephala land parcel at Knossos, but his attempt to buy the land fell through and he died soon thereafter.

Toward the end of the Turkish occupation of Crete, Sir Arthur Evans of Great Britain succeeded in purchasing the Kephala mound. During his first dig in 1900, Evans's men unearthed more than 700 engravings of ancient writing in just three weeks. The script came to be known as Linear-B; its predecessor, Linear-A, is thought to be the first system of writing to represent sounds. Linear-B was later deciphered by the Englishman Michael Ventris; Linear-A remains a mystery, although its contents appear to be the inventory lists of shopkeepers. Evans established the existence of a Cretan culture well before the Mycenaeans, dating it much earlier than the thirteenth century B.C. of Schliemann's hypotheses.

The British archaeologist named this ancestral group the Minoans.

Evans continued his excavation at Knossos until his death in 1941, by which time the site had been almost completely rebuilt in his image of it. Scholars have criticized Evans's liberal use of modern techniques in reconstructing the palace, where he went to great lengths to bring a present-day elegance to features such as the Grand Staircase, the Queen's Hall (Megaron), and the Throne Room. Evans spent $750,000 of his own money in thirty years at Knossos, building a headquarters he called the Villa Ariadne, which today serves as an extension of the British School of Archaeology at Athens. Still, Evans's museum, donated to Greece after his death, remains an important fount of information for students of ancient Crete. Behind a dwelling known as the Little Palace lies an ongoing excavation project that he described as the "Unexplored Mansion," and below the palace ruins, evidence has been discovered of a settlement on the site dating as far back as 3000 B.C.

Other archaeologists made similarly impressive discoveries during this period. To the south of Knossos, the Italian Federico Halbherr led a crew of workers at Phaistos, roughly forty miles to the southwest of Iráklion, along the island's south shoreline. In Gournia, an ancient manufacturing city was uncovered by the American Harriet Boyd Hawes, and in Mallia, French expeditions made still more remarkable discoveries.

World War II brought suffering to the historic island. On May 20, 1941, Nazi Germany's Seventh Parachute Division descended on Crete in pursuit of retreating Allied forces, beginning a three-year occupation of the island. One thousand islanders aided a counterattack that eventually resulted in the retirement of the vaunted Seventh Division, but the Allies' evacuation to Egypt left the island vulnerable to prolonged skirmishes with vengeful German troops. Nazi policy was to murder ten Cretans for each German soldier killed by rebel defenders, and the Cretans endured tremendous hardships through much of the war.

In the postwar years, Crete has become one of NATO's most important bases in Europe. The fierce convictions of Cretans remain unabated: 91 percent of Cretans voted in favor of abolishing the monarchy in Greece in the 1974 national referendum, and in 1979, the islanders took up arms to prevent the loan of art from their museum at Iráklion.

In 1972, Iráklion, now the fifth-largest city and third-largest port in Greece, once again became the capital of Crete; its 244,000 residents make up half of the island's population. It also is the capital of Hērákleion, one of four departments into which the island is divided. Today, its city limits stretch well beyond the medieval city walls. Despite the drawbacks of urban sprawl, many historic features continue to attract visitors to this venerable city. On the site of the old Venetian town square (today known as the Plateía Venizélou), the one-time civic aqueduct, stands the Morosini Fountain. Built in 1638 by the uncle of the military leader who defended Candia against the Turkish onslaught, the fountain combines an Italianate design with friezes of ancient Greek symbolism (it has been missing its pinnacle, a huge statue of Poseidon, since the Turkish invasion). Near Venizélos Square is the Loggia, once a lodge for the Venetian upper class, which has been completely rebuilt after its demolition during World War II. Behind it stands the Iráklion Town Hall, which once served as the Venetian armory.

Located in Plateía Eleutherías (Liberty Square), the social center of the city, is the Archaeological Museum of Iráklion. Housing the world's best collection of Minoan remains, including the Phaistos Disc, the museum consists of twenty rooms providing a chronological portrayal of Crete through the Greco-Roman period. Across town, housed in one of the few extant neoclassical mansions in Iráklion, the Historical and Ethnographical Museum represents Crete's later development, through the Byzantine, Venetian, and Ottoman occupations up to the present day.

Iráklion's St. Mina's Church is actually two churches, one built in 1735 and decorated with Byzantine icons, and the larger portion, employed as the cathedral of Crete, featuring a series of six works by a sixteenth-century master, Mikhailis Damoskinos, who is said to have taught the artist El Greco during his stay on the island. Of the five original gates of the city walls, the Kainouria (New Gate), built in the late sixteenth century, best represents the Venetians' determination to retain their stronghold: the walls at this gate are forty-five yards thick.

Near the Kainouria stands the Martinengo Bastion, the site of the tomb of the revered Cretan writer Nikos Kazantzakis. His novel *Zorba the Greek* gave the outside world a depiction of the quintessential Cretan: sturdy, joyous, and above all, intractable.

Further Reading: *Crete* by John Freely (New York: New Amsterdam, and London: Weidenfeld and Nicolson, 1988) and *Crete* by Robin Mead (London: Batsford, 1980) are both exceptional studies of this remarkable island, covering its roots in Greek myth and its present-day features. *Minoan Crete* by H. E. L. Mellersh (New York: Putnam's, and London: Evans, 1967) and *The Mystery of Minoan Civilization* by Leonard Cottrell (New York: World Publishing, 1971) are very helpful in framing the ancient Minoans' influence on later developments on Crete. Lawrence Durrell has offered a traveler's ode to this fascinating part of the world with the book *The Greek Islands* (London: Faber, and New York: Penguin, 1978). For the best-known fictional presentation of the Cretan people and culture, see Nikos Kazantzakis's *Zorba the Greek,* translated by Carl Wildman (New York: Simon and Schuster, and London: Lehmann, 1952; as *Bios kai politeia tour Alex e Zorma,* n.p.: Eleni Kazantzakis, 1946).

—James Sullivan

Cumae (Napoli, Italy)

Location: Cumae is located about twenty-two miles west of Naples, overlooking the ancient Phlegraean Fields and the Tyrrhenian Sea. It is accessible by car along the Ferrovia Cumana or by bus.

Description: Cumae, perhaps the most ancient Greek settlement on mainland Italy, is set in the volcanic hills that surround the Bay of Naples. Nearby are the ancient Roman ruins of Pozzuoli (now a Neapolitan suburb), Lake Avernus (which, according to the ancients, concealed the entrance to the Underworld), and the remains of the imperial Roman resort town of Baia (now mostly underwater).

Contact: A.A.S.T. (Azienda Autonoma di Cura, Soggiorno e Turismo)
Via Campi Flegri, 3
CAP 80078 Pozzuoli, Napoli
Italy
(81) 5262419 or 5261481

The settlement of Cumae—Cuma to modern Italians, Kyme to the ancient Greeks—is one of the oldest, if not the very oldest, Greek colony on mainland Italy. It was founded sometime before 750 B.C., perhaps as long ago as 1050 B.C. (although this date is questionable), by settlers from the Greek cities of Chalcis and Aeolis. Tradition suggests that Cumae was first settled by emigrants from the large Greek island of Euboea. Thanks to its position over the Bay of Naples and the volcanic Phlegraean Fields, believed to hide the entrance to the classical underworld, Cumae came to play a dominant role in the political history of Greater Greece—Magna Graecia.

The Greek states usually established colonies like the one at Cumae because of overpopulation. In mountainous Greece, good farming soil was scarce, and families were reluctant to subdivide their acreage. At such times they consulted omens and oracles, especially the Delphian Apollo, asking for permission to establish a colony. When the gods approved, city officials selected members of prominent families to lead the venture. Colonists were recruited mostly from citizens, but occasionally outsiders were also invited to participate. Ships were usually provided through city funds, but the actual selection of the site itself was usually left to the priests of the city's primary god or goddess. Settlers from Cumae may also have established other colonies in the area, at Dikaearchia (modern Pozzuoli) and Neapolis, or Naples.

Cumae quickly won a reputation as the major center of Greek culture in Italy. Some authorities have asserted that the major Italic alphabets were derived from a Cumaean Greek standard. The city reached the height of its powers in the fifth century B.C., when the tyrant Aristodemus defeated the Etruscans, a people who had settled farther north on the Italic peninsula. Sometime during the fifth or sixth century B.C., workers excavated a tunnel and chamber under the town's acropolis that would become Cumae's most famous landmark: the Cavern of the Cumaean Sibyl. In Virgil's words,

> Aeneas,
> In duty bound, went inland to the heights
> Where overshadowing Apollo dwells
> And nearby, in a place apart—a dark
> Enormous cave—the Sibyl feared by men.
> (*Aeneid*, VI:12-16; translation by R. Fitzgerald)

The entrance to the cave is described as desolate and deathly:

> The cavern was profound, wide-mouthed, and
>   huge,
> Rough underfoot, defended by dark pool
> And gloomy forest. Overhead, flying things
> Could never safely take their way, such deadly
> Exhalations rose from the black gorge
> Into the dome of heaven.
> (*Aeneid*, VI:331-36)

The Cumaean Sibyl was one of the three great prophetesses of the classical Greek world and the only one in Italy. The other two—the Oracle of Delphian Apollo and the Erythraen Sibyl—were more accessible to the rest of the ancient world. The sibyl herself served as the earthly vessel of the god Apollo, calling the god into her own body at the prayer of earnest petitioners, usually with the assistance of a drug. Virgil explains the process in Book VI of the *Aeneid*, saying that first her face changed expression and color, after which she fell into a kind of fit. The pious took her convulsions to be the presence of the god: "Apollo / Pulled her up raging, or else whipped her on, / Digging the spurs beneath her breast." Not everyone who approached the prophetess expected or received so intense a response. The sibyl also served the petitioners in less dramatic ways, communicating with the dead, interpreting dreams and omens, and offering medical and sexual advice to the lovelorn. Her rituals, known as mysteries, were kept secret for many years and preserved by her priests.

Cumae's position on the Italian peninsula, north of most of the Greek colonies of Magna Graecia, gave the city a special relationship with the native Italian states in Latium and Tuscany. According to legend, one of the Cumaean sibyls offered Tarquinius Superbus—Tarquin the Proud, the last king of Rome—the chance to purchase nine books of proph-

The upper acropolis at Cumae

Archaeological finds near the Cumae acropolis
Photos courtesy of The Chicago Public Library

ecies, called the Sibylline Books, which recorded predictions about the fate of Rome. When Tarquinius complained about the price, the sibyl destroyed three of the books and offered him the surviving volumes at the same price. When he still proved unwilling to pay the price, she repeated the gesture. Finally Tarquinius surrendered and bought the last three books for the same price she had originally asked for all nine volumes. Tarquinius gave the books into the keeping of the Capitoline Temple of Jupiter, where they were preserved for hundreds of years and consulted by the city fathers in critical times. Tarquinius himself was exiled from Rome in 509 B.C., supposedly after one of his sons raped the high-born Lucretia. After a failed attempt by the Etruscan king of Clusium, Lars Porsena, to reseat him on the Roman throne—imaginatively retold by Thomas Babington Macaulay in his *Lays of Ancient Rome*—Tarquinius spent his last years in Cumae.

Unfortunately, the Greek colonists proved just as willing to quarrel with their neighbors as their Greek forebears. An Etruscan expedition was defeated near Cumae in 474 B.C. by Hiero I of Syracuse, the largest and richest Greek colony in Sicily. This victory was celebrated by the poet Pindar in his *Pythian Odes*. Only fifty years later, however, in 421 B.C., the Samnites, a tribe based in the Apennine mountains of central Italy, conquered Cumae and other parts of Magna Graecia. Cumae itself fell into Roman hands in 334 B.C., when the Latins were extending their power down the Italian peninsula.

Cumae itself did not prosper under the Romans. It was eclipsed by nearby Roman towns such as Baia, a famous seaside resort, and the island of Ischia. Baia in particular became so infamous for its debauched and promiscuous night life that the Roman writer Petronius based his *Satyricon* on stories from the town. Julius Caesar established Cleopatra in his summer villa in Baia; she was staying there when news of his assassination reached her in 44 B.C. Emperor Claudius installed his wife Messalina there as well. Furthermore, it was in Baia that Agrippina, the mother of Nero, Claudius's successor, was killed by assassins sent by her son. Cumae and its sibyl paled by comparison.

Cumae continued to decline in significance after the Roman Empire fell. It was sacked by Moslem raiders in 915 and finally destroyed by forces from Naples and Aversa. The last few residents deserted it around the year 1207. Excavations at the site began in 1927 and continue to the present day.

The ruins of Cumae consist of an acropolis and a town farther down the slope. Although excavations are incomplete, they do show that the town contained an amphitheatre and public baths. There was also at least one temple devoted to the worship of Zeus, Hera, and Athena—the three deities later known as the Capitoline Triad, showing connections to Rome rather than to Greece. In Roman times, Emperor Domitian built a road (the Via Domitiana) that ran near Cumae. A triumphal arch spans the remains of the road, the foundations of which are still visible.

Up the hill from the town a road runs to the acropolis, surrounded by a cyclopean wall, almost 250 feet above. At one point workers during the reign of Augustus Caesar drove a tunnel under the acropolis to the sea. Farther along, another, older, carved passage cuts through the hill, tunneling almost 150 yards through the rock. Running on a north-south line, it measures about eight feet wide and seventeen feet high, and includes side corridors looking westward toward the sea. Early Christians buried their dead in crypts along the opposite wall, on the landward side of the passage.

The cavern of the sibyl, where Aeneas' journey to the underworld begins, is still accessible today and forms a major part of the attraction at the Cumaean excavation. In Book VI of *The Aeneid*, the sibyl instructs Aeneas to do two things: offer sacrifices for the souls of his dead comrades, and find the golden bough that will give him access to the underworld. After he has done so, she agrees to lead him to Erebus and back through the nearby entrance at Lake Avernus. In the underworld Aeneas has to confront his own past: he meets his former helmsman Palinurus whom he has not been able to bury (a disgrace for a nobleman), his former lover Dido, Queen of Carthage, who committed suicide when Aeneas sailed away and left her at the end of Book IV, and an acquaintance from the Trojan War, Deiphobus, killed by Menelaus in the sack of Troy. Eventually Aeneas and his guide pass through Erebus and Tartarus, where sinners are tortured, and reach Elysium, where they find Aeneas' father Anchises, the object of their search. Anchises shows his son a pageant of all the Roman line that will spring from him, culminating in Augustus Caesar, Virgil's patron. After this revelation Aeneas and the sibyl make their way to the surface world again, and the Trojan prince sails away north to meet his destiny.

The passage leads to the sibyl's cave (Antro della Sibilla), a square, vaulted chamber with three niches. Virgil—who was in a position to know—said that the sibyl's cave had 100 different openings that carried the god's messages to petitioners. The entrance is marked with marble tablets that recall the dramatic lines of the *The Aeneid*: "Here, as the men approached the entrance way, / The Sibyl cried out: / 'Now is the time to ask / Your destinies!' / And then: / 'The god! Look there! / The god!'"

The road itself continues to the top of the hill. The acropolis is dominated by the ruins of two temples: the Temple of Apollo that Aeneas sought at the beginning of Book VI of *The Aeneid* and the Temple of Jupiter. The necropolis of the ancient city lies north of the temples. Both temples served as churches at times in the early Christian era—dating from the fifth or sixth century. There are also traces of a baptismal pool where prospective Christians received the bath that welcomed them into the church. The sibyl's cave by that time was probably deserted, although tradition states that, until the fourth century A.D., the bishop of Rome himself occasionally trod the stony path to place a petition before the priestess of the pagan god.

Further Reading: No other event in Cumae's history is as significant as its role in *The Aeneid*. Visitors with a literary bent will want to visit the site with a copy of the epic poem in hand—a good modern translation in English is the one by Robert Fitzgerald (New York: Macmillan, 1965; London: Collier-Macmillan, 1968). Those with less ambitious tastes can find a good guide to sites featured in the book in William S. Anderson's *The Art of the Aeneid* (Englewood Cliffs, New Jersey: Prentice-Hall, 1969; Hemel Hempstead, Hartfordshire: Prentice-Hall, 1970). The three-volume *Civilization of the Ancient Mediterranean,* edited by Michael Grant and Rachel Kitzinger (New York: Scribner's, 1988) also provides good background material. J. Boardman's *The Greeks Overseas* (Harmondsworth, Middlesex, and Baltimore, Maryland: Penguin, 1966; revised, New York, Thames and Hudson, 1980; London, Thames and Hudson, 1981) serves as an introduction to the Greek settlements in Italy.

—Kenneth R. Shepherd

Delos (Cyclades, Greece)

Location: Delos lies in the Aegean Sea in the center of the fifty-six-island Cyclades group, so named because the islands were said in ancient times to surround Delos in a circle (*Kyklos* in Greek); to the east of the Peloponnese and southeast of the coast of Attica; just six nautical miles southwest from the better-known island of Mykonos.

Description: Delos, which is also called Little Delos to differentiate it from Rhenea, better known as Great Delos, is a small island known as the sacred birthplace of Apollo and the holy island of the ancient Greeks.

Contact: Greek National Tourist Organization
Town Hall, second floor
Hermoupolis, Cyclades
Greece
(281) 22375

A long and narrow, treeless and hilly island that measures only about four miles long and half a mile wide, Delos was once the "source of Apollonian light," where, on the conical 386-foot-high Mount Cynthos, Leto was said to give birth to Apollo under the shade of a date palm. Today, the island is a vast open-air archaeological museum; according to author Dana Facaros, its "only permanent population are the guardians of the ruins, appointed by the Greek government, and a million or so lizards who play on the broken marble."

According to Greek myth, Zeus began a love affair with a woman named Asteria, who fled her lusty lover in the shape of a quail. Zeus, not one to give up easily, transformed himself into an eagle to more readily pursue her. To avoid capture, Asteria turned into a rock and plunged into the sea. The rock became known as Ortygia (Quail) or Adelos (Invisible One), and floated, unanchored, beneath the sea.

His sexual appetite not quenched, Zeus turned his attentions to Asteria's sister Leto, whom he impregnated while he was in the form of a sword. But Hera, the jealous wife of Zeus, flew into a rage, pursuing Leto and denying her a place to give birth. No island would risk Hera's wrath. Only insignificant Adelos consented to be the birthplace of the future God of Light, and only after being promised that the god born there would forever remain its protector. In gratitude, Zeus caused four adamantine columns to rise from the bottom of the sea and anchor the island. According to another myth, Poseidon touched Delos with his trident, affixing it firmly to the bottom of the sea floor. Leto was carried to the islands, where she gave birth to Artemis, the goddess of the hunt and virginity, and nine days later to Apollo, the god of truth and light. Adelos, the invisible, not under the sun but under the sea, officially became Delos, meaning visible.

Myth aside, why did this barren island propel itself to an importance that equaled Delphi, Olympia, and Epidaurus as a sanctuary to the ancient Greeks? Certainly, the mythology played a major part. But in addition, Delos enjoyed an outstanding geographical location in the center of the Cyclades and was the finest harbor for ships that sailed between the Gulf of Argolis on the Greek mainland and the cities of the Ionian coast of Asia.

The history of the island can be traced to the third millennium B.C., when the island was settled by Carians, Legeles, or Phoenicians. But it was not until 1000 B.C. that the worship of Apollo was launched here by the Ionians, whose rites were mentioned in a Homeric hymn. By the year 700 Delos had emerged as an important center of worship. Delos was also on its way to becoming a commercial port, as pilgrims from all over were drawn to its shores. In the second half of the seventh century B.C. and the first half of the sixth century, the Sanctuary of Apollo was dominated by Naxos and later by Paros. But as the sixth century progressed, Athens began to gradually take control of the region.

With the rise of Athens, the sacred sites of Delos were drawn into the political undercurrent. Indeed, the Athenians took great pains to emphasize the historical and mythological connections between Delos and their city. They pointed out that the first delegation to Delos was led by Erechtheus, the king of Athens, and that Theseus danced around the altar at Apollo after killing the Minotaur on Crete. Not content with these linkages, the Athenians in 543 B.C. produced a Delphic oracle decreeing that the island be purified by removing all graves to Rhenea, thereby severing Delos's connections to its own, more immediate past.

Soon thereafter, the Persian Wars were fought, and the population of Delos fled to Tenos to escape the advance of Persian king Darius I. Darius, according to Greek historian Herodotus, respected—even honored—the sacred site. When the Persian fleet looted the Cyclades in 490, no damage was done to Delos. The Delians were even allowed to return to their home in safety.

The Persians were eventually defeated, and a new era began for Delos. In 478, the first Athenian League was launched, and an Amphictyonic League, or Delian Alliance, was established with Delos at its center. Each island contributed both financial resources and ships to support the navy, which was controlled by Athens. The Delian Alliance was effective; however, in 454, Pericles moved it to Athens, supposedly to protect—more likely to spend—the league's treasury. Not long after, a disastrous plague befell Athens, which was determined to be caused by Apollo's wrath. Instead of purifying Athens, it was decreed, conveniently, that a second purification of Delos was necessary. This time, the edict was exceptionally harsh. In 426 and 425, the Athenians

A view of Delos
Photo courtesy of Greek National Tourist Organization

transferred all the contents of the graves to Rhenea, where they were reinterred in a mass grave (discovered and excavated at the end of the nineteenth century). They also forbade childbirths and burials on Delos, forcing the pregnant and dying to nearby Rhenea.

In desperation, the Delians turned to Sparta for aid during the Peloponnesian War, but the Spartans refused. Nevertheless, Athens punished Delos for its impertinence in consulting Sparta by exiling the entire "impure" population to Asia Minor in 422 B.C. Athenian settlers took the place of the exiled Delians. After a series of setbacks in the Peloponnesian War, Athenians became convinced that their action had offended the gods. To appease the divine powers, they relented and the Delians returned home. Finally, for a brief period between 403 and 394, Delos enjoyed the independence and freedom it had long sought, when the Spartans defeated Athens. A second Delian Alliance was formed, albeit not as powerful as the first. The Delians wished to oust the Athenians altogether from the island, but Philip II of Macedon, the league's leader, refused.

Despite that rebuff, Delos continued to thrive. By the year 314, Delos had grown to become a flourishing religious center in an island alliance under the protection of the Ptolemaic dynasty of Egypt; the year 250 saw it under the guardianship of the Macedonian kings. The influence of Delos continued to grow. The island developed into a cosmopolitan city, full of temples, grand houses, and impressive buildings. By the year 250, Delos was one of the area's leading commercial ports, frequented by merchants from all over the Mediterranean.

When the Romans defeated the Macedonians in 166 B.C., they ceded the island to the Athenians and declared it a free port, hoping to undermine their rival, Rhodes. The Athenians once again exiled the Delians to Achaea, and installed their own colonies. Delos now enjoyed an unbeatable combination: the sanctity of its shrines, the protection of Rome, and the ambition of the Athenians. With the fall of Corinth, Delos was the center of all east-west trade, the most important commercial port in the entire Aegean. People from all over the world—Athenians and other Greeks, Italians,

Syrians, Egyptians, Phoenicians, Jews, and Palestinians—flocked to this thriving island, building their own respective shrines. As many as 25,000 inhabitants settled here. Carpenters set to work constructing new quays and piers to handle the heavy volume of vessels heading toward Delos's shores. The great "melting pot" of the Aegean enjoyed growth that seemed to know no boundaries.

Moreover, its religious influence grew, too. The Athenians organized a festival to honor Apollo each May; in addition, feasts known as the Delia were celebrated every four years, involving music, dancing, athletics, the sacrifice of oxen, and mass revelry. A sacred lake—now dry—was then large enough to stage mock sea battles. Pilgrims arrived in Delos from distances near and far to view the games and processions, or to consult the oracle here.

But the glory of Delos was not to last. While most Greeks supported the war of Mithradates VI Eupator, king of Pontus, against Rome, Delos remained loyal to the Romans. Mithradates took the region and proved a vengeful conqueror; in 88 B.C. his fleet destroyed and sacked many of Delos's treasures, murdering as many as 20,000 people in one day and carrying off women and children as slaves. The island was recaptured by General Lucius Cornelius Sulla of the Roman army, who walled a part of the ruined city and awarded it exceptional privileges, including tax concessions. However, just nineteen years later, Delos was again ransacked by invaders, this time pirates who were connected to the hated Mithradates, and its people were once again dragged off to slave markets. Pirates and scavengers from other islands looted it in search of marble and stone.

The degree of Delos's decline became apparent when Athens attempted to offer the island for sale, but found no takers. In A.D. 363, Roman emperor Julian the Apostate decided to reintroduce paganism on Delos. But an oracle gave a more accurate prediction of the island's future: "Delos shall become Adelos." In the late fourth century, Theodosius I (the Great) banned pagan ceremonies. Pirates took control of the island in the sixth century, putting an end to the small Christian community that had survived there. From that time, house builders on Tenos and Mykonos used Delos for a marble quarry; it gradually became pastureland.

As the Turks began to gain ascendancy, Delos and Rhenea were turned over to the municipality of Mykonos.

After Greece was liberated during the War of Independence in the 1820s, the new Greek state confirmed this ownership.

Excavations of Delos began in 1772 with the work of Pasch van Krienen, a Dutch-Prussian scholar and officer in the Russian military, who sent the artifacts he uncovered to St. Petersburg and Bucharest. In 1829, the members of the Expédition Scientifique de la Morée conducted a small excavation. These were followed in 1872 by a massive project undertaken by the French Archaeological School in Athens. Even today, excavations continue, primarily by French and Greek archaeologists.

Though uninhabited today, Delos still attracts throngs of visitors who come to pay homage to the little island that was a religious and cultural center of the Aegean for 1,000 years. There are many sites that reflect the island's former splendor. Perhaps the most impressive ruin is the terrace of archaic stone lions, which have silently roared at the sun for centuries. Among the many other notable ruins are the Sanctuary of Apollo; the smaller Ionian Sanctuaries of Artemis, Sarapia, and Hera; the Stoas of Philip (built by Philip V of Macedon); the Agora (marketplace) of the Delians; a synagogue built in the second century B.C. by Phoenician Jews; the world-famous glowing mosaics in the House of Dolphins, composed of fragments of precious stones; and the House of Masks, with its mosaics of comical and satiric masks. In addition, the Museum on Delos is filled with sculptures and artifacts from the Archaic, Classical, Hellenistic and Roman periods.

Further Reading: *Land of the Lost Gods: The Search for Classical Greece* by Richard Stoneman (Norman: University of Oklahoma Press, 1987) chronicles the European rediscovery of ancient Greece through the late nineteenth century. Other helpful works on Delos are *Greece: The Cyclades* by Dana Facaros, one of the Cadogan Island Guides (Old Saybrook, Connecticut: Globe Pequot, 1994); *Greek Island Hopping: A Handbook for the Independent Traveller*, also by Dana Facaros (London: Gentry, and New York: Hippocrene, 1979); *Some Greek Islands: The Shores of Light* by Joseph Braddock (New York: Roy Publishers, 1967); *The Greek Islands* by Robin Mead (London: Batsford, 1979); and *Mykonos/Delos: Today and Yesterday* (Athens: Toumbis, 1986).

—Jill I. Shtulman

Delphi (Phocis, Greece)

Location: The ruins of ancient Delphi lie three-quarters of a mile east of the modern town of Delphi, in the Phocis department of central Greece, about six miles from the north shore of the Gulf of Corinth.

Description: Ancient Delphi was the seat of the oracle of Delphi, who made prophecies that affected many events in Greek history; it was an important cultural as well as a religious center. The site was abandoned after the Christian emperors of Rome waged a successful campaign to wipe out pagan practices. Interest in the shrine was renewed in the seventeenth and eighteenth centuries, and its remains began to be excavated in the late nineteenth century. Today visitors can see the ruins of ancient temples and other buildings, and many artifacts are displayed in the Delphi Museum.

Site Office: Delphi Museum
Delphi, Phocis
Greece
(265) 82313

At the heart of Delphi's importance in pre-Christian times was its status as a sanctuary for a divine oracle whose prophecies and advice profoundly influenced the culture and history of the Hellenic world. According to myth and legend, there was from the earliest times at Delphi a shrine of the Earth Mother Ge, or Gaia, whose worship, according to archaeologists, dates back to Ice-Age peoples. Prophecies at the shrine were reported as early as the second millennium B.C. when Delphi was a Mycenaean village. Legends tell of the sibyls, divine prophetesses whose utterances were considered to be revelations from the gods. By the beginning of the first millennium, ancient Delphi was well known to Mediterranean peoples.

According to mythology, a dramatic change occurred at Delphi around 1000 B.C. with the arrival of the Greek god Apollo. It was believed that the oracle of the earth goddess was guarded by a great serpent known as the python. Legend says that Apollo killed the serpent with an arrow, became the lord of Delphi, and was known thereafter as the Pythian Apollo. Apollo was later joined by his younger brother Dionysus, a mystic, who also came to be worshiped at Delphi.

Apollo was the son of Zeus and was considered to be symbolic of the best elements of the Greek spirit. He represented art, music, poetry, beauty, health, political virtue, and moderate behavior. He was unique among the gods for bringing to people knowledge of the thoughts of Zeus. Thus his messages communicated through the oracle provided divine guidance for the best course of human affairs. Delphi was an appropriate place for a temple to Apollo because of the area's physical beauty and the ancient Greeks' belief that it was the true center of the world.

The Temple of Apollo at Delphi, originally built during the eighth century B.C., was a sanctuary where the oracle's activities took place. It contained the sacred omphalos, or stone, which marked the center of the universe. A priestess sat upon a tripod, probably a three-legged chair, to declaim her revelations. Before making a prophecy, the priestess was purified by bathing in the Castalian spring near the temple. She was then placed in a trance induced either by drugs or religious fervor. Questions were put to the oracle either orally or in writing. The answer would be interpreted and often made intelligible by priests who would write out the message if requested. Very important revelations were often engraved on stone. Inquiries from one of the city-states were answered on a sealed tablet and returned to the city by courier.

Not long after the advent of Apollo, the worship of Dionysus was introduced at Delphi. Dionysus was the god of irrationality, ecstasy, and drunkenness, and as such he was seen by the Greeks to provide a balance for the more serene aspects of Apollo. The two gods divided the religious duties of the sanctuary. During the three winter months, Apollo journeyed north to the land of Hyperborea, known in myth as the land of the blessed isles. He was replaced during these absences by Dionysus, although the oracle's activities continued without interruption.

Of the oracles themselves, there are only a few of their pronouncements that have been recorded in history. An oracle around 680 B.C. ordered citizens of Megara to found the city of Byzantium on the Bosporus, which became Constantinople and then İstanbul. Other Greek colonizations were also encouraged by the Delphic oracle. In 547 B.C., Croesus, the king of Lydia, was told that he would destroy a great kingdom if he crossed the Halys River. He did so, was defeated by the Persians, and it was his own kingdom that was destroyed. In 480 B.C., Athens was threatened by Persian military forces. The oracle told the Athenians that they would be unconquerable behind a wooden rampart. This prophecy proved to be true: an Athenian fleet composed of wooden boats defeated the Persians at the decisive battle of Salamis. Many other of the oracle's pronouncements influenced political and economic decisions of importance in the Mediterranean world. In return for the oracle's advice the temple received numerous valuable gifts. The remains of some of these treasures can be seen in the Delphi Museum.

The remains at Delphi are of the third temple on the site, not counting several primitive buildings that may have served as holy places in archaic times. The first substantial temple was built of stone in the late eighth or early seventh

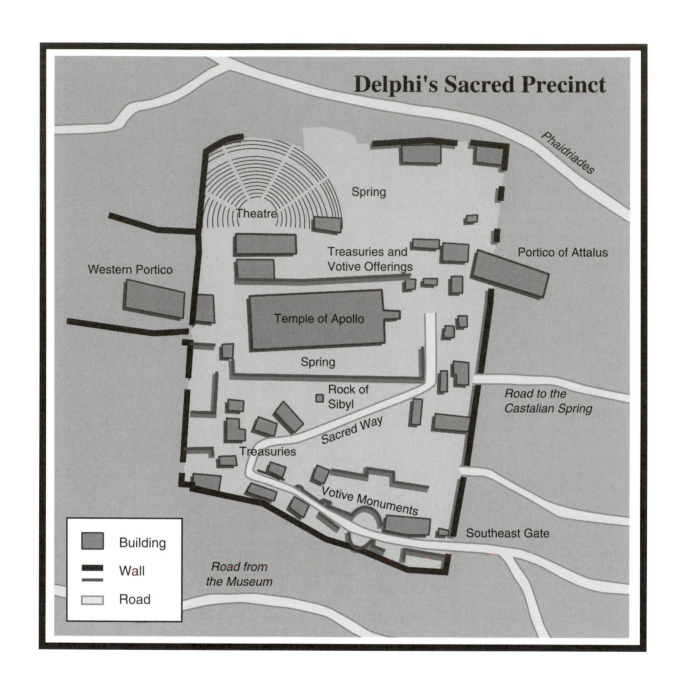

Delphi's sacred precinct
Illustration by Tom Willcockson

century B.C. At the time, Delphi was controlled by the Crissaeans, who levied excessive fees and tolls on pilgrims traveling to the sanctuary. The Delphic priests complained to the Amphyctyonic Council, a sort of mutual-aid group representing twelve Greek cities. The Amphyctyons, with the help of Athens, waged war on Crissa in what became known as the first Sacred War, sometime between 600 and 590 B.C. Crissa was defeated, its territory was dedicated to Apollo, and Delphi became autonomous, with the management of the sanc-

tuary in the hands of the Amphyctyonic Council and the Delphic priests.

This first important temple was destroyed by fire in 548 B.C. An impressive new temple was completed by the end of the sixth century but was destroyed by an earthquake in 373 B.C. Yet another temple was built, which survived until Roman times.

The oracular and religious observances at the temple were accompanied by the Pythian games, held periodically at a stadium built at Delphi about 450 B.C. and rivaling the Olympic Games in importance to the Greek world. The Pythian Games began as musical contests in honor of Apollo; athletic contests were added later. The ruins of the stadium have been excavated together with those of the nearby theatre.

Delphi also was a notable art center. Important Greek historical events were commemorated by paintings and sculptures often given as gifts and offerings to the temple. The most famous and skilled artists in the Hellenic world came to Delphi, making the sanctuary a repository and museum of many of the greatest artistic works of the times. When the artists returned home they spread knowledge of new artistic ideas throughout Greece, creating an unparalleled brilliance in pan-Hellenic culture. The monuments and trophies in the sacred enclosure at Delphi were a record of the accomplishments and artistic achievements of the Greek states. All too many of these artifacts have been lost over the centuries but some still remain to be seen at Delphi.

After the Persian Wars, Delphi, whose oracle's prophecies were often of political significance, became involved in the frequently intense arguments between the Greek city-states. This intervention cost Delphi some of its prestige, and the oracle was even accused of taking bribes. The second Sacred War, about 448 B.C., returned control of Delphi to the Cressaeans, but the shrine regained autonomy by 421 B.C. A third Sacred War, 357 to 346 B.C., was caused by Phocean attempts to cultivate the sacred plain of Crissa: it brought about the fateful intervention of Philip of Macedon. In a fourth Sacred War (c. 338 B.C.), Philip defeated the Athenians and Thebans. This victory paved the way for domination over Greece by Philip and later by his son Alexander, whose conquests were of vast historical consequence. In about 189 B.C. Delphi fell under the control of Roman conquerors and was later pillaged by the emperors Sulla, Nero, Constantine, and Theodosius I. Delphi had enjoyed a short revival under the emperor Hadrian, but Theodosius I, in the name of Christianity, silenced the oracle and in approximately A.D. 390 destroyed the temple and most of the statues and works of art.

About A.D. 363 the emperor Julian the Apostate had attempted to alter the Christian commitment of the empire. His attempts to restore the best aspects of paganism included sending a mission to Delphi with words of encouragement. He received in reply a last despairing message from the temple that the oracle was no longer functioning. Julian's efforts to restore the pagan culture of the Greek and Roman world failed, and most of the pagan artifacts and temples were destroyed in the ensuing years as Christian rulers tried to remove all traces of paganism. Delphi and its environs were destroyed and left in ruins. A small village grew up at the site, which was later named Kastri for a crusader castle that was built by invading Franks near the location of the Delphic sanctuary. Kastri's houses were built of marble taken from the nearby monuments, including the ruined temple. In this manner most of the remains of ancient Delphi vanished. In later years, the sanctuary was buried under earth and rocks falling from the steep flanks of Mount Parnassus, and all knowledge of its precise position and even its name was lost.

By the fourteenth century A.D. western scholars began to rediscover the literature and the artistic beauty of ancient Greece. Among the classic authors who were read again were Pausanias and Plutarch, who described Delphi as it was in its days of importance. In the next century, a traveler named Cyriac found near Mount Parnassus some inscriptions that mentioned Delphi. By the late seventeenth and early eighteenth centuries many European travelers were visiting Greece to investigate the fabulous culture spoken of in the ancient writings. Among the earliest were an Englishman named George Wheeler and a French scholar named Jacques Spon who in 1676 journeyed to the port of Itea. Using the 1,500-year-old description of Pausanias to guide them, they located the ruins of Delphi and described the site accurately and precisely in a book published in 1682. They also described the shabby village of Kastri inhabited by mostly illiterate Greeks who knew little about their ancient history.

The situation in Greece began to change by the beginning of the nineteenth century. After years of rule by the Ottoman Turks, the Greeks began to fight their way back to independence and in the process began to rediscover their heritage. People of other countries also became involved in the renaissance in Greece. In 1812 Byron visited Delphi and described the site in romantic and impressionistic terms. Greece became independent in 1829, and archaeologists began to think seriously of restoring Delphi, one of the most complicated archaeological sites in all of Greece. Early excavators found that Kastri had been built atop much of the site.

The competition for the excavation rights was fierce among France, Great Britain, and the United States. After lengthy and complex negotiations, the Greek government, unable to afford the cost of the excavations itself, awarded the right to the French School in Athens under the leadership of the dynamic archaeologist Théophile Homolle. The village of Kastri was moved at French expense, and on October 10, 1892, the excavations and research began at the Delphic site. They have continued to the present.

The ancient city of Delphi lies immediately to the east of the modern town of Delphi, the former Kastri. It is on the road that runs from the port of Itea twelve miles southwest to the mountain village of Arakhova six miles to the east. Though Delphi is not as well known as Athens, the mystique and historical significance of the famous Delphic oracle continue to make Delphi one of the prime tourist destinations in Greece. The visitor to Delphi will find the layout of the site roughly divided into three parts: the sanctuary of Apollo, the Castalian spring and the sanctuary of Athena, and the museum.

The sanctuary of Apollo, which lies above the main road, can be reached by a path running past the museum and through the remains of an ancient Roman marketplace. The path enters the sacred precinct at the southeast gateway and becomes known as the Sacred Way. The path then leads uphill first to the west, then heads sharply to the northeast, and then north to the entrance of the temple of Apollo. Originally the

Sacred Way was lined with votive monuments erected by Athens, Sparta, Corcyra, Argos, Taras, and other city-states. The monuments have vanished except for their bases. Beyond the monuments are the remains of more than twenty treasuries in which gifts to the shrine were preserved and displayed. Again little can be seen except the foundations, although the Treasury of the Athenians was re-erected early in the twentieth century in the form of a Doric temple.

After the treasuries comes the oldest part of the sacred precinct where cult ceremonies were carried out; it contains the Rock of the Sibyl, thought to be the sanctuary of Ge the earth mother. Only the foundations are left of the third temple, but on the hill above the temple the famous statue of the charioteer was found, preserved when it was buried under a landslide. It is now in the Delphi Museum.

At the northwest corner of the sacred precinct is the theatre built in the fourth century B.C. with room for spectators. Nearby, but outside the boundary of the temple precinct, is the stadion, or stadium, of which many of the tiers of seating survive. These are the sites of the Pythian Games. East of the precinct is the Castalian spring, lying in a gorge between the Phaidriades, where pilgrims purified themselves before approaching the temple and the oracle. Several hundred feet to the east is the sanctuary of the Athena Pronaia with remains of a fourth century B.C. temple of Athena. Finally, the museum is located close to the southern boundary of the ancient city near the main road. The museum contains a collection of finds from the site including an omphalos stone of the Roman era, friezes from the Treasury of the Siphnians, sculptures, and a life-size bull made of silver and gold. Also housed in the museum are numerous artifacts, architectural fragments, and bronzes unearthed during the excavations at Delphi.

Further Reading: Peter Hoyle's *Delphi* (London: Cassell, and New York: Fernhill, 1967) is a knowledgeable, clearly written, and well-illustrated history and description of Delphi. *Delphi and Olympia* by Erik J. Holmberg (Götburg, Sweden: Paul Astrom, 1979) offers a concise and specific discussion of the site with considerable architectural information. *The Delphic Oracle* by T. Dempsey (Oxford: Blackwell, 1918; New York: Benjamin Blom, 1972) is a scholarly study of the history, influence, and fall of the oracle with numerous footnotes to source material. Richard G. Geldard's *The Traveler's Key to Ancient Greece* (New York: Knopf, 1989; London: Harrap Columbus, 1990) contains an informative discussion of Delphi as well as other ancient Greek sites. It is notable for numerous photographs, diagrams, and maps. *Baedeker's Greece* (Englewood Cliffs, New Jersey: Prentice-Hall, and Basingstoke, Hampshire: Automobile Association, 1992) is an outstanding guidebook in the Baedeker series. It includes a descriptive and well-illustrated section on Delphi with detailed maps of the site.

—Bernard A. Block

Didyma (Aydın, Turkey)

Location: On a plateau ten miles south of peninsular Miletus in western Asia Minor.

Description: Site of famous temple and oracle; initially dedicated to a local Anatolian goddess, but later, following the invasion by Ionian Greeks in the first millennium B.C., converted to an oracle of Apollo; endpoint of the Sacred Way, a processional path from Miletus to Didyma; location of the Megala Didymeia, a series of artistic and athletic competitions held every four years in honor of Apollo.

Contact: Tourist Information Office
Yeni Dörtyol Mevkii
Aydın, Aydın
Turkey
(256) 2254145, 2126226, or 2250099

According to Pausanias, a Greek geographer from the second century A.D., Didyma was home to an oracle even before the arrival in Anatolia of the Greek Ionians around 1000 B.C. The original oracle was associated with the worship of a local Anatolian goddess and centered on a temple, sacred grove, and nearby holy spring that gave off a sulphurous vapor.

The invading Ionians replaced this goddess with Apollo, and eventually they came to believe that Apollo had been conceived by Leto and Zeus on the site. By the eighth century B.C. the Ionians had begun construction of a temple there. Remains from this structure, discovered by German archaeologist Heinrich Drerup in 1962, are the earliest yet excavated on the site. The temple is thought to have measured approximately thirty-five by seventy-nine feet. At the end of the seventh century B.C. this structure was enlarged by a twelve-by-fifty-one-foot colonnade.

Work on this archaic temple was probably completed by 560 B.C., as indicated by stylistic similarities to the temples of Samos and Ephesus, both built around this time, and by records of famous gifts being bestowed upon the temple at this date. Additions from this final construction phase include a naiskos, or small temple that housed the famous cult statue of Apollo Philesis carved by Canachos of Sicyon. The naiskos was situated inside the larger temple itself and had no roof, although it was constructed in such a way as to appear from the outside to be enclosed. The circular altar was located east of the temple proper. A well was located to the west of the altar.

Much of the temple was made of tufa (hardened or compacted volcanic debris). The outer edifice, columns of the pronaos (vestibule), and complete upper section of the temple were fashioned from marble. Two ornamental heads and two winged gorgons decorated each point of the temple's pedi-

ment. Other excavated features include a laurel tree (sacred to Apollo), votive statues, and stoas (detached colonnades). Many of these artifacts are now at museums in Berlin and İstanbul.

The temple was approached along the Sacred Way, the route of a procession of crown bearers that began in Miletus ten miles to the north. Near Didyma the route was lined on each side by statues of sphinxes, lions, priests, and priestesses. Some of these figures date to the sixth and seventh centuries B.C.; they now rest in the British Museum. The end of the road opened onto a magnificent esplanade where offerings were made.

The Sacred Way is noteworthy not only for its artistic and religious importance, but for its political significance as well. It functioned as a bridge, extending the political authority of Miletus beyond the traditional boundaries of the polis (city-state). Although the temple was ten miles from the city of Miletus, it was, in fact, Milesian territory. The priest of Apollo was a Milesian official who lived on the temple grounds during his one-year term. Other than the servants of the temple, Didyma had no permanent residents.

The oracle was administered by the Branchidae, a clan of priests who claimed to be descendants of a legendary figure named Branchus. According to the myth, Branchus was originally a shepherd in Delphi, where he caught the attention of Apollo. The god became enamored of Branchus and seduced him, after which Apollo granted the youth the gift of prophesy. Branchus then traveled to Didyma and built the oracular shrine there to Apollo Philesis (Apollo the Affectionate). The Branchidae administered the temple for hundreds of years.

Those wishing to consult the oracle first cleansed themselves in the sacred well and offered a sacrifice on the altar. The client's questions were then put to a prophetess, who entered the sacred spring in the naiskos and became inspired by the sulfurous gases. Her prophesy would take the form of an incoherent utterance, which would be deciphered (and set into hexameter verse) by the priest or his assistants.

By the sixth century B.C. the oracle was famous throughout the ancient Mediterranean and was widely consulted and honored by rulers from across the area. According to Greek historian Herodotus, King Necho I of Egypt, after his victory over the Syrians at Magdolus, offered his battle armor as a gift to the Branchidae; this was the first recorded dedication by an Egyptian ruler to a Greek temple. Necho II is also said to have offered gifts to the oracle. Rulers of closer lands, such as King Croesus of Lydia, made frequent presents to the oracle. Herodotus records that Croesus's gifts of gold were commensurate in style and weight with those he sent to Delphi—indicating that Didyma had truly become a sacred site of the first order.

Remains of the Apollo temple at Didyma
Photo courtesy of Embassy of Turkey, Washington, D.C.

The heyday of Ionian Asia Minor came to an end with invasions by Cimmeria, Lydia, and finally Persia, which conquered the region in 547 B.C. After an unsuccessful revolt by the Ionians in 494 B.C., the Persians under Darius I destroyed the temple. The Branchidae, rather than offering resistance, willingly yielded the temple's treasures to the Persians (according to other sources, the treasures were given not to Darius but to Xerxes I, following the Battle of Plataea in 479). The Branchidae now feared retribution for the betrayal of their office and begged the Persians to take them away. The conquerors agreed, settling the former priests at Sogdiana. Canachos's statue of Apollo eventually was taken to Ecbatana.

The oracle fell silent for almost 150 years, until the coming of Alexander the Great to Didyma. Alexander, ever savvy about the use of local mythology for personal propaganda, employed at this time a philosopher by the name of Callisthenes, whom historians have likened to a press agent. Callisthenes seized the opportunity afforded by the silent oracle; he announced that upon Alexander's entrance into the shrine, the long-dry spring suddenly flowed once more and the voice of the oracle was heard, proclaiming Alexander to be the son of Zeus and predicting his victory over Darius III at the battle of Gaugamela in 331.

The oracle at Didyma was considered divinely active again, and Alexander saw to the temple's reconstruction following his victories in the region. In approximately 330 B.C. (or, according to other sources, 300 B.C.) work began on the temple foundations, the spring of the oracle, and the adyton (the inner shrine of the temple). Architects on the new, Hellenistic temple included Daphnis and Paionios of Ephesus, who had been involved in the construction of the Artemisium at Ephesus. In 300 the stolen cult statue was brought back from Ecbatana by Seleucus I Nicator, emperor of Syria, who assured continued work on the project following Alexander's death in 323. According to legend, Seleucus would have done well to pay more attention to the oracle and less to the temple that housed it. In the early third century B.C. the oracle advised him not to cross to Europe; Seleucus ignored the warning and was assassinated by his former ally Ptolemy Careens.

The Hellenistic temple measured 168 feet by 359 feet, much larger than the Ionic structure that had previously occupied the site. The building was said to be the third largest Hellenic temple in existence, after only the Artemisium and the Heraion of Samos. Initial work on the structure included the addition of a third set of stairs and the elevation of the walls to thirty feet above the base of the peristyle. The naiskos, which has been excavated only in pieces, represents the oldest discovered Hellenistic structure in an Attic style. According to written sources, the temple was home to shrines of five other gods, as well as the sanctuary of Apollo.

Construction stopped at the end of the third century but continued again following the Peace of Apamaeus, which ended hostilities between the Romans and the Syrians in 188 B.C. The Romans enlarged the boundaries of the sacred site and funded the paving of the Sacred Way. Around 200 B.C. Didyma also became the site of the Megala Didymeia, a series of athletic and artistic competitions held every fourth year in honor of Apollo. The athletic games took place in a stadium approximately fifty feet south of the temple; competitions involving music, oratory, and drama took place on the temple grounds themselves.

Work on the temple was again halted at the end of the second century B.C., after which the temple was looted and burned. Under the Roman emperors Tiberius and Caligula, interest developed in the restoration and preservation of the temple (it is said, however, that Caligula wished to install himself as the oracular deity of the temple). By A.D. 46, the door to the temple, damaged during earlier raids, had been repaired. Soon, decorative work was added to the facade, including the fashioning of a frieze. Emperor Hadrian orchestrated further work on the temple between 117 and 137, but the project was never completed.

Goths attacked Didyma in 256, but area residents successfully defended the temple. Ultimately, however, it was religion, and not warfare, that doomed the temple. As Christianity spread, the oracle lost its authority. The temple received a temporary reprieve during the reign of Emperor Julian the Apostate (361–363), but the devoutly Christian Theodosius I officially abolished the oracle in 385 as part of his larger attempt to quash any remaining pagan practices in his empire. The temple was converted to a fortress and Christian basilica, complete with protective wall and castle. Defenses were added until the tenth century, but fires and earthquakes eventually destroyed the structures and forced the area's complete abandonment.

The ruins at Didyma came to the attention of European archaeologists and antiquaries following a visit to the site by Englishman Richard Chandler in 1764. The Society of Dilettanti soon published drawings of the ruins, and Chandler described them in his *Travels in Asia Minor* (1775). The first large-scale excavations were undertaken in 1872–73 by O. Rayet and A. Thomas, and work continued to be done on the site sporadically over the next hundred years. Most recently, work has been conducted under the leadership of German archaeologists. Many of the artifacts uncovered are now in Museums in London, Berlin, and İstanbul.

Further Reading: To learn more about Didyma and the greater Anatolian region, consult *Classical Anatolia: The Glory of Hellenism* by Harry Brewster (London and New York: Tauris, 1933) and *Roman Rule in Asia Minor* by David Magie (Princeton, New Jersey: Princeton University Press, 1950; London: Oxford University Press, 1951).

—Christopher P. Collier and Robert M. Salkin

Diyarbakır (Diyarbakır, Turkey)

Location: Diyarbakır, the largest city in southeastern Turkey, is the capital of the province (*vilayet*) of the same name. It is situated on a bluff in a fertile plain of the upper Tigris River valley, in northern historic Mesopotamia and cultural Kurdistan. The surrounding region is separated from eastern Anatolia by the Taurus Mountains in the north, and from the Mesopotamian plain by the Mardin hills in the south.

Description: Diyarbakır is among the oldest cities in the world. Because of its position at the junction of trade routes linking Asia Minor with neighboring areas, it has been of commercial and military importance since ancient times. Today, Diyarbakır is a predominantly Kurdish and Sunni Moslem city, with many beautiful mosques, most dating to late Turcoman and early Ottoman times.

Site Office: Tourist Information Office
Kültür Sarayı
6, Diyarbakır, Diyarbakır
Turkey
17840-32635

Set in southeastern Turkey in the "cradle of civilization" (the upper Mesopotamia area between the Tigris and Euphrates Rivers), Diyarbakir is considered to be one of the oldest cities on earth. Hence, it has a colorful and violent history. The old town is still surrounded by massive black basalt walls, the city's most prominent feature, originally built in Roman times. The imposing fortifications, said to rival the Great Wall of China in breadth, later gave the city its popular ancient name of Kara ("black") Amid. But the thick walls could not prevent the city, at the strategic head of the Tigris near the Persian frontier, from being a battleground for a long succession of people throughout its roughly 5,000-year history.

Even before the Romans established a colony on the site of Diyarbakır in the third century A.D., the Hurrians, the earliest native race of Anatolia (Asia Minor or modern Asian Turkey), probably founded the first city there, called Hurri-Mitanni. The Hittites conquered the Hurrian kingdom (c. 1400 B.C.), which reemerged in the ninth century B.C. as the Urartian Empire. The Urartes held the city through the second half of the eighth century B.C. until the Assyrians conquered it and called it Amidiya. The Persians and Medes annexed the Urartian state after defeating the Assyrians in 612 B.C. Then came successive periods of Macedonian, Roman and Byzantine, and Sassanid (Persian) domination. A seventh-century Arab group gave Amida its modern name: the Bakr clan renamed it Diyar Bakr—"Place of the Bakr" or "City of the Bakir Tribes." With the decline of Arab influence in the region, the city was subsequently ruled by other Moslem peoples—the Kurds, the Seljuk Turks, the Turcomans, the Mongols, the Persians again—before finally falling to the Ottomans in 1515. This signaled the beginning of a long and relatively peaceful era.

Today, Diyarbakır is predominantly Kurdish, and is Turkey's most Arab city, with distinctive Armenian, Christian, Kurdish, and Arab quarters, a maze of narrow and hidden courtyards. The city has more historical mosques, churches, and other notable religious buildings—most dating from the Ottoman period—than any other Turkish city except İstanbul.

Diyarbakır's classical period began in 331 B.C. when Macedonian king Alexander the Great defeated the Persians. His dynastic successors, the Seleucids, then ruled the eastern Anatolia region. The Seleucid empire soon passed into Parthian Persian hands. The Romans appeared in A.D. 115; over the next several centuries, they and their successors, the Byzantines, fought violently with the Sassanid Persians to maintain control over the town, now called Amida. The city changed hands frequently in wars between the Persians and Romans. The Romans colonized the city in 230, building the first substantial walls in 297. These walls became the main line of defense between the Romans and the Parthian/Sassanian empires of Persia. Roman emperor Constantius II reconstructed the walls in 349. After a lengthy battle in 359, the city was surrendered to Persian king Shāpūr II. Though ceded by Emperor Jovian to the Persians in 363, Amida was soon re-annexed to the Roman Empire. (When Rome fell, its Greek-speaking eastern realm continued as the Byzantine Empire.) Amida was once more taken by the Persians during the reign of Byzantine emperor Anastatius I in 502.

Successive Byzantine rulers, most notably Justinian, expanded and strengthened the walls. The current walls were built over the previous ones by the Seljuks in the late eleventh century. The walls, stretching for three miles around the old city and breached by four huge main gateways, are still in remarkably good shape despite centuries of military assault. Sixty-seven of the original seventy-two defensive towers, some rising to 500 feet, still stand. The walls, gates, and towers are decorated with the designs and inscriptions of conquering armies. Though the old city remains, Diyarbakır has long since spread beyond its confining walls.

Several Arab Moslem groups struggled for possession of the city in the seventh century; some of them also restored the walls. In 638 or 639, during the first great expansion of Islam, the Saracen Arab armies of Khālid Ibn al-Walīd (Sword of Islam) captured Amida, dubbing it Diyarbakır after the Bakr tribe that was granted the town and its area. The city was subsequently held by the Umayyad and Abbasid Arabs, and then the Marwani Kurds before being captured in the early eleventh century by Basil II, Macedonian king of a

Diyarbakır city walls
Photo courtesy of Embassy of Turkey, Washington, D.C.

revived Byzantine Empire. It was taken in 1088 by the Seljuk Turks. Seljuk Malik-Shāh (or Melek Ahmet Sah), prince of Isfahan, completely rebuilt the walls and towers, adding many inscriptions and geometric and animal reliefs. Though the Artukid Turcomans (or Ortokid Turkomans) ruled the Diyarbakır region from the eleventh to fifteenth centuries, the city was sacked by Mongol conqueror Timur (Tamerlane), in the late fourteenth century. Akkoyunlu ("White Sheep") Turcoman tribes retained control in the area throughout most of the 1400s, but Diyarbakır was a Safavid Persian stronghold when the city—and eastern Anatolia—were conquered by the Ottoman sultan Yavuz Selim I in 1515. It became a part of the Ottoman Empire two years later, and was the protected and prosperous capital of a large and important province.

Despite the many fierce struggles for Diyarbakır, it remains Turkey's preeminently Kurdish city. The city's population is overwhelmingly Kurdish, some of them Iraqi war refugees, with a minority of Arabs, Armenians, and even some Afghan refugees. Historically a nomadic mountain people, the Kurds are thought by some to be descendants of the ancient Medes who once ruled southeast Anatolia after the collapse of the Urartian Empire. Others believe them to be descendants of the tribesmen (known as the Kardouchoi or Carduchi by the Greeks and Romans) who occupied the Zagros mountains, and who attacked Greek soldier of fortune Xenophon and the Ten Thousand army in 401 B.C. near modern-day Zakhu, Iraq. In any event, the Kurds are a distinct racial group indigenous to the region traditionally known as Kurdistan, which today straddles the modern borders of Turkey, Iran, and Iraq, reaching to Syria and the former Soviet Union. The Kurds speak an Indo-European language related to Farsi and Peshto, with several mutually unintelligible dialects. Of an estimated 16 million Kurds—a large "minority" indeed—the greatest number are in Turkey, estimated to be between 7 million and 10 million out of a nation of some 60 million. Turkish Kurds are accorded a low status by many other Turks. They are officially referred to as "mountain Turks," the authorities denying them a distinct racial identity, and the use of the Kurdish language is forbidden within Turkey.

The Kurds have never had statehood, partly due to their traditional clan organization, which is further grouped into chieftain-headed tribes; they are further subdivided by religious affiliation, being both Sunni and Shiite Moslem, the proportion varying by region. In the seventh-century Arab conquest of eastern Anatolia, most Kurds accepted Islam; the vast majority of Turkish Kurds are Sunni. The Kurds' various dialects have also been a hindrance to unification. Throughout most of their history, the Kurds did not resist invaders. After the Median empire was replaced by the Persian in 550 B.C., they were nominally ruled at various times by the Seleucids, Parthians, Persian Sassanians, the Arab Caliphate, Mongols, Seljuks, Turcomans, and Ottomans. Under the Ottomans, majority Turkish Sunni Kurds did not enjoy special "nation" status like Christian and Jewish religious minorities.

Because Kurds were known for their military prowess, sultans often employed them as mercenaries in frontier marches. Semi-independent principalities established by the Arab-speaking Hasanwagh and Marwanid Kurds survived in eastern Anatolia into the first half of the nineteenth century, acting as a buffer between the Ottoman-Turkish and Safavid-Persian border; but it was only in the nineteenth century, with the decline of the Turkish empire, that Kurdish nationalism grew. With the collapse of the Ottoman Empire in 1918, an active Kurdish independence movement developed.

The 1920 Treaty of Sevres, drawn up by the Allies at the end of World War I, stipulated that an autonomous Kurdistan should be carved from part of the defunct Turkish empire, a state straddling the current borders of Turkey, Iraq, and Iran. The treaty, however, was never ratified; the Turkish War of Independence had intervened. The May 1919 Greek invasion of Turkey, in Smyrna (İzmir), was the spark that flamed latent Turk nationalism. Led by General Mustafa Kemal, later the president known as Atatürk (Father of the Turks), the Turks defeated the Greeks and proclaimed the secular Republic of Turkey in 1923. While the Kurds had joined Atatürk in warding off the Greek Christians on the west and the Armenians on the east, they were not rewarded with their own nation-state. Instead, the League of Nations granted nation status to Iraq in 1925. In the Turkish Kurd revolts of the 1920s and 1930s, hundreds of thousands of Kurds were executed or deported, and Kurdish culture was suppressed. Numerous short-lived rebellions took place in historic Kurdistan between 1925 and 1945.

The most serious revolt in Turkey erupted in February 1925, while the League of Nations was still discussing the "Kurdish problem." The uprising was due to a number of factors: resentment of Turkish rule, the desire for Kurdish independence, and outrage at the abolition of the religious Caliphate. The organized rebellion was led by Sheikh Said, the hereditary chief of a dervish order. He announced it was time to end the Republic and restore the Caliphate. Insurgents quickly gained control in most eastern provinces, including that of Diyarbakır, and martial law was declared. Turkish troops put down the revolt by April. A special tribunal at Diyarbakır ordered the closing of dervish lodges, and condemned to death Sheikh Said and forty-six other Kurdish nationalist leaders, who were executed shortly thereafter. After the siege was lifted, hundreds of rebels were hanged on the spot in the Diyarbakır area. Sporadic unrest continued in eastern Turkey for many years to come, with major risings in 1930, 1936, and 1939, each followed by executions and deportations. Since the late 1960s, the Turkish government has suppressed Kurdish political and cultural activity, though there has been easing on some forms of cultural expression since 1991.

Diyarbakır currently hosts thousands of Kurdish refugees from Iraq: the first wave fled Iraqi poison gas attacks in 1988, and the second, larger wave came after the end of the Persian Gulf War in 1991. They have joined many Turkish Kurds in Diyarbakır. While some of the Iraqi Kurds are Pesh Merga (Forward to Death) guerillas who have been fighting for an independent Kurdish state in northern Iraq for more than forty years, Diyarbakır today does not appear to be a center for militant Kurdish dissident activity, but in the nearby mountainous region, Kurdish guerilla attacks on the army are still frequent. However, the city, an important Turkish military and NATO base, has not been without its recent tensions. The PKK (Kurdistan Workers' Party) was founded in 1978 in order to establish a Marxist, Kurdish-run state in southeast Turkey. Just one of several Kurdish separatist national movements, the PKK represents the ideological extreme of Kurds' resistance to forced assimilation or marginalization. In the days preceding Turkey's 1980 military coup, Diyarbakır was a hive of separatist activity, and thousands of suspected Kurdish dissidents were locked up in the city's prison, Turkey's most notorious. Since 1984, PKK party activists and other Kurdish rebels have fought a guerilla war with Turkish security forces, mostly in remote southeastern regions. The struggle for an independent Kurdish state continues, with no resolution in sight.

Diyarbakır is a city of modern buildings hiding beautiful basalt-and-sandstone mosques, of old black basalt walls enclosing narrow, dusty winding streets. Roman influence is evident in the old town's rough rectangular grid. The new town, with wider roads and concrete buildings, sprawls outside the walls. The old city's two main streets connect the wall's four main gates, one at each end of the compass point; there are also several smaller gates. The main entrance, to the north, is Harput Gate (Harput Kapisi), once known to the Arabs as Bab El-Arman, or Gate of the Armenians. It is the best preserved of all city gates, a black tower with reliefs and inscriptions in Greek and Arabic.

The Saray, or Palace Gate (Saray Kapisi), accessible from inside the old town, is perhaps the most beautiful; it leads into the Ic Kale, Diyarbakır's military-controlled citadel—possibly the oldest part of the city, dating to the fourth century B.C. Inside the citadel area are the remains of an Artakid Turcoman palace; their animal designs can be seen on the walls. To the east are the Ogrun Gate (Ogrun Kapisi) and the Yeni Gate (Yeni Kapisi), both of which overlook the River Tigris. To the south is Mardin Gate (Mardin Kapi), which overlooks sprawling slums, and to the west is Urfa Gate (Urfa Kapisi), also called Bab El-Rumi. The stretch of wall running between these two gates is the most well preserved, and contains two huge tower bastions added to the fortifications in 1208: Yedi Kardes Burcu (Seven Brothers Tower) and Meliksah Burcu, also known as Ulu Badan (Great Wall), each of which is elaborately decorated with

Seljuk animal motifs. To the northwest is the minor Double Gate (Cifte Kapisi).

Diyarbakır abounds in religious buildings, including twenty-two old mosques with towering minarets, some dating to pre-Ottoman times, as well as some historic churches. Near the center of the old city stands Ulu Cami (Grand Mosque), the oldest place of Moslem worship in Anatolia, and possibly the oldest mosque in all of Turkey. It is also one of the Five Holy Places of Islam. Ulu Cami was the very first of Anatolia's great Seljuk mosques, built by Malik-Shāh in 1091 and 1092, three years after he conquered Diyarbakır. It is modeled on the Great Umayyad Mosque at Damascus—a city that Diyarbakır resembles in some respects. Ulu Cami is claimed as Turkey's oldest mosque because it was built on the site of—and with the masonry from—the Byzantine Syriac basilica of Mar Touma or St. Thomas, at the time of the Moslem Arab conquest of A.D. 639. Though gutted by fire in 1155, the mosque was extensively refurbished and has not been altered since. The large Arab courtyard plan (contrasted with the covered, domed mosques common in Turkey) consists of fountains; Roman/Byzantine columns, capitals, and arcades; Seljuk arches; and Greek and early Arabic inscriptions. Also inside the courtyard is the entrance to Mesudiye Medresesi, the first university in Anatolia, which was built by the Artukids in 1198, and is still used as a Koran school. The city's bazaar and some caravansaries are located in the mosque area.

Another mosque of historic importance is the Citadel Mosque (also called the Süleymaniye Cami, or the Mosque of St. Süleyman), located inside the Ic Kale citadel. It was built by the Artukid Abū-al-Qāsim Ali in 1155 in honor of the twenty-four Moslem martyrs who first breached the city walls during Khālid Ibn al-Walīd's conquest of the city in 639. One of them, Süleyman, was Walīd's son. The Citadel Mosque and many other mosques in Diyarbakır were constructed using alternating bands of black basalt and white sandstone, a favored design of local medieval architects because both stones are common to the region. The largest mosque in Diyarbakır, the Behram Pasa Cami, was built in 1572 by Behram Pasa, Ottoman governor of Diyarbakır. The lovely and gracefully decorated Safa Cami is attributed to Uzun Hasan, the great Akkoyunlu ("White Sheep") Turcoman leader of the fifteenth century. The same clan also built the early sixteenth-century Nebi Cami, or Mosque of the Prophet, and the Kasim Padisah Cami—the last Turcoman mosques before Ottoman domination. The latter mosque's tall, square detached minaret, set on four basalt pillars, is unique in the Moslem world and is known locally as "Dort Ayakli Minare," or "Four-legged Minaret."

Other notable mosques are the Peygamber Cami (1530), Iskender Pasa Cami (1551), Fatih Pasa Cami (1572), and Melik Ahmet Pasa Mosque (1591). Notable Christian churches include the Syrian Orthodox Church of the Virgin Mary and the Surp Giragos Kilisesi, serving the few remaining Armenian families in town. The restored churches of Sts. Cosmos and Damian (Jacobite) and of St. James (Greek), date from Byzantine times.

Diyarbakır, located atop a commanding bluff in the upper Tigris River valley at the junction of trade routes connecting eastern Anatolia and the Black Sea coast with Mesopotamia and the Mediterranean, has been of commercial importance since ancient times. The city has long been famous for its gold and silver filigree work. Diyarbakır is the major market center of southeast Anatolia, and the region's agricultural produce is the best in eastern Turkey. The raising of sheep and goats, as well as animal product processing, are the area's other major economic activities. Its principal exports include woolen and cotton textiles, and copper ore. A railroad from Ankara was built in 1935, and the Diyarbakır University was founded in 1966.

Diyarbakır has experienced rapid population growth since the end of World War II, leading to the creation of the modern town outside the old city walls—and a myriad of social and economic development problems. The recent influx of villager and peasant job-seekers from the surrounding countryside, at the rate of about 50,000 a year, and Kurdish refugee encampments, has only exacerbated the situation. Despite the elegant Moslem architecture, Diyarbakır appears somewhat squalid: it is chaotic, dusty, overcrowded, and overwhelmingly poor. Unemployment is close to 40 percent. It is hoped that the Southeast Anatolia Project, a series of thirteen dams that would convert an England-sized wasteland to irrigated fields, will help alleviate some of Diyarbakır's problems. Many of the town's men have lately converted to Islamic fundamentalism, which, though forbidden by the government, is accepted by authorities because it serves as a substitute for Kurdish dissident activity. Diyarbakır's province is one of twelve in southeastern Turkey still under martial law.

The city's historic strategic importance is underscored by the presence of numerous military installations in the area: the city is home to a Turkish army base (charged with checking regional PKK activities), a Turkish air force base (attached to a joint U.S.-Turkish base used in the Persian Gulf War), and a NATO airbase.

Further Reading: Two books by veteran scholars of Turkey are accessible introductions to the country and its history: *The Companion Guide to Turkey* by John Freely (London: Collins, 1979) and *Modern Turkey* by Geoffrey Lewis (New York: Praeger, and London: Benn, 1974). *Ancient Civilizations and Ruins of Turkey* by Ekrem Akurgal (İstanbul: Haset Kitabevi, 1973) and *The Ottoman Centuries* by Lord Kinross (New York: Morrow Quill, 1977) present more detailed historical surveys, including that of the eastern Anatolian region. By far the most comprehensive modern guidebook is *Turkey: The Rough Guide* by Rosie Ayliffe, Marc Dubin, and John Gawthrop (London: Rough Guides, Ltd., 1994). Solidly written and researched, the nearly 800-page book abounds with historical contexts and contemporary facts.

—Jeff W. Huebner

Dubrovnik (Neretva, Croatia)

Location: Situated on the Dalmatian (Adriatic) coast of southern Croatia, bordering Bosnia-Hercegovina on the east, and Montenegro to the south.

Description: Founded in the seventh century A.D., city called officially by its Latin name, Ragusium, or Ragusa, until modern times (Dubrovnik derived from Slavic word for forest); an independent city-state for most of its history up to 1808; in its heyday, fourteenth to sixteenth centuries, grew wealthy from maritime trade and shipping; one of the most beautifully preserved medieval seaport cities in Europe; popular resort town; heavily damaged 1991–92 by Serb forces after Croatia declared independence.

Site Office: Dubrovnik Tourism Office
Cvijete Zuzoric 1
50000 Dubrovnik, Neretva
Croatia
(50) 26-304 or 25-078

The enchanting city of Dubrovnik is situated on what was a tiny island in the seventh century A.D., when the first immigrants landed there, survivors of the barbarian destruction of their city-state, Epidaurus. Because Epidaurus had once belonged to the ancient Roman Empire, the inhabitants gave their new settlement the Latin name of "Ragusium," or "Ragusa" in its shortened form. Along with Christianity, they brought with them a knowledge of the sea that would provide their livelihood as well as that of future generations for the next 1,100 years.

Ragusa was once an island that belonged to the Dalmatian archipelago of approximately 700 large and small islands stretching 1,000 miles along the eastern Adriatic coast, ending at the bay of Kotor. (The city was joined to the mainland in the thirteenth century.) Its climate is moist, warm, and Mediterranean, with cool nights and hot days fanned by continual sea breezes. Away from the sea, turning inland, lie the mountain ranges of Bosnia-Hercegovina with the highest peak, Mount Orjen, at 6,217 feet, forming a dramatic background to the city.

Protection was the first order of business and one of the highest priorities for the next several hundred years. Ragusans erected wall after stone wall around their growing seaport, with the help of Slavic neighbors on the mainland, who referred to Ragusa as "Dubrovnik," meaning a forested area, with dwellings built of wood. The surrounding region was rich in pine forests, whose heavy scent must have been a relief from the stench of the city's raw sewage decaying in the hot sun, and from the animals roaming freely in the streets.

The town was so well fortified by the late ninth century that in 886 it withstood a year-long siege of Arab forces attacking from the sea. The semi-civilized neighboring Slavs to the north and west also were a constant threat at this time. Nonetheless, increasing numbers of them sensed growing opportunity in the seaport town, and by the thirteenth century, the majority of Ragusa's population was Slavic. Still, Latin and not the Slavic language remained the official tongue, and as late as 1472 a city statute prohibited the use of the Slavic language in government, a reaction to its increasing use among city officials. The Slavs were almost evenly divided between Serbs and Croats until Ragusa achieved its independence from Venetian rule in 1358, when the overwhelming majority of the inhabitants in and around the city were Croats.

The walls, ramparts, and gates of Ragusa formed a complete and extensive ring around the town; today, one can stroll the entire length of the old town on top of these walls and savor the picturesque architecture and sea vistas, marveling that the vulnerable city on a rock survived so many vicissitudes. Throughout the Middle Ages, the most powerful city-state politically and commercially was Venice, which would tolerate no fledgling competitor on the Adriatic; hence from 1205 until 1358, Venice ruled Ragusa with an iron hand, appointing the most important city officials.

Well before Ragusa became a vassal of Venice during the late Middle Ages, it was the richest European city on the eastern Adriatic coast, rivaling Venice in sophistication. Until the period of Venetian domination, Ragusa was for most of its existence an independent city-state, although nominally a part of the Byzantine Empire. Besides carrying on a brisk sea trade, Ragusans did increasing business with the hinterlands of Bosnia and Serbia, with the encouragement of Venice, which hoped to deflect Ragusa from its maritime activity. In 1186, the city concluded a trade agreement with the ruler of Serbia, and in 1189, a similar treaty was reached with Bosnia. Many western goods made their way to the Balkan hinterland in exchange for raw materials, especially ores, while Ragusan merchants served as intermediaries.

With independence from Venice, the Ragusan government established a monopoly on the important salt trade, and salt panning became one of the city's most significant industries until well into the sixteenth century. Ragusan merchants also were not above making fortunes from the slave trade, although the municipal government abolished the practice in 1416 when the Portuguese began surpassing them. Meanwhile, a shipbuilding industry developed and gave Ragusa the reputation of being the "Holland of the east." To foster and stimulate trade relations after 1358, city officials carried out intricate diplomatic negotiations with Ragusa's neighbors.

The Sponza Palace
Photo courtesy of Embassy of Croatia, Washington, D.C.

Achieving independence in 1358 was only one of many milestones in Ragusa's history. In the early thirteenth century, the government ordered the channel separating the city from the mainland (hitherto connected by a bridge) to be filled in. The result was the Placa, Ragusa's main street, from which all other avenues radiated, and the center of town life and entertainment. This act enabled Ragusa to expand beyond the rocky island on which it lay. With each expansion northward, massive stone walls were built for protection. In 1278, the first official archives were established, becoming in time a rich treasure trove of information concerning the city's history and development, its trade relations with dozens of European states, as well as the rise of the Ottoman Turks and their conquests. The majority of these documents were composed in Latin, a written rather than a spoken language. The meticulous Italian archivists and recordkeepers eventually were replaced by equally meticulous Slavs in the fourteenth century.

The Italian influence was overwhelming and visible everywhere, lasting long after Ragusa was freed from Venetian rule by the Peace of Zadar of 1358. Wealthy Ragusans sent their sons to Italian universities; Italian architects were invited to construct churches, fountains, palaces, even the city's all-important aqueduct. The bulk of the grain on which Ragusa depended for its daily bread was imported from Italy. Italian was the language of merchants and the professional classes, even though by the time Ragusa achieved its independence from Venice in the mid-fourteenth century, the overwhelming majority of residents, rich and poor alike, were Slavs, who nonetheless retained "Ragusa" as the official name of their city. Ragusa remained a decidedly western city, partaking in the major cultural and intellectual movements and trends of western Europe, yet always maintaining a very precarious balance between west and east.

By the early fifteenth century, Ragusa was referred to as a "republic." After 1358, Ragusa was nominally a vassal of the Croato-Hungarian kingdom, but in exchange for an annual payment of 500 gold ducats, the rulers of this kingdom agreed to let Ragusa go its own way. For more than a century, Ragusa enjoyed peace, marred by occasional outbreaks of enmity from Venice—attacks on Ragusan shipping and other trade restrictions, with the Ragusans retaliating in kind. In the meantime, neighboring Bosnia and Serbia had fallen to the Turks after fierce resistance. The Ragusans preferred instead to do business with the Turks, who guaranteed their independence and reinstated their salt monopoly in exchange for an annual tribute in gold, a sum that the Ragusan government consistently kept low through persistent negotiation. It was no accident that a city fortress completed in the fifteenth century bore the inscription, "Freedom is not sold for all the gold in the world," although Ragusans apparently were willing to pay a good deal of gold for it.

Independence seemed to go hand in hand with dramatic improvements in the city's quality of life. By the early fifteenth century, stone houses replaced wooden ones, even for the poor. In the same period, Ragusa's government, con-

sisting of the Major Council, the Small Council, and the Senate, initiated city-wide street cleaning and garbage disposal. This entailed hiring and training a team of municipal garbage collectors, and ordering all shop and home owners to sweep the street in front of their property every Saturday afternoon. A sewage system was completed at the same time, which drained all city refuse into the sea. Fresh water supplied all the city's needs by means of an excellent aqueduct system built under the direction of Italian architect Onofrio de la Cava, and completed in the mid-fifteenth century. It was so well built that the city relied on it for its fresh water until the mid-twentieth century. No famine was ever recorded in the city. All grain was imported because of the dearth of arable farmland in and around the city-state. The organization of the grain supply was in the hands of a city commission specifically charged with the import and distribution of the grain, usually barley, millet, and wheat.

Such improvements did not wipe out all unhealthy living conditions. The poor may have lived in stone houses, but these were damp, tiny (one room per floor) and shoddily built. Sanitary ordinances, such as those requiring street sweeping and garbage disposal, were not always taken seriously. Pestilence in the form of bubonic plague ravaged the city, as it did all of Europe, in the 1340s, and reappeared in less dramatic episodes every decade for the next 200 years. The more than 100 doctors registered in the city were unable to cope with serious illnesses. The mortality rate was high, as was the out-of-wedlock birth rate. In view of the latter, an orphanage was built in 1432, where babies were accepted with no questions asked. The population of Ragusa at the height of its wealth in the fifteenth century stood at approximately 40,000, 20,000 less than what it is today.

Despite the fact that Ragusa was a self-governing republic, there was much social stratification, with the city government always in the hands of the same wealthy aristocratic families, until most of them were decimated by the catastrophic earthquake of 1667. Only then did the stranglehold of these patrician families give way to a broadening of the power base. The stipulation that a government official had to be Roman Catholic still held sway, and Eastern Orthodox Christians, while welcomed as visitors, could not even be permanent residents of Ragusa, although Jews and Moslems could.

Ragusa from its beginnings was always a bastion of Roman Catholicism. Today even the serious traveler sometimes fails to notice Ragusa's role as a center of Roman Catholic missionary activity. Dominicans and Franciscans, mainly Italians, vied with each other to spread the Roman Catholic faith among the neighboring, staunchly Orthodox Christian Serbs, as well as among the non-Catholic, non-Orthodox Bosnian Christians, whose "conversion" also was the objective of the Orthodox Church. Islam in time would have the greatest success in converting the heterodox Bosnian Slavs. The Catholic orders also spurred the building of the many beautiful ecclesiastical structures in and around Ragusa. The Benedictine order, first settling on nearby

Lokrum Island in 1023, moved into the city in the fourteenth century and built its own monastery, followed by many other monasteries, convents, and churches. This strong religious presence also spurred the building of the city's cathedral and the famous Church of St. Blaise, the patron saint of Ragusa. Many of these beautiful cloisters and convents still exist. Within the Franciscan compound the first pharmacy in eastern Europe was opened in 1317; it is still in operation.

Throughout the period of its independence, from the mid-fourteenth century onward, the city fathers of Ragusa would not tolerate the building of even a single Orthodox church, so anxious were they to preserve Christian religious homogeneity. By contrast, Jews, who began living freely in their own section of town from 1352 onward, obtained permission to establish their own synagogue in the fifteenth century, judged to be the oldest synagogue still standing in southeastern Europe.

Ragusa at its height in the fifteenth and sixteenth centuries was as wealthy and as cosmopolitan as Venice, with a commercial fleet equal to its Italian counterpart. It was an efficiently governed city constantly improving itself in terms of sanitation, education, and beautification. Two impressive city fountains were commissioned of Onofrio de la Cava, the same architect who had been in charge of the city aqueduct and the Rector's Palace (the seat of the city administration). A lovely canopied square, the "loggia," in the city center was constructed in 1356, able to hold up to 4,000 Ragusans, solely for the purpose of conversation (stone benches were built for this purpose) and leisure activities. A colorful civic pageant was held every year on the feast day of St. Blaise, in early February.

The biggest catastrophe in the city's history until recent times was the devastating earthquake of 1667, which leveled most of the city and claimed thousands of lives. Rebuilding afterward occurred slowly, and on a much reduced scale from earlier days, chiefly because the economic fortunes of the city had declined steeply by the seventeenth century. Powerful new maritime competitors had arisen—the Dutch, French, and British—who were supported by their own wealthy centralized governments. From 1527, Ragusa was technically a vassal of the Ottoman Empire, with which it had nothing—whether religion, language, culture, or custom—in common. The payment of an annual tribute in gold, in return for a precarious independence and security from Turkish attack, was a severe drain on the city's finances. The Mediterranean was no longer the center of European maritime activity: the important trade routes now led to and from the New World.

With a decline in shipping, Ragusa's salt trade and subsidiary industries, especially crafts, went downhill as well. The merchants increasingly were foreigners, and with the weakening of the Ottoman Empire, Ragusa became a pawn of various European powers. Napoléon Bonaparte's conquest of Europe included Ragusa, which fell to French forces in 1808; the city thus lost its independence and became a part of the French Empire. The defeat of Napoléon in 1815 and the subsequent rearranging of European boundaries at the Congress of Vienna merely replaced French control of Ragusa with Austrian. The city remained under Austrian rule until the end of World War I. Only after 1919 and Croatia's joining a Yugoslav federation did the city assume the official name of Dubrovnik.

There may well have been a positive side, aesthetically and architecturally, to the decline of Dubrovnik's fortunes, since the city was too poor in modern times to do much demolition and rebuilding. Until 1991, Dubrovnik attracted many more tourists than residents to the city every year, mainly from Germany, the United Kingdom, and the United States. While most of the city was destroyed during the earthquake of 1667, some of the original buildings had survived the devastation: included among these are the landmark Bell Tower, constructed in 1445, as well as the archbishop's palace, the Rector's Palace, the Sponza Palace (housing the state archives, customhouse and, in the twentieth century, the city museum), numerous Gothic churches, the Franciscan monastery, the two major fountains, as well as most of the city's formidable walls.

With the demise of communism in eastern Europe in 1989, the Yugoslav state began to unravel. One by one, the former states within the federation declared their independence, at the price of civil war between the state and "federal" (largely Serbian) forces. In the summer of 1991, war erupted between Croatia and Serbia, followed by war in Bosnia. Attacked first by federal and then by Bosnian Serb forces (in retaliation for the Croatian government's support of the Bosnian Croats), Dubrovnik suffered heavy shelling from Serb positions in the picturesque mountains overlooking the city. In 1991, the city withstood a two-month siege, and many heavy bombardments afterward, that did not end until late 1992. The beautiful, historic buildings that had survived fire, earthquake, and world wars—the Bell Tower, the Sponza and Rector's Palaces, the monasteries—were heavily damaged. An uneasy peace has been brokered between the Serbs and the Croatian government, and every effort has been made to rebuild Dubrovnik, with generous foreign assistance. Nonetheless, in 1995, tourism was far from making a comeback in what had once been the queen of Dalmatian cities.

Further Reading: While all guidebooks on eastern Europe in English once had lengthy sections devoted to Dubrovnik, this has not been the case in the past several years because of the fratricidal war in the former Yugoslavia. Moreover, books on Dubrovnik in English are few and outdated. A beautifully illustrated guidebook on Dubrovnik, in very good English (which cannot be said of all of them), was published in 1970, *Dubrovnik and Its Surroundings* (Zagreb: Matica Hrvatska). The serious scholar who is interested in the history of Dubrovnik in its heyday, from its institutions to its way of life, literacy, architecture, and education, should consult the highly readable, enjoyable account by Bariša Krekić, *Dubrovnik in the 14th and 15th Centuries: A City Between East and West* (Norman: University of Oklahoma Press, 1972).

—Sina Dubovoy

Durrës (Durrës, Albania)

Location: Situated on Adriatic coast in central Albania, forty miles east of Italy.

Description: Established in seventh century B.C. as Epidamnos; 168 B.C., incorporated into Roman Empire as Dyrrachium; destroyed by an earthquake in A.D. 345; overrun by barbarians and later by crusaders; 1392, ceded to Republic of Venice; Turkish Ottoman forces conquered and annexed area in 1501; remained part of Ottoman Empire until independence in 1913; port completely modernized in 1928 (after 1926 earthquake); in World War II, overrun and occupied by Italian forces; under communist rule, became Albania's second-largest port and tourist center.

Site Office: Alb Seik Travel Agency
RR. Oemal Stafa 1000/1
Durrës, Durrës
Albania
(52) 23595

Durrës is so old that when the communist government undertook an urban renewal project in the 1970s, so many ancient and medieval artifacts were unearthed during construction that residents were scooping them up and using them as decorations in their own homes until the government began fining people for such trespassing. Hence what might be unimaginable elsewhere in the Western Hemisphere is perfectly normal in Durrës, the most ancient city in the Balkans. When the Romans annexed it in 168 B.C., the port city already had existed for more than 400 years.

The fifth century B.C. Greek historian Thucydides was the first person to write of the city's founding, which he surmised occurred in 627 B.C. At that time, immigrants from the Greek island of Corfu settled on the hills overlooking the Adriatic. Corfu was a colony of the Greek city state of Corinth, and tension between the two may have been the impetus for emigration. For the next 450 years, the port city bore a Greek name, Epidamnos, and became an important anchorage on the Adriatic.

Epidamnos grew in population and flourished despite its unhealthy location on the hills directly facing the swampy interior, which separated it from the mainland. Malaria and other fevers were common. Nonetheless, archaeological evidence of an imposing acropolis indicates that Epidamnos was not insignificant. The region also contained precious salt deposits that would be the mainstay of the city's economy in medieval times.

Because of Epidamnos's location on the Adriatic and its natural harbor, the Romans were bent on incorporating it into their empire during their conquest of the Balkans. This finally happened in 168 B.C. For the next several hundred years, the city, renamed Dyrrachium, blossomed as never before or since.

The Romans dubbed the region Illyria; in time, many famous emperors—Constantine, Diocletian, Justinian—were born and raised in this area, which became such an integral part of the Roman Empire. The acropolis either fell to destruction or to reconstruction as an amphitheatre. The amphitheatre, erected in the second century A.D., could seat 15,000 spectators, which would indicate a town of at least that many inhabitants (a population, in that case, that would not be reached again until the mid-twentieth century).

Dyrrachium also became the embarkation point of the famous Via Ignatia—a continuation of the Roman thoroughfare from the Eternal City to Thessalonika. Outside the present-day city of Durrës the stony remains of the famous avenue are still visible. The Romans also erected sturdy fortress walls and made the harbor the principal port on the eastern Adriatic.

For these reasons, Christianity arrived and rooted itself early in Illyria. Scholars have disputed whether St. Paul himself traveled to Dyrrachium; if so, he left no reference to the city in his voluminous letters. However, missionaries contemporary with the great saint did arrive in the first century A.D. and planted the seeds of the new gospel. By the end of that century, Dyrrachium was the seat of a bishopric.

Whatever churches and basilicas may have been erected in the city will remain a matter of conjecture since Dyrrachium fell victim to a severe earthquake in A.D. 345, which leveled the city and its sturdy walls. The timing was unfortunate, since the Roman Empire had fallen into decline. The city would have to recover unaided, and recovery was agonizingly slow. The amphitheatre, the largest in the Balkans, was largely destroyed. The Christian chapel within might have been built afterwards; its mosaics and frescoes are still amazingly intact, although worse for wear.

Unfortunately, the city's walls had suffered damage, offering the city no protection from the hordes of Ostrogoths who pillaged and burned their way through Dyrrachium in 478. By then, the Roman Empire had disintegrated and disappeared in Western Europe. The remaining portion of the ancient Roman Empire in the east, under the influence of Christianity, transformed itself into the Byzantine Empire, centered in Constantinople. The Greek-speaking Christians there followed the Greek rites of Eastern Christianity rather than the Latin rites dictated from Rome.

Illyria was divided culturally between the southern half that was heavily Greek-speaking and Orthodox (Byzantine) and the northern half that remained decidedly Latin and Roman Catholic. Dyrrachium, centered in between these halves, clung tenaciously to Latin Christianity, boasting a

Portion of Durrës's fifteenth-century fortifications

Aerial view of Durrës
Photos courtesy of Alb Seik Travel Agency, Durrës

large Roman Catholic cathedral. Yet there was a sizable Greek Orthodox presence as well.

The Byzantine emperor Anastasius, who during his reign from 491 to 518 extended his firm political control over Illyria, remembered his native Dyrrachium, where he was born and raised. Thanks to his personal intervention and largesse, the city was restored and rebuilt. A triple ring of walls, the remains of which are still visible, surrounded the greatly shrunken town on the hills. No sooner were they in place than wave upon wave of Slavic invaders descended on the city; the walls remained firm, and an increasing proportion of the native population fled to the mountains.

Dozens of rulers and feudal lords, some Albanian, but mostly aggressive European knights and crusaders, invaded the city in the decades and centuries prior to the Turkish conquest in 1501. They found neither wealth nor imposing buildings when they landed, but it was still the starting point of the old Roman highway to Greece and the Levant. Foreigners made up the bulk of the population of the city, whose Latin name the natives in the interior pronounced as Durrës. The errant knights and other European invaders found the native language impossible to pronounce. By 1100, they and other Europeans in the west began referring to former Illyria as Albania, a name used in ancient times as well as by Byzantine authors to refer to this anarchic region of the world.

A German knight visiting Durrës in 1499 described it as a "large destroyed city." Depopulated by recurring bubonic plague epidemics, which began in 1348, as well as by warfare, famine, and the fevers from the nearby swamps, the city could not even guard its own walls, or the bastion that the Venetians had erected in the fifteenth century, a trace of which still is visible. In 1423, the Venetian administration offered exemption from all taxes to attract new residents and retain the ones it already had. Apparently there were few takers, since the tax campaign had to be repeated five years later.

Therefore, it was a simple matter for Ottoman Turkish forces to take the city in the heat of a mid-August day in 1501. There was little resistance. Most of Albania had already succumbed to Turkish conquest by the late fourteenth century. The overlord of Durrës in 1392, who happened to be a native Albanian, had ceded the seaport to Venice that year in the hope that superior Venetian forces could keep the Turks at bay. The Italians eagerly ensconced themselves in Durrës, valuing the salt deposits that would fortify the Venetian monopoly of the salt trade in the Mediterranean. But as it turned out, not even relative security under Venetian rule and tax-free incentives could undo the damage done by centuries of anarchy and pestilence.

The Turks lost no time in transforming the cathedral of Durrës into a mosque. From 1501 to 1512 they set about strengthening and expanding the town walls, remnants of which are still visible to the discerning visitor. The Turks brought few other improvements to Durrës, however. The harbor, by then silted over and practically useless, literally sank into oblivion. Turkish administrators preferred to live elsewhere, away from the swamps and decay of the town, which assumed the character of a military outpost. During an invasion by Italian (Neapolitan) forces in 1606, Durrës suffered enormous damage, especially arson and plunder, from which it took a century to recover.

The famous Turkish traveler, Evliya Çelebi, spent some time in Durrës in 1660, nearly 160 years after the Ottoman conquest of the city. Taking the trouble to count the number of dwellings, he stopped at 150, noting that most of these were wretched huts. He attributed the poverty of the town and the lack of trade and commerce to the rebellious nature of the native Albanians, whom he described as being in constant conflict with the Turks.

This was an exaggeration; the oppressed Christian Albanians, even in coastal Durrës, a traditional stronghold of Roman Catholicism, were converting in droves to Islam and therefore freeing themselves from discrimination and taxation. In fact, Roman Catholicism was prohibited until the sultan issued a special decree, or firman, granting Catholics in Albania the right to practice their religion. Nonetheless, by the end of that century, more than half of the Albanians became Moslem, Durrës being no exception.

Durrës recovered slowly. In the beginning of the eighteenth century, foreign consulates were established in the city, at first French and Dutch, followed by others. The salt trade grew in importance once again. Still, a German visitor as late as the mid-nineteenth century saw little evidence of activity in the harbor, noting only the isolation of Durrës, even from the rest of Albania.

By 1892, an Ottoman source recorded that Durrës contained 550 dwellings, 2 Moslem primary schools and 3 Christian ones, 4 churches and only 3 mosques. While Ottoman sources tended to exaggerate, there is little doubt that the population of Durrës had increased, to a point where for the first time in modern history, it had climbed to nearly 5,000 inhabitants.

Growth continued even after a serious earthquake leveled Durrës in 1905. The Balkan Wars, waged in 1912 and 1913, brought about a redrawing of the map of southeast Europe, and the arrival of political independence for Albania. Briefly, for six months, Durrës even became the nation's capital, until World War I broke out and Italians occupied the city.

Durrës, like the rest of Albania, had barely twenty years to construct a modern nation and forge a government of its own before yet another world war brought Italian conquest, occupation, and postwar totalitarian rule. In those two decades, however, there was a surprising amount of progress. The Durrës harbor was completely modernized after yet another devastating earthquake destroyed much of the town in 1926, and tourism began to bring in income as well as modern roads connecting Durrës to the other coastal cities of Albania and to the nation's capital, Tirana. On the eve of World War II, Durrës was Albania's principal port, a bustling harbor that, unfortunately, was entirely destroyed by aerial bombing during World War II. The unhealthy swamps had disappeared, however, and foreigners were charmed by the beautiful beaches outside of Durrës and the wonderful climate.

The hopeful progress of Albania and its coastal cities was halted for decades because of war and a brutal communist regime. When communism in the Soviet Union began to soften under Nikita Khrushchev in the mid-1950s, Albania's insecure communist party leadership, under Enver Hoxha, would have none of it. They turned instead to communist China for moral support and economic assistance. Poor as China was in those days, the country had an abundance of technicians who began pouring into Albania in the fifties and sixties. When Mao Tse-Tung plunged his country into the frenzied, self-destructive cultural revolution, Albania followed suit. In 1967, the Albanian Red Guard, indoctrinated youth who wreaked havoc throughout Albania, destroyed not only churches but also mosques in Durrës, and publicly humiliated clergy and the religious of all faiths. Minorities were especially sought out and terrorized. Durrës once again slid into cultural and economic oblivion.

Not even Enver Hoxha's death in 1989 brought political relaxation. However, the fall of the iron curtain in Europe and Russia had a rippling effect, even in isolated Albania. The world has since become familiar with horror stories of Alban-ian backwardness and poverty. The ancient city of Durrës has survived far worse. Given political stability and economic opportunities, there may well be a renaissance of the old port. Tourists will be lured by Durrës' beautiful beaches and mild climate; the historically minded traveler will be fascinated by the city's continuous survival over two millennia despite enormous setbacks. While few monuments have survived physically, archaeological excavations can lead to rich restorations, as they have in neighboring Bulgaria. The city's crown jewel, its archaeological museum, is a treasure trove of artifacts attesting to the city's immensely long history and spirit of survival.

Further Reading: The most recent and readable guidebook on Albania, with a comprehensive overview of Durrës' historical sites, is *The Blue Guide to Albania* by James Pettifer (London: Black, and New York: Norton, 1994). A reliable and well-regarded history of Albania in fewer than 200 pages is *Albania: a Socialist Maverick* by Elez Biberaj (Boulder, Colorado: Westview, 1990).

—Sina Dubovoy

Edirne (Edirne, Turkey)

Location: Situated in European Turkey, 10 miles south of Bulgaria, 150 miles west of İstanbul, on confluence of the Maritsa and Tundzha rivers.

Description: Ancient Thracian city called Hadrianopolis when Roman Emperor Hadrian enlarged and renamed it after himself; part of Byzantine Empire after division of Roman Empire in 395; conquered by Ottoman Turks 1362; remained capital of Ottoman Empire until Constantinople fell to Turks in 1453; distinguished by beautiful fifteenth- and sixteenth-century mosques and a famous bazaar.

Site Office: Tourist Information Office
Talatpasa Cali, No.: 76/A
Edirne, Edirne
Turkey
(284) 2255260 or 2121490

The city of Edirne acquired that name when the Ottoman Turks conquered it in 1362. By then, Edirne was already at least 1,200 years old, an ancient city that existed even before Roman Emperor Hadrian enlarged and fortified it in the second century A.D., renaming it Hadrianopolis. (Until the twentieth century, Westerners clung to the name "Adrianople," ignoring its Turkish name). Its sleepy little predecessor, the hamlet of Uskudama, may have existed since the dawn of civilization. When the Romans conquered Thrace, the region in which Uskudama lay, the town's importance was augmented by its role as a vital crossroads between the Danube and the Bosporus.

Under Emperor Hadrian, the Roman Empire was experiencing its golden years. By then a Roman province, Thrace was a rich agricultural region and a gateway to the eastern portion of the empire. Uskudama, nestled in the Maritsa River valley, was a link in the important Via Ignatia, the Roman highway that ran from Dyrrachium (Durrës) on the Adriatic to Thessaloniki. The town was too important to ignore. Fortifying it with a ring of walls and constructing imposing buildings and rectangular streets were in order, turning the former Thracian hamlet into a completely Roman city, an important bulwark protecting this all-important Roman highway. So thoroughly Greco-Roman was its identity that the city adopted the name Hadrianopolis (the Greek rather than Latin name for "City of Hadrian") in honor of its patron.

In present-day Edirne's impressive Old Town, traces of the rectangular street pattern of ancient times is apparent to the discerning eye. In addition, remains of the original walls, rebuilt and expanded since Roman times, are also evident. Roman artifacts are displayed in the city's archeological museum.

The artifacts scarcely do justice to the significance of Hadrianopolis in the waning years of the Roman Empire. Emperor Constantine fought battles with his rival, Licinius, in 314 and 323, just outside the city walls; a Roman emperor, Valens, lost his life in a war with the Goths in 378 that has gone down in history as the Battle of Adrianople, the death knell of the Roman Empire in the west. The Roman outpost in the Maritsa valley lay in ruins. There would follow further devastations from the Slavs and later the Asiatic Bulgars. However, because Hadrianopolis lay squarely within the bounds of the Byzantine Empire—the Roman Empire in the east—the city's strategic importance was not lost on the Byzantine emperors, either, and consequently, after each ferocious wave of barbarian attacks, the city was rebuilt. In Byzantine times, Christian missionaries also made their way to the uncivilized Balkan hinterlands from Hadrianopolis. Often Byzantine armies assembled in the city to do battle with the Slavs and the slavicized Bulgarians.

Ottoman Turks or Osmanlis (an Asiatic tribe that had converted to Islam in the early Middle Ages and taken a name derived from the great Turkish chieftain, Osman) began their holy wars of conquest in the early fourteenth century, steadily eroding the boundaries and finally the heartland of Byzantium, Asia Minor. Fresh from victories in Asia Minor, the Ottoman Turks, under their leader, Sultan Murad I, made their first successful military foray across the Dardanelles when their forces crushed a Byzantine army outside the gates of Hadrianopolis. Capturing the city in 1362, the victorious Turks promptly renamed it Edirne, and quickly turned the leading churches into mosques, whitewashing the interiors to conceal the images so offensive to their Moslem faith. Consolidating their hold over most of Byzantium five years later (Constantinople, the last outpost of Byzantium, did not fall to the Turks until 1453), Sultan Murad turned the city into the Ottoman capital, an honor reserved up until then for the city of Bursa in Asia Minor. Edirne's strategic importance within striking distance of the Balkans and only 120 miles from the impregnable city of Constantinople would not be overestimated.

For the next ninety years, Edirne served as a virtual springboard to further Moslem conquests in the west, too dangerously close to suit western Europeans. A year after the fall of Hadrianopolis to the Turks, Pope Urban V called on the faithful to lead a crusade to recapture "Adrianople." Ironically, those heeding the pope's call were mainly Orthodox Christians and schismatics (Serbs, Wallachians, Bosnians), hitherto mortal enemies of Latin Christianity. In 1364, a sizable Christian army marched unimpeded from Hungary to Thrace, within sight of Edirne's gates. An informant may well have been responsible for the sudden surprise attack of a Turkish force against the crusaders, which annihilated the entire army in the dark of night. Edirne's identity henceforth would be indisputably Turkish.

Interior of the Selim Mosque
Photo courtesy of Embassy of Turkey, Washington, D.C.

Thereafter the armies under Murad I gathered in Edirne often to venture forth on successful military campaigns, resulting in the conquest of Bulgaria in 1366, Macedonia in 1372, and Serbia in 1389. After the fateful battle against the Serbs at Kosovo Polje in 1389, a traitor assassinated Murad I. The new sultan, Murad's son Bayezid I, whose son would commission some of the most distinctive buildings in Edirne, immediately put his own brother to death in order not to prevent a dispute over his claim to the sultanate.

Edirne remained the official capital of the Ottoman Empire until Ottoman forces under Mehmed II conquered Constantinople in 1453, using the largest cannon in the world to breach the walls. For hundreds of years thereafter, Edirne's importance continue to lay in its strategic location, with the Ottoman sultans spending part of each year in the former capital. In the fifteenth and sixteenth centuries, Edirne became beautifully embellished with masterpieces of Ottoman architecture, most of which have survived to this day. There are no surviving monuments to the long centuries of Byzantine and Roman rule, with the exception, perhaps, of the city walls. The famous seventeenth-century Turkish traveler, Evliya Çelebi, was so taken by the city that he wrote a book about Edirne entitled *Edirne in the 17th Century According to Evliya Çelebi.*

Hence Edirne flourished for several hundred years after the Ottoman conquest. While Turkish rule was painful for Christians, who labored under many restrictions, it was a boon to the hitherto persecuted Jewish minority, whose ranks swelled thanks to the well- known tolerance of the Moslem Turks for the Jews. At the city's apogee in the sixteenth century, economic power had passed almost entirely into the hands of Jewish merchants. The native Christian population rapidly converted to Islam, making those who remained Christians a distinct and diminutive minority within the city and region of Edirne.

The city grew to a population of nearly 100,000 at its heyday, but began a precipitous decline as the fortunes of the Turkish Empire waned. By the late nineteenth century, the number of inhabitants stood at barely 30,000, and Edirne had sunk to the status of a provincial frontier post. Because of its location, it suffered repeated invasion and occupation in modern times, especially from imperialist-minded Russians in 1829 and 1878 and nearby Greeks in 1912 and 1920. Modernization passed the city by until the Republic of Turkey was proclaimed in 1923. Thereafter, industry after industry arose in and around Edirne, especially food processing, perfume and soap manufacture, and textiles. Edirne continued to be a major market for the rich agricultural produce of the area, and for regional wines. Today the city is a major railroad terminal in the Balkan peninsula.

Edirne is a gem of Ottoman architecture and as yet, not a major tourist attraction. The small provincial city, located in a loop of the Tundzha River, has a splendidly preserved center. The notable sixteenth-century bridge spanning the Tundzha River, leading into the old town, was designed by Sinan, the most famous architect in Ottoman times. The heart of the old town is the town square, in existence since ancient times, with one surviving tower, the Clock Tower, intact since the second century. A portion of the ancient walls, which originally boasted four clock towers, also has survived the vicissitudes of nearly two millennia. The Greek inscriptions on the Roman walls are in the safekeeping of the city museum, another treasure trove of Ottoman history.

Several splendid mosques lie within short distances of each other. The Mosque with the Three Galleries (Uch Sherefeli Camii), built in 1437, is the earliest example of large-scale mosque building in the Ottoman period. Close by lies the noble Eski Camii, or Old Mosque, which was completed in 1421, taking more than ten years to build, distinguished by an uncommon prayer hall, Roman arches, a portico and pillars. An enormous bazaar or "bedestan," erected in the mid-fifteenth century, is the sole survival of what must have been half a dozen similar bazaars in the city, although this must have been the largest. Begun in the period of Sultan Mehmed I in the mid-fifteenth century, it was renovated in 1896 and consists of one vast building topped by fourteen domes. This bedestan served a dual purpose: as a huge emporium, and as a profit-making venture to support the nearby Eski Camii.

The most famous mosque in Edirne is probably the Selim Mosque, begun in 1569 and completed in 1575, during the reign of Sultan Selim II. It is visible from afar, an integral part of the city's landscape, and one of the masterpieces of the royal architect Sinan. A beautiful, covered courtyard beckons one into the mosque, which is surrounded by no less than four graceful, piercing minarets. One can spend hours gazing at the intricacies of the designs within, and the distinctive architectural embellishments that were characteristic of the Ottoman Empire's most distinguished and renowned architect. The mosque was meant to be the centerpiece of an entire complex of buildings that included schools and other outbuildings. The city's Archeological and Ethnographical Museum is located within the only outbuilding that survived the onslaughts of time.

The imperial palace that housed the sultans when Edirne was still the Ottoman capital, or the Old Palace, was completed in 1450. Much of the graceful building was destroyed in the sieges and occupations of the city in the nineteenth and twentieth centuries, but remains of the old kitchens, as well as the law courts, have survived. After the conquest of Constantinople in 1453, their imperial highnesses would spend only a few weeks a year inhabiting the palace, but it was enough to make Edirne a kind of second capital.

Second in importance to the grand Selim Mosque is its older counterpart, the Mosque of Bayezid II, also consisting of a whole complex of buildings that were completed in 1488, in addition to the house of prayer itself. Some of the buildings consisted of a kitchen and bakery, a hospital and even an insane asylum. According to the inveterate traveler Evliya Çelebi, three times per week, ten musicians played and sang for the unfortunate inmates of the asylum as a way of soothing and comforting them.

Two other small fifteenth-century mosques besides these more famous ones—the Gazi Mihal Camii and the Muradiye Camii—have miraculously survived earthquake and invasion. Not to be overlooked is the Turkish bath or Gazi Mihal Hamami, a part of the mosque of the same name, hence also dating from the same century. The baths separated men and women, with the men's section decidedly larger. The interior is richly decorated and still intact to this day.

Edirne is unequivocally Turkish, small and pedestrian-friendly. It has not been "discovered" by tourists. The few who venture there are charmed by the hospitality of the inhabitants, and the marvels of Turkish art and architecture that, unlike their counterparts in İstanbul, are uncrowded and beckon to the venturesome.

Further Reading: There are many good guidebooks in English on Turkey and İstanbul that cover Edirne in the Ottoman period, but barely mention, if at all, the centuries of history prior to the Ottoman takeover. One has to look at travel books in German to find lengthy descriptions of the present-day city and its historic sites. Of the two most detailed, neither is recent: Marcel Restle's *İstanbul, Bursa, Edirne, İznik*, part of the Reclam series of Kunstführer, or art guides (Stuttgart: Philipp Reclam, 1978) and *İstanbul, Mit Bosporus, Prinzeninseln, Bursa und Edirne* by Günter Wachmeier (Zurich and Munich: Artemis Verlag, 1977). Another excellent source is Godfrey Goodwin's *A History of Ottoman Architecture* (London: Thames and Hudson, and Baltimore, Maryland: Johns Hopkins University Press, 1971). Though not recent, it is readily available in most libraries, extremely comprehensive and detailed, with ample attention paid to Edirne, and fully illustrated.

—Sina Dubovoy

Elba (Livorno, Italy)

Location: A small island in the Tyrrhenian Sea, east of Corsica and six and one-half miles west of the northern Italian mainland; part of Livorno province.

Description: Settled in ancient times by Etruscans and Romans, followed by succession of other rulers; 1814, Napoléon Bonaparte exiled to Elba, where he remained nine months; 1861, united with Italy; heavily bombed and occupied during World War II; island turned to tourism after war, with subsidiary industries in fishing and iron-ore smelting.

Site Office: A.P.T. (Azienda di Promozione Turistico dell'Arcipelago Toscano)
Calata Italia
CAP 57037 Portoferraio, Livorno
Italy
(565) 914671

Elba (from the Latin word "Ilva") is a tiny jewel of an island. Surrounded by the Ligurian and Tyrrhenian Seas and covered from end to end with fragrant *macchia* or island scrub, Elba has balmy weather year round, 100 miles of coastline, white beaches, and scenic mountains. Lying only six and one-half miles from the mainland of Italy, Elba forms part of the Tuscan archipelago that includes the even tinier islands of Gorgona, Capraia, Pianosa, Giglio, and, not to be overlooked, the famous isle of Montecristo, all reachable by ferry. The French-owned island of Corsica is also a short ferry ride away.

Since the collapse of the island's economic mainstay, the steel industry, at the end of World War II, Elba has turned to tourism. There is much for the visitor to behold, from quaint and picturesque villages to Napoléon Bonaparte's still-intact residence; the beautiful harbor of Portoferraio, the island's chief town, still boasts its sixteenth-century walls and gateways; a Roman ruin, thought to be the dwelling of a Roman governor of Elba, bears evidence of a giant-sized swimming pool that had been filled by pipes with heated water. There is even archaeological evidence that Etruscans and Carthaginians had inhabited the island long before it became a Roman colony. Many of their remains were destroyed in the successive waves of barbarian invasions in the wake of Rome's disintegration, when Elba became increasingly deserted, visited only by wandering hermits and ascetics in search of solitude.

Small and remote though it may have been, Elba has been of strategic importance for several centuries. It served as a much-needed stopping point for fleets and armies en route to conquest, and above all Elba has been prized as an island richly endowed with iron ore, one of the few areas in the Mediterranean that contained this mineral coveted for use in weapons production. The earliest name of the island, "Althalos," in fact derives from the Greek words describing smoke arising from burning ore.

Besides iron ore, the Romans, who conquered the island in the second century B.C., were especially interested in its abundance of granite, which they shipped back to their imperial capital. Experts have determined that some of the granite used in the building of the Pantheon came from Elban quarries. Granite was also very important to the city-state of Pisa on the mainland, which was willing to expend large sums for the building of earthen forts near the granite caves to protect them. Both the Romans then and the Italian government in modern times found the island useful as a place of exile for prisoners. Debtors in the Roman Empire frequently were consigned from the dregs of Rome to exile in the beautiful surroundings of Capoliveri (now a tourist center). Thanks to its natural resources, including splendid harbors, the Saracens and, later, Corsair pirates from north Africa struggled with the Italian city-states for control of the island; Elba also became a pawn in the conflicts between the British and the French during the French Revolution and Napoleonic Wars.

Of all the conquerors who set foot on the island, beginning with the Etruscans, probably none was more feared than the Moslem raiders who periodically landed in Elba from north Africa and burned, raped, and pillaged in village after village, enslaving the survivors. The worst of these raiders was the legendary pirate Barbarossa, nicknamed for his red beard, although his real name was Khayr ad-Dīn; appointed by the Turks to command their navy in the Mediterranean, the pirate-turned-Turkish naval admiral spread terror throughout the small island during his attacks. Those who could, fled to the hills and hid.

Elba at that time was ruled by the d'Appiano family, who turned for help to the distant city-state of Florence, whose overlord, Grand Duke Cosimo I de' Medici, was related to the family by marriage. Elba was considered too much of a backwater to attract many settlers from the mainland; nonetheless, its harbors and natural wealth made its protection imperative for the grand duke, who personally oversaw the construction of fortifications and ramparts to guard the main entrance to the island at Portoferraio, or "iron port." The settlement that arose as a result of this protection was dubbed "Cosmopoli" in honor of the grand duke. Not until the seventeenth century did the name revert to the original Portoferraio. Thanks to the modern fortifications and Cosimo's attention to upgrading the island's iron industry, Portoferraio became a major port.

While the Florentines thus limited the effectiveness of the likes of Barbarossa, pirate attacks and raids continued into the seventeenth century. The fortifications were continually

Napoléon's summer home, the Villa San Martino

Napoléon's bedroom in the Palazzina dei Mulini
Photos courtesy of A.P.T. dell'Arcipelago Toscano

improved for the next two centuries, and knowing visitors considered them to be the most advanced in Europe. In 1795, Lord Nelson had another fortress erected to protect the island against invasion from Great Britain's chief enemy, France. In 1801, Portoferraio was able to sustain a year-long siege by the French, although the island did eventually fall to French control on June 4, 1802. By then, the population of Elba had increased significantly, thanks to the addition of thousands of French aristocrats fleeing Jacobin terror during the revolution.

One unexpected boon to Elba's future tourist industry came with the exile of Napoléon Bonaparte to the island as part of his punishment for dragging Europe into more than a decade of bitter warfare. With allied troops already entering Paris in the spring of 1814, the defeated Napoléon had little choice left but to sign the Treaty of Fountainebleau: in exchange for his abdication, the treaty allowed Napoléon to retain a small army of 400 men and officers (eventually, he was permitted to take along 600), granted him a generous pension from the restored Bourbon government (which failed to pay him, however), and allowed him to govern the island of Elba as its sovereign, without, of course, the consent of the Elbans. He set sail for Elba with three allied "observers" (his army arrived later), who were appointed to keep an eye on the warrior's every move.

Napoléon took an instant dislike to his official living quarters in the town hall that had been occupied only ten years before by the Hugo family (the father of the writer Victor Hugo had been one of a series of military governors of the island following the Treaty of Amiens in 1802). He was offended by the stench arising from the raw sewage in the streets, the oppressive heat, and the clamor of carriages on unpaved streets. A one-story house located above the city, with a splendid view of the harbor and sea, the Casa dei Mulini (House of the Mills) took his fancy instead. Several weeks later, he and his entourage moved in, and work commenced on a second story, which was finished in September. Soon his mother, Letizia, joined him, though she lived in a separate, smaller residence nearby. In autumn, his sister Paulina, who had in the meantime deserted her husband (a sculpture of her in the nude is still on display in Napoléon's summer retreat), moved in with him to keep house and act as official hostess. The so-called "Palazzina dei Mulini" thus became the vortex for the town's nightlife and political intrigues. Abetted by his ambitious sister, Napoléon decided to return secretly to France and reclaim the throne, in violation of the Treaty of Fountainebleau. This would mean war, for which he still had an enormous appetite. His hasty departure from Elba on February 25, 1815, was a secret only to the allied commissioners. He bade farewell to his mother and sister and the town's notables, and donated his official residence to the city. After his defeat at Waterloo that spring, Napoléon's second exile turned out to be far less idyllic. This time the allies took no chances and sent him to the distant island of St. Helena, off the coast of west Africa, where he succumbed to cancer (or poisoning) and died in 1821.

During his nine-month sojourn on Elba, Napoléon had taken his governing responsibilities seriously, displaying ability and efficiency. He ordered roads built and refuse collected (fining housewives who dumped garbage in the streets), provided street lighting, and had trees planted along the main avenues. Long-slumbering Portoferraio was slowly modernizing, only to have its progress halted with Napoléon's abrupt departure. He apparently had liked Elba, but chafed under its provincial isolation and, because the promised French stipend of 1 million francs never materialized, felt increasing financial pressure. Meanwhile, the taxes he raised and collected as head of state were barely sufficient for his and his army's upkeep. His sister had given him some of her precious jewels in order to help him renovate a farmhouse in the countryside for a cool summer retreat in hot weather. This would become the Villa San Martino, which, after Napoléon's departure, was sold to a Russian nobleman, who turned part of the summer estate into a museum. The villa and museum are open to the public, as is the partly furnished Palazzina dei Mulini.

Napoléon Bonaparte put Elba forever on the map. Whatever may have been his reputation on the Continent, he left a positive mark on Elba, which in turn became part of the Napoleonic legend. Books have been written about Napoléon's stay on Elba, and the Napoleonic presence is very much in evidence on the island. Thanks perhaps to the need for secrecy and haste to conceal his departure arrangements, Napoléon left behind most of his possessions and his library; his dwellings are still intact, and the garden of the Palazzina is as he left it.

Napoléon's foolhardy comeback was followed by the definitive treaty concluded by the Congress of Vienna in 1815, in which Elba went from French to Austrian control. Accustomed for centuries to being the pawn of European power politics, the inhabitants of the tiny island did not protest. Fishing, iron smelting, and salt panning continued as the island's economic mainstays; tourism was not yet taken seriously, although already a trickle of European visitors, curious to find out more about the larger-than-life Napoléon, made their presence felt. Elban youth were swept into the tide of Italian nationalism, volunteering in the forces of Garibaldi, which were fighting for a unified, sovereign Italian state. Finally, in 1861, the Austro-Hungarian Empire surrendered the Grand Duchy of Tuscany to Italian control. Elba, a part of Tuscany, was ruled for the first time since the Renaissance by native Italians, even if they were not elected by the people.

From then on, Elba has followed the course of modern Italian history. Heavy industry came to Portoferraio in the form of steel mills, which blighted the town's heretofore unspoiled beauty and wholesome air. World War I passed without incident except for a German submarine's strafing of Portoferraio and the harbor, which caused only minor damage. Recession and unemployment followed the war, and Elba was not spared the deep civil unrest that prevailed throughout Italy, and paved the way for Fascism's rise and eventual takeover of Italy in the early 1920s.

World War II passed Elba by until the overthrow of Mussolini in 1943 brought German retribution. Meanwhile the Allies, still considering Italy to be an Axis country, followed the Germans in bombing and finally invading Italy in 1944. The misery and destruction wrought by modern warfare and invasion were unparalleled in Elban history: most of Portoferraio lay in ruins (Napoléon's Palazzina dei Mulini suffered damage, but was not destroyed), and the island's population suffered greatly. As in the days of pirate raids, those Elbans who could, fled to the hills.

The end of the war was followed by the closing down of the steel mills, whose owners deemed them unprofitable. The major source of employment was gone at a time when any employment was hard to come by. What Elbans viewed as a financial and social tragedy for their island, and for Portoferraio in particular, turned out to be a blessing. With the return of peace and the tentative formation of a European community that promised an end to customs and economic barriers, Elbans in 1950 made the historic decision to turn their island into a tourist haven. Major hotel building, restoration projects, and road construction were undertaken; within ten years, the island's tourist initiative had proven successful.

Scientific teams went to Elba in the 1970s to carry out archaeological digging, some of which revealed that Portoferraio rested on the site of what they believed to have been the Roman city of Fabricia; the one major ruin dating from ancient times was determined to have been an immense villa, with a veranda overlooking the sea, complete with what was most probably a heated swimming pool. Napoléon's Palazzina dei Mulini was repaired (it had suffered major roof damage and broken windows) and opened to the public, as was his Villa San Martino three miles away. The Palazzina is largely furnished with Napoléon's belongings, including his library. Just before his ill-fated return to France, Napoléon ordered that the house be turned into a museum. However, when Elba became a part of the Austrian Empire in 1815, the Austrian grand dukes sold the furniture and took most of the books. It has taken nearly a century to restore the interior and recover some of the belongings that were known to have been housed there, and the damage will never be fully undone. Napoléon's summer residence at San Martino contains no original furnishings, but is worthwhile because the building has survived intact, displaying the layout of the house as planned by Napoléon, and containing some interesting artifacts.

There are noteworthy historic sights outside of Portoferraio, all easily accessible via Elba's excellent system of roads and public transportation: Marciana Alto, the oldest continuously inhabited community on Elba, 1,200 feet above sea level; the remarkable Madonna del Monte, 2,000 feet above sea level, with its image of Mary painted on granite that is said to have miraculous powers; the beautiful and popular resort of Marina di Campo with its fortifications built by the Pisans in the thirteenth century; and dozens of chapels and churches, interesting from an architectural and religious point of view.

Further Reading: Of the major guidebooks in English, none makes more than slight reference to Elba, one of Italy's most famous historic places. Moreover, there are no recent historical works in English that focus on Elba. A comprehensive, though not illustrated, book on Elba, is Christopher and Jean Serpell's *The Traveller's Guide to Elba* (London: Cape, 1977). This is extremely detailed and rewarding reading for the serious traveler only. On the lighter, but still informative, side are Andrea Burns's, *Sicily, Sardinia and Elba* (London: New English Library, 1975) and Barbara Whelpton's *Corsica, Sardinia and Elba* (London: Collins, 1967). For students of history and Napoléon, there are Norman Mackenzie's *Escape From Elba: The Fall and Flight of Napoleon, 1814–1815* (Oxford and New York: Oxford University Press, 1982), James P. Lawford's *Napoleon: The Last Campaigns* (Maidenhead, Berkshire: Purnell, and New York: Crown, 1977), and Claude Manceron's *Napoleon Recaptures Paris, March 20, 1815*, translated by George Unwin (London: Allen and Unwin, and New York: Norton, 1968; as *Napoléon reprend Paris (20 mars 1815),* Paris: Forum, 1965).

—Sina Dubovoy

Eleusis (Attica, Greece)

Location: Fourteen miles west of Athens on the Bay of Eleusis, opposite the island of Salamis.

Description: Ancient sanctuary and site of religious rites known as the Eleusian Mysteries performed in honor of Demeter, the Greek goddess of the Earth.

Site Office: Eleusis Museum
Eleusis, Attica
Greece
(1) 554-6019

In ancient Greece, Eleusis was ranked second only to Delphi in religious importance. For centuries the town enjoyed a widespread reputation as the foremost center for the celebration of the mystery cult of Demeter, the earth goddess. Although many of the details about the rites celebrated at Eleusis, known as the Eleusian Mysteries or Eleusinia, are lost, enough has been garnered from archaeological finds and written sources to form a general outline of the beliefs and activities of the cult.

Located just west of Athens at a strategic joining of several roads, the town of Eleusis dates back to at least 1000 B.C. Until about 675 B.C. it was ruled by local monarchs. In the seventh century B.C. Athens assumed control of the region. During the Persian Wars, about 480 B.C., the original temple to Demeter was burned down by an invading Persian army under the command of King Xerxes. The Telesterion, or Hall of the Mysteries, was built in its place by Pericles, the Athenian ruler who also built the Parthenon. During the Gothic invasion of Greece in A.D. 395, King Alaric's army destroyed the Eleusinian sanctuary. Early Christians also played a part in the destruction, building a church near the site, burying their dead in the ruins, and defacing the remnants with crosses scratched into the stone. Although a village existed nearby for many centuries, the dangers posed by pirates in the region eventually forced the villagers to leave. By the seventeenth century the town of Eleusis was abandoned and virtually forgotten.

Demeter was a fertility goddess with apparent origins in the Mesopotamian corn goddess. Her name is an ancient form of the phrase "earth mother." The daughter of Cronus and Rhea, Demeter ruled over not only agriculture, but over the civilization based on agriculture as well. Thus she was the goddess of farming communities and the kind of ordered, permanent settlement, ruled by law, which they represented. She also was the goddess of marriage. Demeter is often depicted as a beautiful maiden with hair the color of ripened grain and wearing a long robe and a veil. Sometimes she is crowned with a ribbon or an ear of corn. In her hand she holds a scepter or ears of corn.

Demeter had two children: Plutos, the god of riches, and a daughter, Kore (Persephone), fathered by Demeter's brother Zeus. The Rites of Eleusis centered on the relationship between Demeter and Kore. The ancient *Hymn of Demeter* relates the story of the two goddesses. In this account, Kore was kidnapped, taken to the underworld, and raped by Hades (Pluto), the god of the dead. For nine days Demeter wandered the world carrying flaming torches and searching for her missing daughter. On the tenth day Hecate advised her to question Helios, the sun god, as to her daughter's whereabouts. After Helios told Demeter of Kore's fate, he further explained that Zeus himself arranged the abduction to provide his brother Hades with a bride.

Overwhelmed with grief and anger, Demeter dressed as an old woman named Deo, or the Seeker, and left Olympus to wander among the mortals. After traveling throughout Greece in despair, during which time she had many adventures, Demeter eventually arrived at Eleusis. Unaware that she was a goddess, King Keleus took Demeter into his household to care for his infant son Demophon. The king's wife, Metaneira, could not persuade the goddess to smile or speak, so lost was she in her grief. But the servant Iambe, a daughter of Pan, made Demeter smile with her lighthearted bantering. Breaking the fast she had maintained since losing Kore, Demeter drank *kykeon,* a mix of water, mint, and flour.

As nurse to Demophon, Demeter did not feed the child the food of mortals but nourished him by breathing on him softly. She anointed him with ambrosia and, each night, placed him in a fire so that little by little his mortal nature would burn away. After a time his parents were amazed by Demophon's rapid growth and godlike bearing. Metaneira spied upon Demeter to discover what secret method the old nurse was employing in caring for the child. When she saw Demeter place the baby into flames, Metaneira cried out in alarm and the child fell to the ground. Demeter angrily explained that, because of Metaneira's meddling, the child would now be mortal.

Revealing her true identity to Metaneira and Keleus, Demeter then demanded that a temple be constructed in Eleusis where her mysteries could be celebrated. Keleus obliged her immediately, and the first temple at Eleusis was built. To show her gratitude for the kindnesses she received from the king and his family, Demeter sent Triptolemus, Keleus's son, to travel the world and teach all peoples the secrets of planting wheat and harnessing oxen. She also gave him the first grain of corn. Triptolemus, traveling in a chariot pulled by dragons, visited all of Greece and Sicily, founding new towns and escaping a host of dangers. When he finally returned to Eleusis, he replaced his father as king.

Overcome with renewed grief over Kore, Demeter retired to her temple in Eleusis and set upon a plan to free her daughter. She caused the world's crops to fail and threatened

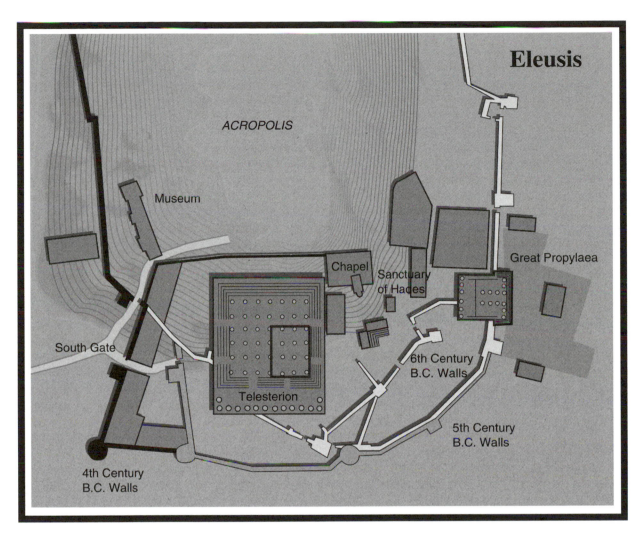

Eleusis
Illustration by Tom Willcockson

all of humanity with starvation unless Kore was freed. One by one the Greek gods visited her to implore her to relent, but Demeter refused. Finally Zeus himself was forced to intervene, arranging with Hades to allow Kore to return to the earth. But Hades tricked his wife into eating some pomegranate seeds before she left. The fruit is a symbol of the marriage union and by eating it, Kore bound herself eternally to Hades.

Zeus again stepped in to free Kore, decreeing that for two-thirds of the year, Kore will live on the surface but must return to the underworld of the dead for the remainder of the year. When in the underworld, she takes the name Persephone; while on the surface, she is Kore. Grateful for the return of her daughter for a portion of the year, Demeter renewed the earth's fertility.

Persephone's journey from the underworld to the upper world and back again mimics the cycle of the seasons and the alternating fertility and barrenness of the soil. Her stay of nine months on the surface corresponds to the fertile period of spring and summer, while her time in the underworld represents the sterile period of autumn and winter.

The rites of Eleusis, then, focused on the fertility of the earth as a cyclical process. The rites were apparently aimed at ensuring the fertility of the soil. In time, they also took on a more spiritual dimension as Persephone's return from the underworld came to symbolize a resurrection from death. The rites were said to derive from certain secret knowledge King Keleus was given by Demeter once Persephone had been returned to her. This knowledge concerned methods of calling upon the two deities whenever the monarch or Eleusis needed assistance.

According to Greek mythology, the Eleusinian mysteries were first formally organized by Eumolpos, a son of

Poseidon. In the war that won Eleusis for the city of Athens, Eumolpos and his son Immaradus were said to have taken part, fighting with the Eleusinians against the Athenian king Erechtheus. When a major battle left both Immaradus and Erechtheus dead, Eumolpos arranged a peace in which the two cities agreed to end their fighting. Although Athens was given dominance over the city of Eleusis, the mystery rites would remain the exclusive property of the Eleusinians. For brokering this arrangement, Eumolpos' descendants were given the hereditary right to supervise the Eleusinian festivals. Later, this right was also extended to those in the Kerykes family as well.

A strict hierarchy ran the Eleusinia, with the priests in charge holding designated posts and wielding specific powers. Organizing the Eleusinia was the work of the Archon Basileus, one of the city's chief magistrates, who was assisted by four epimeletai, or overseers. Two of these epimeletoe were chosen from the Eumolpid and Kerykes families and the other two were elected by the citizenry. The Hierophant, or chief priest, always came from the Eumolpid family. It was he who handled the Hiera, or holy relics, and exhibited them before the faithful. Two Hierophantides, or priestesses, were the Hierophant's chief assistants. They were chosen for life, could marry, and had their statues erected in the sanctuary. Another priestess, the Priestess of Demeter, acted the part of the goddess during the sacred pageant and rivaled the Hierophant in power. The Priestesses of Panageis were a group of nuns whose role at Eleusis is unclear, although they were called "all-holy," could touch the holy relics, and lived in seclusion at the sanctuary. The Keryx, or Herald, was always chosen from the Kerykes family. He called the initiates with a traditional phrase, led them in prayer and served as mystagogos, or guide, to the initiates. The Dadouchos, or Torchbearer, and the supervisor of animal sacrifices were other prominent officials of the Eleusinia.

The rites were celebrated twice each year: the Greater Eleusinia in mid-September (the Greek month of Boedromion) and lasting for about nine days; the Lesser Eleusinia in February (Anthesterion). Candidates were first proposed for initiation into the mysteries by an Athenian citizen who was himself a member of the cult. The neophyte was then admitted to the Lesser Eleusinia held each spring. These rites took place in Agra, located outside Athens on the east bank of the Ilissos River, at a temple to Demeter. This Eleusinia celebrated the return of Persephone from the underworld, although the precise nature of the rites is not known. Purification ceremonies, a period of fasting, sacrifices, and the singing of hymns apparently formed much of the celebration. Hercules was said to have first asked for these lesser mystery rites because he had initially not been allowed to participate in the Eleusinian mysteries at all. Neophytes to the cult were termed *mystae* and were allowed only a small role in the Greater Eleusinia held the following September. Full initiation into the mysteries took place at the following year's Greater Eleusinia.

Every four years the annual Greater Eleusinia festival was celebrated with special fervor and dubbed the *penteteris*.

A special truce lasting fifty-five days was observed throughout the Hellenic world and messengers known as *spondophoroi* were sent to announce the coming festivities. People from all over the Greek world and as far away as Rome would visit Eleusis.

The Greater Eleusinia began when the cult's holy relics were taken from storage at the Telesterion in Eleusis by the Hierophant. The nature of the Hiera is a matter of debate. Some scholars contend that the holy relics consisted of small statues depicting men and women in various sexual positions, and models of human genitalia, but this is not known for certain. It was a common belief among many ancient peoples that human sexual activity stimulated the soil to increased fertility. From Eleusis the relics were transferred in a procession led by the priests to the Eleusinion, a building near the Acropolis in Athens. The relics themselves were carried by the priestesses of the cult.

Three days after the relics were moved to Athens, the initiates were questioned by the cult's priests as to their worthiness to join the festivities. The requirements for admission to initiate status were simple: one must be able to speak Greek because all of the teachings were given in Greek, including certain ritual formulas, and one must not be guilty of murder. Even slaves might participate in the rituals. Only foreigners were definitely excluded from the cult, a restriction enacted after the blasphemous destructions visited upon Greek religious sites during the Persian Wars. This requirement was later revised, after the conquest of Greece by Roman legions, to allow Romans to join. Apollonius of Tyana was denied the right to participate because of his reputation as a wizard. Although labeled "mysteries" and protected by a strict vow of secrecy exacted from each initiate, the rites at Eleusis were routinely celebrated by several thousand initiates and were undoubtedly an open secret to most of the people who lived in the region.

The festival began after the priests determined that all initiates were eligible to participate in the festival. The Archon Basileus cried out "Mystae, to the sea!" and the initiates were led to the seashore where they bathed in the waters to purify themselves. Following this bath, they returned to the Eleusinion where pigs were sacrificed and the blood sprinkled over the initiates. The next stage of the ceremony involved a procession following the Sacred Way to Eleusis. The initiates wore festive clothing and were crowned with myrtle. Many of them carried a *bacchos,* myrtle branches tied into a bundle with wool string or a sack on the end of a stick in which supplies of bedding and clothes were stored. At the head of the procession was an image of Bacchus, one of the names for Dionysus, a Greek god of wine and fertility who was the son of Zeus and either Demeter or Persephone. Bacchus was also believed to have helped the Greeks in their wars against the Persians.

The twelve-mile route was lined with many altars and shrines, which the initiates visited to offer prayers and make sacrifices. At the small bridge leading to the sanctuary, the initiates paused for a brief ceremony in which saffron ribbons were tied around their right hands and left legs, perhaps meant to ward off unwanted visits from malevolent deities. The many stops along the route extended the relatively short excursion into an all-day journey. Carrying torches and singing hymns, the procession reached the Telesterion at Eleusis after dark.

For at least the first day at Eleusis a general fasting was observed, similar to that undergone by Demeter when she was mourning the loss of Kore. The fast was broken by drinking *kykeon* and eating sacred cakes, the meal believed to be the first that Demeter ate when she arrived in Eleusis. An animal sacrifice was offered to Demeter and Persephone, and prayers were said by the Archon Basileus and other priests. After this period of fasting and sacrifice, the initiates were ready to receive the cult's revelation.

The precise nature of the Eleusinian revelation is unknown today, but several contemporary accounts record that the experience left participants feeling awe-inspired, joyful, and in a state of bliss. What is known for certain is that the revelation consisted of three parts: a dramatic presentation of the legend of Demeter and Kore, accompanied by music, singing, and dance; the reciting of sacred teachings during the drama so that initiates would learn the secret mysteries revealed by the story; and the displaying of certain sacred objects by the Hierophant while he stood in a bright light.

The story of Demeter and Kore was reenacted in the Telesterion and at sites within the walled sanctuary where legend claimed that the events had unfolded. While the story was dramatized, the *legomena,* or sacred words, were recited by the Hierophant. This recitation was considered to be an essential part of a successful initiation since the drama alone merely reenacted the familiar legend; the sacred words gave that legend a mystical significance. This was why initiates were restricted to those who knew Greek; without an understanding of what the Hierophant was saying, there could be no revelation.

The holy objects, or *Deiknymena,* displayed by the Hierophant included the Hiera carried during the procession from Athens as well as other objects still a matter of speculation among scholars. Some of these objects were shown only to those who had reached the highest stage of the three degrees of initiation and were known as *epopteia* or seers. It is known that *epopteia* were encouraged to carry with them handfuls of wheat or other agricultural products to show that they participated in the civilized life of the farmer, the way of life that Demeter had brought to humanity. What ceremonies or teachings made up this highest stage of initiation is still a matter debated by scholars.

The last day of the Eleusinia was a time of celebration with much singing and dancing. A ritual involving the pouring out of a liquid onto the soil was also held. The initiates wore new garments brought especially for the occasion to symbolize the change they had undergone during the festivities. On the ninth day, after the festivities were finished, the initiates left for home. Those returning to Athens formed an informal procession, while those traveling elsewhere left directly for home. No initiate was obligated to ever return to the sanctuary, offer sacrifices to Demeter, or live in a particular way. Instead, the Eleusinian experience was meant to enrich an initiate's everyday life from then on.

Although little is known about the actual mysteries revealed to initiates during the Eleusinia, it is evident that the agricultural cycle played a role in the teachings. This cycle of birth, death, and rebirth found in the natural world was linked to the reemergence of Kore from the underworld of the dead each spring. The Eleusinian beliefs were, then, at least partially concerned with the existence of an afterlife.

Because of this apparent belief in an afterlife, some scholars find similarities between the beliefs of the Eleusian mystery cult and the beliefs of the Christians who followed. One sign of this similarity, or at least compatibility, may be seen in the actions of the Roman emperor Valentinian I, a Christian who outlawed all pagan ceremonies held after dark. Valentinian specifically exempted the Eleusinia from this ban. It was only after the rise of the less tolerant emperor Theodosius in the fourth century A.D. that Eleusis was finally abandoned for good.

Since 1882, the Greek Archaeological Society of Athens has undertaken a thorough excavation of the sanctuary of Eleusis. Its efforts have uncovered the damaged ruins of the primary buildings at the site. Only the foundations, stairways, and bases of columns remain of the once magnificent buildings. The rest has been reduced to broken stone.

Among the remnants uncovered are the foundations and broken stone columns of the Telesterion. The building consisted of six rows of seven columns and was surrounded by large tiers upon which up to three thousand people could gather. Broken blocks of masonry also mark the building's location. A well has been discovered where ritual dances were once performed, and two caves uncovered, which initiates believed to be the passages to Hades by which Persephone traveled. In front of one cave lie the ruins of the Sanctuary of Hades. Stone walls, some dating to the sixth century B.C., extend along the eastern edge of the site. Two *propylaea,* or stone entranceways, are also visible, as are a reassembled triumphal arch and a stone courtyard. A museum on the site displays stone statuary and frieze work recovered from the ruins.

Further Reading: *Eleusis and the Eleusinian Mysteries* by George E. Mylonas (London: Routledge and Kegan Paul, 1960; Princeton, New Jersey: Princeton University Press, 1961) is a detailed study of the history of Eleusis as a religious site and an analysis of the rites performed there. *The Rise of the Greeks* by Michael Grant (New York: Scribner's, and London: Weidenfeld and Nicolson, 1987) relates the history of Eleusis to that of other ancient Greek cities, showing how the site became one of the leading religious centers of the Hellenic world. Walter Burkert's *Ancient Mystery Cults* (Cambridge, Massachusetts: Harvard University Press, 1987) is an examination of the factors common to all the ancient mystery cults, including the one at Eleusis.

—Thomas Wiloch

Ephesus (İzmir, Turkey)

Location: Two and one-half miles west of the modern-day village of Selçuk and thirty-five miles south of the city of İzmir, on the northern slopes of Mount Coressus (Bülbüldağ) and Mount Pion (Panayırdağ), immediately south of the Caÿster River, in the province of İzmir. The ancient city commanded a strategic location midway between the Mediterranean and the Dardanelles, slightly northeast of the Greek island of Samos.

Description: Founded either in the eleventh or tenth century B.C.; under the control of a variety of rulers until the Romans made it one of the most important cities among their holdings in Asia Minor; site of many important buildings, especially the Artemisium, also known as the Temple of Artemis (Diana), one of the seven wonders of the ancient world; became an important center in the early Christian church; deserted in the Middle Ages and excavated beginning in the nineteenth century.

Contact: Tourist Information Office
Atatürk Mah., Agora
Carisi, No. 35
Selçuk, İzmir
Turkey
(232) 8926945

According to Greek legend, the Delphic oracle led the Ionian Greeks to Ephesus from their original homeland in central Greece. The oracle advised their leader, Androcles, that "The site of the new town will be shown . . . by a fish; follow the wild boar." Androcles and his men came upon a site inhabited by Lelegians and Carians. As they were cooking, a fish about to be placed over the open fire leaped away, knocking embers into a nearby bush. The fire frightened a wild boar, causing it to flee. Androcles and his entourage chased the wild boar and killed it at Mount Coressus. Androcles then drove the Lelegians, Carians, and Lydians from the upper city and founded the city of Ephesus.

While there is legend, there is no historical record chronicling the foundation of Ephesus. It is believed that the city was founded in the eleventh or perhaps the tenth century B.C. It was dedicated to the goddess Artemis, later known as Diana in Roman times. The original settlement was thought to be at the port of Coressus, 1,200 yards west of the Artemisium. The Ionian settlers found that the Anatolian mother goddess Cybele held sway over the indigenous people of the area. Cybele, with her multiple egg-shaped breasts, was the goddess of fertility and wild creatures. The Ionians combined the worship of Cybele with the Greek goddess Artemis, creating one deity and thus placating the original inhabitants of the city, while also still serving their own religious needs.

The leaders of Ephesus were unsuccessful in making a mark during archaic Greek history. Perhaps this was because the city absorbed so many non-Greek peoples into its society, the Lydians and Lelegians to name but a few. Yet, they were successful in unifying a multi-ethnic settlement behind the goddess Artemis and using her as the glue to hold together the society.

In the middle of the seventh century B.C., according to Herodotus, Ephesus was attacked and seized by the invading Cimmerians. A temple honoring Artemis was burned to the ground. Unlike neighboring Magnesia, Ephesus withstood the attack and soon began to redevelop.

A century later, in 562 B.C., the Ephesians failed to hold back King Croesus of Lydia. According to Strabo, new laws were drafted for the city while under Lydian control by Aristarchus, an Athenian. The city expanded, and a new Artemesium was built.

Lydian rule over Ephesus did not last long. In 546 B.C., the city submitted to Cyrus of Persia. Residents of the city participated in an Ionian revolt against Persia from 499 to 493 B.C. The city also served as a base for an Ionian attack on nearby Sardis. During these six rebellious years the Ephesians massacred Kean survivors from the Battle of Lade. The massacre most likely occurred because of hostilities concerning trade by Kéos, one of the chief commercial rivals of Ephesus.

Ephesus remained under the Persian yoke for almost fifty years, maintaining essentially friendly relations. When the Persian leader Xerxes I failed in his attempt to capture the Greek mainland, he returned home, sacking numerous Ionian shrines in Asia Minor along the way. The Artemisium was one of the few Ionian shrines encountered by Xerxes that was left unharmed.

Gradually, ties to Persia were loosened. Ephesus became a regular tributary of Athens, joining the Delian League after 454 B.C. In 412 B.C., allegiance to Athens wavered during the Peloponnesian War. The Ephesians took part in a widespread revolt against Athens and sided with Sparta for the remainder of the war. In 403 B.C., Ephesus was threatened once more by Persian aggression and thereafter became headquarters for King Agesilaus of Sparta. The Ephesians bolted from the Spartan fold nine years later, however, joining an anti-Spartan maritime league. By 387 B.C., Ephesus was again in Spartan hands. Next, the city came under the rule of Antalcidas, a Spartan sympathetic to Persia, leading to a pro-Persian tyranny that lasted more than half a century.

In 333 B.C., the pro-Persian rulers were stoned to death when Alexander the Great captured the city. After Alexander's death in 323, the city fell into the hands of Antigonus I, one of Alexander's generals. He placed a garri-

Theatre at Ephesus

Hadrian's gate
Photos courtesy of Embassy of Turkey, Washington, D.C.

son at Ephesus, but neglected the city, paying much more attention to his conquests in Egypt and Syria.

The Macedonian general Lysimachus, who had been another of Alexander's commanders, seized Ephesus from Antigonus. He is credited with introducing an age of Hellenic prosperity to the region. He reestablished the city on the northern slopes of Mount Coressus and the southern and western slopes of Mount Pion between 286 and 281 B.C. It is believed the move caused much discord in the city. He ordered the construction of walls encircling the city. The walls enclosed approximately 1,027 acres of land. Colonists from Lebedus and Colophon were introduced. Lysimachus also renamed Ephesus, Arsinoë, in honor of his wife. The name was abandoned following his death at Coroupedion and reverted to its original.

In 280 B.C., Ephesus came under the Seleucid kingdom of Syria's control until their leader, King Antiochus III, the Great, was defeated near Ephesus in 190 B.C. by the Romans. Following the Roman victory, the city was handed over to the king of Pergamum and jointly ruled by both Rome and Pergamum. In September 134 B.C., the king of Pergamum, Attalus III, bequeathed his realm to Rome upon his death—which occurred the next year—since he had no legitimate heirs. Ephesus subsequently became the first city of the Roman province in Asia Minor, over the many objections of Smyrna and Pergamum.

The city was at times difficult to govern. Around 88 B.C., it revolted, along with a number of other Asia Minor cities, against Roman rule. During the uprising many Roman inhabitants of the city were killed. The Ephesians were fined 20,000 talents for the action, which roughly covered the cost of the war.

Ephesus remained at peace, along with most of the other cities in Asia Minor, during the civil war between Julius Caesar and Pompey in 49 B.C. A victorious Julius Caesar came briefly to Ephesus a year later, and announced a favorable new tax policy. The Ephesians showed their gratitude by putting up a statue in his honor. Mark Antony came to the city following the defeat of Brutus and Cassius, Caesar's assassins, in 41 B.C. At Ephesus, he held a meeting to set financial penalties against those who had supported Brutus and Cassius.

The Roman emperors who followed Antony helped to further the development of Ephesus, turning it into the most prosperous commercial and banking center in all of Anatolia, surpassing Miletus, Smyrna, and Pergamum. Numerous public works projects turned it into one of the most impressive Roman cities in the Greek east. Ephesus was spared punishment by Emperor Octavian, unlike many other cities in Asia Minor, for aiding Antony and Cleopatra. This allowed Ephesus to avoid the burden of additional taxation and to continue to prosper. Ephesus managed to safely distance itself from most Roman court intrigues, and also benefited from Emperor Vespasian's policy favoring the development of urban centers and communications within the empire. The road constructed from Pergamum through Ephesus to Smyrna at this time further enhanced the city. Ephesus also became the residence of the provincial governor.

Ephesus's geographical location, at the mouth of the Caÿster River and at the western end of a great trade route from Asia running along that same river's valley, also benefited the city. Easy access to two more bustling trade routes along the Gediz River and the Menderes River further developed the city as an important commercial center. With a population as high as 250,000, Ephesus had outgrown the land upon which it was originally built, advancing all the way down to the coast. So prosperous was the city that it became conspicuous in the Roman world for its abundance of coinage.

The preeminent temple in Ephesus was the Artemisium, an important cult center for the worship of the goddess Artemis and a place of asylum and pilgrimage. The Artemisium built during the reign of Croesus was the largest building in the archaic Greek world, measuring about 55 by 115 yards. All four sides of the Artemisium had double rows of columns, which helped make it appear elongated. The idea to use a double row of columns instead of a single row may have been influenced by construction done in Egypt, where double-columned structures were popular at the time. Assyrian or Hittite influence was apparent in the carved reliefs displayed upon the columns.

A disgruntled and disturbed worshiper named Herostratos set fire to the archaic Artemisium in 356 B.C. When Alexander the Great came to Ephesus, he offered to help in its reconstruction. Appreciative of his offer, the Ephesians declined it with extreme flattery, saying, "It is not fitting for a god to build a temple to a god." Instead, the residents of the city went about rebuilding on their own, following the plan of the previous temple, with a few alterations.

An essential difference was that the newly built temple rested upon a podium approximately three yards high, which was accessed via thirteen steps. The underlying marshy ground and a recurring problem with flooding may have made elevating the platform necessary. The new Artemisium was the largest edifice in the Hellenic world, and was the first building of monumental proportions ever to be built entirely of marble. It was considered the most impressive of the seven wonders of the ancient world.

After the Romans took over Ephesus, they constructed many other public works. An aqueduct system was built to serve the entire city between A.D. 4 and 14. A stadium, capable of seating 70,000 spectators and possessing a track 712 feet long, was completed during the reign of Nero in the mid-first century A.D. Inside the stadium public ceremonies, athletic events, contests, and gladiator fights took place.

The Vedius Gymnasium was erected in 150, by a wealthy Ephesian named Publius Vedius Antoninus, due north of the stadium. Dedicated to the goddess Artemis and to the Emperor Antoninus Pius, it contained a beautifully decorated facade with statues and an interior housing a gymnasium, cold- and hot-water baths, a pool, dressing areas, and a hall. The Harbor Gymnasium and Baths and the Theatre Gymnasium were two additional structures devoted to physical fitness. The former was completed under Emperor Domitian in the late first century and used by the city's athletes as a training ground. It was a massive structure, 360 yards in length. The latter was completed during the second century A.D., also functioning as a small stadium.

Perched along the western slope of Mount Pion was an impressive theatre capable of seating 25,000 people. Originally constructed during the Hellenistic age, the semicircular theatre was altered and enlarged again under Roman rule in the first and second centuries A.D. The theatre had a large stage measuring twenty-five by forty yards, and its auditorium rose thirty yards above the level of the orchestra. A Hellenistic fountain from either the second or third century B.C. occupied the courtyard fronting the theatre.

The Celsus Library, initiated in A.D. 110 by Consul Gaius Julius Aquila and completed in 135, was one of the principal buildings in the eastern section of Ephesus. A reading room three stories high and measuring about eleven by seventeen yards occupied the interior. A two-tiered, horseshoe-shaped gallery ran around the upper two stories on three sides. A courtyard fronted the library, which was destroyed by fire at an unknown date. About 400, the courtyard was converted into a pool. The remains of the library's facade provided an interesting backdrop.

A vast temple honoring the Roman emperor Domitian (reigned 81–96) housed a massive stone statue of the emperor within its walls. It was the first sacred building in Ephesus to be dedicated to a Roman emperor. The temple occupied a central location in the city, opposite the state agora, which hosted a variety of municipal duties.

A small Corinthian-style temple was built by a man named P. Quintilius to honor the Roman emperor Hadrian (reigned 117–138). A distinctive feature of the four-columned temple was the serpent-headed hydra located above the doorway, put there to ward off evil spirits.

Additional buildings of importance included the Prytaneion, the odeon, the temple of Serapis, the Scholastikia Baths, and the East Gymnasium. The Prytaneion was the town hall for the autonomous city of Ephesus. Adjacent to the town hall was the temple of Hestia Boulaia housing a perpetual flame. Built during the reign of Augustus (27 B.C.–A.D. 14), the Prytaneion hosted political business, banquets, and receptions. The building underwent alterations in the third century and was eventually demolished. Its various architectural elements were used in restoration projects throughout the city.

The odeon, erected just after 150, was a small theatre seating approximately 1,400 people. It is believed to have had some type of roof enclosure.

A temple honoring the god Serapis was constructed by Egyptian colonists. Reached by a flight of steps, it was enclosed on three sides by colonnades. Inscriptions in the temple read the rites of the Egyptian cult honoring Serapis. Later, it was converted into a Christian church.

A building referred to as the Scholastikia Baths was erected in either the latter half of the first century or in the first half of the second century. The building originally included a brothel. The two-story structure had public spaces on the ground floor, and rooms for the prostitutes on the second. A Christian woman named Scholastikia extensively renovated the building, no longer housing a brothel, around 400, using some of the remains of the Prytaneion.

Lastly, located on the eastern edge of town was the East Gymnasium. It is believed that the sophist Flavius Damianus and his wife Phaedrina were responsible for its construction.

Multi-story homes of nobility were located throughout Ephesus. Numerous fountains and statues were also spread throughout the city. Many private residences were linked to the various public buildings by a white-marbled road or by the Arcadiane. The marble road was the main thoroughfare of Ephesus linking the Artemisium, the Vedius Gymnasium, the stadium, the main theatre, the Celsus Library, and state agora. The Arcadiane was the primary road leading to the port. Attributed to Emperor Arcadius, the colonnaded street was lined with numerous shops. It was 11 yards wide and 600 yards long, much of it paved with mosaics.

Walls were not the only physical barrier constructed to ensure the city's safety. Access into Ephesus was allowed through only three sets of gates. The most famous of gates was the Magnesian Gateway, constructed by Vespasian (reigned 69–79).

In the early years of the Christian church, following much struggle, Ephesus became an important Christian city. It is believed that the Virgin Mary visited the city with St. John the Apostle sometime between A.D. 37 and 48, following the crucifixion of Jesus Christ. She spent her final days in a small house, today called the House of the Virgin Mary, about three miles out of town. (Pope Paul VI confirmed the site's authenticity in 1967.) A cult of the Virgin Mary eventually replaced the cult of Diana at Ephesus.

St. Paul came to the city and lived in Ephesus from 53 to 56. When he addressed the Ephesians at the local theatre he was driven out of town by an angry mob, led by one of the city's silversmiths named Demetrius. The mob chanted, "Great is Diana of the Ephesians." The incident almost led to a riot. The silversmiths feared Paul's message because it spelled a loss of income for them if Christianity took hold: much of their income was derived from the sale of silver shrines honoring Diana. Paul's address to the Ephesians, the uprising, and the popularity of the cult of Diana are outlined at length in Acts 19 of the Bible. Paul is believed to have written some of his epistles while at Ephesus.

St. John returned to Ephesus in 95; he is said to have written his Gospel and died there. In St. John's Book of Revelation, Ephesus was one of the seven churches addressed. It is also believed that St. Luke may have visited and died in Ephesus.

The coming of Christianity brought additional beautiful buildings to Ephesus and the surrounding area. On a hilltop just outside of the city where St. John once lived stands a church bearing his name. Upon his death, John was buried at this spot. His grave was marked by a memorial that was later enclosed in the fourth century by a church with a wooden roof. Two centuries later, the Byzantine emperor Justinian I erected over the same spot a domed basilica. St. John's grave was believed to be located in the center of the holy structure. Throughout the Middle Ages, St. John's grave was one of the most important shrines in the Christian world. Dust that rose through a small hole over his grave was alleged to have healing powers.

The Church of the Virgin Mary was probably the most important Christian structure built at Ephesus. Situated within an already constructed Roman building, the church was a basilican design completed in the first half of the second century. Originally, the building housing the church was a business and exchange center that was abandoned following an economic slowdown. Measuring 30 yards wide and 260 yards long, the basilica was divided into three sections—a large central nave and two side aisles. The nave and side aisles are separated by two rows of columns. It was the first church ever to be dedicated to the Virgin Mary.

A short distance from Ephesus are the graves of the Seven Sleepers. Christian folklore tells of seven men from Ephesus, all good Christians, who hid in a cave to avoid persecution by the Romans in the third century. All seven men fell asleep and awoke 200 years later, after the Byzantine emperor made Christianity the official state religion. When they died they were buried together at this same site. A church honoring them was constructed upon the graves.

A total of five ecumenical councils were held at Ephesus. The most important of the five held was the third, convened at the Church of the Virgin Mary. It was called by the Eastern Roman Emperor Theodosius II with the approval of Pope Celestine I in 431. At issue were the teachings of Nestorius, the patriarch of Constantinople, who advocated that the Virgin Mary be considered only the mother of Christ, not the mother of God—a teaching that implied Christ was less than fully divine. At the end of the session Nestorius's teachings were condemned. A rival council was then set up under the patriarch of Antioch. The rival council repudiated the condemnation, choosing instead to condemn those who took part in the final decision at the third ecumenical council. Lengthy negotiations finally resulted in the upholding of the condemnation of Nestorius's teachings.

The Ephesian economy was doomed, as were many other cities' along the Asia Minor coast, due to the silting up of its harbor. For centuries the port was dredged in hopes that this would ease the silting problem. The problem continued, however, and caused the local economy to slide into serious decline by the third century. Eventually, the port became useless.

Ephesus was invaded by the Goths in 262. The city was sacked and the Artemisium destroyed. Sections of the ruined temple were stripped away to be used in the building of Hagia Sophia in Constantinople (now İstanbul) and the Basilica of St. John in Selçuk.

Limited construction was completed at Ephesus over the ensuing centuries, with a flurry of activity during the reign of the Byzantine emperor Justinian I in the sixth century. Despite this, however, the city continued a downward slide.

Ephesus found itself increasingly in the hands of adventurers—at one point it was even controlled by a Greek pirate. The thirteenth century witnessed considerable construction in the nearby town of Aya Soluk (now Selçuk) but not at Ephesus. The crusading Knights of St. John managed to briefly capture Ephesus in the latter half of the fourteenth century. Then in 1402, Timur (Tamerlane) obtained possession of Ephesus, along with Aya Soluk, but he considered Ephesus too insignificant to hold, so it passed to Turkey. Ephesus became deserted; its final abandonment could be blamed on malaria, which easily spread into Ephesus from the nearby marshes of the Caÿster River.

In May 1863, excavations began at Ephesus under the supervision of British archaeologist J. T. Wood. Wood supervised work done at the site for eleven years. In May 1869, he came upon the corner of the ruins of the Artemisium. Excavations by other archaeologists continued throughout the nineteenth century. By the century's end, the theatre, the stadium, the odeon, the Prytaneion, and the library had all been exposed. From 1895 to 1913, Austrian scholars took over and continued to find remains of the city from the Roman and Hellenistic times. Excavations have continued throughout the twentieth century.

The ruins from the Roman era at Ephesus are in varying stages of preservation. Some have been found only in fragments. The foundation and a lone column drum are all that remain of the Artemisium. From the temple of Domitian, only two sections of the giant statue honoring its namesake have been salvaged—the head and one of the arms. The Prytaneion is little more than a pile of rubble, but two beautiful statues of Artemis somehow survived and are on display at nearby Ephesus and Selçuk museums. One of the statues was completed during Domitian's reign and the other during Hadrian's reign. The Magnesian gate, the only one of the three gateways still extant, is also in poor condition.

Some other Ephesian ruins are in relatively good shape. Those include the Vedius Gymnasium, the theatre, the marble road, the facade of the Celsus Library, the East Gymnasium, and the city walls. A final group of Roman ruins has yet to be entirely uncovered, among them the stadium and the Harbor Gymnasium and Baths. Almost all visible remains at Ephesus belong to the Roman imperial period. Among the Christian monuments, the Church of the Virgin Mary is well preserved and still standing, as is the Church of the Seven Sleepers. Portions of the Church of St. John are still standing as well.

Further Reading: John Turtle Wood's *Discoveries at Ephesus* (London: Longmans, Green, 1877; reprint, Hildesheim, Lower Saxony, and New York: Georg Olms Verlag, 1975) is a good source of information concerning Ephesian history from the time of Alexander the Great to today. Archaeological excavations are touched upon as well. Guy MacLean Rogers's *The Sacred Identity of Ephesos* (London: New York: Routledge, 1991) is an especially helpful text for anyone seeking information on cult worship at Ephesus.

—Peter C. Xantheas

Epidaurus (Argolis, Greece)

Location: Situated approximately thirty miles from Athens, on the northern side of the Argive Peninsula, on the Saronic Gulf.

Description: Former ancient city, now the site of archaeological excavations; originally settled by Karians, then by Ionians, and Dorians; site of worship for Asclepius, god of healing and medicine; contains one of the best-preserved amphitheatres in Greece, built by architect Polyclitus the Younger; still site of annual theatrical events.

Site Office: Epidaurus Museum and Archaeological Site
Epidaurus, Argolis
Greece
(753) 22009

Epidaurus was not an important city in ancient Greece; rather, it remained an autonomous city-state partly because of its geographical location in hard terrain, and also because of its relationship with dominant Sparta, which helped it to establish good relations with Corinth as well. This, however, did not protect the city from invasion over the centuries.

Although the city is historically most associated with the cult of the god Asclepius, which was introduced to Epidaurus in the sixth century B.C., it was originally settled during the late Bronze Age. Archaeological evidence has shown that another inland settlement also existed that probably controlled the region around it. This site drew pilgrims from approximately the eighth century B.C., when it was first associated with the hunting hero Maleatas, who was later assimilated into the god Apollo. As time passed, the cult of Maleatas ceased to exist as a separate entity, and he was worshiped as Apollo Maleatas, another incarnation of the mighty god Apollo, god of the arts and medicine, and a seer. Archaeologists have found evidence of gifts dedicated to the god around an altar; however, for the most part the shrine of Apollo Maleatas was a simple affair, unlike the temple buildings and sanctuary that arose from the worship of his son Asclepius.

There are several myths associated with the healer Asclepius. According to various legends, the hero-god was born in Epidaurus, Messina, or Arcadia. However, the most ancient and popular belief is that he originated from Tricca, where the oldest sanctuary dedicated to him was situated. According to mythology, Apollo, who was also associated with healing and medicine, fell in love with Coronis (or Aigle), princess of Thessaly, when he glimpsed her bathing in a pool. "Whereas she already had in her womb the fruit of the love of the bright god [Apollo]," wrote the poet Pindar, she fell in love with a mortal and betrayed the god. When Apollo learned of Coronis's betrayal, he sent his sister Artemis to kill the girl with her fatal arrows. However, just as her parents were about to burn Coronis's body on a sacred pyre, Apollo remembered the existence of his son and tore him from the dead girl's womb, carrying him off to be brought up by Chiron, the centaur (half-man, half-horse), a famous teacher who had brought up Jason, Hercules, and Achilles, among others. Asclepius was brought up on Mount Pelion, an area rich with homeopathic herbs, where Chiron taught him the art of healing and medicine. Eventually Asclepius became an even better physician than his teacher and is reported to have accompanied Jason on his expedition to Colchis, and to have had the power to raise the dead. Pluto complained to Jupiter about this resuscitation of the dead, and the latter struck him dead with a thunderbolt. In retaliation, Apollo, angered by the death of his son, killed the Cyclops who had made the thunderbolts. Asclepius received divinity after his death and was worshiped by cults around Greece as a divine healer.

As the Asclepieion at Epidaurus became a famous and popular sanctuary dedicated to the hero-god, another story arose that intimated that Asclepius's birthplace was Mount Titthion, near his shrine at Epidaurus. In this version of the myth, the king of Thessaly took his daughter Coronis to Epidaurus on a trip. Coronis was already pregnant with Asclepius and gave birth to him secretly on a mountain. Reared by a goat and a dog, the child was found by a shepherd who saw a supernatural light emerge from the boy and realized that this was a sign of the child's divine origins. From that day Asclepius became associated with healing and medicine, and the dog and the goat were considered sacred creatures. In the third century B.C. the poet Isyllus dedicated a marble stela to Apollo, and his son, which was placed in the Hieron, the god's supposed birthplace. Asclepius is mentioned in the work of Homer and Pindar among others, and his two sons, Machaon and Podaleiros, can be found in the Homerian epic the *Iliad*.

Asclepius's popularity dates to the sixth century B.C., but his reputation grew slowly. The plague, which decimated the Athenian population around 430 B.C. during the Peloponnesian War, was brought to Epidaurus by the Athenians when they attacked the city. The fact that the city's population was comparatively unaffected by the disease helped to create and promote Epidaurus's reputation as a special healing center. The Asclepieion became known among the infirm and ill for healing, and the festival and games, the Great Asclepeia, held on a four-year cycle between April and June at Epidaurus, helped increase the god's popularity and renown. The event was held nine days after the festival of Poseidon at the Isthmus of Corinth and by the fourth century was attracting pilgrims from all over the pen-

Amphitheatre at Epidaurus
Photo courtesy of Greek National Tourist Organization

insula, who often came from the Corinth games to those held at Epidaurus. Meeting in the city itself, the pilgrims would make their way to the sanctuary, where they would sing paeans (songs) dedicated to Asclepius, the most famous of which was written by Sophocles. Finally in the fourth century, as the international reputation of the sanctuary grew, representatives were sent out by the city to collect money to improve the sanctuary, and architects and artists were brought to the site to erect buildings fitting for dedication to the hero-god. The consequent sanctuary was administered by four financial officials and one priest. Asclepius is often associated with snakes or a serpent. Indeed, large quantities of snakes were originally kept at Asclepieons around Greece. According to mythology, the god himself reared the reptile on Mount Pelion, and it became a symbol of the cycle of life. Asclepius was said to appear in the shape of a serpent to his followers.

In every Asclepieion in Greece, a well or fountain formed the nucleus of the sanctuary. At Epidaurus pilgrims would enter the sanctuary and after being received by a priest would ritually wash themselves in a well (one of the oldest structures in the site at Epidaurus). They would dry and clean themselves spiritually as well as physically, and would then fast. A sacrifice usually involving a bull, or a cow, and a cock would take place. Those in need of healing would sleep on the ground or on a hard stone slab in the Abaton (the Unapproachable Building, specially constructed to house pilgrims), and Asclepius would appear to them in a dream, manifesting himself usually as an animal—most commonly as a serpent. The pilgrims would then tell the priest their dream, and the priest would interpret it as necessary. The priests of the Asclepieions were specially trained in the healing arts. The cured patients would then inscribe their stories on a marble slab, or stela. Many of these are housed in the museum at Epidaurus and give a useful insight into the methods used to cure those visiting the sanctuary. It is reported that Hippocrates, the founder of modern medicine, learned most of his skills from the inscriptions on these slabs.

Before long the visitors to Epidaurus were leaving monetary donations, which helped to finance the construction of further lavish buildings at the site. The Roman consul

Paulus Emilius, who visited the sanctuary in 167 B.C., commented on how impressed he was by the buildings of Epidaurus. However, in 86 B.C. Lucius Cornelius Sulla raided the sanctuary and stripped it of its treasures.

The cult of Asclepius continued to be important well into the Christian era. It was not until the fifth century A.D. that Eastern Roman Emperor Theodosius II forbade the worship of pagan gods; a century later an earthquake demolished much of the beautiful site. Panayotis Kavvadias, sponsored by the Archaeological Society of Athens, was responsible for excavating much of Epidaurus between 1881 and 1928.

Most of the great buildings of Epidaurus were constructed from varying types of limestone, which came from near the city itself and also as far away as Corinth. Indeed, the great Temple of Asclepius was built from huge blocks of Corinthian limestone, which once broken corrodes badly. This, unfortunately, has been the case with the Epidauron site.

The visitor to Epidaurus approaches along the road from Ligourio, and, initially at least, is drawn to the specter of the huge theatre, one of the best-preserved examples of its kind in Greece. However, the original approach for pilgrims to the site was from the northern side of the sanctuary, through the Propylon. As with most other Asclepieions, Epidaurus was situated near water, and to get to the original site the visitor had to cross a stream. The Propylon was the entrance to the site, which separated the sanctuary from the outside world. The original Propylon was a raised platform mounted by six Doric columns, reached by steps and central ramps. There was a square formed by six Corinthian columns and a ridge roof. The inside of the Propylon was formally decorated with inner colonnades, and the building was some 250 feet from the next construction. The effect for visitors would have been a dignified and slow approach through the Propylon before reaching the main buildings in the sanctuary. The architecture of the Propylon suggests that it was erected during the fourth century B.C. Visitors were met with the inscription: "Pure must be he who enters the fragrant temple/Purity means to think nothing but holy thoughts."

The existence of the Temple of Asclepius shows the importance of the god in ancient Greece. The building was Doric in style, designed by the architect Theodotus. It rose from a platform of porous Corinthian stone. Weather has eroded what little remains of the temple, and the platform has had to be closed to visitors. Originally the temple would have been decorated with terra-cotta tiles and marble work. Although it is difficult to imagine what the building originally looked like, according to the fragments and pieces found from the original site, the building must have been quite plain and simple in design, but still an impressive dedication to the god.

The circular Thymele (or Tholos) was noted by Greek traveler and geographer Pausanias, who visited the site in A.D. 160, as the work of the architect and sculptor Polyclitus, who worked in the fifth century B.C. To the south of the temple, only the rings of the foundations of the original building still exist. With a diameter of sixty-six feet, it is much wider than the temple, and its columns were probably taller. It was also a far more extravagant and decorative building than the temple. It would originally have consisted of a circular cella (enclosed chamber), and an external portico of twenty-six Doric columns, mounted by a sculptured frieze; a second inner ring of columns was found within the cella. The building's purpose has been greatly debated. It is situated in the heart of the sanctuary near the temple and the Abaton, which signifies that it was an important building.

The Abaton (or Enkoimeterion) lies near the temple. Pilgrims would come and sleep in this building to await their visitation from the god. It was Ionic in style, and was near the purification well. The building probably belongs to the earliest phase of construction, as the sleeping phase of the healing process was one of the most important parts of this cult. To the extreme east of the Abaton was the sacred font, or holy well in which the pilgrims purified themselves. This was meant to have healing properties.

Situated halfway between the theatre and the sanctuary was the Katagogeion (Hostel), the building used to house pilgrims on the way to the site. It dates from the fourth century B.C. and had roughly 165 rooms spread over two floors. It was roughly 250 feet square and had four internal courtyards. Its existence illustrates the popularity of the sanctuary.

The theatre remains the best-preserved building of its type in Greece. It was used for religious games and contests and was a major part of the cult of Asclepius. The theatre was praised by Pausanias for its beauty and symmetry. Polyclitus is considered to be the architect; construction has been estimated as occurring between 360 and 300 B.C. The structure had originally 34 rows of seats made from grey limestone. These were divided into 12 sections by stairs. It is estimated that the arena could hold up to 6,200 spectators. Around the second century B.C. the theatre was enlarged with a further 21 rows divided into 22 sections, which meant that the stadium could hold up to 14,000 spectators.

Only the foundation of the stage is still in evidence. This arose behind the circular orchestra, which was between sixty-four and sixty-seven feet in diameter. The stage, or proscenium, had a facade decorated by Ionic semicolumns. The stage was renovated over the centuries. The theatre was most probably used for dramatic performances from its beginning, and was built to celebrate the four-year festival of Asclepius.

Athletic events took place in the Stadium, which is located in a natural valley and included a running track. The incline of the natural slopes provided good seating for spectators. Sports were also an important part of the celebration of the deities in ancient Greece.

Apart from the impressive main Asclepieion, the shrine of Apollo Maleatas lies above the main sanctuary, higher up the mountain. It is still being excavated and is not accessible to visitors.

Although Epidaurus was not one of the main cities of the ancient world, it was one of the leading sites dedicated to the healing god Asclepius and his father Apollo. The Theatre of Epidaurus is one of the finest examples of its kind in existence today, and is still a beautiful and tranquil place to visit. Since 1954, between June and August each year, the theatre has hosted plays, operas, and other performances. Maria Callas even sang here. Today more than 10,000 people are drawn to the peace and loveliness of Epidaurus each year. As Henry Miller once commented, at Epidaurus one can hear "the heart of the world beat."

Further Reading: *Guide to Reconstructions of Olympia, Mycenae, Epidauros, and Corinth,* by Niki drossou Paniotou (Congleton, Cheshire: Vision, 1993) is a postcard-sized book with a simple, clear breakdown of the site's history, great photographs, and film reconstructions of original sites. *Epidauros* by T. Papadakis (Zurich: Verlag Schnell and Steiner Munchen, 1975) is an extremely in-depth analysis of the site. *Epidauros* by R. A. Tomlinson (London: Granada, and Austin: University of Texas Press, 1983) offers a good overview of the site's historic background.

—Aruna Vasudevan

Erzurum (Erzurum, Turkey)

Location: Erzurum is the largest city in northeastern Anatolia (Asiatic Turkey) and capital of the province of the same name. The city, 6,400 feet above sea level, is situated at the southeastern edge of a fertile plateau near the source of the Kara Su River. It is surrounded by mountains, which rise in the north to an altitude of about 10,000 feet. Erzurum is 450 miles east of Ankara and 110 miles southeast of Trabzon.

Description: Though the site was probably settled much earlier, the modern city of Erzurum traces its origins to the founding of Theodosiopolis in the late fourth century. Both the Arabs and the Turks knew Theodosiopolis as Arz-ar Rum, variously translated as "Land of the Romans," "Domain of the Byzantines," or "Frontier of Greece," from which its present name is derived. A frontier fortress and garrison town since ancient times, it lay along main trade and military routes in central Asia. It was in Erzurum, in July 1919, that General Mustafa Kemal (later President Atatürk) presided over the first Turkish Nationalist Congress, which led to the War of Independence and the establishment of the secular Turkish Republic in 1923.

Site Office: Tourist Information Office
Cemal Gürsel Caddesi, No. 9/A
Erzurum, Erzurum
Turkey
(442) 2185697 or 2189127

Toward the end of the fourth century A.D. Roman emperor Theodosius I built a frontier fortress on the site of what is now Erzurum and named it Theodosiopolis. Owing to its strategic location along main trade and military routes to Persia, the Caucasus region, and western Anatolia, Erzurum's sovereignty was contested for the next six centuries. Until taken by Turkish tribes in the eleventh century, the town changed hands frequently among the Byzantine emperors, the Persian Sāssānids, and various Arab dynasties, with periods of Armenian rule.

Following the death of Theodosius in 395, the Roman Empire was formally divided into two realms; the eastern part, Greek-speaking and based at Constantinople, came to be known as the Byzantine Empire. Though its possession had been disputed by the Sāssānids, Theodosiopolis fell in 653 to the Arabs, who occupied the town for nearly a century.

The great Arab expansion of the seventh century occurred while the Byzantines were reconquering the Near East. After winning a naval victory in northern Syria against the Arabs in 747, Byzantine emperor Constantine V advanced into Armenia; he razed the walls of Theodosiopolis and transported many of the town's non-Arab inhabitants to Thrace. An able military commander, Constantine checked Arab armies and paved the way for subsequent Byzantine offensives in eastern Anatolia and Armenia. Theodosiopolis also played an important role in the Arab wars waged by Byzantine emperors Leo VI and Constantine VII.

Meanwhile, in the ninth and tenth centuries, a number of Armenian dynasties, ruled by Moslem emirs, had emerged in the eastern Byzantine frontier. The tenth and eleventh centuries, however, saw renewed Byzantine aggression and reconquest in eastern Anatolia, and Theodosiopolis was captured by the Byzantines from an Armenian emirate in 949. Under Basil II, the Byzantine Empire expanded its frontiers into what is now eastern Turkey and Armenia. Theodosiopolis became a Byzantine dependency, called Karin, in 978, but shortly thereafter it was taken by the Bagratid Armenian kingdom of Kars. Basil II began invading Armenia proper in 1021, reannexing the Bagratid kingdom by 1064.

In the eleventh century, a new threat appeared from the east: a Turkoman clan led by Seljukid chieftains. The Turks originally emerged in central Asia in the seventh or eighth century, began migrating westward, and encountered the Arabs in the ninth century. The Arabs had converted the majority of Turks to Islam by the end of the tenth century, and recruited them as warriors. The appearance of the Seljuk Turks in Anatolia set the stage for the decline of Roman-Byzantine-Christian rule.

The Seljuks began raiding Byzantine territory in 1045; a large force led by İbrāhīm İnal sacked Theodosiopolis in 1048. The Seljuks had been carrying out their incursions from bases in northern Persia—until 1071, when Seljuk sultan Alp Arslan, marching north from Baghdad, decisively defeated the demoralized army of Byzantine emperor Romanus VI Diogenes in Manzikert (now Malazgirt), in eastern Anatolia. Manzikert changed everything: "Henceforward," as historian Claude Cahen writes, "there would no longer be any need to return."

After the Battle of Manzikert, the mainly Armenian-inhabited Erzurum fell to the Seljuks, and the way was left open for uncontrolled Turkoman clans—the new frontiersmen of Islam, the creators of a basically Turkish Turkey—to swarm across Anatolia. The Byzantine Empire gradually declined and fragmented over the next four centuries. In the century following Manzikert, the Byzantines and the Turks alternated between warfare and negotiation.

In the twelfth century, eastern Anatolia was composed of the emirates, or principalities, of Erzurum and Erzincan. Erzurum was ruled by a feudal dynasty, the Saltukid clan of the Turkomans; its first known ruler was 'Alī ibn Saltuk in 1103. The Saltuks were descended from him. He died about

Çifte Minareli Medrese, Erzurum
Photo courtesy of Embassy of Turkey, Washington, D.C.

1123. His son, also known as Saltuk, succeeded him, and died in 1168. Coins issued by Saltuk and his son Muhammed indicate that the princes of Erzurum paid allegiance to the Seljuk rulers of Iran—not to the "Roman" Seljuks of Rum, who had established a sultanate just to the west. This allegiance was short-lived, however, as by 1201 Erzurum was securely in the hands of the Sultanate of Rum, which was essentially a small empire in the heart of Anatolia. The Sultanate of Rum evolved into a highly cultured, flourishing state, and Erzurum enjoyed great prosperity, reaching its zenith in the first half of the thirteenth century.

In the 1240s, however, Anatolia was invaded by Mongols: in the winter of 1242–1243, just after the death of Ögödei Khan, son of Genghis Khan, military leader Bayju attacked and sacked Erzurum. On June 26, 1243, Bayju crushed a disorganized Seljuk army under Sultan Keyhusrev II at nearby Kose Dag (or Kuzadag). "In one day," historian Cahen writes, "the course of the history of Asia Minor had been altered beyond recall." With the establishment of the Mongol Empire in the thirteenth century, Anatolia became a major east-west trade crossroads; an imperial highway also linked Tabriz, Persia, the center of the Khanate, to Konya through Erzurum, which grew as a commercial center.

The Mongols' destruction and savagery triggered anarchy and famine in Anatolia. Unpacified, eastern Turkoman tribes overran the lands of the Seljuks and Byzantines, the latter having mostly abandoned Anatolia. The Mongol khan, Mangu, appointed his brother Hülegü to rule over the Seljuks. Following the death of the sultan in 1246, Hülegü restored some order to the Seljuk state, and, for the next three decades, the sultans were puppets of the Mongol khans. After Keyhusrev III defeated the Mongol army in 1276, the Mongols tightened their grip on the Seljuks, though the sultanate lingered until the end of the century.

With the collapse of the Seljuk state, Anatolia fragmented into numerous, mostly Turkoman emirates; Erzurum and northeast Anatolia were ruled variously by Saltuk clan chieftains and the Ilhanid Mongol dynasty based in Tabriz. The rise of the Turkish Osmanli dynasty (later westernized as "Ottoman" and named after clan leader Osman), in the early fourteenth century was interrupted by the invasion of Timur (Tamerlane), the great Tartar warrior who used Erzurum as a staging area for his bloody rampage through Anatolia in 1402. After his death, the Mongol threat soon vanished.

Ottoman Turk ruler Selim I won a decisive victory in eastern Anatolia against the army of Esmā'īl, a Shiite Turkoman clan leader of Persia, in August 1514. In 1515, he annexed the mountainous region from Erzurum to Diyarbakır into the Ottoman Empire. The local dynasties and tribal chieftains in this area submitted to Ottoman suzerainty by 1517. Anatolia was now secure against further invasions from the east. Erzurum flourished for the next three centuries, and, by 1827, was an important trade center with a population of 130,000.

The Turkomans—and, to a lesser degree, the Mongols and Ottomans—left a rich architectural legacy in Erzurum, with a number of Islamic historical monuments dating from the twelfth to fourteenth centuries. The Seljuks especially became renowned for their ornamental tile and relief work. Due to constant warfare and frequent earthquakes, nothing in Erzurum predates the eleventh-century Seljuk conquest—except for the massive walled remains of a raised citadel, which was originally constructed by the fifth-century Byzantines and later rebuilt several times. In Ottoman times, the citadel served as the eastern stronghold of the Janissaries, the sultan's elite guard of warriors. Barely any traces of the old city's medieval (and later) fortifications remain, though the town is still surrounded by an enceinte of earthen ramparts and gates built between 1855, during the Crimean War, and the Russian occupation of 1877. Christian monuments that survived World War I, including the Cathedral, were subsequently razed.

Erzurum's most notable historic buildings are clustered around the citadel in the center of the old city. The Ulu Camii, or the Great Mosque, was founded in 1179 by Muhammed, the third in line of the Saltuk emirs, and later embellished with inscriptions by his grandson, Abu Mansur, which include, among other titles, "King of the Greeks and Armenians." Though its exterior is austere, it has a fine colonnaded courtyard. The mosque's roof collapsed in 1965, but has since been repaired. Erzurum's most famous building is the Çifte Minareli Medrese (Seminary of the Twin Minarets). It was built in 1253 at the height of the Seljuk empire. Though planned as the largest theological seminary of its time, it was never finished. Nearby, one of a cluster of mausoleums called Uc Kumbetler (Three Tombs) dates from the early twelfth century and is said to contain the remains of the first Saltuk emir; the other two *kumbets* are less compelling and date to at least a century later.

Since the Mongols were usually more interested in conquest than constructing monuments, their buildings were rarely completed and were less refined than Seljuk architecture. Nevertheless, Hoca Yakut, the local emir of the Ilhan Mongol rulers of Persia, began the Yakutiye Medrese seminary in 1310; it is perhaps the most fanciful building in Erzurum, with a Persian-influenced facade, turquoise tiling on the minaret, and an intricately colored portal. Ottoman mosques include the classical Lala Mustafa Pasa Camii, supposedly designed by the famous Sinan, architect of İstanbul's greatest mosques, in 1563 (eight years after the treaty with Persia that finalized Ottoman rule in the East) and the seventeenth-century Caferiye Camii.

In the nineteenth century, Erzurum became the main Ottoman fortress against the Russians, who captured and briefly held the city in 1829 and 1878, leading to a decline in the city's importance. It was again occupied by the Russians from 1916 to 1918, during World War I, before being recovered by the Turkish army. The Ottoman Empire's growing rivalry with Russia in the late eighteenth and early nineteenth centuries over Greece and other Orthodox Ottoman territories resulted in the Russians' declaration of war in April 1828. In July 1829, Russian General Ivan Fyodorovich Paskevich took Erzurum; the occupation continued until September 1829, with the Treaty of Adrianople (Edirne). Erzurum was returned to Turkey, though Turkey agreed to give up much land in Georgia and Armenia, as well as to grant autonomy to Serbia and Greece, leaving the Ottomans in a weakened international position.

Russia, desiring Slavic territory in the Balkans, declared war again in April 1877. Erzurum was the scene of a major Turkish-Russian battle in November 1877, but the city was successfully defended by Ahmet Muhtar Pasha against a Caucasian force. The Russians once again occupied Erzurum during the armistice of February-March 1878. The July 1878 Treaty of Berlin returned Erzurum to Turkey and ended the Russo-Turkish conflict, though the Ottomans had to provide for the independence of their Balkan territories and give some eastern provinces to Russia as a war indemnity. This effectively dismembered the Ottoman Empire in Europe.

A severe earthquake destroyed much of Erzurum in 1859, and another in 1901 also caused great damage. (Other major earthquakes occurred in 1939 and 1983.) Armenian nationalist uprisings in eastern Anatolia from 1894 to 1896 resulted in the slaughter of nearly 150,000 Armenians by Turkish troops, with a number in Erzurum. Turkey again ordered the killing of Armenians between 1915 and 1918. The Erzurum Museum contains items removed from two mass graves recently found in the province—the property of some of the more than 1 million Armenians massacred by Turks during this period.

Russia declared war on Turkey in November 1914, at the outbreak of World War I. In February 1916, Russian forces under General Nikolay Nikolayevich Yudenich forced the Turks to evacuate Erzurum's fortress, and established a base there. It had been a brilliantly hard-fought battle, since it took place in winter, over a mile high in altitude. The city remained in Russian possession until 1918, when Russia's forces withdrew after the Bolshevik Revolution. Bolshevik Russia concluded a peace treaty with Turkey and its allies in March 1918, restoring Turkey's lost eastern territories (including Erzurum), as well as frontier districts ceded to Russia in 1878.

The collapse of the Ottoman Empire in October 1918, together with the Greek invasion of May 1919, roused latent Turkish nationalism and resulted in the Turkish War of Independence. In April 1919, local military commander and staunch nationalist General Kiazim Karabekir took control of Turkish troops at Erzurum, which became the main staging center for the war of resistance. He had effectively ruled the city since taking it over from retreating Russian armies in 1918. Erzurum was also the site, in July 1919, where undefeated war hero General Mustafa Kemal, later known as Atatürk, and Karabekir called the first Turkish National Congress; another was held in Sivas in September. The National Pact, drafted and ratified in Erzurum, established modern Turkish borders and eventually led to the founding of the secular Turkish Republic in 1923, with Atatürk as president.

Armenian tradition once had it that earthly paradise was located at Erzurum; the same could not be said today. It has been described as a bleak and somber city, a harsh and wintry place set on a 6,000-foot-high plateau and surrounded by still higher peaks. (One of these peaks is Palandoken Mountain, which has the best ski resort in Turkey.) Since antiquity, Erzurum has been a major fortress town on the route of central Asian armies, and it still serves as an important garrison city vital to Turkey's eastern defenses. The military presence combines with Erzurum's religious conservatism to create a solemn atmosphere.

Destruction wrought by the nineteenth-century Russians, as well as the 1939 earthquake, precipitated the construction of wide new boulevards and tall white concrete buildings. In 1957, Erzurum became the seat of Atatürk University, making it the leading cultural and educational center in eastern Anatolia. The city has just begun to discover mass tourism, with the requisite rug shops and a caravansary, which sells the metalwork of local craftsmen. An important livestock trade center, Erzurum's industries also include sugar refineries, cement factories, meat-processing plants, tanneries, and metalworking.

Further Reading: *A Traveller's History of Turkey* by Richard Stoneman (Moreton-in-Marsh, Gloucestershire: Windrush, and Brooklyn, New York: Interlink, 1993) is a concise, chronological guide to the country's history, from prehistoric times to the present. *Ancient Turkey—A Traveller's History of Anatolia* by Seton Lloyd (London: British Museum Publications, and Berkeley and Los Angeles: University of California Press, 1989) is an accessible account of Turkey's ancient and medieval civilizations. *Pre-Ottoman Turkey* by Islamic scholar Claude Cahen (New York: Taplinger, and London, Sidgwick and Jackson, 1968) presents a detailed, authoritative historical and cultural survey of Turkey from the Seljukid invasion to the rise of the Ottoman dynasty (c. 1071–1330). Though a travel book, *Turkey: The Rough Guide* by Rosie Ayliffe, Marc Dubin, and John Gawthrop (London: Rough Guides, 1994) is a solidly researched and nearly exhaustive guide to the country, full of historical context and contemporary facts.

—Jeff W. Huebner

El Escorial (Madrid, Spain)

Location: Thirty-one miles north of the city of Madrid, in the Madrid province, in the foothills of the Guadarrama Mountains.

Description: The Monasterio de San Lorenzo el Real del Escorial is a monumental monastery, palace, royal tomb, art gallery, and library combined, built by Philip II of Spain in the sixteenth century.

Site Office: Monasterio de San Lorenzo el Real del Escorial
San Lorenzo de El Escorial, Madrid 28200
Spain
(91) 890 59 03 or (91) 890 59 05

Monastery, palace, architectural wonder of the world—the magnificent structure known as El Escorial is all of these things and more. The Escorial, built when Spain was the major political power in the Western Hemisphere whose holdings ranged from the Netherlands to the New World, serves as a monument to both the piety and the ambition of Philip II. Born in 1527, the son of the Habsburg Holy Roman Emperor Charles V (who ruled Spain as King Charles I), Philip inherited his father's Spanish and Dutch holdings and a strong sense of religious duty. Following his victory over the French in 1557 at the Battle of St. Quentin, Philip vowed to construct a monastery in the name of St. Laurence (San Lorenzo), on whose day the battle had been won. When he came to the Spanish throne, he also made it his intention to honor the last wish of his father that a royal pantheon be built to serve as a tomb for the emperor and his wife. Philip envisioned as well a place of meditation and solitude for the kings of Spain, a place of refreshment and renewal for the glory of God and the betterment of the empire.

After a long period of typical indecision, he finally settled upon a site for his great project: a wooded slope of the Sierra de Guadarrama, conveniently close to Madrid but far enough away to provide a secluded natural setting. According to local tradition, Philip surveyed the project from the height of the nearby mountains where a chair-shaped rock is still known as Silla de Felipe II (Philip II's Chair). Today the visitor finds the village of San Lorenzo de El Escorial, the setting for one of the great landmarks of Spanish architecture. Establishing the identity of Spanish Renaissance architecture, it reflects at the same time the many outside influences upon the vast empire, from the classical taste of the Italian Renaissance to the traditional architecture of Habsburg Germany and Austria to the lasting impact of Moorish Spain.

The cornerstone for this ambitious project was laid on April 23, 1563. Philip had chosen as his architect Juan Bautista de Toledo, a Spaniard who had studied and worked in Italy, and whose design was much influenced by Italian

theory. Among his associates was a man from one of the great stone-cutting regions of Spain, Juan de Herrera. When Toledo died in 1567, Herrera assumed the awesome responsibility of completing the project. While his predecessor's plan was followed in the main, modifications became necessary as work proceeded. The Escorial established Juan de Herrera as the greatest architect in Spain.

A journal kept by the historian and observer Padre José de Siguenza records the challenges of organizing the vast army of workers and craftsmen needed for the project. This responsibility fell to Fray Antonio de Villacastin. Craftsmen were sought all over Europe to labor under the exacting requirements of the design: stonemasons, bricklayers, carpenters, cabinetmakers, carvers, metalsmiths, as well as the painters, sculptors, goldsmiths, organ builders, and bellfounders. The new village of San Lorenzo evolved as the artisans set up homes for themselves. Their fine craftsmanship remains impressive to this day, and much of the credit for the result, from the stone facade to the exquisitely crafted towers to the dazzling interiors may be granted to Villacastin. Work progressed for more than two decades. In 1586 the church was consecrated on August 9, the eve of St. Laurence's Day.

Of massive size, the structure is designed in a great rectangle, 676 feet long and 526 feet wide. Inside the rectangular structure is the church, with its towers and great dome, and a complex of smaller rectangles filling in the larger one. This tessellated plan is a Spanish tradition, introduced from the plans of Moorish palaces, commonly built around a central courtyard. Royal apartments and offices occupied the cloisters to the south of the church, with the monastic quarters to the north side. In all, the Escorial includes 16 courts, 88 fountains, 3 chapels, 15 cloisters, 86 staircases, and 300 rooms.

The austerity of the facade, with its towering granite walls little relieved by decoration, makes a sobering impression and signalled a startling break with the lavishness of the Italian Renaissance. The fortress-like walls are eased only by the Habsburgian towers at each of the four corners and by the classical main entry. The steep roofs were sheathed in slate, an innovation from Flanders introduced by Philip himself. The central main entry tower is two stories higher than the rest of the wing, its upper section topped by a pediment, the whole supported by recessed columns. The influence of Palladio is evident in the restrained design. The doorway itself departs from the traditionally prominent Renaissance church entry by effacing itself into the facade.

Herrera found it necessary to make drastic revisions in the plans for the church, the centerpiece of El Escorial. Based on the plan of St. Peter's in Rome, it reflects the more severe lines of the entire structure. The church is designed in

Library at El Escorial
Photo courtesy of Instituto Nacional de Promocion del Turismo

the form of a Greek cross measuring 320 by 200 feet, topped by a dome 60 feet across and 320 feet high. Its interior has been stripped of all ornamentation, leaving the soaring lines and rich art work to dominate. The church facade includes two square bell towers on either side topped by small domes that reflect the main dome. Overlooking the entry from atop the porch are six enormous bronze and marble statues of Old Testament kings. In consequence the approach to the church is called the Courtyard of the Kings.

Inside, the high altar of the church lies behind the intersection beneath the dome and features some of the finest art work in the entire complex. Behind the altar itself are stacked four sections of paintings, separated by columns, with the crucifix appearing in the center of the topmost section, immediately beneath the vault. On either side of the towering altarpiece, or reredos, are bronze statues of saints by Pompeo Leoni, which represent some of the finest sculpture of the Italian Renaissance. At the center base of the reredos, immediately above and behind the altar, is the House of the Sacrament, designed by Juan de Herrera in the shape of a cylinder topped by a dome. It was built of jasper with columns of red jasper, their bases and capitals in bronze, and topped by bronze statues of the twelve apostles.

To either side of the high altar are monuments to Charles V and to Philip II. These also feature statuary by Pompeo Leoni depicting the monarchs and members of their families. The vaults of the church are all embellished with lavish frescoes by such artists as Luca Cambiaso and Luca Giordano.

Philip II's son and successor Philip III began the construction of the royal pantheon, and Philip IV completed it. Both Philip III and IV were devoted supporters of the monastery and lavished much attention upon it. The pantheon was designed by Giovanni Battista Crescenci, who planned it as a sixteen-sided polygon. In the sarcophagi that line the walls are the remains of all Spanish kings, Habsburgs and Bourbons, from Emperor Charles V to Alphonso XIII, with only two exceptions, Philip V and Ferdinand VI. The pantheon is given a royal aspect by the lavish use of bronze and black marble, and a more baroque ornamentation than is found in the church. Since only queens whose sons became kings were entitled to be entombed in the royal pantheon, another pantheon, the Pantheone of the Infantes, was constructed in the nineteenth century for the other queens, princes, and members of the family. Among those laid to rest there was Don John of Austria, son of Charles V and half-brother of Philip II.

The royal apartments comprise about a quarter of the Escorial's total space outside of the church. The apartments of Philip II remain as they were furnished in his day and are a testament to the monastic simplicity in which this ruler of a far-flung empire lived. His private chambers consisted of three rooms totalling approximately ninety square yards. One of the rooms served as his study, another as his bedchamber. The walls are plain white, the furnishing simple; the only decorative element is the beautiful Talavera tiles that extend from the floors to the dado. Here Philip collected some of the books with which he pursued his scientific interests and a few pieces of prized (and priceless) art. A small private chapel connected his bedroom to the church; during his final illness, when he was confined to his bed for fifty days, he was still able to view the high altar. He died on September 13, 1598, at the age of seventy-one.

The throne room and the Hall of Battles both date to Philip's time as well. The throne room, a long hall with a flattened barrel-vault ceiling, reflects the same simplicity as his own apartments, but does include some excellent Flemish tapestries. For the Hall of Battles, Philip commissioned a mural of the Battle on the Higueruela in which John II of Castile defeated the Moors in 1431, and its representation was assigned to a team of painters composed of Niccolò Granello, Lazzaro Tavarone, Fabrizio Castello, and Orazio Cambiaso. The painstakingly detailed painting covers the sixty-foot-long south wall of the gallery and was based upon a fifteenth-

century tapestry. Upon the opposite wall are paintings illustrating scenes from Philip's own military career.

Also dating from Philip's time are three rooms that belonged to his eldest daughter, Isabel Clara Eugenia. A favorite companion of her father's, she often stayed with him at the Escorial. Her rooms are decorated in the same simplicity as her father's, but their furnishings are more in keeping with a royal occupant. From the eighteenth century onward none of the original royal apartments was in use. Alfonso XIII, who reigned from 1886 to 1931, hired José María Florit to furnish them in the style appropriate to Philip's time; Florit was able to do so thanks to a journal kept by one of Philip's servants. When Charles II succeeded Philip IV he too contributed to the enrichment of the Escorial. He founded the Altar of the Holy Host in the church sacristy in 1690; above it hangs a painting by Claudio Coello entitled *The Adoration of the Holy Host by Charles II*, which features superb portraits of the king, Fray Francisco de los Santos, and various nobles and other historical persons, including the artist. The altar was designed by the royal architect, José del Olmo, and was crafted of jasper, marble, and gilded bronze. The crucifix is also gilded bronze. The sacristy itself, where the rare and valuable vessels and vestments used in services are stored, is a long hall 330 feet long with a flattened barrel-vault ceiling. One long wall is fitted with a row of chests of drawers made from precious woods.

Philip II wanted his new monastery to have a suitably impressive library. The large room housing the collection is divided into three sections by two arches. Lavish paintings by the Italian artist Pellegrino Tibaldi cover the vaulted ceiling and the friezes along the upper walls above the shelves. The vault is divided into seven panels where Tibaldi portrayed the liberal arts with their patrons; in the arch above the door in the end wall is a personification of Theology, surrounded by the four Fathers of the Church. Juan de Herrera designed the bookcases with their Doric columns, elaborately carved from precious woods. Although many volumes were destroyed by a terrible fire in 1671, Philip's vision for his library has been well met through the centuries. Today the library contains some 40,000 rare volumes and more than 4,000 manuscripts, nearly half of which are in Arabic. Many of these last came from the collection of Mulei Zidan, sultan of Morocco, donated by Philip III after the collection's capture in 1611. Among the other volumes preserved there are: the prayer book of the emperor Charles V, composed of four beautifully illustrated manuscript volumes on parchment; the *Codex Emilianense*, a copy of a tenth-century manuscript decorated with Mozarabic miniatures; the 'Las Cantigas' Codex of Alphonso X, which contains some of the most important miniatures of the thirteenth century; the *Codex Aureus*, written in gold; and the diary of Teresa of Ávila.

The mathematical elegance and monastic severity of the Escorial were not always appreciated by Spanish monarchs; like other artistic creations, the architecture of the Escorial fell in and out of favor. When the Bourbons came to the Spanish throne the baroque was in full flower, and the early Spanish Bourbons found the Escorial too stark and dreary to live there. It was the resurgence of the interest in the classical style during the eighteenth century that brought it to the attention of Charles III. Charles commissioned Juan de Villanueva, the greatest Spanish architect of the age, to refurbish the palace. Villanueva was both an admirer and a student of the works of Herrera, so the commission was a perfect assignment. Villanueva managed to suit the more elaborate lifestyle of the Spanish Bourbons, construct additional buildings needed for the government, and add two small pleasure pavilions, all the while respecting the overall style of Herrera's design.

One of the most splendid features of the refurbishment was the use of tapestries to cover the plain walls. Many of the tapestries were produced in Madrid at the royal manufactory and were based on designs by the great painters of the day such as Goya, Castillo, and Bayeu. The scenes depicted are of lighthearted country outings and humble peasants. Goya brought a new sensibility of color to his designs, challenging the skills of the royal craftsmen to reproduce it. The fascination of all Europe with the discovery of the ruins of Pompeii is reflected in the Pompeian Drawing-Room, which was filled with wall hangings decorated with Pompeian designs, painted by José del Castillo.

The two small pleasure palaces, the Prince's Pavilion and the Upper Pavilion, provided a place for afternoon gatherings and entertainment. The first was built for the future Charles IV and his wife, the second for the Infante Gabriel, Charles III's favorite son. The Upper Pavilion was designed with music particularly in mind, since Gabriel was a music lover and often hosted concerts there.

As much as it is anything else, the Escorial is one of the world's great treasuries of art. In addition to its architectural painting, the complex is rich with paintings hung, in some places, virtually from floor to ceiling. From the beginning, Philip II had collected an astonishing and rich collection of art from around Europe, and his descendants continued his efforts. Now the Escorial collections encompass a vast range of art, including works by Albrecht Dürer, Gerard David, Hieronymus Bosch, Rogier van der Weyden, Titian, Tintoretto, El Greco, Velazquez, and Luca Giordano. Among the sculpture is the famous crucifix by Benvenuto Cellini, carved from Carrara marble, which hangs in a chapel behind the choir.

Further Reading: For a detailed overview and an abundance of illustrations in both black and white and color, see *The Escorial: the Royal Palace at La Granja de San Ildefonso* by Juan de Contreras y Lopez de Ayala, translated by James Brockway (London: MacDonald, 1963; New York: Meredith, 1967). Both palaces are given thorough coverage.

—Elizabeth Brice

Euboea (Euboea, Greece)

Location: Just off the eastern coast of Greece and separated from it by the Euboean Gulf; usually considered a part of the mainland because of its proximity; with northern Sporades Islands forms department of Euboea.

Description: Approximately 105 miles from northwest to southeast and ranging from 5 to 30 miles in width, the island contains fertile plains as well as mountains and forest regions. In ancient times, seafarers from Euboea established vital trade routes and played a significant role in the development of Mediterranean civilizations. Today the island contains important ruins and artifacts reflecting historic developments stemming back to the Bronze Age. Especially notable are recent archaeological finds at the city of Lefkandi that have challenged the long-held belief in a "Dark Age" occurring between the end of Mycenaean civilization and the emergence of modern Greece.

Site Office: Tourist Police
Chalcis, Euboea
Greece
(221) 83-333, or 24-574, or 22-100

Euboea is the largest Greek island after Crete. At one point, the coastline of its capital city, Chalcis, is less than 100 yards from the Boeotian region of mainland Greece. Running through this narrow channel is the Strait of Euripus, which means "fast current." The island's proximity to the mainland facilitated the construction of a succession of bridges which, since ancient times, helped promote continued migration from mainland Greece. The best-known among these was the Negroponte (Black Bridge), built during the Venetian occupation of Greece in the thirteenth century. For centuries afterward, the entire island was known by the name "Negroponte."

Euboea's earliest known inhabitants were the Abantes, a Thracian tribe that migrated from mainland Greece early in the Bronze Age, which lasted from 3000 B.C. to about 1200 B.C. Succeeding waves of immigrant tribes, including the Ionians, brought to Euboea the comforts and conventions of the brilliant Minoan and Mycenaean civilizations that flourished during this period. Known to archaeologists as "palace cultures," Minoan and Mycenaean societies revolved around elaborate stone buildings whose aristocratic residents governed their cities by hereditary right. Little is known about Greek history from the end of the Bronze Age in 1200 B.C. to about 900 B.C. For many years, historians characterized this as the "Dark Age" of Greece, a mysterious vacuum of some 300 years of diminished prosperity, few achievements, and more primitive society. The archaeological legacy discovered on Euboea in the second half of the twentieth century was to change these perceptions.

The historically significant cities of ancient Euboea included Histiaea, Geraestus, Karistos (also spelled Carystus), Chalcis, and Eretria. The latter two, located on the central-southwestern portion of the island, were ultimately to earn the greatest renown. Chalcis, which reflected many Minoan conventions, was believed to have been settled by Cothus the Athenian sometime following the Trojan War. Eretria, according to the ancient writer Strabo, had once been known as Melaneis and supposedly was settled by Aielus from Athens. These two chief Euboean cities were separated by the fertile Lelantine Plain, named, presumably, for its Lelanton stream, said to have healing powers. Adjoining the stream were mines rich in iron and copper.

Under the Ionian influence, the ancient Euboeans proved to be bold seafarers. They are known to have sailed forth regularly beginning in the eighth century B.C. to colonize and establish trading stations in the Thracian region of Greece and in lands as far away as Italy and Sicily. These Euboean endeavors spurred a commercial and cultural exchange that influenced ancient civilizations through much of the Mediterranean. Additionally, Euboeans voyaged regularly to the ports of Al Mina, Posidium, and Paltus in northern Syria, opening up the Near Eastern world to Greek eyes. They brought back with them gold and silver, then passed these luxury commodities on to mainland Greece and to Euboean markets in Italy. From Al Mina, the Euboeans also purchased fabrics, ivory, metal, and slaves in exchange for Greek wine and oil. The Phoenician alphabet, a precursor of modern Greek, may have come to Euboea from these Syrian cities.

Among the earliest Euboean colonies was the island of Pithecusae (Ischia) in the westernmost waters of the Bay of Naples. Here, the Euboeans traded gold from Syria for Etruscan copper and iron. The tiny island of Pithecusae contained fertile farmlands as well as gold deposits, but lost much of its commercial importance around 750 B.C. when the Euboeans established a second trading station, or emporion, at Cumae, on mainland Italy. Pithecusae was devastated by a volcanic eruption around 500 B.C. Today it is a major archaeological site with the Euboean influence very much evident in ruins preserved by volcanic ash.

The Roman historian Livy called Chalcis "Thermopylae by the Sea" because of its two harbors on the strategic Euripus Strait. The metalworkers of Chalcis, which translates as "bronze," were renowned for their weapons crafted from that metal as well as their artistry with vases and other decorative objects. The city prospered additionally from the

The island of Euboea
Photo courtesy of Greek National Tourist Organization

export of wheat and horses. Horse-breeders, known as *hippobotae,* ruled the city much of the time until the Athenian conquest. Similarly governed were the enterprising citizens of Eretria, a town of impressive scale. Eretrians paved the way for the colonization of Corfu, but retreated in the face of the Corinthian invasion in the eighth century B.C. Eretria was rich both in gold and bronze; its artisans produced elaborately decorated pottery. Additionally, the city reportedly could mobilize 600 horsemen and 3,000 infantry. Eretria's importance as a city is evidenced by the ruins of a temple built around 750 B.C. to honor Apollo Daphnephorus. It is among the earliest known Greek temples.

Led by these two vigorous cities, the island of Euboea enjoyed general prosperity until around 700 B.C., when Chalcis and Eretria became engaged in near-perpetual warfare over ownership of the Lelantine Plain. What came to be known as the Lelantine War triggered alliances and rivalries among other towns in Greece, eventually creating the first known conflict on a national scale in Greek history. The Greek cities of Thessaly and Corinth apparently allied themselves with Chalcis; Miletus and Megara seem to have joined

Eretria. The war had a negative impact on overseas markets as well as the trade flow in the strategic Euripus channel between Euboea and mainland Greece.

It was said that an oracle considered Chalcis supreme among all cities in Greece because of the strength of its armies, a tribute that may reflect the early advantage of Chalcis's bronze breastplates and other armor. These decorative armaments were soon in demand by other warring factions. No clear victor ever emerged from the Lelantine War; instead, years of struggle diminished the power and prosperity of both Chalcis and Eretria. This vacuum opened the way for dictators from Boeotia to impose themselves on Euboean cities for brief periods until the islanders regained self-rule. Eretria eventually built imposing fortifications, some of which remain today. Chalcis continued its colonization endeavors and began minting silver coins engraved with the eagle and serpent of Zeus on one side and a wheel, the symbol of Chalcis, on the other. The Chalcis coin became one of the primary units of commerce in the Mediterranean and gave rise to the "Euboic Standard," which became known throughout the region.

Meanwhile, on mainland Greece, Athens was showing signs of expansionism. Around 506 B.C., Chalcis joined with Sparta and Boeotia to attack the newly threatening city. The allied forces were defeated, and Athens retaliated by invading Euboea and capturing Chalcis. Soon afterward, a new enemy appeared in the form of Persian armies intent on conquests in Greece. In 490 B.C., Persian king Darius the Great demolished Eretria and enslaved its citizenry in revenge for Eretria's aid to an Ionian revolt against Persia nine years earlier. Euboea then joined with Athens and other Greek cities to form the Delian League, an alliance against Persia named for the sacred island of Delos. The league's efforts reduced the Persian threat, but also fed the fires of Athenian imperialism. Euboea soon fell under the domination of Athens, which had become dependent on the island's supply of wheat and cattle.

The Euboeans rebelled against Athens in 446 and again in 411, during the Peloponnesian War. In 371 B.C., Thebes won control of Euboea, making it a Macedonian dependency. In this period, Euboea acquired its most famous resident, the great philosopher Aristotle, who came to the island after the death of his best-known pupil, Alexander the Great. Aristotle died in Chalcis in 322 B.C. The great thinker was said to have thrown himself into the Euripus Strait in frustration over his inability to understand an island enigma: why the powerful current in the strait changes direction several times each day.

In the second century B.C., the island came under Roman domination, but it continued to prosper. Centuries later, it became part of the Byzantine Empire. In 1205, Euboea fell under Frankish subjugation and was divided into three baronies owing allegiance to the king of Salonika. In 1336, the Venetians won almost complete control of the island; it was considered Venice's most significant holding. The island was lost to the Turks in 1470 and became part of Greece in 1830.

Today, the island has nearly 200,000 inhabitants. It still provides bountiful grazing lands, explaining, perhaps, the origin of the name Euboea, which means "rich in cattle." The mountains that once yielded iron and copper today provide magnesite and nickel for export. The city of Karistos exports the green-and-white *cipollino* marble that lent such beauty to Roman temples and palaces in ancient times. The valleys of Euboea produce olives, fruits, vegetables, and grains. Modern Chalcis, with a population of some 36,000, is a major commercial center because of its bridge to the mainland and is an especially popular bathing resort for Greek citizens. The remains of a Turkish fort dating to 1686 are the town's only reminder of the one-time struggle between the Venetians and the Turks.

The historic significance of the island of Euboea is evidenced by its archaeological remains, especially those discovered after 1960. Modern buildings and homes in the capital of Chalcis, of course, have prevented extensive excavations of that region. However, an archaeological museum in this capital city contains an impressive marble sculpture depicting the mythological figure Theseus's abduction of the Amazon Queen Antiope. The work dates to 500 B.C. and comes from the Temple of Apollo Daphnephorus in Eretria. The museum in Chalcis also contains a wealth of Mycenaean pottery along with other artifacts.

More successful have been digs at the site of Eretria, which evidently was rebuilt after its destruction at the hands of the Persians in the fifth century B.C. Archaeological excavations have focused on the ancient West Gate, near which a modern museum is now located. There are fortifications dating to 400 B.C. and ruins of two palatial buildings, a classical theater, a temple honoring Dionysus and two gymnasiums. Of special interest are the ruins of an impressive house dating from the fourth century B.C. The rooms are arranged around a central courtyard. Some of the furniture is preserved as are four pebble mosaics showing mythological scenes including a seahorse, griffins, sphinxes, and floral designs. Visitors to Eretria can also see a small cemetery for warriors and their families dating to 700 B.C. It contains a bronze cremation urn and various weapons buried with the deceased, including a bronze spearhead from the Mycenaean period. Many of the finds from this ancient city are on display today at the Eretria Museum.

The most significant archaeological site on Euboea is the village of Lefkandi, which lies on the once-fought-over Lelantine Plain. Discoveries at Lefkandi have provided archaeologists with new perspectives about what was once considered to be the Dark Age of Greece. Excavations around the city did not begin until the 1960s, when only a small part of the site was uncovered. Among the points of interest was a hillock called Toumba. A cemetery was uncovered there in the 1960s by the British School of Archaeology in Athens. Additionally, trial excavations had uncovered the stone foundations of a large building constructed in the tenth century B.C. Further excavations were clearly warranted. However, the summit of Toumba was owned by a local citizen who wanted to build on the mound.

In 1980, this landowner sent a bulldozer in to clear the way, destroying a third of the historic building before Greek authorities could step in to save the site. The next year, archaeologists uncovered evidence not only of an impressive building but also a vigorous society that flourished from the early Bronze Age to around 700 B.C., right through the so-called Dark Age. Especially revealing was the wealth of rings, bands, and other gold treasures found in tombs dating back to the tenth century B.C. These treasures indicate that Lefkandi once enjoyed a level of prosperity unparalleled in ancient Greek cities other than Athens. The Euboeans of this period had formerly been considered a primitive, secluded people. In fact, their graves contained vases from mainland Greece, bronze wheels from Cyprus, and cut gems and sculptures indicating trade from as far away as Egypt. Euboea's maritime commerce evidently began much earlier than historians previously had thought. One truly unique finding, also from a tenth-century tomb, was a terra-cotta centaur, the mythological creature that is half-man, half-horse. This figure had not otherwise been seen in artwork from that period. Another notable find was a simple vessel with archers painted on it. The earliest known Iron Age depiction of humans on a vase, it also verifies the importance attached to warfare in this period.

The tombs uncovered at Lefkandi revealed a unique combination of burial customs. While corpses were being cremated on a funeral pyre, jewelry, vessels, and personal possessions were tossed onto the fire. Following this ceremony, the ashes were not placed in an urn as was the custom in cities such as Athens. Rather, a few of the burned bones were placed in an open grave with a collection of the deceased's belongings. The remains of those belongings suggest that ancient Lefkandi was one of the first cities to import luxury items from the Near East.

The large, sophisticated structure partly destroyed by the twentieth-century bulldozer contained graves indicating a "heroon," or hero-cult, around 1000 B.C., a phenomenon previously thought to have started two centuries later. One particularly elaborate tomb held the cremated remains of a warrior buried with his iron spear and sword. The ashes of this warrior, now known as the "hero of Lefkandi," were found wrapped in cloth and resting inside a decorated bronze vessel. The cloth is surprisingly well preserved, due largely to favorable humidity levels, and contributes significantly to our knowledge of ancient textiles. Another grave held the skeletal remains of a woman buried with a gold neckpiece, a breast-covering of gold discs, and golden earrings and hair adornment. Archaeologists believe she may have been the warrior's consort. A knife was also found in her tomb, suggesting that she may have been a human sacrifice. A nearby grave shaft held the remains of four horses, also apparently sacrificed by being thrown down the shaft head first.

Today, on the now-peaceful island of Euboea, tourists still come to watch the mysterious changes of current in the Euripus Strait. The riddle that supposedly led Aristotle to despair remains unsolved. But there has been some resolution of a far greater mystery. The so-called "Dark Age" of Greece has proved not so dark after all, as a result of the archaeological legacy found in Lefkandi.

Further Reading: *The Rise of the Greeks* by Michael Grant (London: Weidenfeld and Nicolson, and New York: Scribner's, 1987) includes a discussion of Euboea and its influence on Mediterranean civilization, as do other books by this same author. *Greece: Temples, Tombs and Treasures*, edited by Dale Brown (Alexandria, Virginia: Time-Life, 1994) includes a highly readable chapter on the discoveries at Lefkandi. *The Emergence of Greece* by Alan Johnston (Oxford: Elsevier-Phaidon, 1976) has some brief text and interesting pictures of a few of the island's archaeological sites and treasures.

—Pam Hollister

Évora (Évora, Portugal)

Location: Sixty-eight miles southeast of Lisbon.

Description: A former Roman colony, Arab emirate, and Portuguese royal seat with a wealth of historic architecture; a UNESCO World Heritage site; capital of the district of Évora and the former province of Alentejo.

Site Office: Évora Tourism Office
Pr. do Giraldo
7000 Évora, Évora
Portugal
(66) 226 71

Évora lies in the region of Portugal known as Alentejo, the largest in area, but the least densely populated, of the country's traditionally defined regions. The town is surrounded by wheat farms, vineyards, and groves of cork-oak and olive trees. Into this pastoral setting, however, have come numerous conquerors—Romans, Arabs, and Portuguese royalty—who provided the town with a significant role in national events and a huge concentration of well-preserved historic structures.

Iberos, Phoenicians, Greeks, and Celts preceded the Romans in Portugal, but this last group left the greatest imprint on Évora. In 60 B.C. under Julius Caesar, Liberalitas Julia, as Évora was then known, became a major colony in Lusitania, the Roman province encompassing most of present-day Portugal. The name Lusitania comes from the Lusitani, a Celtic tribe that inhabited the area prior to the Roman occupation. By the first century A.D. Romanization pervaded Lusitania from south to north and a senate was established in Évora. City walls were constructed, and in the second century the fabulous Temple of Diana was erected.

It is not known whether the temple actually was dedicated to Diana, but her name became attached to it because of the probability that the Romans would have perceived Alentejo, a land most suited for hunters and gatherers, as Diana's territory. The temple is bright, simple, and open. Standing on its raised platform are granite Corinthian columns topped with capitals of marble from Estremoz, a small town northeast of Évora. More than in its construction, however, the temple's interest derives from the series of functions it has served—it was a fortress in the Middle Ages and later became a slaughterhouse—and the contrast it provides to the dark, ornate, confined places of worship erected in Évora by Christians during medieval times. Three-fourths of the temple's columns survived an earthquake in 1775, and the remains of the temple can still be viewed today.

When Rome converted to Christianity in the fourth century, so did Portugal, and an important bishopric was estab-

lished at Évora. With the decline and fall of the Roman Empire, however, the next few centuries saw periods of Suevi, Vandal, and Visigoth presence in Évora, until 714–16, when the city was occupied by the son of an Arab governor and converted to a Moslem principality for the next 400 years. With the eleventh-century collapse of the Moslem caliphate, independent regional city-states were formed, and the emirate of Évora, largely controlled by the ibn Wazir family, became one of the more significant and prospered as a university city. The breakup of the caliphate, however, weakened Moslem rule to the point that reconquest by Christians became a real and much-feared possibility. As a defensive measure, Évora's rulers reinforced the Roman and Visigoth walls that surrounded the city. Despite these extensive efforts, the reconquest of Portugal was under way in earnest in the twelfth century. Évora submitted to the Christians in 1159, but was recovered by the Moslems two years later, then was retaken by the Christians in 1166, in a surprise attack led by adventurer Geraldo Sem Pavor, nicknamed Gerald the Fearless. Many legends grew up around Geraldo's attack. It is said that when his soldiers, intimidated by the task at hand, refused to climb twenty feet to the top of the guarded wall, Geraldo scaled the wall on a ladder of spears, beheaded two guards, and returned to his cowering army to show them the evidence of what one man could do. It is also said that he stayed within the walls and fought off hundreds of Moslems with one hand while opening the city gates for his comrades with the other. These legends were preserved in literature by the sixteenth-century lyric poet and historical fiction writer Luíz de Camões in *The Lusiads*. Évora's coat of arms depicts Gerald the Fearless holding two heads in his hand.

Even after the reconquest, the region beyond Évora's city walls was dangerous and unsettled. Therefore, the new Portuguese kings granted administrative power in this and similar regions to religious and military orders that had grown out of the crusading army. The Order of Calatravans was founded in Évora; the order later became known as the Order of Aviz when the town of Aviz, in the Alentejo north of Évora, was bestowed upon it.

In 1383, King Ferdinand I died without a male heir, and his half-brother, John, master of the Order of Aviz, became regent of Portugal. A power struggle began the following year, when King John I of Castile, Ferdinand's son-in-law, invaded Portugal and declared himself king. In the war that ensued, the Portuguese nobility supported John of Castile, while the growing bourgeoisie of Évora, Lisbon, and other important cities supported John of Aviz. John of Aviz also had the assistance of England. He eventually emerged the winner, following a long siege at Lisbon, an outbreak of plague among the Castilians, and a major victory at Aljubarrotta. He was crowned King John I of Portugal in 1385,

Temple of Diana at Évora
Photo courtesy of Portuguese National Tourist Office

beginning the Aviz dynasty. He and his successors made Évora the site of their royal court and residence.

John II, John I's great-grandson, who ruled from 1481 to 1495, held important meetings in Évora, at which he asserted the supremacy of the Crown over the nobility and reformed the justice system. One noble, Fernando, duke of Bragança, who was John II's brother-in-law, particularly objected to the king's

reforms. He launched a conspiracy against the king, was accused of treason, tried and convicted in Évora, and beheaded in the main square. John II confiscated the lands of Fernando and his fellow conspirators, put down another plot, and effectively established royal power over the nobility.

John II also accelerated the construction of the royal palace, which had been started in the first half of the fifteenth century. The palace was probably completed by John III in the next century. Central to the construction were rooms to accommodate receptions, parties, banquets, and other formal events. Festivities celebrating the marriage of Prince Afonso to Princess Isabel of Castile in 1490 drew writer and goldsmith Gil Vicente to the court in Évora. Vicente became court dramatist to Manuel I, John II's cousin and successor, and Manuel's son and successor John III. Vicente wrote farces and tragicomedies for the court and *autos* (plays) to be performed in church; he is credited with the creation of a distinct Portuguese drama. Among his works are *The Portuguese Pastoral Play* composed in 1523 for the Christmas party at court. His last play, *The Forest of Deceits,* was also performed at Évora. Vicente was well liked by the elite, intellectuals, and artists of the day, but his work was censored during the Portuguese Inquisition, which began in 1536, coincidentally the year of Vicente's death. His reputation was not revived until the nineteenth century.

Much construction went on at Évora during the late fifteenth and early sixteenth centuries. A style known as Manueline, so called for King Manuel I, was marked by creative and elaborate decoration, often utilizing nautical images, as this was the golden era of seafaring exploration by the Portuguese. Churches, convents, and monasteries in Évora were built in this style, and many of them still stand today. Among the most significant relics of this period are the Church of St. Francis and the attached Capela dos Ossos (Chapel of Bones). The walls and columns are lined with the skulls and bones of 5,000 people, including monks who died during the plague and soldiers and civilians who were victims of war. The chapel is designed to evoke thoughts on human mortality, but some observers consider it more a reminder of atrocities.

Atrocities did occur at Évora during the Inquisition. In 1543 its first victims were burned at the stake in Évora. The cellars of the fifteenth-century Los Loios monastery became dungeons for accused heretics. The Inquisition in Portugal was generally milder than that in Spain, however, with more burnings in effigy than actual burnings of people. King John III, who initiated the Inquisition, did so more for political than religious reasons; he saw it as a means to increase royal power. The Holy Office of the Inquisition remained a Portuguese institution until 1821, but its power declined sharply under reformist governments in the eighteenth century.

The Aviz dynasty died out in 1580, precipitating an invasion by Spain. King Philip II of Spain annexed the country and was crowned King Philip I of Portugal. Évora was no longer the seat of royalty, but its people were able to influence the course of national affairs. In 1637, Évorans rose up in protest against taxation to finance Spanish military ventures. The Évorans and other Portuguese began to rally behind John, duke of Bragança, who overthrew the Spanish in a bloodless coup in 1640 and was crowned King John IV, restoring Portugal's independence. Lisbon became the royal residence.

Évora never regained its former prominence, and its decline reached its depth with the closing of the university. The university was run by Jesuits, and the Jesuit order as a whole ran afoul of Carvalho e Mello, marquês de Pombal, chief minister to King Joseph Emanuel in the eighteenth century. Pombal, who because of the king's weakness was the virtual ruler of Portugal, believed the Jesuits were opposed to his plans to modernize the country, and accused them of participating in a plot to assassinate the king. In 1758 he closed the university at Évora and the following year expelled the Jesuits from Portugal. Évora's population declined, and little of note occurred in the town until the university was reconstituted in the 1980s.

One positive effect of Évora's years of stagnation, however, was the preservation of its architectural treasures, including many of the religious buildings and private homes constructed during the rule of the Aviz dynasty, as well as remains from the times of Arab and Roman rule—such as the Temple of Diana. Because of this wealth of historic structures, Évora was named a UNESCO World Heritage Site in 1986. Évora also boasts many museums and art galleries, and hosts several festivals each summer.

Further Reading: *A New History of Portugal* by H. V. Livermore (Cambridge and New York: Cambridge University Press, 1966; second edition, Cambridge: 1976, New York: 1977) is a solid, scholarly history that devotes significant attention to Évora, as does *Portugal: A Country Study,* edited by Eric Solsten (Washington, D.C.: Federal Research Division, Library of Congress, 1994; first published as *Area Handbook for Portugal,* by Eugene K. Keefe et al., 1976).

—Patricia Trimnell

Famagusta (Famagusta, Cyprus)

Location: Port city on southeastern coast of Cyprus, facing Syria; forty miles east of inland capital of Nicosia, sixty miles west of the Syrian coast.

Description: Two cities, Turkish Famagusta and adjacent Greek Varosha, together referred to as Famagusta; Famagusta founded in ancient times; became wealthy port under Venetian rule in Middle Ages; conquered by Turks in 1571, after which the Greek population was forced to leave, and settled outside city walls, in Varosha; old city famous for its ancient walls, churches, and ruins; major port city of Cyprus and capital of the district of Famagusta. Since the Turkish invasion of Cyprus in 1974 and its continued occupation of northern Cyprus, Famagusta and the surrounding area has been considered a closed area, patrolled by United Nations troops.

Contact: Tourist Information Office
Leoforos Arch. Makariou III, 17
Agia Napa, Famagusta
Cyprus
(3) 7211796

Throughout its long history, the island of Cyprus has connected East and West, and for centuries the point of contact was Famagusta, an Occidental city from which the coast of Asia Minor is distinctly visible on a clear day. Yet Famagusta is a Cypriot city, not only because of its location but because of its culture—a melange, but not a melting pot, of Greek, Turkish, Italian, Armenian, Syrian, Jewish, and other influences. Its history is incredibly diverse, with antiquities and treasures reflecting the many foreign powers that have dominated Cyprus and left their mark on Famagusta, from the Roman and Byzantine Empires to the French Lusignan family, the Venetians, the Ottoman Turks, and the British.

No one knows definitely when Famagusta was founded. Evidence indicates a settlement was in place as early as 288 B.C., and it is possible that this city was built on the foundation of a still more ancient one. In Greek the town was called Ammokhostos, translated to Famagusta in Latin, and Magusa in Turkish. Important Greek trading colonies dotted the island of Cyprus, and when the Romans completed their conquest of the eastern Mediterranean by the first century B.C., Cyprus became an indispensable part of the Roman Empire. When the Roman Empire was divided in the fourth century A.D., Cyprus fell squarely in the eastern, or Byzantine, portion, and Greek Orthodoxy soon became the dominant religion. (The influence of Christianity had been felt on the island as early as the arrival of St. Barnabas, a disciple of St. Paul, in A.D. 45.)

As always, because of its prime location, the island was a pawn between rival powers. From the seventh to the tenth centuries, the coastal port cities, including Famagusta, were subjected to frequent raids by Moslems from across the sea, the most powerful rivals of the Byzantine Empire. The Moslems did not, however, succeed in capturing Cyprus. Famagusta was an insignificant port at this time, since trade between East and West had diminished drastically because of Moslem dominance of the Mediterranean, which lasted until the Italian city-states successfully wrested control of the sea by the twelfth century.

In the late twelfth century, King Richard I (the Lion-Hearted) of England conquered Cyprus and sold it for a considerable sum to a fellow crusader, Guy de Lusignan. This marked the beginning of the long reign of the Lusignan family over Cyprus. Famagusta, the site of the family's royal palace, grew in wealth as a prominent center of international trade. In the thirteenth century the city was also a launching point for crusading armies setting out for the Holy Land.

Cyprus was ruled as a feudal kingdom, with the Lusignans doling out hundreds of fiefdoms to their loyal Roman Catholic supporters. The new ruling class upset the established order by favoring Roman Catholicism and discriminating against those professing Greek Orthodoxy, creating a tension that was not mitigated until the Council of Florence in 1439 temporarily proclaimed both religions equal. Nevertheless, the Roman Catholic presence explains the appearance of Gothic architecture, still visible in and around Famagusta, as well as the remains of convents and Roman Catholic churches; Roman Catholicism declined precipitously with the resurgence of Orthodoxy following the Turkish conquest of Cyprus in the late sixteenth century.

Henceforth the ruling class spoke French, while, increasingly, Italian traders set up shop or established branches of their business in Famagusta. Mostly Venetians and Genoese, they built many fine buildings and made Famagusta extremely rich on East-West trade. By the fourteenth century, chroniclers of Cyprus recorded that the city had 100,000 inhabitants, no doubt an exaggeration, but an indication of a surge in population stemming from commercial prosperity.

Real political power shifted to those wielding economic control, and by the fourteenth century, those in control were the Genoese and Venetians. The rivalry between the two Italian city-states over control of the Mediterranean played itself out in Famagusta on October 10, 1372. On that day, a new Lusignan king of Cyprus, Peter II, was to be crowned in the cathedral. The respective representatives of the Genoese and Venetian city-states participated in the ceremony, as was the custom, only this time they came to blows, and a terrible melee broke out in the town square. A virtual massacre of the Genoese ensued in Famagusta. In retaliation, the brother of

Famagusta street scene

Famagusta city walls
Photos courtesy of The Chicago Public Library

the Doge of Genoa sailed to Famagusta with an army of 14,000 men, who pillaged the city, burning much of it to the ground. Civil order was not restored for almost a half-century, when Lusignan king James II finally recaptured control of Famagusta from the Genoese in 1464. Lusignan dominance would not last, however; in 1489, the Republic of Venice seized control of Cyprus, and the Lusignan dynasty eventually vanished in genteel exile.

Famagusta never recovered its prosperity, even though commercial activity resumed. Regarding Cyprus as its most valuable possession, Venice used its enormous wealth to rebuild the city of Famagusta, its main port on the island. Venetians aimed to exploit the island's abundant natural resources, especially timber and salt. Cyprus became a colony for the enrichment of the Venetian Republic.

While Venice became rich from the exploitation of the island, the Moslem Turks were conquering region after region of Asia Minor, the Levant, and Egypt. Venetians, even though they were Roman Catholics, were among the main defenders of Constantinople in 1453 and left eyewitness accounts of the fall of the Byzantine capital to the Ottoman Turkish armies of Mehmed II. Still in control of Cyprus, the Venetians neglected to build up the defenses of Famagusta, their major Cypriot port, for thirty-five years, until shortly before the Turks arrived outside the city gates.

Knowing full well that Cyprus would not be spared conquest, the citizens of Famagusta made feverish preparations to repair and fortify their city's defenses. The city's fabled walls were remodeled and strengthened by trained Venetian engineers, and ammunition stores were filled to capacity. The elderly and the children were evacuated to remote villages, leaving a total of 7,000 people within the walled city, according to detailed Venetian reports.

Venetians who survived the assault and siege of Famagusta left graphic written accounts of the ten-month ordeal. Under the command of Lala Mustapha Pasha, the Ottomans had mustered a huge force to conquer Cyprus. Had Famagusta cooperated with the Turks, the city and inhabitants might well have been spared harm. The city's armed resistance against an enemy army numbering at least 50,000 officers and men spelled its doom, however. The port city was besieged for ten months by a Turkish naval fleet numbering 360 vessels, and the inhabitants gradually were reduced to starvation. Lala Mustapha Pasha promised to treat the army mercifully after it surrendered on August 1, 1571, but once deprived of their weapons, the surviving troops and their commander, Mark Antonio Bragadino, were tortured and flayed alive. It was the signal for a full assault on the starving, suffering inhabitants. All young girls and boys were enslaved and shipped to Turkey. On board one of these ships, a young Famagustan girl set fire to the ammunition stored on board, killing herself and her unfortunate companions.

Famagusta's fall became legendary, and stories and plays depicting the bravery and terrible fate of its inhabitants appeared throughout Europe. For the city of Famagusta, the Turkish takeover was a blow from which it never recovered.

In part this was because of the disappearance, through death or flight, of the vigorous merchant class of Famagusta. Then came the expulsion from Famagusta of all non-Turks soon after the conquerors asserted control. Finally, the trade routes had changed, so that by the seventeenth century these routes crisscrossed the Atlantic and the Horn of Africa rather than the Mediterranean.

The story of Famagusta, from approximately the time of the Turkish takeover in 1571 until British control replaced Turkish authority in 1878, is the story of decline and ruin. The once lively, bustling port silted over. The non-Moslems, whom the Turks had expelled from the city, moved immediately outside of the city walls and established the community of Varosha, now indistinguishable from Famagusta. Most of the buildings within the old walled city were heavily damaged or destroyed by the Turkish artillery bombardments. One of these, the Cathedral of St. Nicholas, was repaired and reopened as a mosque, as were several churches in the city. Because of the drastic decline in commercial activity, neither Varosha nor Famagusta blossomed until after World War II.

Nor did the period of decline and stagnation have the side effect, as is sometimes the case, of preserving buildings that might have been torn down for newer, more modern structures. In the mid-nineteenth century the building of the Suez Canal in Egypt led to the razing of many of the splendid old structures in Famagusta, so that their stones could be used in the construction of Port Said. This destruction of architecturally noteworthy buildings finally halted when the British took control of Cyprus in 1878.

The British colonialists were welcomed by the non-Moslem majority of the Cypriot population, even though the new masters of Cyprus paid no heed to the Cypriots' desire for independence and proceeded to exploit the island to their advantage. Nonetheless, Great Britain set about building a modern communications and transportation infrastructure in Cyprus, and lifted the ban on non-Moslems residing in the walled city. The British also proceeded to revive and modernize the port of Famagusta. The latter served the British well during World War II, when ships and submarines left Famagusta's port to spy upon and attack Nazi warships. It was only after World War II, with the return of peace and prosperity to Europe, that tourism developed on a major scale in Famagusta and in Cyprus generally.

Very important for the welfare of Cyprus's historic places, however, was the belated establishment of the Department of Antiquities in 1935. While the British had put a stop to the razing of historic buildings for commercial purposes, little had been done in the way of historic preservation. Under the auspices of the new department, preservation began, continuing well after World War II, with the result that there are many places of historic interest in Famagusta.

One of the most famous sites of Famagusta is the fourteenth-century Citadel, or Othello's Tower, which the Venetians remodeled for the defense of the city in the sixteenth century. No one is sure when and why it came to bear the name of Othello, but it is almost certainly the setting of

Shakespeare's famous play. Besides the Citadel, among the most impressive fortifications in the Mediterranean region is the Martinengo Bastion, which bears the name of its talented Venetian engineer who designed and built the massive fortification in the mid–sixteenth century. The walls surrounding the old town still stand, twenty-six feet thick and fifty feet high.

The former Cathedral of St. Nicholas on the town square, now the Lala Mustapha Pasha Mosque, still is a fine example of Gothic architecture, design, and decoration. In front of it stands a monument to the Turkish poet and nationalist Namik Kemal, who was imprisoned in the former palace of the Lusignans opposite the cathedral. Heavily damaged by Turkish artillery during the siege of 1570–71, the palace was a ruin until the Ottoman Turkish government restored it for use as a state prison. When the British took over, it continued to decay, but was recognized as noteworthy for its antiquity.

Famagusta is famous for its churches, though they are largely in ruins. The fourteenth-century church of St. Peter and St. Paul is said to have been financed by a rich Famagustan merchant with the earnings from a single trade venture with Syria. This massive Roman Catholic structure displays an unusual combination of Gothic and Byzantine features. When the Turks took over the city, St. Peter and St. Paul also was transformed into a Moslem house of worship (the Sinan Pasha Mosque), and it later became the public library of Famagusta.

Other noteworthy churches in Famagusta include the Twin Churches, built in the Middle Ages by the Knights Templars and Knights Hospitallers, two religious and military orders that developed during the Crusades to recapture the Holy Land from the Moslems. The Armenian Church of St. Mary's bears witness to the presence of a sizable Armenian minority during the medieval period; there were originally three Armenian churches, but only St. Mary's has survived. The Nestorian church nearby was built by a merchant of the Nestorian faith (followers of the doctrines of Nestorius, who lived in Palestine in the fifth century) in 1359. There are also several ancient Greek Orthodox churches, a Franciscan shrine, and a medieval convent of the Poor Clares.

In 1960, Great Britain recognized Cyprus as a sovereign republic. Tourism became the primary industry, and thousands of foreign visitors thronged Cyprus's beautiful beaches and strolled through the walled city of Famagusta. Ethnic tensions, however, ran high. Despite the coexistence of a multitude of ethnicities since the Middle Ages, there was little mingling or "melting" of nationalities, especially between the Greek Cypriot majority and the Turkish minority, which accounted for about 18 percent of the population. Inter-ethnic conflict exploded in 1974, leading to the invasion of Cyprus by Turkey, which feared that Greece would annex Cyprus. Thousands of Greek Cypriots fled Turkish-occupied northern Cyprus, and thousands of Turkish Cypriots fled the south. Famagusta (including Varosha) became a ghost town and has remained so in part because of continued condemnation of the occupation of Cyprus by the United Nations, which for several years has been attempting to bring the two sides together to form a new government. Today, United Nations troops patrol Famagusta, but the Cypriot government is not allowed access to the Turkish-occupied portion of the island.

Some day, the fabled beaches of Varosha and the historic splendors of Famagusta may become accessible to foreign visitors as well as to native Cypriots. Until then one can only hope that no damage will befall the monuments to centuries of Cypriot history and culture, since the Turkish occupation authorities have prevented the Department of Antiquities from inspecting or maintaining these ancient historic sites.

Further Reading: An exhaustive study of Famagusta's history, politics, and government, and an illustrated description of all historical monuments and tourist attractions was published in English by the Chamber of Commerce of the city of Famagusta: *Famagusta Town and District: A Survey of Its People and Places from Ancient Times* by Kevork S. Keshishian (Nicosia: Famagusta Chamber of Commerce and Industry, 1985). Absent from its many pages, however, is any reference to the Turkish invasion and continued occupation of northern Cyprus, including Famagusta. The author, an Armenian Cypriot, has a decidedly Greek perspective. Among the most readable, reputable histories of Cyprus are Peter W. Edbury's *The Kingdom of Cyprus and the Crusades, 1191–1374* (Cambridge and New York: Cambridge University Press, 1991), which explores the time when Cyprus was in its commercial and cultural heyday; a lengthy, comprehensive history by Stavros Panteli, *A New History of Cyprus: From Earliest Times to the Present Day* (London and The Hague: East-West Publishers, 1984), which is one of the few all-encompassing histories of Cyprus; and *Cyprus from Earliest Times to the Present Day,* an informative, shorter history by Franz Georg Maier, translated by Peter Gorge (London: Elek, and New York: Fernhill, 1968; as *Cypern: Insel am Kreuzweg d. Geschichte,* Stuttgart: Kohlhammer, 1964).

—Sina Dubovoy

Fátima (Santarém, Portugal)

Location: Central Portugal, ninety miles northeast of Lisbon.

Description: Tiny village on a desolate plateau called Cova da Iria in the Serra de Aire mountains; famous Roman Catholic shrine and the site of pilgrimages since 1917, when three local children claimed to have been visited by the Virgin Mary.

Site Office: Posto de Turismo
Av. Dom Jose Alves Correia da Silva
Fátima 2495, Santarém
Portugal
(49) 531139

The village of Fátima has grown up around one of the world's most important Roman Catholic shrines. The area takes its name from an estate established in the twelfth century, when the warrior Gonçalo Hermigues decided to call his lands "Fátima," the original name of his wife, a Moorish woman who converted to Christianity and then changed her name to Ouriana. The area was populated largely by subsistence farmers in the succeeding centuries. In the early twentieth century, the impoverished inhabitants were to a great degree isolated from the sweeping changes in the Portuguese government and from the developing world conflict.

Portugal's King Charles I was assassinated in 1908, and the liberal republic that was subsequently established was hostile to religion. The Catholic Church in Portugal had been closely identified with the monarchy, was believed to be corrupt, and was considered an instrument of oppression. The republican government abolished the teaching of religion in primary schools, confiscated church property, and sent many nuns and priests into exile. Then during World War I, events occurred at Fátima that led to the rejuvenation of Portuguese Catholicism, not as a tool of the state but as a strictly spiritual phenomenon, making Fátima as important to Portugal's religious identity as Santiago de Compostela is to Spain's.

On May 13, 1917, three peasant children from the village of Aljustrel were tending sheep in the countryside that had once been part of the Fátima estate. According to ten-year-old Lucia dos Santos and her two cousins, nine-year-old Francisco Marto and his seven-year-old sister Jacinta, a bright light appeared above a small oak tree. The light metamorphosed into the figure of a woman. The lady, as the children referred to her, spoke to Lucia. She said that she was the Lady of the Rosary and that she had come to ask them to pray for world peace. She urged them to tell others to pray as well, in particular to say the prayers of the rosary. She told the children that they would face ridicule and suffering, but that their faith would comfort them. Before disappearing, the apparition told the children to return on the thirteenth day of each month.

At first, the children agreed to say nothing about the vision. However, Jacinta was unable to resist telling her parents about it. The families were skeptical. Lucia's parents were particularly distressed and her mother, Maria Rosa dos Santos, threatened her with severe punishment if she did not recant her story. Jacinta and Francisco's parents, Manual and Olimpia Marto, were less strident and in the months to come would provide much-needed support to the children.

The children were resolute and returned to the spot near the tiny oak tree at 12:00 noon on June 13. The woman again appeared to them and when the children asked if they could accompany her back to heaven, the lady replied that she would take Jacinta and Francisco with her in the near future. Lucia was to remain on earth to spread the woman's message.

As word of the children's experiences circulated, increasing numbers of people traveled to Fátima each month. Although none of the pilgrims saw the apparition, many reported seeing bursts of color and light in the air above the oak tree. On July 13, those in attendance heard the children gasp in horror. Afterward, the children said that the woman had shown them a terrible scene of people burning and screaming. This they would interpret as a representation of hell and the fate that awaited those who did not pray.

It was during the July visitation that the woman told the children what has come to be known as the secret of Fátima. The apparition told Lucia to write the secret in a letter and instruct Roman Catholic Church officials that the pope was to reveal it to the world in 1960. However, the Vatican let that year pass without comment. In 1967, Pope Paul VI issued a statement decreeing that the time was not right. The existence of this secret, locked in a Vatican vault, has led to much speculation. Some believe that it foretells the ending of the world. Each pope is believed to have read it. Rumors also have circulated that U.S. president John F. Kennedy, Soviet premier Nikita Khrushchev, and British prime minister Harold Macmillan were shown excerpts by the Vatican and agreed not to reveal the contents.

On the same day the apparition made other predictions that were not kept secret. She told Lucia that the current war would end soon but that if people did not repent, another would start during the reign of a pope to be named Pius XI. The outbreak of this second war would be preceded by a "night illuminated by an unknown light" signaling God's displeasure. Many have interpreted the January 25, 1938, display of the aurora borealis over most of Europe and parts of North America, and the reign of Pope Pius XI during the outbreak of World War II as proof of this prophesy's fulfillment.

Another prediction concerned the mounting influence of Russia and communism. The apparition told Lucia that communism would spread throughout the world until Russia

The Fátima Shrine
Photo courtesy of Portuguese National Tourist Office

was converted to Catholicism. Once that was accomplished, Russia's position as the world's communist power would end. She also asked for the faithful to make communions of reparation on the first Saturday of each month. Some theologians have claimed the dissolution of the Soviet Union in 1991 to be a fulfillment of this prophesy, but it is not the official opinion of the Vatican.

In any event, the growing interest in the events unfolding at Fátima in 1917 alarmed government officials, who were suspicious of the enthusiastic outbreak of religious activity. On August 11, Arturo de Oliveira Santos, the administrator of the council of the nearby town of Ourem, instructed the children's parents to bring them to his office for interrogation. Francisco and Jacinta's father, Marto, refused to surrender the children and instead paid a visit to the administrator himself. Lucia's father, Antonio dos Santos, was still skeptical of the girl's story and decided that a visit to the administrator would "teach her a lesson." As the two men and the young girl stood in Oliveira Santos's office, the administrator threatened Lucia's life if she did not reveal the secret the apparition had told her. Lucia adamantly refused, and the frustrated administrator sent her home.

The next evening, Oliveira Santos visited Aljustrel to question the children again. When he was rebuffed once more, he asked the parents to take the children to the local priory the next morning before they went to Fátima, in hopes that the priest would be able to elicit information from them. Apparently, the administrator did not trust the parents to do his bidding because early the next morning he appeared in the village once again to personally escort the children to the priory. When the children persevered in their refusal to reveal the secret, Oliveira Santos took them to the priory at Ourem. There the interrogations continued, accompanied by threats. Still they did not divulge the information. For a time, they were even placed in the local jail.

In the meantime, a crowd waited impatiently in Fátima. When the children did not arrive by 12:00 noon, those in attendance reported hearing thunder and witnessing flashes of light and many colors. Lucia, Francisco, and Jacinta returned home on August 15. Four days later, the lady appeared to them in Valinhos, located midway between Cova de Iria and Aljustrel.

Eventually, the church hierarchy in Lisbon decided to send a representative to investigate the situation and to interview the children. Reverend Doctor Manuel Nunes Formigao arrived in Aljustrel in time for the September 13 apparition and then returned again in late September. He spent much time with the children, questioning them about what they had seen and what the woman had said to them. Formigao became convinced of their sincerity.

As word of the apparition in Fátima spread, many people pressured the children to ask the woman to show proof of her existence. The children also wanted the lady to tell them who she was. According to Lucia, the woman promised to reveal her identity and to perform a miracle on October 13.

By early October, the tiny village was inundated with pilgrims awaiting the October 13 appearance. The children's homes were practically under siege. Antonio dos Santos's fields were trampled. People were also slowly destroying the small oak tree, taking its leaves, branches, and bark as relics. Lucia's mother, Maria Rosa, was particularly distraught over the upcoming event. She feared that the crowd would become an angry mob if a miracle did not occur. Lucia, however, was very calm. She told her mother not to worry.

By 12:00 noon on October 13, an estimated 70,000 people had gathered at Fátima in a torrential rainstorm. In addition to thousands of believers, the crowd included those who had come to discredit the children's stories and members of the international media who had come to record the event. Once again, the woman appeared to the three children. This time she told them that she was the Virgin Mary. The children also reported seeing Mary, Joseph, and an infant Jesus together as the Holy Family. Lucia said that Mary also appeared to her as the Lady of Sorrows (Mary as she appeared at Jesus' crucifixion) and the Lady of Mount Carmel (Mary as the queen of Heaven). During this visitation Mary also asked that a church be built in Fátima.

As Lucia conversed with the apparition, the throngs of people were astonished to see the sun burst through the rain clouds. Many described it as an orb of silver, dancing in the sky before appearing to plummet toward the earth. Everyone at Fátima that afternoon reported the same experience, even Lucia's parents, who at last expressed belief in her stories.

Worldwide attention was now focused on Fátima. The area was deluged with pilgrims, curious travelers, and relic-seekers. The visitors tore pieces of the tiny oak sapling and scraped wood from the children's homes. Many of the residents resented this intrusion into their lives and blamed the children.

The church press began to publish articles about the controversy. At the same time, the secular press accused the church of fabricating the story in order to regain power. A group of government officials from the town of Santarém arrived with orders to chop down the tree. They then put the tree and other items from the site of the apparition, including crosses, flowers, and coins, on display as objects of ridicule. They had, however, confiscated the wrong tree.

The children continued their routines of prayer and sacrifice. Often, they would refuse to eat or drink. At one point, they tied ropes around their bare skin and wore them as a means of penance. Jacinta was particularly obsessed with the vision of hell that Mary had shown to them and considered her fasting as penance on behalf of those who ate or drank too much, to save them from eternal damnation.

The children also had told their parents that Mary wanted them to go to school and learn to read and write. After the events of October 13, Lucia's mother finally acquiesced and Lucia and Jacinta were enrolled in a local day school. Francisco did not want to attend classes and spent his time praying and waiting for Mary to take him to heaven.

He did not have to wait long. In October of 1918, a powerful influenza epidemic swept through Aljustrel. Francisco and Jacinta were stricken almost immediately. Both believed that this was a fulfillment of Mary's promise to take them to heaven. The sickness waxed and waned over the next several months. On April 5, 1919, Francisco died. He was buried at Fátima.

In July, Jacinta was taken to an Ourem hospital for the treatment of pleurisy. She remained gravely ill, and although she had an open surgical wound that refused to heal, her father brought her home at the end of August. Upon her return, Jacinta told Lucia that Mary had visited her three times. During these visits, Jacinta said, Mary told her that she would be sent to a hospital in Lisbon, would suffer greatly from her illness, and would die alone.

Despite continued pressure from the Portuguese government that church officials not recognize the events in Fátima, Formigao continued his friendship with Jacinta. In January 1920, Formigao made the acquaintance of a Lisbon physician, Dr. Enrico Lisboa. Lisboa had heard of Jacinta's case and offered to provide for her care. By February 10, Jacinta was in Lisbon's Hospital of Dona Stefania for surgery to remove two infected ribs. Her frail condition prevented the surgeons from administering a general anesthetic and the child underwent the procedure with only a local painkiller. After the surgery, she continued to weaken and on February 16 told one of the nurses that Mary was coming for her soon. On the evening of February 20, she asked for a priest in order to make a final confession and to receive communion. The priest heard her confession but said he would return in the morning to administer communion. Jacinta, however, died during the night. Her nurse had left the room momentarily, so Jacinta was indeed alone when she died. The Baron de Alvaiazere, a friend of Dr. Lisboa's who had helped to pay Jacinta's medical expenses, arranged to have the tiny body laid to rest in his family burial vault in Ourem.

On August 5, 1920, Dom Jose Alves Correira da Silva was appointed by Pope Benedict XV to head the new diocese of Leira, which included Cova da Iria and Aljustrel. Ten years earlier, Silva had been imprisoned and tortured by the Portuguese government because of his religious beliefs. He was forced to stand in icy water for long periods of time and was left permanently disabled. Silva fervently believed his prayers to the Virgin Mary had saved his life and one of his first acts as the bishop of Leira was to consecrate his diocese to the Immaculate Heart of Mary. He then turned his attention to the events at Fátima. To protect Lucia from her many detractors and from those who might try to exploit her, he arranged for her to be sent to a school of the Sisters of St. Dorothy near Porto. On June 19, 1921, Lucia left Aljustrel with instructions not to tell anyone where she was going nor to speak of her experiences once she arrived at the convent.

Shortly after the apparition of October 13, 1917, a small, crude chapel had been built on the site. Called the Chapel of the Apparitions, it was constructed of stucco and wood, with its sides left open. On May 13, 1920, the third anniversary of the initial vision, a statue of the Virgin Mary was placed in the chapel. The Portuguese government sent soldiers in an attempt to prevent the installation of the statue, but the pilgrims resisted, and succeeded in putting the statue in place. In 1921, Silva allowed Low Mass to be said for the first time in the chapel and purchased the land around the site from Lucia's father in preparation for the building of a church. The following year, a bomb damaged part of the chapel.

Silva also allowed a cistern to be dug on the site in order to collect rainwater; pilgrims believed the water possessed miraculous properties. The national government continued its opposition to the shrine and issued statements referring to the beliefs concerning Fátima as "this despicable reactionary superstition." In 1927, the government ordered the sealing of the well on the basis that it was contaminated, but desisted in the face of evidence that the water was indeed sanitary. The pilgrimages continued, with the site attracting as many as 300,000 people one day in 1928.

A formal canonical investigation of the events at Fátima was initiated in 1922. On October 13, 1930, the Catholic Church officially established the Shrine of Our Lady of Fátima. Papal indulgences were granted to Fátima pilgrims. Today, the shrine continues to attract numerous pilgrims, many of whom approach the site on their hands and knees.

Construction of Fátima's basilica began in 1928; it was consecrated in 1953. It is of a stark neoclassical design, with giant colonnades and a 215-foot white stone tower. The basilica sits on the far side of a large macadam-paved square that is lined with souvenir stands on the thirteenth day of the month from May to October. To the left and right are two hospitals to house the sick and infirm who come to Fátima

seeking comfort and a cure. Inside the basilica are fifteen altars representing the fifteen mysteries of the rosary, which are based on events in the lives of Jesus and Mary. Bronze carvings over the altars depict scenes from Mary's life. A statue of the Immaculate Heart of Mary stands in an eighteen-foot niche over the basilica's main doorway. It was sculpted by Thomas M. McGlynn, a Dominican priest from the United States.

The two front altars hold the tombs of Jacinta and Francisco. In September 1935, the bishop of Leira ordered the transfer of Jacinta's body to the chapel at Fátima. By the 1950s, the two tombs had been seriously damaged by pilgrims who took fragments as relics. The bishop decided to move the two bodies to the basilica.

In 1925 Lucia was accepted as a postulant in the Institute of the Sisters of St. Dorothy. She became a novice the following year and took vows as a lay sister in 1928. In October 1934, she made her perpetual vows and in the mid-1990s was still living as a Carmelite nun in the cloister in Coimbra in west central Portugal where she is known as Sister Dores.

Several popes have made pilgrimages to Fátima. The most recent was John Paul II, who visited the shrine in May 1991 to give thanks to the Virgin Mary for saving his life after an assassination attempt in 1981. The bullet that pierced his abdomen in that attack was sent to Fátima in 1984 and was inserted into the crown that sits atop the statue of the Immaculate Heart of Mary in the basilica. Pilgrimages continue to fill the country road that lead to Fátima with crowds approaching 1 million on May 13 and October 13.

Further Reading: Most of the literature on Fátima is written by members of the Roman Catholic Church and therefore is spiritual in tone. However, the books still provide fascinating reading and interesting historical background on the children at the center of the events. *Our Lady of Fátima* by William Thomas Walsh (New York: Doubleday, 1954) is immensely readable, as is *Fátima Today* by Reverend Robert J. Fox (Front Royal, Virginia: Christendom Publications, 1983). *Fátima: Pilgrimage to Peace* by April Oursler Armstrong and Martin F. Armstrong Jr. (Garden City, New York: Hanover, 1954) is an account of the Armstrongs' visit to Fátima and presents a sensitive portrayal of the people involved in the events at Fátima.

—Mary F. McNulty

Ferrara (Ferrara, Italy)

Location: Ferrara lies in the heart of the fertile Po Valley; the provinces of Ferrara and Padua meet, marking the northern edge of the Emilia-Romagna region and the beginning of the Veneto region. The city of Ferrara is forty-four miles northwest of Ravenna and forty-five miles northeast of Modena. Bologna is just twenty-nine miles south of Ferrara and Venice is fifty-five miles northeast of the city.

Description: Inhabited from ancient times, along with an important Etruscan settlement in the province, at Spina, Ferrara rose to prominence under the Este family, who held a sumptuous court there from about 1264 to 1598, first as marquises, then as dukes. The fifteenth and sixteenth centuries were the time of the city's great flowering. Vicious despots yet avid patrons of the arts, the Estes attracted great Renaissance artists and poets to their court and contributed much to the revival of the theatre, particularly the comic tradition. The city's fifteenth-century Herculean Addition led to its reputation as the first modern city in Europe and its streets are lined with Renaissance palaces. In the twentieth century, Ferrara was the birthplace of the metaphysical school of painting started by Giorgio De Chirico and Carlo Carrà.

Site Office: A.P.T. (Azienda di Promozione Turistica della Provincia di Ferrara)
Piazzaretta Municipale
CAP 44100 Ferrara, Ferrara
Italy
(532) 35017

Because Ferrara lies near the marshy Po Delta, many early traces of its past have been lost in the mud. It was inhabited from the earliest days, the land being rich and suitable for grain and fruit, with fish in the delta and the river, and game in the surrounding woods. Ferrara itself grew up on a branch of the Po River, which was even then navigable, and until the seventeenth century its economy was based on river tolls, agriculture, and the salt pans of Commachio, thirty-five miles east of Ferrara on the Adriatic. Spina, near Commachio, was a thriving Etruscan port until the fourth century B.C., and the National Archaeological Museum in Ferrara houses pottery from more than 3,000 tombs and an important collection of Attic vases from the fifth and fourth centuries B.C.

Known as Ferraria in Roman times, the town, along with Bologna, was part of the Byzantine Empire at the end of the sixth century A.D. The Byzantine exarchs (viceroys) did some drainage work before the city passed to the Lombard kings; confiscated by Charlemagne, it was then given to the papacy. The popes conferred it as a fief on Tedaldo, Count of Modena and Canossa, a nephew of Holy Roman Emperor Otto I. In 1101 Tedaldo's famous granddaughter, Matilda of Tuscany, besieged the town, after which it was returned to imperial authority. By that time, Ferrara had become a commune, with the Adelardi family as the dominant force. In 1185, Adelardo Adelardi died leaving no son, and Azzo VI d'Este, who was engaged to Adelardi's daughter, Marchesella, took the opportunity to stake the Estes' first claim to Ferrara, a claim they continued to pursue even though Marchesella died in 1186 without marrying Azzo. This claim led to a prolonged feud between the Estes and another prominent family, the Torellis.

Ferrara had developed into a town worth fighting for. It lay on the north-south land route from Venice to Bologna and Tuscany, and the east-west water route connecting northwest Italy with the Adriatic. River tolls and customs duties, plus abundant supplies of grain and salt, brought prosperity. In 1135 Ferrara's new Romanesque-Gothic cathedral was consecrated, replacing the Chiesa di San Giorgio, the church dedicated to Ferrara's patron saint, St. George, which had served as the cathedral since the seventh century. The magnificent facades of the new cathedral were carved by Nicolò, the Lombard master-sculptor trained by Wiligelmo at Modena. Near Ferrara is the Romanesque Pomposa Abbey, founded by the Benedictines in the seventh to eighth centuries. It is said that a monk named Guido d'Arezzo invented the modern musical scale here in the early eleventh century. The church has a Byzantine atrium and fourteenth-century frescoes.

The Este family were the oldest and most distinguished of nobles in northeast Italy, descended from the Obertenghi, great ninth to tenth century Lombard feudal lords who controlled land in Tuscany, Liguria, and Lombardy. They were based in Este, in the Euganean Hills southwest of Padua, but received many grants of land and titles from the Holy Roman emperors, whom they had consistently supported. They also, however, fully recognized papal authority in Ferrara, and this balance between Guelph (supporters of the papacy) and Ghibelline (supporters of the emperor) sympathies, along with their wealth and consequent power, led, in 1196, to Azzo VI's election as podestà of Ferrara at the age of 26. Further problems erupted with the Torellis, who regained power in the town in the next century, but after a three-month siege in 1240, Azzo VII took control for the Estes once more. Obizzo II d'Este became the second marquis of Ferrara in 1264. In an unprecedented move, the town elected him *signore* for life, and the office was soon made hereditary.

In 1243 the Palazzo Ducale (Ducal Palace) was built, while Azzo VII was in control of the city, and it served as the

The Este Castle at Ferrara
Photo courtesy of A.P.T. di Bologna e Provincia

Este residence until the sixteenth century. The first twenty years of Obizzo II's rule were calm, and by 1271 he had virtually gained royal status. In 1278 the papacy secured formal cession of the city from Holy Roman Emperor Rudolf I, making the towns of the Romagna part of the papal state but the papacy was weak and interfered little with the Estes. Beatrice d'Este founded the Monastero Sant'Antonio in Polesine in 1249. Still occupied by nuns today, the group of buildings includes the convent and a church with three chapels. The buildings are also noted for frescoes by the fourteenth-century school of Giotto, and others by fifteenth-century Ferrarese painters.

The Este family continued to consolidate their power throughout the fourteenth century, with Obizzo II succeeded by Azzo VIII; Aldobrandino II; Niccolò I; Obizzo III; Aldobrando III; Niccolò II, a friend of Petrarch; Alberto, who founded the University; and Niccolò III, who came to power in 1393.

Like the similar city-states of Mantua (ruled by the Gonzaga), Urbino (by the Montefeltro), and Milan (by the Visconti, then the Sforzas), the arrangement of rule by a dominant family seemed to work well for all concerned. Ferrara's population was composed of rural aristocrats with town houses in the city, tradespeople and artisans, and lawyers and doctors, none of whom was allowed to become too powerful; at the first sign of influence, the Estes cleverly employed the person within the court. For ordinary people, the courts, its lands, and villas provided employment of all kinds, and some citizens were needed as soldiers to protect the dynasty. The commune's municipal government provided for the day-to-day needs of the city, while the Estes dealt with "foreign" policy and set taxes. Occasionally the Estes abused their power, but rebellion from the people usually forced them back into line.

In April 1333, papal forces making a bid for more control in the province were defeated at Ferrara, and in 1344 the Estensi Signoria, or lordship over Ferrara, was legally proclaimed by papal investiture. In 1382 the plague hit the city, causing much suffering, and in 1385 Niccolò II raised taxes once too often, leading to a popular rising against him in the course of which the minister of finance was killed. The Estes now felt the need to build a fortress to protect them from their subjects, and that year they began construction on Castello Estense (Este Castle), with a passage, still to be seen, connecting it to the Ducal Palace. The castle was designed by Bartolino da Novara, a military architect also responsible for castles in Pavia and Mantua, and was built with four fortified gateways with drawbridges over the surrounding moat. It was completed within two years.

Niccolò III was the twelfth marquis of Ferrara, famous for having fathered countless numbers of children "on both sides of the Po," and the city's first real Renaissance prince. Niccolò was only ten years old when he became marquis in 1393, and he took control in 1402, at the age of nineteen. He is said to have had 800 mistresses and he certainly had 3 wives. His second was Parisina Malatesta, but in 1425 Niccolò discovered that she was having an affair with his favorite illegitimate son and heir, Ugolino. Niccolò promptly imprisoned them overnight in the dungeon of the Estense Castle and then had them beheaded the following day. It was this story that inspired Robert Browning's poem, "My Last Duchess."

Despite this event, Niccolò was known as a humanist. He reopened and revitalized the university and invited Guarino da Verona to the court to teach his children. Guarino and his Mantuan friend. Vittorino da Feltre, were the leading humanist teachers of the day, and prominent people from all over Europe sent their children to be educated by them. Although it was solidly based on Christian principles, the classics, and sciences, the humanist system advocated a balance between mind and body, no corporal punishment, and teaching through play whenever possible. Guarino remained at Ferrara, teaching at the university from 1436, running his school, writing, and translating. His education affected Leonello, the heir, in particular, who became a great friend of his teacher, and who inspired another close friend, Leon Battista Alberti, to write his famous book on architecture, *De re aedificatoria*.

It was during the reigns of Niccolò III's three sons, Leonello, Borso (both illegitimate), and Ercole I, that Ferrara rose to the height of its brilliance. Leonello succeeded his father in 1441 but lived only nine more years. In 1450 Borso, his brother, came to power. Leonello had been a scholar and poet, accomplished in Latin and Greek, a collector of books, and patron of the arts. Borso was showy, popular, lavish, and extravagant. Liberal and fun-loving, he had a reputation for being just. He also wanted to be a duke. In just two years, after much political maneuvering and considerable financial outlay, Borso achieved his aim by becoming the first duke of Modena and Reggio in 1452, the same year in which Italian reformer Savonarola was born in Ferrara. It was not until 1471, however, the year of his death, that Borso became the first duke of Ferrara as well.

It was in Borso's time that the Ferrara school of painting was founded by Cosmè Tura, a native of the city. Before that time, artists in Ferrara had been influenced by Squarcione's Paduan school and then by visitors to Leonello's court, Pisanello and Rogier van der Weyden, who had introduced Flemish oil painting to the city. Mantegna also visited when he was only eighteen years old and became a friend of the young Tura who visited him in Padua and saw the work of Donatello there. Tura became court painter in 1451, and from 1452 on received a regular salary. In 1457 he was given rooms in the ducal residence, and in 1462 the duke gave him his own house. Two years later, Tura bought his own house.

The work of court painter was varied. Tura not only produced portraits and frescoes to adorn the Este residences, he was also called upon to design banners, helmets, tapestries and embroideries, bed quilts and bench covers, and a silver table service.

Borso then took up a project started by Alberto d'Este in 1385: the building and decoration of the Schifanoia Palace. Tura supervised the painting of the Room of the Months, executed between 1469 and 1471, as a celebration of Borso's life. Each month was represented by a painting of three parts, the top dealing with classical mythology, the middle with astrological subjects, and the bottom with fifteenth-century life at court. Each scene featured Borso at work, rest, and play.

The paintings were done chiefly by Francesco del Cossa, Tura's second-in-command at court, who was also influenced by Piero della Francesca when that artist visited Borso's palace. Cossa was assisted by his pupil, Ercole de'Roberti, who was influenced by Giovanni Bellini's work and by the softer Venetian school. Both Cossa and Roberti worked in Bologna after this commission, spreading the influence of the Ferrara school. The younger Lorenzo Costa became Roberti's pupil in Bologna and in 1506 took over as court painter at Mantua upon Mantegna's death. Borso was succeeded by his brother, one of Niccolò III's legitimate sons, Ercole, who had secured the dukedom by fighting and defeating Leonello's legitimate son, Niccolò, while Borso was dying. To ensure his position, Ercole then tried unsuccessfully to poison Niccolò. Later Niccolò invaded Ferrara when Ercole was away, but was defeated. Ercole ordered him captured and beheaded, but then gave him an Este funeral, with all due pomp and ceremony, and buried him in the Este family vault.

Ercole was quite unlike either of his half-brothers and is often considered the greatest Este ruler of Ferrara. Certainly it was he who created the magnificent city of Ferrara that exists today. Ercole I spent money as liberally as Borso, but most of it was spent on the city rather than on the people. He extended the castle, and it then became the ducal residence. He enlarged the university; soon it included 45 professors and nearly 500 students and attracted scholars from all over the world. Copernicus received his *laurea* at Ferrara University in 1504, Paracelso taught there, and it was an important medical center.

The duke maintained a choir for his chapel and employed musicians from all over Europe; he built a theatre and produced Latin and Italian comedies, especially those of Plautus and Terence; he bought 1,000 books for the library and employed 70 painters; his court poets were Tito Strozzi and Matteo Maria Boiardo, who wrote *Orlando Innamorato* there in 1483. Almost every church and convent in the city was rebuilt or redecorated to satisfy the duke's religious inclinations, and three villas were maintained to satisfy his needs for pleasure.

Unfortunately Ercole lost favor with Pope Sixtus IV, due to his tardiness in paying tribute and his having sided with Florence and Milan in their war against Naples and the papacy, the Pazzi War, in 1478. This situation persuaded the pope to go to war with Ferrara in 1482. Sixtus sought and received the assistance of Venice, but eventually he realized that Venice stood to gain far more from the war than did the papacy, and he changed sides, allying himself with Ferrara against Venice. Ferrara lost much territory to Venice before peace was negotiated in 1484. Ercole then turned his attention back to the city, employing Biagio Rosetti to triple its size, an enterprise known as the Herculean Addition, and to build six miles of walls around the new section. This transformation led Jakob Burckhardt to call Ferrara the "first modern city in Europe" in his classic, *Civilization of the Italian Renaissance*.

The highlight of the extension was the Palazzo dei Diamanti, so called because of the 12,600 blocks of diamond-shaped marble stones studding the facade. The name is apt: diamonds were an emblem of the Estes. The palace was built for Sigismondo d'Este, Ercole's brother, who, with the help of half-brother, Rinaldo, had driven the young Niccolò out of the city in his attempt to take it from Ercole. The palace is now the National Picture Gallery.

Ercole married Eleonora of Aragon, the daughter of King Ferdinand I of Naples. It was a happy marriage that produced six children. Two of their daughters, Isabella and Beatrice, became wealthy, powerful women and great patrons of the arts. Isabella married Francesco II Gonzaga, marquis of Mantua, and held a splendid court there. Beatrice married Ludovico Sforza of Milan. A palace was built for him in Ferrara, the Palazzo di Ludovico II il Moro, containing frescoes by Raphael's pupil, Benvenuto Tisi, also known as Garofalo. The palace was never used by Ludovico and now houses the Archaeological Museum.

Alfonso, Ercole's third child by Eleonora, became the third duke of Ferrara, Modena, and Reggio in 1505. Alfonso I's first wife was Anna Sforza, who had died in 1497 in childbirth. His second was Lucrezia Borgia. Alfonso was immediately faced with a conspiracy against him by his brother Ferrante and his half-brother, Giulio, but swiftly had them both put in the castle dungeons where they remained for the rest of their lives. Apparently oblivious to their existence, Alfonso had the castle above their heads extended, building kitchens and a hanging garden for the pleasure of the duchesses.

The next challenge was a new campaign by the pope, in this case Julius II, who was determined to strengthen papal authority in the north. Ferrara and the French supported the pope in his attack on Venice, but again the papacy suddenly changed sides and war ensued. Conflict rather suited Alfonso, whose greatest love was for cannon, but when Julius died in 1513 the war fizzled.

In the brief peace that followed, Titian and Bellini both worked for Alfonso at his court and Ludovico Ariosto became court poet; he wrote *Orlando Furioso* there in 1532. Ariosto also acted as the court's theatre director, responsible for putting on performances of his own comic dramas and those of others for special occasions. In 1531 he was granted a permanent indoor theatre, but in December 1532 it was destroyed by fire, and Ariosto died eighteen months later, broken-hearted. The poet's house can still be seen today, and his tomb is in the Palazzo de'Paradiso, which dates from 1391. It was the seat of the university from the seventeenth century until 1963, and is now the City Library with a complete collection devoted to Ariosto, including a manuscript of his famous epic poem.

While Ariosto was court poet, Dosso Dossi (Giovanni Lutero) was court painter, beginning in 1514. Dosso was influenced by Giorgione, Titian, and the Venetian school. His pastoral scenes reflect the life of leisure led by the ducal family in times of peace: hunting, feasting, and the pleasures of villa life. His great strength lay in rendering the effects of light. Dosso was aided by his brother, Battista Lutero, also a court painter, and together they painted some of the sets for Ariosto's plays.

Lucrezia Borgia died in 1519, the same year that Pope Leo X, a Medici, decided to take Ferrara as his predecessor had intended. The war was continued by Clement VII, who was angered by Alfonso's attempts to regain Modena from the papacy. The Estes succeeded in 1527, but Clement continued to devise plots to murder Alfonso, despite an agreement made in 1531, and the feud ended only with the death of both men in the same year, 1534.

Ercole II reigned until 1558. He had married Renée of France, daughter of Louis XII, in 1528. She was converted to Calvinism at the beginning of the Protestant Reformation and had a chapel built in the castle. She welcomed Rabelais to the court and sheltered John Calvin, who stayed there for some time incognito, much to the embarrassment of her husband.

In 1558 Alfonso II succeeded his father, becoming the fifth, and last, duke of Ferrara, Modena, and Reggio. The following year he built the Palazzina Marfisa d'Este, an elegant single-story residence set in a garden with a grotto and an outdoor theatre. Here, the beautiful if eccentric Marfisa d'Este, an aunt of Alfonso II, would entertain guests, one of whom was the poet Torquato Tasso, who wrote his great epic poem, *Gerusalemme Liberata*, in 1581 on his first of many visits to the city.

Alfonso married Barbara of Austria but they produced no heirs, and when Alfonso died in 1597, Pope Clement VIII took the chance all his predecessors had hoped for and seized Ferrara, absorbing it into the Papal States. The remaining Estes, headed by Cesare, fled to Modena and claimed that city as the new capital of their remaining duchy of Modena and Reggio.

Under the papacy the fortunes of Ferrara rapidly declined. The Civic Museum was founded in the eighteenth century, first at the Paradiso Palace and then the Schifanoia in 1897. In the nineteenth century, a fifteenth-century Carthusian monastery was converted to the city cemetery; it contains the tombs of Borso and Marfisa d'Este. The National Picture Gallery was founded in 1836 in the Palace of Diamonds; it contains paintings and frescoes from the thirteenth century onward including paintings by Tura, Roberti, Garofalo, Dosso, Carpaccio, Guercino, Bellini, and Mantegna.

From 1916 to 1918, two Futurist artists met in Ferrara while in the military service. Giorgio De Chirico and Carlo Carrà were enraptured by the Renaissance art in the Palazzo Schifanoia and the city itself, and the result was the birth of the metaphysical school of painting, Italy's greatest contribution to twentieth-century art. In the Palazzo Massari there is now a Documentary Museum of the metaphysical painters, along with other nineteenth- and twentieth-century art. The university still thrives, with a new advanced chemistry study center specializing in macro polymers. The city is also known for its traditional gastronomic specialties, including Lucrezia Borgia's favorite, *Salama da Sugo,* a sausage cured for one year and served hot in sauce, and the Vino di Bisco continues the wine-making tradition encouraged by Renée of France. Although much of the Este art and book collection went to Modena or Rome, Ferrara and its buildings have been beautifully preserved. Each year at the end of August the traditions of music and theatre are celebrated in the Ferrara Buskers' Festival.

Further Reading: *Ferrara: The Style of a Renaissance Despotism* by Werner L. Gundersheimer (Princeton, New Jersey: Princeton University Press, 1973) begins with the origins of Ferrara and moves through the Renaissance period, concentrating in particular on Niccolò III, Leonello, and Borso. *Dukes and Poets in Ferrara: A Study in the Poetry, Religion, and Politics of the Fifteenth Century and Early Sixteenth Century* by Edmund Garratt Gardner (New York: Dutton, 1903; London: Constable, 1904) includes family trees of the Houses of Este, Gonzaga, Sforza and others, as well as a bibliography. *The Painters of the School of Ferrara* by Edmund Garratt Gardner (New York: Scribner's, and London: Duckworth, 1911) covers the Ferrarese painters from Tura to the Dossis, including the Bolognese painters who followed them. It also boasts a good bibliography. *Cosimo Tura: Paintings and Drawings* by Eberhard Ruhmer (London and New York: Phaidon, 1958) has 118 reproductions of the complete works with a detailed introduction, a chronological list of work, and notes on the plates. *The Painters of Ferrara: Cosmé Tura, Francesco del Cossa, Ercole de'Roberti, and Others* by Benedict Nicholson (London: Elek, 1950) gives the complete story of the Ferrara school of painting. *Dosso and Battista Dossi* by Felton Gibbons (Princeton, New Jersey: Princeton University Press, 1968) provides an account of life in the Este court in general, including Ariosto, literature, the theatre, and the villas. The book also covers the work and lives of the Dossis in detail. *Bellini and Titian at Ferrara: A Study of Styles and Taste* by John Walker (London: Phaidon, 1956; New York: Doubleday, 1957) is a detailed analysis of four paintings done at Ferrara, three by Titian and one by Giovanni Bellini. *The Civilization of the Renaissance in Italy,* volumes one and two, by Jakob Burckhardt (New York: Harper, 1958; London: Muller, 1961) is the classic work on the period. *The Schifanoia Months at Ferrara* by Paolo D'Ancona (Tiranti, Italy: Edizioni del Milione, 1985) concentrates on the Room of the Months painted to glorify Borso d'Este by the school of Ferrara painters.

—Beth F. Wood

Florence (Firenze, Italy): Fiesole

Location: In the region of Tuscany and province of Firenze, four miles northeast of Florence.

Description: Former settlement of the ancient Etruscan civilization, with archaeological remains both of the Etruscan and of the later Roman inhabitants.

Site Office: A.P.T. (Azienda di Promozione Turistica)
Piazza Mino, 37
CAP 50014 Fiesole, Firenze
Italy
(55) 598720

Fiesole lies almost 1,000 feet above sea level, on the slopes and summits of the forested hills of San Francesco and Sant'Appolinare. This location provides spectacular views over the valley of the Arno River and the city of Florence, which stands on its banks. Fiesole is often described as a suburb of that city. In fact, although it has long been economically dependent upon Florence, and was even conquered by Florence in the Middle Ages, Fiesole remains a self-governing commune with a distinct history that precedes the founding of Florence.

Although the hills on which it stands are known to have been inhabited even earlier, the foundation of Fiesole is usually attributed to the Etruscans, who are believed to have arrived here sometime between the eighth and sixth centuries B.C., although some scholars still doubt that they had a permanent settlement at the site. This is just one of the controversies surrounding this people and their civilization. They may have come to northern Italy by sea from the Balkans or Asia Minor, or perhaps overland from northern Europe, settling in the Po Valley and on the coast of what is now Tuscany. They almost certainly founded Fiesole and other hilltop towns as points in their defenses against the Ligurians, who inhabited the region around what is now Genoa, and prospered through trade with the Greek settlements in the south of Italy, whose people referred to the Etruscans as the Tyrrhenoi.

Etruscan kings took control of Rome and central Italy for around 100 years until 507 B.C., when the Romans expelled them and founded the Roman Republic. After the Etruscans' retreat northward Fiesole became even more important to them. By the fourth century B.C. it was protected with fortified defensive stone walls (along the line of the modern highway called the Via delle Mura Etrusche) and provided with numerous temples. At this time it is believed to have had a population of around 10,000, some of whom probably moved to what is now Florence, originally Florentia, when it was founded as a subordinate settlement early in the second century B.C. As the power of the Roman Republic grew, however, that of the Etruscans declined, and the Romans took control of Florence and Fiesole in 90 B.C. Etruscan remains in modern Fiesole include a temple from around the third century B.C., perhaps dedicated to the goddess of wisdom, Menrva (Minerva to the Romans, Athena to the Greeks), and extensive sections of walls around the tops of the hills.

In 82 B.C. Florence was destroyed by the dictator Sulla as a punishment for having supported his enemies in the Civil War, and the Romans colonized Fiesole beginning in 80 B.C. Florence, reestablished by Julius Caesar in 59 B.C., remained dependent on Fiesole, which became the capital of the Roman province of Etruria: its name in Latin was Faesulae. Although it lagged behind while Florence, endowed with easier access to trade by road and river, began to flourish, its status was reflected in the building of an amphitheatre (now known as the Teatro Romano) in the first century A.D. and its expansion to include 3,000 seats for spectators, under Emperors Claudius I and Septimius Severus. Today it is used by the Estate Fiesolana, the arts festival held every summer. Nearby are the Roman baths, established around the same time and later enlarged by Emperor Hadrian, as well as the restored remains of Roman city walls and of a Roman temple, also from the first century A.D.

Florence and Fiesole were attacked by the Ostrogoths in A.D. 406 and by other invaders of Italy as the Roman Empire disintegrated. By 539 Fiesole was a fortress of the Visigoths, who were defeated in that year by an army sent into Italy from Constantinople (now İstanbul) to assert the claims of the Eastern Roman Empire over the region. About this time, the Church of Sant'Alessandro, which stands above the ruins of Etruscan and Roman temples, was constructed. In 570 the Lombards took control of Florence, Fiesole, and the rest of Tuscany, ruling the region at first from the city of Lucca but later, under a series of dukes, from Florence. Charlemagne, the king of the Franks who was to lay the foundations for the Holy Roman Empire, conquered the area in 774, reviving Lucca as the regional capital. From this time onward Florence, Fiesole, and the rest of Tuscany were nominally subject to the Holy Roman Empire and the various marquises of Tuscany chosen by its emperors.

The long rivalry between Florence and Fiesole formally ended in 854, when both cities were placed under a single local ruler, the count of Florence. As Florence continued to develop its role in the trade networks of northern Italy and to raise its political status, succeeding Lucca as capital of Tuscany by 1001, Fiesole was relatively neglected, although it continued to have its own bishops. Their first Duomo, or cathedral, was the Badia Fiesolana, at the foot of the hill on which Fiesole stands. This was abandoned in 1026 and became part of a monastery of the Benedictine order; the next cathedral, the present one, was founded in 1028, about one

Roman amphitheatre at Fiesole
Photo courtesy of A.P.T., Fiesole

and one-half miles away from the Badia. It is dedicated to San Romolo (St. Romulus), who is said to have been the first priest of Fiesole, appointed by St. Peter himself, and to have been martyred on the site of the Badia Fiesolana. After repeated rebuilding, only the nave columns remain from the Romanesque structure of 1028.

During the eleventh and twelfth centuries Fiesole was viewed as an important political and economic rival by the Florentines and was drawn into a series of local wars over trade and political relations. In 1115 Florence became a self-governing commune in opposition to Conrad, marquis of Tuscany, the nominal overlord of both cities. Apparently in the belief that Conrad would use Fiesole as a base from which to recapture the largest and richest city in his region, the Florentines launched an all-out attack on Fiesole in 1123. By 1125 they had destroyed almost everything except the cathedral and the bishop's palace. Florence completed its conquest by formally abolishing Fiesole's independence in that year.

The steepness of the hills in and around the commune of Fiesole discouraged rebuilding for some time after this destruction; indeed, it still helps to protect the town from being completely absorbed into Florence. As a fortunate side effect, the Roman and Etruscan ruins were mostly left undisturbed, under piles of rubble, until scholarly excavations began in the late eighteenth century. Even so, as Florence continued growing and settlement gradually extended up the hills, Fiesole was revived. The cathedral was expanded in 1213, in 1256, and again during the fourteenth century; its treasures include a bust of Bishop Leonardo Salutati and a sculpted altar, both by the artist Mino da Fiesole, and a statue of St. Romulus by Giovanni Della Robbia, installed in 1521. On the same piazza as the cathedral—the Piazza Mino da Fiesole—stands the Palazzo Pretorio, built during the fourteenth century and restored in 1463; it was the headquarters of the town council and of the medieval governors sent to Fiesole by Florence. The founding of new religious establishments is a reliable indication that the population of the area was not only returning to the levels of the years before the Florentine conquest, but was probably increasing to a level capable of supplying resources and labor for these institutions and of sufficiently supporting the burden of rents and taxes to keep them in existence. The monastery and church of San Francesco were built from around 1330 onward on a commanding site at the top of the hill, incorporating in the cloisters parts of the walls of what was probably the Etruscans' fortress.

The revival of Fiesole was probably temporarily checked by the Black Death, which swept through Europe in the mid-fourteenth century and reduced the populations of most areas, but especially of the unsanitary and overcrowded cities. The Villa Palmieri and La Torraccia, two houses near the village of San Domenico di Fiesole, are associated with scenes in Giovanni Boccaccio's *Decameron,* the famous collection of stories set within a framework tale of Florentines seeking refuge from the plague in the hills. After Florence had recovered from the epidemic, it became one of the centers of the Italian Renaissance, a movement in which Fiesole played its part. The Badia Fiesolana, which had been transferred from the Benedictines to the Augustinians in 1439, was extensively rebuilt behind its original Romanesque facade, from 1456 onward, perhaps to designs by Filippo Brunelleschi, under the patronage of Cosimo de' Medici the Elder, ruler of Florence until his death in 1464. He also paid for a number of paintings by Francesco Ferrucci and Francesco di Simone Ferrucci, which can still be seen inside the church, and for its important library, and ordered the construction of a mansion, the Villa Medici (or Belcanto, or the Palagio di Fiesole) just outside Fiesole between 1458 and 1461, to a design by Michelozzo. This house with its garden, which is considered a Renaissance masterpiece in itself, later became one of the meeting places for the so-called Platonic Academy, the group of humanist scholars sponsored by Cosimo's grandson Lorenzo de' Medici, "Il Magnifico" (the Magnificent), who ruled Florence between 1469 and 1492.

Their interest in reviving classical Greek and Roman principles of aesthetics and philosophy played an important part in the Florentine Renaissance.

But the most important of the Renaissance projects in Fiesole, combining religious revival, economic growth, and the ideal of a renewal of classical values, was probably the Monastery of San Domenico di Fiesole, established near the Badia Fiesole around a church built between 1406 and 1435. This became the focal point of the village of the same name, which is still part of the commune of Fiesole today. The monastery is famous as the place where both St. Antoninus and the painter Fra Angelico were trained as monks. The former, whose real name was Antonino Pierozzi, is commemorated by Francesco Conti's painting of one of the miracles attributed to him, inside the church; the latter, whose real name was Guido di Pietro, painted a Madonna with angels and saints, also inside the church, as well as frescoes of the Madonna and Child and of the Crucifixion in the monastery's chapter house. The church contains many other Renaissance paintings and treasures, but has lost several to the Prado in Madrid and the Louvre in Paris over the centuries.

By the time the Church of San Domenico was embellished with a campanile (bell tower), completed in 1613, and a portico, in 1635, both designed by Matteo Nigetti, the Republic of Florence was long gone and the Medicis had become hereditary rulers of Tuscany in alliance with the Habsburgs of Austria and Spain. Throughout the seventeenth and eighteenth centuries, while Tuscany passed from the Medicis to the Lorraine dynasty (in 1737), Fiesole continued to be a place of retreat for the leading families of the city in the valley below, especially in times of plague—as, for example, in 1630 and again in 1633—but, more generally, in response to the widespread but probably mistaken belief that Fiesole was naturally cooler and healthier than Florence. The villas they built among the hills can still be seen today. During the nineteenth century, when visits to Tuscany became fashionable among sections of the British middle class (a fashion affectionately satirized in E. M. Forster's novel *A Room with a View*), those who came to stay in Fiesole included the poet and essayist Walter Savage Landor, who lived in La Torraccia between 1829 and 1835, the Anglicized American novelist Henry James, and the poets Robert Browning and his wife Elizabeth Barrett Browning, who both wrote about Fiesole's landscapes. In 1890 the first electric streetcar was built connecting Fiesole to the center of Florence; this eventually was replaced by buses.

What also attracted visitors to Fiesole from Florence and elsewhere were the Etruscan and Roman remains, which all lie close to one another and which were excavated and restored in stages between 1792, when the theatre was uncovered, and as recently as 1988. Artifacts taken from the sites during the excavations can be found in the nearby Museum Faesulanum, built between 1912 and 1914 in a style intended to emulate that of the Roman temple, and in the Antiquarium Costantini, which opened in 1990 above yet another excavated Roman building, probably a temple, and which specializes in the pottery and other remains of the Hellenistic period in Fiesole and elsewhere.

Fiesole today has a population of approximately 15,000. Its ancient and medieval remains are protected within a special historic zone established by the Italian government, and excavations of the Etruscan sites are still continuing, holding out the promise that eventually some of the mysteries of their lost civilization might be resolved. Since 1976 the town has also been the home of the European University, sponsored by the European Union; the monastic buildings around the Badia Fiesolana church now house most of its activities. Since 1989 the university has also made use of the medieval Villa Schifanoia, on the road below San Domenico di Fiesole. With its own university to add to its older symbols of distinction—the Etruscan walls and temple, the Roman theatre and baths, the medieval cathedral and monasteries, and its special cultural status during the Renaissance and again in the nineteenth century—Fiesole can claim to be significantly more than just another suburb of the city that was once its satellite and later its conqueror.

Further Reading: Whatever its own citizens may think, Fiesole in practice appears under the heading of Florence in most guidebooks and histories, whether of Florence, Tuscany, or Italy, including, for example, Eve Borsook's *Companion Guide to Florence* (London: Collins, and New York: Harper and Row, 1966; revised, New York and London: Harper Collins, 1988), Gene Adam Brucker's *Florence 1138–1737* (London: Sidgwick and Jackson, and New York: Abbeville, 1984) and Christopher Hibbert's *Florence: The Biography of a City* (London: Viking, and New York: Norton, 1993). The chapter on Fiesole in *Etruscan Cities* by Francesca Boitani and others (London: Cassell, 1973; edited by Filippo Coarelli, London: Cassell, and New York: Putnam, 1975) provides an interesting account of what was known of both the Etruscan and the Roman remains at the time that it was published.

—Patrick Heenan

Florence (Firenze, Italy): Piazza del Duomo

Location: The Piazza del Duomo, which surrounds the Cathedral of Santa Maria del Fiore and the Baptistery of San Giovanni, is situated at the northern end of the old city of Florence on the northern banks of the River Arno.

Description: The piazza forms the heart of the traditional religious center of Florence and was created around the baptistery, arguably the oldest surviving building in Florence. In the late thirteenth century, a church dating to Roman times was torn down to make room for the present cathedral, whose dome and campanile dominate the skyline of Florence. The entire complex, but particularly the baptistery dome, was of tremendous importance to the development of Renaissance architecture.

Site Office: Cathedral and Baptistery
Piazza di San Giovanni
CAP 50123 Florence, Firenze
Italy
(55) 294514 or 2302885

Little remains of the ancient history of Florence but the layout of the inner city, which essentially follows the plan of the Roman settlement of Florentia, and possibly the Baptistery of San Giovanni. The Roman origins of this octagonal building are not well documented. Renaissance Florentines believed that it was built as a pagan temple, although modern archaeological evidence suggests that the baptistery, with foundations dating to late Roman times, began as a Christian church. The octagonal plan, which focused on the central baptismal font, and the round-headed arches of the facades and interior were retained through all subsequent alterations, but otherwise the baptistery bears little resemblance now to its earliest incarnation. The baptistery served as the cathedral of Florence until the middle of the eleventh century, when it was replaced in this function by the Cathedral of Santa Reparata, no more than twelve yards away from the baptistery. The new cathedral was built on the site of an ancient Roman temple, possibly dedicated to Mars. Fragments of the Roman temple foundations have been discovered during excavations in the twentieth century.

The twelfth and thirteenth centuries were a time of increasing prosperity for Florence, which became one of the foremost commercial centers of Italy. Several Florentine families also owned some of the most important banks of the country. To reflect the increasing significance of the city, the Piazza del Duomo—then called the Piazza di San Giovanni, a name still borne by a neighboring street—was redesigned in the late thirteenth century to give it an appearance more in keeping with the city's new wealth. The baptistery was clad in green and white marble by Arnolfo di Cambio. The Cathedral of Santa Reparata was torn down at the same time to make room for an altogether more grandiose structure, which was to be named the cathedral of Santa Maria del Fiore. In fact, among the most important specifications for the new cathedral was the stipulation that the new church be larger and more splendid than the cathedrals of Pisa and Siena, Florence's rivals.

Arnolfo di Cambio won the competition for the new cathedral's design, and building was begun under his supervision in 1296. Arnolfo's model, which unfortunately has been lost, called for a Latin cross groundplan, with a simple nave, a huge crossing (to accommodate an unusually large dome), and three apses with ambulatory chapels. As was often the case with buildings of this size and grandeur, construction proceeded slowly and with many, sometimes long, interruptions. Other demands on the city coffers often took precedence, and work ground to a complete halt in the wake of disasters, such as the flood of 1333 and the plague epidemic of 1348. Arnolfo di Cambio certainly did not live to see his work completed, and he was first succeeded in the function of master of the works by the painter Giotto, who made little headway with the main building. However, Giotto's involvement with the project most memorably resulted in the building of the campanile.

The chief purpose of this slender bell tower was decorative: it balances the composition of the chief buildings of the Piazza del Duomo, providing a counterweight to the cathedral. Work on the campanile was begun in 1334 under the direction of Giotto, who died three years later. It was rumored that he died of a heart attack induced by a fear that he had made the base of the campanile too small to carry its weight. Giotto's assistants carried on with the construction and decoration mostly according to Giotto's plan. Andrea Pisano also seems to have had some involvement with the completion of the campanile. Francesco Talenti finished construction of the tower in 1359. Like the baptistery, the campanile is decorated with green and white marble revetments, but its style is otherwise altogether different. Whereas the baptistery is classical in its proportions and harmonies, the campanile, with its vertical thrust, is undeniably in the Gothic style.

Francesco Talenti also made changes in the design of the cathedral during his tenure as *capomaestro*. Adding more ambulatory chapels to the apse and transepts, he diminished the emphasis on Arnolfo's Latin cross plan. As a result, the cathedral looks more like an octagonal church with a very long nave attached to it. The nave, which has two side-aisles, is 295 feet wide and 502 feet long. The emphatic stringcourse immediately under the clerestory strongly stresses the hori-

Baptistery of San Giovanni
Photo courtesy of A.P.T., Florence

zontal elements, so that the eye is irresistably drawn to the sanctuary in the apse. The clerestory windows provide ample illumination, which gives the nave an unusually airy look.

Filippo Brunelleschi, trained as a goldsmith, was first active in the cathedral works in the late fourteenth century, when he created a dome for the baptistery. This shallow, double-shelled octagonal dome carries an octagonal pyramid crowned with a lantern. Although the dome of the baptistery is now overshadowed by the cathedral dome, it was at the time an architectural innovation that stimulated the flowering of Renaissance architecture, most immediately in Florence but also in the rest of Italy.

Brunelleschi then participated in the competition for the design of the bronze doors for the northern entrance to the baptistery in 1401, but he lost to Lorenzo Ghiberti, another goldsmith. The oldest set of doors in the baptistery were cast in the 1330s by Andrea Pisano and are devoted to scenes from the life of John the Baptist. These doors, which have twenty-eight panels, each in a quatrefoil frame, set the parameters for the north and east doors created later. The competition for the north doors of the baptistery was organized by the city government and the most important guilds in the wake of a serious plague epidemic, as a token of gratitude for the city's survival. Seven artists were selected to create a trial panel on the subject of the sacrifice of Abraham, and Lorenzo Ghiberti's was unanimously adjudged the most felicitous. The north doors, with twenty-eight panels in quatrefoil frames, depict twenty scenes from the New Testament, while the remaining eight are dedicated to the four Evangelists and four of the Church Fathers.

Ghiberti worked on the doors for about twenty years, and they now form one of the principal attractions of the baptistery. They are overshadowed only by the east doors facing the cathedral, which are also by Ghiberti's hand. The artist received the new commission about a year after finishing the north doors, and this time the consuls of the Guild of Merchants gave him complete freedom in both design and cost, to "make the door as rich, ornate, perfect, and beautiful as he possibly could," as Giorgio Vasari reports. Ghiberti chose to replace the twenty-eight small panels with ten larger ones, five to each door, each panel depicting a scene from the Old Testament. He worked on them from 1425 to 1452, creating a stunning masterpiece. Each panel, created in a single casting despite its size, is a feat of perspective, of deep and shallow relief, and of delicate detail. Expressing his appreciation for the artistry of these doors, Michelangelo called them the Gates of Paradise, a name that is still in common use. The doors that are now in place in the baptistery are copies, while the originals have been transferred to the cathedral museum.

By the early fifteenth century, the cathedral had progressed as far as the drum, while work on the apse was nearing completion. A competition was organized for the design of the gigantic dome, which Arnolfo had prepared for in the size and foundations of the crossing and the thickness of the crossing walls. Arnolfo's preparations, however, had been made in a spirit of great optimism, as the engineering techniques to build such a huge dome were not available. Arches and domes were then built with the use of wood scaffolding to support the work in progress, which was not stable until it was finished. However, tree trunks long enough to span the Florence cathedral crossing were not available anywhere in Tuscany. Another difficulty was posed by the design of the drum, which was constructed without buttresses to take the side thrust of the dome. As a consequence, the dome had to be unusually light and required a high, pointed design to minimize side thrust. Brunelleschi won the competition in 1418 with a design of a double-shelled dome that he claimed could be built without scaffolding. This design feature initially provoked disbelief and ridicule, and the architect experienced some difficulty persuading his audience that his plan was realistic. The engineering techniques used are still being studied in the twentieth century, and attempts have been made to replicate the project on a smaller scale. In 1420 Brunelleschi and Lorenzo Ghiberti received a joint appointment to carry out the project. After about five years of bickering between the two rival artists, Ghiberti was dismissed, being commissioned at that time to do the east doors of the baptistery. Brunelleschi continued to supervise the construction by himself until his death in 1446.

The dome consists of an inner and an outer shell, which the modern visitor can view from the steep and narrow staircase built in between the shells and leading up to the lantern. The inner shell consists of the primary weight-bearing elements: eight main ribs springing out of the octagonal drum and sixteen minor ribs. The main ribs are repeated on the outer shell and have a decorative effect. The spaces between the ribs are filled with stone in the lower part and with brick, which is lighter, in the top part of the dome. The stone and brick is laid in a herring-bone pattern following Roman practice. The key to construction without scaffolding lay in the order in which the stones (or bricks) were put in place. Brunelleschi proceeded by having his workmen build successive horizontal rings, each strong enough not to require vertical support. As the dome progressed and the opening became smaller and smaller, Brunelleschi became more and more insistent that his directions for the brickwork be followed exactly. Any deviation, he feared, would result in the immediate collapse of the dome.

Brunelleschi designed the buttressed marble lantern to anchor the ribs at the top. Inverted classical corbels were to link the core of the lantern to the ribs of the dome. The architect died just before work on the lantern was to begin, however, and was buried in the cathedral crypt. His successors, under the supervision of Michelozzo, could not believe that the dome could carry the weight of the lantern and were convinced the dome would crumble. They nevertheless followed the architect's instructions, and the dome survived for more than 500 years without incident. Scaffolding erected in the late twentieth century for renovations to repair cracks has partially obscured the frescoes by Vasari and Zuccari in the dome as well as the stained-glass windows by such famous artists as Donatello, Andrea del Castagno, Paolo Uccello, and Ghiberti.

The most recent addition to the cathedral is its facade, which was not completed until 1886 under the architect Emilio DeFabris, in a neo-Gothic style. DeFabris set out to design a facade as close as possible to the one Arnolfo di Cambio would have had in mind, making extensive studies of Arnolfo's work and finally creating a design that is quite in keeping with the rest of the building and the campanile. Besides the green and white marble that is used for the revetments elsewhere, the facade also uses a delicate pink marble.

The Opera del Duomo, immediately behind the cathedral in the Piazza del Duomo, was originally the cathedral workshop during the centuries construction was in progress. It has since then been converted to one of Florence's many outstanding museums. Besides an overview of the cathedral's construction, the museum contains many of the originals of the works of art created for the cathedral, the baptistery, and the campanile. Besides the original baptistery doors, several terra cottas by Luca Della Robbia, and statues from the campanile exterior by Andrea Pisano, the museum displays a Michelangelo *Pietà*.

Further Reading: Howard Saalman's *Filippo Brunelleschi: The Cupola of Santa Maria del Fiore* (London: Zwemmer, and Montclair, New Jersey: Allanheld and Schram, 1980) provides a detailed account of the building of the cathedral dome. *Ghiberti's Bronze Doors* by Richard Krautheimer (Princeton, New Jersey: Princeton University Press, 1971), describes the evolution of Ghiberti's work in exhaustive detail. Among the general histories of Florence during the Renaissance, Gene Bruckner's *Renaissance Florence* (Berkeley and Los Angeles: University of California Press, and Chichester, West Sussex: Wiley, 1969) is very accessible and rich in detail. Christopher Hibbert has written an evocative study of the city in *Florence: Biography of a City* (London: Viking, and New York: Norton, 1993).

—Hilary Collier Sy-Quia and Marijke Rijsberman

Florence (Firenze, Italy): Piazza della Signoria

Location: The Piazza della Signoria lies at the heart of Florence, not far from the Arno River, in the area originally built up by the Romans in the first century B.C.

Description: The Piazza della Signoria is the traditional civic center of the old city of Florence. It is named after the Palazzo della Signoria, better known as the Palazzo Vecchio, which was the seat of the city's government from the thirteenth until the sixteenth century. Cosimo I de' Medici installed himself in the palace in 1537, commissioning the late Renaissance painter Giorgio Vasari to redecorate many of the interiors. From 1865 to 1871, when Florence was the capital of unified Italy, the building served as the lower house of the Italian legislature, and to this day it houses the city council and the mayor's offices. On the south side of the Piazza della Signoria is the Loggia dei Lanzi, now a sculpture gallery, behind which lies the Uffizi Gallery, once housing government offices, but now Florence's most famous art museum.

Site Office: Palazzo Vecchio and State Rooms
Piazza della Signoria
CAP 50123 Florence, Firenze
Italy
(55) 2768465

The earliest firm evidence of a settlement near the present site of Florence is Etruscan. The Etruscans had arrived in Italy, probably from Asia Minor, around 1000 B.C. and had established a flourishing civilization initially more powerful than the Roman settlements in the south of Italy. The Etruscans first settled on a hilltop site above Florence, called Fasula (now Fiesole). They later created a settlement on the banks of the Arno, a little way downstream from the later Roman settlement that grew into modern Florence. The Etruscans floated timber, wheat, wine, and olives from the surrounding hills down the river to the Mediterranean, where they traded their goods for Spanish iron and for jewels and precious metals from Syria, Greece, and Egypt. Thus, the history of Florence as a thriving commercial center can be traced to ancient Etruscan times.

When the expansionist Romans subjugated the Etruscans, Florence passed into Roman hands. In 59 B.C., redevelopment plans for many towns and settlements in Italy, including Florence, were put into effect under the auspices of Julius Caesar. The late medieval Florentine poet Dante Alighieri was thus able proudly to claim that Florence had been re-established through the intervention of Caesar himself—a perspective on their city still dear to the hearts of many modern Florentines. The Romans refashioned the city, now named Florentia, into a military settlement, which was to ensure Roman control of the Via Flaminia, the main road to the north. The city thrived and became a provincial capital by the third century. When Roman dominance began to wane, the Lombards took the city in the sixth century, and late in the eighth century Florence was incorporated into the Frankish Carolingian Empire. Through these changes, however, the Roman town plan survived and continued to form the center of the city.

The city of Florence rose to greater commercial and political prominence under the Frankish Countess Mathilda in the late eleventh and early twelfth centuries. An additional stimulus came in 1125, when neighboring Fiesole was plundered and many of its inhabitants moved to Florence. At this time the city was ruled by consuls who could call on the assistance of the so-called Council of Hundred, made up mostly of wealthy bankers and merchants. By the beginning of the thirteenth century Florence became one of Italy's foremost commercial centers, prominent in both trade and banking. The standard-setting currency of the florin and such commercial innovations as double-entry bookkeeping originated in Florence and testify to its commercial predominance. The city's gains in wealth were accompanied by a loss in self-rule, however, as the Council of Hundred was replaced by a *podestà*, or mayor, who was usually not even a Florentine.

As was the case with many cities throughout Italy, Florence became a scene of ongoing feuding between Guelphs and Ghibellines in the mid-thirteenth century. The Guelphs supported the papacy, while the Ghibellines supported the secular empire, but much more was at stake at the level of local Florentine politics. The Ghibelline party, consisting of the higher nobility, fought to maintain a feudal social organization. The Guelph party, which drew its adherents from the lower nobility, the merchant class, and the guilds, pursued a more democratic form of government. The two political parties fought for control of the city for the better part of two centuries, until the Medici family took over the government in 1434. The conflict first broke out when the Guelphs wrote their own (democratic) constitution in 1250 and appointed a *capitano del popolo* to head a separate government. The Ghibellines meanwhile remained loyal to the *podestà* and the old law, so that the city was effectively divided in two.

In 1255 the Guelphs built the fortress-like Palazzo del Popolo to house the new democratic administration. However, in 1260, at the battle of Montaperti, the Ghibellines routed the Guelphs with the help of Manfred, king of Sicily, putting an end to the democratic interlude. The Palazzo del

Palazzo Vecchio
Photo courtesy of A.P.T., Florence

Popolo was converted into the Palazzo del Podestà, which continued to house the Florentine government until the tower of the new Palazzo della Signoria, now better known as the Palazzo Vecchio, was ready to receive the executive council in 1302. The old town hall, situated just a block or so northeast of the Palazzo della Signoria, then became a prison, called the Bargello.

By 1266 the Guelphs had re-established themselves in Florence, and in 1284 they toppled the Ghibelline government with the backing of the guilds, reinstituting a popular government. They made Florence into a republic, ruled by the so-called Signoria, an executive council. Florence's mercantile prosperity had created an aristocracy of wealth, organized into guilds, whose representatives or priors made up the Signoria. The priors held office for only two months, which

can hardly have contributed to stability. The office of the *podestà* was not abolished at this time, however. Answering to the Signoria, the *podestà* ensured a measure of continuity in the city government.

Before long, the Guelph party broke into two factions, the Whites and the Blacks, over a disagreement concerning the influence of the nobility and wealthiest merchants in the Signoria. By 1293, a major reform measure was passed giving more representation to the minor guilds and artisans and excluding most of the nobility. Ironically, the reform was achieved by a nobleman, Giano della Bella, who had united the minor tradesmen into a powerful force with its own militia companies to outweigh the superior resources of the nobility and wealthier merchants. The reform, known as the Ordinances of Justice, created the new post of the Gonfalonier of

Justice, which was designed to enforce the law against offending nobility. His authority was backed up by a task force of 1,000 men. Although the result was no doubt unintended, the Gonfalonier over time became for all practical purposes the head of the Florentine Republic.

Despite factional conflict, the Guelphs continued to control the city for another decade or so and were able to commission a new town hall to house the Signoria, the Palazzo della Signoria. The commission went to the architect Arnolfo di Cambio, who was involved in the design of many of Florence's most famous late medieval buildings. Construction began in 1299 and continued until 1314. The new building arose on the site of a pre-existing palace, which was razed except for its defensive tower. This tower was incorporated into the design of the Palazzo della Signoria, heightened and topped with battlements. In 1302, the offices of the Signoria were moved to the tower, which also functioned as a watchtower overlooking the Arno Valley. Today, the tower is still one of the most salient features of the Florence skyline, together with the cathedral dome and campanile. The Palazzo della Signoria itself was built on the model of a fortress—and indeed served in that function whenever popular unrest threatened the safety of government officials. Square and rather forbidding, the main block of the town hall, built around a courtyard, consists of three stories of rough-hewn stone topped by the *ballatoio*, crenellated battlements with a covered gallery inside. Arched corbels support the imposing battlements, and frescoes of the city's symbols are almost hidden between the corbels. The doorways are simple and windows are few, particularly at the lower levels, for defensive purposes.

By 1304, construction had sufficiently advanced that the priors were able to take up residence in the building, although workmen still swarmed over the structure. Two years later, the *ballatoio* was finished. Work still continued on the tower as well as on the piazza in front of the Palazzo della Signoria. Apparently, a dispute arose over which facade was to receive the main entrance, a decision that would also affect the layout of the piazza. For the following forty years, the piazza was enlarged as the city bought up adjacent lots from citizens. Today, the piazza extends to the north of the building and arches around to the west, forming one of the larger open spaces in the inner city. By the end of the fourteenth century, the Loggia dei Priori was built on the southern side of the piazza as an outdoor platform for speakers, but it is now noteworthy as a sculpture gallery, called the Loggia dei Lanzi, after the German lancers who were accommodated in the gallery in the fifteenth century. Among the sculptures are Benvenuto Cellini's *Perseus holding the head of Medusa* and Giambologna's *Rape of a Sabine*. Although the initial construction of the Palazzo della Signoria proceeded with unusual speed, then, the piazza itself took about a century to acquire its current look.

During the century in which the physical appearance of Florentine government was so carefully shaped, many changes took place in its organization. By 1300, when Dante became a prior, the split between the White and Black factions of the Guelph party had become unmanageable. Two years later, the Whites were defeated and their more prominent members, including Dante, were driven into exile. This event, and particularly Dante's bitterness over it, are most memorably recorded in the *Divine Comedy*, largely written in exile.

For some twenty-five years, the precarious democracy of the Florentine Republic continued undisturbed. By 1325, however, the city once again fell into the hands of a foreign overlord, Charles of Calabria, who plundered the city treasury to the tune of 200,000 gold florins a year. After he died in 1328, the city turned to constitutional reform in an attempt to reverse the trend toward oligarchy in the city government. The selection of priors as representatives of the guilds was identified as the most important abuse, as the more powerful guilds of the wealthiest merchants dominated the process. A new system of nomination and election gave separate councils to the *podestà* and the *capitano del popolo*. The capitano's council, comprising 300 members, debated and voted on all legislative measures under discussion in the central government before handing them on to the *podestà*'s council, which had 250 members. In addition, there was a cabinet of advisers consisting of twelve *bonuomini* (good men) and the sixteen captains of the militia companies raised in the city's districts.

Although the workings of democracy within the city were at least temporarily improved by these reforms, Florence remained vulnerable to foreign takeovers. The duke of Athens took the city in 1342, beginning a short-lived but intensely autocratic rule. Upon his orders, the west entrance to the Palazzo Vecchio was enlarged and more heavily fortified. When the city managed to expel the duke in 1343, it immediately undertook to remove the fortifications as if to erase all sign of this temporary but ignominious loss of self-government. In 1378, the continuing underrepresentation of the city's poorest craftsmen occasioned the Ciompi revolt, one of the more serious of the many revolts that from time to time forced a recognition of the imperfections in the city's government. For several days, the disenfranchised members of the lower guilds, mainly the wool-washers and -combers, rampaged through the streets against the Gonfalonier of Justice, Salvestro de' Medici.

Florentine democracy permanently buckled under the pressure toward more oligarchic rule in 1434, when renewed turmoil ushered in the Medici era of the city's history. At the end of the fourteenth century, the Guelph party was dominated by the powerful Albizzi, while the Ricci, Alberti, and Medici families—each aspiring to take over the party leadership—formed the opposition and enjoyed the support of the working class. While the city gained in wealth and power, establishing dominion over many other Tuscan cities, the Medici extended their banking empire and managed to secure the much-coveted position of bankers to the Vatican. The Medici influence thus confirmed, Cosimo, known as the Elder and later nicknamed "Pater Patriae," or father of his country, made an unsuccessful bid for leadership in the Guelph party in 1433. His presumption was punished with exile, which caused an uproar among his working-class supporters. By 1434, Cosimo was back in Florence and took control of the government. Espousing a benevolent despotism, he did away with most of the democratic machinery of Florentine government during the thirty years of his rule, forcing the beginning of a new era in the city's history. Cosimo also left a Medici imprint on the Palazzo della Signoria, when, in the 1440s, he initiated a renovation of the building. The internal courtyard was rebuilt under the supervision of the artist and architect Michelozzo to conform more closely to the courtyard of the Palazzo Medici-Riccardi, then the chief Florentine residence of the Medici family. Michelozzo also redid much of the interior decoration of the town hall.

Under Lorenzo de' Medici, also known as Lorenzo the Magnificent, who ruled from 1469 to 1492, Florence again extended its reputation. Already being known far and wide as one of Italy's most important centers of trade and banking, Florence also became the major center of Renaissance art and learning in western Europe. A hotbed of the new philosophy, painting, sculpture, music, and poetry, Florence was increasingly focused on the court of Lorenzo de' Medici, who has gone down in history as one of the most discerning and liberal patrons of the arts. At the very end of Lorenzo's rule, the Medici fortunes took a downturn when the bank failed. This disaster brought another period of instability to the city. Lorenzo's son Piero was unable to repel the invasion of the French king Charles VIII, surrendering the city and fleeing it himself in 1494. The French presence in the city was temporary, and the Dominican monk Girolamo Savonarola soon took control of the city with popular support, instituting a republic of intensely puritanical persuasion. To accommodate the Consiglio Maggiore, the 500-member chief executive council of the republic, Savonarola ordered the construction of a large assembly room known as the Salone dei Cinquecento, on the eastern side of the Palazzo della Signoria. Savonarola's influence crumbled after a few years as did his republic, and by 1498 he was deposed and burned at the stake as a heretic in the Piazza della Signoria.

By 1512 the Medici were back in the city, to be expelled in 1527. They then ruled the city uninterruptedly from 1529 to 1737. First acquiring the title of dukes of Tuscany, the family secured the Grand Duchy of Tuscany for itself in 1569. During the more than two centuries following their return to Florence in 1529, changes to the Palazzo della Signoria focused on further additions to the east end of the building, the interiors, and the courtyard. In 1537, Cosimo I (so called because he was the first Medici Grand Duke of Tuscany) made the Palazzo della Signoria his chief residence, renaming it the Palazzo Ducale, which it remained until the Medici moved to the Palazzo Pitti later in the same century. This move was accompanied by the last name change for the town hall, which became the Palazzo Vecchio. Cosimo I meanwhile commissioned Giorgio Vasari to redo the internal courtyard, to participate in the design of the eastern additions, and to create new interiors for some of Cosimo's favorite rooms.

Cosimo's move into the town hall displaced numbers of government officials, thus creating a need for new office space. Other government offices by that time had found space in sites scattered across the city. To bring all government offices into the same complex, a new building, the Uffizi, was constructed on the south side of the Piazza della Signoria, immediately behind the Loggia della Signoria. Vasari was the architect in charge of its design. Today, the Uffizi is Florence's most famous art museum. Cosimo's successor, Francesco I de' Medici, ordered the architect Bernardo Buontalenti to alter the upper floor of the Uffizi into a gallery to house the growing Medici art collection. The last Medici ruler of Florence, Anna Maria Ludovica, bequeathed the entire collection to the city in 1737. Since then, parts of the collection have been moved to other Florentine museums, but the backbone of the Uffizi Gallery's current collection is still formed by the art treasures of the Medici.

When the last Medici died without heirs, the Grand Duchy of Tuscany passed to the House of Lorraine, who became the next rulers of Florence. In 1799, the Lorraines lost the city to Napoléon, regaining it in 1814. By that time, Florence had lost some of its former importance, but it became the first city of Italy with the campaign for the unification of the country. In 1865 it was made the capital, and the lower house of the legislature was installed in the Palazzo Vecchio, while museums, religious buildings, and palaces were converted into offices, schools, and hospitals— all at municipal expense. When the capital was moved to Rome in 1871, the city could barely afford to convert these public structures back to their original purposes. For the next century, the glorious medieval and Renaissance buildings of Florence suffered from serious neglect. The situation became apparent with the flood of 1966, when the Arno swept through the city, doing great damage particularly to the art treasures and libraries and leaving behind a carpet of mud and debris.

In May 1993, Florence again made international headlines, with the explosion of a car bomb just outside the Uffizi Gallery's west wing. Six people were killed in the attack, which has been attributed to the Mafia, and the building and its collection sustained serious damage. Paintings by Gerrit van Honthorst and Bartolomeo Manfredi were destroyed altogether, as was the painstakingly compiled catalogue of the gallery's holdings. Repairs of the structural damage and restoration of salvageable paintings were immediately undertaken, and the museum reopened after a month to admit once again the flood of tourists who have now become Florence's chief industry.

Further Reading: For a detailed history of the city of Florence from its inception until its golden age, see Ferdinand Schevill, *History of Florence from the Founding of the City through the Renaissance* (New York: Harcourt, Brace, 1936; revised, New York: Frederick Ungar, and London: Constable, 1961). Gene Bruckner's *The Civic World of Early Renaissance Florence* (Princeton, New Jersey: Princeton University Press, 1977) presents a detailed account of the civil government of Renaissance Florence. The same author has also published a very accessible general history of the Renaissance city in *Renaissance Florence* (Berkeley and Los Angeles: University of California Press, and Chicester, West Sussex: Wiley, 1969). Many books have been written about the influence of the Medici on the course of Florentine and Renaissance history, of which J. R. Hale's *Florence and the Medici: The Pattern of Control* (London: Thames and Hudson, 1977; New York: Thames and Hudson, 1978) is among the more balanced. *Florence and the Age of Dante* by Paul C. Ruggiers (Norman: University of Oklahoma Press, 1964) deals with Florentine history of the fourteenth century.

—Hilary Collier Sy-Quia and Marijke Rijsberman

Florence (Firenze, Italy): Ponte Vecchio

Location: South of the center of the city of Florence, the Old Bridge is built on a line running from north-northeast to south-southwest. It spans the Arno River at its narrowest point. Leading from the bridge are the Via Por Santa Maria to the north and the Via Giucciardini to the south.

Description: Triple-arched stone bridge completed in 1345, lined on both sides most of the way across the river with jewelers' and goldsmiths' shops, many projecting at the back over the water. The bridge is surmounted on the east side by the Vasari Corridor, built in 1565. The Ponte Vecchio is open now only to pedestrians.

Site Office: A.P.T. (Azienda di Promozione Turistica)
Via Manzoni 16
CAP 50121 Florence, Firenze
Italy
(55) 23320

The Ponte Vecchio (Old Bridge) is the traveler's emblem of Florence. Spanning the capricious and often violent Arno, which flows east to west on its alluvial bed from the Apennines to the Ligurian Sea just beyond Pisa, the Old Bridge links the thirteenth-century Piazza della Signoria to the Oltrarno, the district "beyond the Arno," which came into its own in the sixteenth century, when the Medici grand dukes of Tuscany took over the palace of their rivals, the Pitti, and their courtiers built houses nearby. Since 1944 and the destruction of the Ponte alle Grazie, the Ponte Vecchio is the oldest of Florence's seven bridges and the only one to have survived both flood and war.

The city's history has always been closely involved with the existence and the vagaries of the Arno, long a source of water and an artery of trade and transport, but always also the "accursed ditch," as Dante called it, with its summer droughts, its savage floods, and the consequent visitations of pestilence on the inhabitants and destruction on their treasures, including the bridges. The Arno at the Ponte Vecchio is about 145 feet above sea level.

Old Florence, which probably existed in the second century B.C. as a colony of the Etruscan city at Fiesole five miles away, was a river settlement. In the first century B.C. an Etruscan ford and market was located near the present Ponte Vecchio. Under the Romans, who had first set up an encampment in 59 B.C. a little to the east of the site of the present bridge, river traffic on the Arno became important. This narrow point of the river must have known bridges from Roman times, but they were wooden structures swept away relatively easily whenever the Arno, swollen by the Sieve, its tributary in the hills, rose in a wild torrent and burst its banks.

Until 1218, when the Ponte alla Carraio was begun, there was only one bridge in the city and it occupied the site of the present Ponte Vecchio. The earliest recorded structure (that of A.D. 972) was of wood. Standing at its northern end was the stone Martocus, a damaged and headless equestrian statue of Mars, the Roman god of war. It had been removed from an earlier temple, and now, regarded as a tutelary figure, it stood facing east, on guard over the town. In the tenth and eleventh centuries the statue on the bridge was crowned with flowers in March, the month of Mars, if the spring was promising, or smeared with mud if the season was bad.

It was at the base of the Martocus that on Easter Sunday in the year 1215 (some sources say 1216) a crucial incident in Florence's violent history took place. A young man named Buondelmonte de' Buondelmonti, who had been betrothed to a member of the Amidei family, decided to marry another woman. As he rode across the bridge on the way to this wedding he was set upon and murdered by the kinsmen of the woman he had rejected. (Dante alludes to the crime in canto 24 of the *Inferno*.) This was the start of a conflict that led to a century and a half of strife in Florence between the Guelph and Ghibelline political factions, the former being allies of the pope, the latter supporting the Hohenstaufen Holy Roman emperors, in the contention for supremacy in Lombardy and Tuscany.

At the southern end of the bridge, in the shadow of the fortified tower houses of belligerent feudal magnates, pilgrims on their way to Rome could stop at a hospice, run by the Knights of the Holy Sepulchre; the structure is still traceable in fragmentary remains. Three more bridges were built over the Arno in the thirteenth century, the first of which, the original Ponte alla Carraia, still made of wood but supported on stone piles, was begun in 1218 and completed in 1220. It was known as the Ponte Nuovo, the New Bridge.

In 1300, the first of many civic disputes brought on by the rebuilding of bridges over the Arno occurred. After being moved during construction work, the Martocus was replaced facing the wrong way, north instead of east. This insult to the god was considered by Dante, among others, as a bad omen for the city, even though by now Florence's patron was not Mars but St. John the Baptist; the incident marked the beginning of the Black/White dissension among the Guelphs. The Martocus was swept away with the nine-arched bridge in the flood of 1333. The memory of the statue remains, however, because of its association with the Buondelmonte murder. According to some, the Martocus's name and guardian function are perpetuated in those of the Marzocco, the heraldic lion in the arms of the city.

By the thirteenth century, shops were already set up on the bridge. In 1206 the Commune of Florence set up the

Ponte Vecchio
Photo courtesy of A.P.T., Florence

Opus Pontis (the Bridge Bureau) to keep the structure in good repair and to rent the mostly wooden shops, chiefly to tanners and makers of purses and other leather goods who needed the water of the river for their trade.

In 1333 the old wooden Ponte Vecchio was lost when another of the Arno's epic floods washed away all the medieval bridges in Florence except the stone Ponte alle Grazie. The present three-arched Ponte Vecchio, of yellow stone with arrow-shaped piers, was finished in 1345. Its builder was probably Neri di Fioravanti, one of the architects of the Duomo and the Bargello, although some attribute it to Taddeo Gaddi, godson and best pupil of Giotto.

The shops and workshops on the new 1345 bridge were now of stone and built into the crenellated walls bordering the bridge. The rents were supposed to be put toward a future rebuilding of the Ponte Vecchio, but in 1378 the Commune privatized the leasing arrangements, farming them out to one of the Medicis in recognition of political services rendered. (The Medicis would rule Tuscany and Florence from the fifteenth century until the Peace of Vienna in 1736.) By the 1420s the already none-too-fragrant tanners and pursemakers had been joined on the bridge by butchers, who gradually ousted their predecessors.

After the temporary banishment of the Medici in 1494, the city authorities sold the shops, although they were still administered by a public body that was charged with overseeing port and tax officials as well as mills, walls, and bridges. Other tradesmen, selling linenware, hosiery, and vegetables, came to occupy some of the bridge shops. In 1593 Grand Duke Ferdinand I de' Medici turned the smelly butchers and the rest off the Ponte Vecchio and replaced them with forty-one goldsmiths and eight jewelers paying twice as much rent, as befitted their loftier occupations. They were not only merchants and craftsmen: many of Florence's greatest artists began their training in these workshops. The demand for local goldsmiths' work and jewelry from churches and wealthy citizens surged with the city's prosperity in the fifteenth century. Strict regulations were aimed at preventing fraud and malpractice in the handling of the precious raw materials. The present-day jewelers of the Ponte Vecchio own their little shops, projecting over the river since the seventeenth century, and still use traditional tools.

The simple stone gallery, the Vasari Corridor, that surmounts and, in some people's opinion, overwhelms the bridge was built in five months in 1565, on the orders of

Cosimo I (the Great) de' Medici, a notable soldier, ruler, and patron of the arts who ruled as duke of Florence from 1537 and grand duke of Tuscany from 1569. The purpose of the gallery was to provide a private covered way between Cosimo I's Palazzo Vecchio offices (now the Uffizi Gallery) just north of the Arno, and the Pitti Palace, on the south side of the river, where his newly married son Francesco was to live with his bride, Joanna of Austria. From the Uffizi, the corridor crosses above the Lungarno Archibusieri and continues elevated on arcades along the north or right river bank, then passes across the top of the east side of the Ponte Vecchio, incorporating the shops beneath, as well as the facade of the Church of Santa Felicità beyond the river. It stretches over some of the houses in the Via Giucciardini and on as far as the Bóboli Gardens, near the Buontalenti Grotto. Three arcades in the center of the bridge are free of shops and allow a wide view of the river and the city. The grand dukes apparently used to "attend" religious services by looking down from a window of the corridor into the Church of Santa Felicità. In 1868 the corridor ceased to be used as a private catwalk, and now, after the repair of structural damage suffered in World War II, it is a gallery of self-portraits, including one by the corridor's creator, Giorgio Vasari.

Vasari, whose name is now chiefly associated with his famous work of biography and art criticism, *Lives of the Artists,* was both painter and architect and studied under Michelangelo. Originally from Aretino, he settled in Florence, where he created a large body of architectural work for Cosimo I, including the Grand Duchy's administrative offices in 1560. He is considered a master of post-1550 Florentine Mannerism.

A bust of Benvenuto Cellini made by Rafaelle Romanelli in 1900 looks down from its plinth on the west side of the bridge, in the open section standing free of both the shops and the corridor arcades. Cellini, another sixteenth-century mannerist artist and a native Florentine of many talents and a colorful personality, was celebrated as goldsmith, engraver, sculptor, swordsman, and author. Like Vasari, Cellini too had Cosimo I de' Medici as a patron; it was Cosimo who in 1545 commissioned Cellini's masterpiece, the bronze statue of *Perseus with the Head of Medusa* for the Loggia dei Lanzi in the Piazza della Signoria. The bust of Cosimo I in the Bargello and *Narcissus* in the Bóboli Gardens are also by him.

The Church of Santa Felicità, just beyond the south end of the Ponte Vecchio, was incorporated into the Vasari Corridor in 1565. The history of Santa Felicità likely makes it the second-oldest church in Florence. It is predated only by San Lorenzo, which goes back to A.D. 393. The little piazza in front of Santa Felicità stands on the site of a fifth-century oratory and of Florence's first Christian cemetery, now commemorated by a fourteenth-century granite pillar. A community of nuns evidently lived in the place as early as the eighth century. The church, consecrated here in 1059 the day after the consecration of the foundation stone of the baptistry, was followed by a fourteenth-century Gothic structure; the present baroque building is for the most part the result of eighteenth-century rebuilding by Ferdinando Ruggiero. Among the most remarkable of the artistic treasures preserved here are two vivid works executed by the early mannerist painter Jacopo da Pontormo during his three-year redecoration, starting in 1525, of the Capponi Chapel: the fresco of the *Annunciation,* which underwent restoration in 1967, and the *Deposition* altarpiece.

The Ponte Vecchio was the only one of the city's bridges to escape destruction during the withdrawal of the German armed forces in 1944. Although the structure itself was spared, streets and buildings at each end were reduced to rubble so as to block the passage.

In the terrible 1966 floods, when water from the Arno was six feet deep in many of Florence's ancient churches and much damage was done to the city's treasures, large quantities of gold were washed away and lost from the shops on the Ponte Vecchio. The shops themselves suffered badly but have since been carefully restored.

During the 1970s the river bed was deepened to relieve pressure on the foundations, which are now owned by the national Ministry of Public Works. The city owns the rest of the bridge, including the corridor, but not the shops.

Further Reading: *The Companion Guide to Florence* by Eve Borsook (London: Collins, and New York: Harper and Row, 1966; revised, London and New York: HarperCollins, 1988) is a meticulous historical account of the city containing much information on the Ponte Vecchio. *Florence and Siena* by John Kent (London: Viking, 1989) has detailed color maps, plans, and drawings that show clearly the architecture and location of the bridge. The helpful text is brief. *The Stones of Florence* by Mary McCarthy (London: Heinemann, and New York: Harcourt Brace, 1959) is a learned and reflective general work on Florence, helpful on the history of the Ponte Vecchio.

—Olive Classe

Gallipoli (Çanakkale, Turkey)

Location: Gallipoli (Gelibolu in Turkish), in the province of Çanakkale, is the chief town on the narrow Gallipoli Peninsula at the northeastern extremity of the Dardanelles at the western entrance of the Sea of Marmara, some 126 miles west-southwest of İstanbul. The narrow strait, 38 miles long and three-quarters of a mile to 4 miles wide, traditionally marks the separation between Europe and Asia.

Description: A major seaport and town in the European part of Turkey, Gallipoli has been an important strategic site since ancient times. It is best known for the World War I campaign in which Allied forces were stopped by a determined Turkish defensive corps. Today, a national park honors the 500,000 soldiers on both sides who lost their lives at Gallipoli.

Contact: Çanakkale Tourism Office
Villayet Konği
Çanakkale, Çanakkale
Turkey
(0-286) 2175012-2173791

Although Gallipoli is perhaps best known as the site of a disastrous Allied invasion in World War I, the remains of the many fortresses and castles that line its beaches are reminders of the strategic importance it has held for thousands of years. The Gallipoli Peninsula, known in ancient times as the Chersonese, was occupied as early as 3000 B.C. The ancient city of Troy defended the Dardanelles Strait from its strategic position to the southwest and on the Asian side. A fortress that could command the strait was built as early as 2500 B.C.

The Greeks knew the strait as the Hellespont. The name derives from the legend of Helle and Phryxus, the children of King Athamas who fled from their stepmother on the back of a flying ram with golden fleece sent by the goddess Nephele. While flying over the strait, Princess Helle fell off the ram into the water. Hellespont was also the setting for the Greek legend of the two lovers Leander, a handsome youth from Abydos, and Hero, an Aphrodite priestess from Sestos. Each night Leander would swim the mile-wide crossing from Abydos on the European side of the Dardanelles to Sestos on the Asian side. One stormy night, he did not emerge from the water. Fearing that he had drowned, Hero threw herself into the sea rather than be without her love. Lord Byron immortalized the story of the two lovers in a poem after he himself duplicated Leander's feat by swimming across the strait.

The name of the Dardanelles is derived from the Roman name of Dardanus, the mythical founder of Dardania,

a city near ancient Troy. The Turks refer to the strait as Çanakkale Boğazı. Gallipoli's written history begins only with the coming of the Greeks, centuries after the legendary siege of Troy.

In the seventh century B.C. Aeolians from Lesbos and Ionian Greeks from Miletus founded a dozen cities on the Chersonese Peninsula. In the sixth century an Athenian colony was established by Miltiades the Elder, who became a tyrant of the Greek colonies on the peninsula. Miltiades' family ruled there until his nephew Miltiades the Younger abandoned the Chersonese to the armies of Darius I of Persia in 492 B.C.

In 483, Darius's son Xerxes marched his Persian army for seven days and nights over a bridge of boats, a feat that impressed the Greek historian Herodotus. After the first Greco-Persian wars, the Gallipoli Peninsula was returned to Athens. It fell briefly to Sparta during the Peloponnesian War but reverted to the Athenians with the end of hostilities. In 338, the peninsula was permanently ceded to Philip II of Macedon by the Athenians. Philip's son, Alexander the Great, accomplished a similar crossing to that of Xerxes, but this time it was a Greek army invading Persia.

In 133 B.C. Gallipoli fell to the Roman Empire, which turned most of the land into state-owned property. In later antiquity a great sea-battle was fought over control of the strait; Crispus, the elder son of Emperor Constantine I, commanded an imperial fleet of 200 ships there and defeated his opponent Licinius and his fleet of 350 ships. During the Byzantine period, Atilla, king of the Huns, reportedly reached Gallipoli and Sestos and is said to have conquered all the fortresses of the area.

In the Middle Ages, Gallipoli featured once again in the designs of Italian city-states such as Venice, which sought to control trade with the East by controlling passage through the strait. In the thirteenth and fourteenth centuries, Venetian traders and European crusaders used it as a starting point for their expeditions to the East. In 1304, a band of Spanish mercenaries seized the fortress at Gallipoli and used it as a base for raids and incursions into the surrounding countryside.

European trade was dealt a serious blow when the Ottomans under Orhan I took possession of Gallipoli after a violent earthquake shook the peninsula in March 1354. The Ottomans quickly transformed Gallipoli into a permanent base from which to expand into Europe. Turkish raiding parties were soon moving into the Balkans and as far north as Adrianople. It was not until Orhan was succeeded by his son Murad I in 1360 that Gallipoli was utilized for permanent conquests. Murad was also one of the first to place tolls on the traffic that passed through the Dardanelles, an action that enraged many trading nations—the Venetians in particular—

Monument to Turkish soldiers at Gallipoli
Photo courtesy of Embassy of Turkey, Washington D.C.

and that involved Gallipoli in numerous conflicts throughout the centuries.

Nor did Byzantium forfeit its claim on the peninsula. In 1366, Amadeus VI, count of Savoy, retook Gallipoli for his cousin John V, emperor of Byzantium, and left a small garrison of Italian mercenaries. But his victory was short-lived; the city was returned to Murad in 1367. Throughout the remainder of the fourteenth century Gallipoli was steadily reinforced as a naval base and arsenal under the Ottomans in preparation for their conquest of Byzantium. Turkish control over this vital link between trade on the Mediterranean and Black Seas caused much consternation in Venice. During the first Ottoman war with Venice (1423–30), Venetian attempts to seize Gallipoli failed. In 1451 the defense of Gallipoli was further reinforced by Mehmed II, the conqueror of İstanbul, when he built fortresses at Kilid Bahr on the European side and a corresponding one on the Asian side at Cimenlik. Remains of both structures still stand. They gave the Ottomans effective control over the strait and encouraged the Ottomans to build a powerful navy that would one day rival that of the Venetians. With the end of the Ottoman-Venetian Wars in the seventeenth century, the Gallipoli Peninsula remained firmly in Ottoman hands.

The strategic significance of the Dardanelles became an issue once again during World War I. Though the Ottoman Empire was traditionally an ally of Britain, the new political leaders of the Young Turk Party had hoped that a rapprochement with Germany would enable them to break the power of the caliphate. Germany, for its part, had encouraged better relations between the Turks and the Germans. On the eve of the outbreak of war, Britain and the Ottoman Empire became embroiled in an argument over ships. Britain had promised to send two ships to İstanbul, but, with the outbreak of war in the fall of 1914, failed to deliver. Germany, taking advantage of the situation, seized the opportunity and sent a battle cruiser, the *Goeben*, thus solidifying a friendship with the Turks. The Turks entered the war in October 1914 on the side of Germany.

In January 1915, an urgent appeal from Grand Duke Nicholas, commander of the Russian Army, prompted the Allies to consider a suitable site for a diversion that would alleviate the pressure on the Russian forces in the Caucasus. British military officials selected the Dardanelles as the best site for a joint naval and military operation. It was later decided that an initial naval attack would proceed without immediate military support while troops would be assembled in Egypt under General Ian Hamilton for the eventual mopping up of the Gallipoli Peninsula.

The naval attack planned for February 1915 was to proceed in three phases: a long-range bombardment, followed by a medium-range bombardment, and, finally, full fire at close range. Minesweepers were to clear the channel to the entrance of the strait under the cover of attack. The long-range bombardment began on February 18, 1915, but had to be suspended because of bad weather and was not resumed until the following week on February 25. Further delays followed,

and the final fleet attack did not take place until March 18. Unknown to British officials, a Turkish steamer had laid a new line of mines in the intervening period. The British quickly lost three of their eighteen battleships, and the attack was abandoned.

Though it had been quickly called off, the attack on Gallipoli caused considerable panic in İstanbul, the capital of the Ottoman Empire. Fears of British occupation led many to flee. The Turkish losses were relatively small, some 118 casualties, but with half their ammunition already spent, it was feared that they would not hold out much longer if the British fleet renewed its attack. Much to their surprise, the British Admiralty decided not to proceed any further without the army. The ensuing month-long delay allowed the Turkish defenses not only to be reinforced but also reorganized by German commander Otto Liman von Sanders, who took charge of the defense of the Dardanelles.

It took over a month for 75,000 men from five divisions, including the Australian and New Zealand Force (ANZAC), the British Twenty-ninth Division, Royal Navy divisions, and one French division, to be assembled and prepared to land on the Gallipoli Peninsula. On the morning of April 25, the main striking force consisting of the British Twenty-ninth and Royal Naval divisions landed at Cape Helles on the southern tip of the peninsula. Another force consisting of the Australians and New Zealanders landed on the Aegean side of the beaches at Gaba Tepe. The French were to create a diversion by landing on the Asian side of the Dardanelles at Kum Kale.

The ANZAC troops did land at Gaba Tepe and managed to establish a foothold on what would later be known as the Anzac Cove but were prevented from proceeding any further by dogged Turkish defenders, who, though outgunned, were rallied by a young Turkish colonel, Mustafa Kemal.

Meanwhile the British landings at Cape Helles fared little better. Like their ANZAC counterparts they were able to establish a beachhead but could not advance further without reinforcement. Even when reinforcements arrived, little progress was made because the Turks had quickly shored up their defenses. Another stalemate, similar to that in western Europe, developed on the Gallipoli Peninsula. The Dardanelles Expedition had become mired in problems.

News of the deadlock in the Dardanelles had serious repercussions for officials in London. The first sea lord, John Abuthnot Fisher, resigned on May 14 when his call to bring an end to the campaign was overruled by the War Council. Winston Churchill, lord of the Admiralty and supporter of the Dardanelles Expedition, was relieved of his post, though he remained on the War Council of the cabinet.

Rather than heeding the call of the first sea lord, the War Council decided to send further reinforcements to Gallipoli. In July 1915 five more divisions were dispatched to the strait and a new strategy devised. The British objective was to capture the hills of the Sari Bair Heights and create a wedge that would divide the Turkish forces. More reinforcements were to be added to the ANZAC positions on the Aegean while a new attack was to be launched from Sulva Bay farther to the north.

The new offensive was launched on August 6, 1915; as with the initial offensive, it proved to be ineffective despite some initial progress. Within a few days, the Gallipoli front was deadlocked once again. By September, Lieutenant General Charles Monro replaced Ian Hamilton as commander. Upon reviewing the situation, Monro was unequivocal in his recommendation that the entire operation be suspended and the troops evacuated. His recommendation was given further support by Lord Kitchener, who also called for the troops to be sent home. In December 1915, the troops from Sulva Bay and Anzac Cove were evacuated under the cover of darkness. One month later, January 9, 1916, the last of the remaining forces at Cape Helles were shipped home, thus bringing to an end one of the most humiliating and frustrating Allied campaigns of World War I. Nearly 214,000 casualties were reported among the British Commonwealth troops that had occupied the beaches. The Turks had a similar number of casualties.

The retreat from Gallipoli had even more serious repercussions for the British military and government than had the setbacks of the previous spring. Winston Churchill left the government to command an infantry battalion in France. The liberal government was superseded by a coalition government. Very soon thereafter, Herbert Henry Asquith resigned as prime minister of Great Britain and was replaced by David Lloyd George.

In Turkey, the effects of the Allies' failure were equally great. Mustafa Kemal, commander of the Turkish Nineteenth Division at Çanakkale, was called the Savior of İstanbul. He further added to his prestige in 1916 when his defenders stopped the southern advance of Russian forces threatening İstanbul. With the end of the war, Kemal embarked on a political career that ultimately led to the founding of the modern Republic of Turkey. Kemal was declared the republic's first president, and in 1933 the Turkish National Assembly granted Kemal the name "Atatürk," Father of the Turks.

Today, the Gallipoli Peninsula is home to a memorial park dedicated to the 500,000 soldiers who died there in World War I. Many of the battle sites, such as Anzac Cove, Cape Helles, and Chunuk Bair, have been preserved for future generations.

Further Reading: A. R. Burns's *The Penguin History of Ancient Greece* (Harmondsworth, Middlesex: Penguin, 1954) provides a good introduction to the strategic importance of the Hellespont in ancient times. Geoffrey Lewis's *Turkey* (New York and Washington: Praeger, and London: Benn, 1955; revised, 1967) provides a good overview of the rise of the Ottoman Empire and the role of Atatürk in the birth of modern Turkey. Alan Moorehead's *Gallipoli* (London: Hamilton, 1956; New York: Harper, 1964) is a popular study of the Gallipoli campaign. Also of interest is Captain Basil Henry Liddell Hart's analysis of the campaign in *The Real War, 1914–1918* (London: Faber, and Boston: Little, Brown, 1930).

—Manon Lamontagne

Gela (Caltanissetta, Italy)

Location: On the southern coast of Sicily, at the mouth of the River Gela, some forty miles west of Syracuse.

Description: Founded by Greek colonists, the city flourished before the beginning of the Common Era, and then ceased to exist. Rebuilt in the Middle Ages, it remained insignificant until oil was found just off the coast in the 1950s.

Site Office: A.A.S.T. (Azienda Autonoma di Soggiorno
e Turismo)
Via G. Navarra Bresmes, 104
CAP 93012 Gela, Caltanissetta
Italy
(0933) 913788

The recorded history of Gela begins with the Greek colonization of Sicily, in the eighth century B.C., but the Greeks were not the first to build a settlement on the low plateau just off the beach. Bronze Age pottery has been found underneath the structures on the later Greek acropolis. The unusual horns, associated with fertility rites, on the bowls identify them as belonging to the so-called Castelluccio culture.

When the Greeks arrived to settle at Gela, they found the plateau uninhabited, however. The native population of Sicels lived in the hills that close off the coastal plain in a semicircle to the north. In ancient sources, the date of Gela's foundation is given as 688 B.C., making it both the last and the westernmost outpost set down during the first wave of Greek colonization. The Gelan settlers hailed mostly from Rhodes and Crete, but they did not maintain strong ties with the mother islands. Like their neighbors at Syracuse, they seem to have set out to subjugate the native Sicels, although all other Greek settlements pursued a policy of peaceful coexistence. The Gelans built fortifications along their acropolis and began a campaign against the Sicels. To maintain their dominance in the area, they also built strongholds at different points in the hills. Greek women being in short supply, the colonists took wives from among the native population, probably by force.

By the early sixth century B.C., Gela controlled the entire coastal area of southeast Sicily, and its citizens founded another city some forty miles to the west in 582 B.C. Before long, this new settlement, Akragas (now Agrigento), became a rival. No exception among the Sicilian colonies, Gela later became more preoccupied with fighting its Greek neighbors than the native population, but in the first 200 years or so relations with other Greek settlements were generally peaceful.

Gela prospered in its early years. The coastal plain proved fertile farmland, and the city quickly grew wealthy from its agricultural exports. One sign of this new wealth was the construction of a temple to Athena on the acropolis in the sixth century B.C. According to custom in Sicily and other parts of Magna Graecia, the temple was clad with painted terra cotta tiles, some of which have been preserved. Gela seems to have been famous for these revetments, shipping them to Olympia when a Gelan treasury was built there.

The Gelans were also the best horse breeders on the island, and some of their wealth may have derived from this source. Whether or not the horses made Gela rich, they undoubtedly made the city famous. Late in the sixth century and early in the fifth, Gelan *quadrigae* were winners of the prestigious chariot races at the Olympic games several times. Horsemen also played an important role in Gela's territorial expansions, as they formed the most important element of the army. A horseman was chosen as the city's emblem when Gela began minting coins early in the fifth century.

Events took a dramatic turn in 505 B.C., the beginning of the time of the tyrants in Sicily, when Cleander took power in Gela. Together with his brother Hippocrates, he enlarged the army and strengthened the city's fortifications, preparing for war. Cleander was not to realize his expansionist goals, however; he was forced to deal with challenges to his own power. He was assassinated in 498 B.C. during a brief civil war. Hippocrates managed to take his brother's place and to assert his power in a more efficient manner. Shortly, he was ready to set out on the campaign that had been planned years before.

Hippocrates' first targets were the Greek settlements on the northern end of the east coast. In short order, his army took Zancle (now Messina) on the straits, Naxos, Leontini (now Lentini), and other, smaller settlements. Taking away much of the riches of these settlements, Hippocrates left them under the watchful eyes of vassal tyrants and marched on to Syracuse, then the only city more powerful than Gela and particularly attractive for its large natural harbor. A battle fought in 492 B.C. at Helorus brought no clear victor in the struggle, and the parties accepted mediation from Corinth. As a result of these negotiations, Hippocrates agreed to desist from his attempt to take Syracuse in exchange for the city of Kamarina, then a Syracusan satellite on the southern coast.

Apparently uninterested in peaceful activities, Hippocrates immediately organized a campaign against the native population in the interior and was killed in battle in 491 B.C. Hippocrates' sons were too young to take over as tyrants of Gela, and his cavalry commander, Gelon, was appointed to act as their protector. Gelon was much like Hippocrates, sharing his ambition and warlike disposition. Born into a priestly family, he was an avid horseman and entered a winning chariot in the Olympic games of 488. Before long, he himself was tyrant of Gela and carried on Hippocrates' campaigns.

In 485, Gelon saw his chance to take Syracuse when a popular uprising brought an invitation for military support

Part of the Capo Soprano, ancient fortifications at Gela
Photo courtesy of A.A.S.T., Gela

from the embattled oligarchs of that city. Rather than restoring the oligarchs to power, he used the opportunity to take power in Syracuse himself. To build up what was now his most important base, he took about half the population of Gela with him, leaving the decimated city under the government of his much less able brother Hiero. He destroyed Kamarina, bringing its inhabitants to Syracuse also, and took Megara Hyblaea, selling its inhabitants into slavery.

Although temporarily in the same family, Gela and Syracuse were not always to be on a parallel course. In 466, many of the inhabitants who had been forced to move to Syracuse returned to Gela, which began a new period of prosperity, witnessed by the wealth of fifth-century terra cottas and sculpture that has been found. It was at this time that the famous Greek tragedian Aeschylus chose to make Gela his home, reportedly writing the *Oresteia* there. Although its whereabouts are not entirely certain, the Gelans had built a theatre, probably on the western edge of the plateau. Aeschylus died at Gela in 456.

In 424 Gela was the scene of a famous congress. Syracuse, then under the leadership of Hermocrates, had made war on Leontini, a small city on the east coast and no match for the superior might of Syracuse. However, Leontini had managed to bring Greece into the conflict, which had sent an expedition in support of the beleaguered settlement. The Greek invasion caused such an escalation of the conflict that all the Greek cities, with the exception of Akragas, entered the fray on one side or another. Gela supported Syracuse, for instance. For once, the Sicilian Greeks decided that war would be fruitless and convened a peace-making congress at Gela.

The peaceful sentiments that had prevailed in 424 did not dominate Sicilian politics for long, however. By 413, Selinus and Segesta were at war over a border dispute, and this conflict eventually brought the end of Gela's prosperity. Again Greece invaded, this time supporting Segesta while Syracuse and Gela supported Selinus. The Greeks were defeated by Syracuse, leaving Segesta without allies, but only temporarily so. This time, the embattled city successfully appealed to Carthage to intervene. The Carthaginians held trading posts on the western tip of the island, but their reasons for sending an expedition to Selinus are unclear. The first Carthaginian invasion of 410 did little to reconfigure the power politics on the island, although it left Selinus in ruins and left Syracuse defeated on the battlefield at Himera.

By a strange twist, Syracuse indirectly provoked a second Carthaginian invasion a few years later, this one to have more far-reaching results. Hermocrates, having been relieved of his command after the battle of Himera, went to Selinus, where he assembled a sizable army and challenged the domination of Carthage over that city. In retaliation, Carthage organized a second invasion and marched through the south of Sicily, taking one city after another. Akragas fell first. Gela was next; the city was destroyed, and its inhabitants were scattered. Kamarina, which Gela had recolonized in the fifth century, followed. Still the Carthaginians advanced to Syracuse, which wisely negotiated a treaty with the enemy in

405. Under this treaty, a large stretch of southern Sicily, including Gela, was ceded to Carthage as an *epikrateia*, or province. The Gelans were allowed to return home, but they were to pay tribute. Building fortifications was forbidden to all tributaries of Carthage.

For more than fifty years, Gela was reduced to an insignificant farming community, until in the mid-fourth century B.C. Timoleon of Syracuse took an interest in the once glorious and prosperous city. He began a program of resettlement with new immigrants from Italy and Greece, and small farming communities were established on almost every hill around Gela. Farther into the interior, the Syracusan tyrant reconstructed a number of Sicel villages. In Gela itself, Timoleon also initiated a program of reconstruction around the original acropolis on the eastern end of the plateau. Gela began growing so quickly once again that it expanded all the way to the western end of the plateau, where, in the late fourth century, the main residential quarters were built on a grid plan. Never before had the city been so large. The western end, for instance, had previously only housed cemeteries and had not been inhabited. Some of the temples destroyed by the Carthaginians were rebuilt. The entire plateau was surrounded with new fortification walls, made out of stone at the foundations and layers of brick at the top, because stone was scarce in the region.

Despite the aid from Syracuse, Gela was officially still in the Carthaginian *epikrateia* and was allied with Akragas, also under Carthaginian domination. When an internal power struggle in Syracuse brought Agathocles to power in 317 B.C., Gela gave refuge to the oligarchs who fled his wrath. Agathocles attacked with a small army, testing the new fortifications for the first time. Apparently, the walls were in fact breached and some of the troops entered the city, but, cut of from reinforcements, they were unable to capitalize on their advances. In the end, Agathocles was able to rescue the troops inside the city but had to call off the attack. He had more success in 311, when, as part of a campaign against Carthage, he captured the city, massacred thousands of its inhabitants, and took much of its portable wealth.

Suffering a defeat at the hands of Carthaginian forces at Ecnomus shortly afterward, Agathocles withdrew to Gela. He restored and elaborated the fortifications of the city to prepare for a possible Carthaginian attack. The walls were also heightened at this time to raise them farther out of the encroaching sands, which were building up everywhere along the southern beaches. Although Carthage never again attacked Gela, its military might finally forced Agathocles to sign a new treaty reaffirming the old boundaries of the *epikrateia*. Even with the treaty, peace and prosperity did not return to Gela. The town was partially destroyed in a Mamertine raid in 282 B.C. The Mamertines, mercenary soldiers who had taken to banditry and operated from a base at Messina, had no interest in anything but booty and did not occupy the city. The surviving Gelans appealed to Phintias of Akragas for aid. Phintias took them to a new settlement he named after himself (now Licata) and completed the destruction of Gela that the Mamertines had begun. A few people may have lived at Gela for the next 200 years or so, but by the beginning of the Common Era it was certainly deserted.

For the next 1,500 years, Gela was little more than a deserted ruin, buried deeper and deeper under the drifting sands. Nevertheless, the hinterland continued to be farmed. Excavations have revealed, for instance, the remains of a huge farmstead complex from Roman times at the mouth of the Gela River. The foundations of the buildings date to the time of Augustus and formed part of an estate that survived intact for centuries and was called the Calvisiana *massa*. Until the thirteenth century, little changed around Gela besides the fact that the farmers paid taxes to a succession of different foreign invaders, from the Romans, to the Arabs, to the Normans.

Then in 1233 Frederick II of Sicily, a Hohenstaufen, began a program of revitalization, seeing that the island's economy had begun a slow decline. As part of this program he resettled Gela, under the somewhat ironic name of Terranova di Sicilia, bringing Greeks and Lombards to create a new community. A modest medieval community developed on the ruins of the ancient Greek city. Some fragments of the ancient structures were incorporated into the medieval buildings, as was the case with the ruins of the temple of Athena, which was used in the Mother Church in the present-day Piazza Umberto I. Since the medieval community was not particularly prosperous, few buildings of any note have survived from that time, however.

The decline of Sicilian agriculture was never really reversed, in part because absentee landlords had little incentive to improve their estates. The area around Gela is something of an exception in this regard, because sometime in the sixteenth century the landowning Pignatelli family built a dam in the River Gela above what was then Terranova. In the eighteenth century, the family still continued their work of maintaining the infrastructure in the area. There are reports that Niccolò Pignatelli, duke of Monteleone and viceroy of Sicily from 1719 to 1722, paid for the upkeep on the river-banks to enable the local farmers to irrigate their land. These improvements saved the local inhabitants from some of the worst poverty suffered elsewhere in rural Sicily in recent centuries.

In 1928 Benito Mussolini devised a project of Italianizing Sicily and effected a number of name changes. Terranova was changed back to Gela. Apparently, the ancient Greek name was more acceptable than the Latin name bequeathed by the Austrian Hohenstaufen. Little else changed in the city, until, in July of 1943, the American army beached its invasion forces at Gela. Finding little resistance, the troops swept through the island in the course of weeks.

In 1948 the ancient fortifications at what is now called Capo Soprano were dug out of the sand, which had beautifully preserved whatever had been left standing by the city's last destroyers. Excavations also laid bare parts of public baths dating to the fourth century B.C., the only baths found in Sicily so far. The most important archaeological finds from Gela and the surrounding area are displayed in the local museum.

In the 1950s oil was found off the shore of Gela, and this finally provided the stimulus for new growth in the city. Within years, Gela once again became a city of note, with a major port and a large petrochemical factory complex situated just off the ancient acropolis beside the river. The juxtaposition of twentieth-century industry with the ancient ruins adequately sums up the strange vagaries of Gelan history.

Further Reading: A brief overview of Gela's history may be found in Margaret Guido's *Sicily: An Archaeological Guide* (New York and Washington: Praeger, and London: Faber, 1967). The first volume of *A History of Sicily*, entitled *Ancient Sicily to the Arab Conquest* (London: Chatto and Windus, and New York: Viking, 1968), by M. I. Finley, places Gela's history in the context of developments in Sicily generally.

—Marijke Rijsberman

Genoa (Genova, Italy)

Location: In northwestern Italy, on the Gulf of Genoa.

Description: A major seaport and industrial center, formerly an independent republic with colonies in the Mediterranean and Black Seas; noted for its art and architecture, including its black-and-white-striped medieval buildings; birthplace of Christopher Columbus.

Site Office: A.P.T. (Azienda di Promozione Turistica)
Via Roma, 11
CAP 16121 Genoa, Genova
Italy
(10) 581407/08

It is at least 2,600 years since a sheltered natural harbor on the northwestern coast of Italy first attracted settlers, the people, now known as Ligurians, who are regarded as the founders of Genoa (Genova in Italian). The city's modern name apparently derives from the Latin word *janua,* meaning "gate": suitably enough, since its merchants began in ancient times to control much of the trade in goods entering Italy from Greece and from Phoenicia (now Lebanon and Syria), and its survival has depended on seaborne trade throughout most of the centuries since.

By the third century B.C. the city of the Ligurians, which surrounded what is now the Castello Hill (from the Latin word "castrum," meaning "camp" or "fortified settlement"), had made an alliance with Rome, already the most powerful of the Italian city-states, and joined in its Punic Wars against Carthage, the North African city founded by the Phoenicians. Genoa was attacked and largely destroyed in 205 B.C., in the course of one of these wars. It was rebuilt and revived as a dependent of Rome, rather than an equal, a status it retained for several hundred years, through the rise and fall of the Roman Empire and even beyond it, serving in a crucial role at the center of several of the main commercial and military routes between the imperial city and its provinces to the north and the east. Like the rest of the empire, Genoa was Christianized during the fourth century, when its first cathedral was built and dedicated to the Twelve Apostles.

Around A.D. 642, nearly 200 years after the collapse of the Western Roman Empire, Northern Italy was invaded by the Lombards, who incorporated Genoa into their short-lived empire. Its disintegration (and absorption into the local population) was followed by years of danger and unrest as the city attracted the attentions of pirates and sought to reestablish its independence and prosperity in rivalry with other Italian cities. The first set of city walls was built in the ninth century, but did not prevent either an occupation by Moslem pirates (known to Europeans as Saracens) during the tenth century or the submission of Genoa to the Holy Roman Empire, based in what is now Germany, soon afterward.

Since Genoa was not located on the main inland routes for pilgrimage and trade in Italy, it had to compete with Pisa and, later, Venice, to develop its role as a focus of maritime trade. Its rise to prominence, beginning in the 1160s, largely depended on its trading post at Galata, a suburb of Constantinople (now İstanbul), the capital of the Eastern Roman (Byzantine) Empire. After the Genoese merchants there were attacked in 1171, apparently by Venetians, the court at Constantinople turned against its traditional ally Venice, punished all of its merchants collectively, and began to favor Genoa instead. Around this time a second set of city walls was built on the ruins of the old walls, from which there still survive one of the towers, the Torre degli Embriaci, and two gates, the Porta Soprana and the Porta dei Vacca, both completed between 1155 and 1157.

The twelfth century also saw the building of the magnificent Church of San Giovanni di Pres, from 1180 onward, and the rebuilding of the city's cathedral in the Gothic style and with a new dedication to St. Lawrence (San Lorenzo). Its tower, not completed until 1522, is covered in the black and white strips of marble that are characteristic of many buildings in the older areas of the city. The cathedral contains the basin, made of quartz, that is said to have held the head of John the Baptist after his execution, and a box, or ark, in which are kept what are said to be his ashes. This box, designed and made by Teramo Daniele and Simone Caldera, is one of the treasures displayed in the Museo del Tesoro di San Lorenzo, in the cathedral's crypt, alongside a cup, the Sacro Catino, which is said to have been first a gift to King Solomon from the Queen of Sheba, and then to have been used by Jesus at the Last Supper.

By the early thirteenth century Genoa had a population of about 50,000 and governed colonies, based on trading posts, not only at Galata but also on the islands of Corsica and Sardinia. Its rivalry with Venice and Pisa continued and intensified, leading to several naval conflicts in the late thirteenth century, which Genoa usually lost, until, having allied itself with the Spanish kingdom of Aragon and helped to finance Aragon's acquisition of Naples and Sicily, it scored its three great victories: over the short-lived Latin Empire of the East, created by Venice during the Crusades, in 1261; over Pisa, at the battle of Meloria in 1284; and over Venice, at the battle of the Curzolani Islands in 1298. These victories confirmed Genoa's strength as at least equal to that of Venice, through its control of outposts on the Black Sea (Caffa and Tana) and on the Aegean (Chios and Lesbos) and its participation in northern European trade in collaboration with Milan. It was also the Genoese, rather than any of their rivals,

La Lanterna, Genoa
Photo courtesy of A.P.T., Genoa

who in 1277 had first used the Atlantic Ocean to trade directly with northern Europe, adding their primacy in that area to their leadership of trade in the Mediterranean and the Black Seas. Yet another set of city walls was completed in 1320, to take in the rapidly expanding suburbs, which helped raise the city's total population to approximately 100,000 by the middle of the fourteenth century.

Unlike its rivals, Genoa did not build its prosperity on government action but on the initiative of its leading merchants who, for example, retained ownership of their fleets and resisted centralization, creating a culture both praised and criticized by later historians for its individualism. The city's political life revolved around the rivalries between factions formed among the noble families who dominated its trade and its rural hinterland. Chronic instability permitted brief periods of direct rule by the Holy Roman Emperor Henry VII, between 1311 and 1313, and by King Robert the Wise of Naples between 1318 and 1335 and culminated in 1339 in the installation of Simon Boccanegra as the city's first Doge (sometimes translated as duke or prince). His supporters claimed to represent the commoners of Genoa but his opponents preferred to submit to Giovanni Visconti, the powerful archbishop of Milan, whose forces soon restored the customary disorder.

It was while Genoa was under Milanese occupation that it received from the poet Petrarch (Francesco Petrarca), who visited the city in 1358, the nickname it still retains: Genova La Superba, Genoa the Proud. Ironically, however, there was less and less for its citizens to be proud of. Wars with Venice continued to absorb much of their energy and wealth, and although they seized control of Cyprus and Tenedos (now Bozcaada) during a conflict that ended in 1353, they and their Venetian rivals alike suffered enormous financial and military losses in the Chioggia War, between 1378 and 1381. Economically declining, militarily and financially weakened and still politically unstable, Genoa could no longer sustain a consistent independence. After submitting to the French between 1396 and 1409, Genoa came under the control of Milanese noble families for most of the fifteenth century, apart from brief periods of independence or French occupation. During these years effective local power passed to the city's creditors, organized as the Banco di San Giorgio. They sold off many assets, including the city of Leghorn (Livorno), which passed to Florence in 1421, but also sponsored many expeditions, such as that of the Genoese explorer Antoniotto Usodimare, who sailed as far as the mouth of the Gambia River in Africa in 1455. Another French occupation began in 1459, although this time it was to last only two years: the power of the Banco was to last much longer.

After the capture of Constantinople by the Ottoman Turks in 1453, neither the Banco nor the city could maintain contact with the Black Sea. The days of Genoa's imperial glory were over, although it retained Corsica for another 300 years. It is telling that the single most famous Genoese adventurer had to seek sponsorship elsewhere. Christopher Columbus was born in Genoa in 1451 and baptized at the Church of Santo Stefano. The building near the Porta Soprana that is now known as Columbus's house was built in the eighteenth century on the probable site of a house that may have belonged to his father and in which he may or may not have lived. Columbus died in 1506, after initiating the conquest of the Americas by Spain rather than by his native city; what are believed to be some of his bones can be seen in the City Hall, the Palazzo Doria Tursi.

Yet another period of French rule lasted from 1499 to 1512, as Genoa became the landing point for the French armies challenging the power of Spain over the Italian peninsula. The Genoese Republic was reestablished in 1528 by Andrea Doria, a member of one of the leading families involved in the Banco di San Giorgio. He arranged for his fellow nobles to hold political positions reflecting their economic power and protected them against the opposition of the lesser nobles and merchants who sought to overthrow him, notably in an attempted coup in 1547. Doria, who died in 1560, at the age of 94, is buried in the Church of San Matteo, long associated with his family. In 1576 the leading families overcame the resentments of the lesser nobles by incorporating them into the political system, giving them council seats and cutting their taxes; the republic was to last more than two centuries longer without any further changes in its constitution.

The renewed republic was founded on a new diplomatic and commercial policy. After centuries of close relations with France—usually friendly, but occasionally involving military occupation—the Genoese bankers had taken the side of France's main enemy, Holy Roman Emperor Charles V, whose possessions almost encircled France. In return, they developed a new role for their city as the meeting point of Charles's northern lands and his Italian possessions, and for themselves as lenders first to him, until his abdication in 1558, and then to his son Philip II, king of Spain and ruler of most of Italy. As these monarchs' German bankers collapsed, the Genoese influence increased, and it was further augmented during the Dutch Revolt, which caused the ruin of their northern rivals, the merchants of Antwerp. Genoa became the most important financial center in Europe (and probably in the world) by building an unchallengeable position in the buying and selling of the silver brought from the Americas to Seville in Spain, while ensuring that each time the Spanish Crown declared bankruptcy, as it did four times in the sixteenth century, their own interests in currency dealing, tax gathering, and landowning were protected.

Political stability and independence had been achieved at last and Genoa was prospering once more. The sixteenth century was another golden age for Genoa, or at least for its leading merchants and financiers, and many of its historic buildings and streets were started or rebuilt with their money and patronage. A new harbor tower, La Lanterna, originally a lighthouse, became and has remained a symbol of the city; the architect Galeazzo Alessi designed a magnificent new church, the Basilica dell'Assunta, and the Porta Siberia, a gateway that marked the entrance to the port: the

Strada Nuova, now called Via Garibaldi, on which the marble palaces of the leading families can still be seen, was laid out; and a new Palazzo Ducale (Doge's Palace) was begun, to grow by accretions over the years and acquire its neoclassical facade in the eighteenth century. The merchants and bankers also patronized the Genoese school of painting, which developed from the end of the century onward, at first under the influence of Peter Paul Rubens, Anthony Van Dyck, and other Flemish artists invited to the city, then through the works of such outstanding Italians as Caravaggio, Bernardo Strozzi, Ansaldo, Carbone, and Gregorio De Ferrari. Many of the most outstanding pictures painted by these artists and their pupils are displayed in the Palazzo Rosso and, facing it, the Palazzo Bianco on the Via Garibaldi. These buildings became public galleries after the Brignole-Sale family gave them to the city authorities in the late nineteenth century. Other Italian and Flemish pictures are displayed in the Galleria Nazionale di Palazzo Spinola.

Along with other northern Italian cities (and, indeed, much of western Europe), Genoa underwent a massive economic decline between 1600 and the 1680s. The volume of its sea trade fell by two-thirds; its silk industry collapsed; and still another French invasion, led this time by King Louis XIV in person, took place in 1684. Yet the republic survived, largely by performing useful commercial and financial services for both the Spanish and French monarchies and for their Italian client-states, and the city continued to grow, notably with the construction of the Strada Grande del Guastato, now the Via Balbi.

During the eighteenth century it became more and more difficult to resist the interference of the great monarchies. A period of occupation by the Austrian Habsburgs, who had succeeded their Spanish cousins as the main rival to French ambitions in Italy, was ended by a revolt in 1746, but Genoa lost its last colonial possession in 1768, when it sold Corsica to France, having found it impossible to suppress the rebellion of Pasquale Paoli, which had made the island ungovernable. Doria's republic was finally destroyed in 1796, when Napoléon Bonaparte led his French Revolutionary army into Italy. Genoa stayed outside the group of cities in northern and central Italy that formed the Cisalpine Republic in 1797; the city's pro-French leaders preferred to create their own "Ligurian Republic." This French dependency survived the destruction of the Cisalpine experiment by Napoléon's enemies in 1799 and Napoléon's creation of a new Italian Republic in 1802 (renamed a kingdom in 1804). By 1808, however, protection had become occupation, Genoa had been incorporated into France, and the anti-Napoleonic blockade had seriously damaged its economy. In 1818, three years after Napoléon's final defeat, the city was incorporated into the Kingdom of Piedmont.

Genoa now became the leading city of the Risorgimento, the movements for Italian unification primarily inspired by Giuseppe Mazzini, who was born in the city in 1805 and studied law at its university. He joined the secret society known as the Carbonari but after a brief imprisonment was

exiled to France in 1831; during his subsequent years of travel around western Europe he established and coordinated units of his own nationalist movement, La Giovine Italia, returning to Genoa in 1833 to organize its first rising (linked with one in Turin), which was scotched by arrests and executions before it could begin. He did not give up his fight even after the failure of the liberal revolutions that had broken out across Italy and most of Europe during 1848. Between 1853, when many of his followers were imprisoned, and 1857 he lived in hiding in Genoa, organizing his Action Party, only to see its energies dissipated in a hopeless adventure, the voyage of a group of rebels from Genoa to Sapri in southern Italy in 1857, which ended in their deaths. Garibaldi's expedition to Sicily and Naples in 1860, which was planned in and launched from Genoa, was more successful: the Kingdom of Italy was established in the same year. The Risorgimento is commemorated by a museum inside the Casa Mazzini, the house in which the nationalist leader was born and to which he frequently returned until his death in 1872.

From the 1890s onward the city that had once been the capital of a commercial empire became, by contrast, one of the centers of protest and organization by the rapidly expanding industrial working class. In 1892 various socialist and anarchist groups met there to create the movement that, at another meeting in Genoa in 1895, became the Italian Socialist Party. In 1900 the workers of Genoa showed their power by mounting a general strike that forced the city's prefect (the representative of the central government) to withdraw his ban on their union federation, the Camera del lavoro. Between the two World Wars Genoa remained the largest port in Italy but also further developed its leading role in heavy industry, especially steelmaking and engineering, sectors dominated by the Ansaldo company, which was later taken over by the state corporation IRI. For five days in January 1944 there was another general strike in Genoa, protesting the shooting of eight prisoners in retaliation for the killing of one German soldier, and on April 24, 1945, Genoa rose against the German occupiers, forcing the officer commanding the 15,000 German troops in the area to surrender the next day and remove the last remaining troops from the port the day after.

In the postwar Italian Republic two more risings showed that for many Genoese the anti-Fascist resistance was not yet a matter of history. In July 1948, after the Communist Party leader Palmiro Togliatti had been seriously injured by a would-be assassin, the local police fled as the city was taken over by angry workers on strike, two of whom died during hours of chaos that continued until the Communist leadership called for calm. In June 1960 marches and riots involving tens of thousands of protesters successfully prevented a proposed meeting in Genoa, organized by the neo-Fascist party, the MSI, at which the former Fascist governor of the city was to speak.

Industrialization has brought both prosperity and unrest to Genoa, but it has also brought enormous redevelopment that has changed the character of the city. The older sections are now separated from the port by an elevated highway; the suburbs have spread, mostly in the form of high-rise apartment blocks, along the coast in both directions; and the air pollution that has damaged many of the historic buildings has only recently begun to be combated. Like Venice, Naples, and other Italian cities, Genoa must now somehow solve the problem of balancing the protection of its heritage with the reform of its political life, tainted by corruption, and of its economy, damaged by the decline of the heavy industry that caused much of the pollution. Then it may once again claim to be Genova La Superba.

Further Reading: There is a great deal of interesting information about Genoa in such general histories of Italy as Giuliano Procacci's *History of the Italian People,* translated by Anthony Paul (London: Weidenfeld and Nicolson, 1970; New York: Harper and Row, 1971; as *Storia degli Italiani,* Bari: Laterza, 1968) and Harry Hearder's *Italy: A Short History* (Cambridge: Cambridge University Press, 1990; New York: Cambridge University Press, 1991), as well as in Alta Macadam's *Blue Guide: Northern Italy from the Alps to Rome* (London: Black, and New York: Norton, 1991), another in the outstanding series of historically reliable, well-written, and comprehensive Blue Guides.

—Patrick Heenan

Gibraltar

Location: At the southern tip of the Iberian Peninsula, overlooking the Straits that bear its name; to the west lies the Atlantic Ocean, and to the east the Mediterranean Sea. Africa is less than fifteen miles to the south. The Rock of Gibraltar is joined to the mainland by a narrow, sandy isthmus and forms the southeast part of a large bay.

Description: An important fortress during the Moorish occupation of Spain; besieged several times during the reconquest before being definitively captured by the Christians in 1462; scene of battle in the War of the Spanish Succession, after which it became a British property; it has remained so, despite numerous sieges and Spanish diplomatic efforts to regain sovereignty. Gibraltar's long history is visible in its Moorish castle, city walls, and, most distinctively, a colony of Barbary apes, possibly introduced by the Moors.

Site Office: Gibraltar Information Bureau
Duke of Kent House
Cathedral Square
Gibraltar
350-74950

The Rock of Gibraltar is a huge limestone ridge that rises 1,350 feet out of the sea. It is joined to the southern tip of the Iberian Peninsula by a narrow sand bank, and—with the Moroccan mountain Djebel Musa—dominates the Straits of Gibraltar, which connect the Atlantic Ocean and the Mediterranean Sea. The ancients believed that the two mountains were the work of the demi-god Hercules, hence the expression "Pillars of Hercules."

For many early eastern Mediterranean peoples the straits represented the end of the world. Stories of the horrors that awaited seafarers who crossed to the other side of the straits may have been exaggerated by traders who sought to protect the routes to the tin of Cornwall and the Scilly Islands (in Britain), or to the mineral-rich city of Tartessos in southwestern Spain. The entrance to the underworld was thought by some to lie in the western face of the rock. Ancient mariners in the area made sacrifices at Gibraltar, which then known as "Mons Calpe."

The Phoenicians established trading posts nearby at the beginning of the first millenium B.C., but they failed to settle Gibraltar. The lack of fresh water, timber for ship supplies, and pastureland for grazing animals made it unsuitable, despite being easily defensible. Instead they founded a city called Carteia, near what is now the town of La Línea in Spain. Carteia later became a Greek colony. The Romans also favored other nearby sites over Gibraltar during their six

centuries of dominance in the Mediterranean. Tingis (modern Tangier) and Saepta Julia (now Ceuta), both on the northern African coast and Portus Albus (now Algeciras, in Spain) became important, as did Julia Traducta (now called Tarifa, which is the southernmost point of the Iberian Peninsula).

Germanic tribes invaded Roman Spain during the fifth century A.D. and struggled for control of the area with the Byzantine Empire over the next 300 years. The Visigothic kings, who reigned from the city of Toledo, were surprised by the arrival of the Moslems in the spring of 711. The first group of invaders was led by Ṭāriq ibn Ziyād. Ṭāriq was challenged by the local Visigoth commander, Theodimir, but eventually succeeded in establishing himself in or around Gibraltar. Mons Calpe became known as Djebel Ṭāriq (Ṭāriq's mountain): the name was corrupted over time, and became Gibraltar. It may have been the Moslems who introduced to Gibraltar the Barbary apes that still live on the upper slopes today.

The Moslems took control of what is now Spain and Portugal, and maintained a foothold until the fall of Granada in 1492. For three centuries Gibraltar formed an insignificant part of al-Andalus—the name given to the large Moorish state based at Córdoba. After the breakdown of centralized power in the eleventh century Gibraltar became part of the kingdom of Seville. But the importance of the rock grew over the next four centuries, when it changed hands several times. It and Algeciras were used as ports by the Almoravids and the Almohads, Moroccan tribes who came to Spain in the tenth and eleventh centuries.

During the fourteenth and early fifteenth centuries the action in the wars of reconquest was centered upon the Gibraltar area. During this time Gibraltar was besieged eight times. At this time around 1,200 people lived in a small, walled town around the Moorish castle, which still exists. The first siege occurred in 1309, when Ferdinand IV of Castile found that his attempt to take Algeciras was hampered by the inhabitants of Gibraltar, who ferried in supplies across the bay. Gibraltar fell to the Christian commander Alonso Pérez de Guzmán, who was accompanied by the archbishop of Seville. But Gibraltar was lost to the Moslems in 1333, and Ferdinand's successor, Alfonso XI, died from the plague during a renewed attempt to capture the rock in 1350. The Moorish rulers allowed the Christian forces to take the body of their king back to Seville in peace. The sixth siege, of 1411, saw the forces of Granada triumph against their fellow Moslems from Fès.

For the next quarter-century Gibraltar flourished as a base for raids into Christian territory. This brought Henry, count of Niebla, to attack the rock in 1436. He was drowned while trying to withdraw, and his body was later hung from the town walls in a basket. Henry's corpse remained there

The Rock of Gibraltar
Photo courtesy of Gibraltar Information Bureau, London

until 1462, when the Christians finally retook Gibraltar under Juan Alonso de Guzmán.

After the end of the reconquest, the conflict between the Guzmán and Ponce de León families—who had taken opposite positions on who should succeed the ailing King Henry IV of Castile—focused on Gibraltar. The conflict was exacerbated by Henry's decision to consider Gibraltar property of the Crown. This move may have been instigated by the Ponce de León faction's influence at court. Henry visited the rock in 1462, and shortly afterward awarded it to Beltrán de la Cueva, a royal favorite rumored to have been the queen's lover. Locals rebelled at this move, and declared for Henry's half-brother, Prince Alfonso de Asturias, as king. Alfonso in turn named Juan Alonso, duke of Medina-Sidonia and a member of the Guzmán family, as governor, causing the ninth siege in April 1466: Medina-Sidonia finally breached the walls and took the keep after a siege of more than a year. With his death and that of Prince Alfonso in 1468, peace was made with the king, and the Guzmán ownership of Gibraltar was confirmed.

This decision was reversed with Queen Isabella's decision to make Gibraltar Crown property once again in 1501. At the same time she granted Gibraltar a coat of arms, whose castle and a key reflect the importance attached to the rock as the bastion of the straits. Isabella's reign also saw the expulsion from Spain of the Jews (1492) and the unconverted Moors of Granada (1499), many of whom crossed to Morocco from Gibraltar. In 1497 the rock was also used as the staging post for the invasion of Melilla, a port on the coast of Morocco that still belongs to Spain.

In September 1540 a Corsair pirate fleet operating from Algiers attacked Gibraltar. The pirates inflicted considerable damage on the town, and took hundreds of people hostage, before their squadron was destroyed by the Castillian fleet near the Alboran Rock (midway between Spain and Morocco, to the east of Gibraltar). Defenses were not improved until 1552, when Holy Roman Emperor Charles V (king of Spain as Charles I) sent the Italian engineer Giovanni Battista Calvi to the rock. Calvi built a wall on the south side of the city, which still exists and is known as the Charles V wall.

King Philip II's reign saw further improvements, under another Italian, Fratino. The period of Spain's imperial greatness also saw the construction of watchtowers in the area to warn against Moslem incursions on the coast. Gibraltar was used as a base for a rapid deployment force created at the end of the sixteenth century to meet the growing threat posed by pirates. Some pirates were Spanish Moriscos—converts from Islam—who were expelled from Spain between 1609 and 1612. Many of them used Gibraltar as a ferry base to reach their new homes in North Africa. English and Dutch pirates also raided shipping in the straits.

Further defensive towers were built during the ministry of Gaspar de Guzmán y Pimental, conde-duque de Olivares, and his concern for the area is shown by the fact that he included a sum for the defense of the straits in his will. Olivares accompanied King Philip IV's party on its visit to the rock in 1624. After arriving from Cádiz, the royal coach proved too large for the gate. To Olivares's dismay, the king had to enter on foot. The governor explained that the gate "was not constructed to let carriages in, but to keep the enemy out."

When King Charles II died without heirs in 1700, Europe was plunged into the War of the Spanish Succession. Louis XIV supported his grandson, Louis-Philippe (known as Philip V of Spain), a claim that was backed by most of the Spanish people. The Austrians, English, and Dutch joined an alliance in support of the Habsburg pretender, Archduke Charles. A combined Anglo-Dutch fleet cruising off the fleet of Spain in the summer of 1704 attacked and took Gibraltar after a brief siege involving the marine regiments under the command of Prince George of Hesse-Darmstadt. The Spanish commander Diego de Salinas had consistently complained to the Madrid government of the weakness of Gibraltar's fortifications, and his garrison succumbed after only four days' fighting. The rock was taken in the name of Charles III of Spain (the Habsburg pretender). Most of the Spaniards expelled from Gibraltar went to live at nearby San Roque.

Almost immediately there were attempts to regain possession of the rock. In the battle fought off Málaga in August 1704 British Admiral George Rooke prevented a fleet of fifty French ships from reaching the Bay of Gibraltar. Spanish troops began to build siege works in the autumn of 1704, and in November a raiding party was led up the steep eastern face of the rock by a goatherd named Simon Suarte. Although they managed to reach the top of the rock, they were discovered by a drummer boy taking food to the lookout and were driven away by the English garrison. The path was subsequently destroyed by English engineers.

The War of the Spanish Succession dragged on until 1713, when the Treaty of Utrecht was signed. Article X of the Treaty deals with Gibraltar, which was ceded to Britain. It begins "The Catholic King does hereby, for Himself, His heirs and successors, yield to the crown of Great Britain the full and entire propiety of the Town and Castle of Gibraltar, together with the port, fortifications, and forts thereunto belonging; and He gives up the said propiety, to be held and enjoyed absolutely with all manner of right for ever, without any exception or impediment whatsoever." Article X states that possession of the rock can only be passed on to the Spanish Crown and also includes provisions to exclude both Jews and Moslems from Gibraltar (both having previously been expelled from Spain). Despite this both religious groups came to flourish in Gibraltar and several Sephardic Jewish families that had been living in Morocco returned to the Iberian Peninsula in this way.

Britain's gains at Utrecht marked a decisive shift in the balance of global power. Historian John Lynch says that the "mere presence of a (British) squadron in Gibraltar was enough to disrupt the Indies trade" on which Spain was so dependent. Not unnaturally, Gibraltar was used as a bargaining piece during the military and diplomatic manuevering that characterized European politics of the eighteenth century. In its many wars against Britain, France could always tempt Spain into an alliance with a promise of help in recovering Gibraltar. The island of Minorca, which had also been taken during the War of the Spanish Succession and threatened the French naval base at Toulon, was used in a similar way. In the second Bourbon Family Compact of 1743, for example, Louis XV of France undertook to win Italian territories for the Spanish crown, to free Spain from the commercial restraints imposed in 1713 and to support it in the reconquest of Minorca and Gibraltar.

The years between the brief thirteenth siege of 1727 and the fourteenth, which began in 1779, were also filled with wrangling over the exact interpretation of the terms of the Treaty of Utrecht. The Spaniards felt that Britain had extended Gibraltar's fortifications beyond their original limit and into the neutral zone of the isthmus. Such disputes continue to the current day, a problem exacerbated by the fact that the area in question is now occupied by Gibraltar's airport, which is the largest in the area. Spain was persuaded by France to join in the war supporting the American colonists against Britain. Initially the threat appeared to be to Britain itself. But this faded, and when, in 1779, the sultan of Morocco—previously an invaluable source of supplies to Gibraltar—was persuaded to withdraw his support, the pressure on the rock became intense. Hostilities began in June, and lasted until September 1782. The Spaniards launched a massive bombardment, known as the Grand Attack, on September 13. But their floating batteries were destroyed by British red-hot shot, and the garrison of some 7,000 men held out. It was at this time that the excellent defensive qualities of the rock were first fully exploited. Siege tunnels drilled into the rock to house cannon during the siege can still be seen. Many of the streets in the town itself show their military origins, such as Upper Castle Road, Cannon Lane, Flat Bastion Road, and Bomb House Lane.

France and Spain again threatened not only Britain, but also the stability of Europe, during the Napoleonic Wars. On October 21, 1805, the battle of Trafalgar was fought off Cape Trafalgar, some fifty miles to the west of Gibraltar. The British under Lord Horatio Nelson destroyed the combined Franco-Spanish fleet commanded by Villenueve, who against his better judgment had left secure moorings at Cádiz. Nelson himself was killed at the moment of victory, and his body was brought back to Gibraltar in a cask of brandy held in the dismasted flagship, *HMS Victory*. (The *Victory* can still be seen in the British town of Portsmouth, while brandy is sometimes referred to as "Nelson's blood" by British seamen). Spain took Britain's side and abandoned the alliance with Napoléon after the French invasion of 1808, which marked the beginning of the Peninsular War, known to the Spanish as the War of Independence. Gibraltar became an important operations base, with numerous sorties directed against the French blockade of Cádiz.

Whereas the first century of British occupation was marked by continual war, the second century was characterized by peace and a prosperity born of the development of commerce. Spain's decline and Britain's colonial dominance meant that any attempt to retake Gibraltar was unthinkable. The Spanish prime minister Leopoldo O'Donnell found his room for manuever was restricted by Britain's refusing to countenance any conquests on the African coast facing Gibraltar.

With the expansion of trade in the Mediterranean, particularly with the opening of the Suez Canal, natives of other ports came to Gibraltar to work. Genoese and Maltese are particularly evident in the nineteenth-century censuses, along with substantial numbers of Britons and Spaniards, and a growth in Gibraltar's Jewish population. Gibraltar became a British Crown Colony in 1830.

By the end of the nineteenth century many strategists felt that Gibraltar was no longer tenable as a fortress and naval base. Weapon technology, they believed, had advanced to such an extent that the rock was vulnerable to attack from large guns concealed in the surrounding land. In view of the situation, proposals were made that Britain accept a long-standing Spanish offer to exchange the colony for Ceuta, a Spanish town in North Africa. However, Gibraltar held such an important place in the psyche of British voters since its capture that no such action was taken.

The naval arms race that preceded World War I saw the construction of several large dry docks in the early 1900s. The were named after Queen Alexandra and the prince of Wales during their visits to Gibraltar in 1905 and 1906 respectively. Another royal, King Manuel II of Portugal, was forced to flee to Gibraltar in the Royal Yacht after the Portuguese uprising of October 1910, and went from there to London. During World War I the Spanish conservatives favored support of Germany, who courted Spain with promise of aid in retaking Gibraltar. Spain remained neutral, however, and Gibraltar's hosptials were used extensively for treating casualties. Armistice celebrations in the colony were muted by funerals for those who perished in the sinking of *HMS Britannia* by a German submarine off Cape Trafalgar.

During World War II General Franco and Hitler cooperated in the Russian campaign, but never reached agreement about an attack on Gibraltar. A total of 16,700 people were evacuated from Gibraltar during that war, most to Northern Ireland, but some to Madeira and Jamaica. Others moved to Tangier and Spain. At this time the British naval group Force H operated from the rock, and was especially involved in running convoys to Malta. General Eisenhower directed Operation Torch from Gibraltar. The rock's new airport was instrumental in ferrying the necessary men and materials needed in the Allies' North African campaign.

The decolonization policy adopted by the United Nations meant that the position of Gibraltar was reviewed by the international community in the postwar years. Spain used the UN to bring pressure on Britain to relinquish the colony, but failed to win significant concessions. Full internal self-government was introduced, perhaps as a way of reducing diplomatic pressure. The 1967 referendum on sovereignty confirmed the status quo. A total of 12,138 residents favored continuance as a British colony, against only 44 in favor of reverting to Spanish sovereignty. The following year a small group of pro-Spanish lawywers and businessmen, called the "doves," were driven from Gibraltar amid considerable violence after expressing their views. Shortly afterward the Franco government closed the frontier.

Spain's entry into both the European Community and NATO led to the reopening of the border in 1985. However, the growth of a large offshore banking sector, together with large-scale tobacco smuggling has led to renewed tension during the mid-1990s. British sovereignty on Gibraltar continues to be an irritant to Spain.

Further Reading: Many of the books written on Gibraltar are biased toward either the Spanish or the British/Gibraltarian position. Even the best all-around work, *The Rock of the Gibraltarians* (Madison, New Jersey: Fairleigh Dickinson University Press, 1988; Grendon, Northamptonshire: Gibraltar Books, 1990), written by the rock's former governor, Sir William G. F. Jackson, announces its viewpoint on sovereignty in its title. J. J. Alcantra's *Medieval Gibraltar* (Gibraltar: Medsun, 1979) provides an account of Gibraltar's tumultuous role in the Christian reconquest.

—Richard Bastin

Gijón (Oviedo, Spain)

Location: On the northern coast of Spain, on the Bay of Biscay, about 235 miles north-northwest of Madrid, in Oviedo province, in the Asturias region.

Description: Industrial city of approximately 250,000 people; birthplace of Enlightenment thinker Gaspar Melchor de Jovellanos; a stronghold of political radicalism in the early twentieth century, leading to its major role in the Spanish Civil War, when it was badly damaged.

Site Office: Oficinas De Turismo
Marques de San Esteban, 1
33206 Gijón, Oviedo
Spain
34 60 46

It was not until the last half of the nineteenth century that Gijón began developing into a large industrial center. This growth was linked to the intensive exploitation of Asturian coal deposits and the processing and export of this mineral. Throughout most of its history, Gijón was a relatively small seaside town and port. Its history is woven into the fabric of the history of Asturias, although occasionally Gijón emerged to play a role in some of the more momentous events in Spanish history.

Very little is known about the site of Gijón before the Roman era in Spain. Artifacts dating from Paleolithic times have been uncovered in the area of Cabo de Peñas (just west of modern Gijón) and in the valley of the Nalon River, at sites southeast of the modern city. The Roman conquest of Spain began during the last decade of the third century B.C., after the defeat of the Carthaginians in the Second Punic War. At this time Spain was inhabited by Basque, Celtic, Iberian, and Celtiberian tribes. By 26 B.C., Roman legions had conquered most of the Iberian Peninsula, with the exception of the region north of the Cantabrian Mountains, which included Asturias. In 26 B.C. Augustus led seven legions through these mountains to subdue the Iberian tribes inhabiting the region. Most of the tribes were defeated within a year, but revolts occurred sporadically until approximately 16 B.C.

Sometime after the pacification of Cantabria and Asturias, the site of Gijón was occupied by the Romans and a settlement known as Gigia was established. It is unclear whether the Romans were responsible for creating an original settlement or whether they merely occupied a pre-existing Iberian village. The village was laid out directly at the foot of Mount Santa Catalina, where the fishing quarter is located today. A map dating from the mid-seventeenth century indicates that for more than 1,500 years the extent of the city remained much as it was during Roman times. Physical evidence of the Roman occupation can still be seen at the Palacio de los Valdés, where ruins of Roman baths dating to the first century A.D. may be viewed in the basement.

Rome's hold on Spain ended during the fifth century, with the invasion of the Alans, Vandals, Suevi, and Visigoths. Of these peoples, the Suevi and Visigoths established lasting kingdoms on the peninsula. The Suevi dominated the western part of the peninsula, including Asturias and thus Gijón, and the Visigoths controlled most of the rest. By 585 the Visigoths, under King Leovigild, had conquered the Suevian kingdom and established themselves as the masters of Iberia.

In 711 the Moorish conquest of the peninsula began. In less than a decade all but Asturias had fallen to the Moors. The Moslem expedition in Asturias was to be launched from Gijón, where in 714 a garrison had been established by Munusa, a Moorish commander. In 722 Munusa dispatched a Moslem force from Gijón under Alaxman to attack the Asturians in their mountain stronghold at Covadonga, about thirty-five miles east-southeast of Gijón. The Asturians, led by their Visigothic chieftain Pelayo, rebuffed the Moors. After Alaxman's defeat, Munusa fled Gijón and was later captured and killed. The battle at Covadonga, which most historians consider a minor skirmish, soon acquired epic proportions and symbolized the beginning of the Christian reconquest. The Moors were never to return to the region, and Gijón was destined to remain the only locale in Asturias that had been occupied by the Moorish invaders.

Facing no Moslem resistance, Pelayo was able to establish an Asturian kingdom that soon came to include parts of neighboring regions. Gijón prospered under the first Asturian kings. In 844 the city's defenders repelled an attack by Norman invaders, who went on to occupy Cádiz and attack Seville. In the tenth century the Kingdom of Asturias became a part of the Kingdom of León. It was under Leonese kings that in the late twelfth century Gijón's port was first developed. Gijón, however, remained relatively small in comparison with the other towns of Christian Spain. By the fourteenth century Salamanca had a population of about 15,000; Valladolid, 25,000; and Barcelona, 48,000. Gijón's population would not reach 5,000 inhabitants until the end of the eighteenth century.

In 1390 King John I of Castile died and left an eleven-year-old heir, who became Henry III of Castile. It is almost a truism of medieval history that trouble ensued when a minor inherited a throne, and Henry's reign was no exception. Gijón surfaced as a base of operation for Henry's chief rival, Alfonso Henriques, who was Henry's uncle. In 1395 Gijón was razed by forces loyal to Henry. The city was not fully rebuilt until 1410.

Roman baths at Gijón
Photo courtesy of Instituto Nacional de Promocion del Turismo

Gijón's harbor was improved significantly in the sixteenth century under Philip II. In 1588 it played a small role in Philip's ill-fated attempt to gain mastery over the English. In July of that year, as the Spanish Armada prepared to round the Galician coast, a violent storm arose and many of the ships were scattered. Two warships found their way to the harbor at Gijón and put in to weather the storm. Gijón's connection to the operation had not ended yet, for in October the city was one of several northern Spanish ports that received remnants of the battered Armada on its return to friendly waters. The port slowly grew in importance, especially after 1778, when trade was opened with the former British colonies in North America.

Meanwhile, in 1744 Gijón witnessed the birth of its most famous native son, Gaspar Melchor de Jovellanos, who became one of Spain's more notable Enlightenment writers and statesmen. Although he spent most of his life outside of Gijón, his connection to the city remained strong. In 1794 he founded the Asturian Institute at the house in which he was born. The institute was an educational center dedicated to providing a progressive education for all people regardless of class. Today the house is a library and a museum displaying works of Asturian artists. It was in Gijón that in 1795 Jovellanos wrote his seminal work, *Informe en Expediente de ley Agraria,* a manifesto for agricultural and mercantile reform. Gijón takes great pride in Jovellanos, and a monument to him is located at the Plaza del 6 de Agosto, a square in the heart of the new part of the city.

Jovellanos was also a Spanish patriot. In 1808 he became a member of the national junta that opposed the Bonapartist regime in Spain. It was also in that year that Spanish deputies departed from Gijón to seek the help of

Great Britain against the occupying French. The British responded by diverting an army, which was meant to go to South America, to Spain to open a campaign on the Iberian Peninsula. Asturian guerrillas, along with compatriots from other parts of northern Spain, effectively tied down most of the French forces who were stationed in the north, thus allowing British troops to land along the coast and conduct raids inland. Some cities, however, fell to the French, including Gijón, which was sacked and plundered.

The pivotal period in the economic and social history of Gijón occurred during the middle and latter parts of the nineteenth century. Of crucial importance was the construction of viable transit lines from the Asturian hinterland to the port of Gijón. These lines allowed Asturian coal to be moved readily to other parts of Spain and to be exported. They also fostered the development of regional industrialization and created a marked shift of Asturias's population from rural to urban areas.

In 1829 the National Mining Office sponsored a study focusing on the development of Asturias's abundant coal reserves. The commission charged with conducting this study suggested that a road be built connecting Gijón with the mining region to the south. It was not until 1838 that construction of a road linking Gijón to the Langreo mining area was begun. Alejandro Aguado, a mine owner, financed the project, which was completed in 1842. Coal shipments through Gijón doubled but transportation costs remained high. What was needed was a railroad linking Langreo with Gijón.

In 1844 the Duke of Riánsares, the husband of Queen Regent María Cristina de Borbon and new owner of the mines in the Langreo Basin, proposed building such a railroad. The duke's wealth, coupled with the political influence of his wife, surely put him in good standing, and a government concession was granted the following year; construction began shortly thereafter. The Langreo-Gijón line began operating in 1855. Much of the coal coming into Gijón was exported or shipped to other parts of Spain. Some, however, was used to fuel the foundries of the region's emerging iron industry, which by the 1870s made Asturias Spain's leading iron-producing region. Gijón's link with the Spanish interior was greatly improved when the Langreo-Gijón line was linked to the rest of the Spanish railway network in 1884. As production of coal and iron increased it became evident that Gijón's port was inadequate. A new and larger outlet, the port in Musel, was opened in 1912.

As Gijón's economy grew, so did its population. In 1900 the city had about 47,500 inhabitants. This represented a more than 900 percent increase over the city's population at the end of the eighteenth century. Politically, the city began to take on a liberal identity. The first socialist organizations in the region were founded by the dockers and metal workers of Gijón in the early part of the twentieth century. By the 1930s Asturias had became known as a radical region, with highly organized socialist, communist, and anarchist groups. Gijón was especially noted for its anarchists.

During the early 1930s the international coal market collapsed and caused great hardship throughout Asturias.

Fearing greater hardship under a conservative government, in 1934 the national leadership of the Socialist Party called for an insurrection to block the inclusion of a right-wing party in the Spanish cabinet. Only Asturias responded, and in October of that year colonial troops were ferried from Morocco to Gijón to suppress the uprising. The Asturian revolt and its brutal suppression solidified the radical parties in the region into a unified bloc.

National elections in 1936 resulted in a left-wing victory. The military, in alliance with other conservative elements, rejected the electoral mandate and revolted in July 1936, thus initiating the Spanish Civil War. Gijón, along with most of the rest of Asturias, remained loyal to the government. However, the nationalists, as the forces opposing the government were known, retained an armed contingent in Gijón. In July and August the workers of Gijón laid siege to the Simancas Barracks, home of 180 members of the nationalist contingent. In an effort to help lift the siege the nationalist cruiser *Almirante Cervera* anchored three miles outside of Gijón and began shelling the city. The shelling was relentless, lasting six hours a day for fifteen days. Much of the city was damaged as entire districts were destroyed. Among the casualties was much of the library of the Asturian Institute, which included drawings by old masters such as Dürer and Velázquez. Despite the shelling, the barracks fell to loyalists in mid-August.

By October nationalist forces had gained control of Oviedo, the Asturian capital. However, nationalist strength in the north was limited, especially after the defeat in Gijón. The tide turned in the north during 1937. Bilbao fell in June, Santander in August. By September Asturias (with the exception of Oviedo) was the only loyalist holdout in the north. On September 1 the nationalists launched an offensive north from Oviedo under General Antonio Aranda, who joined the Navarrese forces of General José Solchaga in early October; these combined forces attacked Gijón, and the city fell on October 21. The fall of Gijón completed Franco's conquest of the north. A cruel repression followed as prominent loyalists were hunted down.

After the Civil War, the city was slowly rebuilt. Its population continued to increase: in 1950 it had about 111,000 residents, and by 1970 it had a population of about 185,000. Along with the reconstruction of the city, the founding of the Universidad Laboral, in 1955, ranks among the city's most significant post-Civil War accomplishments. Gijón also boasts a number of museums.

Further Reading: In the absence of a definitive work on Gijón, one must rely on scholarly histories of Spain as well as travel guides to gain an understanding of the city. Scholarly accounts of Roman and medieval Spain that contain good sections on Asturias include *Roman Spain* by Leonard A. Curchin (London and New York: Routledge, 1991), Harold Livermore's *A History of Spain* (London: Allen and Unwin, and New York: Farrar, Straus, and Cudahy, 1958), *A History of Medieval Spain* by Joseph F. O'Callaghan (Ithaca, New York: Cornell University Press, 1975), and Bernard F. Reilly's *The

Medieval Spains (Cambridge and New York: Cambridge University Press, 1993). A well-researched account of the economic development of Asturias and Gijón during the nineteenth century is contained in an essay by Rafael Anes entitled "Early Industrialization in Asturias: Bounds and Constraints" in *The Economic Modernization of Spain*, edited by Nicolas Sanchez-Albornoz (New York: New York University Press, 1987). As a general reference for Spanish history during the eighteenth, nineteenth, and twentieth centuries, the *Historical Dictionary of Modern Spain, 1700–1988*, edited by Robert W. Kern and Meredith D. Dodge (Westport, Connecticut, and London: Greenwood, 1990), is a valuable guide. Hugh Thomas's *The Spanish Civil War* (London: Eyre and Spottiswoode, 1961; New York: Harper, 1963) remains a classic work on the conflict and contains useful information on Gijón, as does Frank Jellinek's *The Civil War in Spain* (New York: Fertig, 1969).

—Jeffrey M. Tegge

Gordium (Ankara, Turkey)

Location: At the confluence of the Porsuk and Sakarya Rivers, near the town of Polatlı, in Ankara province, about sixty miles southwest of the city of Ankara.

Description: Archaeological site with the ruins of the ancient capital of Phrygia; inhabited from 6300 B.C. to A.D. 189; connected with the legendary Gordian Knot and with King Midas.

Site Office: Gordium Museum
Yassihoyuk Koyu
Polatlı, Ankara
Turkey
(312) 5711422

The remains of the ancient Phrygian capital, Gordium, were discovered near the small town of Polatlı about sixty miles southwest of Ankara. Gordium, whose original name is unknown, was located at the intersection of ancient trade routes, where the Porsuk River flows into the Sakarya River. Known in Turkish as Yassıhüyük (the Flat Mound), it is one of the most important archaeological sites in Turkey. Excavations there by American archaeologist R. S. Young from 1950 to 1974 revealed evidence of occupation from 6300 B.C. to about A.D. 189. Artifacts were found from the early Bronze Age III (2400–2000 B.C.), a time when Anatolia (the western portion of Asian Turkey) gained prominence in the Middle East because it had both the resources and the technology to make bronze items.

Finds dating from the early Hittite period have been excavated in Gordium in the archaeological layer immediately above that belonging to Bronze Age III. The Hittites came to Anatolia around 2000 B.C. and settled in a territory centered on Kussara and Hattusa. Gordium, near the ancient trade routes from northwest Anatolia that led across the straits to Europe, was at the far western edge of the Hittite dominion by the fourteenth century B.C.. It remained under Hittite rule until their empire collapsed about 1200 B.C. The cause of the Hittites' downfall remains unclear, but the empire appears to have been weakened both by internal problems, such as a loss of control over its vassal states, and by external ones, including invasions by the Sea Peoples.

Control of Anatolia passed to the Phrygians, who made Gordium their capital. No sites inhabited by Phrygians between the twelfth and eighth centuries B.C. have been found. The exact origin of the Phrygians is not known, and little was known about their civilization until Gordium was excavated. Phrygian is a name applied to these people by the Greeks, who believed the Phrygians came from southeastern Europe sometime before the Trojan War in the middle of the

thirteenth century B.C. The only source of written history concerning this period comes from the Assyrians, the Phrygians' enemies. The names Muski and Tabal, mentioned in the Assyrian annals beginning about 1160 B.C., are thought to refer to Phrygian peoples and lands. The Hebrew Bible also mentions both Tubal (Tabal) and Meshech (Muski). According to some accounts, their dominion reached as far as the upper Tigris and Euphrates, but no archaeological evidence has been found to indicate their presence in this area. The Phrygians periodically attacked, and were attacked by, the Assyrians, but the city of Gordium developed peacefully.

When the historical record resumed, the Phrygian civilization had taken on an Anatolian character. The Phrygians had established an empire extending from the Taurus Mountains and the kingdom of Urartu (Armenia) on the east to the Lydian border on the west. With abundant resources and an advantageous location, they flourished and developed an advanced civilization. Details of their economy are not known, but decorations on furniture, pottery, and other artifacts show cattle, ducks, geese, and goats. Breeding of livestock, particularly horses, was important, as was the breeding of sheep for wool. There is also evidence of trade with the east, in glass vessels, linen, and cauldrons. It may have been protection of east-west trade routes that sparked some of King Midas's conflict with other powers.

Assyria continued its activities in Anatolia, and about 836 B.C. the Phrygian land of Tabal was defeated and forced to submit to Assyria. In the battles between Assyria and Urartu, Mita, king of Muski, remained loyal to Assyria, but he later joined Tabal and other kingdoms in an alliance against the Assyrians. Tabal was defeated and became an Assyrian province in 712; Mita was defeated but not conquered, the only survivor of the alliance. The historical record is murky about this point, but it seems he retreated northwest out of the Tigris-Euphrates region and out of Assyrian-controlled central Anatolia. He settled in Gordium and became known to the Greek world as King Midas.

Several Phrygian rulers were named Midas, and Gordium is best known today for the legends associated with a mythical King Midas who appears to be a composite of the real rulers. In one such legend, an oracle predicted that a poor man would come to Gordium in an ox cart and become ruler of the Phrygians. Shortly thereafter such a man, a farmer named Midas, entered the town. Since King Gordios had no heir, he named Midas his successor. (Some versions of the story say chariot rather than cart; others say the farmer was Gordios, for whom the city was named after he became king.) Midas placed his cart in the Temple of Cybele, the Phrygian mother goddess, with the reins tied in a knot so intricate that neither end was visible. It was said that whoever could untie

City walls at Gordium
Photo courtesy of Embassy of Turkey, Washington, D.C.

this "Gordian knot" would rule the world. The knot remained for centuries until the coming of Alexander the Great, who cut it with his sword.

There are other familiar legends involving Midas. One story says the god Apollo and the Phrygian river satyr-god Marsyas had a flute competition with Midas as the judge. Apollo was angered when Midas chose Marsyas as winner. He had Marsyas skinned alive and caused Midas to grow donkey ears, which Midas covered with a special hat. Midas's barber, who knew about the ears, was sworn to secrecy, but anxious to tell someone, whispered it to the reeds of the river. (Another version says he dug a hole, whispered the secret, covered it with dirt, and reeds grew spontaneously on the spot.) Forever after, the reeds murmured, "Midas has ass's ears," in the wind.

Another story involves the famous "golden touch." Midas poured wine into a spring, made Silenus the water demon drunk, and held him for ransom from Dionysus. (In another version, Midas entertained Dionysus and gained his favor.) The god granted Midas one wish, which was that all he touched be turned to gold. He soon regretted the choice, when his food and his daughter became gold. He begged Dionysus for a cure and was told to wash his hands in the Pactolus River, which then ran with gold. Some historians believe the early use of brass by the Phrygians may have contributed to the legend; brass (an alloy of copper and zinc) is much shinier and looks more like gold than does bronze (copper and tin), the more prevalent alloy at the time.

The last of the Phrygian kings (probably 725–696 B.C.) was also named Midas. He is mentioned by the Greek historian Herodotus as the first non-Greek ruler to pay homage to the god Apollo at Delphi and dedicate a throne to him. His dynasty was finally toppled by the Cimmerians, who attacked Midas several times before overwhelming his forces in 696 B.C. Gordium was destroyed, and it is said the Cimmerians forced Midas to commit suicide by drinking bull's blood. As a result of the end of Midas's dynasty, Anatolian culture was supplanted by Aegean culture.

The excavations at Gordium revealed hitherto unknown details of Phrygian architecture. The site consists of a citadel atop a low acropolis. The town extended from the foot of the hill to the banks of the Sakarya. At the southeastern edge of the acropolis was found a well-preserved eighth-century B.C. city gate, thought to be the largest pre-classical building in Turkey. The gate was so immense that the ruins, obviously much reduced from the original, are thirty-three feet high. The gate was deeply recessed in the walls and flanked by twin towers from which invaders could be repelled.

The citadel has walls more than forty feet high and a gate reached by a narrow ramp. In the center, facing a series of paved courtyards, is the palace complex of ten buildings. The buildings are in the megaron style, previously identified with Early Bronze Age Anatolian sites such as Troy. The typical megaron was built of mud-brick or stone on a timber frame, with an open vestibule leading to a large rectangular hall with a hearth. Archaeologists are fairly certain that each building had a gabled roof.

The palace had four main megara. The oldest (dating to 760 B.C.) and largest megaron was built on a high stone foundation and topped with a wood, reed, and clay roof. It consisted of a great hall about sixty by fifty-two feet, which probably had a central hearth, and a wooden gallery that ran along three walls, supported by two rows of four wooden posts. There is evidence the gallery may have been used for weaving and storage. Charred remnants of furniture suggest the room was the central chamber in the palace. Geometric borders, depictions of herdsmen driving cattle, and ivory inlays decorate some fragments.

One smaller megaron has a red, white, and blue geometric mosaic floor made of pebbles, representing the earliest example of that particular technique. The design is reminiscent of textile designs and may be an imitation of a carpet, as the largest megaron had an undecorated floor that may have been covered with rugs. Another megaron was probably the Temple of Cybele, home of the Gordian knot. Behind the palace were more megara facing each other along a street leading to the river. Some were probably servants' quarters, as there were cooking facilities, grinding platforms, and weavers' workshops. Others may have housed soldiers.

In the plains below the mound, ninety tumuli (barrow graves) have been located, making the area one of the largest Iron Age burial sites in Turkey. Three, dating to the same period as the main buildings in the citadel, the height of Phrygian culture, were excavated by Young. Each consisted of a timber-frame burial chamber in the form of a house, which was roofed over and then covered with rubble. A protective layer of clay, still intact on many of the tombs, was applied next, then earth added to a height of as much as 200 feet. The earth not only protected the contents, but made the tomb stand out from the plain as a monument. Only one person was buried in each mound, although on occasion Phrygians buried a pair of horses with the dead. One excavated tumulus of moderate size was probably the grave of a princess. In another tomb a young prince was buried with wooden toy animals and other items.

The largest tumulus, Büyük Hüyük, is called the Tomb of Midas (or sometimes Gordios). It is uncertain who was actually buried there, although there is no doubt it was a king. Material in the tomb is from 750 to 725 B.C., which predates Midas, but is too late for Gordios. One of the largest tumuli in Asia Minor, the mound is almost 180 feet high, and

thought to have been about 260 feet before erosion. It is 820 feet wide, and dominates the surrounding area. Its huge size, which probably protected it from grave robbers, also made excavation difficult. The only way it could be accomplished was by drilling a shaft about 150 feet down from the top to locate the cedar- and juniper-framed burial chamber and then tunneling in 230 feet. The grave held the skeleton of a sixty-year-old man about five feet, two inches tall who was buried on a huge bed with twenty rich coverlets.

The tomb contained elaborate furniture inlaid with rare wood in intricate patterns. There were also 178 bronze vessels of different sizes, including 3 giant copper cauldrons on iron tripods. One was plain; one had two busts of bearded men and two female figures on its handle attachments; and the third was decorated with winged female figures known to the Greeks as the Sirens. The Sirens were not part of Phrygian mythology, so the vessel was either imported or copied from an import. Other bronze vessels included two buckets with Urartian or Assyrian designs of rams' and lions' heads, and bronze cups shaped like lotus flowers.

Seventy bronze fibulae (safety pins), a Phrygian invention, were the only items of jewelry adorning the king. Considering the Golden Touch legend and the custom of burying kings with both riches and weapons, the absence of gold, silver, or stone-encrusted swords remains a mystery. Some fibulae and vessels were made of brass rather than the traditional bronze, one of the earliest brass finds.

An alphabetic script, not yet deciphered, was found on five bronze vessels. The letters are inscribed on strips of beeswax affixed to the bronze. This technique is related to discoveries in Assyria dating to the same period, in which inscriptions were found on a "book" with waxed pages. The Assyrian texts were cuneiform rather than alphabetic, but the use of wax suggests it was a prevalent technique in use at the time and may explain the lack of surviving records from the "dark age" beginning in the Middle Bronze Age. When written texts again surface, they are in an alphabetic rather than hieroglyphic or cuneiform system. The fact that the writing found at Gordium was alphabetic rather than cuneiform caused experts to rethink hypotheses about the origin of the Greek alphabet. It was traditionally thought that Greeks visiting ports in the Near East brought back the Phoenician alphabet. Alphabetic finds at Gordium, however, predate any known imports from Greece. Pottery of Greek origin does not appear at Phrygian sites until about 650 B.C., although there is a significant amount of other imported ware. Thus, it is possible the alphabet came to Phrygia first and spread afterward to Greece. This theory is supported by Midas's reported marriage to a princess in Aeolis and the fact that Greek culture at the time was in its early stages and was unlikely to have influenced the stronger, more established culture of Phrygia. There is also one character in the Phrygian alphabet that is not in the Greek alphabet but was in use in some of the Greek cities.

On the plain near Gordium is an imposing Phrygian monument, the Monument of Midas, called Yazılıkaya (Inscribed Rock) in Turkish. It is also known as the Tomb of Midas, but it is not a tomb at all. It consists of a gabled temple facade carved on a vertical rock face and sculpted to look like terra-cotta tiles. (This carving also advances the premise that the Phrygian megara had gabled roofs.) This Yazılıkaya should not be confused with the larger and more famous monument also called Yazılıkaya, a temple site with reliefs carved into the walls of two ravines, near the ruins of Hattuşa.

Following a period of Cimmerian rule in the early seventh century B.C., Gordium was rebuilt as the capital of a smaller, revived Phrygia. The Phrygians were greatly weakened, however, and Lydia gained control of Anatolia and occupied Gordium in 650 B.C.

In the sixth century Persia took over the area. The last sets of ruins beneath the summit of the mound are Hellenistic and Achaemenian (Persian). Southeast of Gordium is a mound, Küçük Hüyük, higher than the citadel at Gordium. It was the site of another fortified town that was destroyed in the Persian invasion. The destroyed town became a tomb for the defeated king of Gordium.

As Alexander the Great marched into Anatolia in the mid–fourth century B.C., he is said to have been enticed by the promise of the fabled Gordian knot, which had stood in Cybele's temple since the time of Midas. He detoured to Gordium, where he wintered in 333. While there, he severed the knot with his sword to fulfill the prophesy—he later ruled most of the western world.

Attacks by the Galatians (Gauls) in 278 B.C. scattered the remaining population of the city. When the Galatians were, in turn, defeated and driven out by the Romans, Gordium was abandoned, not to return to fame until the excavations of the twentieth century. The excavated site offers much to view, and the Gordium Museum is located nearby.

Further Reading: Seton Lloyd's *Early Highland Peoples of Anatolia* (London: Thames and Hudson, and New York: McGraw-Hill, 1967) has a chapter entitled "The Phrygians," which describes the finds at Gordium and includes photographs of the site and some of its artifacts. *The Archaeology of Ancient Turkey* by James Mellaart (London: Bodley Head, and Totowa, New Jersey: Rowman and Littlefield, 1978) contains a chapter about the Phrygians and Neo-Hittites, discussing the history of these peoples as revealed by archaeology; it also includes photographs of Gordium. For background reading and general history, Richard Stoneman's *A Traveller's History of Turkey* (Moreton-in-Marsh, Gloucestershire: Windrush, and Brooklyn, New York: Interlink, 1993) is a worthwhile source.

—Julie A. Miller

Gortyn (Hērákleion, Greece)

Location: On the plain of Mesara along the Lethaios River in the south-central part of Crete.

Description: Archaeological site of the Greco-Roman city that served as the Roman capital of Crete beginning in the first century B.C. Although still not fully excavated, the site has yielded the most significant Roman and early-Byzantine remains on Crete, most notably a series of marble stones inscribed with the famous Law Code of Gortyn. Dating from the fifth century B.C., it is the earliest known legislation in Europe and a wellspring for laws that spread through much of ancient Greek civilization.

Contact: Greek National Tourist Organization
1 Xanthoudidou Street
Iráklion, Hērákleion
Greece
(81) 228-203

According to legend, Gortyn (also spelled Gortyna or Gortys) was founded by the mythical King Minos of Crete, for whom the Minoan civilization is named. This earliest civilization known in Greece began on Crete during the Bronze Age some 2,000 years before the birth of Christ. However, Plato and other ancient writers believed that Gortyn was founded by peoples who migrated south from the Peloponnese Peninsula of Greece, which would have occurred sometime around 1250 B.C. Structural ruins and artifacts later excavated from Gortyn reveal the influence of the Minoans, who built an empire through maritime commerce on the Mediterranean and Aegean Seas. However, little is known about the city until the invasion of a Hellenic tribe known as the Dorians somewhere around the year 1000 B.C. As the Dorians established their dominance, Gortyn developed into the leading city of Crete, reigning supreme for 1,000 years.

According to Greek mythology, it was at Gortyn that Zeus appeared in the guise of a white bull to romance the Phoenician princess Europa. There, she bore him three sons, Minos, Rhadamanthys, and Sarpedon, and thus came to be worshiped as a fertility goddess, ultimately imparting her name to the continent of Europe. Homer mentions Gortyn in both the *Iliad* and the *Odyssey*. In the former work, Gortyn is cited as a walled settlement from which troops of men joined Agamemnon in the Trojan War to recover the fair Helen. In the *Odyssey,* Gortyn is the site of adventures experienced by Nestor and Menelaus on their journey home. Additionally, the Cretan sculptor Daedalus, supposedly the architect of the labyrinth built to imprison the fabled Minotaur of Crete, reportedly took a wife from Gortyn and may have dwelled there himself.

Under the Dorians, Gortyn was a highly militaristic city whose warlike traditions may have been the source of similar institutions in the Greek city-state of Sparta. Sometime near the end of the seventh century B.C., Gortyn replaced Phaestos as the capital of Mesara. Three centuries later, Gortyn was locked in a struggle with Knossos for leadership of Crete and sought an alliance with Macedon. In the third century B.C., Gortyn joined the Koinon, a partnership of the primary cities on the island. In 220 B.C., Gortyn participated in the Lyttos War against Knossos. Later, the city sought the friendship of Rome, but reversed its policy to offer sanctuary to the Carthaginian general Hannibal in 189 B.C.

The nearby town of Lebena served as Gortyn's port for foreign trade via the Lethaios River, which flowed into the Mediterranean. When troops from Gortyn conquered the town of Phaestos in the third century B.C., the city acquired a second harbor at Matala, facilitating greater control of the sea routes between east and west. Plato, who in his *Laws* characterized Gortyn as having one of the best governments in Crete in his time, described the city as prosperous and strong, with a citizenry held in great respect.

However, perpetual warfare among the cities of Crete, combined with Crete's piracy on the high seas, eventually triggered Roman intervention. Between 69 and 67 B.C., the island fell to General Quintus Caecilius Metellus, known thenceforth as "Creticus." According to some accounts, Gortyn sided with Quintus Metellus in his ruthless conquest of Crete, and thus was spared the destruction wrought against other Cretan cities. Crete was combined with Cyrenaica, or Cyrene (Libya), on the north coast of Africa to form a new Roman province, with Gortyn as its capital.

Gortyn, along with the rest of Crete, prospered as Rome put its imperial stamp on the island. Towns were built or expanded, roads carved out, aqueducts constructed, and irrigation projects completed. An imperial mint was located in Gortyn. Among the coins issued from Gortyn in the fourth century B.C. were those depicting the marriage of Zeus and Europa. Zeus is portrayed on some coins in the guise of a bull; on others, as an eagle enveloping the fair Europa. During these prosperous times, Gortyn was especially graced with Greek, Roman, and Byzantine temples and monuments so magnificent that the Italian monk Buondelmonte, visiting their ruins in the fifteenth century, characterized ancient Gortyn as "equal in splendor to our own Florence."

Gortyn also has a strong Christian tradition. St. Paul is said to have founded the Church of Crete, leaving its organization in the hands of his disciple St. Titus, for whom a magnificent domed cathedral in Gortyn was named. By the second century, Gortyn was home to the largest Christian community in Crete, and by the fourth century, the bishop of Gortyn had established his supremacy on the island. St. Nikon

An observer views the Law Code of Gortyn
Photo courtesy of The Chicago Public Library

also is said to have spent a considerable amount of time in Gortyn. A nearby hamlet, or suburb, of Gortyn, called Alonion (also Aloni), was the execution site of the Holy Ten Martyrs of Crete. These resolute bishops were put to death by sword during the reign of the Roman Emperor Decius (249–251) for refusing to consecrate a heathen temple. They were buried in a chapel in the hamlet, and their execution block has been preserved. Their story was told in a hagiographical text written between the sixth and eighth centuries based on an earlier work, now lost.

Gortyn continued its supremacy among the cities of Crete during the early part of the Byzantine era, but suffered considerable damage from earthquakes in the fourth through seventh centuries. At the request of Gortyn's citizens, the Byzantine Empire invested a significant amount of money to repair much of the destruction. However, evidence found in the excavations of nineteen tombs, dated between the fourth and seventh centuries, suggests the citizenry at that time was suffering from poverty, ill health, and low life expectancy. In 824, the city was pillaged and largely destroyed by invading Saracens. The city was ultimately abandoned following the Venetian conquest of Greece and its islands in 1210. Through later centuries, the ruins of Gortyn became overgrown with vegetation, then gradually buried by the soil that washed down from nearby hills. This act of nature was to have fortunate consequences, since it preserved the ruins from total destruction.

Excavations of the ancient city began in the late nineteenth century, revealing a treasure trove for archaeologists, beginning with the Law Code of Gortyn. Discovery of this ancient code, later known as the "Queen of Inscriptions," was an adventure in itself. The Italian archaeologist Federigo Halbherr was lured to the site in the summer of 1884 by reports of recent finds of stone fragments inscribed with ancient writing. He was washing his feet in a millstream when he spotted some writing on the top of a wall, visible only because the water level of the millstream was down at the time. He dug a trench uncovering four inscribed columns. Halbherr reported his discovery to a colleague, Ernst Fabricius. Later that same year, Fabricius visited Gortyn to uncover eight more columns. Fabricius published news of their find in Gortyn late in 1884, sparking widespread interest in archaeological circles.

This ancient set of laws inscribed on marble stones was one of the most sensational discoveries ever made in the Mediterranean region. Consisting of some 17,000 characters, the inscription appears to date to the fifth or sixth century B.C. and may have originally been part of a Prytaneion, comparable to a law court or town hall. The inscription stones were later incorporated into the wall of the Roman Odeon, or covered theatre, sometime in the first century B.C. All of the stones were numbered, presumably to facilitate their reconstruction in a new setting. The letters were carved about an inch high, illegible only in a few places, usually at the juncture between the stones.

The writing spreads over twelve columns, each with fifty-two lines, written in an archaic Doric dialect of eighteen letters, compared to the standard Greek alphabet of twenty-four. The system of writing is known as *boustrophedon,* literally translated "as the ox plows," an easy-on-the-eye style that reads from right to left on one line, then left to right on the following. The writing provides evidence that the civilizations of ancient Crete had a significant role in the development of the Greek alphabet and its Semitic and Phoenician precursors.

The code is said to preserve in writing the laws handed down from Zeus to Minos when the latter was king of Crete. This remarkable piece of legislation evidently served as a model for the constitutions of Athens and Sparta. Rather than being a complete set of laws in itself, the code appears to formalize a comprehensive revision of previously developed law. Nevertheless, through its statutes regarding crime and punishment, class distinctions, family matters, and property rights, it is our chief source of information about the social history of Crete during Hellenisic times.

Under the Code of Gortyn, society was bound by a rigid class structure, headed by the aristocratic ruling class, followed by free people of lower status, then serfs and, finally, slaves. Free men were those belonging to the men's organizations of the city and from whose ranks officials were recruited. Serfs were described as *apetairoi,* meaning those "without comrades." Unlike slaves, serfs had the rights of tenure over their own homes and possessions, although these were considered part of the master's estate; they also could marry and own livestock. A free man who defaulted on a debt could fall under bondage to his creditor, but as a serf, not as a slave. The children of serfs belonged to the master. However, marriages were allowed between a free woman and a serf, and if the man went to live with the free woman, their children were free. If a serf ran away, he lost his ranking and became subject to sale as a slave.

For administrative matters, *kosmos,* or magistrates, were appointed to handle special duties such as collecting produce or dealing with foreigners. The Code of Gortyn mentions a *kosmos hiarogos* (sacral magistrate), who apparently dealt in religious matters. Legal decisions were in the hands of a judge, although no evidence exists as to how he secured the position. The code covered such crimes as theft, rape, seduction, and adultery. The laws were very specific as to the required verdict and penalty in each circumstance, leaving the judge little leeway in exacting justice. Great store was placed on the taking of an oath, generally in the name of a special god. The oath called for severe devastation on anyone who lied, presumably a custom stemming from more primitive times. The scales of justice were tilted heavily in favor of the upper class. Aristocrats paid lower fines for the same offense, and greater evidence was required for their conviction than for members of lesser classes. In some cases, only free men were considered competent witnesses.

The code also regulated the sale of property, the treatment of ransomed prisoners, the rights of children born illegitimately, the payment of debts, and even the recovery for damages done by slaves. But its greatest value lies in its revelations about daily life in the ancient city of Gortyn. Under the code, the father of the family had authority over the children and the disposal of property among them; the mother, however, controlled her own property. Sons had preferential rights in the disposition of property. On the death of the father, town houses and the contents of country houses were turned over to sons. The remainder of the estate was divided with two-thirds going to the sons, one-third to the daughters. When there were no children, the property passed to brothers, their children, or their grandchildren. Other kinsmen were next in line for the estate, followed by any serfs who lived on the property.

The exhaustive rules regarding the disposition of property reflected time-honored rules of wedlock, which involved a system of collective intermarriage. The ancient writer Strabo cites an annual marriage festival in Gortyn. The matrimonial system seems to have relied on tribal endogamy, meaning that cousins of one clan intermarried with cousins of another clan in the same generation. This system reinforced kinships and strengthened social responsibility. Divorce and adoption were allowed under the code, as evidenced by the rules governing property settlements in such events.

The code also reveals much about the rights of women, who evidently fared considerably better in Gortyn than in other cities such as Athens. Should a wife in Gortyn divorce, she was entitled to keep any property she had brought into her marriage, plus half the woven goods she had made for her household. A widow was free to remarry, taking along her own property and that given her by her late husband. She was forbidden to take any property belonging to her children. In a childless marriage, if the wife died, her property went not to the husband but to her clan. The laws were especially specific regarding heiresses. An heiress could be married as early as age twelve. Her husband could be a minor, contrary to the usual custom of marriage only between adults. Should an heiress inherit all her father's estate, she was required to marry his next of kin, with paternal uncles followed by paternal cousins having priority according to age. If she failed to marry the prescribed relative, she had to surrender part of her inheritance.

In addition to the discovery of this fountainhead of Greek law, excavations in and around Gortyn uncovered much evidence of the city's importance under Roman rule. One such find was a temple of the Pythian Apollo, the Greek god of the sun in his role as slayer of the enormous serpent Python, which lurked in the caves of Mount Parnassus. Also uncovered were ruins of public baths and a *nymphaion,* or fountain-monument, as well as some of its piers, arches, and remains of the aqueduct that once fed it with water. Addition-

ally indicative of the Roman influence were the remains of a temple dedicated to the Egyptian gods who were worshiped for a time in Crete following Rome's conquest of Egypt. There is also an amphitheatre, a stadium where chariot races entertained Gortyn's citizenry, and a *praetorium,* the dwelling place and headquarters of Crete's Roman governor.

Near the Roman ruins in the excavation site, the Cathedral of St. Titus is gone, but resting in its place are the remains of an impressive basilica erected during the sixth century and carrying the imperial signs of the Emperor Justinian. The structure features the traditional Byzantine cross-in-a-square design, with a dome over the cross's intersection. The church includes side chapels housed in towers flanking the apse. In one of the chapels are remains of the frescoes that once decorated many parts of the building. After the basilica was severely damaged by the Saracens in the ninth century, the remains of St. Titus and other holy relics were taken to what is now the city of Iráklion in north-central Crete and placed there in a church dedicated to that same saint.

North of the basilica in Gortyn are ruins indicating the site of an agora, or marketplace, where once stood the temple of Asclepius, one of the most famous shrines in Crete during Roman times. A statue of the mythological physician Asclepius, son of Apollo, was once contained in that temple. Now it rests in the Archaeological Museum of Iráklion. Nearby is the Roman Odeon, in whose walls the Code of Gortyn was found embedded. Across the Lethaios River are the ruins of a Hellenistic theatre at the foot of the acropolis hill. On this hill, Italian archaeologists uncovered the foundations of an impressive temple that dates from the seventh or eighth century B.C. and apparently had been renovated several times since then. A sacrificial altar lay inside. The remains of another, more imposing, sacrificial altar were found on a nearby slope. Archaeologists have also uncovered fragments of defensive walls around the city, some dating as far back as Homeric times.

Further Reading: *Everyday Life in Ancient Crete* by R. F. Willetts (London: Batsford, and New York: Putnam, 1969), provides extensive detail about the Law Code of Gortyn and its influence on the social culture of the city. *Crete: Its Past, Present and People* by Adam Hopkins (London and Boston: Faber, 1977), offers historical perspective on Gortyn and the rest of the island as does John Freely's *Crete* (New York: New Amsterdam Press, and London: Weidenfeld and Nicolson, 1988). *Byzantine Crete* by Dimitris Tsougarakis (n.p.: Historical Publications St. D. Basilopoulos, 1988) deals primarily with Gortyn's religious history. Various guidebooks for tourists also contain limited historic and archaeological data about Gortyn as well as useful information for those wishing to visit the site.

—Pam Hollister

Granada (Granada, Spain)

Location: In southern Spain, 250 miles south of Madrid and 160 miles west of Seville. The Mediterranean coast is 50 miles south of the city. The Sierra Nevada mountain range, which lies immediately to the south, contains Spain's highest peak, Mulhacen (more than 11,000 feet), and Europe's highest road.

Description: Granada is Andalusia's fourth-largest city, after Seville, Málaga and Córdoba. There has been human settlement in the Granada area for several thousand years, including a substantial Roman town. Granada was the last part of Moslem Spain to fall into Christian hands during the reconquest, which ended with the surrender of the city to the Catholic monarchs, Ferdinand and Isabella, in 1492. The Alhambra, an extraordinary fortress-palace used by the Moorish kings, has survived virtually intact to this day. On the hill opposite the Alhambra, the Albaicín neighborhood has also changed little over the past 500 years. Ferdinand and Isabella lived for a time in Granada, and are buried in the Capilla Real (Royal Chapel). Next to it is a large classical cathedral, commissioned by the Emperor Charles V.

Site Office: Tourist Information Centre of the Andalusian
Government
Los Tiros House
Padre Suárez Square
Granada, Granada
Spain
(91) 22 10 22

Granada lies at the foot of the Sierra Nevada mountains, presiding over a fertile flood plain, known as the Vega. Its altitude, about 2,000 feet, means that it is considerably cooler than other Spanish cities of the same latitude. Settlement in the area dates back to 5500 B.C. Most of the early inhabitants lived in caves, some of which may have been located in what is now the Albaicín area of the city. In the Bronze Age the first fixed settlements of any size appeared, and people in the nearby coastal areas traded with Greek and Phoenician merchants. By the seventh century B.C. the name Elibyrge was used in reference to what is today the city of Granada.

In the first century A.D. Pliny the Elder used the name Iliberri. Granada became a Roman municipality: the importance of the town can be gauged by the fact that coins were minted with its name. The invasion of the barbarian tribes in the fifth century A.D. had little direct effect on Iliberri: the town was some way from the main highways, and was protected by impressive natural defenses. Later, under the Visigoths who came to dominate Spain, the town flourished, while preserving its essentially Roman character.

In 711 the Moslems crossed to Spain from North Africa. Iliberri was conquered by Ṭāriq ibn Ziyād, the Moslem general who gave his name to Gibraltar. Ṭāriq found next to Iliberis a second town occupied by Jews, Garnatha Al-yehud. Outside the original walls, this area is now occupied by the San Matias neighborhood of the modern city. The name "Garnatha" came to signify the larger town, and this was eventually changed to "Granada." With the arrival of the Berbers in 1010, Zawi Ben Ziri chose Granada as the capital of eastern Andalusia, causing an increase in population to levels not matched again until the fourteenth century. During the eleventh century the control exercised over the city by the prominent Jewish Nagrila family meant that it became known as the "city of the Jews." With the breakup of Moslem Spain and the fall of the city of Córdoba in the eleventh century, new Moslem tribes crossed from North Africa to fight the Spanish Christians. The period of domination by these tribes, the Almohads and Almoravids, saw essential infrastructure fall into decay: it was only with the rule of Ben Ibrahim in 1218 that essential work on Granada's irrigation system, bridges and city defenses was carried out.

With the rapid advance of the Christian reconquest in the first half of the thirteenth century, the governor of Jaén, Muḥammad ben Yusuf ben Nasr, was forced to move his government to Granada in the mid-thirteenth century. The kingdom he created came to extend 200 miles to Tarifa in the west and a further 60 miles in the east to Murcia. Nasrid Granada was strongly Moslem: it gave refuge to other believers from the rest of the peninsula, and Arabic was the only language used. While there were some Jewish residents, the kingdom had no Mozarabes (Christians living under Moslem rule). Although little is known about the Nasrid state, it seems that the second half of the fourteenth century represents the kingdom's cultural zenith. Intensive agriculture and buoyant commerce put the state's finances on a sound footing, allowing the construction of the Alhambra, a large palace-fortress that occupies the principal hill of the city of Granada. It contains some of the finest Spanish Moslem architecture in existence: there are numerous state rooms, and others used by the king and his harem, such as the Patio de los Leones (Lions' Courtyard) as well as impressive gardens known as the "Generalife."

There is no simple answer as to how Granada could maintain its independence for two and one-half centuries after the fall of Seville in 1248. Castile may have accepted the existence of the territory, or found it a useful way of accommodating those of its own Moslem subjects who had become unhappy. Furthermore, the kingdom of Granada was ringed by mountains, and this natural protection was reinforced by

Interior of the cathedral at Granada
Photo courtesy of Instituto Nacional de Promocion del Turismo

hill-fortresses such as that built at Ronda. The Nasrid kings could also count on the support of the Marinid dynasty, who controlled Morocco, though at times this was a threat. Nasrid kings also became tributaries of the Christian Castilians as a way of maintaining themselves in power.

The late fourteenth and fifteenth centuries were characterized by internal disputes, particularly between the proponents of war against the Castilians—such as the Islamic jurists—and the opponents of war, including most of the urban merchants and the governing class. With the fall of Ceuta (on the Moroccan coast) to the Portuguese in 1415 the strategic threat to the Nasrids was increased. In 1462 Gibraltar was taken by the Christians, a sign that the Castilians were once again taking the Granada issue seriously. The situation for the Nasrids worsened when the two major Christian states allied with the marriage of Ferdinand of Aragon and Isabella of Castile in 1469. But it was not until 1481 that the Catholic monarchs were finally convinced of the need to conquer

Granada. This decision followed the Nasrids' attack on the castle at Zahara de la Sierra, and the following year the final portion of the Spanish reconquest began.

Ferdinand initially played one Moslem faction against another, and was quickly able to take Ronda (1485), Málaga (1487) and Almería (1489). The final attack began in 1491, and concluded with the surrender of the city of Granada on January 2, 1492. Legend has it that the Moslem King Muḥammad XI (Boabdil) turned on his journey to exile in Morocco and sighed at the last view of the city he had ruled. The incident is commemorated in the name of a location south of Granada, El Sospiro del Moro—The Sigh of the Moor. It is also said that the weeping Boabdil was scolded by his mother for "weeping like a woman for what he had not been able to defend like a man."

Many of the large estates around Granada were given as prizes to the Christian nobles and military orders that participated in the war. This redistribution of land produced the powerful aristocratic clans that characterized Spain until the nineteenth century. Ferdinand and Isabella lived on in Granada for several months after the conquest, and in 1525 were buried in the city's Capilla Real (Royal Chapel). During their reign Granada had been the scene of not only the successful completion of the reconquest, but also the momentous decision to fund Christopher Columbus's transatlantic expedition.

The terms of the reconquest were lenient at first: the Moorish citizens of Granada were allowed to keep their own religion, customs, law, and dress. But these clauses were quickly erased with the arrival in the city of Archbishop Francisco Jiménez de Cisneros in 1499. He initiated a policy of forcible conversions and mass baptisms. In 1502 this caused the first of a series of revolts in the area: suppression of the uprising was followed by the offer of a choice similar to that given the Spanish Jews a decade earlier—adoption of Christianity or exile. Holy Roman Emperor Charles V (king of Spain as Charles I) issued decrees in 1525–26 that further restricted religious freedom. Moriscos—Moslems who had been baptized Christians—were forbidden to use the Arabic language, wear Moslem costume, or use public baths. Charles V lived in the Alhambra during 1526 and for a time considered transferring the capital of the empire to Granada; he had a new, classical palace built alongside the Moorish rooms. Charles V also ordered the construction of a cathedral in the city. He initially disliked Gil De Siloé's Greco-Roman design for the building because of the contrast it offered with the Gothic of the nearby Royal Chapel. The cathedral houses statues, silver lamps, a lectern, and a series of paintings by the seventeenth-century artist Alonso Cano, who was also responsible for the design of the facade.

Charles's son and successor as king of Spain, Philip II, enforced his father's decrees on religion: the result was a revolt in the Alpujarras region outside Granada, home to numerous Moriscos, many of them engaged in agriculture. The uprising prospered in this remote region from 1568 to 1570, before being put down by Don John of Austria. The Alpujarras Moriscos were deported to North Africa. The rest of the Moriscos in Granada were expelled from Spain in 1611.

Granada went into decline after the reconquest. As early as 1526, a Venetian ambassador visiting the area wrote that the new Christian inhabitants had left the farms of the Moslems to decay. The city was overshadowed by others, particularly the capital Madrid, and the southern metropolis, Seville. As a sign of its backwardness and distance from the center of power after the sixteenth century, Granada was chosen in 1754 as a suitable place of exile for Ferdinand VI's disgraced minister, Zenón de Somodevilla y Bengoechea, the Marqués de la Ensenada.

Granada was occupied by French troops from 1810 to 1812. During this time the Napoléon Theater was inauguarated by Joseph Bonaparte, who had been installed as king of Spain by his brother. In some areas the reforming spirit of the French outlived their occupation: among the numerous administrative changes implemented in the following years was the division of the ancient kingdom of Granada into three provinces, Almería, Málaga, and the modern province of Granada in 1833.

The American writer and diplomat Washington Irving lived in Granada at the beginning of the 1830s. He found a city in decline from the economic and political buoyancy afforded by the royal visits of earlier centuries. The city's state of abandonment was echoed in that of its principal monument: Irving wrote *The Alhambra* while living in the old Moorish palace, which was damaged during the French occupation, and for a time had been occupied by vagabonds and smugglers. Irving's book sold well. Together with artists such as David Roberts, J. F. Lewis, and Gustave Doré—all of whom painted in the city—he aroused new interest in southern Spain, and in particular the Moorish past of Granada.

Spain's prolonged political crisis after the abdication of Isabella II in 1868 saw the country break up into independent states, or cantons. Granada's canton of 1873 was controlled by a Committee of Public Safety of which two members were anarchists: this had the effect of alienating bourgeois support, and the canton quickly collapsed. The social and political frictions that were to tear Spain apart in 1936 were particularly evident in Granada. As irrigated land, the Vega plain of Granada was unusual in not being in the hands of peasant proprietors. Intensive sugar-beet farming was profitable, meaning that rents were as high as 8 percent on the capital value of the land: social conflict resulted because, as historian Gerald Brenan puts it, the laborers were not "the half-starved, down-trodden serfs of the Andalusian Basin, but well-educated and organized socialists, whilst the landowners ... (formed) a compact body belonging, like nearly all Spanish landowers, to the extreme Right." The peasants were well organized in the UGT union, but there was considerable unemployment. Brenan, an eyewitness to the happenings of the first half of the 1930s, says there were frequent riots, and cars were stopped and the owners asked to "make a contribution" to the unemployment fund. He also writes that in the crucial elections of February 1936 the police in the villages around Granada stopped anyone not wearing a collar entering the polling booth: a sign of the power of the right-wing parties in the area. At the time of the Civil War the city had a population of around 120,000, with an adult illiteracy rate of 39 percent. Despite attempts by General Miguel Campins Aura to keep the local garrison from joining the military rebellion that began the war, the uprising nevertheless took hold in Granada in July 1936. There was some working-class resistance in the Albaicín area opposite the Alhambra. One source gives 2,137 as the number of people killed in the nationalist repression, 572 of them during August 1936. Among them was Federico García Lorca, although the exact circumstances of the poet's death are unclear.

The Franco regime did little to correct the economic malaise of the province of Granada, which was one of the five poorest areas in Spain during the 1950s and 1960s. Since the death of Franco in 1975 and the arrival of democracy in Spain, the province of Granada has become one of eight that form the Autonomous Community of Andalusia. It is also the headquarters of Andalusia's own supreme court, and site of one of Spain's most prestigious universities. The population of the city is around 300,000. Tourism is now an important industry. The Sierra Nevada has been developed into a ski resort, and was chosen as site for the 1995 World Skiing Championships: these were postponed until the following year because of lack of snow.

Further Reading: *Historia de Granada,* four volumes, by various authors (Granada: Ed. Don Quijote, 1986) offers a general view of the city. *Reino de Granada,* three volumes, by various authors (Granada: Ed. Ideal, 1992) and Torres Delgado's *El Antiguo Reino Nazarí de Granada* (Granada: Anel, 1971) both provide accounts of the city under the Moslems. Pedro Herrera Puga's *Granada en el Singlo XVI: Asepectos Sociales* (Granada: Imprenta Roman, 1980) is informative concerning the attempts to assimilate the Moslems into Christian Spain and the Moslem revolts of the sixteenth century. Washington Irving's *The Alhambra,* first published in 1832 and available in various editions, provides a romanticized account of the reconquest and a view of Granada in the early nineteenth century. There are few books covering Granada's more recent history, but Gerald Brenan's *The Spanish Labyrinth* (Cambridge: Cambridge University Press, and New York: Macmillan, 1943; second edition, 1950) offers a good idea of the social and economic climate in the area in the years leading up to the Spanish Civil War, and his *South From Granada* (London: Hamilton, and New York: Farrar, Straus and Cudahy, 1957) is a first-hand account of life in a village in the Alpujarra region, near Granada.

—Richard Bastin

La Granja de San Ildefonso (Segovia, Spain)

Location: Fifty miles north of Madrid, seven miles southeast of the city of Segovia, in Segovia province, in the foothills of the Guadarrama Mountains.

Description: Summer palace of the Bourbon kings of Spain, surrounded by gardens and more than two dozen fountains and containing a spectacular collection of Flemish tapestries.

Site Office: La Granja de San Ildefonso
San Ildefonso, Segovia
Spain
(911) 47 00 19

The Royal Palace of La Granja de San Ildefonso has been called the "Spanish Versailles," and with good cause. It was built by the grandson of France's Louis XIV, the builder of Versailles, with the help of French workmen, some of whom came from Versailles. Philip V, French king of Spain from 1700 to 1746, had been raised at Versailles, and he took his French heritage very seriously. And most importantly it was built in a style less inherently Spanish than any other major architectural project of the Spanish baroque, a distinctly European palace that would have seemed equally at home in France or Germany or Italy. Whatever its origins, La Granja de San Ildefonso became one of the most magnificent palaces in Spain, with its beautiful setting, elaborate decor, many gardens, and its array of fabulous fountains.

The site in the forested foothills of the Sierra de Guadarrama at the base of the 8,000-foot Pico de Penalara is indeed a fairy-tale setting for a palace. The woods are thick with elms, oak, alders, ash, and stone-pine. The Valsain and Cambrones Rivers drain the area before uniting to become the Eresma. Deer, wild boar, wolves, and bears once roamed in abundance. In fact it was the region's hunting potential that first attracted royal interest during the Middle Ages when Henry III built a hunting cottage there. In 1450 his son, Henry IV, built a house and a hermitage a few miles away and dedicated the hermitage to San Ildefonso; he credited the saint with saving him from a wild beast. The property was later donated to the monastery of Santa Maria del Parral in Segovia by Ferdinand and Isabella. By the mid-seventeenth century these structures had decayed, but the monks built a summer retreat there, including a brick and granite hostel and a farm or *granja*.

It is here that Philip V enters the story. In 1700 the last Spanish king of the Habsburg line, Charles II, died without leaving an heir and willing his throne to Philip, then duke of Anjou. French king Louis XIV approved his grandson's title to the throne of Spain but refused to cut him out of the French succession, opening the possibility that Philip might one day rule both Spain and France. The reaction of the other European powers to this possibility set off the War of the Spanish Succession, with Philip battling the Habsburgs for his Spanish inheritance. The Treaty of Utrecht in 1713 decided in favor of Philip while depriving him of the Spanish Netherlands and Charles II's Italian possessions. So the grandson of Louis XIV was enthroned in Madrid and, having fought for his Spanish crown, was ready to fight for a French crown as well.

Philip's devotion to his French heritage was thus equal parts nostalgia and ambition. It was supplemented not by Spanish but rather by Italian influence. After his first wife died in 1714, Philip married the Princess Isabella Farnese of Parma. Her influence quickly replaced that of his French courtiers, and her interest in securing land for her children led to Spain's continual involvement in Italian wars.

Obsessively religious, Philip's court was a sedate one but he did indulge the traditional passion of the Spanish monarchy for hunting. In 1720 the king and queen discovered La Granja while on a hunting trip and were at once taken with the remote woodland setting and the charm of the monks' hostel. They purchased the property from the monastery and began to build a manor about the hostel where they might retire from the strain of politics and warfare that surrounded them in Madrid. Philip's chief architect, Teodoro Ardemans, was given charge of the project and produced a European-style country estate house with white plastered bricks and slate-covered spires atop the towers. A church was constructed in the center of the complex, dominating the other buildings and attesting to La Granja's religious heritage and Philip's piety. The church was consecrated in 1723, and the manor was nearing completion under Philip's close supervision, with the king and queen and an abbreviated court frequently in attendance.

In January 1724 Philip astonished the world by announcing his decision to abdicate in favor of his son, Louis. Philip and Isabella moved permanently to San Ildefonso where they lived modestly with a few servants and few luxuries. Philip preferred to devote himself to religious contemplation, and Isabella joined him with apparent willingness. Philip even gained the approval of the pope to establish a collegiate monastery at San Ildefonso and so surrounded himself with clerics. Unfortunately, Louis's marriage to a French princess proved disastrous, and communication and appeals for aid to San Ildefonso were constant; the monarchs' removal was less than complete. Then in August Louis was diagnosed with smallpox and died soon afterward. Philip's second son, Ferdinand, was only ten years old; after much consultation and persuasion, Philip resumed the throne.

With Philip's return to the throne, the purpose of La Granja changed. No longer could it be a modest manor house

Throne room at La Granja de San Ildefonso
Photo courtesy of Tourist Office of Spain, Chicago

for a retired monarch and a few servants; instead Philip transformed it into one of the great palaces of Europe, suitable for the Spanish court. Or, rather, his queen transformed it. Always a taciturn personality, Philip suffered several bouts of mental illness that increasingly incapacitated him. It was Isabella Farnese whose taste guided the refurbishment and expansion of La Granja, with one result being the great baroque facade that faces the gardens. It was probably designed by the Abbe Filippo Juvara, the Sicilian architect who was then also supervising the rebuilding of the Alcazar in Madrid. His pupil, Giovanni Battista Sacchetti, took over the work after Juvara's death.

Philip died in Madrid in 1746 and was interred at his beloved La Granja. Isabella then retired again to La Granja where the building continued. She was said to suffer from a "building mania," for she was also building another palace near Riofrio. This palace, a modest building more in the line of a hunting lodge, is also close to Segovia and now houses a museum of hunting. After her death in 1766 she was also buried at La Granja. It was her son who returned to rule Spain as Charles III and who put the finishing touches upon La Granja. The now magnificent palace served Spanish royalty through the eighteenth and nineteenth centuries. In January 1918 the palace was severely damaged by fire, which gutted much of the royal apartments. Again refurbished, it has become home to a superb collection of Flemish tapestries.

La Granja is made up of the elements of the original monastic structure, the church, the palace, and the gardens. The original building, with its beautifully simple lines and quiet Fountain Courtyard, is the only truly Spanish element at San Ildefonso. The church was built in the form of a Latin cross, with the main chapel, the choir, and two entrances at the four ends. Teodoro Ardemans was responsible for the original design, later modified by Andrea Procaccini, an Italian architect, who adapted the plans to fit the style of international baroque. The central dome and two bell towers dominate the building. Inside, the ornamental screen behind the main altar is flanked by red marble columns with bronze capitals. In a chapel to the left of the altar lies the tomb of Philip V and Isabella Farnese, which features gilt stucco walls, portrait medallions of the monarchs, and sculptures by Hubert Dumandré. The church possesses many pieces of exceptional religious art, including paintings, tapestries and precious metals. Among them is an elaborate sixteenth-century Gothic processional cross fashioned by one of the famous silversmiths of Segovia, Antonio de Oquendo.

The palace of La Granja de San Ildefonso is richly furnished and adorned with all the splendors of which the baroque was capable. Marble, bronze, and gilding are lavishly used. The Marble Room is an example of the height of baroque ornamentation. Columns of Tuscany marble topped with capitals of Carrara marble line the walls. The furnishings and the vaulted ceiling are heavily decorated with gilt, and gilded bronze frames the enormous mirrors. The ceiling fresco was painted by Bartolomeo Rusca. In the Japanese Room, the European fascination with the Orient is in evidence. Here there are columns and panels in intricately worked lacquer-ware designs and furniture in Chinoiserie. Somewhat strangely juxtaposed are some of the finest paintings of the Italian Giovanni Paolo Pannini, depicting four scenes from the Gospels set in a magnificent baroque version of the Temple of Jerusalem.

The throne room is another impressive example with its red walls and draperies, marble tile floor, gilt furniture, and enormous tapestries. Dominating the room are huge gilt and crystal chandeliers from the local crystal factory. The throne itself, a gilded chair thickly upholstered in red velvet, sits before a red and gold tapestry bearing the royal arms.

Among the many paintings by French, Italian, and Dutch artists displayed in the palace are portraits of Philip V and Isabella Farnese by Louis Michel Van Loo. Michel-Ange Houasse, a French master invited by Philip V, is represented by a number of his works that show an Italian influence while documenting early eighteenth-century life. Two paintings depicting the hunt by the seventeenth century Antwerp artist Frans Snyders and an allegorical work by Jacopo Amignoi celebrating the assumption of the throne of Naples by Philip's son Charles are included in the fine collection.

The famous museum of tapestries housed in the former royal apartments includes a stunning collection of this craft, the designs for which often came from the best artists of the age. The work of Raphael in the sixteenth century was said to have transformed tapestries from decorated fabrics into "embroidered paintings," and an Italian work from his design exemplifies that statement. Based on a design from a series for the Sistine Chapel, it was woven in Brussels where the production of tapestries was elevated to a high art. Another important set of tapestries in the collection is the so-called Honour series, brought to Spain as part of the dowry of the Infanta Maria of Portugal, first wife of Philip II. This intricately woven and ambitious series describes the prominent people of the Bible, antiquity, and the Middle Ages. It was woven in about 1520 and was probably designed by Bernaert van Orley.

Perhaps the most distinctive feature of La Granja de San Ildefonso is the sweeping layout of terraced parterre gardens with their fantastic fountains. The terraces, with their geometric layout of paths, parterres, and basins, framed with long cool alleys of shade trees, and set against the backdrop of the green Sierra de Guadarrama, provided a refreshing park for exercise, contemplation, and conversation. The whole was designed by one of the great garden architects of the age, Etienne Boutelou. The garden facade of the palace looks out upon the Great Cascade, a series of terraces and their fountains which stretch into the hills.

The water of the fountains is designed to interplay with the garden statuary, which represent scenes and figures of classical mythology. To produce the statuary for the gardens, Philip V brought a group of French sculptors from his beloved Versailles, including René Frémin, Jean Thierry, Huber Dumandré, Pierre Pitue, and Jacques Bousseau. One of the most famous groups of fountains is that known as "The

Horse-race." This includes the two *Snail Fountains* and the *Fan Fountain*, named for its fan-shaped water sprays; the large fountain portraying Neptune's chariot drawn by horses; then on the next terrace up, the *Apollo Fountain*; and finally the *Andromeda Fountain*.

Perhaps the most superb example of fountain design, surpassing even those at Versailles, is the stunning *Baths of Diana* fountain. It was built by Jacques Bousseau to Boutelou's design in 1742. A large baroque wall with a central niche forms the backdrop; in the niche is the hunter Actaeon, spying on Diana and her nymphs bathing in the basin below. Water is sprayed up from among the bathers against the wall, creating a curtain of mist and a continual cascade down either side of the wall. Philip V, upon first viewing the creation, was only mildly impressed. "You've amused me for three minutes," he said, "but you've cost me three million." If so, every peseta is visible in the result.

In 1723 the city of Segovia deeded over to the king a piece of land for the construction of a reservoir. This lake, built primarily to feed the fountains, forms a beautiful expanse of calm water in contrast to the constant movement of the fountains. It was a popular place for boating and fishing, and the king had a Venetian-style gondola constructed for water parties. In 1867 the consort of Isabella II had a fish-breeding station built, with the result that the lake was stocked with the excellent mountain trout.

As a favorite summer palace of the Spanish Bourbons, La Granja de San Ildefonso witnessed its fair share of history in the making. It was from San Ildefonso that Philip V announced his abdication. A number of treaties were signed in its halls, including that of 1796 which allied Spain to republican France, and it was here that Ferdinand VII revoked the Pragmatic Sanction in 1830. More recently it has served as an important repository of art and as an example of the lavish style of architecture, decorative arts, and garden design called the baroque.

Further Reading: Illustrations of the interior of La Granja and of its gardens are available in *The Escorial; the Royal Palace at La Granja de San Ildefonso* by Juan de Contreras Y Lopez de Ayala, translated by James Brockway (London: MacDonald, 1963; New York: Meredith, 1967). For more about Philip V and the other Spanish Bourbons who lived at La Granja, a useful overview is provided in *The Spanish Bourbons: The History of a Tenacious Dynasty*, by John D. Bergamini (New York: Putnam's, 1974), which traces the dynasty from Philip II up through the current monarch, Juan Carlos.

—Elizabeth Brice

Gubbio (Perugia, Italy)

Location: In the region of Umbria and province of Perugia in central Italy, 24 miles northeast of the city of Perugia and 135 miles northeast of Rome, at the confluence of the Camignano and Cavarello Rivers, on the lower slopes of Monte Ingino.

Description: Town of roughly 30,000 inhabitants with much well-preserved medieval architecture, famed for its artists and religious leaders, and for the Iguvine Tables, a group of bronze tablets that provide insight into language and religious practices in pre-Christian times.

Site Office: A.P.T. (Azienda di Promozione Turistica di Gubbio)
Piazza Oderisi, 6
CAP 06024 Gubbio, Perugia
(75) 9220693

In ancient times, Gubbio was known as Iguvium and was a flourishing town in the region of Umbria. The town lay only a short distance from the Via Flaminia, one of the principal Roman highways leading from Rome northward to the Adriatic Sea. The Romans colonized this route and built fortifications along its length. Iguvium, in time, became allied with Rome, and in 80 B.C., at the end of the first Roman Civil War, it came under a new set of civil regulations in the Roman mode. The ruins of a Roman theatre dating from the republican period and restored under the reign of Caesar Augustus lie just beyond the walls of the old town.

Much of the information about Gubbio and the surrounding region, especially their religious rituals, during the Roman period and earlier comes from several bronze tablets, known as the Iguvine Tables, discovered at Gubbio in 1444. The tablets were subsequently purchased by the town. In 1540, two of the tablets were removed from Gubbio to Venice and later lost. Seven of the tables are still in existence and are housed in the Palazzo dei Consoli, a former government building that is now an art and archaeological museum. The bronze tablets are the only extensive record of the Umbrian dialect that are known to exist to this day. Umbrian is an ancient language in the Italic branch of the Indo-European family of languages. Other languages in this branch include Latin and Oscan. In addition to the value of the tablets as tools for linguistic research, they are also important to the study of pre-Christian religious history in Italy.

The tablets measure from sixteen by twenty-two inches to thirty-three by twenty-two inches and are inscribed on both sides in the Umbrian dialect. Based on linguistic evidence, it is judged that the tablets were inscribed in the period between the fourth century B.C. and the first century

B.C., although some scholars date the tablets to a narrower range of years, 250 to 150 B.C. Four of the tablets and part of a fifth are written using the Umbrian alphabet, with the remaining portion of the fifth and two other tablets written in the Latin alphabet.

The first translation of the Iguvine Tables into a modern language did not occur until 1948. The tablets contain a set of extensive instructions for certain religious ceremonies of a college of priests called the Atiedan Brothers. The Atiedans thrived at Iguvium during the period of the Roman republic. The rituals for sacrifices to various Roman deities, including Jupiter and Mars, are outlined in the tablets, along with sacrifices to the deities of the local Iguvine cult. In addition, the tablets also describe several ceremonies dedicated to Ceres, the Roman goddess of agriculture, to be held in the month of May. One of the ceremonies, called Ambarvalia, was performed in numerous villages and towns. In the Ambarvalia, a bull, a sheep, and a pig were led in procession, three times around the periphery of the planting fields, after which they were sacrificed to the goddess.

The tablets also outline the organization of the Atiedan Brotherhood and the responsibilities of the different positions in the organization. It appears that the brotherhood consisted of twelve members, organized into two groups called *puntis,* each of which contained five priests. In addition to the *puntis,* there was a *fratricus* and an *adfertor.* The *fratricus,* in all probability an appointment rotated among the priests, acted as a presiding officer in the brotherhood and was responsible for obtaining heifers for use in the religious rituals. The *adfertor* was charged with conducting the ceremonial purification of the people, called the *lustration.* The *adfertor,* with the help of an augur, a sort of official diviner but not a member of the Atiedans, read the auspices, or omens, for Iguvium. During the ritual purifications and sacrifice, the *adfertor* was accompanied by two men, robed in purple, called the *prinatur.* As part of the ceremonies, these men, either acolytes or civil officials, joined the adfertor in reciting curses against the enemies of Iguvium.

The Iguvine Tables also reveal information about the structure of society in Iguvium during the Roman period. An interpretation of the tablets indicates the people of Iguvium were initially divided into ten groups, or *decuviae,* but as the population increased, these groups were subdivided into a total of twenty *decuviae.* The function of these divisions is not clear from the tablets. It is known, however, that these divisions played a role in how the citizens arranged themselves in procession for the rituals conducted by the Atiedans. Additionally, groupings other than the *decuviae,* apparently of lower-ranking residents, existed in Iguvium.

Christianity eventually displaced the pagan ceremonies of the Atiedans. A Christian bishop was installed in

The Palazzo dei Consoli
Photo courtesy of A.P.T., Gubbio

Gubbio by the year 413. Then, with the decline of the Roman Empire, Gubbio began to be ravaged by invaders. During the Gothic Wars in the fifth and sixth centuries, Gubbio was virtually destroyed by the Ostrogoths. Subsequent waves of invaders saw the town change hands numerous times until 593, when it was retaken by Rome. Again in the eighth century, Gubbio was invaded and occupied by various groups, until it finally came under the protection of the Roman Catholic Church. Charlemagne reportedly visited the town after his coronation in Rome as emperor of the west.

In the eleventh century Gubbio, dominated by its middle class, became an autonomous commune, or city-state, within the Holy Roman Empire, a status confirmed by Holy Roman Emperor Frederick I's issue of a charter in 1163. This charter was reaffirmed by subsequent emperors. Gubbio's expansion of its territory and its financial power led to conflict with Perugia, resulting in wars in 1151, 1183, 1216, and 1258.

Notable religious figures were associated with Gubbio during the Middle Ages. One of Gubbio's most prominent leaders was Ubaldo, bishop of Gubbio from 1129 to 1160. The citizens of Gubbio believed their bishop marshaled divine support for their forces in the battles with Perugia in 1151; Gubbio's victory is celebrated annually with the Race of the Candles on May 15, the eve of the Feast of St. Ubaldo. Ubaldo was canonized in 1192 and is the patron saint of Gubbio. His remains are housed in the Basilica of St. Ubaldo, located above Gubbio on the heights of Monte Ingino.

St. Francis of Assisi visited Gubbio frequently, and from these visits derives a story that is a recurring subject in the religious art of Italy, that of St. Francis and the wolf. Legend has it that a fierce wolf, in its hunger, was terrorizing the residents of Gubbio. St. Francis, who is reputed to have had the ability to communicate with animals, visited the wolf in its lair and told it of the harm it was doing to the residents of the town. St. Francis struck a bargain with the wolf. If it would stop its attacks, the townspeople of Gubbio would feed and care for the wolf for the rest of its life. The wolf agreed, and in a symbol of good faith and trust put its paw into the saint's hand. This particular scene has been captured repeatedly in art and can be found throughout Italy. In Gubbio, the Church of the Vittorina has a bas-relief over the door depicting the wolf with its paw in the hand of St. Francis.

Gubbio's political loyalties during the Middle Ages alternated between the Guelphs, who were allied with the papacy, and the Ghibellines, who supported the Holy Roman emperors. Guelph battlements can be found sculpted into the cornice of the doorway of the Palazzo dei Consoli. Strife between the two factions continued in the region throughout the thirteenth and fourteenth centuries until, in the late fourteenth century, the citizens of Gubbio asked for assistance from Count Antonio da Montefeltro, who held the duchy of Urbino. Antonio obliged, and the city came under his control. Gubbio's existence as a free town had come to an end. During the late fifteenth century, the family constructed a ducal palace in Gubbio; built to resemble the Montefeltro palace in Urbino, it still stands.

The Montefeltro family controlled Gubbio until their line died out early in the sixteenth century. The duchy of Urbino passed to Catherine de Médicis, then still a child. Taking advantage of the situation, the pope seized the duchy, and for a short time Gubbio became its capital. The duchy was regained by the nobility, but in 1631, it, along with Gubbio, was bequeathed to the papacy and became part of the Papal States.

It was during the late Middle Ages and early Renaissance that Gubbio produced its most famous art and artists. The city's famed miniaturist of the thirteenth century, Oderigi (or Oderisi) da Gubbio, was sufficiently well known to be mentioned by Dante in his *Divine Comedy*. His work is considered the forerunner of the Umbrian school of painting. Gubbio's foremost painter during the Renaissance was Ottaviano Nelli, whose most famous work, the *Madonna del Belvedere,* is in the Church of Santa Maria Nuova in Gubbio. In addition, frescoes by Nelli can be found in Gubbio at the Church of St. Augustine, the Church of St. Francis, and at Santa Maria della Paggiola.

Matteo Gattapone da Gubbio was one of Italy's most important fourteenth-century architects and also worked in Bologna, Spoleto, and Perugia. The Palazzo dei Consoli is primarily Gattapone's, but the facade of the grand entry, including the sweeping stairway from the square, the carved doorway, and its flanking windows, is credited to Angelo da Orvieto.

The town is also justly famous for the Renaissance pottery of Maestro Giorgio Andreoli, who came to Gubbio with his brothers, Salimbene and Giovanni. With the promise of exemption from taxes, Maestro Giorgio, as he is now commonly called, established a factory in 1498 for the production of majolica pottery. Majolica pottery is made of earthenware covered with an opaque tin glaze that is decorated before firing. The result is a beautiful and unusual reddish sheen that gives the pottery the appearance of metal. Although the tradition of ceramics has been maintained in Gubbio, Maestro Giorgio's original factory endured for less than a century, going out of existence when the master's eldest son died in 1576. An example of Maestro Giorgio's work can be found on display in the museum in the Palazzo dei Consoli.

By the eighteenth century, with Gubbio continuing to be part of the Papal States, the power in the town was primarily in the hands of the aristocracy. Most of the common people were powerless and poverty-stricken; these factors fostered support for revolutionary movements, which gained strength in the period following the Napoleonic Wars (in which Gubbio had been annexed for a period to Napoléon's Roman Republic). Gubbio reverted to papal rule after Napoléon's downfall, but revolutionists set up their own for forty days in 1831. The demise of the revolutionary government did not stamp out support for liberalism and nationalism, however. Volunteers from Gubbio took part in the uprisings of 1848–49 and 1859–60, the latter resulting in the creation of a unified Italy.

During World War II, the Germans mounted artillery directly above the town, on Monte Ingino. The town was also a center of anti-Fascist activity, as evidenced by the monument to the forty martyrs, residents of Gubbio, who were killed in retaliation for attacks by resistance fighters.

Today, Gubbio appears much as it did in medieval times. The historic center of the hilly town has five levels of streets, linked by sloping alleys, and is dominated by the Palazzo dei Consoli on the Piazza della Signoria. A fortified wall, portions of which date to the thirteenth century, surrounds the oldest part of the town. A distinctive feature of many of the old houses in Gubbio is the *porta del morto*, or gate of the dead, a narrow doorway through which the deceased, in their coffins, took final leave of their homes. In addition to the Palazzo dei Consoli and private homes, Gubbio's historic buildings include the fourteenth-century Palazzo del Pretorio, begun in 1349 but never completed, which is located across from the Palazzo dei Consoli on the Piazza Grande. This building contains the archives of the commune of Gubbio from its earliest days and a library. It is also the current town hall. The streets above the two palazzi contain another square, on which can be found the town's twelfth-century cathedral, attributed to a builder named Giovanni da Gubbio. Within the cathedral is the Tesoro, or treasury, containing, among other religious items, an embroidered cope, an ecclesiastical vestment given to the cathedral by a former bishop of Gubbio, Marcello Cervini, who in 1555 became Pope Marcellus II. Across the square from the cathedral is the Montefeltro ducal palace.

Tourism plays a major role in the economy of Gubbio, with several festivals held throughout the year. The most famous of these is the Corsa dei Ceri, or Race of the Candles, on the eve of the feast day of St. Ubaldo. In the race, teams of men in medieval costumes representing three different guilds carry large figures of three different saints in procession. The masons, in yellow, carry St. Ubaldo. The merchants, in blue, carry St. George. The peasants guild, in black, carry St. Anthony. Once outside the walls of the old town, the men begin to run, balancing their saints aloft on their shoulders, up the climbing road to the Basilica of St. Ubaldo, atop Monte Ingino. There, the figures remain until the following May. Another popular festival is the Palio della Balestra, an annual crossbow competition with the competitors in medieval costume, held the last Sunday in May.

Further Reading: *The Bronze Tables of Iguvium* by James Wilson Poultney (Baltimore: American Philological Association, 1959) is a scholarly monograph dealing for the most part with a linguistic analysis of the Iguvine tablets, but including an overview of the information they provide concerning religious ritual and the structure of society. *Assisi and Umbria Revisited* by Edward Hutton (New York: McKay, and London: Hollis and Carter, 1953) is a guide to the towns of Umbria, with a chapter devoted to Gubbio. The book emphasizes the religious art and history of the area and also contains descriptions of the interiors of the most prominent secular buildings and churches of Gubbio. *Hill Towns of Italy* by Lucy Lilian Notestein (Boston: Little, Brown, and London: Hutchinson, 1963), another guide to the towns of Italy, also contains a chapter devoted to Gubbio.

—Rion Klawinski

Guernica (Vizcaya, Spain)

Location: Two hundred sixty-six miles north of Madrid near the French border in the Vizcaya region of Spain.

Description: The spiritual home of the Basques, Guernica was destroyed by bombs in an air raid on April 26, 1937, during the Spanish Civil War. It then became the subject of one of Pablo Picasso's most famous paintings and is closely associated with the ravages of twentieth-century war. Only a few structures survived the attack; the rest of the town has since been rebuilt.

Site Office: Turismo Bulegoa
Artekale, 8-E
48300 Guernica, Vizcaya
Spain
(94) 6255892

When most people hear the word Guernica, they think of the famous painting by Pablo Picasso. While that painting may be this little village's claim to fame, Guernica has a long, and recently tragic, history. The village itself is quite small and sparsely populated. Current estimates put the number of inhabitants at around 17,000. It has never been a major tourist attraction and there are really no important structures there. Visitors who do come to Guernica generally do so because of its important place in the history of Spain, first as the seat of the Basque region and second as the first village to be destroyed by air attack in the "total war" waged by the Spanish fascists and their German and Italian allies during Spain's Civil War.

The valley in which Guernica is located has long been inhabited, its rich soil having provided a hospitable environment for both animals and humans for several millennia. Caves in the region contain some of the most significant prehistoric paintings in Europe and testify to the presence of humans in the area during the early Paleolithic period around 15,000 B.C. Recent archaeological work in the region also indicates that the Roman Empire reached these shores—the excavation of coins, glass, and a statuette of the goddess Isis Fortuna support this belief.

The modern town of Guernica was established in the early part of the fourteenth century. Tradition holds that the actual founding by Count Don Tello took place on April 28, 1366. The village grew very slowly and because of its many timber structures experienced devastating fires in 1521, 1537, and 1835. In addition to fires the village was also subject to floods when high tides coincided with periods of heavy rain. Beyond these natural disasters, Guernica experienced continuous conflict with the neighboring village of Lumo. In 1882 the two united to become Guernica-Lumo, now known sim-

ply as Guernica. As a result of this union the village was able to undertake a number of public projects including the creation of water and sewer systems, the construction of schools, and in 1925 the construction of the pelota court where this national game of the Basques could be played. Industry came to Guernica in the early part of the twentieth century, and by 1936 the population had grown to 6,000. Among the most prominent industries in the village were the production of munitions and rope-soled sandals.

During most of Guernica's history it has served as both the spiritual and political center of the Basque population. The Basque assembly, also known as the Batzarrak, met in Guernica as early as the ninth century under an oak tree known as the Tree of Guernica. Meeting under a tree was a political tradition throughout Vizcaya, but the tree in Guernica was the most important. Until 1876 Vizcayan laws were written in the shade of this tree. After Vizcaya came under the power of the Crown of Castile, the Castilian kings had to take an oath to respect the laws of Vizcaya and often traveled to do so under the Tree of Guernica. Remains of an early tree, which is said to have been taken from a sprout of an even older tree, is preserved in a shrine in Guernica, and there is also a new, younger tree planted nearby.

Guernica was a relatively peaceful, out-of-the-way village until 1937. At the start of the Spanish Civil War in 1936, the Basques were granted autonomy, and they elected a president who took office in Guernica. In spite of international support, the Basques had few military resources. The republican government in Spain that had granted Basque autonomy was suffering at the hands of the revolting nationalists, or fascists, in the Civil War; strong support from the Germans and Italians had clearly given the nationalists the upper hand in spite of fierce resistance. With the power of the German Luftwaffe, the northern front in the war appears, in retrospect, to have had little or no chance of surviving the onslaught. However, nothing could have prepared the people of Guernica for what was to happen to them and their village.

Beginning in March 1937, with the aid of the Germans, the nationalists began to bomb the town of Durango, but without great success in hitting their intended targets. In fact, many refugees from Durango had flooded into Guernica during the weeks prior to Monday, April 16, the day of the Guernica bombing. The first day of the week was the traditional day for market in Guernica, and on this Monday, like any other, the people of the city went about the business of buying and selling and then took an afternoon siesta. As the townspeople were preparing for their afternoon and evening activities, the church bells rang out in warning of an air raid. The first attack took place around 4:30 in the afternoon and consisted primarily of large bombs intended to destroy buildings. The second wave was incendiary bombs intended to set

Picasso's Guernica
Courtesy of Museo Nacional Centro de Arte Reina Sofia
Copyright 1995 Artists Rights Society (ARS), New York/SPADEM, Paris

fire to the rubble, and finally the third wave of attack consisted of the machine-gunning of those attempting to flee.

The entire attack lasted some three hours without interruption. The destruction of the center of the village was nearly complete, although the bridge leading in and out of the village, which, it has been speculated, was the true target of the attack, and other structures, as well as the famous oak tree, were left undamaged. The famous Condor Legion of the German army was responsible for the bombing with the assistance of the Italians.

The nationalists in Spain worked diligently to cover up their involvement in the bombing of Guernica and, in fact, closed off the town for five days after the attack so that they could fill in craters created by the bombs and set fire to the ruins in order to make it appear as if a faction of the Basques had actually destroyed the town. They denied involvement in the incident and promulgated the idea that the Basques were responsible in all of their public pronouncements. The nationalists even took pictures of their troops surrounding the oak tree as if they were the saviors of the Basque tradition. In addition, the dead were buried in unmarked and mass graves and lists of the deceased were removed from village registers. The estimates of the number of casualties vary widely, from 200 to 2,000.

The reality of the attack began to slowly emerge through highly contradictory and sometimes greatly exaggerated reports in the press. The bombings made headlines from London to New York. Statements from the mayor of Guernica and the parish priest, who insisted that the town had been bombed by German war planes and that the Basques had not destroyed their own town, began to paint a clearer picture of the events in Guernica. The democratic press within Spain denounced the bombing but it was the artist Pablo Picasso who was able to most effectively capture the horror of the event in his most famous painting.

Several months before the bombing of Guernica, Picasso had been asked to contribute a large mural for display in the Spanish Pavilion at the World's Fair in Paris. Picasso was hesitant to commit and appears to have struggled indecisively with the subject matter for several months. Even at the start of April, just weeks before the fair was to open, Picasso was still making sketches of a rather personal scene—something of a monument to his private life tentatively entitled *The Studio*, showing the love of an artist both for his craft and for his women.

Living in Paris, Picasso was aware of the events in Guernica, and it seems likely that by the end of April he had seen photos of the destruction of the village in the French press. On May 1 Picasso began to develop his ideas for a new mural. What emerged rather quickly, over the next month or so, was not a portrayal of war per se, but rather a striking collection of images, many of which had been explored in Picasso's earlier works, particularly those of the bullfight. A great deal has been written about this painting and all of its possible meanings, but after all is said it remains a powerful indictment of war and cruelty. The image of the screaming mother clutching the dead child and the horse in agony stand in sharp contrast to the flower growing out of the broken spear and the powerful bull, generally believed to represent General Francisco Franco, leader of the nationalists.

This immense painting, some eleven feet by twenty-five feet, was first featured in the Spanish Pavilion in the Paris World's Fair in 1937. The Spanish Pavilion itself opened late and to little fanfare. Dwarfed by the pavilions of the Germans and the Soviets, the Spanish display was essentially a propaganda effort on the part of the republican government, which was suffering heavy losses in the Civil War. Picasso's painting became the centerpiece of the pavilion, which also featured some of his sculpture and works by other notable artists such as Alexander Calder.

At the conclusion of the fair *Guernica* was displayed in several places and was eventually housed at the Museum of Modern Art in New York City. Picasso had proclaimed that the work would never be shown in Spain until after the death of the victorious dictator Franco. In the late 1970s, with the death of Franco and a return to a constitutional monarchy in Spain, the stage was set for the return of the painting. In 1981 it was placed in the Prado in Madrid, and was later moved to the Museo Nacional Centro de Arte Reina Sofia, in the same city. The painting has been called the most important work of art of the twentieth century, a claim that is certainly reinforced by the extreme security measures in place to protect *Guernica* in its current home.

In a rather ironic twist to the history of the bombing of Guernica, one of the structures that was not destroyed was the Unceta factory, which within a week was back in business producing weapons for the nationalists. Today, the Unceta factory employs some 500 and exports guns all over the world.

During the 1940s the village of Guernica began to be rebuilt by the Franco regime through the work of republican prisoners. The overall rebuilding process in Spain was an enormous task; some 300 towns had been effectively demolished during the Civil War. In the reconstruction of Guernica special efforts were made to recreate some of the village's original character. What a visitor to Guernica sees today is a town with little evidence of the atrocities and near-complete destruction of 1937.

Some parts of the village were spared in that Monday bombing raid and are worth noting. The most impressive structure in the village is the Church of Santa Maria, which was begun during the fourteenth century and completed in 1715. The main portal is decorated with sculpture and is crowned by an image of the Virgin Mary. In addition to the original decoration and sculpture on the church, it also houses sculptures by contemporary Spanish artists.

The parliament building, also known as the Casa de Juntas Forales, was built during the early nineteenth century in the neoclassical style. The convent of Santa Maria la Antigua is also housed within this complex. Within the Casa de Juntas is a display on the history of Guernica; just outside, one can see the remains of the famous oak.

In addition to these structures, there are several pieces of sculpture in Guernica that are significant to its history. In the Plaza de los Fueros stands a statue of Count Don Tello granting the charter to the town of Guernica. Erected in 1966, the sculpture was created by Agustín Herranz, a native Guernica artist. In 1986 the *Great Figure in a Shelter* by Henry Moore was placed in the garden area in front of the Church of Santa Maria. This monumental sculpture can be viewed from both within the piece itself and from the outside. To commemorate the fiftieth anniversary of the bombing of Guernica, Eduardo Chillida created the piece *Gure Aitaren Extea (The House of Our Father)*. It is situated near the sculpture by Moore in the garden of the church, and through its window one can see the Tree of Guernica. These sculptures and the town itself have become symbols of peace. In that same spirit, Picasso's painting is perhaps the greatest artistic condemnation of war ever produced. Commenting on this work, and on the social obligation of artists in general, Picasso said:

> I have always believed, and still believe, that artists who live and work with spiritual values cannot and should not remain indifferent to a conflict in which the highest values of humanity and civilization are at stake.

Further Reading: A great deal has been written about Guernica the town and about *Guernica* the painting. *Guernica: The Crucible of World War II* by Gordon Thomas and Max Morgan Witts (New York: Stein and Day, 1975; as *The Day Guernica Died*, London: Hodder and Stroughton, 1976) is a detailed account of the week of the bombing, with numerous personal histories from those who survived. The standard source on the Civil War in general is Hugh Thomas's *The Spanish Civil War* (New York: Harper and Row, and London: Eyre and Spottiswoode, 1961). A good introduction to the period leading up to the bombing can be found in Martin Blinkhorn's *The Emergence of Modern Spain, 1808--1939* (Oxford: Blackwell, 1989). As for the painting, two good texts are Ellen C. Oppler's *Picasso's Guernica* (New York and London: Norton, 1988) and Herschel B. Chipp's *Guernica* (Berkeley: University of California Press, 1988; London: Thames and Hudson, 1989).

—Michael D. Phillips

Guimarães (Braga, Portugal)

Location: Twelve miles southeast of the town of Braga, in the Braga district, northwest Portugal.

Description: Birthplace of Afonso I (Afonso Henriques), first king of Portugal, and site where he came to power, giving rise to the modern Portuguese nation; noted for its well-preserved medieval section, which includes the castle in which Afonso was said to be born, numerous churches, private homes, and museums.

Site Office: Guimarães Tourist Information
83 Largo 28 de Maio
Guimarães, Braga
Portugal
(53) 412450

Guimarães's place in Portuguese history is assured as the birthplace of Afonso I, first king of Portugal, and therefore Guimarães is considered the birthplace of the Portuguese nation as well. The town itself was founded in the ninth century A.D. by one Vimara Peres as part of the resettlement by Christians of lands that had been recaptured from the Moors. The region apparently had been occupied since ancient times, as evidenced by a group of pre-Roman stone huts discovered at Citânia de Briteiros, ten miles from Guimarães. Little of note, however, occurred there until the Middle Ages.

A countess named Mumadona established a monastery at Guimarães in the tenth century, and had a castle built to defend the monastery and town against incursions by Normans and Arabs. By the eleventh century the name Portugal or Portucale was used to refer to the lands between the Minho and Douro Rivers, including Guimarães, in the northernmost section of modern-day Portugal. It was a county dependent on the rulers of Asturias and León. Around 1095 King Alfonso VI of León included Guimarães as part of the dowry when his daughter, Teresa, married the French nobleman Henry of Burgundy. Henry then became count of Portugal. He rebuilt the castle at Guimarães, encouraged foreign trade in the city, and began restoring churches and cathedrals in the area. Henry and Teresa's son, Afonso Henriques, was born around 1109, reputedly in the castle at Guimarães. Henry's cousin, Raymond of Burgundy, had married another of Alfonso VI's daughters, Urraca, and was made lord of Galicia, in what is today the portion of Spain directly north of Portugal.

After Henry died about 1114, Teresa became regent of Portugal. She soon encountered conflicts with local nobles—among other things, they disapproved of her purported love affair with a Galician count, Fernando Pérez de Trava—and she began to believe subjection to Galicia would be a preferable situation. In the end, she managed to maintain a semblance of independence for Portugal and even add to her holdings while under an oath of fealty to Galicia. Then, after Urraca's death in 1126, her son, who had become King Alfonso VII of León and Castile, sent an army to invade Portugal and besiege Guimarães. Teresa and Fernando surrendered, but the terms were unacceptable to Afonso Henriques, now grown to manhood and quite independent-minded. He began to take control of the government, and when Teresa and Fernando sent an army to oppose him, he and his forces won a decisive victory in 1128 at the battle of São Mamede, near Guimarães. His path to victory was undoubtedly smoothed by the widespread internal opposition to Teresa and Fernando's rule. Teresa and Fernando were subsequently exiled to Galicia.

Afonso Henriques spent the next several years trying to expand his territory, an activity that brought him into periodic warfare against his cousin, Alfonso VII of Castile and León. Afonso Henriques also did battle with the Moors, winning a major victory near Santarém in 1139. He declared himself king of Portugal, and Alfonso VII, impressed by his cousin's military might, officially recognized him as an independent ruler in 1143. Afonso Henriques, now King Afonso I, lived and ruled until 1185. During this period he and his army recaptured from the Moors much of the territory that is now Portugal.

Afonso I had made Coimbra his capital beginning in 1139, but Guimarães remained an important center of commerce throughout the Middle Ages. It was a major producer of linen cloth, and a trade fair began to be held there in 1258. In 1357 the English Alliance, a trade agreement between Portugal and England, was signed at Guimarães. Commerce between the two countries flourished, and they became allied militarily as well.

In the fourteenth century King John I, first king of the Aviz dynasty of Portugal, began reconstruction of the Collegiate Church of Nossa Senhora da Oliveira in Guimarães. The original church on the site had been part of the monastery founded by Countess Mumadona in the tenth century; nothing remained of that church, but some portions of a thirteenth-century rebuilding were incorporated into John I's project. The church marked the spot where Wamba, a Visigothic warrior, was proclaimed king of his people in the seventh century. He plunged his staff into the ground and declared he would not accept the royal office unless the staff miraculously grew olive branches. It did, and therefore subsequent religious constructions on the site were dedicated to Nossa Senhora da Oliveira—Our Lady of the Olive Tree. King John I had supposedly prayed to Nossa Senhora da Oliveira during the battle of Aljubarrota, in which he, with the aid of the English, had defeated the Castilian king, who had sought to

Entrance to Nossa Senhora de Oliveira, Guimarães
Photo courtesy of Portuguese National Tourist Office

claim the throne of Portugal. John showed his gratitude by rebuilding the church, which received further alterations in the seventeenth, eighteenth, and nineteenth centuries, and remains one of the dominant features of Guimarães. Adjacent to the church, in the former cloister, is the Museu Alberto Sampaio, featuring many notable artworks and artifacts, including a tunic supposedly worn by John I at Aljubarrota. Guimarães's old town hall also was built in the reign of John I and was reconstructed in the seventeenth century.

John I's illegitimate son, Afonso, count of Barcelos and first duke of Bragança, built an imposing, Norman-influenced palace in Guimarães in the fifteenth century. It

was abandoned a century later (the Bragança family fell out of favor with the rulers of Portugal, then rebounded to provide the country with several of its kings), but was restored in the twentieth century. It now houses a museum, and is the official residence of the president of Portugal when he visits the northern part of the country.

Also in the fifteenth century—about 1465—Gil Vicente, probably Guimarães's second most famous native, was born. Originally a goldsmith—Guimarães is still noted for such craftsmen—he turned to playwriting, became court dramatist to John II and Manuel I, and has been called the father of Portuguese theatre. Some of the work of his first career is evident in Guimarães: a monstrance attributed to him is in the Museu Alberto Sampaio.

Guimarães declined in political importance after the Middle Ages, but the town never lost its sense of its place in Portuguese history, so it has been careful to preserve its medieval monuments. Many of them were restored during and after the Renaissance. The town also has taken an interest in its earlier history. The Museu Martins Sarmento was opened in the early twentieth century to house artifacts excavated at Citânia de Briteiros and other nearby archaeological sites. It was named for the archaeologist who led many of these projects. Part of the museum's collection is displayed in the adjacent fourteenth-century cloister of the Church of São Domingos.

Another notable museum, which also houses an inn, is the former Monastery of Costa, reputedly founded by Afonso Henriques's wife, Queen Mafalda, in 1154. The monastery complex was rebuilt in the eighteenth century and restored after a fire in 1951. Its collection includes many religious sculptures.

Still, the focal point of Guimarães is the castle, on what is known as the Sacred Hill. Even though it is smaller than the nearby ducal palace, the castle takes first place in the hearts of the residents of Guimarães. It is well preserved, despite having been used as a debtors' prison in the nineteenth century; an extensive restoration was undertaken in 1940. It retains its original tenth-century keep, along with many later additions; the keep and its eight surrounding towers inspired the Portuguese coat of arms. A statue of Afonso Henriques stands near the castle, and the adjacent chapel in which he was reportedly baptized draws many visitors.

The castle, medieval churches, and older homes of Guimarães are tightly packed into the town's old quarter. South of this area, in somewhat shocking contrast, are modern, undistinguished apartment buildings and factories producing such products as textiles, shoes, and cutlery. Industry has caused pollution and traffic jams, but has kept Guimarães prosperous, and efforts to preserve the medieval structures of Guimarães are ongoing.

Further Reading: The role of Afonso Henriques in the foundation of modern Portugal is covered in most historical works on the country as a whole, such as *From Lusitania to Empire,* the first volume of A. H. de Oliveira Marques's *History of Portugal* (New York and London: Columbia University Press, 1972) and *The Portugal Story: Three Centuries of Exploration and Discovery* by John Dos Passos (New York: Doubleday, 1969; London: Hale, 1970). The historic sites of Guimarães are described thoroughly in most of the major guidebooks.

—Patricia Wharton and Trudy Ring

Halicarnassus (Muğla, Turkey)

Location: On the Aegean coast of Turkey, in the southwestern corner of Anatolia (Asia Minor).

Description: Ancient city on the site of modern Bodrum. Greek colonists founded Halicarnassus about 1000 B.C. in a region called Caria in Asia Minor. After falling under Persian rule in the sixth century B.C., Halicarnassus was governed by a native Carian dynasty. Under King Mausolus, the city flourished. His tomb, the Mausoleum, was known as one of the Seven Wonders of the World. After Alexander the Great captured Halicarnassus during his invasion of Asia in 334 B.C., the city's importance diminished.

Site Office: Bodrum Tourist Information Office
Barış Meydanı
Bodrum, Muğla
Turkey
(252) 3161091 or 3167694

Halicarnassus, a Greek colony on the Aegean coast of Turkey, is remembered chiefly as the site of the Mausoleum, one of the Seven Wonders of the Ancient World. The spectacular tomb of King Mausolus was built around 350 B.C., during an era of great prosperity for the city. Just seventeen years later, Alexander the Great's conquest of Asia Minor devastated Halicarnassus. Earthquakes and warfare eventually reduced it to ruins; little of ancient Halicarnassus remains today in the coastal resort town of Bodrum.

Although Mausolus's reign represented the height of its civic glory, Halicarnassus had a long prior history as a thriving Greek settlement on the Aegean coast. Its geographic location gave it a natural command of the sea route between the island of Kos and Asia Minor, one factor explaining its importance. Greek colonization of Asia Minor's western coast began about 1100 B.C. The Dorians, a warring people whose origins are unclear, had risen to power in Greece during the twelfth century B.C. The first Greek colonists, the Aeolians, escaped Dorian aggression by crossing the Aegean Sea to the islands of Lesbos and Tenedos. They resettled again in the Troad (land of Troy), beginning the wave of Greek colonization in western Asia Minor. The Ionian people left Greece about a century later to establish several cities south of the Aeolian colonies.

The Dorians then sought to expand their own influence through colonization. They founded six cities along Asia Minor's Aegean coast; three on the island of Rhodes, one on Kos, and two on the mainland (Cnidus and Halicarnassus). Although mountainous and isolated, this region of southwestern Asia Minor, called Caria, was not uninhabited when the Dorians arrived in 1000 B.C. The native Carian people had already settled in the hills surrounding the harbor where Halicarnassus was built. Although initially displaced by the Greek colonists, the Carians returned to the town that had become Halicarnassus and coexisted with the Dorians for centuries. An independent Carian culture remained largely intact despite the city's strong Greek identity. Carian and Greek names, for example, were represented equally among Halicarnassus's residents. Some of Halicarnassus's greatest rulers were Carian satraps (governors) of the Persian Empire later in the first millennium B.C. Not until after the conquest of Alexander the Great were Carians and other native western Anatolian peoples fully Hellenized.

The Greek colonies prospered, and by 800 B.C. they had formed confederations: the Aeolian League (twelve cities), the Ionian League (twelve cities, called the "Dodecapolis"), and the Dorian League (six cities, called the "Hexapolis"). Relations among cities in the Hexapolis were troubled. Halicarnassus was expelled from the league, ostensibly because a Halicarnassian athlete named Agasikles accepted a prize he won at the Dorian games at Triopion instead of offering it to Apollo. A more likely explanation cites resentment at Halicarnassus's domination of the league as the reason for its expulsion by the other member cities.

While the Dorians fought internally, the Aeolian and Ionian Leagues looked toward expansion. In 757 B.C., the Aeolian League city Cyme founded the first Anatolian Greek colony in Magna Graecia (southern Italy) at Cumae. The Ionian League soon led the colonization effort, however, establishing hundreds of cities around the Mediterranean and Black Seas between 750 and 600 B.C. Ionian influence also expanded southward to neighboring Caria; Halicarnassus became completely Ionian by the fifth century B.C.

One of Halicarnassus's best-known natives, the historian Herodotus, was born there between 490 and 484 B.C. and wrote in the Ionian dialect. Herodotus, called "The Father of History," traveled throughout Egypt, Babylonia, Greece, and Asia Minor, recording his observations about regional cultures, geography, history, and folklore. Herodotus's work was part of a larger flowering of Greek civilization in the Ionian cities, which included advances in philosophy, physics, medicine, mathematics, geography, astronomy, architecture, sculpture, and poetry.

Ionian success at colonization, however, aggravated relations with the volatile eastern kingdoms of Lydia and Persia. Threatened by Ionian expansion, King Gyges of neighboring Lydia invaded Ionia about 665 B.C. Within a century, Lydia controlled all Greek cities on Asia Minor's western coast, including Halicarnassus. Lydian government was not necessarily oppressive; the last Lydian king, Croesus, was particularly impressed with Greek culture and was generous toward his subjects. Lydian rule was supplanted in 546

Castle of St. Peter, built from ruins of the Mausoleum
Photo courtesy of Embassy of Turkey, Washington, D.C.

B.C., when Persian king Cyrus defeated King Croesus, bringing all of western Asia Minor and part of eastern Greece under Persian authority. Herodotus, who recorded events of the Persian-Greek war, blamed Croesus for too much ambition in entering military conflict with Persia.

The Ionians fought against Persian rule, beginning with a revolt in 499 B.C. that, despite a few victories in Asia Minor, was crushed five years later. Persia's King Xerxes I led an invasion of Greece in 480 B.C.; he succeeded in demolishing Athens before his naval fleet lost the battle of Salamis. Although Halicarnassus was by then an Ionian city, it was ruled by a Carian queen, Artemisia, who supported King Xerxes. Considered one of Xerxes' closest advisors, Artemisia even joined the battle of Salamis in her own ship. In 479 B.C., the Persian threat to Greece finally ended after Greeks won battles at Plataea and Cape Mycale.

Persian interest in Asia Minor, however, remained a threat to the Greek cities on the Aegean coast. These cities joined the Delian League, a naval confederation chiefly controlled by Athens, in 478 B.C. Halicarnassus became an Athenian naval station before the league disbanded in 404 B.C., after Athens lost the Peloponnesian War to Sparta. A treaty between Sparta and Persia, called the "King's Peace,"

formally relegated all cities in Asia Minor to Persian rule in 386 B.C. Persia dominated the Greek cities for the next fifty years.

The Persian Empire, which spread across Asia Minor and the Near East, maintained power in its distant lands by delegating government to satraps, autonomous military governors. Yet Persian control soon weakened in western Anatolia, a fact symbolized by the evolution of the Carian satrapy into a dynasty. King Mausolus, the satrap who ruled Caria from 377 to 353 B.C., proved to be a powerful and independent leader.

After Mausolus came to power, he quickly gained control of additional territories, including parts of Ionia, Lydia, Lycia, and the islands of Rhodes, Kos, and Keos, greatly enlarging his kingdom. Mausolus then transferred the capital of Caria from Mylasa to Halicarnassus, which he rebuilt into a splendid city. The king forcibly relocated the residents of six Lelegian towns on the Myndus peninsula to his new capital, quadrupling Halicarnassus's population. This move no doubt provided a large work force for the public building projects Mausolus had planned.

Soon after Halicarnassus became Caria's capital in 370 B.C., Mausolus ordered the construction of massive defense walls, with watchtowers at regular intervals, to surround the city. Halicarnassus's harbor was enclosed, and dockyards were built. An agora (marketplace) was located by the harbor. On the summit of the acropolis (now called Göktepe) that rises above the harbor stood a shrine of Ares with an enormous statue; on another hill, near the fountain of Salmakis, a temple was dedicated to Aphrodite and Hermes; and, from yet another summit, Mausolus's palace overlooked the city.

Although technically a mere satrap of the Persian Empire, Mausolus commanded considerable power and influence in southwestern Asia Minor. In a bid for independence, Mausolus and the satrap of Mysia, Orontes, together initiated the Satraps' Revolt of 363 B.C. Within three years Persia's King Artaxerxes II had quelled the rebellion, but Mausolus remained firmly in control of Caria, his sense of power undiminished.

Mausolus's most impressive edifice in Halicarnassus was the shrine he planned as a tomb for himself and his wife (who was also his sister), Artemisia II. The king supervised the beginning of construction before his death in 353 B.C.; the Mausoleum was completed in 351, shortly before Artemisia died. Rising from a point midway up the slopes of the acropolis, the Mausoleum must have been visible from the entire city, perhaps even to ships entering the harbor. Renowned as the largest and most spectacular tomb in the Greek world, the Mausoleum was cited by Philo of Byzantium as one of the Seven Wonders of the World. Its name has passed into the English language as a synonym for a large, stately sepulchre.

Although little of the edifice remains extant, ancient descriptions from writers such as Pliny and Vitruvius supply enough detail to construct an accurate picture of the Mausoleum. Three main sections comprised the royal tomb: a base, a portico, and a pyramidal roof. The tall rectangular base, which served as the structure's basement, measured perhaps 108 by 127 feet, according to one modern estimate. The portico section had an enclosed interior and an outer walkway with thirty-six Ionic columns. Above the portico rose a stepped pyramidal roof in twenty-four levels, crowned by a "quadriga," a sculpture of a chariot drawn by four horses. The entire monument stood about 142 feet high.

The tomb built for a Carian king in a Hellenized city fittingly combined Greek styles with non-Greek forms. The tall base took its shape from Persian structures in Lycia; the colonnade utilized the Greek Ionic order; and the stepped pyramid roof may have been inspired by the Egyptian monuments. Two architects are credited with designing the Mausoleum. Pythius, recognized as one of the most influential architects of his time, wrote books about his architectural designs and theories. His other notable building, still extant, is the Temple of Athena Polias in his native Priene. The second architect of the Mausoleum, Satyrus of Paros, co-authored one of Pythius's books.

Pythius and Satyrus conceived not only a sophisticated combination of cultural styles but also as new approach to design that integrated architecture and sculpture. Scholars estimate that more than 300 statues graced the Mausoleum. The tomb was also decorated with three sculptural friezes, which may have encircled the Mausoleum in horizontal bands, although their exact positions remain unknown. The friezes depicted a chariot race, a battle between Centaurs and Lapiths (the "Centauromachy"), and combat between male Greek warriors and Amazons (the "Amazonomachy"). Both the friezes and the figurative sculptures were created by four renowned sculptors: Scopas, Bryaxis, Leochares, and Timotheus.

In the two years of her reign. Artemisia II faced a serious challenge from Rhodes. Convinced that a kingdom ruled by a woman would be easily captured, the Rhodians sailed their fleet directly into Halicarnassus's harbor in 352 B.C. According to legend, Artemisia's forces surprised their attackers by sailing out of a mysterious "secret harbor" into the main harbor, where they defeated the Rhodian fleet. Artemisia took her revenge by capturing the city of Rhodes.

The period of Halicarnassus's civic splendor and military might, however, soon came to an end. Alexander the Great of Macedonia and his army crossed the Hellespont (Dardanelles) into Asia Minor in the spring of 334 B.C. He immediately entered battle with the Persian army at the Granicus River, on the southern shore of the Sea of Marmara. Despite a battleground that favored the defense, Alexander's forces overcame the Persians, effectively ending their rule in Asia Minor.

Alexander continued his march into Anatolia, where most cities surrendered to the conqueror. Halicarnassus did not succumb so quickly. Under the leadership of Memnon of Rhodes, the Halicarnassians fiercely bombarded the Macedonian army. Unable to stall the invaders any longer, though, Memnon ordered the defenders to abandon their posts. The Halicarnassians set fires in the city before withdrawing to the headlands above the harbor. Alexander's army devastated the city, and the Halicarnassian defenders surrendered to the Carian ruler Queen Ada, Mausolus's younger sister and a supporter of Alexander.

Alexander spared the Mausoleum, but his army destroyed the rest of the city. Halicarnassus's importance as a cultural and political center faded in the centuries following Alexander's invasion. The Macedonians and then the Ptolemies controlled the city until about 190 B.C. Halicarnassus may have been a free city for some time before its incorporation into the Roman Empire's Province of Asia in about 130 B.C. Under the Romans, Halicarnassus suffered losses of wealth and status; it was plundered by Verres in 80 B.C. and later by Brutus and Cassius. As Halicarnassus's fortunes dwindled, the city fell to ruins. The great Mausoleum, which had survived Alexander the Great's invasion, finally crumbled centuries later, probably the victim of earthquakes.

Parts of the Mausoleum survive, however, recycled as building materials for the historic centerpiece of modern Bodrum: the Castle of St. Peter. European soldiers of the crusades built this fortress on a peninsula high above the harbor in the fifteenth century. The Knights of St. John, or "Hospitallers" as they were also known, ruled the southeastern Aegean Sea, maintaining control with fast-sailing ships and strategically located strongholds, including those on Kos and Rhodes and at the mainland port city of Smyrna (now İzmir, Turkey). When their castle in Smyrna fell to Timur (Tamerlane) the Mongol in 1402, the knights relocated to another coastal site—above the harbor of ancient Halicarnassus. The toppled marble slabs of the Mausoleum provided a ready-made quarry for the knights, who incorporated several portions of the royal tomb, such as sculptural frieze panels, into the castle walls. Practically bare of all but rubble, the exact site of the Mausoleum was then forgotten for centuries.

The knights' dominance of the Aegean ended with the advance of Ottoman Turk leader Süleyman I, the Magnificent, in 1522. When Süleyman captured Rhodes, the knights' main base of operation, the crusaders abandoned the Castle of St. Peter and sailed for refuge in Malta. Soon thereafter, the Turks seized the castle, which they used as a prison. With the advent of Turkish rule, the town once known as Halicarnassus became Bodrum, which means "subterranean vault" or "dungeon" in Turkish.

The site of Halicarnassus's "wonder of the ancient world" remained a mystery to the modern world until the nineteenth century. In 1749, a traveler named Richard Dalton managed to gain entry into the Castle of St. Peter, where he noticed panels of a sculptural frieze incongruously built into the walls. Dalton's idea that these were remnants of the great Mausoleum sparked a quest to discover the Mausoleum's location. Scholars suggested several possible sites in Bodrum, but excavations that began in 1856 revealed nothing. A young British archaeologist, Charles T. Newton, evaluated Vitruvius's description and initiated excavations at a hillside location where Ionic architectural fragments were visible in the surface soil. Newton's investigations quickly proved successful. In January 1857, he uncovered a cache of sculptural fragments, and later his team excavated partial foundations, the basement, and the huge staircase and entry stone leading to Mausolus's tomb chamber.

Among the most prized Mausoleum artifacts unearthed by Newton are the figurative sculptures of a bearded man, a veiled woman, and an immense horse. The two human figures, while clearly carved in the Greek style, exhibit characteristics that suggest Persian identities. The man's shoulder-length hair, beard, and mustache mark him as a non-Greek, and the woman's hair is set in a Persian style. Originally, scholars identified the couple as Mausolus and Artemisia themselves and postulated that the king and queen, with the surviving horse, had been part of the chariot sculpture atop the Mausoleum. Modern scholars question the assumption that the figures are Mausolus and Artemisia, and examination has led to a new theory that they were exhibited at a lower height, possibly inside the shrine. As nineteenth-century Turkey did not forbid the export of its antiquities, these sculptures, and many others, were removed by British excavators and given to the British Museum in London.

Today, Bodrum is a wealthy coastal resort town, the main yachting port in Turkey. Very little of ancient Halicarnassus remains visible to visitors—only parts of Mausolus's defense wall and the sparse outlines and scattered seating blocks of a pre-Hellenic theatre, built into the slope of Göktepe. The most historic standing building in Bodrum, ironically, is the Castle of St. Peter, built from the rubble of Halicarnassus's most famous monument. The castle is open to visitors and boasts an excellent Museum of Underwater Archaeology, which displays rare finds from ancient shipwrecks (Bronze Age to medieval) off the Aegean coast.

Further Reading: A number of books address the long, complicated history of Turkey, including significant settlements such as Halicarnassus. Richard Stoneman's *A Traveller's History of Turkey* (Moreton-in-Marsh, Gloucester: Windrush Press, and Brooklyn, New York: Interlink, 1993) presents a span of Turkish history from prehistoric civilizations to the fall of the Ottoman Empire. His account is detailed, thorough, and an excellent source for a broad understanding of Turkish history. John Freely's *Classical Turkey* (San Francisco: Chronicle, and London: Johnson Editions, 1990) relates Hellenic history as it pertains to archaeologically important sites. Sybille Haynes's *Land of the Chimaera* (London: Chatto and Windus, and New York: St. Martin's, 1974) focuses chiefly on the archaeological significance and excavations of historic sites in southwestern Asia Minor.

—Elizabeth E. Broadrup

Hattusa (Yozgat, Turkey)

Location: Central Turkey, approximately 112 miles northeast of Ankara.

Description: Hattusa, with the neighboring sites of Yazilikaya, a sanctuary set in the rock a short distance northeast of the city, and Alaça Hüyük, slightly further to the north, is the major Hittite site in central Anatolia. Set on the hill called Bogazkale, the ruins of the town and its acropolis dominate the modern town of Bogazköy. Although there is a small museum on the site, most of the best-preserved artifacts have been removed to the Museum of Anatolian Civilizations in Ankara. Some of the architectural sculptures—the figure at the King's Gate, for instance—have been removed to Ankara and have been replaced by concrete casts.

Contact: Instanbulluğlu Mah.
Emniyet Cad.
Nizamoğlu 'Konaği
Yozgat, Yozgat
Turkey 5493

When a young French explorer named Charles-Felix-Marie Texier stumbled across cyclopean ruins in the highlands of north-central Anatolia in July 1834, he had no idea that he had found the remains of the capital of one of ancient history's most enigmatic peoples: the Hittites. The Hittite name had been preserved in the Bible: for instance, Uriah, the Israelite soldier whose wife, Bathsheba, King David coveted, was a Hittite. King Solomon had wives who were of Hittite extraction, and his heirs could call upon the rulers of Hatti (Anatolia) for military support. Biblical scholars, however, had little concept of the extent of the Hittites' power and influence, and generally classified them with other Palestinian tribes who opposed the kings of Israel. Some scholars still argue that the Biblical Hittites were actually the Egyptian "sons of Het," a different group of people altogether.

Texier was not the first European to report finds of Hittite artifacts. The Swiss explorer Johann Ludwig Burckhardt described Hittite inscriptions in the ancient Syrian city of Hamath in 1822. But Burckhardt did not recognize the origins of the inscriptions. It was not until 1874 that an Irish missionary named William Wright made the connection between the Syrian inscriptions and the Biblical Hittites. In conjunction with the British philologist A. H. Sayce, Wright published the first book on the subject, *The Empire of the Hittites,* in 1884. Excavations in the early twentieth century uncovered a large cache of inscribed tablets. It was not until 1915, however, that the Hittite language was translated and revealed to belong to the Indo-European family of languages—thus providing the earliest known example of a kingdom run by people who spoke a tongue related to those of modern Europe.

Texier did not know what to make of his discovery. He was actually searching for the settlement of Tavium, a colony of Celtic tribesmen transplanted there during Roman times. Because the native Turks were very suspicious of foreigners, especially Europeans, the French explorer received very little guidance and was forced to ask the local inhabitants about any ruins in the vicinity. When he arrived in Bogazköy, he was directed to the nearby hills. After a difficult climb, Texier came upon the remains of a fortress constructed of gigantic blocks of stone, surrounded by a four-mile fortified wall that enclosed an area of more than three-quarters of a mile. Piercing the wall were two gates, guarded by monumental carvings: on one, lions, and on the other, a man—probably a warrior or a king. "No edifice of any Roman era could be fitted in here," Texier wrote of the site in his 1839 account of his travels. "The grandeur and the peculiar nature of the ruins perplexed me extraordinarily when I attempted to give the city its historic name."

The settlement that became Hattusa was originally founded sometime in the third millennium B.C., probably by native Anatolians. Ancient legends tell of early Assyrian trading parties gathering at trading posts called *karum* and swapping fabric, ready-made clothes, and tin for silver, gold, and copper. In time, local rulers came to dominate the trading centers, and warfare between the centers spread. One of the most powerful ruling houses of the time was that of Kanesh, which, under the rulers Pithanas and Anittas, seized control over most of central Hatti. Anittas took the title "Great King" and enforced his will over Kanesh's rival city-states. Around 1900 B.C., Anittas sacked Hattusa, burned it, and declared it accursed: "I took it by storm in the night," a Hittite scribe records him as boasting, "and where it had been I sowed weed. Whosoever becomes king after me and again settles Hattusa, may the Storm God of Heaven strike him."

Anittas's threat went unheeded. The city was resettled within the next 250 years and a new dynasty was established there, a dynasty with connections to Anittas. Like the rulers of Kanesh, the new settlers tried to control the land of Hatti by conquering neighboring city-states. Labarnas II, the ruler of this new dynasty, was faced with increasing resistance to his plans for empire. He recognized the strategic importance of Hattusa and decided to ignore the curse and to establish himself there. By 1650 B.C. he had renamed himself after his new capital: Hattusilis, "Man of Hattusa."

The heirs of Labarnas were not of Anatolian extraction. They were descended from peoples who had im-

The Lion Gate
Photo courtesy of Embassy of Turkey, Washington, D.C.

migrated to the area from a homeland probably north of the Black Sea, and they spoke a non-Anatolian language. Although they retained their distinctive tongue, they adopted much of the culture of the people they ruled and became known as the Hittites, rulers of the land of Hatti. From approximately 1650 to 1620 B.C., Hattusilis expanded Hittite power from Hattusa across Anatolia and into modern Syria, probably in an attempt to take control over the trade routes to Mesopotamia in the east and to the rich sources of tin in Bohemia in the west. He faced opposition from the warlike states of northern Mesopotamia. At one point in his reign, these states—inhabited by people known as the Hurrians—conquered the land of Hatti itself, forcing Hattusilis back to his fortress of Hattusa. Before his death, however, Hattusilis retook all the territory he had lost, and expanded his borders to the Euphrates River.

Hattusilis also faced problems at home. Around 1640 B.C., he was faced with a dynastic challenge: his chosen heir, his nephew Labarnas, had conspired against him. Hattusilis publicly denounced Labarnas's treachery in a speech that was preserved in the records at Hattusa. In Labarnas's place Hattusilis named as his successor his grandson, who suc-

ceeded him as Mursilis I. Mursilis followed his grandfather's example, expanding Hittite power by pushing his armies south along the Euphrates to Babylon, which he sacked around 1600 B.C. However, the dynastic instability that had plagued Hattusilis also followed Mursilis. After his return from his Babylonian venture he was murdered by a brother-in-law, who then seized the throne. Between 1590 and 1560 B.C., power struggles within the Hattusilis dynasty caused the collapse of Hittite power in Anatolia and the Middle East. By about 1500, the Hittites held sway over only a small part of central Anatolia. At one point Hattusa itself was sacked and burned. However, the Hittites' dynastic problems were simplified about that same time by another ruler, Telipinus, who created a code mandating a legal, hereditary succession. He is considered the last ruler of the Old Kingdom.

The resurgence of Hittite power began approximately half a century after Telipinus. Tudhaliyas I, the founder of Great Hatti, as the Hittite Empire was known to its Egyptian rivals and contemporaries, reestablished control over southern Anatolia and the trade routes to the Middle East. His successors, Arnuwandas I, Hattusilis II, Tudhaliyas II, and Arnuwandas II, were less successful; they proved unable to build on or even to protect the conquests of their forebears. Not until the accession of King Suppiluliumas I, who reigned between about 1380 to 1346 B.C., was Hittite control again extended over most of modern Turkey and a good part of what is now Syria.

Suppiluliumas and his successors maintained a large diplomatic correspondence, preserved at Hattusa, with kings both inside and outside Anatolia. Suppiluliumas himself established relations with Babylon through a marriage to the daughter of the city's king. He created a semi-independent principality from a conquered territory, the former state of Mitanni, to serve as a defensive buffer between his realm and the Assyrians farther east. Even Egypt, one of the most powerful states in the Mediterranean basin, courted the favor of the Hittites. Around 1353 B.C., Queen Ankhesenamen, widow of the boy-king Tutankhamen, wrote to Suppiluliumas asking for a marriage with one of his sons. After some initial hesitation, the king sent the requested son, but the young man fell victim to a palace plot and was killed.

Suppiluliumas was succeeded by his son Mursulis II in 1347 B.C. Mursulis pursued his father's policies, and succeeded in expanding the Hittite hegemony even farther south into Syria and territory that was claimed by Egypt. In 1300 B.C., angry language and the breakdown of diplomacy between Muwatallis, Mursulis's son, and the young pharaoh Ramses II, exploded into open warfare near the Canaanite city of Kadesh, where the Hittites fought the Egyptians to a draw—one of the earliest battles in recorded history. Ramses finally negotiated a peace with Muwatallis's successor, Hattusilis III, in 1284 B.C.; this peace was further cemented about thirteen years later, when Ramses married Hattusilis's daughter. The special relationship established between the two powers lasted for the rest of the century and brought a little-known peace to the Middle East.

The capital of the Hittite Empire was, in the fourteenth and thirteenth centuries B.C., one of the largest and most formidable cities in the world. It covered about 410 acres at the height of its power. The site itself is divided into three distinct parts: the Old City, the New City, and the citadel/acropolis, Büyükkale. The Old City had as its basis the oldest of the settlements at Bogazköy, predating the Hittites. It also formed the Hattusa of the Old Kingdom, the capital of Hattusilis I and his successors. It covered the north slope by the modern village, nearest the major sources of water. In later years it was dominated by the Great Temple, where the Storm God of Hatti, Teshub, and the Sun Goddess of Arinna (appropriate deities for the rough weather of the Anatolian tableland, which can be extremely cold in winter) were ceremoniously worshipped by the kings and queens of Hattusa. The Great Temple also served as a storage place for documents preserved on clay tablets. In its library were records not only of religious myths and legends, but also copies of important state documents.

The New City was separated from the Old City, the citadel, and another fortification on a neighboring hill, called Büyükkaya, by a fortified wall designed to make the progress of potential invaders more difficult. Three major gates cut the fortifying wall of the New City: the King's Gate, dominated by the figure of a man or god carrying a battleaxe, and protected by two flanking towers; the Sphinx Gate, set at the southernmost point on the city's walls; and the Lion Gate, protected like the King's Gate by towers, but decorated with twin lions on its supporting jambs. All these gates look unusual to modern eyes because they are parabolic in shape— they form arcs rather than rectangles or arches, familiar from Roman times to the present. The outer wall was also cut by nine narrow tunnels, or posterns, intended to allow Hittite soldiers a quick way to confront besieging troops. The tunnels were not supported by lintels or arches; they were corbelled, like the *tholos* tombs of Mycenaean Greece, built by piling stones on top of one another, with the top edge of each course projecting out from the course below it until they met, forming a rough diamond-shaped arch. Within the walls were at least four temples, smaller than the Great Temple in the Old City, and houses, some of them large enough to be called castles. Several of the very large houses—Nisantepe, Yenicekale, and Sarikale—may have been alternate royal residences.

Archaeologists have found signs of habitation on Büyükkale itself, the site of the great citadel, dating from about the same time as the Old City. However, this was apparently a separate settlement, occupying the highest and most commanding position. During the Empire period, the main royal residence was in the citadel itself. The second story of the king's house formed a single room, perhaps the reception area in which the king received the tribute and expressions of fealty from vassal states. Clay tablets inscribed with state records and events of dynastic importance were also stored in the building, where archaeologists recovered them thousands of years later.

The Hittite Empire came to a final and violent end around 1200 B.C. Although there is uncertainty about what caused the downfall of Great Hatti, it coincided with the coming of the Sea Peoples, barbaric tribes from southern Europe that arrived in the Mediterranean basin at this time. Sources from Egypt, dating from the time of Ramses III, attribute the Hittites' collapse to the same people who were harassing Egypt's shores. Other sources suggest that the later Hittite rulers lost control over their vassal states and the trade routes that led through them. The Assyrians under Shalmaneser I wrested control of the copper mines in eastern Anatolia from Hittite hands during the reign of Tudhaliyas IV, and the depredations of the Sea Peoples in northern Syria further weakened the empire's economy.

Some scholars believe that events in western Anatolia may also have contributed to the Hittites' decline. In 1924 a Swiss Hittitologist named Emil Forrer proposed a revolutionary interpretation of some of the Hittite records that had been recently translated, dating from the end of the thirteenth century B.C., the traditionally reported time for the Trojan War. Forrer suggested that the western Anatolian kingdoms that appear in the Hittite records were in fact Ilios/Troy (Hittite Wilusa) and the land of the Achaeans/Greeks (Hittite Ahhiyawa). He also discovered, he claimed, a reference to Paris himself, whose kidnapping of Helen, the wife of Menelaus, launched the conflict: a certain Alexandros of Ilios. Another philologist named Ferdinand Sommer disproved many of Forrer's identifications in his 1932 work *Die Ahhijava Urkunden* (The Ahhiyawa Documents). Still, whether the Ahhiyawans can be identified with Homer's Achaeans or not, they seem to have played a part in the destruction of the Hittite empire. Some of the latest texts before the destruction of Hattusa refer to trouble in the west, fomented by the Ahhiyawans.

Although it is unlikely that the Sea Peoples themselves destroyed Hattusa, they may have indirectly weakened the city to the extent that it finally fell victim to some of its nearer, more ancient, enemies. Whoever finally delivered the death blow to the capital of Great Hatti, the site itself was soon reoccupied, probably by native Phrygians, who built their homes and places of worship from the rubble of Hattusa. Cultural descendants of the Hittites, known as neo-Hittites, formed petty kingdoms from the former provinces of the empire. Some scholars believe that the neo-Hittites who settled in Canaan were actually the people that Biblical writers named the Hittites. They maintained their autonomy for another 400 to 500 years before the Assyrians, reaching the apex of their power, absorbed them. The Hittite language and culture fell into disuse and, by about 700 B.C., the last remnants of the empire of Great Hatti had disappeared. The excavated remains of the gates, temples, and other structures, however, stand as evidence of the Hittite civilization.

Further Reading: Any overview of Hattusa and the Hittites should begin with two classic works: *The Secret of the Hittites* by C. W. Ceram, translated by Richard and Clara Winston (London: Sidgwick and Jackson, and New York, Knopf, 1956), and *Hattusha: The Capital of the Hittites* by Kurt Bittel (New York: Oxford University Press, 1970), who led the excavations at Hattusa from 1931 to 1939. *The Empire Builders* by Jim Hicks (Alexandria, Virginia: Time-Life, 1974; London: Time-Life International, 1975), a volume in the Emergence of Man series, provides a good overview of Hittite life and culture. It is profusely illustrated and features as consultants the scholars Oliver Robert Gurney of Oxford University, and Harry A. Hoffner Jr., associate professor of Hittitology at the University of Chicago's Oriental Institute. *The Hittites and Their Contemporaries in Asia Minor* by James G. Macqueen (Boulder, Colorado: Westview Press, 1975; London: Thames and Hudson, 1986), also offers an introduction to Hittite times, but it is intended for a less popular audience. It also provides background information about the peoples who inhabited Anatolia before the Hittites arrived and after Hattusa fell. The companion volume to the BBC television series *In Search of the Trojan War* by Michael Wood (London: British Broadcasting Corporation, and New York: Facts on File, 1985) is beautifully illustrated and gives an idea of the impact the Hittites may have had on the Mycenaean Greek civilizations.

—Kenneth R. Shepherd

Herculaneum (Napoli, Italy)

Location: About five miles southeast of the city of Naples along the Bay of Naples. The partly excavated ruins lie within the modern towns of Ercolano and Resina, at the foot of Mount Vesuvius.

Description: Originally settled by the Oscans, the site was later colonized by the Greeks, Samnites, and Romans. Herculaneum was a quiet, secluded seaside resort for wealthy and distinguished Roman citizens when the town, along with neighboring Pompeii, was completely destroyed by the Vesuvius volcanic eruption in A.D. 79. The buried site was accidentally rediscovered in 1709, and excavations began thirty years later. Herculaneum's ruins, while not as extensive as those of Pompeii, are of major archaeological interest because they are much better preserved than those at the neighboring site. Over 250 years of excavation, which continues today, have revealed a variety of houses—sumptuous patrician villas overlooking the sea, and richly decorated middle-class and plebeian dwellings—as well as many shops and public buildings.

Contact: A.A.C.S.T. (Azienda Autonoma di Cura, Soggiorno e Turismo)
Via San Ciro 15
CAP 80069 Vico Equense, Napoli
Italy
(81) 87 98 343 or 80 15 752

The site of Herculaneum was originally settled by the Oscans, Campania's most ancient people, though Hellenistic culture and commerce progressively supplanted the city's Oscan roots. It was probably settled around 600 B.C. by Greek colonists who, as legend has it, named the town for Heracles (Roman, Hercules). Its name first appears in the historical record as Heracleion or Herakleia, as noted by Greek philosopher and scientist Theophrastus in the fourth century B.C.. In fact, the town plan—a series of main streets intersecting at right angles—supports the belief of early Greek settlement; it was probably first ruled by the Greeks of Cumae and then by nearby Neapolis (Naples), which ruled the Campanian coast and islands. The city was occupied by the Samnites at the end of the fifth century B.C. and then by the Romans at the end of the fourth century. Herculaneum joined an alliance with other Italic towns in the Social War (War of the Allies) of 90 B.C. against Rome; but the town fell to Titus Didius, a lieutenant of General Sulla, in 89, and soon became a Roman municipium.

Herculaneum prospered as a largely residential resort in the first century A.D., with a population of about 5,000.

Unlike Pompeii, which was mostly a commercial city with busy shops, small industries, and agricultural markets, Herculaneum, with one-third the size and one-fourth the population, was a richly cultured retreat for wealthy, distinguished Roman citizens. The city was also inhabited by many artists, craftsmen, and fishermen. Its beautiful and secluded terraced seaside setting lent the town a luxurious, intimate character. With more elaborate private and public buildings, Herculaneum also had a richer artistic life and more architectural variety than Pompeii. There is a clear distinction between noble houses of the patrician aristocracy, which often had inner courtyards and terraced gardens, and those of the lesser classes. Although Pompeii had mostly single-story buildings, many houses of Herculaneum had two or three stories, the dominant features of these houses were atria and peristyles, with much use of wood.

Both Herculaneum and Pompeii suffered severe damage in an earthquake in A.D. 63; the Roman statesman Seneca noted that "the region had never before been visited by a calamity of such an extent, having always escaped unharmed from such occurrences and having therefore lost all fear of them. Part of the city of Herculaneum caved in; the houses still standing are in ruinous condition." The cities were still rebuilding from the earthquake, under the patronage of Emperor Vespasian, when they were buried beneath igneous deposits emitted by the violent Vesuvius eruption of August 24, 79. As Roman historian Tacitus wrote: "the burning mountain of Vesuvius changed the face of the land and said to the sea 'Be thou removed'."

The catastrophe affected the two cities differently because of their locations and the wind direction. Though Herculaneum's ruins are smaller and less striking than those of neighboring Pompeii, its domestic and artistic remains are better preserved. While Pompeii was buried under a pile of pumice-stone and then fine ash, Herculaneum, closer to the volcanic vent, was submerged by a torrent of mud containing lava, slag, and ashes. This semiliquid mass of volcanic matter, fifty to sixty-five feet deep, seeped into every building, gradually hardening into a tufaceous material, thus carbonizing and preserving a variety of organic substances (such as structural timber, wooden furniture, cloth, food, rope, rolls of papyrus, and human bones) that otherwise perished or burned at Pompeii. All but a few of Herculaneum's inhabitants escaped the eruption; a small number of human skeletal remains were found—among the few complete skeletons of ancient Romans ever discovered.

Subsequent eruptions buried the city's ruins under more than 100 feet of volcanic matter. This hard shell of solidified mud completely wiped Herculaneum off the map, but, in contrast to Pompeii, it also prevented the site from being tampered with and plundered. The hard mud, as well as

Ruins of villa garden at Herculaneum
Photo courtesy of The Chicago Public Library

the presence of the modern towns on part of its site, has also hindered excavation over the years.

The ruins remained untouched—and largely unknown—for 1,630 years. While the local population, which eventually settled in the nearby towns of Resina, Portici, and Ercolano, was probably familiar with ancient documents recording the city's existence, no one knew exactly where the buried site lay. The first discoveries were made quite by accident. In 1709, Emmanuel de Lorraine, prince of Elbeuf (Austria) and cavalry commander of the Kingdom of Naples, was sinking a shaft for a well when he came upon the back of the stage of the Herculaneum theatre. Prince d'Elbeuf, however, was more intent on profiting from works of art than on unearthing a Roman town; the decorations and statues he found in the theatre enriched his own villa and were distributed to various museums outside Italy.

Full-scale excavations of Herculaneum were begun in 1738 under the patronage of Charles of Bourbon, king of Naples (later Charles III of Spain), and continued until 1765. These daring and difficult excavations, directed by Karl Weber and Francesco La Bega, were conducted by means of wells and underground tunnels, due to the deep rocky ground, resulting in a maze of corridors and catacombs. Once again,

these excavations were undertaken primarily for the purpose of recovering art treasures—statues, paintings, mosaics, furniture—rather than out of any archaeological or architectural interest. During this rich period of excavation, the theatre, forum, and five temples were discovered. Objects found in these sites are now among the principal treasures of the National Museum in Naples, while paintings and portrait statues found in a building believed to be the ancient basilica of Herculaneum embellish the collection of the nearby Portici Museum.

But the most sensational discovery, unearthed around 1750, was a house outside the town walls, the so-called Villa of the Papyri. One of the largest and most beautiful private dwellings in all antiquity, the house had a facade of over 800 feet and contained a huge indoor colonnaded garden and pool as well as about 60 bronze and marble sculptures. The splendid collection of statuary, also now in the Portici Museum and the Naples Museum, is of a superior quality to any sculptures found at Pompeii. Also found was an entire library of thousands of ancient Greek papyrus rolls, mostly composed of Epicurean philosophical writings. These papyri are now preserved in the National Library of Naples and have only been partly read. The villa was believed to have belonged to Lucius

Calpurnius Piso, a wealthy humanist collector and father-in-law of Julius Caesar. Weber and his archaeologists abandoned the excavation in 1765 because of a dangerous gas buildup in the tunnels.

The Accademia Reale Ercolanse, founded in 1755 to investigate the discoveries, published an eight-volume work on mural paintings and bronze statuary (1757–92); it also published a volume by C. Rosini on the papyri (1797). Excavation outline plans from this period, showing the network of tunnels, were drawn by La Vega, while Weber drew plans of the Villa of the Papyri.

When excavations were resumed in 1823, tunneling was abandoned in favor of working from above the ground, a method that had proven successful at Pompeii. The explorations, carried out in the southern end of town by inexperienced excavators, continued until 1835. They revealed, however, the first houses of Herculaneum, including the peristyle of the House of Argus, one of the city's grandest mansions, with frescoes and a pillared garden. Excavations conducted from 1869 to 1875, after the unification of Italy, were not particularly productive, since the inhabited houses of Resina impeded progress.

An internationally led exploration of Herculaneum, using the most advanced known methods of excavation and preservation, was initiated in May 1927 under the Italian government's archaeological authorities. The excavation was made possible because of the efforts of English archaeologist Charles Waldstein who, beginning in 1904, raised funds from various nations in Europe and America. This modern, systematic excavation, interrupted only by World War II, still continues and has revealed the general plan and elements of the ancient city.

Between one-quarter and one-half of the town's original area of twenty-seven acres has been unearthed, revealing a regular grid of volcanic stone-paved streets with sidewalks, porticoes, and eleven *insulae* (islands, or blocks) of houses, shops, and a variety of public buildings. The newer excavations have disinterred most of the two *decumani* (main roads) that run parallel to the coast—closer to the town than it is today—and three *cardines* (crossroads), which form right angles to the sea. A suburban area outside the town walls, made up of an ancient harbor and beachfront, has also recently been excavated. On the bayside, the town ended on a terraced promontory, with the wealthiest patrician villas enjoying panoramic views of the sea; the *cardines* continued to slope down the steep hillside through the Porta Marina, or sea-gate, to the harbor. It was in this area that most of the intact human skeletal remains were found.

The Decumanus maximus, the larger of the two main roads, located just to the north of the open ruins, forms the boundary between completed excavations and those still in progress; it also appears to mark the division between the town's private and public quarters. The area of the south of the main street, with a strict geometric layout extending as far as the promontory, was a largely residential quarter made up of wealthy and middle-class houses, shops, workshops,

thermae (public baths), and a *palaestra* (sports ground). Most of the public buildings still lie buried (except for the theatre) north of the main street. Excavations continue in the forum area, the actual center of the ancient city, which borders the Decumanus maximus to the north.

Some of Herculaneum's private residences—for example, the so-called Samnite House—date to at least the third century B.C. Houses were mostly remodeled in later years, showing a gradual shift from enclosed central atria to inner courtyards exposed to the sun and air. While most of the houses were single-family dwellings, some patrician mansions had been divided into apartment buildings, while tenements of several stories housed poorer artisans and laborers. In many instances, the original timber used in beams, doors, staircases, partitions, and furniture can still be seen.

Large, luxurious patrician villas on the seaward promontory, at the southern edge of town, include the House of Stags, the House of Telephus, the House of the Gem, the House of the Inn, the House of the Mosaic Atrium, and the House of Argus. Wealthy Romans who resorted in Herculaneum had such dwellings built to specification in order to make full use of the town's charms—sunlight, fresh healthy air, and majestic views of the mountains and the Gulf of Naples; they were designed to distribute the daylight and air through the largest number of rooms. Common mansion features include frescoes, mosaics, peristyles, atria, porticoes, terraces, bay windows, upper floors, and statuary (mostly looted).

The House of the Stags, or Deer (Casa dei Cervi), perhaps the grandest and most beautiful mansion overlooking the bay, was built during the reign of Nero. It contains numerous frescoes and artworks, including a statuette of a drunken Hercules; its magnificently sculptured group of dogs attacking stags is considered among the masterpieces of Herculaneum sculpture.

The House of the Relief of Telephus (Casa del Relievo di Telefo), one of the town's most sumptuous villas, has richly decorated rooms and a spacious, colonnaded atrium garden. It is exceptional for its dazzling marble decorations (rather than frescoes), as well as for a bas-relief depicting the myth of Telephus (a son of Hercules), an example of classical, neo-Attic art of the first century B.C.

The House of the Gem (Casa della Gemma), so named for a Claudian-era gem engraved with the portrait of a woman found there, is a harmoniously elegant, but not extravagant, house with an unusual, finely decorated atrium and a well-preserved kitchen with pots still on the hearth. While Herculaneans, unlike Pompeiians, were not in the habit of creating graffiti, this house's lavatory nevertheless contains the wall-scribbling: "Apollinaris medicus Titi imperatoris hic cacavit bene," or "Apollinaris, physician to the emperor Titus, shitted satisfactorily here."

The House of the Inn, or Hotel (Casa dell' Albergo), once one of the most splendid aristocratic houses in town, was built during the Augustan period and was originally believed to be a public building. However, it was badly damaged by the Vesuvius eruption and further despoiled by the eighteenth-century excavation tunnels of Weber and La Vega. The vast private villa had been divided into modest apartments, hence its somewhat erroneous name.

The House of the Mosaic Atrium (Casa dell' Atrio a mosaico), designed to fully exploit the advantages of its panoramic site, is a fine mansion decorated with frescoes and mosaics. The entrance passage and atrium are paved with the black-and-white checkered mosaic. The house features a solarium, or open terrace, overlooking the sea, and a garden with a marble fountain in the middle. It has an enormous triclinium (dining room), and a wooden door, table, and window frames. The House of Argus (Casa d'Ago), as mentioned earlier, was once one of the finest mansions in town, but was badly excavated in the past. Some vestiges of its former grandeur, however, remain: a peristyle with noble columns, and a garden. It is named for a fresco that no longer remains.

In the residential quarter, rich republican and patrician houses alternated with finely decorated middle-class and more plebeian dwellings. The most notable of these include the House of the Mosaic of Neptune and Amphitrite (Casa del mosaico di Nettuno e Anfitrite), named for the subjects of the best-known mosaic in town; the best-preserved house in either Herculaneum or Pompeii, it contains a wine shop with the ancient wares still displayed on carbonized wooden shelves. The House of the Wooden Partition (Casa del Tramezzo carbonizzato), another unusually well-preserved house, is of Samnite construction (i.e., without a peristyle or colonnaded court). It contains beds, mosaics, frescoes, and a three-doored screen of carbonized wood between the tablinium (living room) and atrium. This is a good example of a patrician dwelling that housed several families. The House of the Partition, or House of the opus graticium, which contained separate apartments, was a fragile plebeian dwelling built of wood, earth, and plaster; it is an example of the type of house that simply disappeared at Pompeii. The House of the Charred Furniture (Casa del Mobilio carbonizzato), a small, elegant pre-Roman house, contains the remains of some carbonized furniture and a painting of Pan discovering a sleeping nymph. The Wooden Trellis House (Casa a Graticcio), named for the wooden trellis framework of the walls, is the only complete artisan's house yet unearthed.

The House of the Fine Courtyard (Casa del Bel Cortile) is one of the most original houses in Herculaneum, with its raised courtyard, stone staircase, and balcony; it contains mosaics, portrait paintings, a marble statue of Eros, as well as many objects of everyday life. The Bicentennial House (Casa del Bicentenario), a patrician residence divided into small apartments, is so named because it was unearthed in 1938, exactly 200 years after digging officially started. One of the upper rooms, which were probably rented to artisans, contains a wooden crucifix in a stucco panel, one of the world's oldest known Christian relics, and evidence that Christians lived here before the eruption. It also has a

huge square atrium and rooms decorated with frescoes and marble. The Samnite House (Casa Sannitica), as its name indicates, is of Italic rather than Roman origin, and is Herculaneum's oldest dwelling. Simply built, it features fresco decorations and a fine atrium surrounded by a gallery with Ionic columns.

A great variety of commercial shops has also been excavated, many of which still contain articles of daily life abandoned during the eruption. The best-preserved of these is the Pistrinum or bakery, owned by Sextus Patuleus Felix, which contains flour mills, storage jars, a large oven, and a stable for asses that turned the mills. Other bakeries, groceries, restaurants, taverns, wineshops, dye-shops, and a jeweler's shop have also been excavated.

Public monuments fully excavated in Herculaneum's residential and suburban quarters include two thermae (bathhouses), the palaestra (stadium-gymnasium complex), and a basilica; in addition, the forum has been preserved in the center of town. The rest of the public buildings are still being excavated, since the demolition of part of Resina and its vineyards in the forum quarter north of the Decumanus maximus and the open ruins; of these, only the theatre has been restored to its original state.

Herculaneum's public baths—one next to the forum quarter in the main excavated area and the other outside the city gate—are in excellent condition. The municipal baths, identified in 1942 and unearthed by deep shafts after 1955, were built in the early Augustan period, about 10 B.C., and were decorated during the reign of Nero; like the larger baths at Pompeii, they had separate men's and women's sections. Various portions of the baths are tastefully adorned with ceiling frescoes, marble mosaic floors, and stucco bas-reliefs; the smaller, simpler, and even better preserved women's baths, especially, have mosaic pavements representing Triton and a labyrinth. The masterpiece of the baths, however, is a piscina (swimming pool), the bottom of which had been painted blue to reflect the blue vaulted-ceiling fresco of sea creatures; bathers stirring the waters had the illusion the pool was teeming with live fish. The small, elegantly decorated thermae suburbane, located below the terrace in the city's extramural southern quarter, were built during a later period. Though they had once been the best-preserved bathhouses known because they were protected against the volcanic flow, they have suffered in recent years due to flooding.

The palaestra, where public games were held, contains imposing colonnades that led eighteenth-century excavators to believe it was a religious building. It is mostly notable, however, for a large bronze fountain in the middle of a 100-foot bathing pool. Discovered in 1952, the fountain is in the form of a five-headed serpent entwined around a tree trunk; the serpent heads served as waterspouts.

The recently excavated basilica, an impressive square hall, was originally a shrine dedicated to town patron deity Hercules, but was later consecrated as the College of the Augustali (Sacello degli Augustali), a meeting place for priests of the Imperial Cult. Though it has fine frescoes, rich statues were looted during eighteenth-century excavations.

The Herculaneum theatre was built during the Augustan period and could seat about half of the town's population, up to 2,500 spectators. While the theatre was largely stripped of its treasures—bronze statues, wall decorations, Corinthian capitals—by Prince d'Elbeuf and subsequent excavators, it nevertheless remains one of the most perfectly preserved examples of a theatre in the ancient Roman world; some marbles, painted columns, and stucco bas-reliefs are still intact. The theatre is reached by going underground; shafts and tunnels bored by Weber and La Vega in the eighteenth century can still be seen.

A number of civic and religious buildings, including several temples and a senate house, are also believed to exist in the forum quarter, either from earlier excavation records or from inscription references. Inscriptions in other buildings tell of a temple of Jupiter, and a temple of the mother of all gods, restored by Vespasian after the earthquake. It is certain that much more will come to light in Herculaneum in the future.

Further Reading: The definitive work on the site is Amadeo Maiuri's *Ercolano* (Novara: Instituto geographico de Agostini, 1932; sixth edition, translated by V. Priestley as *Herculaneum*, Rome: Liberia dello Stato, 1962). A more general account of the excavations and discoveries, with photographs, can be found in *Pompeii and Herculaneum: The Glory and the Grief* by Marcel Brion (London: Elek, and New York: Crown, 1960). Notable from a more scientific standpoint is Charles Waldstein and Leonard Shoobridge's *Herculaneum: Past, Present and Future* (London: Macmillan, 1908). Waldstein initiated the town's twentieth-century excavations. For more recent general guides, *Cities of Vesuvius: Pompeii and Herculaneum* (London: Weidenfeld and Nicolson, and New York: Macmillan, 1971) and *The Art and Life of Pompeii and Herculaneum* (New York: Newsweek, 1979), both by Michael Grant, are to be recommended.

—Jeff W. Huebner

Hierapolis (Denizli, Turkey)

Location: In the Lycus Valley in the center of the Aegean portion of western Turkey.

Description: City called Hierapolis founded in the second century B.C. by the king of Pergamum; later the location of a Roman city and a twentieth-century city called Pamukkale. Site of mineral water pools and unusual mineral formations.

Site Office: Tourist Information Office
Örenyeri
Pamukkale, Denizli
Turkey
(258) 272 2077

Pamukkale, site of a popular health spa, is the modern town on the site of the ancient city of Hierapolis in the Lycus Valley. Although there is no written substantiation, the king of Pergamum is thought to have founded Hierapolis in the second century B.C. to take advantage of the therapeutic waters flowing down from the mountains. For centuries before the town's founding, the local inhabitants knew of the waters and the unique landscape caused by the mineral deposits.

The town sits more than 355 feet above the river valley on a dense mass of calcium, which has been deposited over a period of about 14,000 years, forming a plateau. Calcium-rich carbonated water, originating from an unknown location in the Çal Dagi mountains, bubbles from springs. The hot water (about ninety-five degrees Fahrenheit) flows through the center of town in small limestone-lined canals built up over the centuries to a level six to seven feet above the level of the first known road.

The water then passes over the edge of the plateau and falls in a series of pools to the valley below. As the water cools, carbon dioxide escapes into the atmosphere, while calcium carbonate is left behind. It is estimated that nearly 8,000 cubic yards are deposited each year in the form of constantly changing mineral stalactites and other strange formations. The solidified waterfall is responsible for the Turkish name Pamukkale, which means "Cotton Fortress." Once on the plain, the water is cool enough to be used for drinking and agriculture.

Two theories exist regarding the origin of the name. According to Stephanus of Byzantium, the name Hierapolis, meaning "Holy City," was chosen because of the large number of temples in the area. Coins in circulation up to the reign of the Roman emperor Augustus were inscribed with the name Hierapolis. Archaeologists have unearthed the temple of Apollo and found inscriptions mentioning priests of Cybele, but have not been able to determine what other temples existed in the Roman era. Although multiple gods were worshiped, it was not customary to build temples for all. The second theory, derived from mythology and the custom of naming towns after family members, suggests the town was named for Hiera, wife of Telephus, the mythical ancestor of the Pergamene people.

The site was probably occupied prior to the arrival of the Pergamenes, but any evidence is buried under the thick calcium deposits. Herodotus writes of two other towns in the area, but not Hierapolis. The first written record is a decree inscribed sometime after 183 B.C., honoring Apollonis, mother of King Eumenes II of Pergamum. This suggests that the city was founded by Eumenes very early in the second century B.C. The Lycus Valley was won by Pergamum from the Seleucids of Syria in the Battle of Magnesium in 190 B.C. Anatolia had been under the control of Alexander the Great, but in the turmoil after his death a relatively minor Macedonian general, Seleucus Nicator, took over large parts of the territory, including the Lycus Valley, and became King Seleucus I of the Seleucid Empire. He based his kingdom in Syria and Cilicia, but later moved his capital to Antioch (named after his father Antiochus). Because a Seleucid city, Laodiceia, already existed close to the future site of Hierapolis, there was probably not a Seleucid city on the site.

Pergamum became an independent Hellenistic state in 282 B.C., although it was subject to the Seleucids for a brief time. In contrast to other states at the time, Pergamum had a culture that was almost as sophisticated as those of earlier cities at the height of Greece's prominence or of Alexandria in Egypt. Eumenes II, fearing the power of the Seleucids, aligned himself with Rome. The Attalid dynasty, named for Eumenes II's father, was thus able to rule in peace for fifty years after the Treaty of Apamea, which effectively eliminated most of the other Hellenistic states.

Little is known of the city's history after the Pergamene founding; their Hellenistic city is gone. Archaeological finds indicate the citizens of the city were mainly Greek and Roman, while the local tribes seem to have been Phrygian. The ancient city lies under a thick blanket of minerals, making excavations difficult. The last of the dynasty, Attalus III, was an eccentric who conducted scientific research and experimented with poisons. He died of sunstroke while engaging in another one of his avocations, metalworking, as he worked on a bronze tribute to his mother. He bequeathed his kingdom to the Romans; Hierapolis was now included in the Roman Empire's Province of Asia.

At least four earthquakes shook the area, the most disastrous of which was in A.D. 60 during the reign of Nero. The ancient buildings existing today are all Roman, dating from Nero's rebuilding project, and have been partially buried in calcium.

Cliffs at Pamukkale (Hierapolis)
Photo courtesy of Embassy of Turkey, Washington, D.C.

Roman emperors favored the town, and visits by Hadrian, Caracalla, and Valens resulted in worship of the emperors. Antipater, a distinguished citizen and sophist, tutored the future emperors Caracalla and Geta at the request of their father, Emperor Septimius Severus.

Baths, an important element in Roman society, were built in Hierapolis for its hot mineral waters, thought to be medicinal. The stone baths were very much like their counterparts in Rome, with huge marble-lined rooms, high arches, and barrel-vaulted ceilings. There were exercise rooms and the traditional pools of different temperatures: *caldarium, tepidaria,* and *frigidarium.* One of the baths has been restored and is now the Pamukkale Museum, which also displays items found in the excavations: statues, sarcophagi from the necropolises, and pieces of stone from ruined buildings.

Commerce in the town was varied. Inscriptions found in the ruins mention guilds for metalworkers (nail makers and coppersmiths) and various textile workers (dyers, wool washers, and carpet makers). Wool was an important agricultural commodity, and the textile industry took advantage of the mineral water to set the dyes.

The hot water seeping into the rocks resulted in marble with unique coloring, which proved to be a valuable export for the building of cities, although it was not used in Hierapolis. Other types of stone for the construction in Hierapolis were quarried from the mountainside below the town, resulting in a series of terraces that provided an interesting base for the striking mineral cascades.

In the center of town, water bubbled up from a spring into a sacred pool, the exact nature of which is not known. A common myth among the villagers in the valley held that the pool had no bottom. The pool, near the Shrine of Apollo, was once ringed with columns, which have since fallen into the water along with inscribed marble slabs and other stone debris. Fortunately for archaeologists and historians, the inaccessibility of the stones at the bottom of the pool prevented them from being reused.

The waters and the attendant bizarre calcium formations were not the only fascinations; ancient Hierapolis also had religious attractions. Local gods were merged with the Greek gods and fervently worshiped. Apollo merged with the Phrygian sun-god Lairbenus (who had a temple twenty miles northeast of Hierapolis), and the Anatolian mother goddess, Cybele, became Apollo's mother, Leto. Bozius and Troius became Zeus. The seismic activity and toxic underground gases in the area fostered worship of the Greek gods Poseidon, the earth shaker, and Pluto, ruler of the underworld.

A hillside opening was believed to be the entrance to the underworld and a nine-foot-square chamber was built on the site. Called the Plutonium, the chamber had an arched entrance and a partly paved floor. In the back, a three-foot-wide gap emitted gases and proved deadly to anyone who dared to enter it. According to Strabo, the ancient Greek geographer, a rectangular fenced courtyard surrounded the Plutonium and the gases filled the courtyard with a heavy mist.

Legend holds that the eunuchs of Cybele, under the protection of the goddess, were able to venture a short distance into the underworld entrance, but only if they held their breath. Ancient stories tell of bulls collapsing inside the courtyard and birds dying when released at the mouth of the opening. The physician Asclepiodotus is said to have ventured through the gap in the fifth century, covering his face with a cloak. He tried to find the source of the stream running through the cavern, but after a time, the depth of the water prevented him from crossing and he had to turn back. Curiosity concerning the Plutonium was so great that by the second century A.D. provisions were made for the hundreds of spectators.

The location of this Plutonium eluded early twentieth-century explorers until an Italian expedition unearthed it under the Temple of Apollo. In Strabo's time the temple had not yet been built, and his description of a mist-filled courtyard has proven baffling, as his is the only known report of this curiosity and there is no such mist in modern times. The escaping gases are strong sulphurous compounds, irritating to the eyes and probably harmful, but not as lethal as the ancient legendary gases.

The gas hindered twentieth-century excavators of the Temple of Apollo as it had the ancient builders, who solved the problem by placing two-inch ventilation gaps in the foundation. Built in the third century A.D. from the foundation of a second-century structure, the temple housed eunuch priests of an oracle. The natural rock of the hillside formed the back of the temple, and the remainder was made from blocks of stone, many of them also salvaged from the earlier building. Inscriptions relate to oracles proclaimed during a pestilence, probably an empire-wide plague in the reign of Marcus Aurelius in the second century. The god recommended sacrifices and addressed the city, saying, "you are sprung from me and from Mopsus the protector of your city," implying that Mopsus the seer founded the city. Although Mopsus appears on coins issued in the city, this story of the founding is almost certainly a myth.

On the hillside not far from the temple was the Roman theatre, with inscriptions dating to the second half of the second century A.D. Its entries and corridors were decorated with festoons and foliage in bas-relief. Statues decorated the stage area, and intricate friezes told the life of mythical figures such as Dionysus. A poetic inscription called the city "foremost land of broad Asia, mistress of nymphs adorned with streams of water and all beauty." The attractively restored theatre still stands today. Although damaged by an earthquake, it was never destroyed by humans. Remains at another location revealed that Hierapolis also had another, older theatre.

In imperial times athletic contests were probably held in a stadium on the plain below the town, although no stadium remains have been found. Surviving reliefs in the theatre depict contests between gladiators and wild beasts.

Near the temple was the fountain house, called the nymphaeum. Excavation has yielded some splendid architectural artifacts. It consisted of a large fountain, probably filled with fresh water from a reservoir in the hills, rather than the hot mineral water used in the baths, and piped through a niche in the rear wall. On one side were stairs leading up to the basin. The remaining sides were enclosed by lavishly ornamented walls with pediment-topped rectangular niches, which probably held statues.

The elderly and the ill came to Hierapolis for cures, and many chose to be buried there when they died. As a result, there are four huge cemeteries in the town, making it one of the largest necropolis complexes in Asia Minor. Extending along the road for more than a mile, the necropolis has about 1,200 graves of various types. Some are simple pits; others have multiple sarcophagi sharing a single stone base. Huge Roman tumuli (grave mounds) date from the Hellenistic period. Each of these excavated and restored graves consists of a large round burial chamber covered with a cone of earth and topped with a phallic symbol in stone.

Despite the dire warnings on the tombs of wealthier citizens, grave robbers have been very active. Fortunately, they only plundered valuables and did not destroy the monuments of the tombs themselves. The necropolises were not always so crowded. Epitaphs on excavated tombs indicate that the graves were originally surrounded by gardens or other open spaces with benches for the use of visiting relatives. The spaces were filled with graves at a later date.

Writings from the time reveal an interesting custom. To insure that the deceased would be remembered on a specified date each year, family members made a deposit of money, called a *stephanoticum,* to a guild (or other institution) under an agreement that the interest on the money would finance the placing of a wreath or some other predetermined token or offering.

Christianity was adopted early in Hierapolis, particularly in the sizeable Jewish population, which was receptive to Christian ideology. (There were many Jews throughout Asia Minor; some had been brought from Babylon by King Antiochus III to increase the pro-Seleucid population.) St. Paul is thought to refer to Hierapolis when he mentions the inhabitants of Laodicea who were "neither hot nor cold" in his Epistle to the Colossians. Philip the Apostle lived in Hierapolis until he and his seven sons were killed in A.D. 80. The well-preserved ruins of the Martyrium of St. Philip was another important find in the excavations. At the time of its erection in the fifth century A.D., the Roman government would not permit a Christian church within the town itself, so the Martyrium stands outside of the walls of the city, possibly on the site of Philip's death. It was not a church in the usual sense, but was used for services on St. Philip's feast day. An octagonal central room has doorways leading to the six rectangular rooms that surround it. Around the perimeter are small rooms with outside entrances that were probably used to house pilgrims who came to participate in the feast day celebrations. Philip's tomb was probably moved to the Martyrium, but it has not been located. Missing only its roof, the building survived into the twentieth century with some of its frescoes still visible. As recently as the early 1900s, nomads used the grounds as a campsite, stabling their horses in the building.

Paganism lingered into the sixth century until a zealous bishop ordered the remaining temples destroyed. He then commissioned the building of nearly 100 churches, many of which are still standing.

Although Hierapolis erected a wall to keep out pillaging nomads, sturdier fortifications were not needed during the peaceful period of the *Pax Romana;* and the holiness of the city lent it an additional degree of protection. A straight street, intersected with lesser streets at right angles, ran for more than a mile through the center of town. This main street, lined with columns, public buildings, and, at a later date, two churches, ended about 170 yards outside the walls with a pair of elaborate gates. The huge gate at the north end of the street was built between A.D. 84 and 85 with three arches between round towers. It survives in good condition. The gate at the other end was destroyed.

Toward the end of the Byzantine period, Asia Minor was raided by Turks and Arabs. This turmoil took its toll and the importance of Hierapolis declined. When the Seljuk Turks arrived in the twelfth century, the city was abandoned. The Seljuks took over most of Anatolia, the area that is now the Asian portion of Turkey, and established their Sultanate of Rum (Rome). During this period, a battle of the Third Crusade, led by Holy Roman Emperor Frederick I Barbarossa, was fought at the foot of the falls at Hierapolis. Even after the Mongols conquered Asia Minor, the Sultanate of Rum endured as a vassal state until the Ottoman Turks arose and the Lycus Valley was incorporated into their empire. Ottoman control lasted until the empire was dismantled at the end of World War I. The Republic of Turkey was established in 1923.

The modern town of Pamukkale grew up on the site of Hierapolis. Major excavations of Hierapolis were begun in 1957 by the Italian Paolo Verzone; there had been some limited excavations by a German team in 1898. Since 1957, buildings have been restored and the tourist potential of the area exploited. The sacred pool, with its bubbling spring water, is now in the courtyard garden of the Pamukkale Motel, surrounded by ancient stones. Water has been diverted to the pools of other motels as well. The water is said to cure ailments ranging from heart disease and high blood pressure to nervous disorders and skin diseases.

Further Reading: *Turkey beyond the Maeander: An Archaeological Guide* by George E. Bean (Totowa, New Jersey: Rowman and Littlefield, and London: Benn, 1971) is a traveler's guide to historic excavations in parts of southern Turkey and includes the writer's personal observations. Chapter 20, devoted to Hierapolis, describes the history of important sites and the discoveries made by archaeologists. For a general history of Turkey, Richard Stoneman's *A Traveller's History of Turkey* (Moreton-in-Marsh, Gloucestershire: Windrush, and Brooklyn, New York: Interlink, 1993) is a very readable source. It covers the history of Turkey from prehistory through the 1990s. Although written prior to the major excavations, "The Ruined Cities of Asia Minor" by Ernest Harris, in *National Geographic* (Washington, D.C.), November 1908, provides interesting descriptions of the author's visit to Hierapolis at a time when nomads still camped by the waters.

—Julie A. Miller

Hvar (Split, Croatia)

Location: An island in the Adriatic Sea, off the coast of the Croatian region of Dalmatia; part of the county of Split.

Description: Settled since Neolithic times, ruled by Illyrians, Greeks, Romans, Byzantines, Croats, Bosnians, Venetians, and Austrians in succession, and particularly noted for the Venetian Renaissance buildings in its main town.

Site Office: TZ Hvar
58450 Hvar, Split
Croatia
(58) 741-059

Hvar, an island with an area of about 116 square miles, is one of a group along the eastern shores of the Adriatic Sea that is the upper portion of an underwater mountain range. Hvar shares with these other islands, and with Dalmatia, the mainland region nearby, a complex history of successive invasions and intermingling of cultures, which is indicated by the two main names it has had, each taken from one language into others. Hvar itself is a Serbo-Croat word, derived either from the Greek *pharos,* meaning "lighthouse," or from the Greek island of Paros, or from the ancient Greek settlement of Pharos, now called Stari Grad, on the island; but its Venetian and Austrian rulers knew it as Lesina, which is said to be derived either from a Slav word meaning "forested" or from another Slav word meaning "coming to land."

Neither the Greeks, the Slavs, the Venetians, nor the Austrians were the first to settle on the island. At least 5,000 years ago its inhabitants, Neolithic people whose origins and language are unknown, were already fishing in the Adriatic and trading across the sea with what are now Italy and Croatia. Extensive remains of their settlement have been found in the Grapčeva Špilja (Grabak Cave), near the modern village of Jelsa in the north of the island. These objects, most of which are kept in the archaeological museum in Split on the mainland, include numerous pieces of pottery, stone knives and other tools, and a copper armilla, as well as human and animal bones and seashells. Their discovery has since given rise to the archaeologists' term Hvarska kultura (Hvar culture) to denote Neolithic objects found along the Dalmatian coast and on the islands.

It is possible, though it cannot be proved, that this culture had some connection with those of the Thracians and the Illyrians, the peoples whom the Greeks found inhabiting the Balkans from the eighth century B.C. onward (though they may well have been there earlier). It is also possible that the Illyrians, who settled on Hvar and its neighboring islands as well as on the Dalmatian mainland, were the ancestors of the modern Albanians. They continued to form the majority of the population in the region, including Hvar, for many more centuries, even as the Greeks founded trading posts along the eastern shores of the Adriatic, and later when the Romans extended their empire into the Balkans. On Hvar the Illyrians' main settlement was probably at Pitve—known in ancient Greek as Pityeia—near Jelsa, which was still known as Portus de Pitue (the port for Pitve) as late as the fourteenth century.

Greek traders eventually established two settlements of their own in the south of the island, from around 385 B.C. The legend that the Greeks were sent there from the island of Paros, either on the orders of or with help from Dionysius the Elder, the tyrant of Syracuse in Sicily, may well have some factual basis. Their main towns were Pharos, now Stari Grad, or Starigrad na Hvaru (meaning "old town on Hvar"), which lies at the head of a fjord and was the main settlement on the island from Greek times through to the late thirteenth century; and Dimos, now the town of Hvar, on the southwestern coast, separated from Pharos by mountains. Greeks also settled later at what is now Jelsa. All three settlements probably took part in the maritime trade between Greek centers such as Corinth and Corfu and what is now Bosnia-Hercegovina, where Greek artifacts have frequently been excavated. The Illyrians were apparently forced into the mountainous interior of the island by the new arrivals. It may have been around this time that the massive fortifications were built, of which the ruins can still be seen in Stari Grad, at Tor in the mountains, and at Galešnik. It is still disputed whether these were Illyrian defenses against Greeks, Greek defenses against Illyrians, or the work of local people after the two populations had blended together. The ruins are often referred to as "Cyclopean," in reference to the belief among later Greeks that, like similar walls at Troy and Mycenae, they had been constructed by giants like Polyphemus, the Cyclops mentioned in Homer's *Odyssey.*

In 235 B.C. Agron, the ruler of the Illyrians on the mainland, conquered the island, but his successor Demetrius of Pharos was displaced in 219 when the Romans arrived. Their main interest in the Dalmatian coast and islands was to repel the pirates who used them as bases to attack Roman shipping, but they also took over the settlements on Hvar and elsewhere, as is indicated by mosaic floors that have been uncovered in Stari Grad and at other sites. It was also the Romans who at a later stage brought Christianity to Hvar; the ruins of a Christian baptistery, built in the fifth or sixth century and uncovered in Stari Grad in 1957, may be the site of the island's first cathedral.

By the time the baptistery was completed, the Western Roman Empire had declined and fallen (its last emperor dying in Naples in 476) and Dalmatia was dominated by Ragusa

A view of Hvar
Photo courtesy of Art and Culture Council of America

(now Dubrovnik), Cattaro (now Kotor), and other mainland trading settlements controlled by Italian merchants. The bishops based in these city-states ensured that the mixed Illyrian-Greek-Italian population of Dalmatia would remain loyal to the western, Catholic form of Christianity even after the province was conquered in 535 by the forces of the Byzantine Emperor Justinian I, ruler of the Eastern Roman Empire and head of the Eastern church. Although the two churches were not formally separated until 1054, it is likely that the rivalry between their increasingly separate traditions and practices contributed to the instability of Dalmatia, where Byzantine rule lasted in theory until 1102 but in practice was rarely effective.

As early as the seventh century large areas, including Hvar, were being settled by Slavs as part of the great wave of immigration that had swept down the valley of the Neretva River into what would become Yugoslavia. These new and (as it turned out) permanent inhabitants took no notice of Byzantium until 870, when imperial forces reconquered Hvar

and most of Dalmatia, only to be withdrawn around 886. By around 924 various groups of Slavs became united under Tomislav, now recognized as the first ruler of Croatia, which at that time included only sections of the Dalmatian coast, where its expansion was resisted by the city-states, by the declining power of Byzantium, and by the increasingly powerful republic of Venice. Hvar became part of the Croatian kingdom in the mid-eleventh century during the reign of Krešimir Peter, who expanded his navy and used it to take control of most of Dalmatia. In 1102 Croatia was united with Hungary, under King Kálmán, but retained its own separate administration.

In 1145 much of Dalmatia, including Hvar, was seized by Venice; in 1164 the Byzantine Empire exerted its claims on the region once again; and in 1180 Hungary-Croatia took Dalmatia back. Under the Venetians, construction had begun on the first cathedral in the town of Hvar; dedicated to St. Stephen, it was originally the church for a Benedictine monastery under the control of the first bishop of Hvar. In 1278

Venetian forces took over the island once more and moved its capital from Stari Grad to Hvar. There work began on fortified city walls, while the first palace of the Venetian governors was completed about 1289. In 1331, Hvar, which probably contained most of the island's population, became a self-governing municipality inside the walls, completed around this time, which still surround the medieval quarter, known as the Grad (fortress), and separate it from the newer, lower-lying suburbs. The increasing insecurity of the island is suggested not only by these walls but by the simultaneous building of the arsenal to protect the harbor, as well as by the fortified churches of St. Sebastian at Jelsa and of St. Mary at Vrboska nearby, both constructed in the same period.

None of these measures, however, prevented either the return of Hungarian-Croat rule in 1358 or its removal only about twenty years later. From 1370 onward, the south Slav lands were dominated by Tvrtko I, the ruler of Bosnia, who conquered Hvar, the other islands and much of the Dalmatian coast, as well as parts of Serbia. But in 1389 the Turkish victory over the Serbs at the Battle of Kosovo signaled the beginning of the incorporation of most of these lands into the Turkish Ottoman Empire.

Hvar, however, along with the rest of Dalmatia, remained outside both the Ottoman Empire and the kingdom of Hungary. In 1420 the island was seized for the third and last time by the Venetians, after they had repulsed the forces of two rivals, Hrvoje Vukčić, ruler of Split, and the city-republic of Dubrovnik (Ragusa). The town of Hvar now became important as a naval base halfway between Venice and the island of Corfu. A small harbor was completed there in approximately 1459; it is still known as the Mandrać, from the Greek *mandraki,* meaning "sheepfold." The prosperity of the merchants of the town—mostly Venetians, but probably also including Greeks, Croats, and others—is suggested today by the Palača Gazzari, the Palača Hektorović, the Palača Lučić-Paladini, the Palača Jakša, and other mansions in Gothic and Renaissance styles, which were built into the city wall in the fourteenth and fifteenth centuries, and by the remains of the governor's palace, rebuilt in the Venetian Renaissance style in 1479. The loyalty of the people of Hvar to the Catholic Church is indicated by the construction, initiated in 1461, of a Franciscan friary on the town's seafront and by the rebuilding of the cathedral from 1560 onward, incorporating parts of the Benedictine church that preceded it, as well as a bell tower completed in 1532. It now contains a number of important works of art, such as three Italian portraits of the Madonna, an anonymous portrait of the Madonna and Child from the thirteenth century, and two of the Madonna with saints from the seventeenth century. Other treasures, such as the gilded staff of Pritić or Patrizius, bishop of Hvar from 1509 to 1522, are displayed in a museum inside the Bishops' Palace, next to the cathedral.

Hvar continued to be insecure, since Venice had many enemies, notably the rival Italian Republic of Genoa and the Ottoman Empire. Between 1531 and 1551 the Kaštil or Španjola, the fortress at the top of the hill, 358 feet high, which dominates the old town of Hvar, was built, and the arsenal was refortified and expanded about 1559. The long-expected Turkish attack came in 1571, when a fleet commanded by Ulez-Ali, the bey (governor) of Algeria, who is said to have been a former Christian monk from southern Italy, attacked the arsenal, burned most of the town, and destroyed the friary. The invaders eventually were driven out. A new arsenal, begun in 1579 and completed in 1611, was provided with an arched entrance into its barrel-vaulted main chamber, to permit ships to be brought in directly from the harbor in times of danger.

Venetian rule was not always accepted without protest. An uprising between 1510 and 1514, led by Matija Ivanić, a resident of Vrbaska, expressed the resentment of many of the islanders, and later in the century there was a series of further uprisings on Hvar, culminating between 1610 and 1612 in a major revolt that was eventually put down by the Venetian governor of the island, Pietro Semitecolo. His wife Maria is said to have pacified the commoners with promises of political participation, which were never kept. During the same period Hvar was home to a number of writers whose work, now considered part of the Croatian Renaissance, gave another form of expression to a distinctive identity. The first nonreligious play in the Croatian language, *Robinja* (*The Slave Girl*), was written by Hanibal Lucić (1485–1553) in the town of Hvar; Petar Hektorović (1487–1572), whose poem *Ribanje i ribarsko Prigovaranje,* concerning fishing, has had a major influence on Croatian literature, divided his time between Tvrdalj, a house built around a fish pond at Stari Grad, and a summer residence, now in ruins, in the town of Hvar. Later, Ivan Franjo Biundović, born in the town of Hvar in 1574, had a long career as a Venetian diplomat in England and elsewhere, under the name of Gian Francesco Biondi, and wrote several novels and books on history before his death in 1645. The cultural flowering of Hvar was also aided by the activities of the Kazalište, a theatre, now among the oldest surviving in Europe, which was built inside the Arsenal in 1612.

Hvar continued to be the site of a Venetian naval base, and therefore part of the Venetians' trading network around the eastern Mediterranean Sea and in western Asia, until 1776, when the base was moved to Kotor on the Dalmatian mainland. While the town of Hvar went into decline, the rest of the islanders, surviving as before by fishing and farming, were probably little affected either by this event or by the abolition of the Venetian Republic in 1797, on the orders of Napoléon. Dalmatia, including Hvar, was given to Austria, then Napoléon's ally, and Austrian troops destroyed the Venetian governor's palace, leaving only the loggia, and built a barracks inside the Kaštil. In 1805, after Austria's defeat at the Battle of Austerlitz, Dalmatia was absorbed into Napoléon's Kingdom of Italy, and Hvar was seized by French troops, who built a fortress, Fort Napoléon, above the town of Hvar, which is still used as a military post and radar station. It was bombarded by Russian warships in 1808. In 1809 Dalmatia was incorporated into a new French dependency, the Illyrian Provinces, but this did not outlast Napoléon's empire. In 1812 Hvar was occupied by the British; in 1814 it went, with the rest of Dalmatia, back to Austrian rule.

Although Austria and Hungary were united under the Habsburg dynasty, their possessions were kept separate. Dalmatia was ruled directly from Vienna from 1822 onward while the rest of Croatia was still ruled by Hungary. Even so, what was then called Illyrian nationalism grew steadily throughout the nineteenth century as a movement for creating a single Croatian state, although its leaders and thinkers could not agree on whether to stay loyal to the Habsburgs, to seek independence, or to aim for union with the other Southern Slavs (of Serbia, Slovenia, Bosnia-Hercegovina, Montenegro, and Macedonia) in a new Yugoslavia. In relation to these debates and to European events that were to change its fate, Hvar was relatively marginal and powerless, but could not escape the impact either of conscription (by the Austrians) during World War I or of an Italian occupation just after the war had ended and the Habsburg empire had collapsed. This occupation lasted until 1921; in 1922 Hvar became part of the new Kingdom of the Serbs, Croats, and Slovenes, officially renamed Yugoslavia in 1929.

Hvar was to undergo yet more foreign occupation, however, when the Italians returned in 1941, during Mussolini's conquest of Dalmatia. After the collapse of the Italian Fascist regime in 1943, many of the Italian conscripts stationed on the island joined with the Yugoslav partisans, creating a Garibaldi Brigade, named for their own hero of national unification, to fight the German troops who attempted to take over Dalmatia. Some of them were among the many partisans rounded up and shot by the German occupying force on Hvar before the island was reconquered by partisan forces in 1944.

Beginning in 1944 Hvar was governed as part of the Socialist Republic of Croatia within the Yugoslav federation created by the partisan commander and Communist leader Josip Tito. In 1990 Croatia declared its independence, which was recognized by the international community in 1992. As part of a newly independent Croatia, Hvar has shared with the rest of Dalmatia both the unwelcome side effects of being close to the fighting in other parts of the former Yugoslavia and the hope that some kind of peaceful development and prosperity can be reestablished. Although the farming and fishing that have defined the lives of most of the islanders throughout history continue today, long after the departure of Greek, Venetian, and Austrian rulers alike, and in spite of Croatia's grave economic and political problems, the town of Hvar in particular has been modernized and developed as a tourist center. The Kaštil has been converted into a leisure complex, including bars, restaurants, and nightclubs; the loggia of the Venetian governor's palace has been internally remodeled to serve as a hotel; the friary, rebuilt soon after the Turkish attack of 1571, contains an art museum in what was once its refectory, displaying pictures by Matteo Ingoli, Titian, Tiepolo, and others; and the Arsenal is now partly occupied by a supermarket, while its historic theatre has become a museum. The island's other attractions for visitors, at least in times of peace, include its climate—it has the greatest number of hours of sunshine of any Adriatic island—its traditional cuisine, including local white wines, mullet, lobster, and other seafoods, and the eleven smaller islands near the harbor of Hvar, known as the Pakleni Otoci (Hell's Islands).

Further Reading: Hvar figures prominently in guidebooks to and histories of the former Yugoslavia. However, perhaps the most vivid and evocative description of the island and its past, although now slightly outdated, is to be found in Celia Irving's *The Adriatic Islands and Corfu* (London: Dent, and New York: International Publications Service, 1971).

—Patrick Heenan

İstanbul (İstanbul, Turkey): Blue Mosque

Location: The Sultan Ahmet Camii, popularly known as the Blue Mosque, is located in European İstanbul's Sultanahmet district, the historic heart of the Byzantine and Ottoman Empires. The mosque borders the At Meydani esplanade, formerly the Roman Hippodrome, which lies to the northwest, and is situated about 500 feet southwest of Hagia Sophia, from which it is separated by a small park.

Description: The Blue Mosque, so named for its interior tile decoration, was built for Sultan Ahmed I between 1609 and 1616 by Turkish architect Mehmed Aga. Built partly on the site of the Byzantine imperial palaces and amphitheatre (part of the Hippodrome), it is one of the most beautiful and majestic mosques in İstanbul, rivaling only the Byzantine-era Hagia Sophia and the Mosque of Süleyman the Magnificent. Though the Blue Mosque was constructed after the sixteenth-century Golden Age of Ottoman mosque-building, it is often considered the masterpiece of Islamic Ottoman architecture. It is the only mosque to have six minarets. The porticoed and colonnaded courtyard, which faces the At Meydani, is as large as the mosque itself. Externally, the mosque is characterized by its perfect symmetry: its immense central dome is flanked by four large semi-domes, which, in turn, are surrounded by smaller semi-domes. In the vast interior prayer hall, the most prominent feature is the faience of more than 20,000 İznik tiles decorated with floral designs. The Blue Mosque is one of the largest religious complexes in İstanbul and contains the tomb of the mosque's namesake, Sultan Ahmed.

Site Offices: Sultan Ahmet Camii (Blue Mosque)
Sultanahmet Parki
İstanbul, İstanbul
Turkey

Tourist Information Office
Sultanahmet Meydanı
İstanbul, İstanbul
Turkey
(212) 518 18 02

Built in the early decades of the seventeenth century, the Mosque of the Sultan Ahmet, also known as the Blue Mosque because of its blue-glazed interior tile decoration, is one of the most majestic and beautiful mosques in İstanbul. Indeed, its six towering minarets, its magnificent series of domes and semi-domes, its elegantly geometric proportions and spacious courtyard, its vast prayer hall and 260 stained-glass windows, make the Blue Mosque one of the marvels of the world. Considered by many to be aesthetically superior to the sixth-century Hagia Sophia, the Blue Mosque is one of the masterpieces of Islamic Ottoman architecture, along with the Süleymaniye Camii in İstanbul (built 1550–57) and the Selimiye Camii in Edirne (built 1569–74), both of which were constructed by Sinan, Turkey's most famous architect. It is the only mosque in the world to have six minarets.

The Blue Mosque was built between 1609 and 1616 by the architect Mehmed Aga on the orders of Sultan Ahmed I (ruled 1603–17), hence the mosque's official name, "Ahmet" being a variation of "Ahmed" (foreigners, but not Turks, refer to it as the "Blue Mosque"). Mehmed went to Constantinople (İstanbul) in 1567 to study music. He became interested in architecture and began studying with Sinan, the Royal Chief Architect of Süleyman the Magnificent, who was responsible for building some of the greatest mosques of the Golden Age of the Ottoman Empire. Mehmed based his general design of the Blue Mosque on the Byzantine-era Hagia Sophia, as well as on the work of Sinan. In fact, its ground plan is specifically modeled on the style of Sinan's first commission, Sehzade Camii (1544–48) in İstanbul (also known as the Mosque of Mehmet, son of Süleyman), giving the new structure greater harmony. With the Blue Mosque, architect Mehmed had outdone his master, as well as the architects of Hagia Sophia, and built a mosque that stands as one of the final examples of the great era of Ottoman architecture. Unfortunately, the sultan was able to enjoy his mosque for only a year: he died in 1617 at the age of 27.

The Ottoman Turks ruled most of Anatolia (Asia Minor, or western Turkey) by the end of the thirteenth century, and finally succeeded in capturing Constantinople (formerly Byzantium) in 1453, spelling the end of the Byzantine Empire. By 1516, the Ottomans had also conquered Syria and Egypt, thus achieving political and religious supremacy in the Islamic world.

The Ottoman school of mosque-building became influenced by Byzantine architecture. The Turks were impressed by the supreme architectural qualities of Hagia Sophia, which they immediately converted to a mosque (Aya Sofya Camii), and adopted it as a model for almost every important mosque built in their new capital. Turkish mosque architects also strove for a more pleasing, perfectly symmetrical external appearance. While some earlier mosques, such as Hagia Sophia, had a central dome flanked by two semi-domes, the Blue Mosque's main dome was surrounded by four semi-domes and smaller domes. Though Ottoman mosques were compositionally and structurally Byzantine, their decorative detail was Islamic and Persian in character.

The Blue Mosque is located in the city's Sultanahmet district, the historic heart of İstanbul and the center of both

The Blue Mosque
Photo courtesy of Embassy of Turkey, Washington, D.C.

the Byzantine and Ottoman Empires. It offers a superb view over the Sea of Marmara. Recognizable for miles around, the Blue Mosque is an imposing grey-stoned edifice, situated on a slight hill just south of the Hagia Sophia, to which it was intended to act as a counterbalance. The At Meydani (Place of Horses), a long, narrow, tree-shaded esplanade, spreads out before the mosque to the west. One of the largest religious complexes in İstanbul, the Blue Mosque comprises many outbuildings: a theological school, primary school, mausoleum (including the tomb of Sultan Ahmed), a soup kitchen, and caravanserai.

The complex partly occupies the site of the former Byzantine imperial palaces and the Hippodrome. The Roman Hippodrome, built in A.D. 198 by Emperor Septimius Severus, was the public and cultural center of ancient Byzantium, where chariot races, circuses, and gladiatorial contests were held. The location and dimensions of the present-day At Meydani roughly correspond to the site of the old Hippodrome. Plundered during the Latin crusaders' occupation of 1204–61, the arena had fallen into ruins by the time the Turks conquered the city in 1453. In 1550, part of an arcade of columns at the arena's south end was razed; the entire Hippodrome eventually served as a building-material quarry for

the construction of the Blue Mosque, which sits on the site of the Hippodrome's amphitheatre. In 1890, after being used as a stadium for a polo-like game, the square was converted by the French architect Bouvier into the municipal park.

Among the more noticeable features of the Blue Mosque are its six minarets—towers from which the *muezzin* (mosque officials) call the faithful to prayer. Legend has it that before building began in 1609, religious authorities objected to Sultan Ahmed's plan for a six-minaret mosque because they said it was unholy to rival the six minarets of the great mosque of Mecca; the sultan was later pressured to add a seventh minaret for Mecca, in order to emphasize its pre-eminence. In fact, the seventh minaret had already existed at Mecca, and Sultan Ahmed built six minarets at his mosque only for the sake of harmony—although there were concerns that too many minarets might be too expensive for the declining wealth of the empire and that the imperial palaces had to be destroyed to make way for the mosque's construction. While the four minarets at the corners of the building have three balconies each, the two smallest ones, at the north and west corners of the courtyard, have only two.

Another distinctive feature of the exterior is its series of spherical forms: the mosque has a four-leaf clover shape, with the vast central dome flanked by four large semi-domes, each abutted by smaller semi-domes to the north, south, east, and west; these, in turn, are surrounded by smaller cupolas reinforced on either side by round turrets at the building corners. There are also small cupolas atop the tall octagonal turrets that extend from the height of the four pillars supporting the main dome.

The vast, rectangular courtyard, in front of the building and facing west toward the Hippodrome, is of the same dimensions as the mosque itself. The forecourt has large bronze portals, or gateways, on each of the three sides; sightseers are expected to enter through the graceful west portal, which bears calligraphic inscriptions by the father of the travel writer Evliya Celebi. The closed courtyard, paved with marble, is surrounded by a covered portico resting on twenty-six granite columns and surmounted by a series of thirty small cupolas. At the center of the courtyard stands the *sadirvan,* or ablution (purification) fountain, hexagonal in shape with six columns. Before praying in the mosque—five times daily at set times—believers must perform ablution rituals at the *sadirvan:* washing face, hands (to the elbows), and feet (to the knees).

During the reign of the sultans, the courtyard was also used for all the important religious and political meetings; it was also where the principal Islamic festivals such as the *seker beyrami* (sugar festival) and *kurban beyrami* (sacrificial festival) were celebrated. The forecourt was used as a place of worship for Friday noon prayers when the mosque itself was full. But most important, from the early seventeenth to the mid-nineteenth centuries, the courtyard was the starting point of the pilgrimage to Mecca.

The Blue Mosque's interior is simple, symmetrical, yet grandiose. The great prayer hall, reached by a central door and two side doors, measures 174 feet by 167 feet. The immense central dome, with a diameter of 90 feet and a height of 141 feet at the crown, rises above the hall; the main dome and semi-domes rest, by means of pendentives, on four large arches supported by four massive round, grey marble columns. The pillars, termed "elephant legs," are sixteen feet in diameter and are banded at the waist and ribbed with convex fluting. The two columns at the entrance have fountains built into them. The main dome and pendentives are decorated with calligraphic inscriptions of verses from the Koran, as well as with the names of the Prophet and the first caliphs. On the floor of the great hall is a very fine but worn carpet, upon which are also scattered beautiful old prayer rugs.

Light fills the mosque's prayer hall through five rows of 260 semicircular stained-glass windows in the walls and dome. Some of the windows, modern copies of seventeenth-century originals that were shattered in an earthquake, still have their best-quality Venetian stained glass set in plaster, and form a floral pattern. The doors and window shutters are made of wood inlaid with ivory, mother-of pearl, and tortoiseshell.

The enameled faience wall decorations are the most striking and celebrated feature of the mosque's interior. The bluish-green-shaded painted tiles, which give the mosque its popular name, glitter in the sunlight that streams through the windows. Dating from the late sixteenth and early seventeenth centuries, they cover the walls and the pillars as far as the upper windows, as well as the great dome. The tiles mostly represent stylized flower and tree designs—carnations, cypress, hibiscuses, hyacinths, peonies, roses, tulips—in ornamental patterns with bright, clear colors.

The faience panels on the lower walls and first-floor galleries are especially noteworthy: the tiles, over 20,000 of them, practically exhausted the ceramic kilns in İznik, the ancient Nicaea. Also decorated with floral and geometric designs, this particularly exquisite İznik ware constitutes the last expression of the golden age of sixteenth-century Turkish ceramics. İznik was renowned throughout the sixteenth century as a center of tile and pottery production; skilled artisans were brought from Persia, and by the end of the century the city had more than 300 kilns. The tiles made here decorated public buildings throughout the Ottoman Empire. By the mid-eighteenth century, the local ceramic industry was defunct. The walls and arches of the Blue Mosque's interior are also covered with arabesque stenciling.

The beautiful white Marmara marble *mihrab* and *mimbar* are outstanding examples of seventeenth-century Ottoman sculpture. A *mihrab* is a recess in the wall at the rear of a mosque's prayer hall pointing in the direction of Mecca, to which Moslems turn for prayer. The Blue Mosque's *mihrab,* framed by two great candelabra, contains a piece of the sacred Black Stone from the Ka'aba in Mecca. The *mimbar* is the pulpit where the preacher stands during Friday sermons. The Blue Mosque's *mimbar,* raised and stepped, is a replica of the one at Mecca. It was formerly used only for sermons, or to announce great events. It was from this *mimbar* that the final abolition of the janissaries, the sultan's elite guard force, was announced in 1826. The janissaries had revolted against military reforms in June 1826, and the rebels were eventually burned to death in their barracks in the Hippodrome. The dissolution of the dreaded corps set the stage for real reform in the following years.

In the left part of the prayer hall stands the sultan's loge, or imperial box. On the ceiling under the imperial loge, in the upper gallery to the left of the *mihrab,* is a rare example of early Ottoman decorative style, an array of painted floral and geometric motifs. An exterior ramp on the mosque's northwest corner leads to the *hunkar kasri,* or imperial pavilion, which is connected to the loge by an internal corridor. The ramp and corridor were built so that the sultan could go right up to his box on horseback. The vaulted, underground storerooms and elephant stables at the east end of the mosque have been restored to house the Museum of Carpets and Kilims; *kilims* are woven, decorative rugs used for domestic needs or as prayer mats.

The large square *turbe* (mausoleum), which lies outside the precinct wall to the northwest of the mosque, contains the tomb of Sultan Ahmed I, his wife Kosem, and three of the sultan's sons, two of whom were subsequent Ottoman rulers.

338   INTERNATIONAL DICTIONARY OF HISTORIC PLACES

The mausoleum was built in 1619, two years after Ahmet's death. Also buried here are the Sultans Osman II and Murad IV (both of whom ruled in the seventeenth century), and Prince Beyazit. The mausoleum has a graceful cupola and, like the mosque, is decorated with seventeenth-century İznik tiles. It contains various holy relics, including a turban from the tomb of Yahya Efendi (d. 1570), one of İstanbul's most renowned Moslem saints.

During the Ottoman Empire, the sultans went to the Blue Mosque, often with much ceremony, for the great Friday prayer and for religious festivals. While major Islamic religious festivals and ceremonies are still celebrated at the mosque, it is not otherwise used as a religious facility.

Further Reading: *A History of Ottoman Architecture* by Godfrey Goodwin (London: Thames and Hudson, and Baltimore, Maryland: Johns Hopkins University Press, 1971) is the definitive guide to Ottoman architecture throughout Turkey. Goodwin's *Sinan: Ottoman Architecture and Its Values Today* (London: Saqui, 1992; Brooklyn, New York: Interlink, 1993) is about the famous royal architect of Süleyman the Magnificent, who profoundly influenced Islamic architecture; one of his students, Mehmet Aga, built the Sultan Ahmet Mosque (Blue Mosque). Stanford Shaw's *History of the Ottoman Empire and Modern Turkey,* in two volumes (Cambridge and New York: Cambridge University Press, 1976) is a scholarly analysis of the Turks; volume I deals with the rise and decline of the Ottomans, from 1280 to 1808.

—Jeff W. Huebner

İstanbul (İstanbul, Turkey): City Walls

Location: The main sector of İstanbul is located on the Bulgarian peninsula, on that portion of Turkey that lies to the European side of the Bosporus Strait, south of the inland waterway called the Golden Horn. The walls of İstanbul stretch from the Golden Horn to the Sea of Marmara, four and one-half miles to the south.

Description: The city walls of İstanbul were installed to safeguard the peninsula from ground attacks launched from the European continent. At the north and south ends of these land walls, the bulwarks join sea walls, which keep watch over the shorelines of İstanbul. These links form a circuitous defense of the ancient city and in one form or another, have discouraged enemy approach from all sides throughout much of recorded history. Today, many parts of the walls of İstanbul have crumbled, but sections with remarkably little deterioration remain, testaments to a bygone era in which brute strength was a political necessity.

Site Office: Tourist Information Office
Meşrutiyet Cad. 57/5 Tepebaşi, Beyoğlu
İstanbul, İstanbul
Turkey
(212) 245-6875

For more than 2,500 years, towering stone walls have successfully protected the city of İstanbul through its most perilous moments. Earlier versions of these defenses watched over a rapidly growing international hub. First constructed during the settlement of Byzantium in the seventh century B.C., the walls have been moved outward on several occasions, in each instance accommodating a sizable expansion of the metropolis.

The Roman emperor Septimius Severus was the first leader to enlarge the city by rebuilding its walls farther west. Another Roman emperor, Constantine the Great, destroyed Severus's fortifications and erected a bigger wall to guard the city, which took his name—Constantinople. In the early fifth century, fearing a reprisal of the attack on Rome by the Goths, the administration of Emperor Theodosius II built still greater walls of defense, topped by lookout towers that rose to seventy feet. The walls of Theodosius proved impenetrable for ten centuries until, in 1453, Mehmed the Conqueror, determined to claim Constantinople in the name of Islam, finally managed to breach the ramparts at Top Kapisi (Cannon Gate).

Folkloric history of the city known today as İstanbul begins with its foundation by Byzas, said to be the son of Poseidon, god of the sea. Legend has it that Byzas was instructed by the Oracle at Delphi to settle at the place that lay across from the "city of the blind." Upon arriving at the ideally situated but uninhabited hilltop that lay across the Bosporus from Chalcedonia, a decidedly undesirable location, Byzas determined that the Chalcedonians must be the "blind" to whom the Oracle referred.

The Greeks named the village "Byzantium" after its discoverer, and they built its first walls, enclosing the easternmost tip of the peninsula, Seraglio Point, which overlooks the intersection of the Golden Horn waterway and the Bosporus Strait. The Bosporus was already proving to be one of the world's busiest trade routes, linking Europe to Asia Minor, the Black Sea to the Mediterranean. Recognizing a lucrative opportunity, Byzantium's immigrant Greek merchants prospered, with their successes interrupted only by plundering Persian invaders. The Greeks took back their city in 479 B.C., when the Spartan general Pausanias defeated the Persian army at Plataea. A monument to Pausanias's victory, a serpentine memorial column erected at Delphi, was removed to Byzantium under the reign of Constantine, where it remains today, albeit in dilapidated condition.

Despite infighting between Sparta and Greece, the Greek port city of Byzantium thrived, set apart as it was from the battlefields of the two powers. Many schemers desired control of the blossoming economic center: Philip of Macedon was thwarted from his nighttime raid on Byzantium by an unexpectedly bright crescent moon, the image of which still appears on the flag of Turkey. Alexander the Great brought the city under his rule during the height of his empire; upon his expulsion, the citizens of Byzantium were forced to fend off another attack, this time by the Scythians.

With the influence of the Roman Empire spreading throughout much of Europe, Byzantium was annexed as a Roman province. Swiftly, the city found itself buffeted by political upheavals within the empire. In A.D. 193, it was ransacked by the emperor Septimius Severus, who sought vengeance for Byzantium's support of his rival, the governor of Syria. Over three years, Severus slaughtered the leaders of Byzantium, laid waste to the city, and brought down its original defenses, walls that had been so well constructed they had seemed to be of one stone.

Severus, however, was well aware of the strategic benefits of Byzantium's post on the Bosporus. When he had sufficiently made the city suffer, he rebuilt it into an even more impressive site than it had been in its previous incarnation. New walls extended from the site of the Galata Bridge (which today spans the Golden Horn, connecting the "new" section of İstanbul to the old), continuing past the site of the column of Constantine (the memorial to Pausanias, soon to be imported from Delphi) and the Hippodrome to the Sea of

Remains of the İstanbul walls
Photo courtesy of The Chicago Public Library

Marmara. Along this southerly stretch of Severus's walls ran the bustling square called Tetrastoös; no longer demarcated, the area where the square once lay provides breathing room between two of İstanbul's proudest features, Hagia Sophia and the Blue Mosque.

One hundred thirty years after Severus, during a time of internal power struggles in the Roman Empire, Emperor Constantine the Great defeated his rival Licinius at sea at Chrysopolis (now known as Üsküdar), across from the mouth of the Golden Horn. Claiming he had been instructed in a dream to take his cabinet from Rome, Constantine moved the capital of the empire to Byzantium. Like Severus before him, Constantine avenged Byzantium's support of his rival by tearing down existing fortifications, only to reconstruct them to greater effect. No city was deemed worthy of being the empire's capital unless it sat on seven hills, so Constantine moved the walls of his city nearly two miles to the west of those of Severus, so that they surrounded seven hills. Upon the wishes of his mother, Helena, the emperor officially replaced Roman paganism with Christianity, formally dedicating his new capital city to the Virgin Mary in May 330. Constantine renamed it "New Rome," but it quickly came to be known as Constantinople.

During his reign, Constantine built and/or improved several structures, including his palace, the Forum, the Hippodrome, and numerous churches, all within the vicinity of Seraglio Point. To ensure the safety of his growing city, the emperor employed an estimated 40,000 Goths to erect new walls. When the workers were finished, they took up guard posts along the ramparts, with 5,000 troops stationed between each of seven gates. Constantine's walls were laid out in a series of four rows, the innermost two featuring towers fifty feet apart. Along these walls ran a complex system of cisterns, a waterway that was highly advanced for its time and was instrumental in the development of the city. Built beneath the earth, these cisterns survive today in greater numbers than above-ground structures from the same period. In contrast the walls of Constantine no longer exist.

After Constantine's death in 337, a succession of restless Roman leaders moved the empire's capital elsewhere; not until the reign of Theodosius I toward the end of the fourth century was it restored at Constantinople. Theodosius I is credited with erecting the Golden Gate, the closely guarded entrance to the city at its southwestern corner, a magnificent archway fronted by a moat that would provide the setting for many triumphant returns from battle.

Theodosius I died in 395, leaving Constantinople in the hands of his son Arcadius, who in turn bequeathed the city to his son Theodosius II, seven years of age, in 408. The Goths sacked Rome in 410 and Huns soon threatened the same in Constantinople, whose young emperor was advised by the regent Anthemius to further reinforce the city in 413. The walls of Theodosius II (sometimes called the walls of Anthemius) were built approximately one mile to the west of Constantine's, linking the Golden Gate archway to the shore of the Golden Horn, four and one-half miles to its north. At its northernmost point, the course of this latest line of defense was eventually altered (most likely by the Emperor Manuel I Comnenus in the mid–twelfth century), in order to enclose the imperial Blachernae Palace.

Theodosius's walls were built with limestone blocks, bricks, and concreted rubble. With each segment built separately to allow for varying rates of sinking, significant portions of the walls remain standing today, more than 1,500 years hence, proving without doubt their durability. Above the main wall ninety-six watchtowers rose to heights of sixty feet or more; this wall was thirty feet high, and slightly more than half as thick. Fifty feet in front of the main wall a moat was dug, sixty feet wide and twenty-two feet deep, a daunting obstacle that left invaders vulnerable to counterattack should they attempt to cross, for their initial greeting was a row of archers stationed behind a first, shorter wall. This first wall, ten feet high, was six feet wide at its base, and tapered to two feet at its summit. Various accounts dispute the use of the ditch. Although designed as a moat, there is little record of its being filled, an inconsistency best explained by the perpetual shortage of running water in Constantinople. The walls of Theodosius, the largest undertaking in a long effort to defend the city of Constantinople, would prove to be the city's protector from invasion by land for more than 1,000 years.

In 439, Theodosius II and the Prefect Cyrus completed a circuitous defense of the city, bridging the gap between their completed land walls and sea walls that were partially extant from the times of Severus and Constantine. From the top of the Golden Horn to the tip of Seraglio Point, the thirty-foot-high sea walls included 110 towers and 14 gates. Along the five-mile stretch from Seraglio Point to the Golden Gate there were 188 towers and 12 or more gates. Given the powerful tide of the Sea of Marmara and the difficulty it presented in coming ashore, these last walls were not required to equal the height of their northern counterparts, and they were built to a height of twenty feet.

Reassurance of Constantinople's safety was well timed: throughout the fifth century A.D., the city was inundated with disaster, both natural, such as plague and fires, and man-made—the deadly form of political upheaval that came to be called Byzantine. In 447, a powerful earthquake toppled at least fifty-seven towers along Theodosius's main wall. Threatened by the proximity of the Huns, the city's rival Green and Blue factions joined forces to repair the ramparts in just two months, in the process adding a third line of defense between the main wall and the moat, equipping it

with ninety-two towers, which fell between the towers of the higher wall behind it.

Beginning in 507, Emperor Anastasius I oversaw the construction of a land wall some forty miles to the west of the city. At a length of forty-one miles, this wall connected the Black Sea to the Sea of Marmara. Efforts to safeguard Constantinople from such a long distance proved futile, however, as subsequent attackers were not thwarted until they reached the Golden Gate along the walls of Theodosius.

From the time of the collapse of the Roman Empire, Constantinople—as capital of the Eastern Roman, or Byzantine, Empire—was a site of recurring strife. The city's coveted economy and its evolving spirituality provided a pair of fearsome battlegrounds. Although the walls prevented enemies from taking the city by force, political power plans kept the city in near-constant turmoil. In 622, the walls of Theodosius endured their first enormous threat, as Persians attacked from the east and a coalition of Avars, Slavs, and Bulgars amassed at the gates of the west. The battlements proved insurmountable, and the Byzantines put down an estimated 80,000 enemy troops. In other instances, warring factions of Saracens and Armenians were stopped at the walls of the city, with the latter group of soldiers expressing their anger by camping in the shadows of the Golden Gate, offering human sacrifices to the godly power of the city fortress. In 674, Arab legions were repelled at the walls by the defenders' "fire" (a prototype of napalm). The Arabs never entered the city by force, but continued their attempts until their decisive defeat in 718.

During the winter of 763–64 the sea walls along the Bosporus were damaged when a gigantic ice floe collided with the coast. In 860, feuding Bulgarians and Byzantines were forced to unite in defense against Russians invading from the northeast. Later that century, upon the ascension of the dynastic leader Basil I, a peasant who murdered the reigning emperor, Michael III, the Bulgars renewed their besiegement of Constantinople. Their leader, Symeon, tried desperately to overcome the Theodosian ramparts, in 913 and again eleven years later, but on each occasion he was frustrated. In 1014, in one of the most gruesome spectacles of all the struggles that have taken place at the foot of Constantinople's walls, Basil II (the "Bulgar Slayer") put out the eyes of some 15,000 Bulgarian fighters, sparing one in one hundred so that they could lead the rest home. Upon the soldiers' return, the Bulgarian king died from shock; four years later, his state was annexed to the Byzantine Empire.

From the eleventh century, Seljuk Turks often ruled in Constantinople, aligning themselves with Rome as the Sultanate of Rum but setting in motion a gradual shift from Christian to Moslem influence in the capital city. Meanwhile, Ottoman Turks from the east and Genoese and other immigrants who had settled across the Golden Horn in Galata both eyed the lucrative city with envy. Despite such secular designs on the city, however, it was a religious matter—the rift between Eastern and Western Christianity—that hastened the city's downfall. The Crusades to regain the Holy Land from the domain of the Moslems withered the fortitude of what had become the capital of eastern European culture, as Constantinople's religious diversity made it a target for the Crusaders. In particular, the Fourth Crusade in 1204 was devastating, as Christian soldiers laid waste to Constantinople en route to Egypt, breaking through the sea walls along the Golden Horn just northwest of the still-existent Cibali Gate. Sparing little, the Crusaders damaged many of Constantinople's claims to greatness, looting tombs and vaults for precious treasures and manuscripts.

Greek influence was not restored in Constantinople until 1261, whereupon its leaders were required to combat the increasing determination of the Turks' Ottoman Empire with little help from the west. By the 1420s, the only territory of Asia Minor that the Ottomans did not control was Constantinople. In 1395 the Sultan Bayezid I, nicknamed "Yildirim," or "Thunderbolt," had attacked at the walls of the city but was thwarted. Two years later, he built the fortress Anadolu Hisar across the Bosporus from Constantinople so that he might wrest control of the strait's passage from the capital city.

The stage was set for the fall of Constantinople, and the sultan called the "Conqueror," Mehmed II, three leaders removed from Bayezid, enlisted at least 60,000 men to complete the task. In preparation, in just three months he oversaw the building of the stout Rumeli Hisari, the Castle of Europe, as a base of activity, six miles north of the city on the western shores of the Bosporus. Constantinople had no more than 20,000 defenders, but its troops closed off the Golden Horn by running an enormous chain across it. Not to be outmaneuvered, Mehmed ordered his troops to skirt the blockade by land, and they hauled their ships over the banks of Galata. With fourteen batteries of artillery and four cannon, including the largest one built to date, the Turks mercilessly bombarded the sea walls near the gate of St. Romanos for nearly two months, and on May 29, 1453, their fire finally broke through the fortress.

After more than fifteen centuries of Roman emperors in Byzantium—twenty-two centuries in Rome—and after a succession of more than eighty Christian emperors in the city (the last of whom, Constantine XI, died fighting as Mehmed's men broke through the walls), the city finally succumbed to a permanent new leadership. With their victory, the Moslems eradicated the last vestiges of the Roman Empire.

Mehmed set to work restoring the finery of his new stronghold, quickly doubling its population to 100,000 residents. However, it was not until the rule of Selim I in the early sixteenth century that Constantinople was chosen as capital of the Ottoman Empire. The Ottomans were well aware of the city's success as a hub of global activity, and they permitted a system of autonomous government for various ethnic groups, allowing the Greeks to retain a patriarch and the Jews to remain in their quarter. Each of these groups was in turn required to obey the sultanate. Christians, however, were relegated to an inferior class level, and they were expected to relinquish their sons to the ranks of the Janissaries, the notorious slave army of the Ottomans.

To assure his own safety, in 1457 Mehmed fortified the grand entrance to the city at the Golden Gate, building an enceinte just inside the gate. This added protection came to include seven towers (four Byzantine and three Turkish), lending the structure its name, Yedikule Hisari, or Fortress of Seven Towers. This impressive fortress, with its winding staircase and vaulted chambers, was used as a prison shortly after its completion.

Thus began a period of proud and aggressive Ottoman rule, in which the sultan Süleyman I the Magnificent defined the empire's ostentation and administrated mighty forays onto the European continent. Meanwhile, near the Blachernae Palace, parts of the sea wall along the Golden Horn were destroyed in the earthquake of 1509, a natural occurrence that symbolized the decline of the significance of İstanbul's walls. By the time of the Ottomans' retrogression, the stout defenses of İstanbul were no longer imperative to the survival of the aging city.

Today İstanbul has been stripped of its post as capital of Turkey, but its rich history lingers. As for the walls, large portions of those along the Golden Horn have crumbled or been torn down, but the shorter battlements that line the Sea of Marmara, despite damage incurred during the railway construction of 1871, remain largely intact. Similarly, although the Golden Gate is no longer studded with precious metals, many traces of its intricate sculpture remain, and there is yet much to see along the land walls built by Theodosius. The majestic, oft-restored land walls of İstanbul have been declared by UNESCO as one of the cultural heritages of the world.

Further Reading: *The City of Constantinople* by Michael Maclagan (London: Thames and Hudson, and New York: Praeger, 1968) provides a thorough description of the various walls of İstanbul, putting into perspective their importance in defending this so-often besieged capital city. Both *İstanbul: Tale of Three Cities* by Ryo and Banri Namikawa (Tokyo and Palo Alto, California: Kodansha, 1972) and *A Traveller's History of Turkey* by Richard Stoneman (Moreton-in-Marsh, Gloucestershire: Windrush, and Brooklyn, New York: Interlink, 1993) include good histories of the walls as they detail the development of İstanbul. In addition to these informative sources, colorful portraits emerge in three other books: *The Walls of Constantinople* by Captain B. Granville Baker (London: John Milne, 1910); *Constantinople: The Forgotten Empire* by Isaac Asimov (Boston: Houghton Mifflin, 1970); and the essay on Constantinople written by G. C. Curtis, included in the book *Turkey and the Balkan States, As Described by the Great Writers,* edited by Esther Singleton (New York: Dodd, Mead, 1908).

—James Sullivan

İstanbul (İstanbul, Turkey): Hagia Sophia

Location: On Ayasofya Square in the Old Stamboul section of İstanbul.

Description: Hagia Sophia (known as Ayasofya in Turkish), also known as the Church of Holy Wisdom and the Great Church, was the world's preeminent Christian church for 1,000 years. The present church is the third on the site; the first was constructed in the fourth century, the second in the fifth, and the present one in the sixth. It was converted to a mosque in the fifteenth century and a museum in the twentieth. It remains an architectural and technical marvel and contains some of the most beautiful mosaics created in the Byzantine world. Its name means "divine wisdom"; although popularly called Santa Sophia, it was never dedicated to a particular saint.

Site Office: Hagia Sophia
Ayasofya Square
İstanbul, İstanbul
Turkey
(212) 522-1750

The inspiration for building Hagia Sophia can be traced to the early fourth century and the Roman emperor Constantine I (the Great). Constantine, a convert to Christianity, moved the seat of the Roman Empire from Rome to Byzantium, which was renamed Constantinople. Once in Constantinople, the emperor developed a plan to construct a great church, which would serve as the most important church in the entire Christian world. Constantine's plan was not realized, however, until the reign of his son, Emperor Constantius II. The church was dedicated on February 15, 360, and was called "The Megali Ecclesia" or, in English, "The Great Church." Constantius II's great cathedral, with stone walls and a wooden roof, stood on the highest hilltop in Constantinople, directly above the ruins of an ancient Greek pagan temple. The church did not stand for long. It was burned to the ground on June 20, 404, by a mob protesting the exile of John Chrysostom, archbishop of Constantinople. Emperor Theodosius II ordered the construction of a new church ten years later. It was built on the same site as the original church and followed an identical floor plan. Its dedication was on October 10, 415.

On January 12 and 13, 532, a fire set during the bloody Nika Riots in Constantinople consumed the holy structure. The riots were, at least in part, a protest against the empire's extravagance and unfair taxation policies. Emperor Justinian I successfully put down the revolt, with the help of his wife, Theodora. Work on the new church began only thirty-nine days after the riots, although Justinian may have been planning a larger and grander church even before the fire oc-

curred. Justinian hired the renowned architects Anthemius of Tralles and Isidorus of Miletus to draw up plans for a new church and oversee its construction.

The new church was built on the same site as the previous two. The site was ideal, immediately east of the emperor's palace. To the north were monasteries and mansions for government officials. To the south was a large square named the Augusteum, where official ceremonies were held.

Construction of the new Hagia Sophia was completed in a mere five years, ten months, and four days. The speed of construction could be attributed to the emperor's impatience. Cost seems to have been of little or no concern in the construction of the church. The emperor desired the cathedral to be the most beautiful church on earth, more beautiful and elaborate than even Solomon's temple in Jerusalem. Exotic items were brought from afar to be used in the church. Columns were taken from the Temple of Diana in Ephesus, one of the seven wonders of the ancient world, to be used in Hagia Sophia. Additional fragments from ruins in Athens, Rome, Baalbek, and Delphi were sent along with solid gold ornaments, silver, bronze, expensive stones, porphyry, jasper, and a variety of quarried marble. Wages for 10,000 workers and costs associated with the transport of the various materials used were incurred. Upon completion of the church, Justinian reportedly cried, "My God, I am grateful to you for having chosen me to complete this monument. I am now greater than Solomon." The church would serve as the site of coronations and other imperial functions throughout the Byzantine era.

Justinian visited the construction site almost daily, personally supervising the foremen and workers. His unwillingness to move at a slower pace ultimately led to some construction flaws. Most obvious were problems with settling in the domes, buttresses, and vaulting. These sections of the cathedral were composed of bricks bound in a heavy mortar. To settle properly, they needed time to dry. Ignoring this need so that construction could be hastened, the emperor ordered additional bricks and mortar to be laid atop ones that had not yet dried. The pressure of the massive dome caused the supporting piers, columns, and walls to gradually move out of their proper alignment.

The exterior of Hagia Sophia was meant to be nowhere near as lavish as the interior. Intricate exterior ornamentation was not common during Justinian's time. The powerful mass of the structure—its central dome, many half-domes, and barrel vaults—was meant to capture the eye without the need of further decoration. In this sense, the church followed the lead of its many predecessors. Not until the Middle Ages did the exterior of buildings receive much notice.

It appears that Anthemius and Isidorus incorporated the rectangular ground plan of the Theodosian church into the new structure. The architects also called for the construction

Hagia Sophia
Photo courtesy of Embassy of Turkey, Washington, D.C.

of a massive central dome with half-domes and half-barrel roofs adjoining it. Adopting the Theodosian ground plan, along with an unusual system of serial domes, helped determine the predominantly squarish shape of the exterior. The cathedral measures 245 feet long and 228 feet, 8 inches wide. The length measurement excludes the apse, located on the eastern end of the church, and protruding 20 feet outward. It is the only structural feature to jut out of the building.

The most dominant exterior architectural feature in Hagia Sophia is the vast central dome measuring 185 feet, 8 inches high and 102 feet, 8 inches wide. The dome's current measurement is considerably higher, by some 23 feet, than the original dome, which collapsed in an earthquake shortly after the church was completed. The reconstruction of the collapsed dome was supervised by Isidorus the Younger, a descendant of Isidorus of Miletus. Four massive flying buttresses supporting the western half-dome are believed to have been added to the structure after an earthquake on January 9, 896, or possibly after a tremor on October 26, 896.

There are a number of entrances into Hagia Sophia. Two entrances, one on each side of the apse, are located on the eastern side of the church. The north and south sides have three entrances each. The main entrance, the one used by the

emperor, is on the western side of the church and consists of a series of doors. Originally, to get to the western doors one had to pass through an atrium. The atrium was surrounded on three sides by walls and had a fountain in its center. The walls, fountain, and flooring were made of marble. The fountain was used by the faithful to cleanse their feet before entering the church. The atrium was almost completely destroyed after the Ottoman conquest.

Immediately beyond the atrium site is the outer narthex, also known as the exonarthex. Accessible through one of seven doorways, the outer narthex is a very long, relatively shallow space stretching across the entire western front of the cathedral.

Five doorways spaced across the eastern wall separate the outer narthex from the narthex. The narthex occupies the space between the outer narthex and the main structure of Hagia Sophia. Like the outer narthex, it spans the entire length of the western front, is roofed with cross vaults, and has powerful pillars supporting its many arches. Distinguishing the narthex from the outer narthex are the inner room's double height and depth.

The narthex contains some very beautiful mosaics. Its cross-vaulted ceiling still shines with golden mosaics. Above the Royal Gates, the three central doors of the narthex leading into the nave, is a large mosaic that only recently has been unveiled. It is of Jesus Christ seated on a throne. The Virgin Mary is on his right and St. John the Baptist on his left. Kneeling at Jesus' feet is the figure of an emperor. Two roundels, one a bust of a woman praying and the other of an angel bearing a jeweled wand, are above Christ's throne. An inscription reads, "Peace be unto you. I am the light of the World." The exact date of the mosaic is uncertain.

Just beyond the narthex, Hagia Sophia is split into three distinct spaces. Running along the north and south sides are aisleways that extend the full length of the church. They are accessed through one of six doorways from the narthex, three on the north side and three on the south. The aisleways are subdivided into three sections, or bays, by columnar supports and narrow tunnel vaults. They are vaulted in a variety of forms.

Both the north and south aisles have a central rectangular bay. The four remaining bays flanking the central bays are square. Originally, the southeastern bay was used as a metamorium, space reserved for the emperor during church service. It was from here that the emperor would approach the altar. In the northeastern bay sat the clergy when they were not taking part in the church service. Adjacent to the northeastern bay was the Skeuophylakion, or church treasury, which housed priceless sacramental vessels and utensils.

In the early days, female worshipers were restricted to the upper galleries (the triforium) of the church. Gallery space ran above the north and south aisles, as well as above the narthex. Four labyrinths and four interior stairways permitted access to the triforium. The labyrinths were great winding ramps located at the outer angles of the church. Three of the four labyrinths remain standing. During the tenth century, women were allowed to worship in the middle and western bays of the north aisle. By the fourteenth century, women were again relegated to the upper galleries.

Between the north and south aisleways is the mammoth central nave, accessible from the narthex through the Royal Gates. The Royal Gates are centered on the narthex's eastern wall. The north and south aisleway doors flank the Royal Gates on both sides. The Royal Gates are considerably larger than the doors for the aisleways. The central nave receives abundant sunlight from a ring of windows at the base of the central dome. Additional lighting streams in from a variety of windows located in the walls and above the aisles.

Marble is used extensively throughout Hagia Sophia. A total of twelve different types of marble quarried from across Europe and the Middle East are found in the nave of the church. Eight green monolithic marble columns flank the central nave. Additional green marble columns run down the aisles. At the end of the aisles and in the galleries are ivory-tinted marble shafts. The capitals and moldings on the columns are gold leafed. Four porphyry monoliths made of Egyptian peperino stone support the pendentives. They, too, are sheathed in marble. Marble also covers much of the walls, piers, arches, doorways, and vaults, and frames the windows and doors. The marble used on the walls of the church is cut so that the veining pattern in one slab is reversed in its neighboring slab. Two huge crosses of green Molossian marble flank the Royal Gates on the western wall of the nave. Flooring throughout the structure is Proconnesian white marble. Other colored marble slabs, predominantly in red and green, are used as accents. Unfortunately, a considerable amount of the marble used in the original construction has been shattered by countless earthquakes, is missing in places, or has dulled over the years.

The great pulpit of Hagia Sophia, referred to as the ambo in Byzantine times, is located in the eastern portion of the nave underneath the great eastern arch. The ambo is separated from the rest of the nave by a balustrade. Behind the ambo and between the smaller eastern piers of the basilica is the bema, or sanctuary of the church. The combination of the nave, ambo, and bema back-to-back creates a rectangular interior the length of which is twice the size of its width. Anthemius and Isidorus further visually lengthened the nave by breaking out the east and west arches into half-domes.

Beautiful mosaics can be found throughout Hagia Sophia—directly above the upper parts of the walls, arches, and vaults of the nave and side aisles; in the narthex and outer narthex; and in the triforium. The mosaics are set against a solid gold background. The mosaics do not represent a unified design, but rather are patchwork done over a period of eight centuries. The patchwork effect resulted from restorations necessitated by earthquake damage, by religious upheavals, and by the changing use of the space. A considerable number of mosaics were produced during Justinian's reign. All remaining mosaics attributed to his reign are nonfigural; icon worship did not gain momentum in the region until after his death. If any figural mosaics were done during his reign, they were more than likely destroyed between 726 and 842 by the Iconoclasts, who believed such images promoted idolatry. The Justinian decorations that survive include those of narthex (except for the previously mentioned figural mosaic of Jesus Christ with the Virgin Mary and John the Baptist), much of the abstract decoration of the aisles and galleries, and various designs on the main arches. Principal motifs used in the mosaics include plain or jeweled crosses, eight-pointed stars, squares, roundels, Xs, rhombi, and scrolls. Predominant colors were gold, silver, red, blue, and green.

The Iconoclasts were removed from power in 843, and in the second half of the same century the redecoration of Hagia Sophia with figural mosaics began on a grand scale. In March 867, the image of the Virgin Mary and the Christ child on the half-dome over the apse was ceremonially unveiled. Additional mosaics of the Virgin Mary and the apostles Peter and Paul were completed during the latter half of the ninth century. Work on the mosaics may have been briefly interrupted by an earthquake in 869. Writers from this period are silent about the completion of other mosaics in the church. Yet, there is reason to believe that most of the mosaics seen today, as well as those that were destroyed by earthquakes or by the Ottoman Turks, date from this time.

The classical style of decor had become prevalent in the Christian East when time came to redecorate Hagia Sophia with new figural mosaics. Classical style called for a basilican design devoid of visual barriers, allowing the entire interior to be visible with one glance. In the open space was placed a mosaic pictorial program depicting either the Christian universe or a scheme for salvation. Most important to the pictorial arrangement were the principles of hierarchical arrangement, selectivity, and explicitness. Hierarchical arrangement called for more important Biblical figures to be displayed higher up in the church. The principle of selectivity meant the end of portrayals of lengthy narratives, common before the Iconoclasm, in favor of certain important scenes. Explicitness symbolized victory over the Iconoclasts. Jesus Christ, the Virgin Mary, and the others in the Bible were portrayed with detailed physical characteristics.

The classical style presented unique problems to Hagia Sophia. It was a style that was difficult, if not impossible, to incorporate into the domed cathedral. The structure was simply too massive. In the nave, the area available for mosaics was located above the cornice, seventy-five feet above the floor. At such a height, a figure would have to be at least three times life size to be seen in proper scale. Most areas of the church did not leave adequate space for figures of this size.

Figural mosaics were not only too small, but were also too far removed from one another, making it difficult for them to be seen in the context of a unified story. Finally, Hagia Sophia was a double-shelled structure, unlike the typical single-shelled Byzantine cathedral. Its aisles and galleries constituted an outer periphery, making it impossible to view the entire decorative scheme in a glance.

The solution adopted for the placement of mosaics called for a hierarchical arrangement of single-figure mosaics in the nave and an arrangement of narrative subjects in the galleries. A huge figure of Christ originally occupied the apex of the main dome. It was either a full-length figure of Christ seated on a rainbow, or simply a bust. The original was destroyed after part of the dome collapsed in 1346. A bust of Christ replaced the original, but was most likely destroyed in the sixteenth century, when a Koranic inscription was placed on the dome. In the four pendentives, or supports of the dome, were either four six-winged seraphim or possibly two seraphim and two cherubim. Only the two eastern ones have been preserved, their faces covered by gilded stars.

An enthroned Virgin Mary with the Christ child occupies the eastern half-dome over the apse. Two pale-faced archangels, most likely Michael and Gabriel, stand beside her. Mary, seated upon a jeweled throne, wears blue garments; the Christ child wears a robe of gold. In one of the archangels' hands is a staff that may have been inscribed with the words, "Holy, holy, holy." The other archangel has disappeared except for his wing. The rest of the mosaic is well preserved.

What was originally placed in the great eastern and western arches during the late ninth century remains a subject of controversy. Some have speculated that the twelve apostles were placed in the arches, six in one arch and six in the other.

Others have argued that the Virgin Mary and Christ child were placed in the western arch with Peter and Paul at their side, while all twelve of the apostles were placed in the eastern arch. Both arches collapsed over the ensuing years, the western in 989 and the eastern in 1346. Both arches received new mosaics, which probably bore no relationship to the originals. Fragments of mosaics of the Virgin Mary and child with Peter and Paul were found in the western arch during repairs made from 1847 to 1849. A variety of mosaics were discovered in the eastern arch during this time period. Among those discovered were the Etimasia (a throne symbolizing divine presence), a standing Virgin Mary with child, St. John the Baptist, and Emperor John V Palaeologus.

The decoration of the two tympana was in three registers. The topmost decoration was of a pair of angels or possibly archangels. The middle register was devoted to the prophets; all have been destroyed except for the lower portion of Isaiah. The lowest register consisted of portrayals of fourteen bishops placed in the shallow niches of the south tympana from east to west. Of the fourteen, only four are preserved: St. Ignatius, patriarch of Constantinople; St. John Chrysostom; St. Ignatius Theophores of Antioch; and the mostly destroyed St. Athanasius of Alexandria.

Narrative mosaics were confined to four conical vaults in the gallery. These mosaics did not constitute a narrative sequence, but shared a common theme of theophany, or the visible manifestations of divinity as depicted in the Old and New Testaments. Only bits and pieces of the narrative mosaics remain.

Portraits of the Byzantine emperors are found in the gallery. The extent of the original series is unknown because only three portraits still exist. The oldest, that of Alexander, dates from 895 to 913. Located almost twenty feet above the gallery, it is almost invisible unless viewed under artificial light. Constantine IX Monomachus and his wife, Zoë, pictured making an imperial offering to Christ, make up the second mosaic. The final surviving mosaic, depicting John II Comnenus and family, dates from around the 1120s.

The Deeis, a monumental design from the mid-thirteenth to early fourteenth century, is in the southern gallery of Hagia Sophia. It depicts an enthroned Jesus surrounded on one side by the Virgin Mary, and on the other by John the Baptist. The mosaic is about two-thirds destroyed.

In 1317, Emperor Andronicus II Palaeologus called for the reinforcement of Hagia Sophia after learning that the northeastern vaults were in danger of collapse. The entire structure was restabilized with massive exterior buttressing. In 1344, a terrible earthquake weakened the eastern arch. A period of slow settling and cracking led to its collapse on March 13, 1346, causing much damage to the ambo. The eastern half-dome and the ambo were repaired ten years later, the last significant restoration by Byzantine hands.

May 29, 1453, was the day set by Sultan Mehmed II of the Ottoman Empire to storm the city of Constantinople. Priests, monks, women, and children participated in a procession around the walls of Hagia Sophia in hopes of emboldening the vastly outnumbered defenders of the city. Prayers, the procession around Hagia Sophia, and words of support from Emperor Constantine XI Palaeologus were not enough, however, as Constantinople's defenses crumbled and the city quickly fell. Thousands of the city's residents sought refuge in Hagia Sophia. The Turks successfully stormed the church, chopping down its main door and capturing and killing many of those inside. Pictures, decorations, chains, bracelets, and other holy articles were smashed. When Mehmed entered the church, he was angered by the massacre and ordered it stopped. Calmly, he told those inside to go home, promising them they would not be harmed. (The church had been looted once before, by crusaders who captured Constantinople in the early thirteenth century.)

Contrary to general western opinion, the Turks were quite liberal-minded and considerate in their treatment of Hagia Sophia. They converted the church into a mosque, replacing a cross on the summit of the dome with a crescent. The modified the church's name to djami Aya Sophia Kebir, or "the great mosque of Hagia Sophia." They removed the thrones, altars, and some other features, erecting a mihrab in their place. The mihrab was placed slightly off-center and to the south, indicating to the faithful the direction of Mecca. Four minarets were added to the exterior by Mehmed, Bayezid II, and Selim II. In the eighteenth century Sultan Mahmud I embellished Hagia Sophia the most with a renovation of its gallery and the installation of a library inside the building. Latticework in the new library is a masterpiece of sixteenth-, seventeenth- and eighteenth-century Iznik and Kutahya tiles. A further measure of the Turks' respect for the building is that many mosques were modeled on it.

Another misconception is that the Turks immediately covered all figural mosaics after Hagia Sophia was converted into a mosque. Figural mosaics within easy reach were covered with plaster, whitewashed, or destroyed fairly early, but those not readily accessible remained undisturbed for many years. The decorations in the tympana were not covered until 1573. The mosaic of Christ in the dome remained visible, except for the face, until approximately 1650. Mosaics above the imperial door, in the apse, the great eastern arch, and in the vaults of the north and south galleries remained visible and intact until 1710. Not until the middle of the eighteenth century were all figural mosaics concealed.

A major restoration took place from 1847 to 1849, under the direction of the Swiss architect Gaspare Fossati, during the reign of Sultan Abdülmecid I. The central dome was reinforced with iron rings. Thirteen columns in the triforium were straightened. Damaged nonfigural mosaics were restored. Outer walls, originally spread with plain white plaster, were tinted yellow and marked with brownish-red bands. Both damaged and undamaged figural mosaics were cleaned and covered once more with plaster. Wooden structures around the mosque were torn down to protect it from fire. Abdülmecid took an interest in the mosaics, even pro-

posing at one point that those located outside the place of prayer be left intact and exposed. He was overruled by more conservative forces. All figural mosaics were ordered covered with plaster and a stenciled design imitating the nonfigural designs of Justinian's decoration.

When Fossati discovered the mosaics after all the plaster had been removed, he took time to describe and sketch them. His descriptions and sketches remained unknown for many years. A Prussian architect named Wilhelm Salzenberg also documented the structure around this time. The thirty-seven mosaics discovered and documented by Fossati is almost triple the thirteen remaining today. The disparity between what he found and what remains today is greatest in the tympana, where eighteen mosaics were documented and only three survive. No evidence exists that Fossati's workers destroyed mosaics, except perhaps those that were already seriously damaged. Any destruction that may have occurred probably took place during an interior redecoration following a July 10, 1894, earthquake.

A new era of enlightenment began under the leadership of Mustafa Kemal Atatürk, leader of the modern Turkish republic during the 1920s and 1930s. He permitted foreign archaeologists and scholars to enter Hagia Sophia and begin work uncovering and restoring the remaining mosaics. The uncovering and restoration began in 1931, led by the American Thomas Whittemore, director of the Byzantine Institute of America. Atatürk also surprised many westerners when, on November 24, 1934, he announced the conversion of Hagia Sophia from a mosque into a museum and monument celebrating Byzantine art, ending fourteen centuries of its history as a place of worship. The structure was inaugurated as a museum on February 1, 1935.

Hagia Sophia continues as a museum. Nearby are other historic sites, such as the Baths of Roxelana, remains of an ancient hippodrome, the Blue Mosque, Hagia Eirene, and Topkapi Palace. Hagia Sophia remains an architectural marvel, having stood for almost 1,500 years and sustained more than 1,000 earthquakes. Today, the area of its dome is the fourth largest in the world behind St. Peter's in Rome, the Seville Cathedral, and the Milan Cathedral. Along with the Pantheon in Rome, Hagia Sophia is considered to be one of the two best preserved sacred buildings of antiquity.

Further Reading: Heinz Kahler's and C. Mango's *Hagia Sophia*, translated by Ellyn Childs (New York: Praeger, and London: Zwemmer, 1967; German version, Berlin: Gebr. Mann Verlag, 1967), provides a comprehensive history of the church from the time of Constantine to the Ottoman takeover. It also includes many photographs. Another detailed, illustrated account is Emerson Howland Swift's *Hagia Sophia* (New York: Columbia University Press, 1940).

—Peter C. Xantheas

İzmir (İzmir, Turkey)

Location: İzmir, capital of the province of the same name, is on the Aegean coast of western Turkey, at the end of the Gulf of İzmir, fifteen miles from the sea. The ancient city of Smyrna was situated on a hill near the modern northern suburb of Bayraklı.

Description: İzmir, on the site of ancient Smyrna, is the third-largest city in Turkey and its second leading port, after İstanbul, owing to its spacious yet sheltered harbor. After Ankara, İzmir is the most important city of Asia Minor (Anatolia). Smyrna was one of the most important cities in the Mediterranean world, noted in Greek myth as well as in the Bible; it was once equal to Troy and may have been the birthplace of Homer, the Greek epic poet. Continuously inhabited since the third millennium B.C., Smyrna has been ruled by the Hittites, the ancient Greeks (who raised the city to heights of power and glory in the seventh century B.C.), the Lydians, the Persians, the Macedonian Alexander the Great, the Romans, the Byzantines, the Arabs, the Seljuk Turcomans (Turks), the Genoese, the crusaders (knights of Rhodes), and the Mongols, until becoming a part of the Ottoman Empire in about 1425. İzmir was a key city during the Turkish War of Independence against Greece, 1919–22, and most of the city was destroyed by fire in 1922 as Greek forces retreated; it became part of the sovereign Turkish state a year later. The modern city of İzmir was built largely after the fire; consequently, it retains few vestiges of its ancient and storied past.

Site Office: Tourist Information Office
G.O.P. Bulvari, No. 1/1-D
Büyük Efes Oteli Alti
İzmir, İzmir
Turkey
(232) 4899278 or 4842148

Anglo-Turkish excavations from 1948 to 1951 revealed that the original town of Smyrna was settled at Tepekule, a hill near the modern northern suburb of Bayraklı, some time before 3000 B.C. Its first inhabitants were probably Lelegians, an aboriginal Aegean-Anatolian tribe. The city's name is said to have been derived from an Amazon woman-warrior called Smyrna—an Anatolian name. In ancient times, its advanced culture was rivaled only by that of the fabled first city of Troy. By 1500 B.C. Smyrna fell under the influence of the Hittite Empire, the first major civilization to emerge in Anatolia. Situated between Aeolis on the north and Ionia on the south, Smyrna entered into recorded history when ancient Greeks—the Aeolian settlers of Lesbos (an

Aegean island) and Cyme (an Italic city)—colonized the Bayraklı site. The date of Smyrna's foundation is usually given as 1104 B.C., since Greek pottery of the Protogeometric style found there has been dated to that time.

According to Herodotus, the Greek historian, Smyrna was seized by Ionian refugees from Colophon shortly before 688 B.C.; they barred the gates when the Aeolic inhabitants were at a festival. It became the thirteenth state of the Ionian Federation—even though its epithet, "Aeolian Smyrna," remained long after the Ionian conquest. During the seventh century B.C., when it lay along an important trade route between Lydia and the west, the city-state of Smyrna was at the height of its brilliance and splendor, with massive fortifications and blocks of two-storied houses. The excavations at Bayraklı also unearthed the remains of a temple built to Athena and the wall of the Ionian city. The city's power extended far to the east. Homer was said to have been born in the area at this time, but Smyrna is one of seven cities that claimed the poet. (According to local tradition, Homer was born by the banks of the ancient river Meles, which flowed by Smyrna, and composed poems in a cave at the river's source.)

Meanwhile, the warlike Mermnad kings of Lydia were building a powerful state just to the north. Smyrna was first attacked by King Gyges in the early seventh century, but the city is said to have repelled the invasion; the town's citizens won fame for their exploits against Lydian horsemen. After a long battle, however, Smyrna finally fell around 600 B.C. to Lydian King Alyattes III; the Lydians built an acropolis on a steep peak northeast of the gulf. Two noted Greek elegiac poets wrote of the Ionian struggle against the Lydians: Theognis, around 500 B.C., said that "pride destroyed Smyrna." Mimnermus, some time after 600 B.C., lamented that it was the Smyrnaeans' degeneracy that prevented them from stemming the Lydian advance. Pindar, the Greek lyric poet, also mentioned the fall of Smyrna in an ode fragment written about 500 B.C. In any event, the Lydians destroyed the city's prosperity, and the great power of Smyrna began a long decline. It lost its place in the list of important Greek cities for the next 300 years, though it was commemorated on a silver coin minted in Colophon in the early fourth century. The city did not cease to exist, but its Greek political organization was abolished, and the state of Smyrna was organized into a village system. Smyrna remained little more than a densely populated village through subsequent Lydian and Persian rule. Lydians controlled Smyrna until 546 B.C. when their last king, Croesus, was defeated by Cyrus of the ascendant Persian Empire. The Persians subdued the Greek coastal trading colonies (as well as all of Anatolia) in the course of the next half century.

Around 334 B.C., Macedonian military conqueror Alexander the Great defeated the Persians and conceived the idea of restoring Smyrna to its former Greek glory. Two

The clock tower at İzmir
Photo courtesy of Embassy of Turkey, Washington, D.C.

goddesses worshiped at the oracle of Apollo at nearby Claros were said to have given him the idea in a dream. Alexander ordered the foundation of a new settlement on flat-topped Mount Pagus (now the site of the Byzantine and later Ottoman Kadifekale Castle). According to Strabo, a Greek geographer and historian, Alexander's two top generals, Antigonus and Lysimachus, carried out the plan after his death in 323, placing the city at its current site. Antigonus built the hilltop citadel from 316 to 301, and Lysimachus fortified and enlarged the city from 301 to 281. Smyrna—known for a short time during the Alexandrian era as Eurydicea—soon emerged again as one of the major cities of Asia Minor. In 311, Antigonus, by then ruler of Alexander's Eastern Mediterranean realm, formally decreed that all Greek cities of the region were to be free and autonomous. By 305, Antigonus and Lysimachus were among four self-appointed kings vying for control of Alexander's vast Anatolian empire, eventually dividing it among themselves. Antigonus died in 301 and was succeeded by Lysimachus, who died in 281. In the mid-third century B.C., Smyrna declared its loyalty to Pergamum (the modern Bergama), an influential city-state to the north. When Attalus III, a later successor to Lysimachus, died heirless at the end of the second century B.C., he willed Pergamum to Rome—opening the door to Roman rule in Asia Minor.

Roman influence had been apparent in the area since the second century B.C. As early as 195 B.C., the city built a temple dedicated to Rome. It was a popular place of exile for condemned Roman governors and legates. After Julius Caesar's murder in 44 B.C., his assailants Brutus and Cassius met there for their council of war; around the same time, Trebonius, a Roman proconsul in Asia and a conspirator in Caesar's death, was beheaded in Smyrna by Dolabella, then governor of Syria. Smyrna was chosen by the Roman senate as a site for a temple to Emperor Tiberius in A.D. 26. It soon flourished as a *conventus,* or civil diocese, in what had then become the Roman province of Asia and included south Aeolis; it vied with Ephesus (near modern-day Selcuk) and Pergamum for the title "first city of Asia." The Romans built many impressive edifices and other grand structures in Smyrna, of which only traces remain. The prosperous city of 100,000 people, whose territory spread northward onto the plain, was celebrated for its wealth and beauty, its library, its rhetorical tradition, and its school of medicine, where Galen studied. It was also known for its thermal baths named after the warrior Agamemnon, who is said to have treated his wounded men there. The city's tutelary deity was Cybele, worshiped under the name of Meter Sipylene. Smyrna suffered a catastrophic earthquake in A.D. 178, but the city was rebuilt with the bounty of Emperor Marcus Aurelius. One of the few surviving pre-Ottoman monuments in the city is the Agora, the ancient forum, in the Namazgah Quarter. Though probably originally built during Alexander's time, its current remnants derive from the later reconstruction financed by Marcus Aurelius after the earthquake. The main structures, uncovered by Turkish archaeol-

ogists between the world wars, include a colonnade of fourteen Corinthian columns around a central esplanade and some sculpture fragments. Other statues and antiquities from the site have been removed to İzmir's Archaeological Museum, which contains a wide-ranging collection of artifacts from the İzmir area, including statuary, friezes, tombs, and pottery from many eras. Another ancient site is the imposing fortress of Kadifekale (Velvet Castle) on Mount Pagus, which overlooks the city. While ruins of the original Hellenistic fortress and walls still exist, the present structure dates from Byzantine and Ottoman times.

The Christian religion first spread to Anatolia's great cities just after the crucifixion (c. A.D. 33). It gained a foothold in Asia Minor later in the first century, largely through the extensive journeys of Paul of Tarsus (St. Paul), which are recounted in the biblical book of Acts of the Apostles; his Aegean period roughly occupied the years 53–58, when he wrote his most important epistles. Many Jewish communities were initially hostile to this Roman gentile, but Smyrna's large Jewish population was receptive to Paul and made the city one of the earliest seats of Christianity in the West. Soon, their churches began to be persecuted by the Romans because the Christians resisted imperial rule. When Ephesus, its nearest rival, fell into decline, Smyrna became the site of an important bishopric; Smyrna was one of the "Seven Churches" addressed by St. John the Divine, in the biblical book of Revelation. Shortly after the address, the earliest known written account of a Christian martyr came in a letter from Smyrna: Bishop St. Polycarp of Smyrna was burned there in 155 or 156. The letter said that he was condemned to death by the Roman authorities for refusing to swear "by the genius of Caesar."

Smyrna passed into the hands of the Christianized, Greek-speaking Byzantine realm following the formal division of the Roman Empire in the late fourteenth century, becoming capital of the naval province of Samos. When Constantinople became the government seat, trade between Anatolia and the west declined. Smyrna declined with it. Though Emperor Justinian solidified and strengthened Byzantine rule in the sixth century, Arab raiders ravaged Smyrna throughout the seventh century, sparking several centuries of instability. Then came the Turkish tribes, also called the Turcomans or Turkmen, from the Far East. Originally emerging in Mongolia during the seventh and eighth centuries, they began raiding Byzantine territory in the early eleventh century. Smyrna was repeatedly attacked by the Seljuks, a Turcoman branch, in the late Byzantine period. They first seized Smyrna in 1076, introducing the Moslem religion. A Turkish raider named Tsacha took the city once again in 1090, but it was recovered in 1097 by the generals of Alexius I Comnenus, the founder of a great Byzantine dynasty. Despite growing Seljuk domination in the Asia Minor region, the Byzantines, with the help of the armies of the Fourth Crusade, managed to reoccupy the western third of Anatolia by the mid–twelfth century. The Seljuk state came to terms with the Byzantines in 1176. When the crusaders conquered Constantinople in 1204, the Byzantines established a provisional empire based in Nicaea (modern İznik); Smyrna and its citadel, in ruins, were restored by Byzantine emperor John III Ducas Vatatzes in the period 1222–25.

With the Treaty of Nymphaeum in 1261, Emperor Michael VIII Palaeologus ceded the flourishing port of Smyrna to the Genoese, although it retained Greek ecclesiastical authority. The Byzantines then abandoned Anatolia and returned to Constantinople. This set the stage for rival Turcoman tribes to swarm over former Seljuk and Byzantine lands. Following the collapse of the Seljuk Sultanate early in the fourteenth century, Anatolia fragmented into numerous short-lived emirates, or principalities, established by Turcoman clans and the Mongols. Perhaps as early as 1317, the citadel of Smyrna was taken by (Turcoman) Aydinoglu clan chieftain Mehmet (Mohammed) Bey, founder of the Aydin dynasty. He made his son Umur (or Amur or Omer) Bey the governor of the Aydin emirate of Smyrna, establishing a naval station there in 1329. Soon, the city was recaptured by the Latin crusaders, who placed it in the hands of the Knights of St. John of Rhodes (Greece); though the knights briefly stopped Turkish expansion in the Mediterranean region, they failed to take Smyrna's citadel. As Anatolia was uniting under the Ottoman Turk dynasty in the fourteenth century, Timur (Tamerlane), the last great Mongol Khan invader from central Asia, sacked Smyrna and massacred nearly all of its inhabitants in 1402 and 1403. The conquest was temporary; the Seljuk Aydins took the town over again, briefly, until 1415 when Sultan Mehmet I Celebi incorporated Smyrna into the Ottoman Empire, to which it was formally annexed about a decade later.

Ottoman rule, which was to last almost exactly five centuries, brought renewed stability and prosperity to Smyrna, now also called İzmir, and the city regained its former prominence—even though Mehmet's successors were confronted with repeated Venetian efforts to take the town in the late fifteenth century. Many Jews expelled from Spain after 1492 settled in Smyrna, forming a large and lasting Sephardic community. Despite being damaged by earthquakes in 1688 and 1778, İzmir remained an important commercial and cosmopolitan Ottoman port, and almost every major international mercantile power established a community there. In the eighteenth century, when most of Anatolia saw territorial wars between local feudal lords and pashas (military governors), İzmir replaced Bursa as Anatolia's leading center of eastern trade with Europe. Until the reign of Abdülmecid I (1839–61), İzmir was administered by the *viyalet* (province) of Jezair (the Isles), and not Anatolia. Its governor, however, had less influence over the city than did the Kara Osman Oglu clan of the neighboring Manisa province. By then, İzmir had nearly as many Christian European inhabitants as Turkish: the European makeup was mostly Greek, with sizable Italian, Armenian, British, French, and Jewish communities. Since the city had a non-Moslem majority (before World War I, Greeks alone numbered 130,000 out of a population of about 250,000), the Ottoman ruling class referred to it as "Infidel Smyrna" (Gavur İzmir) where,

it was said, "heathens" enjoyed one of the most cultured lifestyles in the Mediterranean.

World War I and the period immediately following brought turbulence, destruction, and rebirth to İzmir once again. The Ottoman Turk Empire, which had allied itself with Germany in the war, collapsed in 1918, and Greece laid claim to the İzmir *Viyalet* during the Council of Ten, February 3–4, 1919. The claim, however, was never authorized; but since the city was, in theory, under Allied occupation, the Greek mandate was tacitly permitted. A huge contingent of Greek expeditionary forces—aided by British, French, and Italian troops—landed in the harbor and occupied İzmir on May 15, 1919, committing many atrocities against the city's Turkish population. The Greeks pressed inland, clashing with Turkish nationalist democratic forces led by General Mustafa Kemal (who would later become known as Atatürk). İzmir was heavily damaged in the 1919–20 conflict. With the Treaty of Sèvres, August 10, 1920, İzmir and the surrounding Ionian area were to be placed under Greek administration for five years. The Greeks were initially successful against the Turks. But the Kemalists consolidated, and the Greeks lost the support of the British. After the failure of London and Paris peace talks in 1921 and 1922, Kemal's Turkish army advanced to the Mediterranean and drove Greek forces back off the coast; many thousands of Greek refugees from throughout Asia Minor quickly fled the country from İzmir. Kemal's triumphant entry into İzmir on September 9, 1922, was followed by three days of atrocities, as Greek and Armenian inhabitants were raped, mutilated, and murdered. Since the secular republic had not yet been proclaimed, the city's recapture had all the hallmarks of a jihad (Moslem Holy War). Within days, Kemal's forces had "liberated" the town and district of İzmir.

The celebration, however, was short-lived. By the night of September 13, as the Greek army retreated, nearly all of İzmir was in flames. The fires allegedly began in the city's Greek and Armenian quarters—only the Turkish (Moslem) quarter around Mount Pagus would eventually be spared in the two-day blaze. The wind-fanned flames burned wooden houses, and munitions stores exploded. "Without exaggeration," wrote G. Ward Price, a British correspondent stationed aboard the British warship *Iron Duke*, "tonight's holocaust is one of the biggest fires in the world's history." The fires forced thousands of non-Moslems to the quays; Allied ships, strictly neutral, stood idly by, refusing to grant safe passage to the refugees—many of whom leaped into the sea to escape the flames. Finally, an American named Asa Jennings commandeered a number of foreign ships, including a Greek fleet, to take more than 60,000 exiles to Greece; a quarter of a million refugees were eventually rescued. In the end, 70 percent of İzmir burned to the ground, including all the banks, businesses, and consulates in the city's quay-lined European quarter. Thousands—the exact figure is not known—of non-Moslems died in the conflagration. It was, as Kemal said, "the end of an era." But who started the great İzmir fire: the evacuating vanquished or the vengeful victors? Turks have always insisted it was the Greeks and Armenians; the Greeks and Armenians insisted it was Turkish soldiers. In any event, it was for İzmir a true Pyrrhic victory.

With the Treaty of Lausanne, July 24, 1923, which established the borders of modern Turkey, İzmir and the surrounding zone came under Turkish sovereignty. The secular republic, with Kemal as its president, was proclaimed the following October. The treaty also stipulated a great population exchange: the city's large Greek population was sent to Greece in return for ethnic Turks who had been living in Greece and the Greek islands. İzmir's population today barely reflects its rich Greek past. Restoration began almost immediately after the fire. The modern city, on a small delta plain of the ancient Meles River, was planned by René Donje in 1924 and was nearly complete by the 1960s; added were spacious, tree-lined avenues and modern office buildings, apartment houses, stores, and theatres. In April 1928 the city was seriously damaged by another major earthquake. Since World War II, İzmir has grown rapidly; it became NATO's Southeast Command Headquarters (1952) and the center for the large annual International Trade Fair. Historic mosques in İzmir include the Hisar Mosque, largest and oldest in the city, built in the sixteenth century with nineteenth-century restorations; and the Sadirvan and Kemeralti Mosques, both dating from the seventeenth century. İzmir's symbol is the Saat Kulesi, or clock tower; built in ornate late Ottoman style in 1901, it stands in Konak Square, the heart of the city.

İzmir today remains one of Turkey's most popular and prosperous commercial and industrial centers—the country's second largest port and third largest city, after İstanbul and Ankara. A major shipping center with a recently enlarged harbor, it is the principal export center of Turkey. Its port handles the produce of one of the richest agricultural regions of the country. Despite centuries of pillage and plunder, of disastrous fires and earthquakes, of lost ancient monuments and newfound modernity and commerce, this city of nearly 2 million people is still known to Turks as "Beautiful İzmir."

Further Reading: Two books serve as easy-reading introductions to the country and its history, including that of İzmir (Smyrna): *Discovering Turkey* by Andrew Mango (London: Batsford, and New York: Hastings, 1971), and *A Traveller's History of Turkey* by Richard Stoneman (Moreton-in-Marsh, Gloucestershire: Windrush, and Brooklyn, New York: Interlink, 1993). Mango's book, though a bit outdated, takes a largely cultural, region-by-region approach; Stoneman's guide is more chronological, covering points of historical interest from Paleolithic times to today. For more scholarly historical and archaeological surveys on Smyrna and surrounding sites, *Ancient Civilizations and Ruins of Turkey* by Ekrem Akurgal (İstanbul: Haset Kitabevi, 1973), and *Pre-Ottoman Turkey* by Claude Cahen (New York: Taplinger, and London: Sidgwick and Jackson, 1968) are also to be recommended. Though a travel book, *Turkey: The Rough Guide* by Rosie Ayliffe, Marc Dubin, and John Gawthrop (London: Rough Guides, 1994) is a solidly researched, nearly exhaustive guide to the country, abounding with historical context and contemporary facts.

—Jeff W. Huebner

Kanesh (Kayseri, Turkey)

Location: In Eastern Anatolia, in the region of Cappadocia, about twelve miles northeast of Kayseri.

Description: Kanesh, whose modern Turkish name is Kültepe, was first inhabited in the fourth millenium B.C. Early in the second millennium B.C., when Assyrian merchants established a bazaar there, Kanesh became an important center of trade. In subsequent centuries, the city came under Hittite domination. Unrivaled archaeological finds at Kanesh have provided a wealth of information about the ancient peoples and cultures of Anatolia.

Contact: Tourist Information Office
Kağni Pazarı, No. 61
Kayseri, Kayseri
Turkey
(352) 2311190 or 2319295

Kanesh is one of the largest and most significant archaeological sites in Turkey, and it has yielded an unprecedented wealth of information about the Assyrian and Hittite cultures that dominated Kanesh at different times during the second millennium B.C. However, archaeologists have found traces of habitation that date back as far as the fourth millenium B.C. It is assumed that Kanesh was formally settled sometime during the third millenium B.C., just prior to large-scale migrations by Indo-European peoples into the region. This settlement on virgin soil is associated with the fourth (or lowest) building level that archaeologists have unearthed at Kanesh. Evidence that the inhabitants of Kanesh evolved a distinct culture at this time is found in their ceramics. Grouped with other forms of Cappadocian pottery, the earthenware found at Kanesh is decorated with distinctive geometric symbols.

Following raids under King Sargon of Akkad throughout Mesopotamia during the twenty-fourth century B.C., the region of Asia Minor in which Kanesh is located came under the control of the Akkadians, an Indo-European people. The third building level at Kanesh is associated with the Akkadians. There is also evidence of a continued Hattic presence in the city. The Hattians, who do not belong to the Indo-European group, are commonly considered an indigenous culture and maintained their presence at Kanesh well into the second millennium B.C., despite foreign overlordship. Although the areas to the south and west of Akkadian Kanesh experienced upheaval late in the third millennium B.C., the city itself managed to stay out of the conflicts, perhaps in part because of its sheltered location in the mountains.

Early in the twentieth century B.C., Assyrian merchants began to make their appearance in the city, during the reign of Erishum I of Assyria. Erishum pursued a policy of commercial expansionism and opened trade relations with the major Near Eastern cities of the time, including Ur, Nippur, and Akkad. Kanesh became part of this commercial network, in which commodities such as copper, silver, gold, lead, stoneware, ivory, rock-crystal, faience, wheat, and wool were traded widely. Another metal traded at this time was "amutum," probably iron. Amutum was extremely scarce—this being well in advance of the Iron Age—and was worth forty times as much as silver, five times as much as gold. Produce was brought from Assyria and Babylon, usually in exchange for metals.

The Assyrians established a bazaar, known as the *karum,* just outside the city walls. The *karum,* which contained the shops, homes, and warehouses of the merchants, also became a place of residence for metalworkers, who sold their wares to different regions of Mesopotamia through the Assyrian merchants' caravans. The bustling *karum,* which brought a period of extraordinary prosperity to Kanesh, soon overshadowed the city proper. The merchants maintained trade relations with Assur and other cities along established trade routes and also forged new routes, striking out to other cities in Anatolia, such as Diyarbakır, Malata, Urfa, and Adana, that had previously remained beyond reach. Caravans from Kanesh, many quite small but some numbering up to 250 donkeys, also made their way to the Aegean and to the Mediterranean. The Assyrians traded with the indigenous peoples of Anatolia, the Hattians, as well as with the Hittites and the Luwians, both Indo-European groups and relative newcomers to the area.

Operations at Kanesh were often run by junior members or relatives of family firms that maintained their principal base at Assur. Sometimes, families formed partnerships to raise the capital for larger expeditions. It was not only the Assyrians who profited by these ventures, however. Local rulers benefited by the taxes levied on the caravans, and in return they secured the trade routes, safeguarding the caravans from bandits. Local craftsmen were able to expand their markets and sell their wares far and wide. Although the names of nine other bazaars like that of Kanesh are mentioned in ancient texts, only Kanesh itself and the bazaar of Boghazköy have been discovered. Kanesh Karum remained a hub of trade for at least three generations or a period of almost 100 years, lasting roughly from some time in the latter half of the twentieth century B.C. to the mid–nineteenth century B.C.

Under the influence of the prosperous Assyrian merchants of Kanesh Karum, the city evolved a unique culture that is distinct from the neighboring regions. The philosophy, the art, and the way of life that evolved at Kanesh at this time is unusually well understood today because the Assyrians

Ruins at Kanesh
Photo courtesy of Embassy of Turkey, Washington, D.C.

were assiduous record keepers. Twentieth-century excavations have yielded some 15,000 clay tablets from the ancient *karum,* which give an unprecedented record of the Age of the Assyrian Trading Colonies, to use the official archaeological label. These tablets, bearing a cuneiform script, are the earliest known example of writing in ancient Anatolia.

But the tablets are not the only objects from the period of Assyrian dominance that have survived the onslaught of time. Excavations have unearthed the remains of a single wall around an inner city that contains palaces and public buildings, both religious and secular, and the houses of the native inhabitants of the city. The inner city is situated on a hill (hence its modern name, Kültepe, which translates as "Ash Hill"). The lower city, outside the city walls, held the *karum.* The houses both in the city and in the *karum* are quite spacious, usually consisting of three to six rooms in addition to a wide hall. Besides single-story houses, there are also houses with two stories, set on stone foundations and built of mud-brick and wood. The houses have separate living rooms, kitchens, bedrooms, store rooms, and archive rooms. The archive rooms, where the household records were kept, have yielded the most information about the identity of the predominantly Hattic and Assyrian inhabi-

tants. However, the dead were buried within the confines of their own houses in stone cist-graves, and the burial gifts contained in these graves form another rich source of information about Kanesh. Common household items such as cooking utensils, pottery, and statuettes have also been found in profusion.

Although the clay tablets form a record primarily of commercial transactions, they also give insight into the record keepers' language, an ancient Assyrian dialect. In addition, the tablets are a witness to the art and culture that evolved at Kanesh during the Age of the Assyrian Trading Colonies. The tablets are contained in "envelopes" that hold them together. Many of the tablets, particularly those that functioned as legal documents, bear seals on their envelopes to authenticate them. Such seals might belong to lenders, borrowers, witnesses, or parties to an agreement. A variety of seals was in use at Kanesh, in a variety of styles, which indicate not only fashions but also show—through similarities with seals in use elsewhere—the city's connections with other areas in Asia Minor. As a consequence, seals from Kanesh and other sites have been instrumental in the reconstruction of the ancient history of Asia Minor in general and of Anatolia in particular.

Most of the seals are quite small, ranging from three-quarters of an inch to an inch and one-half. In spite of their size, the seals are carved in elaborate patterns. For instance, the cylinder seals have two friezes, one at each end of the seal, which are interconnected with guilloches (a rope pattern). The style of the seal is principally determined by the nature of the friezes. In the Old Anatolian style, the friezes typically consist of rows of animal motifs, either heads or full body images. Birds, hares, and monkeys are particularly common on this type of seal, which originated in Anatolia and has been found in great numbers at Kanesh. The image of a double-headed eagle also occurs frequently and is striking for the degree of realism with which the image is executed. Other seals bear the impressions of gods. A representation of a god accompanied by a lion—which specialists tentatively identify as a god of war—makes a frequent appearance among the Kanesh seals, for instance.

Interestingly, the seals did not function only to validate agreements and authenticate property titles. They were also amulets, shielding the user and the agreements they sealed from misfortune. Their religious function explains the importance of the images depicted on seals, which in turn provide an insight into the pantheon of their makers and users. Besides the Old Anatolian style that is indigenous to the Kanesh region, seals dating to the Age of the Assyrian Trading Colonies discovered at Kanesh display the Old Assyrian style and the Old Babylonian style (the latter evolving from the former) and the Old Syrian style. These different styles bear witness to the cultural influences operating at Kanesh at the time.

The Age of the Assyrian Trading Colonies came to an end at Kanesh in a catastrophic fire sometime around the mid–nineteenth century. What set off the conflagration is uncertain, but some archaeologists speculate that the Hittites were responsible for the destruction of the city. The site seems to have been deserted for about half a century afterward. When the city was rebuilt, around 1800 B.C., the Hittites controlled what by then may have been a city-state.

The Hittites entered Anatolia from the Balkans and the Caucasus in the twentieth or nineteenth century B.C. They settled at various places in Asia Minor and influenced the region's development. They either took part in or were solely responsible for the reconstruction of Kanesh. A bronze spearhead dating to this time bears the inscription, "Pallace of Annita, the king," in an Assyrian cuneiform script. Notwithstanding the fact that the inscription is in Assyrian, it is clear that Annita was the king of Kussara, first Hittite monarch and author of the oldest discovered Hittite text.

The reconstructed city was built on the hill that also held the previous incarnation of Kanesh, and, similarly, had an additional section where the old Assyrian *karum* had been. The section on the hill contained the king's palaces, government buildings, and temples, besides a residential section for the city's wealthier inhabitants. It was surrounded by a single inner wall connected to the fortifications that protected the outer town, which housed the common people. The outer defenses themselves consist of a double wall. The houses are consolidated in tightly configured squares and along a number of streets both in the city on the hill and the lower city. The houses themselves are even more spacious than those associated with the Age of the Assyrian Trading Colonies.

The Hittites used writing, but not nearly as much as the Assyrians before them. Some tablet seals in a distinctive Hittite style have been uncovered, and these show evolving Old Babylonian and other Near Eastern influences. The seals and tablets are not numerous enough to give much insight into the Hittite culture of Kanesh, however, which has left a record of itself predominantly in the form of pottery and sculpture. Pottery finds at Kanesh attest to the fact that the Hittites maintained ties with the greater Mesopotamian civilization, but nevertheless developed a distinctive artistic style. Mostly monochrome, their pottery was wheel thrown and often burnished to a high polish. Depictions of Hittite deities, including a mother goddess, a weather god, and a god of the fields, bear witness to the Hittite religion and to their agricultural orientation. The indigenous Hattic people are also thought to have contributed to the evolving Hittite culture.

After their possibly violent entrance to the region, the Hittites seem to have turned to a peaceful way of life focused on farming and trading. Another period of great prosperity followed the resurrection of the city, but Kanesh did not have great political importance during the Old Hittite and Hittite Imperial periods. Starting in 1000 B.C., however, during the period often labeled as the Neo-Hittite Age, Kanesh gained in political significance, becoming the capital of the Eastern Anatolian region that the Assyrians called the Kingdom of Tabal. Relatively little is known about this period, because the Greeks and later the Romans subsequently destroyed the Neo-Hittite city almost completely. Only some sculpture, painted pottery, and a number of seals dating to this time have been uncovered. However, under Greek and Roman overlordship, the city did retain its importance, continuing to be one of the principal cities of Cappadocia. The city ceased to exist following the Roman period, and today nothing is to be seen on the site besides the excavations.

A Frenchman, Ernest Chantre, was the first in a long line of archaeologists to dig at the ancient site in the 1890s, starting with the hilltop compound. His exclusive interest in tablets, unfortunately, led him to destroy everything else he uncovered. Subsequent digs led by Europeans were scarcely more careful and, after 1925, also included the Assyrian *karum*. In 1948, the first systematic excavations were begun, under the auspices of the Turkish Historical Association and the General Directorate of Antiquities and Museums. Most of the artifacts that have been unearthed since then are on display in museums in Kayseri and Ankara.

Further Reading: Information about the early history of Kanesh, including the Assyrian period, is available in James Mellaart's *The Chalcolithic and Early Bronze Ages in the Near East and Anatolia*

(Beirut: Khayats, 1966). *The Secret of the Hittites: The Discovery of an Ancient Empire* by C. W. Ceram, translated by Richard and Clara Winston (New York: Knopf, 1956; in England as *Narrow Pass, Black Mountain: The Discovery of the Hittite Empire*, London: Sidgwick and Jackson, and Gollancz, 1956; as *Enge Schlucht und Schwarzer Berg: die Entdekung des Hethiter-Reiches*, n.p.: Rowohlt Verlag, 1955) and *The Hittites: People of a Thousand Gods* by Johannes Lehmann (New York: Viking, 1975) both give useful accounts of the development of the Hittite culture and include some references to Kanesh. *Ancient Art in Seals*, edited and introduced by Edith Porada (Princeton, New Jersey: Princeton University Press, 1980) contains a wealth of fascinating information about ancient seals, including those found at Kanesh.

—Christopher P. Collier and Marijke Rijsberman

Kars (Kars, Turkey)

Location: In eastern Turkey, near the Turkish border with both Armenia and Georgia, 178 kilometers northeast of Erzurum, Turkey, on a high, exposed plateau surrounded by mountain ranges. The city is located on the River Kars, a tributary of the Aras, along a section where the river narrows in a deep gorge.

Description: Following a period of Saracen domination, Kars became part of Armenia in the tenth century. Its importance mainly due to its proximity to the east-west trade routes, the city experienced its greatest prosperity under the Armenians at that time. Kars was taken by the Seljuk Turks in the late eleventh century and fortified with a citadel. Destroyed by Timur (Tamerlane) in 1386, the citadel was finally reconstructed by the Turks in the sixteenth century and again in the nineteenth century and is still in use as a military base. Although it is now in Turkish hands and likely to remain so, Kars has been fought over by Armenians, Turks, Persians, and Russians.

Site Office: Kars Tourist Information Office
Ordu (also called Ali Bey)
Caddesi 140
Kars, Turkey
(21) 12724 or 12300

The history of Kars is one of invasion, conquest, and war. Arabs, Armenians, Russians, Persians, and Turks have all, at one time or another, claimed Kars as theirs. Although Armenians have been a constant presence at Kars since its founding, the city, like most of what was ancient Armenia, has been a battleground for other powers. In his *History of the Decline and Fall of the Roman Empire* (1776–1788), Edward Gibbon referred to this area as a "theatre of perpetual war."

The area surrounding Kars has been inhabited since prehistoric times. Tools from the early Stone Age have been found in the hills to the immediate north of the city and late Stone Age implements have been found to its south. Since at least the pre-Christian era, Kars and its surroundings have been associated with the Armenians, an Indo-European people who first appeared in the area around 500 B.C. Strabo, a first-century Greek geographer, called the area Chorzene, and its modern name is believed to be derived from the Georgian *kari,* meaning "gate." After sustaining invasions by the Arabs, Kars came under Saracen rule in the eighth century and grew in size and importance as a result of its location near the east-west trade routes between Asia and Europe.

By the tenth century, Kars was a part of what had become the ancient kingdom of Armenia, ruled by the Bagratid family. King Ashot the Iron, who ruled from 914 to 928, freed Armenia from Muslim raiders, preparing the country for a period of growth and prosperity. Ashot's brother and successor, Abas I, ruled from 928 to 952 and established his capital at Kars. Abas built the Church of the Apostles between 932 and 937, which survives, although in much modified form. The son of Abas, Ashot III, eventually removed the capital from Kars to Ani. It was under these Armenian rulers that Kars reached its greatest prosperity as a center of trade and home to over 100,000 inhabitants. Subsequently Ashot III granted Kars and its surrounding territories to his younger brother, Mushegh, as a separate kingdom. The Armenian Bagratids ruled at Kars until 1064, when the last king, Gagik-Abas, ceded Kars to the Byzantine Empire, although he retained nominal control of the ceded territories.

However, Kars was not in Byzantine hands for long, since reckless Byzantine expansion had left the empire vulnerable to attack. The Seljuk Turks availed themselves of the opportunity and took Kars and its province in 1064, definitively ending Armenia's control of the city. The new rulers were unwilling or, more likely, unable to maintain the conditions that had led to Kars's prosperity, and the city entered a long period of decline and instability. The Turks built a citadel in 1152 but to little avail. An Armenian house called the Zachariads took control of Kars later in the twelfth century, but only under the auspices of the sovereigns of nearby Georgia, whose borders expanded to include the area after the Seljuk dynasty began to break into a number of less powerful principalities. The most notable Georgian ruler was Queen Tamar, who reigned from 1184 to 1212 and was known for her political and military acumen. The power of Georgia waned, however, and in 1236 invading Mongols seized Kars, nearly laying it to waste. Some 150 years later, Timur (Tamerlane), during his military campaigns in northern Persia and the Caucasus from 1386 to 1388, destroyed the citadel and a large portion of the city.

After the ruinous campaigns of Timur, two rival Turkoman dynasties, the so-called White Rams of Diyarbakır and the Black Rams of Van, struggled to control Kars. Eventually, the White Rams were triumphant, but in 1502 they were defeated by Persian forces at the battle of Sharur in Armenia. During the sixteenth century and much of the seventeenth, the Turkoman and Persian Empires fought continuous battles over a frontier that, unfortunately for the local Armenians, always ran across Armenian land. The prize in these contests was control of the Caucasus region, including Kars, which remained of great importance to the Turks because of its strategic location that provided access to Transcaucasia. The Ottoman Turks rebuilt the Kars citadel in 1579, but the fortifications were not sufficient to ensure lasting Turkish control. The Armenians were generally better off

Church of the Apostles at Kars
Photo courtesy of Embassy of Turkey, Washington, D.C.

under the more tolerant Persians, but they were never granted more than a few years of peace at a time.

In the mid–seventeenth century, an expanding Russia also threw itself into the skirmish. Peter the Great managed to make inroads into the Caucasus against both the Turks and Persians in the early eighteenth century, and the (Christian) Armenians settled their hopes on him as a possibly more benign overlord. But the Russians did not get as far as Kars, which remained under Turkish domination. Following the year 1723, the Russians gave up their conquered lands in the Caucasus to Persia, which went on to experience a period of ascendancy under Nadir Shah, who ruled from 1736 to 1749. The citadel of Kars, however, withstood a siege by Nadir Shah in 1731.

In 1807 the city successfully resisted an attack by the Russians, thereby enhancing the citadel's reputation of impregnability. However, attacks upon the city during the nineteenth century saw the fortunes of Kars change. In the Russo-Turkish War of 1828, the Russian general Paskievich captured Kars from the Turks. Paskievich established a tolerant and impartial local civic administration, but the local Armenian population was not to enjoy its benefits for long. By the Treaty of Adrianople (1829), Kars was returned to the Ottoman Empire.

After the war, the Kars citadel was extensively refortified and modernized by the Turks, who correctly continued to consider the city of great strategic importance. A British colonel, Fenwick Williams, who had been involved in the British training of the Turkish army, provided the plans and organization for these improvements. At the beginning of the twentieth century, a British traveler to Kars described the fortress as having a double wall of ramparts, three to five feet thick, sixteen feet apart. The outer ramparts were defended by bastions, or fortified projections in the wall, and the inner ramparts by towers. The circumference of the double wall approached half a mile and its height varied from fourteen to thirty-eight feet. At the northwest corner of the walled enclosure was the inner fortress, which was built from solid stone and overhung the river.

Less than thirty years after the Treaty of Adrianople, the Crimean War broke out, pitting Turkey, Britain, France, and Sardinia against Russia. Kars was again the scene of battle. Initially, the conflict involved only Turkey and Russia, which did well in the Caucasus. After the Russians routed the Turks at the battle of Bashgedikler on December 1, 1853, the remnants of the Turkish army fled to Kars. With the sinking of the Turkish fleet at Sinop, Britain and France found their excuse to enter into the war against Russia. Turkish reinforcements were dispatched to Kars under Colonel Williams, who, since rebuilding the fortress, had been appointed British Commissioner and had been made a pasha, or commander, in the Turkish army. Williams arrived in Kars on September 24, 1854, and directed the further strengthening of the fortifications. One year later, on September 29, 1855, the Russian general Murovyev stormed the city but was beaten back and suffered heavy losses, mainly due to the strategic defensive fortifications hastily erected by Williams. Thus began a two-month siege of the city by 40,000 Russian troops. Williams was in charge of the defense of the city, with its garrison of 15,000 Turkish soldiers. Conditions within Kars were deplorable. The garrison was starving. Colonel Williams, receiving no aid or reinforcements, was forced to surrender the city on November 25, 1855. The surrender provoked a storm of controversy in England: did Kars fall due to an untenable military situation or because of the failure to arrange timely relief for the besieged city? At the close of the Crimean War, the departing Russians destroyed parts of the citadel with explosives. The damage was later repaired by the Turks. During the next conflict, the Russo-Turkish War of 1877–78, Russia once again attacked and occupied Kars. The Turks capitulated to the Russians on January 31, 1878, and on March 3, 1878, the city became part of Russia under the terms of the Treaty of San Stefano.

The nineteenth century had devastated Kars. Its fortress, considered impregnable at the beginning of the century, had been taken three times and was nearly razed. The population of the city—estimated at 50,000 to 60,000 people before 1828—had been reduced to 4,000 by 1907. This included 2,500 Armenians, 850 Turks, 250 Russians, and 300 Greeks. The twentieth century brought Kars little relief.

At the beginning of the century, Turkey experienced the growth of a strong nationalist movement, which was to result in the development of modern Turkey. One of the guiding lights of that movement was Namik Kemal, who, though born in Tekirdag in 1840, was a longtime resident of Kars. (His house in Kars still stands.) Admiring the advances in politics and science in western Europe, Kemal advocated that Turkey adopt them as a model, though adapted to an Islamic context. He translated the works of Rousseau into Turkish and called for freedom and a Turkish fatherland. His ideas disturbed the ruling Ottoman regime, and Namik Kemal died in internal exile at Kéos. However, his ideas influenced a generation of Turks who wished to see a change in Turkey, including Mustafa Kemal, better known as Kemal Atatürk, who was to play a tremendously influential role in the fate of Kars and of Turkey itself.

The Turkish nationalist movement embraced the idea of Pan-Turkism, which demanded the retaking of Kars and its province from Russia. On October 30, 1914, Turkey entered into World War I on the side of Germany. Turkey's proclaimed goal was to regain the territory lost to Russia in the nineteenth century. In the midst of World War I and the turmoil of the Bolshevik Revolution in Russia, the short-lived independent federation of Transcaucasia was formed. The confederation, consisting of Georgia, Azerbaijan, and Armenia, declared itself independent from the Russian Empire and pursued a separate peace with Turkey. To that end, a conference between Turkey and Transcaucasia was scheduled for March 1918 at the city of Trebizond. Before the conference could begin, however, the Treaty of Brest-Litovsk was concluded between Russia and Turkey. By the terms of this treaty, through which Bolshevik Russia withdrew from World War I, the city of Kars and all the territory annexed by Russia in 1878 were returned to Turkey. Nevertheless, the fighting in the area across the Caucasus region continued, and the conference at Trebizond went forward. The Turkish delegation to the conference demanded that the city of Kars be evacuated by the new confederation in return for Turkish recognition of Transcaucasian independence. The newly appointed prime minister of Transcaucasia, a Russian Georgian named Akaki Chkhenkeli, ordered the evacuation of Kars on April 22, 1918.

The evacuation was hasty and confused. Fires were set in the city. On April 25, 1918, Turkish troops entered Kars. Turkey indeed recognized the independent confederacy of Transcaucasia, and the Treaty of Brest-Litovsk was no longer accepted as a basis for negotiations between the Turks and the Transcaucasian confederacy during peace talks held at Batum on May 11, 1918. In that same month, however, the confederation, having declared itself independent, began to disintegrate. On May 26, 1918, Georgia declared her separate independence, and on May 27 Azerbaijan followed suit. On May 30, left with no other alternative by which to pursue a peace treaty with Turkey, Armenia declared its own independence.

The Mudros Armistice, by which Turkey was bound to acknowledge its defeat in World War I, overrode the Batum peace talks. A supplementary clause to the treaty demanded that the Turks evacuate the Kars District. After some delay, the Turks, under pressure from the British, withdrew from Kars in April 1919. At the close of World War I, a separate and independent state of Armenia was recognized by the western European powers, led by Britain. On May 28, 1919, Russian and Turkish Armenia joined to establish a United Armenia, which included the reacquired province and city of Kars.

The Bolshevik Revolution soon spread to the newly recognized state of Armenia, however, and on May 9, 1920, a revolutionary committee was established at Kars. The city was in their hands for two days. Turkey, however, still desired the regions that were once its own, and in September 1920 the Kemalist War began.

Mustafa Kemal, the hero of the Turkish victory at Gallipoli in World War I and now inspector general of Turkey's Ninth Army, ordered that Kars be retaken. On October 24, 1920, the battle for Kars began. In command of the Armenian forces at the fortress of Kars was a General Pirumian. Pirumian was unable to hold the city and, after fierce fighting, the Turks entered Kars. Three days of uninterrupted looting and pillage followed.

The Kemalist War ended on November 17, 1920, and on the basis of the Treaty of Moscow—which was ratified at the Kars Conference during September and October 1921—Kars and its province once more became part of Turkey. The rest of Armenia, along with Georgia and Azerbaijan, was incorporated into the Union of Soviet Socialist Republics, the USSR.

During World War II, Turkey declared war on Germany and Japan on February 23, 1945. In May 1945, the Soviet Union reopened the question of returning Kars to Soviet Armenia. However, no action was taken on this matter at the Potsdam Conference by the United states, the USSR, and Great Britain, who attended the conference and were commonly referred to as the Big Three. The Truman Doctrine, the United States policy to aid any noncommunist country in the face of a Soviet threat, effectively ended the question of Kars's return to Soviet Armenia. The city remains, to this day, part of Turkey.

Kars has a strong folklore tradition, and the area around Kars is famed for both its cheese and honey, in addition to the rough woven carpets, called kilims, made there. The Kars Museum is situated in the new town and has exhibits of artifacts from the Bronze Age, the Greek and Roman periods, and the Seljuk and Ottoman Empires. Part of the museum is devoted to local crafts, including woodcarving and costumes. In addition, the museum houses a good collection of coins. Besides the Church of the Apostles, Kars has another medieval remnant in the Stone Bridge, which crosses the Kars River. The bridge dates from the fifteenth century. It was, like much of Kars, repaired in 1579 and again in 1719, after being damaged in an earthquake.

Further Reading: *Armenia: The Survival of a Nation* by Christopher J. Walker (New York: St. Martin's, and London: Croom Helm, 1980; second edition, New York: St. Martin's, and London: Routledge, 1990) is a detailed history of the wars, the politics, and the suffering endured by the peoples of the Transcaucasian region. *Armenia: Travels and Studies* by H. F. B. Lynch (London and New York: Longmans, Green, 1901) has a chapter devoted to Kars at the beginning of the twentieth century, at which time it was under Russian control. *A Narrative of the Siege of Kars* by Humphry Sandwith (London: Murray, 1856) is a history of the British-directed defense of Kars against the Russians in 1856. The book provides a very complete picture of the politics between Turkey and England in the middle of the nineteenth century and the military situation of Kars, from the point of view of a medical officer in the British Army.

—Rion Klawinski

Kastoria (Kastoria, Greece)

Location: About twenty miles south-southwest of Florina, on Lake Kastoria; capital of the Kastoria department in the western Macedonian region.

Description: Town that was the focus of many medieval power struggles in Greece; today a treasury of Byzantine and post-Byzantine architecture and home to a prosperous fur industry.

Contact: Greek National Tourist Organization
No. 2 Napoleontas Street
Ioannina, Ioannina
Greece
(651) 250 86

The name Kastoria first appears in recorded history at the end of the tenth century A.D., but there has been a town on the site since the sixth century, and there were settlements in the surrounding area much earlier. The ancient Celetrum, a town captured by the Romans in 200 B.C., was probably located on a hill above the current location of Kastoria. Diocletian, Roman emperor from A.D. 284 to 305, established the town of Diocletianoupolis in the vicinity; it was destroyed by barbarian raids, after which Byzantine Emperor Justinian I moved it to the promontory projecting into Lake Kastoria, the site the town occupies today. The historian Procopius says Justinian gave the new town "an appropriate name;" he may have named it for himself, Justinianoupolis, but this is not clear from the records. At any rate, the residents were soon calling the town and the lake "Kastoria," meaning "place of beavers," as those animals abounded in the lake.

Kastoria's history was peaceful until the end of the tenth century, when it became an important base in a number of uprisings and invasions. After the Byzantine Empire annexed Bulgaria in the late tenth century, Samuel, czar of Bulgaria, led a revolt in Macedonia against Emperor Basil II. Samuel captured Kastoria and held it from 990 to 1018, when Basil's forces retook it and restored it to Byzantine rule. Basil's successful campaign brought him the nickname of "slayer of the Bulgars." Samuel is said to have destroyed the Basilica of Agioi Anarghyroi (Sts. Cosmas and Damian). Basil II initiated rebuilding of the church in the autumn of 1018, which he spent in Kastoria, celebrating his victory. The church, still standing today and the oldest in Kastoria, has frescoes inside and out. Its exterior walls are also embellished with a finely detailed pattern of inlaid bricks.

The Norman duke Robert Guiscard, who had already conquered many Italian territories by the late tenth century, turned his attention to the Byzantine realm and occupied Kastoria in 1082. The Byzantine emperor Alexius I Com-

nenus recovered Kastoria shortly thereafter, however; Robert and his successors lost most of their Greek conquests. Robert died of a fever at Cephalonia while planning further advances.

The thirteenth century was very trying for Kastoria. The Despotate of Epirus, which bordered Macedonia, was once part of the Byzantine Empire but became independent in the early thirteenth century. Subsequently launching a campaign of expansion, Epirus captured Kastoria, only to lose it to John III Vatatzes, emperor of Nicea, in 1252, and recover it in 1257 after months of hard battle. It was finally rejoined to the Byzantine Empire in 1259 after a successful military campaign by the new emperor, Michael VIII Palaeologus.

Kastoria then enjoyed many years of peace and prosperity under Michael VIII, but in the next century it was the scene of strife again. In 1320 civil war broke out between Byzantine emperor Andronicus II Palaeologus and his grandson, Andronicus III, who had been stripped of his rights of succession. Andronicus III made war throughout Macedonia, capturing Kastoria, Thessaloníki, and other important cities. Andronicus II, in the face of his grandson's military successes, recognized Andronicus III as co-emperor in 1325, then relinquished the throne to him entirely in 1328. The Serbs, however, had intervened in Byzantium's affairs during the civil war, and once it was over they invaded, under the strong leadership of their king, Stephen Dušan. Stephen took Kastoria in 1334, but relinquished it after the completion of peace negotiations with Andronicus III. The peace lasted until Andronicus III's death in 1341, after which Serbia once again captured Kastoria and much of the rest of Macedonia. Later in the century Albania laid claim to Kastoria. In the 1380s, however, it was taken by a group that would hold it for five and one-half centuries: the Ottoman Turks.

It is unclear whether the Ottomans took Kastoria by force or through a negotiated settlement with the Albanian rulers. The Turks called the town Kesriye. Kastoria retained its majority Christian population under the Moslem rulers, and also experienced an increase in its Jewish community, although there had been some Jews in Kastoria for many years. One of them, the scholar Tobia ben Eliezer, wrote a commentary on the first five books of the Bible during the reign of Alexius I. The Jews who came to Kastoria and the rest of Macedonia during the Ottoman period were generally fleeing less tolerant lands such as Spain and Portugal, where the inquisition was underway. Many of the Jews in Kastoria became involved in overseas trade with Venice and other locales.

Although there were some periods in which the Ottomans persuaded large numbers of both Christians and Jews to convert to Islam, often these conversions were in name only, as the converts continued their original religious and

A view of Kastoria
Photo courtesy of Greek National Tourist Organization

cultural practices. Discrimination against non-Moslems intensified at times when the Ottomans were threatened by outside forces or internal rebellions, and conversions generally increased at these times. Still, both Christians and Jews continued to build houses of worship: at least three Christian churches in Kastoria date from the turn of the seventeenth century.

Kastoria's commercial community continued to center on the fur trade through the years of Ottoman rule. Kastorians' specialty in the fur industry was, and continues to be, the creation of fur garments from remnants left over from foreign workshops that had made coats from entire pelts. Kastorian furriers are said to be able to assemble the pieces so well that their coats cannot be distinguished from those made from a whole pelt. The origins of the Kastorian fur industry are unclear; one theory is that Byzantine officials who fell out of favor were frequently exiled to Kastoria, and

while living there had their old furs repaired by local craftsmen, who through practice developed great expertise.

In the seventeenth and eighteenth centuries many Kastorians involved in the fur trade built large houses, called *archontika,* with workshops on the ground floor and distinctive projecting upper stories that held living quarters. Many such structures still stand in Kastoria, and one houses a Folk Art Museum.

Macedonia revolted against Ottoman rule along with the rest of Greece in the 1820s, but the Ottomans successfully put down that uprising even as most of Greece won independence. The failure of the Macedonian revolt has been attributed to a lack of experienced military leadership and a lack of coordination with the uprisings in southern Greece. Macedonians continued to follow a dream of independence through the rest of the century, however. The Russo-Turkish War of the late 1870s gave impetus to revolutionary forces in Macedonia;

the ownership of the region was one issue to be decided in the peace negotiations between the Ottomans and Russia. The Macedonians hoped a popular uprising would bring about Greek intervention, and forces in Kastoria and other Macedonian cities prepared for insurrection in 1878. Due to indecision on the part of the Greek government, however, assistance from the Greek army was rushed and ineffective, and the revolt failed. The peace treaty between Russia and the Ottoman Empire left Macedonia in Ottoman hands.

The Ottomans' arrest of a Kastorian schoolteacher and Greek nationalist, Anastasios Picheon, in 1887 and subsequent persecution of Greek nationalist activists made Macedonian-Turkish relations deteriorate further. Then, Macedonia began to be contested between Greece and Bulgaria, with armed guerrilla bands of both nationalities active in the area from 1904 to 1908, a period known as the Macedonian Struggle. Eventually tensions in southeastern Europe erupted in the First and Second Balkan Wars in 1912–13. As a result of the wars Macedonia was divided among Greece, Serbia, Bulgaria, and Albania, with Kastoria in the portion going to Greece.

Kastoria endured Italian and German occupation during World War II, and then, along with several other towns in Greek Macedonia, suffered greatly during the Greek Civil War of 1945–49. Communist insurgent forces attacked Kastoria in 1947 but failed to capture it; the natural protection afforded by the lake and surrounding hills made it easily defensible. Still, insurgent activity in the area made it difficult for Kastorians to obtain food and other necessities of life. The tide of the war turned in favor of the Greek government after U.S. assistance and training strengthened the Greek army and the insurgents lost Yugoslavian aid. Kastoria was the base for the final Greco-American campaign against the insurgents in 1949.

Today Kastoria is a prosperous and picturesque city, having remained a leader in the fur industry and retained much of its historic architecture—the latter fact being somewhat miraculous in the face of the invasions and warfare the town has endured. It has more than seventy Orthodox churches; in addition to Agioi Anarghyroi, the most notable ones include the Church of Agios Nikolaos and the Basilica of the Taxiarchoi, both from the eleventh century, and the twelfth-century Church of Agios Nikolaos Kasnitzi. All have intriguing frescoes. Perhaps the most memorable frescoes in Kastoria, however, are in the main church of the lakeside Monastery of the Mavriotissa. The frescoes, mostly from the twelfth century, are crudely executed but arrestingly vivid, especially the one depicting the Last Judgment, full of determinedly avenging angels and a variety of cowering sinners. Some of the portrayals have been interpreted as anti-Semitic, and undoubtedly there is much that is offensive. At the same time, this fresco and others, such as those depicting the Last Supper, the Ascension, and the Dormition of the Virgin, are impossible to ignore. The monastery is also noteworthy as the site of Alexius I's landing when he came to recapture Kastoria from the Normans. Other attractions of Kastoria include the many *archontika* and the variety of cafes and taverns that surround the lake.

Further Reading: *Macedonia: 4000 Years of Greek History and Civilization* (Athens: Ekdotike Athenon, 1983), part of the Greek Lands in History Series edited by M. B. Sakellariou, provides thorough coverage of Macedonian history and Kastoria's role in that history. Although it is in coffee-table format, it contains much solid, scholarly information, dealing with art, architecture, commerce, and religion in addition to the expected political and military history, and it is beautifully illustrated. Some anti-Turkish bias creeps into the sections dealing with the years of Ottoman rule, but overall the book is a valuable resource. Kastoria's extensive Byzantine architecture is also well covered by most of the leading travel guides.

—Jessica M. Bowen and Trudy Ring

Kavala/Philippi (Kavala, Greece)

Location: Kavala is a coastal town at the northern tip of the Aegean Sea in the border region between Turkey and mainland Greece. Philippi, located eight miles northwest of Kavala, provided an outpost for the ancient kingdom of Macedon during its imperialist expansion.

Description: Because of its position as a seaport at the precise point where East meets West, Kavala has enjoyed times of relative prosperity. Today it is the second largest city in the region of Macedonia. Philippi is of interest largely because of the ruins there from Greek, Roman, and early Christian times. It was the site of a major battle between Roman factions in the first century B.C.

Site Office: Greek National Tourist Organization
5 Filellinon Street
Kavala, Kavala
Greece
(593) 228762 or 231653

The histories of Kavala and Philippi have been intertwined since ancient times, when they were known as Neapolis and Krenides respectively. Neapolis, a Greek city-state, was a loyal member of the Second Athenian League during the mid-fourth century B.C., as it had been of the prior Delian League. Krenides, meanwhile, was initially founded by the people of the island of Thasos in 360 B.C. to exploit the mining potential of the area. Within four years, the fate of these two areas became inextricably bound together.

In May 356, Kersebleptes, the Odrysian king, attacked Krenides. Unable to match Kersebleptes' strength, Krenides appealed to Philip II, the great ruler of nearby Macedon (modern Macedonia), for help. Krenides was so rich in natural resources that Philip was eager to help, and after defeating Kersebleptes, Philip took over the colony, renaming it Philippi after himself. The plural form of the name (Philippi), as of its predecessor (Krenides), indicates that the colony was somewhat scattered. Philip now controlled the incredibly productive Asyla mines, which yielded him more than 1,000 talents of silver per year.

After taking control, Philip began to occupy the heavily fortified area that had been settled by Macedonian colonists eager to create and defend their own *polis*. Philip then stabilized the economy by clearing away the forests and swamps from the plains so that the land could be profitably cultivated. According to the philosopher Theophrastus, a disciple of Aristotle, "When the water had been drained off, the land mostly dried out, and the whole territory was brought

into cultivation." From this point forward, Philippi was an invaluable source of income for Philip, while also providing him with a strategic outpost.

Macedon had an antagonistic relationship with Athens, so the Macedonian presence in Philippi was, quite naturally, troublesome for Neapolis, an Athenian ally that was less than a day's march away. Thus, in the summer of 355 the Neapolitans sent two envoys before the Athenian Boule (the lower legislative branch of the Athenian government) to ask for help. The Athenians, however, had problems of their own. While we know from inscriptions that these envoys were received very ceremoniously by the Boule, they were probably unable to obtain much assistance. Meanwhile, Philip's imperialist ambitions continued to grow.

He had set his sights on points farther east, in western Thrace. His goal was eventually to take the wealthy cities of Abdera and Maroneia. Like Neapolis, however, both of these cities were Athenian allies. In 352 Philip set sail from Amphipolis for these two cities. He was blocked, however, by King Amadokos. As Philip was not yet in a position to fight and win, he set sail for home. However, Chares, the Athenian general, was lying in wait at Neapolis to ambush Philip's fleet. That Chares could lay in wait at Neapolis proves that it lay outside of Macedonian control at the time; however, historians believe that the most likely scenario is that Philip took Neapolis some time between 355 and late 354, that Chares then liberated it in spring 353 or earlier, and that Philip had to retake the city around 350.

Philip was fortunate enough to outrun Chares, but he nevertheless learned a valuable lesson: he could no longer risk moving into western Thrace without first securing the territories, including Neapolis, that stood in between. He began to do just that, conquering his eastern frontier one region at a time. The exact date of the Neapolitan fall to Philip is unknown. By 350 B.C., however, Philip had conquered all the surrounding territories. Therefore, even though Neapolis continued to issue its own coinage until 340, it was controlled by Macedon for all intents and purposes after 350. Thenceforth, Neapolis became one of the six port cities supporting the Macedonian economy.

The combined area of Philippi and Neapolis next witnessed major historical events in the middle of the first century B.C., during the famous Battle of Philippi in which the Roman Republican armies of Brutus and Cassius were defeated by the Caesarian armies of Antony and Octavian. With this loss, Brutus's dream of a true Roman Republic was dashed forever, and the Roman Empire became a secured dictatorship.

Two days after assassinating Julius Caesar, Brutus and Cassius fled Rome. Brutus went to Macedonia, and Cassius to Syria, each managing to amass a significant army and,

A view of Kavala
Photo courtesy of Greek National Tourist Organization

between them, control much of the territory from the Adriatic to the Euphrates.

When, one year later, word reached them that Octavian and Antony were leading the Roman forces eastward in pursuit of Caesar's assassins, Brutus and Cassius agreed to meet at Smyrna for a war council. At Smyrna they decided to combine their armies, cross the Hellespont, and proceed along the Via Egnatia westward to meet the Caesarian forces. Antony had sent an advance force of eight legions across the Aegean under the command of Decidius Saxa and Norbanus Flaccus. These legions were traveling eastward along the Via Egnatia (from Dyrrhachium) even as the Republican armies were traveling westward. In the region near Philippi, Brutus and Cassius, assisted by local guides, were able to outmaneuver Decidius and Norbana, eventually coming up behind them on the east side of Philippi.

Norbanus narrowly escaped and was forced to retreat to Amphipolis. Brutus and Cassius purued him as far as two miles west of Philippi, and then took up a defensive position astride the Via Egnatia. They were reinforced by Tillius Cimber, who had just landed a legion on the beach at Neapolis. The Via Egnatia provided the Republicans with an excellent line of communication to Cimber and the naval fleet at Neapolis, while the marshland to the south of the Republican position protected them from being outflanked. They also had the extra advantage of knowing that their naval forces would be able to cut off the Caesarian army's supply lines to Italy—no small matter with winter rapidly approaching. In short, all the advantages were theirs.

Antony and Octavian, meanwhile, had just landed at Dyrrhachium, where Octavian fell ill. With winter coming, quick action was necessary. Antony left Octavian behind and continued on with the rest of the Caesarian army to Philippi. Ten days later, Octavian, the future Augustus Caesar, followed; so determined was he to be present and remind the Caesarian army that they were fighting to avenge the murder of his uncle and namesake that he rode, sick, the entire 350 miles in a litter. And so the two armies stood, separated by a mile or so along the Via Egnatia at Philippi.

Antony saw that the only way to attack the seemingly unassailable Republican position was to build a causeway through the marsh and outflank them. By working almost exclusively at night, Antony's engineers built the causeway in ten days, utterly undetected by the Republicans. Then, on the night of October 22, Antony sent men over who then dug in on the Republican side. The next day Cassius retaliated by building his own causeway that bisected Antony's, thereby isolating the dug-in Caesarians. When Antony attacked in order to reopen his causeway, he left his flank open. Brutus's men, who could see Antony's exposed flank better from their higher ground, attacked the flank, and then sent their main force into Octavian's camp.

After a long day of fighting, Octavian's central rank broke. The decisive charge, led by Messala's legion late in the afternoon, led to a rout of Octavian's camp; in all he suffered 16,000 casualties. However, the news was not all bad for the Caesarians. Antony had been able to break Cassius's line and drive him into retreat. As soon as Antony realized how bad Octavian's defeat was, however, he quit his pursuit and returned to help out. Brutus, returning to find Cassius's camp in disarray, sent troops, though too late to do any good.

Upon seeing those troops approaching, Cassius sent Lucius Titinius to discover whether they were friend or foe. From a distance, it must have looked as if the troops were circling Titinius to capture him, when, in fact, they were simply hailing a friend. When Titinius did not return right away, Cassius committed suicide, believing that the enemy was upon him. Titinius arrived a few moments later and committed suicide himself out of self-reproach for his tardiness. Although the Republicans had lost only 8,000 men, the loss of Cassius was devastating psychologically. Brutus tried to guard against a collapse of morale by sending Cassius's body to Thasos rather than subjecting the men to the ordeal of a public burial. Nevertheless, his men were impatient to dispatch with the Caesarian army.

The Caesarians were, perhaps, in worse shape. Immediately following the first battle, they received devastating news about their fleet: ships carrying crucial reinforcements had been destroyed by Republican forces. This meant that, with winter imminent, the Caesarian army needed a speedy resolution. Every morning, therefore, Antony would line his men up and offer battle to the Republicans. For a little over three weeks the offer was not accepted.

By all accounts, the Republicans should have simply maintained their defensive position. Brutus's men, however, were both impatient and overconfident. Eventually they convinced him to engage in battle; most probably, he realized that

his men's impatience left him very little choice of keeping them together as a fighting unit through the winter.

The second battle took place in mid-November. Once again, Antony's men were lined up to offer battle; this time, however, the offer was accepted. Brutus's infantry, under Messala, attacked the Caesarians' northern flank. The battle lines closed in hand-to-hand fighting, and Octavian's legions were pushed backward. Antony's forces, on the other hand, broke the Republican middle being held by Cassius's men, who subsequently fled in disorder. Brutus tried to retreat and thereby enable the whole Republican force to regroup, but he was pursued by Antony, who had split his cavalry into two columns for the pursuit. When the first column caught the retreating Brutus, they set about capturing the general. In a ploy to buy Brutus time, Lucilius told the advancing Caesarian forces that he was Brutus, allowed himself to be captured, and asked to be taken to Antony. Once brought before Antony, Lucilius told the general that Brutus would never allow himself to be taken alive.

This proved to be true. The next day, Brutus tried unsuccessfully to rally what remained of his troops for one more stand. Once he realized that he was defeated and that his dream for a true Republic would never be realized, Brutus threw himself on his own sword. According to legend, when on the morning of Brutus's suicide he was beseeched to flee, he responded by saying, "Yes, we must fly; but with our hands and not with our feet." Octavian and Antony each reacted quite differently to Brutus's death. Antony had Brutus wrapped in Antony's general's cape for cremation. He let it be known that Brutus alone of the Republicans had acted nobly and unselfishly. Octavian demanded that Brutus's head be cut off and thrown at Caesar's grave. Interestingly, the head of Brutus never arrived in Rome; word circulated that the ship bearing it sank at sea.

Roughly 100 years later, in A.D. 49, Neapolis and Philippi began their transformation into religious centers. In that year, St. Paul sailed from Troad to Neapolis en route to Philippi, where he preached the gospel for the first time in Europe, for which he was cast in jail. He returned in A.D. 55. Nine years later, he wrote his Epistle to the Philippians from a Roman jail cell. Slowly, Christianity established a strong foothold at Philippi. A basilica was built there as early as the fifth century. This church, built on the plan of an octagon inscribed within a square, has been excavated and studied in the twentieth century by Professor Pelekanides. The church was guarded by a large gate. Today one can visit a museum of early Christian mosaics and relics at Philippi. Christianity left its imprint on Neapolis, too; sometime during the Byzantine era the name of the town was changed to Christopolis.

The central feature of the town continued to be the citadel perched on the cliffs overlooking the shore. According to an inscription on a wall of that citadel, the town was burned by Normans in 1185 on their march to Constantinople. The next significant event occured in 1383, when Sultan Murad I amassed an army to march against Manuel II, the Byzantine emperor. In the course of this campaign, Murad beseiged the citadel at Christopolis and ultimately took the town. The sultan did not try to occupy; instead, he recognized the local government, collected the customary poll tax, and installed a small garrison within the city.

Thus from April 1387, Christopolis enjoyed the status of vassalage. However, in 1391, when Sultan Bayezid I marched through Macedonia, the fortress at Christopolis held out against him. When the city did eventually fall, the Turks were so angry that they leveled the city to its foundations, scattering its inhabitants. With the city all but annhilated, the Turks installed a garrison there for strategic purposes.

In 1425, the Venetian admiral Michiel attacked the Turkish garrison at Christopolis, killing forty-one (among them the leader, Ismael Bey); unfortunately, Michiel left behind a garrison of eighty men under the leadership of a captain named Dolfin without leaving a galley moored on the beach. Subsequent attack by a Turkish force of 10,000 men routed the Venetians; after that, Christopolis began to decline.

We know from the written accounts of a mid-sixteenth-century French naturalist named Belon that Kavala was colonized on the site of old Christopolis following the war between Turkey and Hungary—either in 1527 or 1528—and that Ibrahim Pasha (grand vizier from 1523 to 1536) repaired the Roman aqueduct so that water could be brought to the city. Toward the end of the fifteenth century, moreover, Jews who had been persecuted in Europe migrated into Macedonia; by the mid-sixteenth century, Kavala had become a Jewish center. As East-West trade increased in the late fifteenth and early sixteenth centuries, Christians too began to leave their mountainous refugee areas and return to the urban centers. This influx of both Jewish and Christian populations breathed new life into Kavala.

The city grew rapidly to five districts, each containing around 500 two-story houses, as well as lead-roofed mosques, imarets (poorhouses), seminaries, inns, and warehouses. Despite the waves of migration, the inhabitants remained largely Greek, and the language tended to be Greek as well. Nevertheless, the town was Turkish. It belonged to the Turkish *eyalet* (province) of the Aegean Islands, falling under the command of the *kapudan pash*, or chief admiral. During times of war, the *bey* (lord) of the city was expected to provide two galleys for the Turkish campaign. The military presence at Kavala was large—more than 2,000 men—because the threat of attack by the Venetians was very great.

As the Turkish Empire slowly diminished, the situation in Kavala became increasingly more unruly and anarchic. During the Austro-Turkish war, and even after the treaty of Passarowitz in 1718, disorder reigned. The Albanian *armatoli* (private militia) made things even worse, causing Sultan Ahmed III to abolish the institution in 1721 (in fact, the practice continued illegally for the next 100 years). After the end of the Russo-Turkish war in 1774, virtual anarchy reigned in Macedonia, while Albanian terrorism increased. Order was not restored to the region until the establishment of the Greek state in 1830 (although Kavala remained Turkish unilt 1912).

Kavala is remembered as the birthplace in 1769 of Muhammad ʿAli Pasha, ruler of Egypt from 1805 to 1848. Ali made many gifts to the city such as schools, roads, waterworks, and other public utilities. Still remaining are the copper-domed buildings of the Imaret, an old-people's home he constructed and endowed in 1817. The town has commemorated Ali with a statue of him riding a horse with his sword drawn; the statue stands in a prominent position within the city. Other points of interest for the modern-day visitor are the remains of the Byzantine walls and sixteenth-century castle and the aqueduct constructed by Süleyman I (the Magnificent).

Further Reading: For a very readable and informative account of Neapolis and Philippi during ancient times, see J. R. Ellis's *Philip II and Macedonian Imperialism* (Princeton, New Jersey: Princeton University Press, 1986; London: Thames and Hudson, 1987). *Mark Antony: His Life and Times* by Alan Roberts (Upton-upon-Severn, Worcestershire: Malvern, 1988) gives a very clear account of the battle, free of the confusing jargon that so often muddles military history; however, his slant is certainly in favor of the Caesarians. For a more pro-Republican account see Max Radin's *Marcus Brutus* (London and New York: Oxford University Press, 1939). Finally, for a clear and comprehensive account of Kavala throughout the Middle Ages and beyond, see A. E. Vacalopoulos's *History of Macedonia: 1354–1833,* translated by Peter Megann (Thessaloníki: Institute for Balkan Studies, 1973).

—Lawrence F. Goodman

Kayseri (Kayseri, Turkey)

Location: In the steppes in east-central Anatolia near the extinct volcano Erciyeş Daği.

Description: Capital of the Kayseri province of Turkey; also known as Mazaka, Kaysarīya, Eusebia, and Caesarea Cappadociae; former capital of the historic region of Cappadocia and Seljuk Danişmend emirate; an early Christian center; also noted for extensive Seljuk and Ottoman architecture.

Site Office: Tourist Information Office
Kağni Pazarı, No. 61
Kayseri, Kayseri
Turkey
(352) 2311190 or 2319295

Kayseri (also known as Kaysarīya or Caesarea), originally called Mazaca (Mazaka), is a large oasis in the bare steppes of east-central Anatolia. Just northeast of the city is the snow-capped peak of Erciyeş Daği, a 12,800-foot extinct volcano. It is a spot where ancient trade routes from Persia to Europe crossed those connecting the Black and Mediterranean Seas. Nothing is known of Kayseri's beginnings, although people have lived in the area since the fourth millennium B.C., the earliest being the Hatti.

Kayseri is in Cappadocia (Kappadokia), an area that today consists of a triangle with its points at Kayseri, Nevşehir, and Niğde. Known as Katpatuka in Assyrian and in Greek, the name derives from the Hittite for "land of the well-bred horses." The distinctive geology of Cappadocia played a role in the development of the area. It consists of a huge mesa made up of tuff (tufa), volcanic mud and ash covered by lava flows from eruptions of Erciyeş Daği and Hasan Daği. The effects of erosion on these layers of dramatically different densities, as well as earthquakes, subsidence, and oxidation have produced a singular landscape. Tall cones (often called "fairy chimneys"), needles, pyramids, and other shapes in a rainbow of colors cover the surface. This forbidding area was used for defensive purposes, first by the Hittites, who carved fortresses in the soft cliffs, and then by the Persians and subsequent invaders, who maintained them.

Sometime around 2000 B.C. the Hittites arrived, most likely from the Caucasus, and settled in Asia Minor. They supplanted the Hatti and established a powerful kingdom that covered an area bounded by the Euphrates, the Black Sea, and the Mediterranean. Centuries later the Hittite kingdom dissolved, weakened by invasions and internal troubles, and the Phrygians came to power in the area. Kayseri became an important town under the Phrygians.

Until excavations at Gordium, little was known about the Phrygians. The ancient Greek version of their origin says they came from Europe before the thirteenth century B.C. Assyrian records from as early as 1160 B.C. refer to tribes called the Muski and Tabal that were probably Phrygians. Kayseri is thought to have been the nucleus of the land of Tabal, which grew to be one of the most powerful kingdoms in central Anatolia. Tabal was subjected to battles involving Urartu (Armenia), the Muski, and Assyria, and in 712 became an Assyrian province for a time. Later, when Phrygia fell to the Cimmerians, Lydia moved into Anatolia.

Lydia, the Cimmerians, and the Assyrians alternately controlled the region until the middle of the sixth century, when Cyrus II (the Great) of Persia defeated King Croesus of Lydia. Cappadocia remained under Persian rule until the coming of Alexander the Great in 333 B.C., when it became independent. About 150 B.C., Mazaca (Kayseri) was renamed Eusebia and became the capital of the Kingdom of Cappadocia under Ariarathes V Eusebes Philopator (the Pious Father-Loving).

The nearby Kingdom of Armenia rose to be a great power under Tigranes the Great, who ruled from 95 to 55 B.C. He invaded Cappadocia more than once and in 77 B.C. destroyed Kayseri. When Cappadocia became a Roman province in A.D. 17, the Romans called its capital Caesarea Cappadociae. Caesarea was taken and held briefly by the Persians after their success in the Battle of Edessa (Urfa), but it was soon back in Roman hands.

The Cappadocians were converted to Christianity by St. Paul, and from Cappadocia the faith was disseminated throughout eastern Anatolia. King Tiridates III of Armenia, the first to establish Christianity as the official religion, was converted in 314 by St. Gregory the Illuminator, who came from Caesarea.

Caesarea was centered on an ancient acropolis on the slopes until the fourth century A.D. A new city was built on the plains around a church and monastery built by St. Basil the Great, bishop of Caesarea. Basil, born in Caesarea in 329, was called "master of the holy." He was one of three Cappadocians (the others were St. Gregory of Nazianzus and St. Gregory of Nyssa) whose writings were said to be second only to the scriptures in formulating the theology of the early Christian church. As the founder of eastern monasticism, St. Basil encouraged the adoption of an ascetic way of life. His teachings led to the formation of Anchorite monastic groups, in which monks lived in cave dwellings and farmed the fertile volcanic soil on small plots between the rock formations, thus allowing for solitary meditation alternated with communal work.

During the Roman persecutions (and the much later Arab invasions), Christians carved refuges out of the rocks,

Ruins near Kayseri
Photo courtesy of The Chicago Public Library

building entire cities, including churches and chapels, beneath the surface. Hermits still live in some cones and chimneys today. The rock churches, many of which feature elaborate frescoes, remain tourist attractions, as do the restored underground cities.

After the division of the Roman Empire, Caesarea came within the eastern, or Byzantine, realm. Byzantine Emperor Justinian I enclosed Caesarea within walls and built a citadel of black volcanic rock in its center, covering an area about 875 yards by 220 yards. The city was of strategic importance and needed strong fortifications. It was attacked countless times over the centuries. Although the city walls are in ruins (one section remains partially intact), the citadel was later rebuilt and restored and now serves as a modern shopping center.

Despite an "Eternal Peace" at the end of Justinian's first war with the Persian King Khosrow I in 532, another war and dubious peace followed eight years later. When Phocas seized the Byzantine throne in 602, Khosrow II took advantage of the chaos and attacked, taking the eastern provinces and occupying Cappadocia. It was Heraclius, who came to the Byzantine throne in 610, who decisively defeated Persia and regained all of the Byzantine territory.

When Arabs began their attacks, the ancient empires of Byzantium and Persia, spent by wars and internecine struggles, could not mount strong resistance. Warriors of Islam poured out of Arabia after the death of Muhammad in 632 in search of booty and in an effort keep potential threats to their kingdoms at bay, with little concern for annexing territory. Their raids occurred almost annually and took their toll on Asia Minor. They reached Caesarea as early as 647. There was a brief respite from 656 to 663, when the Arabs were preoccupied with their own dynastic struggles, but their attacks subsequently resumed. Caesarea was captured in 726, but in 740 Emperor Leo III drove the Arabs out. Leo's son Constantine V, who reigned from 741 to 775, continued the war. He attacked the Arabs in their own lands, taking advantage of the fact that the Arabs had for various reasons turned their attention eastward. His victories were so impressive that the mere threat of his attack is said to have turned back an Arab army intent on invading Cappadocia.

The peace was only temporary, however, and in 797 Arabs again invaded Cappadocia. On more than one occasion, Byzantium had to pay tribute to the Arabs. The Byzantine Empire had the power to win a war with the Arabs, but they were hampered by internal intrigues and dissension, and preoccupied with religious debates. Emperor Nicephorus I, who reigned from 802 to 811, refused to pay tribute, but was defeated and forced to pay by a huge Arab army in 806. Internal troubles in Arabia in 809 stopped the raids until 890. The Byzantine triumph over the Arabs came under Nicephorus II Phocas, who was from a Cappadocian family. His Army of Caesarea proclaimed him emperor, and Theophano, widow of the previous emperor, Romanus, married him and acknowledged him as ruler. He reigned from 963 to 969.

Turks were the next group of invaders. Their coming marked the beginning of the Kayseri and Turkey that exist today. Roman writers as early as the sixth century knew of the Turks, whom they grouped with the "Huns" of antiquity. In the sixth century the ancestors of the Seljuk Turks moved south from Mongolia and the Lake Baikal region of Siberia into Uzbekistan. It was a time of great migrations driven by many factors, including population growth and shifts in climate.

After serving as *gazis* (warrior horsemen of Islam) for the Persian caliphs, the Seljuks under Khan Toghrïl Beg took control of northern Persia in the eleventh century. In 1055 Tohrïl and his *gazis* enlisted the aide of *mamluks* (slave soldiers) and forced the caliph to recognize Tohrïl as sultan of Persia and Mesopotamia. The resulting dominion was called the Great Seljuk Sultanate.

Under Kiliç Aslan II, Seljuks conquered and sacked Caesarea in 1067 and made it the capital of the Danimend emirate, which included Cappadocia, Sivas, and Amasya. In 1071, Tohrïl Beg's successor, Alp Arslan, defeated the Byzantine army at Manzikert and set up the Seljuk capital in the coastal city of İznik (Nicaea). With the help of the crusaders, the Byzantines took back their territory temporarily in 1097, but Caesarea soon returned to Seljuk control. The Seljuk tribes joined to found the Sultanate of Rum (Rome) to rule over all the Turkish principalities. Formerly nomadic, they made the central Anatolian town of Konya their fixed capital. Despite the brutality, bloodshed, and forced conversion to Islam that were part of the Turkish conquest, some of the population welcomed the Turks and voluntarily converted; Byzantine rule had also had oppressive aspects, including high taxes. Towns were destroyed and looted, but the Seljuks rebuilt them in their own style.

Much of the noteworthy architecture that survives today in Kayseri is of Seljuk origin. The characteristic Seljuk building was fashioned to resemble a tent, its roof supported by posts. Seljuk buildings were decorated with mosaic tiles, wood and masonry carving, and carpets, all in uniquely Turkish designs that continue to be used today.

Islam discouraged extravagant private houses and fostered concern for community welfare and the glorification of Allah. Surviving buildings, therefore, are mainly mosques, schools, soup kitchens, and baths. In addition, sultans, princesses, and viziers competed in the building of impressive tombs.

The Ulu Camii (Great Mosque) was built in 1142 by a Danişmend emir, Melik Mehmet Gazi, in a style known as "Anatolian basilica." The marble capitals of four rows of stone pillars, many salvaged from older buildings, support the central dome. Its minaret is cylindrical and decorated with a tile mosaic. Distinguishing features are its original finely carved wooden *mimber* (pulpit) and a fountain near the entrance.

Before Western Europe had a medical school, Anatolia had its first, built in 1205 in Kayseri, the Giyasiye Medrese. Both the school and the adjoining hospital, the

Sifdahiye Medrese, were built with funds left by a sultan's daughter. The Seljuks also rebuilt Justinian's citadel, in 1224. Nearly ten feet thick with eighteen towers, the walls are interrupted by a single gate, which is topped with two typically Seljuk lions. The citadel was further strengthened under the Ottomans in the sixteenth century.

The first mosque complex built in Turkey was the thirteenth-century Honat Hatun in Kayseri, consisting of the Honat Hatun Camii (mosque), *medrese* (Koran school), and *türbe* (mausoleum). The mosque is in the basilica style with a decorated entry. Its *medrese,* now the ethnographic museum, has two rooms with vaulted ceilings opening into a courtyard. It is considered a Seljuk masterpiece. Important both for its architecture and for its decoration, the octagonal *türbe* is ringed by a double ribbon of geometric designs. The sparse interior holds the Sultana Mahperi's white marble coffin decorated with exquisite calligraphy, and the coffins of two princesses.

Throughout Kayseri, including peculiar places such as the middle of busy highways, are many other twelfth- through fourteenth-century *kümbets* (mausoleums). Because of the nomadic heritage of the Seljuks, the *kümbets* are normally circular, to resemble yurts (tents). They are of two stories, with a sumptuous coffin placed in the upper story. In 1247, three *kümbets* were built by viziers: the cylindrical Sırçalı (Crystal) Kümbet, which no longer has its crystal tiles; the octagonal Çifte (Twin Vault) Kümbet, with a cone-shaped roof; and the Ali Kafer Kümbet. Also notable is the Döner (Turning or Revolving) Kümbet, the twelve-sided tomb of a princess, which has a conical roof that rests on a dainty tower so that it appears the roof could revolve with the slightest touch. It is decorated with characteristic Seljuk designs: arabesques, the Tree of Life, two-headed eagles, and lions. Other interesting buildings of the period are the 1249 Haci Kılıç Camii and Haci Kılıç Medrese, with impressive gates, and the Sahabiye Medrese (1267), decorated with what are considered to be among the finest examples of late Seljuk geometric carving.

In 1243 Kayseri was under Mongol rule for a short time. The coming of the Mongols, successors to Genghis Khan, significantly affected the power structure of the area. They brought great destruction to the countryside and forced the Seljuk rulers to submit to the Mongol khan. With the subsequent arrival of the Ottomans, the Seljuks lost control of Anatolia. The Ottomans went on to forge an empire that would last, with temporary interruptions, until World War I.

Various Turkish sultans ruled Kayseri. The Ottoman Bayezid I, called Yıldırım (Thunderbolt) for his swift movements in battle, took control of the city in 1397. At Angora in 1402, the Tartars under Timur (Tamerlane) defeated Bayezid. He was taken prisoner and displayed in an iron cage by Timur's army. The Tartars ruled the Ottoman territory until Timur's death in 1405. After the Tartars, Kayseri was occupied by Karaman Turks. The Egyptian Mamluks then ruled for a period until Ottoman power revived. Except for the city of Constantinople, all of the former Byzantine Empire had fallen to the Ottomans by the 1420s. They finally captured Constantinople in 1453, and changed its name to İstanbul. Although Kayseri did not officially become part of the Ottoman Empire until 1515, it had an Ottoman ruler for several years prior to that.

Within Justinian's citadel, the Ottomans built a fountain and a small mosque, the Fatih Camii. Also of Ottoman origin is the five-domed Kurşunlu Camii (1580), its name derived from its lead-covered cupola. It is said to have been built from designs of the famous Ottoman architect Sinan Pasha, although it is not considered representative of his work.

An important trading center since the days of the Assyrian merchants, Kayseri has three ancient covered markets where goods are still sold. The Bedesten (Turkish for a covered market hall) was built as a cloth market in 1497. The well-preserved caravanserai, the Vezirhani, was built in 1727, and the covered bazaar was built in 1859 and has since been restored.

Following the dismantling of the Ottoman Empire at the end of World War I, Kayseri became part of the new Republic of Turkey. Since then Kayseri has grown significantly, and the modern town extends well beyond the remains of the old city walls. Modern apartment buildings contrast with the great monuments left by the Ottomans, Seljuks, and Byzantines.

Further Reading: *Turkey: The Rough Guide* by Rosie Ayliffe, Marc Dubin, and John Gawthrop (London: Rough Guides, 1994) contains sections about Kayseri and Cappadocia, as well as information about Turkish culture, customs and sights. *A Traveller's History of Turkey* (Moreton-in-Marsh, Gloucestershire: Windrush, and Brooklyn, New York: Interlink, 1993) by Richard Stoneman presents an overview of the history of Turkey.

—Julie A. Miller

Keos (Cyclades, Greece)

Location: Westernmost of the Cycladic Islands, thirteen miles southeast of the Attic promontory, Cape Colonna (Sounion). Keos is fourteen miles long from north to south, ten miles wide from east to west, and peaks in the center, at Profitis Ilias. As in ancient times, Keos's principal cities and population are concentrated in the north, in Chora (ancient Ioulis), and Livadi (ancient Koressos). The site of ancient Poieessa and Karthaia are on Keos's southwestern and southeastern coasts.

Description: Colonized as early as the fourth millennium B.C., the island was home to a flourishing Bronze Age settlement. During classical times, Keos's four autonomous city-states merged twice to form a federation. In 480 B.C. Keos contributed to the Athenian forces at the battles of Artemisium and Salamis and joined the Delian League and the Second Athenian League. Around 200 B.C. Ioulis absorbed Koressos, and Poieessa fused with Karthaia; Ioulis later incorporated Karthaia and became the island's principal town. When the Byzantine Empire was partitioned in 1207, four Venetian adventurers acquired Keos; in 1566 the island came under Ottoman rule. Keos participated in the Greek Revolution of 1821, and in 1912 it joined the kingdom of Greece. Today, tourism, manufacturing, animal husbandry, and agriculture are important Keian industries.

Site Office: Archaeological Collection
Village Hall
246 00 Chora
Keos, Cyclades
Greece
(288) 22079

Keos's history, like that of the other Cyclades, was closely intertwined with developments in the wider Greek world. Keos (alternatively spelled Ceos, Kéa, and Tziá) was known even in ancient times for its four city-states, fine harbor and mineral resources, famed lyric poets, and stringent laws. The Keians have played important roles in Greek politics, from the classical period to the formation of the modern Greek state.

According to legend, the island was initially inhabited by nymphs who were frightened away by a lion and fled to Carystus; a promontory on Keos called Leon marks the event, and a colossal stone lion reclines near Ioulis. The nymphs taught the Keian hero Aristaios shepherding, cattle raising, and beekeeping, industries important to Keos to this day. Bacchylides' *Ode I* celebrates King Minos's romance with the Keian princess, Dexithea, by whom he had a son, Eu-

xanthios; when Minos departed, he left half of his soldiers to protect her. Kallimachos, who retold in his *Aetia* Xeomedes's fifth-century B.C. chronicle of Keos's mythology, mentions the hero Keos, son of Apollo and Melia, and one of Euxantios's descendants, the hero Acontius.

Acontius was celebrating the festival of Artemis at Delos when he fell in love with Cydippe, an Athenian noblewoman. He tossed in her direction an apple on which he had written, "I swear by the sanctuary of Artemis to marry Acontius." She read aloud the inscription but threw the apple away. When Cydippe's father arranged to give her to another man in marriage, she fell ill shortly before the wedding. After this happened three times, he consulted the Delphic oracle, who revealed that Artemis was punishing his daughter for her conduct. Cydippe told her parents of the incident at the festival, and Acontius won her hand in marriage.

The pear-shaped island's mountainous terrain is broken by fertile spring-fed valleys and rises to a central promontory, Profitis Ilias; in antiquity, its abundant water supply earned it the name Hydrousa. The spacious Bay of Ayios Nikolaos in northwest Keos and the island's arable land, favorable location for trade, and deposits of iron, lead, silver, copper, and *miltos* (red ochre) attracted colonists early. Kephala on the northwest coast preserves Late Neolithic houses and a small cemetery dating from 3500 B.C., the oldest discovered graves of the Aegean.

During the Bronze Age, trading relationships with mainland Greece, Crete, and other Aegean islands contributed to Keos's growth and prosperity, but Keos was not a Minoan colony; Keians borrowed these foreign elements selectively and imbued them with their own traditions.

Ayia Irini, a cape on the Bay of Ayios Nikolaos, was home to a flourishing Middle to Late Bronze Age settlement. A temple discovered there contained parts of more than fifty terra-cotta female figurines sporting Minoan-style flaring skirts, short jackets, and garlands (or necklaces); the statues, some of which are nearly life-sized, were apparently cult images of the divinity worshiped there. The temple itself is unique, since few such structures are thought to have been built during the Mycenaean Age; it was destroyed by an earthquake around 1300 B.C. but was rebuilt and remained in use for three more centuries. Around 700 B.C. the head of one of the statues was placed on a new stand in the temple, where it was revered for two more centuries; an inscription on one of its offerings refers to the male fertility god Dionysus, prompting some to suggest that worshipers mistook the head's pointed chin for a beard.

In classical times (early fifth to late fourth centuries B.C.) four independent poleis (city-states) dominated Keos. Koressos, a principal port overlooking the modern village of Livadi on the Bay of Ayios Nikolaos, was the center of the

A view of Keos
Photo courtesy of Greek National Tourist Organization

island's smallest polis. Built in the seventh century B.C. atop a double-humped ridge, the ancient, fortified town was a thriving seaport in classical times. The area has yielded remains of residential sectors, fine Attic pottery, a temple of Apollo and a temple of the Dioscuri, and kouroi, statues of young men that were sometimes used as funerary monuments.

The city of Poieessa (now, Poisses), whose territory was slightly larger, overlooks a valley on Keos's southwestern coast. Between the two polis centers sit the temples of Apollo Smintheus and Athena Nedusia; according to Greek geographer Strabo, the latter was founded by Nestor on his way home from Troy.

Karthaia (modern Poles), on the southeast coast, is practically deserted today, but the massive walls and nearby temples of Athena and Apollo bear witness to its former prosperity. The latter commemorates Epeius, the legendary Keian artist who built the Trojan horse and sailed with other Cycladic islanders to Troy.

The centrally located Ioulis (present-day Chora) claimed the largest territory and continues to be Keos's main city. Simonides, the premier poet of victory odes, and his nephew Bacchylides were natives of Ioulis. A fierce rivalry grew between Bacchylides and Pindar, dubbed the Keian Nightingale and the Theban Eagle, respectively; sometimes both poets composed victory odes in honor of the same athletic event.

The circumstances surrounding the formation and demise of the four city-states are sketchy, but Karthaia, Koressos, and Ioulis were all producing their own silver coins as early as the archaic period (seventh to early fifth centuries B.C.).

Keos contributed a contingent of soldiers in the 480 B.C. battles of Artemesium and Salamis and joined the Delian

League, which was formed in 478 B.C. to protect Athens and its allies from Persian aggression and to liberate those under Persian control. The financial contribution of Keos's four city-states (paid together as a single unit) was larger than that of the average league member. Possession of a strategic harbor and mineral resources had enabled the tiny island to grow comparatively wealthy.

In the fourth century B.C., Keos's city-states showed increasing signs of unification, with Ioulis emerging as the headquarters of a federal government that may not have encompassed Poieessa. Evidence for this federal government can be found in an inscription on the Sandwich Marble stele of Athens recording a loan the city-state of Delos made to the Keians as a group before 377 B.C., and two treaties Keos established around 364 B.C. with Histiaia and Eretria in Euboea. The treaties outlined the acquisition of citizenship rights for immigrants from the various cities; because the criteria for enrolling citizens in the individual Keian cities were incompatible with each other, the treaties carefully spelled out procedures for inducting foreigners into an island-wide federal state. This federation may have been a vestige of earlier Eretrian supremacy in Keos, when the Eretrians revolted from Athens and set up a unified state in Ioulis in 450 B.C.

Around 376 B.C. Koressos, Ioulis, and Karthaia joined the Second Athenian League, the purpose of which was to safeguard Athens and allied states from Spartan imperialism; Poieessa may have joined separately. The Keians attempted unsuccessfully to secede from the league and in the process murdered an Athenian envoy. The Athenian general Chabrias quelled the uprising, and negotiations culminated in a surprisingly mild settlement: in return for the Keians' loyalty, Athens pledged to refrain from taking reprisals for the rebellion, to readmit Keos in full standing to the league, and to allow disaffected Keians to live in any of the allied states. Shortly afterward, Athens mandated that the Keians resume governance by individual city-states.

Athens sought to reestablish control by monopolizing Keos's *miltos* production. The natural blend of red ferric oxide, clay, sand, and other substances imparted a red color that was highly prized by Athenian vase painters; it was thought to possess medicinal properties as well. Decrees posted in 350 B.C. in Koressos and Ioulis granted Athens exclusive rights to Keos's exceptional *miltos* and stipulated that those two cities, along with Karthaia, accept all Athenian legislation pertaining to the matter. Harsh penalties awaited transgressors; wrongdoers in Ioulis, for instance, would suffer confiscation of their property, half of which would go to the people of Ioulis and the other half to the informer.

The Keians were as renowned for their character and norms as they were for their natural resources. Aristophanes punned the distinction between the forthright Keians and wily Chians. And the Keians had a maxim, "Whoso cannot live well shall not live ill," embodied in a law related by Claudius Aelianus in his *Varia historia:* "Those among them who are very old summon each other as if it were for an interchange

of presents or some ceremonial sacrifice. Having gathered together, they drink hemlock with a wreath on their head, having realized that they have become useless for the activities that serve the fatherland when their mental faculties already are in decline." The prescribed age for euthanasia was sixty; Strabo remarked that in the wake of a siege by Athens, Keians voted that the oldest citizens commit suicide, prompting the Athenians to raise the siege.

Self-killing on behalf of the polis was considered praiseworthy; nevertheless, ancient authors were struck by the alacrity with which some individuals imbibed the "Keian Cups." In his *Factorum et dictorum memorubilium (Memorable facts and sayings)* Valerius Maximus told of a Keian woman over ninety years old who was determined to fulfill her "last resolution" before a group of Romans. After distributing her effects and exhorting her eldest daughter to preserve her memory and sacred possessions, she made libations to Mercury and invoked his assurance of a safe passage to the best part of the underworld, then she "took with a firm hand the beaker, in which the poison had been mixed." She described how the poison was taking effect and asked her daughters to perform the final act of closing her eyes. Maximus added, "She made us Romans leave, however, stunned as we were by the unprecedented spectacle, streaming with tears."

External influences impacted even more profoundly on the Keians during the Hellenistic and Roman periods. Keos contributed soldiers to the League of Islanders under Ptolemaic control and became a naval base for the Ptolemaic general Patroklos; officials installed a resident governor at Koressos, which they renamed Arsinoë after the late queen. The Egyptians were actively involved in the island's affairs, dispatching judges to settle internal disputes at Karthaia and Arsinoë; Ptolemaic control apparently did not extend to Poieessa, which was experiencing Macedonian encroachment.

Sometime after 244–43 B.C. Keos reconstituted a federal government in the wake of another threat, Aetolian pirates who raided the Cyclades for slaves and booty; a treaty protected Keians from attack by Aetolians and persons claiming Aetolian citizenship. The individual city-states continued to issue distinct coins, organize themselves differently, and function independently—sometimes discordantly—but they continued to enjoy the benefits and rights conferred by their federal status, as well. This federation dissolved around 196 B.C., when the Aetolian threat receded, and Koressos reverted to its original name.

Keos's increasing subjection to Athenian control had more far-reaching consequences. After the Battle of Philippi in 42–41 B.C., Mark Antony presented Keos and other islands as gifts to Athens, a prominent example of a trend that enabled wealthy families to amass large estates at the expense of smaller landholders. The population around Koressos began to contract, and two of the city-states permanently lost their autonomy: around 200 B.C. Ioulis absorbed Koressos, and Poieessa merged with Karthaia. The latter merged with Ioulis in late Roman or Byzantine times. According to archaeologists J. F. Cherry, J. L. Davis, and E. Mantzourani, these syntheses involved "the surrender of local autonomy, the transfer of local institutions, and the relocation of populations which had previously been more widely distributed."

The next major turning point occurred in A.D. 1207, when Keos and six other Greek islands incorporated in the Byzantine Empire became spoils of war divided among four Venetians: Domenico Michieli, Pietro Giustiniani, and the brothers Andrea and Geremia Ghisi. After a brief respite, during which Keos was restored in 1278 to Byzantine rule, the Ghisi reclaimed their possessions in 1296.

For the next three centuries the Spanish, Venetians, Genoese, Franks, and Turks jockeyed for control of the island, now called Tziá. Recurrent conflicts, especially the Turco-Venetian War of 1463–79 and the 1537–38 campaigns of Khayr ad-Dīn (known to Europeans as Barbarossa) ravaged Keos and its Aegean neighbors, decimating their populations. In 1538 the Duchy of the Archipelago wrested control of Keos from the Turks, but by 1566 the island was firmly under Ottoman rule, cemented by Sultan Selim II's appointment of his loyal follower, the Portuguese adventurer Joseph Nasi, as duke of the Archipelago.

Under Turkish hegemony, which lasted for two more centuries, Keos became known as Murtat Adasi (Egg Island), and its mostly Greek population remained largely self-governing and concentrated at Chora (ancient Ioulis). Both the Turkish and the Venetian Empires exacted tribute from Keos, which by the mid–seventeenth century had become a major exporter of wine, silk, barley, and acorn caps used for tanning and dyeing processes.

Keos became involved in sporadic Greek revolts aimed at dismantling the Ottoman power, in particular a 1647 attack by Francesco Morosini on Keos and a brief foray of Russian forces across the island under the command of Lambros Katsonis in the late eighteenth century.

Keos was among the first of the Greek islands to participate in the Greek Revolution of 1821, and in 1912 it joined the kingdom of Greece. Throughout the nineteenth century the island continued to be a major provisioning point for ships, and demand for its exports heightened.

When the *archons,* members of the Keian ruling elite, went abroad to participate in the politics of the new Greek state, other Keians seized the avenues for social, political, and economic advancement that were opening up for them; the traditional *voli* system of land tenure broke down, facilitating farmers' access to land. New industries developed, including a coaling station at Kobba and an enamel factory at Livadi, and the latter emerged as a key city. Out-migration soared in the 1940s and 1950s, however, when acorn exporting declined and the coaling station and factory ceased operation.

Today, Keos manufactures aluminum kitchenware for the Greek mainland, and animal husbandry and almond, honey, and barley production have gained new ground. Most of Keos's population lives in the northern part of the island, where tourism is a booming business. Visitors may view the stone lion near Chora, remains of the ancient city-states and their temples, and a twenty-four-meter-high stone tower at Ayia Marina. The Archaeological Museum of Keos at Chora houses many of the island's antiquities, including the statues recovered from the Temple at Ayia Irini, and the Numismatic Museum at Athens conserves Keian coins.

Further Reading: *Landscape Archaeology as Long-Term History: Northern Keos in the Cycladic Islands from Earliest Settlement until Modern Times,* edited by J. F. Cherry, J. L. Davis, and E. Mantzourani, volume sixteen of the Monumenta Archaeologica Series (Los Angeles: Institute of Archaeology, University of California, 1991), comprehensively overviews Keos's history, addressing in depth the formation and demise of the Keian city-states. Anton J. L. van Hooff's *From Autothanasia to Suicide: Self-Killing in Classical Antiquity* (London and New York: Routledge, 1990) discusses the Keian prescription regarding suicide. For more information about Keos's prehistory and relationships with Athens, see *The Temple at Ayia Irini: The Statues* by Miriam Ervin Caskey, with John L. Caskey, Stella Bouzaki, and Yannis Maniatis (Princeton, New Jersey: American School of Classical Studies, 1986) and *The Second Athenian League: Empire or Free Alliance?* by Jack Cargill (Berkeley and Los Angeles, and London: University of California Press, 1981). *The Latins in the Levant: A History of Frankish Greece (1204–1566)* by W. Miller (London: Murray, 1908) meticulously recounts the Venetian-Turkish conflicts.

—Maria Chiara

Kestel/Goltepe (Niğde, Turkey)

Location: In southern Turkey, sixty miles north of Tarsus.

Description: Ancient mining and tin smelting site, dating from 3200 B.C. The site was excavated, beginning in 1990, by a team led by Aslihan Yener, assistant professor at the Oriental Institute of the University of Chicago.

Contact: Tourist Information Office
Belediye Hizmet Binasi 3.
Kat C Blok
Niğde, Niğde
Turkey
(388) 2323393

A rich mining area in southeastern Turkey has provided important sites for archaeologists studying the origins of the metal industry in the Mediterranean region. One of these sites is Kestel, the location of a Bronze Age tin mine, about sixty miles north of Tarsus. Nearby is Goltepe, an ancient miners' village, that with the mine has provided researchers important new insights into the development of the tin industry. Kestel and Goltepe are situated in the southernmost edge of the Cappadocia in central Turkey, in a region renowned for its volcanic landscape. Rock formations in this area were carved into lunar-landscape-like places of worship during the early Christian period.

Several cities in the region are part of frequented tourist routes, but the two Bronze Age sites are far removed from them, in the central Taurus Mountain range about three miles west of the small, isolated town of Camardi in Niğde province.

Kestel and Goltepe date from the beginning period of copper and bronze use in southwest Asia, between 3200 and 2000 B.C. They are situated in a strategically important location, along a north-south fault zone. The area links central Anatolia to the Cilician plains to the south and to the Mediterranean coast.

The sites were excavated by Aslihan Yener, assistant professor at the University of Chicago's Oriental Institute, beginning in 1990. Her work provides important evidence for the ways in which tin was mined in the Bronze Age and how it was smelted so it could be combined with copper to produce bronze. Her discoveries firmly established for the first time that tin was produced in Anatolia to be used in the making of bronze. The mine and village were a full-blown industrial operation that represents how, by the year 2000 B.C., tin production had come to play an important role in the development of civilization in Anatolia.

Tin's economic role in the metal technology of the time is perhaps akin to that of oil in industry today. It was a key ingredient in the high-technology metal of its age: bronze. Bronze is an alloy made by combining copper with as much as 5 to 10 percent tin. Because it is more easily cast in molds and harder than copper, bronze replaced copper in the production of tools, weapons, and ornamental objects.

Yener's work has the potential to change established theories about economic and metallurgical developments in the Bronze Age Mediterranean world. The Bronze Age, which began about 3000 B.C., was a time of great economic expansion throughout the Middle East. During this time, great city-states such as Troy rose in Anatolia and empires developed in Mesopotamia. The Bronze Age lasted until 1100 B.C., when iron became the most important metal in manufacturing.

Despite the importance of bronze and the role tin played in its production, scholars have long believed that tin was not readily available in the Middle East. Cuneiform texts on clay tablets speak of sources to the distant east, and researchers have believed that Afghanistan was perhaps the only likely location of tin mines. Yener's discovery shows that tin came from local sources, and was not exclusively imported from afar.

Goltepe and Kestel are in a highland resource zone that provided materials for more complex societies elsewhere in Anatolia as well as in Syria, the Aegean region, and Mesopotamia. The resource zone played an important role in the industrial revolution that helped develop the Mediterranean area during ancient times.

In their work at Goltepe and Kestel, researchers initially used relatively primitive techniques to determine that the area had sufficient traces of tin to make the mining of tin during antiquity a possibility. Bryan Earl, an expert on tin from the Cornwall area of Britain, and the Turkish Geological Survey both conducted operations in local stream beds in which they panned for tin with a shovel. Those efforts found traces of alluvial cassiterite, or tin ore, in large enough percentages to justify further exploration.

The Kestel tin mine is made up of a variety of mining chambers, all of which have been depleted. Yener did preliminary work at the site in 1987 and 1988 by running five soundings, or exploratory digs, into the mine to sample and date it. That exploration recovered cassiterite, as well as evidence, such as ceramics, dating the mine to the Bronze Age.

Researchers believe that the mine was first used as an open-pit mine. Later, chambers were developed and workshops set up at the mine's entrance. The mine produced hundreds of tons of tin, which was probably transported by donkey caravan.

Yener's work indicates that children were the miners at Kestel. The underground mining system, which measures

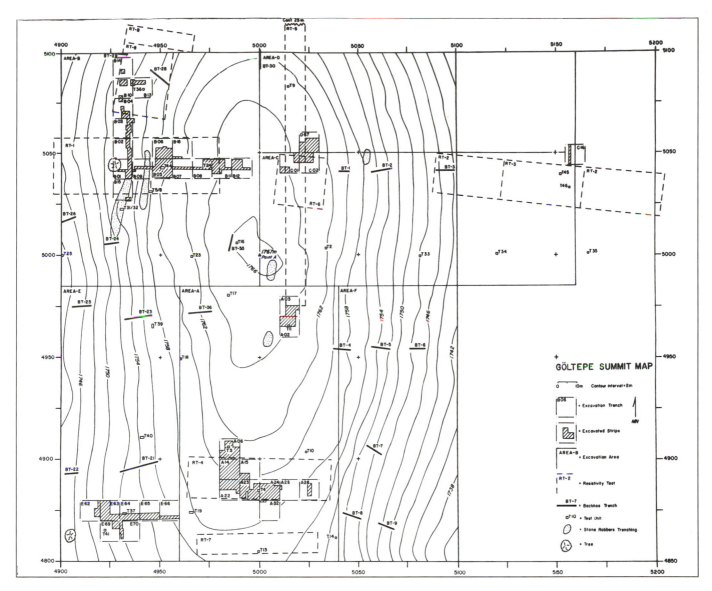

Diagram of the excavations at Kestel and Goltepe
Illustration courtesy of University of Chicago

more than two miles, is made up of shafts that are about two feet wide, too small for adults to enter easily. Yener and her colleagues also have recovered a burial site inside the mine with the skeletons of a number of twelve- to fifteen-year-olds, a finding that supports the view that the miners at the site were children. By examining the skeletons further, it will be possible to determine if they died of mining-related illness or injury.

The mining was done with stone tools and fire. Miners would light fires by the ore veins, thus making it easier to batter away the ore. The mine probably produced about 5,000 tons of ore during its 1,000 years of operation.

Although the researchers had found tin ore at Kestel, some skeptics thought that there was not enough tin to prove that the mine was actually a tin mine. Working with tin experts from Cornwall in southwestern England, an area famous for its tin deposits, Yener discovered industrial debris

at the mining village of Goltepe, about a mile from Kestel. Those deposits provided clues about how the tin was smelted. Instead of evidence of only low-grade tin, one ton of tin-slagged crucibles with 30 percent tin content was discovered at Goltepe. This establishes beyond doubt that tin-metal was being produced and was the motivation for the mining and smelting industry.

After mining, the ore was first washed, much the way panners in the early American west recovered gold. The ore was ground and then smelted in covered crucibles, into which workers blew air through reeds. Droplets of tin were encased in molten slag, which was ground out, rewashed and re-smelted in a labor-intensive, repetitive process. This intensive use of stone tools to crush not only ore but also slag to release tin globules explains why more than 50,000 stone tools were found at the site. It is an unparalleled collection of such materials.

Radiocarbon dating indicates that Goltepe was occupied from 3290 B.C. to 1840 B.C. It developed over several phases and began as a community of pit-houses built into the soft sedimentary bedrock. Later, a walled community developed.

Exploration at Goltepe began in 1990 with the intention of reconstructing the tin metal production process, as well as the village context within which the activity took place. Pottery provided important information about the trade connections of the site. The fragments recovered at Goltepe are similar to those produced in central Anatolia, the Cilician coast and the Aegean, Syria, and Mesopotamia. The fragments date from various points in the Bronze Age.

People in the village lived in open-pit homes, which apparently were built by cutting through bedrock and then putting up stone walls. The researchers recovered painted plaster, suggesting that the stone facings were plastered. The houses apparently had interior posts that supported a wattle-and-daub superstructure. In one of the subterranean homes in Goltepe was found a hearth and groundstone tools. The ashes in the hearth were dated to about 3200 B.C. Also recovered at the site were examples of earthenware painted in a purplish-red color.

Also unearthed were subterranean workshops and a well-built north-south wall that could have been part of a wall around the village. The wall, which was more than a yard high in some places, was made of large, irregularly shaped stones, and was sealed with blocks.

A discovery of particular archaeological importance at Goltepe consisted of hundreds of crucibles and crucible fragments. Analysis of the crucibles as well as further research on the Kestel mining site yielded further important information on tin production in Anatolia during the Bronze Age. Perhaps most important is Yener's discovery that tin can be smelted at relatively low temperatures in crucibles. Other tin sites known to exist throughout the Mediterranean area also could have been sources of tin through the labor-intensive smelting processes the team recreated.

Several hundred people probably lived on the site. It is unclear whether the food consumed by the community came in as trade or if some form of agriculture took place locally. Terraces on mountain slopes suggest that some crops may have been raised there.

Researchers also want to learn of the division of labor among men, women, and children for the various tasks of mining, smelting, grinding metallurgy, and agriculture.

Yener, a leader in the expanding field of isotope research, hopes further work may also connect tin produced at Goltepe with artifacts found elsewhere in the Middle East. Because mining sites produce metals with characteristic isotopes, she will be able to determine where the metal used in particular artifacts came from. Yener's work originated from a study she began in 1980 to identify sources of metals used in the production of weapons and other objects in the ancient Near East. Yener, an American of Turkish descent, began the work as a member of the faculty of Bosporus University in İstanbul.

Further Reading: More information on Yener's findings is available in her article "Tin Processing at Goltepe, an Early Bronze Age Site in Anatolia" in the April 1993 issue of *American Journal of Archaeology* (New York), volume 97, no. 2.

—William Harms

Konya/Çatal Hüyük (Konya, Turkey)

Location: Konya is in central Anatolia on the Konya Plain in the Republic of Turkey. Çatal Hüyük is about thirty miles southeast of Konya.

Description: Konya is Turkey's eighth-largest city, called Iconium in Greek and Roman times; capital of the Seljuk Sultanate of Rum and home of the Whirling Dervishes, a tribe of mystics. Çatal Hüyük has remains of the first known city, which dates to the seventh millennium B.C.

Site Office: Tourist Information Office
Mevlâna Cad., No: 21
Konya, Konya
Turkey
(332) 3511074

Konya, in central Anatolia, is Turkey's eighth-largest city, with a population of more than 500,000. Nearby is Çatal Hüyük, one of the most important archaeological sites in the Near East. Although Neolithic remains had been discovered in parts of Turkey near the Fertile Crescent, most archaeologists did not believe that similar civilizations had existed on the Anatolian plateau. In 1958, British archaeologist James Mellaart discovered such a civilization at Çatal Hüyük, about thirty miles southeast of Konya. In 1961 he began excavation of the thirty-two acre site, made up of two flat hills or tumuli (*Hüyük*) roughly in the shape of a fork (*Çatal*).

Folklore holds that the Konya plain was the first dry land to appear after the Old Testament flood. In fact, a large shallow lake covered the plain in 16,000 B.C. When the waters receded, a fertile plain emerged, watered by a river flowing from the Taurus Mountains. The terrain, with grasslands for pastures, marshes teeming with game, and forests to provide wood for houses, could support a large human population. About 7000 B.C., on the banks of the river, where the lake bed adjoins the clays of former swamps, the city of Çatal Hüyük emerged, thought to be the very first city. Before the Çatal Hüyük finds, small rural villages were the standard and experts did not expect to find seventh-millennium-B.C. urban settlements. Çatal Hüyük exhibited the earliest evidence of ordered city life, and the oldest known mirrors, pottery, textiles, wooden vessels, and frescoes. There were 5,000 to 6,000 people, living in family units in about 1,000 houses. Until the discovery of Çatal Hüyük, Jericho, only one-fourth its size, was the largest and the oldest known Neolithic city.

Although the site has not been fully explored, a wedge-shaped portion extending down to the level of the earliest settlement has been excavated. There are fourteen layers from about sixteen feet below the ground to more than fifty-five feet above. Radiocarbon dating estimated the site to be from 6250 to 5400 B.C., while tree-ring evidence indicates 7100 to 6300 B.C. The town was never destroyed by war and experienced almost 1,000 years of relative peace, during which each successive city was built atop the previous one. Then, for unknown reasons, the inhabitants deserted the site and built a new settlement across the river.

Each layer held tightly packed mud-brick houses, with no streets to separate them, just occasional courtyards for refuse and sanitation. Each house had two rooms: one rectangular, with sitting/sleeping platforms, and a kitchen area; the other a storehouse. There were no doors; a ladder led from the flat roof down into the kitchen area. Exterior walls were blank, with a few openings for light near the roof, making city walls unnecessary. Every third or fourth house served religious purposes as shrines with religious decorations and cult objects.

Interior walls of the houses had plastered panels. Some were painted with simple geometric designs in red paint, others with more elaborate multi-colored scenes framed by pilasters. Paintings in the shrines were particularly extensive, but even these only served as decoration for about a year. The custom was to apply new coats of plaster frequently; one house had 120 layers. Fortunately for archaeologists, the new plaster preserved many extraordinary paintings. Some mimicked the patterns of local textiles, while others depicted scenes of everyday life, including deer hunts, and scenes of nature and wildlife. One picture documented the eruption of the volcano, Hasan Daği, with the city in the foreground. Scenes of vultures pecking at headless corpses provided corroboration of theories about local burial customs. Skeletons buried under the sleeping platforms in homes appeared to have been exposed and only moved to the house after the bones had been picked clean.

Since each family rebuilt its house on top of the old one, archaeologists have had the opportunity to study 1,000 years of generations of the same families. While most citizens were of the same stock, there was enough genetic diversity to assure a healthy and expanding population. Overpopulation was probably responsible for the spread of Neolithic culture through a large area as people migrated to establish new settlements.

Finds in successive layers at Çatal Hüyük demonstrated the development of technology. The earliest layer showed limited use of the first known pottery, but with each successive layer the pottery became more prevalent and more decorative. Abundant resources in the area gave Çatal Hüyük the leisure to advance culture and art. Although engaged in some hunting and gathering, the inhabitants also kept domesticated animals (cattle, sheep, goats) and cultivated several types of grains (wheat, barley, peas, lentils), making them one

Whirling Dervishes at Konya

The Mawlana Museum
Photos courtesy of Embassy of Turkey, Washington, D.C.

of the best-fed communities in the ancient Near East. Although some clothing was made from fur and animal skins, most was from cloth adroitly woven on upright looms from plant fibers, wool, and animal hair. Antler toggles for men or pins at the shoulder for women fastened the clothes.

They made tools and weapons such as arrowheads and sickle blades from locally available obsidian, and ceremonial daggers from flint imported from northern Syria. Two volcanoes, Hasan Daği and Karaca Daği, provided the hard black obsidian, so smooth it was used for the first mirrors found in some women's graves. The use of raw materials from distant areas, including ores imported as early as 7000 B.C., for the first primitive metalworking, suggests the existence of extensive trade routes.

As shown by the many shrines, religion was important from the beginning. In addition to religious wall paintings, the buildings held stone and pottery statues, some carefully made presentations, others crude, probably for offerings rather than permanent display. Most were female, with animal heads (usually bulls) substituted for male offspring. Two- to four-inch statues of the mother goddess, with large breasts and hips, crouching to give birth, were prevalent. These were probably connected with the Phrygian goddess, Cybele, and the later Artemis.

Çatal Hüyük was abandoned toward the end of the seventh millennium B.C. and Anatolian civilization shifted to other sites. About thirty miles away, the town that is now called Konya developed in a well-watered spot in the arid steppes. It too had ancient beginnings, as Neolithic remains have been found in the area around the Alâeddin Tepesi, site of the original *tepe* (acropolis). Much later, during the tenure of the Hittites (late second millennium B.C.) it was an important provincial trading center, called Kuwanna, on the main road between the steppes and the mountains.

Subsequent occupants were Phrygians, Greeks, Romans, and Byzantines, who called the town Iconium. Iconium assumed importance in the Christian church, which convened one of its earliest councils there in A.D. 235, and was the site of sermons by both St. Paul and St. Barnabas. With the downfall of the Byzantine Empire, however, the entire area declined. Iconium of the Greeks, Romans, and Byzantines had vanished. The best of the pre-Turkish artifacts are now in the Museum of Anatolian Culture in Ankara, although some remain in the Archaeological Museum in Konya. The town that survives today dates to the Seljuk Turks who first reached Iconium in A.D. 1068.

Chinese records from 200 B.C. suggest the Turks originally came from Mongolia and the area south of Lake Baikal (now in Russia). Due to climate changes, overpopulation, and other influences, a general migration from central Asia occurred in the sixth century that included the ancestors of the Seljuks. By the tenth century they had converted to Islam and were headquartered in Bukhara (Uzbekistan) under a khan called Seljuk. His successors went to Persia to serve the caliphs as warrior horsemen or gazis of Islam. In 1055 the Seljuk khan, Tuğrul Bey, backed by an army of gazis and *Mamluks* (slave soldiers), forced the caliph to acknowledge him as sultan of Persia and Mesopotamia, thus establishing the Great Seljuk Sultanate. Tuğrul Bey's successor, Alp Arslan, followed looting gazis into Anatolia, trying to maintain control, and defeated the Byzantines in the Battle of Manzikert in 1071.

For a brief period after Manzikert the Seljuks tried to establish their capital at İznik (Nicaea) on the coast, but were forced inland by a Byzantine-Crusader army in 1097. Byzantines reestablished rule for a time, but by the 1140s battles and internal strife weakened the empire sufficiently that the power of the Turks revived. The Seljuk Sultanate of Rum (Rome), a consortium of tribes, ruled all the Turkish self-governing states. Breaking with tradition, the nomadic Seljuks had a permanent headquarters at Konya.

The Turkish conquest of Anatolia was frequently brutal and cities were often looted and destroyed. Some inhabitants fled, but many, tired of excessive taxes, welcomed the Turks. Some voluntary conversions came about because of Moslem free education and health care, but this goodwill did not stop a century of massacres, forced conversions, and other atrocities.

The more progressive of the Seljuk sultans, particularly Alâeddin Keykobad I, who ruled in the early thirteenth century, combined harsh Turkish rule with the refinements of Persian culture, turning Konya into a great city, where literature and the arts flourished. Many buildings surviving today were built under his rule. The tall pointed arches, roofs supported by poles and light-colored stone of typical Seljuk buildings were reminiscent of tents. This period also produced distinctive designs in tile, wood, masonry, and carpets that became prototypes for the typically Turkish designs produced under the Ottoman Empire.

The oldest building in Konya, the Alâeddin Camii (mosque), shares the hill (partly natural, partly remains of prior settlements) with the ancient acropolis. The mosque was begun by Sultan Mesut I in 1130 and completed in 1221, re-using columns and masonry from classical buildings. The three-domed Iplikçi Camii, built in 1230, is the oldest working mosque in the city.

Büyük Karatay Medrese, an Islamic school dating from 1251 and now a museum of ceramics, mixes Arab and Greek elements to produce a typical Seljuk portal. Quotations from the Koran and names of caliphs encircle a dome of gold, blue, and black tiles representing the heavens. The decorations include Christian and pagan designs and, despite Islamic prohibitions, representations of animals, birds, and even angels. Another religious school, the Ince Minare (Slender Minaret) Medrese was also built in 1251 and its minaret severely damaged by lightning in 1901. It has a typical Seljuk doorway wreathed with Arabic phrases that is even more decorative than the one at Karatay. As at Karatay, reliefs show angels and animals. The medrese is now a museum displaying Seljuk design elements and carvings from ruined buildings, and drawings of lost Seljuk palaces. Modern-day Konya's symbol, a two-headed eagle, is reportedly derived from some of these drawings.

Sahip Ata, a complex built around a mosque in 1283, noted for its brick and stone entrance and its tiled mihrab (pattern of tile pointing toward Mecca); the stone walls of a palace, now protected by a concrete canopy; and the Sırçalı Medrese, named for its crystal-like blue tiles, are other surviving Seljuk remnants.

In the thirteenth century, Konya became the home of the order of the Whirling Dervishes or Mevlevi. A Sufic mystical order within the Islamic religion, it was founded by Jalāl ad-Din ar-Rūmī, called Mawlana (Our Master). Although the order was disbanded during President Atatürk's 1925 modernization, dervishes still whirl; the ceremony is now officially considered cultural rather than religious. Mawlana's tomb remains a place of pilgrimage.

Mawlana was born in 1207 and soon a wandering dervish (literally, a "beggar" in Turkish) foretold his future eminence. After receiving a warning from God when he was twenty years old, Mawlana and his father fled their central Asian home of Balkh just before the Mongols massacred the populace. They moved to Konya where the sultan welcomed them warmly. Mawlana completed his multi-volume masterpiece of Persian devotional poetry, the *Mathnawi*, in the 1250s and was recognized as the leading mystic of the time. His influence is still felt in modern Turkey's Islamic culture. The date of his death in 1273 was called the Wedding Night, as he satisfied his earthly yearning for union with God (the One Friend) on that day.

Dervishes achieve communion with God through their whirling ceremony. After the leader of the order whispers a private message (possibly from Mawlana's works) to each, the dervishes begin to whirl slowly with folded arms. As the pace increases, they each enter a trance and unfold their arms, receiving blessings in the upturned palm of one hand, transmitting them to earth with the down-turned palm of the other. In a circle, skirts swirling, the entire group makes a full revolution of the room, then falls to the floor. After they have performed the ceremony three times in succession, they pray. A special choir and orchestra accompany the ceremony and a simple flute, the ney, provides a low, eerie sound to signify a soul seeking union with the One Friend.

Under a dome of turquoise tile in the former *tekke* (monastery) of the dervishes (now the Mawlana Museum) is Mawlana's tomb, covered with a cloth of woven gold weighing ninety pounds. As a sign of respect, a Turkish son traditionally stands when his father enters a room, but according to myth, Mawlana's father's tomb rose when Mawlana was buried and now stands upright. The museum preserves the traditions of the dervishes with recorded ney music and displays of rugs and illuminated Korans. However, its most important display is the original illuminated manuscript of the *Mathnawi*.

With the coming of the Mongols, successors to Genghis Khan, in 1241, the balance of power shifted again. The Seljuks continued to rule, but as Mongol vassals. Survivors of the Byzantine Empire arose, took advantage of chaos and

famine brought on by the brutal Mongol invasions, and recaptured Constantinople in 1261. The new emperor, Michael VIII, solidified his position by marrying one of his daughters to the Mongol khan. Final Seljuk attempts to regain power came in 1276 when Keyhusrev III, with the help of the Egyptians, defeated the Mongols and took Konya. Soon, however, the coming of the Ottoman Turks finished the ambitions of both the Byzantines and the Seljuks. After the end of the Sultanate of Rum, Konya became the seat of the kingdom of Karaman. The Ottoman Murat I marched on Karaman with an army of Christian mercenaries, marking the first time a Moslem used Christians against other Moslems, and the first use of cannon and muskets by a Turkish army.

The founder of the Ottomans probably came to Anatolia in the general migration after the Battle of Manzikert. A mythical version of the founding says that Süleyman Şah left Persia, fleeing the Mongols in the middle of the thirteenth century, and leading a gazi army of "400 tents" came to Rum to serve the sultan. The sultan allowed him to keep any territory he grabbed and held. Under his grandson, Osman, many more loot-hungry gazis joined the tribe of the Osmanlis (Ottomans), giving Osman sufficient power to be named sultan. He was forced to pay tribute to the Mongols, but when Mongol power declined, his descendants established a powerful state centered at Bursa (Prusa).

The defeat of the Ottomans by the Tatars under Timur (Tamerlane) was merely a temporary setback that ended with the death of Timur in 1405. By the 1420s the Ottomans controlled all of the Byzantine Empire but Constantinople itself. In 1453, the Turks took Constantinople and renamed it İstanbul. All of Anatolia then belonged to the Turks.

After the rise of the Ottoman Empire, Konya declined until it was little more than ruins by the nineteenth century. It was only with the coming of the railroad at the end of the century that the town began its renaissance, becoming the vibrant town it is today. When the Allies dismantled the Ottoman Empire after World War I, the intended partition of Anatolia was never accomplished. Instead, the Allies recognized Turkish rights to most of the territory that makes up modern Turkey, and the Republic of Turkey was born in 1923, with Mustafa Kemal (Atatürk) as president.

Atatürk's attempt to outlaw the dervishes and his other reforms have had little effect on the importance of Konya as a religious center. Pilgrims still come to Mawlana's tomb and the dervish ceremony continues, and, although the practice is officially discouraged, many women wear veils. Konya remains a conservative Moslem town, observant of Islamic law.

Further Reading: From prehistoric Anatolia to modern Turkey in the 1990s, *A Traveller's History of Turkey,* by Richard Stoneman (Moreton-in-Marsh, Gloucestershire: Windrush, and Brooklyn, New York: Interlink, 1993) traces the history of Turkey in an informative, yet interesting and easy-to-read style. For details on the archaeological digs at Çatal Hüyük, the authoritative source is Chapter I, "Çatal Hüyük: The Neolithic Revolution" of *The Archaeology of Ancient Turkey* by James Mellaart (London: Bodley Head, and Totowa, New Jersey: Rowman and Littlefield, 1978), the archaeologist who discovered and excavated the site. Similar information, and details of life in the Near East (from the beginnings of time through the first century B.C.) can be found in the beautifully illustrated *The Cultural Atlas of Mesopotamia and the Ancient Near East,* by Michael Road (Oxford: Equinox, and New York: Facts on File, 1990). This book includes photographs of some artifacts found at Çatal Hüyük. *Turkey, the Rough Guide,* by Rosie Ayliffe, Marc Dubin, and John Gawthrop (London: Rough Guides, Ltd., 1994), a general travel guide to Turkey, provides further information on the dervish ceremonies and details on Seljuk architecture.

—Julie A. Miller

Kos (Dodecanese, Greece)

Location: The Greek city of Kos is located on the island of Kos, at 112 square miles one of the largest of the Dodecanese Islands situated off the coast of Asia Minor (Turkey). The capital of the island, Kos is situated on its northeastern coast.

Description: Kos, a thriving harbor city, was founded in 366 B.C. and has fallen under the rule of the Ptolemies of Egypt, the Romans, Venetians, and Turks. It is rich in Roman artifacts and is near the site of the Asclepieion, the medical school that grew up after Hippocrates' death in the mid–fourth century B.C.

Site Office: Greek National Tourist Organization
Information Desk
Akti Kountourioti
Kos, Dodecanese
Greece
(661) 37520

Excavations beneath the modern city of Kos have revealed human settlements there from the Minoan and Mycenaean periods. The island was colonized by peoples from the mainland of southwest Asia Minor and later by Dorians from mainland Greece. It remained relatively unimportant until the city of Kos was founded on the northwest tip of the island in 366 B.C.; its sudden significance at about this time owes largely to the island's most famous son and the father of medicine, Hippocrates, whose fame spread throughout the ancient world from the mid-fourth century on.

A native of the island, Hippocrates (c. 460–c. 377 B.C.) instituted a type of medical practice that contrasted sharply with the accepted methods of treatment and has endured into our own time. Hippocrates is the first known doctor to record case histories and other medical data in an empirical manner, and the first to document such illnesses as epilepsy. His writings and those of his followers form the *Hippocratic Corpus*, which includes such treatises as *Prognostic* and *Regimen in Acute Diseases*.

Hippocrates' influence as a teacher in Kos drew many disciples, but it was his reputation after he died that played an important role in the founding of the city in 366 B.C. Students and philosophers continued to arrive there, and around the mid–fourth century construction began on the famous Asclepieion, or Sanctuary of Asclepius (god of healing and medicine), two and one-half miles outside the city of Kos. Situated on the site of a temple to Apollo Kyparessios, the Asclepieion was built on three terraces connected by a grand staircase. There physicians following the guidelines of their great teacher established a medical school.

Kos quickly became an important maritime and trading center in the Aegean and eastern Mediterranean regions.

In 336 B.C. Alexander the Great occupied the island, which then passed to the Egyptian Ptolemies after Alexander's death in 323. Ptolemy II Philadelphus, who was later to gain maritime supremacy in the Mediterranean, was born on the island in 308 B.C. The Asclepieion enjoyed wide prosperity throughout its early years under Ptolemaic rule and continued to attract the interest of post-Hellenic rulers.

Throughout much of the Hellenistic period under Ptolemaic rule, the city of Kos prospered. Advances among Greek cities in the areas of law and individual liberty were felt on the island, and it was there as well as in cities on the Greek mainland that nonpolitical clubs and associations began to appear after 300 B.C. These associations, usually private, were fairly small, with no more than 100 members. Often they were formed for the sole purpose of religious worship, but many appear to have been primarily social in intent. The clubs provided their members with a few privileges; when a member died, for instance, the club would sometimes pay for his burial, but that was about the extent of its largesse because most clubs were constantly in need of money. The religious associations would meet on designated days and rent out their facilities during the rest of the year. Associations bearing the name of the trade practiced by its members began to form; the most prominent in Kos was comprised of physicians. These associations are considered a precursor of the professional trade guilds that emerged during the Roman Empire.

Kos flourished until at least 88 B.C. The growth of international trade, new clubs, luxuries such as silk clothing, and improved technology such as better constructed private homes, municipal roads, and facilities all enhanced the quality of life. As evidence of this growing prosperity, certain clubs were said to honor new members by crowning them with crowns of gold instead of the traditional crown of leaves. Despite the riches, however, the poor within the city were not well provided for; individuals were expected to fend for themselves. While wealthy citizens would often make donations to the city for the purpose of building bridges or other public facilities, no organized means of providing aid to the poor existed.

The wealth of the city of Kos during the Hellenistic period was also reflected in the abundance of festivals held there, typically characterized by dancing and music. Over time the festivals became very elaborate and costly city functions, with professional performers commanding high fees. Additionally, the production and trading of raisins and wines made Kos famous. The importation of silkworms also influenced an emerging silk industry, which had a great effect on the economic prosperity of the city. Silk from Kos became highly sought after throughout the Mediterranean region.

The Ptolemies, whose influence on Kos had diminished between 88 B.C. and 30 B.C., did not oppose the Roman

Mandraki Harbor at Kos
Photo courtesy of Greek National Tourist Organization

Empire due to their own internal problems. During the second century A.D., Kos saw much of its prosperity decrease under Roman rule.

After a Roman civil war in 49 B.C. between political party leaders Caesar and Pompey, special privileges were granted by the victorious Caesar to certain cities, such as Kos, that had supported him against Pompey. Between 41 and 33 B.C., the city had a popular Roman ruler named Nicias, who brought prosperity to the city, although not on a level with that experienced during the Hellenistic era. After the first century, the Romans added baths and statuary to the Asclepieion.

The city of Kos prospered during the Byzantine era, becoming the seat of a bishop. Several Christian basilicas dating from the fifth and sixth centuries have been unearthed there. In 554, however, an earthquake destroyed the Asclepieon and many of the buildings in the city center. (Additional damage may also have come during raids from Asia Minor.)

The Treaty of A.D. 992 between the Greeks and the Venetians formed a union that influenced trade through Kos. The Greek government, wanting to increase trade and shipping to meet its financial needs, granted special consideration to Venetian merchants. Venetians gained the opportunity to increase trade and to smuggle items that were banned by the Byzantines, who then dominated the region. They also received reductions on import and export duties as well as certain guarantees against the powerful Byzantines. The two governments entered into several additional treaties after their initial treaty in 992. Because Kos was a gateway to Asia Minor, it was an essential trading location that saw much Venetian commerce. Eventually the Venetians became too ambitious for the Greeks, who retaliated by granting the same trading and shipping privileges to the Genoese as they had granted to the Venetians; Kos then witnessed a great influx of Genoese trading.

By the eleventh century, however, Kos was under constant threat from the Saracens as well as from pirate raids. The island came into the hands of the Knights of the Order of St. John around 1306, and they occupied the city of Kos in 1315. During their tenure the Knights of St. John took stone from the ruins of the Asclepieion site for the construction of buildings in Kos.

In 1522 the Turks, after raiding the island in 1457 and 1477, took control of Kos and the rest of the Dodecanese, maintaining control until 1912.

While under Turkish rule, the inhabitants of the Dodecanese were given special privileges and treated fairly well, although that was not necessarily the case for the rest of

Turkish-occupied Greece, where corruption by Turkish rulers was fairly widespread. A Turkish judge resided in Kos, but the rest of the islanders were responsible for their own affairs and electing their own government. Kos was able to collect local taxes, duties, and funds for schools, doctors, and even a municipal pharmacy. During the Greek War of Independence, Kos suffered financially, but it enjoyed many of its privileges until 1867.

During the Cretan insurrection in 1868, Turkey attempted to assimilate Kos and the other Dodecanese inhabitants into its government and kingdom. Kos maintained its fiscal autonomy until the Turkish revolution in 1908. In 1909, because the Turks wanted to establish uniformity in all parts of the Turkish kingdom, Kos's privileges were eradicated, the Turkish language was imposed, and conscription was enforced. The latter meant starvation for the families who needed all of their members to assist in sponge-fishing in order to provide income. The harsh measures imposed by the Turkish government were met with much protest by the residents of Kos. In 1912, the Turkish government attempted to soften its policies by imposing a one-tax-for-all measure.

The city of Kos and its island were brought under the control of the Italians while Italy was at war with Turkey over Tripolitana during 1912 and 1913. Under the Treaty of Laussanne, Kos and the rest of Dodecanese were to remain under Italian occupation until Turkey agreed to withdraw from Tripolitana. The Dodecanese were viewed as a convenient foothold to North Africa and to the Aegean Sea, as well as a strategic location for military bases to control the Mediterranean.

At first, Kos viewed the Italians as their liberators from Turkish domination but soon realized that would not be the case. Kos took part in the Dodecanese rally to become reunited with Greece; however, Italian occupation dampened their efforts. In 1912, a general assembly consisting of all the Dodecanese as well as the city of Kos drew up a declaration asking that they be reunited with Greece. In the interim, the islands wanted to be named the "Aegean Commonwealth." The Italian government retaliated by forbidding the display of the Greek flag and the holding of public gatherings. From 1912 until the beginning of World War II, the population of Kos, along with the rest of the Dodecanese Islands, decreased by 50 percent due to inhabitants leaving either voluntarily or being expelled by the Italians. In 1947, Kos was given back to Greece, thus officially ending the Italian occupation.

Greece and Turkey continue to vie for control of the Dodecanese Islands. Turkey argues that the islands cannot belong to Greece because of the intervening high seas between them and the Greek mainland, and further argues that Greece wants complete control of the Aegean. Turkey wants to take the matter to international courts in hopes of having boundaries drawn, but Greece adamantly objects to the idea. Although the city of Kos has had a long history of belonging to many peoples and being dominated by many cultures, it views itself as a Greek city and part of Greek society.

Today, Kos is a well-preserved historical city only slightly affected by tourism. Its principal monuments include the fifteenth-century Castle of the Knights, begun in 1450 by the Venetian Fantino Guerni, then governor of Kos, and completed in 1478; the Archaeological Museum with ancient artifacts from the Hellenistic and Roman eras; and the fourth-century B.C. Asclepieion.

Further Reading: Important sources on the history of Kos and the Dodecanese include *History of Greece from the Beginnings to the Byzantine Era* by Hermann Bengston (Ottawa: University of Ottawa Press, 1988); *Hellenistic Civilisation* by Sir William Tarn (London: Arnold, 1927; third edition, London: Arnold, 1952); *A History of the Greek People 1821–1921* by William Miller (London: Methuen, and New York: Dutton, 1922); and *A Short History of Modern Greece 1821–1945* by Edward S. Forster (London: Methuen, 1941; second edition, 1946).

—Jessica M. Bowen

Kotor (Montenegro, Federal Republic of Yugoslavia)

Location: Southern Dalmatian coast of the former Yugoslavia, at the southeastern end of the Gulf of Kotor (Boka Kotorska); bounded on the west by the Adriatic Sea, east by Mount Lovcen (the highest of the Njegusi Mountains); approximately fifty miles south of Dubrovnik and thirty miles north of the Albanian border.

Description: A jumble of narrow streets and small buildings jammed onto a rock that juts into the Bay of Kotor like a finger, Kotor is one of the most scenic cities in Europe. It has been damaged several times by earthquakes (most recently in 1979), but most of its medieval streets and structures remain much as they were in the thirteenth century. The town is encircled by city walls that were begun in the late Byzantine era, then rebuilt by Venice during the Renaissance; notable structures in the town include the Church of St. Luke, a small box-like building in a tiny square, which is among the first examples of Serbian architecture.

The most striking visual features of Kotor are its medieval architecture and its dramatic isolation. The city, whose appearance has changed little since the Middle Ages, sits in the shadow of Mount Lovcen, the highest peak of the Njegusi range. Kotor is encircled by walls, and the only access to the city is through three gates or by sea. Over the centuries, Kotor has prospered from maritime trade and also has suffered from foreign invasions and natural disasters—the latter being primarily earthquakes, with major ones occurring in 1563, 1657, and 1979. After the 1979 quake, the town was reopened to visitors only gradually. The quakes are due to a thrust fault (so called because it is formed when rock slabs press against each other, thrusting one side over the other) that runs in a line from north Africa, and turns north through Italy, east, and then south down the Adriatic coast, directly under Kotor.

If the rock under the Kotor streets is shaky, the political affairs are shakier still. But this is nothing new in this land. Through the centuries, Kotor has been preserved by three natural advantages: its valuable harbor, its fortified position, and its isolation. But the town's advantages have also invited foreign domination, and on this coast, few foreign powers have waited for an invitation to invade.

The site of Kotor was occupied by Greeks, perhaps as early as 800 B.C., who were able to displace the native Illyrians along the coast, if nowhere else. Later, the Romans supplanted Greece, establishing a presence along the coast in 227 B.C.; unlike the Greeks, the Romans made their way inland, conquering territory they came to call Illyricum. The coast, by differentiation, was called Dalmatia, as if it were a separate land.

In contrast to many other towns on the Dalmatian coast, Kotor has retained few traces of Roman rule. The town, which the Greeks called Acurion, was renamed Acruvium by the Romans, but judging from the amount of artifacts that have been found, the more important Roman town on this part of the Dalmatian coast was Budva, a few miles south. The coastal territory provided the Romans with a gateway to trade in eastern Europe and a stepping-stone to conquests inland in Illyria and Macedonia. Eventually, the area became attractive to wealthy merchants, and country villas were built by the sea, like an ancient version of the Riviera. The greatest impact on Kotor's development, however, came after the division of the Roman Empire into eastern and western portions in the fourth century A.D. Kotor came under the rule of the eastern, or Byzantine, realm, but it took several centuries for Byzantine religion and culture to pervade the town.

The Roman Catholic Church was well established in the area by the eighth century, ahead of the Eastern Orthodox Church headquartered in Constantinople. In 809, the town built its first Catholic cathedral (dedicated to St. Tryphon) and also founded its sailors guild. The still-extant Guild of the Boka Sailors is one of the oldest in the world—its bylaws are among the longest surviving codes of law on the sea.

Byzantine influence finally took hold—strongly—in the latter half of the ninth century. In 864, two Eastern Orthodox missionaries named Cyril and Methodius traveled from Constantinople into the rugged Serbian countryside. Their mission was successful on two fronts: the Orthodox Church was solidly established in Kotor and its environs, and the empire was politically strengthened in an area that could serve to contain the Avars. This warring tribe had been moving south from what would become modern Hungary to parts of the Adriatic coast that were uncomfortably close to Constantinople. In 867, Kotor (called Dekateron or Decadron under Byzantine rule) was invaded by Saracens, who burned and looted the town, Constantinople responded by building a new system of city walls that came to extend far beyond the boundaries of the town. Byzantine rule turned out to be home rule, of a sort, for Kotor, which was granted independent commune status under the empire. Kotor had its own senate and councils of nobles, and minted its own coins. The town was partially shielded from incursions by the difficult mountain terrain, the walls, and natural protection of the harbor. Its people were sailors and shipbuilders, which made them different from their mountain neighbors. The influence of the Roman and Byzantine empires was stronger in Kotor than in the Slavicized mountains. But frequent battles inland, above the port, caused many of those neighbors to move in.

A view of Kotor
Photo courtesy of The Chicago Public Library

The walls were enough to stop invasions from the mountains for as long as Byzantium had the power to bolster them. That power had mostly evaporated by the early part of the eleventh century, when the Bulgar king Samuel, who had established his capital south of Kotor in Macedonia, swept into town and established south Slavic rule in Kotor for the first time in 1002. Neither this shift of power, nor the next, in 1102, when the Hungarian king Könyves Kálmán became the ruling monarch over the southern Dalmatian coast, had as much effect on Kotor as the establishment of an independent Serbian state nearby in 1159.

Until then, the Serbs for centuries had been nominal subjects of the Byzantine Empire, practicing the Eastern Orthodox religion and paying tribute to Constantinople, but they had to a great degree managed their own political affairs, with tribal leaders being the chief authorities. In the mid-twelfth century, under Grand Zupan (clan leader) Stephen Nemanja, the Serbs renounced their allegiance to the empire, made a series of military conquests, and formed their independent state. In 1186 Kotor was incorporated into the Serbian realm. In 1196, Stephen Nemanja's son, Stephen Nemanja II, was crowned the first king of Serbia by his younger brother, Sava (who later was canonized an Orthodox saint).

Though Stephen's capital was at Ras, more than 100 miles away, his coronation provided the impetus for the building of a new church in Kotor. The Church of St. Luke, begun in 1196, was one of the first buildings in a truly Serbian style of architecture. A small box of a building with a central dome that looks like a lopsided hilltop, it is much larger on the inside than an outside observer would expect. Over the next hundred years, Serbian architects built on the foundation of the box and dome, a native form that mimicked the surrounding mountains by stressing the upward thrust of the dome over the room below it. Walking inside is, in a sense, like walking inside a mountain cavern.

During the period of Serbian rule, Kotor enjoyed wealth and importance it has never attained since. Many of the buildings lining the narrow streets and facing the tiny squares were built during this time. Kotor was the main port of the Serbian kingdom, and a major commercial rival of both Ragusa (later Dubrovnik) and Venice. Perhaps it was the jealousies of local people during this period of prosperity that later caused other Yugoslavians to look upon the seafaring people of Kotor as too clever to be trusted, and as argumentative as lawyers. One saying quoted by an ethnologist centuries later was "They are quick-witted enough to shoe a flea and to split a hair into nine strands." As to competition with the Venetians, Venice was quick with a remedy.

The kingdom of Serbia reached its height in the middle of the fourteenth century under Stephen Dušan, who ruled a portion of central Europe that extended as far south as central Greece, but several states broke off after Dušan's death in 1355. The kingdom was also threatened by foreign invaders, including Ottoman Turks and Venetians. Independent Serbia was finished in all but name on June 28, 1389, when Serbian Prince Lazar Hrebeljanovic was defeated at the Battle of Kosovo by Turkish troops led by Sultan Murad I. Kotor remained Serbian, but by that time, the strategically located city had been troubled by foreign powers for at least 19 years. The city had asked Hungary/Croatia for protection from Venice in 1370; in 1378 it was briefly taken by Venetians commanded by Vettore Pisani. Kotor asked the king of Bosnia, Tvrtko I, for help in 1385, but Bosnia had its own problems with invaders. In 1391, Kotor voted to go it alone; the the city proclaimed itself an independent republic, like Dubrovnik up the coast, and like Venice across the Adriatic.

Unlike the situation of Venice, however, Kotor's independence was constantly threatened by advancing Turkish armies. The lands surrounding the new independent republic were rapidly diminished. Even tributary towns on the bay, such as Perast, were actively assisting the Venetians against Kotor. Weighing the pros and cons of subjugation by Turkey or by Venice, Christian Kotor's choice was clear. In 1420, Kotor asked Venice for protection. Venice responded by rebuilding and extending the city walls, introducing its culture to Kotor, and establishing a shipping base in the city. Once Venice's rival, Kotor, known to the Italians as Cattaro, had become its vassal.

Venice established ports all over the Dalmatian coast, but where it did not or could not, enemies flourished. While Venice tried to stay out of the conflicts between Christianity and Islam (unsuccessfully, most of the time), religious affairs were only a pretense for plunder by the pirates of the Adriatic. Caught between Turkish pirates (Barbary corsairs) and Christian pirates (from the "pirate city" of Segna on the northern Dalmatian coast), Venetian ships (many of them built in the shipyards at Kotor) were often captured by the smaller, faster, pirate craft, and emptied. By the 1570s, with a Turkish presence already established on the north shore of the Bay of Kotor, having withstood at least one Turkish siege (in 1539), and still digging out from under a 1563 earthquake, Kotor was bedeviled on both land and sea. Friendly ships were being captured by Turkish pirates at a rate sometimes as high as twenty-five a month. Then, in 1572, the plague came to Kotor.

Trials such as these have decimated other cities; they certainly did not do Kotor any good. Yet, despite the success of the Turks elsewhere in Serbia, the city remained outside the Turkish sphere. Part of this is due to the harsh battles put up by the people up in the mountains in the neighboring kingdom of Montenegro, established after the demise of the Serbian kingdom. While most of the rest of Serbia and nearby states were subjugated by empires—first by the Ottoman Empire, then by Austria, France, England, and back to Austria—the little fortress state up on the Black Mountain retained its independence.

Montenegro has been called the "Serbian Sparta," but the people are descended from a mixture of nationalities (like the rest of Yugoslavia and the former Yugoslavia, whose populations, paradoxically, are as strongly insistent on their hereditary differences as any country in Europe). To this day, traditional Montenegrin dress includes a black band, a symbol of mourning that pays tribute to the defeat at the Battle of Kosovo.

Montenegro was first linked with Kotor in the early fourteenth century, as both were part of the Serbian kingdom. If Montenegro were located at sea level, its border would not be much more than a stone's throw from the city. But because of the hard climb up the rocky face of the mountain, Montenegro remained a world away. Even today, Kotor natives say that snow in Montenegro stops at the old border, halfway down the mountain.

Other borders in the region frequently shifted, however, and the upheavals kept a steady flow of Serbs coming to Kotor. Venice was Catholic and many of the Serbs were Orthodox, but the two faiths managed to coexist. Their common enemy, Islam, however, was continually denied entrance. Another Turkish siege in 1657 failed to break Kotor. In 1667, another earthquake shook the city. The city rebuilt.

Both Venice and Kotor were in the midst of decline. Austrian armies under the House of Habsburg drove the Turks out of central Europe at the end of the century. By the time the Habsburg navy docked in the bay in 1797, Venice's power was a memory and little fight was left in Kotor. These factors did not stop other nations from fighting in, and over Kotor. Over the next seventeen years, three other nations laid claim to the port.

First France, in the midst of Napoleonic expansion, took control of the entire Yugoslav hinterland. France occupied Kotor from 1807 to 1813. Then, very briefly, England took over. Montenegro claimed Kotor for its own in the same year, and from late 1813 into 1814, managed to have access to the sea for the first time since Stephen Dušan ruled the kingdom of Serbia. The access was short-lived. Austria returned and united the coast for the first time in centuries, by joining the old republic of Dubrovnik together with Dalmatia and the Bay of Kotor—all under Austrian rule.

The events that led up to the assassination of Austrian archduke Francis Ferdinand by a Serbian nationalist on June 28, 1914 (525 years to the day after the Battle of Kosovo) were to affect the entire world, and involved in no small part the ages-old problem of access to trade on the Adriatic. With Austria controlling the coast, the Habsburgs also controlled access to overseas markets. South Slavs and Serbian nationalists rebelled against the arrangements, which were draining their resources to Austria's benefit, in the "Pig War" in the late nineteenth century (so called because Austria levied a customs embargo on livestock exports). The embargo mostly served to strengthen Serbian nationalism. In 1912 Montenegro started the First Balkan War. Two years later, the world went to war.

In Kotor, the Austrians had done one thing no other conquering state had done—they built a military road up the face of the mountain to Montenegro. This road is cut into the hard stone of the mountain face, zigging and zagging among the boulders, and has been called one of the most dramatic drives in the world. It was used by Austria in World War I in 1918 to briefly subdue Montenegro at a time when the Austrian hold on Kotor had become more precarious than ever before.

In February of that year, the Dalmatian sailors manning the Austrian ships anchored in the bay heard that the Russian czar had been driven from power. The sailors tried to do the same with their Austrian captains. For a short time, the red flag of revolution flew on Austrian ships, but the mutiny was crushed, and the ringleaders—three of them native to the bay—were shot. From then on, the Austrian navy was hampered in the area, doubtful of the loyalty of its crews. With the Allied victory, both Montenegro and Kotor became joined again in the new state of Yugoslavia.

Modern Yugoslavia was not created by the Treaty of Versailles, nor were any of the multitude of border disputes that had arisen over centuries settled by naming the national entity Yugoslavia. The treaty has been called a recognition of something that had already been established, with a mix of confusion thrown in, because large parts of Croatia were thrown under the rule of Italy, Yugoslavia's worst enemy. Between the wars, Yugoslavia was a mess. It had no effective administration, and several unhappy minorities. Its king, Alexander, was assassinated in 1934; his successor, Prince Paul, favored the Croats over the Serbs. Yugoslavia tried to stay out of World War II, but this fragile conglomeration of bickering states was destroyed by Hitler, whose forces invaded on April 6, 1941. Germany began partitioning Yugoslavia on April 17.

During the war, the inhabitants of Yugoslavia fought the Germans, and each other, as never before. The Soviet Red Army eventually liberated the land, and communists, led by Josip Tito, declared the country a Federal People's Republic in 1945. In 1948, Russia invaded Yugoslavia, but this time Yugoslavia's hard-won independence would not be lost. Yugoslavia thereafter had an independence of Russia unknown by any other Soviet satellite in Europe. Tito's communism seemed to be working until the 1960s, when the country experienced severe economic difficulties and came to the brink of civil war.

There is no evidence that the civil war that has ripped through most of the former Yugoslavia in the 1990s has had much of an effect on Kotor. Contraband and gasoline is smuggled to the Serbs from Albania through Montenegro; it is possible that some items from abroad used in the ongoing civil war are also going through Kotor, but this is speculation.

Kotor, a town now too small for a post office, is joined to the land, but is more like an island, isolated between the sea and the mountains, a narrow rock that its people have been clinging to for more than 2,000 years. Nothing has shaken them off yet.

Further Reading: *Serbian Legacy* by Cecil Stewart (London: Allen and Unwin, and New York: Harcourt Brace, 1959) is a fine blend of history, both political and architectural, and personal travelogue by the author, who writes both authoritatively and well. It is also, unfortunately, out of print. *The Yugoslav Coast* by Lovett F. Edwards (New York: Hastings House, and London: Batsford, 1974) is very much the same sort of book, only with less emphasis on architecture and more on travelogue. This, too, is well written by an author who knows his subject thoroughly. Both include sections of Kotor and the Bay of Kotor, and both provide excellent descriptions. *The National Question in Yugoslavia: Origins, History, Politics* by Ivo Banac (Ithaca, New York, and London: Cornell University Press, 1984) is an exhaustive, meticulous attempt to sort out the various ethnicities in Yugoslavia.

—Jeffrey Felshman

León (León, Spain)

Location: On the Bernesga and Torio Rivers in northwestern Spain.

Description: Former capital of the medieval kingdom of León, now famous for its Gothic cathedral and other church buildings.

Site Office: Tourist Information Office
Plaza de la Regla, 3
León, León
Spain
(987) 23 70 82

Like so many cities in western Europe, León began as a fortress of the Roman Empire, becoming the headquarters of the Seventh Legion under Emperor Servius Sulpicius Galba in A.D. 68, approximately 150 years after the Romans had first landed on the Mediterranean shores of the Iberian Peninsula. The city's modern name may well derive from the word "legion." For the Romans, northern Spain was a frontier zone: difficult to defend, underpopulated, and far from their major cities. Only 200 years after its foundation the settlement was overrun by invaders (probably the Germanic people known as the Suevi), and in the fifth century the region was again invaded by Suevi, Alans, and Vandals, as the Roman Empire contracted and collapsed. Little is known of life in León under the Romans or under the invaders, although remnants of the Roman period are displayed in the city's archaeological museum, located in a medieval building that used to be the chapter house of the Monasterio de San Marcos.

In 586 León was conquered yet again, this time by the Visigoths, who had begun their invasion of the Iberian Peninsula eighteen years before. It is possible that the city's modern name derives from that of Visigoth leader Leovigild, rather than from the Roman "legion." Their rule was marked by frequent rebellions and disputes among members of the royal family and powerful nobles. It lasted until 711, when a Moslem army of Arabs and Berbers swept into Iberia from North Africa during yet another civil war and drove the Christian rulers into the Asturias, the region to the north. After being repulsed from France in 732, the Moslems, known to medieval Europeans as the Moors (in reference to Morocco), established a political structure throughout the peninsula. By tolerating Christians and Jews, they averted discontent for centuries. But they were unable to dislodge the Christian state established in the Asturias, whose forces gradually moved southward. In 740 the Christian king, Alfonso I, seized León from the Moslems and held it for some years; in 792 it became, as it has remained, the seat of a Catholic bishop.

The city was briefly lost to the Moslems again but was retaken by the Asturians, led by King Ordoño I, in 850. In 914 the city was designated the capital of a small but growing Christian kingdom, later named León, by King Ordoño II, replacing the former headquarters of the Christian forces at Oviedo. The tomb of Ordoño II, who died in 923, is inside León's cathedral, and portions of the church that he ordered to be built in the city survive in the fabric of the Church of San Salvador de Palaz del Rey (the Holy Savior of the King's Palace). From around this time the city was symbolized by the figure of a lion rampant, now part of the coat of arms of the kings of Spain, which forms the centerpiece of the national flag.

The Moslem caliphate, based at Córdoba in the southwest, did not give up its northern frontier towns without a fight. León was badly damaged during an attack by Moslem forces led by the chief minister and general al-Manṣūr (or Almanzor) in 996. Six years later the Moslems were driven out, and King Alfonso V initiated the reconstruction of the city's walls and main buildings.

In 1035 the province of Castile broke away from the kingdom of León to become a separate state, under King Sancho III Garcés of Navarre, with its capital at Burgos. Only two years later the two were reunited by Sancho's son Ferdinand I, who married a princess of León and then defeated her brother Bermudo III, his predecessor as king of León, in battle. It was Ferdinand I of Castile-León who began the building of the Church of San Isidoro el Real (St. Isidore the Royal), which was to house the remains of that saint, brought from Seville in the seventh century; of San Vicente (St. Vincent); and of Ferdinand himself and other rulers of León.

The narthex, the section of the church that contains the royal tombs, is known as the Pantéon (an oddly pagan term for a Christian building) and consists of two Romanesque chambers, on the western side of the church, completed in 1063. The statues on the entrance to the tombs may be the first to have been erected anywhere in Spain since Roman times; the interior, however, harks back to the traditions of the Visigoths, with leaves and vines carved into the columns and wall paintings depicting the Last Judgment and symbols for the twelve months. The rest of the church in its present form, however, dates mainly from about 100 years later, having been rebuilt on the orders of King Alfonso VII, starting in 1149. By then the influence of the Moslem culture that the kings were dedicated to destroying had begun to affect architecture, including the so-called Moorish arches, shaped like horseshoes, which can be seen throughout the church.

By then, also, King Alfonso VI of Castile-León, having once more reunited the realm after his father Ferdinand I's disastrous division of it among his five children, had in 1082 taken the title Emperor of Hispania, reflecting his successful reconquest of central Spain. León was now one of the leading cities of a steadily expanding kingdom, in which the growth of trade in both food and luxury goods encouraged further growth

The Pantéon at the Church of San Isidoro el Real
Photo courtesy of Instituto Nacional de Promocion del Turismo

in the city's population. However, expansion brought with it discontent, both among the merchants of León and other cities, who resented the taxes and subsidies demanded by the nobles, and among the peasants, who did not share in the new wealth. After Alfonso VI's death in 1109, the accession of his daughter Urraca, the first queen of León, and her marriage to the king of Navarre provided pretexts for these resentments to erupt into yet another civil war, which lasted seven years, and during which the city of León was in chaos. The victory of the nobility in 1117 was followed by mass executions of citizens, and León was relatively quiescent from then on.

The Empire of Hispania proved to be short-lived, and in 1157 Castile and León were separated once again. In 1168 the Leonese royal court sponsored the creation of the Order of the Knights of Santiago (St. James), a group of military monks prominent as warriors against the Moslems, as landowners, and as guardians of the pilgrims who passed through

the city on their way to the tomb of St. James at Santiago de Compostela. The scallop shell that symbolized the pilgrims still appears in the decorations of the Monasterio de San Marcos (monastery of St. Mark), which the knights founded in León between 1168 and 1173.

It was also in León that the first meeting of the Cortes, the representative assembly of the kingdom, took place, in 1188, under King Alfonso IX, who needed to raise new funds to continue his war against King Ferdinand III of Castile, his own son. Although this was among the earliest parliamentary bodies in Europe, it would be misleading to call it (as some have) "democratic." As in Aragon, England, and other medieval kingdoms with similar assemblies, its members were handpicked by the monarchs to represent the three groups without which government would have been impossible: the landowning nobility, the church authorities, and, with greater weight because of their financial power, the leading merchants

Ark containing the remains of San Isidoro el Real
Photo courtesy of Instituto Nacional de Promocion del Turismo

of the cities. Nevertheless, the ordinance that this Cortes passed, proclaiming the (theoretical) right of every householder to kill anyone, even the monarch, who trespassed on his home, has played much the same symbolic role in Spanish history as the Magna Carta of 1215 in British history.

After several decades during which the progress of the Christian Reconquista (Reconquest) of the peninsula was repeatedly disrupted by disputes among the five Christian kingdoms—León, Castile, Aragon, Navarre, and the former Leonese dependency of Portugal—the first two of these were united once again under Ferdinand III of Castile, whose hostile father Alfonso IX of León died in 1230. From now on the joint kingdom would again be known as Castile-León, and increasingly just as Castile.

By the time of Ferdinand's death, in 1252, the Christians had taken Córdoba, the former Moslem capital, and Seville and had achieved control of all of modern Spain and

Portugal apart from Granada in the southwest. Their victories were marked by the building of new cathedrals in Burgos, Toledo, and León. The new cathedral in León, called Santa María de Regla, was built between 1258 and 1303 on the site of three earlier cathedrals and in a style that shows the influence of French church architecture as seen at Chartres or Reims on its builders, Maestro Enrique and Juan Pérez.

The cathedral houses the remains of San Froilán, the patron saint of the city, which are kept in a silver reliquary on its high altar. Its 125 distinctive stained-glass windows, covering a total of about 19,375 square feet, were installed at various dates between the thirteenth and twentieth centuries. The windows are dominated by yellow, red, and gold—the royal colors of the old kingdom—and give its interior a brightness that contrasts strongly with the relative gloom of most medieval church buildings in Spain. Other additions to the original design include choir stalls from the fifteenth century; a spire on the south tower, designed by Joosken van Utrecht and completed in 1472; and a staircase and a trascoro (screen), both designed by Juan de Badajoz in the sixteenth century.

From the time of Ferdinand III, the kings of Castile and León held court more often at Burgos than at León, which underwent a relative decline. Nevertheless it continued to be a stopping point on the pilgrimage route to Compostela, and it remained important enough for King Alfonso XI to order another rebuilding of the city walls, still following the line of those built by the Romans and under Alfonso V. It is these that partially survive today as a boundary for the historic center around the cathedral, although most of the bastions have been demolished over the centuries.

León was largely unaffected by the intermittent civil wars that marked Castile-León's history from the death of Ferdinand III onward, as the nobles resisted the attempts of the monarchs and their ministers to exert control over them. Between 1476 and 1479, however, Queen Isabella I of Castile-León succeeded in repelling an invasion by the Portuguese and created a joint monarchy with her husband, King Ferdinand II of Aragon. Their successful efforts to impose a centralized administration on their subjects included suppressing the excessively independent military orders. Thus in 1493 (one year after the final defeat of the Moslems at Granada and the unification of what is now Spain), Ferdinand became grand master of the Knights of Santiago, and their property, in León and elsewhere, passed into the hands of the royal family.

In 1516 Ferdinand was succeeded as king of Spain and as grand master of the order by his grandson Charles I, who was to become Holy Roman Emperor Charles V. During the Revolt of the Communidades, the uprising by most of the cities of Castile-León against Charles's attempts to extort a new *servicio* (subsidy) from them in 1519–20, León was one of the few cities whose delegates voted in favor of the subsidy. It was perhaps partly in recognition of this loyalty during the first test of his rule that Charles later placed his gigantic and complex coat of arms on the roof of the Monasterio de San Marcos when its reconstruction, which had begun under Ferdinand II in 1513, was completed in 1549. Below his shield the building's front is decorated with a series of busts representing numerous heroic personages admired at the time, ranging from the Roman emperor Augustus to Queen Isabella herself. Charles's reign, which ended with his abdication in 1556, also saw the construction of the Church of San Isidoro, next to the monastery, with a facade decorated with sculpted scallops; and of the arcades that still surround the Plaza del Mercado, the site of the city's main open-air market until recent times, laid out in the district that had traditionally been inhabited by Jewish residents until their expulsion from Spain by Ferdinand and Isabella. The revival of León's fortunes in this period is also indicated by the building of the city hall and the ornate Palacio de Los Guzmanes, which both stand on the Plaza de San Marcelo. The city's most famous resident in these early years of the newly unified Spain was Juan Ponce de León, who led the Spanish expedition that discovered Florida and claimed it for Ferdinand and Isabella in 1513. He and his family, which had been prominent in the city for centuries, are commemorated by the Torre de los Ponces, a medieval tower that stands to the south of the cathedral.

In the seventeenth and eighteenth centuries León, once a royal capital but now a small city far from the center of events, played little part in the history of Spain, which declined from being the most powerful state in western Europe to being one of its poorest and least influential. In 1808, during the French invasion of Spain, the Church of San Isidoro was attacked and damaged; many of the medieval books in the church's library were destroyed, and the twenty-three tombs inside the Pantéon have remained unrepaired ever since. Later in the century, as the Spanish economy developed, León prospered too. Its cathedral was restored between 1868 and 1900, and one of its most notable buildings, the Casa de Botines, designed by the famous Catalan architect Antonio Gaudiy Cornet, was completed in 1894.

In 1920 the city still had only about 22,000 inhabitants. Its sixfold growth since then is mainly attributable to its new status as the headquarters of the companies created to exploit the coal and iron discovered to the north of the city. In 1934 there was unrest among the coal miners, in sympathy with the major uprising by the miners of Asturias, which was suppressed by government troops led by General Francisco Franco. When Franco initiated the Spanish Civil War with his own uprising, starting in Morocco on July 17, 1936, his supporters among the army garrison stationed in León seized control of the city just two days later. The city was thus spared from attack or bombardment both during the Civil War and World War II (during which Spain was neutral) and continued to grow to its current population of more than 144,000.

In 1978, following the death of Franco in 1975 and the return of parliamentary democracy after nearly forty years of dictatorship, León became part of the autonomous community, or region, of Castile-León (not to be confused with the much larger medieval kingdom of that name). The Monasterio de San Marcos, declared a national monument under Franco, is now a luxury hotel, although its church and its museum are separately managed. These buildings, the cathedral, the other churches, and the palaces in the historic quarter of the city still evoke the long-gone glories of the medieval kingdom of León. Their beauty and their historic importance deserve to be appreciated even by those who recall, uneasily, that the Reconquista that justified and paid for their construction also involved the suppression of the Moslems and Jews of Spain and later inspired Franco's Fascist regime.

Further Reading: The background of the golden age of León is comprehensively and fascinatingly narrated in a recent book by Bernard F. Reilly, *The Medieval Spains* (Cambridge and New York: Cambridge University Press, 1993), which is refreshingly free from the hostility to Islam that has been characteristic of too many works on Spanish history.

—Patrick Heenan

Lucca (Lucca, Italy)

Location: In the Tuscany region and Lucca province, on the left bank of the River Serchio, fourteen miles northeast of Pisa.

Description: Once a Roman colony and site of the formation of the First Triumvirate of Caesar, Crassus, and Pompey; in the Middle Ages, became the seat of the Margravate of Tuscany and was involved in many struggles between the papacy and the Holy Roman Empire. Noted for medieval and Renaissance art and architecture; birthplace of the architect Matteo Civitali and the composer Giacomo Puccini.

Site Office: A.P.T. (Azienda di Promozione Turistica)
Piazza Guidiccioni, 2
CAP 55100 Lucca, Lucca
Italy
(583) 491205

Lucca, from an Etruscan word meaning "city in the midst of a wood," was a minor Etruscan city, falling in 584 B.C. to the possession of the Ligurians, who controlled it for nearly three and one-half centuries, at which time it fell to the Romans after a prolonged siege. Lucca was declared a Roman colony in 177 B.C. Julius Caesar was proconsul in Lucca for a time, and the city was the site of a conference that resulted in the formation of the First Triumvirate of Caesar, Crassus, and Pompey in 60 B.C. The Romans were impressed with the military abilities of the city's residents and recruited many of them for service in the army. Much construction was carried out in Lucca during the period of Roman rule, and the city's residents are said to have been the first in Tuscany to convert to Christianity.

After the barbarian king Odoacer deposed Western Roman Emperor Romulus Augustulus in 476, Lucca lost one-third of its territory. Theodoric, king of the Ostrogoths, murdered Odoacer in 493. After the death of Theodoric, Byzantine Emperor Justinian I sent Belisarius and later Narses to conquer Italy. When Teias, last of the Ostrogothic kings, fell at Sarno, most of Etruria surrendered, except for Lucca. The city held out against a long siege by Narses until it, too, succumbed and became part of the empire. The barbaric wars weakened the territories of northern Italy, thus enabling the Roman Catholic Church to assume an increasingly important role in economic and civil affairs.

In 568, the Lombards, a northern tribe gifted in the skills of warfare and little else, invaded northern Italy. One account places their locus of power at Pavia near Milan; another at Lucca. Although the Lombards are considered the least civilized people to ever overtake Italy, they did not impose their culture on others. Instead, they adopted much of the local culture: their leaders converted to Christianity, they took on Roman styles of dress, and they absorbed many Roman legal practices into their code of law.

To counteract a tightening of Lombard control and increased demands for tribute from the papacy, Pope Stephen II asked the Franks to invade northern Italy. They did so in 755 and returned in 773 under the leadership of Charlemagne, who replaced Lombard rule with Frankish rule. Charlemagne was crowed emperor of Rome by Pope Leo III in 800, thus laying the basis for the Holy Roman Empire. This event was to have great influence on Lucca and medieval Europe as a whole, as the struggle for power between the empire and the church informed the course of events that followed. Many Holy Roman Emperors believed they had power over the church; the church hierarchy thought differently.

Under Frankish rule, Lucca remained a center of power and flourished. By the year 900, for example, Lucca had at least fifty-seven churches. The Franks made the surrounding area a margravate, and the first margrave established his seat in Lucca. In 1027 the Margravate of Tuscany was conferred upon the Canossa family. The most renowned member of this family was the Countess Matilda, known as the Great Countess.

Matilda, born in 1046, was the daughter of Boniface III, margrave of Tuscany, and Beatrice, daughter of the duke of Lorraine. Boniface was assassinated in 1052 by the soldiers of Holy Roman Emperor Henry III; the emperor resented the power Boniface had accumulated. Beatrice was left with three children, Frederick, Matilda, and Beatrice (or Beatricio—the genealogists differ on the gender). The widow married Duke Godfrey II of Lorraine and arranged an engagement of the child Matilda to his son from another marriage, Godfrey the Hunchback. They were eventually married by proxy after the death of the elder Godfrey in 1069, but young Godfrey was hesitant to come to Lucca until 1073, and was immediately recalled to Germany by Holy Roman Emperor Henry IV, a close friend of his. Matilda disliked her husband, as she was devoted to the church and would side with it in its struggles against the empire until her death. Her husband died in 1076, as did her mother, leaving Matilda to inherit all the riches and power of the house of Canossa, which included Tuscany, Spoleto, and Lombardy. This was just at the time when the disputes between the emperor and the church were beginning to flare.

Although Matilda did intercede for Henry IV, who was also her cousin, during his humiliation and excommunication before Pope Gregory VII in 1077, she eternally angered him by donating all of her lands to the church that same year. Power struggles between the church and the empire continued; Matilda's army, fighting the empire, was defeated

The Holy Face crucifix
Photo courtesy of A.P.T., Lucca

at Milan in 1080. Lucca returned to the church's sphere of influence when Matilda's troops bested the imperial forces near Sorbara in 1084. In 1089 Pope Urban II, Gregory VII's successor, persuaded Matilda, in the interest of the church, to marry Guelph d'Este, son of the duke of Bavaria; she was forty-three, the groom nineteen. When Henry learned of the marriage he led his forces into Italy against Matilda and, for a time, was victorious. Matilda then persuaded Henry's son, Conrad, to rebel against his father, and when Conrad was crowned king of Italy, he conferred upon Matilda the title of vice regent. Matilda died in 1115 at the age of sixty-nine.

The constant wars between the church and the empire influenced the rise of independent urban republics, or "communes." The growth of a prosperous mercantile class was another factor in this movement. The communes often were formed by an alliance of the old noble military families and the new mercantile elite (whose supporters were known as the *popolo*), who banded together to counteract the power of the empire or the church. The commune leaders formed councils, which are first mentioned in the histories of Lucca in 1088. In Lucca, the council had five members, one for each parish, and they were usually nominated by the emperor or the margrave of Tuscany until 1162, when Holy Roman Emperor Frederick I granted the Lucchesi the right to elect their own representatives.

Communal government did not always function smoothly; the nobility sometimes fought among themselves or banded together against the *popolo*. Unity sometimes resulted when the city was threatened by a neighboring commune, such as in the bitter feud between Pisa and Lucca. Communal loyalty was often divided between the empire and the church: factions that sided with the empire were called Ghibellines or Whites; those that sided with the church were Guelphs or Blacks, but the adoption of such names was probably more a means to draw support locally than a sign of deeply held beliefs. From the eleventh to the thirteenth century, there was a confusing mess of battles between cities, Guelphs and Ghibellines, between autonomous communes and imperialists (church or empire, it made little difference), and between citizens, often distinguished by their class affiliation. Lucca and Pisa waged war upon each other, simultaneously attempting to curry favor with the emperor and the pope. It is odd, for example, that Lucca became for a period a refuge of the Guelphs, since the emperor had bestowed upon it many privileges. During the invasion of Tuscany in 1260 by Manfred, son of Emperor Frederick II and king of Naples and Sicily, Lucca took in many of the fleeing Guelphs. Unable to stand up to Manfred's forces alone, however, Lucca surrendered to him, and a Guelph-Ghibelline coalition was formed to govern the city. But when the imperial forces were finally defeated and Manfred killed at Benevento in 1266, the Lucchesi expelled the Ghibellines from power.

When the Ghibelline leader Castruccio Castracani, aided by Pisan troops, sacked Lucca in 1314, hundreds of the city's residents fled, including a good many of its artisans,

who took with them the skill of fine silk-weaving. Then Castruccio became popular in Lucca, so that the Pisan leader, Uguccione della Faggiuola, became jealous and sentenced him to death. Uguccione himself set out for Lucca to see the sentence carried out, but tarried along the way. During this period, the Pisans revolted; news of the revolt reached Lucca before Uguccione could, and the citizens freed Castruccio and named him lord of Lucca on the very day of his scheduled execution.

In Castruccio, Lucca had a brilliant warrior and leader. The neighboring town of Sarzana elected him its lord as well, and he built a strong fortress there. He also took Pontremoli and Fosdivino from the marquess of Malespina. He built a citadel, the Augusta, at Lucca, from which the succeeding lords of Lucca would rule until it was destroyed in 1370. He conquered the Florentine-held Altopascia and then Pistoia; later Pisa fell, and he was made duke of Lucca and Pistoia, of Volterra and of Luni, and vicar of Pisa. The Florentines retaliated, lest he should conquer all of Tuscany; his forces and the Florentines visited many atrocities upon one another. During the campaign, Castruccio became sick with a fever and died in 1328, at the age of forty-seven.

After his death, Lucca was bought and sold, invaded and besieged, until Pisa gained control in 1342. According to Lucchesi accounts (dating from after the occupation), the Pisan government was brutally oppressive. However, for at least part of the period of Pisan occupation, Lucca was allowed to conduct its own internal affairs, and while under the watchful eye of the Pisans, enjoyed more autonomy than the Lucchesi would later admit. Through the intercession of the papacy and a payment of up to 500,000 florins, Lucca persuaded Holy Roman Emperor Charles IV to declare the city independent in the late fourteenth century. One of the Lucchesi's first acts after their liberation was to destroy the citadel that Castruccio had built, simply because Pisan officials had occupied it.

Lucca remained a relatively free republic until 1799, with few interruptions. By now a reduced state, both in size and importance, Lucca survived by trying to remain on good terms with as many of its neighbors as possible. It came for a time under the rule of the aristocratic Guinigi family. In 1430 it was nearly captured by expansionist Florence, but it escaped when the Visconti family of Milan, to whom the Lucchesi had turned for help, sent military commander Niccolò Piccinino to rout the Florentines. The Florentines then besieged Lucca in 1437, and to make peace Lucca had to give up a great deal of its territory. Under a new treaty a few years later, however, Lucca regained some of the lands it had lost.

By the early sixteenth century, the nobility had obtained supreme power in Lucca. They imposed regulations on the silk trade that resulted in less income for the weavers but greater profits for the nobles who handled much of the buying and selling of silk. In 1531, the people revolted. The ragged revolution was the first in the city to challenge the authority of the aristocracy. With the help of a priest, the revolt was quickly put down. In succeeding years the nobles intensified their power by passing laws prohibiting any save the aristocracy from becoming members of the city's governing body.

The power of the nobles remained secure until the French Revolution of 1789 created turmoil throughout Europe. In 1799, French forces entered Lucca and imposed a democratic constitution on the city. This government was to last just six months, however, and disappear in the war between France and Austria.

In 1805, following a period in which Lucca was overrun and bankrupted by, in turn France and Austria, Napoléon installed his sister, Maria-Anna (also known as Élisa) and her husband Felice Baciocchi as princess and prince of Lucca. Though the Lucchesi were not exactly enthused about this, they found in Élisa a tireless reformer and a diplomat, who smoothed over some of the harsher pronouncements of her brother. She established schools, reformed prisons, eased the divestiture of church property, expanded hospitals, and provided for the immunization of children against smallpox.

After Napoléon's downfall, the 1814 Congress of Vienna made the Bourbons of Parma dukes of Lucca. María Luisa and her son Charles Louis did much for the betterment of the territory. They built dykes to tame the Serchio, good roads and bridges, and an aqueduct to bring fresh water to the city. In 1847 the Bourbons ceded Lucca to the grand duke of Tuscany, Leopoldo II, and in 1860 it was incorporated into the newly united Italy.

Lucca is where the English writer and critic John Ruskin undertook his study of architecture in the nineteenth century. It is also the birthplace of architect Matteo Civitali (1436–1501) and the operatic composer Giacomo Puccini (1858–1924), known for such works as *La Bohème, Tosca,* and *Madama Butterfly.* Today, it cannot compete with Pisa (and the Leaning Tower) for tourist trade, and its economy is based on agriculture—products include olive oil, fruit, and wine—and light industry. However, it has several sites of interest.

One of Lucca's oldest attractions is the Holy Face crucifix (Volto Santo), brought here from the east in the eighth century. Many legends surround this image of the crucified Christ with a highly evocative face. One such legend claims it was carved by Nicodemus, who fell asleep before he could complete it. When he awoke he found that an angel had done the work for him. It is housed in the Cathedral of San Martino, built in the sixth century by St. Frediano, an Irish bishop. Pope Alexander II, while bishop of Lucca in the eleventh century, initiated expansion, and Guidetto of Como

added the facade with its balconies and decorated columns in 1204. Nicola Pisano contributed the Nativity and the Deposition. The cathedral also houses the greatest work by Jacopo della Quercia, a tomb for Ilaria del Carretto, the wife of Lucca's one-time ruler Paolo Guinigi. The cathedral's pulpit, two beautiful fonts, and a couple of other tombs are by Matteo Civitali, Lucca's foremost master architect, who also built the Palazzo Pretoria, near the house where Puccini was born.

Villa Guinigi, built in 1418 for Paolo Guinigi, was presented to the city upon his fall from power. It is the home of the Museo Nazionale Guinigi, with Etruscan, Roman, and Hellenistic artifacts, as well as art masterpieces by Berlinghiero Berlinghieri, Ugolino Lorenzetti, and Fra Bartolommeo.

Lucca has several other old palaces and noble estates. Another home associated with the Guinigi family, the Palazzo Guinigi, has a tower where oak trees were planted. The effect is startling. The Palazzo Mansi is now a museum displaying the works of Tintoretto, Veronese, Bronzino, and Andrea del Sarto. Villa Reale, just outside the city, was splendidly furbished with gardens, a pool, fountains, and statues when Napoléon's sister Élisa took it as her summer residence. María Luisa de Bourbon added an observatory, a greenhouse, a coffee house, a library, and a study. A music festival is held there each year. Of its Roman heritage, Lucca retains part of its street plan, one wall preserved inside the church of Santa Maria della Rosa, and the Piazza Amfiteatro, which follows the outline of the arena of the old Roman amphitheatre, with houses built where the seats once stood.

Further Reading: The only book dealing thoroughly with Lucca was published in 1912 and reflects the biases of the time: *The Story of Lucca,* by Janet Ross and Nelly Erichsen (London: Dent, and New York: Dutton, 1912). It contains the complete history of the turbulent period of Guelph–Ghibelline struggle. *A Guide to Tuscany* by James Bentley (New York: Viking, 1987; Harmondsworth, Middlesex: Penguin, 1988) contains a better-than-average tour of the city, while *A Concise History of Italy* by Christopher Duggan (Cambridge and New York: Cambridge University Press, 1994) places the characters of Lucca's story in the context of Italian History. Christine Meek contributes two academic works on Lucca in the fourteenth century: *The Commune of Lucca under Pisan Rule, 1342-1369* (Cambridge, Massachusetts: Medieval Academy of America, 1980) and *Lucca: 1369–1400: Politics and Society in an Early Renaissance City-State* (Oxford and New York: Oxford University Press, 1978). For a discussion of Napoléon's Italian empire, see *The French in Italy: 1796–1799* by Angus Heriot (London: Chatto and Windus, 1957).

—Gregory J. Ledger

Madrid (Madrid, Spain)

Location: Situated in the exact center of the Iberian
Peninsula in central Spain.

Description: The capital of Spain and center of government,
commerce, communications, higher education,
and administration. At a height of 2,119 feet,
Madrid is Europe's highest capital city. The
Manzanares River runs through the southwest
section of the city before joining the Jarama, a
tributary of the Tagus River.

Site Office: Ayuntamiento de Madrid
Plaza Mayor, 3
2801 3 Madrid, Madrid
Spain
(13) 66 48 74

Between the first and the tenth centuries A.D., Madrid was
settled by several groups. The first were the Celtiberians, who
were followed by Romans, and then Visigoths. However, the
earliest authentic historical mention of the town, then known
as Matrít, occurs in the surviving Arab chronicle. Muhammad
Ibn 'Abd ar-Rahman V, emir of Córdoba, put up walls between 852 and 856, and built a defensive castle or Alcázar on
the present site of the royal palace. The region was then taken
by the Moors, and the expansion of the Christian communities was checked during the tenth century. The Moorish
emirate of Córdoba disintegrated into a score of independent
units during the first half of the eleventh century, and by the
end of the century, Christians had reconquered the Tagus
River area. It was not until 1083, in the reign of Alfonso VI,
king of León and Castile, that Madrid was finally taken from
the Moors.

During Alfonso VI's rule, Madrid saw the birth of the
city's future patron saint: San Isidro Labrador (the Laborer).
Isidro was a serf of the wealthy Vargas family and is said to
have performed numerous agricultural miracles (it is held that
angels plowed his fields when he was busy praying). He is
particularly famous for miracles involving water, most notably the rescue of his infant son after his wife accidently lost
the boy down a well (she was nevertheless canonized herself
as Santa María de la Cabeza). According to the legend, Isidro
saved his son by miraculously raising the well's water level,
thereby floating the boy to safety. Because of the saint's close
association with water, his remains were traditionally paraded through the streets of Madrid in times of drought.

The first *cortes,* or courts, of Castile were called by
Ferdinand IV in 1309 and were gathered in Madrid to prepare
for war on Granada. Later *cortes* were gathered there in 1329
and 1335 by Alfonso XI, who endowed Madrid with plenary
jurisdiction in 1339 and created the *Ayuntamiento,* or town

council, in 1346, to be governed by *homes sabios,* or "wise
men." In the same year, Alfonso XI granted the town permission to found a school of grammar studies. Henry IV frequently resided in Madrid, made it into a hunting set, was
married there, and in 1454 bestowed on it the titles of Muy
Noble and Muy Leal (Very Noble and Very Loyal). In 1476
Isabella and Ferdinand confirmed the privileges of the town,
and personally heard cases and administered justice in the
Alcázar. It was not, however, until the mid–sixteenth-century
reign of Charles 1 of Spain, later Holy Roman Emperor
Charles V, that Madrid came into the limelight of European
history.

After defeating the armies of Francis I of France at
Pavia, Charles held Francis captive in the Alcázar until the
treaty of Madrid releasing him was signed on January 14,
1526. In the absence of Francis from France, his mother,
Louise of Savoy, acted as regent, and it was she who faced
the task of negotiating the conditions of her son's release from
his Madrid captivity. Charles was extremely demanding and
required Francis to surrender two of his sons as hostages, to
give up Burgundy and all the other territories that had been
held by Charles the Bold at the time of his death, to pay an
indemnity, and to surrender the villages of Thérouanne and
Hesdin. The peace between France and Spain was to be sealed
by a marriage between Charles's niece, Mary, and the dau-
phin. However, the marriage never took place. Eventually,
Louise conceded Burgundy and a peace treaty was drawn up
agreeing to Charles's conditions and granting Francis his
freedom and permission to marry Charles's sister, Eleanor of
Portugal.

When Philip II moved his court from Toledo to Madrid in July 1561, he named the town *unica corte* or "one and
only court," without declaring it his capital. Nobody thought
the move would be permanent, but except for six years during
the reign of Philip III, from 1600 to 1606, the court never
moved again.

From 1561 until the beginning of the seventeenth
century, the city underwent extraordinarily accelerated
growth, and, in 1579, the first municipal decrees were dictated. Miguel de Cervantes died at his home in Madrid in
1616. He would later become Spain's most celebrated author,
primarily through *El ingenioso hidalgo Don Quijote de la
Mancha.* Three years later, in 1619, the Plaza Mayor was
completed and, in the following year, the festivities for the
celebration of the beatification of the peasant Isidro were held
there. In 1622 there were again huge festivities when St.
Isidro was canonized.

During the reign of Philip IV, which began in 1621,
Madrid's growth continued and the size of the city quadru-
pled. In 1631, Philip began in the heart of the city the
construction of a new palace with extensive and very beauti-

Botanical gardens in Madrid
Photo courtesy of Instituto Nacional de Promocion del Turismo

ful gardens. It later became the Retiro Park. The park contains a large lake and magnificent sculptures, including the only public statue of the devil in Europe, *Angel Caido* (Fallen Angel), created in 1885 by Ricardo Bellver. In 1635 the construction of Philip's palace was completed, and Madrid was enclosed with a twenty-foot stone wall, pierced by five principal gates and eleven doorways. Only three structures remain where once these openings stood: the late–eighteenth-century Puerta de Alcalá and Portillo de San Vicente, and the nineteenth-century Puerta de Toledo. In the same year in which the palace was completed, playwright Lope de Vega (el Fénix) died in Madrid and was buried in the church of St. Sebastian amid spontaneous public mourning.

The reign of the melancholic Charles II in the late seventeenth century was a restless period in Spanish history that left the kingdom in a state of near-ruin and anarchy, and gave rise to the War of the Spanish Succession. With the new century, the new Bourbon dynasty came to the Spanish throne and continued to rule from Madrid. Philip V made great efforts throughout his reign to lift Spain from its decay, and it was a

blow to his capital when the Alcázar burned down in 1734. The palace was destroyed together with the treasures it contained, including many very early paintings by Velázquez.

Charles III implemented plans for the regular cleaning of the streets and in the same year he created the *Cuerpo de la Milicia Urbana,* the predecessor to the Municipal Guard. By 1765, a system for lighting the streets at night had been developed. Charles ran not only the country but also the city, instituting there and elsewhere schemes for the construction of roads, bridges, and drainage works. His undertakings included the restructuring of many of the city's streets and loveliest walks, and the building of some of its finest monuments, as well as the foundation of hospitals and the botanic gardens. By the end of his reign, Charles had transformed Madrid from what had been an uninspiring city to a town with outstandingly beautiful public buildings and monuments; clean, lighted streets; and a feeling of civic pride.

The reign of his son Charles IV saw less reform and construction, but nonetheless some interesting innovations appeared, such as the instigation of "diligent coaches," the

Convento de las Descalzas Reales
Photo courtesy of Instituto Nacional de Promocion del Turismo

earliest taxicabs. More significant, however, was the great fire that swept through the Plaza Mayor in 1790 and made its total reconstruction necessary. Then in 1808, Napoléon Bonaparte arrived in Madrid with his troops and on May 2 there was a Spanish uprising against the French invaders. Nevertheless, Charles was forced to abdicate in favor of Bonaparte, who gave the throne to his brother Joseph. Joseph's reign as king of Spain was brief, but the reforms that the city underwent were intense. Joseph ordered the demolition of numerous convents and replaced them with plazas. In 1813 the French retreated from Madrid, and in 1814 Ferdinand VII was restored to the throne and bestowed on Madrid the title of *Muy Heroica*.

The 1834 cholera epidemic severely depleted the city's population, but by that date the city was beginning to look much as it does today. By the end of the nineteenth century, gas installations, electric and horse-drawn streetcars,

electric street lighting, hydraulic lifts, and public telephone services were common. Alfonso XIII, who was born king in 1886 after his father's death, inaugurated in 1910 the work on the Gran Vía, today the greatest thoroughfare of the city. Alfonso's most notable innovation was the construction of the "University City" in 1927. Madrid soon had an underground train service, opening in 1921, and an air link with Barcelona, and was in every way a modern capital city. From 1930 to 1940 the capital had a sad and difficult decade, starting with the Second Republic, which ruled from 1931 to 1936, and was followed by the Spanish Civil War, from 1936 to 1939. At the end of the war, Madrid was faced with the task of reconstructing much of the city at the same time that the population was swelling at an enormous rate with immigrants from the countryside and smaller towns.

The Fascist regime of General Francisco Franco, which began at the end of the Civil War, was greeted by extremes of popularity and hostility. However, the opposition was in the minority, and it was generally recognized that harmony needed to be restored. Franco's coup in 1936 had failed to take Madrid, which remained republican, and the Civil War ended only when Madrid fell to Franco on April 1, 1939. His government, already recognized by Italy and Germany in 1936, achieved recognition by France and the United Kingdom in 1939. This approval was short-lived and the whole of the United Nations condemned his regime on February 9, 1946. However, there were huge demonstrations of support for Franco in Madrid, and it suited him to consolidate his rule of Spain in comparative political isolation from the rest of the world. After Franco's death in 1975 and the restoration of the monarchy, now limited in power, Spain has gradually become less isolated, joining NATO and the European Economic Community (now the European Union) in the 1980s.

Spain's new political and social freedoms brought a wave of creative revitalization, particularly in Madrid. The city became a magnet for artists, musicians, thinkers, and socialites from across the country. Painters such as Féderico Amet and Miguel Barcelo and sculptors such as Jorge de Oteiza, Pablo Serrano, and Miguel Ortiz, among many others, fed the cultural renaissance. In the later 1980s, the explosion of creative energy became known as the *movida madrileña* (Madrid Happening).

Madrid today is a modern capital city known for its welcoming population with an enormous appetite for life and fun. It is famous for its splendid collection of museums and art galleries, was named cultural capital of Europe in 1992, and is widely claimed to be the European city with the largest area of green spaces. Especially famous is the Retiro Park, the former palace gardens, an enormous and painstakingly kept stretch of woods and flower gardens in the center of the city. A large and beautiful lake is surrounded on one side by a monument reminiscent of an ornate Greek temple, which was once the setting for theatrical productions. The old quarter of Madrid retains the same pattern of winding sloped streets that it had in the days of Philip II, and the layers of the city's growth are clearly visible to any casual stroller.

The Salamanca district is of particular interest as the first development outside the early city limits, although it is now in the very heart of the city. It was carefully planned around the Retiro Park and the neoclassical Prado museum to house the wealthy citizens outside the cramped city center.

The wide, straight streets, laid out on a grid pattern, are lined with trees and imposing eight- or nine-floor blocks. This is also the luxury shopping area, where the leading fashion houses have their boutiques.

In sharp contrast to the spacious elegance of the Salamanca is the Latina, the zone of dark and twisting streets that is typical of the medieval town. Not far from the Plaza Mayor, this is the original heart of Madrid, today known for its shabby but charming buildings and the Rastro, the famous and extremely popular flea market of the capital. Officially held on Sundays, the Rastro consists of hundreds of "stalls," often simply wares laid out on the pavement of the winding streets, which traditionally sell antiquities, oddities, and other secondhand goods.

Of the hundreds of museums and art galleries in Madrid, two are particularly noted for the superb quality of their permanent collections. The most important is the Prado, one of the world's major art galleries, whose principal facade, with a large imposing portico of six Doric columns, faces a statue of Velázquez. It was built from 1785 to 1816 as the Museum of Natural Sciences. The north facade has four Ionic columns reached by stairs to compensate for the variations in the ground level, and faces a statue of Spanish artist Francisco José de Goya. The south wing looks onto the botanic gardens. The museum has undergone great alterations and enlargements, but its exhibition space is still too small and it has taken over the nearby building, el Cáson del buen Retiro, to house the Spanish school of the nineteenth century. The second noteworthy collection is in the Centro de Arte Reina Sofía built by Charles III in 1776 on the site of a building erected in 1566 by Philip II as the General Hospital of St. Charles. The collection includes Picasso's *Guernica*, which the artist had forbidden to be housed in Spain until Franco's death.

Further Reading: Madrid's general historical background is excellently covered by Raymond Carr in *Modern Spain, 1875–1980* (Oxford: Oxford University Press, 1980; New York: Oxford University Press, 1981) and *Spain, 1808–1975* (Oxford: Clarendon, and New York: Oxford University Press, 1966; second edition, 1982), and by A. Launay and M. Pendered in *Madrid and Southern Spain* (London: Batsford, and New York: Hippocrene, 1976). The best introductory guide in Spanish is *Guia de Madrid* by José del Corral (Madrid: El Pais, 1990).

—Clarissa Levi

Mafra (Lisboa, Portugal)

Location: Sixteen miles northwest of Lisbon in the Lisboa district of west central Portugal.

Description: One of Portugal's oldest established communities, on an abrupt ridge over the steep river valley and flood meadows of the Tagus, best known for its colossal palace-convent built 1717–35 in fulfillment of vow by King John V.

Site Office: Comissão Municipal de Turismo
Avenida Oliveira Salazar
Mafra, Lisboa
Portugal
(61) 52023

The ancient settlement of Mafra spent most of its history as a tiny village of 100 residences, a castle that changed hands with political regimes, and a church dating to the twelfth and thirteenth centuries. Located in the notably cultivated lowlands of the Estremadura region, Mafra remained largely uninvolved in Portuguese national affairs, despite the village's proximity to the capital at Lisbon and to the port of Ericeira on the Atlantic Ocean, as the nation became a major international trade and colonial power from the fifteenth century onward. Then when John V succeeded to the throne in 1706, Mafra lost its anonymity as he chose that pastoral hamlet for the great monument to the prosperity of an evolving Portugal. The palace-convent he built at Mafra was to become the most extravagant, imposing work of Portuguese architecture and one of the largest structures in Iberia.

The village of Mafra was recovered from the Moors in 1147 and was shortly thereafter presented to the Order of Calatrava by Afonso Henriques. Sancho III of Castile, founder of the order, made repairs to the Moorish castle. A parish was established and the Romanesque-Gothic Church of Santo André was erected in the late twelfth to early thirteenth century. In 1276 the parish priest Petrus Juliani, an astrologer, doctor, and theologian also known as Petrus Hispanus (Peter of Spain), became Pope John XXI. The pleasant village and environs of Mafra were intermittently favored by the nobility, particularly for summer homes. The estates of Thomaz de Lima Noronha e Vasconcellos, the Visconde (Viscount) de Vila Nova de Cerveira, an aide to John V's queen, were well established at Mafra by the late seventeenth century, a significant time in Portugal's history.

In 1693 gold was discovered in Portugal's colony of Brazil. Other sources of wealth from this location included silver, emeralds, diamonds, and rare hardwoods, of which the royal coffers claimed a one-fifth share. John V became king in 1706 and began at once his consumption of the new Portuguese fortune. His reign was marked by the construction of many grand architectural monuments, but the palace-convent at Mafra was the most unrestrained and expensive.

John and his wife, Maria Ana, initially had difficulty producing an heir; the couple's first son died in infancy. An administrator of the king confided John's desire for offspring to Frey Antonio de San José, a Franciscan friar of the Order of Arrabida. Frey Antonio said that a vow from John to erect a convent dedicated to St. Anthony in the locality of Mafra would ensure God's delivery of children to the royal couple. John promised, and his daughter Maria Barbara de Bragança, a future queen of Spain, was born in 1711. (A son, Joseph, who would be John's successor, followed in 1714.) Prior to Maria's birth, John formally declared his intention to build a convent for the Arrabidan friars, but the location as a prescription of Frey Antonio is debatable. The Visconde de Vila Nova de Cerveira offered his estate in Mafra, and perhaps implored the queen to persuade John to take his offerings. The village sheltered the poorest community of friars then extant in Portugal as well. Lastly, Mafra was the first segment of the Portuguese coast Maria Ana beheld when she arrived from Austria. By consideration of one or a combination of these circumstances, though he never accepted the viscount's tribute, John decided to build in Mafra.

The original design and intentions for the palace-convent were humble in comparison to the monument completed in 1735, for the period of building coincided with a rebirth of the arts in Portugal. A national Portuguese style had not survived the union with Spain, which lasted from 1580 to 1640. John IV tried to import the baroque style that was popular throughout Europe, but native architects resisted. Continuing conflict between Spain and Portugal diverted time and money from other architectural commissions. Peace for the two countries followed the signing of the Treaty of Utrecht in 1713, however, and with the immense riches derived from Brazilian mines, there was a resurgence of political and financial security unknown to Portugal since the reign of Manuel the Fortunate. Royal patronage of the arts was liberal and available when João Frederick Ludovice arrived from Italy, and the country was ripe for the revitalization of architectural development. Introduced to this political and cultural climate, the architectural genius of Ludovice brought forth a renaissance and the distinction of the palace-convent at Mafra.

Ludovice was born Johann Friedrich Ludwig in Swabia in 1670. He learned the craft of the goldsmith from his father before he left for Augsburg at age nineteen to join the imperial armies at war with France. Eight years later he proceeded to Rome, where he sought work and living accommodations with a Jesuit community. He Italicized his name to Ludovice, recanted his Lutheran faith, and married an Italian. While in Rome Ludovice attended lectures by famous

Mafra's convent
Photo courtesy of Portuguese National Tourist Office

architects and found employment as a sculptor of altarpieces at the Church of the Gesù, where he enjoyed the company of distinguished masters. Whether or not he had acquired a rudimentary knowledge of architecture in his native village or in the army, by 1701 Ludovice had obtained an informal degree of architectural expertise heavily influenced by papal architects and architecture.

Ludovice received a commission to work as a goldsmith for Jesuit institutions in Portugal from 1701 to 1708. Ludovice eventually broke his contract by taking non-Jesuit commissions as well, but apparently did not fall from favor with the Jesuits. By the time the palace-convent in Mafra was conceived, John V could have afforded to import a genuine Italian architect, but the modest scale of the original plan required little more than an architect who could reproduce an Italian design. Ludovice's architectural expertise was untried, but his familiarity with Roman architecture and his German roots (which he held in common with the queen) were factors in his favor. The complex at Mafra was to make his reputation.

In its enormity, the monument at Mafra rivaled Philip II's El Escorial outside of Madrid when finished. The Spanish convent-palace antedates the palace-convent at Mafra but was not its prototype. Mafra's architecture was derivative of contemporary monuments of Rome with a Latin basilica, unlike that of El Escorial, which had a Greek-cross plan. The physical arrangement of the palace, basilica, and convent was influenced by German and Austrian architecture. The scope of the project grew with John V's increased interest in the design and his extravagant nature. The labor required for the enormous venture was conscripted and guarded. A town with a hospital was erected to house and care for the workers, who numbered about 45,000 at the project's peak. After the monument's dedication ceremony in 1730, construction continued for almost six years, keeping as many as 15,000 craftsmen, stone masons, and laborers employed at the site.

The quarries of Pero Pinheiro between Mafra and Sintra were mined for stone. Bricks were fashioned all over the country and shipped to the site by way of a network of new roads built for this purpose. Industrial smoke from cement furnaces blanketed the region around Mafra. Pink, white, and grey marble was obtained from every quarry in Portugal, and Carrara marble was imported from Italy.

The complex became a vast expanse of masonry occupying 290,000 square feet. The English writer William Beckford once commented that it appeared a giant resided at

Mafra. Eleven chapels comprise the basilica featured at the center and forefront of the ornate 71-foot facade. It is flanked by the king's and queen's apartments, German-style pavilions, in the middle of the palace. A total of 5,200 doors and 2,500 windows provide entrance and light to 880 rooms. Six Ionic columns form three arches beneath two towers and a dome, and the belfry contains a carillon of 114 Flemish bells. The library holds 35,000 volumes, and in the royal apartments one passage alone measures 812 feet in length. The convent at the rear of the structure at one time housed 300 monks of Arrabida, the order at whose suggestion the palace-convent was originally constructed.

In 1720 John V instituted a Royal Academy of History and endowed Mafra with its magnificent library. In contrast to the darkness of most of the complex, the library is light, of a rococo style with white-painted bookshelves and a white plaster barrel roof. It has many study alcoves and a flat central dome surrounding a rayed sun. Among the books, most of which are two and three centuries old, is a collection of illustrated fifteenth-century manuscripts. At the same time John commissioned a library at the University of Coimbra; although the latter is not generally considered as attractive as the library at Mafra, the two are considered among the most beautiful rooms in the world.

For the laying of the complex's foundation stone in 1717, John traveled to Mafra for the first time and celebrated the occasion with religious ceremonies and banquets. He returned intermittently to oversee the construction. In October 1730 he consecrated the basilica with one of the most extraordinary ceremonies of his reign. Vast numbers of infantry and cavalrymen were on parade, clergy were honored by royalty, and John posed dramatically on his throne. The four-day celebration culminated on the king's birthday with a solemn presentation of the church's relics to the bishop of Leirìa.

John died in 1750; his son, King Joseph I, bestowed upon Ludovice the title of Grand Architect of Portugal, which recognized the more than four decades the architect served John V. Joseph mandated that the office granted the architect would die with him. At his death in 1752 Ludovice was saluted for the monument he built at Mafra, for his other architectural and creative work, and for his instruction to other architects. He had developed a distinctive Portuguese architectural style that reflected his Roman experience, the baroque design that was fashionable but had been unattainable in Portugal, and the influence of local constructions.

In 1753 Portugal's first School of Sculpture was established at Mafra by the Italian Alessandro Giusti. A student of the painter Conca and the sculptor Maini, Giusiti came to Portugal from Italy to assemble a chapel in Lisbon. He settled in Mafra to found and direct the school. The marble reliefs at the Mafra palace-convent are attributed to him and his students. In 1756 Joaquim Machado de Castro became his assistant and disciple, and later became director of the school. Machado de Castro developed into one of Portugal's greatest sculptors. His work in the church of Mafra's palace-convent attests to his excellence.

The Lisbon earthquake of 1775 had little effect on the palace-convent's solid frame, but the School of Sculpture was removed to Ajuda and activity at Mafra decreased significantly in the nineteenth century. King John VI sometimes sought refuge at Mafra from his wife, Queen Carlota, who had plotted to overthrow him, and in 1807 he contributed an organ to the church. The palace-convent fell into disrepair as popular antagonism toward the church mounted, however, and the repairs made by Ferdinand, consort to Queen Maria II, in the middle of the century did not restore the monument to its original glory. Most of the palace-convent was converted into a military barracks in 1886. The monarchy last utilized the vast royal quarters at Mafra on October 5–6, 1910. Manuel II and his mother Amelia fled from Lisbon's Necessidades Palace and spent their last night in Portugal at the palace-convent at Mafra. For fear of the encroachment of a mob from Lisbon, Manuel descended from Mafra to the port at Ericeira on the morning of October 6 and sailed to Gibraltar, then to England, where he died in exile.

The palace-convent was designated a national cultural monument in 1939. Since 1911 the monument has contained the Religious and Profane Art Museum. It also houses Mafra's city hall and an elementary school. Army regiments are still garrisoned there, and a church remains a part of the structure. The palace-convent is regarded as monotonous and monstrous by some modern observers; size is the edifice's most outstanding feature and obliterates its beauty. The contemporary Portuguese author José Saramago wrote *Baltasar and Blimunda*, a historical novel of the palace-convent at Mafra. Saramago celebrates the absurdity of the Mafra institution in this comical novel, satirizing John V's vanity and the extravagance of his patronage to the arts and the church.

Further Reading: Robert C. Smith Jr.'s lengthy article "João Frederico Ludovice: An Eighteenth Century Architect in Portugal" in *The Art Bulletin* (New York), volume 18, number 3, 1936, is the most comprehensive of all writings on the village of Mafra. The article is filled with historic details, placing the Mafra complex in its social and political context. General histories of Portugal provide helpful background information. Some of the better ones are H. V. Livermore's *A New History of Portugal* (Cambridge and New York: Cambridge University Press, 1966; second edition, 1976) and Cedric Salter's *Portugal* (New York: Hastings, and London: Batsford, 1970).

—Patricia Trimnell

Málaga (Málaga, Spain)

Location: The capital of the province of the same name in the region of Andalusia, situated on the south coast of Spain at the foot of the Montes de Málaga, skirted by the provinces of Cádiz to the west, Granada to the east, Seville to the northwest, Córdoba to the north, and by the Bay of Málaga on the south.

Description: One of the oldest of the Mediterranean ports, established by the Phoenicians about 1100 B.C. The second-largest city in Andalusia, after Seville, and one of the principal ports of Spain, exporting olive oil and almonds, as well as sweet Málaga wine, famous for centuries. Modern Málaga forms part of the Costa del Sol, a popular tourist area.

Site Office: Málaga Tourism Office
Pasaje de Chinatas, 4
29016 Málaga, Málaga
Spain
(952) 21 34 45 or 22 89 48

The city of Málaga was founded by the Phoenicians as a trading post in the twelfth century B.C., the name deriving from the salt fish trade for which it was well known (*malac* means "to salt"). Subsequent invaders included the Carthaginians, Romans, Vandals, Visigoths, and Moors. For several long periods in its history (with the notable exception of the later sixteenth to the eighteenth centuries) the city has been an important port. The Spanish reconquest of Málaga and southern Spain in 1487 resulted in the expulsion of the Moslems and later the Moriscos (Moslems who had been converted to Christianity). Soon after the discovery of America in 1492, the cities of Cádiz and Seville took precedence over Málaga as a port, until, at the beginning of the nineteenth century, it became popular with English visitors. The Civil War from 1936 to 1939 led to the destruction of many of Málaga's ancient buildings and monuments. Surviving buildings and monuments of architectural interest include the Alcazaba (ninth century), the Gibralfaro (thirteenth century), the church of Santiago el mayor (1490), the Iglesia del Sagrario (late fifteenth century), the Cathedral (sixteenth to eighteenth century), the Museo de Bellas Artes (c. 1550), the Fuente de Neptuno (1560), the lighthouse (1588), the Museo de Tradiciones y Artes Populares (seventeenth century), the English Cemetery (1830), the Plaza de Toros (1874), and the Ayuntamiento (town hall, 1911–19). Recently the city has become a highly popular tourist resort, with areas entirely devoted to large hotel complexes and tourist trade amenities.

In the time of the Phoenicians, the city of Málaga formed a part of the kingdom of Tartessos. This kingdom covered approximately the area of modern Andalusia, and is mentioned in the Bible as "Tarshish." The wealth of Tartessos was based on the mineral resources of Río Tinto and Sierra Morena, and the kingdom flourished during the eighth century B.C.

Around 636 B.C., the Greeks established themselves at Mainake, near Málaga. By this time the economic dominance of the Phoenicians was in decline. Remains of the Greek settlement have been discovered near Torre del Mar, some twenty miles along the coast from the modern city of Málaga. By 500 B.C., the Carthaginians, earlier called upon the by Phoenicians for protection, were breaking up the region of Tartessos and taking southern Spain from the Greeks.

In the First Punic War (264 to 241 B.C.) Rome took several Italian territories from the Carthaginians and in the Second Punic War (218 to 201 B.C.), the Roman troops of Scipio Africanus extended their realms to cover Spain. Thus, Málaga became a Roman municipium. It may have benefited from its lack of resistance to Roman rule in that it did not have to put up with frequent Roman interference as did other towns. The Romans continued to take parts of the peninsula until about A.D. 50. Julius Caesar was the governor of the region during the first century B.C., at which time it was the most prosperous part of Spain. It became known as *Baetica,* stemming from *Baetis,* the ancient name for the Guadalquivir River, with Málaga situated in the east of the province. The later name of Andalusia probably has its origins in *Vandalusia,* from the subsequent fifth-century Vandal occupation. By 478 the Visigoths, fierce Teutonic warriors, were in control of Andalusia and much of the rest of Spain, though between 522 and 621, Málaga, Cartagena, and the Balearic Islands formed a province called Spania, held by the Byzantines as an outpost of the Roman Empire.

A small intellectual revival occurred in Andalusia in the sixth and seventh centuries, but discontent arose at the level of power allowed to the Catholic Church, which made the Visigoths unpopular with the majority of the people. In 711, 7,000 Berbers, led by Ṭāriq ibn Ziyād and sent by Mūsa of Damascus, defeated the Visigoths and killed King Roderick, successor to the throne of Toledo. The Arab conquerors were on the whole welcomed by the people as rescuers from the warrior band of Visigoths who had encouraged overtaxation and the persecution of Jews. The first emir of al-Andalus (The Isle of the Vandals) was Mūsa's son, 'Abd al-'Aziz. Before long he was overthrown by the forces of the new caliph, who had dethroned Mūsa in Damascus. A new governor was sent in his place in 716. Although now a minority, the Christian population was allowed to retain its own laws, and in exchange for tax payments, was also protected by the laws governing the province. These people became known as *mozárabes* (meaning Arab-like). Jews were allowed the same freedoms, and began to return from their enforced or self-im-

The Alcabaza, or fortress, of Málaga
Photo courtesy of Tourist Office of Spain, Chicago

posed exile effected under the Visigoths, many settling in Málaga. Because of the tolerant attitude of its rulers, al-Andalus was the only European region in which Jews now settled in significant numbers.

In 756 'Abd ar-Raḥmān I, having fled Damascus in the wake of a massacre of descendants of the Ummayyad dynasty, named himself the independent emir of al-Andalus, which at that time stretched well beyond modern Andalusia. Málaga, along with Seville and Córdoba, was a great city in this now-Moslem kingdom. The emir proved a successful ruler, partly because he created a professional army and continued to encourage tolerance among the different religious and ethnic groups.

In the ninth century, 'Abd ar-Raḥmān II and Muḥammad I fostered a prosperous al-Andalus, though it ran into trouble briefly at the end of the century when Christian riots against the widespread Islamic influence destroyed the peace. 'Abd ar-Raḥmān III, of mixed Arab and Spanish parentage, with an Arab name but the fair coloring of a European, seemed to bridge the gap. Religious adherences were observed, but during this golden age of al-Andalus, despite some fighting between Moslems and Christians, fanaticism was not rife. It was fairly common in Málaga and other cities for Moslems and Christians to intermarry, and to have family and other links with those of the other faith.

In 929, the emir took the proclamation of independence even further than did his namesake of two centuries earlier by naming himself caliph and sole ruler, and declared that he was not answerable to any higher authority, any religious or military leader. During this period of Málaga's history, agrarian production in the area included cotton, rice, dates, sugar, and oranges, all of which had been introduced by the Arabs. Irrigation processes were refined and learning in the sciences and arts developed, especially in the main cities. Málaga itself was famous for its singers. Highly civilized al-Andalus was starting to make Christian Europe look positively backward.

With the dawning of the eleventh century, however, dissatisfaction was in the air, and the number of rebellions rose sharply. Nine caliphs ruled between 1008 and 1031; by the latter date the region had been divided into a number of smaller principalities known as *taifas*, each headed by a ruler with a particular political sympathy. Málaga and Ronda were populated mainly by Arab Berbers and their descendants. As there were numerous factions, state rule was an uneasy and often short-lived affair. In the later years of the eleventh century, Alfonso VI, king of Castile and León, demanded tribute from the *taifas* and attempted to extend his monarchist rule to al-Andalus (he took Toledo with the aid of El Cid in 1085). By 1110, amid religious and economic unrest, the Almoravids (African Berbers originally brought in by the *taifa* governors to counter the influence of Alfonso VI), under the leadership of Yūsuf ibn Tāshufīn, conquered al-Andalus for themselves.

Almoravid rule abolished the *taifas* and lasted some fifty years. Another Berber tribe, the Almohads, led by Muḥammad ibn Tumart, denounced the Almoravids as pagans; their dominance in al-Andalus was already dissolving in widespread rebellions by the time the Almohads took control in 1147. Fighting broke out frequently as the Almoravids retained control of the Balearic Islands and other regions of Spain. A famous victory over Alfonso VIII of Castile and the Christians at Alarcos in 1195 was not decisive; Moslem al-Andalus was breaking up under the consolidated strength of Christian Spain. The battle at Las Navas de Tolosa in 1212 effectively brought Almohad control to an end, uniting the armies of Castile, Aragon, and Navarre under Alfonso VIII in a common purpose.

Although by the middle of the thirteenth century most of Spain was under Christian rulers, an area including Málaga and stretching from Gibraltar to Almeria was established as the Emirate of Granada from 1238 until as late as 1492. This situation came about because the Arab conqueror of Granada, Muḥammad ibn-Yusuf ibn-Nasr of the Nasrid tribe, was a follower of Ferdinand III, king of Castile and León. Granada

still had to pay tribute to Castile's Christian king. Many of Spain's Moslems, displaced from their former homes elsewhere in Spain, found a haven during the next century in the Emirate of Granada, which was enjoying an economic and cultural revival. Málaga and Almeria traded with Italy and North Africa; the land was cultivated successfully to produce a variety of foodstuffs; the silk trade was a major asset; arts, crafts, and poetry flourished in the comparatively peaceful state. Civil wars among Christian factions of Spain meant that Christians were fully occupied with internal affairs, especially in Castile, and the reconquest of Málaga and southern Spain was delayed for some years.

Moslem rule was in theory handed down from father to son as in Christian kingdoms; however, uprisings against the heir often meant that this passing of the gauntlet did not take place. The ruler (caliph, sultan, emir) would appoint a chamberlain; or a man with strong military backing took the position for himself. After the *hajib*, as he was known, the next step in the hierarchy was a state council (*diwan*). Local administrators (*walis*) governed provinces or cities, and military rulers and their warriors occupied the borders. Taxes were imposed on individuals, products, rents, and imports and exports, and formed the bulk of the state revenue. Educational and health institutions were maintained from a public welfare fund supplied by donations.

When the Christians conquered most of Spain by the mid–fourteenth century many of the surviving Moslems left the country altogether, or converted to Christianity. Only the Emirate of Granada, which included Málaga, remained an enclave of Moslem control. In other parts of Spain, North African influences were so ingrained that they could not be shaken off. They can be detected in the philosophy and literature of the period, the cultivation of crops such as sugarcane (plantations still exist just outside Málaga), apricots, and almonds, and in irrigation practices. Numerous Islamic laws and institutions were retained, and certain features of the "typical" Andalusian house are Moslem in origin. Peter, king of Castile and León from 1350 to 1369, had a Moorish palace built for himself in Seville; a century later Henry IV of Castile had adopted Moorish customs and was running his court along the lines of Moslem rulers. In 1469 Henry's half-sister Isabella, now queen of Castile, was married to Ferdinand, her cousin and the heir to Aragon. Ferdinand took the throne in 1479, five years after Henry's death, and a new imperial era approached for Málaga and southern Spain.

The Emirate of Granada posed a potential danger to the Christian monarchs, not only in itself, but as a direct route to Spain for Moslem invaders from North Africa. Castile and Aragon combined forces against Granada in a war beginning in 1481, resulting in the fall of Málaga in 1487 and the surrender of the entire kingdom of Granada by 1492. Despite subsequent rebellions, Moslem forces never regained control.

Ferdinand and Isabella are often considered in a heroic light as the founders of modern Spain. Christopher Columbus journeyed to America under their patronage. They unified a country full of ambitious local rulers and battling factions. However, in addition to these honorable achievements, they were also harbingers of poverty on a more local level. Their rule was certainly not advantageous for Málaga and Granada, which, during the next seventy years, were brought to their knees in economic terms. The silk industry was ruined by the imposition of crippling taxes; formerly lucrative exports were halted. The Jews were expelled and Moslems were forcibly converted by the Inquisition, despite the original conditions of surrender, which allowed them to remain and practice their religion in peace. Money, property, and possessions were confiscated from Moriscos by members of the Inquisition whenever they could find a scrap of evidence in the Moriscos' disfavor. This backtracking on the surrender agreement understandably produced riots; those near Granada in 1500 led to the exile of all non-converts. Likewise, an uprising in 1568 was used as a weapon against the Moriscos in order to justify their expulsion, hindering the development of Málaga and other regions where they had become important to the economy. Those who did convert and were permitted to remain were prohibited from contact with North Africa; transgression was punishable by death.

While Málaga and other cities in Andalusia should have been among the wealthiest regions in the world in the sixteenth century, with riches flooding in from America and the Indies, much of this wealth was absorbed by the monarchs and nobles to finance their wars. Andalusia was no longer required by Spain—and was no longer able—to survive on the fruits of its own labors; farming and culture went into a decline and the region was run on almost feudal lines, while thousands of miles away, the exploitation of the recently discovered continent continued apace. Málaga's position as a foremost port of Spain was transferred to Cádiz, where cargo from the New World was imported and carried to Seville and the north.

Although losing much in self-sufficiency and status, Málaga benefited from Spain's imported wealth in that the now-famous landmark, the Cathedral, was built (or at least begun) in 1528, on the site of an earlier mosque. Designed by Diego de Siloé, it was continued by Diego de Vergara the Elder and the Younger in the mid– to late sixteenth century. Construction continued on and off until the end of the eighteenth century (an earthquake in 1680 destroyed part of the building) when the facade was completed. One of the two towers is stunted, having never been built to completion. This unbalanced look gave rise to its nickname, La Manquita, meaning one-armed or one-handed. The nearby palace, which now houses the Museo de Bellas Artes, was also constructed in the sixteenth century, and the Fuente de Neptuno (Neptune Fountain), dates from 1560. The lighthouse was begun in 1588, with an extension made in 1900. Other similar projects were under way all over Spain at this time as an architectural boom took place under the Christian monarchs.

During the next 200 years, while much was happening in Spain as a whole—the Thirty Years War; fighting against the Netherlands; loss of territories to France and other coun-

tries; the War of the Spanish Succession; the Treaty of Utrecht; the Seven Years War; the reign of the reformer Charles III; and war with France and Britain—very few events of note occurred in Málaga's own history, or indeed in that of Andalusia. Like the rest of Spain, it was occupied by the French during the Napoleonic Wars from 1808 to 1814. After the expulsion of the French, Málaga began to develop as a vacation spot, especially for the English. By 1830, there were enough English visitors to Málaga to justify the building of the first Protestant cemetery, by William Mark, the British consul and an enthusiast of the city. It is near the bullring, and referred to as the English Cemetery. Among those buried there is the poet Jorge Guillén, who lived from 1893 to 1984.

The next indication of a small tourist industry for Málaga was the building of a hotel in the 1840s. English residents numbered some 120 by 1850, drawn mainly by the still-prosperous wine trade. A further 300 visitors every year meant that English ways began to be adopted, anglicizing the area, to the delight of the visitors, eager for the comforts of home.

On the political front, Republicanism was making inroads into Málaga and other Spanish cities in the 1830s. In 1831 General J. M. Torrijos led a liberal uprising in Málaga but it was quashed by General Moreno. Torrijos and his fifty-odd followers were executed in the city. The revolutionary south of Spain found itself in an unusual position of sympathy with certain intellectuals of Madrid, and hosted tours by Republican speakers. By 1837 Málaga was an important Republican center. By the 1860s economic problems had escalated. Rural laborers usually employed in farming were jobless because of a bad harvest and turned south for work, to Málaga and to northern regions. Several days' rioting resulted when the government ceased payments to the unemployed in 1868. The Civil Guard, a repressive police force established in 1844, was a source of fear and anger for Andalusians who saw it as an organization whose main purpose was to protect the interests of the wealthy landowners. Poverty and desperation led to rural Andalusians as well as urban dwellers throughout Spain to embrace the anarchist philosophy espoused by Mikhail Bakunin of Russia. By the 1870s, discontent in Andalusia and elsewhere took the form of armed revolt and political secession. Independent cantonal republics were set up, but these could not hold out for long against the national government. Only Málaga and Cartagena remained strongholds for a few months rather than weeks.

In the next decade, municipal government also had its problems. Much of the money intended for town improvements in Málaga found its way into private pockets, as municipal governors allocated it to family, friends, and other favored individuals. The streets were badly constructed, dark, and dirty as a result of bills left unpaid by the inefficient local governors. Cities and towns became more crowded after population increases in the 1900s. Although thousands of Andalusian peasants crossed the Atlantic to seek employment, population continued to increase in Málaga and other provincial cities. In 1900 Málaga was one of six towns in

Spain with a population exceeding 100,000; by 1931, it was one of eleven.

During and after World War I, emigration to America slowed to a trickle, and the move to the towns increased, further aggravating the employment problem. Strikes ensued in Andalusia, with demands for better working conditions and higher wages. In some towns 50 percent of the workforce was jobless by 1936. In the 1920s and 1930s much of the land in provinces like Málaga was owned by a wealthy minority, with peasant farms occupying a tiny area. It was argued that the variable quality of the land made it impractical to enforce redistribution. Landowners rented out areas of their land to farmers at exorbitant prices despite the falling price of crops, breeding dissatisfaction.

Religious unrest was also abroad. An antireligious minority of radical socialists burned down twenty-two churches and religious buildings on May 11 and 12, 1931. In other cities similar actions took place as the Civil Guard looked on, having received instructions from the national government not to become involved. The bishop of Málaga was expelled for criticizing the role played by the government in these incidents.

The south in the 1930s was home to a large number of revolutionaries and anarchists, who spoke to the laborers as no intellectuals had done before, going among them as friends, feeding their desire for social justice, and absorbing their concerns into the anarchist ideal. Membership of the national union, the Confederación Nacional del Trabajo (CNT), established by the anarchists in 1910, exceeded that of the older socialist organization, Unión General de Trabajadores (UGT). So strong was the anarchist support in Andalusia that by 1932 the government began to fear a national rebellion. They were prepared when this occurred in early 1933, and managed to suppress it. However, by 1936 they had lost control of much of the region, and in Málaga churches and convents were again desecrated by extremist anarchists.

The same year saw the beginning of the Spanish Civil War, in which the anarchists made an uneasy alliance with the Republican government against the conservative insurgents, supported by the aristocracy, the church hierarchy, and the military, and led by General Francisco Franco. In February 1937 Franco's forces took Málaga, and Republican supporters were executed by the thousands—about 21,500 in the eight-year period during and immediately following the Civil War. The poets Emilio Prados and Manuel Altolaguirre were both from Málaga; both went into exile for the duration of the war.

Spain remained neutral in World War II, but many areas were already crippled from the effects of the Civil War. Conditions were very bad in Andalusia, forcing many people to migrate to the large cities farther north.

The Falangists, a nationalist political party with the support of Franco, competed with the church for popular support. One of its young members, José Luis Arrese, developed a plan for a new Málaga in the early 1940s, but nothing came of his vision. The bishop of Málaga, Angel Herrera, then started his own planning program, but was blocked by the Falangists. The government finally took on the task of rebuilding the city and began to construct blocks of flats for workers to replace the existing slums.

In the 1950s the reviving postwar economy of Europe encouraged travelers to look farther afield, and in the 1960s, under Franco's Ministry of Tourism, the south coast along the province of Málaga was reborn as the Costa del Sol. At the cost of millions of pesetas to government and private industry, Málaga and the surrounding area were transformed by the late 1960s. The Costa del Sol now incorporates the resorts of Estepona, Fuengirola, Nerja, San Pedro de Alcántara, and, perhaps most famously, Marbella and Torremolinos. A tourist boom during the 1970s made these and other Spanish resorts extremely popular, especially with British tourists, and in recent decades supply has been hard pressed to match demand. This growth of the travel trade coincided with rapid industrial growth in the city and the decline of dependence on agriculture. Unfortunately, modernization and popularity have brought the common toll of a sharp increase in crime and the summer deluge of visitors; the ancient buildings that survived the Civil War now look onto the tourist hotel blocks and modern offices, and for many months of the year, beaches are crowded with tourists.

Further Reading: Among the respected scholarly works on the history of Spain are *Modern Spain, 1875-1980* (Oxford: Oxford University Press, 1980; New York: Oxford University Press, 1981) and *Spain, 1808-1975* (Oxford: Clarendon, and New York: Oxford University Press, 1966; second edition, 1982), both by Raymond Carr. The books provide excellent coverage of Málaga's politics in the nineteenth century, the economics and agriculture of the region, and an extensive chronological table of major events in Spain during the period. *Spain* by George Hills (London: Ernest Benn, and New York: Praeger, 1970) contains excellent bibliographical details. The book is divided into "The Making of Spain," "The Making of Modern Spain," and "Modern Spain." It is informative especially on political power struggles and economics in Málaga and Andalusia, with plenty of figures and statistics as supporting evidence. *Southern Spain: Andalucia and Gibraltar* by Dana Facaros and Michael Pauy (London: Cadogan, and New York: Macmillan, 1991; revised, 1994) is a cut above the run-of-the-mill travel guides, tracing the history of southern Spain with a contemporary eye. Colloquial and often subjective, it is nonetheless a good condensed rundown of events in the region from prehistory to 1992. Another good travel guide is the *Blue Guide: Spain,* fifth edition (London: Black, and New York: Norton, 1989), which provides a brief history of Málaga, including some interesting facts and figures, and name-drops of personalities connected with the city. *A Traveller's History of Spain* by Juan Lalaguna (Moreton-in-Marsh, Gloucestershire: Windrush, and Brooklyn, New York: Interlink, 1990; revised, 1994) is of interest to the general reader with useful subtitles and headings. It is especially good on Moslem al-Andalus of the eighth century onward and includes maps and drawings, chronologies of rulers and monarchs, and major events in Spain.

—Sarah M. Hall

Malta

Location: The Republic of Malta is located in the Mediterranean Sea approximately 58 miles south of Sicily, 180 miles north of Libya, and 180 miles east of Tunisia. Of the five major islands that compose the republic, only the islands of Malta, Gozo, and Comino are inhabited.

Description: Strategically located in the central Mediterranean between Europe and Africa, Malta has been an important harbor, naval base, and cultural crossroads for thousands of years. An independent country within the British Commonwealth since 1964, Malta has been dominated for most of its history by a variety of outside powers, creating a unique society flavored by each of the cultures that once called the islands home. Although its historical fortunes were largely dictated by the volatile state of affairs in the Mediterranean, modern Malta is very much an independent state. Its economy relies chiefly upon tourism, basic manufacturing, and a growing high-tech manufacturing sector, and Malta's 350,000 residents enjoy a standard of living comparable to that of Greece, Ireland, or Spain. Moslem influences abound in Malta, but the country is decidedly European, with English and Maltese being the official languages and well over 90 percent of residents belonging to the Roman Catholic Church. Malta is also the third most densely populated country in Europe, after only Vatican City and Monaco.

Site Office: The National Tourism Organization
280 Republic Street
Valletta, CMR 02
Malta
224-444

Covering only 122 square miles spread out over the five islands of Malta, Gozo, Comino, Comminotto, and Filfla, the tiny island nation of Malta has always played a significant role in the often fractious history of the Mediterranean Basin. But not until 1964— and some would say, 1979, when Britain removed the last of its naval personnel from its base at Grand Harbour—did Malta achieve the independence necessary to write a history of its own.

Although pivotal, Malta's primary role in the evolution of the European balance of power has been as a strategic rest stop for the great military forces of the ages. Phoenicians, Carthaginians, Romans, Normans, Arabs, Turks, and Britons have all left their stamp on Malta. Even the French managed to hold onto the islands for two years during the Napoleonic era. Virtually every nation to dominate the Mediterranean

claimed Malta as its own for as long as it could, and the capture of Malta by an up-and-coming power usually signaled a change in the status quo.

Few nations can claim a culture as unique as the one that has evolved in Malta. Cobbled together over the centuries from the models provided by whoever had their ships anchored in the harbor at the time, Maltese culture is an exotic blend of European, Middle Eastern, and North African influences. European traits dominate the mix; Catholicism is the primary religion, and English and Italian are widely spoken. But no matter where its components may have originated, Maltese culture is distinctly Maltese.

Like many small nations, Malta has had to take drastic steps in recent years to diversify its economy. Malta enjoyed an easy affluence for centuries from an artificial economy based on providing goods and services to foreign occupiers. Following independence and the later withdrawal of British naval forces from Grand Harbour in 1979, the Maltese government began a series of five- and seven-year plans to create a self-sustaining economy based on industry, agriculture, and tourism. Not surprisingly, tourism—providing goods and services to foreigners, mostly Britons—has been very successful, employing almost one-third of the country's labor force. After a slow start in the 1960s, manufacturing has grown rapidly in recent years. Maltese skills in cross-cultural wheeling and dealing, honed for centuries, have given the country a valuable edge in international trade and finance. The Maltese currency, the Maltese lira, is among the world's strongest and most stable, and the tiny country's merchant fleet, at 10.6 million tons, is the twelfth largest on earth.

Agriculture has been another matter. Malta has always been dependent upon imports of food for its survival. The limestone islands themselves are notoriously short of arable soil, so much that the Maltese government has to declare that whenever soil is removed from a construction site it must be transported to an agricultural area. Naturally occurring fresh water is also scarce, and 70 percent of the nation's water supply comes from government-run desalination plants. Modern farming methods have dramatically increased the country's agricultural output, but Malta remains a net importer of food.

It is generally agreed that the first inhabitants of Malta sailed to the island from Sicily in about 3800 B.C., during the Neolithic Age. The pottery that has been found on the island both dates these people and links them to Sicily by virtue of its distinctive design. Very little is known about the cultural characteristics of these earliest residents, although they did leave behind a sizable collection of impressive stone relics. Sturdy, yet easily carved, the distinctively golden limestone of which Malta is composed has been pivotal to every culture that has called it home. Modern Maltese cities, often reckoned

Aerial view of Malta
Photo courtesy of Malta National Tourist Office

among the world's most beautiful, are constructed almost entirely of limestone.

Among the artistic and architectural artifacts left by the earliest settlers is the oldest freestanding building in the world, at Skorba. The enormous Neolithic temples found at Hagar Qim, Ggantija, Mnajdra, and Tarxien are about 1,000 years older than the famous Egyptan pyramids at Giza. Statues representing a goddess of fertility have also been found throughout the islands. Interestingly, no metal artifacts from the era have been found anywhere on the islands, leaving the builders of the massive temples with only primitive flint and obsidian tools with which to cut the huge stones and carve their intricate designs.

Near the temple at Tarxien in the modern city of Hal Saflieni lies the Hypogeum, the most prominent of Malta's archaeological wonders. An enormous underground temple constructed sometime around 2400 B.C., the Hypogeum's 1,600 square feet reach 40 feet into the earth and consist of three levels complete with an oracular shrine and a complex series of chambers and tombs. The Hypogeum, hewn from solid rock with primitive tools, is an excellent echo chamber equipped with a "microphone" in the form of a small opening in one wall. When spoken into, the opening transmits the speaker's voice

throughout the Hypogeum, imparting it with an inhuman quality sure to impress even the most stubborn of nonbelievers. The Hypogeum was discovered by accident in 1902 by a developer who had purchased the land above the structure as a site for new houses. In the course of making the necessary excavations for foundations, the Hypogeum was unearthed. However, the builder waited until after completing the houses before reporting his discovery to the authorities.

The people who built the Hypogeum disappeared mysteriously around 2000 B.C. Theories regarding their disappearance point to either an epidemic or, more likely, an invasion by a more advanced culture armed with metal weapons. Maltese history becomes sketchy between 2000 B.C. and 850 B.C. Fragmentary remains indicate the short-lived presence of a southern Italian cutlure immediately following that of the builders of the Hypogeum. Tools and relics of unknown origin from the Bronze Age and Iron Age have been found on the islands, however, substantiating the theory of continuous habitation.

In approximately 850 B.C., the Phoenicians became the first of many Mediterranean powers to colonize the strategically important islands. It was probably the Phoenicians who gave Malta its name, from their word *maleth,* meaning haven or hiding place. They used Malta primarily as a convenient way-station for seaborne trading, particularly during the winter months when the islands' sheltered natural harbors provided much needed refuge between Europe and Africa. The most important artifact of the Phoenician era is the Cippus, a marble column dedicated to the god Melkart. The inscription on its base, written in both Greek and Punic, played a pivotal role in helping archaeologists unravel the mysteries of the first Phoenician alphabet.

Phoenician rule gave way to a brief period of Carthaginian domination, as evidenced by findings of coins, inscriptions, and Punic-style tombs. This era was to be fleeting, however. Soon Rome rose to dominate the Mediterranean, and, naturally, claimed Malta as its own in 218 B.C. during the Second Punic War.

Originally used strictly for naval purposes and under the Roman praetorship of Sicily, the islands' status within the empire was eventually upgraded to that of *municipium,* or allied city. Municipal status allowed local leadership to control domestic affairs, to coin money, and to send ambassadors to Rome.

Roman rule in Malta was relatively benign. The locals provided goods and services to the occupying legions, just as they were destined to do for the next two millenia. Rome left its stamp on Malta in the form of grandiose architecture and a legal system that enshrined such notions as democracy and individual rights, though only a limited segment of the population enjoyed these benefits. The remains of impressive Roman baths can be seen today at Ghajn Tuffieha.

The first chink in the Roman armor came in A.D. 60 when, according to legend, St. Paul was shipwrecked on the island of Malta and began to convert the Maltese to Christianity. Paul proved his credentials immediately, by surviving a snakebite unharmed, and he was offered lodging at the house of Publius, the "chief man of the island." During his stay, Paul healed Publius's ailing father. Publius later became the first bishop of Malta. The overwhelming dominance of Christianity in Malta today is ample proof of his evangelical success.

When division within the Roman Empire dictated in 395 that Roman territories be split in two, Malta was ceded to the Eastern Roman Empire, dominated by Constantinople. Malta actually lay well within the Western Empire, but its independent geographical status made it a negotiable territory.

Malta enjoyed a relatively long period of near-independence following the schism. As the Roman era waned, Malta remained a colonial outpost mostly in name and was closely tied to its nearest neighbor, Sicily, with which it established close social and economic ties that have remained in place ever since.

In 870 Saracens swept through the Mediterranean, capturing Sicily and, almost by default, Malta. Surprisingly, very little is known about the Arab occupation of the islands. Their impact upon the Maltese language is undeniable, and much that is Maltese, from architecture to cuisine, is endowed with an unmistakable Arab flair. Many Maltese place names, such as the aforementioned cities of Hagar Qim and Mnajdra, are quite obviously of Semitic origin.

The Arabs' most enduring legacy in Malta is the city of Mdina. Originally settled during the Bronze Age, Mdina was fortified by the Phoenicians around 1000 B.C. and named Malet. In order to make the city easier to defend, they split the city into two cities—Mdina and Rabat—by digging a deep moat. They then walled and heavily fortified the central city of Mdina. In Arabic, Mdina can be translated as "walled city," while Rabat means "suburb." Mdina's basic layout and street plan have not changed since the Arabs made these modifications.

The Saracens also introduced valuable crops like cotton, lemons, oranges, and pomegranates to the islands. Many of the islands' terraced farms and irrigation systems have been dated to Arab rule.

Following the Norman Conquest of Sicily, which began in 1060, Malta fell to the Normans under Roger de Hauteville. Roger established a formal feudal system on the islands, which meant that for the first time, average Maltese citizens were actually included in a rigid system imposed by an occupying foreign power, as opposed to being expected only to provide necessary goods and services. The system was fortified by fiefs, titles, taxes, and rights derived from the count. Maltese history is undecided about Roger. His feudal hierarchy may have added a measure of "civilization" to the islands, but it also curtailed freedoms and invited abuse. Nobles expected tribute, but when the islands were raided in the succeeding centuries by competing Moslem and Christian pirates, they also hid away in their well-fortified homes and let the beleagured peasants defend themselves.

Disputes over the succession of Roger's heirs led Malta to be passed around from one Mediterranean power to another, beginning with Charles of Anjou, brother of Louis IX of France, in 1266. In 1409 the Maltese and Sicilian portion of the kingdom passed to the Crown of Aragon, which also included portions of present-day France and Spain.

Malta was transformed forever in 1530, when on March 24, Charles V, Holy Roman emperor and king of Spain and Sicily, ceded the islands to the Knights Hospitallers of the Order of St. John of Jerusalem, who then took the name of Knights of Malta, by which they are remembered today. In return for the islands, the knights were required to pay a nominal rent of one falcon or hawk each year to the king of Sicily, as Maltese falcons were highly prized. (Falcons can no longer be found in the wild on Malta, but it is this treaty that provided the basis for Dashiell Hammett's famous mystery novel *The Maltese Falcon* and the Humphrey Bogart film of the same name.)

The oldest of the ecclesiastical orders of knighthood which flourished throughout Europe during the early portion of the second millenium, the Malta order operated as a quasi-autonomous society under the auspices of Rome. All of the orders had a dual military and religious function, and most had specialties, too. The Order of St. John specialized in the construction and administration of hospitals, which were not at all common in 1530.

The order first built a hospital for Christian pilgrims in Jerusalem early in the eleventh century. When Jerusalem was recaptured by the crusaders in 1099, the hospital was under the auspices of a man named Gerard. When the Crusades ended, Gerard reorganized his institution into the Hospitallers of St. John and drafted a constitution that was approved in 1113 by Pope Paschal II. A few years later the knight Raymond du Puy gave the order its military function. Initially, this function was limited to escorting pilgrims and providing protection for the sick and poor. But a series of strictly military battles against the Moslems in conjunction with other allies expanded the military function and turned the order into a full-fledged mercenary organization at the disposal of the pope.

Despite its historical fame, the order's membership never grew beyond approximately 500 to 600 sworn knights. Bureaucrats and hospital personnel accounted for significantly more of the order's population than did the knights for whom the order is remembered. The order was composed of members of noble families of England, France, Spain, Germany, and Italy and was devoted to the protection of Christianity and, in reality, the bedevilment of Islam, in particular the Ottoman Turks. Knights took vows of celibacy and poverty and swore allegiance to their elected grand master, who operated under authority of the pope.

For the knights, Malta provided a suitable base of operations in their efforts to stem Turkish piracy in the Mediterranean. Every six months, the order sent a naval expedition along the Mediterranean coast to do battle with Ottoman pirates and participate in the anti-Islamic crusades of various nations of Christian Europe.

Obviously, the knights were not successful in their efforts to overthrow Islam. But they were effective enough to raise the ire of the Ottoman Süleyman I (the Magnificent), who in 1565 ordered the invasion of Malta. Süleyman sent an armada of more than 200 ships and 40,000 troops to the islands, intent upon destroying the order. Süleyman had not counted upon the skills of Jean Parisot de Valette, then grand master. With a motley force of knights, men-at-arms, Maltese soldiers, volunteers, angry citizens, and last-minute assistance from Sicily, de Valette withstood the siege, turning back the Turks within four months and declaring victory on September 8, 1565. The Turks never again attempted to penetrate the western Mediterranean.

Malta's success against the numerically superior Turks was heralded throughout Christian Europe as an important victory in the centuries-long struggle against the "infidel." However, it was also indicative of the problematic nature of the relationship between the Maltese and the Order. Although sworn to poverty and celibacy, the knights were anything but poor or, it is safe to assume, celibate. Jean de Valette is remembered for his victory against the Turks, but without the help of 7,000 Maltese, victory would not have been possible. Furthermore, had the order taken up residency elsewhere, Malta never would have been invaded with such ferocity in the first place. In 1565 the Maltese were still pleased with the daring exploits of the newest occupiers, but it was not long before the Knights of St. John degenerated into a lazy aristocracy of questionable worth.

It was the knights, however, who left the most significant legacy on the islands. Their wealth and influence allowed them to import the most significant artists and architects of the day, and modern Maltese city centers are still composed primarily of the magnificent limestone buildings commissioned by the Order. The city of Valletta, the modern capital, was created by the Order following the success of the Great Siege. Built on the peninsula of Sciberras and named for Jean de Valette, Valletta was built on a grid pattern and planned from the very beginning. During its construction between 1566 and 1571—funded in large part by donations from Christendom outside Malta—garbage disposal and drainage systems were built into the plan. A planning department akin to a modern zoning board oversaw the construction and established strict rules regarding the use of specific properties within each district.

Naturally, Valletta was heavily fortified, and a moat was dug between Grand Harbour and Marsamxett Harbour to separate Valletta from the mainland. The moat, more than 3,000 feet long, 65 feet wide, and 60 feet deep, was the most important of the various fortifications established at the new city.

By the late eighteenth century the order had grown into an anachronistic, bloated, and corrupt aristocracy that treated the Maltese badly when it bothered to notice them at all. In 1775 a group of Maltese clerics staged an unsuccessful public revolution against the oppressive rule of the order. Following the French Revolution, the order went bankrupt as the French republican government seized the order's important revenue-producing estates in France. When Napoléon Bonaparte landed at Malta in 1798 on his way to Egypt, the islands were his for the taking. Under the guise of seeking fresh water supplies, he landed at Valletta on June 12 and declared Malta to be a department of France. The useless knights made no attempt to stop him and, after 268 years, their rule on Malta came to an ignominious end. The islands were ceded to Napoléon without a shot being fired.

Napoléon stayed in Malta for six days, writing the laws and issuing the proclamations necessary to transform the islands into a standard department of France. It did not work. The Maltese revolted, successfully blockading the French troops behind the walls of Valletta and the nearby Cottonera for two years as they appealed for help from other Mediterranean powers, including Naples, Portugal, and Britain. It was the British fleet led by Admiral Horatio Nelson that came to the aid of the Maltese, and by 1800 the French garrison had been expelled. Malta became a de facto British Crown Colony. In 1802 the Treaty of Amiens returned the islands to the Knights of St. John, but the Maltese protested and insisted upon the continued protection of Britain, subject to a Declaration of Rights. Britain agreed and the relationship was formalized by the Treaty of Paris in 1814.

The British transformed the islands, introducing sweeping reforms in government, law, public health, transportation, and education. They also dramatically expanded existing naval facilities at Grand Harbour, renaming the area the Dockyard. Before the end of the nineteenth century the Dockyard had become Malta's economic mainstay. Malta served Britain well in its various nineteenth-century imperial wars and benefitted greatly from the opening of the Suez Canal in 1869. In World War I Malta served as a "hospital island" and earned the nickname "nurse of the Mediterranean."

Malta suffered another great siege during World War II, in the form of Axis air bombardment, and was one of the most heavily bombed countries of the conflict. Bombing began in June 1940, before the island or its British garrison was ready to fight, and continued for the next two years in combination with attacks on merchant vessels attempting to reach the choked-off nation. British fighter planes and Maltese ground troops, however, held an actual invasion of the islands at bay. The Maltese burrowed into the hills, digging caves and tunnels for refuge from the bombs. Royal Navy convoys attempting to bring supplies to the islands were routinely sunk, and by August of 1942 Malta stood alone against the Axis. In that same month, however, one tanker and four cargo ships of an original convoy of fourteen vessels sailed into Grand Harbour with food and fuel for the Maltese. In recognition of their heroism, King George VI awarded the Maltese Britain's highest civilian honor, the George Cross. This was the first time such an honor had been bestowed upon an entire population. American President Franklin D. Roosevelt also paid tribute to the Maltese when, in 1943, he presented the people with a Presidential Citation and a plaque for "valorous service far above and beyond the call of duty."

Although constitutional rule in Malta was attempted more than once, constitutions were suspended from 1936 to 1947 and from 1959 to 1962, and power was vested exclusively in the hands of the colonial governor. At the heart of the problem were demands for special treatment from the English, Italian, and native Maltese.

On September 21, 1964, Malta became an independent state within the British Commonwealth, although the government negotiated a ten-year agreement with Britain to provide for defense and financial aid.

In the 1970s, the island opted for nonalignment, successfully negotiating various agreements with an unlikely assortment of allies, including the United States, Libya, Tunisia, Italy, the Soviet Union, and China. Malta pursued especially close relations with Libya, the military component of which was eventually curtailed in favor of improving ties with the United States. The government also renegotiated Malta's treaty with Britain, receiving substantially higher rental payments for use of the Dockyard naval base by British and other NATO warships and agreeing to the complete withdrawal of foreign troops by 1979. On December 13, 1974, Malta declared itself a democratic republic, ridding itself forever of the last vestiges of colonialism. The British withdrawal in 1979 completed the cycle, and Malta has been self-governing ever since.

Although Western in culture, language, and religion, Malta has steadfastly maintained its nonaligned status. In 1988 four British warships stopped briefly at Malta during a routine patrol of the Mediteranean. A general strike was called throughout the island in protest of the visit because the ships were believed to be carrying nuclear weapons, in violation of Malta's nonaligned status.

In 1989 Malta was the site of the first summit meeting between American President George Bush and Soviet Premier Mikhail Gorbachev. The summit was supposed to have been held aboard two warships anchored off the islands, but high seas forced talks to move to more stable ground. The summit quickly became known in newspapers around the world as the "seasick summit." Malta is expected to join the European Union sometime around the end of the twentieth century.

Further Reading: Information on the Knights of Malta abounds. Of particular note are *Malta of the Knights* by Elizabeth Wheeler Schermerhorn (New York: Houghton Mifflin, and London: Heinemann, 1929) and *The Ecclesiastical Orders of Knighthood* by James Van der Veldt (Washington: Catholic University of America Press, 1956). The *Europa World Book* (London: Europa Publications, 1994; as *Europa World Yearbook,* Detroit: Gale, 1994), published annually, is an invaluable source of information regarding little-known countries, such as Malta. Malta is also well-served by travel guides.

—John A. Flink

Mantua (Mantova, Italy)

Location: Seventy miles southeast of Milan.

Description: Capital of the province of Mantova in Lombardy, ruled for centuries by the Gonzaga dynasty, renowned for their patronage of the arts.

Site Office: A.P.T. (Azienda di Promozione Turistica)
Piazza Andrea Mantegna 6
CAP 46100 Mantua, Mantova
Italy
(376) 350681

Mantua (Mantova in Italian) is located on a peninsula, surrounded on three sides by the Mincio River, which widens into Lago Superiore (Upper Lake), Lago di Mezzo (Middle Lake), and Lago Inferiore (Lower Lake). The first settlement on the site, founded in the tenth century B.C., was one of two major towns—along with Felsina, which was to become Bologna—among a group of twelve in the valley of the Po River that were founded by the Etruscans. The town was highly developed for its time, with a flourishing trade in pottery and agricultural products. Its geographic position was advantageous not only because it gave access to trade along the Mincio River, but also because it was militarily defensible. Even so, Mantua was not spared by the Celtic invaders who arrived in the region during the sixth and fifth centuries B.C. and who went on to conquer most of northern Italy.

The destruction of the Etruscan civilization in its heartland left Mantua and many other settlements in the area underpopulated and undefended, and therefore easy for the growing power of Rome to conquer and control. Mantua accordingly became subject to the Roman Republic in approximately 220 B.C. Under Roman rule the city and its hinterland were revitalized through programs of reclamation of the marshes and swamps around the three lakes, and agriculture flourished once more. Mantua became famous throughout the Roman Empire through its association with the life and works of the poet Virgil (some of whose writings form the major source of information on the early history of the city). Virgil was born in Andes, a small village nearby, in approximately 70 B.C. His connection with the city is remembered today with a monument in the Piazza Virgiliana, which was created at the beginning of the nineteenth century.

In the fifth century A.D., much of Italy was overrun by a series of invaders, starting with the Visigoths in 410, the Vandals in 455, and the Ostrogoths in 476. As the Western Roman Empire disintegrated, control over Mantua and other cities changed hands several times, both as a result of these invasions and of the subsequent conflicts in the early sixth century between the Ostrogoths, who, unlike the other invaders, settled on the land, and the forces of the Byzantine

emperors, who sought to impose their claims as heirs to Rome. In 568 another eastern Germanic group, the Lombards, completed their conquest of what has since been known as Lombardy, the region that includes Mantua. In 774 Charlemagne, ruler of the Franks, conquered the part of Lombardy that included Mantua. During these upheavals Mantua had become a religious center, with both a cathedral and a Benedictine monastery. Sometime in the ninth century it also became known for an important sacred relic, the Preziosissimo Sangue, believed to be the "most precious blood" shed by Jesus during his crucifixion, which was preserved inside the monastery church (later to become the church of Sant'Andrea, or St. Andrew). In 945 Lothar II, a Frankish King of Italy, granted the bishops of Mantua the right to mint coins that would also be used in the towns of Verona and Brescia, an indication of the economic and political power of the bishops versus the claims of the local nobles, the counts, who ruled these three towns.

Mantua became a possession of a greater noble, Boniface of Canossa, marquis of Tuscany, during the eleventh century. His family retained control, usually in alliance with the bishops, until the death of his descendant Matilda of Tuscany in 1115. By then a movement toward communal autonomy had begun to develop in northern and central Italy, generally as a result of the growth of a prosperous merchant class that collaborated with military families (Consorteria) to curb the powers amassed but often abused by the local bishops. After several years of negotiation, Mantua, too, became a commune in 1126, under consuls chosen by the leading merchants. For many towns, including Mantua, this system brought conflicts with local military families, for the constant threat of war among the various city-states increased the influence and ambitions of these families. Mantua had already been attacked in 1121 by the forces of Verona and in 1125 by those of Reggio and Modena, and Mantua went to war against Verona five more times until 1150. Each side sometimes resorted to brutality, as on the notorious occasion when the Mantuans cut off the noses of around 3,000 Veronese prisoners. The rivalries among the cities of northern Italy were further complicated after 1137, when the succession to the throne of their nominal overlord, the Holy Roman emperor, was disputed between Konrad von Hohenstaufen (Conrad III) and Henry of Bavaria; each city took sides, either with Conrad's supporters (the Ghibellines) or with Henry's (the Guelphs).

In 1167 Mantua joined the Lombard League, a group of northern Italian towns allied against the policies of Emperor Frederick I (Barbarossa). Ghibelline and Guelph now became labels used respectively for those loyal to the emperors and those who resisted them, either in the name of local autonomy or in support of the popes. The Palazzo della

Painting by Giulio Romano in the Palazzo del Tè

Mantegna fresco in the Camera degli Sposi at the Ducal Palace
Photos courtesy of A.P.T., Mantua

Ragione, the town hall of Mantua built between 1198 and 1250, with its bell tower and its balcony from which the city's leaders addressed crowds of supporters, can be seen as a symbol of Mantua's Guelph allegiance.

The ruling families of Mantua faced challenges within the city as well as without. During the early thirteenth century, new religious ideas flourished throughout Italy, in the form of various sects, including, for example, the Franciscan friars. In Mantua the radical Cathars, who rejected the authority of bishops and popes and protested against what they saw as neglect of the poor, developed a significant following. By 1260, however, most of the Cathars had been eliminated by the inquisition, which the Guelph elite, led by the Este and San Bonifazio families, welcomed to the city. Only fifteen years later, following further civic upheavals, Mantua passed into Ghibelline hands, with the election for life of Pinamonte Bonacolsi as Captain of the People. He and his family ruled Mantua as despots from 1275 to 1328, displacing even the power of the nobility. Growing resentment among the old landowning families eventually led to the downfall of the Bonacolsis, however; Passarino Bonacolsi was assassinated by his brother-in-law Luigi Gonzaga and his followers on August 16, 1328, and Luigi was elected captain in his place. Mantua was to be ruled by the Gonzagas, one of the most powerful political dynasties in northern Italy, until the eighteenth century.

Realizing that they could not afford to militarily or commercially challenge such larger and richer rivals as Milan and Venice, the Gonzagas were generally shrewd and cautious in their policies, preferring to increase their wealth by selling their soldiers' services as mercenaries and to secure their political position through arranged marriages into the ruling families of Italy and Germany and through appointments within the Catholic Church. Nor did they neglect defense: one of the greatest of the dynasty, Francesco I, signore of Mantua from 1388 to 1407, ordered the building of the Castello San Giorgio to protect Mantua against any possibility of attack from the lakes; the structure was completed in 1406.

Gianfrancesco I took control of Mantua in 1395 and was given the title of Marquis by Emperor Sigismund in 1433. He concentrated on developing the city's role as a center of culture and learning. Although the city did not acquire a university, the school he helped to found in 1423, headed by the Humanist scholar Vittorino da Feltre (Vittorino Ramboldini), became an influential part of the network of institutions promoting Renaissance ideas and attitudes. Gianfrancesco's son Lodovico III, who succeeded him in 1444, continued to patronize the arts and education. In the summer of 1459, he hosted the Congress of Mantua, a series of meetings of delegates from the major powers in Christendom, called by Pope Pius II to discuss a possible crusade against the Turks, who had conquered Constantinople six years before. Some of the delegates became ill with fever, for the lagoon and rivers were infested with malarial mosquitoes. Lodovico's response was to expand his sponsorship of schemes for land drainage and for urban development, including new buildings, paved streets, new piazzas and, to show his interest in what was then new technology, a zodiacal clock in the town hall.

Some of the best artists and architects of the period worked on these schemes. Andrea Mantegna, a leading painter of the Padua School, arrived in Mantua in 1460. His works can be found in many buildings, including the frescoes in the Camera degli Sposi in the Ducal Palace, on which he worked from 1465 to 1474. The frescoes depict the court life of Lodovico and his German wife, Barbara of Brandenburg. The decoration of the vaulted ceiling of the chamber, also painted by Mantegna, is one of the earliest examples of *trompe l'oeil*. Other pictures by Mantegna, several by his son Francesco, and Mantegna's tomb can all be seen in the Church of Sant'Andrea on the Piazza Marconi, rebuilt in 1472 to designs by the Florentine architect Leon Battista Alberti, who also designed the Church of San Sebastiano. This has been expanded and restored many times since but retains his Greek-cross plan, resembling a plus sign, rather than the much more common Catholic Church pattern based on the elongated nave. Another important architect brought to Mantua by Lodovico III was Luca Fancelli, who began the building of the Reggia dei Gonzaga, which incorporates the grand Palazzo Ducale (Ducal Palace).

Lodovico III's reputation as a patron of the arts, though considerable, is exceeded by that of Isabella d'Este, the wife of his descendant Francesco II and, after his death in 1519, a leading adviser to their son, Federigo II. For nearly fifty years she associated with the great artists and humanist thinkers of the time, and in addition governed Mantua for long periods when her husband and then her son were away fighting. Isabella and Federigo together employed Giulio Romano (Pippi de'Giannuzzi), a pupil of the great painter Raphael, as both painter and architect from 1524 onward. Romano's work includes the interior of the Bishop's Palace, the rebuilding of the Cathedral of Santi Pietro e Paolo (Sts. Peter and Paul) after a fire in 1545, and the Palazzo del Tè, Federigo's country house, which was built from old stables in 1525. They also added new buildings to the Reggia dei Gonzaga complex, one of the largest palaces still standing in Italy, with more than 450 rooms, 12 courtyards and gardens, 3 piazzas, and a church. Today it houses many fine museums, among them the Greco-Roman Museum, which contains the ancient artifacts collected by the Gonzagas.

The Gonzagas' rank was raised from marquis to duke in 1530, after Lombardy, which had come under intermittent French military occupation during the Italian Wars (from 1494 to 1559), was reconquered by its hereditary overlord, Holy Roman Emperor Charles V. During Charles's visits to the city in 1530 and 1532 he stayed at the Palazzo del Tè. Isabella d'Este died in 1539, but her example of patronage continued to influence the Gonzagas for at least another century. Vincenzo I extended the tradition to include music: the first performance of Claudio Monteverdi's opera *Orfeo* took place at his court in 1607.

By then Mantua's population had greatly increased, perhaps approaching 80,000 and probably exceeding this level in the twentieth century. However, a crisis in food production and a number of outbreaks of the plague, especially between 1630 and 1633, seriously depleted the populations of all the cities of northern Italy, including Mantua. The city also suffered directly from armed conflict for the first time in centuries, with the outbreak of the War of the Mantuan Succession (1628–31), after the direct, Italian line of the Gonzaga family died out. In 1630 Mantua was attacked and largely destroyed by Austrian troops, but in 1631 it was relieved by French forces, sent to secure the duchy for the Gonzaga-Nevers family, a branch of the dynasty long settled in France. In return the city was obliged to join with France and Savoy in a war against Spain that lasted from 1635 to 1659 and involved huge expenditures for a city that had already lost its former prosperity. The Duchy of Mantua (the city with the surrounding countryside) thus became an informal French dependency years before its sale to the French king Louis XIV in 1701, and the death of the last of the Gonzagas in 1708.

In the same year that he acquired Mantua, Louis XIV found the major powers of Europe ranged against his plans to make his grandson king of Spain. Although the resulting War of the Spanish Succession did not prevent this plan from being realized, it did deprive both France and Spain of many territories, including the Duchy of Mantua and most of southern Italy, which passed to the Habsburg dynasty of Austria under the Treaty of Rastatt, signed in 1714. The Austrians strengthened Mantua's defenses, making it, along with Peschiera, Verona, and Legnago, one of the key points of their "quadrilateral" defensive line, thereby exposing the city to involvement in further conflicts between the Habsburgs and the Spanish Bourbons for control of Italy, notably in the form of a Spanish attack in 1734.

That attack was repulsed, but two generations later, following the decline of Spain and the eruption of revolution in France, French troops, led by Napoléon Bonaparte, laid siege to the city between June 4, 1796, and February 2, 1797, and attacked it again in 1799, occupying it until 1814. Napoléon arranged for many of the great artworks owned by the Gonzagas to be taken to France, and heavy taxation led to an uprising in Mantua and other Italian cities in 1809. The leader of the uprising, Austrian Andreas Hofer, was executed in Mantua in 1810. A memorial to him can be seen in the Citadel on the north side of the Lago di Mezzo.

Mantua returned to Austrian control in 1814, but like other Italian cities it was affected by the growth of Italian nationalism—the movement known as Risorgimento—and there were frequent protests and uprisings against foreign rule. The crypt of the Church of San Sebastiano now contains the tomb of a group of rebels known as the Martyrs of Belfiore, executed by the Austrians after the major uprising of 1851–52. Mantua, Venice, and other Austrian territories in the peninsula joined the Kingdom of Italy in 1866 following Austria's defeat in the Austro-Prussian War (Seven Weeks War). Although the countryside around the city suffered from a decline in agricultural production and a rise in unemployment, which led to waves of strikes and protests eventually suppressed by the army in 1884, Mantua itself was able to develop into a major industrial center with the expansion of the railway network. The city absorbed much of the rural population and became involved in the food processing, paper making, ceramics, and furniture industries that have dominated its economy during the twentieth century. The city suffered from heavy bombing in World War II, which caused severe damage to some of its medieval buildings, although, fortunately, many of its portable art treasures were saved because they were stored in the Ducal Palace. Since the war, Mantua has also been the site of a growing petrochemical industry, on the outskirts of the old town, which, unfortunately, has caused serious pollution of the lakes around the city. In spite of these changes, the population of Mantua, at around 60,000, is still less than at the height of its glory under the Gonzagas. The city has retained its medieval walls, its historic buildings, and its many cobbled streets.

Further Reading: The Longman History of Italy series, including John Larner's *Italy in the Age of Dante and Petrarch* (London and New York: Longman, 1980) and Eric Cochrane's *Italy, 1530–1630* (London and New York: Longman, 1988), provides a comprehensive account of the background, and many of the main events, of Mantua's history. Further details can be found in Selwyn John Curwen Brinton's *The Gonzaga: Lords of Mantua* (London: Methuen, and New York: Brentano, 1927) and in the two volumes of Mrs. Julia Mary (Cartwright) Ady's biography *Isabelle d'Este, Marchioness of Mantua 1474-1539* (New York: Dutton, and London: Murray, 1903), although both these works are somewhat outdated and may be difficult to obtain.

—Monique Lamontagne

Messenia (Greece)

Location: Southwestern part of the Peloponnese region in Greece. Bordered by Elis in the north, Arcadia in the northeast, and Laconia in the east. The Mediterranean Sea surrounds this region on the west and south.

Description: Not to be confused with its ancient capital, Messene, Messenia is a historic region in the Peloponnese and a department of modern Greece with a history that dates back as far as 2600 B.C. It is known for its excellent arable lands and boasts of its mountains as one of the world's best natural fortifications. Messenia was plagued by many wars and invasions including the First, Second, and Third Messenian Wars, which indirectly led to its independence after 500 years of Spartan control.

Site Office: Tourist Police
Kalamata, Messenia
Greece
(721) 23187

The archaeological record of Messenia indicates some form of civilization in the region as early as the second millennium B.C. While the history of the western and northern areas of the Peloponnese, which included Messenia, is difficult to reconstruct, archaeological excavations show that in the Early and Middle Helladic periods (extending roughly from 2600–1600 B.C.), Messenia was one of the most active areas in Greece. The discovery of tholos tombs, which held the bodies of deceased royals and their possessions, was made in Messenia. There is evidence of more than 100 settlements during the Bronze Age, and tholos tombs were dispersed throughout the region; such a matrix suggests that Messenia was divided into a number of small principalities. The population grew in these numerous principalities, which included a fortified site on the hill called Ano Englianos, eventually to become the center of a large dynasty.

By the middle of the Late Helladic or Mycenaean period (1500–1100 B.C.), most of the properties on the hilltop were destroyed, and in their place a palace of considerable size was erected. This palace, better known as the Palace of Pylos, was home to the legendary royal family of Nestor. In the Homeric poem the *Iliad,* Nestor is identified as a wise counselor to the Achaeans during the Trojan War. This is the earliest mention of Messenia in Greek literature.

According to one tradition, springing from the second-century Greek historian Pausanias, Messenia was held by the descendants of Nestor for two generations after the Trojan War. Another tradition testifies to the control of Messenia by Nestor's descendants up to the Trojan War, when the region fell into the hands of Orestes, son of another Homeric hero, Agamemnon. Modern archaeological evidence suggests that after having been exchanged by many hands, Messenia, along with the rest of the Peloponnese, fell under the control of the Dorians as early as 1000 B.C.

In the eighth century B.C. another foreign invasion came, this time from the Spartans, who were becoming one of the more sophisticated peoples on mainland Greece. In the spirit of expansionism, Sparta looked upon the verdant and fertile plains of Messenia as a potential tributary. What ensued between the Messenians and the Spartans was the beginning of a lengthy trilogy called the First, Second, and Third Messenian Wars, which spanned from 743 B.C. to 460 B.C. The cause of the First Messenian War has been a matter of dispute among historians, as has been the whole historical sequence of Messenia's culture and tradition.

Some attribute the war to an altercation between the Spartan Euaephnos and the Messenian Polychares, the champion at the fourth Olympiad in 764 B.C. Polychares, wronged by Euaephnos, took his grievances to the Spartans. When the Spartans ignored him, Polychares struck out randomly against Lacedaemonians, neighbors of Messenia under allegiance to Sparta. Spartan kings Alkamenes and Theopompus demanded the surrender of Polychares, but the Messenians and one of their kings, Antiochus, refused. Androcles, the other Messenian king and brother of Antiochus, strongly urged his people to surrender the Olympic hero. The dissension took an ugly turn, becoming an internal war among the Messenians; the outcome was the death of Androcles.

While the Messenians were distracted by internal conflicts, the Spartans of Lacedaemonia attacked the small bordering town of Ampheia. The Lacedaemonians continued their rampage, assaulting other insignificant towns across the Messenian territory. Antiochus's son and successor, Euphaes, met this encroachment with quick retaliation. In the fifth year of the battle, however, the Spartan forces asserted their greater strength and forced the Messenians to take refuge on the fortified Mount Ithome. There, Euphaes died of his wounds from battle. His successor, Aristodemos, committed suicide not long afterward.

Without a leader, the Messenians were forced to surrender or flee. Those fortunate enough to escape fled to other parts of Greece such as Messenia's ally and neighbor, Arcadia. Others sought refuge overseas, especially southern Italy and Sicily. Those who had surrendered became helots, or Spartan serfs, especially the people of the Stenyklaros plain in northern Messenia and Makaria near the Messenian Gulf.

The Spartans divided their new lands into *kleroi,* or land-lots. The Messenian helots cultivated the Spartan *kleroi* in the Stenyklaros and Pamisos Valley, enduring this way of life for approximately fifty years until an uprising on their

Ruins of city walls at Messene, Messenia's ancient capital
Photo courtesy of Greek National Tourist Organization

part led to the Second Messenian War, which lasted roughly twenty years (670–650 B.C.). Their rebellion came about partly because of the Spartans' defeat by Argos in 669 B.C. Despite their helot status under Spartan rule, the Messenians had a considerable amount of autonomy. By comparison with their counterparts, the Laconian helots, the Messenians lived farther from Sparta and had, as a natural defense, a formidable mountain barrier. The Messenians also were able to detect the growing internal dissensions among the Spartans and saw an opportunity to regain their lands.

The hero of the Second Messenian War, Aristomenes, led the Messenians on a campaign against Sparta. The Messenians were successful in enlisting the sympathies of the Arcadians, Argives (inhabitants of Argos), Pisatai, and Eleans, while Sparta maintained an alliance with the Corinthians and the people of Lepreon. Internal disunity, one of the main reasons for the defeat of the Messenians in the first war, was again a cause for their defeat in the second. The treachery of the Arcadian king, Aristocrates, in receiving bribes from Sparta and his subsequent flight from his allies, resulted in the Messenians fleeing to Mount Eira. This fortified mountain, on the banks of the Nedon River, and close to the Ionian Sea, was the Messenians' stronghold for more than a decade.

However, similar to their ancestors who were left without a leader at Mount Ithome during the first war, the Messenians were abandoned by Aristomenes on Mount Eira when he could no longer maintain his position. Thus, the Second Messenian War ended with the complete subjugation of the Messenians to Spartan rule.

From the end of the war to the beginning of the fifth century B.C., no revolt was effective in de-stabilizing its enemy. Sparta divided the conquered territory to be worked by helots and a number of principalities. The conquest of territories such as the Stenyklaros plain, the lower west portion of the Pamisos Valley, and the section at the base of the Arkritas peninsula provided the Spartans with a handsome economic resource. Without doubt, such a conquest aided Sparta in becoming a formidable presence in the Peloponnese.

Revolts on a larger scale occurred as early as 489 B.C. Overseas, Messenian refugees offered their new allegiance to Anaxilas, tyrant of Rhegion (modern-day Reggio di Calabria) in southern Italy. As a reward for capturing the commercial city of Zankle, Anaxilas presented the city to the Messenians, and it was renamed Messene. On the mainland, Sparta was engaged in the Battle of Plataea in 479 B.C. during the Persian invasion of Greece. The helot population had grown to the degree that there were 35,000 Messenian and Laconian light-armed troops at the battle, the equivalent of seven helots for each Spartan.

Animosity toward Sparta among its former allies grew both during and after the Persian invasion. The helots saw the growing enmity as an opportunity to host an uprising. However, the occasion to revolt further intensified with an earthquake that devastated Sparta in 465 and reduced its population considerably.

The Messenians' plan to attack Sparta itself was thwarted by the Spartan king, Archidamus. The Messenians retreated to Mount Ithome, and Sparta enlisted the service of the Athenians, considered experts in overtaking fortified enemy stations. When the Athenians were unable to produce successful assaults on the inhabitants of Mount Ithome, the Spartans suspected their ally of harboring sympathies toward the Messenian helots. With much disgust, the Spartans dismissed the Athenians. After a stalemate of almost ten years, the Messenians were forced to surrender, but the surrender was made on the terms that the Messenians were allowed to leave the Peloponnese. If they ever set foot in the territory again, they risked becoming slaves of the person who captured them. Thus ended the Third Messenian War.

The expelled Messenians were received by the Athenians, who by this time had become hostile toward the Spartans. The Messenians were settled at Naupaktos on the Gulf of Corinth. Their gratitude toward the Athenians was displayed in their military service during the Peloponnesian War (431–404 B.C.). The Messenians served with the Athenian naval forces and captured Pheia on the coast of Elis, and a couple of years later aided the Athenian general, Nikostratos, to spread pro-Athenian democratic ideals in Corcyra (Corfu) in 428. Their campaign again took them to Aetolia with Athenian general Demosthenes in 426. The Messenians' fidelity drew the approval of Demosthenes so much that he enlisted a great number of them to serve under his command to fight Eurylochos and his Spartans. After a victory over the Spartans in Sphakteria in 421, the Messenians were again expelled from Naupaktos and surrounding areas as a condition of the peace treaty with Sparta. The homeless Messenians fled back to Sicily and Cyrene.

Sparta's determination to preserve its dominion over the Peloponnese was met with an equal desire by the Thebans. At the request of the Arcadians, Thebes sent an army to attack Sparta; the Theban army was successful in defeating the Spartans in the Battle of Leuctra in 371. The army then penetrated Laconia, a Spartan stronghold, and devastated the area.

Seeing the strength of the Spartans begin to wane, Epaminondas next led his army and allies into the Spartan territory of Messenia. The procession awakened the serfs who belonged to the old Messenian race, and they eagerly joined in the march to the slopes of Mount Ithome, where General Epaminondas founded the city of Messene in 369 to become the prototype of newly liberated cities. The

Messenians, who had been without a home for so long, were finally called to reclaim what had belonged to them and their ancestors.

The site of the new city, Mount Ithome, was chosen deliberately by Epaminondas as the strongest natural fortress in all of Greece. Epaminondas hoped that the strategic location of this new state would cut off Sparta from its economic resources. With Mount Ithome as its acropolis to the west, the new city of Messene began the work of populating itself. Former Messenian helots as well as helots from other regions, exiled Messenians, and others who had aided the Thebans were welcomed to settle in the allotted land. Among the settlers were the repatriated Messenian members of ancient priestly families who revived the cult worship of Zeus Ithomatas, for whom the mountain is named. Up until the end of the fourth century, the city was called Ithome but was later appropriately named Messene.

The inhabitants of Messene took all measures to secure themselves from further altercations with Sparta and its allies. The city was heavily fortified by a circuit wall five and one-half miles long and ten feet wide that to this day is considered an example of phenomenal military structure of the third and fourth centuries B.C. The people began to forge diplomatic relations with the Arcadians, Argives, and Thebans. Messenia also won a seat among the military powers after fighting in the Battle of Mantinea in 362. In 365, Messenia was successful in securing an alliance with the Athenians protecting it against future Spartan aggression.

For the next 150 years, Messenia took a position of neutrality while the Macedonians and the Achaean League conducted warfare. Its neutrality was challenged as Messenia was attacked by various forces, primarily Sparta. However, warfare came to an end with the Peace of Naupaktos in 217. This peace, acknowledged by all the affected parties, recognized a greater problem: the threat of Roman power.

Messenia broke off its position of military neutrality and allied itself with Rome, under whose rule it lived in relative prosperity. While the greater part of Greece achieved economic stability during Roman occupation, the wealth was much more concentrated in the Peloponnese. This may be attributed to the efficient provincial government management and favorable weather conditions that gave and continues to give its people rich, arable soil and crops.

Messenia, and in particular its capital Messene, experienced economic prosperity unlike its neighbors. Documents reveal that the inhabitants not only paid the Roman government property tax but also supplied soldiers and slave-sailors. The citizens of Messene were thus able to pay their assessments and furnish resources to the Romans with considerable ease.

After the fall of the Roman Empire, Messenia, like the rest of the Greek world, was swallowed up by a series of foreign invasions, and was part of the Byzantine Empire for nearly seven centuries (500–1200 A.D.). Messenia then became part of the Achaia province under Frankish control from 1205 to 1432. The Ottoman Empire held almost all of Messenia under its dominion from 1460 to 1821, with a minor interruption by the Venetian occupancy from 1685 to 1715.

In 1821 the Greek War of Independence was declared. The formal declaration of the revolution of the Greeks against the Turks was made in Kalamata, present-day capital and chief port of Messenia. The declaration was made in the church of Agii Apostoli, which was erected in the thirteenth century. Kalamata was sacked by the Turks during the War of Independence, captured by the Italians during World War II, and damaged by an earthquake in 1986, but it retains many historic structures, including the Agii Apostoli church, a thirteenth-century Frankish castle, and Byzantine-era churches. The region has many other historic towns, such as Koroni, with its Venetian castle, and Methoni, with a Venetian fortress and Turkish baths. The Palace of Pylos, also known as Nestor's Palace, lies near Hora, which has a museum displaying antiquities from the palace and elsewhere. At the site of the ancient capital, Messene, one can view remains of the defensive walls, a theatre, a stadium, and temples.

Further Reading: *The Minnesota Messenia Expedition: Reconstructing a Bronze Age Regional Environment*, edited by William A. McDonald and George R. Rapp, Jr. (Minneapolis: University of Minnesota Press, and London: Oxford University Press, 1972), offers one of the best archaeological accounts of Messenian sites. Likewise, J. B. Bury's *A History of Greece: To the Death of Alexander the Great* (London and New York: Macmillan, 1900; fourth edition, edited by Russell Meiggs, London: Macmillan, and New York: St. Martin's, 1975) and George Grote's *A History of Greece; From the Earliest Period to the Close of the Generation Contemporary with Alexander the Great,* Volume One (London: John Murray, 1862), provide thorough readings of Messenia's political and military engagements. Readers will find the historical relationship between Messenia and its long-time enemy well documented in Paul Cartledge's pro-Spartan book *Sparta and Lakonia: A Regional History 1300–362 B.C.* (London and Boston: Routledge and Kegan Paul, 1979). M. Rostovtzeff's *The Social and Economic History of the Hellenistic World,* Volume II (London: Oxford University Press, 1941) will provide an erudite study of Messene's (Messenia's ancient capital) standard of living, as well as the currency exchange, during the Roman occupation.

—Hyunkee Min

Metapontum (Matera, Italy)

Location: On the Gulf of Taranto, thirty miles southwest of the city of Taranto.

Description: Ancient coastal city of roughly fifteen square miles, situated between two rivers, the Brandano and the Basento. Burial site of the philosopher Pythagoras; temporary headquarters of the Carthaginian general Hannibal during the Second Punic War. Ruins remain of the large Doric temple Tavole Palatine, as well as a city center (Agora), a grand theatre and, most recently, unearthed, a spring-site temple to Zeus. The Antiquarium museum houses a well-organized collection of maps, religious statues, vases, and jewelry.

Site Office: E.P.T. (Ente Provinciale per il Turismo)
Viale delle Sirene
CAP 75012 Metapontum, Matera
Italy
(835) 741933

For 2,300 years the dead of Metapontum concealed a mysterious secret. Buried undisturbed beneath marshy mud for centuries, skeletons in a Metapontum cemetery revealed evidence that would change how scientists viewed the history of Europe. Excavations at the site in 1992 offered new answers to one of medical history's most ignominious questions: how did syphilis spread into Europe during the fifteenth century?

Formerly, historians had blamed the continent-wide epidemic on inhabitants of the New World. The ship crews of Columbus and other explorers were thought to have contracted the disease through sexual intercourse with native Americans and brought it back to Europe. But at this southern Italian site, etched into the bones of sixteen excavated Greek skeletons, scientists found worm-like markings left by treponemes, the bacteria causing syphilis, at least 1,700 years before Columbus sailed. Scientists now postulate that the disease was present in isolated pockets throughout Europe, possibly mistaken for leprosy.

In total, archaeologists found 320 undisturbed skeletons at Metapontum, the largest cemetery from an ancient Mediterranean society ever uncovered. Evidence of venereal disease was not the only discovery they made. The site has been a treasure trove for agricultural archaeologists as well. Four miles from the city walls, near a spring, archaeologists discovered remarkably preserved ancient seeds of olives, figs, blackberries, alfalfa, grapes, and beans. Some grape seeds were found inside actual grapes, with intact leaves and stems. Ancient farmers are thought to have left the seeds as votive offerings in terra-cotta cups at the spring site, near a

temple to Zeus Aglaios. The Metapontum seeds were so well preserved in mud near the spring that their DNA did not erode. Since ancient farmers chose the best seeds for their next year's crops—thereby imposing an artificial evolution on the plants—molecular biologists at Italy's University of Camarino are comparing the seeds' DNA codes with today's domesticated crops and wild varieties to see how DNA codes have shifted over 2,300 years.

While Metapontum was renowned in antiquity, the city's origins are disputed. The historian Eusebius wrote that the Greek colony was settled by Achaeans in 773 B.C., although modern historians consider that date too early. The city was built in an easy-to-defend position between the mouths of two rivers, the Bradanus (now Brandano) and the Casuentus (Basento). Metapontum was destroyed by the Samnites in the middle of the seventh century B.C., but was said to be later recolonized by the Achaean leader Leucippus, whose head appeared on Metapontine coins during the fourth century B.C. The town was built by the Achaeans as a buffer between their city of Sybaris and the hostile people of Tarantum.

Metapontum was surrounded by fertile farmlands. As Greece lost arable land to erosion and overpopulation, Metapontum became a valuable trading colony exporting corn, wheat, and other grains in exchange for olive oil and wine. The earliest coins from Metapontum, dating from 550 B.C., depict ears of corn. The city had its own treasury at Delphi, filled with valuable objects. Games were held in the city dedicated to the Greek river-god Achelous. In its prime, Metapontum had at least 20,000 inhabitants as well as 10,000 neighboring farmers.

Near the Casuentus (Basento), inhabitants of the area built an artificial harbor (now silted in and lying 1,000 yards inland from the coast). As the town grew, wide east-west avenues were built, crossed by smaller streets to form a grid. Outside the town, early Greek surveying skills were evident in neatly spaced allotments 380 yards apart separated by drainage ditches every 224 yards.

The agricultural city was the final refuge and burial site of the philosopher Pythagoras, whose cult continued to flourish there after his death in the later sixth century B.C. Pythagoras and Xenophanes of Elea were the two philosophers who shifted the focus of philosophical development during the sixth century from Greece to southern Italy.

Pythagoras himself never wrote a word, but his practices caused an angry backlash both in Greece and Magna Graecia, as the Greek colonies in southern Italy were called. It was not so much his mystical theories about the immortality of the soul and the importance of numbers that angered authorities in Greece and Italy as the secret societies formed by his followers. The Pythagorean cult stressed withdrawal

Greek temple at Metapontum
Photo courtesy of E.P.T., Metapontum

from society for the refinement of the soul and the pursuit of intellectual life. Its members were vegetarians who led generally ascetic lives. While this behavior was perfectly acceptable in India and other contemporary Asian cultures, the Pythagorean stance directly contradicted the principles of the Greek polis, which demanded participation as a citizen. This early conflict between individual and state caused a struggle that shook the Greek world.

After being driven from Greece, Pythagoras settled in Croton some 100 miles south of Metapontum along the coast, where he formed a brotherhood of 300 members who shared vows of secrecy and allegiance to his principles. Other clubs formed in cities nearby. But the behavior of the cult members angered the populace, who attacked the building where the Pythagoreans met, setting it on fire and killing many members. Similar mob attacks ensued on Pythagorean groups throughout Magna Graecia.

Pythagoras' tomb was erected in Metapontum, where he is said to have starved himself to death after being chased by angry mobs across the bottom half of the Italian peninsula. The cult carried on after the philosopher's death, first travel-

ing to Greece and then reappearing later in Magna Graecia. The center at Metapontum was kept alive by the philosopher's pupil Aristeas, who appeased the locals by encouraging adherents to worship Apollo with reverence.

The Metapontines became more warlike in the centuries that followed, throwing in their lot with neighboring participants in various conflicts. Unfortunately, they could not pick a winner: in 413 B.C. they allied themselves with the Athenians during the siege of Syracuse, Sicily; the Athenian force was devastated. In 332 B.C. they joined Alexander of Epirus in the fight between the Tarantines and a Lucanian-Brutti coalition. Alexander was killed and buried near Metapontum. When the Metapontines appealed to the Spartan invader Cleonymus to help them fight the angered Lucanians, he conquered the town himself in 303. Later, they backed the Tarantines against Rome when Pyrrhus of Epirus brought his fearful elephants onto the peninsula. When Pyrrhus later fled to Sicily in 278, the Romans marched into Metapontum, conquering the city.

But the Metapontines' greatest military gaffe was siding with Hannibal during the Second Punic War. After the

Carthaginian general slaughtered 90,000 Roman soldiers at the battle of Cannae, the Romans decided to hole up in Rome rather than meet Hannibal on the open field. Hannibal settled in Metapontum for a time. Plutarch tells us that the Metapontines tried to lure the Roman general Fabius into a trap: "counterfeit letters came to [Fabius] from the principal inhabitants of Metapontum, with promises to deliver up their town if he would come before it with his army." Fabius was about to fall into the snare when he consulted the "omens of the birds" and decided that "Hannibal . . . had laid an ambush." When Hannibal abruptly left the Italian peninsula in 207 B.C. to defend Carthage, the citizens of Metapontum evacuated the city rather than face Roman revenge. Although not completely deserted, the bustling town of thousands diminished, never to recover. Later it was sacked by the slave leader Spartacus.

As the town's harbor filled with silt and the surrounding land eroded from prime farmland into marshy sand, Metapontum dwindled. By the mid-second century A.D., the town was in ruins. These ruins lay buried for centuries in the marshy wetland of southern Italy, uninhabitable because of malaria-carrying mosquitoes. During the twentieth century, when national drainage programs began after World War II, Metapontum rose from the mud to reveal its former glory. Among the monuments at the site are the Tavole Palatine or Knight's Table, a huge Doric temple probably originally dedicated to Hera. Fifteen of its thirty-two large columns are still standing, making it one of the better preserved monuments in Magna Graecia. The Antiquarium nearby houses fine Greek vases, bronze statues, a sixth-century B.C. Achaean alphabet, and fine jewelry. Remains of a second temple to Apollo Lykeios stand near the city center. A theatre in the town center is an early architectural example of an auditorium constructed from an artificial mound supported by a retaining wall. The Temple to Zeus Aglaios, near which the preserved seeds were found, is being reconstructed just outside of town.

Scholars of Magna Graecia often lament the devastating effect of malaria in southern Italy. The disease is said to have wiped out the populations of many coastal towns during the Middle Ages. Yet scholars have disagreed about exactly when malaria became critical on the coast, some maintaining that it had been present during the Greek colonies' heydays. "The customs of the Sybarites seem to prove that they had some acquaintance with marsh fever and tried to guard against it," writes early twentieth-century traveler Norman Douglas. "'Whoever would live long,' so ran their proverb, 'must see neither the rising nor the setting sun.' A queer piece of advice, intelligible only if the land was infested with malaria."

Modern excavations have proved Douglas right. As well as the bone scars for syphilis, scientists found evidence indicating that malaria was present. In fact, from what excavators have uncovered, life in ancient Metapontum was far from healthy. Scientists concluded that the population was anemic and malnourished. The inhabitants were small of stature and had a life expectancy of thirty-two years. In spite of the presence of a medical school in the city, researchers found that many skeletons had broken bones that had healed naturally, without any attempt to set them. This find could explain the vehement curses inscribed into lead tablets found in one Metapontum tomb. The inscriptions revile fifteen local doctors by name for ineffective medical treatment: "possibly an early medical malpractice suit being appealed to some higher court in the hereafter," writes *New York Times* reporter John Noble Wilford.

Twentieth-century excavations at Metapontum have revealed findings of critical scientific and historic importance. The ancient city is sure to divulge more of its secrets in the years to come.

Further Reading: For a comparative study of the rise of philosophical ideas during antiquity, see William McNeill's *The Rise of the West: A History of the Human Community* (Chicago and London: University of Chicago Press, 1963). McNeill compares contemporaneous philosophical movements from Europe and Asia and explores their mutual influences. For a complete report of the medical findings in the Metapontum tomb excavation, see National Geographic's *Research and Exploration* magazine (Washington, D.C.), November 1992. For an early twentieth-century antiquarian's travels through Italy south of Rome, including a visit to Croton and a geographical interpretation of Pythagoras' philosophy, see *Old Calabria* by Norman Douglas (Boston and New York: Houghton Mifflin, and London: Secker, 1915).

—Jean L. Lotus

Metéora (Trikkala, Greece)

Location: On the summit of distinctive rock formations in the Cambunian Mountains in Thessaly, Greece, just north of the town of Kalambáka, also known as Kalabáka or by the ancient name Stagoi.

Description: A group of Byzantine monasteries established in medieval times in a two-square-mile mountain refuge that offered sanctuary from warfare and religious persecution. The monasteries of Metéora are famed for their remote location atop perpendicular rock masses and for their brilliant frescoes and icons preserving an art style unique to the late Byzantine era. The Metéora is one of the most popular tourist attractions in northern Greece.

Contact: Tourist Police
10 Hatzipetrou Street
Kalambáka, Trikkala
Greece
(432) 22-813

As a monastic retreat, Metéora is ideally located in the Cambunian Mountains in the northern Greek region of Thessaly. Here, rock masses long ago eroded into a forbidding labyrinth of contorted mounds, crags, and monoliths, some rising more than 2,000 feet in the air. These gritstone and conglomerate rocks may have been formed from deposits of soil, sand, and stone at the bottom of a primordial sea that once covered the region of Thessaly. Water currents over the centuries gradually carved the rock masses into fantastic shapes. Later, when the sea had retreated, the exposed rocks were gradually smoothed by winds and rains over millions of years. Another theory attributes the origin of this mysterious rock formation to large salt deposits from a primeval sea.

The word "Metéora" has been translated as "between heaven and earth" or, similarly, "middle of the air." Either translation is an appropriate description for both this geological phenomenon and the sanctified life style sought by the monks who retreated to this strange but magnificent setting. As early as the ninth century, hermits and ascetics began seeking seclusion amid the pinnacles and caves of Metéora for safety from brigands and to achieve closer communion with the divine. The earliest known monastic colony was built on a rock called Dhoúpiani, under the leadership of the Superiors of the Ascetics of Stagoi. Several other small colonies were built in the next few centuries. By the 1300s, the need for such an inaccessible refuge became paramount as war broke out between Serbia and the Byzantine Empire, spreading violence throughout northern Greece. Serbia won considerable lands from the Byzantines during the struggle. In 1346, the Serbian Stephen Dusan assumed the title of

emperor over territories including Thessaly, bringing some semblance of peace to the war-torn region.

About 1350, another settler came to Metéora. He was Athanasius Koinovitis, from the Greek monastic community of Mount Athos, known as the Holy Mountain of Orthodoxy. According to one legend, he was carried to the top of one of the monoliths by an eagle. Another story claims that Athanasius originally settled on a lower crag but fled upward when he saw demons hovering over his cave. In any event, he ascended some 1,800 feet to the top of the broadest rock. There, he built the first structures of what was to become the Great Meteoron Monastery. In this early settlement, Athanasius lived with nine brethren observing very strict monastic rules, including one forbidding the presence of women. According to legend, Athanasius, famous for his abhorrence of the female sex, had once been approached by the widow of a Serbian ruler seeking his blessing. He not only refused to bless her, but he condemned her heartily for being a woman and issued the frightening warning that she would die soon. She died, in fact, within three months.

Athanasius and his early followers climbed the sheer cliffs of Metéora by wedging timbers into rock crevices. As more members joined the community and more monasteries were built, long rope ladders provided access. Some of these ladders stretched up to 130 feet and could be folded and drawn up. Ropes coiled around hand-powered winches were used to draw up both provisions and visiting brethren, who huddled in rope nets or baskets as they violently lurched and spun on their way up. This hair-raising ascent could take up to half an hour and was a test of faith in itself.

Among Athanasius's disciples was John Uros Palaeologus, son of the Serbian emperor Symeon. John Uros abdicated his throne and became the monk Joasaph, eventually succeeding Athanasius as leader of the monastic community, which continued to attract new members. The Serbian rulers of Thessaly encouraged monastic development, and the monasteries of Metéora benefited from donations of land and property in Walachia and Moldavia. In 1388, Joasaph used the financial resources provided by his father to enlarge the Meteoron, the richest and most commanding of the clifftop monasteries. The hermit-king Joasaph was forced to flee to Mount Athos for a time, possibly as a result of the Ottoman invasion of Thessaly in the late fourteenth century. He returned to Metéora as abbot in 1401. He died there in 1423 and was buried next to his spiritual guide, Athanasius. Joasaph is known as the "Father of the Great Meteoron."

Following the Ottoman conquest of Thessaly in 1393, monastic communities had become asylums for Hellenism and Orthodoxy. The number of monks at Metéora rose into the hundreds. Each monastery contained cells for its inhabitants, a church or two, a refectory, and sometimes a library to

Metéora
Photo courtesy of Greek National Tourist Organization

preserve manuscripts. Early in the sixteenth century, the saintly Larissis Bessarion II became revered throughout Thessaly for reviving the Orthodox faith in those harsh times and for looking after the welfare of persecuted mountain inhabitants. It was during this period, and perhaps under his influence, that some of the major monasteries of Metéora were built or expanded. Eventually, Metéora became one of the most powerful strongholds of Christianity, with a community of twenty-four monasteries decorated by the leading artisans of the time. Second only to Mount Athos, the monastic community of Metéora became known as Orthodoxy's second Holy Mountain.

Influenced, perhaps, by their harsh natural surroundings and by the violence of warfare in the outside world, the ascetic monks of Metéora often went to extremes in practicing the mortifications they believed essential to their salvation. The founding brothers of the Varlaám Monastery imposed especially harsh privations on the monks living under their rule. Community members lived on a single meal each day consisting of only bread, beans, and water. They prayed through much of every night. As a form of additional mortification, one of the founding brothers was known to wear an iron chain under his habit bound tightly around his waist directly against the skin. Just before his death, he stretched himself out on his bed and arranged his arms in the form of a cross.

As Metéora served as a preserve for Orthodoxy, so it also served as a refuge for learning and art. Following the Turkish conquest, Greece had descended into a state of intellectual stagnation. Education declined drastically, and the Greek language itself seemed headed for extinction. While documented evidence is scanty, it is believed that monastic communities, particularly those at Metéora and Mount Athos, provided a haven for students and scholars who had not fled to the West. Thus, the Byzantine tradition of education never wholly disappeared. Similarly, art was greatly affected by the Turkish occupation, particularly after the capture of Constantinople, the heart of the Byzantine Empire, in 1453. Traditional Byzantine art declined. With the disappearance of patronage for works on a grand scale, a type of folk art developed instead. This simpler, more naive style became known as "Cretan," although it had not originated in the island of Crete but had come there from Constantinople via Thessaloníki in northern Greece and Mistra in the Peloponnese. Among the most renowned Cretan painters of the 1500s were Theophanes and Damascenos. Theophanes became famed for his frescoes, while Damascenos specialized in movable icons. The works of both men can be seen in the monasteries of Metéora, and their influence is evident in the artwork of the Italian Renaissance.

The wealth and power of Metéora were also its curse. In the sixteenth century, the riches contained in the monasteries generated widespread rivalries and corruption. At one point, the Great Meteoron fell under the rule of an abbot who accepted bribes from the Turks, a crime for which he was eventually banished. In another incident of declining morals, a monk tried to bring into the community two women disguised in monks' garb. Fraud and clandestine sales eventually robbed the monasteries of many treasures.

Because copyists were few and most monks illiterate, manuscripts were rare. The monks of Metéora devotedly collected written materials in the dark age of learning in Greece, but they were often unaware of their value. Outsiders commonly sought out the innocent monks and bought their manuscripts at bargain prices. A notorious scoundrel in this effort was the Cypriot monk Athanasius. Known as "the Orator," this slick-tongued impostor traveled through Greece on the secret instructions of Cardinal Mazarin of France. As a result of his devious transactions, many of the manuscripts once housed in Metéora now rest in France's Bibliothèque Nationale.

In this era of fraud and corruption, the once-noble monasteries of Metéora suffered considerable deterioration. During the eighteenth and nineteenth centuries, they were looted on several occasions. The monasteries were later to sustain severe damage during World War II and the subsequent Greek Civil War. This downward turn of fortune was finally reserved in 1960, when the structures were restored to ensure preservation of the remaining manuscripts and frescoes. A highway built in the early 1960s paved the way for mass tourism. However, the influx of visitors put an end of Metéora's reclusive appeal and drove away most of the monks. Only a few of the monasteries are still inhabited. The individual monasteries of Metéora can be reached by bridge or steps that were cut in the rocks in the 1920s. A few are inhabited by caretaker Orthodox monks, who now use motorized winches to bring in provisions.

The Great Meteoron is the oldest and largest of the monasteries and still houses a small monastic community. Its crowning glory is the sixteenth-century Church of the Transfiguration, or Metamorphosis, which contains a large narthex supported by four columns. The church has beautiful frescoes depicting the Nativity, the Transfiguration, the Crucifixion, and the Resurrection, as well as the cruel deaths inflicted on Christian martyrs during the Roman persecution. Across from the narthex is a chamber filled with the skulls and bones of deceased monks from times past. The Great Meteoron contains full-length portraits of its founders, Athanasius and Joasaph, with their long black beards. Here, one also can see exceptional examples of post-Byzantine icons, including several depicting events in the life of Christ and another showing the heroic St. Demetrius thrusting a lance into an invading Bulgar. A domed refectory contains a museum displaying icons, manuscripts, and etchings. Visitors to the Great Meteoron also are afforded a spectacular view of titanic boulders and spectral rocks tapering into the shape of cones.

Separated from the Great Meteoron by a frightening abyss, the Varlaám Monastery is the second oldest at the site and is also still inhabited. It once housed such sacred relics as the supposed finger of St. John and the putative shoulder blade of St. Andrew. The buildings here, extending directly from surrounding rock, appear to have been constructed haphazardly. The most distinctive attraction is the Church of Agion Panton, or All Saints, built in 1542 by two brothers, Nectarios and Theophanes Apsarades, from nearby Ioannina. The frescoes of Agion Panton offer a fearsome depiction of the Apocalypse, complete with hermits, soldiers, flashing swords, and martyrs in colorful red and purple vestments and surrounded with golden haloes. The frescoes were painted by Frangos Castellanos and Georgios of Thebes in 1565 and restored in the late nineteenth century. The one-time refectory is now a museum and library containing a Book of Gospels once belonging to the Byzantine Emperor Constantine VII. At Varlaám, one also can see a demonstration of the net-and-pulley system the monks once used for access.

While the St. Stéfanos Monastery building complex is comparatively small, its interior appears lighter and more spacious than those of the Great Meteoron or Varlaám and is the best preserved in all of the Metéora settlement. St. Stéfanos is now a nunnery. Because of its easy accessibility from the road, it is also one of the most popular stops for tourists. The church within St. Stéfanos has, in a silver container, the sacred head of St. Charalambos, who was believed to possess supernatural powers to ward off pestilence. The church also has an elaborate wood-carved iconostasis and bishop's throne.

Looming directly above the town of Kalambáka, the Holy Trinity Monastery was built on the summit of one of the narrowest monoliths and is distinctive for its balconies and arcades. Fans of James Bond may remember Holy Trinity as the backdrop for the climactic chase scene in the 1981 movie *For Your Eyes Only*. The original foundation of the monastery was built by the monk Dometius in 1476. Its chapel contains inexpert wall paintings added 200 years later. Holy Trinity is especially picturesque because of the gardens between its separate structures and the view it affords of the snow-capped Pindos Mountains in the distance.

Originally founded in the fourteenth century, the Rousanou Monastery, also known as St. Barbara since its conversion to a nunnery, was expanded by two brothers from Ioannina, Joasaph and Maximos, who came to Metéora around 1545. The monastery is beautifully situated and the most visible of the Metéora structures from the valley below. Consequently, it is the building most often depicted on posters and in guidebooks about this region of Greece. Because of deterioration in its walls, which had been built directly into rock formations, Rousanou was abandoned in the 1980s. A small church, dedicated to St. Barbara, is decorated with excellent Cretan frescoes depicting Orthodox martyrdoms of unmatched brutality.

The last of the monasteries accessible to the public is St. Nikolaos, also called Anapafsa. Founded in 1388 and expanded in 1628, this monastery appears very deteriorated and is not open to the public on a regular basis. However, it contains the most beautiful frescoes of the Cretan master Theophanes, including scenes of the Last Judgment and Paradise.

The small town of Kalambáka, at the foot of the rock formations of Metéora, offers restaurants, hotel accommodations, and bus service for the tourist trade. The town, itself, was once famed for its medieval architecture, but its historic buildings were almost totally destroyed during World War II. Kalambáka's most distinctive sight is the Byzantine Church of the Dormition of the Virgin, built in the eleventh century on the ruins of a fifth-century basilica. The church contains frescoes by the Cretan monk Neophytos, son of Theophanes, but they have been severely blackened by candle smoke and incense burning over the centuries.

Adding a lighter note to the solemn atmosphere of Metéora is a local custom practiced at the Church of St. George, built in a shallow cave not far from the Great Meteoron. Colorful scarves can be seen hanging across the face of the cave. Once a year, on St. George's Day, young men from nearby communities hazard the difficult climb to take down the old scarves and hang new ones as signs of favor to their chosen ladies. The custom is said to have evolved from a long-ago incident during which an injured man's leg was miraculously cured by the scarf wrapped around it.

Further Reading: The definitive work on the subject is *Meteora: The Rock Monasteries* by Donald M. Nicol (London: Chapman and Hall, 1962; second edition, 1975), but the book may not be easily available. A concise history of Metéora and descriptions of the major monasteries can be found in both *The Companion Guide to Mainland Greece* by Brian De Jongh (Englewood Cliffs, New Jersey: Prentice Hall, and London: Collins, 1983; revised by John Gandon, New York and London: HarperCollins, 1991), and *The Berlitz Travellers Guide to Greece* (Zurich, Switzerland: Berlitz, and New York: Macmillan, 1992). A more intellectual discussion of Greek monastic communities, with frequent references to Metéora, is available in *The Greek Nation: 1453–1669* by Apostolos E. Vacalopoulos, translated from the Greek by Ian and Phania Moles (New Brunswick, New Jersey: Rutgers University Press, 1976).

—Pam Hollister

Milan (Milano, Italy)

Location: Milan is located in the Po Basin of northern Italy, bordered on the north by the Alps and on the south by the Po River. It is the capital of the province of Milano and of the Lombardy region.

Description: Today Italy's second-largest city, Milan is thought to have been founded by Celtic tribes in the seventh century B.C. Following the Roman conquest of the area in the late third century B.C., it was transformed into a Roman colony called Mediolanum. The city went on to become a major commercial and religious center and capital of the Western Roman Empire. The Middle Ages and early modern era brought centuries of warfare and foreign domination, but Milan occasionally rose to dominate the region, as in the signorial period from 1277 to 1447. Historically overshadowed by Rome, Milan became the economic powerhouse of Italy following World War II. Milan is one of the world's primary centers of fashion and design, home to the bulk of Italian manufacturing prowess and the de facto capital of northern Italy. Milan is also the headquarters of the Northern League, a political party advocating greater autonomy from Rome for regions and provinces.

Site Office: A.P.T. (Azienda di Promozione Turistica del Milanese)
Via Marconi, 1
CAP 20123 Milan, Milano
Italy
(2) 870016

The Lombardy region of Italy has been inhabited from at least 3000 B.C. Relatively little is known of the earliest inhabitants, although they did leave behind a mysterious legacy of thousands of symbols carved in the rocks in the region of Val Camonica, near Lake Iseo. Even more mysteriously, these inscriptions continued to be made well into the Roman era, despite the fact that Lombardy was most likely home to a long succession of otherwise different tribes during this period. Evidence also exists of Bronze Age societies that lived in dwellings constructed on pilings in the water at the Lakes Orta, Varese, and Ledro.

By 400 B.C. the Lombardy region was populated by several Celtic and Ligurian tribes, in contrast to the rest of the Italian peninsula, which was home to numerous loosely-related Indo-European tribes now known as the Italics. Because the residents of the north bore little resemblance to anyone else in Italy, early Roman forays into the area marked the beginning of the growth of the empire. In a particularly relevant piece of foreshadowing, the Romans didn't consider Lombardy to be a part of Italy. They called the area Cisalpine Gaul, for "Gaul this side of the Alps."

By 283 B.C., the Romans managed to conquer the entire peninsula. They immediately founded colonies in the Lombardy region, including Brixia (Brescia), Comum (Como), and Mediolanum (Milan). Lacking much by way of wealth, Lombardy weathered the first centuries of empire relatively easily. The north developed rapidly during the Pax Romana. Mediolanum grew into an important commercial and trading center. Roads linked the cities of the north to each other and to the south. The numerous lakes of the Lombard plain became popular vacation spots for well-to-do Romans, bringing both capital and prestige to the region.

By the third century A.D. the Roman Empire had fallen into decline and faced increasing threats of invasion. Seeing a need for rapid, fundamental change, Emperor Diocletian (reigned 284 to 305) split the empire into western and eastern halves for ease of administration and defense. Emperors of the western region kept their army, court, and administrative headquarters at Mediolanum because it allowed for easy defense of the borders on the Rhine and Danube. Furthermore, as a thriving commercial city, Mediolanum was worth defending.

While Rome withered, Mediolanum flourished, taking on more and more administrative responsibilities and, for a time, arguably outshining Rome. Diocletian's successor Constantine brought further fame to the city with the historic Edict of Milan in 313 that legalized the religion of Christianity. This act also removed the common threat of repression that had united the Christians, and a schism soon developed between early Catholic orthodox believers and followers of the Egyptian bishop Arius, who denied that Christ was of the same substance as God. Milan at first came under Arian bishops but became orthodox when Ambrose (later patron saint of the city) was elected bishop in 374. Legend says that when Ambrose, then a consular governor, attempted to calm the crowd gathered during the election of a new bishop, a child shouted "Ambrose Bishop!" The crowd seconded the motion, and Ambrose found himself with the job. He had not even been baptized.

Ambrose is credited with numerous noteworthy accomplishments during his tenure from 374 to 397. Legend holds that bees had flown into his mouth as an infant, apparently attracted to the honey of his tongue; such was his reputation for eloquence. Ambrose is also credited with converting St. Augustine and forcing Emperor Theodosius I to do penance for ordering a civilian massacre in Thessaloníki before allowing him to enter church. Today, native-born Milanese are known as Ambrosiani and the Milanese church still celebrates Mass according to the Ambrosian rite.

Milan's cathedral
Photo courtesy of The Chicago Public Library

At the beginning of the fifth century Milan began what was to be well over 1000 years of occupation by various foreign powers, some benevolent, some tyrannical. Given its position just south of the Alps, Milan was occasionally sacked simply because it lay on the path to Rome, undoubtedly feeding the anti-Rome tendencies of even the earliest Milanese.

The Visigoths under Alaric I invaded in 410, and Attila, king of the Huns, sacked and plundered the city in 452. Apparently, Attila had no grand plans for Milan; the Milanese were able to reconsecrate their cathedral only a year later. By 470 all of Italy was in the grip of Gothic general Odoacer, who six years later deposed the western emperor and crowned himself king of Italy. He ruled from the new capital at Pavia.

Ostrogoths led by Theodoric invaded Lombardy in 489, ostensibly under orders from the eastern empire headquartered at Constantinople. By 493 Theodoric had murdered Odoacer and crowned himself king. Theodoric proved to be an effective ruler; the arts and letters flourished, and Milan entered the sixth century in much better shape than it had been in the fifth.

In 535, however, Emperor Justinian decided to reincorporate Italy into the Byzantine Empire, touching off the Gothic War. Milan was occupied by Justinian's troops in 538, but the Byzantines were ousted the following year by an invading coalition of Goths and Burgundians. Byzantine forces finally gained control of the peninsula in 553, only to levy oppressive taxes against the people. Disease, starvation, and border skirmishes with Franks and Alamanni only exacerbated the situation.

In 568 the Byzantine leader Narses, one of the generals responsible for the success of the Gothic War, invited a small group of Lombard mercenaries into the region to help him deal with remaining pockets of Gothic resistance. After surveying the land, the mercenaries returned in 569 under the leadership of their king, Alboin, conquering the major cities of the Po Basin within a matter of months. The population of Milan dwindled as residents headed for the comparative safety of Genoa, still in Byzantine hands. The Lombards kept the north until 774, giving their name to the region and forever altering its cultural foundations, thereby contributing greatly to the divide that has existed ever since between north and south.

When the Lombards set their sights on Rome in the 750s, Pope Stephen II, and later Pope Hadrian I, turned to the Franks for assistance. Frankish leader Charlemagne besieged the Lombard capital at Pavia for ten months, finally occupying the city on June 7, 774. The Franks took over the administration, and their elite entered into the region's nobility alongside the Lombards, who remained in the area. Milan was given free rein during Charlemagne's rule, and in many ways the city again outstripped Pavia, the official capital, as the political and economic center of the empire. Charlemagne even allowed the Milanese to continue practicing their Ambrosian rites while he introduced standard Roman liturgy everywhere else in his empire.

Charlemagne's mostly absentee administration was characterized by petty internal squabbling among minor aristocrats and lords. The situation quieted when Charlemagne was crowned western emperor in 800, but conflicts flared again after his death in 814. In Milan the archbishops assumed political power during this turmoil, providing a measure of stability. Archbishop Anspert (868-881) even defied the papacy itself, leading the resistance to a French Carolingian whom the pope wished installed as king of Italy. Archbishop Manasse was given permission to mint coins in 949. Meanwhile, Hungarian invasions of Lombardy had become increasingly destructive. Defensive walls and castles were constructed in the area, and in 960 Archbishop Walpert traveled to Germany to enlist the aid of the Saxon king Otto, who was crowned Holy Roman Emperor Otto I at Rome in 962. Perhaps because of the archbishop's role in his accession, Otto was particularly generous to Milan. New buildings were erected and older ones restored.

Emperor Henry II was crowned king of Italy at Pavia on May 14, 1004, but quickly moved his court to Milan following a revolt against his rule. In 1026 his successor Conrad II was crowned king at Milan, making the de facto center of the empire once again its official capital.

This era also saw the birth of the commune, or free city-state, in Milan. The decision to crown Conrad king in 1026, for example, was made not strictly by the nobility, but by a popular assembly established in 1024. The assembly was composed primarily of nobles and church officials, but it vested authority in more than one person or even a handful of people. In 1042 a coalition of artisans and merchants staged a rebellion, forcing archbishop Ariberto da Antimiano and his allies among the nobility out of Milan. Eventually, dissatisfaction with the rule of the archbishops and nobles led to a reform movement and the reduction of the archbishop's authority. The commune took another step forward in 1044 when the various classes came to a cooperative agreement regarding civil administration. But by 1045 the situation had deteriorated into intermittent civil war that would last for most of the remainder of the century.

By the end of the eleventh century, however, Milan's situation improved. In 1097 the various social classes formed the commune civitatis, a truly inclusive democracy not unlike the modern parliamentary model. The people elected their own representatives, and the representatives in turn elected consuls (administrators) to run the city. Furthermore, Milan's location on a route to the Holy Land during the First Crusade (1097–1130) provided a steady flow of Christian soldiers, creating an economic boom for the industrious Milanese.

But like everything else in Milanese history, the boom was not to last. The need to maintain safe passage into and out of Milan for the crusaders led to war with the smaller rival towns of Cremona, Lodi, Pavia, and Como, all of which were beaten by powerful Milan. The vanquished cities appealed to Emperor Frederick I Barbarossa for help, and in 1158, long after the crusade had ended, they received it. Frederick besieged Milan in August 1158, and the city fell in a month's time. When the Milanese revolted four years later, Frederick undertook a second full-scale siege, which destroyed the city and garnered the swift surrender of everything that was left standing.

Frederick's intervention in the north made it plain to the cities of Lombardy that the freedom of their communes would never be guaranteed unless they put aside their differences and presented a united front to the common enemy. They had a strong ally in Pope Alexander III, who opposed Frederick's Italian excursions. In April 1167 the communes of Mantua, Cremona, Bergamo, and Brescia formed the first Lombard League, with Milan joining shortly thereafter. Together, the league repelled Frederick several times, and in 1176 the Milanese militia destroyed his army in a surprise attack, forcing Frederick to flee to Venice alone and make peace with the pope. In 1183 the liberty of the communes was recognized in the Peace of Constance. Although highminded, the treaty really only bound the communes to each other when presented with a common enemy. When unthreatened from without, they were free to fight with each other.

The next great era in Milanese history was the signorial period, from 1277 to 1447. The signori, remembered today as either lords or despots, maintained the integrity of the communes but ran them as personal fiefdoms, sometimes for better, and sometimes not. The Milanese signori developed their base of power over the course of successive generations; eventually, they controlled Mortara, Novara, Vercelli, Vigevano, Pavia, Piacenza, Tortona, Bergamo, Como, Alessandria, Verona, Padua, Pisa, and Siena. In 1395 the signori were invested with the title of duke.

When Filippo Maria, the last duke of Milan, died without an heir in 1447, a new entity called the Ambrosian Republic (in honor of St. Ambrose) was proclaimed, but it lasted only three years. In 1455 a federation of Milan, Venice, Rome, Florence, and Naples became the Italian League, which created something like unity and persisted for forty years.

An ill-considered marriage between Milanese nobility and the French court in 1389 came back to haunt the city in 1498 when, after a complicated series of intrigues and untimely deaths, the French king Louis XII proclaimed himself duke of Milan, backing up his claim with an invasion of the city a year later. French domination lasted from 1500 to 1512, during which time many of Lombardy's greatest works of art were sent to France. The Milanese regained control of their city from 1512 to 1515, but the city returned to French rule following Francis I's victory at Melegnano. Francis met his match, however, in Charles I of Spain, crowned Holy Roman Emperor Charles V in 1519. Deciding that the thriving city of Milan would make a good base for central European operations, Charles invaded, finally beating the French at Pavia in 1525. Charles left an all-powerful Spanish viceroy in Milan, and the city remained under Spanish control for nearly 200 years.

Milan prospered in the early years of Spanish rule, but declined as Spain declined, losing markets to foreign competition. The War of the Spanish Succession of 1713 distracted the Spanish long enough for the ruling Habsburgs of Austria to cross the Alps and claim both Milan and Mantua. Luckily, the Habsburgs were a benevolent clan, as despots go, and instituted important reforms that successfully turned around the stagnant economy. More importantly, Austrian rule gave the north a significant boost over the sleepy south, allowing Milan to take the economic lead it has held ever since.

Austrian rule continued uninterrupted until 1796, when Napoléon crossed the Alps and proclaimed Italy a vassal state of France. He renamed Lombardy the Cisalpine Republic and issued a full slate of new laws and edicts before departing. In 1799 the French were forcibly removed from Italy by an Austro-Russian army and British naval power, but the determined Napoléon came back with a vengeance in 1800, winning a major victory at Marengo and crowning himself king of Italy at Milan. Napoleonic rule lasted until 1814, when the Austrians returned and were greeted enthusiastically by the northerners, chafing under fourteen years of high taxes and political oppression. Austrian sovereignty was recognized by the Congress of Vienna.

A series of revolts between March and July of 1848 shook Austrian rule, but did not defeat it. In 1849 King Charles Albert of the Piedmontese House of Savoy, who had been a prime mover in the earlier revolution, declared war on Austria. Piedmontese forces managed to gain control of Lombardy and Tuscany, putting Milan under Italian control once again, but full Italian unification would have to wait until 1860.

The north and its de facto capital of Milan played host to many of the most fundamental changes to follow in the wake of Italian unification. The already-prosperous north boomed through new trade links with the rest of the Italian peninsula, much of which had been closed to commerce. The new wealth also brought with it new labor problems and new approaches to politics and the role of the state. In 1882 the first Italian socialist party, the Partito Operaio Italiano, was founded in Milan. Fascism also made its debut in Milan when Benito Mussolini established his inflammatory *Popolo d'Italia* newspaper in the city and organized the first fascist organizations in the turbulent period between the world wars.

Milan suffered heavy allied bombing during World War II and was further tormented by continued German occupation even after the Allies had captured the south in 1943. An effective resistance movement in the north wrought havoc with German designs on the region's industrial capacity and was responsible for capturing Mussolini and his mistress during an attempted escape to Switzerland in 1945. After shooting both, the Partigiani, as the resistance was called, hanged Mussolini from his toes at a nondescript gasoline station in Milan, where unconfirmed legend says passers-by were invited to beat the body with a stick if they chose.

Blessed by the Marshall Plan in the years following the war, Milan took the lead in the economic rejuvenation of the 1950s. Unlike Britain, France, and Germany, Italy was primarily pastoral prior to the war (with the exception of the industrialized north). Postwar rebuilding resulted in a Milan-centered northern economic powerhouse of more than 60,000 businesses, from giants like Alfa-Romeo to small, family businesses that have come to be known as models of "Italian capitalism," a phenomenon characterized by small, closely held businesses whose futures lie securely in the hands of only a handful of people. In 1986 Italy announced that it had eclipsed Britain to become the fifth-largest economy on earth. Italians hailed this event as Il Sorpasso, even after it was revealed that there had been more than a little creative accounting on the part of government number crunchers in arriving at the final figure.

That Milan today ranks among the world's great cities is undeniable, as it is in ample possession of those things commonly associated with greatness: significant economic power, industrial and technological prowess, intellectual sophistication, and a large, cosmopolitan populace willing and able to perpetuate them. But it lacks the fundamental credentials of sovereignty and political influence. In Italy, Rome makes the rules and sets the pace.

Certainly there are other great cities in a similar position. Milan is a rarity because it neither graciously accepts its status as second city nor blithely ignores the edicts promulgated in the city that outranks it. It fights with Rome each and every day, and with increased vigor in the postwar years. Opinions vary on exactly how Milan is to reach its goals, and, indeed, even what these specific goals are—not that such simple vagaries would slow down the Milanese.

Despite the lofty rhetoric dreamed up by both sides, the conflict between Milan and Rome, north and south, revolves around money. In this, Milan has the high ground. Wealth created in Milan represents 10 percent of the country's gross national product, even though the city's population of 1.4 million accounts for slightly more than 2 percent of Italy's total population of 60 million. Milan is the world's largest exporter of fashion clothing (much-vaunted Paris is only in sixth place) and is home to many of Italy's most recognizable businesses, including such familiar names as Alfa-Romeo, Olivetti, Pirelli, and Gianni Versace. Per capita income in Milan is 38 percent higher than in Rome,

and the Milanese pay 25 percent of all national taxes. This last statistic, more than anything else, is at the root of Milan's grievance with Rome. As a saying popular in the north puts it, "Milan works and Rome eats."

This perpetual frustration with "enslavement to the south" has found its voice in the postwar era in the form of several regionalist political parties with platforms ranging from greater autonomy from Rome, to jingoistic anti-immigration rhetoric, to an all-out cry for secession from the republic. Generally grouped under the banner of the Northern League, these parties made their first significant showing in Italian politics in local government elections held in May 1990. In June of 1993 the Northern League took control of Milan, legitimizing Milan's status as the de facto capital of northern Italy. What the future holds for Italy's "secret capital" remains to be seen.

Although Milan is certainly an urban, modern city geared to industry and fashion, visitors can do more than shop at the exclusive fashion boutiques on Via Borgospesso and Via della Spiga. The city's cathedral (Duomo) is the fourth-largest church in the world; the domed roof contains 135 spires and 3,200 statues. Inside is stored a nail supposedly taken from Christ's cross. Among the many other architectural attractions is La Scala, Milan's opera house, completed in 1778 and reopened in 1946 after the repair of serious damage from the bombing raids of World War II. The city is also home to several outstanding museums. Perhaps the most famous work of art in the city is Leonardo da Vinci's *Last Supper,* housed in the Vinciano Refectory.

Further Reading: Materials in English devoted exclusively to Milan are relatively rare, but the city is covered in great detail in many histories of Italy. *History of the Italian Republics in the Middle Ages* by J. C. L. Simonde de Sismondi (London: Routledge, and New York: Dutton, 1906) gives exhaustive coverage to each of the prominent republics during one of the most important periods in Italian history. *The Cadogan Guide to Lombardy, Milan and the Italian Lakes* by Dana Facaros and Michael Pauls (London: Cadogan, and Old Saybrook, Connecticut: Globe Pequot, 1994) is an excellent overview of Milan and Lombardy, past and present.

—John A. Flink

Minorca (Baleares, Spain)

Location: The northernmost island of the Balearic Islands chain, 27 miles northeast of Majorca and about 140 miles southeast of Barcelona in the Mediterranean Sea.

Description: The second-largest of the Balearic Islands, with an area of 264 square miles. The landscape consists of a plateau divided into small farms; a central hill, Monte Toro, that rises 1,100 feet; and cave-studded cliffs that drop to the sea. Pine forests cover about fifteen percent of the island, and more than 1,000 ancient stone monuments dot the countryside.

Site Office: Oficina de Tourismo
Plaza de la Constitución
07001 Mahón, Minorca
Spain
(71) 36 37 90

Minorca has a long history of habitation. Archaeological evidence, including carved antelope horns, flint implements, pottery fragments, and rock art, indicates that the first inhabitants of Minorca were Neolithic people from the east who had settled the island by 4000 B.C. Between 1800 and 1500 B.C., the culture was characterized by the use of coastal caves for homes and tombs. These early Minorcans also constructed artificial caves; at Cales Coves, the square openings of about 150 such structures still remain in the rock.

Sometime between 1400 and 1200 B.C., a new culture developed on Minorca. The so-called sea peoples, mysterious conquerors from Asia Minor, began to push westward in the Mediterranean, founding a nation on the island of Sardinia. Scholars theorize that colonists from Sardinia reached Minorca, driving out or absorbing the native inhabitants.

The new settlers brought their culture with them, characterized most dramatically and enduringly by the construction of large stone monuments called taulas, talayots, and navetas. Taulas (tables) are T-shaped towers constructed of two huge stones weighing several tons. The supporting stone averages sixteen feet tall, and the cross stone averages thirteen feet long and five feet wide. Most taulas are surrounded by a circle of upright stones with an opening that faces south. Talayots are roughly conical towers with a base of sixty-five to eighty feet in diameter. Some are completely solid while others contain a small chamber. Because archaeologists have found no relics inside the talayots, no one is sure of their use or significance.

Navetas (boats) are so called because most resemble the upside-down hull of a long boat. About sixty of these structures still stand in the more remote areas of the island. Navetas are characterized by a narrow chamber supported by pillars. Some, in addition, have a stone ceiling topped by an upper chamber. Archaeologists have found skeletal remains inside navetas, indicating that they served as community graves.

In approximately 1500 B.C., the Phoenicians settled the island, naming it Nura, and founding the cities of Maghen (now Mahón) and Jamma (now Ciudadela) as centers of trade. Ciudadela eventually became the island's capital. In about 500 B.C., the Greeks established a presence alongside the Phoenicians, using the island as a point in their trade route through the Mediterranean. Around 400 B.C., the Carthaginians drove out the Greeks and Phoenicians. The islanders joined the Carthaginian armies as mercenaries. These warriors were prized for their skill with the sling. In fact, many scholars believe that the name of the island group is derived from the Greek *balein,* for sling.

The Balearic sling-throwers also used their talents in the practice of piracy, staging offshore attacks on passing ships, which they pelted with stones. When the Minorcan sling-throwers attacked the fleet of the Roman commander Caecilius Metellus around 123 B.C., the Romans drove the attackers into hiding in the talayots, eventually slaughtering many of them. Minorca and the other islands surrendered to Roman rule.

For several hundred years, life in Minorca as a frontier Roman province was peaceful and uneventful. Although construction of stone monuments and the use of caves for burials continued, Roman culture gradually supplanted the traditional ways. It was under Roman rule that Minorca got its present name, from the Latin for "smaller one" (as compared to the larger Majorca). The name was first mentioned in a text by St. Hippolytus in the third century A.D. The discovery of the ruins of several churches with exquisite mosaics dating from the late fifth century indicates that Christianity was firmly established by that time.

The Vandals, Christian marauders from northern Europe, overran Minorca around A.D. 421. Their rule lasted until about 530, when the Byzantine Empire under Justinian I conquered the island. Persistent raiding by Arab pirates chipped away at the Byzantine hold on power. Finally, by the late eighth century, Moors—Moslems from North Africa—had more or less completed their conquest of Minorca and the other Balearic islands, though some scholars place the final cementing of Arab control as late as the beginning of the tenth century.

The Moors built a grand fortress on Mount Santa Agueda in the north-central part of the island around the year 1000 and ruled Minorca in a spirit of religious tolerance for the next 200 years. However, during the eleventh century, Castile, Aragon, and other Christian kingdoms of northern mainland Spain began to wage campaigns to extend their rule southward into Moorish territory. This pressure eventually

A taula and talayot on Minorca
Photo courtesy of Instituto Nacional de Promocion del Turismo

led the Moors to initiate forced conversions of Christians on Minorca, one of their last Spanish strongholds, during the early thirteenth century.

The king of Aragon, James I, succeeded in seizing control of Minorca from the Moors in 1232. Aragonese control of the island was slack, however, and Moors continued to govern the island. James I died in 1276 and left Minorca, along with Majorca, to his younger son, who ruled as James II. James II was involved in a bitter rivalry with his older brother, Peter III, who had inherited the Aragonese throne. When Peter III died in 1285, his son Alfonso III succeeded him and overran James II's realm.

Alfonso III was responsible for one of the most tragic episodes in Minorcan history. In 1287, he planned an invasion of Moorish North Africa. The Moorish governors of Minorca foiled his campaign, however, by warning their brethren in North Africa. Alfonso reacted with extreme brutality, seizing all property on Minorca and murdering or selling nearly all of the inhabitants into slavery.

Alfonso's public works projects included the construction of the Church of Santa Maria in Mahón. But his control over the island proved short-lived; the Treaty of Anagni restored Minorca and Majorca to James II in 1295. As a result of this period of Aragonese rule, most present-day Minorcans trace their descent from the Catalonians who settled there during the thirteenth century.

Between 1300 and 1700, a variety of misfortunes took their toll on Minorcans. The population had been nearly wiped out by Alfonso's terror, and what few people remained suffered from periodic outbreaks of cholera, smallpox, and plague. Incessant pirate attacks weakened them further. In the fifteenth century, the island was used as a place of exile for criminals from the Spanish mainland. The population thus became increasingly lawless. A rebellion in 1451 left the island leaderless as well. In 1469, the marriage of Isabella I to Ferdinand II joined the kingdoms of Castile and Aragon, and Minorca and the other Balearics became part of a united Spain.

During the reign of Holy Roman Emperor Charles V in the sixteenth century, Fort St. Philip was constructed. This military installation was to play an important role during the French-British rivalry over the island over the next 200 years. But even with the fort, Minorca enjoyed no relief from marauding corsairs. In 1535, Khayr ad-Dīn, the notorious red-bearded Arab pirate known to Europeans as Barbarossa, plundered the port of Mahón, murdering or enslaving half its people. He pillaged Ciudadela even more devastatingly three years later.

The War of the Spanish Succession (1701–1714) brought new rulers to Minorca. King Charles II of Spain had died in 1700, and his heir, Philip of Anjou, a French duke, took the throne as Philip V. An alliance consisting of England and Austria and other members of the Holy Roman Empire opposed Philip's assuming the throne of Spain and declared war on France. The alliance sought to install Archduke Charles of Austria as King of Spain, and Minorca along with the other Balearic Islands supported this cause. In 1708, forces led by Major General James Stanhope landed on Minorca, and Great Britain took over the island. When the war ended with the defeat of the French in 1714, Minorca officially became a British possession under the Treaty of Utrecht. In 1722, Minorca's capital was moved from Ciudadela to Mahón.

Although tensions between the devoutly Catholic Minorcans and the Protestant British simmered, the island prospered after the war under the prudent direction of the British governor, Richard Kane. He spearheaded agricultural reforms and building projects, including a road, which still bears his name, from Mahón to Ciudadela on the other side of the island. He also donated the clock on the front of the town hall, which had been built in 1631. Although succeeding governors did not match Kane for just and intelligent leadership, the island's significance grew as European powers increasingly came to regard it as an important strategic point in the Mediterranean.

Britain and France had been engaged in conflicts over commercial control of colonies in India and North America. In 1756, this rivalry led to the outbreak of the Seven Years War in Europe. France believed that capturing Minorca from the British would give it control over the Mediterranean. On April 17, 1756, a French fleet of more than 175 ships and 12,000 troops under the Duke of Richelieu attacked Minorca. The British forces, under the eighty-four-year-old General Blakeney, fended them off for over two months from Fort St. Philip, but a shortage of troops, the dilapidated condition of the island's defenses, and the inability of British Admiral George Byng to provide reinforcements led to the British defeat and Byng's eventual court-martial and execution. The French occupied the island for the next seven years. One of the remnants of this occupation is a French-style church in the village of San Felipe with an inscription to Louis XV above the door. When the war ended in 1763, the victorious British regained Minorca under the Treaty of Paris. During this phase of British rule, a new town, George Town (now Villa Carlos) was built.

In 1782, Britain, once more at war with France, found its defenders on Minorca besieged. After withstanding the siege for six months, the British finally surrendered the island to the joint Spanish and French forces.

But Minorca's revolving-door occupation was to continue. In 1793, Britain and France were yet again at war, this time over Napoléon's aggressions toward neighboring countries in Europe, and Spain was now an ally of Britain. Admiral Horatio Nelson took Minorca in 1798. In 1802, Britain peacefully handed Minorca to Spain under the Treaty of Amiens.

In the beginning of the nineteenth century, the United States established a presence on Minorca. Its trade had been, along with that of other European powers, continually disrupted by the extortion practiced by the Barbary rulers. President Thomas Jefferson sent the U.S. Navy to defend Mediterranean shipping in 1801, and used Mahón as its winter base until 1805. In 1815, the navy established a permanent base there (its first foreign naval base), and for several decades notable American naval commanders received their training there.

The nineteenth and early twentieth centuries were relatively uneventful on Minorca. The striking Teatro Principal, the oldest opera house in Spain, was built in 1824. A prosperous shoe-manufacturing industry, still an important source of income for Minorca, was developed in Ciudadela. Minorca and the other Balearic Islands remained neutral during World War I.

During the Spanish Civil War (1936–1939), Minorca went its own way in supporting the Republican cause, while the rest of the Balearics supported the Fascist rebels under Francisco Franco. Minorca managed to avoid takeover by the rebels throughout the war, partly thanks to an agreement between Great Britain, who sided with the Republicans, and Italy, who supported the Fascists, prohibiting Italy from landing there. However, shortly before the end of the war, on the eve of the final Fascist victory, Franco, in a deal brokered by the British, forced the surrender of Minorca.

World War II, which erupted five months after the end of the Spanish Civil War, and its aftermath caused economic hardship on Minorca and the other Balearics. After the war, tourism had the potential to become an important industry on the islands. However, as punishment for Minorca's support of the Republicans during the Civil War, Franco suppressed the development of the tourist industry on the island. The Minorcans' work ethic and resourcefulness enabled them to prosper at other industries, including the manufacture of shoes, cheese, and costume jewelry. The tourist industry's relatively slow growth resulted in less garish development on Minorca than that suffered by many other popular tourist destinations. The first charter-flight tours to Mahón began in 1958.

In 1983, Minorca along with the other Balearics became an autonomous province of Spain. Since then, the Castilian language, imposed during Franco's regime, has gradually been replaced by the traditional dialect called Menorquín, which is based on Catalan.

Today, Minorca is a prosperous island community of a handful of towns generally known for their cleanliness. Mahón has many surviving buildings from the British colonial period, especially Georgian houses with sash windows, not found anywhere else in the Mediterranean.

Further Reading: *The Balearic Islands* by L. Pericot Garcia, translated by Margaret Brown (London: Thames and Hudson, 1972; New York: Praeger, 1973), is a scholarly archaeological study of Minorca and the other Balearic Islands from Neolithic times through the Roman conquest. *Mediterranean Island Hopping: The Spanish Islands* by Dana Facaros and Michael Pauls (New York: Hippocrene, and London: Sphere, 1981) provides a readable, well-researched introduction to the island, its history, people, and contemporary ambience. *The Balearics: Islands of Enchantment* by Jean-Louis Colos, translated by Christine Trollope (Skokie, Illinois: Rand McNally, and London: Allen and Unwin, 1967), contains more than 130 striking, fully annotated black-and-white photographs of historic sites on the islands.

—Lisa Klobuchar

Mistra (Laconia, Greece)

Location: About three miles west of Sparta in the Peloponnesian Peninsula (ancient Morea) of Greece.

Description: Site of the walled, hillside city that was the cultural and political center of the Byzantine Empire in the fourteenth and fifteenth centuries during an era of political weakness in Constantinople and the resurgence of Hellenism. Mistra is chiefly notable for its intellectual influence on the Italian Renaissance and the fine developments in architecture and wall paintings that reflect the last phase of the Byzantine Empire. The ruins of Mistra's Frankish castle, its Palace of the Despots, and its architecturally elaborate churches can be viewed today.

Site Office: Tourist Police
Mistra, Laconia
Greece
(731) 93-315

Mistra's establishment as a city resulted indirectly from the Fourth Crusade. In 1204, the Frankish Crusaders and their Venetian allies diverted from their original destination of Egypt to pillage Constantinople and apportion Byzantine lands among themselves. Peloponnesus was awarded to the Frankish de Villehardouin family, which set up its political base on the open plains of Sparta. In 1249, searching for a better fortified base against Greek insurrection, William de Villehardouin looked to the nearby mountain range of Taygetus, on whose desolate slopes the ancient Spartans had once abandoned sickly infants to die. William built his fortress castle atop a steep, rocky hill known as Myzithra, presumably for its resemblance to a conical-shaped cheese made locally. The name evolved into "Mistra." There, under William's rule, young noblemen from all over Christendom came to be initiated into the chivalric traditions.

Frankish rule of Mistra, however, was short-lived. As Byzantine armies sought to reclaim their territories, William was captured by Emperor Michael VIII Palaeologus and forced to relinquish his stronghold at Mistra as part of his ransom in 1262. Mistra then became the capital of a despotate ruled by sons and brothers of the Byzantine emperor and the base from which to drive the Franks from Peloponnesus. Mistra flourished as a center of Byzantine art and scholarship for nearly two centuries, eventually numbering 50,000 inhabitants. Six centuries later, Goethe chose Mistra as the setting for Faust's meeting with Helen of Troy, awakened after her thirty-century sleep in Sparta, and the place where their son Euphorion was born and died.

Despite its forbidding terrain, Mistra grew rapidly under Byzantine rule as Greeks fled the plains of Sparta, still under Frankish dominance, to live under a ruler of their own religion. Under the brief tenure of the Franks, the original city had developed in three sections: the castle atop the hill surrounded by a wall; a larger section extending down to the bottom of the hill, also walled; and a third, unfortified section on flatter ground, where the Greeks built their earliest churches. The upper section of the city contained mainly the homes of officials and courtiers clustered around the royal court. Elsewhere, homes of rich and poor stood side by side among shops and churches.

The first governor of distinction in Mistra was Andronicus Palaeologus Asen, son of the ex-king of Bulgaria and nephew, through his mother, of the Byzantine emperor Andronicus II Palaeologus. Serving from 1315 to 1321, Andronicus Palaeologus Asen achieved many victories against the Franks in Peloponnesus, inducing lesser Frankish lords to accept Greek domination and the Orthodox religion. Following a period of disorder on the peninsula exacerbated by civil war, Emperor John VI Cantacuzenus appointed his younger son Manuel to the Mistra Despotate in 1349. Shortly before leaving for Mistra, Manuel married a noblewoman of Latin descent, a move that may have influenced his policy of friendliness toward Latins in the Peloponnese. With their help, Manuel was able to subjugate the rebellious Greek lords of the region. Also, through alliances with the Venetians and the Latin governor of Achaea, Manuel scored a series of victories against the Turks, reducing their raids in the Peloponnese.

In the meantime, Manuel converted Mistra into a renowned center of learning by creating facilities for study, encouraging the collection of books, and commissioning the illumination of precious manuscripts. He also promoted the construction of churches that were to become Mistra's most lasting legacy. It was under Manuel's rule that the fabulous Palace of the Despots was built in the upper city, providing the royal court far more splendid accommodations than the Frankish castle. With the scarcity of level ground prohibiting broad avenues or open squares in Mistra, the palace lands served as a ceremonial parade ground.

Respected as a firm but compassionate ruler, Manuel died in 1380. After three years of squabbles within the Byzantine royal family, the governorship of Peloponnesus was awarded to Theodore, a younger son of Emperor John V Palaeologus. Theodore I arrived in Mistra in 1383 hoping to increase his holdings. But years of failed alliances, fruitless warfare against Latin princes, and encroachments by the Turks left him in despair. At the turn of the century, he agreed to sell Mistra to the Knights Hospitaller, one of the three great military orders with Peloponnesian holdings, and retire to the

Medieval ruins at Mistra
Photo courtesy of Greek National Tourist Organization

seacoast town of Monemvasio. But when the knights entered Mistra, the rebellious inhabitants rose against them and almost strung up the knights' delegates. Theodore repudiated his agreement, returned to Mistra, and settled accounts with the knights by refunding their payment. Although not a popular ruler, Theodore I had done his share to encourage the arts and enhance the architectural glory of Mistra. He died in 1407, a few days after taking monastic orders.

The Byzantine emperor, now Manuel II Palaeologus, Theodore's brother, appointed his twelve-year-old son, also named Theodore, to the Mistra Despotate. Manuel visited the city himself to ensure the competency of the child's ministers. Theodore II grew to be a considerable scholar and superb mathematician, but had little interest in power. A marriage was arranged for him with Cleope Malatesta, from the royal family of Rimini, Italy. She arrived in Mistra and was wed to Theodore, but initially he shunned her. Only when he recognized that she shared his intellectual fervor did he come to admire her. His courtiers became devoted to her as well. During Theodore's thirty-six years as governor in Mistra, he earned the goodwill of the leading Greek scholars, and philosophy and literature flourished for the last time in the Byzantine Empire. With the aid of his imperial brothers, he cleared the peninsula of Latins. Agriculture and commerce prospered. In 1443, Theodore II departed from Mistra to settle the religious problems of Selymbria on the Sea of Marmara. He left the despotate to his brother Constantine.

As Mistra had grown as a political stronghold, the city had attracted many of the best artists from war-ravaged Constantinople and many of the most renowned scholars as well. Chief among them was the eminent philosopher Gemisthus Plethon. He came to the city around 1407 and later established a school of thought to overturn the prevailing Aristotelianism, with its emphasis on logic, in favor of revived interest in Plato's teachings. Plethon stressed the mystical side of Platonism, and developed a system uniting Greek mythology with Greek logic. Through his visit to Florence and that of his pupil Bessarion of Trebizond, who later became a cardinal of the Roman Catholic Church, Plethon greatly influenced the philosophical texture of the Italian Renaissance.

Another distinguished thinker, a young Greek ecclesiastic named Isidore, studied under Plethon, wrote extensively while in Mistra, and later became metropolitan of Kiev and head of the Church of Russia. Others attracted to Mistra included Gennadius II Scholarios; John Eugenikos, brother of the metropolitan of Ephesus; Charitonymous Hermonymous; and George the Monk. Another, John Moschus, was to follow Plethon as Mistra's chief philosopher in residence.

While Plethon's Platonism became popular with Renaissance thinkers, his religious views were not at all appealing to the Greeks. Plethon set down these controversial views in a book called *On the Laws,* of which only a contents outline and related notes have survived. In this work, Plethon proposed a new religion based primarily on the teachings of Zoroaster, and also cited thinkers such as the Brahmins of India. He entrusted his manuscript to the despot of Mistra, but eventually it found its way into the hands of the patriarch of Constantinople. Shocked by its heretical contents, the patriarch had the work burned. Except for a year-long sojourn in Italy, Plethon spent the rest of his life under the patronage of the imperial family in Mistra, dying there in the 1450s around the age of ninety. He was buried in Mistra, but his body was removed from its tomb in 1465 when Venetian troops under Sigismondo Pandolfo Malatesta of Rimini made a brief incursion into Mistra. The Italian leader took the body of the great philosopher home to Rimini and placed it in an elaborate sepulchre, a more fitting tribute to the man who had contributed so greatly to the Italian Renaissance.

While intellectual achievement played a major role in Mistra's history, the city is best known for the many churches reflecting the last phase of Byzantine architecture at a time when the empire itself was largely under Islamic domination. The churches of Mistra were well-proportioned and graceful, combining the Roman-style domed basilica with the Greek-cross floor plan. This design was complemented by recessed windows, detailed wood carving and elaborate frescoes. The churches often housed women's galleries and extra chapels, an innovation allowing space for the tombs of Mistra's rulers and clerics. Some churches had open porticoes and belfries reflecting Western influences.

The oldest church surviving today is the Agioi Theodoroi, completed in 1295 under the observation of the learned cleric Pachomius. Today, it contains the earliest surviving frescoes in the city. Here Pachomius founded a monastery, the Brontochion, and won from the Byzantine emperor the singular distinction of removing his monastery from local jurisdiction to that of the patriarchate of Constantinople. Another early church is the beautiful Aphendiko, with tall buttresses, cupolas, billowing apses, and inclined roofs. Surviving today in its sanctuary are frescoes depicting full-length figures of the better-known bishops of the Eastern Church. In the Metropolis, or the Cathedral of St. Demetrius, another of Mistra's great churches, are splendid frescoes depicting the life of St. Demetrius, the miracles of Christ, and the Last Judgment. The sanctuary contains a marble beam and arched frames providing fine examples of Byzantine stone carving. Among other notable religious structures surviving today are the Peribleptos and the Pantanessa Convent, both built around the middle of the fourteenth century. The former features the simplest architecture among Mistra's churches, and frescoes of such gospel scenes as the Nativity, the Transfiguration, and Christ's entrance into Jerusalem. The seven-domed Pantanessa contains frescoes of painstaking detail, one of which is considered the most dramatic surviving depiction of the raising of Lazarus. Complementing Mistra's architectural achievements is the castle fortress atop the hill. Though now in ruins, it still reflects its Frankish origin combined with Byzantine and Turkish restorations.

Mistra enjoyed its last taste of Byzantine glory under the governorship of the despot Constantine, who ruled from 1443 to 1449. As the ever-threatening Turks were occupied by yet another crusade advancing into the Balkans, Constantine led an army northward in 1444 to capture Athens and Thebes. But he lost these regions to the Turks again two years later, and his army was left decimated. An invasion of the Turkish army into Mistra was prevented only by winter weather, which made mountain travel difficult. Meanwhile, the death of Emperor John VIII Palaeologus in 1448 left the imperial throne to the despot Constantine, who became Emperor Constantine XI Palaeologus. With other claimants actively seeking the crown, Constantine, before leaving Mistra, hurriedly held his imperial coronation in the Cathedral of St. Demetrius in January 1449.

The new emperor decreed that two of his brothers, Demetrius and Thomas, would rule the empire's Peloponnesian holdings, with the Mistra Despotate falling to Demetrius. Unfortunately, the brothers quarreled and provided little aid to the emperor when the Turks launched an invasion of Constantinople. The imperial city fell in 1453, and Emperor Constantine was killed, to the great sorrow of his former subjects in Mistra. His brothers now faced rebellion in their own armies. For nearly a century, the imperial troops in the Peloponnese had recruited Albanians, who made good fighters but remained aloof from the Greeks. With Constantine dead, the Albanians, who felt little loyalty to his two brothers, revolted. The two brother despots had to call on the Turkish sultan to restore order in the ranks, leaving them obligated to pay annual tribute to their one-time enemy. This soon led to more trouble.

By 1458, the annual tribute had fallen three years behind and, fearing Western interference in the Peloponnese, the Sultan Mehmed II led a great army into Greece. Their own armies being no match for the Turks, Demetrius and Thomas were forced to cede much of their lands to the sultan, although Mistra was initially spared. But the brothers continued quarreling over their remaining territory, each seeking help from the West. In the spring of 1460, the sultan summoned Demetrius to Corinth. When he failed to appear, the sultan brought his army to Mistra and persuaded Demetrius to surrender the city in exchange for lesser holdings. By 1461, virtually all of the Peloponnesian Peninsula was under Ottoman control, and a Turkish governor came to Mistra to live in the Palace of the Despots.

Although its Byzantine era had ended, Mistra prospered during its next two centuries under Ottoman rule, in large part because of the silkworm farms in Mistra's valley, which attracted considerable foreign trade. At the beginning of the Ottoman era, Mistra had more than 40,000 inhabitants, and the establishment of a Jewish colony in the city was indicative of its status as a commercial center. But in 1684, when war broke out in the Peloponnese between the Turks

and the Venetians, Mistra was again threatened. Two years later, with their army threatened by plague, the Turks abandoned the city. Under the new Venetian rulers, Mistra's Greek inhabitants suffered higher taxes and the deliberate erosion of their silk industry to prevent competition with Italy.

In yet another reversal of fortune, the Turks launched a counteroffensive, taking from Venice nearly all its possessions in the Peloponnese. Mistra along with the rest of the province returned to Ottoman rule. Half a century later, however, Russia declared war on the Turks and sought allies among the Greek inhabitants of the Aegean Islands and the Peloponnese. In 1770, Russian troops in the Peloponnese were welcomed by the metropolitan of Lacedemonia and soldiers from Mistra and surrounding areas. The small Turkish garrison in Mistra surrendered; however, the metropolitan prevented a general massacre of Turkish residents of Mistra. But the Russian army soon retired to the coast, leaving Mistra largely defenseless when the Turkish pasha and his Albanian troops attacked later that same year. Under the pasha's orders, the once great city was set afire and parts of it were left in ruins. When order was restored, the population had dropped to 8,000. The resilient Greeks of Mistra restored some of their structures and prospered modestly over the next several decades with the revival of the silk industry in that region. But in 1824 Ibrāhīm Pasha destroyed the city during the Greek War of Independence. With little will to rebuild, the surviving citizens abandoned Mistra to settle around Sparta. The city that had sprung from Spartan roots ultimately lost its residents to Sparta.

Today, Mistra is inhabited only by nuns of the Pantanessa convent. Although Greek authorities began considerable restoration in the late nineteenth century, the city still reflects the great destruction caused by centuries of war. Byzantine artistry remains forever mingled with bullet holes and saber scars, a mute tribute to the dramatic history of this once glorious city.

Further Reading: *Mistra: Byzantine Capital of the Peloponnese* by Steven Runciman (London and New York: Thames and Hudson, 1980) is a thorough examination of the city's history, complete with maps, illustrations and photos. *The Living Past of Greece* by A. R. and Mary Burn (Boston: Little, Brown, and London: Herbert, 1980), *The Companion Guide to Mainland Greece* by Brian De Jongh and John Gandon (New York: Harper, 1989), and *Greece Observed* by André Barret, translated from the French by Stephen Hardman (London: Kaye and Ward, and New York: Oxford University Press, 1974; as *Voir la Grece,* Paris: Hachette, 1971) all discuss Mistra's history and architecture, along with other notable sites in Greece.
—Pam Hollister

Modena (Modena, Italy)

Location: City in northern Italy, twenty-five miles northwest of Bologna, thirty-five miles southeast of Parma, and forty-five miles southwest of Ferrara; in the province of Modena and the region of Emilia-Romagna.

Description: Settled from the earliest days, Modena first flourished in the eleventh and twelfth centuries; the city and surrounding province have much Romanesque architecture and sculpture. The city reached its peak in the seventeenth century under the Este family. Now a thriving city, one of the wealthiest in Italy, it is rich in museums and libraries, is the headquarters of automakers Ferrari and Maserati, and is the home town of two famous opera singers, Luciano Pavarotti and Mirella Freni.

Site Office: (A.P.T.) Azienda di Promozione Turistica della
Provincia di Modena
Corso Canalgrande, 3
CAP 41100 Modena, Modena
Italy
(59) 220136

Among the oldest objects to have been found in Italy is the *Venus of Savignano* (now in the Museo Pigorini, Rome) dating from late Paleolithic times, discovered in what is now the province of Modena, just southeast of the city, indicating that the area was inhabited very early. The Apennine culture succeeded the earliest settlements, flourishing from the fourteenth to twelfth centuries B.C. The Protovillanovan people took over the area about 1200 B.C. There were early Iron Age Villanovan settlements from the ninth century, and the Etruscans occupied the area from about the eighth century, with Bologna, then called Felsina, as their capital. The Celtic Gauls took over the whole area between the Apennine Mountains and the Alps, and in the southwest of the province of Modena there are some Celtic huts to be found, made of sandstone with plaited rye-straw.

By the third century B.C. what is now Modena had been conquered by the Romans and was known as Mutina. In 187 B.C. the Romans built the Via Emilia, the road linking Rimini to Piacenza, and which even today is the main road of the small city of Modena, running right through its center. In the years following the downfall of the Roman Empire, the area was almost destroyed by wars and became increasingly marshy as drainage was neglected, but by the sixth century A.D. a new town had arisen, originally known as Cittanova or Città Geminiana, named after its patron saint, San Geminiano, whose feast day, January 31, is still celebrated in Modena.

In approximately 1000 the town came under the rule of the counts of Canossa when walls surrounding it were built

in the shape of a pentagon; now there are broad avenues where the walls stood. It was in the eleventh and twelfth centuries that Modena began to develop, under the Countess Matilda Canossa, who was the patron for the building of its Duomo, or cathedral, begun in 1099 and completed in 1184, just nine years after the founding of Modena's university. The architect of the cathedral was Lanfranco, a Lombard, and it is one of the most important examples of Romanesque sculpture in Italy. Much of the sculpture was done by Wiligelmo, who founded a school of Romanesque sculpture in Modena, and there is also sculpture by the Maestri Campionesi (Masters of Campione), the Lombard master-masons who, together with their rivals, the Maestri Comacini, created and spread the Romanesque-Lombard style all over northern and central Italy. Inside the cathedral is the tomb of San Geminiano.

The whole province of Modena contains many examples of Romanesque buildings and sculpture, including more work by Wiligelmo and his school. There is a large number of churches, but the most important building is the Abbey of Nonantola. Originally founded in 752, it was destroyed and rebuilt several times until its final reconstruction in the Romanesque period, after an earthquake in 1117. Additions and alterations were made over the centuries. Between 1914 and 1917 much work was done to restore the abbey to its original Romanesque state, including a pre-1117 crypt that had been covered over in the fifteenth century. Some time later a cycle of Romanesque frescoes was revealed. Nonantola is just to the northeast of the city of Modena.

The symbol of Modena, the Torre Ghirlandina (Ghirlandina Tower), was built beside the cathedral serving as both its bell tower and as a defense post, housing the municipal charters and other important civic documents. Although both the cathedral and the bell tower are sinking a little, making the latter lean somewhat, it stands nearly 300 feet high; the base dates from the same period as the cathedral while the upper part, with its spire and distinctive bronze-garlanded weather vane, dates from 1261–1319 and is Gothic in style. A second symbol of Modena is a statue known as *la Bonissima,* representing, according to popular legend, a rich and noble woman, famous for her generosity to the poor.

The twelfth and thirteenth centuries were troubled times, and in 1167 Modena joined the Lombard League in resisting Holy Roman Emperor Frederick I's attempts at colonization in northern Italy. In 1226 Frederick II renewed imperial claims to the region, but this time Cremona, Parma, Reggio, and Modena backed him, and in 1235 Mantua and Ferrara joined in support of the emperor, helping him in his victory at the Battle of Cortenuova in 1237. This change of sides was not unusual in these times, when cities chose whether to back pope or emperor not only on an ideological

444

Modena's Duomo
Photo courtesy of A.P.T. di Bologna e Provincia

basis but as a result of ongoing blood feuds between noble families, with their ever-changing alliances, and inter-commune rivalries. After Frederick II was defeated in 1250, conflict continued, focusing on the nobles. In Modena it was the Savignano and the Guidoffi families. Many communes created organizations called the *popolo*, composed of wealthy non-nobles who aimed to control the nobles' power, but in Modena there were no great bankers, tradesmen, or industrialists; most businesses were run only with the support of the nobility.

One such family in the region was the Este family of Ferrara, just northeast of Modena. When Azzo VII d'Este died, in 1264, his illegitimate grandson, Obizzo II, was accepted by the people of Ferrara as an absolute monarch, and in 1288 Modena offered Obizzo lordship of its territory, as did Reggio in 1290. By this time people were tired of constant conflict and hoped that unity under the Este family would bring peace. The local nobles supported the move, seeing it as a means of curbing any further growth in the strength of the commune and any developing *popolo*. So the Marquis Obizzo II became lord of Modena and Reggio. In 1289 a fortified manor was built for the lord in the northern part of Modena, later to become the Ducal Palace.

In 1325, however, there was a war between Modena and Bologna, part of an ongoing conflict between the two cities. A trophy of this war was a bucket stolen by Modena from Bologna. The bucket came to be housed in the Ghirlandina Tower, where it can still be seen. A long, mock-heroic poem was written about this bucket by Alessandro Tassoni in 1622, and a statue honoring the poet stands near the tower.

The middle of the fourteenth century was a difficult time for the whole of Italy, already weakened by wars and famine, when the Black Death struck. In Modena it lasted through much of 1348, and was followed by another period of slow recovery. However, at this time the town did produce Tommaso da Modena, a painter whose frescoes can be seen in the chapter house of San Niccolo in nearby Treviso.

In the early fifteenth century feudal relationships were still strong in Modena, with a constant struggle between the counts Cesi of Gombola, who held Modena as an imperial fief, and the commune, or municipal government, which was supported by the Estes of Ferrara. In 1450 Borso d'Este succeeded his brother, Leonello, as marquis of Ferrara, and spent much money and time on persuading Holy Roman Emperor Frederick III that he should make Borso a duke—of Modena and Reggio. He succeeded in 1452 and the Duchy of Modena and Reggio was created.

With the Este family in power in Modena, the duchy and the city quickly grew in stature. There was a setback when Alfonso I d'Este lost Modena to Pope Julius II, who was seeking to expand papal territory, about 1510. In the early 1520s Alfonso antagonized the new pope, Clement VII, a Medici, by trying to regain the duchy, but the Estes' efforts were eventually successful, and by 1531 they had secured Modena and Reggio once more. Alfonso was succeeded by

Ercole II, son of Alfonso's marriage to Lucrezia Borgia, and on Ercole's death, Alfonso II became duke. When Alfonso II died in 1597, leaving no heir, the pope promptly evicted the Estes from Ferrara, reclaiming it for the Papal States. The remaining Estes, under Cesare, fled to Modena, which they chose as their new capital for the Duchy of Modena, Reggio Emilia, and Carpi, encompassing the territory that remained to them. This new status immediately triggered a rise in the fortunes of Modena, which reached its peak during the seventeenth century.

Some significant building had already begun before Modena became the dukes' capital. This included the Church of San Pietro, dating from 1476, which has a grand composition, *Canaan Nuptials*, painted on its entrance door by Modenese artist Ercole Setti. The Palazzo della Ragione (Palace of Justice) and the Palazzo Civico (Civil Palace) were begun in the sixteenth century, then attached to each other in the seventeenth century to become the building now known as the Town Hall. In the Sala del Fuoco (Fire Room) within the Town Hall the walls were decorated by another local painter, Niccolò dell'Abbate, a Mannerist painter who became part of the Fontainebleau school in 1552.

Cesare d'Este and the rest of his family soon ordered work to be done on his new capital to make it worthy of the title and to provide the Estes with a suitable setting for the life to which they were accustomed. The city walls were extended and streets were widened. In 1634 work began on the Palazzo Ducale (Ducal Palace), ordered by Francesco I d'Este. This was designed by Bartolomeo Avanzini and built on the site of Obizzo II's castle. The huge baroque palace, with three large towers and three stories, took several centuries to complete. Eventually the palace became a military academy; its cadets are now considered to be some of the country's top infantry and cavalry recruits. A small palace (Palazzina) was also built, designed by Emilia's most important seventeenth-century architect, Gaspare Vigarani, adjacent to the formal Ducal Gardens.

Avanzini, architect of the Ducal Palace, also designed the Collegio San Carlo (College of St. Charles). Several churches in Modena also date from the seventeenth century. These include the Church of San Giorgio, another of Vigarani's designs; the reconstruction and enlargement of the old Church of San Agostino, ordered by Alfonso IV's wife, Duchess Laura Martinozzi, in 1663; and the Church of San Vincenzo in which, eventually, sixteen dukes of the Este family were buried. The Church of the Madonna del Voto was built by the community in thanks for the end of the plague of 1630, and the Church of San Bartolomeo was built for the Jesuits in grand style.

Modena continued to be the capital of the Este duchy throughout the eighteenth century, despite suffering somewhat, both socially and economically, as a result of the Wars of Polish and Austrian Succession. Modena, however, was the only seigneurial state that not only survived, but also enlarged its territories and political influence in the eighteenth century. There had been a crisis when Duke Rinaldo

I had lost Modena to the French, from 1733 to 1736. Francesco III, his heir, pulled out of the alliance with the Holy Roman Empire to join the Bourbons in order to regain the dukedom, but it was threatened again from 1742 to 1748 by the Austro-Sardinian occupation, during the War of the Austrian Succession. However, the 1748 Treaty of Aix-la-Chapelle, which ended the war, returned the Este territories to the duke and in 1755 Francesco III was, in addition, appointed governor of Lombardy by Empress Maria Theresa. Pro-Habsburg again, Modena then became a center for the Jesuits, who were under attack elsewhere. The duchy extended its boundaries thanks to a marriage between Duke Francesco's son, Ercole, and Maria Teresa Cybo, who was heiress to the duchy of Massa.

With such strong ties between the Estes and Austria, both political and then through marriage, Francesco was influenced by the reforms being made by Maria Theresa and her son Joseph II. He set up a Council of State to advise him on policy, appointed three secretaries of state to oversee various parts of the duchy, and established a new judicial administration. He also produced his *Codice di Leggi e Costituzioni* in 1771, which, while consisting largely of traditional Roman law, did include some significant changes, in particular the establishment of a Supreme Council for Justice, which became the appeals court and provider of guidance for judges and lawyers. Between 1753 and 1755 he managed to gain control of much ecclesiastical property, and in 1788 a new Land Register was produced and the whole tax system revised. These reforms, designed to relieve poverty by redistributing land, caused the church to lose far more property and privilege than the nobility of the area did.

Francesco's concern for the poor manifested itself in two buildings as well, a public hospital called the Grande Ospedale Civile degli Infermi, built from 1755 to 1762, and a poorhouse called Grande Albergo dei Poveri, built from 1767 to 1771 but which in 1788 became an art gallery. In 1883 the municipal authorities acquired the building, which is now the Palazzo dei Musei, a magnificent museum and art gallery.

The building contains the two most important collections gathered by the Estes—the Biblioteca Estense (Este Library) and the Galleria Estense (Este Gallery). The library is one of Italy's most important and valuable collections of books, letters, and manuscripts, composed of four basic components: the Este Collection, begun in the eighth century; the Muratori Archives; the Music Section; and the Campori Foundation. The library includes several illuminated manuscripts, the two most famous being *De Sphaera*, with work by Lombard miniaturists, and the world's most extensively illuminated book, the *Bible of Borso d'Este*, 1,200 pages with more than 2,000 miniatures, heraldic emblems, and decorations.

The collection now in the Este Gallery was begun in 1598 by Cesare d'Este, who took many works of art from Ferrara to Modena when the family was expelled. Francesco I, his successor, added greatly to the collection and the tradition continued. There are many works by Emilian and Tuscan painters, including Cosmè Tura, Dosso Dossi (Giovanni Lutero), the Carracci family, Guido Reni, Giovanni Francesco Barbieri (Il Guercino), and the Venetians Tintoretto and Paolo Veronese. There is a bust of Francesco I by Gian Bernini, and until 1992 there was a Velazquez painting of the same subject, sadly stolen along with Corregio's *Campori Madonna*.

Other collections in the Palazzo dei Musei include the d'Este Lapidary Museum, consisting of most of the Roman remains found in the area; the Archivio Storico della comunità (Community Archives), with documents on Modena dating from 969 onward, and two more museums, the Museo Archeologico (Archaeological Museum) and the Museo di Storia e Arte Medievale (Medieval Art and History Museum).

Modena's university buildings were designed by Tarabusi and dated from about 1750, when they became a permanent home for the university established in the twelfth century. The campus contains zoology, anatomy, and mineralogy museums. Another important Modena institution is the Muratoriana Studies Center and Museum, adjacent to the Committee of National History Library. The center holds the writings of Ludovico Antonio Muratori, one of Modena's most famous citizens, considered the father of Italian history. He was librarian of the Este Library for some time. His tomb lies in the fourteenth-century Church of Santa Maria Pomposa, a church he restored, which is next to the center and museum named for him. Muratori's reputation as a scholar made eighteenth-century Modena a center of intellectual development.

In 1796, the Este Duchy of Modena effectively ceased to exist, with the establishment of Napoléon's Cisalpine Republic. Ercole III was deposed, losing his duchy, but on his death in 1803 his brother-in-law, Ferdinand of Habsburg-Lorraine, who was married to Maria Beatrice d'Este of Modena, took the title. Ercole's grandson became Duke Francesco IV in 1814 when Napoléon's republic collapsed. The Italian nationalist movement, or Risorgimento, deposed Francesco briefly in 1831; the duke actually had already fled the city before the nationalists made their announcement on February 9. However, Austrian intervention reinstalled Francesco and subdued the uprising. One of the main instigators, Ciro Menotti, was hanged, so becoming a hero of the Risorgimento. His statue stands opposite the Ducal Palace. Francesco V became the last duke of Modena in 1846, but was deposed in 1859 as the nationalist movement pressed on. In 1860 the duchy became part of a unified Italy.

The twentieth century saw the rise of socialism with Emilia-Romagna as its stronghold. There was an upsurge of nationalist feeling after World War I and the socialist Benito Mussolini gradually transformed his ideas into Fascism. During World War II the city and province of Modena formed one of the most active centers of resistance to Nazism and Fascism. There are several memorials in the area to those who fell during this struggle, including the 136 citizens of the village of Monchio who were killed by German troops in 1944.

Both the city and the province of Modena have prospered in the late twentieth century. The city has the highest per capita income in Italy. The province is home to the makers of Ferrari and Maserati cars. Ferrari has opened a company museum near its factory in the town of Maranello, just south of the city of Modena. The museum displays one of the world's largest collections of Ferraris. Modena is also rich in theatres, music, and cinema, perhaps an encouragement to two of its most famous citizens of the twentieth century, the tenor Luciano Pavarotti and the soprano Mirella Freni, both of whom have returned to their home town to sing at the City Opera House.

Further Reading: *Power and Display in the Seventeenth Century: The Arts and Their Patrons in Modena and Ferrara* by Janet Southorn (Cambridge: Cambridge University Press, 1988; New York, Cambridge University Press, 1989) has 200 pages arranged in three parts, the first of which is *The Este in Modena (1598–1658)*. The other two parts deal with Ferrara. Two volumes in the *Longman History of Italy* provide significant insight into events in Modena. They are *Italy in the Age of Dante and Petrarch* by John Larner (London and New York: Longman, 1980) and *Italy in the Age of Reason* by Dino Carpanetto and Giuseppe Ricuperati, translated by Caroline Higgitt (London and New York: Longman, 1987; as *Italia del Settecento*, Bori: Laterza, 1986).

—Beth F. Wood

Monaco

Location: On the Mediterranean Sea in the southeasternmost corner of France, approximately nine miles east of the city of Nice and five miles west of the Italian border. Monaco has its own coastline and territorial waters in the Mediterranean, but is otherwise completely surrounded by France.

Description: The Principality of Monaco, as it is officially known, is the second-smallest sovereign state in Europe (only Vatican City is smaller). Its total area of approximately seven-tenths of a square mile is the result of landfill and reclamation projects that increased the principality's surface area by 20 percent. With a population of approximately 30,000, Monaco ranks as the most densely populated country in the world. Its main industries are banking, manufacturing, and gambling and tourism. The Grimaldis, Monaco's ruling family, have been in power almost continuously since 1297; although the country has adopted most Western democratic institutions, they reign today as the last supreme monarchy in Europe.

Site Office: Direction du Tourisme et des Congres de la
   Principauté de Monaco
2a, Bd des Moulins
Monte Carlo MC 98030
Monaco
92 16 61 66

Compared to the sweeping and often random events that have shaped its larger neighbors, the history of the tiny Principality of Monaco appears sharply defined and logical in its unfolding. The characters evolve from one generation to the next, but the backdrop stays the same until well into the nineteenth century, when gambling made its debut, and has remained essentially the same ever since.

Today, Monte Carlo, Monaco's central gambling and nightlife district, is probably better known than the name of the country that surrounds it. Monte Carlo has become synonymous with luxury, and the Monte Carlo Grand Prix is one of auto racing's most important events. But there is also a side to Monaco not seen by the world at large. Small to medium-sized manufacturing companies and growing high-technology firms in the newly reclaimed district of Fontvieille now account for one-quarter of the country's economic output. The service sector, led by international banking, has grown rapidly in the latter half of the twentieth century. Although less than half the size of New York City's Central Park, Monaco has grown into a thriving, modern country.

The Rock of Monaco, the heart of the country and site of the royal palace, juts out into the Mediterranean, creating an easy-to-defend fortress surrounded on three sides by sea and cliffs. The rest of the principality sits on a densely clustered set of hills; almost all of this area has been developed since the advent of gambling and the growth of the tourism industry. Monaco took its name either from the temple of Hercules Monoecus, which was rumored to have existed on the rock prior to Greek or Roman knowledge of the territory, or from the Ligurian tribe the Monoikos, who inhabited the rock about 600 B.C.

Before the rise of the Grimaldis, Monaco's ruling family, the rock changed hands regularly, but none of the occupying groups established cultural roots. After the Ligurians, the Greeks, Phoceans, Carthaginians, Phoenecians, Saracens, and Romans all occupied the rock prior to 1297. Hannibal used its natural harbor for his fleets in 221 B.C. while fighting the Romans. Julius Caesar anchored his ships there in 69 B.C. during his Greek campaign. The Lombards took over after the fall of the Roman Empire in 476, but were driven out by Charlemagne in 774. Shortly thereafter, the Saracens from North Africa swept through the area. Many important peoples passed through Monaco, but few of their influences remained. The rugged peninsula may have had significant strategic value, but its lack of arable land and scarce natural resources made it less than inviting for long-term occupation.

By the twelfth century Monaco had become an important fortress in the ongoing feud between the Guelph (pro-papal) and Ghibelline (pro-emperor) parties of Genoa. In 1174 the Guelphs created a charter granting themselves sovereignty over the small outcropping in order to build a proper fortress upon it. This measure had undoubtedly been considered by others but never been accomplished. The stated purpose of the fortress was "for the defense of the Christians against the Saracens and for the use of imperial troops in case of war between the [Holy Roman] Empire and the Provençals." In 1190 Holy Roman Emperor Henry VI made the act official, awarding Monaco to the Guelphs. The fortress was built in 1215.

The Grimaldis did not rise to prominence in Monaco until 1297. They had, however, begun to play an important role in Genoese politics as early as 1133, when Otto Canella was a consul in the city. His son Grimaldo Canella gave his name to the family. The clan's reputation, even into modern times, has ranged from brilliant efficiency to ruthless greed. Before long the Grimaldis were among the leading families of the Genoese Republic. Politically, they were allied with the Ghibellines.

The family achieved the height of its prominence in the late thirteenth century under the leadership of Rainier

The Prince's Palace

Casino at Monte Carlo
Photos courtesy of Monaco Government Tourist and Convention Bureau

Grimaldi, namesake of the current ruler. Rainier served in the navy of Charles II, king of Naples, with whom the family was allied. His palace in Genoa was reputed to rival the king's.

Rainier's wealth and power fueled the jealousy of his nephew François Grimaldi, known to history as François the Spiteful, who became desperate for a way to enhance his standing within the family. In 1295 he commanded an army of Ghibellines in the successful rout of enemy Guelphs from Liguria, forcing them to take refuge on the Rock of Monaco. Two years later he finished the assault when, according to legend, on the cold night of January 8, 1297, François led a small contingent of followers up the steep and rocky slope to the very gate of the fortress. With his men safely hidden in the shadows, François, disguised as a monk, knocked on the gate, asking for shelter for the night. Taking pity on the destitute holy man, the soldiers let him in. Once inside the fortress, François pulled his sword from under his cloak, and, calling to his comrades outside the fortress, made quick work of the half-asleep Guelph soldiers who dozed inside. The event is commemorated in the principality's coat of arms, featuring two monks brandishing swords and the motto "Deo Juvante," meaning "With the Help of God."

Continuing attacks by the Genoese Guelphs forced François to flee the Rock four years later. But the Grimaldis had stayed long enough to become impressed with the Rock's potential as a base for their sea trading business, from which the family fortune derived, and they resolved to regain the territory.

By 1338 the family had earned enough money and respect within certain circles to actually purchase the Rock outright from the Genoese Republic, which had held title to the land since 1190. Charles Grimaldi, son of Rainier, became the sole lord of Monaco and the Rock, then home to a simple fortress and several narrow rows of houses in which lived the 200 or so inhabitants, who came to be called Monégasques. The outbreak of the Hundred Years War disrupted Charles's plans for expansion of the settlement, and in 1349 he left Monaco to serve as admiral of the French fleet under Philip VI.

During the six-year absence of their newest ruler, the Monégasques prospered as fearsome land-based pirates. The vantage point and defenses supplied by the Rock made piracy easy, and the region's lack of natural resources made piracy it necessary. Upon his return from the war, Charles put an end to the piracy and began to realize his prewar dream of building the Rock into a true principality. Among his first priorities was to increase the size of his tiny domain, which at the time consisted of about three miles of coastal land and the Rock itself. The principality was barely a scratch in the southern flank of France, reaching a maximum of half a mile inland, with northern boundaries similar to those of today. Having been handsomely rewarded by Philip for his service to France, Charles was able to purchase the neighboring towns of Roccabruna and Mentone (later to be known as Roquebrune and Menton). The resulting increase in population and associated economic activity created something like

a viable township, complementing the military function of the fortress.

Unfortunately, Monaco's former owners, the Genoese, felt threatened by Charles's newfound friendship with France, and in 1355 they attacked the principality from land and sea with troops totaling 4,000. Hopelessly outnumbered, Charles surrendered. In 1357 he died, and his son, Rainier II, fled the principality, never to return.

Monaco eventually passed to the kingdom of Aragon, and in 1419 Rainier II's three sons, Ambroise, Antoine, and Jean, who had done well in the family trading business, repurchased the domain in the name of Grimaldi. The longest-lived of the three brothers, Jean, left a will stipulating that the eldest Grimaldi son would inherit the principality. If there was no male heir, the eldest daughter could inherit, but only if her husband adopted the name and coat of arms of the Grimaldis, thereby assuring the future of the dynasty. This edict has stood essentially unchanged ever since, with one amendment made in 1487 by Lambert Grimaldi that excluded from succession family members who had entered the religious life.

The sixteenth century was a turbulent period for the Grimaldis. In 1505 Jean II was murdered by his brother Lucien, who then succeeded him. Lucien and his brother Augustin were then murdered by their nephew François. In 1604, after a relatively long reign of fifteen years, Lucien's grandson Hercule was stabbed to death in an alley by a group of Monégasque men whose daughters he had molested.

Hercule left a seven-year-old son, Honoré II, as heir. The boy-ruler and his sister, Jeanne, were hidden beneath the palace by members of the family until their maternal uncle Federico Landi, prince de Valdetare, arrived from Milan and was declared regent. Federico installed Honoré II, calming the Monégasques. He then returned to Milan to sign an agreement with the Spanish government, which had overrun Milan, to allow Spanish troops to be garrisoned at Monaco. Honoré II spent the next ten years in Milan with his uncle. Monaco finally became a principality in 1612 when Federico prepared a document declaring Honoré II to be "regnant, prince and seigneur by the grace of God of Monaco, Menton, and Roquebrune."

When Honoré II returned to Monaco in 1615, nearly twenty years old, he may have been the first prince of the Rock, but he was prince of an occupied rock. The Spanish garrison had virtually overrun the small principality, with Spanish soldiers marrying Monégasque women and effectively controlling Monaco's economy and trade. Honoré's primary objective during his reign was to rid his principality of the occupying Spaniards.

In 1630 he sought help from Cardinal Richelieu of France, chief minister to Louis XIII. It took five years, but eventually Honoré succeeded in obtaining a pledge of protection from France. Unfortunately, the Thirty Years War intervened, calling French troops away to battle and putting Monaco's woes at the bottom of their list of priorities. Honoré tried twice to dislodge the Spanish garrison, the first time without success. On November 17, 1641, however, Honoré commanded another attack on the Spaniards, successfully overpowering them and freeing Monaco from their occupation. Even though help from France was no longer needed, the brave act caught the eye of Louis XIII, who commanded Honoré to travel to France for an audience.

The invitation was to be a fortuitous one, for the association with the French court would be vital to Monaco's future security and wealth. Louis XIII immediately invested Honoré with the Order of the Saint Esprit and made him a duke and peer of the realm, complete with title to the duchy of Valentois in Provence. The revenues generated by the property, which was ten times larger than Monaco, were his to keep. He was also awarded the title of "duc et pairs étrangers," given to foreign princes who owed loyalty to France. In 1651, only seven princes held the title, including the Prince of Monaco. The title put Honoré and his descendants near the top of the French hierarchy at court, behind only the king, queen, heir apparent, legitimate male descendants of current or former kings, and princes of the royal blood. In a court estimated to include 10,000 people, this was a very high position.

Nearly two centuries of relative peace and prosperity ensued on the Rock following Honoré's recognition by Louis XIII. Marriages between the princes of Monaco and their heirs to the daughters of important noble families helped Monaco flourish economically.

The outbreak of the French Revolution changed this state of affairs. Monaco may have been a sovereign state, but its rulers spent more time in France than they did in Monaco. Furthermore, Monaco's status was not taken seriously by all, and many viewed the strategic Rock as a sliver of France that had somehow been lost in the pages of history. Besides, the ambitious and sometimes greedy Grimaldis were exactly the type of aristocrats who had suddenly fallen out of favor in France.

The year 1788 found Honoré III without money, and he was forced to travel to France in 1790 to seek aid. Unfortunately, he found Louis XVI and his family imprisoned and unable to release funds. Honoré stayed in Paris, determined to find a solution to his problems among old friends. By 1793, however, revolutionary fervor had reached even the Rock. On January 13 of that year, the People's Councils of Monaco, Menton, and Roquebrune, which had been strictly advisory bodies, voted to create a conventional assembly. They also voted to put an end to Honoré's reign and ask for union with France. Their request was granted only hours after a French revolutionary tribunal of 749 members voted to execute Louis XVI. On March 4, 1793, the former Principality of Monaco became part of the French department of the Alpes-Maritimes and was renamed Fort Hercule.

For the next twenty-one years Monaco would be administered as a French city. Its own revolutionary elements, known as the Société Populaire, decreed that all churches must close at night, that church bells could not ring, and that priests could not hear confession. Otherwise, life went on as before for most Monégasques. Unlike the case in France, where the bloated aristocracy had fueled the flames of class warfare, the only class division in Monaco was between the Grimaldis and their subjects. After the dynasty had been declared null and void, there was little additional cause for revolt.

Honoré and his family were temporarily imprisoned in France and Honoré's daughter-in-law Françoise-Thérèse, wife of his youngest son, Joseph, went to the guillotine as the very last victim of Robespierre's Reign of Terror in 1793–94. The Grimaldis retained their prestige, if not their power, the same way they always had: by keeping the company of the right people. In 1800 Honoré IV married the sister of Joachim Murat, one of Napoléon's most capable generals.

Following the fall of Napoléon, the Grimaldis were returned to their throne on May 30, 1814, by the Treaty of Paris. Another Treaty of Paris, signed on November 20, 1815, placed the principality under the protection of the king of Sardinia. This protectorate was officially established by the Treaty of Stupinigi, signed on November 8, 1817. Life on the Rock went on much as it had. The Monégasques went about their ancient trades, Menton and Roquebrune made the principality agriculturally self-sufficient, and Monaco was, for the most part, left alone.

In 1848 dissent once again rose from within when the towns of Menton and Roquebrune declared themselves independent of the central government. The towns provided much to the principality in terms of food and services, but received little in return. Topography separated them from the day-to-day workings of Monaco. While both towns wished to remain under the protection of Sardinia, neither wanted to remain subservient to Monaco. Prince Florestan of Monaco never ceded his rights to the breakaway towns, and the treaties with Sardinia were never finalized. In practice, Menton and Roquebrune became independent towns, with neither Monaco nor Sardinia taking a leading role in their administration. In 1849 Charles II, son of Florestan, attempted to win back the towns by making a surprise appearance in Menton, complete with prearranged "supporters" and a flashy display of the Grimaldi colors. Unfortunately, he was nearly torn apart by a hostile crowd and was imprisoned in nearby Villefranche for five days.

In 1860 Sardinia transferred its rights to Menton and Roquebrune to France. Charles III, now prince of Monaco, vehemently objected to the deal. On February 2, 1861, however, he agreed to formally relinquish all claim to the two towns in return for 4 million francs and an agreement from France of perpetual sovereignty for the principality. Monaco's already diminutive territory was reduced by 90 percent and its population by 80 percent, but its perpetual independence was guaranteed. Except for reclamation projects reaching southward into the Mediterranean, the principality's borders have remained the same ever since.

One condition of the treaty with France was not made public until 1918. It stipulated that the Grimaldis could not cede any portion of the principality to any country other than

France, nor could they accept a protectorate from any other country. This condition was expanded in later years.

The most fundamental change to take place in modern Monaco was initiated in 1856, when, shortly before his death, Florestan approved the establishment of a gambling and hospitality facility on the Rock. With the secession of Roquebrune and Menton the principality was desperately in need of a new economic engine. The Riviera was attracting more visitors every year, and tourism seemed the industry of choice.

Gambling served two purposes for Monaco. First, it gave travelers a reason to visit the otherwise unattractive Rock, which at the time was neither comfortable nor well supplied with tourist facilities. Second, and more important, gambling promised to do an excellent job of separating foreigners from their money. Such gambling revenue entered the Monégasque national economy from without; no internal resources were expended to generate it. The foreign origin of this revenue was guaranteed by laws prohibiting Monégasque citizens from entering the casinos except as employees.

The gambling concession was originally let to a consortium of entrepreneurs who called themselves the Société des Bains de Mer, the Sea-Bathing Society. Their initial foray failed. However, in 1863 the concession was purchased by entrepreneur François Blanc, who, at the time, had been running the enormously successful Bad Homburg casino in Hesse-Homburg, a small, sovereign state in central Germany similar in many ways to Monaco. Anticipating the impending illegalization of gambling in Hesse-Homburg, Blanc was eager to set up a new operation in Monaco. On April 1, 1863, Blanc established the new Société des Bains de Mer et du Cercle des Étrangers (SBM), or, the Sea-Bathing Society and Foreigners' Club.

Blanc built the casino on the plateau of Spelugues, renamed Monte Carlo in 1866 after Charles III. He also invested heavily in hotels, restaurants, roads, and other improvements necessary to make the venture successful. No expense was spared. The casino opened with great fanfare on New Year's Day, 1865. Monaco was transformed from a curious backwater to a glittering resort almost overnight.

Some problems remained. Accessibility to Monaco was still poor, requiring a four-hour carriage ride over the hills. Such problems were largely solved by the arrival of rail service in 1868. In 1869 the new railway brought more than 170,000 tourists to Monaco. The SBM was a success, and the principality was revitalized. The original agreement between SBM and the principality stipulated that in return for the concession the prince receive 50,000 francs per year, a personal allowance of 2,000 francs per week, and 10 percent of net profits generated by SBM's operations. By 1869, Prince Charles III had abolished all taxes on Monégasque citizens.

During the development of the casino a customs union treaty was negotiated with France that went into effect in 1865. The treaty eliminated all border formalities between the two countries, allowing people, goods, and services to flow freely between them. Monaco was still allowed to levy its own taxes and write its own laws, but frontier patrols and tariffs were completely eliminated.

Following the death of Charles on September 10, 1889, Charles's son Albert came to the throne, becoming Prince Albert I. An avid scientist and adventurer, Albert undertook scientific expeditions around the world, ranging from the rain forests of Brazil, to the Arctic isles of Spitzbergen, to the American West. He founded Monaco's famous Oceanographic Museum, which, during the mid-twentieth century, was administered by French underwater explorer Jacques Cousteau. The "scientist prince" also founded the Anthropological Museum, the Exotic Gardens, and the International Commission for the Scientific Exploitation of the Mediterranean, and played a leading role in the creation of the Institute of Human Paleontology in Paris.

Prince Albert also endowed Monaco with its first constitution. The document, ratified on January 5, 1911, provided for a National Council elected by universal male suffrage and created a public domain separate from the prince's holdings. Specific accounts were created for public purposes, including health, education, and infrastructure. Previously, public services on the Rock had been provided at the discretion of the prince and, officially, from the prince's pocket.

Albert did not enact the constitution solely out of altruism. Discontent was high in Monaco at the beginning of the twentieth century. Of the population of 20,000, only 2,000 were Monégasque. Gambling and tourism may have created a viable economic engine for the principality, but they also required expertise that could only be provided by foreign specialists. Albert also spent most of his time either in Paris or on scientific expeditions funded with public money. Although respected, he was not popular.

Approximately 600 protesters marched on the palace on April 4, 1910, threatening revolt if Albert did not hear their demands for democratic representation. Fearing violence, Albert appealed to France and Britain for help, explaining that the security of French and British citizens was at stake. French troops in Marseilles were alerted, and 300 British sailors on a warship anchored at Villefranche were given leave to go to Monaco. Guns and ammunition were packed in wine cases and stored in the Hotel de Paris and the sailors were under orders to arm themselves in defense of the prince, if necessary. The constitution was, therefore, a means of preventing armed revolt.

Monaco remained neutral through both World Wars. However, its strategic importance provoked yet another bilateral treaty with France. The Franco-Monégasque Treaty of 1918, not made public until after the signing of the Treaty of Versailles in 1919, stipulated that France would take responsibility for the principality's defense and territorial integrity if, in return, Monaco would formulate its policies in the best interests of France. Furthermore, heirs to the Monégasque throne could only be of Monégasque or French nationality.

Aside from a decline in tourism, Monaco weathered World War I easily. During World War II, however, the Rock was not so fortunate; it was occupied first by Italian and, later,

German troops. Where Albert had tried to intercede for peace in the first war, his son, Louis, cooperated with France's Vichy regime, a puppet of the German government, in the second, dramatically eroding his support in later years.

Louis died on May 9, 1949, having left power in the hands of his twenty-five-year-old grandson, Rainier III, four days before. Rainier was the son of Prince Pierre, comte de Polignac, and Princesse Charlotte de Monaco, daughter of Louis. To the world outside, Rainier is perhaps most famous for marrying American actress Grace Kelly, whom he met while she was shooting a scene for Hitchcock's *To Catch a Thief* in Monaco. The wedding took place on April 19, 1956, in a ceremony watched around the world, and the new princess became popular as a patron of charity and the arts. (The world's attention would turn back to Monaco when Grace was killed in a motor accident in 1982.)

The birth of Rainier's son Albert-Alexander-Louis-Pierre on March 14, 1958, provoked familiar power struggles within the Grimaldi family. The son of Rainier's sister, Antoinette, had previously been heir to the throne; with those prospects shattered, she attempted to limit her brother's power by throwing her political support behind Jean-Charles Rey, the most influential member of the National Council and an advocate of constitutional reform.

Fearing for the future of his dynasty, on January 29, 1959, Rainier suspended the constitution, dissolved the National Council, and abrogated the rights of political assembly and demonstration. On Monégasque radio he declared, "I cannot tolerate any pressure whatsoever which might undermine my complete rights." He went on to say that "For a year, the National Council has hindered the administrative and political life of the country [and] a certain council member has been intriguing ceaselessly for many years for the purpose of furthering his own ambitions." Rainier was called a dictator in the foreign press despite his popular wife's statements that six months of princely rule would ease Monaco's difficulties. In fact, Rainier had always been a dictator; his actions, while rash, were legal under the Constitution of 1911, which supported the notion of divine right and left ultimate power in the hands of the prince.

But Rainier had not counted upon the reaction of Charles De Gaulle, then president of France. In October of 1959, De Gaulle gave Rainier six months to solve his constitutional crisis or risk the future of the guarantee of friendly relations with France. For De Gaulle, the crisis served as a convenient way to apply pressure on Rainier to correct the real problem of a tax-free zone so close to France. Many of France's wealthiest citizens had successfully avoided paying French income taxes simply by depositing their money in Monégasque banks and claiming to be residents of Monaco. In April 1960, Rainier capitulated, allowing French citizens to be subject to French taxes while in residence in Monaco. Almost immediately, two-thirds of the 16 billion francs deposited in Monégasque banks by French citizens was trans-ferred to Switzerland. (A similar tax ordinance applying to French corporations was adopted in 1963.)

On December 17, 1962, a new constitution was adopted, abolishing the principle of divine right and giving the vote to all Monégasque citizens aged twenty-one or older. Final executive authority still rests in the hands of the prince, but now through a minister of state, a French civil servant selected by the prince from a list of three candidates supplied by the French government. The Council of Government, consisting of three members appointed by the prince, assists the minister of state. The National Council, composed of eighteen members, is elected directly by the people.

The National Council has the authority to pass laws, but only the prince can promulgate them.

The 1962 constitution also further refined the relationship with France, stipulating that Monaco would become an autonomous region in the Alpes-Maritimes if ever the hereditary ruler should die without an heir. Because adoption of an heir is permitted, the future of the Grimaldi dynasty seems secured.

The modern Monégasque economy is nothing short of a miracle given the principality's paucity of indigenous resources. As of 1990, industry centered in Fontevielle accounted for 27 percent of the principality's economic activity. Chemicals, building and public works, transportation, plastics, and electronics have been successful contributors to the industrial sector. The tourism industry contributed another 25 percent. The service sector, led by international banking and real estate, makes up most of the remainder. Monaco's population of 30,000 enjoys full employment, and the country even employs 19,000 non-resident workers from neighbors France and Italy.

Historic sites in Monaco include the 1878 Opera House, designed by Charles Garnier, architect of the Paris Opéra; the Monaco Cathedral, where Prince Rainier and Princess Grace were married; the Prince's Palace, of which a limited portion is open for public tours; and the museums founded by Prince Albert I. Many visitors come for the Monte Carlo Grand Prix and another major event, the Monte Carlo International Television Festival, founded by Rainier and considered television's version of the Cannes Film Festival.

Further Reading: *The Grimaldis of Monaco: Centuries of Scandal, Years of Grace* (New York: Morrow, and London: Harper, 1992) by Anne Edwards is a complete historical overview of the Grimaldis and their domain. The *Europa World Year Book* (London: Europa, and Detroit, Michigan: Gale), published annually, provides up-to-date information on economics, politics, and demographics. The *World Bibliographical Series* (Oxford: Clio, and Santa Barbara, California: ABC-Clio, 1991) edition on Monaco, compiled by Grace L. Hudson, provides an exhaustive listing of books and articles relating to the principality. Modern Monaco is also well covered by a wide variety of travel guides.

—John A. Flink

Monreale (Palermo, Italy)

Location: Situated on the slope of Monte Caputo at the edge of the fertile plain of the Conca d'Oro, five miles southwest of Palermo.

Description: Monreale is the site of the Benedictine Abbey and Cathedral of Santa Maria la Nuova built for the Norman King William II in the twelfth century. The cathedral and abbey are premier examples of the eclectic building style that evolved during the period of Norman domination over Sicily. The village of Monreale itself owes its existence entirely to the religious complex.

Site Office: A.A.P.I.T. (Azienda Autonoma Provinciale per l'Incremento Turismo)
Piazza Duomo
CAP 90142 Monreale, Palermo
Italy
(91) 6402448

Monreale is a relative latecomer to the stage of Sicilian history. The island had seen centuries of rich Greek cultural development, followed by periods of Roman, Byzantine, and Arab domination. By 1060, when the Normans conquered Sicily, the principal religion was Islam and the language most frequently spoken was Arabic. However, pockets of Greek culture had managed to survive, as well as small Christian communities that hearkened to the Greek Orthodox Church. The Normans brought with them a French and an Italian influence (the latter because they also owned territories in southern Italy) and Roman Catholicism. They pursued a policy of peaceful coexistence with the Arab and Orthodox populations that included religious tolerance, and the result was an extraordinary mixture of cultural influences.

In 1174, the Norman King William II founded a Benedictine abbey on a mountain slope overlooking the Conca d'Oro southwest of Palermo, with the intention of making it the second archbishopric of Sicily as well as a royal mausoleum. The king was so magnanimous in his endowment of the abbey that the Monreale abbot became the second largest landowner in Sicily after the king. A few years after the founding, its possessions included castles, villages, a town in Apulia (Italy), grain mills, a sugar cane refinery, and a tunny fishery, as well as a long list of privileges, such as exemption from all taxation, fishing rights, free pasturage, and free hospitality anywhere on the island. Historians speculate that William's intention, at least in part, was to diminish the political influence of the archbishop of Palermo. If so, he went to some effort to conceal his motivation. Initially, the foundation did not have a name of its own but was referred to as "super sanctam Kiriacam," after a surviving Greek Orthodox chapel. These references thus attached—if spuriously—the new institution to an already existing tradition, so as to suggest that nothing new was being established with the creation of the richly endowed abbey and archiepiscopal hopeful in such proximity to the Palermo archbishopric. The pope in fact granted Monreale archiepiscopal status only a few years later, in 1183. Following that date the royal complex was referred to only by the name of Monreale (or Montis Regalis, in Latin).

A building campaign of unusual intensity was undertaken immediately after the founding, and a mere two years later, in 1176, construction of the cloister had advanced sufficiently far that 100 monks from the Benedictine monastery of La Cava, near Salerno, Italy, were able to move in. As early as 1183, the remains of William I and his wife were transferred to Monreale, another testimony of the extraordinary progress of the construction. By 1185, the bronze door in the main portal of the cathedral was installed. Little additional information about the history of the construction of the complex has survived, but it would appear that by the time of William II's death, in 1189, the architectural elements had been more or less completed. The decorations may have taken somewhat longer, but they too were finished at an early date. In the beginning, the complex was made up of the Cathedral of Santa Maria la Nuova, a monastery with a cloister, and a royal palace. Walls, which have since crumbled into near-oblivion, surrounded the complex along its exposed western and northern sides, and a defensive structure was built on the mountain top.

The flagship of the complex was, and still is, the cathedral, a mostly Romanesque structure with a basilican nave and three apses at the east end. The exterior decoration of the building, particularly the profusion of polychrome-interlaced blind arches of the east end, is clearly of Islamic origin. The painted wooden ceiling in the nave is similarly decorated with Islamic motifs. The mosaics in the apses and along the walls of the nave, by contrast, are undoubtedly of Byzantine workmanship and inspiration. For instance, the cathedral's principal mosaic, in the top of the central apse, is a Christ Pantocrator, with his right hand raised and his left holding the Book of Life, possibly influenced by the Pantocrator at the Cathedral of Cefalù, on Sicily's north coast. The Pantocrator at the Church of the Dormition at Daphne, Greece, however, was undoubtedly a model to both the Sicilian mosaics. A Catalan and Provençal influence is evident in the cathedral's sculpture, while the bronze doors in the western and northern portals hail from Pisa and Apulia. Although it certainly cannot be said that a synthesis of these elements was achieved, the unusual confluence of divergent traditions is of breathtaking and unique beauty. Little remains of the rest of the original Norman complex but the monastery's cloister, with its polychrome-inlaid columns and Moorish fountain.

Interior of the Cathedral of Santa Maria la Nuova
Photo courtesy of The Chicago Public Library

In addition to its other riches, the Benedictine abbey acquired four rural districts in its vicinity in 1182. About 2,000 people, mainly Moslem farmers, thus came under the jurisdiction of the abbot. In spite of religious tolerance, the abbey undertook to Christianize the local population, a plan that produced considerable resentment and led to revolt by the end of the century. In the Moslem insurgency that began in 1197 and spread across large parts of the Sicilian northwest, local farmers took much of the abbey's landholdings. Norman rule had disintegrated by then, and no central authority was in place to put down the rebellion. Adding to the chaos, the abbot is reported to have entered into conflict with his own monks a few years later. It was not until the 1220s, when Frederick of Hohenstaufen asserted his rights and authority as King Frederick I of Sicily (he was also Holy Roman emperor as Frederick II), that the abbey's holdings were restored and the insurgents punished. The Benedictine monks were granted a series of charters that enabled them to claim Arab farmers as part of their work force.

The next 100 years or so saw a massive emigration of Arab Sicilians, who no longer felt welcome on the island. As a consequence, much of the Sicilian countryside became deserted, and the wealth of the Monreale abbey diminished considerably. Whereas in 1182 there had been fifty hamlets in Monreale's domain, by the early thirteenth century only twelve remained, and it is by no means certain that these had been continuously inhabited. Nevertheless, it still was one of the richest archbishoprics in the West, and the office of archbishop of Monreale (who was usually also the abbot) was one of the most coveted. Over the course of the centuries, the most prominent Italian families, including the Medici, Farnese, Borgia, and Barberini, managed to secure the office for their sons. (No native-born Sicilian ever was archbishop.) As a consequence, the archbishop was rarely resident in Sicily, and little came of William's original intention to make Monreale the religious focus of the island.

Another consequence of Monreale's having been frequently headed by absentee abbots/archbishops was that exceptionally few changes in the Norman complex were made over the course of the centuries. One of the earliest changes took place late in the fifteenth century, when Giovanni Borgia had a sacristy built along the southern side of the cathedral's sanctuary. In 1547, a portico was built along the northern side of the nave. More drastic changes took place a few decades later, when the cathedral finally received an inlaid floor. The most significant changes made at this time did not involve the cathedral, however, but the monastery. As soon as Monreale had been made an archbishopric, the east wing of the monastery had been converted into an archiepiscopal residence. In the 1570s and 1580s this section of the monastery saw extensive reconstructions that gave it a much more Italianate appearance. The new design was based on ancient Roman villas, particularly the Villa d'Este at Tivoli, and was laid out with loggias, terraces, and gardens. The reconstruction must have seemed rather odd alongside the older Romanesque buildings in the complex. However, it is difficult to judge the appearance of the complex at this time, since later modifications in the nineteenth century have almost completely obliterated the sixteenth-century archiepiscopal palace. Only drawings of the ground plan and elevations remain to give an impression of its appearance.

The village of Monreale in the meantime was beginning to come into its own in the sixteenth century. The town acquired a printing press in 1554, one of the first on Sicily. Its first book, predictably, concerned the religious government of the Monreale archbishopric. The Jesuits established a public school in the town in 1551. One of their first pupils was Antonio Veneziano, a local boy who, after a life of adventure, was to become the most famous poet of Sicily of the sixteenth century, often referred to as the Sicilian Petrarch. After completing his education at Palermo and Messina, Veneziano returned to Monreale and was accused of double murder. He then began a life as a bandit in the Monreale area. Later he was captured by Algerian pirates, was ransomed, and returned once again to Monreale. He died in prison, in 1593, at the age of fifty, for having written a satire against the Sicilian viceroy.

In the late sixteenth century, the abbey finally recognized the fact that Monreale had not played much of a role in the public or personal lives of the Sicilian kings since the death of William II. Sicily had passed from Norman into Austrian hands and then, starting in the late thirteenth century, had come to play a fairly marginal role in Mediterranean history as a possession of the House of Aragon. When the island passed into the Habsburg empire in 1516, Monreale's position remained entirely unchanged. In 1589, then, under Archbishop Ludovico II de Torres, the mostly unused royal palace was made into the Archiepiscopal Seminary. Architectural changes involved a modification of the northern facade in a Renaissance style visible particularly in the windows. Ludovico also published the first description of the cathedral, *Historia della Chiesa di Monreale,* under the name of his secretary, Giovanni Luigi Lello. The archbishop also added a chapel to the cloister and another one to the south side of the cathedral. The cathedral chapel, the Capella di San Castrense, is the most important example at Monreale of Italian Renaissance architecture. The monastery underwent further renovations in the mid-seventeenth century, when the dormitories and refectories in the west wing were modernized.

Disaster struck the abbey in 1647, when a revolt that broke out in Palermo spread to the surrounding countryside, inspiring the local population to attack the complex. The rioting had begun over the food shortages that presented a serious hardship to the common people of Sicily throughout much of the sixteenth and seventeenth centuries. As a wealthy landlord exempt from taxation and from the secular courts, the abbey, with its fabulous riches, must have presented itself as the most natural target to the starving poor. A lack of organization and unity among the rebels, however, enabled the army to put down the revolt in a matter of months. The archbishop of Monreale was wise enough to absolve the rebels in a public ceremony, but he failed to address the political realities that had provoked the revolt. Apparently determined to hang on to the abbey's wealth, he chose to exorcise the demons and witches that he identified as the cause of the uprising. About this time the abbey began leasing out its landholdings, possibly in an effort to obscure its role as the villain in the drama of the peasants' misery.

Late in the seventeenth century, under Archbishop Luigi Alfonso de los Cameros, the cathedral's east end underwent a complete rearrangement. The Capella di San Giovanni Battista was torn down, while the mosaic of John the Baptist was moved to the south wall of the nave. Cameros instead built the Capella del Crocifisso along the east end. The sumptuous decorations in this chapel, which by and large have survived, make it one of the more important examples of the baroque style as it evolved on Sicily. Cameros was also responsible for the baroque altars in the side apses.

From 1701 until 1714, the Austrians and Spanish fought over the succession to the Spanish throne. One consequence of this conflict was that the Piedmontese Victor Amadeus II had the opportunity to take Sicily in 1713. Although before long it became clear that Victor Amadeus had no chance to hold onto the island, who would manage to take it from him remained an open question until 1720. Both the Austrians and Spanish laid a claim to Sicily and took the occasion to continue their fighting, this time on Sicilian soil. In the middle of the conflict, the Monreale abbey's holdings were sequestered by orders from Madrid, but they were restored in 1720, when the Austrians finally won the conflict. The Austrian Habsburgs did not hold Sicily for long, only until the death of the Emperor Charles VI in 1734, when Sicily once again became part of the kingdom of Spain under the Bourbons. During the Austrian period, the third of the east end chapels was built, the Capella di San Benedetto, dedicated to St. Benedict, founder of the Benedictine order. This chapel later in the same century was fitted out with additional sculpture, of which the altarpiece is the most significant. Under Bourbon rule, in the mid-eighteenth century, the Archiepiscopal Seminary was altered again, this time to conform to the rococo style. A third floor was also added to the structure then.

The Bourbon Ferdinand I, ruler of Sicily from 1759 to 1825, made a general effort to diminish the influence of the Catholic Church on the island, dissolving several smaller monasteries and abolishing the archbishopric of Monreale. Ferdinand's intention was to stop the drain of wealth from the island into the coffers of the non-resident archbishops, some of whom never even bothered to maintain the buildings properly. Before long, however, the archbishopric was re-

stored, through the influence of Archbishop Francesco Testa at the Vatican.

In 1773 another revolt, originating in Palermo, spilled over into the countryside and resulted in what has been called "gang warfare" at Monreale. Again the rebellion was suppressed, and again life went on as usual at the abbey. A new wing was built onto the monastery in 1780 on the south side of Cathedral Square. This wing has since been converted to the Convitto Guglielmo. But the nineteenth century saw more and more upheaval in Sicily, with a revolt in 1820 and another in 1848. On both occasions, the rioting spread to Monreale, and in 1848 the local peasantry ended up playing a major role in the revolt, under the leadership of the peasant di Miceli. He gathered a band of farmers, armed them, and led them to Palermo, where they joined in the street fighting against the army and the police. The rebels successfully stormed the Palermo jail—always the first target in the Sicilian uprisings—but as before the revolt was doomed to failure. Lacking a vision of social change, the rebels were exclusively focused on the immediate relief of their hunger. Di Miceli himself was content to become a tax collector and official in the Sicilian coast guard in 1849, and he subsequently became a wealthy landowner in the Monreale district.

When Sicily finally became a part of Italy in 1861, the religious houses were abolished, and the Benedictine monastery at Monreale was dissolved. This situation did not prevent the recurrence of rebellions, however, and in 1866 another uprising broke out, this time starting in Monreale and moving to Palermo. Ironically, di Miceli was again at the head of the rebels, but this time he found little profit in the venture. This uprising was to be the last popular revolt on Sicily, and it was the last time that the peace at the abbey complex was disturbed. Monreale is a monastery once again, and the cathedral has become a major tourist attraction. The town of Monreale is now a modest market center, where the citrus fruit and other produce of the Conca d'Oro is traded.

Further Reading: Among the wealth of material on the architecture of the Monreale royal complex, Wolfgang Krönig's *The Cathedral of Monreale and Norman Architecture in Sicily* (Palermo: Flaccovio, 1965) is both the most thorough and the most accessible. For readers of Italian, a general history of Monreale is available in S. Spinnato's *Monreale* (Palermo: Spinnato, 1962). Denis Mack Smith has written an excellent general history of Sicily since the Middle Ages in *Medieval Sicily: 800-1713* and *Modern Sicily After 1713*, volumes two and three of *A History of Sicily* (London: Chatto and Windus, and New York: Viking, 1968).

—Marijke Rijsberman

Mount Athos (Chalcidice, Greece)

Location: On easternmost (third) prong of the Chalcidice Peninsula, in the Greek part of Macedonia; Mount Athos is highest point, 6,670 feet, on peninsula.

Description: In Greek, the "Holy Mountain," an autonomous theocracy 35 miles long, 131 square miles, governed by Holy Council of twenty monks; Kariai, administrative capital; spiritual center of Orthodox Christianity since ninth century A.D.; consisting of twenty monasteries, seventeen Greek, one Serbian, one Russian, one Bulgarian, and their dependencies, connected to one another by footpaths; Athos off limits to all women, children, and casual tourists; accessible only by bus or boat.

Contacts: Ministry for Foreign Affairs
Directorate of Churches
2 Zalokosta Street
Athens, Attica
Greece
(1) 362-6984

Ministry for Northern Greece
Directorate of Cultural Affairs
Platia Diikitriou
Thessaloníki, Thessalonike
Greece
(31) 270-092

Mount Athos, "the Holy Mountain," in Greece, is a close-knit monastic republic that has existed since the disintegration of the Roman Empire. Its atmosphere of intense spirituality and peace stands in stark contrast to that of its neighbor Bosnia. Mount Athos and Bosnia represent the extremes that have always characterized the history of the Balkan Peninsula, where democracy was first born, where Islam and Christianity confronted each other in mortal combat, where radical religious sects flourished, and where a terrorist act ignited World War I—all taking place in a geographic area the size of Texas. It was from the Balkans that Orthodox monks headed eastward, bringing Christianity, the alphabet, and Byzantine culture to Russia.

Mount Athos, located in what was once part of the Byzantine Empire, was already old by the time of the formal split between the Eastern and Western churches in 1054. Up until then, Roman Catholic monasteries, although in a minority, existed side by side with their Eastern Orthodox counterparts on the Holy Mountain. The Catholic monks departed and headed west after the schism, making reconciliation even on spiritual Athos, let alone in the world of power politics, impossible.

According to one legend, Athos had its beginnings when the Virgin Mary was suddenly blown by high sea winds on to the top of the beautiful mountain. How long she stayed there to recover from the shock is uncertain, but shortly afterward, the first hermits and other ascetics started to arrive. They did not live in a community, however, but in caves or rude huts (then, as now, Mount Athos was heavily wooded), seeking spiritual perfection on their own.

Circumstantial evidence and tradition point to the existence of a religious settlement on Athos hundreds of years before 843, when the first recorded document that has been preserved, composed by monks, indicates clearly the existence of an established religious community on Athos. This was at the height of the "iconoclastic" controversy in the Byzantine Empire, ignited by the Islamic proscription of images in worship. The repudiation of images challenged Orthodox Christians in particular, perhaps because of their geographic proximity to Islamic regions, to justify their veneration of icons and statues. Some Orthodox clergymen and even one Byzantine emperor desired to "purify" Christianity of profane images, even if this meant creating a schism within their own ranks, and deepening their differences with the Latin church, which is exactly what the leaders of Islam desired. The monks on Athos sided with traditional image worship, and their opinion prevailed. From then on, Athos became the seat of ultimate authority in Orthodox spiritual matters, similar to the papacy in the Roman Catholic Church. Unlike the pope, the patriarch of Constantinople was merely a figurehead, in large part because the patriarchate lacked the territorial and political independence of its counterpart in Rome.

Another milestone in Athos's history occurred in 885, when Byzantine Emperor Basil I formally recognized the self-governing independence of Mount Athos. Perhaps in response to a request from Mount Athos, the empire issued an edict in 1060 that forbade women and almost every female animal, except hens, sows, and cats, to reside in or enter the territory of Athos. Despite this 1,000-year ban on females, change, even on Mount Athos, has seeped in at a glacial pace; many of the monasteries have telephones, and a few have electricity. These changes may well be a harbinger of greater concessions to females in the centuries to come.

By the time Athos received its charter, the formal split between Eastern and Western Christianity had taken place, and Athos's greatest monastery was 100 years old. Built in 963 (erroneously designated by the Greek government in 1963 as the anniversary of the "founding" of Mount Athos, the Great Lavra owes its existence to the leadership of one of the most remarkable of Athos's monks, St. Athanasius. Until his arrival at Athos, the monastic community lived by and large according to the spiritual rules of St. Basil, the father of Eastern monasticism. His rules were flexible, making allowance for those monks who wished to live a hermit's existence

Zográphou Monastery on Mount Athos
Photo courtesy of The Chicago Public Library

apart from the community, yet giving them a place in the community. Those who chose to live in the community were encouraged to divide their days between work and prayer, and to build churches around which their lives revolved.

A monastic movement eventually arose in Byzantium that had the effect of a tidal wave in remote, changeless Mount Athos. Athanasius the Athonite in the mid-tenth century introduced "Studite" monasticism, which prescribed for monks a life of strict discipline and subordination to the authority of the *igumen,* or abbot. Athanasius, canonized after his death in 1000, spearheaded the building of the Great Lavra, over which he became abbot, and where he imposed his rule. Other monasteries on the Holy Mountain followed suit, but not without much tension between the two "rules"—tension that led to numerous clashes.

In the end, the stern discipline of Athanasius's rule won out, although a determined minority continued to follow the rule of St. Basil. Allowance was made for those whom St. Basil himself had tried to discourage—the extreme individualists, the hermits in their isolated caves and huts; still, even they bowed to community pressure to attend services, at least on high holy days. Clearly even the formidable Athanasius

was no dictator. Perhaps his strict rule triumphed because the times were chaotic, and the need was paramount for Eastern Christians to appear as a unified, disciplined army against the growing threat of Islam. Today, in contrast, the fastest growing, most "modern" monastery (in terms of such creature comforts as electricity, telephones, and motorboats) practices the rule of St. Basil: not surprisingly, it attracts the youngest members to its portals.

Nearly 200 years after the death of Athanasius, another remarkable monk, St. Sava, patron saint of Serbia, made his mark on Mount Athos. In addition to establishing Hilandar Monastery in 1197, of which he became abbot, he convinced the monks of the need for an independent monastic order, according to which they could live on their own and do the necessary farmwork and other chores important to the daily survival of monastic life on Athos. From then on, farming, crafts, and even artistic communities arose that were composed of men who were monks, who did not take holy orders, and who devoted themselves the better part of each day to labor rather than to the liturgy. They could choose to live by themselves, or to form small communities, which served particular monasteries as dependencies.

This pragmatic departure from the prevailing Athanasian rule ensured the survival of Mount Athos through the centuries. By the fifteenth century, there were forty monasteries and 20,000 monks living on Athos. Monasteries had their own treasuries, and the majority of them had generous endowments from patrons who donated money as well as income-producing property outside of Athos, located mostly in Asia Minor. When the Turks completed their conquest of Asia Minor in 1453 (the year Constantinople fell to them), the monasteries slowly began losing income as well as members, since an important donor and patron had been the Byzantine state. The new rulers, however, did not confiscate most church property, nor did they interfere in the life of Athos, which retained its autonomy as long as it paid an annual poll tax that all non-Moslems owed the Turkish Ottoman government.

There were also negative milestones in Athos's history: numerous fires have damaged every one of the monasteries, with the exception of the Great Lavra, and fire still continues to be the greatest natural hazard. For a period in the Middle Ages the crusaders set up a "Latin Kingdom," which included Thessaloníki and Mount Athos in the north. The Roman Catholic archbishop of Thessaloníki and his allies approved the sacking of many monasteries on Athos, robbing them of all their gold, silver, and jewels. The plight of the monks was so great that Pope Innocent III felt compelled to put a stop to the pillage by extending his official protection to Athos, and threatening with excommunication anyone who harmed the Holy Mountain. The depredations ceased, but the priceless objects that had been pillaged were never recovered.

In the early nineteenth century, many Greek monks joined in the Greek War of Independence against the Ottoman Turks, which brought the wrath of the Turkish government upon Athos in the form of burdensome taxes. While Greece won its independence in the Treaty of Adrianople of 1829, modern-day northern Greece continued to be part of the Ottoman Empire until 1912; thereafter, for the next decade, there was constant warfare: first the Balkan Wars, which ended in 1913; World War I; followed by war between Greece and Turkey, which lasted until 1922.

The drain on monastic finances and the declining numbers of monks during the nineteenth century were gradually offset by the arrival, by the thousands, of Russian monks and the building of a pretentious, heavily endowed monastery, dubbed the "Rossikon," which symbolized the Russian government's drive for dominance of Mount Athos. Nonetheless, czarist patronage, coupled with increased Serbian and Bulgarian support of their monasteries after their respective nations became independent, saved Mount Athos from impoverishment.

World War I brought new difficulties to Athos, as allied troops, mainly British and French, used Greece as a base for attacking the Ottoman Empire in Turkey and the Middle East. French troops occupied Athos and remained there for more than a year after the war officially ended in the fall of 1918. Their occupation resulted in the pillaging of treasures and artwork that exceeded even that of the thirteenth century; this time, however, papal authority could not put an end to the stealing, and the monasteries never recovered their losses. In World War II, when Nazi German troops passed through the Holy Mountain without committing a single act of theft or sacrilege, the monks of Athos remembered them with gratitude long after the war ended, though they learned that the rest of Greece had not fared nearly so well.

The postwar period brought with it a drastic decline in the numbers of monks, which threatened the future viability of the Holy Mountain. Gone was the support of the Russian government after communism took over and adopted "official atheism" for the next seventy years. By the 1960s, the number of Russian monks in the enormous Rossikon monastery had dwindled from six thousand in its heyday to no more than eight. Similar drastic reductions occurred when communism asserted itself in Yugoslavia and particularly in Bulgaria, after World War II .

Yet, positive events have to a large extent offset the trials and tribulations of the twentieth century on Mount Athos. The Treaty of Lausanne in 1923, signed by the former allies of World War I as well as Turkey, recognized northern Greece, including Athos, as belonging to Greece; three years later, the Greek government guaranteed the autonomy of Athos in a special charter. Decades later, when Greece officially joined the European Community in 1980, the community recognized the special status of Athos as a Byzantine treasure, whose charter was exempt from community rules. Seemingly, Athos's status as the seat of Orthodox Christianity is secure and its treasures are safe. However, violent conflicts in that volatile region of Europe threaten peace and stability, even for Mount Athos.

In the 1980s, Orthodoxy has undergone a revival that has resulted in a resurgence of monasticism. Barely 1,200 monks lived on Mount Athos in the mid-1960s; thirty years later, the number is closer to 2,000. In part this religious revival may have been a reaction to the excessive materialism of the Western world. Certainly the fall of communism in Russia, the largest Orthodox nation in the world prior to the Bolshevik takeover in 1917, has much to do with the Orthodox revival, and with the growing numbers of Russian Orthodox monks on Athos.

The number of visitors to Athos has also grown in the years since World War II. In the 1970s, the crush of tourists, many of them casual and curious, prompted the council that rules Mount Athos (consisting of one representative from each monastery) to request the Greek government to institute a permit system for visitors to Athos. Permits are issued by the Ministry for Foreign Affairs in Athens and the Ministry for Northern Greece in Thessaloníki. Although Greeks are exempt from the necessity of a permit, only the serious-minded among them, and no children, are allowed to visit the Holy Mountain. For all others, planning a visit to Athos must be done months in advance, with a maximum initial stay of four days, although the stay can be extended. One does not have to be Orthodox to visit Athos, although preference is given to visitors who are, and in some monasteries the non-Orthodox are more restricted in terms of what they are allowed to see.

Only boats and buses reach Mount Athos, although neither to this day has a regular schedule. Upon arrival, the visitor must expect long treks from monastery to monastery, usually through beautiful dense woods that can be full of snakes. Visitors arrive in the town of Kariai, administrative capital of Mount Athos, with a few small shops and restaurants, and devoid of women and children. From there, the traveler decides his itinerary. During a short stay it is impossible to view all the monasteries, and some, perched high up on the mountainside, require a vigorous climb. Along the way one may encounter "beggar" or "vagabond" monks who live in the woods and beg from the monasteries and their dependencies as a way of life; one may also come upon the lonely, rude hut of the hermit. There is at least one artist colony situated on the isthmus whose monks are devoted to religious art and iconography.

The world of Mount Athos revolves around the liturgy, and has more in common with medieval Byzantium than the modern world. Few monasteries have electricity, although most have telephones that connect only to the "capital." Time is measured according to Byzantine time (hence sunset is considered midnight) and the calendar is the Orthodox Church calendar, thirteen days behind the calendar used in the rest of the world. The day begins at dawn with vespers, followed by a main noon service, a nocturnal service, and often a sleepless vigil for those monks choosing to stay awake all night in prayer. The monasteries are repositories of ancient manuscripts and precious icons, some rescued from the depredations of ravagers and occupiers; such is the supposedly miraculous icon of the "Virgin Guarding the Gate" in the huge Iviron monastery. In the most "modern" monastery of all, Vatopedi, with electricity and telephones (and the only monastery that adheres to the "modern" calendar), one can still find the original bronze portals that had belonged to the Hagia Sophia church in Constantinople, before it was turned into a mosque in 1453. Most observers agree that the most impressive frescoes are to be found in the oldest and largest monastery, the Great Lavra.

Further Reading: While most guidebooks on Greece cover Mount Athos, few, if any, do justice to Athos in its spiritual context. One that comes close is *The Real Guide: Greece* by Mark Ellingham (Englewood Cliffs, New Jersey: Prentice-Hall, 1992) There are no serious or comprehensive books on the history of Athos available in English. One of the best is in German: Erich Feigl's *Athos: Vorhölle zum Paradies* (*Athos: From Limbo to Paradise*) (Vienna-Hamburg: Zsolnay Verlag, 1982).

—Sina Dubovoy

Mount Helicon (Boeotia, Greece)

Location: In the department of Boeotia, between Lake Kopais and the Gulf of Corinth, central Greece. At the foothills of Mount Helicon lie the cities of Thespiae and Lebadea.

Description: Mount Helicon is known for the inspiration it imparted to the poet Hesiod, whose work served to systematize what became the basis for Greek mythology. He wrote of the nine Muses, sister goddesses of ancient Greek religion. Today, remnants of buildings used in worship of the Muses dot the mountain and surrounding area.

Contact: Greek National Tourist Organization
2 Amerikis Street
 P.O. Box 1017
Athens, Attica
Greece
(1) 322-3111

Mount Helicon in ancient times was known for its beauty and its religious associations. At one time a lushly forested area, it has been inhabited since the Neolithic period. Its greatest fame, however, came in the Greek classical age when it was a center for worship of the Muses and inspiration for Greek mythology.

The mountain range in which Mount Helicon is located made it possible for communities to be isolated from one another. Except for small, narrow areas, there are very few plains in the region, creating a natural barrier. The inhabitants were segregated into three distinct groups: the herdsmen, the farmers, and the traders. There were frequent conflicts between the disparate groups.

During Neolithic times, when society shifted from hunting to farming, many people were drawn to the fertile pastures near Mount Helicon. The second-century geographer Pausanias described Mount Helicon as "the most fertile in Greece." Some of the larger and lower pastures near Mount Helicon eventually became the home of wealthy cattle barons who were rich enough to arm their own cavalries during battles.

It was common in rural areas for tracts of land belonging to citizens to be cultivated by hired tenants. The writings of Hesiod, himself a farmer in Ascra, tell of "thetes," a class of farm laborers who received wages for their services to landowners. Despite this, farmer-tenants took much pride in cultivating their land on Mount Helicon and within the surrounding rural areas. Agriculture was viewed by many as a noble calling. For them, any other occupation, including the military profession, albeit a popular career, was secondary to farming.

The Boeotian region is characterized by intense and significant religious devotion; it is also known for its superstitions. The Greek philosopher Heracleides described the religious fervor of the Boeotian area as "astonishing for the hope it puts in life." Mount Helicon was one of the most important religious sites in ancient Greece.

Hesiod believed that he was inspired to write by the nine Muses, divine goddesses of ancient Greece who were said to inhabit Mount Helicon. His version of the origins of the gods and men in the poem Theogony represents a synthesis of legends that helped establish Greek mythology. Theogony contains the earliest references to Pandora and tells of how Prometheus stole fire from the heavens. Another of Hesiod's poems, *Works and Days*, was written as a letter of reproach to Perses, his brother. Perses was supposedly lazy and unproductive, scheming against Hesiod to gain control of their father's property.

The mythology surrounding the Muses is complex. There were nine Muses: Clio, Euterpe, Thalia, Melpomene, Terpsichore, Erato, Polhymnia, Urania, and Calliope. Calliope was the leading Muse. Their mother was Mnemosyne, or Memory. Although the Muses were believed to be virgins, or at the very least, unmarried, they were also recorded as mothers of several Greek gods, including Orpheus. They were patronesses of the arts and sciences: Clio was the Muse of history, Erato of love and poetry, Urania of astronomy.

Many popular statues of the Muses depict them cradling scrolls or lyres in their arms. A statue of the god Apollo shows him dressed in the flowing robes common to musicians of the era, leading the choir of Muses on Mount Helicon. It is from the Muses that the word "music" is derived. The religious sanctuary of the Muses, in a sacred grove near Mount Helicon, was adorned with statues of the Muses, of Dionysus, and of the great lyric poets.

Religious festivals in honor of the Muses were held every four years at Thespiae, a town at the foot of Mount Helicon. It is likely that in the beginning the festivals were attended only by the patrons of well-known poets of the time. Later, as the festivals grew in significance, persons involved in other aspects of the arts, as well as the sciences, also participated.

Another festival held on Mount Helicon every four years was called Erodita and was held in honor of the god Eros. This god, armed with flaming torches, originally symbolized sexual vigor and virility. It was during the Hellenistic age that Eros was sentimentalized as the son of Aphrodite. During the Roman era, he became Cupid. At the festivals, every bride would offer a lock of her hair and a piece of clothing to Eros as a symbol of her youth and virginity.

Near Mount Helicon were two fountains dedicated to Aganippe and Hippocrene. Legend claims that the Hippocrene

Valley of the Muses
Photo courtesy of Consulate General of Greece, Chicago

fountain was created from the imprint of the hooves of Pegasus, the winged horse belonging to Bellerophon, grandson of Sisyphus, one of the legendary rulers of Corinth.

Mount Helicon and the surrounding area have witnessed numerous historic events. Thebans fled to Mount Helicon during the Trojan War when the valley was overrun by northern tribes. Thebes subsequently headed up the Boeotian League, formed in the early sixth century. The republic of Boeotia and other league members opposed the growing imperial power of Athens, but about 457 B.C. Athens succeeded in breaking up the league and bringing all its members except Thebes into the Athenian-led Delian League. Boeotia rebelled against Athens in 447 B.C., and in a battle near Mount Helicon inflicted heavy losses on Athenian forces sent to quell the rebellion. Athens signed a treaty promising to respect Boeotian autonomy.

In 371 B.C. the Spartans invaded Boeotia in order to attack the Thebans. Both groups had a long history of fierce battles against one another. Despite the fact that the Spartan population had been dwindling in number for at least 100 years, King Keombotos of Sparta led a surprise march through the Mount Helicon region against Epaminondas, the leader of the Thebans. In the brutal fighting that followed, known as the battle of Leuctra, the Spartans lost 400 of their 700 soldiers. Boeotia, which had assisted Thebes, became an important political force in Greece and remained so for more than thirty years, until the conquests of Macedonian kings Philip II and his son Alexander the Great brought Macedon to pre-eminence in the 330s. Later, the area came under Roman rule. The Roman emperor Constantine I (the Great) took statues and other materials from the religious buildings at Mount Helicon to adorn his new capital, Constantinople, in the early fourth century A.D. Even as Christianity (Constantine was the first Christian emperor) supplanted pagan religions, however, Mount Helicon retained some religious significance; the remains of several small Christian chapels have been found in the area.

Descriptions of Mount Helicon by Pausanias helped modern archaeologists determine how the area looked in classical times and led them to discover the remains of several structures. These include a theatre in which the contests in honor of the Muses were held; a monumental altar; and the ancient well that was part of the Hippocrene fountain. A nearby spring may have been the source of the

Aganippe fountain. The sacred grove is almost treeless now, devoid of the mystery and beauty that attracted the ancient Greeks, but its historic associations are still able to evoke strong emotions.

Further Reading: *A Traveler's History of Greece* by A. R. Burn (New York: Funk and Wagnalls, and London: Hodden and Stough-ton, 1965) is a good source for detailed and in-depth information on many aspects of Greek history, including Greek culture and mythology. *Ancient Greece* by Peter Green (New York: Viking, 1973; as *Concise History of Ancient Greece*, London: Thames and Hudson, 1973) provides detailed and obscure information on historical events. It also includes excellent overviews of ancient sites in Greece.

—Jessica M. Bowen

Mycenae (Argolis, Greece)

Location: Approximately halfway between historic Argos and Corinth, in low-lying hills between the peaks of Mount Zara and Mount Profitis Ilias, overlooking the Argive plain and the Dervenaki pass in the northeastern section of the Peloponnese region.

Description: One of prehistoric Greece's most important cities; center of the Mycenaean civilization, which flowered during the Late Bronze Age (1400–1100 B.C.).

Site Office: Mycenae Archaeological Site
Mycenae, Argolis
Greece
(731) 66585

Greek legend contains many colorful stories about the early years of Mycenae. The city was supposedly founded by Perseus, slayer of Medusa and son of Zeus (king of the Greek gods) and Danae, a mortal. The Perseid dynasty provided Mycenae with many of its rulers in these legends; the last of the dynasty was Eurystheus, famous for the labors he imposed upon Hercules. After Eurystheus, the Mycenaeans chose Atreus as king.

The house of Atreus was one of the most tragic in all of Greek mythology. Atreus hated his brother Thyestes for seducing his wife; as revenge, Atreus tricked Thyestes into eating his own children, whom Atreus had secretly killed and prepared as a feast. Thyestes cursed his brother and his children, a curse that was to play out in future generations. First, Atreus's son and heir, Agamemnon, was forced to kill his own daughter to appease the gods and generate a fair wind for the Greek fleet sailing to Troy. The war had begun at the behest of another of Atreus's sons, King Menelaus of Sparta, whose wife, Helen, had been stolen and taken to Troy. After leading the Greeks to victory, Agamemnon returned to Mycenae only to be murdered by his wife, Clytemnestra, and her lover Aegisthus, the son of Thyestes. Agamemnon's children, Orestes and Electra, continued the cycle of revenge by killing their mother and her lover. Ultimately, during the rule of Orestes' son Tisamenus, the descendants of Hercules returned and triumphantly claimed the Mycenaean throne.

While less colorful than these legends, the archaeological record of Mycenae is long and varied. Excavations have uncovered an Early Bronze Age (Early Helladic) settlement and succeeding settlements from the Middle Bronze Age (2000–1550 B.C., also known as Middle Helladic Period). Most of the Early and Middle Helladic settlements have been lost due to rebuilding during the Late Bronze Age (Late Helladic). Excavations from these first Mycenaean settle-

ments have revealed little about the origins of Mycenaean culture, so these origins remain a mystery. One theory, unsubstantiated, is that contact occurred between Mycenae and Egypt around 1500 B.C., leading to the colonization of Mycenae by Egyptians. Supporters of the theory point out the strong Egyptian influence on the funeral customs and furnishing of grave shafts at Mycenae. Whatever the origins of the Mycenaeans, however, it is quite obvious that they were heavily influenced by the Cretan Minoan society, the Egyptians, and other peoples of the eastern Mediterranean.

The term "Mycenae" has been used throughout history to describe not only the city of Mycenae, but also mainland Greece and associated islands, with the exception of Crete, during the Late Bronze Age. Mycenaean civilization throughout the eastern Mediterranean experienced accelerated growth and prosperity beginning in the Middle Bronze Age around the seventeenth century B.C. At that time the Minoan civilization, centered on the island of Crete at Knossos, was the preeminent power in Greece. Initially, Minoan control of the seas confined Mycenaean growth to mainland Greece. With the collapse of the Minoan civilization around the fifteenth century B.C., Mycenae experienced unprecedented growth and a marked increase in both wealth and population. Mycenae took over Minoan foreign trade and numerous Minoan colonies in the Aegean, and even occupied the island of Crete. It was in the Argive plain, where the city of Mycenae was located, that the growth and prosperity were most impressive. Between 1400 and 1100 B.C., Mycenaean civilization reached its greatest heights economically, politically, and culturally.

Mycenaean settlements spread across the Greek mainland into southern and central Greece, to the northeast as far as Thessaly, and, to a lesser degree, into northwestern Greece and Macedonia. An expanding presence on the mainland was matched with an increased Mycenaean presence throughout the eastern Mediterranean. Mycenaean pottery dating from 1550 to 1400 B.C. has been uncovered on many of the Cycladic islands, especially Naxos, attesting to Mycenaean activity in the region. Additional Mycenaean pottery remains, dating from the fourteenth to the twelfth centuries, have been unearthed at sites such as Troy, Aeolis, Clazomenae, Ephesus, Miletus, Sardis, Kéos, Samos, Lesbos, and Cyprus, indicating trade and the establishment of some colonies. Farther afield, pottery has been found in inland Syria, Palestine, and Amman, Jordan. To the south, archaeological excavations in Egypt have revealed Mycenaean vases and wall paintings depicting foreigners bearing gifts to the pharaoh. The gifts shown in the paintings have a decidedly Mycenaean appearance. Discoveries of Mycenaean pottery have also been made to the west in Sicily, and what may have been a Mycenaean settlement has also

The Lion Gate at Mycenae
Photo courtesy of Greek National Tourist Organization

been found near Taranto, Italy. Amber and other materials rare in the Mycenaean world have been uncovered at many Mycenaean cities, leading to speculation that trade may have occurred between the Mycenaeans and the peoples of interior Europe.

The Mycenaeans exchanged their beautifully painted jars of varying styles and sizes for a number of items. From Egypt came ivory and gold. Copper, one of the most important commodities of the Late Helladic Period, undoubtedly came from Cyprus. Imported foodstuffs included a variety of herbs and spices from the Middle East. Wine came from several regions, albeit in very small quantities, since most wine consumed at Mycenae was homegrown.

The items that Mycenae exported, as well as those it used domestically, were created in numerous workshops. Goods produced, in addition to pottery, included jewelry, ornamental blue glass, construction materials, textiles, perfumes, metal goods, archers' bows, and components for ships. Many of these items were created specifically for export; at some point Mycenae's exports may have expanded to include the shipment of timber to the treeless lands of the Nile.

The Mycenaeans also had a well-trained and well-equipped military. The Mycenaeans were knowledgeable of metal technology and skilled at exploiting local ores. Their elite army, drawn from the aristocracy, was one of the first forces to effectively use the war chariot and the long sword. The war chariot was a lightly built two-wheeled vehicle pulled by two horses and possibly used for the rapid deployment of heavily armed troops. The variety of weapons unearthed from the graves of aristocrats attests to their ability to overpower their foes. The graves housed heavy thrusting spears, short double-edged swords, daggers, and bows; heavy linen tunics, possibly reinforced with bronze; and helmets, made of, or reinforced with, bronze.

It is conceivable that Mycenae's expanding trade and colonization constituted an attempt to accommodate a burgeoning population. Mycenaean Greece was heavily populated and far from being economically self-sufficient. It has been estimated that there were as many people living in Greece during the Mycenaean heyday of the Late Bronze Age as there were some 900 years later in classical Greece. The most populous region was the Argive plain, which included Mycenae. This is demonstrated by the disproportionate number of major Late Helladic monuments and gravesites concentrated there. Archaeological surface surveys have indicated that the population on the Argive plain most likely reached its peak in the thirteenth century B.C.

Little is known about the social and administrative system at Mycenae. Most of what is known has come down from documentary evidence on tablets uncovered at Pylos, a Mycenaean settlement in the southwestern quarter of the Peloponnese. Much of the evidence is vague, making it dangerous to generalize, and may be only specific to the Pylian state. It appears, however, that Mycenaean society vested a great deal of power in the king, who directly oversaw economic activity. He most likely had a variety of subordi-

nates assisting him. There was a middle class made up of farmers and some artisans. Beneath the middle class was a large number of workers. There may or may not have been a distinction between slaves and free workers.

Scattered evidence has been found suggesting that the Greek world during the Late Helladic Period may have been politically and culturally unified under Mycenaean control. This evidence has been hotly contested. Those supporting the theory of a united Greece point out similarities in artistic expression, common language, a Mycenaean-led expedition of various Greek states against Troy, and a well-developed system of roads radiating outward from Mycenae. Those opposed to the theory contend a unified Greek state would not be made up of individual locales sealed off from each other by fortified walls.

Until the 1950s, little information was available about the language of Mycenaean Greece. What was known came from symbols painted upon vases. All that changed with the excavation of two houses at Pylos, "the house of the oil merchant" and "the house of the wine merchant." Excavations at the two sites revealed many clay tablets written in a syllabic script known as Linear B. When the two buildings burned the clay tablets were hardened, allowing them to survive. Linear B was first discovered at Knossos. It was a modification of the Minoan Linear A script, a language that has since been identified as being non-Greek. Linear B, however, was revealed to be the most ancient form of Greek when it was deciphered in 1952 by Michael Ventris, an amateur cryptographer from England. The script was thought to have originated on Crete but subsequent analyses showed it was not native to the island. It was, however, widespread on the Greek mainland. The contents of the documents found at Mycenae were unimportant; they listed perishable goods received or issued from palace stores, along with some personnel information that provides a bit of insight into the kingdom's administrative organization. The script appears never to have been used for display, and literacy seems to have been reserved only for the administrative class.

The origins of classical Greek polytheistic religion can be found in the Mycenaean age. Familiar ancient Greek deities such as Zeus, Poseidon, Hermes, Ares, Dionysus, Hera, Artemis, and Athena are mentioned on the Pylos tablets. It can be inferred that Hephaestus and Apollo were also worshiped. Along with names of familiar gods, unfamiliar deities are mentioned, most prominently a goddess known as Potnia. The unfamiliar gods and goddesses may have represented religious cults that were later absorbed into, rather than expelled from, ancient Greek religion. The various deities were served by priests and priestesses. The most frequent offerings to the gods and goddesses were agricultural, especially olive oil, honey, and wool. The sacrifice of animals sometimes occurred. Human sacrifice is also suspected, but in very few cases.

The Mycenaeans were active farmers. Chief crops harvested from the surrounding fertile plain included wheat, barley, olives, figs, and many types of fruit. From grapes they produced wine. Flax was also grown, probably to make linen. Animals such as horses, oxen, sheep, goats, and pigs were domesticated. The oxen pulled plows, and their hides supplied leather. Small horses were used to draw light chariots. Sheep were raised for their wool, goats and pigs for meat. Hunting may have been a popular sport. Documents mention hounds and refer to the hunting of deer.

Mycenaean society began to come apart around the thirteenth century B.C. Over the course of one hundred years Mycenaean culture experienced rapid decline. Specific reasons for the decline and eventual collapse remain a mystery. Several hypotheses have been put forth to explain it, including climatic changes, destruction by earthquake, economic collapse, internal strife, and invasion. Some historians and archaeologists believe Greek tradition, which maintains that Mycenaean society fell to the invading and less civilized Dorians, who conquered both the mainland and the Aegean Islands. Other historians and archaeologists subscribe to the theory that the Dorians did not invade but rather moved into lands already in the throes of decline.

There is reason to believe that during the apogee of Mycenaean civilization, between 1425 and 1230 B.C., fears of invasion arose and many cities were fortified. Defensive walls around the city of Mycenae were built on the summit of a hill around 1340 and were extended a century later. The Perseia Fountain enclosed by the walls ensured a continuous source of water within the citadel. The fortified walls were built of huge stone blocks, of varying shapes, put together without the use of mortar. So large were the stones that legend declared them built by giants called Cyclopes, who were invited from Asia Minor for the construction—hence the term "Cyclopean walls." The citadel was triangular in shape, measuring approximately 400 yards across from east to west and 200 yards from north to south. It commanded a south-facing view of the Argive plain all the way to the sea. It was placed so as to also control a northward pass through hills connecting the Isthmus of Corinth to central Greece. This allowed the Mycenaeans to maintain control of the Argive plain and access to the sea. Very large sections of the fortified walls are still standing today.

The citadel's main entrance was from the southwest through an opening called the Lion Gate. It was constructed about 1250 B.C. The gate was closed off by double wooden doors sheathed in bronze and secured by a wooden bar. Above the double bronzed doors was a massive stone sculpture of two lions, the oldest example of monumental sculpture in Europe. The sculpture survives to today with only the heads of the two lions missing. Access to the gate was via a graded road that led up to a ramp. There were three additional entrances into the citadel, the North Gate and two auxiliary gates.

Many important buildings were located within the citadel. Most important, occupying the upper sections, was the palace complex, consisting of multiple buildings that were expanded many times over the years. It was a combination royal residence, storehouse, court, accounting office, and religious and social center. The oldest section of the complex was at the very top of the hill. An important later addition to the palace was a grand staircase, consisting of about forty steps, erected in its southwest corner. Painted stucco floors and frescoes on the walls provided for lush surroundings. Many workshops and storage areas were in the eastern section of the palace. A three-story building referred to as the House of Columns was built along the fortification walls and at one point may have been part of the main palace. To the north, flanking a court, were two houses, today dubbed the Gamma and Delta Houses. Off to the southwest were two structures, the Granary House (also known as the Citadel House) and the Tsountas (named for archaeologist Christos Tsountas). Together, the two may have encompassed an important shrine area for cult worship. The Granary House was so named because of the carbonized barley, wheat, and vetches found in its basement. Workshops were also in this area. To the northeast are traces of at least three buildings. Lead objects, fine ivories, and pottery have been found within the ruins. On the north and northwestern slopes is an area originally consisting of a series of buildings, terraces, passages, and courts. Bronze works, female figurines, vases, and jewelry have been found among the rubble. Outside the citadel was a section that is today referred to as the Lower Town. The Lower Town spread across approximately 250,000 square yards and housed most of the population. The most important buildings of the Lower Town were located closest to the citadel.

The city of Mycenae was attacked—perhaps by the Dorians—around 1250 B.C. The attack destroyed buildings located outside the citadel. Between 1150 and 1100, the city was sacked many time; the palace was destroyed and never rebuilt. Some surviving sections outside the citadel continued to exist thereafter as a small city-state. Many Mycenaeans fled the city over the course of the invasions. They fled to Achaea, Cephalonia, Ithaca, Euboea, Epidaurus, small sections of Argolid and Attica, Kéos, Naxos, Crete, and Cyprus.

Mycenaean society slipped into an age of feudalism, great poverty, and much violence, lasting almost 400 years. Numerous sites throughout Greece were abandoned. The most dramatic area of abandonment occurred in the southwestern Peloponnese where 150 known sites in the thirteenth century were reduced to a mere 14 sites one century later. Laconia, Argolid, Corinth, Attica, Boeotia, Phocis, and Locros also experienced dramatic losses. The population of Greece plummeted. Artisans lost their royal patrons, causing arts and craftmaking skills to be forgotten. Literacy also declined. Stone structures were often replaced with ones of timber and mud. Elsewhere in Greece, some advanced cultures may have survived in this era, however, as evidenced by finds on the island of Euboea.

At the start of the Greek classical age, Homer immortalized the long-gone Mycenaean civilization in the *Iliad*. The remainder of Mycenae's history did not match its earlier glory. In 480 B.C., the city sent nearly 400 men to fight alongside troops from other Greek city-states against the Persians at Thermopylae. The following year Mycenaean soldiers participated in fighting at Plataea. Argos besieged Mycenae in 470 B.C. and destroyed it two years later. When the Roman geographer Pausanias visited the site in the second century A.D. it was almost totally deserted.

Mycenae came under the care of the Archaeological Society of Athens in 1837. By 1840, they had cleared away an opening at the Lion Gate. In 1874, Heinrich Schliemann began conducting archaeological excavations at the site hoping to uncover the Treasury of Atreus, a huge tomb believed to contain the remains of Agamemnon. The Treasury of Atreus was eventually uncovered by a Greek archaeologist named Stamatakis in 1878. It was found empty, having already been raided and robbed of its treasures. Schliemann managed to discover the palace, two homes, the Perseia Fountain, and many tombs. The high point of his excavation came when he unearthed a circular stone chamber called the Tholos Tomb, containing skeletal remains and unplundered goods from the sixteenth century B.C. The items included gold-faced masks, gold and silver cups and jewelry, embellished bronze swords, crystal knobs, decorated daggers, ivory hilts, and pottery. It has been surmised from the number of empty holes in the walls and from bronze nails found in the tomb that bronze plates originally covered the tholos tomb's walls.

Alan Wace began work at Mycenae in 1920 and continued excavating until 1923, exposing the Granary. In 1939, he returned and began excavating gravesites. Work was interrupted by World War II and the Greek Civil War, and did not resume until 1950, under the supervision of the Greek archaeologist Dr. John Papadimitriou and an American professor named George Mylonas. The two found Mycenae's second tholos tomb in 1951, outside of the citadel's walls. The second tomb contained fourteen royal and twelve private citizens' graves. It was slightly older than the first tholos tomb and contained fewer riches.

Some sections of Mycenae were destroyed by invaders, and others altered by construction undertaken during the Hellenistic Age. Still, the surviving sections of prehistoric Mycenae remain a testament to the greatness of the Late Bronze Age in Greece.

Further Reading: Maitland A. Edey's *Lost World of the Aegean* (New York and London: Time-Life, 1975) is a well-written account of both the Minoan and Mycenaean civilizations. Special emphasis is placed on how the Minoans influenced the later Mycenaean society. Richard Hope Simpson's *Mycenaean Greece* (Park Ridge, New Jersey: Noyes, 1981) provides an extremely detailed account of all known Mycenaean sites. Frank H. Stubbings's *Prehistoric Greece* (London: Granada, 1972; New York: Day, 1973) provides much insight into the archaeological discoveries at Mycenae. V. R. d'A. Desborough's *The Greek Dark Ages* (London: Benn, and New York: St. Martin's, 1972) is an excellent source on the decline and fall of Mycenaean society.

—Peter C. Xantheas

Naples (Napoli, Italy)

Location: On the southwestern coast of Italy, to the west of Mount Vesuvius.

Description: One of the largest cities in Italy, formerly the capital of the Kingdom of Naples and Sicily; famous for its Roman, medieval, and Bourbon heritage and notorious for its slums and organized crime.

Site Office: E.P.T. (Ente Provinciali per il Turismo)
Piazza dei Martiri, 58
CAP 80121 Naples, Napoli
Italy
(81) 405311

Like Cumae, twenty-five miles west along the coast, the city that was to become Naples (Napoli in Italian) was founded by Greeks from the island of Rhodes, who began settling along the shores of Southern Italy around 1000 B.C. This new city, Parthenope, became a refuge for people from Cumae after its defeat and destruction by the Etruscans in 524 B.C., but after 474 it became known as Palaeopolis ("Old City"), in contrast with the nearby Neapolis ("New City"), founded around that date. In about 400 both these cities came under the control of the Samnites, who were displaced in their turn by the Romans after a siege that lasted from 329 to 326. It was they who amalgamated the two Parthenopean settlements into the single city that has since developed into Naples.

Up to the collapse of the Western Roman Empire, Naples was a center of Greek-speaking culture and learning, attracting, among many others, the Emperors Augustus and Nero and the Roman poet Virgil, who wrote his major works in the city, including the *Georgics,* his cycle of verses on rural life, and the *Aeneid,* the epic tale of Aeneas, the mythical founder of Rome. Naples later became a focus of Christian ritual and pilgrimage, based on the veneration of San Gennaro (St. Januarius), a bishop of Naples who was murdered in nearby Pozzuoli in A.D. 305. What is said to be his blood is contained in two vessels now kept inside a silver bust of his head, made in 1306, in the seventeenth-century chapel named for him inside the Duomo (cathedral). This mysterious substance is believed to return to a liquid state every May, September, and December. The saint's body is said to lie inside the city's catacombs, now also named for him, which were started around the second century A.D.

Romulus Augustulus, the last of the Western Roman Emperors, fled from Rome in 455 during an attack by the Vandals and died in Naples in 476. The city then passed under the nominal control of the Eastern Roman Emperors at Byzantium (now İstanbul), but in practice it was governed first by the Ostrogothic king Theodoric, up to his death in 526; by the Byzantine general Belisarius until 543; and by the Ostrogoths again until 568, when the first of a series of local rulers, often known as dukes, created an effectively independent state. These rulers included bishops of Naples (from 754 to 800 and again from 876 to 898) as well as feudal nobles. Meanwhile, the merchants and craftsmen of the city gradually established a reputation as producers and traders of linen and other textiles, making contacts with Christians and Moslems alike throughout the eastern Mediterranean and beyond and raising their prosperity to the point that the city was able to make and use gold coins.

By the eleventh century, however, Naples, Amalfi, and other southern cities were faced with increasing competition from such northern ports as Genoa and Pisa, and with the problem of trying to assimilate the Normans, the descendants of the Scandinavian Vikings who came to Italy as mercenaries in its numerous wars and settled in great numbers in the south. In 1059 Pope Nicholas II made one of their leaders, Robert Guiscard, duke of Apulia and Calabria and ruler of most of southern Italy. The Normans steadily expanded their holdings and political influence, and in 1139 Naples, too, was incorporated into their Kingdom of Sicily by Roger II, a nephew of Robert Guiscard. In 1194 this kingdom was inherited by Holy Roman Emperor Henry VI (Heinrich von Hohenstaufen), who planned to use it as a base for increasing imperial power over Italy and decreasing the power of the papacy. His son Frederick II reasserted this policy by founding the University of Naples in 1224 as a rival to the popes' University of Bologna. The Hohenstaufen family retained control of southern Italy until the death of King Manfred in a battle against the pope's ally Charles of Anjou, a brother of the French king Louis IX, in 1266.

Charles of Anjou was king of Naples until his death in 1285, three years after the outbreak of the Sicilian Vespers, the popular revolt that was to lead to Sicily's becoming a separate kingdom under Manfred's son-in-law Peter III, king of Aragon. Charles had already made Naples his capital, in preference to Palermo, and begun building the Castel Nuovo—also known as the Maschio Angioino—and the Duomo, the cathedral of Naples. The best known of his successors, the Angevin kings, is perhaps Robert the Wise, who ruled from 1309 to 1343 and revived the city's reputation as a cultural center by inviting Boccaccio, Petrarch, and other poets, as well as Giotto and other painters, to his court there. His memorials include the Church of Santa Chiara, a Gothic building completed in 1328 and since frequently restored after earthquakes and warfare, most recently after World War II; the Castel Sant'Elmo, a palace built on the Vomero hill above the old city; and, next to it, the Certosa of San Martino, a Carthusian monastery that he sponsored. Naples under the Angevins was a city of around 30,000 people, its port

Staircase in the Palazzo Reale, Naples
Photo courtesy of The Chicago Public Library

and markets dominated by Catalans, northern Italians, Germans, and Jews, and its finances by the great banking houses of Florence, and the capital of a kingdom that, suffering from chronic economic decline and rural depopulation, gradually collapsed into a number of competing regions led by warring nobles.

In 1442 the Angevin king René I was overthrown and replaced by Alfonso V of Aragon, who reunited the kingdoms of Naples and Sicily as Alfonso I and rebuilt the Castel Nuovo in Gothic style, providing it with a triumphal arch, the Torre della Guardia, as its gateway in 1467. Although his successor Ferdinand II of Aragon (who ruled Naples as Ferdinand III) managed to suppress a revolt by the leading nobles in 1484, the kingdom remained disunited and many of the former rebels welcomed an invasion by the armies of King Charles VIII of France in 1495. After nine years of struggle between the armies of France and Spain, Naples became the seat of Spanish rule in the south and intervention in the north of Italy. The Italian Wars continued under King Francis I of France and Holy Roman Emperor Charles V, but these did not alter the political position of Naples, which benefited economically from its new role. The Spanish were careful to incorporate the local nobility into government by allowing their traditional royal parliament and city council to continue, and to sponsor building projects, such as the Palazzo Reale (royal palace), begun in 1600, and the main avenue of Naples, which, until the twentieth century, was named for the first Spanish governor, Pedro of Toledo.

The increasing burden of taxes on the rural population and the high level of consumption among the nobility encouraged the growth of Naples, which became one of the largest and most overcrowded cities in Europe. Between around 1500 and 1547 its population rose from perhaps 100,000 to at least 240,000. Expansion brought with it both plague, as in 1576, and social unrest, as in the unsuccessful uprising against the Spanish, which broke out in 1585 after some years of economic depression. These problems continued and intensified as Naples underwent economic decline through the course of the seventeenth century, sharing with other Italian commercial centers in the collapse of their markets in the face of English and Dutch competition. In July 1647, riots against new taxes led to the creation of a "Parthenopean Republic" under Tomasso Aniello, or Masaniello, a fisherman who was only twenty-four years old when he seized power for nine days in October. His removal and execution by the Spanish was followed by the arrival of the French duke Henri de Guise, who proved unable to organize effective resistance to the return of Spanish rule in August 1648. This was generally welcomed by the nobles, who had been badly frightened by the widespread support for the revolt in the countryside. Eight years later yet another outbreak of the plague killed more than half of the 450,000 inhabitants of the city.

In 1707 the Kingdom of Naples and Sicily was detached from the Spanish throne and awarded to Charles VI, a Habsburg archduke from Austria. The branch of the French royal family, the Bourbons, that had taken over in Spain did not accept the arrangement. In 1734 the Infante Charles of Bourbon (later Charles III of Spain) seized both parts of the kingdom as Charles IV and confirmed the establishment of yet another dynasty by his victory over the Austrian army in the battle of Velletri ten years later. His home, the Capodimonte Palace, became the Royal (now National) Museum after his death. Since 1777, when King Ferdinand IV inherited the art collections of the Farnese family, it has housed their ancient sculptures and art objects from Rome and Herculaneum alongside pictures by Caravaggio, Botticelli, Titian, and several Neapolitan painters. The museum now also displays many of the statues and mosaics taken from the excavations at the site of Pompeii, the Roman city near Naples that was destroyed in an eruption of Mount Vesuvius in A.D. 79.

By the end of the eighteenth century, the city's population had reached 400,000, and rebellion was in the air once again. In 1799 another short-lived Parthenopean Republic was created after the arrival of a French Revolutionary army, but the city was swiftly reconquered by a peasant "Army of the Holy Faith," led by Cardinal Fabrizio Ruffo, which killed hundreds of republicans while a British fleet under Horatio Nelson ensured that there was no escape by sea. In 1806 Ferdinand IV was driven out of the city by another French invasion, but this time the kingdom was given to Joseph Bonaparte, an older brother of Emperor Napoléon I. Only two years later he was transferred to Spain and replaced by the French general Joachim Murat, who, having taken part in the disastrous attack on Russia in 1812, broke with Napoléon in 1814. After only one more year, however, Murat returned to Napoléon's camp in the hope that his restored empire would last. After Murat's execution the former Ferdinand IV returned as Ferdinand I, ruler of what was to be known from 1816 as the Two Sicilies, and founded the Memorial Church of San Francesco di Paola in 1817.

Ferdinand's repressive policies were opposed by secret societies, mainly the Carbonari, which rose against him in 1820, forcing him to announce a written constitution and accept liberals as his ministers. Only with the aid of the Austrian army, which arrived in 1821, was he able to restore the absolute monarchy. Ferdinand II, who succeeded him in 1830, was equally determined to suppress any revolutionary tendencies. His reign saw the opening of Italy's first railroad, from Naples to Portici, in 1839; the draining of marshes, which allowed the expansion of the city to the east of the historic center; and the foundation of the Camorra, the criminal organization whose name is as closely linked to Naples as that of the Mafia is to Sicily. It originated in 1842 as a branch of a Spanish robbers' network, formally known to the authorities as the Confraternita della Guardugna (brotherhood of robbers), but it soon acquired the name by which it is now known (derived from a Spanish word meaning "extortion money") and seized control of most prostitution and gambling, and even of daily life inside the city's jails.

Although many of the poorest and most desperate Neapolitans came, willingly or not, under the influence of the Camorra, others tried to continue the city's well-established tradition of unsuccessful revolutions. In January 1848 yet another rising, first in Palermo and then in Naples, was to be the first of the many to break out across Europe and give that year its nickname, the Springtime of Nations. Twelve years later the inexperienced Bourbon king Francis II, who had succeeded Ferdinand II in 1859, was expelled from Naples, not by its own citizens but by a rebel army led by Giuseppe Garibaldi, which crossed from Sicily to seize Naples in September 1860 and thus complete the absorption of the Two Sicilies into the new Kingdom of Italy.

Naples was now a provincial city geographically and socially distant from the centers of political and economic power. (Indeed, it is striking how many of its most famous natives, from the opera singer Enrico Caruso to the film actress Sophia Loren, have had to go elsewhere to find success.) The city nevertheless continued to expand in the years following unification. Many of its medieval buildings were destroyed in the 1870s and 1880s to make way for what was intended to be a modern and sophisticated city center, symbolized by the Galleria Umberto I, a complex of arcades completed in 1883. The disastrous cholera epidemic of 1884 was followed by further limited programs of slum clearance, new building, and improvements to the water supply, as well as the intendedly symbolic renovation of the frontage of the Duomo in a neo-Gothic style. Yet the Forcella, the area around the main railroad station, became and has remained one of the poorest, unhealthiest, and most crime-ridden slums in Europe, acquiring the nickname "kasbah" (probably in the 1940s) in reference to its continuing role as the city's main location for the open-air buying and selling of everything from bread to heroin, from cigarettes to medicines.

In 1922 the March on Rome, the event that symbolized the seizure of power by the Fascist dictator Benito Mussolini, was preceded by a rally of his armed followers in Naples in October, which he himself came to address. Although the Fascist regime managed to suppress the activities of the Camorra, at least to the extent of driving it further underground, its network was fully in place to benefit from the black market that developed in the fall of 1943. An intensely destructive German occupation lasted from September 8 of that year until its defeat by a popular rising, which began on September 28 and was followed by the arrival of British troops on October 1. With the disruption of utilities, which had always been inadequate, typhus spread through Naples, claiming even more lives than had the German army. This was followed in 1944 by the most recent eruption of Mount Vesuvius.

In the postwar years the Camorra expanded further, benefiting from the return of Neapolitan criminals expelled by the United States and the absorption of Tangiers into Morocco in 1961, which allowed the Camorra to seize control of the cigarette-smuggling operations formerly based there. Since the 1960s it has diversified into heroin smuggling and speculation in real estate, while keeping up its traditional business of extorting "protection" money from stores and companies in and around the city, but its hold on Naples has been significantly weakened, partly by increased government action against its leaders but mainly because of internal feuding between its two main factions, the Nuova Camorra Organizzata (the new organized Camorra) and the Nuova Famiglia (the new family). The legitimate alternatives to the Camorra, including the new and expanding metal, chemical, and electrical industries, attracted many families from the surrounding countryside and helped to raise the general standard of living, at least by comparison with the chronic poverty of much of Naples. Yet fifteen years after the war, of the 1,170,000 people then living in the city, around two-thirds had no regular employment; the older textile, food, and leather industries were drastically declining; and a large proportion of the funds provided by the Cassa del Mezzogiorno, the government fund for southern Italy created after World War II, had been embezzled, used to subsidize inefficient and short-lived industrial developments, or recycled as cash for political campaigns.

The population of Naples reached 2,700,000 in 1971. After yet another outbreak of cholera in 1973, mass movements of protest among the unemployed and the homeless, who refused to pay electricity bills, blocked traffic, and occupied public buildings, led to the victory of a leftist coalition led by the Communist Party in the city elections in 1975. They were no more successful at solving the city's problems than their Christian Democrat predecessors had been, as was shown in 1980, when a large part of the relief supplies sent to the city after the earthquake of November 23 was stolen by the Camorra.

The history of Naples has been marked by a succession of foreign conquests and citizen uprisings, epidemics and natural disasters, and the building of magnificent palaces and churches in the midst of the overcrowded and unsanitary homes of the poor. The cultural achievements of Neapolitans under Roman, Angevin, Spanish, and Bourbon rule—including, not least, the invention of the pizza some time in the seventeenth century—have unfortunately never been accompanied by the achievement of sustained prosperity, political stability, or social justice for most of the city's residents. The proverbial instruction to visitors, "See Naples and die," may well seem less a matter of praise for the city's vitality and fascination than a cruel joke at its expense.

Further Reading: Naples has often been described by its visitors, who have included the German poet and playwright Johann Wolfgang von Goethe, the Italian diarist Giovanni Casanova, the French novelist Stendhal (Marie-Henri Beyle), the British admiral Horatio Nelson, and his mistress Lady Emma Hamilton. In particular, Norman Lewis, a British writer who served in military intelligence during World War II, gives a vivid and memorable account of the city in his book *Naples '44* (London: Collins, and New York: Pantheon, 1978).

—Patrick Heenan

Nauplia (Argolis, Greece)

Location: On the Greek Peloponnesian peninsula, near the head of the Gulf of Argos, forty-four miles southwest of the Isthmus of Corinth.

Description: Important port and trading post; occupied by a variety of conquerors during the Middle Ages; important site in Greek War of Independence; capital of Greece from 1828 through 1833.

Site Office: Tourist Information Office
Iatrou Square
Nauplia, Argolis
Greece
(752) 24444

Nauplia's historical significance derives principally from its role in the Greek War of Independence and its status as the capital of Greece in the first few years after the nation broke free of the Ottoman Empire. Nauplia is one of the most picturesque cities in all of Greece, with well-preserved neo-classical architecture and striking military fortresses. Nauplia is located on the northern shore of the western edge of a rocky peninsula that juts out into the Gulf of Argos; two hilltop citadels rise above the city and dominate the rocky southern shore of this peninsula. The citadel Its-Kale sits on the smaller of the two crags, 280 feet above sea level; this site has been fortified since ancient times. The Palamidi, an imposing Venetian fortress built in the eighteenth century, is perched on the larger of the two crags at 700 feet above sea level.

Paleolithic and Neolithic remains have been found in the vicinity of Nauplia, and the site may have been a naval station during the era of the Mycenean civilization in the second millennium B.C. Greek legend holds that the city was founded by the mythical Nauplios, the son of Poseidon and the Danaid Anymone. Nauplios's son Palamedes, for whom the higher of the two crags, and its fortress, are named, is credited with the invention of lighthouses and dice, as well as the creation of several letters of the Greek alphabet. According to the epic poet Homer, Palamedes was killed by his fellow Greeks in the Trojan War when he was mistakenly charged with treachery after playing a trick on the great warrior Odysseus. In the twelfth century B.C., during the time of King Agamemnon's Trojan expedition, Nauplia was a self-governing city-state.

In 625 B.C. Nauplia was invaded by the neighboring city-state of Argos, and thereafter played little part in classical Greek history. Nauplia, whose name is derived from the Greek for "naval station," became an important port again for the Byzantine Empire in the Middle Ages. Its calm coastal waters and location in the eastern Mediterranean on the crossroads of Europe and Asia made it an ideal harbor and trading post.

The Byzantine Empire declined after Constantinople was sacked in 1204 by crusaders, who set up the Latin Empire of Constantinople. Nauplia remained a Byzantine possession for a time, but in 1210 it fell to the crusaders, who made Nauplia part of the Duchy of Athens in their empire.

In 1388 Venice purchased Nauplia. The Venetians saw Nauplia as a strategic site for protection of their sea trade routes. They named the city Napoli di Romania and immediately made efforts to protect the harbor by repairing the walls and further fortifying the hilltop citadel of Its-Kale. In 1471, the Venetians constructed an additional castle, the Castle Pasqualagio, on the tiny islet of Bourdzi, located in the harbor about 500 yards off the coast.

By 1453, the Ottoman Empire had overrun Constantinople (a Byzantine possession again), along with most of the other former Byzantine lands, and conquered almost the entire Peloponnese peninsula. Its-Kale, however, proved to be nearly impenetrable to frequent Turkish attacks; the Venetians fought off two significant attacks on Nauplia by the Turks in 1470 and 1500. In 1502 the Venetians further strengthened the city's defenses by enclosing it in walls. In 1540, however, not even the walls could protect the city; the persistent Turks captured Nauplia and made it the capital of the Turkish region of Morea.

In 1686 Venice recaptured Nauplia from the Turks. In 1711 the construction of an impressive complex on the difficult terrain at the top of the Palamidi was begun. When it was completed in 1714, it had three independent fortresses connected by ramparts and enclosed within solid and ornate walls. In 1715, only one year after its completion, the Turks attacked the fortress and were able to drive the Venetians out in only eight days. The fortress was well built, but it had not been garrisoned with sufficient troops for its defense.

The Ottoman Empire had a feudal-like economic system that relied upon the practice of bringing in uneducated Albanians to work as servants. During the time of Turkish rule, relations among the Greeks, Albanians, and Turks were strained, and there were frequent revolts and disturbances. After quashing one particular uprising by the Albanians, the Turks led a group of these servants up to the Palamidi and hurled them off the cliffs to their death along the shore of the Gulf of Argos below.

In the eighteenth century, the balance of power in Europe shifted from the city-states such as Venice toward such powers as France, Britain, and Russia. Greek merchants increasingly began to establish ties with these emerging powers; reciprocally, the European governments were interested in Greece because of its strategic location on the fringes of the Ottoman Empire. When Catherine the Great of Russia

The Bourdzi castle
Photo courtesy of Greek National Tourist Organization

invaded the Peloponnese briefly in 1770 and later wrested control of the Black Sea from the Turks, the relationship between Russia and Greece was strengthened. Russian leaders gave the Greeks the privilege of trading under the Russian flag and several years later appointed Ioánnis Kapodístrias, a Greek originally from the island of Corfu, to be a joint secretary of state for the Russian government. In addition, Napoléon made no secret that he considered making some of the Greek isles and the Peloponnese a French colony.

The events of the French Revolution at the end of the eighteenth century stirred the national consciousness of the Greeks and helped to plant the seeds for a Greek cultural renaissance and independence movement. The Greeks were painfully aware that while they had their own language and their own church, they were lacking a nation to call their own. In 1814, a secret organization known as the Philiki Etairia, roughly translated as the Society of Friends, was founded by three Greek merchants in Odessa, Russia, for the purpose of discussing how the Greeks could best obtain their independence. Within a few years similar organizations were formed all over Greece and in some cities elsewhere in Europe.

The foundation of the Philiki Etairia coincided with the rise of a significant movement in Europe known as philhellenism. Greek merchants in Russia, Germany, France, Italy, and England developed ties with the people of these countries and helped to spread information about Greece throughout Europe. These nations felt an affinity with the Greeks because the Greeks were Christians and descendants of the founders of western civilization. Philhellenic committees formed all over Europe to offer the Greeks moral and economic support in their struggle to free themselves from Ottoman rule. The European governments, however, did not support the Greeks as unconditionally as did the general public because they were justifiably concerned about the intense internal rivalries that divided the Greeks. Various Greek factions were unable to reach a consensus on the path that their road to independence should follow.

In 1821 the Greeks caught the Turks by surprise and began the War of Independence by taking over several small fortresses in the Peloponnese. Later in this same year, the Greeks convened the first National Assembly in Epidaurus, a few miles from Nauplia, to draft a provisional constitution and to appoint a titular government. In the late summer of 1822 the Greeks captured Nauplia; this would have been a major victory had the Greeks been able to resolve their internal differences. These differences were highlighted when, after the capture of Nauplia, the titular government tried to convene the second National Assembly in Nauplia. However, Theódoros Kolokotrónis, the commander who had taken Nauplia, did not care for the government and refused to allow the meeting to take place; the meeting site was changed to Argos. While the Greeks fought among themselves, they had little time to concentrate on fighting the Turks.

The Greeks probably would have been crushed by the Turks if it were not for the wave of philhellenism rising across Europe. By 1824, the philhellenic movement had become a great romantic crusade to free the Greeks from the control of the sultan. Most notable among the philhellenes were Lord Byron, Victor Hugo, and Goethe. Young men from all over Europe volunteered to fight in Greece. In 1824, Lord Byron helped the Greeks secure British financing. Byron, desiring a more active role, came to Greece to fight, but died of a fever not long after his arrival.

Even with the support of the European community, it appeared for some time that the Greek independence movement would not succeed. In September 1824, shortly after Lord Byron's death, the first installment of the British loan, some £50,000, arrived in Nauplia on the ship *Florida*. Very little of this or subsequent installments, however, went for military purposes; most of it was spent for government payoffs to individuals. In 1826, after the arrival of another European loan, the provisional Greek government was forced to seek refuge at the Bourdzi castle because two rival factions carried out a gruesome feud on the streets of Nauplia to decide who would control the distribution of the recently arrived money.

Finally, in 1827, after European military intervention under the leadership of Sir Edward Codrington and Sir Richard Church, a cease-fire was approved and the Ottomans recognized a Greek independent nation that consisted only of the Peloponnese and a few Greek islands. Ioánnis Kapodístrias was selected to be the first president of the Greek nation, and in 1828 he arrived in Nauplia, having moved the provisional capital there from Aegina. On the day of his arrival in Nauplia two factions were bombarding each other from different sides of the city. This was indicative of the difficulties that his presidency would face. His tenure saw some positive accomplishments, especially in the realm of public works, but he encountered opposition from liberals who wanted such reforms as press freedom and from other opponents who simply desired additional power or wealth.

On Sunday, October 9, 1831, Kapodístrias arrived at the Church of St. Spyridian in Nauplia for an early morning service. As he entered the church, he was assassinated by two of his political enemies; each of the assassins delivered a mortal wound, one with a dagger and the other with a pistol. One of the murderers was caught immediately and lynched on the spot and the other was caught and executed twelve days later. After the assassination, two rival governments were set up and civil war broke out.

The great powers of Europe—France, Great Britain, and Russia—had protecting power over Greece, but were divided over which government to support. Finally, they agreed that Otto, the seventeen-year-old son of King Louis I of Bavaria, should become the king of Greece. In 1832, the National Assembly ratified the selection of Otto in the Church of the Annunciation just outside of Nauplia, in the suburb of Pronoia.

Otto and his 3,500 Bavarian troops arrived by ship in the harbor of Nauplia in January 1833. The troops disembarked several weeks before the young prince to prepare for the February 6 crowning of the new king. The philhellene George Finlay wrote the following about the day that Nauplia waited for their king to disembark among his new people:

> The scene itself formed a splendid picture. Anarchy and order shook hands. Greeks and Albanians, mountaineers and islanders, soldiers, sailors and peasants, in their varied and picturesque dresses, hailed the young monarch as their deliverer from a state of society as intolerable as Turkish tyranny. The music of many bands in the ships and on shore enlivened the scene . . . and the sounds of many languages testified that most civilized nations had sent deputies to inaugurate the festivals of the regeneration of Greece.

Otto's court was located in a villa enclosed within the walls of the Palamidi. His Bavarian troops cut approximately 900 steps into the rocks on the northwest side of the hill to allow access of the Palamidi from the old town. Each day, until the Greek capital was moved to Athens in 1834, Otto's ministers and staff climbed this flight of stairs to get to work. The stairs still exist today and visitors can still make the grueling 700-foot climb.

The significance of Nauplia declined after the capital was moved to Athens. An uprising of soldiers in Nauplia, however, helped bring about the abdication of King Otto in 1862, and the town was a British evacuation point in World War II. The evacuation, made under cover of darkness, resulted in a shipwreck in which several people were killed.

The major points of interest in Nauplia today, though, are the many reminders of the independence movement and the early years of modern Greece. The Ministry of War building, used by both Kapodístrias and Otto, is located on the central Syntagma (Constitution) square in a preserved Venetian mansion. At the entrance of the Venetian St. Spyridian Church, constructed in 1702, a bullet hole from the assassination of Kapodístrias can still be seen. The Church of the Transfiguration, another Venetian church, was used as a mosque by the Turks during their rule of Nauplia. Otto returned this church to the Roman rite and allowed a memorial monument to be constructed inside that lists the names of the many philhellenes who died in the independence movement, including Lord Byron. Nauplia also features a marble statue of Kapodístrias, a marble obelisk honoring the French who died in the war, and several statues of the leaders of the Philiki Etairia (one of these statues is located in front of the building that housed the first high school of liberated Greece, begun by Otto in 1833).

Most of the Turkish structures in the city were destroyed after the Greeks gained their independence. However, there are several converted mosques and public Turkish fountains that can still be seen today. Two of the most prominent converted mosques are located in the central Syntagma square and are easily recognizable because they are aligned to face Mecca instead of being placed at right angles with the rest of the square. The former mosque of Vouleftikó, which housed the first Greek parliament, is restored as a historical landmark while the other former mosque on the square has been converted into a movie theatre.

Nauplia's three fortifications, Its-Kale, the Palamidi, and the Bourdzi castle, are extremely well preserved, and visitors are permitted to enter all three of these historical sites. It is even possible to stay in a hotel within the grounds of the Its-Kale. The Its-Kale fortress is especially interesting because the original walls, with their polygonal masonry, can still be seen. The remains of additions made to the fortress by successive conquerors display the architectural styles of their respective eras.

Nauplia's famous Archeological Museum is located on the central Syntagma square, in a former Venetian mansion that served as the naval arsenal in the fifteenth century and later as the Venetian governor's palace. The museum houses a collection of Mycenean pottery, several remnants of Minoan frescoes found at nearby Tiryns, and the famous "Linear B" tablets that expanded the understanding of Greek history. The museum also displays the only perfectly preserved Mycenean suit of bronze armor, discovered in an excavation at Dhéndra in 1960.

After being a site of struggle for centuries, Nauplia today is very peaceful. The town is the center of much publishing and artistic activity. Its narrow streets are lined with old houses with colorful balconies and shutters, and its abundant stores and cafes are frequented by both the city's residents and its many tourists. Apart from its own interest as a city and its historical landmarks, Nauplia's location and modern facilities make it a popular point of departure for other sites in the Pelopponese.

Further Reading: *A Short History of Modern Greece* by C. M. Woodhouse (New York: Praeger, 1968; as *The Story of Modern Greece,* London: Faber, 1968) summarizes the history of Greece since A.D. 324. Of particular interest concerning Nauplia is the outline of the problems that Kapodístrias faced as the first president. *The Greek Struggle for Independence, 1821–1833* by Douglas Dakin (Berkeley and Los Angeles: University of California Press, and London: Batsford, 1973) provides an extremely detailed account of the events that led to the establishment of the modern Greek state. *Blue Guide: Greece* by Robin Barber (London: Black, and New York: Norton, 1987) is a comprehensive tour guide that features excellent descriptions of historic sites in Greece.

—Mark D. Hanafee

Naxos (Cyclades, Greece)

Location: Naxos is the largest of the Cycladic islands—a group of islands in the middle of the Aegean Sea. Lying 100 miles southeast of Athens, Naxos sits in the heart of the Cyclades.

Description: Picturesque island, home to the ancient Cycladic civilization, and later under Roman, Byzantine, and Venetian rule; part of an independent duchy in the late Middle Ages; has many archaeological sites and historic buildings. Chief town also called Naxos; the island also has many smaller villages.

Site Office: Archaeological Museum
Naxos, Cyclades
Greece
(285) 22725

Naxos is not only the largest of the Cyclades, but also the lushest and, by many accounts, the most beautiful. It has been called "the pearl of the Aegean" and "the garden of the Cyclades." Because the island is so much larger than the surrounding islands, it enjoys a natural water supply that far exceeds those of the other Cyclades. As a result, irrigation is possible on Naxos, whereas it is not on the other islands. Naxian agriculture is, therefore, more productive.

In ancient Greek times the island was famous for its wine and was known as the home of Dionysus, the god of wine. Coins on the island were even stamped with Dionysus's image. According to legend, Naxos is also where Dionysus met his bride, Ariadne. After she helped Theseus defeat the Minotaur on Crete, Ariadne and Theseus fled to Naxos. There, while Ariadne slept, Theseus deserted her on the beach. Dionysus came upon Ariadne after she awoke and had begun to weep. He fell in love with her and married her. Naxos's historical significance, however, begins thousands of years before the Greek era.

Cycladic culture began around 3000 B.C. and was characterized by bronze sculpture, the growing of cereal grains, the cultivation of vines and olive trees, and trading in obsidian. Traditionally, three divisions are made within Cycladic culture: Early Cycladic (3000–2000 B.C.), Middle Cycladic (2000–1500 B.C.), and Late Cycladic (1500–1100 B.C.). On Naxos, archaeologists have confirmed thirty-three Early Cycladic sites, but only four Middle Cycladic sites. When offering explanations as to Naxos's apparent prosperity during the Early Cycladic period, archaeologists suggest that trade in marble may be, in part, responsible; Naxian marble is the best available in the Cyclades. Despite the large number of Early Bronze Age sites, this period of prehistoric Naxos has been relatively underexplored. There has, however, been some study of a fortified settlement at Panormos, on the southeastern coast, that dates from the Early Cycladic (2300–2100 B.C.).

There is an excavated site at Grota, near the modern town of Naxos, on the west-central coast that contributes to archaeologists' understanding of the Late Cycladic phase of prehistory. Excavation was begun here in 1949 by professor Nikolaos Kondoleon of the University of Athens, and continued intermittently until his death in 1974. Subsequently one of his students continued his work there. Material found at Grotta, and later on Crete, is consistent with the theory that Naxian culture migrated to Crete and provided the basis for the Minoan culture. As the Ionian Greeks began to populate the Aegean in the tenth century B.C., Greek culture began to work its way into the Cyclades.

Theodore Bent, one of the nineteenth century's foremost experts on the Cyclades, argues that Naxos (and the Cyclades in general) are important for the study of ancient Greece for three reasons. First, unlike mainland Greece, the Cyclades were never subject to any invasions by barbarous tribes; hence much of the cultural record has been preserved, whereas on the mainland many artifacts have been destroyed. Second, the Latin influence that literally overtook the mainland does not seem to have penetrated the Cyclades very deeply (and here he uses Naxos as his example). Finally, the Turks, who also overran much of the mainland at one time, never seemed to have much to do with the heart of the islands. In short, the Cyclades offer an undistorted picture of ancient Greek culture.

The Ionian Greeks were quick to settle on Naxos and even quicker to prosper there. By 735 B.C. Naxos had become so prosperous that it had already established a colony, also called Naxos, on Sicily. The first true ruler of Naxos, however, did not appear until the sixth century B.C. At that time, a man named Lygdamis helped Peisistratus establish himself as tyrant of Athens. Peisistratus returned the favor by setting up Lygdamis as tyrant on Naxos. Lygdamis, in turn, set up Polycrates as tyrant on Samos (an island in the far eastern part of the Aegean), establishing a triple alliance that spanned the Aegean. As historian Charles Frazee writes, "Lygdamis . . . made Naxos the most powerful city of the Cyclades, forcing islands from Andros to Paros to pay him tribute."

Then, in 501 B.C., Persia invaded the Aegean. The Persian forces were repelled by an 8,000-man army at Naxos; however, the Persians returned eleven years later, and this time they were successful. Naxos fell to forces commanded by Datis and Artaphernes. Cunning leadership, though, freed Naxos from Persia ten years later. In 480 B.C. the Persian king Xerxes I (the Great) took ships and crewmen from the Cyclades in order to equip himself to win his fight against Sparta and Athens. At the battle of Salamis, Demokritos, commander of the Naxian troops, deserted Xerxes and joined

A typical old stone house on Naxos
Photo courtesy of Greek National Tourist Organization

Sparta and Athens. This was enough to turn the tide against Xerxes and give the victory to Athens.

As a result, Athens was able to form a "league" in the Cyclades. Naxos joined this league; however, they soon discovered that Athenian rulers were no less oppressive than the Persians had been. In 471 B.C., Naxos tried to leave the league but was suppressed. The Athenians established military colonies, called *cleruchies,* on Naxos to ensure its allegiance. Naxos and the rest of the Cyclades ceased being strong, independent entities.

All of the Cyclades came under Roman rule after 146 B.C. Within two centuries, Christianity had been brought to the island. According to tradition, St. John was exiled to Patmos where, having served the term of his exile, he decided to preach the Gospel on Naxos. From there, the new religion eventually migrated to the rest of the Cyclades. When the Roman Empire was divided, the Cyclades, logically, went to the eastern realm and became the possession of the Byzantine Empire.

In 815, Naxos's quiet prosperity was interrupted by a Moslem invasion. Fifteen thousand Moslems had fled Spain for Egypt after a failed revolt; however, Egypt denied the refugees access. Consequently, they settled on Crete and used the island as a base for piracy. They were led by a man named Abu-Hassan. Naxos was attacked by these pirate forces in 831 and again in 904. The Moslems were eventually expelled from Crete by the Byzantine emperor Nicephorus II Phocas. Frazee believes that either Phocas, or his immediate successors, built the fortress of Apalyrou on Naxos. It controlled the access to the island's interior.

By the late eleventh century, the Cyclades were once again prosperous. In 1083 Emperor Alexius I Comnenus established an independent metropolitan in the Cyclades, with the title of Paro-Naxias. Alexius also placed several key governmental figures on Naxos, indicating (as did the name of the metropolitan) the central importance of Naxos in the region.

The Fourth Crusade during the early thirteenth century resulted in the Latin conquest of Constantinople. In 1204 Byzantine emperor Alexius V fled, and his empire was divided between the Venetian Crusaders and the Latins. This marked the beginning of the Latin Empire (which would rule until the Byzantines reconquered Constantinople in 1261). Baldwin I was named the first Latin emperor, and he and the doge of Venice drew up a partition treaty in order to clarify who received which lands. With respect to the Cyclades, Baldwin was given Tinos, Venice received Andros, and the rest of the islands were left unassigned. This created a tremendous opportunity in the Cyclades, and a Venetian named Marco Sanudo took advantage of that opportunity.

Marco Sanudo, who had taken part in the Fourth Crusade, was the nephew of Enrico Dandolo, the doge of Venice. The way in which he established himself as ruler of the Cyclades attests to his cunning as well as his skill. After the partition treaty, the Venetians knew they had to begin occupying their new territories quickly before their archrival, Genoa, could move in and stake a claim. When word came to Venice that the Genoese fleet had already put to sea, Sanudo equipped eight galleys (at his own expense) and went to engage the Genoese. He heard that they had taken up in the Naxian fortress at Apalyrou, besieged them there, and once he had defeated them, claimed the island in Venice's name.

Sanudo knew, however, that Venice was unable physically to occupy all the land awarded to it in the partition treaty. He cleverly asked for, and was granted by the Venetian Republic of St. Mark, the right to possess the Cyclades and run them as a private venture. Realizing that the Cyclades were not Venice's alone to grant, Sanudo then appealed to the newly crowned Latin emperor, Henry of Flanders. In 1207 in Constantinople, Henry granted Sanudo the Cyclades.

The Aegean during the Middle Ages was infested with pirates, and Sanudo knew he would need protection from them if his rule were to succeed. Nevertheless, he also sought a high degree of autonomy. He believed that these two needs would be best met by becoming a vassal of the Latin Empire, rather than by becoming yet another governor of the Republic of St. Mark. He appealed to Constantinople for vassal status and was successful. In 1210 Flanders gave Sanudo the title of Duke of the Archipelago.

Sanudo proved to be a very effective ruler, simultaneously prosperous and sensitive to the plight of native Naxians. In contrast to the Venetian practice in Crete, Sanudo did not confiscate the land of the Greek nobility on Naxos. Instead, he split the island into fifty-six fiefs, and distributed them in such a way that there was plenty left over for the *archontes,* the Greek nobles.

One of Sanudo's first acts was to build a castle in the town of Naxos, on the same spot where Lygdamis had built his castle seventeen centuries earlier. While building his castle, Sanudo commissioned a Catholic cathedral dedicated to Mary's Annunciation. In 1208, Sanudo asked Pope Innocent III for, and received, a Catholic bishop. Marco Sanudo died in 1228 and was succeeded by his son, Angelo. In direct succession, the Sanudi ruled from 1207 until 1383; they were followed by the Crispi, who ruled for another 183 years. While the Sanudi rule was orderly and relatively free from incident, the Crispi rule was neither of these.

That the Crispi came to power at all was due to an act of treachery. In 1371, Niccolò Sanudo became the Duke of the Archipelago. Turkish aggression had escalated in the Aegean, and Venice considered it necessary for there to be a strong western ruler in the Cyclades. Niccolò had already proven himself to be a weak and selfish ruler. Hence, the Venetian authorities contacted Francesco Crispo (well known for managing a personal fleet of pirates, and married to a Sanudo niece) and let him know that they would not mind a change in rulers. In March, Crispo came to Naxos, supposedly for a hunting expedition with the duke, his cousin. Under some pretense, Francesco left the hunting party; several hours later his attendants returned with the news that the duke had been ambushed by a band of rebels and had died falling off his horse. Crispo, under the guise of stemming a revolutionary plot, assumed control. Thus began the Crispi reign.

As the Byzantine Empire gradually fell to the Turks, the dukes of the archipelago relied increasingly on Venice for protection. As Venetian power began to wane in the Aegean, the duchy experienced a similar decline. In 1416 Mehmed I, sultan of the Ottoman Empire, insulted that Duke Giacomo did not congratulate him upon assuming the sultanate, ordered a naval attack on the Cyclades; Naxos was besieged. The Venetians eventually came to the rescue, but attacks by the Turks continued throughout the fifteenth century. By 1494 the citizens of Naxos had become so discontented that they poisoned Duke Giovanni. His heir, Francesco III, was too young to rule, so the Naxians appealed to Venice. The republic sent three governors to Naxos who ruled there until 1500, at which time they turned over the rule to Francesco. In 1510, however, Francesco apparently went insane. After several months of strange behavior, he murdered his wife and attempted to murder his son before fleeing to Rhodes. The Venetians captured him, took him to Crete (where he died), and reinstalled rule by governors. This lasted until 1518, when Francesco's son, Giovanni IV, had become old enough to rule.

Giovanni was the twenty-first and last duke. In 1537 Khayr ad-Dīn (known to Europeans as Barbarossa or Redbeard), of Sultan Süleyman I's fleet, demanded that Giovanni pay the sultan tribute and accept vassal status. Giovanni had watched the Turks ravage the surrounding islands and knew that he had no choice. He paid 6,000 ducats and agreed to pay an additional 5,000 per year. He also wrote to the pope, apologizing for the deed.

Giovanni proved to be a dissolute and incompetent ruler. Utterly dissatisfied with him, the Naxians sent a secret envoy to Sultan Selim II; Giovanni did not discover this until too late. Selim gave the duchy to a Jew named Joseph di Nasi, who, realizing that the Greeks hated Jews, sent a Spaniard, Francis Coronelli, as his ruling agent. Coronelli married a niece of the Crispi and took her name. Once Joseph died, the sultan took back the duchy and ruled through his dragomans.

In 1770 the island was occupied by the Russians; it was regained by the Turks in 1774. Naxos finally joined the Greek kingdom in 1830 at end of the Greek War of Independence. During World War I all the Cyclades were occupied by the British; during World War II, Naxos was in the possession of the Italians, and later of the Germans.

Tourism began to develop on Naxos in the 1970s. The chief town of Naxos and several smaller villages contain notable historic sites. The town of Naxos has the Venetian castle, Byzantine churches, the remains of the Grota settlement, and an archaeological museum. Another archaeological museum is in the seaside village of Apiratho. Near the village of Apolona is an ancient marble quarry.

Further Reading: Charles Frazee's *The Island Princes of Greece: The Dukes of the Archipelago* (Amsterdam: Hakkert, 1988) is perhaps the most historically informative source on Naxos. For physical and cultural description, see J. Theodore Bent's *The Cyclades; or, Life among the Insular Greeks* (London: Longmans, Green, 1885); J. Irving Manatt's *Aegean Days* (London: Murray, 1913; Boston: Houghton Mifflin, 1914); and Eric Forbes Boyd's more recent *Aegean Quest* (New York: Norton, and London: Dent, 1970). Charles Stewart's *Demons and the Devil: Moral Imagination in Modern Greek Culture* (Princeton, New Jersey: Princeton University Press, 1991) also includes interesting examinations of Greek culture. For information about the prehistoric significance of Naxos, see R. L. N. Barber's *The Cyclades in the Bronze Age* (Iowa City: University of Iowa Press, and London: Duckworth, 1987).

—Lawrence F. Goodman

Nora (Cagliari, Italy)

Location: Twenty-two miles southwest of the city of Cagliari, two miles outside the town of Pula, in the Cagliari province and Sardinia region.

Description: Semi-submerged abandoned ruins of a port city, Nora is situated on a promontory ending in a sharp cliff. Six acres have been excavated. A Roman city is built on top of Carthaginian and Phoenician ruins, about a quarter of which have been slowly engulfed by the ocean. Uncovered ruins include indigenous Sardinian towers called *nuraghi,* Roman and Phoenician mosaic floors, a Roman amphitheatre, a nymphaeum, several temples, a municipal plan mapping out the city, and a sewer system.

Contact: E.P.T. (Ente Provinciale per il Turismo)
Piazza Deffenu, 9
CAP 09125 Cagliari, Cagliari
Italy
(70) 654811 or 651698

The now-abandoned port city of Nora makes an appearance in the ancient world's earliest travel guides. Pausanias, author of the ten-volume *Itinerary of Greece,* written in the second century A.D., described the then-thriving city as having divine origins. According to Pausanias, Nora's founder was none other than Norax, son of Hermes.

These mythic origins befit modern Nora. Hermes was, after all, the god who conducted shades of the dead into the underworld. Today's Nora seems to lie somewhere between the worlds of the living and the dead. Set on a rugged isthmus on the coast of southern Sardinia, Nora's vacant, windswept ruins form a watery ghost town sinking into the sea. Cemeteries, temples, and roads are partly submerged. All were buried for centuries until excavated in 1954.

But for one week each year, the abandoned town springs to life when Sardinians host the colorful festival of St. Efisio, Sardinia's patron saint. The four-day pilgrimage party starts in Cagliari and ends at the saint's burial site, outside of Nora. Experts consider the festival the most authentic folklore event in Europe, with colorful costumes, music, and a parade of brightly decorated ox-drawn carts.

It is now known that the town was settled by Phoenicians around 700 B.C. Nora's location on a tongue of land stretching into the sea made the city an excellent stopover for Phoenician ships as they transported silver, metals, and luxury items across the Mediterranean. The Spanish influence was probably present because ships from the Phoenicians' richest colony in Tartessus, southwest Spain, no doubt docked in Nora's harbors.

The isthmus was also defensible from the land where Sardinia's rugged indigenous tribes inhabited the mountains. Little is known about the Bronze Age Proto-Sards who roamed the island from 1500 B.C. onward. They too had a settlement at the site which pre-dates the Phoenician colony.

Beaten down by wave after wave of invaders, the Proto-Sards left no written record but they did leave axes, arrows, and ceramics, as well as hundreds of bronze statuettes. By the time the Romans invaded, the Proto-Sards had disappeared. All that remained were their unique stone towers, shaped like truncated cones and called *nuraghi,* which means "height" or "habitation." More than 7,000 *nuraghi* dot the island, including one on Nora's site. They are thought to be defensive structures, serving as retreats for villagers during times of attack.

The sturdily built *nuraghi* outlasted many newer Roman and Phoenician buildings because they were made of basalt blocks taken from extinct volcanoes, and did not decay. Like the Egyptian pyramids, the cones were constructed by rolling stones up an earth ramp until the tower was finished and then digging out the structure. Typically, the towers measured roughly thirty feet around at the base and were constructed of two-foot mortarless blocks of stone. Inside cracks were sealed with mud. A single doorway, almost always facing south, was attached to a guard-niche. Inside, a second story was reached via a staircase frequently interrupted by defensive gaps of five feet or more. At Nora, the Romans propped up part of their aqueduct with the remains of the local *nuraghi,* by then in ruins.

When the Phoenicians arrived to colonize the promontory, they focused their efforts along the shore and did not settle very far inland. Nora may have been colonized during the height of the Phoenician sea power—around the forty-five-year era between 709 and 664 B.C. when the seafaring merchants dominated the Mediterranean.

The Greeks coined the name "Phoenician" from the Greek word for "blood-red" because the Phoenician sailors had sunburned skin. Phoenicians inhabited the coasts of Syria and Lebanon from the fifteenth century B.C. onward and were an offshoot of the Semitic peoples. They considered themselves Canaanites and their king, Sidon, was said to be descended from Ham in Genesis. Unlike the Semites farther inland, they became sailors and traders in luxury items.

Prowess at sea was useful when dealing with their powerful enemies. The Phoenicians were subjugated repeatedly by neighboring empires, including Assyrians, Egyptians, and later, the Persians. They would hire out their navy to their conquerors and thus spare themselves from slavery or destruction.

Perhaps the constant state of tyranny fueled a desire to form their own colonies elsewhere. The Phoenicians

Roman theatre at Nora
Photo courtesy of E.P.T., Cagliari

started colonies in Sicily, Africa, Malta, Sardinia, Corsica, and parts of Spain. Ties of race and sentiment ran strong in colonies. Some, like the city of Carthage, later overshadowed their predecessors. To their colonies Phoenicians brought a written language, the prototype which later became the Greek alphabet, as well as luxury goods imported from Egypt and Babylon.

Excavations at Nora revealed evidence of a wealthy city with a thriving import business: Egyptian and Greek artifacts were found including gems, terra-cotta objects, vases, glass and ivories. The Cagliari museum now displays most of these objects.

Typically narrow Phoenician streets and the foundations of buildings were excavated as well as mosaics from the period which seem primitive compared to later Roman examples.

By around 560 B.C., Nora was under the control of invaders from Carthage. From their base in Nora, the Carthaginians eyed the island's interior. Subduing the native Sards took some doing, as the cave-dwelling indigenous people were expert guerilla fighters. However, the Carthaginians prevailed and imposed harsh penalties on the natives. Remaining inhabitants were driven into the mountains and exiled from agriculturally fertile areas.

Carthage ruled the island, exporting grain from the fertile interior and mining the mountains, until the end of the First Punic War. After peace was made in 241 B.C., Rome took advantage of Carthage's weakness and demanded the island as tribute. In 238 B.C. the Roman takeover was complete, but it took several campaigns to subdue the native population. Locals mounted a strategy of brigandage and low-scale guerilla mischief, beginning a tradition that Sardinians uphold to this day.

In Nora, gruesome evidence of the turmoil remains. The Phoenicians often sacrificed children during times of great distress or to avert disaster. A *tofet,* or burial ground, for cremated children has been uncovered by archaeologists in Nora. The ashes of bodies were placed in large jars buried underneath carved stone slabs and a statue of the goddess Tanit.

Under Roman rule, Nora was the capital of Sardinia. In 27 B.C. it became a municipium. Principal structures in the city were built during the following century including a theatre, an aqueduct, private villas, baths with elaborate drainage systems, and a temple of Juno. A Roman road linked Nora to Cagliari and Sulcis.

Nora—and for that matter all of Sardinia—maintained a low historical profile. Sardinia became the grain belt of Rome, supplying harvested grain for the empire. Romans still struggled with the natives and viewed Sardinia as a backwater and a place of exile.

Over the next centuries, Sardinia was converted to Christianity, thanks to Nora's most famous citizen, St. Efisio, who was martyred by the Romans in A.D. 303. St. Efisio looms large in Sardinian legend, but of his actual history scholars are skeptical. This is partly because the saint's legend is extremely similar to that of another saint of the same era, St. Procopius. The legend of St. Procopius made its way into southern Italy when it was read at the Council of Nicaea in A.D. 787 and translated by Anastasius the Librarian.

St. Efisio's legend was recorded by a historian priest and written in a history called *Analecta Bollandiana* during the ninth century. According to the legend, the Efisio was a Roman officer during Diocletian's persecution of Christians. He was purportedly born in Jerusalem (then called Elia) and raised as a pagan by his mother, a pagan lady of nobility named Alexandria, and her Christian husband, Christopher. Alexandria took Efisio to Antioch to meet the emperor and to recommend him for service in the military. Diocletian, taken with the young man's zeal toward the heathen gods, gave him an army commission and sent him to seek out and punish Christians in southern Italy.

Efisio set out to carry out his orders, but, like St. Paul, he experienced a road-to-Damascus conversion. An earthquake shook the land in a place called Vrittania and the young officer had a vision of a hovering silver cross and the voice of Christ compelled him to speak the word of God. When Efisio arrived in Gaeta, he asked a silversmith named John to forge a cross similar to the one in his vision. By displaying the crucifix, Efisio drove away hoards of Saracen invaders and then proceeded to Sardinia. From Cagliari, the new Christian wrote to the emperor and his mother announcing his conversion.

According to legend, the angry emperor dispatched an officer called Julicus to Sardinia who imprisoned Efisio in Nora and cruelly tortured him. Julicus got his comeuppance when struck suddenly by a fatal fever, while Efisio's wounds miraculously healed. Soldiers sent to destroy the prisoner were converted and his mother, in turn, converted after arriving to visit her son.

At this, the emperor sent a second minion, a judge named Flavianus, to punish Efusio's insubordination. Torturing ensued including such ancient standbys as red-hot skewers, whipping with thongs of oxhide, burning hot coals applied to the back, wounds rubbed with salt and a rough hair-cloth. Efisio survived the ordeal calmly, refusing to retract his beliefs. Flavinanus, dumbstruck, finally sentenced him to execution by beheading.

Regardless of the legend's veracity, Efisio became the patron saint of the island. The saint's relics were hidden by the local Sards when coastal attacks by Vandals drove the population inland during the fifth century. The saint's small church was built in nearby Pula in 1089. It features a narrow nave supported by uncharacteristically wide pillars, which divide it into two aisles. Nora was abandoned after the Vandal attacks and the ruins were left untouched until the sixteenth century, when Philip II of Spain had a tower built on the isthmus which is now used as a lighthouse.

In 1647 the first festival of St. Efisio was held, and the observance continues to this day. The saint's relics are taken out and paraded from Cagliari to Pula during the festivities from May 1 to 4. Island militia members march in traditional embroidered scarlet vests, peasant men and women wear their region's costumes, musicians perform on bagpipe-like luneddas. Festivities also include feasting and breathtaking horse races through the village. The festival ends with an enormous fireworks display, the largest in Italy.

But when the festival ends, Nora returns to its quiet, craggy solitude. Other than a few visitors, the site remains frozen in time—neither alive nor dead, but somehow caught forever in between.

Further Reading: For a colorful modern travel diary, including personal reminiscences and interviews with sundry countesses and marquesses, see Virginia Waite's *Sardinia* (London: Batsford, and New York: Hippocrene, 1977). A more scholarly history of Sardinia, including guides to excavations, is found in Margaret Guido's *Sardinia* (New York: Praeger, 1963; London: Thames and Hudson, 1964). For the complete legend of St. Efisio, and hagiography in general, see Hippolyte Delehaye's *Legends of the Saints* (South Bend, Indiana: University of Notre Dame Press, 1961; London: Chapman, 1964).

—Jean L. Lotus

Ohrid (Macedonia)

Location: In the southwest corner of Macedonia on Lake Ohrid at an altitude of 2,300 feet.

Description: An ancient city, established approximately 400 B.C., originally called Lichnidos until the ninth century when Greek missionaries Cyril and Methodius began their Christianizing efforts and the city became known as Ohrid. Cyril and Methodius developed the first Slavic alphabet here and established Ohrid as a religious and cultural center. In the eleventh century, Ohrid served briefly as the capital of the independent Macedonian Empire. The city was conquered by Ottoman Turks in 1395 and transformed into a Turkish city, with numerous Turkish landmarks still in evidence.

Site Office: Fond za Razvoj na Turistička Propaganda
Sobranie na Grad Ohrid
9600 Ohrid
Macedonia
(389) 96 22 375

Ohrid is one of the oldest cities in the world and is situated on one of oldest lakes in Europe. Fifteen miles long and seven miles wide, shimmering Lake Ohrid lies in both Macedonia and Albania. Only a small portion of the much-contested region of Macedonia is an independent republic; the rest lies in Bulgaria and Greece. The language spoken in Ohrid and in the Former Yugoslav Republic of Macedonia (its official name) strongly resembles Bulgarian, but the region's past is inseparable from Greek history. The earliest inhabitants of Ohrid, the Illyrians, were completely Hellenized in culture, and the upper class spoke Greek.

Ohrid was once the capital of an independent Macedonian Empire in the eleventh century; by then, the city was already more than 1,000 years old. Its arresting beauty, healthful climate, and abundant lake fish made it attractive to settlers and conquerors alike. Like the rest of the Balkan Peninsula, of which Macedonia forms a part, Ohrid was destroyed over and over again, invaded by Greeks, Romans, Goths, Byzantines, Bulgarians, Serbs, crusaders, and Turks. In 1919, it was annexed to the Yugoslav Kingdom, which attempted strenuously to impose the Serb language and culture on its people.

Since 1990, the former Yugoslav state of Macedonia has been yet another "breakaway" republic, with Skopje as its capital. Prosperous Ohrid, the most popular tourist attraction in Macedonia, continues to draw visitors from around the world to its pristine lake and beautiful old town. Every summer since 1961, the Ohrid Summer Festival has taken place, featuring outstanding musicians, dance, and theatre. Ohrid boasts its artists' and writers' colonies, and its many museums. It is in fact a living museum city, with thirty picturesque churches, some originating in the fifth century A.D., a towering medieval fortress dating from the eleventh century, mosques, and historic Balkan homes.

One thousand years before the advent of the Slavs to the Balkan Peninsula came the Illyrians, at first a primitive tribal people who quickly fell under the influence of the Hellenic civilization. Not far from present-day Ohrid, in Trebenishte, archaeologists unearthed a burial place containing the ancient tombs of wealthy Illyrians. The objects found in the graves bear witness of the extent to which the settlers, at least the wealthy among them, identified themselves as Greek. They called their settlement by the lake Lichnidos, a Greek word. Before Philip of Macedon conquered the area and annexed it to his Macedonian kingdom in the fourth century B.C., the Illyrians had called their region Dessaretia.

The Romans easily overran Macedonia and Greece and proceeded to annex it in 148 B.C. In fact, their empire in the east was defined by the Danube River, beyond which lay wilderness and savage tribes. So long as the Roman Empire remained stable, the border was impregnable.

Under Roman rule, Lichnidos expanded, its streets were paved, and some imposing buildings were erected, as the remains of a classical theatre attest. The industrious Romans proceeded to build a paved road that ran from the imperial capital itself eastward along the Dalmatian coast, through Lichnidos to Byzantium (later known as Constantinople and İstanbul). This was the Via Ignatia, and because Lichnidos lay astride this important route, the city attracted commerce and industry.

In the twilight of imperial Rome, Christianity began to spread throughout the region, abetted by the system of roads that enabled a missionary to traverse the length and breadth of the empire. By the second century A.D., missionaries had established a foothold in Lichnidos, well before the advent of the most famous local missionaries of all, Cyril and Methodius. By the fifth century, Lichnidos was the seat of a bishopric and an attractive haven for missionaries on their way east.

Several medieval churches in Ohrid were built on ancient foundations. The cathedral of St. Sophia, built in the eleventh century, is one of these churches, with little trace of the original fifth-century basilica that preceded it. Near the Imaret Mosque lie the remains of a fifth-century church that was never used as the foundation of a later structure, and whose name has been lost. It was a large church and the richness of its surviving floor mosaics attest to the town's wealth when it was built.

Lichnidos fell squarely into the eastern portion of the Roman Empire when Emperor Constantine divided it administratively in the early fourth century, with Constantinople the

The Birth of the Virgin, twelfth-century frescoes from St. Panteleimon's monastery

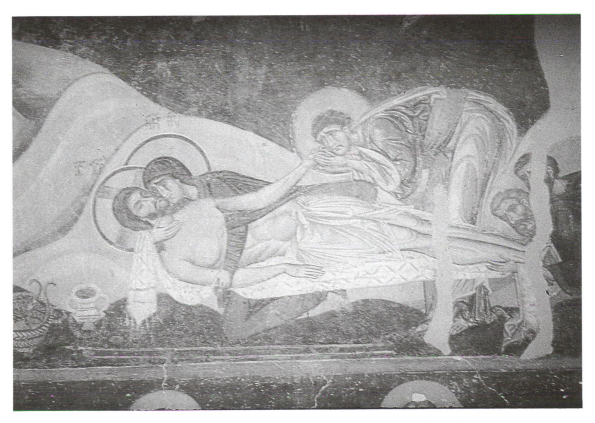

The Lamentation of Christ, twelfth-century frescoes from St. Panteleimon's monastery
Photos courtesy of Ministry of Information, Macedonia

capital of the eastern half. The two portions became more and more distinct, separated by language and cultural differences. As the Roman Empire weakened, barbarians streamed across the Danube, overrunning the Balkan Peninsula. Lichnidos was not immune from attack, despite its solid walls. Slavic tribes crossed the Danube in the sixth century, and one of these tribes, calling itself the Brsjaci, settled in the vicinity of Lichnidos.

By the ninth century, Macedonia was overwhelmingly Slavic. "Ohrid," the settlement's new name, reflected this ethnicity. "Vo hrid," or "on a cliff," were Slavic words denoting the location of Lichnidos—on the crest of a hill. A church document emanating from Constantinople in 879 is the oldest existing record of this new Slavic name. Since Constantinople was a Greek city and the writer was probably Greek, his use of "Ohrid" implies that the change of name may have occurred well before then.

In the ninth century, the missionary brothers Cyril and Methodius stopped in Ohrid on their way to the heathen Slavic peoples beyond the Danube. Among their disciples were Clement and Naum, who accompanied them all the way to Moravia. Cyril and Methodius devised the first written alphabet for the Slavic language, based on Greek characters and dubbed "Glagolitic." Eventually, rising tensions between Eastern (Orthodox) and Western (Latin) Christianity led to the expulsion of Cyril and Methodius (and their disciples) from Moravia, which to this day is Roman Catholic.

Clement left Moravia for Ohrid, arriving in 886, a year after the death of Methodius in Rome; fourteen years later, in 900, Naum arrived. Hitherto, literacy was confined to those who knew Greek (or Latin). The two men, Clement in particular, set about simplifying the Glagolitic alphabet into a version known since then as Cyrillic, after the great saint, reflecting the sounds of the Slavic language as spoken in Macedonia. This became the alphabet of the south Slavs (Serbs, Bulgarians, Macedonians). When Bulgarian missionaries carried out their Christianizing efforts in Kiev, Cyrillic also was adapted to the Slavic language spoken in that region, as well as in Russia; until the nineteenth century, Cyrillic was even the alphabet of the non-Slavic but Orthodox Christian Romanians.

The legacy of Cyril and Methodius was so profound that even in the days of strict communism in Bulgaria, their feast day was a major state holiday. The influence of Cyril and Methodius on Clement and Naum was profound. Above the grave of St. Naum is a fresco (painted hundreds of years after his death) depicting the two missionaries standing side by side next to his bier, seemingly for eternity.

Both Clement and Naum carried on the educational work of their saintly mentors in Ohrid. Clement founded a school in the monastery he established in 893, St. Panteleimon's, and the musically inclined man even organized a choir to sing the liturgical music he composed. He also undertook the monumental task of translating the Bible from Greek into Slavonic, and wrote at least one literary panegyric of St. Cyril.

Upon his arrival in Ohrid in 900, Naum established the beautiful Monastery of St. Michael the Archangel, lo-cated on Lake Ohrid (renamed Sveti Naum after his death in 910), some fifteen miles outside the city. In due course, the monastery became a center of culture and literacy. In 1802, a fire destroyed the archives and the library, but the oldest biography of St. Naum, written in the first half of the tenth century, was preserved elsewhere, in a fifteenth-century copy. St. Panteleimon's and Sveti Naum were the first monasteries in the Slavic lands, both of which trained missionaries to carry on the work of Christianizing the Slavs farther east, especially in Bulgaria and the Kievan principality. Of the two monasteries, only Sveti Naum survives to this day. Partly in recognition of the achievements of these two men, Ohrid was elevated to an archbishopric in 893, a seat it retained until 1767.

After the division of the Roman Empire into Eastern and Western portions, Macedonia remained under the hegemony of Byzantium. Linguistic and ethnic differences, rather than cultural ones, created constant tension between the southeastern Slavs and their Byzantine rulers. Even a common foe did not unify them. In the tenth century, Macedonia was conquered by the Bulgarian state. A violent revolt against Bulgarian rule in the mid-tenth century ended with the establishment of the first independent Macedonian state. The leading rebel's son, Samuil (Samuel), ascended the Macedonian throne as the kingdom's first czar in 976, and made Ohrid his capital in 1001. He ruled for thirty-eight years, steadily expanding his kingdom until it stretched as far east as the Adriatic Sea and as far north as the Danube and Sava Rivers.

Meanwhile, the Byzantine state had regrouped its forces and set out to reconquer both the Macedonians and Bulgarians. The Byzantine emperor Basil II personally took charge of the campaign, and in 1014 his army defeated the Macedonian force commanded by Czar Samuil, deliberately and cruelly ordering Samuil's defeated army to be blinded. Just that year, Ohrid had been made the new imperial residence of the czar, with an impressive new fortress built on the highest ridge overlooking Ohrid. Samuil never took up residence there, having apparently committed suicide. The conquering Byzantine army set fire to the fortress and razed the city walls. Byzantium loosely controlled Macedonia until 1334. Weakened by Moslem conquests of its territories, Byzantine rule over Macedonia gave way to the control of the neighboring Serbs, who took advantage of the political void, until they, too, succumbed to Moslem conquest in 1389.

Despite the frequent turmoil, the city of Ohrid attained its cultural apogee in the Middle Ages, thanks in part of the legacy left by its famous missionaries. The monastery school of St. Panteleimon became renowned for its copying and illustrating of religious books. Churches were also built in the town and filled with valuable frescoes and icons. In the tumultuous eleventh century, the cathedral of St. Sophia was constructed on the foundation of a former basilica. In the early 1950s, reconstruction and restoration of St. Sophia's was completed. Layer upon layer of whitewash was removed to reveal some of the finest frescoes in eastern Europe, among them, a whole gallery of female saints. The earliest known portraits of Cyril and Methodius are among the precious surviving paintings.

The little church of St. Nikola Bolnichki nearby was founded by the Serbian czar Dushan in the fourteenth century; its western facade depicts him, his wife, Jelena, and their sons St. Sava and Simeon. St. Sava went on to found the Monastery of Hilandar at Mount Athos; his father, Czar Dushan, eventually relinquished the throne and joined his son there as a monk. The frescoes of Czar Dushan and his family were the last to be painted in Ohrid before the Turkish takeover. Right before this happened, St. Clement's remains were transferred from St. Panteleimon to the Church of Bogorodica Perivlepta (the Virgin of Perivlepta), renamed St. Clement's. This church originally was built in 1295 on the foundations of a far more ancient one. Within is a separate gallery of Ohrid icons, a property of the national museum, containing over 100 icons, some created in the twelfth century.

Another landmark is St. John the Divine's, situated in the little fishing village of Kaneo, a few miles from Ohrid. The church and its breathtaking location on a jagged cliff overlooking the lake are often depicted on postcards and in guide books. It was constructed of baked brick in the thirteenth century, and not a single record survives of its history. Remarkably and quite inexplicably, it possesses a portrait fresco of the Renaissance humanist and Roman Catholic scholar and priest, Erasmus of Rotterdam. Besides the portrait, there are many icons that were uncovered during the 1963–64 restoration of St. John's. Although it suffered much damage from fire and neglect, it is among Ohrid's most distinctive religious monuments.

Because of Ohrid's location on a major commercial route, many outsiders were attracted to the city and settled within its walls. Much of the city's trade and commerce in the Middle Ages was in the hands of Jews, who had lived in the area since pre-Christian times (Greek-speaking Jews were referred to as "Romanioti"), and whose numbers augmented during the Middle Ages. Many Yiddish-speaking Askenazi Jews from western Europe, primarily Germany, fled persecution and found somewhat greater tolerance and acceptance in the Balkans. In the fifteenth century many Spanish-speaking Sephardic Jews fled to Macedonia to escape the inquisition in Spain. From the twelfth through fourteenth centuries, Ohrid flourished as a Jewish cultural center of some note, where Jewish scholarship and an active religious life were carried out. Among the most famous names from this period were the rabbi Judah Moskoni, and the linguist, traveler, doctor, and author, Moshe Moskoni. There was at least one synagogue in Ohrid, which survived intact into the nineteenth century.

The conquest of Ohrid by Ottoman Turks in 1395 undoubtedly played some part in the Jewish community's success. Despite their reputation for ethnic and religious intolerance, the Ottomans welcomed Jews to their empire, and the new arrivals found safety, security, and many opportunities hitherto denied them. Ironically, it was the further expansion of such opportunities that ended Ohrid's role as a Jewish cultural center. When the Turks conquered Constantinople in 1453, the Jews began to leave the small cities and towns of the Balkans for large urban areas. By 1930, there were only eleven Jews remaining in Ohrid. Today, the presence of this once-vibrant community and its contributions to the city's commercial and cultural life are largely forgotten.

Ohrid remained under Turkish rule until 1912. Although all non-Moslems were considered second-class citizens by the Ottoman Turks, interference in the domestic affairs of the non-Moslem provinces of the empire was kept to a minimum. Life in multi-ethnic Macedonia went on as before.

Nonetheless, the Turks discriminated blatantly against non-Moslems. The oldest and largest surviving church in Ohrid, St. Sophia, was in due course transformed into a mosque; the same fate befell the ancient monastery of St. Panteleimon's. With liberation from Turkish rule in the twentieth century, strenuous efforts were made by the Macedonian and Yugoslav governments to turn back the clock and reconvert these Turkish monuments to Christian structures. By this time, there was a sizable Moslem minority, who were now discriminated against.

When Turkish rule came to an end, Macedonia was partitioned among Greece, Bulgaria, and the newly established Yugoslav kingdom. Ohrid was in that part of Macedonia dominated by the Yugoslav Serbs. From 1919 until World War II, the Yugoslav government made every effort to eradicate the Macedonian language and culture, which it referred to as "south Serbian," while Macedonia was designated as "southern Serbia."

After World War II, ethnic conflicts and identities were suppressed and submerged in the name of state socialism, although Macedonia was given autonomy and republic (or state) status. In fifty years of communist rule within the former Yugoslavia, Ohrid was transformed from a poor agricultural backwater to Macedonia's most popular tourist attraction.

It is somewhat ironic that the former communist regime of Yugoslavia took the greatest care to restore Ohrid's Christian heritage. In a region where the church was identified so closely with national feeling and national identity, this solicitude is surprising. There was no comparable attention paid to non-Christian monuments, such as the centuries-old synagogue and the Turkish buildings.

Thus far, even in the independent Macedonian state of the 1990s, only the Christian, Slavic part of Ohrid's history has been emphasized. However, the visitor will discover that many foods, beverages, buildings, and even customs and clothing are legacies of the centuries of Turkish rule. Today, one can view the famous Imaret (or Imperial) Mosque, standing on the site of the former St. Panteleimon monastery. Like the monastery, the mosque encompassed an entire compound consisting of a house of worship, a hostel, and a soup kitchen for the poor of every ethnic background. A Moslem dervish monastery also was built in Ohrid, and the tomb of one of its most distinguished and revered abbots has survived in the courtyard. The multicolored Ali-pasha Mosque graces the center of the city.

Further Reading: All guidebooks focusing on eastern Europe adequately cover Ohrid, although as of this writing, none focuses on Ohrid in particular. In 1977, a fifty-page, illustrated guidebook on Ohrid appeared in translation entitled *Ohrid: A Guide to the City and Its Surroundings* (Ohrid: Biljana Tourist Society, 1977). It is extremely detailed, but biased (not a single mention is made of any minority group's influence, unless it is negative). Readers interested in Macedonia's historical context should consult R. Malcolm Errington's *A History of Macedonia* (Berkeley: University of California Press, 1993). A fascinating little-known aspect of Macedonian history is explored by Alexander Matkovski in *A History of the Jews in Macedonia* (Skopje: 1982), with numerous references to Ohrid.

—Sina Dubovoy

Olympia/Elis (Elis, Greece)

Location: Olympia is located in the western Peloponnese region in southern Greece, on the north bank of the Alpheus River. Elis is located some thirty-five miles to the northwest. Both are located in the modern department of Elis.

Description: The site of both the Olympic games held in ancient Greece and the most important sanctuary to the god Zeus, Olympia was an important center of Greek culture. The nearby city of Elis oversaw the games held at Olympia and was a complex of stadiums and temples rather than a city in its own right.

Site Offices: Olympia Archaeological Museum
Olympia, Elis
Greece
(624) 22529, 22742, or 22717

Museum of Olympic Games
Olympia, Elis
Greece
(624) 22544

Olympia's early history is lost in legend. Archaeological findings dating back to 1500 B.C., mostly ruined stone buildings, have been discovered on the site. Legend claims that the first sanctuary built at Olympia was called Gaeon and was dedicated to Gaea, the Earth deity. Later, Olympia became the focus for the worship of the god Cronus and an altar was constructed on the Hill of Cronus where the faithful would offer prayers and sacrifices to the deity. Cronus was in turn replaced with Zeus and the other familiar deities of the Greek pantheon.

The origin of the Olympic games themselves is also found in legend. Several conflicting traditions have come down to modern times. The first holds that Zeus was raised by five brothers, known as the Kourites, in a small cave, named Idaion Andron, near Olympia. The Kourites held athletic contests among themselves to see who was best at wrestling, boxing, and other sports. Victors received wreaths made from a wild olive tree. Soon the gods became interested in competing in these contests, too, and so the games were more formally organized. Every five years, to honor the five original Kourite brothers, games were held at Olympia. During these early contests Zeus himself was said to have bested Cronus at wrestling, while Apollo beat Hermes in the foot race.

A second tradition traces the competition's origins to Hercules. After cleaning stables for Augeas, the king of Elis, Hercules was angered when Augeas refused to hand over some oxen, as promised. In retaliation, Hercules destroyed the city of Elis. He then traveled to Olympia to pay tribute to Zeus and organized a series of athletic contests to celebrate

his victory. A third version holds that the games began following a victory by Pelops over Oenomaus, the king of Pisa, in a chariot race presided over by Zeus. By winning the race, Pelops won the hand of Oenomaus's daughter, the princess Hippodamia.

Whatever their origin, athletic contests were held at Olympia in honor of Zeus and the other Greek gods for many centuries before Greece entered the historical record. What is known for certain is that the games were reorganized by King Iphitus about 884 B.C. Seeking a solution to the constant civil wars that threatened the stability of the Greek city-states, Iphitus consulted the Delphic oracle. Through the oracle, Apollo told Iphitus that to bring the warring Greeks together, the Olympic games of legend must be revived. Therefore, Iphitus, Lycurgus, the leader of Sparta, and Cleisthenes of Pisa signed a treaty establishing the Olympic games and declaring that Elis and Olympia were sacred sites. No armed men were allowed to visit either location and, during a three-month period before and after the Olympic games, a truce existed between all the Greek city-states.

This Sacred Truce, as it came to be called, held great power over the otherwise independent Greek city-states. The Olympic Parliament was empowered to levy fines on all those who violated the Sacred Truce and, if these fines were not paid, the Delphic oracle would no longer answer that city's questions and the city's athletes could not participate in future contests. The Sacred Truce of the Olympic games became an almost religious observance among the Greeks. When the Sacred Month of the Games was announced by the Spondophoroi, or traveling heralds, all inter-city strife was ended and the Greeks turned their attention to the athletic contests at Olympia. Once, during the Peloponnesian Wars, Sparta violated the Sacred Truce and was heavily fined by the Olympic Parliament.

The Olympic games were held every forty-ninth and fiftieth months during the full moon of midsummer, or during the current months of July and August. The period of the full moon was chosen so that the games could last into the night. The four-year period between games was called an Olympiad, and each of the Olympiads was named for an athletic victor at the previous contests. The Olympiads were commonly used by the Greeks as a way to establish historical chronology. Greek historians placed important events in time by noting the Olympiad in which they occurred.

The first recorded winner of an Olympic contest appeared in 776 B.C. Coribus from Elis won a 200-meter foot race. To honor him, Coribus's name was given to the first Olympiad. From that point on, a list of victors was kept for the historic record, causing many historians to date the beginning of recorded Greek history to this date.

At first, the games at Olympia were overseen jointly by the two nearby cities of Elis and Pisa. The growing

Remains of the Temple of Zeus at Olympia
Photo courtesy of Greek National Tourist Organization

importance of the games throughout the Hellenic world, however, soon drove the two cities into sometimes violent confrontation over this arrangement. About 472 B.C., this rivalry ended with the destruction of Pisa and the complete takeover of the Olympic games by Elis.

Elis not only controlled the Olympic games but the religious sanctuary at Olympia as well. The sanctuary's priests were traditionally chosen from among the members of Elis's most prominent families. The ten members of the Olympic games jury also resided in the city.

Strict rules governed the Olympic games. Contestants had to be Greek citizens not accused of murder or sacrilege. Training had to begin in an athlete's home city a year before the games and, one month before the games, athletes were obliged to move to Elis or Olympia for final training.

Spectators to the games were held to standards as well. They had to be free men, not convicted of a sacrilege. Women were not allowed to watch the nude contests between men, although some historians believe that maidens could attend. Any woman who violated this rule could be thrown from nearby Mount Typaion, although there is no record of this punishment ever being meted out. One woman, wishing to see her son compete in a foot race, reportedly slipped into the games in disguise. When her son was victorious, she ran out to him, revealing her secret. Because of her family's prominence in the games—her husband, three brothers, nephew, and son had all won wreaths over the years—she was pardoned for her offense.

The Olympic festivities drew people from all over Greece, who saw the games as a chance to sell their goods, recite poems for money, negotiate business deals, or even sign a peace treaty with a neighboring city. Since Olympia had limited permanent residential quarters, the throngs of visitors slept in the open air or set up tents for themselves. At night there was feasting as musicians played their instruments and poets recited the epics.

The early Olympic games consisted of a simple foot race, but over the centuries the games became more numerous and complicated. At their peak, a total of fourteen contests were held over a four-day period. Two classes of games were held at Olympia: those physical contests held in the stadium where the athletes competed in the nude, and those on the racetrack with chariots. The original foot race measured nearly 200 meters. In 724 B.C., a foot race twice that length was added to the games, and in 720 B.C. a race twenty-four times the length of the stadium was created. The pentathlon was introduced in 70 B.C.

This consisted of a series of five contests: distance jumping, throwing the discus, throwing the javelin, a foot race, and wrestling. Additional contests included boxing in 688 B.C.; the pankration, a combination of wrestling and boxing events, in 648 B.C.; and foot races for oplites, or those in full armor, in 520 B.C. Competitions for boys were also added: boxing in 616 B.C. and the pankration in 200 B.C.

Olympic chariot racing also grew more elaborate as time passed. In either 740 B.C. or 680 B.C. (Greek historians differ on the exact date) a four-horse chariot race was introduced. In 648 B.C., single-horse races began, and in 384 B.C., chariot races with four colts.

The games began with a procession from Elis in the evening. The judges (known as Hellanodicae), athletes, and trainers passed from Elis to the fountain of Piera to offer sacrifices. Piera marked the boundary between Elis and Olympia. From there they traveled via the Sacred Way to Olympia. Thousands of spectators from every Greek city-state greeted them with cheers of support. The Theoroi, officials from the city-states participating in the games, sat in reserved seats from which they could observe the contests.

The first day of the games began with sacrifices to the gods, for the games were meant as religious tributes. Then the athletes vowed on the Great Altar of Zeus that they were eligible to participate in the games and that they would obey the Olympic rules while competing. All judges, trainers, and even the athletes' parents made similar pledges. Trumpeters, who announced the beginning of races with a blast of their horns, and heralds, who introduced athletes by calling out their names to the crowd, were chosen on the opening day of the festival.

The second day of the games began early with the chariot races, followed by the horse races. Up to forty chariots participated in a nine-mile race around two stone posts set in the ground of the hippodrome, or race course. The chariot race was a free-for-all; few of the starting charioteers made it to the finish line. The turns at each post were so sharp to make, especially at high speed, that chariots routinely toppled over or crashed into one another. Because the olive wreaths of victory were given to the owner of the chariot and horses and not to the driver, it was not uncommon for one man to have several chariots entered in the same race to increase the likelihood of winning. Following the races, the spectators then hurried to the stadium to watch athletes compete in the pentathlon. These contests lasted until evening and marked the end of the first day of the festival.

The third day of the Olympic games began with a sacrifice to Zeus, the most solemn moment of the entire festival. After a procession to the Great Altar of Zeus, the assembled priests slaughtered a herd of 100 bulls. The bulls' thighs were burned as a sacrifice while the rest of the meat was used in the victory banquet at the end of the games. The boys' events took up the remainder of the third day.

On the fourth day the various foot races were held, as well as the wrestling and boxing events, and the pankration. This day marked the end of the athletic events of the Olympic festival.

On the fifth day the victory banquet was held. It began with a procession to the Temple of Zeus, where each winning athlete received his wreath of olive branches. When the athletes emerged from the temple, the crowd showered them with flowers. At the banquet, meat and wine were feasted upon until late at night. The following morning, the athletes, trainers, and their families left for home.

Victors in the Olympic games won far more than wreaths from a wild olive tree. Statues were erected in their honor in the Altis area of Olympia. In their hometowns, victors were paraded in a magnificent chariot, songs and poems were written in their honor, they were given choice seats at all public spectacles, statues were carved in their image and placed in prominent locations in the city, and they were exempted from paying all taxes. In Athens and several other cities, generous cash rewards were also common. In some Greek cities, a section of the city wall was torn down and the victorious athlete led in through the opening. This ritual signified that any city with strong citizens had no need to defend itself from enemies with a wall. So great was the renown of Olympic victors that even politicians such as Alcibiades of Athens entered the chariot races in order to win popular support.

One sign of Olympia's importance in the Hellenic world is the profusion of buildings, statues, and other archaeological remnants found at the site. Perhaps the most impressive of these structures is the Temple of Zeus. Some 90 by 210 feet in size, the temple was designed by the architect Libon of Elis and built from the spoils Elis won after it defeated and destroyed Pisa. In fact, much of the building material came from the ruins of Pisa. Dedicated in 457 B.C., the temple is decorated with carvings depicting Zeus and his children. In one such sculpture, Zeus oversees the chariot race between Pelops and Oenomaus. In another, Zeus's son Apollo mediates a peace between the Lapiths and the Centaurs, warring factions in a great battle. Inside the temple, the twelve labors of Zeus's son Hercules are depicted. The most stunning artwork, now lost, was the central statue of Zeus himself by the renowned Greek sculptor Phidias. Made of chryselephantine (gold and ivory) and standing some forty feet high, the statue depicted Zeus seated on a regal throne and holding a scepter in his left hand and the figure of Victory in his right. So beautiful was this statue that it was considered to be one of the Seven Wonders of the ancient world.

Other buildings at Olympia included the Bouleuterion, the meeting place of the Olympic senate; the Heraion, a temple dedicated to the goddess Hera; and the Leonidaion, a residence for visitors to Olympia. The Altis, a walled open area about 650 by 520 feet in size in which the temples of Zeus and Hera were located, also housed a number of treasuries, or small temples, built by individual Greek cities. Olympia's stadium, capable of seating 40,000 spectators, surrounded an open grassy space with stone markers to show the starting and finishing lines.

Following the battle of Cheronia in 338 B.C. and the conquest of the Greek city-states by Alexander the Great, the

Olympic games greatly expanded as the Macedonians began sending athletes to the events. The Macedonians also built the Philippieion at Olympia, a round central building surrounded by eighteen Ionic columns and housing sculptures by Leochares. The building commemorated the victory of Alexander's father, Philip II, over Athens and Thebes.

The widespread popularity of the Olympic and related Greek athletic festivals, and the numerous benefits to the victors, eventually forced out the amateur athletes the original games were meant to attract. Young men became professional athletes who specialized in a particular event. The development of body, mind, and spirit that the original games hoped to encourage was forgotten. By the fourth century B.C., Sparta forbid its citizens to engage in specialized, and thus unbalanced, training. Such training, it was argued, produced bad soldiers. Consequently, Sparta's citizens ceased competing in the Olympic games.

The Roman conquest of Greece in 146 B.C. marked the beginning of the decline of the Olympic games. During the reign of Nero, from A.D. 54 to 68, the overbearing emperor entered all the contests and was repeatedly crowned the victor in each of them. Although he twice fell off his chariot, which was pulled by ten horses instead of his opponents' four, Nero easily won the rigged chariot race as well. New games were added for his benefit, such as singing and musical competitions, so that the vain emperor could show off his negligible skills. In A.D. 67, eager to win yet more honors, Nero organized the games two years earlier than scheduled.

In A.D. 394, Theodosius I abolished the Olympic games altogether. He saw them as a pagan relic that Christian Rome did not wish to continue supporting. The great statue of Zeus was taken from its place at the temple at Olympia and transported to Constantinople, where it was later lost in a fire. In 420, Theodosius II burned the Temple of Zeus. Earthquakes during the sixth century destroyed many of the remaining structures. Centuries of landslides from the nearby Hill of Cronus and periodic flooding from nearby rivers eventually buried Olympia in twenty feet of mud.

Olympia lay in ruins, virtually forgotten, until the early nineteenth century when France sent troops to the area to help Greece gain independence from the Turks. The French force had a scholarly contingent, known as the Expédition scientifique de Morée, which began excavations at the foot of the Hill of Cronus in hopes of finding some remains of ancient Olympia. They soon discovered the metopes, stone carvings, from the Temple of Zeus depicting some of the twelve labors of Hercules. These metopes are now housed in the Louvre. Later French excavations were conducted under an agreement with the Greek government that any artifacts uncovered would remain in Greece.

In 1875 a new dig was organized at the site, this time by German archaeologists from the German Archaeological Institute in Athens. Chancellor Otto von Bismarck had urged the establishment of the institute as a way to increase German prestige in classical scholarship, and a similar institute had been founded earlier in Rome to excavate Italian sites. The Greek project was led by Ernst Curtius, a German archaeologist who had long wanted to dig at Olympia. After twenty-three years of effort, Curtius finally convinced the German government and Crown Prince Frederick to finance his project.

Between 1875 and 1881, the dig at Olympia unearthed most of the site's complex of temples, city treasuries, and outbuildings. Among the structures excavated were the Temple of Zeus, the Temple of Hera, and the Philippieion. Some 130 sculptures, 1,000 terra-cottas, and 13,000 bronze objects were also uncovered. After the project was completed, a five-volume description of the findings was published.

The rediscovery of Olympia inspired many Western Europeans to discuss a revival of the Olympic games of old. The leading spokesman for this idea was the French baron Pierre de Coubertin. On June 16, 1894, Courbertin presided over the International Congress of Paris, a meeting of representatives from thirteen nations interested in reviving the Olympic games. From this meeting came an international agreement to hold the first modern Olympic games in Athens in 1896. Since that time, the games have been held, except for wartime interruptions, on a regular basis in cities around the world.

The 1936 Olympic games, held in Nazi Berlin, were the first to feature a torch lit at Olympia and carried by runners to the games site. This has since become a symbolic ritual at every Olympic competition. As a propaganda gesture, Nazi leader Adolf Hitler donated a large sum to reopen excavations at Olympia, especially the unearthing of the original stadium. During World War II, when Nazi troops occupied Greece, German archaeologists worked on this excavation project. Following the war, this massive task was finally completed by the German School of Archaeology in 1960.

Today, Olympia is the site of two museums, the Museum of Olympic Games and the Olympia Archaeological Museum. Housed in the museums are collections of sculpture dating to the fifth century B.C., metopes and pediments from the Temple of Zeus, and works by the Greek sculptors Paionios and Praxiteles. A large collection of bronze armor and weapons, including the helmet worn by Miltiades, an Athenian general who fought the Persians, is also on display. A monument to Courbertin, founder of the modern Olympics, stands near the International Olympic Academy.

Further Reading: *Olympia: Brief History and Complete Archaeological Guide* (Olympia: Sp. Fotinos, 1971), translated into English by Evi Pawloff-Valmas, provides a detailed look at the archaeological remains at Olympia today. *The Olympics: A History of the Games* (Birmingham, Alabama: Oxmoor House, 1992) by William Oscar Johnson gives an overview of the modern Olympics festivals as well as a brief account of the ancient Greek games. *Olympic Games in Ancient Greece* (New York: Harper, 1976; London: Harper, 1978) by Shirley Glubok and Alfred Tamarin provides a detailed description of the athletic contests for younger readers. *Discovering the World of the Ancient Greeks* (New York: Facts on File, 1991) by Zofia Archibald contains a useful summary of the German excavations at Olympia and a guide to the site today.

—Thomas Wiloch

Orvieto (Terni, Italy)

Location: On the western edge of the Umbrian region of central Italy; on an enormous square rock 640 feet above the junction of the Paglia and Chiana Rivers; fifty-two miles from Perugia and sixty-eight miles from Assisi.

Description: An ancient town first settled by the Etruscans; known primarily for its thirteenth-century cathedral built to house the relics from a miracle that occurred in the nearby town of Bolsena. Orvieto is also known worldwide for its wine and pottery.

Site Office: A.P.T. (Azienda di Promozione Turistica dell'Orvietano)
Piazza Duomo, 24
CAP 05018 Orvieto, Terni
Italy
(763) 41772

Orvieto stands on the site of an early Etruscan settlement called Volsinii. The Etruscans arrived in Italy from Asia Minor sometime between the tenth and eighth centuries B.C., building sophisticated towns with streets often as wide as fifty feet across and paved with stones. Arches and barrel vaulting characterized the homes, which also featured water pipes, fountains, and sewers. The region was rich in minerals and the Etruscans mined the earth for copper, tin, lead, and zinc. Expert craftsmen, the Etruscans developed a flourishing export trade in bronze figurines.

The town's location, on the surface of a rock 640 feet above the ground, provided a natural defense against marauders. Nevertheless, the Romans were able to reach Volsinii in 283 B.C., kill its inhabitants, and destroy its buildings. The remains of Etruscan tombs with their characteristic wall frescoes and terra-cotta portraits can still be found in Orvieto.

The Romans rebuilt the town and named it Orvieto. One popular legend holds that Orvieto took its name when a tribe of barbarians attempted to cart off a cache of metal chalices. The cups suddenly turned to gold, prompting the thieves to exclaim "aurrum vetitum," which means "forbidden gold." Another story, which likely originated with the area's wine growers, links the Latin name with the golden color of the region's wine.

In the early Middle Ages, Orvieto was the seat of a Lombard duchy and a Tuscan countship. It grew into a bustling market town inhabited by farmers, shopkeepers, and artisans. Although primitive in comparison to the cities of northern Italy, the towns of Umbria did function as cooperatives. The residents shared the rights to surrounding pastures and forests, and they took turns guarding the walls and providing for their upkeep.

After Holy Roman Emperor Charles III (the Fat) was deposed in 887, anarchy and civil wars prompted many people to seek refuge in walled cities such as Orvieto. The influx of people brought an increase in trade, and merchants acquired wealth and authority. Craft guilds, run by the various families of the town, ran the government on an ad hoc basis. However, the lack of a formal government left the town vulnerable to frequent shifts in power. Competing families used force and alliances with supporters to gain control. In spite of the feuding, by the twelfth century, Orvieto had achieved the status of an independent commune.

The town's alliance with Rome remained strong. When the struggle for power between the Guelphs, who supported the pope, and the Ghibellines, who supported the German Holy Roman Emperors, intensified in the thirteenth and fourteenth centuries, Orvieto joined with Florence, Montepulciano, and Bologna on the side of the Guelphs. It was a rivalry fueled by passion and anger, resulting in thousands of deaths. The larger city-states of Florence, Milan, Naples, and Venice took possession of smaller cities in their efforts to acquire the upper hand, although none was strong enough to conquer the others. A few smaller cities, like Orvieto, were able to survive by joining in strategic alliances. In fact, cities frequently changed sides, fighting or aligning with their neighboring cities as it suited their needs. In 1448, Orvieto became part of the Papal States, a territory that divided northern and southern Italy.

The Palazzo dei Papi was built in the thirteenth century as a summer residence for the popes to escape the heat of Rome. The popes also occasionally used Orvieto to hide from their enemies. Pope Clement VII, for example, retreated there in 1527 when the armies of Holy Roman Emperor Charles V invaded Rome. It was from the papal palace in Orvieto that same year that Clement issued his refusal to allow England's king Henry VIII to divorce Catherine of Aragon. This led to Henry's break with Rome and the founding of the Church of England.

In 1263 an event occurred that led to the construction of Orvieto's most famous landmark. A Bohemian priest in the town of Bolsena, southwest of Orvieto, had a remarkable experience after he expressed doubts about his faith. The priest had questioned the veracity of transubstantiation, the belief that the body and blood of Christ are transmuted to bread and wine once they have been consecrated. One day, while preparing the Eucharist in the crypt of the Church of St. Cristina he saw that drops of blood were falling onto the altar cloth. This event was recognized as a miracle by the Roman Catholic Church, which commissioned a cathedral to honor the event and to provide a depository for the relics. Pope Nicholas IV laid the cathedral's cornerstone in 1290.

Orvieto's Duomo
Photo courtesy of A.P.T., Orvieto

Orvieto's Cathedral, known in Italian as the Duomo, was constructed over three centuries and was finally completed in 1600. However, some of the mosaics on the facade were installed as late as the nineteenth century. The Duomo is a spectacular example of the Romanesque-Gothic architectural style. More than 100 architects, sculptors, painters and mosaic artists contributed to the project. The facade was designed by Lorenzo Maitani, a native of Siena. On the lower parts of the pillars Maitani created reliefs that depict scenes of the Last Judgment and hell in grisly detail. Although the scenes of paradise are somewhat more serene, they are also remarkable for their vividness. These scenes recall the penchant of the ancient Etruscans for portraying religion as a fearsome force.

Andrea Pisano, Andrea di Cione (Orcagna), and Michele Sanmicheli also contributed to the cathedral's facade. The rose window is the work of Orcagna. Inside, frescoes in the Capella Nuova, a chapel near the High Altar, were created by Fra Angelico and depict the Prophets and Christ in Glory. The chapel is also noted for another striking scene of the Last Judgment, this one done by Luca Signorelli. Opinion varies on the possible influence of Signorelli's Last Judgment on Michelangelo's work in the Sistine Chapel: many critics believe that Michelangelo drew on Signorelli's work in the Duomo, while others point out that Michelangelo called Signorelli's writhing figures "sacks stuffed full of nuts."

In the Capella del Corporale the relics of the Miracle of Bolsena are kept in a gold and enamel reliquary. The cloth and the Host are displayed on the Feast of Corpus Christi, instituted by Pope Urban IV after the miracle occurred, and at Easter. Since 1337 Orvieto has held a parade each June to celebrate the Miracle of Bolsena. A pageant featuring costumes of the fifteenth century was added to the festivities in 1950.

Another notable Orvieto structure, the Church of San Domenico, was built between 1233 and 1264 and holds the thirteenth-century tomb of Cardinal de Braye. The tomb was created by Arnolfo di Cambio, a favored sculptor of popes and princes. Encrusted with colored marble and inlaid mosaics, the tomb is noted for the two life-like deacons that Arnolfo sculpted in the act of drawing curtains to shield the cardinal's body from view.

During the Renaissance, Orvieto absorbed villages and countryside until its boundaries reached south to Lake Bolsena, west to the coast, east to the Tiber River, and north to Lake Trasimeno. Neighboring cities, such as Siena and Perugia, also were trying to grab as much land as possible. The men of Orvieto were ready at any time to abandon their work and rush into battle. When necessary, the independent Umbrian cities stopped fighting long enough to join with the centrally located town of Perugia to protect the Via Apruntina, the region's major trade route, against the cities located on the Old Roman Way.

But, as before, the town continued to prosper in spite of the frequent skirmishes. Farmers harvested apples, cabbages, olives, grain, and grapes for wine. Ironworkers forged armor and weapons for the armies and intricate railings for the homes of nobility. The town's carpenters became renowned for their wooden toys.

Today, Orvieto's most famous craft is pottery, an industry that began with the Etruscans. In the thirteenth century, Orvieto was the center for the production of majolica, a tin-glazed earthenware. The Gothic-style pots and urns feature designs with a Middle-Eastern quality. Manganese and copper are used to create the predominant colors of purple and green.

The clay for Orvieto pottery is still mined from the nearby rivers and then stored in the Etruscan-era cellars. These cellars once connected each house with stores for food and water when the town was under siege. In the sixteenth century, Pope Clement VII de' Medici was concerned that the town's water supply would be cut off during an invasion so he ordered the digging of a well. Designed by Sangallo the Younger, St. Patrick's Well is 203 feet deep. It boasts two concentric staircases flanked by seventy-two windows. Sangallo's ingenious design allows the staircases to spiral without ever meeting.

Orvieto's reputation for producing white wine is as old as the town itself. Indeed, Signorelli requested that part of his payment for work on the Duomo be made in wine. The volcanic tufa stone that provides the base for Orvieto is rife with caves that are used to ferment the region's Trebbiano grapes. The wine is still made in the original manner, without chemicals.

Other noteworthy aspects of Orvieto include such architectural landmarks as the eleventh-century Church of Sant'Andrea, the thirteenth-century People's Palace, and a fourteenth-century fortress that has been converted to a public garden. The former Palazzo dei Papi now houses the Archaeological Museum. The Church of San Francesco, built in 1266, is notable for the tomb of Boniface VIII, the pope who canonized Louis IX of France.

Further Reading: Although Orvieto is rich in historical significance, it usually only merits short chapters in travel guides on central Italy. Of these, *Italy: A Travel Survival Kit* by Helen Gillman and John Gillman (Hawthorn, Australia, Berkely, and London: Lonely Planet, 1993) and *Italy At Its Best* by Robert S. Kane (Lincolnwood, Illinois: Passport Books, 1989) give the most lively descriptions of Orvieto's charms. Elizabeth Lambert's article entitled "An Insider's Orvieto" in *Architectural Digest* (Los Angeles), January 1994, is also interesting for its discussion of Orvieto's artisans.

— Mary F. McNulty

Ostia Antica/Portus (Roma, Italy)

Location: Ostia Antica, or Roman Ostia, used to lie at the mouth of the Tiber (the town's name was derived from the Latin word *ostium* meaning "mouth"), on the west coast of Italy, and is some fifteen miles southwest of Rome. The old Roman road, the *Via Ostiensis*, used to be the major artery connecting the two. A little to the north of Ostia stands the site of Portus, which was linked to Rome by the *Via Portuensis*. Since the shore has advanced more than two miles since Roman times, the modern Ostia Marina and Lido di Ostia are to be found on the coast, the Lido now a popular resort close to Rome, often called the Lido di Roma.

Description: A fifteenth-century castle and a little village comprise today's Ostia, close to the vast expanse of ruins known as Ostia Antica, which truly looks like a great town destroyed. The remains of innumerable buildings reach into the distance, several with magnificent mosaics. Between 42 and 106 A.D., two new ports were built slightly to the north of Ostia; this location, which became known as Portus, gradually acquired more importance than Ostia. Remains of the quays of the Portus harbor were found during the construction of Rome's international Leonardo da Vinci airport, which is situated here. Vestiges of a great cemetery serving Portus lie to the south of the town in an area known as the Isola Sacra, made into an island by the cutting of a canal.

Contact: A.A.S.T. (Azienda Autonoma di Soggiorno e Turismo)
Viale dei Volsci, 8
CAP 00049 Velletri, Roma
Italy
(6) 9630896 or 9633367

Two legends obscure the evidence of the real beginnings of Ostia. First, Roman writers including the epic poet Quintus Ennius were claiming even in the third century B.C. that Ostia had been established as the first Roman colony by the fourth legendary king of Rome, Ancus Marcius, in the seventh century B.C. Second, the poet Virgil wrote that Ostia was founded at the place where Aeneas, hero of the *Aeneid*, landed after his adventures following the Trojan War. In addition, the historian Livy relates that the Romans defeated the Veii to win the Maesian forest, thereby extending their control to the coast; they later, according to Livy, established a settlement at the mouth of the Tiber to maintain and protect a supply of salt for Rome. No archaeological evidence has been unearthed to prove any of these versions.

The earliest archaeological evidence for a settlement in Ostia comes from the remains of a rectangular Roman fort, called the *castrum*, dated to the mid-fourth century B.C. This structure appears to be roughly contemporaneous with a number of others along the coast of Latium and to have formed part of a new line of Roman defenses. In the middle of the century, Gauls and Greeks were plundering the area, and in 353 B.C. a force from the southern Etrurian town of Caere even raided Roman territory. Russell Meiggs, in his book *Roman Ostia*, speculates that the Ostian *castrum* may have been erected in rapid response to Roman failure to fend off a Greek attack in 349 B.C. Other historians have suggested the *castrum* dates from between 338 and 317 B.C. While Antium (modern Anzio) to the south is reckoned to have been established as a colony in 338 B.C., Ostia is commonly referred to as Rome's first colony.

The *castrum* at the mouth of the Tiber covered about five and one-half acres and defended both coastline and river. Roman commerce was already expanding: trade agreements with Carthage dated to the middle of the fourth century B.C. The site of Ostia was probably also establishing itself as a trading post. Rome, at this time beginning to organize a naval fleet, in 311 B.C. appointed naval overseers to Ostia.

Evidence of Ostia's greatly increased importance to the Romans came in 267 with the appointment of one of four Roman *quaestors* to manage the Ostian fleet. The First Punic War against Carthage, from 264 to 241 B.C., involving the fight for Sicily, led to the creation of a powerful Roman navy. By the end of the war, Ostia harbored a permanent fleet, which would come into action in the Second Punic War from 218 to 201 B.C. against the Carthaginian general Hannibal.

By this time, Ostia was no doubt Rome's main naval base. Fleets arriving at and departing from the area are recorded through the period. General Scipio Africanus Major's soldiers left Ostia with supplies to go fight Hannibal in Spain in 217 B.C. The following year, massive supplies of grain for the Romans arrived from Hiero of Syracuse. The Ostian navy's main tasks were to protect the coastline—seemingly maintaining a fleet of thirty vessels—and to receive vital food shipments. In 215 B.C., it was temporarily sent to Tarentum to guard against possible danger there after communications between Philip V of Macedon and Hannibal had been revealed.

In the second century B.C. Ostia developed considerably as a trading port, importing food and luxuries. Rome's population and needs were growing rapidly. Victorious campaigns abroad encouraged foreign trade. Ostia shared with Puteoli (modern Puzzuoli) on the eastern coast Rome's import trade. The historian Strabo recorded large vessels bearing goods "to Puteoli and Ostia, Rome's harbor." Many of the Spanish, African, and Sicilian goods probably arrived by

Columns of Ostia Antica's theatre

"Maska" sculptures at Ostia Antica
Photos courtesy of Italian Government Tourist Board-E.N.I.T.

Puteoli, but the main corn imports appear to have been shipped to Ostia. The story recorded by Polybius of the Syrian prince Demetrius's elaborate escape from Rome via Ostia in 162 B.C. to go claim the Syrian throne indicates that ships also traveled east. The Ostian streets were lined with supply stores for restocking vessels. Timber was also transported from Ostia to the capital.

It is difficult to know exactly how Ostia expanded at this time; the town was almost completely rebuilt in the second century A.D., but evidence of growth in the second century B.C. has been found in many places, in particular in some use of concrete (probably not employed until after the Second Punic War) in certain new walls. The corn trade was doubtless boosted in Ostia as in Rome by Gaius Gracchus's laws encouraging cheaper distribution. It was probably also in this period that travertine boundary stones were erected along parts of Ostia's riverside to set limits to development and the unloading of cargo. That the Ostian *quaestor's* main duty had shifted from supervision of the fleet to supervision of corn imports shows the importance to Ostia and Rome of trade in the second century B.C.

Ostia and its role in the corn supply assumed strategic importance in the civil war between rival generals Sulla and Marius at the beginning of the first century B.C. Marius made the capture of Ostia his first objective upon returning from Africa, and despite a Roman garrison stationed there, he was able to take and plunder the town with ease. A few years later Sulla issued orders for his army to take Ostia if Rome did not yield. New walls were erected around Ostia, perhaps under the direction of Sulla. An inscription declared that the walls were donated by the Roman senate and people. The enclosed area was thirty times greater than that of the fourth-century B.C. *castrum*.

Piracy remained a major concern for Ostian trade and security, however. An organized fleet of plunderers entered the Tiber estuary in 67 B.C., demolished a Roman fleet, and pillaged Ostia. Pompey was given the task of clearing the seas of such marauders, which he did swiftly following this incident.

After Octavian became the first Roman emperor as Augustus in 27 B.C., a praetorian cohort was posted to Ostia, from where Augustus may have embarked on his journeys. A plot against him by Fannius Caepio and Varro Murena in 23 B.C. failed, but Fannius Caepio managed to escape and was taken down the river to Ostia in a chest. Due to a food crisis the following year, Augustus assumed responsibility for the administration of the supplies and dispatched his stepson Tiberius to serve as *quaestor* in Ostia. Augustus would set up an efficient administration that would benefit Ostia.

In the Roman imperial period, Ostia, while closely linked to Rome, became a strong town in its own right. Important building initiatives, many of them monumental or religious, changed the face of the city, and although most structures would be either buried under or transformed by late imperial developments, the original forms of several important constructions such as the Temple of Hercules or the four small temples on a single podium east of the *castrum* remain discernible. Leading Roman families built sumptuous dwellings in Ostia similar to those in Pompeii.

Archaeologists cannot precisely date Ostian buildings of the early empire. The theatre probably originates from Augustus's reign and is thus one of the first permanent theatres in Italy; it was completely reconstructed in the second century A.D. Unlike the sumptuous new buildings in Rome employing much marble, even the Ostian theatre was constructed of tufa at this time. Three temples are reckoned to date from the Augustan period.

By this time Ostia stood out as the most important town on its stretch of coast. The poet Virgil chose the mouth of the Tiber as the supposed site of Aeneas's new Troy. Its increased prominence during the reign of Augustus was due partly to the massive corn imports from Egypt; much corn was eventually sent to Puteoli on Italy's eastern coast rather than to Ostia's overtaxed harbor.

During the early empire, Ostia was clearly important to Roman rulers, who visited there frequently. Numerous inscriptions honor them. The town's forum at the heart of the former *castrum* may have been set out during Tiberius's reign. An important temple to Rome and Augustus, with travertine columns used in its construction, was built to face the Capitolium. Warehouses, called *horrea*, were mushrooming; the *horrea* of Hortensius, also employing travertine, was of grand proportions. Gradually bricks began to take the place of tufa in the town's building, facilitating higher construction, and marble became more common for grand buildings. An aqueduct was built sometime in the early empire, as were new baths. Archaeologists have speculated that there was even an imperial palace at Ostia at this time.

Work began on the new port of Ostia under Claudius in A.D. 42. This was to prove the most fundamental change to Ostia during the Roman Empire, a massive enterprise from which Claudius's architects tried to dissuade him. But the silting up of the Tiber's mouth, the dangers of famine, and the difficulties of maintaining a steady flow of supplies throughout the year caused Claudius to insist on developing the harbor. The corn supply was too important to security to be left to chance.

The harbor was dredged from the Tiber's north bank. Parts of the quays were unearthed in the twentieth century during the construction of Rome's international airport of Fiumicino. An inscription notes the digging of canals from the river to the sea, to help connect the harbor with Rome and protect against flooding. Suetonius wrote that the digging of the canal to drain Lake Fucine alone took 30,000 men eleven years to complete. The harbor was already in use in the early 60s. Tacitus a little later wrote of a storm destroying 200 boats there. Claudius also tightened the corn bureaucracy, replacing the senatorial *quaestor* in Ostia with a procurator reporting to the *praefectus annonae* in Rome. A cohort was also assigned to Ostia by Claudius to fight fires.

Claudius frequented Ostia more than probably any other emperor. He surveyed much building at the site, and he departed from there for his travels outside the peninsula, as when he went to accept honors for the conquering of Britain.

He even organized a hunting party to kill a whale stranded in the harbor. The *Grandi Horrea* (great warehouses) are thought to have been built during his reign, as are the baths under the *Via dei Vigiles*, but with the manpower required for the port there was probably little other construction at this time.

Nero too had elaborate plans for a canal from Ostia into Rome to facilitate the transport of corn inland. The project could also have been used to drain marshes on the coast. Work began, but the scale of the task and the political situation in Rome made continuation too difficult. Even wilder plans of Nero's to expand Rome's walls to incorporate Ostia were recorded.

After Vespasian had restored order following a civil war upon Nero's death, new building in Ostia probably resumed under Domitian. The buildings were of a new type: they were raised three and one-quarter feet above the old, in part to protect the town against floods, in part to support foundations for the greater number of stories necessary to help house the expanding population.

Ostia's major expansion, the one that has left its clearest marks on the archaeological record, occurred during the reigns of Trajan, Hadrian, and Antoninus Pius, through the first half of the second century A.D. The major event in Trajan's period was the creation of a further port, or great hexagonal basin dug out behind Claudius's harbor. It could receive even the largest ships transporting corn from Alexandria. The basin dramatically increased Ostia's trading importance. Ostia, already the major Roman gateway to the west, now surpassed Puteoli as the main receiver of goods from the east. New *horrea* sprouted around Trajan's port. A new road, the *Via Portuensis*, was built to serve the Claudian and Trajan ports. Although these ports stood at some distance from Ostia, the urban center remained in control of the port area, known as *Portus Ostiensis*. The trading guilds remained based in Ostia.

In the town, new buildings were rising now to three, four, even five stories, these blocks being known as *insulae*. Brick was becoming the preferred building material; its use allows for easier dating of buildings as the bricks tended to be stamped with a consular date. The dwellings in the *insulae* were not destined only for manual workers; some were luxurious, and some blocks became early forms of a garden city. With whole quarters being redeveloped, planning was carefully considered. Porticoes ran down the major streets lined with shops.

The middle class grew as trade spread wealth across the citizenry. Ostia became a cosmopolitan town; people from many other regions began to settle there, as did freedmen in large numbers. The trade guilds, supported by rich patrons, assumed great importance and built impressive headquarters. Magistrates and town council members were increasingly likely to be middle-class traders. Many new public buildings were erected, including no fewer than eight public baths, such as the well-excavated Baths of Neptune and Forum Baths, with sumptuous mosaics. The forum was extended, added to by the curia and basilica, and covered in marble, as was, most impressively, the grand *capitolium*. Much richer decoration characterized this period, with particularly fine marble and mosaic work, sculptures, and inscriptions.

It is unwise to read too much detail into the inscriptions, which praised the emperors of this Flavian period, who were clearly generous in helping ennoble Ostia. Hadrian promised 2 million sesterces for the building of the Baths of Neptune. Roman senators also became patrons in Ostia, while two particularly powerful Ostian families, the Fabii and Egrilizi, had members who attained the status of consul.

Along the coast to the south of the town, an area already favored by wealthy Ostians and Romans during the late Roman Republic, more and more suburban villas were constructed. The Ostian shore was the setting for the second century Christian apologist Minucius Felix's famous dialogue between a pagan and a Christian, *Octavius*.

The traditional Roman cults maintained their power at this time. Just about all the traditional temples were restored in the second century, the most impressive work being on the Temple of Vulcan, while under Hadrian a grand new temple was erected to the Capitoline triad. Each year a Roman *praetor* would attend the Ostian games to Castor and Pollux. But with the influx of foreigners, foreign cults also established themselves, in particular those to Cybele, Isis, and Serapis. Buildings for these cults were not grand, and it seems followers came from the working, rather than the governing, classes. The greater popularity of Mithraism, a religion based on a Persian deity that gained a strong foothold in the Roman world, would follow from the mid-second century.

New developments in Ostia reached their apogee under Hadrian. Yet Ostia was still much smaller than the most important towns of Africa, Spain, and Gaul, though with its new *insulae* it was possibly more densely populated. It was not a suburb of Rome, even though it attracted Roman patronage. With few exceptions, its citizens played little part in Roman affairs. Russell Meiggs describes it as essentially a middle-class town. Under Antoninus Pius, the building program was less frenetic, but some important works of his time remain, such as the *Horrea Epagathiana et Epaphroditiana* and the Forum Baths. The so-called "Imperial Palace," an ornate, splendid edifice outstanding for the refinement of its detail, is attributed to this period. The only evidence to date for its having any imperial connection is the name of an imperial princess marked on a number of water-pipes. Estimates of Ostia's population indicate perhaps 20,000 inhabitants at the end of the republic and 50,000 by the end of Antoninus's reign.

The second half of the second century witnessed slow decline at Ostia in conjunction with the slowing of the imperial economy. The guilds remained strong, however, and the popularity of Mithraism led to a mushrooming of temples. For a time Ostia was given the name of "colonia felix Commodiana," perhaps because of the predilection of Commodus (son of Marcus Aurelius and emperor from 180 to 192) for the imperial Laurentine villa not far off, or because of his concern for the corn supply. The theatre was rebuilt in his time, according to the evidence of the brick stamps, but the inscription is dedicated to Septimius Severus.

Septimius Severus appears to have been particularly popular in Ostia judging by the number of inscriptions to him.

The *Grandi Horrea* were expanded, the Piccolo Mercato and Vigiles barracks altered, and a large semicircular emporium was erected close to the river. By the time of Septimius Severus's death, there were purportedly seven years' worth of reserves of corn. Few signs point to much building activity in the years following Septimius Severus, except for an extraordinary, original round temple, very possibly an imperial project given its grandeur, perhaps dating from Severus Alexander's or Gordian rule. Roman patronage seems to have dwindled at the turn of the century.

The third century was for Ostia a time of deterioration, reflecting the decline of the Roman Empire. Rome would import fewer materials from overseas. Local powers of government were eroded. The role of *quaestor* became increasingly burdensome, and Roman authority appeared more distant and uninvolved. After the middle of the century, the Roman *praefectus annonae* took direct charge. With the brick industry collapsing, builders reverted to the use of tufa mixed with brick, and marble was no longer used liberally. Plague had wrought its effect on the population. Throughout this decline, the port area survived much better than Ostia itself.

It is difficult to judge when the new ports outside of Ostia developed into an urban center in their own right. The clearest archaeological proof for a large population growing around the ports in the second century comes from a large cemetery on the Isola Sacra, stretching around one and one-quarter miles south of the canal. Local traders of all descriptions are buried here, indicating a thriving community. The second-century physician Galen wrote of knowing "the harbor or the city near the harbor which they call Ostia," referring to both as "populous centers." The change in the name of the newer port area also indicates how it emerged as a separate city. Initially referred to as *Portus Augusti Ostiensis* or *Portus Ostiensis*, after Trajan's basin was added it became known as *Portus Augusti et Traiani felicis* or *Portus Uterque*. By the end of the second century, the harbor and town around it was called simply Portus. For some professions, the two centers shared a guild, but some, such as the shipbuilders, divided into separate organizations. Portus had its own temples from the second century as well.

In the third century, workers probably began moving from Ostia to Portus, which was taking the economic upper hand. The Ostian warehouses, which had held massive reserves, were abandoned. Trade was still carried out on Ostia's Piazzale delle Corporazioni. But contemporary mention of Portus, or Portus Romae even, no longer referred to Ostia.

However, Ostia developed along its pleasant seashore. Although new buildings for trade were no longer built in third-century Ostia, theatre and baths flourished. Fountains and nymphaea were extremely popular and numerous, and curved walls came into vogue. For the prosperous, the living in Ostia at this time appears to have been easy.

Two religions were of nearly equal importance in the region during the third century. Fifteen shrines to Mithra unearthed in Ostia, the earliest evidence dating from 162, attest the widespread importance of this cult, the appeal of which rested mainly with the working people. The Ostian upper classes would in time be won over to Christianity, but the Christian community was slow to build. A large number of Christian lamps dating from before the end of the century have been found. Later records also indicate a Christian martyr, Aurea, in Ostia in the middle of the century.

Under Constantine I, who officially changed the ports area into an independent town in the early fourth century, Portus benefited from imperial support at Ostia's expense, although the building during his reign of a Christian basilica in Ostia is recorded. It may be that Ostian sympathies lay with Constantine's rival Maxentius, who moved the imperial mint there when Aquileia was under threat in 312. Such a move may have been seen as an honor, but the mint was closed under Constantine. Through the fourth century, a great many laudatory inscriptions were carved to the emperors. The lavish words contrasted with the enfeebled empire, now under attack by Germans and Goths. The rich thrived in Ostia still. Some second-century houses, such as the Fortuna Annonana, were now beautifully redone, as indeed were some commercial buildings turned into houses, such as that of Psyche and Eros. But the *insulae* were not maintained in the same manner: much of the general craftsmen's work was shoddy, and while services were kept in operation, restoration was patchy, planning was largely neglected, and old material taken from other sites was employed, so that in one new public lavatory, for instance, seats were found made from inscribed cemetery marble.

The *praefectus annonae* possibly spent part of the year in Ostia, some imperial officers doubtless lived there, and rich merchants may have come there for peace from the busy Portus. The middle classes, however, had suffered. Local government and the previously active guilds had come to a standstill. Inscriptions to public careers or of guilds disappeared almost entirely in the fourth century. Social responsibilities were neglected, the rich apparently trying to avoid public office. Only one official, the *praefectus annonae* Vincentius Ragonius Celsus, stands apart, receiving a spate of honors around 390 for his services to Ostia and Portus, in particular restoring certain buildings such as the Forum Baths and the theatre. But the revival was probably very brief, and Celsus held the post for less than four years.

From 314 there is proof from attendance records of the Council of Arles that Portus had its own Christian representatives, later bishops; Ostia too had its own bishop. Religious tensions may have been pronounced in Ostia, possibly exacerbated by the gap between rich and poor, and one Christian source refers to persecution of Ostian Christians. St. Augustine stayed here while waiting to return to his native home in Africa with his mother, Monica. They both shared a powerful experience of religious ecstasy in Ostia, recorded in St. Augustine's *Confessions*, before Monica died in the town.

From the fifth century, Ostia quickly declined into obscurity. When the Visigoth Alaric invaded Italy to sack Rome in 410, he took Portus first, but ignored Ostia. The poet Claudius Rutilius Namatianus, traveling back to Gaul, noted how all that remained of the place was the barren pride of having once welcomed the legendary Aeneas. In 455 the Vandal Gaeseric led a further invasion of Italy. Portus was pillaged, and no doubt Ostia too. By the end of the fifth century, the latter's public water supply ceased to operate; people lived in partially collapsed buildings; the dead were buried in random graves. Portus revived: Cassiodorus describes a busy harbor scene at the turn of the century. In 573, the Gothic leader Vitigis realized he had to seize the well-defended, walled Portus to stop Justinian's general Belisarius from getting supplies through to Rome. Belisarius did use the Tiber and Ostia once Portus had been taken, but after this the town fell further and further into disuse.

Saracens attacked in the ninth century. The Ostians who remained appear from late makeshift fortifications to have put up meager resistance before retreating inland. Pope Gregory IV decided to build a new settlement there, possibly in part to commemorate Aurea, the Christian martyr of third-century Ostia. The fortified spot was first named Gregoriopolis and was slightly smaller in size than the fourth-century B.C. *castrum*. The inhabitants did not stay to defend it when Saracens attacked again in mid century, simply locking the gates and fleeing. The Saracens occupied it while they pillaged Portus, which they also found abandoned. Pope Nicholas I added gates and towers to the Ostian fortifications after the middle of the ninth century, but the life had been drained from the place. The river mouth was becoming too silted up to continue shipping, and the few remaining inhabitants had reverted to the work from which Ostia originated, exploiting salt beds. The marshland to the east then flooded, and the large lake that remained became a source of trade for fishermen. Fruitless litigation continued for a long time between the salt farmers and the fishermen, while malarial mosquitoes ravaged the population. By the twelfth century, Ostia counted only a handful of inhabitants. Its antique marble was destroyed for use as lime, as a Papal Bull of 1191 attests. Richard the Lionhearted landed at the mouth of the Tiber at the end of the twelfth century and saw the immense ruins of ancient walls.

These ruins were greatly exploited for good building materials by the Pisans through the eleventh and twelfth centuries. Ostian stone can be found in Pisa's cathedral, in the Florence Baptistery, and possibly in Amalfi's cathedral. Ostian marble went into the magnificent fourteenth-century facade of Orvieto's cathedral. By the fifteenth century, the Romans as well began using Ostia for building resources.

The fifteenth century brought some revival. The present standing walls were erected, together with a church and an impressive castle ordered by Cardinal Giuliano della Rovere, later Pope Julius II. The castle guarded the Tiber's route to Rome against invaders, but in a dramatic flood of 1557 the Tiber's course was shifted, so that the castle lay almost half a mile from the river.

Renewed interest in classical culture led to a different kind of plundering during the Renaissance: collectible inscriptions and art, in particular statuary, became highly prized. The passion for collecting spread across Europe. A Scottish painter, Gavin Hamilton, was collecting Roman works for Lord Townsend in 1774, defying the risk of malaria for the booty. Ostia soon became a stopping point on the Grand Tour of Europe. An Englishwoman traveling there in 1805 describes the galley slaves working to produce salt for papal advantage and praises the area's fruit, reputed in early Roman times.

At the start of the nineteenth century, private foraging in the ruins was halted thanks to the Director General of Antiquities, Carlo Fea. Official excavations began, conducted by Petrini, who organized the first haphazard digs in 1802 and 1804. Working for the pope, his role was to reveal as much of the ancient city as possible, record his findings, and hand the artifacts to the Vatican. Further excavations from 1824 to 1834 were conducted more on Cardinal Pacca's personal behalf; he financed the digs and retained the finds.

Much more serious excavations started in 1855 under Visconti, working for Pope Pius IX, although again finds were destined for the papal collection and building material for works in Rome. Visconti's passion led to the creation of a museum at Ostia, and his records were essential in piecing together Ostia's history. After 1870 the excavations were under the direction of the new Italian government. Digs continued intermittently until 1907, when they were made more systematic. In 1912, the first detailed work on Ostian history by Paschetti was published. In 1938 excavations were speeded up to reveal a far larger area of the site in preparation for a grand international organization planned to be held in Rome in 1942. Despite the war, a great deal of work was carried out in haste. For many years the excavations were headed by Guido Calza.

The coast, distant now from the classical sites, was developed as a resort in the twentieth century. A monument by Pietro Consagro to the provocative author and film director Pier Paolo Pasolini was erected in 1993, close to where he was murdered. Many of the sites of Portus are not open to visitors, although a museum of Roman naval ships was inaugurated in 1979. Roughly two-thirds of Roman Ostia have been excavated, the finds divided between five regions into which it was apparently once split. The historian Gaston Boissier said of Ostia and Pompeii that: "These two towns are the best-preserved relics of Roman antiquity." Further discoveries are still to be made.

Further Reading: The best book on the history of Ostia and Portus is written in English. *Roman Ostia* by Russell Meiggs (London and New York: Oxford University Press, 1960; second edition, Oxford: Clarendon Press, 1973; New York: Oxford University Press, 1974) is a fascinating work, gathering together archaeological and literary evidence to form a picture of Ostia and Portus's history. Meiggs makes clear his debt and gratitude to two earlier Italian scholars of Ostia, Guido Calza and Italo Gismondi. Calza authored, among other publications, an informative, useful booklet, *Ostia* (Rome: Liberia dello stato, 1931; ninth edition, Rome: Istituto poligrafico e zecco dello stato, 1974), which is available in a quirky translation. The bibliographies in these titles refer to further specialized reading. Meiggs's bibliography is very detailed.

—Philippe Barbour

Padua (Padova, Italy)

Location: Part of the Veneto region and Padova province in northeast Italy, in the Venezia Euganea region, twenty-two miles west of Venice; twenty-one miles southeast of Vicenza; forty-nine miles east of Verona.

Description: One of Italy's most important art cities, known in Italy as Padova, the birthplace of Mantegna, housing Giotto's masterpiece in the Scrovegni Chapel and Donatello's great sculpture, *Gattamelata*. There is also work by Altichiero, Menabuoi, the Lombardos, Il Riccio, Guariento, Giorgione, Giovanni Bellini, Tintoretto, Titian, Costa, Veronese, Longhi, and Falconetto. Also a pilgrimage center, with the remains of St. Anthony of Padua and St. Prosdocimus. A university town; "Il Bo" was the second university to be founded in Italy, in 1222, giving the city the nickname "La Dotta"—the Learned. Home of one of the first botanical gardens. Padua is now the economic capital of the Veneto.

Site Office: A.P.T. (Azienda di Promozione Turistica)
Riviera dei Mugnai, 8
CAP 35137 Padua, Padova
Italy
(049) 8750655

Legend has it that Padua was founded in 1185 by Antenor, a hero of the Trojan War, who is said to have founded a state in the north of Italy. Livy, the famous Roman historian, would have it so, but he was a Paduan, born in the village of Teolo, just outside Padua in the Eugane Hills. In Antenor's Square, the Piazza Antenore, there is a sarcophagus containing bones that some believe to be those of Antenor, though others say they belong to a tenth-century Hungarian warrior. Nearby is another monument, erected on the 2,000th anniversary of Livy's birth.

Padua was a small fishing village in the fourth century B.C., a Roman colony known as Patavium in 89 B.C., and a municipium by 45 B.C. A few traces of the Roman period can still be seen; the Old Castle, La Torlonga, retains sections from A.D. 99. There are ruins of the amphitheatre and forum, remnants of the harbor and a few Roman bridges built over the River Bacchiglione, or River Medoacus as the Romans called it. The Civic Museum holds more artifacts. The layout of the old town, with its narrow, cobbled streets, is of Romans design.

From the fourth century, Padua was the seat of a bishop, the first being St. Prosdocimus, whose relics lie in the fifth-century Chapel of the Virgin (Sacellum di Prosdocimus), within the sixteenth-century Basilica di Santa Giustina.

In A.D. 602 Patavium was virtually destroyed by the Lombards, led by their king, Agilulfo. Under the Lombards, and then the Carolingians, the area was divided into dukedoms and eventually several feudal establishments evolved, some of which developed into self-governing towns and then cities. Padua was one of these, and in the ninth century new Romanesque walls were built, though little remains of them today. At the end of the ninth century the town's oldest church, Santa Sofia, was begun, but not finished until the twelfth century. Meanwhile, Padua had retained its bishop, and as trade and craftsmanship developed, so did its prosperity, so much so that by the twelfth century it had become a commune (independent city-state).

In 1107 Padua joined three other communes—Verona, Treviso, and Vicenza—in the Veronese League. In 1164 the league became the Lombard League and expanded to include most of the cities in the area to defend themselves against Holy Roman Emperor Frederick I (Barbarossa), who made several attempts to take the territory. He was defeated in 1176, and in the 1183 Peace of Constance the cities were granted their own constitutions. The oldest private house in Padua, 19 Via Santa Lucia, was built just prior to that point, around 1160, for Ezzelino I, grandfather of the dreaded tyrant Ezzelino III da Romano.

The thirteenth and fourteenth centuries were a time of great prosperity for Padua and the Veneto. The First Crusade (1097–1130) had proved very profitable for the area; many crusaders passed through Venice, and with money came a surge of building and expansion. In 1201 the commune of Padua built a canal from the medieval spa town, Battaglia Terme in the Euganean Hills, to take building materials down to the city.

The Eugane or Euganean Hills, southwest of Padua, are volcanic hills whose rich soil produces the local Montecchia DOC wines, and in which there are many hot mineral springs. Three well-known spa towns are situated near Padua: Battaglia Terme and Montegrotto Terme, which draw on the same springs and were used by the Romans, and Abano Terme, six miles south of Padua. All three still provide therapies, including mud baths, believed to treat several ailments, particularly respiratory and rheumatic problems.

The University of Padua, Italy's second university, was founded in 1222. The first was Bologna. Its more familiar name, Il Bo, comes from that of an inn that once stood on the site, "Bo" meaning "ox" in Venetian dialect. During the thirteenth and fourteenth centuries the university rapidly developed as a prehumanist center, participating in the revival of classical literature. The establishment of a university was perhaps the most significant product of the city's desire to create buildings and institutions that would reflect its affluent status.

Frescoes in the Scrovegni Chapel, painted by Giotto di Bondone
Photo courtesy of A.P.T., Padua

In approximately 1215 work began on the building of the commune's law courts, Palazzo della Ragione, to house the city's parliament or *podestà*. In the following century a huge upstairs hall, the *Salone,* was added. It is one of the largest medieval halls to survive, measuring 250 by 89 feet, with a ceiling 85 feet high. It is said to have been frescoed by Giotto and his school, but the present frescoes date from the fifteenth century and were painted by Nicolo Miretto and Stefano da Ferrara. There are over 300 scenes including cycles on the Work of the Months, the Liberal Arts, the Trades, the Signs of the Zodiac, the Apostles, and the patron saints of Padua, of whom there are six: Prosdocimus, Daniel, Anthony, Francis of Assisi, Giustina, and Louis of Anjou.

The commune was also mindful of the continuing uncertain political situation. Frederick II, the grandson of Barbarossa, was both emperor and king of Sicily, and he built a group of supporters, known as Ghibellines, who favored state economic control and religious and intellectual tolerance, and who tended to include the rural nobles, a class that had developed from the *seigniories* during the ninth and tenth centuries. The *seigniories* were strong family units, with a tradition of fighting each other. During the thirteenth century, these rural magnates moved into the cities, where they felt safer, establishing quite dazzling courts there, and it was these families who developed into the famous Renaissance potentates.

Padua, however, supported the opponents of the Ghibellines, the Guelphs, who pledged loyalty to the pope. The Guelphs favored religious orthodoxy, the liberty of the commune, and the interests of the emerging merchant class. Frederick II's ambitions in the region led him to send his lieutenant, the tyrant Ezzelino III da Romano, on a march on Verona in 1232. By 1236 Ezzelino had gained possession of Verona, Vicenza, and Padua for the emperor and king. Padua remained under Ghibelline control until Charles of Anjou finally defeated Conradin, Frederick's grandson, in 1268, after which Charles was supreme in Italy until 1282.

These turbulent times are reflected in the building of a new set of defensive walls and nineteen gates. The commune began them and Ezzelino continued the work. Two of the gates remain, Porta Altinate and Porta di Ponte Molino. The old Castle, La Torlonga or La Specola, built in 1237 by Ezzelino, reinforced by the Carraresi, and transformed into an observatory in 1767, is now used as a prison.

However, the most outstanding legacy of the thirteenth century is the Basilica di San Antonio, commonly known as Il Santo. It was begun in 1232 and has been a great pilgrimage center ever since, particularly on June 13, St. Anthony's Feast Day. St. Anthony the Hermit was born in Lisbon, Portugal, in 1195 to a noble family, and after entering the Lisbon monastery of the Canons Regular of São Vivente, he was inspired by the lives of Franciscan missionary martyrs. He succeeded in being sent to Morocco to follow in their footsteps, but on his return voyage to Portugal a storm blew him off course and he landed (or was shipwrecked) in Sicily. Receiving ordination from the Franciscans there, he soon

became known as a great preacher and traveled in northern Italy and southern France, converting heretics. In 1221 St. Anthony preached daily in Padua throughout Lent and in 1223 St. Francis of Assisi made him the first professor of theology for his friars. He spent the last years of his life just north of Padua and died on his way to the city center on June 13, 1231, at the age of 36. Pope Gregory IX canonized him in the following year and the Paduans immediately began building the basilica to contain his body, which lies in the chapel of his name. St. Anthony has become the patron saint of miners, lovers, pregnant women, and the poor. Many also believe he is able to find lost property.

The basilica is a Romanesque-Gothic extravaganza, reminiscent of San Marco in Venice. The interior is Byzantine in character and is an art-lover's paradise, containing work by painters such as Altichiero and Giusto de'Menabuoi, and by sculptors such as Donatello, the Lombardos (Pietro, Tullio, and Antonio), Jacopo Sansovino (or Tatti), and Il Riccio. The Antonian Library, which has a fine collection of rare books and manuscripts, is also housed in the basilica.

In 1318 Padua appointed Jacopo da Carrara to the governorship and so began the generally benign rule of the da Carrara family or the Carraresi, under which the city reached its peak economically, culturally, and politically. The Carraresi lost the city briefly to the dalla Scala or Scaligers of Verona in 1328 but regained it nine years later, when they were admitted into the Venetian nobility. They would hold power until 1405. They had a splendid court, and the royal palace, or Reggia, attracted great artists and craftsmen from all over the country, including the brightest star of the century, Giotto.

Giotto di Bondone arrived in Padua in 1303 at the peak of his powers and spent the following years there painting scenes from the life of Christ and the Virgin Mary in the Scrovegni Chapel, which has since become known as his masterpiece. The chapel lies in the Arena Gardens, site of the ruins of Patavium's Roman amphitheatre, hence the other name of the chapel, the *Madonna dell'Arena*. The chapel was built at the instigation of Enrico Scrovegni to expiate the sins of his father, Reginaldo, a usurer and miser so avaricious that the church refused to bury him and whom Dante condemned to hell in canto 17 of the *Inferno*. The master cycle of frescoes consists of just under forty panels, perfectly preserved.

Adjacent to the Scrovegni Chapel is the Church of the Anchorites, which once held a series of frescoes by a young Mantegna, a native Paduan, in the Ovetari Chapel. Much work has been done to restore the old Augustinian church, built between 1276 and 1306, but tragically only a small portion could be saved of Mantegna's work, though some frescoes by another Paduan, Ridolfo Guariento, remain. The church also holds several fourteenth-century tombs, including those of Jacopo da Carrara and Umbertino da Carrara.

St. George's Oratory (Oratorio di San Giorgio) features a cycle of fourteenth-century frescoes by Altichiero and his pupil, Jacopo Avanzo, illustrating scenes from the life of St. George.

Another fourteenth-century building, the Museo Civico agli Eremitani, which was built as an Augustinian convent, has been restored and now acts as the Civic Museum and Municipal Gallery. The museum contains Egyptian, Etruscan, pre-Roman, and Roman archaeological finds, while the gallery contains a Giotto crucifix, Giorgione's *Leda and the Swan,* and works by Giovanni Bellini, Tintoretto, Titian, Costa, Vivarini, Veronese, Guariento, and Longhi. The baptistry of the sixteenth-century cathedral next door contains a cycle of frescoes painted by Giusto de Menabuoi between 1376 and 1378 and a polyptych by the same artist. The cycle was painted for Fina, wife of Francesco il Vecchio da Carrara.

Petrarch was a friend of Francesco, and the magnate gave the writer some land in Arquà, the most attractive village in the Euganean Hills. Petrarch spent the four years preceding his death in 1374 in Arquà. The visitors' book in his villa bears many famous names, including Byron, Shelley, and Mozart. Petrarch's body lies near the village church.

The early fifteenth century brought more change to Padua when in 1405 the city was one of the first to be incorporated into the Venetian Republic, although it managed to maintain a large amount of autonomy. Venice lost its dominant trading position with the east and turned to its newly acquired mainland regions to provide wealth. Agricultural land was created through drainage and irrigation, producing vineyards and orchards, fields of maize and rice, and mulberries for the raising of silkworms. In 1469 books began to be printed in Venice, and by the end of the century a considerable paper-making industry had developed at Padua to help to supply the new printing capital of Europe.

Venice also made use of Padua by sending its young aristocrats to be trained at the university, which continued to expand. In this century the university was one of the greatest centers of humanism in Italy. In 1463, the School of Philosophy was joined by a chair of Greek, the first professor being Demetrius Chalcondyles, who later went to Florence University. By 1497 a chair for the study of Aristotle in Greek had been founded. Many Englishmen of the time studied there, including John Tiptoft, earl of Worcester, who lectured there from 1471 to 1509; and Thomas Linacre, the founder of the Royal College of Physicians in London and personal doctor to Henry VIII, who took his degree as doctor of medicine in 1484. Padua's greatest art master, Andrea Mantegna, was born in approximately 1420 on the Isola di Carturo and was apprenticed to Francesco Squarcione in his excellent Paduan workshop or *bottega.* Since the destruction of his frescoes in the Church of the Anchorites, during World War II, Mantegna's best work is to be seen in Verona. However, Donatello's great statue of Gattamelata, a huge equestrian bronze that influenced Renaissance sculpture, remains in front of Il Santo. Gattamelata (originally Erasmus da Narni) was a *condottiere* or mercenary general who fought valiantly for Venice in 1432. Donatello of Florence was the best sculptor of his day, and he was in Padua from 1444 to 1453 creating the *Gattamelata* and the altar for Il Santo.

The following century brought yet another renewal, particularly at the university, which reached new heights of importance. Padua University led Europe in both teaching and research of anatomy and medicine. Giovanni de Monte translated the works of the Greek physician Galen and lectured there in the early seventeenth century. He was followed by a great number of equally brilliant teachers and researchers including the Belgian Andrea Vesalius; Eustachio, for whom the eustachian tubes in the ear are named; Columbo, who discovered the pulmonary circulation; Ingrassias; Aranzi; and Fabrizio, who discovered the valves in the veins and who taught William Harvey at Padua, the Englishman who discovered the circulation of blood. The famous Anatomy Theatre, established in 1594, is thought to be the oldest in the world.

Another of the university's innovations was the Botanical Garden. In 1533 the Venetian government instituted a chair of simples (medicinal plants) at Padua and was then persuaded to establish the Botanical Gardens specifically to provide plants for research into medicine, desperately needed in a time when no effective remedies existed to combat diseases such as typhoid, malaria, the plagues, and syphilis. Here, in 1565, the first lilac in Europe was grown and, in 1590, the first potato. Goethe visited the Gardens in 1786, and it is said that this is when he finally accepted a theory of evolution, sitting under the tree now known as Goethe's palm.

Galileo held the chair of mathematics at Padua from 1592 to 1610, and while there he invented a thermometer, a proportional compass, and a refracting telescope. The latter led to his astronomical discoveries that were to cause him so much trouble. The lectern from which he taught is preserved.

For a time, theatre companies were developing in the city, and Padua, with Siena and Venice, was a center for comic playwrights in particular. The Paduan dramatist Angelo Beolco created Ruzante, a comic peasant character who may have inspired the later Commedia dell'Arte, and outdoor performances of his plays were given in the Loggia Cornara, which, with the Cornara Odeon, was designed by Falconetto of Verona.

Building continued throughout the century, in particular a new defense system, built between 1509 and 1557 after the Siege of Padua in 1509 by Emperor Maximilian I. Falconetto designed two of the gates. For the Benedictines, Il Riccio designed the Basilica of St. Justinia, a pink brick church reminiscent of Il Santo and containing the *Martyrdom of St. Justine* by Veronese, painted in 1575.

In the late sixteenth century in the Square of the Seigniors, the Palazzo del Capitanio was built for the Venetian rulers of Padua between 1599 and 1605, replacing the Carraresi Palace. The same square boasts a clock tower by Falconetto with a great astronomical clock, conceived by Jacopo Dondi. In the School of St. Anthony there are eighteen sixteenth-century frescoes describing the life of the saint, three of which are by Titian.

The seventeenth century began badly as the Veneto suffered from plagues and warfare. But the situation improved as the great age of villa-building began in the valley of the River Brenta, running between Padua and Venice. It was in 1678 that a woman first received a degree in Italy; Elena Cornaro Piscopia earned a Ph.D. at Padua and her statue is in Il Bo. By the beginning of the eighteenth century Italy's population had grown to 13.5 million, with the highest density in the north. The university continued to flourish, and Oliver Goldsmith's name is registered there for a period between 1765 and 1767. Bartolomeo Cristoforo built the first piano in Padua early in the eighteenth century.

Padua's largest square, the Field in the Valley, or Il Prato, dates from 1775. Previously a marshy field, it was the site of the Roman theatre, and in the Middle Ages tournaments took place there. In 1775 the Podestà had the field drained and the huge square (953,939 square feet) now has gardens surrounded by canals crossed by four baroque bridges, and eighty-seven statues of famous Paduans including Galileo, Petrarch, and Ruzante. On April 30, 1797, the first tree of liberty was planted here, and on August 22, 1808, the first hot-air balloon took off from Il Prato.

In 1797 Napoléon abolished the Venetian Republic in the Treaty of Campoformio and Padua was under the authority of Austria until 1866. During the 1848 revolutions, Padua University was one of the principal centers of rebellion. Dissidents and liberals, fighting for a free and united Italy, would meet in the extraordinary Caffè Pedrocchi, designed by Giuseppe Jappelli and named after the first owner, Antonio Pedrocchi. Originally, the neoclassical building was open twenty-four hours a day, had no doors, and every university student was entitled to a free newspaper and glass of water.

Padua was annexed to the new kingdom of Italy in 1866, leading to a resurgence of development in all areas. On November 3, 1918, a treaty was signed at Villa Giusi, which finally freed the Veneto from Austria. While the northern part of Padua suffered from bombing in World War II, the center of the city retains its historic character.

Since 1946, when a national referendum made Italy a republic, Padua has gone from strength to strength commercially. The city boasts a huge trade fair complex and hosts an annual trade fair in June. The city is a commercial and trade link between Italy and northeast Europe, the Near and Middle East, and, with Verona, the agricultural center of the Veneto.

Further Reading: *Venice and the Veneto* by James Bentley (London: Arum, 1992) is a reliable, informative, and readable account of the area, with a twenty-five-page chapter on Padua and its environs. *The Veneto: Padua, Vicenza, Verona* by Dorothy Daly (London: Batsford, and New York: Hastings, 1975), though not as reliable factually, contains a more colorful and personal account of the city by an English woman who made her home in Padua. There is a twenty-eight-page chapter on Padova, and separate sections on the Euganean Hills and the villas of the Brenta. *Northeast Italy* by Dana Facaros and Michael Pauls (Chester, Connecticut: Globe Pequot, and London: Cadogan, 1990) is an excellent guide to the whole of northeast Italy, with a good background history and chronology and a useful directory of artists, written in an entertaining style. *Mantegna (1431–1506) and the Paduan School* by Reginald H. Wilenski (London: Faber, 1974) and *The Frescoes of Mantegna in the Eremitani Church, Padua* by Giuseppe Fiocco with an introduction by Terisio Pignatti (Oxford: Phaidon, 1978) are both good starting points for learning about Mantegna and those he influenced. *Giotto: The Arena Chapel Frescoes* by James H. Stubblebine (London: Thames and Hudson, and New York: Norton, 1969) and *Giotto: La Capella degli Scrovegni* by Carlo Carra (Milan: Pizzi, 1945) give accounts of Giotto's masterpiece in Padua. *Giotto and Some of His Followers* by Osvald Siren, translated by F. Schenck, in two volumes (Cambridge, Massachusetts: Harvard University Press, 1917; London: Milford, 1918), gives a broader account and includes details of some of the Giottoesque painters present in Padua.

—Beth F. Wood

Paestum (Salerno, Italy)

Location: Fifty miles south of Naples, one-half mile from the present-day coastline near the southern end of the Bay of Salerno; enclosed by a three-mile-long fortification wall with northern, eastern, southern, and western gates; bounded on the south by the Temples of Hera I and II and on the north by the Temple of Athena. The sanctuary of Hera at Foce del Sele is located five and one-half miles north of Paestum; the Sanctuary at Santa Venera lies one-third mile east of the south gate.

Description: Founded around 600 B.C. by colonists from Sybaris, Poseidonia was one of Greece's most important colonies. It retained its Greek character even after it fell under native Lucanian (Samnite) dominion in the fourth century B.C. In 273 B.C., the Romans established a colony there and renamed it Paestum. Three centuries later, the site became a malarial swampland, deserted except for occasional settlements of Christian worshipers. The rediscovery of Paestum's architectural wonders in the eighteenth century fueled debates about the evolution of European architecture. The Museum of Paestum preserves the antiquities found there.

Site Office: A.A.S.T. (Azienda Autonoma di Cura, Soggiorno e Turismo)
Via Magna Graecia, 151
CAP 84063 Paestum, Salerno
Italy
(828) 811016

Paestum played a key role in the political, commercial, architectural, and religious life of the ancient Greek west, but for centuries it was literally obscured from view. The inscriptions found at the site and classical historical and literary references to Paestum are tantalizing, but few; Ovid, Virgil, and Martial, for instance, paid tribute to the roses that bloomed there twice a year (and do to this day). The archaeological evidence provides a wealth of information on the city's history and the lives of its inhabitants. Paestum, with its well-preserved Doric temples and Greek sanctuaries coexisting with Roman public buildings, offers the visitor a unique glimpse into the past.

The site that became Poseidonia, later, Paestum, was occupied during the Paleolithic, Neolithic, Copper, and Iron Ages. The nearby Gaudo tombs, unearthed by American soldiers during World War II, point to a large Bronze Age settlement dating from about 2400 B.C., and Mycenaean potsherds and remains of Late Bronze Age structures almost lend credence to the legend recorded by Greek geographer Strabo that the Thessalian explorers Jason and the Argonauts landed at the mouth of the nearby River Sele, where they erected the sanctuary dedicated to Argive Hera.

Poseidonia was an outgrowth of the great westward expansion of the Greeks beyond the Aegean area, which took place between the eighth and sixth centuries B.C. Increased contact with Egyptians and Middle Easterners, and population growth, drought, and other internal pressures stimulated a renaissance of Greek culture, an intellectual ferment that spawned coinage, new philosophical schools and styles of music, and innovative techniques in architecture, metallurgy, pottery, and sculpture. Greeks from the mainland, Asia Minor, and the Cyclades traveled the Mediterranean in search of new lands, trade routes, and new ideas. They confined their western colonies mainly to Sicily and southern Italy, from the Bay of Naples on the west coast to the Gulf of Tarentum (Taranto) on the eastern side; the network of independent city-states became known even in ancient times as Magna Graecia (Great Greece).

One of the Greeks' most prosperous and legendary Italian colonies was Sybaris, established on the peninsula's southeast coast around 720 by the Achaeans of the Peloponnese and the Troizenians of the Argolid. The Sybarites were renowned for their luxurious lifestyles; according to one ancient observer, they piped wine directly from the vineyards into the city and banned roosters from within the city limits to avoid waking people too early. Aristotle recorded that when Sybaris's founders quarreled, the Troizenians were expelled from the town, fled west, and founded Poseidonia. Poseidonia may have received another influx of Sybarites, after their town was sacked by neighboring Croton (Crotone) in 510. An Archaic Greek tablet found at Olympia lauds their close connection: "The Sybarites and their allies and the Serdaioi are yoked together in terms of true and guileless friendship forever, the guarantors are Zeus, Apollo, and the other gods, and the city of Poseidonia."

The landscape offered no natural defenses, so early settlers built a wall around the polygonal limestone shelf that dictated the city's outline; pottery dated between 625 and 600 corroborates Strabo's claim that this event occurred toward the end of the seventh century. Poseidonia's prime location on a fertile agricultural plain on trade and communication routes with other Greeks and with the Romans, Etruscans, and other Italic peoples contributed to its rapid ascent. By the mid–sixth century, Poseidonia was striking its own silver coins, stamped with the image of Poseidon, and receiving goods imported from Egypt and the Greek east.

The oldest Greek materials found there come from the Sanctuary of Hera at Foce del Sele, located about one mile south of the Sele River and five and one-half miles north of Poseidonia. In 1934, archaeologists Paola Zancani-Montuoro and Umberto Zanotti-Bianco unearthed a sanctuary complex

Temple of Hera I
Photo courtesy of A.A.S.T., Paestum

that included altars, a temple, and a Treasury built between 570 and 560 B.C. Many of the Treasury's thirty-three surviving sculptured sandstone metopes (frieze panels) portray epic cycles, a pictorial breakthrough in Greek architecture; these include the Trojan War and its aftermath and the heroic exploits of Hercules (Herakles), among them his capture of the Kerkopes, gnomes who had robbed him while he slept. Initially Hercules bound the Kerkopes to a pole and slung them over his shoulder, but he eventually let them go because, in the words of John Griffiths Pedley, "they made him laugh with scurrilous remarks about his hairy backside which they were uniquely well situated to observe."

The low, double-roofed Underground Shrine, uncovered north of the Roman Forum in 1954 by archaeologist Pellegrino Sestieri, housed eight bronze vessels brimming with a honeylike substance. A lion-shaped handle decorates one of the *hydriai* (water jars), and an Athenian black-figured *amphora* (storage jar) depicts Hercules' apotheosis on one side and a Dionysiac scene on the other. This building, constructed between 510 and 500, may have been an offering to Hera or a cenotaph built to honor an important political figure.

The impressive Doric temples bear witness to Poseidonia's period of greatest prosperity, the Greek occupa-

tion of the sixth and fifth centuries. The western Greek architects drew from their Achaean roots but conceived a highly imaginative design, a testament to their propensity for transforming their culture wherever they journeyed.

Around 550, they began constructing the imposing Temple of Hera I (also known as the Basilica), which faces east and measures 80.4 by 178.1 feet. Front and back facades both feature nine columns; each side has eighteen columns. A central seven-column colonnade divided the *cella* (main interior room) into two spaces, possibly for a double cult; sixth-century objects found throughout the area, including an inscribed silver disk proclaiming "I am sacred to Hera; strengthen our bows" and sculptures depicting the marriage of Hera and Zeus, support this theory. The exterior and interior columns, three of which front an interior porch on one end of the colonnade, have the same dimensions—another departure from convention. The Doric columns show a marked *entasis,* a convex swelling of the shaft that dispelled the illusion of concavity that would have resulted if the columns had been perfectly straight. Sculptured leaves, lotus flowers, rosettes, petals, and other floral motifs ornament their capitals.

The much smaller (108.2 by 47.5 feet) Temple of Athena (dubbed the Temple of Ceres), built around 500 in the

Detail of Tomb of the Diver
Photo courtesy of A.A.S.T., Paestum

city's northern section, also faces east and displays a bold mixture of eastern and western Greek architectural styles. The twelve side and twenty-six flank columns are Doric, but the eight columns forming the interior porch are Ionic, a first in the history of architecture and, along with its eastern Greek counterpart, the Temple of Athena at Assos, a probable inspiration for the Hephaisteion and Parthenon.

The massive Temple of Hera II (alternatively, the Temple of Neptune) was built around 450 alongside the Temple of Hera I. Six flank by fourteen side columns define its 197- by 82-foot perimeter. The fourteen inner columns dividing the cella into two aisles support a tier of columns that once held up the roof. Its classical "Doric refinements" (e.g., a stucco coating that gave the columns the appearance of marble) suggest an architect trained in mainland Greece.

The Greek temples and sanctuaries yielded no definitive traces of a cult of Poseidon, as might be expected in a city that bore the sea god's name and impressed his image on its coins. From its foundation through the Roman era, Poseidonia's protectors were female divinities. The Sanctuary of Hera at Foce del Sele, for instance, yielded 25,000 terra-cotta figurines, many of them popular depictions of Hera holding a child or a pomegranate, potent symbols of fertility. Later sanctuaries featured representations of a warlike Hera brandishing a weapon (Hera Hoplosmia) and Hera Hippia, protector of horses, a reference, perhaps, to an avocation of aristocrats, who alone could afford four-horse chariots. And the Temple of Athena yielded sixth- and fifth-century terra-cotta figurines of both a nurturing and a bellicose Athena and pottery fragments inscribed with "MENERVA," the goddess's Roman counterpart.

Fifth-century Poseidonia saw the emergence of a democracy, as indicated by the remains of an agora zone with its *bouleuterion,* meeting place for the town council (or *ekklesiasterion,* the assembly meeting place) and the simple, uniform burials found outside the city. The painted Tomba del Tuffatore (Tomb of the Diver), discovered one mile south of Paestum in 1968 by archaeologist Mario Napoli, is a noted exception. All four sides and the lid of the rectangular stone coffin, which dates from around 480 B.C., celebrate the pleasures of aristocratic life. On one of the longer side panels, for instance, five reclining male figures enjoy a banquet; two caress and glance admiringly at each other while a third looks on, another flings the dregs of his wine at a target, and the fifth figure sits alone holding his cup. The lid slab shows a young man plunging from a diving board into water, marking, perhaps, the transition from life to death. The lavish tomb melds Greek and Etruscan traditions.

Poseidonia's proximity to foreign influences accounted for its brief tenure as a Greek city-state. Around 400 the native Samnites, whose local name was "Lucanian," seized control of Poseidonia. Except for a brief interlude, when Alexander I (the Molossian), king of Epirus, accepted an invitation from the city of Tarentum (Taranto) to relieve the Greek colonies of their Italic oppressors and captured Poseidonia in 335, the town remained in Lucanian hands during the fourth century. The Lucanian occupation was, in Pedley's words, "more a fusion of peoples than a conquest," since the city retained its name, coinage, and Greek language, and the Greek temples, sanctuaries, and other traditions remained in use, though with important modifications.

Hera assumed new attributes: Hera as Woman-Flower and the kneeling Hera Eilytheia, who assisted women in childbirth. Figures of men and women carrying piglets and containers—devotees of the usually exclusively female cult Demeter Thesmophoros (Bringer of Laws)—also made their appearance.

Poseidonian artists produced their red-figured wares locally but emulated Corinthian and Athenian vase painting traditions. Two famed vase painters of the late fourth century, Assteas and Python, signed their vessels in Greek and decorated them with scenes from Greek mythology, tragedy, and comic theatre; one shows robbers besetting a miser and his treasure chest.

The numerous graves dotting Poseidonia's territory are striking indicators of the population growth, intensive agricultural production, and social and political change that characterized the period. Extravagantly painted tombs dating from 370 to 290 portray lively funeral scenes, including chariot races, boxing matches, soldiers on horseback, and processions of mourners. The richness of the tomb paintings and graves of wealthy individuals suggests that members of the Lucanian elite were tightly integrated into the Greek town, and perhaps even formed an oligarchy.

The Romans, too, were poised to capitalize on the town's wealth and strategic location. In 273, the encroaching Roman army established at Poseidonia a Latin colony composed of Roman and other Latin peoples, and renamed the city Paestum. Paestum aligned itself politically with Rome during the Carthaginian wars and was expected to provide ships and sailors when needed; Emperor Vespasian planted a colony of naval veterans there, apparently to cement the city's allegiance to Rome. In return for its fealty, Paestum remained relatively autonomous, striking its own bronze coins during the reigns of Augustus (31 B.C.–A.D. 14) and Tiberius (A.D. 14–37).

Whereas the Lucanians had comfortably amalgamated Greek traditions with their own, the Roman conquest of Poseidonia heralded a more complete, deliberate break with the past. Religious, political, social, and architectural conventions all took on a Roman character.

The transformation was most conspicuous in the grid-like building plan of the city, begun shortly after the Roman colonization; instead of orienting buildings in relation to the natural landscape, as the Greek architects had done, the Roman engineers designed structures with reference to each other.

In constructing the Forum, the heart of the Roman town, they expediently employed Doric-style architecture and capitals from the Greek buildings, but chose a site different from that of the Greek agora, as if to reassert their dominance. The Forum and its public buildings—circular, stepped *comitium,* where public meetings took place for the election of magistrates; *curia,* the senate assembly house; *aerarium,* the treasury or jail; *Lararium,* home of the Lares, the divine protectors of the city; and amphitheatre—formed a complex that bisected the town from its northern to its southern borders.

The Romans built their shops, *macellum* (public food market), and basilica (where legal and business transactions took place) on top of a sixth-century sanctuary of Hera. Nearby stood the *capitolium,* the temple of Jupiter, Juno, and Minerva; its Corinthian capitals buttressed a Doric frieze embellished with running women and warriors striking battle poses. Or the temple may have honored Bona Mens, the goddess of good sense, moderation, wisdom, and memory, whose cult took hold in Rome during the third century B.C.

A large third-century-B.C. gymnasium complex features a *piscina,* or swimming pool, which may have been used for ritual bathing of the cult statue of Venus Verticordia (Venus Who Turns the Heart) and Fortuna Virilis.

Private homes with airy courtyards, modest wall paintings, and floor mosaics were arranged in rectangular blocks resembling those found at Pompeii; the houses had, in fact, been refurbished after being blanketed with volcanic ash from the A.D. 79 eruption of Vesuvius.

Gone were the glorious tomb paintings of the Lucanian era; Paestans favored cremation, then reverted to inhumation in simple rock-hewn burials with meager grave goods—a lamp, a bronze coin, a glass ointment bottle, an occasional flash of jewelry—suggesting a leveling of status.

The Romans' propensity for incorporating Greek institutions is most evident in Paestum's religious sanctuaries, especially the Sanctuary at Santa Venera, located roughly one-third of a mile outside the city's south wall. Pellegrino Sestieri, Mario Napoli, and other investigators excavated a major religious complex initially constructed by the Greeks in the early fifth century B.C. and equipped with a sophisticated drainage system.

Inside the rectangular Oikos, where sacrifices were held, an unusual faceted circle of blocks touches three of its four inner walls. Greek and Roman votives representing the female world, such as female figurines, pots, bone hairpins, an alabaster ointment flask, a bone cosmetic box shaped like a duck, amber pendants, and ceramic loom weights, littered the area.

The large Rectangular Hall contained animal bones, votive objects, and a marble base inscribed with a dedication to Venus. Its five column drums, surrounded by omega-shaped niches, may have served as dining tables, seats for worshipers, or supports for cult objects. A nearby *piscina,* installed in the second century A.D., had once sheltered fish or eels kept for aesthetic, ritual, or economic purposes.

The Romans embarked on an ambitious renewal of the sanctuary in the second century B.C.; a later inscription found there records, "Sabina, wife of Flaccus, saw to the construction of the goddess' shrine from the ground up, to the decoration with stucco work, and to the provision of seats and pavements; she paid for it with her own money, and she approved the work." Sabina and another benefactor, Valeria, may have presided over a cult of Venus; Greek figurines of Aphrodite and Roman draped marble statuettes of Venus were found there.

Apart from the few luxury items, benefactions of wealthy individuals, and classical literary references, the picture that emerges from the Roman graves and houses is that of a city whose days of glory were numbered. The 133 B.C. construction of the Via Popilia through the Vallo di Diano east of Paestum diverted trade from the city; soon afterward, Paestum suffered a steady decline. Strabo reported that as the Sele River became silted up, sand and mud choked off streams, and the area became unhealthy. People migrated from this malarial swampland, and the city took on a lackluster appearance.

Early medieval settlements grew up around the Temple of Athena, then converted into a Christian church, and the Church of Santa Maria dell'Annunziata, which had been built from dismantled Roman columns and capitals. Papal records document repeated attempts to establish a bishopric at Paestum, but the unhealthy environment and Saracen invasions of 877 prompted the city's residents to transfer the bishopric to neighboring Capaccio. For a time, Capaccio flourished; recent excavations have explored traces of medieval dwellings, including a baronial residence, and a twelfth-century cathedral, Santa Maria del Granato (St. Mary of the Pomegranate), whose iconography inherited Hera's attributes.

Pottery remains and coins attest to a brief burst of industrial activity in the twelfth and thirteenth centuries centered on the Temple of Athena zone. But the revival did not last: the scant written records of the succeeding centuries described the landscape and its inhabitants as hostile.

Paestum occasionally appeared on sea charts, and its buildings, half-hidden by vegetation, attracted the occasional notice of scholars and travelers, among them the eighteenth-century architect Ferdinando Sanfelice, who sought unsuccessfully to decorate structures at Capodimonte with columns from Paestum. Travelers and antiquarians largely remarked on the glories of Naples, the discovery of Herculaneum in 1738, and the 1748 excavations at Pompeii. Paestum remained virtually lost and forgotten until the middle of the eighteenth century.

The Enlightenment movement reawakened interest in the collection and interpretation of ancient art and architecture. Between 1745 and 1750, Count Felice Gazzola, a Neapolitan courtier who participated in hunting parties in the forests close to Paestum, commissioned some of Naples's finest artists to draw Paestum's monuments. The 1784 publication of Antonio Magri's drawings in *Rovine della citta di Pesto detta ancora Poseidonia* by P. A. Paoli, Gazzola's heir, ignited intense interest in Paestum's history and architectural splendor.

Thomas Major, a Londoner, engraved drawings of Paestum's monuments by Magri. The French architect J. G. Soufflot and draftsman Robert Mylne, supplemented them with historical and numismatic comments, and published *Ruins of Paestum otherwise Poseidonia in Magna Graecia* in 1768, in order to demonstrate "the state of Grecian Architecture in its Infancy, that Elegance, Grandeur, and Magnificence which has been the Admiration of the succeeding Ages." Like-minded thinkers challenged the old notion, embodied in the works of Vitruvius, that the simpler Doric Order was a pale imitation of Roman architecture, the latter tracing its roots to Etruscan structures. Prominent among them was the German theorist Johann Winckelmann, whose landmark publication *Anmerkungen zur Baukunst der Alten* (Remarks on the Architecture of the Ancients) overturned Vitruvius's scheme and proposed instead that the Romans had based their architecture on that of the Greeks. Lively debates over the possible Roman, Etruscan, or Greek origins of Paestum's structures ensued, inspiring new theories concerning the form, function, evolution, and laws of design of Greek architecture and their application to contemporary building programs.

Leading architects and artists, among them Giambattista Piranesi, now included Paestum in their itineraries, and they strove to produce accurate renditions of its monuments. Yet the guidebooks and travel diaries of the day perpetuated the "lost city" image. Literary giants including Sir Walter Scott and Percy Shelley undertook the pilgrimage, and their travel diaries resonate with an aura of mystery; traveling to Paestum was a metaphysical journey akin to the nineteenth-century explorers' search for the source of the Nile.

Archaeological excavations of Paestum's urban centers and necropolises were sporadic and antiquarian in flavor until Vottorio Spinazzola's systematic exploration of the Temple of Hera I and Forum areas in the early twentieth century, followed by the meticulous work of Paula Zancani-Montuoro and Umberto Zanotti-Bianco, Mario Napoli, and recent investigators, including archaeological teams from the Universities of Michigan and Perugia.

Today, visitors may view the Greek temples and altars, *bouleuterion/ekklesiasterion,* Underground Shrine, and foundations of the two extramural sanctuaries. Most of the visible structures, however, are Roman: the houses, Forum and other public buildings, gymnasium and *piscina,* and the western part of the amphitheatre. The cobbled Roman highways and fortification walls appear much as they did two millennia ago.

The Archaeological Museum, inaugurated in 1952 and located next to the Church of the Annunziata, exhibits votive figures, ivory plaques and statuettes, coins, gold and silver jewelry, pottery (including some of Assteas' and

Python's works), the bronze vases from the Underground Shrine, Greek and Lucanian tomb paintings, architectural members from the main temples and sanctuaries, and other antiquities. Principal attractions are panels from the Tomb of the Diver and a facsimile of the Foce del Sele Treasury frieze, which displays its sculptured metopes.

Further Reading: Michael Grant's *A Guide to the Ancient World: A Dictionary of Classical Place Names* (New York: Wilson, 1986) and Margaret Guido's *Southern Italy: An Archaeological Guide* (London: Faber, 1972; Park Ridge, New Jersey: Noyes, 1973) briefly introduce Paestum's history and landmarks. *Paestum: Greeks and Romans in Southern Italy* by John Griffiths Pedley (London and New York: Thames and Hudson, 1990) is a comprehensive, clearly written, and beautifully illustrated account of the city's history and archaeology. Pedley's in-depth analysis brings to life the customs and innovations of its western Greek founders and highlights the continuity between the Greek, Lucanian, and Roman eras. "Excavation at Paestum 1982" in *American Journal of Archaeology* (Boston) volume 87, number 3, 1983, pages 293–303, by Werner Johannowski, John Griffiths Pedley, and Mario Torelli, focuses on recent work at the Sanctuary at Santa Venera. Pellegrino Claudio Sestieri's *Il nuovo Museo di Paestum* (Rome: Istituto Poligrafico dello Stato, 1964) describes the antiquities exhibited in Paestum's National Archaeological Museum. The works detailing Paestum's central influence on debates about the evolution of Western architecture appear in *Paestum and the Doric Revival, 1750–1830,* edited by Joselita Raspi Serra (Florence: Centro Di, and New York: National Academy of Design, 1986). For more background on Paestum's wider cultural contexts, see *Megale Hellas. Nome e immagine: Atti del ventunesimo convegno di studi sulla Magna Grecia, 2–5 Ottobre, 1981* (Taranto: Istituto per la storia e l'archeologia della Magna Grecia, 1982), T. W. Potter's *Roman Italy* (Berkeley and Los Angeles: University of California Press, and London: British Museum Publications, 1987), and A. G. Woodhead's *The Greeks in the West* (London: Thames and Hudson, and New York: Praeger, 1962).

—Maria Chiara

Palermo (Palermo, Italy)

Location: Situated on a bay along the northern coast of Sicily, some forty miles east of the westernmost point of the island. Palermo is surrounded by the fertile plain known as the Conca d'Oro (Golden Shell).

Description: Palermo started in the ninth century B.C. as a Phoenician trading post and later was heavily fortified during the protracted wars between Carthage and the Greek cities. It came under Roman rule during the First Punic War (264-241 B.C.). In the ninth century A.D., invading Arabs brought prosperity to Palermo by making it their administrative seat, and the Norman kings of Sicily added to its glory in the eleventh and twelfth centuries by establishing a flourishing court there. Subsequently, Palermo and Sicily declined under centuries of foreign domination, which continued despite several revolts and revolutions. Sicily was made a part of Italy in the 1860s, and Palermo became the Sicilian capital in 1948, when the island finally gained regional autonomy.

Site Office: A.A.P.I.T. (Azienda Autonoma Provinciale per l'Incremento Turistico)
Piazza Castelnuovo, 35
CAP 90100 Palermo, Palermo
Italy
91 586122

In comparison to the magnificent ancient Greek cities on Sicily, Palermo had a humble beginning. When Syracuse, Selinus, and Akragas (now Agrigento) vied with Athens for pride of place in the Mediterranean world in the sixth century B.C., Palermo (then known as Panormus) was one of a number of Phoenician trading posts at the western end of the island. Whatever wealth was generated by the Phoenicians' trade with the native Elymians and Sicans was taken elsewhere. When Carthage intervened in the internecine warfare of the Greek cities in the early fifth century B.C. (thus beginning more than two centuries of intermittent war with the Greeks), Palermo acquired some military significance. However, it did not truly become a city until after the Romans took Palermo in 253 B.C., in the course of the First Punic War. But even under Roman rule, the eastern end of the island remained dominant and saw the most economic development.

In 31 B.C. Palermo received the favored status of Roman *colonia*, thereby acquiring a greater degree of self-rule and becoming exempt from some of the heavy taxation imposed on Sicily by the Romans. The city also saw an influx of Roman veterans, who made Palermo their new home and

altered the face of the modest town. This era of prosperity lasted for several centuries until the break-up of the Western empire. Subsequent urban development has virtually effaced the Roman presence, however. In the fifth century A.D., the Vandals briefly gained control of Sicily and then sold the island to the Ostrogoths, under whose administration Palermo began a long decline. The Ostrogoths eventually lost Sicily to Byzantium in 535, but the city's fortunes failed to revive under the faltering dominion of the Byzantine Empire. Palermo had been Christianized by the fifth century, and a struggle for influence ensued between the Eastern and Western churches. Byzantium eventually won this struggle in Palermo, as elsewhere on the island, with the result that the city became chiefly Greek Orthodox, although it also had sizable populations of Jews, Italians, and Moslems.

Prosperity finally returned to Palermo when the Arabs, profiting by the weakness of the Byzantine Empire, landed an invading army on the western shores of Sicily in 827. The invaders captured the western half of the island in short order; Palermo, after falling to the Arabs in 831, was made the administrative center of the Sicilian Emirate. The Arab conquest, essentially complete by 878 with the fall of Syracuse, brought improvements in administration and agriculture as well as a tremendous boost in commercial activity. Sicily fared well under its new rulers. The Arab administration pursued a policy of religious tolerance, and Palermo, which saw rapid expansions over the next century, continued to support an extremely diverse population. Nevertheless, Palermo became a predominantly Islamic city, with mosques springing up in all quarters. Arabic, of course, became the official language. By the tenth century, Palermo, in the estimation of contemporary visitors, became a city second only to Baghdad in wealth and culture, surpassing the great cities of Europe.

Internal power struggles in the emirate weakened Arab control over Sicily in the eleventh century. The Normans took the opportunity to invade, landing near Messina in 1060 and sweeping westward through the island over the next decade. By 1071, the Normans, under Roger I, had advanced to the walls of Palermo. After a blockade of five months, the city surrendered and became the capital of the new Norman kingdom of Sicily. However, the Norman takeover did not immediately bring major changes. While the new rulers brought Roman Catholicism with them and quickly established an archbishopric at Palermo (with local endowments for the Benedictine and Cluniac orders), they were at the same time tolerant of Sicilian Arabs, in fact incorporating them into the Norman army and administrative system, which was left virtually unchanged.

The Normans also brought the Romanesque style of architecture, as can still be seen in churches in and around

Church of San Giovanni degli Eremiti
Photo courtesy of E.P.T., Palermo

Palermo dating to that time. However, they employed Arab, Byzantine, and Italian craftsmen to carry out the decorative programs for these buildings according to their native traditions. The Martorana and the Palatine Chapel in Palermo are among the most beautiful examples of the eclecticism fostered by the Normans. Moorish wood ceilings are juxtaposed with Byzantine mosaics within the general setting of northern Romanesque architecture. The Church of San Giovanni degli Eremiti, on the other hand, looks strikingly like a mosque, although it was built in the 1130s, three generations after the Norman Conquest. The remarkable confluence of cultures sustained under the Norman domination of Sicily focused on the court at Palermo, which became a renowned center of science under Roger II in the early twelfth century. Poetry also flourished under his patronage. However, most of the city's wealth derived from its textile production and from the fact that it was a trade nexus between Europe and the East.

By the mid-twelfth century, however, Norman rule weakened. Roger was succeeded in 1154 by his son William I, significantly nicknamed "the Bad." Upon William's death in 1166, his wife, Margaret of Navarre, became regentess on behalf of their minor son, William II. During the six years of Margaret's regency, Walter, archbishop of Palermo, con-

trolled the government. When William II came of age in 1172, Walter at first functioned as his advisor. After a few years, however, William set to work to diminish the influence of the archbishop, establishing a second archbishopric at Monreale, a mere five miles southwest of Palermo. Well on his way to consolidating his position as the supreme authority in Sicily, William died childless in 1189, and the kingdom passed out of Norman and into German hands.

More than thirty years of chaos followed William's death. Henry VI of Hohenstaufen was unpopular with the Sicilians, and his arrival in Palermo in 1194 sparked a rebellion, which was brutally repressed. By this time, the delicate balance between the Christian and Moslem populations had been lost. In 1197, Henry died, leaving a three-year-old heir, Frederick II, and in the ensuing disorder, the Germans in Palermo began seizing Moslem property wherever they could. A massive Moslem uprising, originating in Palermo, spread to the countryside, which was still predominantly Islamic. Order was quickly restored in the city itself, although the rebels held the surrounding country for more than twenty years. The thirty-year reign of Frederick II of Hohenstaufen (1220–50) was marked by relative stability and a revival of culture and the arts at the Palermo court. Frederick's death, however, brought more than a century of upheaval, during which Sicily declined as one of the main agricultural suppliers in the western Mediterranean. Its capital declined as a center of culture and commerce.

In 1260 the pope sold Sicily to England, and a year later he resold it to Charles of Anjou, without, of course, the consent of the legitimate king of Sicily, Manfred of Hohenstaufen. The ensuing conflict over the crown escalated into outright war in 1265. The Angevins carried the victory in 1268, but they were extremely unpopular, and in 1282 a revolt broke out in Palermo, which later came to be known as the Sicilian Vespers. Much romanticized as a heroic bid for independence on the part of native Sicilians, the revolt has been memorialized in such works of art as Verdi's opera, *I vespri siciliani (The Sicilian Vespers)*. At an open-air celebration on Easter Monday, 1282, a French soldier was killed in what apparently began as nothing more than a brawl with the Palermitan crowd. The incident sparked an outburst of violence against the hated French, who were massacred wholesale. Although no exact figures are available, it was said that by the end of the day thousands of Frenchmen lay dead in the streets of the city. Before long, this expression of resentment of foreign overlordship turned into a full-fledged social revolution against feudalism and the autocratic (and generally rapacious) barons. A movement for municipal independence made itself felt, although it lacked organization and forethought.

Soon after the massacre, a number of prominent Palermitans convened a parliament and declared the city an independent republic. They sent armed bands to cities and villages all over western Sicily to incite popular rebellion there as well. Anti-feudal sentiment was rife virtually everywhere, and, joined with resentment of the French, it proved too much for the established order. In fact, by the end of April, all of Sicily had joined the revolution, and the French had been thrown out. Palermo first made an appeal to the Papacy for help in securing independence from Anjou and in putting down the baronial forces. When the pope proved unwilling to support the cause, the rebels turned to Peter of Aragon. After months of delay (as well as consultation with the barons), Peter landed an army at Trapani on the west coast in August 1282, and days later he was crowned king of Sicily at Palermo. Peter quickly made accommodations with the barons, and the outcome of the revolution was that Sicily had exchanged one foreign ruler ready to support the feudal state for another of similar disposition.

Until 1377 Aragonese dominance of Sicily was contested by Naples, then an Angevin possession; for almost 100 years the island was subject to armies ranging over the land and feeding off the people. Depopulation and destruction of farmland and forest resulted. Palermo escaped these ravages relatively unscathed, but the city's former glory was altogether lost. The Aragonese ruled in absentia, and with the disappearance of the royal court the artistic and cultural life of Palermo simply faded away. Commercial activity began to decline as well. In 1412, Sicily went from Aragon to Castile,

but this change was of little consequence to the island. Except for a brief interruption in the eighteenth century, Sicily was ruled from Palermo for the following 400 years by a long succession of Spanish viceroys, whose principal interest was in collecting the desired taxes for the Spanish crown. The island continued its slow decline during this period, but Palermo did not, on the whole, share in this fate. It is true that Palermo's industry was on a downward slope, as is demonstrated, for instance, in the gradual disappearance of its sugar refineries and textile industry. But the city had a flourishing and growing artisan class. By 1385, there were already forty guilds or *maestranze,* and these played a significant political role in the city. The decline in agriculture brought large numbers of destitute peasants to Palermo, resulting in extensive slums.

The growth of an urban proletariat in Palermo was balanced at the other end of the social spectrum by an influx of the landed gentry, who managed to avoid taxation by taking up residence in the capital. The old center of Palermo thus saw a proliferation of wealthy aristocratic residences. In addition, the Spanish viceroys undertook an aggressive program of urban embellishment to symbolize the power and glory of the Spanish crown. One result of this politically motivated ostentation is the magnificent Quattro Canti, the square at the intersection of the Via Toledo and the Via Maqueda, designed and constructed in the early seventeenth century. The three-story screen facades at the four corners of the square, modeled on the Quattro Fontane in Rome, are filled with statuary. The first story holds sculptures symbolizing the four seasons, while the second story displays sculptures of Spanish sovereigns. The third story is taken up with figures of Palermitan saints. The decorative program thus lays out an ideal political order with nature at the bottom, secular power at the center, and religion at the top.

Such visual representations, however, could do little to improve the realities of the economic and political mismanagement that was the consequence of a government uninterested in the general prosperity of the island. By the sixteenth century, "Palermo was a parasite town that consumed much of the revenue of the island and yet despised the country districts which made its luxurious living possible," as historians M. I. Finley, Denis Mack Smith, and Christopher Duggan describe it in their *History of Sicily.* Neither the Spanish viceroys nor the Palermo parliament did anything to diminish the barons' feudal stranglehold on the countryside and the majority of the cities. Sicily's economic condition had deteriorated so drastically that in the seventeenth century the common people suffered from chronic food shortages. A bad harvest in 1647 aggravated the misery of the urban poor of Palermo to such an extent that a bloody revolt broke out. Before long, the rebellion spread to the countryside. From May through September of that year, the people rioted, breaking into prisons to free the inmates and into monasteries to lay hold of the food supplies stored there. Although many of the guilds had been sympathetic to the rebellion at first, it was they in the end who broke its back. By September they were persuaded to act as a militia, restoring order and returning Palermo to the viceroy and the barons without effecting any changes to alleviate the economic misery of the overwhelming majority of Sicilians.

Over the course of the eighteenth century, the population of Palermo doubled to 200,000, and most of the new arrivals swelled the ranks of the poor. In 1773 the starving people rose up again, and this time they controlled the city for more than a year. But again the *maestranze* acted as a police force and eventually put down the rebellion without concessions. A few years later, in 1781, Sicily received a viceroy who seriously undertook reforms, perhaps as a result of civil unrest. Domenico Caracciolo developed a constitution and attempted to strengthen parliament so as to make it more than a tool of the barons. Predictably, he was unpopular with the latter, who managed to force him out in 1786, before he was able to consolidate his reforms.

In 1799 Napoléon invaded Naples, where the Spanish king Ferdinand III then held court. Ferdinand fled to Palermo with his court, which inspired high hopes in the hearts of many Palermitans. It was expected that the city would profit tremendously by being the home of a royal court. However, Ferdinand was not at all interested in Palermo or Sicily, apparently considering his stay there as a most unfortunate exile to a backward and barbaric part of his realm. He left Palermo again as soon as he thought possible, in 1802, but he was premature. Returning to Palermo in 1806 and finding it in much worse shape than he had left it, he called in the British to protect him against a possible invasion of Sicily by Napoléon. The British, surveying the economic and political shambles on the island, attempted to force the reforms that Caracciolo had been unable win. A reformed parliament was convened and a constitution ratified, with the reluctant consent of Ferdinand. Unfortunately, the British presence formed the backbone of the new government, and when they left in 1815 the reforms were again in jeopardy. In 1816 Ferdinand returned to Naples, annulled the constitution, and abolished parliament.

A mere four years later, this perfidy resulted in a peasant revolution that was concentrated, as before, in Palermo. The Neapolitan army sent to subdue the revolt managed to isolate the city, and after heavy bombardments the rebels surrendered. But the situation remained explosive. By the 1840s there was renewed unrest, and this time the disenfranchised found support for their cause with factions of the middle class and even some aristocrats finally influenced by Enlightenment thought. A riot in Palermo on January 12, 1848, once again brought the peasants up in arms against the government, which fled while the police were assassinated. A full-fledged revolution developed, and an organizing committee in Palermo called for a new constitution and reconvened parliament, holding out for independence from Naples. Throughout much of the spring and summer of 1848, peasant squads fought the militia led by the conservative aristocracy. By September the Spanish crown had mobilized an army at Naples and sent it to reinforce the Sicilian militia. After several months during which it wiped out the revolution in the rest of the island, the Neapolitan army laid siege to Palermo. Six months later, the city agreed to an armistice and surrendered.

In 1860 another insurrection in Palermo pitted the peasants against the radical reformers. Garibaldi seized the opportunity to land an invading army at Marsala, on the west coast. After taking the island, he introduced a series of reforms and swayed popular sentiment toward the annexation of Sicily by Piedmont in Italy under Cavour, which was realized in 1861. With the unification of Italy, this move became inconsequential, because Sicily would probably have become part of Italy even without the annexation. A mythology manufactured mostly in the nineteenth century represents Sicily as essentially and always Italian, but the truth is that there had been little national sentiment throughout the island's history and what there was of nationalism had little to do with Italy. In fact the Sicilians, largely unhappy with their incorporation into a unified Italy, began a long campaign for regional autonomy.

Palermo itself did not fare well by the incorporation, declining as the first export center of the island until, by 1900, it was outstripped by Catania on the east coast. Although Palermo's population doubled between 1861 and 1921 and trebled between 1861 and 1961, the city lacked the economic base to support the growing population. Serious housing shortages were exacerbated by Allied bombardments of the city in 1943 and not remedied until the 1950s and 1960s.

By the 1960s, Sicily had finally gained regional autonomy, and Palermo had been made the island's bureaucratic center. The city benefitted greatly from its status as capital of the region, and an upturn in the economy with new industrial development occasioned a building spree that unfortunately seriously defaced Palermo's cityscape. Today the old center, with its architectural treasures preserved from the Norman past as well as from the fifteenth and sixteenth centuries, is surrounded by an ugly urban sprawl that has swallowed up many of the charming ancient communities in the vicinity of Palermo.

Further Reading: A general history of Sicily is available in M. I. Finley, Denis Mack Smith, and Christopher Duggan's *A History of Sicily* (London: Chatto and Windus, 1986; New York: Elisabeth Sifton Books-Viking, 1987; originally published in three volumes, London: Chatto and Windus, and New York: Viking, 1968). *The Norman Kingdom of Sicily* by Donald Matthew (Cambridge and New York: Cambridge University Press, 1992) is an in-depth study of developments on the island under the Normans. Steven Runciman supplies an excellent history of the Sicilian Vespers in *The Sicilian Vespers: A History of the Mediterranean World in the Later Thirteenth Century* (Cambridge and New York: Cambridge University Press, 1958).

—Marijke Rijsberman

Palestrina (Roma, Italy)

Location: Twenty-three miles east-southeast of Rome along Via Prenestina; seventeen miles south of Tivoli.

Description: Hillside town, originally known as Praeneste, set amid the ruins of the enormous ancient Temple to Fortuna Primigenia. Originally a wealthy Etruscan city, Praeneste became a Roman *municipium* and later a Roman summer resort. Ruins remain of three separate towns constructed in tiers descending Monte Ginestro. Nearby are remains of a grand villa, reputedly built by Emperor Hadrian. A medieval cathedral, with intact bell tower, dates from 1100. The seventeenth-century Palazzo Barberini, built inside the original temple walls, now houses a museum.

Site Office: Prenestino National Archaelogical Museum
Palazzo Barberini
CAP 00036 Palestrina, Roma
Italy
(6) 95 58 100

When Allied bombers attacked the Italian city of Palestrina during World War II, they unearthed the remains of ancient Praeneste. The hillside city mentioned by Livy and Horace was famous in history for its mountain breezes, its unusual Roman dialect, and its excellent roses and nuts. But Praeneste was a city of extremes. In the shadow of the enormous Temple to Fortuna Primigenia, Praeneste's fortunes swung between bloody warfare and luxurious prosperity.

The year of Praeneste's founding is unclear, but the city existed in some form as long ago as the late eighth century B.C. Although it lay south of the traditional boundaries of Etruria, the triangle between the Tiber and Arno Rivers, Praeneste was originally an Etruscan settlement. Like other Etruscan city-states, Praeneste was allied with the Latin League, a loose federation of independent cities.

Also, like other Etruscans, Praeneste's residents adopted Greek culture and gods. Its foundation myth was similarly Greek in character. According to the legend, Praeneste was founded by Telegonus, the well-meaning illegitimate son of Odysseus and Circe; while attempting to find his father, Telegonus was mistaken for an intruder and killed Odysseus with a sword his mother had given him. It was also under this Greek influence that inhabitants built what later became the enormous Temple to Fortuna Primigenia, a pyramid of terraces scaling the hillside, visible from Rome and from the sea.

The inhabitants of Praeneste seem to have had a higher standard of living and finer appreciation of the arts than their Etruscan cousins elsewhere on the peninsula. A heightened taste was evident in the contents of the Barberini

and Bernardini tombs from the sixth and seventh century B.C. While Etruscan cemeteries in Caere to the north were filled with everyday, simple, and functional items, Praeneste's tombs were stocked with luxurious objects, jewelry, and ivories. Delicately carved bronze cists (burial chambers) were supported by lions' claws and adorned with miniature Greek statuettes. Symmetrical, Asian-influenced faces and patterns decorated bronze finger bowls. A gold belt clasp found in the tombs is decorated with the tiny details of 131 creatures such as lions, sirens, and chimeras, each sprouting a tiny human head from its back. According to archaeologist Luisa Banti, these were "the personal belongings of individuals who loved being surrounded with art objects, people who wanted to enjoy the sight of beautiful, finely wrought objects in their afterlife."

The inhabitants of Praeneste sought out fine art overseas. The city is thought to have imported foreign craftsmen to Praeneste to set up local schools. Local ivory artists, for example, worked in Oriental styles, but employed a homemade woodlike finish not found in the contemporary Greek or Oriental pieces.

The industrious, seafaring merchants were surrounded by a far less advanced and more aggressive local population of Villanovans and other tribes not that far removed from the Bronze Age. In spite of hostile neighbors, the league of Etruscan city-states prized their independence so much that they could not cooperate to fend off attacks. In that way, the cities of Etruria "resembled medieval Tuscany," writes Banti, "where geographically close cities . . . were separated not so much by distance as by suspicion, hate, old rancours and struggles."

Praeneste's governors chose to ally themselves with their increasingly aggressive Roman neighbors. According to Roman historian Livy, Praeneste sided with the Romans in the battle at Lake Regillus (499 B.C.) against other states in the Latin League. The city was also attacked for years by a primitive tribe of the Aequi, entrenched in the nearby Alban Hills.

Praeneste scrapped with Rome several times, taking advantage of Rome's weakened position after the Gallic invasions. By the time the Latin League was destroyed, losing to Rome in the great Latin War (338 B.C.), Praeneste was punished with the loss of some territories. But rather than be incorporated into the Roman state, the two cities forged a treaty whereby Praeneste was guaranteed independence.

During the relative peace that followed, the cult of the goddess Fortuna Primigenia flourished in Praeneste. The cult is known to have been flourishing by 241 B.C., but the goddess's temple is thought to have been built even earlier. In Praeneste, the goddess Fortuna was called Primigenia (First-born or First-bearing). She was especially revered by matrons and was represented suckling two babies, thought to

Detail of the Nile Mosaic
Photo courtesy of E.P.T., Rome

be Juno and Jupiter. Her widely consulted oracle, known as the Praenestine Lots, was practiced by priests in the sanctuary who inscribed oracles onto sacred pieces of oak wood. The practice continued until the temple was closed in Christian times, first by Constantine and later by Theodosius.

The original temple was constructed at what is now the second-lowest tier of the complete ruin. Built inside a natural grotto, the temple floor was decorated with detailed mosaics depicting a sea scene of Poseidon's temple adorned with swimming fish. The famous *Nile Mosaic* was found in the nearby two-story basilica. This remarkable work depicts the flooding of the Nile, with details including a shepherd's hut, mummies, ibises and Roman warriors. Even the Greek philosopher Carneades, the famous skeptic and founder of the anti-Stoic New Academy in Athens, was impressed by the temple when he visited Rome in 156–155 B.C.

Praeneste's economy boomed during these years. The city provided contingents to the Roman army and possessed the right of exile: banished Romans were permitted to live in peace there. Nuts and roses were cultivated and said to be the finest in Italy. Praenestines even developed their own Latin pronunciation—shortening and softening words—which was sometimes ridiculed by other Romans.

A slave revolt was put down in 198 B.C., but even greater trouble was brewing nearby. By 90 B.C., Praeneste, like many of its neighbors, had become a Roman *municipium*; political upheaval in Rome, less than twenty-five miles away, inevitably spilled into the city.

In 87 B.C., the aging dictator Marius, who had been in exile in Carthage, returned in glory with his ally Cinna and marched into Rome, wresting power from Sulla, Marius's former friend and general, in a bloody coup. (It is said that Marius's guards stabbed everyone whom the returning dictator did not salute.) Marius appointed himself co-consul, but eighteen days into his reign he died of pleurisy at the age of 71.

In 82 B.C., at age 27, Marius the Younger, adopted son of the dictator, was made consul and was promptly attacked by his father's old nemesis, Sulla. The cunning old general defeated young Marius in battle near Sacriportus, and Marius fled to Praeneste, which Sulla besieged. In the city founded by Telegonus, Marius may have felt an affinity for the outcast son of another historically dominating father, Odysseus.

But Marius could not live up to his father's reputation. Six years before, when the elder Marius was imprisoned and a soldier entered to put him to death, the dictator had bellowed, "Man, darest thou murder C. Marius?" causing the frightened executioner to drop his sword and flee. In contrast, Marius the Younger, upon receiving news of Sulla's further victories at Rome, crumpled under the pressure and killed himself.

Praeneste suffered deeply for defying Sulla's army. Troops massacred 12,000 male residents and razed the city, dismantling it stone by stone. Only the temple was left untouched. Sulla moved the remaining inhabitants to a lower altitude and built an army barracks on the site of the former settlement. Sulla delegated the affairs of Praeneste to consul M. Terentius Varro Lucullus in 73 B.C., and ten years later, much of the land had been purchased by proprietors.

Fortuna's temple soon benefited from the new regime. During Sulla's reign, the temple was expanded magnificently by master builders who, for the first time, used concrete mixed with volcanic mortar. The revolutionary technique enabled architects to build larger and stronger vaults, which, in turn, could support larger roofs. The design of the new temple retained traditional Greek elements such as colonnades, an axial layout, open terraces, and arcaded walls, but executed them on an enormous scale made possible by the new construction techniques. New terraces were added with so many steps that the city was later nicknamed Polystephanos, or "many wreathed." Steep ramps led up to a large enclosure roofed by porticoes. At the pinnacle sat an enormous theatre used for religious performances. When construction was finished, the temple filled the entire hillside.

Meanwhile, the resort qualities of Praeneste remained, especially during the hot summer months when mountain breezes cooled the town. Wealthy Romans quickly studded the city with villas and amused themselves watching gladiatorial contests. Roman poet Horace compared Praeneste favorably to Timur and Baiae, although these resorts were more fashionable.

Emperors Augustus, Hadrian, and Marcus Aurelius Antoninus vacationed there, the latter two constructing villas. Praeneste was also a stop in the self-imposed wanderings of Emperor Tiberius, who (as Tacitus reports) detested Rome and wished to "indulge his sensual propensities in private." Writer Pliny the Younger and statesman Symmachus also stayed in Praeneste.

By 313 A.D., the temple of Fortuna had been closed by order of Emperor Constantine, and oracle reading was forbidden. A bishop was appointed and Christianity was imposed on the city. During the following 800 years, the medieval town dwindled to a small village, now called Palestrina, whose population easily fit inside a single terrace of the temple. A cathedral was constructed there in 1100.

But Praeneste's quarrel with Rome was not over. In 1297 the Colonna family, who owned the city, revolted from the pope. The city was crushed ruthlessly by the papal army. According to eighteenth-century historian Edward Gibbon,

> in the use of victory [the papal armies] indulged the meaner passions of jealousy and revenge; and instead of adopting the valour, they trampled on the misfortunes of their adversaries. The captives, in their shirts, with a rope round their necks, solicited their pardon: the fortifications, and even the buildings . . . were demolished.

The town was rebuilt but demolished again less than 200 years later by the cardinal and papal general Vitelleschi in 1437. According to Gibbon, when Stefano Colonna returned from exile to view the smoldering ruins of his family's city, a man asked him, "where now is your fortress?" Colonna is said to have laid his hand on his heart and replied, "Here." Colonna rebuilt and fortified the town in 1448.

Palestrina's most famous modern resident was Giovanni Pier Luigi da Palestrina, the renowned sixteenth-century composer of 105 masses and many madrigals, magnificats, and motets. Palestrina—a true musical visionary who invented polyphonic harmony—was employed by the Vatican during his tumultuous career. He was thwarted and manipulated by the caprices of six different popes and their bureaucracies. One of his compositions was even credited to a pope living centuries earlier.

In 1630, after Palestrina's death, the town was purchased by the Barberini family, who built a palace along the semicircular lines of the original temple.

Although the town slipped into obscurity and poverty during the ensuing centuries, the area grew in importance as an archaeological site. In 1738, the famous Ficoroni Casket, depicting the arrival of the Argonauts in Bithynia and the defeat of Amycus by Pollux, was excavated there. In 1771, a Roman calendar created by grammarian M. Verrius Flaccus was discovered after having been incorporated as building material in the Church of San Agapitus. By 1855, excavation was underway on the necropolis at the foot of the hill. Objects in the oldest graves date to the seventh century B.C. and include Phoenician-style silver cups. Bronze and Etruscan artifacts were also uncovered.

The pace of archaeological excavations at Palestrina increased markedly after Allied bombing during World War II uncovered the true scope of the original temple complex, which extends far into the plains below the modern town. The Palazzo Barberini has been converted into a museum to house the wealth of objects found in the area. Although art thieves stole several bronzes in 1991, the museum's chief attraction, the *Nile Mosaic,* remains.

Further Reading: *Etruscan Cities and Their Culture* by Luisa Banti, translated by Erika Bizzarri (Berkeley and Los Angeles: University of California Press, and London: Batsford, 1973; as *Il mondo degli Etruschi,* Rome: Biblioteca di Storia Patria, 1968) is a comprehensive study of theories regarding Etruscan life, art, religion, and politics. For a historical treatment of technological advances and other elements of the ancient lifestyle, see *The Oxford History of the Classical World,* edited by John Boardman, Jasper Griffin, and Oswyn Murray (Oxford and New York: Oxford University Press, 1986).

—Jean L. Lotus

Palma (Baleares, Spain)

Location: On the southern shore of the island of Majorca, which lies about 100 miles off the eastern coast of Spain in the Mediterranean Sea.

Description: The capital and largest city of the Balearic Islands, with buildings dating from the Middle Ages and the Renaissance, surrounded by zigzagging streets called the Avenues. Moorish walls still stand on the side facing the sea. Principal buildings include the Gothic cathedral; Almudaina, the royal palace; and Bellver Castle, a Gothic structure located in the hills overlooking the suburb of Terreno to the west of the city.

Site Office: Oficina de Tourismo
Jaime III, 10
07001 Palma, Baleares
Spain
(71) 71 22 16

Palma was founded in 120 B.C. by Caecilius Metellus, a Roman naval commander, and was probably named Palmeria (Latin for "triumph") in honor of his conquest of the islands. Historical information on Palma and the islands in general during Roman rule is sparse, suggesting that life was peaceful and comfortable. Palma has little in the way of Roman ruins. However, archaeologists have noted surviving hints of Roman influence, such as the layout of the streets and the style of the gate of the Almudaina. Whatever structures remained after the Vandals drove out the Romans in about A.D. 450 was destroyed, primarily during the Middle Ages.

The Vandals ruled until 534, when Majorca became part of the Byzantine Empire. During the next few hundred years, though Christians lived peacefully side by side with Moslems, Christian kingdoms struggled with Moors for control of the islands. It was during this time that the first wall around Palma was built. It enclosed a small area that now contains the Almudaina, the cathedral, and the land about 300 feet to the north. A second wall went up in the 800s. This one enclosed the area bordering what is now the Borne and the Avenida General Mola, extending about 600 feet north of the cathedral and palace and about 1,500 feet east to the site of the Moorish fortress of Gomera on the southeast side of the city, at the end of what is now the Calle del Sol.

In 902, the Moorish Emirate of Cordova (which later became the Caliphate of Cordova) established sole control of Majorca. In 1015, the wali of Denia, who served as the governor of the island, declared Majorca an independent emirate. Early in the twelfth century, the Moors constructed a third wall to protect Palma.

During this period of Moorish rule, Palma was called Medina Mayurka, and the walis ruled from their palace there,
which stood at the site of the present Almudaina Palace. Along with international trade, Moorish skill at irrigation and other farming techniques, construction, and industrial crafts such as pottery making and jewelry making were key factors in the prosperity of the region. Moslems worshiped at the magnificent mosque, which later became the great cathedral of Palma. Under the wali of Morthada, who wrested control of the Balearics from the wali of Denia, Majorca became a base for pirate raids against ships from neighboring Christian kingdoms. In 1114, largely to put an end to piracy, the Catalans and Pisans joined forces in a quasi-crusade to drive the Moors from Majorca. They captured Palma, tore down the wall, plundered the city, murdered nearly every Moslem, and freed about 30,000 Christians imprisoned by the Moors during the siege.

However, the Moors were still powerful throughout Spain, and in 1115 the Christian conquerors withdrew from the city. In 1116, Palma's walls were rebuilt, and the city came under the control of the Almoravide Moors, a dynasty of North African Berbers. The fanatical Almohades took over in 1203. Although the Almohades believed in strict adherence to Islamic teachings and were intolerant even of less devout Moslems, to preserve Majorca's prosperity they encouraged Christian traders to use Palma as a distribution center for their goods. On the other hand, the Moors still engaged in acts of piracy, presumably against the vessels of these same merchants.

Moorish domination of the Balearics came to an abrupt end in the first half of the thirteenth century. Abu Yahya, the Almohade leader at the time, had captured two Catalan ships. James I, the king of Aragon, demanded their return. Abu Yahya consulted Christian merchants in Palma about how he should respond, and they told him that James, only in his early twenties at the time, had little power. He refused to return the ships to James.

James responded to this insult vigorously. He embarked for Majorca in September 1229 with a fleet of 150 ships. The voyage was stormy, and James vowed that if St. Mary granted him safe passage, he would dedicate a church to her. His fleet landed safely at San Telmo on Majorca's west coast. The next day they engaged the Moors at Santa Ponsa, near the coast about ten miles southwest of Palma. This battle resulted in the rout of the Moors, with 1,500 killed. However, James suffered serious losses as well—two of his most valued commanders died in the battle. In the days following, the Aragonese king besieged Palma, and both sides hurled projectiles with their various forms of catapults. The Moslems tried to drive out James's army by cutting off the spring that supplied water to their camp. James ordered his troops to drive the Moors from the spring; the fallen Moors were decapitated and their heads flung over the city walls. The

The town hall at Palma
Photo courtesy of Tourist Office of Spain, Chicago

Moors answered this outrage by hanging live Christian prisoners, crucifixion style, along the section of the wall that was under attack by the Christian forces. With encouragement from the prisoners and the blessing of his clerical advisers, James continued the offensive in spite of the danger to the prisoners. According to accounts, however, the Christian prisoners were miraculously untouched by the bombardment, and at nightfall the Moors took them down.

Over the next few months, the two sides carried on peace negotiations but were never able to agree on terms. The battle flared into increased ferocity at the end of December, as the Moslems desperately tried to drive off the Christians. Finally, on December 31, James's forces began a fierce assault. They blasted an opening in the wall, and the cavalry and foot soldiers poured through. In the hand-to-hand fighting that took place in the streets of the city, 20,000 Moslems were killed. For the next eight days, the city was ransacked. For a month thereafter, prisoners and valuables were sold at auction. Abu Yahya was among the captured, and though he was initially promised mercy, some evidence exists that he was tortured to death.

James secured most of the island by the spring of 1230. The same year, he instituted a constitution for the city of Palma, the Carto de Poblacio, or Charter of the People, which guaranteed equal rights for ordinary citizens, clergy, and nobility alike. The development of the middle class as a result of this bill of rights was one important factor in the development of Palma as a great commercial center in the Middle Ages. The Catalan language that was introduced at this time is still spoken by the islanders in a dialect called Mallorquin, which incorporates French and Arabic words. In 1230, in fulfillment of his promise at sea the previous year, James began construction of Palma's great cathedral on the site of the Moslem mosque.

James I died in 1276, and his kingdom was divided between his two sons; the Balearics went to his younger son, who ruled as James II, and Aragon went to the elder, Peter III. The brothers engaged in a bitter dispute over power in the region, and it was not until the Treaty of Anagni in 1295 that James could finally install himself at Palma as undisputed king of Majorca.

James II continued his father's legacy, instituting measures that contributed to the continued blossoming of Palma as the center of the prosperous Majorcan kingdom. He founded the city's weekly market in 1302 and fortified the island against Moorish pirates. In the first years of the fourteenth century, he began construction of Bellver Castle, which he used as a vacation spot, in the hills above what is now the Palma suburb of Terreno. James's administration was headquartered at Almudaina Palace. Among his remodeling projects there included the addition of the rooftop garden and zoo, and bathrooms equipped with hot water heaters.

James II's reign also witnessed a flourishing of culture, epitomized largely by the birth of the great period of Catalan Gothic architecture. Numerous churches went up, including the Church of San Francisco, which James II commissioned in honor of his eldest son, who gave up his claim to the throne to enter the church. James II was also the patron of one of the most accomplished medieval scholars and missionaries, Ramon Llull.

Llull was born sometime between 1232 and 1235 into one of Palma's noble families. The story of his conversion from a young libertine into a pious scholar is poignant. A devotee of the courtly pursuit of love that was in fashion at the time, Llull was infatuated with Ambrosia de Castello, the virtuous wife of a wealthy Palma merchant. He composed courtly poems extolling the beauty of her breasts and followed her everywhere; to get her attention he once even barged—on horseback—into a mass she was attending at the Church of St. Eulalia. After this spectacle, Ambrosia presumably could take no more, so she agreed to a rendezvous. Llull

arrived, believing his passion would soon be fulfilled. But when she undid her bodice to show him the breasts he had coveted, he was horrified to see that they were covered with cancerous tumors.

Shock and remorse drove him to give up his hedonistic ways and to seek enlightenment. His wish apparently was granted to him in meditation on top of Mount Randa about twenty miles west of Palma, where he had a vision in a bush that revealed to him the single system that united all the sciences of the world and the mathematical formula that underlay the spirit and the church. Llull became a linguist, theologian, and scholar and wrote more than 200 books on nearly every academic discipline that existed at the time. He zealously took up wandering missionary work and learned Arabic to better communicate with those he sought to convert. Llull lived to be about 81, when he was either martyred in North Africa or died of illness in Majorca. He is buried in a chapel in the Church of San Francisco in Palma. The Catholic Church awarded him beatification, but petitions by Majorcans over the centuries for canonization have gone unheeded.

James II died in 1311, and his second son, Sancho, succeeded him. Sancho's reign coincided with the peak of Palma's glory as a trade center. Many wealthy merchants built mansions in Palma during this period, and a noteworthy—albeit anonymous—school of painting developed. Sancho was an able ruler, but was hampered by asthma attacks and lived for many years away from Palma in the mountains of Majorca and on the mainland. He had no male heir and was succeeded upon his death in 1324 by his nephew, James III.

James III's lack of diplomatic skills was a key factor leading to his loss of the kingdom to his brother-in-law, Peter IV of Aragon, in 1349. As part of the kingdom of Aragon, Palma's fortunes suffered. Aragon waged war on various other kingdoms, exposing Palman trading vessels to attack, and Peter began to impose heavy taxes on Majorca's citizens to finance this warfare. As the wealthy were squeezed by the dwindling of trade, they passed the burden onto their poor tenants in the form of higher rents and taxes.

This hardship eventually led to the tragic pogroms in Palma, beginning in the fourteenth century. Although Jews did not enjoy all the same privileges of the Christian population—during the time of James I's conquest, for example, they were forbidden to carry arms and since 1285 had been prohibited from living outside their ghetto—they had long held positions of influence, serving as doctors, merchants, scholars, cartographers, and translators. However, their role in tax-collecting contributed to their unpopularity with Majorca's peasant population. In 1391, the peasants attacked the ghetto, killing 300 Jews, then turned their wrath on Palma's governor, who fled along with 800 Jews to Bellver Castle. The peasants were able to win cancellation of the debts they owed to Jews and were allowed to keep whatever valuables they had looted from the ghetto. A number of Jews were forcibly converted.

Persecution of Palma's Jews continued. In 1413, restrictions that had long been in force were elevated to the status of law when the king handed down a decree making it illegal for Jews to live outside the ghetto, eat with Christians or sell food to them, or, among other things, carry arms. Punishments ranged from public flogging to confiscation of all property.

These persecutions culminated in 1435, when the Jewish population of Majorca was officially eradicated. During the Easter Holy Week that year, three Jewish leaders were arrested for staging a mock crucifixion in Palma and were sentenced to be burned alive. They were offered the less ghastly option of hanging if they would convert. They agreed, and when the rest of the Jewish population heard the news, they appeared to request baptism themselves. A mass conversion was staged at St. Eulalia. The three prisoners were ultimately pardoned.

Although the Jews had ostensibly converted to Catholicism, many continued to practice their old religion and prejudice against them continued. Even those who embraced the Roman Church were not trusted to blend with the "true believers" and were forced to remain apart. They became known as the Chuetas, and even today they form a unique subgroup of Majorcan Catholics and worship in the Church of Montesion (Mount Zion) located in the former Jewish ghetto.

In 1488, the Inquisition, which had raged in Aragon since the first half of the thirteenth century, struck Majorca, and Palma's plazas became the setting for many public executions of Jews and other non-Catholics for the next 200 years. These executions, known as *autos-de-fé*, often took the form of all-day public holidays during which a mass was celebrated and the condemned were tortured and burned alive.

In the mid–fourteenth century, Palma's prosperity reached its peak. Its port was the center of trade between Europe and Africa, and the merchants of Palma, who owned approximately 900 ships, grew rich, building magnificent palaces in the city. In 1456, the Lonja, one of the Mediterranean's grandest stock exchanges, was completed in Palma. Located in the southwest corner of the city, the Lonja once perched directly on the waterfront, and merchant vessels could unload their cargo of gold, gum arabic, honey, leather, and other commodities directly inside.

But even amid this splendor, the seeds of Palma's social and economic decline were already germinating. The persecution of the Jews, which had essentially neutralized that vital and productive workforce, hastened the decline in ways that must have been difficult for the agents of the Inquisition to recognize. Other events of the fifteenth and sixteenth centuries contributed further. Natural disasters, such as earthquakes, outbreaks of plague, and devastating floods in 1403, took their toll. In 1453, the Turks conquered the Byzantine Empire, and eastern markets collapsed. With the discovery of the Americas in 1492, the attention of Europe shifted westward, and Queen Isabella prohibited the Balear-

ics from trading with them. Finally, Portuguese navigator Vasco da Gama's discovery in 1498 of the sea route to India around the southern tip of Africa meant even fewer ships would call on Palma's harbor.

Although the wealthy had ways of staving off the effects of the economic recession and managed to maintain their luxurious lifestyle, the poor continued to suffer. The animosity of the rural peasants that brought about the 1391 pogrom became an enduring fact of life. In 1521, peasants and artisans, led by a Palman named Juan Crespi, staged an uprising against the ruling classes. They drove Palma's viceroy from the island, killed nobles, and took control of the city. The fighting went on for two years throughout Majorca until the viceroy, supported by reinforcements from the mainland, returned. Crespi agreed to let Emperor Charles V mediate their dispute. In exchange for mercy, Crespi and his followers surrendered to the viceroy. The promise proved empty, however, and the viceroy ordered Crespi and about 420 fellow rebels to be drawn and quartered.

During the next few hundred years, Palma's fortunes were largely dictated by events on the Spanish mainland and abroad. From the mid–fifteenth through the sixteenth century, the Turkish conquest of the Aegean, as well as persecution of Moors in Spain, led to increased threats from piracy. Palma responded in 1560 by beginning restoration of the old Moorish walls, a program that continued into the eighteenth century. Palma saw sporadic action in various wars and conflicts, including the War of the Spanish Succession, the Seven Years War, and struggles between liberals and traditionalists in the Carlist Wars. Palma's leaders tore down several old buildings outside the city walls in 1771 to improve the city's defenses. However, by the late nineteenth century, defense took a back seat to the need for beautification and expansion of the city, and work on tearing down the walls began. As a result, the lovely view of the city from the harbor that visitors enjoy today was first revealed.

The groundwork for Palma's future role as a major tourist destination was laid in the nineteenth century when Archduke Luis Salvador published a number of books about Majorca. In 1910, the government founded the Fomento de Turismo on Majorca to stimulate tourism.

The Spanish Civil War was a painful period for Palma, as it was for the rest of Spain. Early in the fighting, Majorca fell to the nationalist forces of Francisco Franco, and Palma was occupied by his Italian allies, who carried out bombing raids against the Spanish mainland from its airport. During that time, the Ramblas, a popular promenade like the Borne, was renamed the Via Roma.

Today, Palma remains a beautiful and prosperous city, though now its revenue comes chiefly from tourism rather than trade. Many of the historic buildings now house museums and government offices rather than noble families. The Almudaina, for instance, is a government building, the Lonja hosts art exhibitions, and Bellver Castle is home to a historical museum that exhibits Roman mosaics and other antiquities.

Further Reading: *Majorca* by Arthur Foss (London: Faber, and Levittown, New York: Transatlantic, 1972) is a comprehensive study of the island's history, with abundant material on the significant historical sites of Palma. *The Balearic Islands* by L. Pericot Garcia, translated by Margaret Brown (London: Thames and Hudson, 1972; New York: Praeger, 1973), is a scholarly archaeological study of Majorca and the other Balearic Islands from Neolithic times through the Roman conquest. *Mediterranean Island Hopping: The Spanish Islands* by Dana Facaros and Michael Pauls (New York: Hippocrene, and London: Sphere, 1981) provides a readable, well-researched introduction to Majorca and Palma, their history, people, and contemporary ambience. *The Balearics: Islands of Enchantment* by Jean-Louis Colos, translated by Christine Trollope (Skokie, Illinois: Rand McNally, and London: Allen and Unwin, 1967), contains more than 130 striking, fully annotated black-and-white photographs of historic sites on the islands.

—Lisa Klobuchar

Pavia (Pavia, Italy)

Location: On the northern bank of the Ticino River, five miles from its junction with the Po, about nineteen miles south of Milan.

Description: An important city in Roman times; reached its pinnacle in the Dark Ages as the capital of successive barbarian kingdoms; later ruled by Italian noble families. Famed for its university and its medieval and Renaissance architecture.

Site Office: A.P.T. (Azienda di Promozione Turistica del Pavese)
Via Fabio Filzi, 2
CAP 27100 Pavia, Pavia
Italy
(382) 27238 or 27706

Pavia, a university and industrial city, enjoyed the high point of its long and varied history as one of the most powerful capitals of the barbarian kingdoms of the so-called Dark Ages. Its significance dates to Roman times, however; Pavia's location on the northern bank of the Ticino River, about five miles from where it meets the Po, made it a strategic and commercial center for the Romans and their successors. Ticinum, as Pavia was then called, was a large and prosperous Roman colony. The Roman historian Cornelius Nepos was born here in the first century B.C., and here Emperor Augustus and his wife Livia Drusilla welcomed Tiberius back from his successful campaigns in Germany.

In A.D. 352 Ticinum was the site of one of the last battles between the armies of Constantius II and Magnentius for control of the empire; Constantius emerged victorious. In 408 the Roman general Stilicho assembled his army here to prepare to fight the Visigoths, but Emperor Honorius, who suspected Stilicho of conspiring against him, ordered the army to kill Stilicho and many of his officers. The soldiers pillaged the town while the Visigoths bypassed Ticinum on their way to sack Rome. Ticinum escaped pillage by Attila the Hun forty-four years later by voluntarily surrendering many of its valuables.

In 476, generally considered the date of the fall of the Western Roman Empire, Orestes and his son Romulus Augustulus, the last emperor, fled to Ticinum only to meet the barbarian conqueror, Odoacer, who killed Orestes, deposed Romulus, and became the king of Italy. By that time, the Catholic Church was becoming wealthy and influential in politics: the brave and eloquent local bishop, Epiphanius, was able to persuade Odoacer to put a stop to the burning and plundering of the city. Odoacer sought the position of representative of the Eastern Roman Empire but was refused; he therefore continued as king of Italy, bringing several years of peace.

Odoacer was not a particularly strong ruler, however, and he was not on good terms with the Byzantine emperor Zeno. With encouragement from Zeno, Theodoric, king of the Ostrogoths, invaded Italy from the north in 488. In 493, Theodoric overtook Ravenna, seat of the Roman aristocracy and Senate; there he killed Odoacer. Pavia became Theodoric's headquarters and protected his family and noncombatant Ostrogoths when the king turned to other campaigns. Then Theodoric developed Pavia, where he built a palace, baths, and an amphitheatre. He divided his time between Pavia, Ravenna, and Verona.

Theodoric was an effective ruler, enjoying support from both the Goths and the former Roman subjects of Pavia. He maintained Roman legal traditions while weeding out some of the corruption that had marked the waning years of the empire. His reign was not without its share of intrigues, however. Boethius, a philosopher, high-ranking senator, and one-time friend of Theodoric, was suspected of conspiring with the Byzantine emperor against the king, and in 523–24 Theodoric had him imprisoned on an invented charge of witchcraft in the tower of Pavia, which stood until 1584. There Boethius wrote the great work, *The Consolation of Philosophy.* He was executed by strangulation in 524.

Boethius had been a supporter of Roman traditions, and the divisions between these traditionalists and the Goths deepened after his death. Theodoric's popularity helped hold the kingdom together, but after his death in 526 the factions became polarized and the kingdom weakened. In the unrest that followed, Byzantine emperor Justinian I sent his army to invade northern Italy, beginning the Gothic Wars. Pavia was one of the last cities to hold out against the Byzantines. It was still occupied by Gothic soldiers in the early 540s, but eventually it fell as well.

Byzantine control, however, did not last long. The invasion of the Lombards from what is now Hungary in 568 proceeded without opposition until they laid siege to Pavia, the first city to resist the forces of the Lombard king Alboin. Pavia finally fell, however, and subsequently became the capital of the Lombard kingdom of Italy, which was to last slightly more than 200 years.

Pavia was not the first choice of the Lombards as their central administrative seat. Alboin preferred Verona, and Agilulf, who succeeded Alboin's son Cleph and grandson Authari, liked Milan. But Pavia did finally become preeminent by the 620s during the reign of Arioald. Cunipert, who reigned from 679 to 700, further expanded the role of Pavia. The city was filled with grand buildings, including the palace and numerous churches, and learned men, including grammarians, lawyers, and notaries.

Lombardy reached the height of its power under Liutprand, who became king in 712. He revised several laws,

Small cloister at the Certosa de Pavia
Photo courtesy of A.P.T., Pavia

drawing upon Roman tradition and integrating it smoothly into the Lombard legal framework. His military ventures were nearly always successful, so that by 743, the year before his death, Lombard military power was unrivaled in Italy. Later in the century, however, the papacy, which had begun to perceive Lombard strength as a threat, invited the Franks—who heretofore had been on friendly terms with the Lombards—to invade northern Italy. They besieged Pavia in 754 and forced the Lombard king Aistulf to accept a treaty wherein he ceded Ravenna, which he had captured in 751, to the papacy. He broke this treaty in 756; again the Franks invaded, and the treaty was restored.

Aistulf died in 756, and his successor, Desiderius, the last Lombard king, attempted to reassert Lombardy's power. He attacked papal territory in 772–73. Pope Hadrian II called upon the Franks, led by Charlemagne, to invade Italy. They besieged Pavia and drove Desiderius from power; Charlemagne claimed the territory as his own. Under Charlemagne and his successors in the Carolingian dynasty, Pavia remained the administrative center of Lombardy. The Carolingians and their successors, the Holy Roman Emperors, visited the city infrequently, however, and had little power to control their Italian territory.

By 1024, when Holy Roman Emperor Henry II came to Pavia, the situation had deteriorated to the point that the citizens attacked and burned the royal palace; the people of Pavia felt that maintaining the royal seat had become a financial burden and no longer brought the advantages it once had, when rulers took a more personal interest in the city. Conrad II, Henry's successor, attacked Pavia two years after Henry's death; meeting much resistance, he agreed to move the royal residence to a location outside Pavia's walls and the place of coronation to Milan. Pavia was no longer in any sense a capital.

In the ongoing medieval struggles between the Holy Roman Empire and the papacy, Pavia always sided with the empire, making it the enemy of Milan and the Lombard League, which sided with the papacy. Pavia was often the place the emperor would begin or end a military campaign. It was faithful to Frederick I (Barbarossa) when he attempted to regain control over Lombardy in the eleventh century, and it became his rallying point in the battles against Milan, which headed the Guelph (pro-papal) league. Pavia also stood by Frederick II in his thirteenth-century wars with Pope Gregory IX; this support had as much to do with Pavia's rivalry with Milan as with its imperial affiliation. Pavia was ruled briefly in the late thirteenth century by Mastino I della Scala of Verona, who had assisted Frederick II's grandson, Conradin, in his attempt to capture Sicily; the venture did not succeed, but Conradin was sufficiently grateful to Scala to make him lord of Pavia. Pavia eventually rebelled against Scala, then submitted to Holy Roman Emperor Henry VII in 1310.

The powerful aristocrat Matteo I Visconti, however, asserted his rule over Pavia in 1313, two years after he took possession of Milan. Visconti rule left an impression on Pavia: Matteo's grandson Galeazzo II built the great Castello de Pavia (also known as the Castello Visconti), hosted the great poet Petrarch there, and founded the University of Pavia. The court at Pavia was said to be more brilliant than the Visconti seat of power, Milan. Galeazzo's son Gian Galeazzo, also a patron of the arts, lived at the Castello while beginning construction of the famed Certosa de Pavia, a family chapel and monastery just outside the city. At Gian Galeazzo's death in 1402, his son Filippo Maria was made count of Pavia, and at his death in 1447 Pavia passed to his son-in-law Francesco Sforza.

The Sforza family held lavish family weddings and other festivities in Pavia, and also engaged in a number of intrigues. In 1494 Francesco's son, Lodovico Il Moro, was serving as regent for his nephew Gian Galeazzo Sforza, but wanted to seize power for himself. He feared reprisal by the kingdom of Naples, however, and invited Charles VIII of France to invade Naples. Lodovico entertained Charles at Pavia in a lavish style. Within a week of the royal interview Gian Galeazzo died; he had been ill for some time, but even so Lodovico, who had kept his nephew a virtual prisoner in the castle, was suspected of poisoning him. After Gian Galeazzo's death, Lodovico crowned himself duke. The Naples invasion did not succeed; Charles returned to France. Lodovico, having turned against Charles, then made it likely that he would be deposed by France; the mission was carried out by Louis XII, Charles's son, with the help of Swiss mercenaries. It is said that Lodovico showed the powers of Europe how easy Italy was to invade, enabling them to decide their contests on the chessboard of Italy.

After Louis XII died, France and Spain both claimed northern Italy. Struggle ensued, and Pavia saw a quick succession of French, then Spanish overlords. The Spaniards helped to install Francesco, son of Lodovico, as duke of Milan. However, the castle at Milan was occupied by a French garrison, so Francesco ruled from the Castello de Pavia. Francis I of France, with the help of the pope and Alfonso I d'Este, invaded Italy in 1524. At the Battle of Pavia, France was overwhelmingly defeated, and Francis gave up any claim to Italian territory.

Under Spain, Pavia experienced nearly two centuries of neglect and impoverishment. Italian nobles, allied with France, attempted to capture the city in 1655, without success. With the War of Spanish Succession (1701–14) and Austria's subsequent possession of Lombardy, the situation began to improve. Austrian archduchess Maria Theresa suppressed the Spanish Inquisition, put an end to other abuses of power by the clergy and the judiciary, developed agriculture, patronized the arts, and helped restore and modernize the famous University of Pavia. Her son, Holy Roman Emperor Joseph II, carried on her work.

When Napoléon invaded toward the end of the century, Pavia, loyal to its Austrian rulers, resisted. Napoléon, however, overwhelmed the city, and once he became emperor Pavia accepted his rule. Napoléon further endowed the university and improved the canal between Pavia and Milan. After the downfall of Napoléon and the later union of Italy, the university continued to thrive; it is still one of the country's top educational institutions.

Modern Pavia is overshadowed by Milan, but the university remains a source of distinction, and has helped the city's economy to flourish. Industry and agriculture also are important to Pavia, as are tourists, many of them making side trips from Milan. Pavia's long and varied history is apparent throughout the city. The street plan is the lasting legacy of its days as the Roman city of Ticinum. The Duomo, or cathedral, begun in 1488, is located at the intersection of the city's two main streets. Many architects worked on it, including Giovanni Antonio Amadeo, Leonardo da Vinci, and Donato Bramante. The cathedral received additions until 1930; despite the many hands that worked on it, the cathedral has the appearance of being not quite complete. Pavia's most important church, the Basilica di San Michele, was founded in 661 and rebuilt in the twelfth century after being struck by lightning. Charlemagne was crowned in the original church, and Emperor Frederick I in the rebuilt one. The church's golden sandstone exterior bears an odd relief of intricate friezes depicting the Apocalypse. Other detailed carvings decorate the interior.

The University of Pavia, with a student population of about 22,000, stands in the northeast part of the city. It is particularly renowned for its medical and law schools. In the middle of the university campus is the crypt from a seventh-century Arian church, Sant' Eusebio.

The Castello Visconti was damaged in the Battle of Pavia, but its remains house the Museo Civico, displaying Roman artifacts, Lombard and medieval carvings, and twelfth-century mosaics. Behind it is a twelfth-century church with the evocative name of San Pietro in Ciel d'Oro—St. Peter in the Golden Sky. It takes its name from its gilded ceiling, which is mentioned in Dante's *Paradiso*. It houses the relics of St. Augustine, retrieved from Carthage by Liutprand, who is buried there. Also entombed is the philosopher Boethius.

About five miles north of Pavia is the Certosa di Pavia, an architectural jewel. Its foundation was laid by Gian Galeazzo Visconti in 1396. It is the work of many medieval masters, but especially of Amadeo, who with his pupil, Bergognone, devoted thirty years to its sculptural decoration and design of its elaborate facade. It served as a monastery until suppressed by Napoléon, but was reoccupied by the contemplative Cistercian order in the late twentieth century. A few of the Cistercians there have been released from their vows of silence so they can guide visitors around the complex. It houses many fine works of art and the tombs of Gian Galeazzo Visconti, Lodovico Il Moro, and his wife, Beatrice d'Este.

Further Reading: For a reverential look at Pavia, see *Lombard Towns of Italy* by Egerton R. Williams Jr. (London: Smith, Elder, and New York: Dodd, Mead, 1914). It devotes much space to the history and architecture of Pavia. Chris Wickham explores the Dark Ages, when Pavia was the seat of power for consecutive barbarian kingdoms, in *Early Medieval Italy: Central Power and Local Society 400–1000* (Totowa, New Jersey: Barnes and Noble, and London: Macmillan, 1981). For the medieval period, see *Society and Politics in Medieval Italy: The Evolution of Civil Life 1000–1350* by J. K. Hyde (New York: St. Martin's, 1973) and *Lords of Italy: Portraits of the Middle Ages* by Orville Prescott (New York: Harper and Row, 1972). Mrs. H. M. (Katherine Dorothea) Vernon's *Italy from 1494 to 1790* (London: University of Cambridge Press, 1909) details Pavia's long period of decline and some degree of renewal.

—Gregory J. Ledger

Pergamum (İzmir, Turkey)

Location: In the İzmir province of southeastern Turkey, 175 miles southeast of İstanbul, 25 miles east of Aegean Sea.

Description: Following an obscure early history, this city became the capital of the ancient Greek province of Mysia and was variously called Pergamon, Pergamos, and Pergamum. It was taken over by Persians in the mid-sixth century B.C., reverting to Greek control in 334 B.C. Expanding to the largest city state in Asia Minor, Pergamum reached its apogee under Eumenes II (197–159 B.C.), who established a huge library, an impressive theatre and many monuments. The Romans took control of the city in 133 B.C. After the fall of Rome, Pergamum was incorporated into the Byzantine Empire. Ottoman Turks conquered the area in the fourteenth century, changing the name of the city to Bergama.

Site Office: Tourist Information Office
Zafer Mah., İzmir Cad., No.: 54
Bergama, İzmir
Turkey
(232) 6331862

This thoroughly Turkish city, only recently become a major tourism center, is located in a fertile plain of southeastern Anatolia, in the heart of Asia Minor. The history of the earliest settlements on the site is not clear, but by the time the Hellenic armies laid siege to nearby Troy, the Greeks found a Mysian settlement there. This settlement, possibly called Teuthrania, seems to have belonged to King Teuthras of Mysia. The local inhabitants resisted the invaders and also aided King Priam of Troy in his doomed struggle. The Greek Pergamos, presumably attracted by the fertility of the soil and the abundance of water, is reputed to have conquered the city a short time afterward.

The Greeks built sturdy fortified walls around the newly renamed Pergamum, perched on a steep height a thousand feet above the valley of the Caicus River. It was sheep-grazing and vineyard country, and cypress trees lined the long road from the valley to the walled enclave above. Some of these original walls have survived the ravages of time. The lush region might have been idyllic had it not been for the earthquakes and the incessant marauding armies crisscrossing the nearby plains of Anatolia. Despite these dangers, the population of the town grew.

However, around 546 B.C., the city fell to the Persians, under whom it seems to have experienced some decline. By the fifth century B.C., the Greek philosopher Xenophon described the city as a "small fortified town on the summit of a hill," referring to the acropolis, which as yet had no magnificent buildings adorning it. However, Pergamum was more than a fortified city. Whoever controlled it also was in control of the surrounding territory, the province of Mysia, and so it attracted a new conqueror. The Persians stayed in control until 334 B.C., when the armed forces of Alexander the Great marched into the area. After he scored a decisive victory against the Persians in the valley, he bequeathed Pergamum to one of his officers, Lysimachus, a native of Mysia. Lysimachus became tremendously rich from Alexander's bounty and the booty acquired from other conquests. Because Pergamum was so heavily fortified, he considered it to be a safe haven in which to deposit his wealth.

The eunuch to whom Lysimachus entrusted his fortune eventually gathered an army and rebelled against him. Pergamum became an independent city-state, with the eunuch, Philetaerus, in command. Upon his death, his nephew Eumenes succeeded him. Eumenes turned out to be a brilliant warrior and considerably expanded Pergamum's domain. His son and successor, Attalus I (reigned 241–197 B.C.), was the first ruler of Pergamum to declare himself a king.

By the time Eumenes II succeeded his father Attalus I, the city state of Pergamum stretched over almost all of western Asia Minor; the city itself, poised on its hilltop pedestal, was crowded with buildings of every sort, including the king's sprawling palace. Artisans from the Greek mainland, in particular jewelers and sculptors, flocked to wealthy Pergamum to set up shop. All the necessary resources, from money to talent, were handy when Eumenes embarked on a frenzied building campaign, which lasted for thirty-nine years, until his death.

The population of the terraced city must have been sizable, considering that the outdoor theatre, which Eumenes commissioned, consisted of eighty rows and 15,000 seats. It was by no means the only theatre in Pergamum: the hospital, for which the city was renowned, contained its own 4,500-seat theatre for the entertainment of its patients (entertainment being considered essential convalescent therapy). Miraculously, the city theatre has survived intact to the present day and still seats an audience of 15,000. Its architects designed it to complement the mountainous landscape, situating it dramatically on a steep incline of the acropolis: it was the steepest theatre in the ancient world. It even came equipped with a mobile wooden stage, which the Romans later replaced with a permanent one of marble. The other marvel that Eumenes II had erected was the largest altar to Zeus in the ancient world.

Eumenes II's greatest achievement, however, was the construction of one of the largest libraries in the world at that time. It housed over 200,000 volumes and threatened to outshine the famous library in Alexandria, Egypt. The Pergamum library attracted scholars and philosophers throughout the

Theatre at Pergamum
Photo courtesy of Embassy of Turkey, Washington, D.C.

Hellenic world. The Ptolemaic rulers of Egypt, jealously guarding the reputation of the Alexandrian library, prohibited the export of papyrus, out of which books were made. Far from depressing the book trade, this prohibition stimulated experimentation with an alternate source of writing material. Dried sheepskin turned out to be an ideal medium for writing, more durable than papyrus and in abundant supply around Pergamum. Pergamum soon became the manufacturing hub of "parchment," a corruption of the city's name.

At his death in 159 B.C., Eumenes II left a beautiful city, of which modern archaeologists have constructed models, conveying concretely what the city must have looked like at its apogee. These models show a terraced and congested urban center, crowded with impressive and elegant classical buildings, libraries, schools, and hospitals. Water was collected at various points in the city and piped to individual households. Pergamum was then by far the most important city in Asia Minor, and one of the leading cities of the ancient world.

Eumenes II was the last enlightened and effective ruler of independent Pergamum. The last of the independent rulers of the city-state, Attalus III, was by contrast a sadistic despot, who mercifully ruled only a few years, from 138 to 133 B.C. Bowing to the inevitable, Attalus III bequeathed his city state to the fast expanding Roman Empire, a sensible act that was meant to spare the city the destruction inevitably attending a military confrontation. However, there were patriots in Pergamum who took issue with this bequest and rose up against Rome. In all, it took two years for the Romans to subdue the disgruntled rebels. From then on, Pergamum was one of Rome's most important and richest provinces and a transmitter of much of the culture of the Hellenistic world. By that time, Pergamum consisted of four levels, connected, in that pedestrian heyday, by stairways. The Pergamum library had became a sort of university, which might have left an enduring legacy had it not been for the unfortunate political vicissitudes of that era.

One of the Roman Empire's disgruntled leaders, Mark Antony, marched his army into Asia Minor, as had so many before him. Having fallen in love with Cleopatra of Egypt, the Roman general, without consulting anyone, ordered his army to dismantle the library and to ship all of its precious cargo to Egypt as a gift to Cleopatra. In this drastic way, Mark Antony hoped to console Cleopatra for the recent loss of the invaluable Alexandrian library to a destructive fire. Mark Antony's action was a staggering blow to Pergamum, however, which cost the city its intellectual promise and its pre-eminence as a leading center of learning in the ancient world.

Except for the terrible loss of its library, the subsequent Pax Romana was a boon to Pergamum, which prospered under Roman rule. Besides being a leading manufacturing center of parchment, other major industries in Pergamum were the production and export of silver vases, brocades, perfumes, and wines.

During the first and second centuries A.D., Pergamum became widely known for its hospital, or sanatorium, just outside the city gates. Literally a compound that was first constructed in the fourth century B.C., the hospital gained in importance when the Romans turned it into a major shrine to Asclepius, the god of healing. There was a temple to the god inside the compound, and pilgrims to the city regularly made their way to the hospital, dedicated to Asclepius, where they drank the healthful mineral waters that poured out of a spout in the form of a lion's mouth. The hospital catered to the wealthy with their real and imagined illnesses, providing a luxurious dormitory for their stay. It did not refuse services to the poor, but they had to find accommodations elsewhere. The hospital prided itself on the fact that it never lost a patient, although in that era the mortally afflicted would usually have been too ill to make their way into the hospital to upset its spotless record. The therapeutic regimen included dieting, purging, mudbathing, sunbathing, and imbibing the mineral waters. The hospital also sought to encourage a cheerful and positive outlook, as is evident from the presence of the hospital theatre, whose repertory consisted mainly of comedies. The Roman Emperor Marcus Aurelius, who reigned from 161 to 180, is known to have been cured of tuberculosis at the hospital. His reign coincided with the activities of the most famous physician in the Roman Empire, Galen. The celebrated scholar-physician was based at the Pergamum hospital, which boasted an excellent library. Galen wrote at least 500 medical treatises in the hospital library. The renown of the hospital continued for some time after the deaths of Galen and Marcus Aurelius, until a ferocious earthquake in A.D. 253 heavily damaged the city.

Under Roman rule, Pergamum declined to the status of a provincial capital; nonetheless, the Roman passion for grandiose architecture found an outlet in the city. Many monumental buildings were erected, the largest being the amphitheatre, which seated 50,000 spectators, and an aqueduct. As the population of Pergamum increased during the Roman Empire's Golden Age, the congested city could expand only in one direction—downward, into the valley below. Much of Pergamum's monumental architecture on the hilltop and in the valley succumbed to damage in the earthquake of 253.

Recovery was slow, and when the city was back on its feet times had changed. Christianity had been making inroads since the time of St. Paul in the first century A.D. Pergamum, with its renowned monumental altar to Zeus and its many pagan temples, was slow to adapt to the sea change. In fact, persecution of Christians in this pagan stronghold was especially vicious: a local bishop, Antipas, was made a martyr and died after suffering hideous tortures. Eventually, however, Christianity won out and the pagan temples were converted to churches. The Temple of Asclepius within the hospital compound was turned into a basilica.

Following the breakup of the Roman Empire in the fourth century, Pergamum was slowly transformed from a Roman enclave into a Byzantine city, Greek in culture and conforming to the Eastern Church in religious matters. At about the same time the city changed from a hilltop citadel to a low-lying village. The enormous Greek Orthodox Church of St. John, popularly referred to as the "Red Basilica" (this name deriving from the reddish hue it has acquired over the centuries), was constructed in the early Byzantine era. It defines the city's profile today, being the largest building in the city.

Despite its relative proximity to Constantinople, Pergamum remained vulnerable to invaders, in part because of its exposed location and in part because its thriving agricultural economy made it a rich source of booty. Less than a hundred years after the death of Muḥammad, the founder of Islam, Arab armies began their forays into Asia Minor, bent on conquering the monolithic Byzantine Empire. In the winter of 716–17, Arabs attacked and looted Pergamum, setting fire to it as they left. The Byzantine Empire was then still strong enough to beat back such hostile incursions, but not to put an end to them. The walls surrounding Pergamum were rebuilt, the builders carelessly quarrying their construction materials from the city's ancient monuments—fallen buildings, pillars and the like. However, the reconstructed walls failed to keep at bay crusading armies from Europe, Seljuk Turks, and, later, Ottoman Turkish forces. By 1336, the Ottoman Turks subjugated Pergamum and the surrounding territories. The victors proceeded to transform most of the churches into mosques, and later the Turks cut up many ancient statues in the process of building their cemeteries. This once thoroughly Greek city in time underwent a total transformation in culture and identity.

The Turks renamed the city Bergama and, neglecting the ancient acropolis, continued building in the valley. The ancient city and its citadel fell into greater ruin and decay, a process that continued until 1872, when a German engineer, working on a road paving project in Bergama, noticed a peasant hauling what appeared to be a block of ancient limestone to a kiln to be burnt. Acting on a hunch, Karl Humann, the engineer, relieved the astonished peasant of his burden and sent the stone to the state museum in Berlin, where experts determined it to have been part of an ancient statue of Zeus. Intrigued by this find, the German government agreed to finance an archaeological expedition to the ruins of Bergama, the first of many. In the course of the excavations, archaeologists unearthed or identified whole buildings. One of these was the home of what most likely had belonged to a Roman consul, complete with a central courtyard, a floor paved with mosaics, and wall paintings. A great many artifacts, including whole pillars, were sent off to the state museum in Berlin, rather than being left in place. Consequently, what earthquakes and political vicissitudes did not destroy, German archaeologists managed to carry off, though at least these treasures were not destroyed. The Ottoman and post-Ottoman governments that allowed this despoliation to happen may have been inspired by Turkey's deep discomfort with the extent and sophistication of Anatolia's Greek heritage. The Greek population in Turkey was not forcibly expelled until 1923 (at the same time that Greece expelled its Turkish citizens). Hence for centuries, in Bergama as elsewhere, the two peoples lived side by side, their cultures intermixing. Most of the Greeks eventually converted from Greek Orthodoxy to Islam and thus relieved themselves from the many burdens—one of which was heavy taxation—imposed on non-Moslems. The converts became Turkish in every respect, masking the historic Greek presence.

Bergama began attracting a few European tourists in the nineteenth century, mainly religious persons traveling the route of the early Christian missionaries, particularly St. Paul, who was a native of Asia Minor. These travelers braved the isolation, the poor roads, and the many bandits in the area to marvel at the pagan ruins and to meditate on the transience of earthly things. Some of them, such as the American missionary D. L. Miller, sought out the few Christians on their peregrinations through Turkey to give them encouragement and to pray with them for the conversion of the Turks. Since most of the early visitors of the ancient ruins were steeped in the classics, their travel narratives are often insightful and informative. The trickle of tourists turned into a deluge after World War II, however, and they now form the basis of the town's major industry. The ruins on the acropolis rival many in Greece itself, and the beauty and drama of the landscape provide an enchanting backdrop to this archaeological wealth.

Further Reading: Bergama is well covered in most guidebooks, although there does not seem to be a separate book on the city and its ancient sites except for Elisabeth Rohde's beautifully illustrated German study, *Pergamon, Burgberg und Altar* (Berlin: Henscheverlag, 1976). A fascinating and sensitively written account of one woman's explorations of Turkey, which delves into the history of Pergamun at considerable length, is Harriet-Louise H. Patterson's *Traveling through Turkey, An Excursion into History and Religion* (Valley Forge, Pennsylvania: Judson, 1969). Finally, an unsurpassed study of Asia Minor in its heyday is by historian and classicist David Magie, *Roman Rule in Asia Minor to the End of the Third Century after Christ* (New York: Arno, 1975), which contains references to Pergamum.

—Sina Dubovoy

Phigalia/Bassae (Elis, Greece)

Location: East of the Ionian Sea at the Nedha River gorge in the department of Elis on the Pelopennese peninsula.

Description: Phigalia was an important market town in ancient Greece. In the nearby locality of Bassae, the Phigalians erected a temple to Apollo Epikourious, atop Mount Cotilion; the temple is today one of the best-preserved structures of its kind.

Contact: Tourist Information Office
43 Ethnikis Antistaseos Street
Tripoli, Arcadia
Greece
(71) 239392

In ancient times, Phigalia played a pivotal role in the development of Greece by connecting the pastoral district of central Peloponnesian Arcadia to the western portion of the peninsula, and especially the ancient states of Elis and Messenia. (Today's boundaries put Phigalia and Bassae in the department of Elis, next to the modern department of Arcadia.) It served as an important market for agricultural products originating in the fertile center of the Peloponnese. The town was also an important trade center for the people of the Aegean Sea region on the other side of the peninsula, who used the Nedha River for transportation instead of the mountainous overland route across the Pelopponese.

Phigalia played a pivotal role in the development of what was to become the first great maritime empire of the Mediterranean region. Not as expansionist as the Romans who were to follow, the Greeks were neither as materialistic nor as self-serving, preferring instead to foster mutually beneficial relationships among the various tribes of their islands as well as with seafaring peoples in neighboring countries.

The market town was a staple of ancient Greek life, and the center of all commercial, intellectual, and even political activity for most of the Greeks' Golden Age. Even Athens, the first great capital of Europe, was really a market town at heart, differing from such provincial outposts as Phigalia more in population and physical size than anything else.

The market town was so important to the political and social structure of ancient Greece that the Greeks even established carbon-copy market towns throughout their empire, and particularly in Italy. For example, Caulonia, Italy, which remains a thriving seaside center today, was founded by the Greeks when Rome was little more than a collection of thatched huts.

The market at Phigalia was also pivotal to a trading arrangement with nearby Elis, a sovereign state located in the western region of the Peloponnese, just west of Arcadia and south of Delphi. Goods could be loaded aboard seagoing vessels in the Nedha for their short voyage west to the Ionian Sea.

The region was also an important meeting place for a host of intellectual and political leaders from the Golden Age of Greece. The native Phigalians, who were not great fighters, preferred discourse to war. However, because of its strategic importance in the heart of the peninsula and its valuable natural resources, the town changed hands more than once.

Phigalia and its surrounding territory were the subject of great dispute between the Arcadians and the Spartans from neighboring Laconia. Sparta captured the town in 659 B.C. but was driven out shortly thereafter by Arcadia, due largely to the presence of 100 elite mercenary soldiers from Oresthasion, another important Arcadian town. The Delphic oracle had told the Phigalians that they would recover their town only with the help of the Oresthasians. The mercenaries were fierce and self-sacrificing warriors; all of them died in the battle. For many years thereafter the Phigalians honored their saviors by placing gifts at the Oresthasians' tomb.

In the late fifth century B.C., the Phigalians built the monument for which they are remembered today. The temple at Bassae, which means "the glens," is said to have been built as a shrine to Apollo Epikourios (the Succorer, or the Helper), who was credited with delivering the Phigalians from a plague during the Pelopennesian War (431–404 B.C.). The Greek geographer Pausanias claims the temple was built about 430 and designed by the architect Ictinus, who also designed the Parthenon. Others dispute Pausanias, however, saying the plague in question probably took place about 420, as an earlier one, in 430–29, was confined to Athens. They also question the attribution of the temple to Ictinus. The temple was probably built between 420 and 417 B.C., on the foundations of an earlier temple. No matter what the facts surrounding its construction, the temple stands today as an outstanding and unusually well-preserved example of provincial classical Greek architecture. Built almost entirely of locally quarried limestone, it is notable for being the first structure to use Corinthian columns.

The fifth-century temple to Apollo was a departure from traditional Greek style in many other regards. One significant aspect of the shrine, which is a partially open-roofed structure, is the fact that the length of the building runs north to south, rather than east to west, the contemporary norm. The traditional east-west direction would place the front door of the structure directly in line with the rising of the sun on the horizon, while the Temple of Apollo, which at first glance seems to ignore this important sociological and architectural rule of thumb, includes an interior door that faces eastward to compensate for this directional aberration. The temple's north-south orientation may have been dictated

Temple of Apollo at Bassae
Photo courtesy of Embassy of Greece, Washington, D.C.

by the less-than-ideal topography of the rugged and rocky Mount Cotilion. Another theory suggests, however, that the temple was planned to face neighboring Mount Ithome, the site of a shrine built in honor of Diana and Demeter. This shrine was excavated early in the twentieth century, and the remains of the statuary found at the site now resides in the National Museum, in Athens.

The temple to Apollo measures approximately 125 feet in length and 47 feet in width. Although it is appended with six columns at each end, in keeping with the architectural norm of the period, its north-to-south flanks consist of fifteen columns each, as opposed to the usual thirteen. At the center of the temple is the cella. The interior of the cella, in another departure from standard procedure, is not divided in the usual fashion into three separate aisles, but is partitioned by five short walls extending toward the center of the structure. Each wall ends with a three-quarter Ionic column, differing in style from the thirty-eight Doric columns which form the main support system of the temple. A lone Corinthian column stands at the center of the temple, near where the statue to the patron god is usually located. The column is the oldest known Corinthian column in Greece.

It was long thought that the floor in the center of the cella, which is roofless, had been hollowed out in order to collect rainwater. This theory was disproven, however, when serious archaeological study revealed that the ancient floor had merely settled into the earth, creating a basin-like appearance in no way connected to the structure's original design or purpose.

The portion of the cella behind the lone Corinthian column was fitted with a roof, creating a temple-within-a-temple that most scholars believe made use of the remains of an earlier temple built at the site. The old temple was situated in the standard east-west position, and the area reserved for the obligatory statue seems to have stayed in situ even while the later, more impressive edifice was constructed around it in nonconforming style.

Legend has it that the temple was originally home to a colossal statue of Apollo wrought of bronze. This impressive icon, unfortunately, was lost to the city of Megalopolis and put up in that city's market square. Although its visibility assured its inclusion in many contemporary accounts of the time, its display at this site also hastened its destruction by weather and vandalism. Excavations at the temple at Bassae have unearthed fragmentary remains of a large marble statue, presumably placed in the temple to replace the original bronze occupant.

Phigalia remained an important commercial area for several centuries after the building of the temple. It became part of another territorial dispute between Corinth and Achaea during the Achaeo-Aetolian contests in 221 B.C. It is believed that during this time period the Phigalians erected a wall approximately three miles in circumference around the city. Most of the wall still stands. The Aetolians remained securely and peacefully established in Phigalia until driven out by the drunkard Macedonian king Philip V in the early third century B.C. Eventually, the area became largely deserted; the mountainous region was known only to a few shepherds. Phigalia became, as it remains today, a tiny village.

The temple was discovered accidentally in 1765 by an archaeological team headed by a French explorer and architect named Joachim Bocher. Much of the temple was standing at the time, but a portion had collapsed, probably due to an earthquake. The temple's remote location, however, prevented its being plundered for other structures, as antiquities in heavily traveled areas often were. At worst, the sheepherders and assorted provincials who were aware of its existence pilfered many of the metal straps and bits that held together the more precarious pieces of stone.

In 1902 the Greek archaeologist Kavvadias led a major reconstruction effort that restored almost the entire structure by by 1906. Missing from the reconstruction, however, were the twenty-three tablets of the ornate 102-foot frieze that had originally encircled the cella atop the columns. The frieze depicts battles between Greeks and Amazons,

Centaurs and Lapiths. The prize pieces had been removed in 1812 by a team of mercenary antiquarians that included British explorers, the Austrian vice-consul to Greece, and an Estonian baron. With the aid of approximately sixty men, the team secretly spirited away the tablets under cover of darkness. Four years later they surfaced in Corfu and were purchased by the British Museum. Some additional pieces of the frieze were discovered in 1961.

The walls of Phigalia are also well-preserved. Just less than three miles in total circumference, the walls include several gates, posterns, and both square and round guard towers. The walls vary in thickness from approximately six feet to ten feet and also in style of brickwork, indicating that the walls were either built at differing points in time or haphazardly repaired over the centuries until they were no longer needed. The region is also home to the temple of Dionysus Akratophoros and the tomb of the Oresthasians, both of which are situated in the hills overlooking Phigalia.

Further Reading: Phigalia/Bassae, and especially the temple to Apollo, are covered in many survey works devoted to ancient Greece. The Life World Library edition covering Greece (New York: Time, 1963) contains a good overview of the site and the various modern expeditions to it. The *Baedeker's Greece* (Leipzig: Baedeker, 1909) edition from 1909 incorporates a surprising amount of detail in its descriptions of the area and its reconstruction which was, in 1909, timely news. *Greece and the Aegean Islands* by Philip Sanford Marden (Boston and New York: Houghton Mifflin, 1907) is of the same era, but somewhat broader in scope.

—Dellzell Chenoweth

Pisa (Pisa, Italy)

Location: Capital of Pisa province in the Tuscany region of Italy. The city is situated in a flat plain on both banks of the Arno River.

Description: Famous for its art treasures; Piazza dei Miracoli (cathedral square) consists of the cathedral, the Campanile (Leaning Tower), the baptistery, and the Camposanto (cemetery). The structures contain a blend of Romanesque and Gothic influences. The cathedral houses sculpture, mosaics, and paintings. The baptistery is rich in Gothic ornamentation, containing an octagonal marble baptismal font and a pulpit. On the corridor walls of the cemetery are Gothic windows decorated with frescoes by fourteenth- and fifteenth-century artists. The Leaning Tower, still an important historical and cultural landmark, personifies the spirit of Pisa.

Site Office: A.P.T. (Azienda di Promozione Turistica)
Lungarno Mediceo, 42
CAP 56100 Pisa, Pisa
Italy
(50) 542344 or 541800

Pisa, a town of 100,000 in central Italy near the Ligurian Sea, has a proud history as both a city-state and as a great maritime power. During its heyday in the Middle Ages, Pisa rivaled such other maritime city-states as Genoa and Venice. To most people, however, Pisa is the home of the famous Leaning Tower and birthplace of one of the most renowned scientists in the history of the Western world, Galileo Galilei.

Etruscans from the eastern Mediterranean settled in the area between the sixth and third centuries B.C. Gradually the community developed, and Pisa grew into a small fishing village. In 180 B.C. it became a military colony and naval station of Rome. Excavations later revealed a considerable Roman presence, which included a forum and an amphitheatre. While shipbuilding and trade brought further growth to the region, Pisa's fate relied in large part on the stability of Rome. Thus the subsequent decline of the Roman Empire in the fourth and fifth centuries A.D. led to general upheaval and rule by various peoples, including the Lombards, a Germanic tribe, and the Franks.

During the early medieval period, the city's reputation as a maritime power gradually grew. Pisa's relatively remote, seaside location proved an advantage since the Pisans were far enough away from Rome to maintain independence from both church and state. The sea too laid open the riches of the world, and thus a new class of wealthy and influential merchants developed. Between 1016 and 1052, the Pisans conquered Sardinia and Corsica. They also participated in the First Crusade, which helped the republic consolidate its position in the Near Eastern ports of call, particularly in Constantinople. The victory over Amalfi, a major commercial and naval power, in 1136 solidified the city's reputation as a maritime and military stronghold. Commercial expansion accompanied this growth in military stature. Pisa's prosperity was based, in part, on wool and leather. By the late thirteenth century an aggressive industrial base began to add to the city's increasingly complex infrastructure. When the population began to shift from country to town, Pisa faced mounting social problems such as overcrowding, unsanitary living conditions, and, most devastatingly, the Black Death in 1348.

Pisa's economic prosperity coincided with a remarkable artistic flowering between the eleventh and thirteenth centuries. Art and architecture flourished in Pisa, and a distinctive style that combined Romanesque and Gothic influences with Byzantine flourishes, known as Pisan Romanesque, emerged. This style had great influence on nearby towns and as far away as Sardinia. The greatest Pisan sculptor of this era was Nicola Pisano.

When the Pisans defeated the Saracens in 1063, they returned home with such treasures that they were able to begin construction of a cathedral. The Cathedral of Pisa (Cathedral of Santa Maria Assunta) forms part of a series of masterworks of architecture and art located in the Piazza dei Miracoli. The other monuments consist of the Baptistery, the Campanile, or leaning tower, and the Camposanto, or cemetery. The baptistery, which has a circular ground plan, was dedicated to St. John the Baptist. In 1155 construction began on the Walls, a method of fortification intended to protect these great religious monuments. Although work on the Camposanto began in 1278, it was not completed until the fifteenth century.

Undoubtedly the most famous of these symbols of power and prosperity is the Leaning Tower. Work on the Leaning Tower began in 1173. The bell chamber at the top was completed about 1350. The tower is the final monument of the ensemble of buildings constructed in the Piazza dei Miracoli. The cylindrical, Romanesque tower is sheathed in white marble and consists of eight levels of round-arched arcades and a winding staircase of 294 steps.

Construction was suspended in 1174 when it appeared that the tower had begun to lean: uneven settling of the ground coupled with a shallow foundation caused the building to shift toward one side. Work did not resume until 1275 when three additional levels were added, ostensibly to correct the problem. The long delays were intended to allow the foundation to settle.

The structural problems of the tower have been of major concern ever since. In 1840 a panel of experts met to discuss the troublesome issue. In 1935 cement was injected into the foundations as a method of waterproofing, but this

The Leaning Tower of Pisa
Photo courtesy of A.P.T., Pisa

measure proved futile. Another committee met in 1990 to decide on measures to safeguard the famous structure. At that time a suggestion was made to place 600 tons of lead on the north side of the foundation. Whether this latest attempt to correct the problem will be sufficient is, at the moment, uncertain. However, for safety and structural reasons, the Leaning Tower was closed to the public in 1990.

One of the most famous Pisans, Galileo Galilei, was born there in 1564. He studied medicine and mathematics—the latter with a private tutor—at the University of Pisa before becoming a lecturer of mathematics at the university at age twenty-five. Galileo was a demanding scholar with a violent temper and, like many great intellects, he did not suffer fools gladly. His incessant questioning was said to exasperate his teachers at the university. Although he was offered a chair there, Galileo instead moved to the University of Padua where he remained nearly two decades. He spent his last years in Florence, by which time his controversial theories about the nature of the universe had led to religious conflict. Galileo relied on the good graces of the powerful Medici family for protection from the wrath of the church.

Galileo reportedly conducted public experiments from the top of the Leaning Tower. He was thus able to prescribe the law of uniform acceleration for falling objects. Further, Galileo, based on his studies, came to agree with the Copernican theory that the planets revolve around the sun. The Catholic Church, however, insisted that the earth was the center of the universe. Consequently, Rome condemned Galileo for teaching a doctrine contrary to Scripture. In addition to his controversial theories, Galileo greatly developed the telescope and discovered the existence of craters on the moon and the satellites of Jupiter.

In the days before Italian unification, citizens of the Italian peninsula identified themselves with a city rather than a nation. Out of the violence and chaos of medieval Italy emerged the notion of the self-sufficient community or city-state. Power lay not with a centralized government but with powerful families and craft guilds. Every element of intellectual or cultural activity originated, they believed, in the town. But the city also had its dark side as personal vendettas and family feuds erupted and community battled against community.

The thirteenth and fourteenth centuries saw Pisa's supremacy threatened by Genoa. Continual rivalry with other provinces along with internal factions began to take their toll. In 1284 the Genoan fleet defeated Pisa decisively at the Battle of Meloria. The fleet suffered heavy losses, and at least 10,000 Pisans were imprisoned. While Pisa never fully recovered from this devastating loss, it did achieve a number of significant successes, winning victories against its Tuscan neighbors of Lucca in 1314 and Florence in 1315.

The city then entered a period of relative peace and prosperity until it was taken over by Milan in 1392. Within a decade, Pisa's nemesis, Florence, gained control in 1405. Although volatile, the Pisa-Florence connection proved beneficial to both parties: Pisa developed as a cultural and intellectual center under the ruling Medicis, while the acquisition of Pisa and other ports increased the wealth and power of Florence. As Florence prospered, so did the Medici family.

The Medicis improved relations between Pisa and Florence, developing the port of Pisa, rejuvenating its moribund shipbuilding trade, and even reviving Pisa's famous university, which had fallen on hard times. Under the Medicis, too, something of a cultural renaissance occurred in Pisa, and major public works, such as the Aqueduct of Asciano in 1601 and the Canal of the Navicelli in 1603, were financed. In addition, a bit of Pisa's ancient maritime prowess was recalled when the Order of the Knights of St. Stephen formed in 1561.

Over the centuries, the roles of pope and emperor, of church and state, caused friction in Italy. The pope's followers were known as Guelphs; advocates of the emperor were called Ghibellines. In 1241 Pisa was excommunicated by the pope for its support of imperial policies and its seizing of clerics who were traveling to Rome to take part in a major council. Less than two centuries later, in 1409, the Roman Catholic Church assembled in Pisa to end the Great Schism, which had divided the church between the see of Rome in Italy and the see of Avignon in France. The council summoned Pope Gregory XII of Rome and Pope Benedict XIII of Avignon, but both refused either to attend or to send representatives. The delegates at Pisa then declared both popes heretics and elected in their place one pope, Peter of Candia, the Gregorian cardinal of Milan, as Pope Alexander V. Most of Europe accepted the new pope. However, Gregory and Benedict refused to relinquish their authority. Thus, Europe was divided. Hungary and areas of Italy and Germany followed Gregory while Spain and Scotland supported Benedict. This chaotic situation led finally to the Council of Constance in 1415, which ended the schism when all three popes either abdicated or were deposed and a new one elected. The Council of Pisa is important because it was here that the "conciliar theory," a doctrine that claimed the supremacy of councils over the pope, found widespread acceptance.

In 1494 French armies under Charles VIII invaded Italy. Pisa briefly regained its independence at the end of the fifteenth century and into the early sixteenth century until falling again to Florence in 1509. In 1527 Rome was sacked and the French king, Charles V, placed Italy under the rule of the House of Habsburg. During the naval battle of Lepanto in 1531, Pisa joined a coalition of European powers, including the papacy, Spain, Venice, and the House of Savoy, to defeat the Ottoman Turks. In 1559 Italy fell to the Spanish branch of the Habsburg family, remaining until the turn of the century when the French Bourbons replaced the Spanish line of succession. At the Peace of Utrecht in 1713, major changes were again made as most of Italy came under Austrian Habsburg rule. Austria dominated the country until Napoléon invaded Italy in 1796.

With the fall of Napoléon, the Congress of Vienna in 1815 separated Italy into seven states: Lombardy-Venetia, Tuscany, Parma, Modena, the Kingdom of the Two Sicilies, Piedmont-Sardinia, and the Papal States. In 1860 Pisa joined the Kingdom of Sardinia, and the following year Italy was unified under King Victor Emmanuel II.

During World War I Italy supported the Allied cause. However, in 1922, the fascist dictator Benito Mussolini assumed control of the country. When World War II began in 1939, he banded with Germany, although Italy did not actually enter the conflict until June 1940. Pisa was the target of heavy bombing raids during the war. Casualties ran high, and the old churchyard as well as the marble walls of the Piazza dei Miracoli were seriously damaged. The Allied invasion of 1943 forced Mussolini from power, and the country was eventually liberated by Allied forces. A new era in modern Italian history began when, in 1946, Italy declared itself a constitutional republic.

Although no longer the power it once was, Pisa remains an important commercial center of Italy. Moreover, its rich architectural and cultural heritage continues to attract thousands of visitors each year. In addition to the monuments of the Piazza dei Miracoli, other important structures in Pisa and surrounding area include the Museo Nazionale di San Matteo, which houses collections of medieval pottery and twelfth- and thirteenth-century Pisan paintings, and the Medici Arsenal and Fort, which still bears the Medici coat of arms. Pisa continues to commemorate its historical traditions with several colorful festivals, including the Luminara di San Ranieri (candlelight festivities) and the Historical Regatta of St. Ranieri. Both are held in June. The Regatta of the Ancient Maritime Republics, in which boats representing Pisa, Amalfi, Genoa, and Venice compete, is held every four years.

Further Reading: While Italian guidebooks are plentiful, obtaining an all-purpose, up-to-date history of Pisa in English can be difficult. Available information tends to be sketchy and is usually scattered throughout pages of general history books of Italy. For a brief and readable introduction to Italian history, see Vincent Cronin's *A Concise History of Italy* (New York and London: American Heritage, 1972). More specific to Pisa is Christopher Hibbert's *The House of Medici: Its Rise and Fall* (New York: Morrow Quill, 1974; as *The Rise and Fall of the House of Medici,* London: Lane, 1974), an insightful account of the famous banking family that ruled Florence, controlled the papacy, and influenced the history and culture of Italy for generations. Giuliano Valdes's *Art and History of Pisa* (Florence: Bonechi, 1994) is an illustrated history of Pisa and its monuments. J. H. Plumb's *Horizon Book of the Italian Renaissance* (New York: American Heritage, 1965; as *Penguin Book of the Renaissance,* Harmondsworth, Middlesex: Penguin, 1965) explores the long-lasting effects of the Renaissance on Italian life and culture.

—June Skinner Sawyers

Pompeii (Napoli, Italy)

Location: On the Bay of Naples in southern Italy, in the region of Campania and the province of Napoli, on the southwest slope of Mount Vesuvius at the mouth of the Sarno River.

Description: Pompeii is the best known of the ancient Roman cities that were destroyed in the eruption of Mount Vesuvius in the year A.D. 79. The buried cities gradually were forgotten, until in the eighteenth century a series of discoveries brought them back to light, revealing invaluably descriptive remnants of ancient life.

Site Office: A.A.S.T. (Azienda Autonoma di Cura, Soggiorno e Tourismo)
Via Sacra #1
CAP 80045 Pompeii, Napoli
Italy
(81) 8632401

In the past few centuries, the city of Pompeii has been resurrected not only from the ashes of Mount Vesuvius, but from the ashes of human memory as well. Demolished in A.D. 79 by a volcanic eruption that buried the city under as much as fifteen feet of debris, Pompeii was all but forgotten by generations of Neapolitans until a series of accidents brought the long-entombed city and its neighbors Herculaneum and Stabiae to the startled attention of historians and curators. Today, after more than two centuries of wildly disparate levels of commitment to the investigation and preservation of these ancient Roman cities, Pompeii once again lies exposed to the sun in southern Italy—a monument both to human achievement and to the uncontrollable forces of nature.

Traditionally, Hercules is credited with the foundation of Pompeii as he traveled abroad to complete the labors that Hera demanded of him. Evidence of human endeavors on the site of Pompeii can be traced to the eighth century B.C. By the sixth century B.C., the Oscan tribe had established itself on the site. Scholars are uncertain whether this ethnic group was indigenous to the region or whether it emigrated from another; what is fairly certain is that the Oscans built the first walls in Pompeii, and that the city's name has its origin in their language.

During the first half of the sixth century B.C., Greek merchants from Cumae, recognizing the lucrative possibilities of Pompeii's strategic position on the Sarno River, took the city and established a trade route linking inland Italy to the sea. These settlers instilled a lasting Hellenic influence on the region's culture and customs, and they built some of the oldest remaining structures in Pompeii, including the Doric Temple of the Triangular Forum and the Temple of Apollo. The Greeks shared worship of Apollo with the Etruscans, who

conquered the city and governed it from 550 to 474 B.C. In 474, the Cumaeans won a decisive battle at sea over the Etruscans and regained control of Pompeii.

Thus began a period of impressive growth for the Campanian coast, as prosperous Greek merchants developed the foundation for the city of Pompeii in the walled, gridded design that it still retains. The Greeks were obliged to share the fruits of the region, however. In 420, Campania was besieged by the Samnites, a contentious tribe from the neighboring Italic hills who held the coastal area for more than a century. Greek Naples fell in 326; later, in 314, the Samnites were driven from Campania's metropolitan centers by Roman legions. Until 283, the Samnites waged periodic campaigns on the southern coast, but they were repeatedly driven off by the Romans. The last threat to Rome's rule over Campania occurred between 90 and 80 B.C., when the peoples of the region united against the Romans in the Social War, an uprising decisively quashed by Roman general Sulla.

At the base of Vesuvius, the city known to the Romans as Cornelia Veneria Pompeianorum had for some time functioned as a summer resort; now, at the urging of Sulla's nephew, it was used as a military colony as well. A local senate was created, and Pompeiians prospered commercially from sales of wine and oil. In 73 the legendary warrior Spartacus led a band of 70,000 slaves against the Roman state, taking refuge in the crater of Mount Vesuvius (which was assumed to be extinct) before being put to death by the Romans. Further political unrest during the rule of Emperor Claudius I settled into relative stability under Nero, until, in A.D. 59, a riot erupted at the city's amphitheatre between Pompeiian citizens and Nucerian colonists. Reacting to the incident, Roman lawmakers stripped the townspeople of their amphitheatre privileges, barring them for ten years from the gladiatorial battles and theatrical events that had occupied their leisure time.

Inexplicably, Pompeiians gathered in the amphitheatre on February 5, A.D. 62, when a massive earthquake toppled much of the city. Their sophisticated mountain-spring waterworks were flooded, and the Temples of Isis, Jupiter, and Venus Pompeiana (the latter erected to honor the protectress of the city) were demolished.

For seventeen years, the people of Pompeii worked to rebuild their city in the aftermath of the earthquake. In August 79, seismic tremors shook the region, a warning to the citizenry that was compounded when their waterways mysteriously stopped flowing. On August 24, Mount Vesuvius erupted in flames, igniting lightning storms and earthquakes. Pompeii was deluged by quarter-inch pumice stones and eight-inch rocks alike, followed by ground-surge deposits—a mix of hot ash and gas traversing the volcano's surroundings at sixty miles per hour. Because no red-hot lava flowed from

Pompeii, with Mount Vesuvius in the background
Photo courtesy of The Chicago Public Library

Vesuvius, the city was not ruined, but it was buried under as much as fifteen feet of debris. Of approximately 20,000 residents, it is estimated that 2,000 perished in the disaster, mainly due to asphyxiation by the volcano's noxious ground surges. The neighboring cities of Herculaneum and Stabiae were wiped out as well. Rome's Emperor Titus declared a state of emergency and contributed his own money to the aid of victims.

Of the disaster's casualties, the best known was the Roman leader Pliny the Elder, who was working at the naval base in Misenum when word came of the cloud issuing from Vesuvius. Intending to study the phenomenon from his ship in the Bay of Naples, he was compelled to offer his service to the distressed coastal dwellers whose only escape route was by sea. According to a letter written by his nephew Pliny

the Younger to the historian Tacitus, the elder Pliny succumbed to poisonous fumes as he and his colleagues sought to escape the hailstorm of ash, stone and mud with pillows atop their heads.

By the third century A.D., a "second Pompeii" was begun at Civita, to the north of the buried city, but half-hearted efforts to settle the area were finally abandoned in the eleventh century after the debilitating effects of more earthquakes and eruptions, as well as invasions. According to the historian Procopius, a Vesuvian eruption in the year 472 carried ash as far as Constantinople; in 512 another eruption was reportedly witnessed from across the Mediterranean.

Gradually, Pompeii faded from the collective memory of the people of Italy, until no one could say exactly where the buried city lay. As a perpetually disputed territory, southern Italy had other concerns, subdued as it was by invading Saracens, Germans, Angevins, and Spaniards. In the fifteenth century, the European Renaissance heightened awareness of the continent's ancient history, but Pompeii was as yet remembered only vaguely, as "la Città."

Around the year 1594, a wealthy Neapolitan named Count Muzzio Tuttavilla ordered a rerouting of the Sarno River for the irrigation of his country villa Torre Annunziata, overlooking the Bay of Naples coast. Diggers on the project unearthed inscriptions reading "Pompeii," but Tuttavilla, assuming the artifacts referred to the ancient leader Pompey, had them tossed aside. In fact, his unwitting crew had carved through the length of Pompeii—from the amphitheatre to the Temple of Isis to the Forum to the necropolis of Porto Ercolano—but the city remained unrecognized. In 1631, the second-worst eruption of Vesuvius on record took place, claiming 18,000 lives and pushing back the shoreline at the Bay of Naples by one-half mile.

Not until 1709 did another discovery merit renewed interest in the lost civilizations of Italy. In the town of Resina, which had been built atop the solid rubble of the Vesuvius lava flow in Herculaneum, workers digging a well for a local monastery came upon sections of marble seats from the Herculaneum Theatre. This time, someone recognized the significance of the find, not for its historical implications, but for its marble content. At nearby Porcini, the Austrian Prince d'Elbeuf organized a tunneling project to mine the region for more valuables, but his enthusiasm waned and the undertaking was scuttled by 1716.

In 1738, the remnants of the Herculaneum Theatre were positively identified, encouraging a new investigation of the natural vaults at the foot of Vesuvius. Charles III, the first of the Spanish Bourbons and successor to the expelled Austrians, had assigned Rocco Gioacchino de Alcubierre, colonel of engineers, to the site four years earlier. Alcubierre proved to be a much better treasure-hunter than an archaeologist; to be fair, methods of archaeology at the time were primitive. Even Johann Joachim Winckelmann, the so-called father of archaeology, agreed that extracting valuable goods was more important than "disengaging" the city "merely to satisfy the idle curiosity of a few." In this spirit, the tunnels of Herculaneum were refurbished as galleries, and scores of ancient articles were shipped to the Royal Palace at Portici, eventually coming to the National Archaeological Museum in Naples. When the well-trained Swiss architect Karl Weber arrived at Resina to institute a semblance of order to the excavations, a resentful Alcubierre requested and received permission to relocate the dig to la Città.

In March 1748, Alcubierre's crew of twenty-four uncovered the Temple of Fortuna Augusta, but it was not enough to satisfy the restless colonel, who returned to Herculaneum. In consequence, the ancient city of Pompeii was not definitively identified for another fifteen years. Weber, lacking sophisticated tools and methods to combat obstacles such as carbonic gases and insufficient light, met further complica-

tions at the hands of his jealous colleague, who sabotaged Weber's scaffolding. Undaunted, in 1750, Weber discovered the Villa of the Papyri just outside Herculaneum, an 800-foot-long dwelling still considered a monumental find for its unparalleled collection of ancient bronzework. During World War II, the precious Papyri statues fell victim to Hermann Göring's pillaging in Italy. At war's end the collection was restored to the National Museum in Naples.

For fifteen years, Weber's remarkable excavations attracted increasing global attention. Among his most important finds was the first complete library extant from ancient times, and on August 20, 1763, digging crews at la Città cleared away an inscription reading "res publica Pompeianorum," "commonwealth of Pompeiians"—indisputable certification of Pompeii's existence on the site. In 1765, the new project manager, Francesco La Vega, removed his staff from Herculaneum to concentrate on the spectacular finds at Pompeii. New discoveries included the Great Theatre (1764), the gladiators' barracks behind the theatre (1766), the Temple of Isis (1767), and the Villa of Diomedes (1771). By now, growing circles of the intellectual and aesthetic elite were smitten by Italy's carefully preserved classicism. Designers imitated the romantic engravings unearthed at Pompeii, which perhaps most visibly influenced Thomas Jefferson's detailed appointments at Monticello.

Upon the departure of the Bourbons, the French leader Jean-Antoine Étienne Championnet took the reins in the Kingdom of Naples, sufficiently supporting the excavations so that in 1799 La Vega named a villa for him. Despite La Vega's vehement protests, Championnet's successor, King Ferdinand IV, revoked funding and called off the excavation project. In 1806, Joseph Bonaparte reinstated the archaeological work, and fifty crew members disinterred the Basilica, the House of Sallust, and the Temple of Fortuna Augusta. Bonaparte's successor, Murat, acceding to the wishes of his wife, Queen Caroline, an enthusiast of classical history, financed a staff of 500 and helped publish François Mazois's influential study of the ruins. When the Kingdom of Naples fell in 1815, war veterans were stationed at the ruins to guard against pillagers, and a disconsolate Caroline left southern Italy, never to return to her beloved Pompeii.

Ferdinand's return to the throne again endangered the excavation process, which was suffering from so many inconsistencies. Ferdinand had the city's flagstones removed, to facilitate carriage passage through the ruins. (In the city's heyday, these flagstones lined the streets so that residents could skirt the mud and sewage of the city's plumbing conduits and rainwater. After Ferdinand's reign ended, the stones were replaced.) Ferdinand's romantically inclined son, Francis I, upon his ascension, was more sympathetic than his father to the fragile history that lay under his jurisdiction, but then there followed one of the most inappropriate in a long line of dubious developments in the work at Pompeii. Giuseppe Garibaldi, general of the Italian Nationalists, and his Red Shirts, financed by the French author Alexandre Dumas père, in their successful Expedition of the Thousand against the Austrians, gratefully awarded the unqualified Dumas with the directorship of excavation. Though excitedly committed to his new task, Dumas reportedly lodged at the Royal Palace, feasting and drinking at the taxpayers' expense.

Bringing an overdue sense of determination to the project, a newly independent Italian government assigned Giuseppe Fiorelli, a well-respected scholar, to the excavations as it sought to capitalize on the country's vast artistic heritage. Fiorelli, a political outcast from the times of Austrian rule in Italy, had been summoned to disprove the authenticity of some "preserved" corpses offered for sale to the king's brother, the count of Syracuse, at Cumae. In gratitude, the count granted Fiorelli the overseer's duty at Pompeii. With 500 workers at his disposal, Fiorelli advanced the techniques of archaeology considerably, instituting three simple but rigid rules of procedure: all wall markings were to be traced and gates identified, work at one site was to be completed before the crew moved to another, and drainage systems were to be cleared as faithfully as roadways. During his career, Fiorelli's efforts helped formalize the excavations at Pompeii, Herculaneum, and Stabiae, as well as those in Egypt, Greece, Iran, and Mesopotamia.

Fiorelli's approach was antithetical to prior methods: rather than dig from street level into the Pompeiian buildings, leaving chaotic piles of earth along the way, he entered houses through their roofs, clearing away debris and ensuring the safekeeping of artifacts and murals, which, until this time, had been removed from their resting places to museums. In 1871, Pompeii produced 3,329 inscriptions of graffiti; that figure doubled in thirty years, as Fiorelli's successors upheld his meticulous methods. Roofs and columns and furniture were repaired, and backyard gardens were restored to their original greenery with new plants.

In 1907 and 1911, two necropolises built by Pompeii's pre-Roman inhabitants were unearthed, a discovery that fed the debate regarding the region's ethnic history. By 1926, Professor Amedeo Maiuri had proven the existence of the Italic Oscans on the site of Pompeii at the dawn of recorded history. Coinciding with these breakthroughs at the beginning of the twentieth century, Mount Vesuvius put on a display of might for the modern world: culminating a period of activity that had begun in 1905, Vesuvius erupted in April 1906, spewing an enormous "Pine Tree" cloud that reached 30,000 feet at its highest point. Naples was covered in two inches of ash and the volcano claimed 100 lives. Ominously, Vesuvius also erupted in March 1944, during World War II, six months after an Allied bombing run targeted the Italic Oscan ruins.

By most indications, at the time of its demise Pompeii was something of a middle-class town; much of its aristocracy had moved to the country as the successful merchant class began to broaden its influence. In contrast, Herculaneum, smaller than Pompeii (55 acres to the latter's 163), was also less frantic. Its residents were not traders but fishermen and artisans, and its private residences were generally built in an architectural style superior to that of the faster-paced Pompeii. Some of Pompeii's larger homes were divided to accommodate greater numbers of tenants. Some featured front rooms designed for trading; streetside shops that opened to the layout of the typical home (an atrium, followed by a receiving room); and finally a peristyle consisting of the private rooms (bedrooms, dining room, and the family garden). Most houses featured decorative mural painting, important artwork differentiated by scholars as the "first through fourth" Pompeiian styles, which evolved rapidly from simple colors to spatial illusions to a primitive sort of impressionism and, lastly, an increasing flair for the baroque.

Today, the uncovering of Pompeii is virtually complete, while work at Herculaneum is still under way. Inside the main entrance to Pompeii (the Porta Marina) lie the Temple of Venus and the Basilica, prototypes in design of the Christian church. The Basilica, the judicial and financial center of the city, opens onto the Forum, creating a public square suited for elections and announcements. Outstanding examples of Pompeiian architecture include the elegant House of the Faun, the Villa dei Misteri, and the House of the Vettii.

With the passing of the director Amedeo Maiuri in 1963, the work at Pompeii was left yet again to deteriorate into weeds and disrepair. Not until Giuseppina Cerulli Irelli, the first woman director of the project, was appointed did its prospects turn for the better. In 1979, a comprehensive exhibit of art and artifacts called "Pompeii A.D. 79" toured major museums in Europe, Japan, and the United States. One year after this triumphant international showing, however, an earthquake wreaked extensive damage on the collection at the National Museum in Naples and the ruins themselves. Both sites were temporarily closed for repair, as nature seemed once again to assert its indignation over human efforts to restrain it.

Further Reading: *Pompeii and Herculaneum: The Glory and the Grief* by Marcel Brion, translated by John Rosenberg (London: Elek, and New York: Crown, 1960; as *Pompéi et Herculanum,* Paris: Michel, 1960) is an oversized book containing extensive reprints of Pompeiian artwork and a detailed history of the excavation process. *In Search of Ancient Italy* by Pierre Grimal (New York: Hill and Wang, and London: Evans, 1964) is a good source for early Pompeiian records, as is *Naples and Southern Italy* by Edward Hutton (London: Methuen, and New York: Macmillan, 1915). During the twentieth century, archaeologists' attentions have to some extent reverted to the smaller towns buried under the ash of Vesuvius; as a result, two of the better books for information on the history of the digs are primarily devoted to Herculaneum: *Buried Herculaneum* by Ethel Ross Barker (London: A. and C. Black, 1908), and *Herculaneum: Italy's Buried Treasure* by Joseph Jay Deiss (London: Thames and Hudson, and New York: Harper and Row, 1985).

—James Sullivan

Priene (Aydın, Turkey)

Location: In western Turkey (Asia Minor), in the Menderes (ancient Maeander) River valley, about ten miles inland from the Aegean Sea.

Description: Ancient Greek/Hellenistic city-state; member of the Ionian League and later the Delian Confederacy; known especially for its gridded plan and Temple of Athena Polias, considered a paradigm of the Ionic style of architecture.

Contact: Tourist Information Office
Liman Cad. No.: 13
Kuşadası, Aydın
Turkey
(256) 6146295

Writings by Strabo, Greek geographer of the first century B.C., and Pausanias, Greek traveler and geographer of the mid-second century A.D., describe the genesis of Priene as a Greek polis or city-state founded by Ionians and Thebans at the beginning of the first millennium B.C. This archaic Priene was a precursor to a new city that thrived as part of the great Hellenistic civilization that flourished in Asia Minor even as Athens declined.

The archaic site, lost under an extensive delta created by centuries of silt carried by the Menderes River west to the ocean, was reported to have been on the Aegean coast. The site of the later city, built in the fourth century B.C., is on Mount Mycale about ten miles from the sea. This gradual westward shift of the coast was calamitous for the Ionian cities whose economies depended at least partly on their ports; Priene's loss of its two viable seaports undoubtedly had an impact on the city's stature, in terms both of trade and of strategic importance to whichever power was ruling the region.

Along with Miletus, Myus, Ephesus, Colophon, Lebedos, Teos, Clazomenae, Erythrae, Phocaea, Samos, and Chios, Priene belonged to the Dodecapolis, a loose federation of twelve cities that was formed before 800 B.C. Miletus and Priene ranked among the most important members of this Ionian League, which had its religious center, the Panionium, on Priene's territory. Tradition held that the Aipytidai dynasty that ruled in Priene was descended from Neleus, who had been the *oikistes* of Miletus during the Bronze Age.

Relations among the league members were ambivalent, with city-states occasionally going to war with one another, usually over territorial disputes. For example, Samos and Priene maintained a hostile relationship for centuries, contending repeatedly for disputed territories. On at least one occasion Samos sought the assistance of Miletus to defeat their mutual rival Priene in the sixth century B.C.

Priene faced a new peril during the seventh century B.C. when the Lydian usurper Gyges included the Ionian city-states in his expansionist plans. He was thwarted when the Cimmerians suddenly invaded Asia Minor from the north. This band of raiders menaced much of Anatolia throughout the seventh century, and only during the reign of Gyges' son Ardys did the Lydians successfully expel the Cimmerians. Back on track with their own agenda, the Lydians promptly sacked Priene.

In the sixth century B.C., Priene revived and regained prosperity under Bias, one of the so-called Seven Wise Men of Greece. Astute leaders such as Bias of Priene contributed to the city's authority within the Ionian League; responsibility for the Panionium, for example, carried enough status to suggest that it took more than Priene's geographic proximity to win control of the site. Priene remained a popular target for greater powers in the sixth century, and Bias became adept at negotiating favorable terms for the much-invaded city.

In the mid-550s, however, Priene and the other cities submitted to Croesus, the last king of Lydia. That situation was short-lived, as the Persian king Cyrus II (the Great) envisioned his empire expanding westward to the Aegean. With the Persians on the verge of conquering the Greek cities one by one in 547-546 B.C., the sage Bias advised the leaders of the Ionian League to remove their respective cities' populations to Sardinia. He envisioned a safe haven there, an Ionic island with one capital city. His radical advice went unheeded, however, and Priene and the rest of the coastal area became subject to Persia. Indeed, Priene was all but annihilated when the Persian generals overthrew Croesus and his kingdom of Lydia while conquering Asia Minor. After capturing Priene, the Persians enslaved the population.

The Persians introduced regional governments ruled by satraps, but in practice the cities maintained a modicum of autonomy. The reduced Ionian cities continued to be ruled by their Greek tyrants, for example, with the Persian king's approval. But that was no substitute for self-determination, and in a rare display of political unity, in 499 B.C. the cities, under the leadership of Aristagoras of Miletus, expelled their pro-Persian tyrants and established democracies. The cities persevered in this Ionic Revolt for years, a significant feat considering Persia's power at the time. In 494 B.C. Priene sent twelve ships to join the Ionian rebels at Lade, an island near Miletus, but the assembled fleet was no match for the Persian navy, which crushed the rebellion for good.

The vanquished Greeks appealed to Athens for protection, leading to the formation of a league under that city's control, ostensibly to keep the coast of Asia Minor free from Persia. This so-called Delian Confederacy comprised most of the cities along the Aegean coast, not only those of Greek

Athena temple at Priene
Photo courtesy of Embassy of Turkey, Washington, D.C.

heritage, but non-Greek cities too. Priene made modest contributions to the confederacy in 450–442 and 427–425 B.C.

Rivalries within the league surfaced again soon enough. In 442–440 Samos and Miletus fought over Priene; Athens intervened in Miletus's favor. Further turmoil ensued when the Peloponnesian War between Athens and Sparta ended with the latter's victory in about 400 B.C. The Delian Confederacy fell under Sparta's control, but unlike Athens, Sparta was inexperienced in maintaining overseas properties. Sensing a new opportunity, the Persians again made advances toward the coast, and through the King's Peace in 386 B.C. recovered Priene and the other Greek cities.

Priene was rebuilt in the fourth century B.C., apparently on the site of Naulochus, one of the two harbors that served the original city. The well-preserved ruins have provided one of the best records of the built environment during Hellenistic times. Unlike other sites in Asia Minor, where Roman culture largely superseded the Greek, there were notably few interventions at Priene during the Roman period.

Priene offers an unusual example of a Greek city built according to so-called Hippodamian principles based on a rectangular grid plan. Unlike the gridded city of Miletus, sited on low ground adjacent to its harbor, Priene's grid was applied unconventionally to the steep site on the southeastern slope of Mount Mycale. The city faces south, with the main avenues running east-west on the level. The north-south cross streets, in contrast, are stepped, transversing the four tiers of the natural shelf on which Priene sits.

Aside from its meticulous plan, Priene is known mostly for the Temple of Athena Polias designed by the architect Pythius in 340 B.C.; he also was responsible for the Mausoleum at Halicarnassus (one of the Seven Wonders of the Ancient World). The Roman architect and writer Vitruvius referred in the first century B.C. to a now-lost book by Pythius, in which the latter described the Temple of Athena Polias as being the paradigm of the Ionic style in architecture. Additionally, the temple had the distinct honor of being dedicated in 334 B.C. by Alexander the Great, who was passing through the area. (An inscription with his name on one of the walls of the front porch resides in the British Museum.)

The Temple of Athena was peripteral—a colonnade wrapped around the entire building—and hexastyle, with six columns at the front and rear, and eleven on the sides; a pair of columns situated *in antis* (inside the projectile walls) graced each of the front and rear porches. The rear porch was called the opisthodomus, and this temple was the first Ionic example to borrow this feature from the older Doric temples. Relying on the Temple of Athena Polias as a model, all large Ionic temples built after this date featured such an opisthodomus as well as the specifically Ionic vocabulary established at Priene. In fact, the temple is an especially trustworthy model for deducing the ideal Ionic proportions and measurements. The rectangle of the plan, for example, is in a precise 2:1 ratio (60 by 120 Ionic feet); all further measurements—width of columns, spacing of columns, proportions of the cella and pronaos, and decorative features—are precisely related to one another by certain ratios.

The theatre, whose auditorium for 5,000 spectators was hollowed out of the hillside, is striking for its horseshoe-shaped orchestra. The theatre at Priene is the best surviving example of this orchestra type. The horseshoe shape developed in Hellenistic times as the stage became broader and began to occupy space that once was part of a full circular orchestra. The generous seating capacity in the auditorium suggests that the theatre entertained residents of the larger Prienean territory in addition to those within the city itself.

Following the street grid between the civic center and Priene's west gate stand several blocks of private houses, also from the Hellenistic period, which rank among the best-preserved examples of residential construction from the period. One of these houses is particularly notable: it is believed to have been the Sanctuary of Alexander the Great, possibly commemorating his visit in 334 B.C. The Prienean houses varied in size, but many were sited only four to a block, allowing for spacious rooms and generous interior courtyards within the private walls.

South of the theatre and just two blocks east of the Temple of Athena Polias was Priene's civic center, which straddled the main east-west avenue. The Temple of Zeus Olympius and the agora, both from the third century B.C., were on the south side of the avenue. Facing this complex across the street on the north was the Sacred Stoa, whose northeast side opened up to two key buildings of the Greek polis, the bouleuterion (council house) and the prytaneum (state dining room and council committee building), both from about 150 B.C. The bouleuterion is especially intact, with tiers of seats rising on three sides of the nearly square room; also surviving are the remains of an altar at the center of the room. Surrounding the agora on three sides were stoas, with the agora's north side open for access to the main avenue. The eastern stoa adjoining the agora adjoined a temenos that housed the Temple of Zeus Olympius, a small Ionic temple with four columns along the front portico.

In 283 Lysimachus, the Macedonian general who had taken control of much of Asia Minor, passed down the decision in yet another border dispute between Priene and Samos; in particular, Samos claimed a *peraia* on Mount Mycale which the Prieneans felt encroached on their territory. Priene received Dryussa, and Samos won Batinetis. In 281 Priene became part of the Seleucid kingdom when Seleucus I (ruled 306–281 B.C.) gained western Asia Minor after defeating Lysimachus.

In about 277 B.C. the Celtic Galatians invaded Priene's territory, taking prisoner the citizens of the unprotected countryside and treating them savagely. Priene offered refuge within the city walls to those who could escape the pillaging. Drawing on the city's residents, one Prienean named Sotas organized defensive forces that finally drove the invaders away.

Priene remained part of the Seleucid Empire until 246 B.C., when it was ceded with the rest of southern Asia Minor to Ptolemy III of Egypt (ruled 246–221 B.C.) in the aftermath of the Laodicean War. In 196 B.C. the town reverted to Seleucid rule, however, when Antiochus III (ruled 223–187 B.C.) recovered Asia Minor for the Seleucid Empire.

The Romans entered the area in the early second century B.C., defeating Antiochus III in 190 B.C. The Treaty of Apamea in 188 B.C. ceded Asia Minor to Rome, and Cn. Manlius Volso began organizing Asia Minor for the empire. Within the new province of Asia, Priene was designated as a free confederate of Rome.

In about 155 B.C., in Oropherne's War, Ariarathes V of Cappadocia and Attalus II of Pergamum waged war against Priene. Just a quarter of a century later, in 130 B.C., Ariarathes VI of Cappadocia founded the Sacred Stoa at Priene. This stoa was an imposing structure indeed, 525 feet long and 39 feet wide under the portico, which featured 40 Doric columns along the front. Twenty-four Ionic columns that supported the wooden roof also divided the inner space longitudinally. Enhancing the monumental impact was an open promenade almost 21 feet wide set between the building and the main avenue. Erected at about the same time as the Sacred Stoa were a gymnasium and stadium at the southernmost point of the city. Several Doric stoas highlighted these two buildings.

After 27 B.C., the Temple of Athena was rededicated to the goddess Athena Polias and the emperor Augustus; in general, Priene lapsed into insignificance under Roman rule—Miletus was geographically much more strategic with its active harbors—and did not merit the sort of monumental structures that the Romans superimposed on many other classical sites. This lack of building may have been a drawback for Priene at the time, but it was this very inactivity that preserved the Hellenistic integrity of the site. The Romans doubled the depth of the stage at the theatre, but even that did not affect the basic structure.

With the spread of Christianity, Priene was made a bishopric subject to the archbishop of Ephesus. Priene otherwise had become increasingly peripheral, and lapsed into disarray during late antiquity. Many of the public buildings were converted to residential use, filled with small houses that were built without regard for the orderly city plan. By the late seventh century, invasions of Asia Minor had devastated many of the Hellenistic cities there; Prieneans moved from the vulnerable lower city to the acropolis, a high, fortified location.

During the eleventh to thirteenth centuries, the area below the acropolis—by then known as the town of Sampsun—was reoccupied. In 1204 Sampsun was the center of an *episkepsis*, a property-based fiscal division within the Byzantine system of administration. That same year the town became the capital of the ephemeral state of Sabbas Asidenos, a local ruler in Anatolia who was active from 1204 to 1214; he assumed power in Sampsun and in the lower Menderes Valley when the Fourth Crusade conquered Constantinople. The Greek-ruled empire based at Nicaea absorbed his territory, possibly as early as 1205. Remains indicate that Asidenos's capital consisted of a fortress on the acropolis that was rebuilt in the twelfth and thirteenth centuries, and a small fort in the lower town with scattered houses outside its walls.

The Turks began moving into Asia Minor, and by 1300, during the reign of the Byzantine emperor Andronicus II Palaeologus (1282–1328), Priene and all of the territory that had been Ionia were solidly part of the Ottoman Empire. This was not an easy transition, as the weakened Byzantine government sporadically attempted to drive the Turks out in the latter part of the thirteenth century. In the midst of this struggle, the determined Turks devastated Priene. Turkish settlers eventually supplanted Christianity with their Islamic culture, leaving the traces of Hellenistic culture and Christianity to be discovered by others.

English traders from Smyrna discovered the ruins of Priene in 1673. In the eighteenth and nineteenth centuries, the Society of Dilettanti—a group of young, wealthy Englishmen whose main interest was classical antiquity—conducted expeditions to Asia Minor; their first visit to Priene was in 1764. These students concentrated mostly on the Temple of Athena Polias, and as late as 1869 they found the temple walls still in place to about the height of a human being. However, local inhabitants subsequently carted off many of the blocks for house building and other construction.

Despite the English forays into the ruins at Priene, concentrated archaeological intentions surfaced only at the end of the nineteenth century, and it was the Germans who took the initiative in uncovering the entire site. In October 1894, the director of the antiquities collection at the Berlin Museum, R. Kekulé von Stradonitz, and the man who had discovered the great altar at Pergamum, Carl Humann, on a visit to Priene in the course of excavating nearby Miletus and Didyma, decided to set up operations at the site on Mount Mycale, too. Humann was an engineer who had chanced upon archaeology while helping the Ottoman government build railways, and to facilitate access to Priene, he had a nine-mile road built to the nearest town, Söke. He also erected the expeditionary headquarters at Priene.

Humann began the first excavation on September 18, 1895. Soon Theodor Wiegand joined him at the site, and it was Wiegand who took over the expedition when Humann died the next year. Wiegand concluded this initial series of archaeological campaigns on April 24, 1899, and within five years published the discoveries as *Priene: Ergebnisse der Ausgrabungen und Untersuchungen in den Jahren 1895–1898 (Priene: Results of the Excavations and Investigations*

in the Years 1895–1898, with H. Schrader, Berlin: Reimer, 1904). Despite the fact that parts of the excavation had not been extensive enough, and that some building descriptions—especially of the private houses—were too brief or not even included, this work was important for presenting in one source an entire picture of an ancient Greek city.

The modern town of Sampsun continues to occupy the site below Priene, and the villages of Turunçlar and Güllübançe also are located nearby. From Güllübançe the ancient site can be entered through the northeast gate. Five columns of the Temple of Athena Polias' peripteral colonnade were re-erected in 1964, adding drama to the already impressive ruins.

Further Reading: After Wiegand's important work (noted above) was published in the early twentieth century, much of the scholarship on Priene has appeared in German, including F. Hiller von Gaertringen's *Inschriften von Priene* (*Inscriptions of Priene*, Berlin: Reimer, 1906), A. von Gerkan's *Theater von Priene* (1921), K. Regling's *Münzen von Priene* (*Coins of Priene*, Berlin: Schoetz, 1927), and the wider-ranging *Ruinen von Priene* (Berlin and Leipzig: Gruyter, 1934) by Martin Schede. There is an excellent, painstaking summary (twenty pages of small type!) by G. Kleiner in August Friedrich von Pauly's *Realencyclopädie der classischen Altertumswissenschaft* (Stuttgart: Metzler, 1894–1919). Sources in English that have appropriated the most pertinent information include *Greek Architecture* by Arnold Walter Lawrence (Harmondsworth, Middlesex: Penguin, 1968; fourth edition, 1983) in the Pelican History of Art series and *The Architecture of Ancient Greece* by William Bell Dinsmoor (London and New York: Batsford, 1950) both of which provide substantial discussions of Priene's urban and architectural development. In *Archaic Greece: The City-States, c. 700 to 500 B.C.* (London: Benn, and New York: St. Martin's, 1976), L. H. Jeffery presents a thorough, well-annotated survey that places Priene and Ionia in the context of the entire Greek world during the archaic period. *Ionian Trade and Colonization* by Carl Roebuck (New York: Archaeological Institute of America, 1959) is volume nine in the Monographs on Archaeology and Fine Arts series from the Archaeological Institute of America and the College Art Association of America—and reflects those credentials. This specialized study focuses on economic development in Ionia in the earliest centuries, before Persian domination began in the sixth century B.C. *Greeks in Ionia and the East* by J. M. Cook (New York: Praeger, 1963; Aylesbury and Slough, England: Watson and Viney, 1962) is a very readable account organized both chronologically and thematically, with excellent illustrations. *Persia and the Greeks: The Defence of the West, c. 546–478 B.C.* by Andrew Robert Burn (London: Arnold, and New York: St. Martin's, 1962) is an exhaustive study of the tensions between the two ancient powers. Of particular interest concerning Priene are chapters about Cyrus the Great and about the Ionic Revolt. *Anatolia: Land, Men, and Gods in Asia Minor*, volume one: *The Celts in Anatolia and the Impact of Roman Rule* by Stephen Mitchell (Oxford: Clarendon, and New York: Oxford University Press, 1993) offers a scholarly look at developments in Asia Minor in the centuries following Priene's apogee.

—Randall J. Van Vynckt

Pula (Istria, Croatia)

Location: In northwestern Croatia, at the southwestern tip of the Istrian Peninsula, about fifty-five miles south of the Italian city of Trieste.

Description: An ancient city well-known for its Roman ruins, especially its amphitheatre, which in size and state of preservation rivals the Colosseum in Rome. The city also has several structures from the Byzantine and Venetian eras. Situated at the site of a well-protected natural harbor, the city's commercial and naval shipping, and the industries associated with these, have long held an important place in its economy.

Site Office: TZO Pula
Istarska 13
52000 Pula, Istria
Croatia
(52) 24-062

Throughout most of its history Pula existed on the fringes of great empires and unless it figured strategically, it was generally left to its own devices. Among the many masters of the city were Romans, Byzantines, Venetians, and Habsburgs. The Romans and Venetians, who together accounted for more than 1,300 years of rule in Pula, indelibly stamped a Latin character on the city. In fact, its Italian name, Pola, appears throughout numerous historical texts. The city remained largely Italian even after Venetian rule lapsed in the eighteenth century, when Pula passed to the Habsburgs. It was not until the mid-twentieth century, after World War II, that the city lost its Italian ambiance when most of the city's Italian residents relocated or were expelled. Today, for the first time in more than two millennia, Pula is a Slavic city, populated mainly by Croats. Its Latin legacy survives only in its architecture.

According to tradition, Pula was founded by the Colchians who came here in their quest to retrieve the Golden Fleece. Based on the legend of the Argonauts, this theory has never been substantiated. It is known, however, that the Istrian peninsula was well–populated by Illyrian tribes as early as 500 B.C. Excavations near Pula have uncovered evidence of burial grounds dating from between 500 and 300 B.C. By 200 B.C. the city had developed into a prospering settlement and a refuge for people fleeing Celtic invaders.

In response to raids by Illyrian tribes on Roman enclaves at the head of the Adriatic, Roman legions campaigned in the region and captured Pula in 178 and 177 B.C. The Romans soon established a military and naval outpost. The city was destroyed by the Romans in 39 B.C. while pacifying a rebellion of Illyrians and Dalmatians. It was later rebuilt under Augustus and named Pietas Julia.

It was under Augustus and his successors that the city witnessed true prosperity as evident in the large building projects undertaken and completed. The Triumphal Arch of Sergius, built in 27 B.C., was the first of many such projects. Another major project, the Temple of Augustus, was constructed from 2 B.C. to A.D. 14. It still stands today as a proud testament to the Augustan legacy. Outshining all, however, was the amphitheatre, which was completed during the first century A.D. It encompasses three tiers and seventy-two arches and holds a capacity crowd of 23,000. Like most Roman amphitheatres, it was the site of many gladiatorial combats between man and beast. The amphitheatre is still used today as a venue for dramatic performances, concerts, and film screenings.

After the fall of the Western Roman Empire in 476, Pula, like all of Istria, came under the control of the Ostrogoths. For the next two centuries, control of Pula was to be contested by several powers. In 539 Istria was captured by the Byzantine forces of Justinian. It was from Pula that in 544, Belisarius, the great Byzantine military commander, assembled a fleet and launched his expedition to oust the Ostrogoths from Italy. Shortly afterward Ravenna was captured and made the seat of Byzantine rule in Italy. It was from Ravenna that Pula was ruled. During the seventh century the Lombards began establishing a foothold on the Italian peninsula. At the beginning of the seventh century Istria, including Pula, was devastated by the Avars and Slavs, acting in alliance with the Lombards. Ravenna fell in 751, completing the Lombard conquest of northern Italy. Venice and most of Istria, including Pula, remained outside the realm of Lombard control and steadfastly clung to their Byzantine protectors.

In 774 the Lombard kingdom fell to Frankish forces and came under the rule of Charlemagne. Seven years later the region was awarded to Pépin, Charlemagne's son. In 788 the Byzantine empress Irene sent a force to Italy to unseat the Franks but it failed miserably, ending in 798 with the loss of Byzantine possessions in Istria. Frankish rule in Pula lasted until 811 when it, along with all of the former Byzantine holdings in Istria, were returned in exchange for Charlemagne's right to use the title of emperor.

Throughout the ninth century, Byzantine rule in the upper Adriatic slowly weakened. It was during this period that Venice, though technically remaining under Byzantine rule, began to develop as a strong regional power. During the last half of the tenth century, Slavic pirates raided with impunity throughout the Adriatic. In the absence of Byzantine protection, Pula along with many other coastal towns of Istria and Dalmatia appealed to Orseleo, the doge of Venice, for aid. Perhaps wanting to extend Venetian influence or, more likely, to secure the Adriatic as a safe trading route, the doge chose to come to their assistance and during 998 and 999 succeeded

Pula's Roman amphitheatre
Photo courtesy of Embassy of Croatia, Washington, D.C.

in clearing the region of most of these Slavic corsairs. Pula welcomed the Venetians as liberators and pledged 2,000 pounds of olive oil yearly as tribute to the doge. In essence Pula became a protectorate of Venice, which by the year 1000 had become independent of the Byzantine Empire. Thus began a period of Venetian rule in Pula that would endure for nearly 800 years.

As Venice expanded its hold on the region, rival seafaring powers emerged and began to test Venetian resolve in the Adriatic. Pula became the site of several challenges to Venetian hegemony. In 1145 a Pisan force made an attempt on the city. War was narrowly averted through the efforts of the papacy. Less than half a century later a Pisan squadron returned to Pula and occupied the city. Venice prepared for war but the Pisans withdrew peaceably.

Pula remained on the fringe of Venetian control and somewhat vulnerable to forces desiring to weaken Venice's growing power. In 1240 a Guelph (pro-papacy, anti-imperial) coalition, which included Venice, laid siege to Ferrara, an important possession of the Holy Roman Emperor Frederick II. In response to the capture of Ferrara, Frederick sought a reprisal and chose to strike at Istria and Dalmatia, the most vulnerable of Venetian possessions. Imperial agents incited the residents of Pula and Zara into rebellion. Venetian forces,

however, responded decisively and put down the insurrection, razing the walls of Pula.

The foremost challenger to Venetian control of the Adriatic, however, was Genoa. In 1378 a Venetian flotilla under Vettore Pisani scored a great victory over the Genoese near Antium (Anzio). Upon the arrival of winter, Pisani's squadron put in at the harbor of Pula. In May 1379 a Genoese fleet under Luciano Doria appeared outside Pula's harbor and challenged Pisani to leave the harbor and engage in battle. At first Pisani balked, feeling that his defensive position was preferable, but his pride got the better of him when he was accused of cowardice by his crew. In the ensuing battle off the coast of Pula, the Venetians suffered a terrible defeat. Despite the defeat at Pula and the near capture of Venice itself, Venetian rule was eventually consolidated and Pula remained firmly under Venetian control.

Throughout most of the period of Venetian control Pula languished as a backwater. Venetian disregard for this Istrian port was evidenced in the late fifteenth century when members of the Venetian senate nearly succeeded in implementing a plan to dismantle the city's amphitheatre and bring it to Venice. They had already carted away most of Pula's sixth-century Byzantine basilica. Venice was also indirectly responsible for decimating the city's population with the plague, which Venetian traders had brought back from trading voyages to the east. By the early seventeenth century, plague and other epidemics had taken their toll and the once-thriving city had a population of only 300. Some progress, however, did occur during the Venetian period. In the seventeenth century significant improvements were made to the city's cathedral and a fortress overlooking the harbor was constructed.

Venetian rule of Pula came to an end at the close of the eighteenth century at the hands of Napoléon Bonaparte. Napoléon's victories at Arcole and Rivoli in late 1796 sealed the defeat of the First Coalition and the loss of Austrian possessions in northern Italy. French armies completed the conquest of northern Italy in the spring of 1797 with the capture of Venice. Weary of war, in October of 1797 Austria agreed to the terms of the Treaty of Campo Formio. In this treaty Austria acquired most of the Venetian Republic, including Pula. In exchange Austria recognized French primacy in northern Italy, agreed to the French occupation of the left bank of the Rhine, and ceded the southern Netherlands (Belgium) to France. Austrian rule of Pula was interrupted from 1806 to 1814 when French forces held the city. It was returned to Habsburg control as a result of the Congress of Vienna in 1815.

Pula, with its fine harbor, represented a tremendous prize to the virtually land-locked Habsburg Empire. As a result of its strategic importance, the city witnessed prosperity it had not enjoyed under the Venetians. By the middle of the century Pula had become a major imperial naval base. It developed rapidly during the second half of the nineteenth century and new shipyards, warehouses, and factories were built. In 1876 a railway link to the Habsburg hinterland was completed.

The importance of the naval base at Pula was reflected in its defenses. By the outbreak of World War I the harbor was protected by a coastal artillery battery and an extensive mine field. During the war the main fleet of the Austrian imperial navy was based here. The imperial navy was assigned somewhat limited objectives; chiefly to defend the coasts and protect sea lanes for supplying armies operating in the southern theatre. The Austrian fleet was successful in fulfilling these objectives. However, under somewhat bizarre circumstances, Pula was not spared an attack by the Allies.

At the close of the war, in October 1918, Admiral Miklós Horthy de Nagybánya, chief officer of the imperial navy, gave the command to surrender the fleet at Pula to the Yugoslav National Committee, acting on behalf of the Allied armies. Shortly afterward, unaware of what had transpired, two Italian commandos swam into the harbor at Pula and attached mines to the *Viribus Unitis*, the flagship of the imperial navy. They successfully destroyed the ship, killing fifty-one seamen. In the following month Italian forces seized Pula and other parts of Istria in accordance with what had been promised them by the Treaty of London. Pula was officially awarded to the Italians in 1919 by the Treaty of St.

Germain and the city became a part of the Italian province of Venezia Giulia.

During World War II Pula was under Italian military control until 1943, when German troops displaced the Italians. In 1945 Yugoslav forces entered the city as occupiers. The city was officially transferred to Yugoslav control in 1947. Most of its Italian residents either fled or were forced out. By the end of the war the city's shipyards, which had suffered extensive damage, were operating at only one-fifth to one-third of their prewar capacity. Under the Yugoslavs, Pula's shipyards were rebuilt and developed to specialize in the construction of tankers. The city's strategic importance was recognized by the Yugoslav government and Pula became a naval district, functioning as a part of the coastal defense.

The city grew rapidly under the Yugoslavs. In 1948 it had a population of 20,741. By 1961 it had grown to nearly 37,000 and by 1991 to about 62,000. Unlike many other parts of Croatia, the city's population was somewhat homogeneous, mainly made up of ethnic Croats. Its ethnicity proved to be advantageous for it remained unscathed during the civil war between Serbs and Croats in Croatia following the collapse of Yugoslavia in the early 1990s. The city retains many of its Roman structures, including the Triumphal Arch of Sergius, Temple of Augustus, and its crown jewel, the amphitheatre; it also has numerous buildings from the Byzantine and Venetian periods, including a thirteenth-century town hall and seventeenth-century cathedral.

Further Reading: Many sources on Pula are available but no truly definitive work in English exists. The dedicated researcher must seek out general works on specific periods in Pula's history, concentrating on histories of the empires that ruled the city, and hunt for pertinent details. For information on Pula before the Roman period, John Alexander's *Yugoslavia before the Roman Conquest* (New York: Praeger, and London: Thames and Hudson, 1972) provides a good background. *A Dictionary of Greek and Roman Geography*, edited by William Smith (London: Murray, 1872) provides an extensive, though somewhat dated, outline of Pula during the Roman period. There are many good sources for Pula under the Byzantines and Franks. Among these are *The Rise of the Saracens and the Foundation of the Western Empire*, volume 2 of *The Cambridge Medieval History* edited by J. B. Bury, H. M. Gwatkin, and J. P. Whitney (London: Cambridge University Press, and New York: Macmillan, 1913; second edition, 1966) and volume IV from the same series, *The Byzantine Empire*, part 1, *Byzantium and Its Neighbors*, edited by J. M. Hussey (London: Cambridge University Press, and New York: Macmillan, 1913; second edition, 1966). Both contain good background material and annotated indexes to guide the reader to germane sections. The mammoth two-volume *Venetian Republic* by W. Carew Hazlitt (London: Black, 1913; New York: AMS, 1966) provides invaluable information on the city under the doges. Also of some use are the lively accounts found in Francis Marion Crawford's two-volume *Gleanings From Venetian History* (London and New York: Macmillan, 1905).

—Jeffrey M. Tegge

Ravenna (Ravenna, Italy)

Location: Capital and principal port of the province of Ravenna, in the Emilia-Romagna region of northeastern Italy, about sixty-five miles northeast of Florence. The city is on a low-lying plain near the confluence of the Ronco and Montone Rivers, six miles inland from the Adriatic Sea, with which it is connected by canal to the harbor and seaside resort of Marina di Ravenna, formerly the Porto Corsini.

Description: Originally inhabited by native Italic peoples before being occupied by the Etruscans and Gauls. The Romans established the port of Classis in the first century B.C., and Ravenna became the most important military seaport on the upper Adriatic. The city achieved its greatest glory as the capital of the Western Roman Empire beginning in the early fifth century; it was later (sixth and seventh centuries) capital of Italy's Ostrogothic kings. As seat of the Exarchate of Ravenna, the city was the center of the Byzantine government in Italy until the Lombard invasion of the mid-eighth century. Due to the influence of both Rome and Constantinople, Ravenna's art and architecture reflects a unique blending of East and West. Its fame rests primarily on the fact that it was the birthplace of, and is still the richest depository for, early Christian art; its fifth- to eighth-century churches and monuments contain the most magnificent Byzantine mosaics in the world. After being ceded to the States of the Church in the eighth century, Ravenna declined in importance, though it was a free city-state in the twelfth and thirteenth centuries. In subsequent centuries, it came under the rule of the Papal States, the local nobility, and the Venetian Republic. Ravenna became part of the Kingdom of Italy in 1861.

Site Office: A.P.T. (Azienda di Promozione Turistica)
Via San Vitale, 2
CAP 48100 Ravenna, Ravenna
Italy
(544) 3575516

The peaceful, provincial-looking city of Ravenna, one-time capital of the Western Roman Empire and famous for its early Christian monuments and the greatest mosaics in western art, was probably first inhabited by native Italic peoples who migrated south from Aquileia (near Trieste) about 1400 B.C. In ancient times, the Adriatic Sea lay closer to Ravenna, which was a small island settlement on coastal lagoons that have since silted up. It entered history as an Umbrian town, and was later occupied by the Etruscans. In the sixth or fifth century B.C. the Gauls, a Celtic people, established Cisalpine Gaul, located between the Apennines and the Alps (at the time, "Italy" referred only to the peninsula proper).

Around 187 B.C., the Romans built Via Aemelia, a long highway running straight from the Adriatic port of Rimini to what is now central Italy (the old road still exists, as Via Emilia), establishing colonies along the way. Ravenna soon became an important Roman center, owing to its access to the sea and its strategic position along the network of highways, which eventually served as trade and conquest routes from Rome to Dalmatia and the Danube.

Ravenna was the base from which Julius Caesar negotiated with the Senate in 50–49 B.C. before setting out on his famous crossing of the Rubicon River, thus invading Italy and starting the civil war. Emperor Augustus, Caesar's great-nephew, built the harbor of Portus Classis, which was soon linked to the old city (three miles away) by the Via Caesarea canal. Surrounded by impassable marshes, the city had clear military advantages. By the first century B.C., Ravenna had become a prominent Roman port, the base for Augustus's naval fleet in the Adriatic Sea, capable of sheltering as many as 250 warships. Almost nothing is now visible of the ancient Roman city and port.

The Roman Empire was formally divided between East and West in A.D. 395 following the death of Theodosius the Great. The Eastern (Byzantine) Empire was based at Constantinople, the Western at Rome and then Milan. Ravenna's period of greatest glory began in 402, when Theodosius's son, Western Emperor Flavius Honorius, threatened by barbarian Goth invasions, moved his court from Milan to the easily defensible Ravenna—thus making it the capital of the Western Roman Empire. The relocation came just in time: Alaric the Goth sacked Rome in 410. During the reign of Honorius, as well as that of his half-sister, the beautiful and strong-willed Galla Placidia, Ravenna was embellished with many magnificent churches and monuments, and the art of mosaic-making flourished. Alaric had captured Placidia in Rome; when the Goths attacked Ravenna, she saved the city by marrying Alaric's brother-in-law, Goth King Ataulphus. She accompanied him everywhere, even in battle. After his murder, she married one of Honorius's generals, Constantius, who became Emperor Constantius III, ruling jointly with Honorius. He died seven months into his reign, however, and Placidia began a love affair with her half-brother, resulting in scandal and her exile to Constantinople. She returned to Ravenna after Honorius's death in 423, ruling what remained of the Western Empire as regent during the minority of her son, who became Emperor Valentinian III.

The last western emperor, Romulus Augustulus, was forced to abdicate in 476, and Constantinople became the seat

Christ as the Good Shepherd, mosaic from Galla Placidia's tomb

Christ on the way to Calvary, mosaic from Sant' Apollinare Nuovo
Photos courtesy of Provincia di Ravenna

of the sole Roman emperor; the popes became virtual emperors in Italy. With the fall of the Western Empire, the Germanic general Odoacer crowned himself the ruler of Italy, making Ravenna his capital. In 488, Byzantine Emperor Zeno commissioned Ostrogoth king Theodoric the Great, who had been brought up in Constantinople, to invade Italy; the Ostrogoths quickly occupied the peninsula, but the impregnable Ravenna held out for three more years. Theodoric tricked Odoacer into sharing Italy with him—but then murdered him in 493; Theodoric made Ravenna the capital of his Ostrogothic realm that same year. He built a royal palace and ushered in another era of great building activity, including several Arian churches. Although Theodoric stabilized his lands, ruling through Roman officials, he failed to establish a Gothic kingdom because his authority was checked by a jealous emperor and suspicious Catholics: during the first barbarian invasions, the Roman pope and local bishops assumed great power, filling the vacuum left by the collapse of the Roman system; the church and the emperor opposed Italy's strong government, as well as the Christianized barbarians' adoption of Arianism, which, heretically, denied the full divinity of Christ. The last small renaissance of Latin letters occurred under Theodoric's reign.

Following Theodoric's death in 526, as well as the murder of his daughter Amalasuntha nine years later, the leadership in Constantinople decided to assert its force. Byzantine Emperor Justinian I (the Great), who reigned from 527 to 565, sought to revive imperial power in Italy in order to vanquish the Goths and restore Roman institutions and the Catholic Church. He sent an invasion force of Roman troops who were, ironically, largely foreign and Greek-speaking. Tough Ravenna was spared destruction in the terrible wars between the Goths and the Eastern emperor for control of Italy, which lasted until the Goths were finally subdued in 552; the city was eventually conquered in 539–40 by the young and brilliant general Belisarius. Justinian would leave a permanent mark on Italian civilization; he introduced Byzantine art, which would reach its highest expression under his reign in Ravenna. Byzantine art became the great formative influence of Italian painting.

Ravenna became capital of the Exarchate of Ravenna, or Exarchate of Italy, an imperial governorship created in 584 under Eastern Emperor Maurice. As such, Ravenna became the military and administrative center of the Byzantine government in Italy. In the early seventh century, the exarchate included a diagonal strip of territory extending from north of Ravenna to south of Rome, the peninsula's southern extremities, and various coastal enclaves. Under the exarchs, Greek-speaking viceroys, Ravenna saw a third period of prosperity, growing in power and splendor, until it was conquered by the Lombards in the mid-eighth century. The exarchs would reign over increasingly smaller areas of Italy for the next 500 years.

One of the leading cities of western Europe for centuries, Ravenna enjoyed great political, economic, intellectual, and artistic activity from the time of Honorius to the coming of the Lombards, Italy's Dark Ages. Ravenna was the birthplace of western Christian art, and is still the world's richest treasure-house of early Christian art. Its fame rests on the quality and quantity of its fifth- to eighth-century churches and monuments, which, in many instances, surpass even those of Rome. As capital city of the Western Roman Empire for two and one-half centuries, and as a major port of entry for the Eastern Empire, Christian Ravenna's art and architecture reflects a blending of west and east, of Roman architectural forms and Oriental decoration, especially in its Byzantine mosaics. Its churches are adorned with the finest mosaics ever made; it was mostly Greek craftsmen from the Constantinople court who made mosaics the new medium of public art in fifth- and sixth-century Ravenna. The fusion of eastern and western techniques, designs, and imagery have resulted in a highly individual art unrivalled anywhere else in the western world.

The small Tomb of Galla Placidia and the great Church of San Vitale—which stand on the same grounds in the northwest part of the city—contain the most celebrated and elaborate mosaics in Ravenna. Placidia's mausoleum, built around 440, about a decade before her death, is the oldest complete monument in the city. Besides governing the declining empire, Placidia was a devout Christian who endowed many churches and priests; archbishop at the time, St. Peter Chrysologus, built several projects with Placidia's support. The building technique of the simple red-brick mausoleum is western, but its Latin cross composition—with barrel-vaulted roof and a central dome—is eastern in nature. The interior's entire upper surface—the dome, roof, and walls—is covered with glittering mosaics of gold, green, and blue, a triumph of decorative art: represented here are a cross, symbols of the Evangelists, eight of the Apostles, animals, and a depiction of Christ the Good Shepherd above the doorway. There are three marble sarcophagi in the mausoleum, containing Placidia's second husband Constantius III, and her son Valentinian III. Curiously, the remains of Placidia are not believed to reside in the tomb; she died in Rome in 450, and is probably buried near St. Peter's.

The Church of San Vitale, an octagonal basilica built of marble and topped by a great terra-cotta dome, is the masterpiece of Byzantine art in Ravenna and one of the most splendid examples of Byzantine architecture and decoration in western Europe; its magnificent, brilliantly colored mosaics are among the last great works of art in the ancient world. The church is not the last work of the Romans, but the first work of the Romanesque. Begun in 525–26 during the reign of Theodoric by Bishop Ecclesius, at the place where Ravenna patron St. Vitalis suffered his martyrdom, the church was not consecrated until after Justinian ruled Ravenna, in 547–48, by Archbishop Maximian. Unlike those in the Tomb of Placidia, San Vitale's decorations are exclusively Byzantine in style. The famous mosaics, probably made in local workshops and influenced by similar work in Constantinople, are all in the presbytery. They include portraits of Justinian (with Maximian) and his wife, Empress Theodora, with their respective courts; Old and New Testament scenes and figures; and a depiction of Christ the King enthroned on a globe of the world, flanked by archangels and by St. Vitalis and St. Ecclesius, the founder of the church. Christ is also shown with the Apostles and the sons of St. Vitalis.

Other extant monuments dating to the time of the Arian emperor Theodoric include his tomb (Mausoleo di Teodorico), the Church of Sant' Apollinare Nuovo, the Orthodox (or Neonian) Baptistery, and the Arian Baptistery.

The Tomb of Theodoric, located in the northeast part of Ravenna and built around A.D. 520, is a strange, half-barbaric structure: a two-storied decagonal rotunda built with huge, square limestone blocks is capped by a monolithic, single-slab 300-ton limestone dome. The limestone was brought by the Goths from the region of the former Yugoslavia. The interior ornamentation is sober and austere, showing a clear Germanic influence. When the city was captured by Justinian's general Belisarius in 540, Theodoric's bones were scattered and the mausoleum was converted into a church.

The Church of Sant' Apollinare Nuovo, a basilica built by Theodoric around 519, on the east side of town, was originally an Arian cathedral but became a Catholic church around 570; it was dedicated to St. Apollinaris, the first bishop of Ravenna (and a reputed martyr), in the ninth century when the saint's remains were moved here from his other basilica at nearby Classis. Although parts of the church were remodeled in the sixteenth and eighteenth centuries, it has a well-preserved early Christian interior, with three naves separated by twenty-four Corinthian marble columns from Constantinople. The Arian mosaics—second in brilliance to those of San Vitale—were begun by the Ostrogoths and finished by the Catholics; they are of great artistic and theological interest because they may be the oldest such representations in existence. Slightly older than the works in San Vitale, they show the Roman port of Classis, with ships, on the left wall; the town of Ravenna with its churches and Theodoric's palace, on the right wall; above them are represented saints and prophets. Depicted above the windows are compositions from the New Testament, thirteen on each side: the teachings and miracles of Christ, shown beardless (on the left), and the passion and resurrection of Christ, with a beard (on the right). Above the arcades of the presbytery the Byzantines added other figures: the Magi, followed by a row of twenty-two holy virgins leaving Classis and offering crowns to the enthroned Virgin Mary and Child, on the left side; and, a row of twenty-six martyred saints with wreaths leaving Theodoric's palace and moving toward Christ enthroned with attendant angels, on the right side.

The Orthodox Bapistery (later called San Giovanni in Fonte), also known as the Neonian Baptistery because it was decorated by Archbishop Neone, who served from 449 to 452, is an octagonal brick monument located southwest of the town center. The fine mosaics in the dome ceiling, some of which have been restored, are also among the oldest in Ravenna; they depict a scene of the baptism of Christ, around which are portraits of the Twelve Apostles. The Byzantine reliefs over the arches of the rotunda represent the prophets. The Arian Baptistery (later called Santa Maria in Cosmedin),

northeast of the town center, dates from the early sixth century; the mosaics in the dome, also restored, are a more simplified version of the Neone mosaics, with the Apostles arranged around a Baptism scene. The lower band of mosaics shows the throne of God adorned with the cross.

In open country a few miles south of Ravenna stands the plain brick Church of St. Apollinaris in Classis (Sant' Apollinare in Classe Fuori). The last great Byzantine monument in Ravenna, it is also the largest and best preserved basilica in the city, and the only remaining relic of ancient Classis. The church, which contains twenty-four Byzantine columns and the marble sarcophagi of Ravenna's fifth- to eighth-century archbishops, was begun about 535 and consecrated in 549 by Maximian; it was completed with a tall, cylindrical campanile in the ninth century. The church was restored in the eighteenth century. The apse and triumphal arch are adorned with beautiful sixth- to seventh-century mosaics, portraying green-and-gold symbolic representations of Christ the Savior and the Transfiguration; below, St. Apollinaris is shown praying amidst a flower-strewn landscape between rows of sheep, signifying the Apostles. Along with these depictions are Old Testament sacrifice scenes.

Other monuments of historical interest include the Cathedral and the Archbishop's Palace. The Cathedral of Sant' Orso, located just southwest of the town center, stands on the site of the oldest church in Ravenna. Originally built from 370 to 390 and dedicated to its founder, St. Ursus, the church was destroyed by an earthquake in 1733 and rebuilt from 1734 to 1743. Only the crypt, the pulpit, and a campanile survive from the earlier structure.

The Archbishop's Palace, now the Episcopal Palace Museum (Museo dell'Arcivescovado), stands behind the cathedral. It contains the so-called Throne of the Archbishop Maximian, a sixth-century masterpiece carved in Alexandria, Egypt, and adorned with rich ivory reliefs. The throne, thought to be a gift from Emperor Justinian, is considered the finest piece of ivory work in existence. The palace's Archiepiscopal Chapel (Oratorio di Sant' Andrea), built around 500 during the reign of Theodoric, also contains fine sixth- to seventh-century mosaics in the vaults and arches; they show portraits of the saints, as well as other early Christian animal representations and symbols.

The death of Emperor Justinian in 565 led to an increasingly unstable Byzantine Italy. The Lombards, Germanic invaders from the north, overran the Po River Valley in 568. They established a kingdom in the north, initially sharing Italy with the semi-independent Byzantine dukes, the new city of Venice, and the exarchs of Ravenna. Gregory the Great, a Roman noble who became pope in 590, was a crucial figure in the development of the papacy and the western church. Determined to protect Italy from barbarian rule, he persuaded the Lombards to spare Rome and Ravenna. During the seventh century, however, different parts of the exarchate were conquered by Lombard dukes. In the early eighth century, Eastern Emperor Leo III, the Isaurian, aroused the wrath of big Italian landowners with a new system of taxation.

Injury was added to insult in 726, when Leo signed a decree banning images of Christ and the saints, ordering their destruction. Italy revolted against the Byzantine government under leadership of the papacy, which was gradually assuming rule of Rome and its duchy. The revolt seriously threatened the exarchate, with the Lombard kings siding with the emperor and rebels. After the suppression of the revolt, around 750, the southern dominions of the exarchate were annexed to Sicily. Though Ravenna had been seized several times by the Lombards in their struggles with the Byzantines, the city finally fell in 751 to the ruthless Lombard king Aistulf, who drove out the last exarch. Byzantine power in Italy was a thing of the past.

The popes, more threatened by the Lombards than by the absent Byzantines, sought help from across the Alps. At the invitation of Pope Stephen II, Frankish King Pépin III (the Short) captured Ravenna in 754; he drove the Lombards from Ravenna and restored the lands of the exarch, passing the city to the papacy in 757 (archbishops, however, retained princely powers). Pépin's son, Charlemagne, defeated and captured his father-in-law, Lombard King Desiderius, in 773–74. He assumed the "Iron Crown of Italy," renewing the exarchate's grant to the Holy See. In 800, Pope Leo III crowned Charlemagne emperor of the Roman people; Rome was once again seat of the empire. The Holy Roman Empire would last 1,000 years. Ravenna's fortunes declined slowly and gracefully in subsequent centuries. Venice overtook Ravenna as the leading port on the Adriatic as the Classis harbor gradually silted up and was abandoned in the ninth century.

Otto I the Great, a German of the Saxon dynasty, invaded northern Italy in the 950s and was crowned Holy Roman Emperor in 962. The strong government of Otto and his Germanic successors allowed trading cities to expand their power and influence, at the expense of the church and nobles alike. The papacy went into decline. Many important cities of northern Italy became independent communes, or free city-states. Beginning in 1154, recently elected Emperor Frederick I Barbarossa crossed the Alps five times, determined to reassert imperial power in Italy. His activities marked the beginning of a renewed conflict between the papacy and the empire. Frederick sacked the free cities of the north; his destruction of Milan (1162) led to the formation of the Lombard League, a unified front of cities stretching from Ravenna to Asti. The League, with the support of Pope Alexander III (whom Frederick had exiled from Rome), along with the support of Sicily and Venice, defeated Frederick's knights in 1176; the emperor was forced to recognize the pope and the free cities. Ravenna became a commune in 1177, but its independence was short-lived.

Pope Innocent III (who reigned from 1198–1210) laid the foundation of the States of the Church, which would include the territory of Ravenna. Although he had persuaded Frederick I to respect the States of the Church, Innocent's death led to a further rift between the popes and the emperors. Beginning in the thirteenth century, during the brilliant reign of Frederick II, Ravenna, like all of Italy, became embroiled in the strife between the Guelph and Ghibelline factions. Broadly, the Guelphs, led by the popes, supported religious orthodoxy, the liberty of the communes, and the interests of their emerging merchant class. The Ghibellines, led by the emperor, supported state economic control, religious and intellectual tolerance, and the interests of the rural nobles. The death in 1250 of Frederick—who had been excommunicated twice by the popes—marked the decline of imperial power in Italy. In 1276, in return for papal recognition as Holy Roman emperor, Rudolf of Habsburg recognized the papacy's rights over the States of the Church and ceded Romagna (which included Ravenna) to the popes again.

In subsequent centuries, Ravenna was contested by the Holy See, the local aristocracy, and the Venetian Republic. From 1297 to 1441, the city was ruled by the house of da Polenta, a noble Guelph family of Romagna famous for offering refuge to the exiled Dante. The Italian poet finished the *Divine Comedy* and died in Ravenna in 1321. Dante's Tomb (Sepolcro di Dante) is located in the town center, next to the large tenth-century Romanesque Church of St. Francis (San Francesco). The tomb's neoclassical exterior was built in 1780; the interior sarcophagus contains the poet's remains. Dante Alighieri was exiled from his native Florence, and took refuge first at Verona and then at Ravenna, at the court of his friend Guido Novello da Polenta, lord of Ravenna.

Venice had become a powerful maritime state in Italy by the end of the fourteenth century, following the defeat of its archrival Genoa in 1379–80. Its efficient government, great commercial activity, and trade rights with the east led Venice to colonize many lands throughout the fifteenth century, including those of northern Italy, the Dalmatian coast (in the former Yugoslavia), many Aegean islands, and areas in the Levant. Ravenna came under Venetian control in 1441, enjoying a brief period of renewed prosperity. Despite the French invasion of northern Italy in 1494–95, Venice annexed more mainland possessions. With the League of Cambrai in 1508, however, various powers—including the papacy, the empire, and other Italian states—that had suffered loss of territory at Venice's hands tried regaining their lands; they temporarily defeated the Venetian Republic in 1509. League member Pope Julius II, determined to take full control of the States of the Church again, recovered the towns in Romagna that had been seized by Venice. Ravenna was formally returned to the Papal States in 1509.

The Battle of Ravenna, which occurred in 1512, was one of the major engagements of the Italian Wars against foreign domination (France, Spain, Germany) in the late fifteenth and early sixteenth centuries. Julius II, aiming to expand the Papal States and to unite all of Italy, dissolved the League of Cambrai and formed the Holy League (1511) in which Venice, Spain, and the papacy joined to drive the French out of northern Italy. After successfully defending Bologna in early 1512, Louis XII's French army, led by the twenty-three-year-old Duke Gaston de Foix along with 5,000 German mercenaries, besieged Ravenna on Easter Sunday. The League army was led by Ramon Cardona, Spanish viceroy of Naples. Though Foix was killed during the final charge against retreating Spanish forces, the French and Germans decisively won the battle, which became famous for foreshadowing modern artillery tactics. Ravenna surrendered to the French the following day, and was plundered. Only a few months later, however, the city was recaptured without difficulty by Julius II; following his death in 1513, the French army, weakened by the loss of its leader and opposed by a strengthened Holy League, withdrew back across the Alps.

Once again, Ravenna belonged to the Papal States until 1859, except for another period of French domination under Emperor Napoléon Bonaparte, from 1797 to 1814. The Emilia-Romagna region was one of the first to join forces with King Charles Albert of Sardinia-Piedmont and Garibaldi, military commander of the Roman Republic, in the quest for a unified Italy, in the 1840s. The city proclaimed its union with the Kingdom of Sardinia in 1860, which became the Kingdom of Italy in 1861. Italy became a Fascist dictatorship in 1925 under Benito Mussolini, though the anti-Fascist resistance movement was born in Emilio-Romagna. The region and Ravenna suffered heavy damage by the Fascists and their Nazi allies during World War II. Ravenna was taken by the Allied forces on December 5, 1944.

Today, after considerable postwar rebuilding, Ravenna is a modern, industrious city. It continues to be the principal port of Emilia-Romagna, with a newly constructed ship channel and harbor. The discovery of large offshore natural gas deposits in the 1950s led to the building of one of Italy's largest petrochemical plants near Ravenna—and to the growth of a large industrial quarter on the outskirts of the city. Yet Ravenna still remains a major international center for the study and production of mosaics.

Further Reading: Though slightly outdated, *A Short History of Italy,* edited by H. Hearder and D. P. Waley (Cambridge and New York: Cambridge University Press, 1963), is a succinct, scholarly introduction to Italian history, from Classical times to the present day. *History of the Italian People* by Giuliano Procacci, translated by Anthony Paul (New York: Harper, 1970; Harmondsworth, Middlesex: Penguin, 1973; as *Storia degli Italiani,* Bari: Laterza, 1970), presents a more in-depth view from the year 1000 to the twentieth century. *Wonders of Italy* by Giuseppe Fattorusso (Florence: Fattorusso, 1925) teems with photographs of Ravenna's early Christian and Byzantine art, architecture, and mosaics. *Venice and North-Eastern Italy* by Eric Whelpton (London: Hale, and New York: International Publications Service, 1965) is a travel book that contains descriptions of Ravenna's numerous historic sites. Of the numerous travel guidebooks available on Italy and its regions, *Cadogan Guides: Northeast Italy* by Dana Facaros and Michael Pauls (London: Cadogan, and Old Saybrook, Connecticut: Globe Pequot, 1990) contains intelligent, often witty historical accounts of the Emilia-Romagna region and the city of Ravenna.

—Jeff W. Huebner

Rhodes (Dodecanese, Greece)

Location: Rhodes is a member of the Dodecanese, a group of twelve islands located off the southwestern tip of Asia Minor. Rhodes is the largest of these islands.

Description: Both the island and the town of Rhodes have ancient roots that stretch back to the medieval Knights of St. John and earlier to classical Greece. Major sites include: Acropolis and other ancient ruins; medieval fortifications that surround the old city; Palace of the Grand Masters; the Knights' Hospital and Archaeological Museum, and, beyond the city limits, Kameiros, the remains of a Doric-style temple, and the Acropolis of Lindos. Rhodes was also the former site of the famous Colossus of Rhodes, one of the Seven Wonders of the World.

Site Office: Greek National Tourist Organization
5, Archbishop Makarios and Papagou Streets
Rhodes, Dodecanese
Greece
(241) 23-655 or 23-255

Rhodes is the name of an island in the Aegean Sea off the coast of Asia Minor (Turkey) as well as the island's chief city. It is the largest of the Dodecanese Islands. With a population of just under 90,000, the modern city of Rhodes, located on the northern tip of the island, is a major port that exports olives, wine, fruit, vegetables, honey, and other items. It is also an important historic site with ancient roots that date to the classical era of Greece.

Rhodes's earliest inhabitants were Carians and Phoenicians, but around 800 B.C. the island was taken over by Dorian Greeks who had come from northwestern and central Greece. In 490 Rhodes fell under Persian rule before joining the Delian League as an ally of Athens. In 411, however, the Rhodians decided to support Sparta in the Peloponnesian War. In return, Sparta recognized Rhodian independence.

In 408 B.C., the three Doric city-states of Ialysus, Lindus (now Lindos), and Camirus united into one state with Rhodes as the capital. This period is often referred to as Rhodes's Golden Age, during which Rhodes became an important political, commercial, and religious center. Rhodians founded colonies in adjacent areas and along the coasts of Asia Minor and Europe. This golden era lasted from the fifth to the third century B.C.

Rhodes became the most important maritime center in the Mediterranean. Its trade flourished with Italy, Greece, Macedon, Asia, and Africa. Around this time, Rhodes assumed its position as a major naval presence in the Aegean Sea, establishing the earliest code of naval law, the Rhodian

Sea Law. Rhodes's reputation as a naval power was so respected that the great Roman emperor, Augustus Caesar, professed admiration for the island. Rhodes reached its height as a commercial and naval power in the third century B.C.

Rhodes's reputation, though, rested on more than just its military and maritime prowess. It was also a great center for the arts. A major school of sculpture was established there as was a famous school of rhetoric, founded by the great orator Aeschines. Such renowned and highly respected Romans as Cato, Cicero, and Julius Caesar attended the school. Rhodians avidly pursued the finer arts in all their variety.

Under a succession of Greek, Roman, Byzantine, and Turkish rule, Rhodes always managed to maintain its independent spirit. Rhodes came under the control of Alexander the Great, but when he died in 323 B.C., the Rhodians once again proclaimed their independence. This did not please Demetrius I Poliorcetes, king of Macedon from 294 to 288 B.C. At that time Macedon was the most powerful and largest kingdom in the Greek world. In 304 B.C. Demetrius laid siege to the town, but the stubborn Rhodians were not the type to give up easily. They fought a valiant and what must have seemed interminable struggle. The famous siege, which lasted some twelve months, forced the Rhodians to promise the city's slaves a chance at freedom if they would help defend the town against Demetrius.

Demetrius created a fearsome weapon called the *helepolis,* which surpassed in both strength and ingenuity any weapon of war in existence at the time. The movable structure standing on eight large wheels contained soldiers who, safely protected from harm's way by shutters, could hurl their weapons from the porthole-type openings located in front. Not surprisingly, these giant and seemingly impenetrable siege machines frightened the Rhodians. During the Crusades, later generations would use variations of Demetrius's helepolis, such as the battering ram, the wooden tower, and, perhaps the best known of them all, the catapult. Yet Rhodes managed to withstand the constant onslaught. The siege finally ended when King Ptolemy of Egypt, who was allied with Rhodes, sent a fleet to assist the Rhodians.

Demetrius left Rhodes so impressed with the Rhodians' courage and bravery under fire that he gave them the siege weaponry as a token of his admiration. The Rhodians in turn sold the equipment and from these sales built an immense statue, the Colossus of Rhodes, dedicated to the sun-god Helios or Apollo. It became known as one of the Seven Wonders of the World.

Historians are still not certain of the exact location of the statue. They do know, however, that it stood about 120 feet high. The sculpture was built in the image of a naked Apollo, gazing out at the ocean, wearing a gold crown and

Rhodes harbor, with medieval fortress in background
Photo courtesy of Greek National Tourist Organization

holding a torch that served as a beacon to seafarers. The sculptor was Chares of Lindus, a native Rhodian. The statue itself consisted of plates of bronze that were held together by iron braces and blocks of stone. The mammoth work took some twelve years to complete and, according to some accounts, cost the modern-day equivalent of $75 million. In 225 B.C., an earthquake knocked it off its base. It fell to the ground, its massive head buried deep in the soil. Only the legs remained intact.

By the second century B.C., Rhodes had allied itself with Rome. During the Roman civil wars, Rhodes sided with Caesar. As a result, in 43 B.C., the Roman General Cassius attacked the city, devastating the once mighty Rhodian fleet and forever ending Rhodes's status as a major naval power. The island was eventually incorporated into the Roman Empire. With the division of the Roman Empire into East and West in A.D. 395, Rhodes then became a part of the Byzantine Empire. Over the centuries the city was often attacked by looters.

The Colossus stayed in the place where it had fallen until the seventh century, when the Saracens attacked Rhodes and reportedly sold the last remnants of the famous statue to a Jewish merchant. It was cut into strips and moved to the mainland. The Colossus of Rhodes was never seen again.

Rhodes was taken briefly by the Saracens in the seventh century and by the Genoese in the thirteenth. In 1309 the island became the headquarters of the Knights of St. John of Jerusalem (also known as the Knights Hospitallers). The Knights Hospitallers, later known as the Knights of Malta, built many of Rhodes's medieval structures, including the Palace of the Grand Masters. The Knights of St. John were refugees from Jerusalem and Cyprus. Originally they were a charitable brotherhood founded to care for the sick and the poor. The Hospital of St. John of Jerusalem was established around 1048 by merchants from the Italian province of Amalfi as a refuge for pilgrims. The Hospitallers took vows of poverty and chastity. Their patron saint was St. John the Baptist. By the middle of the twelfth century they had assumed their military function.

For more than a century and a half, the Hospitallers as well as the Knights Templars and the Teutonic Knights defended Christianity in Asia Minor. The Hospitallers formed their own naval fleet and played a major role in the Crusades, medieval military expeditions organized by the Church to liberate the Holy Land from non-Christian, mostly Moslem control.

Rhodes became the headquarters of the order, which was then divided into eight national units, or tongues: Provence, Auvergne, France, Italy, Aragon, England, Germany, and Castile. The French, who outnumbered the others, took over administrative functions. Each tongue had its own "inn," or residence. The entire order was ruled by the grand master.

During the fifteenth and sixteenth centuries, the Turks attacked Rhodes numerous times. In 1480, for example,

under Grand Master Pierre d'Aubusson, the knights were able to resist the relentless assaults of Sultan Mehmed II. A little more than forty years later, however, a third and more deadly siege under Sultan Süleyman the Magnificent dealt the final blow to the Hospitallers.

Süleyman, aware that the Hospitallers were helping the Christians raid the coast of Asia Minor, assembled in 1522 an army of about 150,000 men and transported them in some 300 ships. The knights, with only about 5,000 men on their side, received the help of various groups, including 200 Genoese, 50 Venetians, and around 6,000 natives of Rhodes. This siege lasted six months. When the Turks blockaded the town from the sea, the knights, weakened by starvation and disease and betrayed by a defector, were finally forced to surrender. Rhodes was set on fire. Only 180 knights survived the long siege. On the night of January 1, 1523, the remaining knights and Grand Master Villiers de L'Isle-Adam left the island forever, eventually settling in Malta.

The Turks ordered that the knights be treated with utmost respect and courtesy. Rhodes then fell under Turkish control. Thus from 1522 to the early years of the twentieth century, the Dodecanese Islands were part of the Ottoman Empire. In 1912 the Italians took over, having captured Rhodes after a brief siege.

When the Italian dictator Benito Mussolini came into power, he made it his mission to restore the order's palaces to their original grandeur. The Italians improved roads, authorized construction of numerous hotels and other buildings, and restored many of the ancient monuments. Italian rule over Rhodes did not last long. During World War II, Italian control gave way to German, but British and Greek forces liberated the island in 1945. After the war, the Dodecanese Islands were officially reunited with Greece.

Many classical and medieval remains still exist on the island, including the Doric Temple of Athena in Lindos and the Acropolis with temples and theatre in the city of Rhodes proper. Also in the vicinity are a restored stadium (c. third century B.C.) and a small theatre. There are also ruins of the temples of Athena Polias and Pythian Apollo. Today two bronze deer, symbolic of Rhodes, stand at the entrance where, according to tradition, the Colossus of Rhodes probably stood at the city's port.

The modern city of Rhodes consists of two distinctive parts: the Old City is surrounded by the magnificent military walls built in the fourteenth, fifteenth, and sixteenth centuries. The New City refers to the rest of the city that has been built since the Italian occupation of 1912. Today it consists of a resort community of hotels, shops, and nightclubs.

Although the Knights Hospitallers left centuries ago, they, perhaps more than any other group during Rhodes's long history, remain a major presence in the town, even to this day. The fortress walls they built still exist, as do their palaces. The walls are as much as forty feet thick.

The Palace of the Grand Masters, or Knights' Palace, was completed in the fourteenth century. The palace and the fortress still dominate the town. The palace is perhaps the most durable symbol of Rhodes. It was damaged by earthquake twice—once in 1481 and again in 1851. When the Turks took over, they converted it into a prison. After being severely damaged by an explosion in 1856, the Italians later restored it to its former grandeur. The palace, which is said to resemble the Papal Palace in Avignon, France, contains ancient Hellenistic and Roman mosaics as well as sixteenth- and seventeenth-century Western furniture.

Work on the Knights' Hospital and Archaeological Museum started in 1440 and was completed in 1505. The two-story former hospital now houses the Archaeological Museum, which contains collections of coins, pots, and Mycenaean and Roman sculpture as well as the ancient masterpiece, Aphrodite of Rhodes (c. third century B.C.).

Knights Street (also called Street of the Knights) is a cobblestoned street that recalls the era of the Knights Hospitallers, lined as it is by fifteenth- and sixteenth-century Gothic buildings. The knights used to reside in these inns. The street has been restored to resemble a typical street scene in medieval times.

Despite its latter-day reputation as an international holiday resort, Rhodes continues to be a historic site of major international significance, with monuments that date from the golden era of ancient Greece to the medieval era of the Knights Hospitallers.

Further Reading: General studies of Rhodes written for a contemporary audience are difficult to locate. Most references to Rhodes are buried in general histories of Greece. A good introduction to the world of ancient Greece is Robert G. Kebric's *Greek People* (Mountain View, California: Mayfield, 1989). N. G. L. Hammond's *A History of Greece to 322 B.C.* (Oxford: Clarendon, 1959; third edition, Oxford: Clarendon, and New York: Oxford University Press, 1986) is a modern interpretation of Greek culture and ideas. Similarly, Antony Andrewes's *The Greeks* (London: Hutchinson, and New York: Norton, 1967) examines Greek society in the classical period from 750 to 350 B.C. Robert Payne's *The Splendor of Greece* (New York: Harper, 1960; London: Pan, 1964) is a serviceable introduction to the Greek isles. A more scholarly approach is offered by Michael Grant in *The Rise of The Greeks* (London: Weidenfeld and Nicolson, and New York: Scribner's, 1987). *The Knight in History* by Francis Gies (New York: Harper and Row, 1984; London: Hale, 1986) traces the development of the medieval knight, including the Knights of the Hospital of St. John of Jerusalem, examining the significance and longevity of the knightly figure in Western culture.

—June Skinner Sawyers

Rimini (Forlì, Italy)

Location: In the Forlì province in the Emilia-Romagna region of northeastern Italy; on the Riveria del Sole of the Adriatic and northeast of Mt. Titano and the republic of San Marino.

Description: Historic town and seaside resort on the Adriatic Sea, founded as a Roman colony in 268 B.C.; ruled by the Malatesta family from the thirteenth to sixteenth centuries.

Site Office: A.P.T. (Azienda di Promozione Turistica)
P. le Indipendenza, 3
CAP 47037 Rimini, Forlì
Italy
(541) 51101

Rimini became a Roman settlement in 268 B.C., when the Romans gave it the Latin name Ariminium, from which its modern name derives. According to the Greek geographer and historian Strabo, the site was originally inhabited by an Umbro-Etruscan civilization. After its founding as a Roman colony, then a municipium, Ariminium became important in Roman wars and politics because of its strategic location on the Adriatic coast; it served as a base for naval and military operations during the Second Punic War (218–201 B.C.). Both Julius Caesar and Augustus favored the site; the latter in 27 B.C. had a triumphal arch built at the intersection of the Roman roads Via Flaminia and Via Aemilia, where the settlement continued to grow. Augustus also caused a bridge to be built over the Marecchia River, which flows into the Adriatic at the foot of the Apennine Mountains near Rimini; the bridge was completed by Tiberius in A.D. 21.

The Council of Rimini in A.D. 359 attempted, unsuccessfully, to resolve a dispute arising from the teachings of the Alexandrian priest Arius concerning the divinity of Christ. Thereafter, with the decay of the Roman Empire, Rimini was captured by the Goths before falling under Byzantine rule from the sixth to the eighth centuries. The Lombards ruled briefly over a region made up of Rimini, Pesorro, Fano, Singalia, and Anacona—all situated on the Adriatic coast and together came to be known as the Pentapolis. Rimini then fell into the hands of the Franks, under Frankish king Pépin III, who in 756 gave the city, along with other territories, to the church in an act that laid the foundation for the Papal States.

In the twelfth century Rimini became an independent commune, and in the thirteenth century the city and surrounding region became the domain of the famous Malatesta family, which ruled there until the sixteenth century. The papal-allied Guelph chief, Malatesta da Verucchio, was elevated to the status of mayor in 1239, but other members of the Malatesta family clamored for power and were eventually recognized as lords of the town in 1334. Further familial strife occurred when Gianciotto Malatesta (the Lame) murdered his wife, Francesca da Rimini, and her lover, Gianciotto's brother, Paolo. This example of family brutality was recorded in Dante's *Divine Comedy.*

The most notorious of all Malatesta lords, however, was undoubtedly the fifteenth-century Sigismondo Malatesta of Rimini. One of the more extreme instances of brutality at a time when greed and treachery characterized many of the mercenaries, called *condottieri,* who tried to gain power throughout Italy, Sigismondo remains something of an enigma to modern historians. While he inspired the wrath of many, including Pope Pius II, he was also a respected man of letters who held a keen interest in the arts and learning, and carried out duties as a soldier as well. His anti-papal stance was as strong as his love for learning, and it is significant that among the remains enshrined in the church Sigismondo built in Rimini, those of the Byzantine scholar Gemisthus Plethon (which he brought back to Italy after capturing Sparta from the Turks) are believed to be present, while no remains of any Christian saint are enshrined there.

Sigismondo's hatred for the pope, combined with the numerous acts of violence for which he was accused (murdering two wives, molesting his sons and daughters, just for starters), led Pope Pius II in 1461 to an extraordinary measure, perhaps never undertaken by any pope before or since. In a strange act of reverse canonization, Pius publicly consigned Sigismondo Malatesta to hell with the words, "By an edict of the people, he shall be enrolled in the company of Hell as comrade of the devils and the damned." A year later, Sigismondo was excommunicated and his effigy was burned in the vicinity of St. Peter's before a large audience. Sigismondo, himself still among the living throughout these public humiliations, was amused by the proceedings. Eventually, after a number of defeats, Sigismondo was forced to reconcile with the church; Pius accepted his repentance and revoked the excommunication. In return, Sigismondo turned over the majority of his land—although not Rimini—to the pope.

Sigismondo's most lasting historical legacy is not his barbarous temperament, however, but the church he built, now known as the Malatesta Temple, Rimini's most celebrated monument. Taking over an unfinished thirteenth-century Franciscan church, Sigismondo commissioned Leone Battista Alberti to redesign the exterior, with assistance from Agostino di Duccio on the interior reliefs in an attempt to create a personal monument that would serve as well to declare his love for Isotta degli Atti (who also inspired a number of his poems). The temple comprises a mix of strange, often undecipherable, elements, with nothing remaining of

The Malatesta Temple
Photo courtesy of Agenzia Turistica Provincia, Rimini

the original Christian iconography. Thus it is rather odd that the temple has since 1809 been used as Rimini's cathedral. The interior contains Roman arches, classical and pagan decorations (sybils, the planets and the moon, flower-laden dancing children, marble elephants symbolic of the heraldic badge of the Malatesta family), as well as the intertwined initials of S and I, for Sigismondo and Isotta, nestled in numerous places among the rare marbles, carvings, and sculpture. A painting of Sigismondo by Piero della Francesca hangs in the Malatesta Temple, showing, in profile, a most handsome and charming man. Through lack of funds, Sigismondo abandoned construction of his personal temple in 1461.

Sigismondo was succeeded by his illegitimate son Roberto, who cleverly managed to prevail over other, legitimate heirs. He ultimately proved himself to be a successful commander of the papal army and later reconciled himself with the pope. Roberto's own son, Sigismondo, was defeated in the defense of his lands against the oppressing faction of Pope Alexander VI, thus ending the rule of the Malatesta family. In 1509 Rimini found itself in the hands of the Papal States.

During the Napoleonic Wars, Rimini fell briefly under French rule. At this time the only Italian army still in existence was led by Gioacchino Murat, brother-in-law of Napoléon Bonaparte. With the defeat of the Kingdom of Italy, Murat found himself the only surviving Napoleonic ruler in Italy. The Congress of Vienna began in 1814, but Murat's representatives were not invited, and while he plotted with the remnants of a Napoleonic party in Italy, Napoléon himself fled from Elba to France in March 1815. Murat tried to gain support once more in a proclamation issued from Rimini and calling for eventual independence and the unification of Italy. However, the promise of a constitution remained vague, and Murat was not a credible candidate. On May 2, Murat's army was finally defeated and forced to travel south.

Rimini reverted to papal rule and was eventually annexed in 1860 to the Kingdom of Italy. This did not mean that Rimini was no longer to court controversy. In 1872, the town held the first ever Italian Socialist Congress. Thirty-five delegates met to craft a break with Karl Marx's International Working Men's Association and, as a result, created the Italian Federation.

With Rimini's difficult past behind it, the attractive town became a major beach resort. During the nineteenth century, Rimini was already beginning to enjoy its popularity as a tourist favorite, and after 1920 a number of suburban seaside towns began to emerge to the city's south. However, during World War II, Rimini suffered considerable destruction from intensive Allied bombing; more than 90 percent of the houses in the city were destroyed. Once the damage was mended, Rimini was again able to bask in Italian sunshine.

In addition to the Malatesta Temple, important monuments in Rimini include Augustus' triumphal arch, the oldest of the city's monuments, dating from 27 B.C.; the remains of a castle Sigismondo Malatesta built in 1446; the restored Palazzo dell'Arengo of 1204; the picture gallery; the civic library; and several medieval and Renaissance churches.

Further Reading: *The Malatesta of Rimini and the Papal State* by P. J. Jones (Cambridge, London, and New York: Cambridge University Press, 1974) gives a detailed history of Rimini's ruling family from the mid-thirteenth century to the early sixteenth; Jones' research is exhaustive, and his narrative, while focusing on the Malatesta lords, provides insightful glimpses into the city's history during this period. A broader, if less detailed, treatment of the city's history is found in *Cadogan Guides Italy* by Dana Facaros and Michael Paulis (London: Cadogan, and Old Saybrook, Connecticut: Globe Pequot, 1988). H. V. Morton's *A Traveller in Italy* (London: Methuen, and New York: Dodd Mead, 1964) also contains some useful passages on Rimini, as does *A History of Italy 1700–1860* by Stuart Woolf (London and New York: Methuen, 1979).

—Nicolette Loizou

Rome (Roma, Italy): Capitoline Hill

Location: The Capitoline Hill (in Italian *Il Monte Capitolino*) is just west of the center of the old city. To the southeast it looks over the Roman Forum toward the Colosseum; to the northwest, on the side from which it is now most easily approached, are the busy Piazza Venezia and Via del Teatro di Marcello.

Description: The center of the ancient Roman world, and the place from which, notably in the United States of America, numerous national and local legislative buildings, and sometimes the hill on which they stand, have taken their name. The Capitoline is one of the Seven Hills on which ancient Rome was built. It is about 150 feet high and consists of the two summits, the southern one originally the more important, together with the ground between them and the slopes. The Capitoline Hill or Capitol contains buildings and artifacts dating from three main periods—antiquity, the Middle Ages, and the Renaissance. The main architectural features defining the modern appearance of the Capitoline Hill are the Piazza del Campidoglio (Capitoline Square), designed by Michelangelo and surrounded on three sides by Renaissance palaces—one still serving as Rome's town hall, the others now museums housing Roman antiquities and art—and the sixth-century Church of Santa Maria in Aracoeli. Two flights of steps and a roadway, of different periods and designs, lead up to the church and the square. The huge modern white marble monument to Victor Emmanuel II of Savoy, the first king of Italy, encroaches on the northwest slope of the hill.

Site Office: Museo Capitolino
Piazza del Campidoglio
CAP 00184 Rome, Roma
Italy
(6) 678 2862

The Capitoline Hill and the buildings on it, in their various religious and political forms and functions, were originally distinct entities, but they are all now often referred to collectively as the Capitol. Like certain buildings and institutions of ancient Athens, some of those in ancient Rome, such as the Pantheon and the Colosseum, have given their name, or a form of it, to similar features in other cities, ages, and cultures. The word "Capitol" is one of these legacies from Rome, inherited most famously by Capitol Hill in Washington, D.C., the site of the buildings housing the United States Congress.

As early as the Bronze Age (2000–1000 B.C.) there were people living on what was to become the Capitoline Hill. It is the smallest of the Seven Hills on which Rome was founded (the other six were the Palatine, the Aventine, the Caelian, the Esquiline, the Viminal, and the Quirinal Hills), but ancient Rome's religion and politics centered on it. In Rome the actual hill is sometimes alluded to just as the Capitoline. It takes its name from the Latin noun Capitolium, used to denote the temple of the supreme god Jupiter Optimus Maximus (Best and Greatest), built late in the sixth century B.C. on the highest point of the hill's southern summit. Jupiter was considered to be Rome's tutelary deity. The name Capitolium arose, according to tradition, because the *caput* ("head" in Latin) of a certain mythical Tollius was turned up by men digging the foundations. The corresponding adjective was *capitolinus,* the Capitoline Hill, or Hill of the Capitolium; the manifestation of the god who was honored there was called Jupiter Capitolinus.

The temple to Capitoline Jupiter, parts of which have been revealed by archaeological excavations, was begun, according to tradition, by the fifth of Rome's legendary kings, Tarquinius Priscus; it was finished by the seventh and last, Tarquinius Superbus, and dedicated, after his expulsion, in 509 B.C., the year in which the republic was set up.

The Capitolium was the largest of the city's early temples. Here were kept the Sybilline Books, brought from Cumae to be consulted before great decisions of state. The temple was almost square—about 200 feet long and 185 feet wide. It was dedicated to the Etruscan triad of gods: Jupiter, Juno, and Minerva, each of the three having his or her own *cella,* or chamber, in the splendid interior. Jupiter's *cella* was the central one. It contained a primitive terra-cotta statue of the god, later replaced by one of gold and ivory. The senate held its first meeting of the year here. The chamber dedicated to Juno was on the left of Jupiter's; Minerva's was on the right.

Outside there were three rows of six Tuscan columns in the front and four columns along each side. The entablature was made of wood, later gilded, and the pediment was of terra-cotta; above it was a terra-cotta statue of Jupiter driving a chariot drawn by four horses.

The Capitolium was burned down in 83 B.C. during the civil wars, and again in A.D. 69 and A.D. 80, but it was always rebuilt with increased magnificence; with its bronze doors and gilded roof it must have shone like a beacon. By 69 B.C. marble columns had replaced the original wooden ones.

New consuls came to the temple to make their vows and offer sacrifices, and the triumphal entries and processions of victorious generals finished here. Domitian's new temple of A.D. 82, fronted by six Corinthian columns, lasted until the fifth century, but was pillaged during the barbarian invasions.

In 390 B.C., in a famous incident, the *Arx* or citadel that occupied the steep northern summit was saved from a

Fragments of the colossal statue of Constantine the Great
Photo courtesy of Italian Government Tourist Board-E.N.I.T.

night attack by the Gauls when Juno's sacred geese, kept on the hill, honked in alarm and alerted the former consul Manlius to the presence of the enemy, who were flung back down the rocky cliff. In 344 B.C., on the same north summit, where later the Roman mint would stand, the temple of the goddess Juno Moneta was built. It was on this spot that Augustus is supposed to have received a sybilline prediction of the coming of Christ. The Church of Santa Maria in Aracoeli now occupies the site of Juno's temple.

Between the summits and the temples, the 78 B.C. Tabularium, or state record office, was built into the hillside on the Forum side. Its grey tufa podium, recently restored and impressively visible from the Forum, serves as base for the Palazzo Senatorio, and remains of its facade are built into the Palazzo del Museo Capitolino.

To the south side of the Capitoline Hill, at its highest point and above a cliff, is what is generally thought to be the Tarpeian rock, the infamous place of execution from which traitors were thrown to their death; a site on the north side of the hill has also been suggested as holder of this doubtful honor.

In addition to the damage done to buildings and precious objects by invaders, the hilltop and steep slopes were gradually altered by removal of construction materials for reuse, by weather erosion, and by neglect. All the ancient buildings were in ruins by the Middle Ages. In time, like other once-grandiose sites, the Capitol became home to a market. By 1534, prior to a visit to Rome by Holy Roman Emperor Charles V, Pope Paul III commissioned Michelangelo to restore the historic hill. Michelangelo accordingly designed the Cordonata stairway, the Campidoglio Square to which it leads, and the facades of the surrounding buildings, two of which already existed. Though he did not live to complete the plans himself—the work was not finished until the seventeenth century—the Piazza del Campidoglio was built essentially as designed by Michelangelo.

In ancient Rome the top of the Capitol was accessible only by a zigzag path climbing from the Forum. When the modern city developed during the Renaissance, the focus of citizens' activity had changed, and so the new complex on the hill was oriented to face northwest, away from the Forum and toward St. Peter's.

There are now three ways up the Capitoline Hill from the Via del Teatro di Marcello. On the left, 124 steep marble steps rise to the facade of the Church of Santa Maria in Aracoeli. The Aracoeli staircase was constructed in 1348 and funded by money given with prayers or thanks to the Virgin for deliverance from a plague. On the right there is a winding nineteenth-century roadway, the Via delle Tre Pile. In the middle stands Michelangelo's imposing 1536 Cordonata stairway, enclosed by stone balustrades continued to the right and left along the open townward side of the square. Various archaeological fragments and pieces of statuary mark all three access routes, and the spaces between are landscaped with gardens and shrubberies.

At the bottom of the Cordonata are two granite Egyptian lions; halfway up on the left, at the spot from which he made a famous speech to the people of Rome, stands a statue of Cola di Rienzo, the patriot, political reformer, and rebel against the papacy. At the top, on either side of the stairway, raised on pedestals and each wearing his legendary eggshell cap and accompanied by his white horse, are giant marble late-Roman statues of the mythological twin heroes Castor and Pollux, the Dioscuri or sons of Zeus.

The Piazza del Campidoglio is one of the organized complexes of palaces, open squares, and flights of steps that characterized urban planning in Rome during the baroque period (seventeenth and first half of the eighteenth centuries). The square on the Capitoline Hill is enclosed on the left by the Palazzo del Museo Capitolino or Palazzo Nuovo (New Palace), added to the plan by Michelangelo in 1536 to balance the two existing palazzi, at the back by the Palazzo Senatorio, and on the right by the Palazzo dei Conservatori (Palace of the Custodians). The creamy colors, matching neoclassical lines, and symmetrical proportions of the three facades combine into a majestic unity.

In the center of Michelangelo's trapezoid pavement, designed in concentric geometric stars, stands a copy of the Capitoline Museum's famous bronze statue of the philosopher-emperor Marcus Aurelius. The original, kept indoors in one of the galleries, is the only equestrian statue surviving from imperial Rome; originally it was gilded. It was made during the subject's lifetime and later piously saved from destruction because it was thought to represent Constantine, the first Christian emperor of Rome. During the Renaissance it served sculptors as the pattern for equestrian statues. It was brought from the Lateran Palace in 1538 and given its high plinth by Michelangelo.

At the back of the square, in the middle, rises the graceful pilastered facade of the Palazzo Senatorio, which was built above the ruins of the Tabularium as, before it, was a medieval fortress where from the twelfth century, it is thought, Rome's senate met. Michelangelo redesigned this latter building, and the present Renaissance palazzo implements his plans, as executed and modified by Giacomo della Porta and Girolamo Rainaldi in 1592. In front of the double stairway whose flights converge on the raised entrance, there is a fountain with two second-century statues, representing the Rivers Tiber and Nile, on either side of a porphyry goddess of Rome. The 1582 bell tower above the Palazzo Senatorio, the equestrian statue in the middle of the Campidoglio pavement's central star, and the midpoint of the Cordonata's width are aligned with each other. Rome's present-day city council meets in the Palazzo Senatorio; in the council hall there is a second-century statue of Julius Caesar. The remains of the Tabularium beneath the Senatorial Palace are visible *in situ* and in a special gallery built in 1938 below the square and the palace; this gallery contains extensive material unearthed by modern archaeological excavations.

The Capitol's main galleries of Roman antiquities and artifacts, known collectively as the Capitoline museums, are

The Piazza del Campidoglio
Photo courtesy of Italian Government Tourist Board-E.N.I.T.

housed in the buildings to the left and right of the Piazza del Campidoglio. The Capitoline collection was founded by Pope Sixtus IV in 1471, when he gave the city his important group of bronze statues recovered from the city's ancient ruins. It was the world's first collection of classical sculpture, though the Vatican's is now the largest. The status of sculpture was then very high and long remained superior to that of painting. Sixtus IV placed the pieces he had given in the Palazzo dei Conservatori and decided that henceforward Rome's ancient art treasures should be kept there on the Capitol. Subsequent popes, especially Clement XII and his successor Benedict XIV, both in the eighteenth century, added to this collection; it was Clement XII who opened it to the public in 1734. After 1870, when Rome became the capital of a united Italy, the city itself became a donor, augmenting the number of exhibits with objects brought from archaeological sites on municipal land. Many of the ancient Roman exhibits in the Capitoline museums are copies of Greek originals.

On the left of the piazza is the Palazzo del Museo Capitolino or Palazzo Nuovo. Incorporating what is left of the Tabularium's facade, it was started in the middle of the sixteenth century and completed in about 1654, when it began to be used as a museum. It houses the Museo Capitolino or Capitoline Museum, which was extended into the other buildings in 1876 when more space was needed to display the results of new discoveries.

At ground level there is a pillared portico and an inner courtyard with a fountain by Giacomo della Porta and a reclining colossal statue of a river god. This personage (Marforio) was one of the city's group of what were called the talking statues, on which from the early sixteenth century it became the custom for wits and cynics to affix their written comments and countercomments (called *pasquinades,* from the name of the supposed originator of the tradition) on current Roman scandals. The ground-floor galleries house artifacts relating to oriental cults practiced in imperial Rome, as well as carved sarcophagi of diverse ancient provenance.

The first-floor galleries contain many famous statues, including the bronze *Dying Gaul* (at one time thought to represent a gladiator), which is a Roman copy of a third-century B.C. Greek original; the red marble *Laughing Silenus,* an imperial Roman copy of a Hellenistic bronze; the *Young Boy with a Duck* (or Goose), a replica of a second-century Hellenistic bronze; two grey marble centaurs in Hellenistic style from the villa of Hadrian at Tivoli; and a replica of the Greek fifth-century B.C. *Wounded Amazon.* There are also numerous portrait busts of Greek and Roman cultural celebrities (philosophers, poets, emperors, statesmen). Another room is called the Hall of the Doves after two Greek-inspired exhibits: a mosaic from the floor of Hadrian's Villa depicting a vase with four of the birds drinking from it, and a statue of a young girl holding a dove. In the so-called Cabinet of Venus is the Capitoline Venus, in Parian marble, probably copied from a second-century B.C. Greek original.

To the right of the piazza is the Palazzo dei Conservatori, where Rome's magistrates met in the Middle Ages and where Sixtus IV housed his original collection. It was rebuilt in the fifteenth century, redesigned in the sixteenth by Michelangelo, and rebuilt between 1546 and 1568 by Giacomo della Porta. The exhibits are mainly sculptures, shown in the Sale (Rooms) dei Conservatori and, also on the first floor, in the Museo del Palazzo dei Conservatori. On the second floor there is a picture gallery, the Pinacoteca Capitolina. There is an annex at the southern end of the Palazzo dei Conservatori, the Palazzo Caffarelli, housing the Museo Nuovo. It was built in 1580 and at one time was used as the German Embassy.

In the inner courtyard among other statue fragments are conserved the huge head, hand, and foot of a colossal statue of Constantine the Great. On one of the staircase landings on the way up to the Sale there are relief panels from triumphal arches honoring Marcus Aurelius. Among the classical and post-Renaissance treasures housed above is the magnificent marble-floored Hall of the Horatii and the Curiatii, with its recently restored frescoes by the Cavaliere d'Arpino (Giuseppe Cesari) depicting the legendary story of patriotic combat between two families of brothers at the time of the Roman kings. As in the Museo Capitolino, world-famous antiquities abound in the Sale. Among the bronzes presented by the founder, Sixtus IV, there is the *Spinario,* or *Boy with the Thorn in his Foot.* In a room named after it stands the late sixth- or early fifth-century B.C. Etruscan bronze *Wolf of the Capitol,* or *She-Wolf of Rome,* with its little Romulus and Remus added underneath by Antonio Pollaiulo late in the fifteenth century.

Next, at the same level, is the Museo del Palazzo dei Conservatori, largely devoted to public dignitaries of ancient and post-Renaissance Rome. One hall, though, contains relics of archaic cultures. In another there are statues of fourth-century officials; in others, exhibits relating to Rome's Christian martyrs. The world's oldest signed Greek vase, the Aristonothos Krater, dating from the seventh century B.C., is here. The seven rooms of the new wing, the Braccio Nuovo, contain ancient Roman statues, monuments and sarcophagi.

Upstairs the Pinacoteca Capitolina, founded by Pope Benedict XIV in 1749, contains paintings by Italian and other artists of the fourteenth to the seventeenth centuries, including Titian, Veronese, and Caravaggio.

The Palazzo Caffarelli, which is used when open to exhibit ancient Greek and Roman sculptures, is built over the remains of the Capitolium, the sixth-century B.C. Temple of Jupiter Capitolinus from which the hill received its name. Parts of the podium and walls can still be seen.

To the left of the museums and the square, on the northwest summit and highest point of the Capitol, the Church of Santa Maria in Aracoeli maintains a 2,500-year tradition of religious worship on the site. The church is named, according to an old belief, after the altar erected by Augustus on this spot to mark the Tiburtine prophecy of Christ's coming; this altar is said to lie beneath the pavement in the seventeenth-century chapel of St. Helen in the middle of the left (north) transept.

There was a Christian church here in the fifth century. The Benedictines took it over in the tenth century, and in 1250 Franciscans rebuilt it in brick in a mixture of Romanesque and Gothic. It is oriented roughly east-west and has an apse and transepts.

The rich but disciplined interior is divided into three naves by twenty-two columns taken from ancient sites. The gilded ceiling created in 1575, decorated with naval emblems, commemorates the 1571 battle of Lepanto, when the son of the Holy Roman Emperor Charles V, Don John of Austria, commanding the fleets of Spain, Venice, and the pope, won a victory over the Turks. The inlaid floor contains numerous interesting tombs and memorials, including a monument to Cardinal d'Albret by Andrea Bregno (1465), Donatello's tomb of Giovanni Crivelli, and the Bufalini Chapel, where there are frescoes (c. 1486) by Pinturicchio of the martyrdom of St. Bernardino and of St. Francis receiving the stigmata. On pilasters in the central nave, there is a pair of *ambones* (pulpits for the reading of gospel and epistle) from approximately the twelfth century, ornamented with mosaics by members of the Cosmati family. The main chapel has a baroque altar with a twelfth-century Byzantine Madonna, and the third chapel in the left-hand nave has a mid–fifteenth-century St. Anthony of Padua by Benozzo Gozzoli. In the sacristy, but brought out every year at Christmas for the Nativity scene, is preserved an olive-wood figure of the infant Jesus, much revered locally.

Further Reading: *A Traveller in Rome* by H. V. Morton (London: Methuen, and New York: Dodd Mead, 1957) supplies a colorful context of historical and literary facts and speculations, drawn from the author's intimate knowledge of Rome. *Rome: The Biography of a City* by Christopher Hibbert (Harmondsworth, Middlesex: Penguin, and New York: Norton, 1985) also sets the history of the Capitoline Hill in context; the manner is more academic than Morton's, but again the effect is authoritative and colorful. *Eyewitness Travel Guide: Rome*, edited by Ros Belford and Rodney Palmer (London and New York: Dorling Kindersley, 1993), though in guidebook format, contains much unhackneyed fact and is unusually well illustrated, in color.

—Olive Classe

Rome (Roma, Italy): Colosseum

Location: In the heart of what is left of the ancient imperial city, to the northeast of the Palatine Hill, the Colosseum stands at the eastern end of the modern Via dei Fori Imperiali, where several busy avenues converge.

Description: The massive stone and concrete remains of a huge freestanding oval amphitheatre built in the first century A.D. It is one-third of a mile in circumference and covers an area of 6 acres or 2.4 hectares. It is 615 feet long, 415 feet wide, and 164 feet high, measured externally overall. It is least damaged on the northeast side. The outer facade presents three superposed tiers of eighty arches each, plus an attic story. Within there were originally rows of seats rising nearly to the top of the outer wall and descending to a central arena. Beneath ground level are the extensive remains of the chambers, cages, etc., used in the preparation and organization of the spectacles mounted in the arena.

Site Office: Colosseum
Piazza del Colosseo
00184 Rome, Roma
Italy
(6) 735227

The Colosseum is the most famous building surviving from imperial Rome (27 B.C. to A.D. 476). The Anglo-Saxon historian and theologian, St. Bede the Venerable, quotes the traditional saying:

> While stands the Colosseum, Rome shall stand;
> When falls the Colosseum, Rome shall fall;
> And when Rome falls—the World.
> (translated by Lord Byron)

It is difficult, though, to admire its durability and its magnificent form and proportions untroubled by thoughts of the purpose for which it was built: it was destined and used for the organized, brutal, and spectacular killing in public of humans and animals, in order to give pleasure to callous crowds of spectators.

It was originally called the Flavian Amphitheatre when construction began between A.D. 70 and 72, and was part of an extensive program of building by Emperor Vespasian, founder of the Flavian dynasty. He chose to build it in a marsh, which had earlier been turned into an ornamental lake adorning the grounds of the fabulous Golden House belonging to Emperor Nero, of whose extravagance the sober Vespasian disapproved. The engineering problems presented by the waterlogged site found a lasting solution in the skill of the Colosseum's architects and the enduring strength of Roman concrete. Vespasian's amphitheatre was finished by his eldest son, Emperor Titus, who consecrated it in A.D. 80.

The edifice was built by prisoners taken in the Jewish war conducted by Vespasian and Titus. Titus inaugurated the new stadium with ceremonies that included a 100-day series of shows and games, among them fights between men and wild beasts; 5,000 animals were killed on the first day. The Latin satiric poet Martial won equestrian rank with the small book of epigrams he published in A.D. 80 to celebrate the consecration.

Further work on the building was conducted after Titus's death in A.D. 81 by his brother, Emperor Domitian, the last of the Flavians. In about A.D. 230 Emperor Severus Alexander restored the amphitheatre and added the fourth level. Some earthquake damage occurred in the fifth, thirteenth, and fourteenth centuries.

The main frame of the structure and its facade are of travertine, a light-colored calcareous rock quarried in Tivoli. The secondary walls are of tufa and brick. The interior was originally faced with different kinds of colored marble, but this is gone, pillaged over the centuries. The inner bowl and the arcade vaults are Roman concrete.

The inner of the two main circumference walls is less well preserved than the outer facade. Its crumbling arches leading into the auditorium seem to be impregnated more deeply with the sinister function of the place, but they too impress by their grandeur and harmony.

Outside, the columns that embellish each of the three rows of arches are of the three Greek orders, their degree of ornateness increasing with their height above the ground: Doric order columns at street level, Ionic for the tier above, Corinthian for the top row of arches. The design and ornamentation of the facade exerted a great influence on Renaissance architecture. In the original building the second- and third-floor arches were occupied by statues, while the lower ones were rented out to vendors. The last, topmost story, added by Severus Alexander in the third century, is without arches but has Corinthian pilasters. At each level, behind the facade, corridors under a mixture of barrel and groin vaulting lead around inside the building and beneath the seating structure, giving access through the inner circumference wall to the auditorium.

When the Romans built their amphitheatres, of which the Colosseum is the largest and the best known, they aimed at maximum spectator capacity and optimum viewing facility. The focus was on the sanded central arena (the word "arena" means "sand" in Latin; sand gave a foothold and absorbed the blood spilled in the hunts and combats). Amphitheatres were often built into a hillside to utilize this natural support, but the architects and engineers of the Colosseum, with Roman concrete technology available, were able to make this huge structure self-supporting. Iron clamps, the marks of whose

A section of the Colosseum
Photo courtesy of The Chicago Public Library

position can still be seen, were used to reinforce the joints; the stability of the structure was reduced when the clamps were stolen in medieval times. Besides the eighty entrance arches, the building had four main entrances, the widest, in the northeast, reserved for the emperor.

The Flavian Amphitheatre later came to be known as the Colosseum; the earliest recorded mention of the name is in Bede. The new name probably alluded to a nearby gigantic (in Latin, *colosseus*), 100-foot statue of Nero rather than, as one might suppose, to the stadium's own size. Mussolini ordered the demolition of the gilded bronze statue of Nero in 1936. Its brick base, more than twenty feet square, sits between the Colosseum and the Via dei Fori Imperiali.) "Coliseum," another form of the word, is sometimes used to refer to other large stadiums and theatres.

The ellipse-shaped auditorium or *cavea* provided numbered seats for 50,000 paying spectators, with free standing room at the top for thousands more. Some authorities say the capacity was more than 80,000 in all. The arena was separated from the auditorium by a protective wall fifteen feet high. On top of this wall on the northern side, with a ringside view but out of the direct sun, was the podium, a wide terrace where the emperor reclined and where special seats were reserved for senators, ambassadors, and other dignitaries, including the Vestal Virgins, whose duties included attendance at the games.

The main seating area was divided into three tiers, one above the other, the level of the places in inverse proportion to the social level of their occupants. The common people sat in the topmost seats. An official made sure spectators stayed in their designated sections. The blocks of seats were wedge-shaped, with open stairways leading up to the exits on the landing of each floor. Special dress regulations prevailed, the ordinary citizens of Rome wearing the white toga and functionaries their more colorful official robes.

The seats, now missing like the wooden floor of the arena, were shaded from the sun by awnings. The awnings were fixed to masts resting on corbels in the attic story, where, from a narrow platform, specially trained sailors controlled these contraptions as best they could by means of a system of ropes and pulleys; the unwieldy canopy flapped and roared if there was any wind. Beneath the auditorium and the arena were underground chambers and passages, and stone dens for the animals brought for the games from their menagerie on the nearby Caelian Hill. Their cages were dragged under the building and raised by pulleys to the sanded arena.

The games usually went on for several days. The so-called sports that brought such great crowds to the Colosseum had already long been popular in Rome. In the mornings there was the slaughtering of wild animals. Lions, bears, hippopotamuses, crocodiles, and elephants were trapped for that purpose in the countries of the empire. Combats between gladiators were first seen in Rome in the middle of the third century B.C.; they were outlawed in A.D. 407, but the fights with wild beasts went on until A.D. 523. Until A.D. 326 criminals could be sentenced to fight with animals in the

Colosseum. The professionals, called *bestiarii* or *venatores*, who fought against animals here were trained in showy techniques at special schools in the city.

This horrible show was followed by the afternoon's more refined pleasures—fights, often to the death, between gladiators from imperial Rome's four official training schools and some private ones, using swords, lances, and tridents to wound and kill, and nets to entangle and bring down. A successful gladiator could become famous, popular, and rich.

The gladiators paraded around the arena before the fight, shouting when they came before the emperor's box, "Hail, Caesar! those about to die salute you!" Adversaries were paired by selection or lot. Fighting began at a signal from the emperor. The men's efforts were urged or forced on by the trainers and accompanied by the playing of trumpets and horns, and the raucous cries of the crowd. Sometimes the loser appealed to the emperor to spare his life. If the crowd gave the thumbs up sign, Caesar would make the gesture that would save him, but if the people shouted "Kill him" and gave the thumbs down, that was the end.

Between the morning and the afternoon shows, numbers of condemned criminals were forced to take weapons and kill each other. Popular legend holds that Christians too were sent into the arena, to be torn apart by wild beasts, but no evidence exists of Christian martyrdoms taking place in the Colosseum.

At first, few women, only the small group of Vestal Virgins and the empress, were allotted official places in the Colosseum. Later, women were allowed to occupy seats at the top of the tiers with the lower-class men and even occasionally took part in fights against each other.

Private promoters, such as businessmen or politicians, could buy popularity by hiring the Colosseum and putting on games at their own expense. Mock military battles were also staged there, and the arena could be flooded for naval displays using water brought from a special reservoir on the Caelian Hill. Machinery for producing these special effects, in addition to that for raising the wild animal cages to arena level and clearing away the corpses of dead combatants, was housed at and below floor-level.

The popularity of the most savage of these entertainments waned as Christianity spread in Rome and the empire, especially after the conversion of Constantine the Great in about A.D. 312. The last recorded games in the Colosseum were in 523.

By medieval times the place was sometimes used as a bullring and for plays or parades. The structure was turned into a fortress in the thirteenth century by the Frangipani family. In 1312 the Holy Roman emperor gave the building to the city of Rome. By the fifteenth century it was serving as a quarry, furnishing travertine and marble for Renaissance churches and palaces, including St. Peter's, the Farnese Palace, the Palazzo Venezia, and the Palazzo della Cancellaria.

Its fabric plundered by vandals for building materials and encroached on by nature to picturesque effect, the Colosseum became a compulsory item on the things-to-see lists of artists, poets, and tourists visiting Rome. In 1749 it was declared sacred by the Roman Catholic Church because of the Christians who, it was believed, had met a martyr's death in the arena.

From the nineteenth century the Colosseum increasingly attracted the interest of archaeologists, becoming the object of excavation and preservation work and rising, with the other long-neglected antiquities of Rome, to the status of national and world heritage. The 1,000-year process of heedless destruction of the Colosseum was stemmed when several of the nineteenth-century popes, notably Pius VIII, began to conserve and restore it instead, though the huge main gap in the walls remains.

In 1864 the foundations were excavated in a search for the Frangipani treasure believed buried there; nothing of the sort was found, but the underground construction of the building was uncovered and examined. In the 1890s and 1930s the ancient site was cleared of later accretions, and modern restoration work has been going on since 1973.

The Colosseum is open to the public, restoration work permitting. There is a collection below ground of archaeological exhibits (the Antiquarium), with architectural and sculptural fragments recovered from the site, together with equipment used for the spectacles. A scale model of the original complete building can be seen in the upper part of the building.

The triple Arch of Constantine (A.D. 315) stands nearby, put up to mark the newly converted Constantine the Great's third and final victory over Maxentius in the battle of the Milvian Bridge near Rome three years earlier. This achievement concluded Constantine's successful struggle for the title of Emperor of the West. From 325 until his death in 337 he reigned over the whole empire. Under his rule Christianity became the state religion. Though Constantine and Roman culture in general were still far from enlightened by modern standards, the heyday of the cruel games was passing.

Further Reading: *A Traveller in Rome* by H. V. Morton (London: Methuen, and New York: Dodd, Mead, 1957) has the personal enthusiasm and interesting local and historical detail one associates with this author. *Blue Guide: Rome and Environs* by Alta Macadam, fourth edition (London: A. and C. Black, and New York: Norton, 1990) provides an excellent description of appearance, construction, and layout of the building at all periods.

—Olive Classe

Rome (Roma, Italy): Forum

Location: The Forum in Rome lies in a valley roughly 1,700 feet from the left bank of the Tiber River. Originally two small public squares, the Forum spread to its outermost boundaries at the Capitoline and Palatine Hills over the course of 900 years of expansion and development.

Description: Since the beginning of civilized Rome, the Forum was the city's public square, offering a meeting place used for merchandising, legislation, litigation, worship—all manner of social activity. Resting on an oblong parcel of land measuring 300 feet by 200 feet, the Forum is home to some of the Vestal Virgins, the Basilicas Aemilia and Julia, the Curia (Senate House), and the Arches of Titus and Septimius Severus. At the height of Roman influence, eight roads led from European provinces directly to the center of the Forum, where distances from the capital to outlying replicas of the Forum were marked by columns of gold and marble. For centuries after the fall of the Roman Empire, the site was neglected and abused. Excavations were begun in the late eighteenth century at what is now one of the best-known archaeological sites in the world, and to this day the Forum dig continues to yield the impressive remains of ancient Rome.

Contact: E.P.T. (Ente Provinciale per il Turismo)
Via Parigi, 11
CAP 00185 Rome, Roma
Italy
(6) 4881851

The ruins standing today in Rome's ancient Forum are little more than rubble. The monuments, temples, and meeting-houses of the Forum were plundered throughout the Middle Ages for their wealth of precious ornaments and building materials, and for centuries the area was an urban cow pasture buried under layers of refuse. Yet the visitor to the ruins still senses the site's greatness; many of the most ambitious efforts of ancient Roman society took place here, and dramatic episodes of history seem to come to life in the rarefied atmosphere of the Forum.

The oldest remnants of organized urban activity in the area of the Forum date to 753 B.C., the year given as the birth of Rome by the writer Varro in 47 B.C. By this time the area, an ideal setting for commerce due to its proximity to the Tiber River, was already being used as a marketplace and meeting place. Both inhumations and cremations have been unearthed ten to fifteen feet below the original ground level of the Forum; burials apparently continued there until just before the establishment of the first buildings in the square. The establishment of the Comitium, the circular outdoor meeting space used for important pronouncements and trivial gossip alike, formalized the use of the area for the first time. About a century later, with Rome quickly urbanizing, the first structures were erected in the Forum, establishing its two small squares as the city's most important common area.

As early as 900 B.C., traveling Greek traders were arriving, enticed by the richness of the Italian land and its commercial opportunities. By the eighth century, the adornments of burial vaults under and near the Forum were beginning to betray a disparity in the wealth of the entombed, indicating an emergent class system in Rome, but the Romans were also evolving as a religious and social people given to a certain agreeability known as "clementia."

According to legend (and the writings of the Augustan historian Livy), the abandoned twins Romulus and Remus, rumored to be the sons of Mars, god of war, were suckled by a she-wolf until they were discovered on the banks of the Tiber. Later ascending to leadership in the city, the brothers engaged in a power struggle culminating in the murder of Remus at the hands of his brother on Palatine Hill. Romulus then initiated the so-called Rape of the Sabines, a nearby Italic tribe, to remedy a scarcity of women in his city. The intervention of the Sabine women halted an ensuing war between the Romans and Sabines, and the two tribes united to bolster the population of Rome. On the outskirts of the Forum, the remains of the Vulcanal, or Altar of Vulcan, purportedly date to the time of Romulus; in its day it was most likely a holy sanctuary adorned with lotus trees and statues.

The Etruscan king Tarquinius Priscus ruled the city during the years 616 to 579 B.C., but Rome was fast becoming home to a diverse culture, infused with the lineage of the Sabines and with a Latin heritage that aligned the city with other settlements of the Latium region of ancient Italy. Inscriptions in both Latin and Etruscan have been discovered from this period, indicating a peaceful coexistence in the growing city.

Tarquinius's visionary public works projects included the inception of the Cloaca Maxima, the drainage system that served the marshy area in and around the Forum, which had been built on a landfill; the Temple of Jupiter on Capitoline Hill; and the sprawling Circus Maximus facility. In 1898, the Italian archaeologist Giacomo Boni unearthed the Forum's original layer of paved surface, dating to the time of the Tarquins, beneath nine other layers. Boni also discovered the famous Lapis Niger, or Black Stone, the ancient monument inscribed with the warning that begins, "Whosoever defiles this spot, let him be forfeit to the spirits of the underworld"— early proof of the Forum's unparalleled sanctity in the minds of the Romans. Resembling the Greek alphabet, the inscription is the oldest known example of Latin writing. Some

Arch of Septimius Severus
Photo courtesy of The Chicago Public Library

scholars claim that the Lapis Niger covers the Tomb of Romulus.

Rome's rapidly developing system of government, an aristocratic republic, was significantly advanced by the plebians' revolt begun in the early fifth century B.C. The Roman Republic, founded in 509 B.C., had two tribunes: one of the patriarchs and another of the commoners. Bronze plaques inscribed with the Twelve Tables, an early benchmark of Roman law, were put on display in the Forum, and the motto of the republic, "Senatus Populusque Romanus," heralded the "Senate and People of Rome." In 367 B.C., the Temple of Concord was erected to signify the accord between the plebians and their senators. With its crumbling fragments cut off from the Forum by a modern roadway today, the temple gamely continues to commemorate the Licinio-Sextian laws, named after the filibustering tribunes who secured more extensive representation for the lower classes.

Around this time, many of the republic's posts of authority were being established: they would be held by public servants who came to dominate the daily discourse of the Forum with their proclamations, opinions, and hearsay. The Romans' innovative system, based on the leadership of

two magistrates (consuls) elected yearly, would owe little to the Greek model of public representation. The positions of the censors, administrators of the public's accountability to taxation and military service, were created in 440 B.C., followed by the quaestors (financial administrators), praetors (court officials) and aediles (administrators of public works and affairs). The pontifex maximus, overseer of the sacred law, was charged with the caretaking of the Vestal Virgins, and the augurs were charged with the decoding of portentous omens such as weather changes and the flight patterns of birds.

In addition to its function as a primary meeting place for such dignitaries, the Roman Forum also served as a gathering spot for the city's less dignified residents. Prostitutes, moneylenders, fortune tellers, and gamblers could be found commingling with state authorities, and in 338 B.C. the consul Gaius Maenius saw fit to ban the produce sellers and butchers of the Old Shops in favor of "less aromatic" booksmiths and silversmiths. Florists, perfumers, and jewelers regularly plied their trades along the route known as the Sacred Way (Via Sacra).

In 484 B.C. the Temple of Castor and Pollux was dedicated at the intersection of the Sacred Way and Tuscan

Street. Built to honor the twin brothers of Helen of Troy, the temple is survived by three Corinthian columns measuring to a height of forty-one feet atop a twenty-three-foot platform. According to legend, the Heavenly Twins, who had brought news of the Roman victory at Lake Regillus over the forces of the deposed Tarquin dynasty on the wings of their white horses, were first spotted watering their mounts at the Fountain of Juturna, the Forum watering hole that doubled as the focal point for the dispensation of the news of the day.

The earliest erected building to house the Senate, the Curia Hostilia, had been in existence since the reign of the Etruscan king Tullius Hostilius. This earliest model of the Curia featured the famed Goddess of Victory statue near its podium; the statue, much disputed among the Christian and pagan factions of later years, was removed by law in A.D. 357, returned after protests thirty-five years later, and then vanished permanently two years after that. In 328 B.C., the Romans engaged in a furious sea battle with the soldiers of Antium; the Roman victory was commemorated by the assembly of the prows of captured ships in the Forum, adorning the speakers' podium outside the Curia, later to be called the Rostra (which translates as "prows").

The last incident of intrusion on Roman development before the city's great advancement occurred when the Gaulic Celts sacked the city in 390 B.C. The city's quick recovery and subsequent building of a defensive wall added to its growing prestige, and Roman leaders shortly began to harbor imperial designs. Fourth-century skirmishes with the Samnites, the Campanians, and the Volsci led to the enlargement of the Roman sphere of influence, and a series of indecisive battles with the Greek leader Pyrrhus in the year 280 further cemented the dominant status of the Italian city in the world's theatre of war. Pyrrhus's ambassador to Rome returned to his master with this description of the emergent city's glorious Senate: "It is an assembly of kings."

By the time of Julius Caesar, Rome was universally recognized as the dominant power in the Western world. The ambitious Caesar returned to Rome from his ambassadorship in Asia Minor, holding the posts of aedile, pontifex maximus, and finally praetor. Adding to his fame and fortune, Caesar wed the daughter of a future consul, and he waged successful campaigns abroad, against the Gauls and the Brits. Returning triumphantly, Caesar marched on Rome to ensure his ascendance to leadership; less well received than he had expected, he seized the state's treasury at the Forum's Temple of Saturn and declared himself dictator.

Near the foot of Capitoline Hill, the Romans had honored Saturn, their god of agriculture. The Romans symbolized their association of that discipline with wealth by housing their state treasury in the Temple of Saturn. Here was centered the six-day celebration of Saturnalia, the pagan Christmas, during which social status was temporarily set aside and townsfolk exchanged candles and clay dolls. In the room ensconced beneath the stairway leading to the Temple of Saturn's fourth-century portico, Caesar plundered 15,000 bars of gold and 30,000 of silver.

With his well-funded regime, Caesar sought to win the support of the Roman people by bestowing upon them public works of unprecedented magnificence. Reconstructing the Forum and reorienting it toward the northwest, Caesar crowned his lavish efforts with a banquet for 22,000 people held in the square in 45 B.C. He restored the Curia (for the second time since 80 B.C.), built his own Forum behind the Curia, disassembled and rebuilt the Rostra, and built the Basilica Julia.

Begun in 55 B.C., the Basilica Julia was incomplete upon Caesar's assassination eleven years later; his successor, Augustus, saw the project through to its completion. Flanking the basilica were two main thoroughfares leading from the riverfront into the Forum: the Vicus Jugarius to the northwest and the Vicus Tusculus to the southeast. Faced with marble, the imposing basilica was the central courthouse, featuring thirty-six columns surrounding four chambers that competed with each other for acoustic space, as orators' voices bellowed throughout the echoing halls of the building.

To the north of the basilica lay the Temple of Venus Genetrix (the Mother), the goddess of horticulture and love from whom Caesar claimed to be descended. Outside the temple, Caesar erected an equestrian statue of himself, the horse having been pilfered from an earlier statue of Alexander the Great. Inside the temple, Caesar depicted Venus wearing pearls, and he had a likeness of his lover Cleopatra fashioned from gilt bronze. Reconstructed during the fourth century A.D., part of the Temple of Venus can be seen in the Forum today.

Caesar's murder at the hands of conspirators at a Senate meeting held at the Curia Pompeia (the Curia Hostilia was then undergoing repairs) was followed by the legendary speech given by Mark Antony on the Rostra. To honor his fellow consul, Mark Antony erected a monument to Caesar, the "Glorious Father of the Country," in the Forum. Despite the abolition of Caesar's dictatorship, the deposed emperor was declared a god, and his grand-nephew and adopted son Octavian (later Augustus) rose to the crest of Roman leadership.

The reign of Augustus produced a lasting peace (his rule initiated the Pax Romana, which lasted 200 years) and a time of remarkable Roman prosperity. An ardent supporter of urban grandeur, he left an important mark on the city, changing it from a "city of brick into one of marble." In 29 B.C., he erected a new Temple of Caesar on the site of his adoptive father's cremation, and he again rebuilt the Rostra, this time to face the Capitol. By the end of Augustus's tenure, Rome had seen the restoration of every building in the Forum, with the exception of the Tabularium, the record-keeping office on the square's west end dating to 78 B.C., and the Carcer, the prison situated next to the Curia. Seeking to further enhance the city's bold impression, Augustus built his own Forum, as had Caesar; in years to come the Roman emperors Trajan and Vespasian would similarly grace the city with their own commons.

The Forum's original basilica, the Aemilia, built in 179 B.C. by the consul Marcus Aemilius Lepidus, was restored during the Augustan age after a fire, prompting Pliny the Elder to call it one of the most beautiful buildings in the world. This basilica was fronted by a two-floor portico overlooking the Forum; inside, the hall was used for the dual purpose of commerce and justice. The basilica was built on the squatting site of the Forum's ever-busy butchers and moneychangers, who simply moved their trade to a position along the outer wall of the new building. The basilica's portico was an early and impressive example of the use of the Romans' formula for concrete; its design, a high-roofed hall lined with colonnaded aisles, has been the classic model for Christian churches for centuries. The marble pavement of the basilica still reveals the glimmer of gold coins burned into its surface as startled merchants fled the invasion of the Goths in the year 410.

Around the time of Augustus, "pit sand," a brownish-red volcanic residue, began to be used in making concrete. Roman public works enjoyed the availability of the building material, which proved most effective until the advent of Portland cement in the nineteenth century. However, the city was not invincible; almost a century after Augustus's work in Rome, a devastating fire swept through the city for six days beginning on July 18 in the year 64. Only four of the city's fourteen districts were spared. When the blaze was contained, the Emperor Nero began the daunting task of repairing the city. Cut short by his suicide, the repairs were enjoined by Vespasian, who embossed his currency with the phrase "Roma Resurgens," indicating his commitment to a complete revivification of the city. Vespasian's own Forum was dedicated in the year 97, eighteen years after his reign ended.

After yet another fire, in the year 191, Emperor Septimius Severus resumed improvements to the original Forum with his restoration of the Temple of Vesta, the circular house of worship credited to the ancient regime of the second Roman King Numa; the structure was originally designed from thatchery to resemble the earliest huts of ancient Romans. Its stone remains, and those of the House of the Vestal Virgins, can be seen today in some of the best-preserved states of the buildings of the Forum. The House of the Vestal Virgins, built around a central courtyard ornamented with ponds and gardens, was the resplendent home of the sacred virgins, of whom there were at first four, and later six or seven. Men discovered in the house were punished by death; virgins who were despoiled were punished by live burial. Chosen from between six and ten years of age, the virgins served as many as thirty years in their religious roles, experiencing a curious mixture of comfort and hardship. Their pagan order existed until A.D. 394, well after the advent of Christianity in Rome. Their elevated social status, complete with financial benefits, prestigious carriage rides, and the entrustment of important documents, was tempered by the requirements of their position, among them the keeping of an eternal flame and the daily fetching of holy water in enormous jugs that were not to be set down.

For his contributions to the splendor of Rome, Septimius Severus was rewarded with the construction of his great arch, erected in 203 upon the ten-year anniversary of his accession. This monument featured the names of Septimius and his sons Caracalla and Geta; after Caracalla ordered the murder of his brother, Geta's name was removed from the face of the arch. Adorned with elaborate bas-reliefs of the emperor's battles and reclining gods (later copied for ornamental use throughout Europe), the Arch of Septimius Severus was chosen by the Holy Roman Emperor Charles V in the sixteenth century as the focal point of his triumphant procession; creation of his route resulted in the destruction of two churches, hundreds of houses, and parts of some of the ancient monuments of the Forum.

The monument to Septimius Severus, portions of which remain standing, was a stunning example of the Roman arch, a design introduced in A.D. 81 with the Arch of Titus. Located near the Forum's main entrance to the southeast, this earlier arch depicted the capture of Jerusalem eleven years earlier. Its unique design, used in the Forum for free-standing monuments and copied extensively, was made possible by the invention of thrustless concrete and was eventually incorporated into bridges and aqueducts throughout the world. The treasures of the Roman raid on Jerusalem were kept in the Temple of Peace, probably until the sack of the Goths in 410. Some believed the Visigoth King Alaric was buried with those spoils, others that they were looted by the Vandals. The Jews believed the treasures were scattered in the Tiber for eternity.

The buildings of the Forum fell into decline, along with the Roman Empire. The Forum had already suffered when the imperial palace on Palatine Hill came to command the attention of Rome's clinging aristocracy, who increasingly challenged the traditions of the square. Defying the accepted notion that the Forum should honor immortal figures only, Domitian erected an equestrian statue of himself in the Forum proper. Later, the reign of Diocletian saw no less than seven columns honoring living citizens erected in the Forum. The Curia, already twice in need of restoration, burned in A.D. 283, to be rebuilt by Diocletian. In A.D. 608, the Forum's last monument was erected, to honor the Byzantine interloper Phocas, who had awarded the Pantheon to Pope Boniface IV.

Although Rome suffered successive invasions by the Visigoths, Ostrogoths, and Byzantines, Rome retained enough elegance in the year 800 to host the coronation of the Frankish king Charlemagne. The *Einsiedeln Itinerary,* a guidebook written about this time by a Swiss monk, relied on a fourth-century map to describe (for the most part accurately) the ancient buildings of the Forum and its surroundings. A century later, earthquakes toppled some of the city's ancient monuments, which were starting to be obscured by rubbish and debris.

Throughout the Middle Ages, the frequent plunder of Rome led to the deterioration of the Forum landmarks, as precious stones and metals, statues, and fixtures such as gates and columns were removed for use elsewhere. Advances realized through the caretaking of the Renaissance were temporarily eradicated with the Napoleonic Wars, but at the beginning of the nineteenth century Pope Pius VII instituted a comprehensive, systematic plan for the preservation of the ruins. Upon Rome's nomination as the capital of a united Italy in 1872, some of the world's best scientific minds began to work in earnest to restore the splendor of ancient Rome.

The Forum has seen a measure of neglect even in the twentieth century. For instance, the Curia, preserved since the Middle Ages in its second function as a Christian church, was destroyed during the 1930s and has been resurrected to resemble a simpler, much earlier version of the venerable hall. But ongoing excavations continue to reveal new clues to the puzzle of ancient Rome. In recent years, professors from Colgate University and the University of Pisa have unearthed the sacred boundary line ascribed to Romulus: a thirty-foot-wide, nine-foot-deep gully lined with a wall demarcating the "pomerium," the hallowed ground given to the omen-reading of the ancient augurs. The ditch and wall lay beneath the northeast slope of the Palatine Hill, at the origin of the Sacred Way, the boulevard that led proudly into the middle of the Forum in the heyday of Rome. The historian Tacitus recorded his tale of this most holy aspect of the ancient city centuries ago; these recent discoveries lend a certain truth to long-held "myths" about the birth of Rome.

Further Reading: Christopher Hibbert's *Rome: The Biography of a City* (London: Viking, and New York: Norton, 1985) and Georgina Masson's *The Companion Guide to Rome* (Englewood Cliffs, New Jersey: Prentice-Hall, 1965) are outstanding sources of information for the history of Forum development. The Time-Life book *Rome: Echoes of Imperial Glory* (Alexandria, Virginia: Time-Life, 1994), part of the Lost Civilization series, provides a concise, evocative chapter on the Forum. Roman Italy by T. W. Potter (London: British Museum Publications, and Berkeley: University of California Press, 1987) includes some important notes on the rise of Rome not found elsewhere. Three other books worth noting are *The World of Rome* by Michael Grant (New York: World Publishing, and London: Weidenfeld and Nicolson, 1960); *Roman's World* by Frank Gardner Moore (New York: Biblo and Tannen, 1965); and *Traveller in Rome* by H. V. Morton (New York: Dodd Mead, and London: Methuen, 1957).

—James Sullivan

Rome (Roma, Italy): Pantheon

Location: In the Piazza della Rotonda, east of the Piazza Navona, on what was the Campus Martius, which at the time the Pantheon was built still lay outside the city walls.

Description: The Pantheon is the best preserved of Rome's ancient buildings. It is a stone, brick, and concrete edifice of great beauty and imposing size, dating mainly from the second century A.D. and essentially in its original form. A porched rotunda, its overall design combines three geometrical shapes—rectangle (the portico), cylinder (the main body), and hemisphere (the dome). Its construction represents an enormous feat of architectural and constructional skills, exploiting the Roman technique of using stone- and brick-faced concrete to maximize the possibilities of the arch, the vault, and the special case of the vault, the dome. In the early years of its existence it was a temple dedicated to the Roman gods, but later it became a Christian church (Santa Maria Rotonda) and a burial place for noted Italians.

Site Office: E.P.T. (Ente Provinziale per il Turismo)
Via Parigi, 5
CAP 00185 Rome, Roma
Italy
(6) 488 3748

The Pantheon's name comes from *Pantheion,* the Greek word for a temple dedicated to all the gods (from *pan* = "all" and *theios* = "pertaining to the gods"), and the Roman Pantheon was originally dedicated to the seven planetary deities. As well as being a religious edifice, it was also intended as a monument to the greatness of emperors and empire. Similarly, in modern times the building is both a place of worship and a shrine for national figures.

The Campus Martius was a flat, open, partly marshy area in a bend of the Tiber. It was named after the temple and altar to Mars that had been built there with many other early temples and cult constructions. During republican times in the late sixth century B.C., the Campus Martius was state property, used by the citizens for voting, military training, and athletics. From 55 B.C. the Campus, with the important Via Flaminia running through it from north to south, was developed as a prime site for public buildings. It was as part of this development that what is now the Pantheon first came into being.

The first Roman temple on this precise site was erected—probably in the shape of a rectangle bordered by a colonnade on which rested a gabled roof—in about 27 B.C., during the third consulate of Marcus Vipsanius Agrippa.

Agrippa was Emperor Augustus's son-in-law, general, and minister, and the purpose of the building was to commemorate the naval battle of Actium in 31 B.C., which saw the momentous victory of Octavian (the name by which Augustus was then still known) and Agrippa over Marcus Antonius (Marc Antony) and Cleopatra. At much the same time as Agrippa was erecting this temple, Augustus was using ground on the Campus Martius for a crematorium and mausoleum for himself and his family.

Agrippa's original, south-facing temple was probably of travertine, a cream or light brown local limestone, much used in imperial Rome, that hardened under exposure. Agrippa constructed it as part of a whole building complex, in association with his baths. These, started in 25 B.C., were the first public baths in Rome, and remains of them can be seen nearby to the south, in the Via del'Arco della Ciambella. Between the Pantheon and the baths, Agrippa built another temple, the Basilica Neptuni.

Together with the rest of his civic buildings on the Campus Martius, and their lawns and parks, Agrippa's temple was damaged in a disastrous fire in A.D. 80. It was restored by Domitian, who ruled from A.D. 81 to 96. In A.D. 110 the temple was struck by lightning and burned down. A few years later it was completely rebuilt by the emperor Hadrian.

The new temple, which faced north this time, was of stone, brick, and concrete, with marble adornments. The work started in A.D. 118 or 119, Rome's grandeur being then at its height, and the consecration took place between 125 and 128, when Hadrian, who traveled a great deal over the empire, was present in Rome. A century had elapsed since Agrippa's temple was built.

Hadrian was an admirer of the Greeks. He was also a great ruler, organizer, and patron of the arts, who left other grand still-surviving monuments to bear witness to the boldness of his vision. The most notable is Hadrian's Wall, which defended the border between Roman Britain and Caledonia.

The large dedicatory Latin inscription running along the pediment of the Pantheon below the sculpted tympanum states that the edifice was built by Marcus Agrippa during his third consulate, that is in about 27 B.C. This inscription is said to have come from Agrippa's original building, and the stamp-marks on bricks used in other sections prove that the major part of the present temple is not Agrippa's but Hadrian's. It was designed, perhaps by Hadrian himself, to a different plan from Agrippa's and on a larger scale. It was Hadrian who gave the magnificent new building the name Pantheon and, thanks to its solid construction and periodic conservation and restoration over the centuries, it is Hadrian's temple that has survived, substantially as he made it. The mausoleum, also circular and durable, constructed by Hadrian from 130 B.C. for himself and his family, forms the base

The Pantheon
Illustration courtesy of Italian Government Tourist Board-E.N.I.T.

of the Castel Sant'Angelo. Some columns and remains of the walls of the Hadrianum, Antoninus Pius's memorial to Hadrian, are still to be seen just to the northeast of the Pantheon in the Piazza di Pietra.

Hadrian's Pantheon was subjected to some superficial alterations in the early part of the third century A.D. by the emperors Septimius Severus and his son Marcus Aurelius Antoninus. They may have used materials from Agrippa's temple in work on the portico; an inscription on the architrave records their labors.

The pagan Pantheon was closed by the first Christian emperors of Rome in the fourth century. Left untended, it suffered in the fifth century at the hands of barbarian invaders, but it was still the property of the emperor. During the divisions, decline, and decay of the remains of the ancient Roman Empire, Christianity had been gaining ground, and some of the ancient imperial civic buildings were taken over and consecrated to the new religion.

It is possible that the column of the Byzantine emperor Phocas in the Forum was erected in acknowledgment of his gift of the Pantheon in 608 to Pope Boniface IV. In 609 the Pantheon became the first Roman temple to be turned into a Christian church; it was dedicated to St. Mary and all

martyrs (Santa Maria ad Martyres). According to legend, Christian martyrs' bones were transported here by the wagonload from the catacombs on the Appian Way.

In A.D. 667, the rapacious Constans II Pogonatus, another Byzantine emperor, visited Rome and stole, among other treasures, the Pantheon's gilded bronze roof tiles; in 735, Gregory III installed a lead roof. In 1270 a square bell-tower was added on to the Pantheon behind the pediment; a 1534 engraving shows it still there.

While the popes were at Avignon, from 1309 to 1377, the building was used as a fortress in the Colonna family feud with the Orsinis. The tide turned back in favor of the Pantheon's fame and dignity in the early 1430s when Pope Eugenius IV freed it of accretions, including a lean-to poultry market beside the portico. It became an architectural model as well as a revered possession of the pontiff and a repository for relics and city treasures. A fishmarket in the piazza, however, was not moved until 1847.

In 1563 Pius IV repaired and largely recast the Pantheon's bronze double door. The Barberini pope Urban VIII commissioned Gian Bernini to add two open square-domed turrets at the front; these objects of derision were removed in 1883. It was also Urban VIII who took the bronze from the portico's ceiling and used it to make the baldaquin in St. Peter's and a large number of cannon, some say eighty, for the Castel Sant'Angelo, inspiring the satirical comment that the Barberinis had finished off the job the barbarians had started.

Between 1662 and 1666 Pope Alexander VII had the collapsed left-hand side of the portico restored, and the level of the piazza lowered. In 1747 Paolo Posi restored the interior for Benedict XIV.

The Pantheon is entered from its north side through a portico or pronaos, which underwent restoration in 1974. Originally the pronaos was raised above several steps, and there was a much larger open space in front. Making the architectural transition between the portico and the interior rotunda is a rectangular feature as wide as the pronaos and the same height as the rotunda up to the level of the second row of coffering of the dome. This transition is not always admired but, unlike Greek monumental buildings, Roman ones concentrated the visual interest on the interior rather than on the exterior. The Pantheon's combination of pronaos and rotunda is important in the history of architecture.

The portico, with its gabled roof and triangular pediment supported by columns, is conventional in design but impressive in scale and atmosphere. It is about 112 feet wide and 50 feet deep. Sixteen monolithic columns, of the Corinthian order without flutings, support the roof. Some are of gleaming reddish granite, some of gray. The columns are more than forty feet tall and measure about fifteen feet around, with white marble capitals and bases. With the change of orientation, Hadrian's front row of eight columns stands on the foundations of the rear wall of Agrippa's temple. The other columns behind the front eight are arranged in four rows of two each, forming three aisles. The middle aisle, the widest, leads to Pius IV's twenty-four-foot-high bronze double door; the other aisles lead to the giant semicircular niches where statues of Augustus and Agrippa may have once stood. The ceiling was originally covered with bronze.

The Pantheon's interior is spectacular. The rotunda's grandiose design posed formidable constructional problems, and the techniques used to meet the challenge aimed at and achieved a combination of simplicity, grandeur of scale, richness of ornament, and durability. The dominating forms in the harmonious overall pattern of this architectural figuration of the divine cosmos are the sphere and the circle with their derivatives. The rectangular shapes that embellish the floor, the walls, and the cupola are placed so as to coexist urbanely with the all-encompassing curves.

Between marble floor and coffered dome, the cylindrical temple chamber of the rotunda is formed by a circular wall, or rather a double shell, about twenty feet thick, into which, evenly spaced, there are built seven huge niches. Six of these are framed by a pair of tall Corinthian columns, and then an equally tall pilaster on each side; the seventh, the apse, round-arched like the entrance portal opposite, has two freestanding columns. Against the inner wall, at regular intervals between the niches, there are eight pedimented aedicules or shrines; each

pediment rests on two Corinthian columns of granite, yellow marble, or porphyry. Though smaller than the niches, the aedicules still dwarf human visitors. Bejeweled antique statues of the gods originally stood in the niches and aedicules.

Much of the colored marble paneling applied to the walls of the rotunda still remains; the floor, also of colored marble, has been restored to its first design. Marble decoration came into use in the buildings of Rome in the first century A.D. Large amounts of the colored varieties were imported in response to showy Roman taste.

Many of the Pantheon's ancient niches and aedicules are now adapted to Christian worship, serving as chapels and displaying Christian devotional and memorial artifacts. The apse contains the high altar, above which there is a seventh-century icon of the Virgin and Child.

From the sixteenth century onward, numerous famous people have been buried in the rotunda. The tomb of Raphael, who lived from 1483 to 1520, is situated in the third aedicule on the left, with De Fabris's nineteenth-century bronze bust of the artist, and on the altar stands a statue of the Madonna made to Raphael's design by Lorenzetto. To the right, under an empty niche, is an epitaph to the dead fiancée beside whom Raphael wished to be buried.

The remains of other Italian artists lie in the Pantheon, including those of Baldassare Peruzzi (1481–1536) and Annibale Carracci (1560–1609). The first king of united Italy, Victor Emmanuel II (died 1878), his son Umberto I (died 1900), and Italy's first queen, Margaret of Savoy (died 1926), are also entombed there.

At the next highest level, above the columns of the niches, a carved cornice runs around the vast chamber. Above this is the attic stage, mainly a restored version dating from 1747, with grille-covered rectangular openings corresponding to the niches and shrines below. Between the rectangular openings, red marble pilasters once alternated with sets of three marble panels; some of this can still be seen above the niche on the right of the apse. A Renaissance stucco frieze, added just below the dome, runs around above the attic story.

Domed structures had been known since before 1500 B.C., but the hemispherical dome of the Pantheon was the largest in the world until the twentieth century and the availability of reinforced concrete and modern steel. With an internal diameter of 142 feet, it is more than a yard wider than the dome of St. Peter's in the Vatican. Its height, or radius, and its diameter are equal to those of the cylindrical-shaped chamber below, so that it forms half of the complete sphere, thus representing the cosmos. From the inside, the dome seems to start from the top of the attic stage, but in fact it begins at the top of the highest cornice on the outside.

The dome is unsupported—that is, it is held up by its own fabric, without additional supports such as pillars or columns. It is ribbed, and decorated with five concentric rows of coffers, diminishing in size as they rise toward the round oculus, or little eye, a bronze-edged circular opening at the top of the dome, which lights and ventilates the otherwise windowless building. It is likely that the coffers originally contained gilded bronze rosettes and moldings. The exterior of the dome is stepped, the five divisions corresponding to the rows of coffering within. The oculus is approximately twenty-seven feet across; it admits a round or elliptical shaft of light that moves with the changes of the sun over the interior. Although glass windows were in use in Rome from the first century B.C., the oculus is not glazed. Any rain falling through the opening drains away at floor level into the channel built by Agrippa to draw away water from the marshy parts of the Campus Martius.

It is not known exactly what method was employed to construct the great dome, or what the proportion is of brick to concrete. It stands without the support of arches except at its lowest level. Roman concrete, the most common building material in the city from the first century B.C., was made particularly strong and adaptable by an admixture of local volcanic ash. The Pantheon rotunda and dome exemplify these qualities. For the dome, the types of aggregate material mixed into the concrete were graded, the heaviest employed at the bottom and the lightest—pumice—at the top around the center of the vault. The upper part of the dome is four feet thick; the lower is much thicker, but its fabric was made lighter by the coffering.

A staircase rises from the portico to the dome, from which vantage point one has a wide view of the city. Outside the Pantheon, in the piazza, an elaborate sixteenth-century stone fountain stands, incorporating an Egyptian obelisk that dates from the third century B.C., but was placed on its present site in 1711.

Further Reading: *A Traveller in Rome* by H. V. Morton (London: Methuen, and New York: Dodd, Mead, 1957) contains interesting facts about the Pantheon, including some out-of-the-way detail, and is enlivened by the author's personal impressions and reflections. *Roman Art and Architecture* by Sir Mortimer Wheeler (London: Thames and Hudson, and New York: Praeger, 1964) is terse and authoritative on the subject of the Pantheon and has well-chosen illustrations. *Rome: The Biography of a City* by Christopher Hibbert (Harmondsworth: Penguin, and New York: Norton, 1985) covers 3,000 years of Roman history, in which the story of the Pantheon is colorfully situated. Information about construction methods used in the Pantheon can be found in *A History of Technology* (Oxford: Clarendon, and New York: Oxford University Press, 1956), vol. II, chapter 12, "Building Construction," by Martin S. Briggs.

—Olive Classe

Rome (Roma, Italy): Ponte and Castel Sant'Angelo

Location: In the Vatican City on the right bank of the Tiber.

Description: Former tomb of the Roman Emperor Hadrian, later used as a fortress and a prison, now a national museum.

Site Office: Vatican Information Office
Vatican City-State
(6) 6982

During the reign of the Roman emperor Hadrian, which lasted from A.D. 117 to 138, many magnificent buildings were erected throughout the city of Rome, including the Pantheon, in 125, and the Temple of Venus, in 135. The last of these was the tower begun in 135 and intended as a mausoleum for the emperor and his family. This was completed in 139, under Hadrian's adopted son and successor Antoninus Pius. It consisted of four sections: a square base faced with marble and decorated with columns and statues; a circular structure, made of travertine and peperino stone, which was about sixty-six feet high; a tumulus (mound of earth) planted with trees on top of the tower; and, crowning the structure, a statue of Hadrian driving a quadriga (four-horse chariot). The mausoleum continued to be used for burials of members of the imperial family until the interment of Emperor Septimus Severus in 211, and it was connected with the city center by a bridge across the Tiber River known as the Pons Aelius (Hadrian's full name was Publius Aelius Hadrianus). This bridge was designed by the architect Demetrianus; of its eight original arches, the three central ones are still standing. These are the two structures that, much rebuilt, were to become the Castel and Ponte Sant'Angelo—the castle and bridge of the holy angel.

Between 271 and 275 Emperor Aurelian ordered the construction of what is now called the Aurelian Wall, a system of fortifications consisting of sixteen gates connecting twelve miles of wall on the left bank of the Tiber, surrounding and protecting the city. On the right bank the mausoleum, embellished with added towers, became a fortress, not part of the wall itself but central to the system. It could be reached through one of the gates, then known as Porta Aurelia but known in the Middle Ages as the Porta Sancti Petri or Gate of St. Peter, which gave access to the Pons Aelius.

By the end of the fourth century the official acceptance of Christianity had given a new prominence to the area around the first Basilica of St. Peter, built for the bishops of Rome between 320 and 329. As it rapidly became a focus for pilgrimage, its presence gave the fortress a new and increasingly significant role in the life of the city. It overlooked, and could therefore be used to monitor movements across, the Pons Aelius, the only remaining bridge providing easy access for pilgrimages to the Vatican Hill, on which the basilica stood. For several centuries the bridge also continued to be an important part of the sacred route for papal processions.

During the fifth century the fortress and the end of the bridge nearer to it were enclosed in a new wall in such a way that anyone coming off the bridge onto the right bank had to pass through a walled-off area and through the fortress to reach the basilica. This reinforcement was probably a response to the increasing insecurity of the times, which were characterized by the invasions of the Visigoths in 410, of the Vandals in 455, and of the Ostrogoths in 476. This last challenge to the city directly caused the collapse of the Western Roman Empire. When the Ostrogoths' leader Theodoric the Great made himself king in 493, the mausoleum became a prison as well as a fortress, and additional ramparts and chambers were built onto it.

In 532 Rome was captured by the forces of Byzantine emperor Justinian I. During a battle in 536 statues from the mausoleum were broken and the pieces used as ammunition for catapults fired by the Byzantine defenders against the Ostrogoths. War continued over the next sixteen years, during which the Ostrogoths briefly regained control of the city, until the final victory of the Byzantine forces in 552. However, the citizens of Rome were to suffer even more destruction with the invasion of the Lombards in 568, which was followed by widespread starvation and an outbreak of the plague. Such was the devastation that in 590 Pope Gregory the Great led the people through the streets of Rome in prayer against the plague. According to legend, as they approached Hadrian's Mausoleum, the image of the warrior archangel St. Michael appeared, waving his sword. The pope believed this to be a sign that the city would be victorious in its struggle against the disease. To commemorate the alleged miracle he had a chapel built on top of the fortress, above which was placed the marble statue of the angel, which gave the fortress its modern name. This statue was destroyed by lighting in 1497 and was eventually replaced by a bronze figure during the reign of Pope Benedict XIV, in the middle of the eighteenth century.

Although Rome was under the nominal authority of the Byzantine emperor, the papacy, under Gregory and his successors, developed a powerful and economically efficient local regime with its own army, able in practice to function as an independent authority in Rome. The imperial officials in Constantinople (Byzantium) may well have wished to challenge this development but were normally too busy dealing with the problems created by successive Arab and Lombard invasions of the emperors' various territories. In 753, when Rome itself was once again attacked by the Lombards, Pope Stephen II turned to the Frankish king Pépin III, the Short, instead of the Byzantine officials, to oust the Lom-

The angel atop Castel Sant' Angelo
Photo courtesy of Italian Government Tourist Board-E.N.I.T.

bards, and three years later Pépin, upon receiving the papal blessing as "Patrician of the Romans," returned all confiscated land to the Catholic Church, thus giving the papacy a new form of legitimate power over what became the Papal States. Under Pépin's son Charlemagne, who succeeded his father in 768, the church received extensive financial support and the city of Rome was rebuilt and refortified while the Caroliginian Empire continued to expand. In 800 Charlemagne was crowned Emperor of Rome by Pope Leo III, thus laying the basis for the Holy Roman Empire, fully established by Otto I, the Great, in 962. These events form the background to the history of the Castel and Ponte Sant'Angelo from the time of Leo III onward, with the development of the district near the Vatican Hill known as the Borgo—from the old term burgus Saxonum, the borough of the Saxon (or German) inhabitants—around the Castel Sant'Angelo. As the imperial city became the papal city, so the imperial mausoleum became the papal fortress.

In 846 Rome was attacked yet again, this time by the mixed Arab-Berber Moslem forces known as the Saracens, who destroyed much of the area outside the Aurelian Walls, including the Basilica of St. Peter. In 847 Pope Leo IV ordered the construction of twelve yards of defensive walls, which incorporated the Castel Sant'Angelo and were completed six years later. The area within these walls, parts of which can be seen today, included the Burgo and became known as the Città Leonina (Leontine City). Rome now consisted of two walled districts, the old imperial city within the Aurelian Wall, on the left bank of the Tiber, and the Leontine City (roughly equivalent to the modern Vatican City) on the right bank, with the Castel Sant'Angelo and the Ponte Sant'Angelo as the only convenient and supervised point of access between the two, giving it a militarily strategic position in any future power struggles.

By the beginning of the tenth century the papacy was no longer in alliance with the Carolingian Empire, which had been divided among Charlemagne's descendants, and therefore found itself isolated in its power struggles with the wealthy ruling families who eventually took control of Rome and of papal appointments. They then led the city's resistance to the attempts of the Holy Roman Emperors to impose their control on the city. A revolt against Otto I in 964 failed when the Romans who had sought refuge inside the Castel Sant'Angelo were captured by imperial forces. In 974 Roman supporters of the powerful Crescenzi family murdered Pope Benedict VI, who had been chosen by Emperor Otto II, inside the fortress, but imperial forces were able to regain control. In 998 the Crescenzi led yet another attempt to seize the Leontine City but this, too, ended in failure after a brief battle with the forces of Otto III, who laid siege to the Castel Sant'Angelo until the rebels surrendered.

During the eleventh century the popes themselves began to seek ways of asserting their authority against the demands of the Holy Roman Emperors. In 1083 Emperor Henry IV tried to repeat the successes of the three Ottos by attacking the Vatican, but failed to capture Castel Sant'

Angelo, where Pope Gregory VII and his followers held out for a year before being rescued by the Norman duke Robert Guiscard. His troops did not stop at victory but went on to plunder the city. The papacy continued to be a source of conflicts among the emperors, the ruling families, and a growing number of Roman radicals. In 1188 Pope Clement III obtained the loyalty of many ruling families and of rebel leaders through reforms that eventually enhanced the Catholic Church's political and economic power. Significantly, the Castel Sant'Angelo, which had proved to be an essential military asset, became papal property at this time. But the conflicts did not cease; papal elections came more and more under the influence of the kings of France, and from 1309 to 1377 the popes lived not in Rome but in Avignon, France, in what the Romans called the "Babylonian Captivity." To add to the political and economic decline of Rome, which was a side effect of this period, the city suffered a major earthquake and an outbreak of the plague in 1348. In the hope of securing a miracle similar to that claimed for Pope Gregory in 590, Pope Clement VI marched with the people of Rome along the Pons Aelius toward the Castel Sant'Angelo. It has been suggested that the names Castel Sant'Angelo and Ponte Sant'Angelo became commonly used by the people of Rome after this event.

Between 1378 and 1417 there was simultaneously one succession of popes in Avignon and another in Rome; beginning in 1409 there was even a third pope, who resided in Pisa. This division and the growing corruption of the church reduced Rome to an economic backwater and led many desperate Romans to give their support to Pope Boniface IX, who brought political and economic stability to the city. During his reign large parts of the Vatican area were restored and the fortifications of the Castel Sant'Angelo were improved. Successive popes continued to improve and alter the appearance of the castle with more fortifications during the fifteenth century. In 1410 John XXIII—now regarded as an antipope—started the construction of a covered passageway along the wall, later known as the Passetto Vaticano, which links the castle to the Vatican palaces. Its purpose was to give the popes quick access to their fortress when the city came under attack.

In 1450, during a celebration held by Pope Nicholas V at St. Peter's Basilica, the congestion created by the papal procession and those shopping at market stalls on the Ponte Sant'Angelo became so great that 172 people died, being trampled or drowned when panic broke out. Although more space was designated in front of the bridge on the city side, between 1451 and 1453 the bridge itself was rebuilt, the old gates at each end were removed, and two large towers were built on the side of the castle. In addition, two small chapels were built on the left bank of the Tiber (the city side) in order to shorten the bridge. From 1492 Pope Alexander VI saw to the completion of the Passetto Vaticano, which he used in 1494 during the threat of a French assault, and to the reconstruction of the fortress around four bastions, named after the four Evangelists (Sts. Matthew, Mark, Luke, and John, the writers of the Gospels). In 1527 Rome was attacked by unruly soldiers of the Holy Roman Emperor Charles V; the soldiers had not been paid for months. During their Sack of Rome, as it became known, churches, hospitals, and tombs were pillaged and numerous sacred treasures were stolen. Pope Clement VII fled through the Passetto Vaticano to the Castel Sant'Angelo along with nearly 3,000 people who found refuge with him in the fortress for several months. Among them was the sculptor Benvenuto Cellini, who had earlier been imprisoned in the castle, and it is his memoirs that are the main source of information about these months. Only after a ransom had been paid by Clement was the city restored to order in 1528. Clement later installed papal apartments where Hadrian's tumulus had been, the section in which he took refuge. Ever since then these apartments have been kept ready in case of invasion.

Restoration of the city's infrastructure dominated the rest of the sixteenth century and the early seventeenth century. In 1530 Pope Clement ordered the erection of the statues to Sts. Peter and Paul, which still mark the entrance to the Ponte Sant'Angelo on the left bank, replacing the two small chapels that had been damaged in 1527. Under Pius IV and Urban VIII the fortifications were extended to include outer ditches and defensive walls. By the mid-seventeenth century, Rome was in a position to renew its role as a center of culture, and great architecture and sculpture were encouraged, especially the work of Gian Bernini. Among his many commissions from Popes Clement IX and Clement X, Bernini was asked between 1667 and 1672 to design the ten angels, two of which he sculpted himself, which still line the Ponte Sant'Angelo, displaying symbols of the sacrifice of Jesus to visitors passing on their way to the Vatican. The bridge as it looks today was largely rebuilt and completed by Bernini's pupils in 1688, after the artist's death in 1680.

French Revolutionary troops took possession of Rome in 1798, occupying the Castel Sant'Angelo. There they painted the bronze angel red, white, and blue and placed a Phrygian cap, symbol of their revolution, on its head. Under the puppet Roman Republic that they established, relations between Pope Pius VII and Napoléon improved to the point that French soldiers were withdrawn in 1800, but they returned to take possession of the castle once again, in 1808, when Napoléon created a dependent Kingdom of Italy. After Napoléon's final defeat in 1815, Rome was directly governed by the popes, as part of the Papal States. In 1849 the threat of a republican rising led by Giuseppe Garibaldi and Giuseppe Mazzini forced Pius IX to ask the French president Louis-Napoléon Bonaparte (Emperor Napoléon III from 1852) to maintain order. French troops were to occupy the Castel Sant'Angelo for the next twenty-one years, but they were eventually ousted by Prussian troops sent to aid the nationalist movement, which, having unified the rest of Italy in 1861, took over Rome and made it the national capital in 1871.

Between 1880 and 1901 the castle served as a prison, but then it was decided it should be restored. Major excavations of the Ponte Sant'Angelo were carried out in 1892, and its end arches were altered to fit with the embankments on the Tiber that were built between 1892 and 1894. In 1925, under the Fascist government of Benito Mussolini, the Castel Sant'Angelo became a museum of art and military history, the Museo Nazionale di Castel Sant'Angelo, and the Ponte Sant'Angelo was pedestrianized. Excavations were also carried out around the castle in the months before its restoration in 1933–34. In 1929 the Lateran Treaties between Mussolini and Pope Pius XI created the independent state of the Vatican City, guaranteeing the pope's possession of several properties, including the Castel Sant'Angelo. Today the castle and the bridge have become important historical monuments and tourist attractions in their own right.

Further Reading: Christopher Hibbert's *Rome: The Biography of a City* (Harmondsworth, Middlesex: Penguin, and New York: Norton, 1985) provides an excellent, detailed account of the city's long history. Mark S. Weil's *The History and Decoration of the Ponte S. Angelo* (University Park, Pennsylvania, and London: Pennsylvania State University Press, 1974) is an expert academic study of the bridge.

—Monique Lamontagne

Rome (Roma, Italy): Vatican

Location: The Vatican City-State occupies approximately 109 acres within the city of Rome. It lies on the west bank of the Tiber River and is defined for legal purposes by the ancient walls that surround it on all sides except at St. Peter's Square, which is open to Rome.

Description: Vatican City is the world's smallest sovereign nation, consisting of just seventeen one-hundredths of a square mile and completely surrounded by the city of Rome. The Vatican is the titular seat of the Roman Catholic Church and, as such, occupies a unique niche among the world's countries and religions. The Vatican belongs to many international organizations in its own right, and has observer status at the United Nations. Yet, it is still a strictly religious entity, having acquired the trappings of nationhood only to guarantee the pope's freedom from secular constraints in his administration of the church. The Vatican is also the world's only elective monarchy, as the pope is elected by the college of cardinals, but is invested with supreme power until death, not unlike a king.

Site Office: Vatican Information Office
Vatican City-State
(6) 6982

The Vatican is an anomaly. Geographically, it is not only completely surrounded by another country, but by another city. It is almost entirely encircled by ancient walls, and has neither an airport nor international waters to call its own. Its total surface area is most accurately measured in acres, rather than square miles. Its population hovers around 800 and the country, for that is what it is, has an annual birthrate of essentially zero.

Yet in the pantheon of worldly power and influence the tiny state on the banks of the Tiber ranks with the capitals of earth's most important nations, regularly intoned in sentences that also include references to Washington, London, Paris, and Moscow. As the spiritual home to more than 800 million Catholics worldwide, the Vatican wields great power within its peer capitals, and its influence in matters reaching from the ballot box to the battlefield is undeniable.

By all generally accepted precedents of international law, the Vatican is a sovereign state, free to do as it wishes within its borders. For practical purposes, however, the Vatican is the seat of the worldwide Roman Catholic Church, and holds sway over its international flock in much the same manner that parishes dictate to churches, dioceses dictate to parishes, and archdioceses dictate to dioceses. In a manner of

speaking, the Vatican is the only rectory in the world to qualify for a seat in the United Nations. (Vatican officials opted for mere observer status, instead.)

But it is exactly this dichotomy that gives the Vatican its unique brand of influence. Shrouded in secrecy and mystery, the inner workings of the papal state are known to only a select few, imparting the tiny nation with an otherworldly aura that other capitals cannot hope to achieve. Worldly capitals rise and fall, but the Vatican—the Roman Catholic Church—appears invulnerable.

It is odd then that the instrument that gave the Vatican its legal sovereignty did not materialize until 1929. Called the Lateran Treaty, after the Lateran Palace, the former official residence of popes, the document was promulgated by no less than Fascist dictator Benito Mussolini, a figure remembered today primarily for his arrogance and ineptitude. Still, 1929 hardly qualifies as ancient, and even Washington, D.C., has been in existence as a legally incorporated entity for a much longer period of time. However, the Lateran Treaty really only codified a state of existence that had been accepted for centuries.

The Lateran Treaty, signed on February 11, 1929, defined a legal distinction between the Vatican City-State and the Holy See. The Holy See, which refers strictly to the office of the pope in his capacity as leader of the worldwide Roman Catholic Church, was defined as a separate and pre-existing entity that had done its business around the world for centuries. The Vatican City-State, in simplest terms, is merely a sovereign piece of real estate upon which the pope can make decisions regarding the administration of the church, free from the restraints of a secular government.

The Vatican City has a total surface area of just less than 109 acres and is surrounded on all sides except for the piazza of St. Peter's Square by walls built primarily between 1540 and 1640. The Vatican's acreage is covered in near-equal parts by buildings, pavement, and gardens. The 160 acres of sovereign papal holdings outside the Vatican and included in the Lateran Treaty are actually larger than the state itself. These holdings include several sites in Rome, including the Lateran Palace, and the papal summer residence of Castelgandolfo in the Italian countryside.

Despite its diminutive geography, the Vatican has its own embassies abroad, called Papal Nuncios and officially accredited to the Holy See, that are accorded the same diplomatic privileges as the embassies of any other state. About 100 Swiss Guards serve as the Vatican's independent and purely defensive army. The tiny state is bound by the Lateran Treaty to remain outside the temporal disputes of the secular world, including Italy, and thus is barred from maintaining an offensive military apparatus. The government runs its own television and radio stations, publishes a daily newspaper, and

Detail of Raphael's Transfiguration in the Picture Gallery of the Vatican Museums
Photo courtesy of The Chicago Public Library, used by permission of the Vatican

maintains a prolific press. There is a Vatican railroad line and a heliport for use by high-ranking Vatican officials and visitors.

The Vatican is also unique in that it is the world's only elective monarchy. Upon the death of a pope, members of the college of cardinals elect a new pope, who is then invested with supreme power over the Holy See, Vatican City, and the worldwide church until death. The pope appoints a pontifical commission and a president to handle the day-to-day administration of the city. As of 1995 the pope was John Paul II, Polish-born Karol Wojtyla, elected October 16, 1978, the first non-Italian pope since the death of Adrian VI, who was Dutch, in 1523. John Paul II is the 266th Roman pontiff in an unbroken line that began with St. Peter immediately following the death of Christ.

Although commonly known as His Holiness, the pope's full title is even more imposing: His Holiness Pope John Paul II, Bishop of Rome, Vicar of Christ, Successor of the Prince of the Apostles, Supreme Pontiff of the Universal Church, Patriarch of the West, Primate of Italy, Archbishop and Metropolitan of the Province of Rome, Sovereign of the Vatican City-State, Servant of the Servants of God. At the Vatican, there is little doubt who is in charge.

The cardinals may elect the pope, but the pope creates the cardinals. The number of cardinals worldwide has fluctuated over the centuries, but stood at 145 in the mid-1990s. Cardinals residing in Rome and serving as the pope's im-

mediate advisors are known as cardinals in curia. The Roman Curia, as the collective is called, serves as the primary administrative body of the church and as the papal court. A cardinal secretary of state manages the Vatican Secretariat of State and the Council for the Public Affairs of the Church. The Secretariat of State was divided in 1988 into two divisions, one for General Affairs and another for Relations with States. Cardinals also manage other offices, tribunals, and commissions throughout the church hierarchy. Cardinals not stationed in Rome manage archdioceses around the world. Some archdioceses may cover tens of thousands of square miles and contain dozens of individual dioceses and hundreds of parishes.

Relatively few people are born to Vatican citizenship. Between the signing of the Lateran Treaty in 1929 and the year 1980, for example, only fourteen children were born within the Vatican walls. Normally, Vatican citizenship is reserved for those few authorized to live within its boundaries, although the families of lay employees, such as the Swiss Guard, are usually not granted citizenship even though they reside within the city walls. Citizenship is also extended to all cardinals, resident or not, and those assigned to diplomatic service abroad with the Holy See. Most Vatican citizens hold the distinction only for the duration of their duty to the state, after which time their legal status reverts to that of their home state. Any person who would be left stateless is automatically granted Italian citizenship.

The Vatican also issues its own coins and postage stamps. Stamps are good for use only on mail originating in the Vatican, and are an important source of revenue as they are prized by collectors and rarely used after they have been purchased. The stamps are printed by the Italian state printing office. Coins are struck by the Italian mint and are of the same sizes, denominations, and compositions as Italian coins. They are legal tender throughout Italy and, since 1932, the Republic of San Marino, as well.

Vatican City is home to some of the world's premier museums, galleries and archives, many of which, such as the Egyptian Museum, display a surprising interest in non-Christian religions. Others include the Etruscan Museum, Pio Clementine Museum, the Gallery of Tapestries, and the Missionary-Ethnological Museum.

Most famous of the Vatican's attractions is the Sistine Chapel, named after Pope Sixtus IV who ordered its construction in 1477. The chapel was completed in 1480 and originally featured the works of many of the premier artists of the day, including Pier Matteo d'Amelia, who originally covered the ceiling with a simple, starry sky. The ceiling was made famous by Michelangelo, who began his famous fresco in 1508 under orders from Julius II. The ceiling was completed in October of 1512 and inaugurated on All Saints' Day (November 1) of the same year. Michelangelo went on to paint *The Last Judgment* on the altar wall of the chapel, a project that took five years, from 1536 to 1541, and covered more than 240 square yards and represented 391 separate figures.

The area now covered by the Vatican has been put to many uses over the centuries, and some that would seem to be mutually exclusive. The Vaticanum was one of the fourteen original administrative regions of Rome established by Emperor Augustus around the time of the birth of Christ. Of the fourteen, only the Vaticanum was on the west bank of the Tiber River. The Vatican was not always the high-rent district it is today, either. Tacitus described the district as "infamous" due to its marshy and malarial nature. The satirist Martial, critiquing the local wine, said "Vaticana bibis, bibis venenum," or "Drink Vatican and you drink poison." Pliny said its snakes were large enough to swallow children whole.

Still, the Vaticanum was sufficiently lush to attract large numbers of wealthy Romans, many of whom built expansive villas at the site. The area was further developed by the emperors Caligula and Nero. The former built a chariot track called the Gaianum and an artificial lake called the Naumachia to be used for mock sea battles.

Caligula never lived to see the completion of his grandest project at the Vaticanum, the Vatican Circus. Intended as a private arena for gladiatorial games, the Vatican Circus was completed by Nero sometime immediately prior to A.D. 64 and stood adjacent to the current St. Peter's Basilica. In the center of the circus stood the obelisk brought from Egypt by Caligula. The obelisk was destined to stand in its initial location for over 1,500 years, long after the circus was destroyed. Eventually, St. Peter's Basilica was con-

structed only a few feet away, and in 1586 Pope Sixtus V ordered the obelisk to be erected in the center of St. Peter's Square, where it stands today.

The reason the Vatican stands where it does begins with the great fire that destroyed the city of Rome in July of 64. The city east of the Tiber was lost, but as the only significant district of Rome on the west side of the river, the Vaticanum and its circus were spared. Emperor Nero, who had been at the seaside during the conflagration (nobody really knows if he had been fiddling or not) immediately announced plans for a new and grandiose Rome to rise from the ashes of the previous, primarily wooden city. Such was his enthusiasm that many luckless Romans privately wondered if Nero had ordered the fire just to clear the way for his dream of a new and improved capital city.

Needing both a scapegoat and a distraction, Nero blamed the fire on the new sect known as Christians and announced a grand day of games at the Vatican Circus scheduled for October 13. The populace, weary of rebuilding their city, responded by filling the circus on the fateful night, there to be treated to an evening of chariot racing and Christian-slaughtering. Some of the hapless Christians were killed by wild animals, while others were burned at the stake and still others were crucified. According to the historian Tacitus, St. Peter, the disciple charged by Jesus Christ with the task of "confirming the faith of the brethren," was among the Christians crucified at the Vatican Circus that day. His body was buried in a crevice in a nearby necropolis. Shortly thereafter, a small shrine appeared on the site. The shrine became a church and the church eventually grew into St. Peter's Basilica, the largest church in the world. St. Peter's and the city-state that surrounds it together mark the grave of St. Peter, the first bishop of Rome.

It should be stated that the events of St. Peter's crucifixion and burial, although probable, have never been conclusively proven. There is, however, enough circumstantial evidence in support of the legend to satisfy Vatican authorities, who have never wavered from their assertion that St. Peter's Basilica stands above the tomb of the Prince of the Apostles.

To begin with, in order for Emperor Constantine to build the first church on the site in 312, over a million cubic feet of dirt had to moved from the Vatican Hill. This was done despite the fact that the already-leveled surface of the former Vatican Circus was only a matter of a few feet away. The hill location had already been venerated by the faithful since immediately after Peter's death. This would have been a very compelling reason, Catholic historians say, for Constantine to insist upon the site he chose.

That the site served as a Roman necropolis had always been known. The ground beneath the basilica is the resting place of hundreds, and possibly thousands, of people. This includes the 142 popes interred there, although most of the dead are unknown. That the Vatican rests upon the grave of somebody is beyond dispute.

But do the countless remains include those of St. Peter? Archaeological evidence was not sought until 1939, and even then the decision was made by accident. Pope Pius XI had in his life expressed a desire to be buried near to his predecessor, Pius X. Upon the death of Pius XI, his successor Pius XII ordered a renovation of the burial grottoes beneath the basilica in order to honor his predecessor's request. However, soon after excavation had begun workmen inadvertently discovered ancient layers of building material and sarcophagi below the known grottoes, prompting Pius XII to order a full-scale, scientific excavation of the area beneath the altar.

The year 1939 was a bad year to begin sensitive archaeological excavations anywhere in Europe, and in Italy in particular. However, work continued even as World War II raged throughout the continent and, in its latter years, found its way to Rome.

Excavations quickly revealed evidence that the site had been in use for much longer than had been thought. Layer upon layer of passages and burial sites stretched downward into the earth, startling in their complexity and impressive in their attention to detail. At the lowest level of the ancient Roman necropolis there lies a "street" of burial chambers designed to resemble the houses in which the dead spent their lives. Frescoes and mosaics abound, including the oldest known mosaic of Christ, the *Christ Helios.*

Directly beneath the altar the remains of a small shrine were discovered, presumably the one erected in the second century in honor of St. Peter. At the foot of the shrine was a slab set into the ground, like a gravestone. Upon removing the stone, a grave was found, but it was empty. A cache of bones found nearby proved in 1956 to have come from a variety of livestock and several humans, including a woman.

However, a scholar from the University of Rome, Margherita Guarducci, pursued the case. The remaining walls of the original shrine had been covered with graffiti left by early Christian pilgrims, much of which proved to be decipherable. Among the scribblings, which required six years to thoroughly analyze, Guarducci found the snippet "Petros eni," meaning "Peter is within." The discovery made headlines.

There was still the problem of a lack of bones at the site. Had they been looted in an earlier century or, as Guarducci and others surmised, relocated in order to prevent their theft? Further study of the nearby Red Wall in which the first cache of bones had been found revealed a small ossuary lined with marble, indicating a tomb of relative importance. The niche was empty. Acting on input from workmen involved in the earlier excavations, Guarducci searched through storage rooms at St. Peter's, double-checking the boxes of bits and pieces taken from the grottoes in earlier excavations until, in a box marked "graffito" she found bones removed from the ossuary ten years before. Apparently, a monsignor responsible for making daily inspections of the excavations had mistakenly taken the bones from the ossuary and deposited them in storage.

Scientific analysis of the bones proved that they were from a man of between sixty and seventy years of age, bearing out traditional accounts of St. Peter at the time of his death.

Furthermore, among the bones were found bits of valuable purple and gold cloth. The bones had also been deposited in what seems to be the only ossuary built by Constantine when he erected the first basilica at the site. Although the conclusions drawn by these discoveries continue to be disputed by scholars, it satisfied Vatican authorities. On June 26, 1968, Pope Paul VI announced to the world that the bones of St. Peter had been found. They were reinterred and, officially, the matter has been put to rest.

The story of the construction of the Vatican as a place central to Christianity begins in A.D. 312 when Emperor Constantine I did a religious about-face following his victory over Maxentius at the Milvian Bridge outside Rome. Crediting his victory to the new Christian god, Constantine not only legalized the religion but did everything he could to promote it. Interestingly, Constantine himself remained unbaptized until 337, when he lay near death.

To begin his conversion, Constantine ordered the construction of a basilica for the bishop of Rome, then Pope Miltiades. The basilica was built on a piece of imperial property at the eastern edge of the city known as the Lateranum, after its original owners, the Laterani family. Other buildings at the site were given to the bishop and over the years the Lateran grew into the administrative center of Christendom. The Lateran Cathedral remains the cathedral church of Rome to this day.

A few years later, Constantine made his biggest contribution to the new religion by ordering the construction of what is now known as Old St. Peter's at the site of the shrine erected over the grave of the saint. Officially, Old St. Peter's was known as a martyrium, in honor of the martyr. The site continued to be used for burial purposes, with layers of the new faithful buried atop the Roman dead from earlier centuries.

The official residence of the popes was still at the Lateran, but the area around Old St. Peter's grew quickly. Buildings were constructed to house pilgrims and clergy assigned to the basilica. Housing was even provided for popes and their entourages in case they needed the space. Facilities to administer to the poor were also constructed. A thriving commercial district known as the Borgo grew up between Old St. Peter's and the nearby Tiber to serve the needs of the constant flow of pilgrims, adding greatly to the wealth of the church and of the surrounding area.

The Vatican began to take its modern physical shape in 847 following an attack by Saracen pirates the year before. Rome proper resisted the attack, protected by its immense walls, but St. Peter's was not so fortunate, prompting Pope Leo IV to order the basilica and its associated precincts to be encircled with a protective wall. The thirty-nine-foot-high wall held four gates and was fortified with forty-six towers for surveillance and defense.

The new city was officially inaugurated on June 27, 853, and named the Leonine City. Many branches of papal government and administration were transferred to the new, fortified city, although the Lateran remained the seat of the papacy. The Leonine City's walls were not fixed, however, and were enlarged and modified by various popes as the church bureaucracy grew, effectively ending with the walls completed by Pope Urban VIII in 1640.

The significance of the Vatican grew as the centuries passed, with more and more functions being transferred there from the Lateran. In 461, Pope Leo I was the first pope to elect to be buried at St. Peter's. Beginning with Pope Leo III in 795, popes were crowned at St. Peter's and made their way to the Lateran to take possession of the seat of their power. Leo III also presided over another leap in the importance of the Vatican when, on Christmas Day in the year 800, he crowned Charlemagne emperor of Rome, setting a precedent to be followed by rulers throughout the West. In fact, when Henry IV was excommunicated by Pope Gregory VII in 1084, he attacked Rome and occupied the Leonine City. He even chose his very own pope, called an "antipope," Clement III, to perform his crowning in St. Peter's. His successor, Henry V, invaded in 1111 and abducted Pope Paschal II, forcing him to proceed with the crowning in St. Peter's, even though the pope agreed to do it only in secret.

For several centuries the popes presided over a vast swath of Europe known as the Papal States, which they administered as secular rulers in the era long before the advent of democracy. Temporal authority over the Papal States grew slowly, beginning in the seventh century when the popes were faced with the inability of rulers headquartered in Constantinople to provide adequate defense of Italian lands against the hostile Lombards of the north. In league with local nobility, the popes served as the only on-site leaders the otherwise fractious landed gentry could agree on. Their power was formidable, but it was not written in black and white.

Papal authority was legitimized with assistance from another "illegitimate" leader, Pépin III, the Short, who had ruled the Frankish kingdom for four years without any formal authority. Needing papal authority to legitimize his throne, Pépin in 751 sent an emissary to Pope Zacharias to ask for help. With papal blessing, Pépin was anointed king by the bishops of Francia the same year.

Still living under the constant threat of Lombard domination, Pope Stephen in 754 appealed to then-King Pépin III to return the favor by forcing the Lombards to return several provincial cities to the republic and guarantee the safety of the Papal States. Pépin agreed, invading Italy in 755 and forcing the Lombards to return Ravenna and several other cities. Upon Pépin's retreat, the Lombards backed out of the deal and in 756 invaded Rome. Pépin came to the rescue, this time forcibly removing the Lombards from the city and all disputed territories. The keys to the cities of Rimini, Ancona, Fano, Pesaro, and Senigallia were placed on the altar at St. Peter's, legitimizing papal authority.

The Papal States endured, sometimes in name only, until 1870, when Rome was declared the capital of the unified Republic of Italy. Rule was only interrupted three times: in 1798 when a Roman republic was proclaimed in the wake of the French Revolution, in 1808 when Napoléon temporarily annexed the Papal State, as it was then known, and in 1848–49 when revolution created a second and equally short-lived Roman republic.

In 1871 the government of the recently united Italy unilaterally passed the precursor to the Lateran Treaty in the form of the Law of Guarantees. Under the law, the popes were to have sovereignty over the Leonine City, Lateran Palace, and Castelgandolfo, and the popes would enjoy equal sovereignty while in Italy. Ironically, the Law of Guarantees did not guarantee much, and did nothing to promote the idea of papal sovereignty anywhere outside Italy. Pope Pius IX despised the law and lived out his reign as a voluntary prisoner in the Vatican Palace, as did his three immediate successors. It was not until the Lateran Treaty was signed in 1929 that the so-called Roman Question was finally settled to the satisfaction of all parties.

Another glitch in the history of the "Eternal City" came in the fourteenth century when the popes were forced to reign from Avignon, in what is now the Provence region of France. An association with Charles of Anjou intended to prop up both Charles's empire and that of the papacy quickly went sour when Sicily fell to the Vespers in the waning days of the thirteenth century. With Charles supplying the military muscle necessary to recapture the island, the popes were left with no choice but to bow to French interests.

However, after forcing the ouster of his predecessor, Celestine V, Pope Boniface VIII took over and immediately adopted an autocratic style intended to put the power back into the papacy. In 1301 he made the mistake of denouncing the French king, Philip the Fair, for his policy of taxing the clergy, threatening to excommunicate him, if necessary. By 1303, Boniface VIII was dead, but the rift between France and the Holy See was severe.

Following a short interlude in the form of Pope Benedict XI, who reigned from 1303 to 1304, the college of cardinals elected Clement V, the archbishop of Bordeaux, as the next pope. Clement V was crowned in Lyons and established his seat of power at Avignon, where it was to stay until 1376. The papacy remained subservient to French interests during its entire tenure in France.

In 1377 Pope Gregory XI returned the papacy to Rome, establishing himself at the Vatican because the Lateran had fallen into disrepair during the long absence. So had the city of Rome. Reduced to a population of 17,000 and stripped of its wealth and glory, Rome was a dangerous city in desperate need of rejuvenation. Gregory died in 1378, and, under intense public pressure, the cardinals elected Urban VI, a Roman aristocrat who had every intention of keeping the papacy firmly rooted in Rome.

Unfortunately, Urban proved to be not only incompetent, but violent as well. The college of cardinals, terrified of a pope who threatened them with physical violence if they did not see things his way, fled Rome and elected a Frenchman, Clement VII, to be their new pope. This action ushered in the Great Schism, which was to last until 1417. Urban and his supporters stayed in Rome, while Clement and the college of cardinals returned to Avignon, where they received support from the French government. An attempt at reconciliation in the General Council of 1409 backfired terribly when it declared null and void both existing popes and elected its own, resulting in three popes instead of two. A second General Council held from 1414 to 1417 resulted in the election of Pope Martin V who, in 1420, returned the papacy to Rome and, specifically, the Vatican, from where it has originated ever since.

Despite the proliferation of offices, museums, and galleries that take up most of the Vatican's limited space, it is the towering St. Peter's Basilica that anchors the city-state both to its location and to the minds of the people of the world. The rest of the Vatican can only be seen from within its walls, but St. Peter's, with its balcony overlooking the piazza, is a familiar sight to anybody who watches television news.

The current St. Peter's was commissioned in 1506 by Pope Julius II and consecrated in 1626, 1,300 years after the consecration of Constantine's original basilica. Many of the most famous architects and artists of the day participated in its construction, and the basilica remains to this day the largest church in the world.

Old St. Peter's had already undergone several minor renovations, most either strictly structural or doomed to incompletion by the death of their champion. However, Julius was determined to make good on the attempts of his predecessors. First, the old basilica was in dangerous condition. Second, and probably just as important in Julius's eyes, the brilliance of the Renaissance had begun to take hold in Rome and elsewhere in Italy, making the old cathedral a poor relation when it should have been a guiding light.

Julius ordered the architect Donato Bramante to design a new St. Peter's. Bramante based his plan on a Greek cross, with naves of equal length crossing beneath a hemispherical dome. His plan remained largely intact despite the embellishments added by later architects. Bramante died in 1514, at which time the threesome of architects Fra Giocondo, Giuliano da Sangallo, and Raphael took over and changed the plan to a Latin cross with a long central nave. Due to a series of political, military, and religious crises, work on the basilica came to a virtual halt until 1539 when demolition of the western portion of the old basilica was completed and a foundation was laid for the new apse. The architect Sangallo (Giuliano's nephew) supervised the proceedings.

Michelangelo took over the project upon Sangallo's death in 1546, immediately demolishing most of his predecessor's work and reverting to Bramante's original plan. Given a free hand by Paul III to make whatever changes he felt necessary, Michelangelo was determined to finish enough of the massive project before his death that nobody could make major changes to it. By 1558 he had completed the first phase of his project in spite of opposition from the supporters of Sangallo's plan.

Several teams of architects labored on the new basilica, essentially honoring Michelangelo's plan, until 1605 when Pope Paul V ordered the demolition of the remainder of the Constantinian nave to make way for a long central nave and a return to the Latin cross design that Michelangelo had so despised. The new nave was designed by Carlo Maderna, and incorporated Michelangelo's columnar facade, although the decision to lengthen the nave, easily defensible for purposes of elaborate ceremony, had the effect of making the grand dome visible only from a distance.

In 1656 Pope Alexander VII ordered the architect Gian Bernini to design an appropriately monumental piazza in front of the new basilica to formalize the square that had stood before it for centuries. Bernini had earlier contributed several interior structures to the basilica, including the impressive altar that stands above St. Peter's tomb. Bernini designed a dual colonnade consisting of 284 Doric columns arranged in four rows, enclosing the square in enfolding arms. Above the colonnade he placed statues of 96 saints, looking down at the assembled pilgrims below.

Further Reading: Material on the Vatican is voluminous, covering both its history and the collections in its museums. *Vatican City State*, part of the World Bibliographical Series compiled by Michael J. Walsh (Oxford: Clio, and Santa Barbara, California: ABC-Clio, 1983), contains an exhaustive listing of research materials, mostly in English. *The Vatican*, text by various authors (New York: Vendome, 1980), is a valuable source of the coffee-table variety, featuring color photographs and fact-filled captions and essays. *The Vatican: Spirit and Art of Christian Rome* by the Metropolitan Museum of Art (New York: Abrams, 1982) is essentially similar, although more technical in its explanations of the Vatican's art and architecture. The Vatican is also well documented in travel guides covering Rome. Individual components of papal history, such as the Papal States, are covered in many standard reference collections.

—John A. Flink

Salamanca (Salamanca, Spain)

Location: One hundred thirty miles northwest of Madrid, on the banks of the River Tormes.

Description: A famous university town, among the most prestigious in all of Europe for some 400 years. Many of the important structures in Salamanca are built of "golden" sandstone, and the town is renowned for its Plateresque architecture.

Site Office: Tourist Information Centre
Plaza Mayor 10
37002 Salamanca, Salamanca
Spain
(923) 218342

Although it is famous as one of the great university cities in the Western world, Salamanca and its university are both well past their days of glory. Still, the city is one of the most beautiful in all of Spain if not Europe and has aged quite gracefully. The sandstone used in the construction of many of its buildings gives the city an impressive appearance, and this stop along the Route of Silver is often referred to as the "golden filigree." Indeed, one traveler from early in this century referred to the color as that of "a Gloire de Dijon just before it drops its first petal." In addition to the beautiful color of its buildings, Salamanca is also known for the Plateresque style of its architecture. This highly detailed sculptural style is well suited for the sandstone of the buildings. The soft stone does not weather well after years of exposure to the elements, but its color becomes increasingly richer. The city was declared a World Heritage Site by UNESCO in 1988, and, indeed, its long history and historical structures are testament to the rise and fall not only of Salamanca and its university but of the power and influence of Spain itself.

Originally settled by a tribe called the Vettones, Salamantica, as it was known in ancient times, was captured by Hannibal in 217 B.C. Legend holds that the invading army disarmed the Salamantine men but did not think to search the women. They were therefore able to re-arm their men, who fought off the captors and escaped into the hills. Thus, Salamantine women have traditionally been known as shrewd and cunning. It is said that Hannibal was so impressed with the ingenuity of the escaped Salamantines that he allowed them to return to the city unharmed. Later Salamanca became an important Roman city and was a station on the Vía de Plata, which stretched from Mérida to Astorga. In 715 it was captured by the Moors, and it was not retaken until the eleventh century, after some 300 years of intermittent warfare.

The modern history of the city begins in 1085, when Alfonso VI gave Salamanca to his daughter and son-in-law. In the early thirteenth century, sometime around 1220, Al-

fonso IX founded a university there. Later, Alfonso X (the Learned) donated his substantial collection of more than 100,000 books and manuscripts to the library of the university. He also founded the university's law school.

The University of Salamanca, like the city itself, flourished for several centuries. It was considered one of the world's four great universities, the others being those of Paris, Oxford, and Bologna. In the fifteenth and sixteenth centuries it could claim 10,000 students and 25 colleges. It had numerous faculty members of high repute, including one woman, Beatriz de Galindo. The university was not averse to progressive ideas—in the sixteenth century it taught the Copernican theory of the universe, which other universities had shunned as heretical. Columbus's plans and discoveries were debated at the University of Salamanca; the universities in Mexico and Peru were modeled on it; the Council of Trent resulted from its philosophies; and the concept of international law was first considered within its walls. It was particularly well known for its distinguished faculty of law.

The glory days were numbered, however. The independent streak that has always characterized the people of Salamanca culminated in an insurrection in 1521, which was crushed by Holy Roman Emperor Charles V (king of Spain as Charles I). Shortly thereafter, ultra-clericalism took hold in Salamanca and both the city and the university began a long decline. Finally, the city was overrun by the French in 1811, during the Peninsular War. Much of Salamanca was damaged or destroyed in the fighting between the French and the British, and during the conflict a munitions storage area was ignited by unknowing smokers, inflicting particularly heavy damage to one section of the city.

During the 1930s, Salamanca served briefly as the headquarters of the nationalists in the Civil War. Throughout the period of the Franco regime, Salamanca was essentially neglected and continued its slow deterioration. Beginning with Spain's economic boom in the 1980s, much of Salamanca was restored and rejuvenated, and it is now a much-visited tourist attraction. Its university and colleges continue to function, but with nothing comparable to the prestige they held in the fifteenth and early sixteenth centuries. Where it once ranked among the greatest universities in all of Europe, Salamanca is no longer even considered to be one of the most important universities in Spain. The university does run a well-known language program, and during the summer months the city is filled with students, especially from the United States.

The last great professor at the university was Miguel de Unamuno, who spoke out forcefully against both sides in the Civil War and died in Salamanca in 1937 while under house arrest. A statue of Unamuno now stands in one of Salamanca's many squares, a constant reminder of the country's troubled past. Other notable scholars who have

Plateresque facade at the University of Salamanca
Photo courtesy of Instituto Nacional de Promocion del Turismo

taught at the university include the Biblical commentator Fray Luis de León. Legend holds that one day, in the middle of a lecture, he was arrested and imprisoned by the inquisition. When he was finally released five years later and returned to the classroom, it is recorded that he began speaking precisely where he had left off, with the words "As we were saying yesterday. . . ."

With its long academic history, many traditions developed at the university, most of which have now faded into memory. The walls of the city are covered with names of successful candidates for the doctoral degree who, upon completion of their exams, had their names and honors inscribed on the public walls of the city in a mixture of bull's blood and olive oil. It was then considered polite to arrange a bullfight for fellow students; until the mid-nineteenth century, these were staged in the grand Plaza Mayor.

The university itself is a fine example of Plateresque architecture, and in front of the particularly ornate main facade stands a statue honoring Fray Luis de León, who seems to look on approvingly. The arms of Ferdinand and Isabella are visible on the main entrance with the inscription "The Kings to the University and the University to the Kings." One of the popular activities of visitors to the university is trying to find the small frog that sits atop a skull in the ornate carvings on the facade. The entire sculptural presentation is an allegory on virtue and vice.

Within the university one can visit classrooms including the classroom of Luis de León, which has been preserved as it was when he taught there, with the original benches and pulpit still in place. One can also visit the university chapel. Unfortunately, the library, with its important collection of documents concerning the exploration of the New World, is not open to the public.

Beyond the university, the city has many famous sites, most notably the Plaza Mayor. Most cities in Spain were originally laid out around a central square; Salamanca's is among the most picturesque and quaint in the entire country. It is a wonderful place to sit and dine at an outdoor cafe or stroll through the bookstores and shops. The plaza is Spanish baroque in style, with three-story buildings on all sides. It measures 243 by 269 feet and was originally laid out in 1729 to the designs of Alberto de Churriguera. It was completed in 1755 with the construction of the baroque town hall on the north side of the plaza.

The Cathedral in Salamanca is unusual in that it is actually two cathedrals side by side. In most areas of Europe a new cathedral was built in place of a cathedral that had been destroyed by fire, or elements of the old cathedral were incorporated into the new structure. In Salamanca, however, the new cathedral was built next to the existing cathedral, which was never torn down. The newer of the structures was begun in 1513 but was not completed until 1733. It can be classified as late Gothic, although many styles were incorporated during the long building process. The church has a beautiful facade and one of the most highly decorated domes in all of Spain. The tympanum above the main portal is richly carved with scenes of the adoration of the shepherds and the magi. Unfortunately the stained glass is in poor repair, with some areas quite severely damaged. Inside, the cathedral contains a distinctive and very ornate altarpiece with a Romanesque crucifix said to have been carried by El Cid on his campaign.

The older cathedral dates from the twelfth century and is a much simpler structure, designed in the Spanish Roman-esque style. The interior of the church is quite interesting and beautifully adorned. Of special interest is the retable with its fifty-three panels depicting scenes from the life of Christ and of the Virgin. Perhaps the most unique work of art in the cathedral is a thirteenth-century statue of the Virgen de la Vega, patron saint of Salamanca. This sculpture is made of copper and covered in Limoges enamel. The cloisters are filled with the Gothic tombs of past bishops as well as some fine pieces of sculpture. The exterior of the church is graced by the Torre del Gallo, a Romanesque-Byzantine dome that once supported a weathercock, hence its name.

One of the most unusual structures in the town is the House of Shells, which was built as a private residence in the fifteenth century by a doctor. It has been restored and is now open to the public. The intricate facade of the building is made up of hundreds of simulated scallop shells.

One of the most visited sites in Salamanca is the Convent of Las Duenas. Inside the convent one must climb the stairs in order to view the carved capitals of the architectural columns. These are covered with demons, saints, and animals of every type. Among the figures are many from Dante's *Divine Comedy*, including a portrait of Dante himself.

For one of the best views of the Salamanca, one must travel a short distance outside the city along the banks of the River Tormes to the massive Roman bridge that spans the river. More than 1,300 feet long with twenty-six arches, nearly half of the structure is original. It is believed to have been built as early as the second century A.D.

Many important figures in Spanish history spent time in Salamanca, among them St. John of the Cross, Columbus, Cervantes, and Ignatius Loyola. All of them were attracted to this city for its intellectual life and, it appears, were soon captured by its magical atmosphere.

Further Reading: Unfortunately, there are few books dedicated solely to Salamanca either written in or translated into English. In Spanish there are many books, but the best information in English comes from two sources. The first is an ongoing, multivolume series entitled *The History of Spain*, edited by John Lynch (Oxford and Cambridge, Massachusetts: Blackwell, 1989–). In addition to the standard travel guides, there are some dated, but rather charming, accounts of travel in Spain such as *The Cities of Spain* by Edward Hutton (London: Methuen, and New York: Macmillan, 1906). In this book, Hutton dedicates a chapter to his experience in Salamanca, which he calls "the rose of the desert." A recent article entitled "A University Town Cast in Gold" by Malcolm Bradbury in *New York Times Magazine*, May 16, 1993, is a brief but very approachable introduction to this wonderful city.

—Michael D. Phillips

Salamis (Famagusta, Cyprus)

Location: On Famagusta Bay, on the east coast island of Cyprus, directly north of the city of Famagusta.

Description: The most important Cypriot city of ancient times; scene of several battles during the Persian Wars; visited by St. Paul and St. Barnabas; site of Jewish uprising in A.D. 116; damaged by earthquakes, then rebuilt and renamed Constantia by Eastern Roman Emperor Constantius II; abandoned after destruction by Arabs in 648; remains of numerous Roman and Byzantine structures remain visible today; since the Turkish invasion of Cyprus in 1974 it has been occupied by United Nations troops and closed to visitors.

Contact: Tourist Information Office
Leoforos Arch. Makariou III, 17
Agia Napa, Famagusta
Cyprus
(3) 721796

According to the poet Homer, Salamis was founded after the Trojan War in 1193 B.C. by the archer Teucer, who came from the Greek island of Salamis in the Saronic Gulf of the Aegean Sea. In addition to giving the city its name, Teucer is said to have brought Greek religion to the island; the Cypriots worshiped Aphrodite as their primary goddess. The word Cypriot is derived from a root meaning "wanton or licentious in nature," thereby providing a connection to the goddess of love.

However, Homer's history has been known to be more fanciful than faithful, and it is generally agreed that references to Teucer, in reality, describe the occupation of the island by the Sea Peoples, who settled many of the islands in the eastern Mediterranean in the millenium before the birth of Christ.

Whatever its origins, Salamis soon began to prosper. It was an important Mediterranean seaport, exporting the ores of neighboring mines and the distinctive local pottery. It also was a leader in naval and merchant shipbuilding. Salamis's many advantages, however, also made it a point of contention among many powers throughout history. Whoever controlled the eastern Mediterranean also controlled Salamis, whether by design or by default.

Salamis and the rest of Cyprus therefore witnessed numerous waves of immigration and, often, invasion. There was significant Phoenician migration to Cyprus about 800; the Phoenicians were fleeing the expansion-minded Assyrians. Cyprus may have come under some degree of Assyrian control as well. Egypt, whose power rose about the same time Assyria's collapsed, became interested in Cyprus in the late seventh and early sixth centuries B.C. When the Egyptian king Ahmose II invaded Cyprus in the early sixth century, he met with the leading Cypriot prince, Euelthon, the ruler of Salamis. Euelthon became a viceroy to the Egyptian ruler and agreed to pay tribute. The tribute most likely was not large, as Egypt wished to maintain friendly trading relations with Cyprus. Egyptian control lapsed with the death of Ahmose in 526 B.C.

A new power was on the horizon, however: Persia. Cyrus II, the Great, had already asserted himself in the area before he died in 529, and his son, Cambyses II, conquered the island in 525. The Persians interfered little with Cyprus: their land-based empire needed the island's sea access. However, the Cypriots were taxed heavily, and little of their tax money came back to them in the form of public works. The people of Cyprus therefore joined the Ionian Greek cities—which were also under Persian rule—in revolting against Persia in 499.

The king of Salamis, Gorgus, was urged by his brother Onesilus to join the revolt. Gorgus refused, and was exiled from the city. Onesilus took control and rallied other Cypriot cities to his cause. He was killed battling the Persians at Salamis, his courage gaining him a permanent place in the national pantheon of heroes.

Fighting between Greece and Persia, with Cyprus as a pawn, dragged on for many years; there was a major Greek naval victory at Salamis in 449. The Greeks eventually won the war, but Cyprus remained under Persian domination, and the Phoenicians of Salamis, who had been allied with the Persians in the war, asserted themselves as rulers of the city. Cyprus remained culturally Greek, however, and continued to trade extensively with Greece.

In 411 B.C. a Greek Cypriot prince named Evagoras, who claimed descent from the founder of Salamis, ousted the Phoenicians and proclaimed himself ruler of Salamis. He further Hellenized Cyprus, using the Greek alphabet on its coins, and he brought many Greek artists and scholars to his court. He was made an honorary citizen of Athens for assisting that city against Sparta in the Peloponnesian War. He also tried to unite Cyprus against Persia, but was unable to gain the support of all the cities on the island. He was defeated by Persia on land at Citium and at sea off Salamis. He died in 374. His successors were vassals to Persia, although they engaged in occasional periods of revolt.

Peace and a brief period of independence came to Cyprus with the rise of Alexander the Great, who defeated the Persians in 333 B.C. at the Battle of Issus. Because the Cypriot kings had allied themselves with Alexander and had helped defeat Persia at the siege of Tyre, Alexander granted the island autonomy. His death ten years later spelled the end of independence, however, and the possession of Cyprus was contested for several years until it was won in 294 B.C. by

Remains of Roman gymnasium at Salamis
Photo courtesy of The Chicago Public Library

Ptolemy I of Egypt, one of Alexander's generals. He founded the Ptolemaic dynasty, which controlled Cyprus for 250 years.

The Egyptians brought a more centralized government to the island than had previous conquering powers. Salamis, which had been capital of the island, relinquished that role to Paphos, but Salamis was the only city allowed to retain its own mint. The city also managed to make cultural contributions during Ptolemaic rule; among its distinguished natives was the historian Demetrius of Salamis. It remained a center of commerce for a time as well, but its harbor was gradually silting up, and its role as a port would soon be outstripped by the new city of Famagusta. Also at this time, Jewish communities began to grow in Salamis and elsewhere in Cyprus.

The Ptolemaic dynasty, suffering from internal conflicts, eventually declined in power, and the Roman Empire was able to gain control of Cyprus in 58 B.C. The Romans governed the island competently, constructed many public works there, and allowed the Cypriots to maintain their religious and cultural practices. Salamis was essentially destroyed by an earthquake in 15 B.C., but was rebuilt by Augustus as a proper outpost of the empire. Some remains of

the Roman marketplace remain visible at Salamis today, as do portions of a Roman gymnasium.

The apostle Paul and his companion Barnabas landed at Salamis about A.D. 45 to spread Christianity throughout Cyprus. This was a homecoming for Barnabas, a native of Salamis. They converted the Roman proconsul, Sergius Paulus, and Cyprus became the first Christian province of the Roman Empire. The island was not entirely Christianized, however; Barnabas is said to have been stoned to death by the Jews of Salamis several years later. He was buried with a copy of the gospel of St. Matthew, and his tomb was discovered more than 400 years later by the archbishop of Constantia, as Salamis came to be known.

The destruction of Jerusalem by the Romans in A.D. 70 brought many Jewish refugees to Cyprus. During this period the Jews found themselves alienated both from the pagan Romans and from the Christians. In A.D. 116 Jewish discontent culminated in open revolt in Cyprus, Egypt, and Cyrene. The revolt was particularly violent in Salamis, where many non-Jews were killed and many buildings destroyed. The revolt was quashed by emperor-to-be Hadrian with the help of the Libyan consul Lucius Quinctus. Following the revolt all Jews were expelled from the island.

In the fourth century, Cyprus suffered from drought and earthquakes. Salamis, in particular, was devastated by earthquakes in 332 and 345. Constantius II rebuilt the city and renamed it Constantia, after himself. After the division of the Roman Empire toward the end of the century, Cyprus became a remote, neglected province of the eastern realm.

This was a period of peace, however, and the cities of Cyprus began to regain some prosperity. They also were intensely religious. Constantia became the site of an archepiscopal see, and a great basilica, the remains of which have been excavated, was built there in the fourth century. It also became, once again, the capital of Cyprus.

Constantia's new prosperity lasted for roughly two centuries. Then the city, which had seen so many invaders, saw yet another: the Arabs. About 648 an invasion fleet of 1,700 Syrian vessels sailed to conquer Cyprus, the lone holdout in an Arab empire that encircled the eastern Mediterranean from Egypt through Palestine and north to Syria. Salamis-Constantia was looted and destroyed, with the plunder being divided between the Syrian emir and his Egyptian allies. Most of the population was killed. Arab attacks on Cyprus continued for another three centuries, and despite the island's eventual recapture by Byzantium, Salamis-Constantia was never rebuilt. Today only ruins from the Roman and Byzantine eras provide evidence of the ancient power and glory of the city. The area is not open to foreign visitors, because of the Turkish invasion of Cyprus in 1974 and the subsequent occupation by United Nations troops. At some point, however, it may again be possible to view the reminders of Salamis's past.

Further Reading: *Cyprus* by H. D. Purcell (New York: Praeger, 1968; London: Benn, 1969) is a comprehensive overview of the entire history of Cyprus. *Cyprus in History* by Doros Alastos (London: Zeno, 1955) is similar in scope. *Cyprus: A Place of Arms* by Robert Stephens (New York: Praeger, and London: Pall Mall, 1966) tells the story of Cypriot history from the perspective of conquest and struggle.

—Dellzell Chenoweth

Salerno (Salerno, Italy)

Location: On southern Italy's western coast, in the Campania region, some thirty miles southeast of Naples.

Description: Salerno is the principal city and capital of the Salerno province. Once a Lombard principality, Salerno was conquered by the Norman Robert Guiscard and later fell under Hohenstaufen rule. An Allied invasion was launched from Salerno Beach in World War II. The city is home to Europe's oldest medical school and the Norman Cathedral of St. Matthew, built in the eleventh century by Robert Guiscard.

Site Office: E.P.T. (Ente Provinciale per il Turismo)
Via Velia, 15
CAP 84100 Salerno, Salerno
Italy
(89) 224322 or 224539

The Italian coastal city of Salerno, known to the ancient Romans as Salernum, takes its name from the word for sea salt (*sal*) and the name of the nearby river, Irnus (now Irno), located to the east of the city. In 197 B.C., the Romans founded the colony of Salernum on the site of an earlier town, Irnthi, which is thought to have been established by the Etruscans. Excavations indicate that a center of Etruscan activity existed near the Gulf of Salerno, but very little is known of the city's Etruscan origins.

Salerno became part of the Lombard Duchy of Benevento in A.D. 646 and served as the capital of an independent Lombard principality from 839 to 1076, when the city fell under the control of the Norman conqueror Robert Guiscard. Ironically, the Lombards of Salerno had actually invited their oppressors into southern Italy. The Norman descent into this part of the world was chronicled in the eleventh century by a monk named Amatus (sometimes called Aimé) who lived at Monte Cassino. According to his account, in 999 a group of forty young Normans stopped in Salerno on the way back from Palestine and were graciously received by the city's Lombard prince, Gaimar IV, who ruled Salerno from 999 to 1027. During their stay, Saracen pirates attacked the city. The Normans immediately took up arms against them. Their courage inspired the local populace to join them in repelling the marauders. Impressed by their valor, Gaimar IV offered the Normans rich rewards if they would remain in his court. However, the men were eager to return to their homeland after a long absence and refused the offer. They promised, however, to propose the idea to their equally courageous friends in Normandy. They left Salerno accompanied by the prince's envoys, laden with gifts that were calculated to lure the Normans southward: almonds, lemons, pickled nuts, gilded iron instruments, and fine clothing.

Word of the Lombard invitation spread quickly throughout the towns and manors of Normandy. Unfortunately, a disreputable element—adventurers, gamblers, professional fighters, and their accompanying riff-raff—headed south, responding to the promise of easy money. In the summer of 1017, a party of such fortune-seekers crossed the Garigliano River and joined the forces of the Lombard Melus, who led them into battle against the Byzantine forces that threatened southern Italy. The alliance prompted Byzantium to send great reinforcements, and the joint Lombard and Norman forces experienced a dismal defeat. Not only did Melus fail to free his people, but Lombard independence was forever crushed.

Southern Italy was at this time a land torn asunder by the continuous conflict among numerous rebellious states and cities and by the interests of three different religions and four different races. A man with fighting skills could always find employment. As mercenaries, the Normans served a variety of masters in the years following Melus's defeat. Many young Normans came to serve Gaimar in Salerno, while others turned to his rival and brother-in-law, Prince Pandulf of Capua.

In spite of pressure from Pope Leo IX to stem the tide of liberators turned oppressors, Lombard ruler Gaimar V of Salerno refused to turn on his Norman allies. Gaimar's long and mutually beneficial alliance with the Normans had been strengthened by the marriage of his sister to Drogo, a brother of Robert Guiscard. On June 2, 1052, Gaimar was murdered by his four Lombard brothers-in-law, all sons of the count of Teano, the eldest of whom proclaimed himself Gaimar's successor. Gaimar's brother, Duke Guy of Sorrento, escaped capture by the insurgents and immediately sought help from the Norman forces, which were already mobilized between Melfi and Benevento. Only four days after Gaimar's death, the Norman army arrived at Salerno's walls. The four brothers did not stand a chance against the Norman forces. The usurpers barricaded themselves in the citadel, taking Gaimar's young son, Gisulf, with them. However, their own families had been taken hostage, and Guy was able to obtain his nephew's release. Guy recognized Gisulf as the rightful heir and successor of Gaimar, and although they would have preferred to see Guy assume the throne, the Normans joined him in declaring themselves for Gisulf. Gisulf confirmed all the existing territorial possessions of his Norman rescuers. Demonstrating a moral sensibility unusual for their era and position, Gisulf and Guy promised to spare the lives of Gaimar's murderers. However, as the prisoners left the citadel, the Normans took it upon themselves to slay the four ringleaders, along with thirty-six others—one person for every wound on Gaimar's body.

Salerno's refractory Norman allies turned into conquerors with the advent of Robert de Hauteville, who won the

Salerno's cathedral
Photo courtesy of E.P.T., Salerno

surname of Guiscard—the Cunning. Robert was born in Normandy in 1015 into a family of knights, one of the many sons of Tancred de Hauteville who left their mark on history. It is said that Robert rode alone from Normandy and arrived penniless in southern Italy in 1046. He began his career in southern Italy as the leader of a band of marauding Norman horsemen. In the mid-1050s, the Guiscard led a highly successful campaign against the southern territories of Gisulf of Salerno, during which Cosenza and some neighboring towns fell to the Normans. Robert's rise to power gained great momentum upon the death of his brother, Count Humphrey, in the spring of 1057. Humphrey had entrusted Robert with the guardianship of his young son, Abelard, but the Guiscard characteristically seized for himself all his nephew's lands. Robert was elected duke of Apulia by the Norman counts of Melfi, and in August 1057 the Normans formally acclaimed him as his brother's successor, adding all of Humphrey's personal estates to the Guiscard's already extensive possessions. In just eleven years, Robert became the greatest landowner and most powerful figure in southern Italy.

Gisulf of Salerno, against the advice of his uncle Guy of Sorrento, had been determined to oppose the Normans almost from the day of his accession. His policy proved short-sighted and counterproductive, since he did not have the resources to contain the Normans. Gisulf would have done well to follow his father's policy of goodwill and cooperation for the sake of maintaining Salerno's independence. Instead, Gisulf invited hostility: Robert and William de Hauteville relentlessly pared away Salerno's outlying territory until little was left but the city itself.

Robert Guiscard soon found it necessary, however, to find a way to reconcile the Lombards to their Norman overlords. Marriage alliances traditionally resolved such problems, but Robert was already married to Alberada of Buonalbergo, and his son, Bohemund, was little more than a baby and not marriageable, even by the day's standards. However, Robert discovered a pretext to annul his marriage to Alberada on the grounds of consanguinity. Finding himself legally a bachelor, Robert proposed to marry Gisulf's sister, Sichelgaita of Salerno. Although Gisulf had never hidden his hatred for the Normans, who had taken from him almost everything he possessed, he was in desperate need of an ally strong enough to hold William de Hauteville and Richard of Capua in check. Gisulf gave his consent to the marriage on the condition that Robert first bring William to heel. Robert agreed and was successful in curtailing his brother's activities. The marriage then took place in 1058 or, according to certain historians, 1059. Sichelgaita was reputed to be a perfect wife for the Guiscard: she scarcely ever left her husband's side, least of all in battle, her favorite activity. Anna Comnena, daughter of Alexius I Comnenus, a Byzantine emperor who faced Robert's invading armies, reported that "when dressed in full armor the woman was a fearsome sight."

Through his marriage to Sichelgaita, the Guiscard finally acquired prestige and regard in Lombard eyes. Those who had previously obeyed him simply out of compulsion now did so with respect for ancient customs, remembering that the Lombard people had long been subjects of Sichelgaita's ancestors. In 1060, Robert set his sights on Sicily, and he took the island from the Arabs in a matter of years. Once Sicily was securely in Norman hands, however, Robert returned to southern Italy to consolidate and expand his possessions there.

Robert's position in Italy was strengthened after Amalfi took the opportunity to place itself voluntarily under his protection in 1073. Gisulf had never forgiven the Amalfitans for their role in the murder of his father twenty-one years earlier, and although never strong enough to occupy the city, he took every opportunity to make life difficult for its people and to brutally torture hapless Amalfitan merchants who fell into his clutches. Robert loathed his brother-in-law and would have attacked Salerno long before, had it not been for Sichelgaita. The loyalty of Amalfi, however, facilitated this task when the opportunity to move against Gisulf arose.

Robert's acquisition of Amalfi greatly disturbed Pope Gregory VII, who excommunicated Robert and began to raise an army against him that included Pisans and Salernitans. Humbly, and in hope of reaching an agreement, Robert and his forces marched to meet the pope in Benevento. Fortune smiled on the Guiscard once again: dissension broke out in the pope's ranks. The Pisan contingent found itself face to face with Gisulf, against whom the Pisans bore a grudge for the pirate-like conduct of his ships. Soon the entire papal army was split, and the pope faced a terrible humiliation.

Robert and his Norman rival, Richard of Capua, joined forces, and the Normans turned their attention to Salerno, where Gisulf grew more insufferable by the day. His arrogance and irrational behavior had alienated his allies, and his few remaining friends, including the pope and Sichelgaita, warned him to moderate his actions, but to no avail. In the summer of 1076, the Norman siege of Salerno began. Gisulf was in a hopeless position. All of the other Lombard states of southern Italy had already been absorbed by the Normans, and he had no allies except for the pope, who had no army. The Salernitan army was greatly outnumbered. The besieging force was made up not only of the most seasoned Norman troops, but of important contingents of Calabrian and Apulian Greeks, as well as Saracens from Sicily.

Having long foreseen the possibility of a siege, Gisulf had ordered every inhabitant of Salerno to keep a two-year stock of provisions, which should have kept the populace from starving when the Normans appeared at their gates. As soon as the besiegers had taken up their positions, however, Gisulf seized one-third of every household's supplies for his personal use and later requisitioned what little remained. His people began to starve. Amatus wrote that at first the Salernitans ate their horses, dogs, and cats, but soon even these were gone. Emaciated corpses lined the streets, but Gisulf pretended not to notice. Few complained, for Gisulf was known to blind and maim those who raised their voices.

Salerno held out for six months, and then on December 13, 1076, through treason from within, the city's gates were opened to the Normans. The starving garrison capitulated without a struggle. Gisulf and one of his brothers retreated with a few others to the citadel, inaccurately known as the Castello Normanno (it was built by the Lombards). Its ruins may still be seen today on the heights northwest of the city. In May 1077, Gisulf was forced to surrender. Robert demanded all the territorial possessions of the prince of Salerno and those of his brothers, Gaimar and Landulf, along with the city's most treasured relic, a tooth of St. Matthew. Robert had long coveted this holy relic and knew that Gisulf kept it with him in the citadel. Gisulf handed over a fake, the tooth of a recently deceased Salernitan Jew. Robert had the tooth examined by a priest who was familiar with the relic, and the deception was discovered. The message was sent to Gisulf that if he did not present the genuine tooth the very next day, all his own teeth would be torn out. Gisulf then promptly handed over the relic and was allowed to leave Salerno.

Robert had now captured the greatest and most populous city south of Rome, the seat of Europe's earliest school of medicine. Making Salerno his capital, he restored the faltering city to its former glory and began a new era of magnificence. To mark the city's rebirth, Robert built the Cathedral of St. Matthew, whose crypt houses the sacred relic, on the site of an earlier church from the era of Constantine. The building was drawn up in the Norman Romanesque style, but Robert employed Sicilian workmen, both Greek and Arab, to carry out much of the work. As a consequence, the cathedral is strongly reminiscent of the eclectic architecture of Norman Sicily. It is particularly rich in mosaics of Byzantine inspiration. Parts of the cathedral were rebuilt in subsequent centuries, but the belltower is the original Romanesque construction, topped with an eighteenth-century addition. Today the building also houses a museum, the Museo del Duomo, which, among other treasures, holds the Salerno ivories. These are a set of exquisitely carved medieval ivory door panels illustrating Old and New Testament scenes. Preserved almost intact, the Salerno ivories are a rare surviving example of a popular medieval art form. Pope Gregory VII, who spent the last year of his life exiled in Salerno, consecrated the cathedral in 1085. He died that same year and was buried in its south apse, where his tomb may still be visited. Today, the austere Norman cathedral is almost lost among the bustling city's modern buildings. However, a close look reveals an inscription on the facade: "Built by Duke Robert, greatest of conquerors, with his own money."

Salerno's medical school also continued to flourish under the Normans. Brian Lawn, in his book *The Salernitan Questions,* describes the school as "the birthplace and nursery of what has been called the scientific renaissance." The Salernitan masters began an intense study of Aristotle's scientific and medical texts in the late twelfth century, and so they stimulated the development of a scientific approach to natural phenomena that had not been seen in the west since the ancient Greeks. Anticipating Arab scholarship that came to dominate Western scientific thought during the following centuries, the Salernitans formulated a set of "research" ques-

tions in physics—known as the Salernitan questions—that guided scientific discovery for centuries afterward.

Robert's hard-won conquests slipped out of de Hauteville hands when, in the next century, William the Good of Sicily committed a blunder of statesmanship, accepting an alliance with the powerful King Frederick Barbarossa of Germany through a marriage—that of his aunt Constance (Robert's niece) and King Frederick's son, Henry VI of Hohenstaufen. When William died in 1189 without heirs, the door was opened wide to German rule. Henry was unable to claim his Sicilian throne immediately. In his absence, an illegitimate de Hauteville, William's cousin Tancred, was crowned king. Upon his return, Henry drove out the usurpers, and in 1194, Salerno surrendered to him without a struggle and was sacked by Henry's forces.

Henry VI was succeeded by his son, King Frederick II, who was also known as the *Stupor Mundi,* or "wonder of the world." Frederick did much to improve Salerno's famous school of medicine, a secular institution staffed by married lay men and women whose preeminence was recorded by Thomas Aquinas. Frederick decreed that no one practice medicine in his realm without a university degree from Salerno's medical school, which he licensed in 1224 as the only school of medicine in the Kingdom of Naples.

Upon Frederick's death in 1250, Pope Urban IV sold the Sicilian crown to Charles of Anjou, brother of the French king, Louis IX. A Salernitan, Giovanni da Procida, played a major role in the Sicilian Vespers, an uprising against the French feudal lords in 1282. Subsequently, da Procida returned to Salerno and was responsible for the expansion of its harbor. He also established an annual fair to encourage trade. Nevertheless, in the fourteenth century Salerno was outstripped by Naples, and the city subsided into a comfortable provincialism. The medical school and university continued to be prominent, however.

Salerno once again made its appearance on the stage of European history during World War II. Salerno Beach served as the site of Allied landings during the campaign to drive the German army out of Italy. On September 9, 1943, one day after Italy had officially surrendered, General Mark Clark's United States Fifth Army landed. Officially known as Avalanche, this invasion was soon called the Avalanche of Errors. German panzer attacks nearly pushed the invaders back into the sea, and military expert J. F. C. Fuller has described this campaign as one of the war's most strategically useless. Thousands of American and British lives were lost, and the city suffered much destruction before the arrival from Calabria of the advance guard of the British Eighth Army, which forced the Germans to withdraw. American paratroopers reinforced the beachhead, and the Allies were finally able to move on to Naples by October 1.

Today, Salerno thrives as a university town with an excellent beach and an active seaport. The city's main industrial products include textiles, machinery, construction materials, ceramics, and wrought ironwork. Ruins of the Castle of Arechi, prince of Benevento, and the remains of a palace and gate from the Lombard period are still in existence. The city's principal monument still is the Cathedral of St. Matthew.

Further Reading: *The Other Conquest* (New York and Evanston, Illinois: Harper and Row, 1967; as *Normans in the South,* Harrow, Essex: Longman, 1973) and *The Kingdom in the Sun* (New York and Evanston: Harper and Row, 1970), both by John Julius Norwich, are engagingly written guides to the Norman history of southern Italy. *Lords of Italy: Portraits of the Middle Ages* by Orville Prescott (New York and Evanston, Illinois: Harper and Row, 1972) is also an intriguing work and provides many references to the history of Salerno. *Salerno* by Hugh Pond (Boston and Toronto: Little, Brown, and London: Kimber, 1961) is a full-scale study of the World War II Allied invasion. Brian Lawn's *The Salernitan Questions: An Introduction to the History of Medieval and Renaissance Problem Literature* (Oxford: Clarendon Press, and New York: Oxford University Press, 1963) gives an account of Salerno's role in the development of medieval scientific thought.

—Caterina Mercone Maxwell and Marijke Rijsberman

Samothrace (Evros, Greece)

Location: Northernmost island in the Aegean Sea archipelago, situated between the Thracian coast of Greece and Gallipoli, Turkey.

Description: Volcanic island sixty-nine square miles in size, composed of eroded granite and surrounded by rough seas; known for its exceptional scenic beauty and the ruins of an ancient sanctuary. In Homer's *Iliad*, Poseidon watched the Trojan War from Mount Fengari, which, at 5,250 feet, is the island's highest point, and the highest in the Aegean. The famed "Winged Victory" sculpture, now in the Louvre in Paris, was found here.

Site Office: Samothrace Archaeological Museum
Samothrace, Evros
Greece
(551) 41474

The island of Samothrace was originally occupied by Thracians during the Neolithic and Bronze Ages. Of Indo-European stock and language, the Thracians were noted warriors often viewed as primitive because of their preference for open-air villages. However, they were also talented poets and musicians, and evidence of a Thracian silver monetary system has been found. On Samothrace, the Thracians built a sanctuary for the worship of the Great Gods, a pre-Greek Chthonic cult. The Great Gods included the Kabeiroi, a little-known group of powerful and wrathful deities; the Thracian mother goddess, Axieros; and others associated with her such as Axiersos and Axiersa of the underworld, and Kadmilos, a vegetation god. These dieties were worshiped as protectors of nature and of seafarers.

Excavated coins from the seventh century B.C. imprinted with the face of the Greek goddess Athena pinpoint the arrival of Greek settlers. The Greeks built the town of Palaipolis and expanded the sanctuary. The Great Gods were merged with the Greek gods Pluto, Persephone, and Hermes. The mother goddess was first associated with Cybele, a deity of Asia Minor, and ultimately with Demeter. In spite of the Greek influence, the Thracian language was retained in the cult rituals and was preserved in inscriptions on stone and pottery.

Cults and mysteries were not unusual during the Hellenistic age. However, several characteristics set the sanctuary at Samothrace apart from other ritualistic cults. Initiation was open to all men and women, regardless of nationality or social standing. Initiates could pass from the first level to the second level without the lengthy waiting period demanded by some cults; many at Samothrace went through both stages on the same day. The Samothrace cult was also unusual for its rite of confession and absolution. Initiation could take place at any time of year.

The sanctuary was built outside of town on a ridge of the hill of St. George between two streams. Only two of the buildings were used for the initiation rituals: the Anaktoron (Hall of the Lords or House of the Masters) and the Hieron. The rectangular Anaktoron, with an interior of ninety by forty feet, was the site of the first degree of the rituals. Along the inner walls, limestone bases supported wooden benches from which members of the cult watched the proceedings. At the southeast corner was the sacristy. The northern end, known as the Holy of Holies, was hidden by a partition.

Within the Hieron, the second stage of initiation was enacted. One hundred and thirty-one feet long with a double colonnaded entrance, the Hieron is noted for its similarity to a Christian basilica. Its interior was also lined with benches. At the south end are the remains of an apse that likely was backed by a curtain. Marble sculptures of the goddess of victory were added to the roof corners sometime between 150 and 125 B.C. One has survived, albeit without its head, and is on view in the local museum.

Other buildings included the Hall of Votive Gifts and a necropolis from the sixth century B.C., and the Altar Court from 200 B.C. with a colonnade that once formed the stage wall of the theatre. The Hall of Votive Gifts, on the west side of the Hieron, was graced with a Doric facade and a marble mosaic pavement. Steps once led to a large altar above which a theatre was built into the face of the hill in approximately 300 B.C. The theatre, with its characteristic outer Doric columns and inner Ionic columns, may have been the site of drama competitions during the summer festivals.

Many of the original buildings were lavishly refurbished with marble during the Ptolemy dynasty, a line of Macedonian rulers of Egypt who reigned from 323 to 30 B.C. The first original marble building erected was the Temenos, an open-air walled shrine constructed between 350 and 340 B.C. It was noted for an Ionic entranceway that at one time was engraved with a frieze of female dancers. Fragments of the frieze are on display in the museum. It is thought that famous statues of Aphrodite and Pothos sculpted by Skopas may once have stood within the Temenos.

Ptolemy II Philadelphus and his second wife, Arsinoe II, were particularly immersed in the cult activities at Samothrace. The latter commissioned the Arsinoeion, built between 289 and 281 B.C. on the site of an earlier cult building southwest of the Anaktoron. (Excavations have unearthed the remains of a seventh- or sixth-century altar and terrace as well as walls of a fourth-century double precinct.) The Arsinoeion, the largest surviving example of a roofed rotunda from Greek antiquity, has an outer diameter exceeding 393 feet and foundations approximately 8 feet wide. On the exterior, a wall of

Temple of the Great Gods at Samothrace
Photo courtesy of Greek National Tourist Organization

Thasian marble supported a Doric entablature. Inside, the marble wall held a gallery of Corinthian half columns with an Ionic cornice. In the center of the tiled domed roof, a hole allowed smoke from sacrificial fires to escape. Historians presume that the Arsinoeion was used to stage dramatic performances and processionals to welcome foreign visitors to the annual summer festival.

The Ptolemaion, a ceremonial entranceway to the sacred area of the sanctuary built for Ptolemy II Philadelphus from 285 to 280 B.C., now lies in ruins. Constructed of Thasian marble, the gateway's six Ionic columns faced the city to the east; Corinthian columns faced the sanctuary. In the foundation, a barrel-vaulted tunnel diverted water from seasonal heavy rains.

Opposite the Ptolemaion are the foundations of a stepped circular arena dating from the fifth century B.C. that may have been used for sacrifices. Near the steps are bases on which at least twenty statues once stood. To the north are the remains of a small building that Alexander IV (son of Alexander the Great, born after the latter's death) and Philip III (Alexander the Great's successor) dedicated to the Great Gods.

The rituals of the cult have remained a mystery throughout the ages; historians can only surmise the activities from the buildings and the inscriptions carved on them. It is likely that initiates entered the Anaktoron from the sacristy whereupon they were clothed and crowned. A priest probably led them into a hall for a water purification rite. Libations were often poured into a pit as an offering to the Great Gods. The initiate was then placed on a circular wooden platform as members of the cult performed ritualistic dances. He or she was then led to the Holy of Holies, the inner sanctum where further rituals were likely performed and a priest displayed symbols of the mysteries said to signify heaven and earth.

Similar rituals were performed at the Hieron for the second stage of initiation, with the addition of the confession rite. The existence of marble stepping-stones near the entrance of the building has led some to speculate that the initiate stood there to make a confession and receive absolution before entering the building. Numerous stones with holes in their centers have been found in the ruins; archaeologists believe they were used as sconces for torches. Thus they think it likely that the rituals were carried out in the evening by torchlight.

Many important names, both historical and mythical, have been linked with Samothrace. Among the initiates were the Greek historian Herodotus; King Lysander of Sparta; the Roman scholar Varro; and Piso, Julius Caesar's father-in-law. Philip II of Macedon and Olympias of Epirus (the parents of Alexander the Great) were said to have met and fallen in love while visiting the sanctuary in the fourth century B.C. Dardanus, the founder of Troy, was born in Samothrace. In the *Iliad,* Homer tells of Poseidon, the Greek god of the sea, watching the Trojan War while perched on Samothrace's Mount Fengari. Aristarchus, who edited Homer's works, was born on the island in 155 B.C. Hadrian is said to have visited Samothrace, as did St. Paul on his trip through Macedon in about 55 B.C.

While the presence of the sanctuary allowed Samothrace to remain relatively independent of the fighting among the Greek city-states, the island nonetheless suffered frequent conquests. It was used as a naval base by Athenians, Thracians, Ptolemies, Seleucids, and Macedonians. In the fifth century B.C., the Aegean islands were under Persian rule. Athens led the battle to free them, and afterward, when Athens demanded a tax to provide protection against further Persian aggression, the Delian League was formed for that purpose. Samothrace was one of the members of the League that paid a hefty tax to Athens in exchange for protection. The treasury was originally kept on the island of Delos but was eventually moved to Athens during that city's drive for more power.

Macedon gained control of Samothrace in 340 B.C. and held it for the next two centuries. In 170 B.C., after the Battle of Pydna, in which the Romans annihilated the Macedonian forces, Perseus, the last king of Macedon, attempted to hide on the island but was captured by the Romans. The Romans then made Samothrace a free state and disbanded all leagues. In 84 B.C., pirates from the Barbary Coast laid siege to the island, but it was quickly rebuilt by the Romans.

By the fourth century A.D., as Christianity spread through the region, the cult that had existed at Samothrace for centuries was obsolete, even though some followers continued to gather there. The island continued to be inhabited, as evidenced by the remains of an early Christian basilica and several small Byzantine churches.

In 1414, Palamede Gattilusio, a nobleman from a prominent Genoan family, arrived on Samothrace and built three forts from antique stone found on the island. The ruins of two towers from these medieval structures still stand. Excavations have revealed the remains of a harbor water break and a Byzantine church beneath the towers. Turkey occupied the island in 1457, controlling it until 1913, when the Ottoman Empire ceded Samothrace to Greece.

In 1863, Champoiseau, then serving as the French consul at Adrianople, visited Samothrace and came upon a valuable archaeological find. After climbing the hill behind the sanctuary, Champoiseau entered a large fountain-house. Perched on the prow of a marble ship, which lay diagonally across a pool, was a majestic sculpture of Nike, the winged goddess of victory. Known variously as the "Winged Victory" or "Nike of Samothrace," the sculpture had been commissioned to commemorate the naval victory of Antigonus II of Macedon over Ptolemy II in 258 B.C. Unfortunately, archaeologists have not been able to determine exactly who created the Winged Victory, but it has been ascribed to both Lysippus and Leochares. Champoiseau brought the sculpture back to France, where it can be seen in the Louvre. Although its removal to France saved it from certain theft or destruction, and although its mounting in the Louvre is by all accounts impressive, nothing could have matched the sculpture's intended position sailing out majestically over the fountain's reflecting pools.

Three years later a French archaeological mission arrived on Samothrace to map the site. Austrian expeditions followed in 1873 and 1875, at which time the Ptolemaion was excavated. Marble recovered at the site was supposed to be divided between Austria and Turkey. However, the Turkish shipment never arrived in İstanbul. Further excavation was undertaken by Sweden from 1923 to 1925. An American team from the Institute of Fine Arts at New York University, led by Karl Lehmann and Phyllis Williams Lehmann, began a systematic excavation in 1938. The Lehmanns are credited with providing much of what is known about the sanctuary.

Today Samothrace is primarily a tourist attraction with visitors coming to view the ruins of the sanctuary and to enjoy the natural hot springs.

Further Reading: Most general tourist guides give an adequate description of the island of Samothrace. *Blue Guide: Greece* by Robin Barber (London: Black, and New York: Norton, 1990) includes a fairly detailed discussion of the island's history. *Baedeker's Greece* (Englewood Cliffs, New Jersey: Prentice-Hall, and Basingstoke, Hampshire: Automobile Association, 1992) does an adequate job of describing the origin of the Great Gods and the work of the excavation teams. For a more in-depth treatment of Samothrace and other historic Greek sites, read *The Living Past of Greece* by A. R. and Mary Burn (Boston: Little, Brown, and London: Herbert, 1980).

—Mary F. McNulty

San Marino

Location: Completely surrounded by Italy. Located in the north central portion of the country, San Marino is sandwiched between the Romagna and Marche regions of Italy, twelve miles west of the Adriatic Sea.

Description: With an area of approximately twenty-four square miles, the Most Serene Republic of San Marino, as it is officially known, is the third-smallest independent country in Europe after Vatican City and Monaco, and the oldest republic in the world. Situated on the slopes of Mount Titano, San Marino's geography has blessed the tiny republic with ease of defense and easily verifiable borders; a combination of qualities that has played a pivotal role in the republic's constant struggle to maintain its independence. San Marino was founded in the early fourth century A.D. and, despite the turmoil that has raged around it over the centuries, has maintained its independence almost continuously ever since.

Site Office: San Marino State Tourist Office
Palazzo del Turismo Contrada Omagnano, 20
San Marino
(378) 882400

Far from an administrative curiosity, the diminutive Republic of San Marino is the world's oldest republic, with a proud history reaching all the way back to the fourth century A.D. When most of the rest of post-Roman Europe was populated by tribes whose territory was defined by borders that could change at any time, San Marino had already established a sovereign territory, a crude but effective democratic government, and the beginnings of a body of national tradition that has survived to this day.

San Marino's political and geographic position is unique within Western Europe. Even Monaco, while much smaller, possesses its own territorial waters and international port on the Mediterranean Sea, while the small semi-states of Andorra and Gibraltar are both ruled by outsiders: the former by the president of France and the bishop of Urgell, and the latter by the British government. Except for relatively brief periods of occupation by foreign powers, San Marino has always stood alone, and stood well.

Modern San Marino possesses amenities comparable to those of Western Europe in general, and its standing as an independent state is recognized by the entire international community. San Marino has its own diplomatic corps, calendar (For San Marino, time begins at 301 A.D., the year of the republic's founding) passports, and militia, and belongs to several international organizations in its own right. Unemployment in San Marino is negligible and social services are excellent.

Despite the trappings of modern society, San Marino's history is on prominent display throughout the republic and the national heritage is very much alive and well. The most obvious reasons for this are that San Marino was spared most of the destruction wrought in other parts of Europe over the centuries, leaving a great deal of the oldest parts of the republic intact, and that the people of San Marino cling to their heritage with greater tenacity than most, as small countries like theirs were at one time not so rare.

Archaeological excavation in San Marino has revealed that the region was inhabited by humans since at least the ninth century B.C. This evidence has been gleaned primarily from discoveries of small hand tools, weapons, coins and day-to-day human refuse, such as picked-clean animal bones. There is, however, no evidence that the area had permanent settlements until at least the latter half of the third century A.D.

It was at this time that San Marino the republic began to take shape. According to local legend, San Marino was officially founded on September 3, 301, by Marino (or Marinus), a stonecarver from the island of Rab in Croatia. Marino had left Rab in the company of several colleagues to assist in the restoration of the Italian seaside town of Rimini, located on the Adriatic shore just east of Mount Titano, which had been severely damaged by a series of barbarian invasions. It is not clear whether Marino came to Rimini of his own volition or whether he was forced to contribute his skills to the effort because of his status as a Christian. However, nearby Mount Titano served as Marino's quarry for the building stone he needed to do his work in Rimini, and he often traveled there.

During his residence in Rimini, Marino devoted his time not only to stonecutting, but to Christian penance and the conversion of pagans. According to the legend, Marino left Rimini following an encounter with a woman believed to be possessed. After attempting in vain to seduce him, the woman declared that he was really her husband and publicly accused him of spreading the illegal Christian religion.

Taking the hint, Marino left Rimini and settled on the slopes of Mount Titano, there to dedicate his efforts to meditation and a simple life free from the pressures of evangelism. Unfortunately for Marino, wandering shepherds discovered his hiding place within a year, and the mad woman of Rimini followed shortly thereafter. Upon seeing her, Marino shut himself up in a cave on a cliff. After six days of prayer and fasting, he emerged and exorcised the spirits that had possessed the woman. She then returned to Rimini to spread the word of the holy man on the mountain.

Marino's successful exorcism greatly enhanced his status with the people of Rimini, many of whom became his followers and helped him to build a temple to St. Peter on the

One of San Marino's three towers
Photo courtesy of Republic of San Marino

highest peak of Mount Titano. His fame eventually reached the ears of Felicissima, the Riminian woman who owned Mount Titano. While vacationing on the mountain, Felicissima's son Verissimo made the mistake of aiming an arrow at Marino as he stood next to the church, and was immediately paralyzed. Felicissima begged Marino to take pity on her misguided son. Marino healed the boy and, in deference to the awesome power of the new religion, all fifty-three members of Felicissima's family converted to Christianity. Felicissima even gave her mountain to Marino, to which he was appointed deacon by the bishop of Rimini soon after. It is generally accepted that Marino died on September 3, 366. He was later canonized, becoming St. Marino, or, in Italian, San Marino.

From the fourth to the tenth centuries life in San Marino was peaceful and uneventful, creating the environment of freedom and tolerance for which the tiny republic would later become known. A letter written by a monk named Eugippio in 511 describes the territory, already known as San Marino, as being inhabited by monks, shepherds, hunters and woodsmen, artisans and peasants, denoting the rapid growth of the community.

It was not until 885 that San Marino's political and legal independence was first enshrined in a legal document. Called the "Placito Feretrano" and signed on February 20, 885, this document was written to settle a dispute between Deltone, the bishop of Rimini, and Stefano, abbot of the monastery of San Marino. Deltone accused Stefano of stealing church lands on the west side of Mount Titano. In the ensuing negotiation Stefano managed to prove that the church had never owned the disputed lands and that, furthermore, it had not gained anything from them in "40, 50 or 100 years." The small territories of Casole, Erviano, Fabbrica, Fiorentino, Flagellaria, Grinziano, Laritiniano, Petroniano, Pignaria, Ravellino, and Silvole were ceded to San Marino.

Oddly enough for a document of such pivotal historical importance, the original has long since been lost. A copy of the the Placito Feretrano made in the eleventh century but with the original dates intact was found in the republic's national archives in 1749.

San Marino continued to grow at a steady pace through the Middle Ages. At the time of the Placito Feretrano the republic was barely two and one-half square miles in size. By the end of the thirteenth century it would reach nearly sixteen square miles, more than half its present size.

The dawning of the republic's prescient system of democracy is generally dated to December 12, 1244, the date on which another document found in the archives explains that the citizenry had established a two-man city government headed, at the time, by Filippo da Sterpeto and Oddone Scarito, both of whom were lawyers. The Arengo, or council of the heads of all the families of San Marino, also came together sometime in the thirteenth century. This probably happened informally at first, as an exact date is not available.

Existing documents from 1253, 1295, 1302, 1352–1353, 1491 and 1505 trace the evolution of San Marinese law, much of which is still valid today. Originally drafted by the members of the Arengo and the municipal administration, these statutes are notable for their simplicity, fairness, and faith in the idea that with individual freedom there must also come individual responsibility. There were no legal distinctions between class. Service in the diminutive military was mandatory for San Marinese men between the ages of fourteen and sixty.

It was illegal to walk the streets at night without a light and guards and night watchmen were held responsible for the losses of their employers if they could not apprehend the culprits. Hospitality was the rule when dealing with commoners, but the wealthy were widely distrusted. In fact, wealthy and powerful people were not allowed to buy land or build houses within a one-mile radius of the capital city. Even then, the San Marinese knew that keeping a low profile would be the only way to maintain their independence in an often hostile part of the world.

Treason and murder carried the sentence of capital punishment, public swearing resulted in 150 days in prison, and anyone victimized by a crime was free to retaliate in kind 10 days after the criminal's conviction if the offender had not yet made just compensation. It was against the law to interrupt speakers during government proceedings, littering carried with it a fine, and families living within the central city were forbidden to own more than three goats. Country dwellers were free to own up to six.

By the sixteenth century most of San Marino's political structures were also in place. Based upon nothing so much as common sense, the system of government created by the San Marinese is an apt reflection of their historically ingrained disdain for vesting too much power in any one person or group.

San Marino is divided into nine parishes: Acquaviva, Borgo Maggiore, Chiesanuova, Domagnano, Faetano, Fiorentino, Montegiardino, Piere, and Serraville. Prior to February 22, 1977, San Marino had no diocese of its own, despite being a sovereign state, and jurisdiction over the nine parishes was shared by the neighboring sees of Montefeltro and Rimini. Currently, Montefeltro, Rimini and San Marino are ecclesiastically united under a common bishop.

For purposes of political representation the republic is divided into ten manors, each one headed by a magistrate entitled "lord of the manor," who serves as the government's representative to the people. The ten manors—some of which also serve as parishes—are Acquaviva, Chiesanuova, Domagnano, Faetano, Fiorentino, Fratta, Guaita, Montegiardino, San Giovanni Sotto la Penne, and Serravalle.

Th republic is also divided into nine townships, again with some overlap amongst manors and parishes. These are the capital of Citta di San Marino, Borgo Maggiore, Castello di Fiorentino, Castello di Chiesanuova, Castello di Faetano, Castello di Montegiardino, Castello di Montecerreto (or Acquaviva), Castello di Serravalle, and Castello di Domagnano (or Montelupo). All of the townships are formed around their own principal towns, hence the name "township," but not all possess castles or other fortifications, despite the name "castello."

The national government of San Marino is composed of six separate bodies: the Captains Regent, the Great and General Council, the State Congress, the Council of Twelve, the State Mayors and the Arengo. Additionally, there are nine town councils that administer the nine constituent townships of the republic.

At the top of the administrative ladder are the Captains Regent. Elected to a single six-month term each April 1 and October 1 by the Great and General Council, the two Captains Regent hold supreme executive authority over the other branches of government and serve as commanders-in-chief of the small military. The twice-yearly ceremonial investiture of the Captains Regent has been carried on without interruption since the policy was first formalized in 1244.

The office of Captain Regent is neither all-powerful nor strictly honorary. The Captains-Regent have significant authority, but, as in most democracies, their power is checked by the other branches of government. They can issue emergency or ordinary decrees, although these can be overturned by the Great and General Council. At the end of their six-month term, all Captains Regent must be judged by the Sindacato; a citizen tribunal wherein the people have the opportunity to judge "deeds done and deeds not done."

The Great and General Council is composed of sixty consiglieri and holds the greatest legislative power. Members serve five-year terms, during which time they appoint the Captains Regent, the three secretaries of state, and the nine portfolioed government ministers. The council also appoints diplomatic agents and consuls, confers citizenships and honorary titles, and approves all treaties and agreements. The Great and General Council also creates and implements laws and bylaws, approves the annual budget, and handles government administration. The council must meet in public unless the subject matter to be discussed is deemed to be a matter of national security.

Councillors are elected by proportional representation of all citizens eighteen years of age and over. Candidates must be literate, over twenty-five years of age and not hold any consular, diplomatic, or ecclesiastical post for a foreign state. They must also not belong to the police force or the gendarmes.

The State Congress is composed of the two Captains Regent, the secretaries of state, and the nine portfolioed government ministers. Collectively, the congress was a consultative body until 1945, when a new law gave the congress executive and administrative powers.

The Council of Twelve handles civil, penal, and administrative functions. It is elected by the Great and General Council for the full duration of a five-year term and acts as the judicial body for certain civil matters regarding property and other civil law.

The two State Mayors are elected by the Great and General Council from within its own ranks for a five-year term. The mayors represent the interest of the state in legal, administrative, and financial matters.

The Arengo is composed of the heads of all the families of San Marino and, as in most democratic systems, holds ultimate authority by virtue of holding all the votes. Formerly a sovereign body with specific powers, the Arengo has retained only the right to petition the Captains Regent and the Great and General Council. On the first Sunday after April 1 and October 1, members of the Arengo are welcome to petition the government in a public forum. The Captains Regent and Great and General Council are legally bound to discuss all petitions heard within the ensuing six-month period.

The Arengo was abolished in 1600 when the rapid growth of the republic made governing through such a large body of people impractical. However, the Arengo was reinstated by the people through a referendum held in 1906, at which time its powers were altered.

Due to the republic's small size and almost familial familiarity, legal authority is placed in the hands of foreign judges, usually brought in from Italy. Putting the power in the hands of foreigners, the San Marinese believe, is the only way to maintain impartiality in such a small, close-knit society. All judges are nominated by the Great and General Council and one, called the legal commissioner, is obliged to live in San Marino for the duration of his service. Magistrates and jurisconsults are also nominated from outside the republic's citizenry. The commander of the police force is also foreign, although the officers themselves are San Marinese.

San Marinese law is composed of a patchwork of ancient and modern statutes, common law and tradition. For example, the criminal code was not formalized until 1975 and civil marriages did not exist in San Marino until 1953. Again, San Marinese custom and law are firmly rooted in consensus and common sense.

There is much more to San Marino's legacy of freedom than nationalistic jargon. In 1296, when the tiny republic still had to prove its independence from time to time, the chief magistrate of neighboring Montefeltro attempted to impose taxes on the citizens of San Marino. Naturally, the San Marinese argued that they were not under the jurisdiction of the magistrate of Montefeltro. They eventually appealed to Pope Boniface VIII, who assigned two lawyers of the papal court to deal with the matter. They, in turn, delegated responsibility for the case to a Master Ranieri, who just happened to be abbot of St. Anastasio Monastery in Montefeltro.

Unimpressed by the documentary history of the republic, Ranieri insisted upon direct consultation with its residents. With the liberty of their homeland at stake, the San Marinese arrived for the trial en masse, all prepared to answer Ranieri's question "Quid est libertas?" or, "What is freedom?" Their answers, many of which were recorded, were consistently unwavering. Ugolino di Guiduccio, a resident of Casole, pointed out that "The people of San Marino have never paid tributes to anyone, nor have they any intention of starting now." Martino of Montecucco said that "Man is born free and for what he owns he owes thanks only to Our Lord Jesus Christ." Local politician Gianni di Biagio, when asked how he knew the San Marinese were a free people, answered simply that "They have never been subject to anyone." The exact outcome of the 1296 debate is not known, but it can be guessed with reasonable accuracy, given San Marino's continued status as a sovereign state.

From 1460 to 1463 the last armed conflict to take place on San Marinese territory was fought between the house of Malatesta and the rulers of Urbino. San Marino sided with Pope Pius II, and Frederick, the count of Urbino. The alliance won the dispute, which settled, among other things, the territorial integrity of San Marino. The size of the republic's territory has remained unchanged since Pius II's Papal Bull of 1463.

In 1503 San Marino's independence was briefly interrupted by the intrusion of Cesare Borgia, son of Pope Alexander VI, duke of Romagna and Machiavelli's prototypical "Prince." Borgia conquered the republic and held it for six months. In 1543 Fabiano of Monte San Savino, eventually to become Pope Julius III, allied himself with the lord of Rimini and attempted to invade San Marino. They left Rimini with more than 500 soldiers on the night of June 4, 1543, with the intent of attacking under cover of darkness; the force lost its way in a dense fog, effectively ending the campaign.

San Marino experienced a long period of internal decay during much of the seventeenth and eighteenth centuries. The law of asylum, however noble, meant that the republic played host to a relatively large population of miscreants and ne'er-do-wells. Government became corrupt and ineffectual, and the system of public education, in which the San Marinese took rightful pride, broke down completely.

In 1691 a priest named Ascanio Belluzzi used his own money to fund a school for the youth of the republic. This should not have been exceptional, as the Statutes of 1600 declared that "Children must be instructed in habits and customs as well as learning."

Perhaps the biggest mistake made during this period was to succumb to the temptation to draw a line between the classes. In very un-medieval fashion, San Marino had never observed a legal distinction between nobility and peasantry. These distinctions certainly existed, but San Marino was unique in the age for not enshrining them in law. However, on October 20, 1652, a decision was made to reduce the number of members in the Great and General Council from sixty (it had been as high as eighty-five) to forty-five, but with thirty members to be chosen from among the upper classes of the city of San Marino and the remaining fifteen from among the peasantry. Not only was this mandated elitism a step backward for the progressive republic, it encouraged power plays by the aristocracy.

On October 17, 1739, Cardinal Giulio Alberoni, legate of Romagna, crossed into San Marino in the company of an abbot and two lawyers at the behest of some members of San Marino's corrupt aristocracy. As a representative of the pope, Alberoni took it upon himself to stage a show trial of two alleged criminals, Marino Belzoppi and Piero Lolli, to show the people that their little republic was worthless without the support of Rome.

The peasantry did not take kindly to this imperial notion, prompting Alberoni to seek assistance in the form of Ravenna's chief of police, who arrived in San Marino accompanied by forty-seven constables and an executioner. Over the next few days, they were joined by 500 troops, forcing the San Marinese to hand over the keys to the city. Alberoni immediately set about punishing those who would not swear loyalty to the authority of the church, going so far as to arrange a public ceremony during which leaders of all the noble houses were to swear their allegiance to the new papal government. Only three manor houses, Serravalle, Faetano, and Montegiardino did so, while the rest of the nobility stood in defiance.

The San Marinese pleaded their case directly with the pope, eventually winning the dispute and their freedom from Alberoni. The country was liberated on February 5, 1740, which to this day is celebrated in San Marino as St. Agatha's feast day, and is a national holiday.

When Napoléon invaded Italy in 1796 the officials of San Marino were understandably nervous. Would their official neutrality be recognized? On February 7, 1797, a Napoleonic representative named Gaspare Monge was welcomed into San Marino by the Captains Regent with a surprising answer. Finishing a speech on behalf of the Emperor, Monge said:

"I come from General Bonaparte, on behalf of the French Republic, to give to the Republic of San Marino the assurance of peace and an everlasting friendship. Citizens, changes may be brought about in the political constitution of your neighboring countries. If one of your frontiers should be contested, or if some part of the neighboring states should be absolutely necessary to you, I have been charged by the general to beg you to inform him. He will be only too pleased to put the French Republic in a position to demonstrate its sincere friendship for you. As far as I am concerned, citizens, I am happy to be responsible for a mission which must be agreeable to both republics and which gives me the possibility to express the admiration that you inspire in all the friends of freedom."

The San Marinese were delighted with the offer of friendship from the most powerful man in the world, but rejected his offer of additional territory, directing Monge to tell his leader:

...the Republic of San Marino, quite satisfied with its own small size, doesn't dare to accept his generous offer, neither does it entertain the ambitious wish to enlarge its territories in fear that its independence will be compromised in the future.

San Marino's independence was confirmed at the Congress of Vienna in 1815, at the end of the Napoleonic Wars.

In 1849 San Marino played host to Italian patriot Giuseppe Garibaldi who, in the company of several thousand men, successfully maneuvered his way between French and Austrian armies from Rome all the way to the neutral territory of the republic following a French siege of that city that Garibaldi could not hope to fend off. Pursued at the end by an Austrian army, Garibaldi and his followers had no choice but to enter San Marino without explicit permission from the Captains Regent.

Nonetheless, Garibaldi was welcomed into the republic. His residence in San Marino lasted less than a day, as he, in the company of several advisors and 162 troops, slipped out of the republic under cover of darkness, fooling the watchful Austrian army. Despite his short stay in the republic, Garibaldi was later granted honorary San Marinese citizenship.

Noble or not, neither the Austrian government nor the papal government appreciated San Marino's show of support for the underdog Garibaldi. On June 24, 1851, San Marino was invaded by more than 4,000 Austrian troops and a contingent of papal soldiers, sparking a civil war that lasted until 1854. The fighting only came to an end when the Grand Duke of Tuscany attempted to formally take over the republic, resulting in a stern rebuke by Napoléon III, who, upholding the guarantee of his ancestor, came to San Marino's aid.

With the establishment of a united Italy in 1860, San Marino's independent status was finally assured. A formal pact of friendship was signed between the two countries in 1862, and the new Italian state opened an official consulate in San Marino in 1874.

U.S. President Abraham Lincoln was also honored with San Marinese citizenship as a token of respect in honor of his efforts to maintain the Union. Touched by the gesture, Lincoln replied to the Captains Regent on May 7, 1861:

Although your dominion is small, your state is nevertheless one of the most honored in all history. It has by its experience demonstrated the truth so full of encouragement to the friends of humanity that a government founded on Republican principles is capable of being so administered as to be secure and enduring.

San Marino managed to stay out of the limelight for several decades following the unification of Italy, sending only humanitarian aid to its larger neighbor during World War I. However, the tiny republic could not avoid the spread of fascism. In the election of 1923 the Fascist Party of San Marino won an outright majority in the Great and General Council, and two fascists, Giuliano Gozi and Filippo Mularoni, were elected as Captains Regent in April of the same year. In practical terms, the advent of fascism in San Marino meant only that the police force was given somewhat broader powers. Even though very little recognizable as fascism was actually instituted in San Marino, the fall of Italian fascism on July 28, 1943, is still celebrated as a national holiday.

San Marino maintained its independence and neutrality throughout World War II, but it could not stay out of the war completely. On June 26, 1944, British planes mistakenly dropped 243 bombs on San Marino, despite the fact that the republic's borders had been clearly marked with enormous white crosses easily visible from the air. During the course of the war, the tiny republic of less than 20,000 offered shelter to more than 100,000 refugees.

In 1945, only two years after the fall of fascism, San Marino experimented with another political system: communism. The citizens were inspired by a Romanian promoter named Maximo Maxim, who told them that they could solve their postwar financial problems by opening a state-run casino, operated, of course, by Maximo Maxim.

In 1949 the casino was opened, and was an instant success. This deeply troubled the Italian government, because Italian money was leaving the country and filling the coffers of the San Marinese government. When San Marino legalized divorce, which was not legal in Italy at the time, the Italian government could stand no more and took economic action. Severe restrictions were placed on trade with San Marino, roadblocks and purposely tedious border inspections and blockades were instituted, and San Marino was effectively isolated. Maxim left the country in 1951 to try his luck anew in Switzerland. Like fascism before it, which had hardly changed the day-to-day lives of the San Marinese at all, communism quietly faded away in 1957, never having implemented anything close to actual Marxism.

Modern San Marino is a prosperous nation-state. It still has no border formalities with Italy, although determined travelers can have their passports stamped upon entry, if they so desire.

Seeing the need to secure an economic base for the future of the republic, the San Marinese government has targeted the convention and tourism business. In 1988 the San Marino Arte Auction was created, bringing a prestigious annual event to the republic. New convention centers have opened to take advantage of the new influx of business, and the Ministry for Tourism and Sport has redoubled its efforts to turn San Marino's "marginal tourists" (day-trippers from Rimini) into long-stay vacationers. Visitors to San Marino can view such historic monuments as the three towers that guard the small nation from the heights of Mount Titano; the Basilica del Santo, a nineteenth-century church that replaced a much older one, and houses some of the relics of the saint for whom the country is named; and the Palazzo del Governo, the seat of San Marino's government. The small republic also has several museums, including one dedicated to Garibaldi and another to philatelic and numismatic displays (San Marino's postage stamps are much prized by collectors).

Further Reading: Information about San Marino is not widely available in English, although a few sources do exist. The best single source in English is *The Republic of San Marino: Historical and Artistic Guide* (San Marino: Azienda Tipografica Editoriale, 1981), printed and distributed by the Ministry for Tourism and Sport and available by writing to the San Marino State Tourist Office. Written by Nevio Matteini, an obviously devoted believer in the values held dear by San Marino, the book is an excellent chronicle of the republic's history, covering its development in great detail from prehistoric to modern times. Additional information can be found in works dealing with the people and institutions that had an impact on San Marino, such as Garibaldi, Napoléon, and the Roman Catholic Church.

—John A. Flink

Şanlıurfa (Urfa, Turkey)

Location: In southern Turkey, near the Syrian border; between the Tigris and Euphrates Rivers, at the foot of the Taurus Mountains. Sector where "Oriental" Turkey meets Western Turkey.

Description: Capital of the province of Urfa in the Republic of Turkey. City built around a hilltop fortress; includes a cave said to be the birthplace of Abraham. Called Edessa by the Macedonians, Urfa by the Turks, and renamed Şanlıurfa ("famous" or "glorious Urfa") in honor of the town's opposition to French occupying forces in 1919.

Site Office: Tourist Information Office
Asfalt Yol Cadesi, No. 4/D
Şanlıurfa, Urfa
Turkey
(414) 2152467 or 2157610

For thousands of years Urfa has occupied an important crossroads. North-south trade routes from Anatolia into Egypt and southern Mesopotamia, and east-west routes from the Mediterranean to Persia, met in the fields at the foot of the mountains in Anatolia. On these rich pastures, the town grew to be an important regional center. Its position on the edge of the Western world, looking east, enhanced its importance for the Babylonians, Persians, Seleucids, Parthians, Sāssānids, Romans, Byzantines, Arabs, and crusaders, all of whom occupied the town at various times.

The first known settlers were the Hurrians, an advanced Indo-European people who came from the Caucasus Mountains and from Persia as part of a general migration. Aided by their early use of chariots, they built a great civilization centered on Urfa. A huge jagged rock dominates the town, the site of successive citadels. The Hurrians probably built a fortress on the summit around 3500 B.C., although there is no tangible evidence for such a structure. At some point in their history, they may have lived in caves, since they were called Hurri, which comes from the Babylonian word for "cave." The name of the town evolved from the name of the people: Hurri became the Aramaic Orhai, and ultimately the Turkish Urfa.

Urfa is closely associated with the Old Testament figure Abraham by both Moslems and Jews, although the local stories differ from those in the Bible and the Jewish Midrash. The Babylonian King Nimrod (Nemrut), who ruled Urfa, had a dream that foretold the birth of a child who would be greater than himself. Nimrod killed all the male infants (as Herod later did) and separated men and women to assure there would be no more babies. About to give birth, Abraham's mother went to a cave under the mountain. Abra-

ham was born in the cave and remained hidden there for ten years, suckled by a deer. Local Moslems and Jews believe Abraham was living in Urfa when God called him to take his family to Canaan. One tradition says he needed a sign before he believed. Following God's instructions, he baked four kinds of birds in a pastry and threw it in the air. The birds flew away and Abraham believed he had been visited by God.

Although Abraham's father, Terach (Terah), made and sold idols, Abraham hated polytheism and destroyed the graven images. His father took him to Nimrod, who sentenced him to be burned alive for his sacrilege. Abraham refused to renounce monotheism in order to save himself. A fire was lit at the top of the cliff, but since it was so hot no one could get near it, Abraham was spared. Nimrod then had Abraham thrown over the cliff by catapult, but a sacred spring broke his fall. Stories differ on whether God created the spring to break Abraham's fall, or whether there was already a spring, sacred to many gods, at the foot of the rock. The cinders from the fire fell into the spring and became the holy carp that are still revered. (Other versions of the story say the fire was below the hill and that the spring came forth to put it out.) Urfa has two other suggested biblical connections. One local tradition says the prophet Job lived in Urfa for a time; the other makes the claim that the Garden of Eden was nearby.

Today the Ibrahim Halilullah Dergâhi (*Dergâh:* "prayer place"), said to be the birth cave of Abraham, is surrounded by mosques, religious schools, and charitable institutions. It is entered from the courtyard of the Mevlid-i-Halil-Camii (Mosque) Şerifı. In back of the mosque enclosure is another cave, said to be the home of Abraham's deer-nurse. Gölbaşı, "at the lakeside," is the name given to the area at the foot of the citadel, with three sacred carp pools, including Abraham's Fish Pond, the Birket İbrahim. It is a cool, green spot in otherwise parched Urfa.

Above Urfa is the *kale,* or citadel, atop a large rock, the possible site of a Hurrian fortress and an attraction for all invaders since. Although each new group captured and further fortified the citadel, other invaders always followed and conquered it. What survives today was most likely the fort built by crusaders in the twelfth century. At the top of the rock are two Corinthian columns inscribed "Nimrod's Throne." Despite the name, they are most likely remnants of an early Christian church.

The Hittites, who entered Anatolia in the same migration that brought the Hurrians, destroyed the town in the fourteenth century B.C. After the Hittites, the Assyrians controlled the area, although there are no remains of their town. The Assyrians ruled until the coming of the Macedonians under Alexander the Great.

The founder of the Seleucid dynasty, Seleucus Nicator (one of Alexander's generals), was given control of

Hamil ur-Raman Camii (mosque) and complex
Photo courtesy of Embassy of Turkey, Washington, D.C.

Babylonia. The Seleucids named the town Edessa, after a town in Macedonia. Edessa subsequently became the capital of the kingdom of Osrhoëne, founded in 132 B.C. by an Iranian, Osrhoës, who had probably been a governor under the Seleucids. He was succeeded in 127 B.C. by an Arab, Abdu bar Maz'ûr. Osrhoëne was a vassal kingdom under the kingdom of Parthia, which controlled Mesopotamia.

Rome and Parthia had once been allies but the Roman general Gnaeus Pompeius Magnus (Pompey) took steps to bring Parthia under Roman rule. He turned the Parthian territory of Osrhoëne over to a local sheik, Ariamnes or Mazares (Maz'ûr) who was officially named Abgar II. Pompey counted on Abgar's assistance when he marched into Mesopotamia.

Contention between Parthia and Rome continued into the reign of Abgar VII of Edessa. Abgar tried to remain neutral, but ultimately submitted to Rome (under Trajan) to save his kingdom. In A.D. 116, however, he joined in a revolt involving the Parthian army and the conquered provinces, which extended all the way to the Persian Gulf. Rome won Mesopotamia, sacking and burning Edessa; historians believe Abgar may have been killed in the fighting. Southern Meso-

potamia became a Roman vassal state under a Parthian, King Parthamaspates, who was appointed by Hadrian. In 123 the local dynasty in Edessa was restored briefly, subject to Parthia, but by the year 163 it was once again under Roman rule. In 164 there was again fighting between the Romans and the Parthians all around Edessa, and King Mannus Philorhomaios VIII, who had been driven from the throne by the Parthians, was restored to his position by the Romans. In 195, however, Osrhoëne declared its independence from Rome. As punishment its status was changed to that of a mere Roman province, but later the same year Rome needed its troops in the west and restored the Abgar dynasty to the throne.

About A.D. 200 the people of Edessa became Christians. (Native tradition attributes the conversion to one of the seventy-two disciples of Jesus, but that suggestion results in a conversion date at odds with known chronology.) Abgar VIII, who reigned in the late second and early third centuries, was the first Christian king; his ancestor, Abgar I, is said to have carried on written communications with Jesus. The Christian church in Edessa used Aramaic, the language of Jesus, in its worship, and its successor, the present-day Syrian

Orthodox Church, uses Syriac, which is derived from Aramaic. Records show the existence of a Christian "temple" in Edessa as early as 201. Edessan missionaries soon expanded the church into Iran.

Osrhoëne remained a Roman client kingdom under Abgar IX, who built his palace at the foot of the citadel in Edessa. In 216 the status of Edessa changed when the Roman Emperor Marcus Aurelius Antoninus (Caracalla) invited the kings of Armenia and Osrhoëne to visit him. They were imprisoned, and the area incorporated into the Province of Mesopotamia. Caracalla set up winter headquarters in Edessa, intending to do battle with Parthia. Although there were no battles, his presence stirred up rebellion and he was assassinated near Edessa in 217.

In 260, Edessa was besieged by the Sāssānid Persian Emperor Shāpūr I. The Roman Emperor Valerian was unable to attack with his ailing and demoralized army. He tried to negotiate instead, was taken prisoner, and died in captivity. Edessa was ruled by Persia until Shāpūr was defeated by the kingdom of Palmyra, a Roman vassal. Edessa remained under Palmyra during the reign of Queen Zenobia, who renounced obedience to Rome. She was eventually captured by the Roman Emperor Aurelian and led away in golden chains.

From its early Christian beginnings, religion remained important in Edessa, which became the center for preaching and catechism in Aramaic/Syriac. In the fourth and fifth centuries, it became a center of both Syriac religion and literature. St. Ephraem Syrus, a Syriac writer noted for his commentaries on religious subjects, founded his theological "School of the Persians" in Edessa. When Nestorius, patriarch of Constantinople from 428 to 431, promulgated his doctrine of the humanity of Jesus, the Edessans adopted this belief and became an important Nestorian center. Nestorius was sentenced to torture and deportation as a heretic by the Council of Ephesus, and his writings were ordered burned. Because of its Nestorian tendencies, the School of the Persians was closed and the theologians deported. Edessa was also the site of the martyrdom of three confessors in 309 and 310, under the Roman Emperor Diocletian.

Between 502 and 506, the town of Edessa was damaged greatly during a series of Persian wars against the Byzantine Empire, to which the town belonged following the division of the Roman Empire. It was later restored by Justinian I, although the wars continued as Justinian tried to defend and extend his Byzantine Empire. In 532 he and the Persian king Khosrow I achieved an "everlasting peace." Justinian paid significant tribute to Khosrow, to induce him to guard the eastern edge of the empire against invading nomads. When the Armenians revolted against Rome, Khosrow started another war, taking advantage of Justinian's having been occupied with the Goths. Khosrow invaded the Mediterranean region, destroying Antioch and laying siege to Edessa in 544. From 607 through 609, Khosrow II took over Armenia, Mesopotamia, Syria, and northern Asia Minor. Diverted by continuing battles between Orthodox Christianity and another heretical sect, the Monophysites, Byzantium

was prevented from defending itself. (Edessa had converted and become a center of the Monophysites.) Khosrow II was finally killed by his subjects, who were exhausted by wars. In the peace that followed in 628, the Byzantine Emperor Heraclius, a native of Edessa, restored Orthodox beliefs. He also attempted to revive his empire, but was prevented from doing so by the coming of the Arabs.

Caught between the Sāssānids and Byzantium, Edessa had suffered greatly in the sixth and early seventh centuries. The coming of the Arabs in the mid-seventh century finally brought some peace. The heavily taxed natives, with Semitic roots, felt a kinship with the Arabs and welcomed them. The town of Edessa declined in prominence after it came under Arab rule and became part of the Diyar Mudar (abode of the Mudar tribe) along the Euphrates River.

The Arabs stayed until the eleventh century, although battles between the Byzantines and Arabs continued. The Arabs called Edessa "Al-Ruhā," a version of the original name that evolved into the Turkish "Urfa." In 942–43, a Byzantine army came to Al-Ruhā and carried away the mandylion, a gold image of Christ. In exchange they promised permanent peace with Edessa and returned 200 Arab prisoners. Promises aside, a Byzantine general occupied Edessa again in 1030. Although the Seljuk Turks, under Alp Arslan (the Valiant Lion), raided Edessa for booty in the 1060s, it remained under the Byzantines until 1083, when it came under the control of the Armenian King Sembat. Later it was part of the Armenian Kingdom of Cilicia founded by Philaretus, a Greek leader of the Armenians. When the first crusaders appeared, the Christian Armenians were their natural allies. The County of Edessa, a minor European feudal state under Count Baldwin of Boulogne, was established and fortifications were built on the citadel. Ruins of the castle and other crusader buildings remain today.

After an initial period of peace under the crusaders, the Turks rallied their forces. Al-Ruhā was taken by the Turkish leader Zangī in 1144. The negligent and unambitious crusader count Jocelyn II and his fortified but poorly defended town were no match for this impassioned defender of Islam. The defeat represented the beginning of the total defeat of the crusader states. The pope called the Second Crusade in response, but the crusaders never again reached Edessa. Although Edessa was restored to Jocelyn II for a few days in 1146, Zangī's son, Sultan Nureddin, conquered Edessa, pillaging the town. The people were enslaved, and Jocelyn II was taken away in chains. In 1182 Edessa came under the rule of Moslem hero Saladin, who fought off Christians in the Third Crusade and founded the Ayyūbid dynasty. Except for a conquest by the Seljuks that lasted for four months in 1234, Edessa remained under Saladin's successors.

The Ulu Camii (Mosque), patterned after the Grand Mosque of Aleppo in Syria, was built during this period (twelfth century) and remains today. Its minaret is the belfry of an earlier Byzantine church. Also surviving is the Halil-ur Rahman Camii, built in 1211, near the sacred pond of Abraham. It contains symbols representing the diversity of religious beliefs in the area, including the sun and moon (the planets were worshiped in the Urfa area until the eleventh century). Surviving artifacts from Neolithic times through the Seljuk period are displayed in Urfa's Arkeoloji Müzesi, opened in 1988.

Edessa was ravaged by the Mongols under Hülegü in 1260 and partly restored at the end of the fourteenth century. After suffering further decline under Timur (Tamerlane), it was then ruled by the Egyptian Mamlūk dynasty until the Mamlūks were defeated by the Ottoman Sultan Selim I in 1516. The name of the town was changed to the Turkish "Urfa," and its ownership was contested among rival Moslem princes. It finally became part of the Ottoman Empire in 1637 under Murad IV. Except for a brief occupation by Muḥammad Alī Pasha's Egyptian army in 1837, Urfa remained in the Ottoman Empire until World War I.

The late-sixteenth-century Ottoman caravansary, "Gümrük Hanı," still forms a large part of the bazaar known as the *Kapalı Çarşı,* which exists today. Goods are both sold and made in the covered bazaar, more accurately illustrating Turkish traditions than any other surviving market.

Surviving from the seventeenth century is the Abdürrahman Camii and Medrese (religious school), near the largest of the sacred pools. It replaced a twelfth-century mosque, using its minaret. Also near the pool is the Rizvaniye Camii, built by an Ottoman governor in 1716.

Much of Urfa's history in the nineteenth century revolved around Turkish treatment of the Armenians. In the mid-1890s there were 60,000 people living in Urfa, 20,000 of them Armenians. In October 1895 hundreds of Armenians were killed during a two-day massacre by the Turkish and Kurdish populace. The Armenian quarter of the town remained under siege for two months, with no food or water allowed in. The Armenians were falsely accused of having a cache of weapons, which the Turks demanded in return for lifting the siege. Rumors also circulated that the sultan had ordered extermination of the Armenians. In December, a crowd of Turkish soldiers and civilians invaded the quarter, massacring thousands of dazed and starving Armenians. Three thousand survivors took refuge in a church, normally considered a place of refuge under Islamic law. Troops entered, however, and burned the church, killing all the occupants. They went on to plunder and burn everything in the Armenian quarter. In 1915, there were two further massacres in Urfa, and about 550 Armenians were killed.

After World War I, the French occupied Urfa for two months before they retreated, leaving it to the Turks. As a result of heroic battles with the French, the city was renamed Şanlıurfa, "Glorious Urfa." In 1923 the Republic of Turkey was formed with Kemal Atatürk as president.

The twentieth-century city of Şanlıurfa, with a population of 300,000, serves as the commercial center for an area that extends from thirty miles north of Şanlıurfa all the way south to the Syrian border. Yet it remains a quiet community that still has flocks of sheep in the streets. There are a significant number of Kurds in Şanlıurfa and, although Turkish is the official language, Kurdish and Arabic also are spoken there. Religion remains important in Şanlıurfa, as it has become a center of fundamentalist Islamic beliefs.

Further Reading: *Eastern Turkey: a Guide and History* by Gwyn Williams (London: Faber, and Levittown, New York: Transatlantic Arts, 1972), devotes part of a chapter to Urfa, giving a brief history and tour of the town. A short section on history, travel, and sights in Urfa is included in *Turkey: The Rough Guide* by Rosie Ayliffe, Marc Dubin, and John Gawthrop (London: Rough Guides, 1994). Richard Stoneman's *A Traveller's History of Turkey* (Moreton-in-Marsh, Gloucestershire: Windrush, and Brooklyn, New York: Interlink, 1993) is a source for general history of Turkey, although it does not cover Urfa specifically.

—Julie A. Miller

Santiago de Compostela (La Coruña, Spain)

Location: In the region of Galicia, in the northwest of Spain, on the Atlantic coast.

Description: Founded in the ninth century, Santiago is a pilgrimage city where the bones of James the Great (St. James, or Santiago in Spanish) were allegedly found on a hillside in the Galician countryside by a hermit attracted to the site by stars. The site became known as Campus Stellae (Field of Stars), which became corrupted to Compostela. Santiago de Compostela is a beautiful city, mostly hewn out of golden granite. Because of the west Galician weather, most of the city is covered in vegetation.

Site Office: Oficina de Turismo
Rua del Villar, 43
15705 Santiago de Compostela, La Coruña
Spain
(81) 58 40 81

Santiago de Compostela is a beautiful granite city situated in an area of Spain once known as "finis terrae" (the end of the earth), as it lay on the western edge of the then-recognized world. Founded in the ninth century, the city's history and development has been rooted in the popular belief that the remains of St. James lie there. As a result, Santiago became a popular place of pilgrimage, and thousands of people have traveled to the site along "el camino de Santiago" (the Pilgrim's Way or the Way of St. James).

James the Great, according to myth, came to Spain from Palestine to preach the gospel. Although he later returned to his native land, on his death his body was taken by two of his disciples and brought back to Spain. At the end of the remarkably short seven-day voyage that ended at the Roman port of Iria Flavia, the disciples, having encountered a belligerent queen, a dragon, and two wild bulls, used the bulls to lead them to the place where James the Great's remains would be buried. The animals led the disciples to the forest Liberum Donum, where the body lay for approximately 800 years.

Despite references to it by the Irish St. Adelhelm in the seventh century and the Spanish monk Beatus de Liebana in the following century, the tomb remained untouched until the ninth century. Between 813 and 842, Pelayo, a hermit, noticed a great star burning in the sky above an area on the hillside near San Fiz de Solvio. He informed Theodemir, bishop of Iria Flavia, who on examining the site discovered a grave. Theodemir found three bodies and claimed that one of them was the remains of James—although whether this was James the Great (the brother of St. John), or James the brother of Jesus is unclear. Soon after this the site became known as the "field of

the star" (Campus Stellae, soon corrupted to Compostela—although some historians argue that Compostela comes from Compositum, which in Low Latin has a funereal connotation, or Compostum, meaning "placed together").

The discovery of the tomb came at a fortuitous time. Spain had been besieged by the Moors since the seventh century, and the warriors of Islam fighting their jihad (holy war) had captured most of the Iberian Peninsula, apart from Asturias in the extreme north. The Asturian king, Alfonso II, built a small chapel on the site of St. James's tomb, which later became the site of the city's famous cathedral, although al-Manṣūr, king of the Moors, destroyed the building in 997 and took the chapel bells to a mosque in Córdoba. The bells were later returned, and the building was re-erected in the last quarter of the eleventh century.

The legend of St. James the Great grew when Alfonso's successor, Ramiro I, claimed that St. James rode at his side at the Battle of Clavijo, and helped him defeat Abd ar-Raḥmān II. Subsequently St. James was taken up as a symbol for the conquest of the Moors, and the Christian war cry in the reconquest of Iberian land was "For St. James and a united Spain!" St. James (Santiago) became the patron saint of the country.

Alfonso III, building on the saint's reputation, erected a basilica in the "field of the star," and enlarged the site by donating sections of land to the saint. Over the centuries various kings, bishops, and dignitaries visited Santiago de Compostela. In addition, the site became popular with ordinary people who sent gifts and requests for help to the shrine. As the shrine's popularity increased, Santiago de Compostela began to increase in size. Bishop Sisnando built a moat and city wall to strengthen the town against Norman invasion in the second half of the tenth century.

In the eleventh century the town expanded southward and toward the northeast. During this period, which coincided with the Crusades, Santiago de Compostela became a popular destination for pilgrimages. As pilgrims (jacobeo/jacobita) flooded into the town, merchants, tradesmen, shopkeepers, and innkeepers followed. Most of the immigrants settled in the new, growing suburbs, on the pilgrimage route. Although the commercial district was semi-autonomous, control of the town lay in the hands of the bishop, a fact that led to a revolt in the early twelfth century. The pilgrimage was also responsible for the growth of small towns along the route through the Pyrenees and along the northern peninsula. The popularity of the pilgrimage route also gave rise to a series of "tourist literature." The Clunaic order was producing guides to recognized routes and itineraries from the eleventh century.

During this period Bishop Diego Gelmírez worked to improve the profile of the city, whose religious eminence was being challenged by Toledo. Worried that the town would lose

The cathedral of Santiago de Compostela

the income generated in part by the pilgrims coming to visit St. James's tomb, Gelmírez campaigned to have Santiago de Compostela turned into an archbishopric. In 1120 Gelmírez became an archbishop and a papal legate for most of western Spain and Portugal. Gelmírez was also responsible for the construction of a number of impressive buildings, including the bishop's palace, several churches, and the lovely Romanesque cathedral. Today Santiago is a mixture of Romanesque and baroque architecture, for the most part instigated by the church.

The city suffered from famine and disease during the Middle Ages, a fact that retarded the growth of a middle class and allowed the church to maintain control. While the population starved, the church still invested large amounts of money in construction, and some of the loveliest churches and houses in the city were built during this period. By 1600, however, the effect of the church's domination of city affairs had led to the demise of commerce and industry, and the emergence of a genteel class of lawyers, churchmen, and the less wealthy aristocracy.

By the end of the seventeenth century, Santiago de Compostela's importance as a place of pilgrimage was declining. Pilgrims and the city's inhabitants despised each other—the French were so disgusted with the treatment of visitors to the city that their king, Louis XIV, passed a law forbidding his subjects to participate in the pilgrimage. During this period the credibility of the story of St. James's visit to the Iberian Peninsula and the authenticity of the remains found in the field of the star were called into question. The remains were later exhumed and examined in the nineteenth century and the authenticity of the remains was announced in a papal bull by Leo XIII. The tomb was also examined in 1946, and Bishop Theodemir's remains were found. However, the discovery of a necropolis from the first century, which some historians claim contains the remains of Galician nobles, has also fueled the controversy surrounding St. James's remains.

Internationally, Spain's economic and financial reputation was in decline by the seventeenth century. A series of regional revolts in Portugal, among other places, in the 1640s and the extreme poverty and famine that most of the people were experiencing exacerbated the sense of hopelessness felt by most Iberians. Thus, the ruling elite and church began to look for another saint who might raise the spirit of the nation; the recently canonized St. Teresa was a popular choice. Supported by King Philip IV and the count-duke of Olivares, who carried around the saint's actual heart set in diamonds, St. Teresa was made co-patron by the pope in 1627. This aroused such disgust among supporters of St. James that the decision was reversed in 1629.

In the succeeding centuries Santiago de Compostela became an intellectual and cultural abyss. The Age of Enlightenment largely passed over the city. The once-respected university became an enclave for the aristocracy and clergy. The dominance of the conservative church adversely affected the economic and intellectual progress of the city and its people. Once a humanist center, the city slowly stagnated in the eighteenth and nineteenth centuries.

The city suffered further adverse effects because of its support of the losing faction in the Carlist Wars, which arose from a disputed right of succession for the Spanish crown. On the death of Ferdinand VII in 1833, Don Carlos, the late king's brother, supported by the conservatives and the church, fought for the crown with Ferdinand VII's daughter Isabella, whose support lay with the army and liberals. The subsequent war lasted six years and ended with Isabella II's succession after she came of age in 1843. Unfortunately, because Santiago de Compostela gave its support to Don Carlos, the new queen made La Coruña, a liberal city, the capital of the region and the center of justice and administration. As a result, Santiago de Compostela's economy and population suffered. The city remained at a population of approximately 20,000 from the eighteenth century until the beginning of the twentieth century.

During the twentieth century Santiago de Compostela again became a thriving city. Its economy improved dramatically with the development of industry and a cattle market. The university again became a respected institution, and now draws students from all over Spain. Tourism also became an important source of income for the city, and pilgrims still come from all over the world to visit the famous cathedral. Although there were originally four major pilgrimage routes as described in the 1140 Codes Calixtinus by Aymery Picaud (the Toulouse route from Arles to Col de Somport; the Le Puy route to Roncesvalles; the Limouisin route, from Vézelay to Roncesvalles; and the Tours route, from Paris to Roncesvalles), this has expanded to include routes through Spain, for example from La Coruña, Barcelona, and Seville, and routes through the Pyrenees. Today the pilgrim can choose from a multitude of ways to get to the city.

The cathedral itself lies at the center of all of the routes into the city, in the Obradoiro Square, which lies in the middle of the city. The square is also the site of the Palacio de Arzobispo Gelmírez (the Palace of Archbishop Gelmírez) and the Colegio de San Jerónimo. The structure of the building is a mixture of Romanesque and baroque architecture. Most of the building dates only from the eleventh and twelfth centuries, as much of the original cathedral was desecrated during a raid by the Moor leader al-Manṣūr in 997. Although he took the cathedral bells as a trophy, he left the tomb of St. James alone.

Bishop Diego Peláez was responsible for beginning reconstruction of the building in 1075. He employed Master Bernardo (the elder) to perform the work. The Chapel of El Salvador was built during this period. Further work occurred in the last decade of the eleventh century, under Master Esteban, assisted by Master Bernardo (the younger). The building was completed in 1128.

The work done under Master Esteban received its stimulus from the Abbey of Cluny. Esteban's influence and the close ties with France due to the popularity of the pilgrimage meant that the architecture was not only Romanesque but was directed by French style too; this type of design became known as Compostel-Romanesque.

The crypt was erected under Gelmírez and later reinforced by Master Mateo. This acts as a base for the Portal de La Gloria, one of the best features of the cathedral, built under the direction of Master Mateo sometime between 1168 and 1188. Originally the west front of the building, it is now situated inside the cathedral. It depicts scenes from the Old Testament and of the Glorification of Christ according to St. John. Mateo's work is clearly influenced by French Gothic style; the expressions of his figures' faces are more realistic and less static than Romanesque sculpture. Almost 200 granite sculptures form the portal. The Portal de La Gloria replaced the Romanesque facade and forms the atrium of the Basilica.

Master Mateo features in the building itself. An image of him can be found kneeling at the foot of a central column bearing a statue of St. James. The marble column on which St. James rests contains five holes in which millions of pilgrims over the years have come to place their fingers. The indentations have occurred as pilgrims praying in front of the apostle have rested with their hand pressed against the Tree of Jesse, which is found on the column.

From the portal the visitor can get a prime view of the seventeenth-century baroque chancel. This forms the site of one of the most important aspects of the pilgrimage. The chancel houses a twelfth-century statue of St. James, which the pilgrim can embrace. The chancel was renovated in the seventeenth century by Domingo de Andrade and Peña de Toro. They used marble and gilded wood, and solomonic columns to enhance the baroque style. The famous Botafumeiro, the huge censer, is attached to the octagonal fifteenth-century dome. As the cathedral originally housed the pilgrims, the censer served as a way of fumigating the building. Today the censer is used during important services and raised by a gigantic pulley situated on the front of the altar. It takes eight priests to swing the censer, which gravitates in an arc of 80 to 100 feet. St. James's remains can be found in the crypt situated below the altar. They had been misplaced after being moved for safety during an English invasion, but were rediscovered in 1879.

Among the many beautiful chapels found in the cathedral is the impressive Chapel of Las Reliquias, which since 1535 has housed the Royal Pantheon. Among the valuable possessions found here is the silver bust of Santiago Alfeo. Bishop Theodomir's tomb can be found in the vestibule here. The Chapel of Las Reliquias together with the equally magnificent Chapel of San Fernando and the plateresque portals form part of the area around the cloisters.

The cathedral has several interesting facades. The Facade de las Platerias is the southern Romanesque facade, which dates from between 1078 and 1103. The fourteenth-century La Berenguela (clock tower) is one of the most beautiful baroque towers still in existence today. The northern Azabacheria facade is named after jet sellers who lived near here. It was renovated by Ventura Rodríguez and is neoclassical in style. The Puerta del Perdón (Holy Door) is only opened when St. James's Day, July 25, falls on a Sunday. It dates back to 1611 and was the work of González Araujo and Fernándo Lechuga.

The cathedral of Santiago de Compostela is one of the most beautiful buildings in Europe, and is certainly one of the main reasons why the city was declared a UNESCO World Heritage site in 1985. But Santiago de Compostela also houses several other magnificent buildings. The Palacio de Arzobispo Gelmírez lies to the north of the cathedral. It is Romanesque and was commissioned by Gelmírez in 1120 following the destruction of the Archbishop's Palace during riots objecting to the domination of the clergy in Santiago. It was renovated in the eighteenth century, when its present facade was also created.

Although pilgrims were originally housed in the cathedral, various hostelries opened in the city. The Hostal de los Reyes, today an extremely expensive hotel, lies in the northern section of the square. Its chapel is built from Coimbra (Portuguese) stone.

In the southern area of the square lies the College of San Jerónimo, which was founded by Archbishop Alonso III de Fonseca, and opposite the cathedral lies the Palacio de Rajoy (Rajoy Palace), which was built in the mid–eighteenth century as a confessor's seminary. It has a neoclassical facade and contains a sculpture of St. James. Today it is the base of the Council of Galicia and the town hall.

The squares that lead off the cathedral are also worth noting and visiting. The largest is the Plaza de Quintana, which houses the oldest monastery in the city—the Monastery of San Pelayo de Antealtares, which dates back to the ninth century and was originally meant to house St. James's tomb. The Plaza de Azabacheria was once the financial center of Spain, and the Plaza de las Platerias was the silversmith's center. The cloisters of the cathedral once housed several silversmith shops.

Further Reading: *Pilgrimage Route to Compostela: In Search of St. James* by Abbe G. Bernes, G. Vernon, and L. Balen (London: Robertson-McCarta, 1990) is a thin but extremely comprehensive guide to the pilgrimage route. *Cities of Spain* by David Gilmour (London: Murray, and Chicago: Dee, 1992) presents detailed narratives on the history, legends, and customs of major Spanish towns. *A History of Spain and Portugal* by Stanley G. Payne (Madison: University of Wisconsin Press, 1973) is a dry but accurate textbook.
—Aruna Vasudevan

Sarajevo (Bosnia-Hercegovina)

Location: In the Miljacka River Valley, at the foot of Mount Trebević approximately 70 miles inland from the Adriatic coast of the former Yugoslavia, 140 miles southwest of Belgrade.

Description: Capital of the Republic of Bosnia-Hercegovina, one of six republics (becoming independent states after 1991) comprising the former Yugoslavia. Major administrative and military center under Turkish Ottoman rule; became part of Austria-Hungary in 1878; site of the assassination of Archduke Francis Ferdinand of Austria in 1914, the event triggering World War I. A siege on Sarajevo beginning in April 1992 has resulted in the worst fighting in Europe since World War II.

For five centuries as part of the Ottoman-controlled Balkans, Sarajevo was one of the most important and most beautiful cities in that historically vexed region. Its residents, at least until recently, have thought of their city as a place of unique diversity; many foreign observers have confirmed their claim. It was a city that stood between East and West, acting more as a bridge joining the two cultures than as a wall separating them. It bore splendid, harmonious traces of its important status under Ottoman and later Austrian rule. Since its founding, it has been inhabited by people from all the major Western religions: it was the seat of a Roman Catholic archbishop, a Serbian Orthodox metropolitan, and the Reis ul-Ulema, Yugoslavia's highest Moslem religious authority. Its citizens—Croats, Moslems, Jews, Serbs—lived and worked together in an environment marked by cultural, linguistic, and religious intermingling quite unlike any other city in the region, perhaps anywhere. But after the disintegration of Yugoslavia within months of the fall of the Berlin Wall in 1989, the atmosphere of pluralism in Sarajevo turned terrifyingly uncertain as its citizens—a third of them Moslems—found themselves subject to a new wave of violent Serbian nationalism that began to work its way through Croatia toward their region. Since then, uncertainty has given way to a daily horror in Sarajevo almost impossible to imagine. A Serbian siege of the city that began in April 1992 has destroyed much of it and largely isolated its citizens from any outside assistance.

If the future of Sarajevo in mid–1995 remains uncertain, the early history of the area of which it is the capital is also obscure, at least by comparison with what we know of its Balkan neighbors. Only late in the twelfth century do sources begin to speak of Bosnia in any great detail. It was under the sway of Hungary after Kálmán annexed Croatia in 1102. After the Byzantine army defeated Hungarian forces at the battle of Zemun in 1167, Bosnia, which had provided men for the Hungarian cause, fell under Byzantine rule. Yet the area was distant enough from Constantinople that it was more or less left to its own affairs under the direction of local nobles during the brief period of Byzantine rule, which lasted probably until about 1180. The Dalmatian coast, along with southern Croatia and Bosnia, was again taken by the Hungarians after 1180, but they too appear to have taken little interest in directly managing the affairs of these territories. The central part of Bosnia, including the area that later became known as Vrhbosna, now the city of Sarajevo, by all evidence remained independent throughout this period.

Prior to that, not much is known about the site of modern Sarajevo; indeed, little of consequence accrues to the city's history until the fifteenth century, when it became a major administrative center under the Ottoman Empire. The Illyrians, ancient peoples who inhabited the Balkan region from the third millennium B.C., established settlements nearby, and during the Roman era the region including Bosnia and Hercegovina was named Illyricum. The Romans established a resort near Sarajevo at Ilidža, the source of the Bosna River; a sulfurous spa still exists there. Goths, followed in the latter half of the sixth century by Slavs, colonized the region.

It was the Slavs who erected a fortress, which became known as Vrhbosna, on a hill east of Sarajevo. This served as a trade center into medieval times, although it received relatively little attention until the early fifteenth century. Exactly when the Turks captured Vrhbosna and established a town below is a matter of some dispute. Scholars have often held that the Turks captured a number of Bosnian settlements, including Vrhbosna, in 1435 or 1436, and some sources indicate that it was as early as 1429. However, recent scholarship demonstrates that the Turks took Vrhbosna and other towns in eastern Bosnia at least a full decade later, between 1448 and 1451. Whatever the date, it is certain that the town quickly evolved into an important location for the expanding Turkish Empire, and that Sarajevo for all practical purposes first came into being at their hands. The Turks renamed the settlement Bosna-Serai, the Turkish word *serai* meaning "palace," from which the first two syllables of Sarajevo derive. Sarajevo remained a Turkish city for nearly five centuries.

The city's founder is considered to be Isa-bey Ishakovic-Hranusic, a Turkish vizier in charge of the empire's western regions during the middle of the fifteenth century. Isa-bey built a bridge across the Miljacka River, thereby joining three or four small medieval settlements. On the left bank he built a *serai* with residences for his family and his suite of soldiers and administrators. In addition, he put up government buildings to serve the needs of the developing province, as well as the city's first mosque, the so-called

Views of Sarajevo
Photos courtesy of Amira Dzirlo

Emperor Mosque (Careva Džamija), the first public baths, and the Hippodrome (at-mejdan).

By joining the existing settlements and developing a central market area for the new town, Isa-bey laid the groundwork for the site's rapid expansion. By the end of the fifteenth century the town had already spread to the mouth of the Koseva stream, where there had stood a mosque along with an inn and a few shops. The portion of the town that encompassed these buildings, until recently still called Hiseta (*hise* means a share of the whole), recalls the parceling of land in those days, done for the purpose of stimulating private building.

Without a doubt the greatest period of building in Sarajevo occurred under the governorship of Gazi Husref-Bey from 1521–1541. During this period, Sarajevo became the commercial, administrative, and military center of the Turkish presence in the Balkans. Gazi Husref-Bey devoted nearly his entire life to the development of Sarajevo. In addition to the region's principal mosque, the Gazi Husref-Bey Mosque, or Begova Džamija (completed in 1530), he supervised the building of numerous educational and municipal structures including schools (among them a special school of Sufi philosophy), libraries, inns, public baths, a free kitchen, a large covered market, caravanserai, and about 300 shops. While the principal development at this time obviously benefited the city's ruling Turkish-Moslem population, Gazi Husref-Bey's stewardship was marked by an openness to non-Moslem cultures that was largely responsible for Sarajevo's multi-ethnic character in the centuries to come. Franciscan monasteries in Visoko and Fojnica were rebuilt, as well as the Catholic church in the city's Latin Quarter (an area originally developed by merchants from Dubrovnik). According to local legend, he also helped establish the Eastern Orthodox Church in Sarajevo, located near his mosque. Under Gazi Husref-Bey, Sarajevo jumped from the classification small town (kasaba) to large city (seher), and became a model for the development of other Bosnian settlements.

The latter half of the sixteenth century saw further expansion with the construction of a number of mosques, synagogues, monasteries, and Orthodox churches. In the center of the city, on the left bank of the Miljacka River, was the *serai*, the residence of the Bosnian governors, while on the right bank grew up a large *carsia* or business and administrative center. The rapidly growing population began to spread out into the surrounding mountains in groups known as *mahalas*, small settlements with about fifty houses each, a mahala mosque, and an elementary school.

By the seventeenth century, Sarajevo had more than 100 mosques, many of them lead-covered. Crafts and trade provided the city's economic basis, and Sarajevo consolidated its role as a major trading center located on the primary routes connecting East and West.

Along with its prominent position within the Ottoman Empire, Sarajevo was also the scene of revolts that came about as a result of that empire's internal struggles and gradual weakening, and also through the sharpening of differences between the Turks and the Bosnian Moslems. A peasant revolt of 1682 stemmed from disputes among craftsmen and guilds, and especially among craftsmen and rich traders. Tired of being taken advantage of by traders who were able to buy goods cheap in Sarajevo to sell them at extraordinary profit in the West, peasants from the mountains surrounding the city along with a number of poor inhabitants within Sarajevo raided and demolished many of the municipal buildings as well as several grand houses. Such events, along with several natural disasters including flood, fire, and plague, left Sarajevo severely weakened for much of the seventeenth century.

This gradual decline reached its nadir with a catastrophic Austrian attack led by Prince Eugene of Savoy in 1697. After nearly all of Sarajevo was burned to the ground, the Turks included in their massive rebuilding program a more rigorous plan to strengthen Sarajevo's defenses. The area on Vratnik Hill was fortified with walls and gates beginning in 1727. The fortified area could be entered through one of seven gate-towers, three of which, the Sirokac, Ploca, and Visegrad gates, survived into the late twentieth century. Bastions for cannon were also built, and the famous Yellow Bastion was not finished until 1819.

Its efforts mostly focused on rebuilding what the Austrian forces had destroyed, Sarajevo did not exert itself to any great extent beyond its own confines during the entire eighteenth century. After Bosnia-Hercegovina passed from the Turkish Empire to the Austrians in 1878, representatives of all the nation's peoples—Moslems, Serbs, Croats, Jews—formed a Peoples Government, a collective body of organized resistance. Embittered by 400 years of Turkish occupation, the Bosnians were further angered when they discovered that the liberty they expected to result from the peace treaty of San Stephan had in fact been replaced by a decision between the Turkish and Austrian powers at the Berlin Congress to hand Bosnia over to Austria-Hungary. The Peoples Government's armed resistance to Austro-Hungarian soldiers at the same time as they withdrew obedience to the Turkish Sultan had little effect, however.

For the Austrians, the occupation of Bosnia-Hercegovina was the first step in a program of colonial expansion toward the East. The new administration wanted to demonstrate its culture-carrying mission in the capital of the occupied land. During their forty years of rule (1878–1918), the Austrians built many new public structures in Sarajevo; they founded institutes there and laid out parks. Still preserving and developing its function as the political and cultural center of the province, Sarajevo, formerly a commercial and crafts center, began rapidly industrializing. Although some traces of European architecture had already appeared under the Turks in the nineteenth century, after the Austro-Hungarian occupation a marked Viennese influence was felt; some remarkable neo-baroque and Secession style buildings have survived into the 1990s, although the ongoing siege of Sarajevo has left their condition, along with everything and everyone else in the city, in question.

Sarajevo leapt to the world's attention on June 28, 1914, when a nineteen-year-old Serbian nationalist named Gavrilo Princip assassinated Archduke Francis Ferdinand (heir apparent to the Austrian throne) and his wife while the two were in Sarajevo on an official visit to observe Austrian military maneuvers just outside the city. This event played a significant role in the start of World War I.

Many historians consider that the Austro-Hungarian Empire set up this episode as a pretext for declaring war on Serbia. Serbia (which included Macedonia) and Montenegro were independent states in 1914; Slovenia, Croatia, and Bosnia-Hercegovina were part of Austria-Hungary. Serbia, behind a movement for the unification of all these republics and spurred by the Mlada Bosna (Young Bosnia) of which Princip and his young collaborators were members, bitterly opposed the Austrian presence. Whether the war party in Vienna had actually anticipated the outcome of the archduke's visit is a matter of speculation, but it is clear that after the assassination Austria wasted no time in seeking to put an end to any further Serbian resistance. On July 23, 1914, Austria sent an unacceptable ultimatum to the Serbian leaders, and five days later they declared war on Serbia. Shortly afterward, the declaration spread beyond this local conflict and escalated into world war.

Following the war in 1918, the "Kingdom of the Serbs, Croats, and Slovenes" was proclaimed; in 1929 it became the Republic of Yugoslavia and in 1931 adopted a parliamentary constitution. The new republic, with a centralized government in Belgrade, eliminated many historic borders. Sarajevo became part of the Drina region, composed of eastern Bosnia and parts of western Serbia.

On April 6, 1941, the German air-force bombed Sarajevo and several days later occupied the town. The National Liberation Movement soon organized resistance to the German, Italian, Ustase, and Cetnik troops who had turned Sarajevo into their base of military operations. The city was liberated in April 1945, and later that year Yugoslavia, with Bosnia-Hercegovina as one of six constituent states, was made a communist republic.

In the 1990s Sarajevo has endured undoubtedly the greatest trial of its 600-year existence. In October 1991, in the wake of the collapse of communism and the fall of the Berlin Wall, Bosnia-Hercegovina declared itself a sovereign republic, following similar moves by Slovenia and Croatia after the first multi-party elections in the then-crumbling Yugoslavia. The Serb-controlled Yugoslav federal army in Belgrade took immediate action to arm Serb extremists under Radovan Karadžić in an attempt to take any territory containing Serbs and add it to the state of Serbia, "expelling" all non-Serbs. After moving through Croatia in late 1991, destroying the towns of Dubrovnik and Vukovar among others, the Serbian army began its attack on Sarajevo in April 1992. By early 1993, less than a year after the war began, 200,000 civilians in Bosnia-Hercegovina had been killed, and 2 million forced from their homes. Sarajevo, which was (and, in mid-1995, remained) completely surrounded by Serb troops,

counted more than 10,000 civilians dead, about one-quarter of them children. These numbers have continued to mount dramatically ever since.

But the numbers killed by shell-fire hardly tell the story of the conditions of life in Bosnia-Hercegovina and its capital. By August 1993, the outside world began to learn of concentration camps, containing mostly Moslems and Croats, in Bosnia-Hercegovina. "Ethnic cleansing," a phrase little heard since the end of World War II, was discovered to be a Serb motive in their move to create a "new Yugoslavia." Atrocities including the widespread rape of Moslem women were being committed by Serb forces. In Sarajevo, electricity, water, food, all were cut off, and outside humanitarian assistance has been severely hampered, despite the presence of UN forces, by Serbs controlling the city's airport. Trees have gradually disappeared from Sarajevo's parks and cemeteries to provide fuel.

On May 16, 1995, after a shaky cease-fire during which foreign diplomats were still prevented from entering the city, the heaviest fighting in Sarajevo for two years broke out. While Western leaders were trying to propose a plan to Serb leader Slobodan Milošević that would ease sanctions on Serbia in return for the latter's recognizing Bosnia's borders, the citizens of Sarajevo were once again bracing themselves against the shells that have fallen on every section of their city, leveling much of it.

Nearly all of Sarajevo's buildings have been severely damaged or destroyed. The Gazi Husref-Bey Mosque, once the principal mosque of the former Yugoslavia, has taken more than 100 hits. The destruction of this great monument is perhaps a suitable metaphor for the loss of ethnic diversity its namesake so admirably desired. The words of the seventeenth-century Sarajevo poet, Mohammed Merkesja, who, far from his native city, reflected longingly on its beauties, must strike the residents there now as a voice from some irretrievable past: "Clear, running water escaping from the sunlight, eddying through the shade of trees in that rose garden of the world. . . . The high mountains that surround Sarajevo embrace the sky, and with their snow-covered and misty tops look like dignified elders. . . . It is impossible to count the beauties of this exemplary town and mention all its miracles."

Further Reading: *The Late Medieval Balkans: A Critical Survey from the Late Twelfth Century to the Ottoman Conquest* by John V. Fine (Ann Arbor: University of Michigan Press, 1987) is the second of two indispensable studies Fine has written on the Balkans from the Slavic colonization of the area to the Turkish invasion. It provides limited but useful information on the founding of Vrhbosna, the site that became Sarajevo. *History of Yugoslavia* by Vladimir Dedijer et al. (New York: McGraw-Hill, 1974; Maidenhead, Berkshire: McGraw-Hill, 1975) gives important details relating to trade conditions in Sarajevo from the fifteenth century on, and offers a balanced analysis of the events surrounding the assassination of Archduke Francis Ferdinand and the Austrian response to it leading to World War I. Two shorter studies by Fred Singleton, *Twentieth-Century Yugoslavia* (New York: Columbia University Press, and

London: Macmillan, 1976) and *A Short History of the Yugoslav Peoples* (Cambridge and New York: Cambridge University Press, 1985), both provide information about Sarajevo in the larger national context. A number of recent books on the fall of Yugoslavia, and more specifically the siege of Sarajevo, have appeared since 1991; while on the whole unscholarly and ahistorical, these provide some sense of current conditions in Bosnia-Hercegovina and its capital. Particularly noteworthy among this group are *Sarajevo: Exodus of a City* by Dzevad Karahasan, translated by Slobodan Drakulić (New York: Kodansha, 1994; originally published as *Dnevnik selidbe,* Zagreb: Durieux, 1993) and *Sarajevo: A War Journal* by Zlatko Dizdarevic (New York: Fromm, 1993). The former title, written by a professor at the Academy of Theatrical Arts at the University of Sarajevo, offers a rare and highly sophisticated treatment of the situation within the city. Karahasan writes compellingly about Sarajevo as a "closed" system, drawing comparisons to the closed systems of the Nazi concentration camps. Dizdarevic gives a concise overview of the political conditions leading to conflicts in the former Yugoslavia. Two first-hand accounts written by foreign journalists worthy of consulting are *The Fall of Yugoslavia: The Third Balkan War* by Misha Glenny (London and New York: Penguin, 1992) and *A Paper House: The Ending of Yugoslavia* by Mark Thompson (New York: Pantheon, and London: Vintage, 1992). *Letters from Sarajevo: Voices of a Besieged City* by Anna Cataldi, translated by Avril Bardoni (Longmead, Shaftesbury, and Rockport, Massachusetts: Element, 1994; originally published as *Sarajevo: Voci da un assedio,* Milan: Baldini and Castoldi, 1993) makes no claims at scholarship but offers graphic accounts of a besieged Sarajevo from the point of view of its citizens.

—Amira Dzirlo and Paul E. Schellinger

Segovia (Segovia, Spain)

Location: North of the Sierra de Guadarrama at an altitude of just over 3,280 feet; on a narrow triangular ridge of rock above the turbulent confluence of the Eresma and Clamores Rivers. Southeast of Castile and León, one of Spain's fourteen state-size "autonomous communities"; sixty-two miles northwest of Madrid.

Description: Once an important Roman city, then a prosperous Arab stronghold, and after that, the city in which many of the kings of Castile resided. Now an industrial city, producing textiles and paper, Segovia is famous for its historical buildings and is celebrated in Spain for its traditional gastronomy, particularly its roast suckling pig. It is the capital of Segovia province and the episcopal see of the archdiocese of Valladolid.

Site Office: Oficina de Turismo
Plaza Mayor, 10
4001 Segovia, Segovia
Spain
(11) 43 03 28

According to legend, Segovia, a city mentioned by Pliny, was founded by Hercules before being destroyed and later rebuilt by the Celts in the eighth century B.C. All that the Celts left to the area they inhabited was the root of its name: "Seg". Segovia's better documented history begins in the year 96 B.C., when it was taken by the Romans for use as a pleasure resort. It had the added advantage of being strategically situated to control access to the Duero valley sixty-two miles to the north. For this reason the city acquired considerable military importance and is surrounded by walls erected by Alfonso VI in the late eleventh century.

After the Romans, the Visigoths came to the province of Segovia, leaving Visigothic names to several towns. From the eighth to the tenth centuries, the area was a zone of almost constant fighting, a semideserted province through which Germanic tribes passed on raiding expeditions. The city was repeatedly taken and lost by the Moors from 714 until 1079, when it was conquered by Alfonso VI, who began the great task of re-population. Alfonso showed equal regard for Jews and Muslims, although he principally wanted to entice Spaniards down from Asturias and Galicia in the north. Following Alfonso's capture of the city from the Arabs, medieval industry and commerce began to flourish, and the great period of Segovia's prosperity commenced. During the Middle Ages, Segovia was an important wool town with celebrated markets outside its walls, and in the fifteenth century the city entered what is now known as its golden age, with a population that rose to well over 60,000.

Until 1479, the Iberian Peninsula consisted of five independent kingdoms, with Castile and León together occupying sixty-two percent of the whole. In 1469 the two branches of the royal house of Trastamara were united by the marriage of Isabella, daughter of John II of Castile, to Ferdinand, son of John II of Aragon. They were known as the Catholic monarchs and presented an unchallengeable claim to the Castilian throne. At the end of the fifteenth century, the last of the Trastamaras made their court in Segovia and took the thirteenth-century Alcázar castle, still one of the city's principal monuments, for their palace. It was here that Isabella I was proclaimed queen of Castile on December 13, 1474.

The Jewish community was of great economic and cultural importance to the medieval city, with each Jew over fourteen years of age having to pay the bishop "thirty pieces of gold," a calculated allusion to the payment to Judas. The community was much persecuted and lived in what are today known as *las juderias* (the Jewries) until their expulsion from Spain by the Catholic monarchs in 1492. The Jewish quarter of Segovia, where the poet Antonio Machado lived for thirty years during the 1800s, retains many of the original names of its narrow streets.

When Charles I of Spain, later Emperor Charles V, arrived from the low countries bringing his Flemish court and favorites, he outraged his Spanish subjects with the new levies he imposed. Led by Juan Bravo from Segovia and Juan de Padilla from Toledo, an uprising took place known as the rebellion of the *comuneros* or the "war of the communities." Segovia was selected for special punishment, and the revolt, put down in Villalar in 1521, ended with the execution of the two leaders in Segovia. The uprising is often regarded as the most glorious moment in the history of the town.

Segovia is celebrated for its three great historical monuments, the most remarkable of which is undoubtedly the colossal Roman aqueduct, one of the finest remaining and, incredibly, still functional. Built under the Roman emperor Trajan, it is known as *el puente del diabolo* or devil's bridge since, according to tradition, it was constructed by the devil in a single night. Made from great blocks of the local yellow granite brought from the Sierra Guadarrama, it is unlined and uncemented, but put together with such artistry that it has survived the centuries and still performs the function for which it was designed, carrying the city water down from the Rio Acebeda. The structure is of two superimposed tiers of 118 elegant arches. It measures 2,372 feet across, with a height of up to 92 feet where the ground beneath it is lowest. The thirty-six arches destroyed by King Almamun of Toledo in the eleventh century were restored by the fifteenth, and at the beginning of that century the aqueduct again became functional. Today, it towers over the Plaza del Azoguejo, which still holds its traditional markets adjacent to the main Via Roma leading through the city from Ávila to Valladolid.

The Alcázar of Segovia
Photo courtesy of Instituto Nacional de Promocion del Turismo

Segovia's cathedral dominates the Plaza Mayor in the center of the town. The original cathedral was razed during the *comuneros* revolt. Emperor Charles V undertook the reconstruction of the cathedral and chose the present site. He had the original cloisters painstakingly removed stone by stone from the old site in the gardens of the Alcázar. The cathedral's massive and attractive design was executed in granite that glows golden pink in the sun. Because the Gothic style was replaced by Renaissance architecture in the six-teenth century, Segovia's cathedral was the last of its kind in Spain.

The inside of the cathedral is spacious and elegantly simple, lit through spectacular stained-glass windows. Magnificent fifteenth-century flamboyant Gothic choir stalls, removed from the original cathedral, lie between the wide aisles. Each of the chapels, separated from the nave by delicately wrought-iron screens, has fine works of religious art, including a 1571 baroque altarpiece by Juan de Juni.

Other important works include a silver arch with the relics of San Frutos, the patron saint of the city, and a Gothic image from the fourteenth century of Nuestra Señora de la Paz (Our Lady of Peace), given by Henry IV of Castile. The cathedral also houses an important museum where the Corpus Christi monstrance that is paraded through the streets on the feast day is displayed. The chapterhouse displays a fine collection of Brussels tapestries illustrating the short reign of the third-century Queen Zenobia of Palmyra.

Each of the three most celebrated historical treasures of Segovia is associated with a different epoch. If the great aqueduct recalls the past-but-not-forgotten pagan world, and the cathedral a time of Christian religious fervor, the Alcázar of Segovia is a reminder of a period of war and chivalry in the city's history. In this Muslim palace, kings were born, tournaments took place, and courts were held. Isabella came out of the Alcázar to be proclaimed queen of Castile. Philip II and Anne of Austria were both married within its walls. For many years the Alcázar, always a defensive edifice, served as a state prison, although its beautiful Gothic and Moorish towers and spirals with Renaissance ornamentation appear too whimsical for the scenes of violence they have witnessed. The palace stands on the tip of a cliff, providing it with marvelous views over the city and the Sierra Guadarrama, and making it a striking landmark.

The Alcázar started as a primitive fortress during the reign of Alfonso el Sabio (the Wise) in the thirteenth century. The homage tower was added in the fifteenth century. The Catholic monarchs and, later, Philip II contributed greatly to its modification, although it was Charles III who converted it into the Royal Academy of Artillery. The conversion nearly caused its destruction by fire in 1862. It was restored twenty years later, and the general military archives were installed within it. The interior, notably the throne room, is still filled with period furnishings and pictures. Several of the magnificent earlier ceilings have surprisingly been retained, providing attractive settings for the display of medieval armor and works of art that it houses.

Segovia is famed for much more than its aqueduct, cathedral, and Alcázar. It has a number of superb Romanesque churches. Eighteen of the original thirty still remain, and of those within the city walls, one of the earliest, built in the eleventh century, is generally considered the finest: San Juan de los Caballeros (St. John of the Knights). It has a robust tower that today houses a splendid collection of ceramics by Ignacio Zuloago, although its most outstanding feature is its famous portico, originally from the later church of St. Nicholas, covered with decorative carvings of flora and fauna with magnificent heads.

The most remarkable of the Romanesque churches in Segovia is the last, San Estobán, dating from the thirteenth century. It has a very striking five-story tower 184 feet high with arched windows. Inside there is a *calvario* from the thirteenth century taken from an earlier church and a Gothic polychrome statue of Christ carved in wood similar to one in the church of Christo de la Vega, formerly Santa Leocadia, in

Toledo. A delightful square in the center of the old walled city's aristocratic section, the Plaza San Martin contains a statue of Juan Bravo and a fine fifteenth-century house known as Casa de Juan Bravo, although he did not live there. The Plaza also contains the twelfth-century Romanesque church of San Martin, founded in the tenth century as a Mozarabic (Moorish-Arab) temple. Its portal is a fine example of Romanesque architecture in Segovia.

Some of the finest Romanesque churches are, however, to be found outside the city walls, in what are known as the *arrabales*, or outskirts. The early twelfth-century church of San Millan is erected on the site of an earlier church whose original Mozarabic tower it incorporated. It sits in the center of a large quare, which allows room for a good view of the simple primitive Romanesque structure as it contrasts perfectly with the Moorish transept.

The most intriguing of the Segovia churches outside the walled city is undoubtedly the Vera Cruz (True Cross) chapel, formerly the church of the Holy Sepulchre. It is believed to have been constructed by the Templars, who inscribed a dedication to the Holy Sepulchre on the polygonal building dated April 13, 1246. The church later passed to the order of Malta, which still owns it and celebrates religious festivals there, notably on Good Friday, when members of the order dress in the original black choir habits of the knights and parade at night by candlelight along the road to Zamarramala. The circular interior of the church is modeled on the mosque of the rock in Jerusalem, and has two floors consisting of one small chamber each where secret ceremonies were conducted. To one side of the circular corridor, forming a ring around the chambers, is a tall, square, four-story tower affording a superb view of the city and the Alcázar. To the right of the main entrance is the small Lignum Crucis (Wood of the Cross) chapel with an ornate flamboyant Gothic altar where for centuries relics of the True Cross were venerated.

The well-preserved residences of the old aristocracy of Segovia were generally built with towers to add to the city's defenses. It is largely due to these fascinating houses in the narrow uneven streets of the old city that Segovia has such a fairy-tale appearance when viewed from outside the city walls. The district known as the Cloisters, whose residents enjoyed special privileges, contains many Romanesque town houses and used to have a gate that was closed at night, separating the quarter from the rest of the city. Among the most notable of the secular domestic buildings is the sixteenth-century Casa de los Picos, built as a fortress to guard the Puerta de San Martin and covered with diamond-shaped stone spikes from top to bottom. The Casa de los Picos has generated its own legend. According to the story, the first owner had the spikes added in order to avoid having his house associated in the usual way with his own name, which was infamous because he was the city's executioner. Like so many legends generated by the buildings and residents of Segovia, ecclesiastical even more than secular, this one has been shown to be unfounded.

Further Reading: As with most other historic Spanish towns, there is nothing available in English except for tourist guides, atlases, gazetteers, and encyclopedia articles, many of which are informative and helpful. Further reading is normally to be found in ordinary, mainly political histories, and in histories of art and architecture. Other important works include *Islamic Spain, 1250–1500* by L.P. Harvey (Chicago: University of Chicago Press, 1990) and *Enrique IV and the Crisis of Fifteenth-Century Castile, 1425–80* by William D. Phillips Jr. (Cambridge, Massachusetts: Mediaeval Academy of America, 1978).   —Clarissa Levi

Selinus (Trapani, Italy)

Location: On the southern coast of Sicily, some fifty miles southwest of Palermo.

Description: Ancient Greek city that became extinct in the third century B.C., a victim of the First Punic War.

Contact: A.A.S.T. (Azienda Autonoma Provinciale per l'Incremento Turistico)
Via V. Sorba, 15
CAP 91100 Trapani, Trapani
Italy
(0923) 27273-27077

Selinus, named for the wild celery that still grows in the area, was settled by Greek colonists some time in the seventh century B.C. Ancient Greek historians disagree about the exact date of Selinus's founding: Thucydides gives 628 B.C., while Eusebius and Diodorus Siculus both give 650 B.C. Pottery shards found during excavations in the twentieth century support the earlier date.

Greek colonization of Sicily had begun on the east side of the island about 100 years earlier. Centuries of deforestation and soil erosion have turned large stretches of Sicily into arid desert, but at the time the island was highly fertile and extremely suitable for agriculture. Wheat, wine, olives, fruit, nuts, and wool formed the main staples. The eastern colonies prospered, and the Greeks began to push westward along Sicily's southern coast and into the interior. Greek colonists from Megara Hyblaea, an earlier Greek settlement on the eastern coast of Sicily just north of Syracuse, joined with new immigrants from Megara in Greece to found Selinus, a farming community that, for a while, formed the westernmost outpost of Greek settlement.

Situated in the swamps between the mouths of the Rivers Selinos (now Modione) to the west and Cottone to the east, Selinus depended on the rich farmland farther into the interior. The city itself was built on two hills rising out of the marshlands. The southern hill adjoining the beach became the site of the pear-shaped acropolis, while the residential part of the ancient town was built immediately to the north, on the second hill. The two river mouths formed small natural harbors, enabling communications with the Greek world and with the Carthaginian and Phoenician traders. Although several small temples were built on the acropolis itself, the most important early sanctuary lay outside the city proper. Built in the valley to the west, the sanctuary of Demeter Malophoros (apple-bearing Demeter, goddess of fertility) originally enclosed only a number of open-air altars, but by 600 B.C. the Selinuntines were wealthy enough to build a stone hall there. Later a small sanctuary to Hecate Triformis and a temple to Zeus Meilichios were added to the complex.

Selinus flourished and grew extraordinarily wealthy as a result of its exports of agricultural products to both Greece and North Africa. As a consequence of the colony's success, it became one of the earliest cities in Sicily to mint silver coins, at about 550 B.C. These coins bear an image of the wild celery leaf, the city's emblem. Selinus's wealth was translated into grand religious architecture, as was usual in the Greek world, where a city's riches were most easily measured by the size of its temples. In the mid-sixth century the Selinuntines began building their greatest temples in the valley to the east of the city. These temples, now known as temples E, F, and G, took so long to build that they show a transformation of style from archaic to classical. Temple G was, in fact, never finished at all, despite the work having continued for more than 100 years. Although the general mold of Selinuntine architecture is Greek, following Doric models, Selinus is the only ancient Greek settlement on Sicily that shows a distinct style in its architectural decoration and sculpture. The terra cotta revetments and carved metopes, for instance, are unlike anything else produced in Magna Graecia and seem to have been famous at the time. The most interesting surviving examples are on permanent exhibit at the Museo Nazionale Archeologico at Palermo.

Late in the sixth century, the acropolis was ringed with walls, and a defensive gate was built at the narrow northern end to give access to the residential area. The area within the walls was laid out along one of the earliest-known grid plans, along a north-south axis. The residential area was given a grid plan at a slight angle to the one on the acropolis. Roads connected the various temples to the acropolis hill and the residential sector. The harbor was enlarged at about the same time and quays and storehouses were built along the waterfront. Probably all its produce left Selinus by boat, either directly to Greece or to Mazara, a little farther west on the coast, where the Selinuntines seem to have traded with Carthaginians and Phoenicians.

Since Selinus was dependent on farming and controlled an area stretching for miles in every direction, the acropolis fortifications served to defend the city from attack by the two local tribes, the Sicans to the east and the Elymians to the west, into whose territory Selinus continued to expand. The Carthaginian and Phoenician traders, who maintained trading posts on the western tip of the island, had no interest in territorial expansion and would not have been considered a threat at the time. Before long, Selinus became embroiled in border disputes with its neighbors, particularly Akragas (modern Agrigento) and Segesta, a hellenized Elymian settlement some forty miles to the north. At various times, Selinus also found itself at odds with Syracuse. Ultimately, this internecine warfare became Selinus's downfall.

Temple ruins at Selinus
Photo courtesy of Italian Government Tourist Board-E.N.I.T.

Little is known about Selinus's skirmish with Akragas. Early in the fifth century B.C., Akragas laid claim to Heraclea Minoa, a small settlement between the two cities that belonged to Selinus. Some years later the dispute came to an end, with Akragas emerging victorious.

The year 480 brought a massive Carthaginian invasion against the majority of the Greek settlements, at the invitation of the cities Himera and Messina. For unknown reasons, Selinus seems to have remained outside the Greek alliances and to have supported the invasion. At the decisive battle, however, Selinus was not present, for reasons unknown. Although Syracuse resoundingly defeated the Carthaginians at Himera, Selinus escaped the fate usually reserved for defeated enemies and came away with its independence, perhaps because it successfully persuaded the victors that it had been neutral in the conflict.

Although the city continued to prosper, by the mid-fifth century B.C. the first reports surfaced of malaria epidemics. Sand had been drifting up along large stretches of the southern beaches of Sicily, choking the river mouths at Selinus and enlarging the swamps surrounding the city. According to legend, the city called upon the philosopher Empedocles, who reportedly solved the problem by draining the marshlands. If he did so in fact, his solution must have been temporary.

By 416, the permanent conflict between Selinus and Segesta flared into open warfare. Selinus secured the powerful support of Syracuse, prompting Segesta to call upon Carthage and, when the first appeal came to nothing, upon Greece. A Greek force invaded Sicily in 415 but was defeated by Syracuse two years later. This, however, did not end the hostilities between Selinus and Segesta. Segesta seems to have had little hope of defending itself and again called on Carthage in 410. This time, Segesta's appeal found a willing ear, and in 409 an expedition led by Hannibal laid siege to Selinus. The city and its allies were unprepared for the attack, and, after only nine days of fighting, the walls were breached, the city plundered, and many of its inhabitants massacred.

The Carthaginian army then marched to Himera, where it met and defeated the forces of Syracuse. Upon this victory, Hannibal and his army returned home, apparently satisfied with the results. The tragedy of Selinus was then compounded by Hermocrates, a former commander of the Syracusan fleet. He had assembled a small army of mercenaries to force Syracuse to reinstate him. When the effort failed, he took his band to Selinus, gathering more refugees along the way, and ensconced himself in the ruins of the old city. By these actions, Hermocrates directly challenged Carthage's overlordship of the area and so precipitated the next Carthaginian invasion. In 406, Carthage laid siege to Akragas. After seven months of fighting, Akragas was abandoned to the Carthaginians, a defeat for Syracuse, then the most powerful force on the island. The defeat was at least in part due to dissension among the allies. Gela was the next city to fall to the Carthaginian expedition, and it was followed by Kamarina, the last obstacle between Carthage and Syracuse.

Syracuse was not ready to engage the enemy at its own walls, however. In a treaty signed in 405, Syracuse ceded a vast stretch of southern Sicily, including Selinus and Akragas, to Carthage as an *epikrateia*, or province. The arrangement enabled the Africans to exact tribute from the cities in their domain. Those Selinuntines who had fled the carnage several years earlier were allowed to return, but they had to pay tribute to Carthage and were not permitted to rebuild the city's fortifications. Selinus never recovered, although it survived for another 150 years.

Between 350 and 300 Syracuse made several unsuccessful attempts to recapture Selinus and other territories from Carthage, but the Carthaginian hold over the area was unbroken until early in the third century B.C., when a conflict between Akragas and Syracuse brought another Carthaginian army to the walls of Syracuse. It was in 278, at this time of chaos, that Pyrrhus, following his proverbially exhausting victories in Italy, arrived in Sicily, enthusiastically welcomed by the Greek cities. The Greeks were rarely loath to call for foreign invasions, and this one must have seemed their best chance to get rid of Carthage. Selinus was one of the cities to support the new invader and immediately began rebuilding its fortifications. Pyrrhus indeed quickly drove the Carthaginians out of southern Sicily, only to have his forces driven out almost as quickly by the next invaders, the Ro-

mans. The Roman expedition provided Carthage with an opportunity to reestablish itself in most of the area previously lost to Pyrrhus.

The Roman invasion was the beginning of the First Punic War, a conflict fought in large part on Sicilian soil. After several years of skirmishes between the two foreign powers, Carthage decided to give up its base at Selinus in 250 B.C. By that time, the city was a small farming community, apparently not a sufficiently significant possession to fight over. Carthage evacuated all inhabitants to Lilybaeum (now Marsala) and destroyed the never-to-be-rebuilt city. Malaria, associated with the marshes surrounding the site, is sometimes given as the primary cause for the failure to rebuild this briefly prosperous and powerful city. A hamlet, Marinella di Selinunte, too small to have a recorded history, grew up a little way to the east, but otherwise the area is deserted to this day.

In the eighteenth and nineteenth centuries, Selinus became a favored stop on the Grand Tour of Europe. Visitors came for the archaeological interest of the magnificent ruined temples (toppled by a medieval earthquake) to the east of the city. The main attraction to the nineteenth-century tourist may have been that the ruins offered an edifying—and at the same time romantic—spectacle of human ambition come to nought.

Excavations on the acropolis in the twentieth century have uncovered the ancient fortifications, the northern defensive gate, the remains of a few small sanctuaries, and the grid plan. The temple of Demeter Malophoros has also been partially preserved. A small museum provides information about the ancient city, but the most interesting archaeological finds are displayed at Palermo.

Further Reading: Margaret Guido's *Sicily: An Archaeological Guide; the Prehistoric and Roman Remains and the Greek Cities* (New York and Washington: Praeger, and London: Faber, 1967) has a section devoted to Selinus that contains both a brief review of Selinus's history and a readable account of its remains. The first volume of *A History of Sicily*, entitled *Ancient Sicily to the Arab Conquest*, by M. I. Finley (London: Chatto and Windus, and New York: Viking, 1968) places Selinus's history in the larger context of the history of Sicily.

—Marijke Rijsberman

Seville (Sevilla, Spain)

Location: Southwestern Spain, on the Guadalquivir River 60 miles from the Atlantic Ocean and 320 miles from Madrid; capital of the province of Sevilla, in the region of Andalusia.

Description: Spain's third-largest city, which for long periods of its 3,000-year history was the largest and most important city in the country. There are substantial Roman remains at Italica, five miles to the north. In Seville itself, the entire *casco antiguo* ("old city," within the old city walls) follows the street plan of the Moorish occupation and within this area most buildings date from the nineteenth century or earlier. Seville has frequently played a key role in Spanish history and has inspired many works of art, music, and literature.

Site Office: Tourist Information Centre
Avenida de la Constiución, 21
Seville, Sevilla
Spain
(95) 422 14 04 or 421 81 57

Seville lies some sixty miles from the Atlantic Ocean in what was once a fork in the Guadalquivir, which is Spain's only navigable river and was for centuries the city's prime source of income. The mild climate of southern Spain made it attractive to prehistoric people, and there is evidence of early settlements throughout the province of Seville, particularly in the Aljarafe to the west and around Carmona to the east of the city: artifacts from these and numerous other local sites can be seen in Seville's Archaeological Musuem.

Although the Celtic migrations probably reached southern Spain by the seventh century B.C., early external influences came almost exclusively from the eastern Mediterranean. Some historians state that the Phoenicians founded the coastal city of Cádiz as early as 1100 B.C., and while this may be an exaggeration there is certainly evidence of Phoenician influence in the Seville area by the seventh century B.C. Remains of the trading post that grew up near what is now the Plaza de la Alfalfa is the first evidence of settlement in the city itself. This find coincides with the legend that the city was founded by Hercules, who in the form of the god Melkart was worshiped by the Phoenicians.

Herodotus writes of Greek traders reaching the Straits of Gibraltar, and although their influence was far greater on Spain's northeast coast, it seems probable that there was at least some kind of Hellenic contact with the Seville area by the seventh century B.C. For both the Greeks and the Phoenicians the principal attraction of Spain lay in the country's considerable mineral wealth, particularly the silver deposits in the Sierra Morena mountain range to the north of Seville.

Excavation at Rio Tinto, thirty-five miles northeast of Seville, shows that mining was taking place there as early as 3000 B.C. Until its destruction by the heavy open-cast work of the nineteenth century, a hill called Cerro Salomon existed in the area, leading some commentators to speculate that Rio Tinto could have been King Solomon's mines, and Andalusia the mythical Elysian Fields. There is also speculation over the location of the city of Tartessos, which is mentioned in the Bible as Tarshish, the intended destination of Jonah. Tartessos, it seems, grew rich on the mineral deposits, before clashing with the Phoenicians and their Carthaginian successors. Tartessos may have been located alongside the lake at the mouth of the Guadalquivir, which has now dried up to leave the marshlands of the Doñana Natural Park, but excavations have failed to provide any concrete evidence to back these ideas. The most spectacular relic of Tartessian civilization is the Carambolo treasure. Found near Seville in 1958, the treasure's gold necklace and armbands are thought to date from the sixth century B.C. and are the most important exhibit in the Archaeological Museum.

The Carthaginians increased their interest in the area after the second treaty with Rome (348 B.C.) and again after the First Punic War (264–241 B.C.). Much of the Second Punic War (218–201 B.C.) was fought in Spain, where Scipio Africanus successfully cut Hannibal's supply routes in the turning point of the conflict. Scipio's victory at Ilipa (possibly Alcala del Rio, fifteen miles north of Seville) in 206 marked the end of Carthaginian settlement in Spain, and the beginning of six centuries of Roman dominance. In the same year Scipio built homes for his veterans and in a new town he called Italica. The first wholly Roman settlement in Spain, Italica was to become an aristocratic city and administrative center, supplying Rome with two of its greatest emperors, Trajan (ruled A.D. 98–117) and his successor Hadrian (ruled 117–138), who constructed a new residential area in the city. The ruins of Italica can be seen in the village of Santiponce, five miles north of Seville. Among them are a 25,000-seat amphitheatre, the third largest in the Roman world. The Archaeological Museum now houses many of the best mosaics and sculptures, while others form part of private collections housed in the Casa de Pilatos and Palace of the Duchess of Lebrija.

Seville itself was a thriving trading center under the Romans. Known as Hispalis, the Roman town occupied what is now the southeast area of the "old city." Much of the modern-day street plan follows the Roman blueprint, centering on the Calle Abades, with the Plaza de la Alfalfa area serving as forum. Large columns that once formed part of a Roman temple can still be seen in the Calle Marmoles, and a further two were taken to the Alameda Square in the sixteenth century and crowned with statues of Hercules and Julius Caesar. Caesar defeated the sons of Pompey at the battle of

Seville's cathedral
Photo courtesy of Instituto Nacional de Promocion del Turismo

Munda in 45 B.C. and then came to Seville to chastise the citizens for not having supported him in Rome's civil war. Despite this he made them free citizens of Rome and reconstructed the city walls.

The decline of the Roman Empire brought with it the arrival of migratory "barbaric" tribes. The Vandals took Seville in 426 but stayed only three years before heading south to forge a new African empire. Another Germanic tribe, the Visigoths, took over, and their dominance of Spain lasted for three centuries. The Byzantine Empire was a constant threat to Visigoth rule in the south during the sixth century, and at some stages Seville was probably a border town between the two domains. The scholar-saints Isidore and Leander were both archbishops of Seville during the period of Visigothic dominance, and the Catholic prince Hermenegild made his last stand in the city against the forces of his father Leovigild, who was an Arian Christian.

The Visigothic kings were displaced by the Moslem invasion from North Africa in 711. Seville, then the most important city in Spain, fell to the Moslem forces the following year, and remained under Moslem rule for 537 years. For much of this period Seville, known as Isbiliya, was ruled from Córdoba. However, with the fragmentation of Moslem Spain

after the death of the dictator al-Manṣūr in the early eleventh century, Seville became the most powerful of the independent *taifa* states and at some stages ruled even Córdoba itself. The growth of Christian power in the north threatened the Moslem states, who looked to their Islamic brethren from North Africa for support. In 1090 the Almoravid tribe crossed to Spain. But the Almoravids took power themselves and deposed the Sevillian al-Mut'amid, who is remembered fondly in the city as "the poet king." A century later, with the Almoravids now gone, southern Spain again came to be dominated by a Moroccan tribe, the Almohads, who applied Islamic law more strictly than did their predecessors—it is said that wine-drinking had previously been widely accepted in Seville.

The southward push of the Christian Castilians proved irresistible after the battle of Navas de Tolosa in 1212, and in 1247 King Ferdinand III had encircled Seville, which was refused help by the Moroccan states. The following year the city fell. Ferdinand divided up the city into areas for the Jewish, Moslem, and Christian populations. The largely peaceful cohabitation was disturbed by attacks on the Jewish population, the worst of which occurred in 1391. (The Jews were expelled from Spain by the Catholic monarchs in 1492, and the Moslems in 1502.)

King Ferdinand was later made a saint, and his body is periodically displayed in the cathedral. The cathedral occupies the site of the massive twelfth-century central mosque, which was used for Christian services by the conquerors but suffered earthquake damage and was demolished in 1401. Only the ablutions courtyard (the "orange-tree patio") and the magnificent 300-foot alminar tower, known as La Giralda, survive. Completed in the early sixteenth century, the Cathedral of Seville is the third largest in the world, after St. Peter's in Rome and St. Paul's in London. It contains numerous works of art, the world's biggest altarpiece, and the tomb built for Christopher Columbus—although it is not clear where exactly the explorer's remains are. Columbus's 1492 expedition left from Palos de la Frontera (sixty miles west of Seville, in the province of Huelva) but he lived for some time in the city, which benefited enormously from his breakthrough.

In 1503 Seville became the seat of the Casa de la Contratacion customs house and was granted a monopoly on trade with Spain's new colonies. The city boomed: its population surged from 40,000 in 1480 to nearly 130,000 by 1588, making it one of Europe's largest cities, outstripped only by Paris, London, and possibly Naples.

Seville's large population of Jews and Moslems, the influx of foreigners at a time of religious upheaval, and the Spanish Crown's desire to prevent heresy reaching the colonies meant that the inquisition was particularly active in Seville. Victims were burned at *autos-da-fe* in the Plaza de San Francisco, next to the Town Hall. In some years as many as 2,000 people may have died this way. The last execution took place in 1781.

The opportunity of making enormous fortunes and the transitory nature of the population meant that Seville in the sixteenth and seventeenth centuries became—in the words of contemporary playwright Lope de Vega—a "new Babylonia" and a paradise for tricksters and confidence men of all sorts. Writers including Cervantes (who was imprisoned in a building that still stands in the Calle Sierpes), Quevedo y Villegas, and Mateo Alemán made the name of Seville synonymous with the "picaro" figures that characterize the Spanish literature of the period. The wealth from the colonial trade also attracted artists such as Murillo, Velázquez, and Zurbarán, much of whose work reflects the religious zeal of their patrons.

Seville experienced both Spain's sudden rise to power in the sixteenth century and its rapid decline in the seventeenth. Gold and silver receipts dropped off, nearly half the population died in the plague of 1648, a working-class rebellion began shortly thereafter, and the river began to silt up. In 1717 Seville finally lost the monopoly on trade with the Americas to Cádiz. Some compensation for this loss came with the arrival of King Philip V and his court in 1729; they stayed for four years, making Seville the Spanish capital once again, if only briefly.

Seville did make some cultural contributions in the eighteenth century, becoming a center for flamenco music and for bullfighting: construction of the Real Maestranza, Spain's first round bullring, was begun in 1760.

Napoléon's armies invaded Spain in 1808, and the city was occupied from 1810 to 1812. The Peninsular War marked the beginning of more than a century of constitutional crisis in Spain in which Seville was seldom directly involved but was nevertheless disrupted significantly. Reflecting the change in regimes who tried to stamp their authority on the town, the Plaza de San Francisco had its name changed seventeen times in 140 years. The city also lost some of its art treasures to foreigners: Napoléon's general Nicolas-Jean de Dieu Soult took with him some of the best Murillo paintings in 1812, and many more were bought from monasteries by Baron Taylor, creating a boom in Spanish art in the Paris of the 1830s.

Seville continued to influence writers and artists of the nineteenth century, however. The writer Prosper Mérimée was inspired to write *Carmen* by his experiences in Andalusia. *The Alhambra* author Washington Irving was a resident of Seville for a time, while Lord Byron found inspiration in Tirso de Molina's (Gabriel Téllez) much-reworked classic *Don Juan* while living in the city. Together with novelist-politician Benjamin Disraeli, Hispanist Richard Ford, Bible salesman George Borrow and painters Gustave Doré, David Wilkie, and David Roberts, they made Seville the capital of a romantic and often falsified Spain that captivated the imagination of travelers. The composers of operas saw the city the same way. *The Force of Destiny, Carmen, Don Giovanni, The Marriage of Figaro, The Barber of Seville,* and numerous other operas, from the eighteenth and nineteenth centuries, are all set in the city.

The nineteenth century also saw the construction of Seville's first bridge, an impressive iron structure named for Queen Isabella II, but known locally as the "Triana bridge" after the neighborhood it joined to the central city. The Guadalquivir River, which had flooded the town sixteen times in only sixty-three years during the seventeenth century, broke its banks several more times in the 1890s. Because of the threat of flood, the Alfonso XIII channel was dug at the beginning of the twentieth century and the river was blocked, only to be reopened in 1992.

The city's five-mile medieval walls were torn down in the 1860s to make way for a highway. Some sections still exist to the north of the city center in the Macarena district. The city was also shaped by the arrival of the railroads and, most importantly, by the decision to hold the 1929 Iberoamerican Exhibition in the city. Entire neighborhoods such as Porvenir and Heliopolis date from this period, as do the buildings around the María Luisa Park, including the huge Plaza de España and the Hotel Alfonso XIII, both designed by Anibal González.

The exhibition was not a commercial success, and plans to stimulate foreign trade were frustrated by the economic downturn of the 1930s. The Depression was also an aggravating factor in the political upheaval in Spain, which eventually degenerated into the Civil War of 1936–39. Left-wing political activism in Seville had earned the city the nickname of "Sevilla la roja" (Seville the red). But the right-wing military coup succeeded in the city as a result of swift action by military governor Quiepo de Llano, who later claimed that he had taken the city with only 180 men. Quiepo de Llano became famous for his bloodcurdling radio broadcasts, and is thought by some to have ordered the death of poet Federico García Lorca in Granada. Llano's coup in Seville was to prove decisive in the war: General Franco's rebel troops flew up from Morocco into Seville's former Tablada airfield, which was also later used by Hitler's "Condor Legion." The Francoist crackdown in Seville is thought to have been particularly severe: estimates of victims range from 9,000 to 49,000 people.

The 1940s saw substantial population growth in Seville. The total number of inhabitants rose from 281,000 to 376,000, most of it accounted for by immigration from the surrounding countryside. A further population boom occurred after the death of General Franco and the return to democracy in 1975. Population increased from around 640,000 to 725,000 in the decade before 1992. This trend, together with a tendency for residents to move out of the old city center, meant—for most practical purposes— the incorporation into Seville of the villages of the Aljarafe hill.

After the election of Seville-born Felipe González as president of Spain in 1982, the ruling socialist party favored projects centering on the underdeveloped south. The biggest of these was the Expo '92 World Fair, which attracted more than 100 participating nations, 43 million visitors and more than $8 billion in infrastructure spending in the area. The exposition was held on reclaimed marshland on the Island of the Cartuja, part of which is now used as a theme park, and the rest for business and scientific research. Road systems and other communications were revolutionized in preparation for the event. Seven new bridges were constructed and the AVE, a $4 billion, high-speed train (capable of going 180 miles per hour), was built, making the 320-mile journey to Madrid a 150-minute trip. Seville today is well known for its festivals, particularly the processions of the Semana Santa (Easter Week) and the Feria (April Fair), each of which are attended by more than 1 million people every year.

Further Reading: The University of Seville's *Historia de Sevilla* series covers all of the city's history, and is far more reliable and less romanticized than many other works on Seville. Some of the books in the series are *La Ciudad Antigua* by Antonio Blanco Freijeiro (1984); *La Sevilla del Siglo XVII* by Antonio Dominguez Ortiz (1986); *Sevilla en el Siglo XX* by Alfonso Braojos, Maria Parias, and Leandro Alvarez (1990) and *Siglo XVIII* by Francisco Aguilar Piñal (1982). There is no good, general account of Seville in English, but the classical origins of Andalusian culture are covered well, albeit sometimes in rather speculative fashion, in Allen Joseph's *White Wall of Spain* (Iowa City: University of Iowa Press, 1983).

—Richard Bastin

Side (Antalya, Turkey)

Location: On the southern coast of Turkey, off the Gulf of Antalya. Between the mouths of the Eurymedon and Melas Rivers, Side is located approximately 40 miles east of the city of Antalya, in Antalya province.

Description: Ancient city colonized by Greeks during the seventh century B.C. after the Trojan War; later under Roman rule. Has extensive and well-preserved ruins, primarily from the Roman period, including a theatre, city wall, main gate, aqueduct, temples, and agoras.

Site Office: Tourist Information Office
Side Yolu Üzeri
Side, Antalya
Turkey
(242) 7531265

As both vacation resort and classical site, Side is one of the best known in Turkey. In addition to its scenic location off the Gulf of Anatylya, Side boasts a rich history dating back to the seventh century B.C. Side's monuments of historical significance include its theatre, agoras, Apollo Temple, Roman baths, fountain, and city wall. The theatre, built on colonnaded arches, seated 15,000 people in its time, and is one of the largest ancient theatres in Asia Minor. The two agoras once hosted slave auctions; one of them is now filled with souvenir shops. The famous Roman baths, supposedly used by Antony and Cleopatra, now form a museum, housing some of the most significant artifacts found by archaeologists throughout Turkey. In the courtyard off the bathhouse lie many other ruins of ancient Side. What is left of Side's imposing city wall can be seen near the main entrance to the city.

During early Greek settlement in Asia Minor, Herodotus wrote that in this region "the climate is the most beautiful in the world and there is no other region more blessed." Side was settled by Greeks in the second half of the seventh century B.C. as one of the five cities making up the region of Pamphylia. From Pamphylia, a word meaning "the land of all tribes," there arose Attaleia, Perga, Sillyum, Aspendus, and Side. Many of Side's immigrants came from Cyme, an Aeolian Greek city just north of Smyrna. The Greeks, who were skillful and adventurous sailors, mingled with the native population. The exact character of the natives, and whatever existing settlement the Greeks found, remain unknown.

Despite widespread Greek immigration, Side maintained its "barbaric" or "pre-Greek" dialect long after its neighboring cities had shown evidence of a Hellenistic form of language. The native Sideten tongue, most likely of An-atolian origin, is of considerable historical importance not only because of its anomalous character, but also because it has allowed historians to postulate a genealogical relationship between the ancient Hittites and Anatolians. It has not been determined if there was any difference between the language or ethnicity of those in the city as opposed to those living on the outskirts. However, Side maintained this dialect until the third century B.C., despite its remaining unintelligible to inhabitants of any of the other Pamphylian cities.

Although the word Side translates as "pomegranate," a symbolic reference to fertility, there is some question as to the specific derivation of the city's name. The word Side is considered an atypical translation of the word pomegranate, leading scholars to assume that this word also had early Anatolian origins. The same name was found in many locations around Asia at this time. Another explanation of the city's name connects it to a woman named Side, daughter of Taurus and wife of Cimolus, but the woman was probably mythical. Side's currency remained emblazoned with a pomegranate up to the fifth century, with the image taking various forms.

In its early centuries Pamphylia came under Lydian and then Persian rule. There is little knowledge of Sideten history, beyond the history of the region as a whole. However, the inhabitants of Side did not have the best of reputations; the great harpist Stratonicus, when asked who were the most wretched of all men, replied, "The men of Side."

In the fourth century B.C., Alexander the Great conquered southern Asia Minor; Side readily surrendered to Alexander. The city's official language then became Greek by Alexandrian decree. During this time, Athena, the city's principal deity, was displayed on its coinage holding in her hands a pomegranate. Yearly games began to be held in Side in her honor, and she was worshiped in the temple dedicated to her.

Upon the death of Alexander the Great in 323, Antigonus I, who had been one of Alexander's generals, took control of Pamphylia. Then, during the third century B.C., the region fell under the rule of the Ptolemies of Egypt and the Seleucids of Syria. Side apparently remained unaffected by these changes and uninvolved in military conflicts. In 190 B.C. Side was the location of a naval battle between forces of the Seleucid leader Antiochus III, led by Hannibal, and the Rhodians, who were fighting on behalf of Rome. While Side sympathized with the Roman cause, it stayed out of the battle. Side was friendly to the Romans, and during the second and first centuries B.C. the city enjoyed its first great period of prosperity as a result of this relationship. Some time during this period the elegant city wall was erected.

The wall, constructed with regular breccia block, acted as a splendidly decorated fortress protecting the penin-

Side's theatre
Photo courtesy of Embassy of Turkey, Washington, D.C.

sular city from attack during wartime. The remains of this wall, still visible near the main gate, show the interior configured in a three-tier fashion. The third story, in partial preservation, has a series of windows. Reinforced by towers placed at irregular intervals with loopholes at the middle tier used to project missiles upon military aggressors, this wall provided protection from both land and sea attacks. The bottom story supported the two upper stories with archways and pillars. The wall at Side is one of the best such structures on the southern coast of Turkey.

During the early second century, the harbor of Side became the major port for pirates from the Cilicia region to sell their captives into the slave trade. The inhabitants of Side developed a reputation as "shameless pirates" as well, as they allowed their agoras to be used for the auctioning of slaves and booty. A Roman naval expedition led by Servilius drove many of the pirate vessels out of eastern Lycia and western Pamphylia; these regions then became a Roman province. Side remained independent, however, for some time after its neighbors had come under Roman rule, and the slave trade continued until Pompey cleared the sea of pirates in 67 B.C.

Side eventually did come under Roman rule, and it flourished, especially in the second and third centuries A.D.

Legitimate trade replaced the slave market, and Side's most ornate city buildings were erected, along with its main gate, a city fountain, aqueduct, colonnaded streets, and its theatre. Although the city had four gates, one from the countryside and three from the sea, the main gate, similar to the one found at Perga, was constructed for the purpose of better securing the city from attack. The main gate consists of four doors surrounded by a convex, semicircular wall. Across from the main gate, also probably constructed at this time, was the city fountain. Once three-tiered and decorated with statues, the fountain now retains only its lowest tier, basin, and holes where water once flowed. Its statues and other artifacts found during excavation of this site are on display in Side's museum. Also built during this period was Side's aqueduct. Once bringing water from the nearby Melas River, all ten sections of the aqueduct remain standing. West of the main gate, there still exists a hole in the city wall where a pipe bringing water into the city was once situated. Two colonnaded streets added to Side during this time led into the city from the main gate, one to the north and the other to the south. Although the southern road is overgrown with vegetation, the northern road is still in use today, having been well preserved.

Side's immense theatre, seating 15,000 people in 48 rows of seats, was erected during this period and is now one of the best preserved in all of Asia Minor. The theatre was designed by a native Anatolian architect, and although a minor Greek influence can be detected in its exaggerated curve, a predominantly Roman style is evident in the theatre's barreled substructures. The walkway to the theatre is paved in a mosaic, and at the theatre's foot is a marble kiosk decorated with the twelve zodiacal signs around its ceiling. Theatregoers reached their seats by way of archway, passageway, and stairs. The stage area was decorated with columns, niches, statues, and reliefs. Behind the stage were representations of deities such as Apollo, Ares, Hermes, Asclepius, Hercules, Hygeia, Nike, and Nemesis—all now in rather poor condition.

Originally, the theatre was used for the regular performance of plays. In late Roman times, however, an additional wall was constructed between the two-tiered stage and the audience to facilitate gladiatorial fights involving wild animals. Around the fifth or sixth century A.D. the theatre was used as a church; inscriptions have been found on certain seats ceremoniously reserved for priests.

As Roman authority waned in Side and its region during the third century, there came a resurgence of piracy, most notably an attack on Side by Scythian buccaneers from the Black Sea region. Although not escaping damage, Side emerged from this siege victorious. Also during this time, incursions by the Isaurian people from the mountains to the north contributed to the overall state of disarray at Side. Due to the various invasions and the general decline of the Roman Empire by the middle of the fourth century, Side and its neighbors had become impoverished. One result of this in Side was the construction of a wall cutting the size of the city in half. This left the northeastern part of the peninsula deserted.

Under Byzantine rule after the division of the Roman realm, Side expanded well beyond its original limits during the fifth and sixth centuries. The theatre and other buildings show evidence of repair at this time, and outside the main gate a forum was erected. Under the patriarch of Constantinople, Side came to be at the center of a thriving diocese. Also in the sixth century, Side native Tribonian became chief legal minister to Emperor Justinian I and helped draft Justinian's law code.

By the seventh century, Side again fell into disarray as a result of Arab attacks. Eventually the city came under the rule of the Arabs. Despite brief periods of control by the Byzantines, during which there were minor improvements, Side continued to decay and was largely in ruins by the tenth century. After a fire, what was left of its population came to inhabit Antalya. During the twelfth century, the Arab geographer Edrisi described what was once the city of Side as a "burnt Antalya."

In 1811, the English naval officer Captain Francis Beaufort came upon Side during his survey of ancient Anatolia. Sailing into Side aboard his ship, *Frideriksteen,* Beaufort described the site as being ovrrun by vegetation and in ruins but, at the same time, filled with magnificence. Upon the completion of his journey Beaufort wrote and published a detailed description of the ancient sites all along Anatolia's southern coast.

While Beaufort had praised Side, Sir Charles Fellows, who came to Side during the 1830s, was considerably less impressed by what he found. Fellows reported that upon his arrival in Side, he came across not one residence or housing establishment—only tents inhabited by nomads, scattered randomly about the ancient city. To Fellows's dismay, most of the ruins appeared to be Roman, not Greek (Greek antiquities were held in greater esteem). Later archaeological findings at Side have proven Fellows's assessment to be correct. Several scholars have explored the area, and from 1947 to 1966, the archaeology department of the University of İstanbul worked on the restoration of ancient Side.

Over the past few decades Side has grown considerably. To the objections of some, Side has become not only a place of historical value, but also a booming tourist site. With restaurants overlooking the sea, vendors along the old colonnaded street, and a number of hotels to choose from, Side now attracts sightseers year-round.

Further Reading: Works that provide helpful information on Side include Harry Brewster's *Classical Anatolia: The Glory of Hellenism* (London and New York: Tauris, 1993), George E. Bean's *Turkey's Southern Shore* (New York: Praeger, and London: Benn, 1968), and John Freely's *Classical Turkey* (London: Viking, and San Francisco: Chronicle Books, 1990).

—Christopher P. Collier

Siena (Siena, Italy)

Location: Central Italy, 43 miles inland from the Mediterranean Sea, 44 miles south of Florence, and 143 miles northwest of Rome; in the province of Siena and region of Tuscany.

Description: A small medieval city, 1,115 feet above sea level, whose ancient ramparts enclose approximately two and one-half square miles of narrow streets running between tall medieval mansions and palaces. Eleven streets lead into the magnificent scallop-shaped Piazza del Campo, built between 1297 and 1310. Siena is famed for its architecture and art, as well as for the Palio delle Contrade, a dangerous race around the square between horses from ten of the seventeen *contrade,* or wards, run each year on July 2 and August 16.

Site Office: A.P.T. (Azienda di Promozione Turistica)
Via di Città, 43
53100 Siena, Siena
Italy
(577) 42209

Siena, built on three clay hills after which the color "burnt sienna" is named, was a medieval commercial and banking center and is still dominated by the splendid Gothic architecture of its public and domestic buildings. The city is equally renowned for the artistic treasures of these buildings' interiors.

The emblem of Siena is the same as that of Rome: a she-wolf suckling Romulus and Remus. As Romulus was purported to have founded Rome, so Remus's sons, Senius and Aschius, were postulated by legend as the founders of Siena. In fact, the city's origins can only be conjectured from a few Etruscan tombs found outside the Porta Camollia, its northern gate, until it came under Roman rule as Saena Julia. Only scraps of evidence make it possible to trace the city's development, in some ways similar to that of many other towns, toward emergence as a democratic republic. Siena, at first a settlement governed by a civic personage and then by a bishop, came under the dominion of an aristocratic oligarchy from whom were drawn three to a dozen consuls.

As early as 1147, the nobles were constitutionally forced to accept fifty ordinary citizens, along with a hundred from their own ranks, into the city's general council. As the period of Siena's cultural ascendancy began in the thirteenth century, the citizens were in open rebellion against the privileged aristocratic exemption from taxation, and they sought to remove the nobles from power altogether. The resulting compromise created a supreme magistracy of twelve nobles and twelve *populani* known as the Government of the Twenty-four. Rivalry with Florence over the control of intervening territory led Siena to take the part of the Ghibellines,

as those siding with the emperors came to be known, against the Guelph cause, essentially more democratic and anti-German than pro-papal, to which Florence committed itself.

Following a cycle of intermittent hostilities and hasty reconciliations between the towns, the Sienese and their allies emerged victorious over a much stronger Florentine force at Montaperti in 1260. Clement IV, pope from 1265 to 1268, gave the Sienese the right to collect tithes from the Holy Land, although they were soon to be usurped as papal bankers by the Florentines. The benefits of the Montaperti victory, however, were soon eroded by the crushing defeat of pro-imperial Siena at Colle di Valdelsa by the pro-papal Florentines and French in 1269, and by the augmented strength of popular, pro-papal sentiment within what had become a city-state. By 1277, the popular party had excluded all nobles from the supreme magistracy, whose numbers had been increased in 1270 from twenty-four to thirty-six, all of whom sided with the pope against the emperor. In 1280, the new constitution was confirmed when the supreme magistracy was reduced to fifteen. The constitution was definitively sanctioned in 1285, after which point the city was ruled by a council of only nine, all non-noble, which achieved prosperity within the city for some seventy years.

By the early fourteenth century, the city's principal buildings had been erected. The Black Death arrived at Siena in May 1348 and by October had reduced the city's population of 52,000 by two-thirds. The death toll was greater than that of London or Paris, and not until the twentieth century did Siena's population climb back to its pre-plague level. The accompanying economic crisis meant the suspension of building in the city for two generations, and the plague also affected Siena's cultural life by claiming several of its best artists, including the brothers Pietro and Ambrogio Lorenzetti. Work by the brothers and other artists had brought Siena a reputation as a major cultural center. The city's affection for the arts, however, survived the plague, and it would produce other great painters during the Renaissance.

Meanwhile, the nobles and populace had each come to resent the exclusivity of the social group from which the magistracy was drawn. Sporadic unrest continued until Charles IV, king of Bohemia and Holy Roman Emperor from 1355, overthrew the government and substituted a council of twelve drawn from the lowest social class. The unrest intensified, and Charles IV was for a time held captive before virtually selling his captors an imperial patent. Siena's mostly oligarchical constitutional arrangements had continued the medieval tradition and differed from those of most cities to its north, in which power came to be concentrated in the hands of a single prince even if, as in Florence, the constitution itself was republican. Siena had had to pay the price of sporadic but serious disturbances and blood-letting.

Nave of Siena's cathedral

Pulpit of Siena's cathedral
Photos courtesy of The Chicago Public Library

In addition to political strife, the late fourteenth century saw much economic distress in Siena. There were several years of bad harvests, resulting in food shortages, and the city's streets and water system were in a state of disrepair. There also were struggles between workers and employers, culminating in an uprising by wool workers, in anger over poor wages and working conditions, in 1371. The workers seized stockpiles of grain from various houses and attacked the Palazzo Publicco. The revolt was put down, but the power of the traditional aristocracy remained tenuous for ten to twenty years thereafter. The aristocracy itself was changing as well; whereas Siena's elite had once been active in banking and commerce, it now was adopting more of a feudal model, like the rest of Europe, in which wealth was gained renting out lands they owned or from holding church or civic office. The result was further decline in Siena's once vibrant economy.

At this time Siena also had to contend with invasions by mercenary bands bent on pillaging the city. It became so costly to either defend the city or buy off the mercenaries that from 1399 to 1409, Siena, as a protective measure, accepted the suzerainty of Milan.

Siena remained involved in conflict with its neighbors. In the early fifteenth century, Siena allied with Florence against Pope Gregory XII, and then broke with Florence in a territorial dispute involving Lucca. When Florence sided with Venice and Eugenius IV, Siena was allied with Milan and the future emperor, Sigismund. A peace treaty was signed in 1433, although relations with Florence remained tense, and sometimes broke into open warfare. Siena enjoyed a long period of prosperity and cultural renewal toward the end of the fifteenth century, with the city's aristocrats appearing to be firmly in control.

The aristocratic families actually were feuding and factionalized, however, and in 1524 one of the leading families, the Petrucci, were thrown out of the city by their political opponents and the government was overhauled. Then in 1527 the sack of Rome had the effect of further depressing trade in Siena, and the economic difficulties led to a series of revolts by the lower classes which, combined with factionalism among the nobles, plunged the city into chaos. Imperial troops were called in to restore order in 1530, after which a permanent garrison was installed in Siena.

There followed many years of attempts to reform and stabilize the government of Siena, without any lasting success. The city also became increasingly dependent on, and dominated by, the soldiers and emissaries placed there by Holy Roman Emperor Charles V, a Spaniard. While their power came to be resented, Siena remained loyal to the emperor until 1552 when, with the help of the French, the Sienese launched a rebellion. They drove out the imperial troops, but the empire soon regrouped. It failed to retake the city in 1553, but in 1554 began a long and debilitating siege, resulting in Siena's capitulation in 1555. The siege reduced the population to a size smaller than it had been after the Black Death. The emperor gave the city to his son, King Philip II of Spain; two years later Philip sold it to Cosimo I de' Medici, duke of Florence, who had at first abetted the rebels and then switched to the imperial side. Cosimo formally incorporated the city into his Grand Duchy of Tuscany in 1559. Cosimo and his successors, who ruled Siena until 1737, suppressed Sienese culture and any sense of independence, wishing the citizens to grant their primary loyalty to Florence. The city still produced gifted artists, but their work was little known outside Siena, indicating that the rest of the world did not view the city as any sort of leader. Some bright spots in this period were advances in education and medicine; in the mid-eighteenth century, for instance, the doctors of Siena's Santa Maria della Scala hospital were among the first to practice widespread inoculation against smallpox.

The Medici line of dukes died out in 1737, after which Siena passed to the dukes of Lorraine, who were driven out by Napoléon's revolutionary armies in 1799. After the fall of Napoléon, Siena was restored to Lorraine, whose rulers inspired little enthusiasm among the Sienese. This is evidenced by the fact that in 1859 Siena was the first Tuscan city to vote for a unified Italy under the monarchy of Victor Emmanuel II.

By the nineteenth century Siena had become a popular tourist destination, especially for writers and other intellectuals. Their writings often described Siena as a decaying but picturesque city. The Sienese reaction was to restore their city and take a new pride in their medieval heritage, a pride that persists to this day. Siena in the twentieth century has remained aloof from political and cultural upheavals, preferring to glorify its past, one that is indeed rich.

Art historians speak imprecisely of a "Sienese" school of architecture, painting, and decoration, which is commonly related to such other manifestations of the city's culture as the spirituality of St. Catherine of Siena (1347–80), who played an important part in bringing the papacy back to Rome from Avignon, and St. Bernardine, born in 1380 and notable for his preaching and reformatory endeavors. The essential characteristic of the Sienese school is variously said to be a reliance on color, and on blending clearly Byzantine structures, motifs, and impersonal, religiously determined stylization into innovative western Gothic decorative features.

In sculpture, the artists chiefly associated with the school include Nicola Pisano, famous for grafting Romanesque design onto increasingly emphasized Gothic features in the thirteenth century, and for replacing classical with Christian figures. His son, Giovanni, deepened the dramatic impact of Gothic styles by further incorporating monumental statues into the architectural framework. Nicola Pisano is responsible for the strong sense of movement in the figures on the pulpit of Siena cathedral, and Giovanni for the figures on the cathedral facade as far up as the gables over the three recessed doorways. The upper portion of the facade dates from the middle of the first half of the fourteenth century, and is by master-builder Giovanni di Cecco.

In painting, the principal names associated with the Siena school include Duccio di Buoninsegna, who, in the late thirteenth and early fourteenth centuries, deepened the Byzantine features of Sienese art by strengthening color and energy within the new Gothic framework and produced the *Maestà* (Virgin in Majesty) altarpiece now in the cathedral museum. His pupil, Simone Martini, accentuated the Gothic elements again in his use of silhouette and patterns of color and whose 1315 *Maestà* is in the Globe room of the Palazzo Publico. The Lorenzetti brothers were connected with the tradition introduced at Florence by Giotto and developed the naturalistic tendencies of Duccio with solid figures and expressive forms; both brothers worked at Siena. Pietro is the creator of the frescoes in the church of San Francesco, and Ambrogio of the allegorical frescoes of good and bad government in the Palazzo Pubblico.

Duccio's earliest work was for Florence; Simone Martini moved early from Siena to work in Naples, Pisa, and Orvieto before returning to Siena and then moving to Avignon; and the Lorenzetti brothers also worked outside Siena. Therefore, the "Sienese school" as a category used by art historians is relevant for work well beyond that executed in or for the town of Siena.

In Siena, the Gothic brick Palazzo Pubblico on the Piazza del Campo was built between 1288 and 1309. The elegant 289-foot tower at its side, the Torre del Mangia, named after a medieval bell-ringer, was added from 1338 to 1348. The accompanying chapel, built as a public offering after the end of the 1348 plague, was erected between 1352 and 1376. Inside the Palazzo, the major works of art include Simone Martini's *Maestà* and his portrait of the Sienese general Guidoriccio da Fogliano, and the frescoes of Ambrogio Lorenzetti. The chapel has frescoes by Taddeo di Bartolo di Fredi, impressive inlaid fifteenth-century choir stalls, and an altar painting by Giovanni Sodoma. The decoration dates from the early fourteenth to the early fifteenth centuries. It is, nonetheless, only when viewed as a power center of the city's late medieval administration in its historical context rather than merely as an interesting building with fine interior decoration that the power of the Palazzo ensemble achieves its greatest effect.

The cathedral, Byzantine in its decorative style, is impressive chiefly for the breathtaking quality of its interior, which measures 289 feet in length and 174 across the transept. Originally Romanesque in conception, the exterior was fundamentally changed by a later addition of the upper Gothic portion of the facade. Later still, in 1339, a plan was begun to incorporate the present cathedral into a huge new one (the present nave would have formed the transept), but work had to be abandoned because of the Black Death. Built partly in black and white marble, it was erected on foundations laid in the mid-twelfth century and was consecrated by 1255, although the aisles were still unroofed, and the dome had still to be built. The slim campanile, constructed of alternating layers of white and black marble, has remarkable pillared windows whose width increases with their height.

From 1258 to 1314, construction was controlled by the Cistercians of San Galgano, who finished the dome in 1294. The Romanesque lower part of the facade, mostly designed by Giovanni Pisano, was built between 1284 and 1296, but the cathedral was then fundamentally replanned with an extension toward the east and the construction of the baptistry. The Gothic upper portion of the facade by Giovanni di Cecco was inspired by the facade of the cathedral at Orvieto, and the whole facade appears compressed between the two massive side towers with their delicate stone tracery. Many of the original statues have now been removed to the cathedral museum, leaving reproductions in their place.

The interior, in which a tall nave is separated from the lower-roofed side aisles by clustered pillars faced in marble and supporting massive round arches, is dominated by the effect of the alternating bands of black and white and the profusion of rich but heterogeneous decorative features. The bands of color relate the cathedral to the Romanesque architecture of Pisa and Lucca, while the main structure is derived from Lombard models and there are specifically Cistercian features like the decoration of some of the capitals. Of exceptional interest is the partly mosaic paving of the cathedral floor, which depicts various allegories and classical and biblical scenes. Domenico di Pace Beccafumi, inheriting the bright colors of the Sienese tradition, is responsible for some of the paving and the paradise frescoes in the chancel. Pinturicchio also worked on the paving and created the fresco of the crowning of Aeneas Piccolimini with the poet's laurel wreath by Frederick III in the Piccolimini library adjoining the cathedral. An intense *John the Baptist* by Donatello can also be seen in the chapel.

Other notable artists whose work remains in the cathedral or has been removed to the cathedral museum include Bernini, Ghiberti, and Jacopo della Quercia. The magnificent inlaid woodwork stalls of the chancel were carved at various dates between the fourteenth and sixteenth centuries. The inlaid panels were added in 1503. The marble pulpit was commissioned from Nicola Pisano to be richer and more elaborate than the one he had just completed in Pisa. Among his assistants were his son, Giovanni, and Arnolfo di Cambio, who was paid a sum nearly equal to that paid to Nicola. Di Cambio probably died in 1302, and is best known as the architect of Florence's cathedral.

The Pinacoteca contains an exceedingly rich concentration of the finest Sienese painting from the thirteenth to the sixteenth centuries. A score of other buildings and churches in this small town are not only marvels of architectural splendor, but also contain within their walls a whole catalog of artistic masterpieces. The town, however, does not give the impression that it is a mere museum catering to the cultural tourist. The street flagstones reaching to the base of the patrician houses that line them increase the impression of narrowness and bustle, and the appurtenances of the tourist industry do not altogether swamp the manifestations of a vigorous local culture as they do in some other Italian cities.

Modern Siena preserves a continuity with its medieval past. Even the Palio delle Contrade is based on the medieval division of the city into the wards of Città, Camollia, and San Martino, originally divided into fifty-nine *contrade*. Seventeen still exist, and the *palio* starts with a procession from each, carrying banners, with the chariot of liberty, or *carroccio*, and the municipal guard armed with crossbows. Each year, ten *contrade*, chosen by lot, send a horse and rider to compete in the race of three circuits of the sand-strewn course around the edge of the square. The *palio*, or standard of the Virgin, Siena's patron, is awarded to the winner. The race, at the heart of a public festival, replaced the bullfights of the fifteenth and sixteenth centuries and the mounted buffalo races from the late sixteenth century.

Further Reading: A mass of work exists on the art and architecture of Siena, on its principal artists, and on its history. For the general non-Italian-speaking reader, the most important modern works include *Painting in Florence and Siena after the Black Death: The Arts, Religion, and Society in the Mid–Fourteenth Century* by Millard Meiss (Princeton: Princeton University Press, 1951; Oxford: Oxford University Press, 1952); *Siena: A City and Its History* by Judith Hook (London: Hamish Hamilton, 1979); *A Medieval Italian Commune: Siena under the Nine (1287–1355)* by William M. Bowsky (Berkeley: University of California Press, 1981); and *Siena and the Sienese in the Thirteenth Century* by Daniel Waley (Cambridge and New York: Cambridge University Press, 1991).

—Claudia Levi

Sinop (Sinop, Turkey)

Location: Seaport on the southern coast of the Black Sea, in north-central Turkey. The chief town of the Sinop province, it lies on a low isthmus linking the Boztepe Peninsula to the mainland. The town is isolated from the Anatolian Plateau by a high, forested mountain range.

Description: Sinop (historically Sinope) is the oldest city and only natural harbor on the Black Sea coast. It was an important port for the second millennium B.C. Hittites, but was destroyed by the Cimmerians. Sinop was refounded in the seventh century B.C. by Ionians from the Aegean kingdom of Miletus, and by the fourth century B.C. had become the most flourishing Greek port on the Black Sea and the center of a wide-ranging maritime power. In the second century B.C., Sinop became the capital of the Pontic kings. It continued to prosper until being colonized by the Romans in the first century B.C. and declined further during the Byzantine period due to the diversion of traffic to other ports. It was part of the Greek Empire of Trebizond in the Middle Ages, but was taken by the Seljuk Turks in the thirteenth century. After the Mongols weakened the Seljuk state, Sinop became a principality under the Turcoman Candar (or Isfendiyar) Dynasty in the early fourteenth century, which it remained until being annexed by the Ottoman Empire in 1458. The Crimean War (1853–56) began when Russia destroyed much of Sinop in 1853.

Site Office: Tourist Information Office
Hükümet Konaği, Kat
Sinop, Sinop
Turkey
(368) 2615207-2618301

Sinop is the oldest city and only natural harbor on the Black Sea coast. According to legend, it was founded by the Amazons, who named it for their queen Sinova (or Sinope), the daughter of a minor river god. Sinope attracted the interest of Zeus, who offered her any gift she desired in return for her favors. She requested everlasting virginity, thus foiling his amorous intentions. Zeus complied, and she lived happily ever after. Sinop's ancient inhabitants, however, ascribed the city's origin to Autolycus, a companion to Hercules.

Sinop's fine natural setting on an easily defended peninsula led to a late Bronze Age settlement in the second millennium B.C. It served as a port for the Hittite "Old Kingdom" (c. 1700–1200 B.C.) capital of Hattusa, south of Sinop; the Hittites were the first major civilization to emerge in Anatolia. The town was destroyed by the nomadic Cimmerians of south

Russia about 700 B.C. Around 630 B.C. Sinop was founded as the largest Ionian colony of the distant Aegean kingdom of Miletus. The Miletians imported their Greek culture; in the fifth century B.C., Sinop received an Athenian colony and, by the fourth century, the city had become the most prosperous Greek settlement on the Euxine (Black) Sea. Sinop extended its maritime authority over a wide range of country; it established a number of Greek colonies along the coast to the east, and had the most dominant naval fleet on the Black Sea. Sinop also flourished because it was the terminus of a great caravan route from the Euphrates River to the sea, over which traveled the products of central Asia and Cappadocia. The ruins of a second-century B.C. Hellenistic temple still stand in the center of town.

In 220 B.C. the Ionians successfully defended Sinop from an attack by Mithradates III, king of Pontus, a region in northeastern Anatolia bordering on the Black Sea. Pontus, a former district of Cappadocia, emerged as a strong political realm at the end of the fourth century B.C. after Persian nobleman Mithradates, who claimed descent from Darius the Great, founded a small kingdom and dynasty there in 337-36. In 183 B.C. King Pharnaces I finally took Sinop by surprise, annexing it as the capital of the Pontic monarchy, which combined Persian and Hellenistic traditions.

Pontus reached its zenith as an expansionist power during the long reign of Sinop native Mithradates VI Eupator, also known as "the Great" (115/120–63 B.C.). Mithradates built his birthplace into a splendid city of marble avenues, gymnasiums, and colonnaded agoras, with naval arsenals and well-built harbors. Virtually no trace of these monuments remains, except the city walls. The defenses, originally built by Sinop's seventh-century colonists, were fortified by the Pontic kings, who added the isthmus-straddling walls. The city walls were strengthened by subsequent occupiers. Though Sinop's ruined citadel had originally been built by the Hittites, its present layout dates from the second century B.C., when Mithradates IV ordered its reconstruction. The citadel, too, was rebuilt by later rulers.

Mithradates VI came into conflict with Rome when he tried to take the neighboring kingdoms of Bithynia and Cappadocia, thus beginning Rome's three Mithradatic Wars in the early first century B.C. Though Mithradates ceded his conquered territory—except for the overseas province of the Crimea—to Roman general (and later dictator) Sulla in 85 B.C., Sulla's lieutenant Murena nevertheless initiated a second war against the Pontic king in 83, losing in 82. When Rome annexed Bithynia in 74, Mithradates went to war with Rome for the third time. General Lucius Licinius Lucullus expelled Mithradates from Pontus; decisively defeated at Nicopolis by lieutenant Pompey, Mithradates took refuge with the king of Armenia in 72. Lucullus finally captured

Harbor at Sinop
Photo courtesy of Embassy of Turkey, Washington, D.C.

Sinop in 70, nearly burning it to the ground, and the city was annexed to Rome by now-general Pompey in 63–62, becoming part of the Roman province of Bithynia-and-Pontus.

Mithradates VI—Rome's greatest enemy in Asia Minor—was a remarkable, almost romantic historical figure. He was renowned for his stature, strength, courage, despotic traits, and political and strategic skills. He modeled himself after Alexander the Great, and thought of himself as a Hellenistic liberator. He was also known for his intellect; Roman historian Pliny the Elder said he had mastered twenty-two languages, the number spoken in his dominions. Perhaps most remarkably, Mithradates is said to have so thoroughly saturated his body with various poisons, as well as their antidotes, that he was immune to their effects. After Lucullus took Armenia in 69 B.C., Mithradates tried in vain to poison himself; instead, he had himself stabbed to death by a Gallic mercenary in 63. His body was sent to Pompey, who buried it in Sinop's royal mausoleum.

While Sinop became a Roman colony under Julius Caesar, it began to lose much of its prosperity due to the diversion of shipping trade to Amisus (Samsun) and Ephesus, the Roman Empire's main port in Asia Minor. Sinop declined through the Byzantine period, when it was part of a region called Paphlagonia. Attempts to revive its fortunes in the sixth and seventh centuries were thwarted by frequent Persian and Arab raids.

Sinop became subject to the Greek Empire of Trebizond (the modern Trabzon) in the Middle Ages. The city of Trapezus, as the ancient Greeks called it, had been founded as a seaport colony by Sinop's Ionian settlers. But it began to assume greater importance as a Black Sea power when it was recaptured from the Seljuks by Alexius I Comnenus, Byzantine emperor from 1081 to 1118. After the Comnenus dynasty was driven from the Byzantine throne, two of Alexius's great-grandsons, Alexius and David, established a stronghold here in 1204, the year Constantinople was sacked by Latins during the Fourth Crusade. Alexius became Trebizond emperor Alexius I. The most successful of breakaway Byzantine states, Trebizond existed until 1461, when it became part of the Ottoman Empire.

Trebizond's golden age lasted through the thirteenth and fourteenth centuries. The state, which had a largely Greek culture but a diverse population, managed to survive the transition from Byzantine to Turkish influence by fighting, forging alliances with, and paying tributes to a succession of neighboring domains not in the coastal empire proper: the Seljuk Turks, Mongol khans, Turcoman chieftains, as well as distant traders like Genoese and Venetians. Sinop, a remote but coveted port in a "march" (frontier) province of the empire, was constantly a source of contention.

Seljuk sultan Kay-kaus, wishing an outlet to the Black Sea, acquired Sinop in October 1214 in exchange for the release of Trebizond emperor Alexius I (called Grand Comnenus), who had been captured by frontier Turcomans. As a result, Trebizond was made a vassal kingdom to the Seljuks, and Alexius was compelled to pay tribute to Kay-kaus and to assist the Turks in military matters. Sinop's defenses were restored, a Turkish garrison was installed, and a number of churches were converted into mosques. The town was commanded by Hethoum, a possibly Islamized Armenian; it was felt he was better qualified than a Turk to deal with the port's predominantly Christian merchants.

Sinop's Seljuk domination was short-lived. The Mongol successors of Genghis Khan swept into Anatolia in the 1240s, and Seljuk sultans became puppets to the Mongol khans. Sinop was then a principality ruled by the *pervane* Mu 'in al-Dīn Süleyman; *pervane* (butterfly) was the title Seljuk sultans conferred on their viziers, or administrative ministers. Though Trebizond briefly recaptured Sinop in 1254 or 1259, the town was retaken by Seljuk sultan Keykubat before 1262. At the time of Mongol leader Hülegü's death in 1265, the *pervane* Süleyman paid homage to the new chief of the Il-Khans, the western dynastic branch of the Mongols, and received the town of Sinop as a personal possession. A virtual autocrat (he had proclaimed himself "King of the Emirs and the Viziers"), Süleyman immediately began rebuilding Sinop as a solidly fortified Moslem base, but was dispatched by the Mongols in 1278. Though a Byzantine attack on Sinop was repulsed by Turcomans in 1280, the principality of Sinop was still in the hands of either the son of the *pervane*, Muhammed Beg, or his grandson Mas 'ud (Beg or Chelebi) at the end of the century.

In the early fourteenth century, when Sinop was in the hands of Ghazi Chelebi, believed to be the son of Mas 'ud, the Sinop-Kastamonu region was conquered by a Turcoman dynasty variously known as Candar, Jandar, or Isfendiyar. The dynasty took its name from Semseddin Yaman Candar, who had served in the army of the Seljuk sultan in the late thirteenth century and was awarded a region west of Kastamonu (just south of Sinop) for his services. Candar's son Süleyman, or Sülayman Pasha, had captured Kastamonu by 1314, accepting the suzerainty of Abu Sa 'id, ruler of the Il-Khans. Süleyman conquered Sinop sometime after 1324, since Ghazi Chelebi's activities had been recorded at least until that date. In any event, the death of Abu Sa 'id in 1335 hastened the demise of Il-Khanid power, leaving Sinop firmly in the hands of the Candar-Turcoman state.

The Genoese fleet—one of Europe's most powerful at the time—had begun trading in the Black Sea in the late thirteenth century, and the local Candar emirs allowed the Genoese to use Sinop as the center of their Black Sea trade throughout the fourteenth century. But its trade with Sinop was often disrupted by piracy, which characterized the early fourteenth-century reign of Prince Ghazi Chelebi. The Gen-

oese and the Seljuks were the last to reconstruct the town's ancient hill-top castle.

Dynastic struggles led to the division of the Candar principality into two realms around 1380. The Kastamonu branch accepted Ottoman (Turk dynasty) suzerainty and was annexed by Sultan Bayezid I in 1391; the Sinop branch remained under Candarid rule. In 1402, however, Central Asian Tatar conqueror Timur (or Tamerlane) briefly interrupted Ottoman expansion and restored the entire region to Candar rule again. Dynastic rivalries following Timur's death (1405) led to the division of the principality again in 1417, with one branch falling to Ottoman influence. Following Sultan Mehmed II's campaign along the Black Sea shore, Sinop was incorporated into the Ottoman Empire in 1458; the rest of the principality, as well as Trebizond, followed suit in 1461. Sinop then became a secondary port to Samsun, which has since emerged as the largest Turkish port on the Black Sea.

Sinop did not again rise to prominence until the late Ottoman era. Increasing vulnerability to foreign intervention led to eighteenth-century military reforms and to the creation of a naval base in Sinop in the 1770s. Renewed Russian claims to authority over Christian (Eastern Orthodox) territory in Ottoman lands kindled the Crimean War of 1853–56. During a diplomatic standoff, the Ottoman army crossed the Danube into Russian territory in October 1853; on November 30, 1853, Russian vice-admiral P. S. Nakhimov, commanding a powerful Black Sea fleet, responded by dramatically attacking the main Ottoman fleet anchored in the harbor of Sinop. Most of the Turkish warships were either sunk or run aground, and wind-fanned fires spread from the harbor to the town, reducing much of it to ashes. Only one Ottoman vessel, under an English captain, escaped destruction. Britain and France, outraged by the "Sinop Massacre," joined the war in March 1854, quickly forcing a Russian withdrawal. While Turkish troops had initially invaded the Crimea, the remaining two years of the war were fought largely between Anglo-French forces and Russia; the Ottomans were hardly involved. The Ottoman Empire had prevailed.

Sinop again rose to prominence during the Cold War. The port's proximity to the Soviet Union (the northernmost point in Anatolia, Sinop is less than 100 nautical miles away from the Crimea) precipitated the building of a U.S. military base in Sinop's citadel area. Most recently, the base was used as a secret NATO listening post, which is in the process of being dismantled. Today, the town's main activities include tourism, fishing, and agriculture. Corn, flax, and tobacco are grown in the well-drained valleys and fertile coastal strip.

Few ancient monuments remain in Sinop save the continually rebuilt city walls and citadel, some columns, and the sparse ruins of the second-century B.C. Hellenistic Temple of Serapis (the Egyptian form of Apollo), which was excavated in 1951 and stands in the museum grounds near the Municipal Park in the center of town. While most of the present city walls date from Byzantine, Genoese, and Seljuk times, some of its inscribed stones date from the early Greek and Roman periods. The Balatlar Kilise, a ruined seventh-century Byzantine church, still retains some badly damaged frescoes, although the church's finer icons have been removed to the city museum.

The oldest mosque in town is the Alaeddin Keykubat Camii, built by the Seljuks in 1214 and named for the prominent sultan; its splendid original *mihrab* (prayer niche) is displayed in Istanbul's Museum of Islamic Art. Behind the mosque is the Alaiye Medrese (or Pervane Medrese) seminary, built in 1262 by Keykubat's *pervane,* Mu 'in al-Dīn Süleyman; the most notable feature is its marble-decorated portal. The building now houses a museum with some interesting tombs. The Saray and Fetih Baba mosques were both built in 1339; the Cifte Hamam (Turkish bath) was built in 1332. The fourteenth-century Sayid Bilal mausoleum, on a hilltop at the edge of town, offers a good view of Sinop and the sea.

Further Reading: The definitive *Pre-Ottoman Turkey* by Claude Cahen (New York: Taplinger, and London: Sigdwick and Jackson, 1968) is a general survey of Anatolian culture and history from 1071 to 1330. Lord Patrick Balfour Kinross's *The Ottoman Centuries* (New York: Morrow, 1977) and Stanford Shaw's *History of the Ottoman Empire and Modern Turkey,* two volumes (Cambridge and New York: Cambridge University Press, 1976–77) are authoritative accounts of the Ottoman era (c. 1280–1918). Richard Stoneman's *A Traveller's History of Turkey* (Moreton-in-Marsh, Gloucestershire: Windrush, and Brooklyn, New York: Interlink, 1993) and Seton Lloyd's *Ancient Turkey: A Traveller's History of Anatolia* (London: British Museum Publications, and Berkeley and Los Angeles: University of California Press, 1989) present concise, readable accounts of the country's history. *Trebizond: The Last Greek Empire* by William Miller (London and New York: Macmillan, 1926) may also be of interest.

—Jeff W. Huebner

Sintra (Lisboa, Portugal)

Location: North and northwest of the Tagus River, approximately eighteen miles northwest of Lisbon and eleven miles east-northeast of Cabo da Roca, the westernmost point of the European continent.

Description: Small town at foot of and including Serra de Sintra, for centuries the favorite summer residence and vacation spot for royalty, nobility, and distinguished visitors from abroad; abundant with palaces and lush vegetation; renowned for semitropical climate and ethereal beauty, the subject of a profusion of writings, paintings, etchings, and photographs second only to Lisbon among Portuguese cities; modern Sintra made up the historic center at Vila Velha, the contemporaneous district of Estafãnia, and the village of San Pedro.

Site Office: Sintra Tourist Information Office
Praça da República, 23
2710 Sintra, Lisboa
Portugal
(01) 923 11 57 or (01) 924 17 00

Sintra, a pretty hill town near Lisbon, was an ideal summer distraction for Portuguese kings of the thirteenth to eighteenth centuries seeking respite from the tensions of office and life in the sweltering, plague-ridden capital. To the Romantic writers and wanderers who later visited the spot, Sintra was idyllic, a fairy-tale land complete with castles, arid purple plains, and craggy mountains partially obscured by mist and fragrant with giant magnolias and lemon trees, lavender, periwinkles, and maidenhair. Robert Southey, who traveled there in 1800, asserted that, "for beauty . . . all existing scenery must yield to Cintra." While Sintra was rarely the stage for decisive political or military events, it was a platform for ideas and schemes, where notables passed idle days developing fantasies that often became real campaigns. From Sintra, John I germinated the Portuguese Empire and Manuel I awaited the return of the expedition to India. Summering in Sintra, Portugal's royalty abided and incited events affecting the world outside of the realm Byron called a "glorious Eden."

The Celts are thought to have established the first settlement on the site of Sintra. They were followed by Romans, who established the stronghold named "Mons Lunae," and later by the Alanni, Goths, and Moors, who gave the site its present name and held Sintra as a garrisoned outpost of Lisbon when it was occupied by Christian forces in 1093 during the Christian Reconquest of Portugal. At that time the fortifications at Sintra were surrounded by tilled fields. The region known to the Moors as Balata was considered a marvel of fertility, as it contained the rich lands of the Tagus River and the towns of Lisbon, Sintra, and Santarém. Alfonso VI, king of León, maintained his 1093 occupation of Sintra briefly; an army of Yūsuf ibn Tāshufīn retook the town for the Moors in 1095. Henry of Burgundy, count of Portugal, reinstated Christian command from 1109 to 1111, and after Lisbon was conquered by Afonso I (Henriques) in 1147, Sintra was isolated and surrendered prior to attack. According to legend, when the Christians entered the Moorish castle, all defenders had evacuated through two subterranean passages. Walls still remain of the Castelo dos Mouros, which dates from the early Moorish occupation in the eighth century; beneath the edifice are storage vaults, and Southey observed from his uncle's hilltop cottage in 1800 a cistern within the walls, out of which bubbled a spring of pure water. Along with six other great Moorish citadels conquered in the twelfth and thirteenth centuries, the stronghold in Sintra is represented on the arms of Portugal.

Afonso I granted the property in Sintra to the knight-monk Gualdim Paes, grand master of the Order of the Templars. After the suppression of the Templars, holdings were passed to the Order of Christ by King Dinis, and the knights gave it back to Dinis's wife, Queen Elizabeth, who was later made a saint. In the early fourteenth century, Elizabeth established and maintained the first royal residence in Sintra, a small Gothic palace, more a hunting lodge, mentioned in a minor opus composed by the "Poet-King" Dinis, and possibly built on the site of the Moorish commander's residence. Beatriz, wife of Afonso IV, successor to Dinis, later spent a summer in Sintra.

In 1387 King John married Philippa, the daughter of John of Gaunt, duke of Lancaster, and Sintra, with its cool air, mist, and double the rainfall of any surrounding municipality was joyfully reminiscent of England for the queen. In the early fifteenth century, John expanded the lodge at Sintra into the Paço da Vila or Paço Real (National Palace), enhancing the accommodations to be enjoyed for many summers by their family of four sons and two daughters, in light of two trends predominating in the late fourteenth and early fifteenth centuries. From across the Pyrenees came a new style of residential elegance, customized in the French court and indicative of a less violent period, emphasizing comfort rather than security. At the same time, domestic architecture on the Iberian Peninsula had become dominated by the Mudéjar style, which employed Moorish elements such as horseshoe arches.

With the completion of John's renovations, the palace included two floors, a tower, twenty-six rooms, detached kitchens, terraces, and a grand hall with two noteworthy attached chambers. Several decorative additions made by John commemorate significant incidents that had occurred at

Palácio da Pena
Photo by Evelyn Heyward, courtesy of Portuguese National Tourist Office

his court. The marriage in 1429 of John's daughter Isabella to Philip III, duke of Burgundy, occasioned an assembly of prominent guests to the palace at Sintra. Among them was Jan van Eyck, an envoy in Philip's train, who greatly influenced the Portuguese artist Nuño Gonçalves. To solemnize the marriage, the so-called Swan Room was festooned with painted swans, the emblem of Bruger, Isabella's future home. The decorations of the Magpie Room came about following a less formal royal union. According to legend, the queen one time caught her husband in flagrante delicto with one of her ladies-in-waiting. When the courtiers began to spread the rumor, John asserted his innocence and shunned their gossip by ordering the painting of this palace chamber ceiling with twittering magpies. The birds hold in their claws the red rose of the House of Lancaster and in their beaks a pennant bearing *por bem,* "for good," or "no harm done." Decorative additions also included tiled walls and other Mudéjar accents fashioned by artisans from the community of Moors still living untroubled under Christian rule.

While at retreat with his family in the palace, John decided to crest his prosperity as king, and the great Portuguese Empire was conceived. John's original suggestion of a tournament was rejected by his sons, who proposed he consider taking the coastal town Ceuta in Morocco instead. In 1415 he sent an expedition under the vanguard of his son Henry, later King Henry the Navigator, which invaded and captured the port.

During the reign of Manuel I (the Fortunate) in the early sixteenth century, another series of national episodes left their marks on the palace at Sintra. Having traveled through Granada to court his first wife Isabella, daughter of the king and queen of Spain, Manuel greatly admired the well-preserved Moorish Alhambra, and he transcribed aspects of the architecture to the palace at Sintra upon his return. His greatest additions to the exterior were the twin conical Moorish chimneys that tower above the palace. He also added tiled, orange-scented patios; the palace contains the oldest and largest number of colored glass tiles (known as *azulejos*) in Portugal. The Gothic Arabesque architecture exemplified in the Paço da Vila flourished in Portugal from 1505 to 1520 and was appropriately designated as the Manueline style.

Manuel's major political move to reentitle the country's nobility inspired the decoration of a vaulted ceiling in the palace. The nobility had been suppressed under the rule of Manuel's predecessor, John II. Among the many noblemen killed during the campaign was Manuel's brother Diogo, duke of Viseu. When Manuel took the throne, he pardoned the nobles who had been exiled and organized them into a court of ministers. As a tribute, he placed the coats of arms of the seventy-two noble families on the ceiling of the Sala dos Brasões in the Paço da Vila.

Portuguese exploration begun by John I was at its height under Manuel I, and the summer of 1499 saw two years of anticipation rewarded with news of Vasco da Gama's successful voyage to India. The king spent the summer hunting in the Serra de Sintra overlooking a port on the outskirts of town while watching the sea for the fleet's return. On July 10 he spied the advance of Nicolau Coelho, a sailor and emissary of Gama, who had remained in Goa, on the western coast of India. To acknowledge Portugal's good fortune, Manuel immediately vowed to erect a chapel on the peak from which he witnessed Coelho's arrival. The Palácio da Pena would be built on the ruins of this chapel in the mid-nineteenth century.

In the mid-sixteenth century, the palace at Sintra was home to one of the more colorful Portuguese kings, Sebastian, who succeeded to the throne at the age of three in 1557. The young king established his reputation as an unbalanced individual in 1569 when after three sultry summer months during which thousands of citizens died miserably from plague in Lisbon, he opened the graves of former kings buried in Alcobaça so that he could meditate upon the bones of his ancestors. According to tradition, while residing in Sintra as a youth, Sebastian sat on the Patio of the Negress, one of the *azulejo*-bedecked terraces introduced by Manuel I, and listened to Portugal's national poet Luiz Vaz de Camões read from his *Os Lusíadas,* his epic glorifying the era of Portuguese discoveries. Sebastian was inordinately inspired by these sessions and was deluded by a grandiose yearning to obtain victory over the Moslems in Africa. A royal council convened in Sintra in February 1578 in an effort to dissuade the king from personally conducting his hopeless endeavor, but the council was unsuccessful. Sebastian was lost, presumably killed, in Africa; his ill-fated enterprise subjected the country, already weakened by its overextended empire, to Spanish domination under Philip II in 1580.

Another addition to the royal lore associated with the Paço da Vila was contributed by Afonso VI. Delinquent in his kingly duties, then manipulated by the nobility, Afonso in 1668 relinquished control of the government to his brother Peter, later King Peter II. The Cortes (parliament) decided that Afonso should be exiled but that his queen, with her sizable dowry, should remain in Portugal; once her marriage to Afonso was annulled, she married Peter, her former brother-in-law. Afonso was retrieved from exile to frustrate rumors of a coup meant to reinstate him and was guarded in a small room in the palace. The tiled floor in the room of his imprisonment is worn through where the deposed king paced in despair and misery for nine years until he died in 1683. Though never in disuse, the palace was no longer a favorite residence of the court by the end of the eighteenth century. Damages incurred in the earthquake of 1775 were clumsily repaired, and a tremendous influx of wealth from Brazil was allocated to the construction of other palaces.

In response to the 1808 invasions of Napoléon, Portugal's "oldest ally," England, sent military aid in the person of Arthur Wellesley, duke of Wellington. The French advance was successfully aborted by Wellington, and the Convention of Sintra, concluded and ratified in August 1808, provided for the evacuation of the defeated French from Portugal aboard Wellington's ships, but did not force the French to surrender their arms or booty. The astonishing leniency of the convention enraged the Portuguese, and in November 1808 William Wordsworth wrote *The Convention of Cintra,* a tract in which he expressed the disgust over the convention shared by the British public. (Despite its title, the convention was most likely signed in Lisbon rather than at Palácio de Seteais, the estate of the Dutch consul Daniel Gildemeester, as some histories suggest.)

Several decades later, Prince Ferdinand of Saxe-Coburg-Gotha, consort to Queen Maria II, began perhaps the most eccentric addition to Sintra's architecture when he decided to transform the partially ruined sixteenth-century monastery of Nossa Senhora da Pena atop the Serra de Sintra into a new royal residence. The result was a fantastic high Romantic mixture of Moorish, Norman, Gothic, and Bavarian styles. In 1839 Ferdinand purchased the monastery and collaborated on the erection of the palace with the German military architect Ludwig Von Eschweg, who stands in effigy on a nearby peak. A bizarre construction of fantasy similar to the Paço da Vila less than three miles beneath it, the Palácio da Pena is a Disney-like facade: the Arab minaret contains no stairs to the top, the Norman drawbridge does not draw, and the books inside are reputed to be only jackets. The palace's greatest authentic treasure is the alabaster altarpiece carved for John III by sixteenth-century Norman sculptor Nicolau Chanterène. It is placed in the commemorative chapel originally constructed by Manuel I, around which the palace was built.

By this time Sintra was a summer destination not only for Portuguese royalty, but for the Romantic writers of the late eighteenth and nineteenth centuries. To these travelers, most of whom came from England, Sintra's lush flora, historic buildings, and refreshing climate seemed to be Romantic beauty materialized. Among the earliest of these Romantics to visit the city was William Beckford, in 1787. He was followed in 1800 by Robert Southey, who described the town as "the most blessed spot on the whole inhabitable globe," and eleven years later by Lord Byron, who called it

"a new Paradise." When Danish writer Hans Christian Andersen visited in 1866, his guide and reference in Sintra were lines from *Childe Harolde's Pilgrimage,* Byron's extended narrative poem of a traveler in Europe. Not all of the Romantics visiting Sintra were foreign, however. The town was a favorite of Portugal's own Almeida Garrett, who said of Sintra, "here is spring enthroned."

In 1887 a railway line was opened between Lisbon and Sintra, and with it came a new group of summer visitors: middle-class tourists. In time, such travelers became a mainstay of Sintra's economy. Following the end of the Portuguese monarchy in 1910, the royal palaces were acquired by the Provisional Government for the people of Portugal and now are the settings of concerts, exhibitions, pageants, and official receptions. The Palácio de Seteais was acquired by the Portuguese Board of Tourism and is now a luxury hotel. Several of the eccentric country houses once belonging to minor royalty have been purchased by wealthy foreigners, including many British and Japanese expatriates.

Further Reading: Although H. V. Livermore has written two other works on Portugal since *A History of Portugal* (Cambridge: Cambridge University Press, 1947), this initial work is the most comprehensive regarding Sintra. Hans Christian Andersen's *A Visit to Portugal, 1866,* translated, with notes and commentary, by Grace Thornton (Indianapolis, Indiana, and New York: Bobbs-Merrill, and London: Owen, 1972) contains the famous author's observations about the city, as well as commentary on the observations of authors who had preceded him to Sintra. Andersen's account chronicles minor elements of interest lost by more recent histories, though his research is a compilation of undocumented folklore; Thornton's additions render the work a bit more academic. For poetic descriptions of Sintra's famous beauty, one should read Robert Southey's letters or Lord Byron's *Childe Harold,* both of which are available in a variety of editions.

—Patricia Trimnell

Skopje (Macedonia)

Location: In Macedonia on the banks of the Vardar River, 220 miles south-southeast of Belgrade.

Description: Also known as Skoplje, or Turkish Üsküp (Üsküb) under the several centuries of Ottoman rule, or ancient Scupi. Principal city of the geographic area historically called Macedonia. Capital of the Yugoslav Republic of Macedonia from 1945, and capital since 1993 of the independent republic of Macedonia, a landlocked, mountainous country that lies nestled between Serbia, Bulgaria, Greece, and Albania; transportation and trade center.

Site Office: Turistički Sojuz na Skopje
Dame Gruev, Gradski zid bl. 3
91000 Skopje
Macedonia
(389) 91 11 84 98

Since medieval times, Skopje has been central to Macedonian politics and culture, which—unlike those of the powerful, ancient Macedons such as Alexander the Great—have been determined and generally subsumed by the area's various conquerors. Predominant in Yugoslav Macedonia (as opposed to Greek Macedonia to the south, whose capital is Thessaloníki) have been the Bulgarians, and indeed, until the rise of a specifically Macedonian nationalism in the twentieth century, the residents were generally considered to be of the same stock as the Bulgarians. In any case, the natives of Skopje and the Vardar valley came to prefer Bulgarian overlordship over the occasional Serb incursions from the north. For several centuries, however, when the Ottomans ruled the entire Balkan region, the political interests of both the Serbians and the Bulgarians were kept at bay in Skopje.

Ancient Scupi, as Skopje was called prior to the Middle Ages, was the site of an Illyrian tribal center, but permanent settlement did not occur until the Roman emperor Domitian (ruled A.D. 81–96) founded a colony there, in the district of Dardania in the Roman province of Upper Moesia. This isolated location in the far south of the province may have been chosen for the relatively peaceful conditions there, as well as for proximity to the neighboring province of Macedonia. The Roman settlement—originally called Domitiana—was actually somewhat to the west of modern Skopje, on a hill at the confluence of the Vardar and Lepenac Rivers. As with many other Roman colonies, the settlers were military veterans, in this case mostly from the Seventh Legion.

From the fourth century, the town not only served as capital of Dardania but was the bishopric for the diocese based there. The first known bishop of Scupi was Paregorios,

who participated in the Council of Serdica in 342–343. Two basilicas of the late fourth century have been discovered. At about the same time, the ancient Roman theatre at Scupi ceased to be used for performances, and the site was put to residential use.

The Huns under Attila (ruled 434–453) devastated Scupi in the fifth century, causing the population to flee into the nearby hills. When the town was leveled by a major earthquake in 518, the Romanized residents remained in the more secure enclaves, having adapted to the ways of surviving in the countryside.

The emperor Justinian I (ruled 527–565) was born near Scupi in 483, and as part of his campaign to restore areas that had been destroyed by invaders from the north, he had Scupi rebuilt. The Byzantines eventually constructed several fortresses in the area.

In the sixth and seventh centuries, the large-scale slavicization of the Balkan peninsula took place, with various tribes of Slavs migrating from north of the Danube through Macedonia and all the way to the Greek coast. The concentration of Slavs was heaviest along the roads established by the Romans, avenues for invasion as well as for commerce. One branch of the so-called Orient Express route proceeds through Skopje on the way to Thessaloníki and Athens. The increasing Slavic settlements led to tensions, as the Byzantines and Slavs disputed rights to the city.

Bulgarians invaded Byzantine imperial territory upon the expiration in 846 of a thirty-year peace treaty, and they probably entered Macedonia west of the Vardar River at that time. The Byzantines interfered, under Basil I (ruled 867–886). It was not until the Bulgarian-Byzantine Treaty of 904 that the transfer of most of Macedonia to the Bulgarians was recognized by Byzantium, and the Bulgarian Simeon (ruled 893–927) centered his empire there.

Later, the Bulgarian king Samuel (ruled 980–1014) made Skopje the capital of his short-lived empire. The town frequently served as a center of anti-Byzantine revolts, and the Byzantine emperor Basil II (ruled 976–1025) sent his forces to Skopje to take the town from the Bulgarians in 1004; this successfully put an end to Samuel's expansionist intentions. The Byzantines rebuilt the fortress walls during the Comnenus dynasty late in the eleventh century. The Serbs under King Stephen Nemanja, a clan leader (ruled c. 1167–96) under Byzantine suzerainty and founder of the Serbian state, captured Skopje in 1189; this initial Serbian incursion was brief, however, as the Byzantines promptly regained the town.

In the thirteenth century Skopje was contested by Bulgaria, Serbia, Epirus, and Nicaea. From 1282 the town again was in Serbian hands, this time under King Stephen Uroš II Milutin (ruled 1282–1321), who marched upon the

Daut Pasha baths

Roman acqueduct at Skopje
Photos courtesy of Republic of Macedonia Ministry of Information

town in the very first year of his reign. In about 1299 a Byzantine mission headed by the diplomat Theodore Metochites arrived at Skopje to negotiate the politically expedient marriage of the five-year-old princess Simonis (daughter of Emperor Andronicus II Palaeologus) to Milutin, a move that would bring the Serbs into the Byzantine fold.

The Serbs were not content to be vassals of the Byzantine state, however, and the Serbian king Stephen Dušan (ruled 1331–55) seized from the Byzantine Empire not only Macedonia, but Albania and parts of Greece, too; he made Skopje the capital city of his fledgling empire. On Easter, April 16, 1346, leading clergy—the archbishop of Serbia, the archbishop of Ohrid (the religious center in Macedonia), and the Bulgarian patriarch of Trnovo among them—joined Dušan in a huge convocation he had called at Skopje. The assembly elevated the archbishop of Serbia to patriarch of the Serbs; he in turn promptly crowned Stephen Dušan as emperor or czar of the Serbs and Greeks. The new czar's realm included the Bulgarians and Albanians as well.

Skopje enjoyed a brief period of independence under the Serbian prince Vukašin Mrnjačvević, who acquired the town by 1366. By then, the Turks under Sultan Murad I (ruled 1360–89) were moving into Macedonia, but they let Vukašin govern Skopje as an independent principality within the Ottoman Empire. Under Sultan Bayezid I (ruled 1389–1402), the Turks finally took Skopje in 1391, renaming the town Üsküp. As with much of the southern Balkans, the longevity of Ottoman control ultimately determined the enduring oriental nature of the city, accented by caravanserai and mosques. As a provincial capital for the Ottomans, Skopje developed into a major commercial center.

Despite the Ottomans' initial tolerance of Christianity, Turkish settlers contributed to the Islamization of Skopje, spurring a proliferation of marble mosques. In the fifteenth and sixteenth centuries the Turks tried to force conversions in Macedonia, with little success. In Skopje the churches had to maintain a low profile behind high walls and without steeples. More likely to gain converts was the simple fact that Moslems obviously enjoyed economic and political privileges that were denied nonbelievers.

The sixteenth century was a time of turmoil in Skopje. In 1520 the town was devastated by another major earthquake. Uprisings in Skopje against the Turkish suzerainty occurred in 1572, 1584, 1585, and 1595. In 1689 Austrians under Count Silvio Piccolomini burned Skopje to the ground, either to combat a cholera epidemic or because the town was Turkish; the latter option was certainly possible, as the Austrians and Turks were actively engaged in a struggle for southeastern Europe. In any event, Piccolomini expressed regret that he had to destroy a town with so many fine mosques and beautiful gardens.

Skopje, while continuing to serve its commercial and administrative purposes for the Turks, generally declined throughout the eighteenth century. With the construction in the nineteenth century of the Belgrade-Thessaloníki railway, however, the town began a revival. As with the roads that had first made Skopje a market town, the direct railway access increased the city's commercial prospects.

By the end of the nineteenth century, opposition to Turkish oppression was building throughout the empire, with a concentration of activity in the southern Balkans. As the main city in Macedonia, Skopje was bound to play a role in any significant political movements. Indeed, when the Internal Macedonian Revolutionary Organization (VMRO) formed in 1896 in neighboring Bulgaria, with the nationalist and socialist Georgi "Gotse" Delchev in charge, Skopje was designated as the center of one of six revolutionary target districts. The VMRO originally agitated for removing the Turks and joining Bulgaria (to the dismay of Serbia, which considered Macedonia to be part of "Old Serbia"), but eventually developed a notion of Macedonian autonomy. Delchev led an anti-Turkish insurrection in 1903, and though Skopje itself was not the most active front, it is estimated that more than 1,000 revolutionaries battled the Turkish forces in the city. Delchev died in the attempt, and the Turks crushed the uprising. In general, politics and propaganda have colored discussions of these complex political activities throughout the twentieth century, making it difficult to sort out pro-Bulgarian, anti-Serbian sentiments from purely pro-Macedonian activities.

Skopje's liberation from the Ottoman Empire finally came in 1913, when the city was given to the Serbs at the conclusion of the Balkan Wars of 1912–13. Leon Trotsky's graphic reportage during the conflict included accounts of atrocities in Skopje, mostly against Albanians. Quite contrary to the budding interest among many residents of the area in forging a Macedonian identity, Serbia immediately subsumed the long-lost "Old Serbia" as South Serbia. It was this expanded Serbia that dominated the new Kingdom of the Serbs, Croats, and Slovenes (renamed Yugoslavia in 1928) that was formed in 1918.

The next chance for any sort of Macedonian self-awareness came during World War II. As elsewhere in the Balkans, Skopje's communist partisans opposed the German occupation of the city that began in April 1941. However, the Macedonian Communist Party opted to ally itself with its Bulgarian counterpart over the objections of Tito's communists in Belgrade. The Macedonian party repeatedly failed to adhere to the Yugoslav communist line; finally the presence of Bulgarian troops—under German auspices—wore thin in Skopje, especially as the Bulgarians too were interested in suppressing Macedonian nationalism in favor of a Bulgarian identity. The communists in Skopje eventually got around to backing Tito's partisans over the Bulgarians, ensuring a future for the region after liberation in 1944 as the Federal Macedonian People's Republic, one of Yugoslavia's six internal entities.

Within liberated Yugoslavia under Tito (ruled 1943–80), Skopje remained the capital of the new Macedonian Socialist Republic. Befitting a capital for a people forging a national identity for themselves, in the subsequent five decades Skopje became the center for a specifically Macedonian culture embodied in a national Macedonian university (1949), an autocephalous Macedonian church (1967), and concentrated efforts at developing a distinct language and literature.

Natural disaster obliterated the Macedonian capital when an earthquake of magnitude nine on the Richter scale struck the city of 228,000 inhabitants at 5:17 a.m. on July 26, 1963. More than 1,000 persons died, fully half of the population was left homeless, and three-quarters of the city lay in ruins. Prior to the catastrophe, business and industry in Skopje represented 38 percent of Macedonia's economy, and the consequences of this loss would have been dire indeed.

Fortunately, the international response was immediate, with seventy-eight countries eventually providing more than $300 million in aid to the city. The goodwill that had even the United States and Soviet Union cooperating earned Skopje the sobriquet "City of International Solidarity." Reconstruction began immediately, as the components for prefabricated houses were flown in; the result was a suburban sprawl of such dwellings covering about sixty square miles on the west bank of the Vardar. More substantial steel construction supplanted many of the emergency structures along the new, wide streets, but in some of the suburbs the streets still are lined with small, prefabricated houses.

While the initial inclination of some experts was to move Skopje from its troubled site, which earthquakes had destroyed now for the third time, the city's leaders decided to rebuild a less earthquake-prone city at the same location. Engineers drilled 100,000 holes under Skopje to determine where the terrain would be able to support new buildings, even in the event of another earthquake of the same magnitude. A number of sites would support twenty-story buildings, while many other sites that were unsafe for any construction were transformed into parks.

Skopje burgeoned throughout the 1960s and 1970s as structures utilizing the latest in glass and concrete technology were erected in the city center. The activity attracted many new residents to the city, nearly doubling the population. Skopje ranked as the third-largest city in Yugoslavia, just behind Belgrade and Zagreb, with a substantial industrial base, especially in steel. In the wake of the earthquake, the city's oriental character was significantly reduced. A few medieval Turkish structures—notably the Daut Pasha's baths, the Kale fortress high above the city, and the caravanserai—survived, as did part of the quaint *aršija* shopping district in the old town. But the new buildings eclipsed the once-prominent minarets of the few remaining mosques.

As Yugoslavia disintegrated in the 1990s, Skopje was bound to share in the strife. The breaks to independence of the northern republics of Croatia, Slovenia, and finally Bosnia-Hercegovina, where violence and genocide became the rule, set the stage for Macedonia's declaration of independence. The main contention seemed to be not so much the existence of the country but the fact that it insisted on calling itself Macedonia. This was an affront to the Greek government, which considers Macedonia to be an appellation suit-

able for regions or provinces, but not for countries; both the Bulgarian and the Greek governments fear a desire by their own Macedonian minorities to seek unity with the newly independent Macedonia at the expense of the parent governments.

Contributing to the woes of President Kiro Gligorov's government was the lack of foreign aid or investment in his largely unrecognized country, continuing to subject the population of 2 million people to economic turmoil. As well, by the mid-1990s ethnic tensions were surging with the unrest among the substantial Albanian population in and around Skopje. The demands of this minority for Albanian-language institutions directly challenged the fervent Macedonian nationalism that inspired the nascent republic, and portended further crises in the capital.

Further Reading: *History of Yugoslavia*, by Vladimir Dedijer et al. (New York: McGraw-Hill, 1974; Maidenhead, Berkshire: McGraw-Hill, 1975) provides an informative, detailed summary of the turbulent history of the nations and states that have determined Skopje's regional milieu. Addressing more specifically the national heritage of which Skopje has long been the heart are two books. *Macedonia: Its People and History* by Stoyan Pribichevich (University Park and London: University of Pennsylvania Press, 1982) requires discretion, as the author lets his ethnic biases run rampant; its wealth of detail still makes the book useful. *A History of the Macedonian People*, written by members of the Institute of National History in Skopje (Skopje: Macedonian Review Editions, 1979), obviously has its own agenda, but the basic history seems solid enough and offers depth. *The Early Medieval Balkans: A Critical Survey from the Sixth to the Late Twelfth Century* (Ann Arbor: University of Michigan Press, 1983) and *The Late Medieval Balkans: A Critical Survey from the Late Twelfth Century to the Ottoman Conquest* (Ann Arbor: University of Michigan Press, 1987), both by John V. Fine, provide a comprehensive, clear analysis of the complex politics among the various powers—including the Byzantines, Serbs, Bulgarians, and Turks—that struggled for centuries for control of Macedonia. Specialized studies include *Pannonia and Upper Moesia* by András Mócsy (London and Boston: Routledge and Kegan Paul, 1974), one in a series of scholarly volumes called *History of the Provinces of the Roman Empire*. Two highly readable personal narratives are *Black Lamb and Grey Falcon: A Journey through Yugoslavia* by Rebecca West (London: Macmillan, and New York: Viking, 1941), an incisive commentary—including a chapter on Skopje—on conditions in pre-World War II Yugoslavia by the grande dame of Balkan travel, and *A Paper House: The Ending of Yugoslavia* by Mark Thompson (New York: Pantheon, and London: Vintage: 1992), a first-person journalistic account including Thompson's visit to Skopje as hostilities threatened to spread to Macedonia in the early 1990s. Specialized studies of the nascent revolutionary politics are *The Events of 1903 in Macedonia as Presented in European Diplomatic Correspondence*, edited by Basil C. Gounaris (Thessaloníki: Museum of the Macedonian Struggle, 1993), with documents in original English, French, or German, and *The Politics of Terror: The Macedonian Revolutionary Movements, 1893–1903* by Duncan M. Perry (Durham and London: Duke University Press, 1988).

—Randall J. Van Vynckt

Sofia (Sofia, Bulgaria)

Location: Southwest Bulgaria, in the heart of the Balkan Peninsula; surrounded by the Balkan and Rhodope mountain ranges; at foot of Mount Vitosha.

Description: Largest city in Bulgaria, established in the fourth century B.C.; Roman administrative center first century A.D.; destroyed many times during barbarian invasions; conquered by Ottoman Turks in 1382; by fifteenth century, city called Sofia after its most important church; capital of Bulgaria since 1879; famous for its mineral baths and as important winter sports site; Bulgaria's oldest and most important administrative, cultural, and political center.

Site Office: Committee for Tourism
1, Sveta Nedelya Square
Sofia, Sofia
Bulgaria
84 131

Sofia is one of the most ancient cities in Europe, established well before Rome became an empire. Archaeological diggings in modern times have revealed that even before Thracian nomads settled on the site of the present-day city, others had inhabited the area at least 30,000 years ago. The attractions of the area were the same then as now: mineral springs, a temperate climate, fertile soil, and a beautiful location. Two major mountain chains, the Balkan and Rhodope ranges, come together in Sofia, capped by brooding and beautiful Mount Vitosha on the city's outskirts. Historically the mountains have been the city's playground: a place of refuge, a winter sporting haven, and a hiker's paradise. To this day, Sofia and its immediate vicinity are renowned for their healthful mineral springs. The oldest baths in Sofia have been functioning since Roman times.

The mountains of Bulgaria have posed no barrier to the invasions of peoples and conquering armies. The highest point in and around Sofia is only 7,000 feet, and the famous Iskŭr mountain pass is less than 10 miles away. By virtue of its location as a crossroads between Europe and Asia, Sofia was overrun and dominated by the Thracian, Roman, Byzantine, and Ottoman Empires, and until recently the city was the capital of a Soviet satellite state. Each successive empire left its mark—from beautiful Thracian artifacts to Roman statuary, Byzantine and Roman churches, Turkish mosques, and the dreary totalitarian architecture that dominates downtown Sofia's skyline today. If not exactly a melting pot, Sofia became a colorful mosaic of cultures.

The city has not always been called Sofia. When the Serdi Thracian tribe settled at the foot of Mount Vitosha in the fourth century B.C., they named their city after themselves: Serdica, a name that was retained for the next 1,000 years. In 339 B.C., Philip II of Macedonia destroyed the town, the first of many instances when it was laid waste.

The Romans under Emperor Trajan established the boundary of their empire in Europe at the Danube River, the northern border of present-day Bulgaria. Serdica became a Roman town as early as A.D. 29, and with Roman domination came paved streets, piped water, and major public buildings such as temples, baths, a theatre, a city council building, and a mint. The Roman administration regarded the city as important enough to mint its own coins, many of which are on display at the Ethnographical Museum. On the outskirts of the city walls lay the villas of the elite who governed the area, just as there are today. Serdica's inhabitants of Thracian descent were referred to as Illyrians.

In the fourth century A.D., Emperor Constantine divided the Roman Empire into its western Latin-speaking and eastern Greek-speaking portions. Constantinople became the capital of the Eastern Roman Empire. Serdica fell squarely within the eastern part. When Christianity became legal at this time, there was at least one Christian church already in Serdica, the church of St. George. A small rotunda surmounted by a dome is all that remains of the fourth-century church compound that once covered one square block. Archaeologists in modern times have discovered an even older structure underneath, believed to have been a Roman mineral bath. So solidly rooted was Christianity in Serdica that in 342 the city was chosen as the site of an important church council attended by 170 bishops.

The deterioration of the Roman Empire at this time was deeply felt in Serdica; wave upon wave of barbarian tribes fought their way past the ill-guarded Danube border and swept down the mountain passes. Between 441 and 447 Attila, king of the Huns, easily crossed the border with his army, cutting a wide swath of destruction. Serdica was decimated. Survivors fled into the mountains, and the city lay in ruins for the next hundred years.

The Byzantine Empire gained strength during the reign of Emperor Justinian I in the mid–sixth century. Attempting to wrest Rome from barbarian control, he and his armies crisscrossed Thrace. Justinian recognized Serdica's strategic importance and ordered it rebuilt, with protective walls twenty feet thick. At this time, the Church of St. Sophia, after which the city would be renamed some 800 years later, was built on the shattered remains of two older churches. Simple and austere, free of any decorative art, the church stood on the highest point of the city, outside the walls, and became a focal point of religious devotion.

The name Sofia was not adopted for the city until 1376. In the interim, the city acquired a different name from

Rila Monastery
Photo courtesy of Embassy of Bulgaria, Washington, D.C.

yet another group of conquerors. Shortly after the rebuilding of the city under Justinian, Slavic invaders swept through Thrace, spreading terror and destruction. By the mid–seventh century, the Slavs outnumbered the area's original inhabitants. They settled in the region and became farmers, adopting Christianity in the process. Serdica's name eventually became the Slavic "Sredets," meaning "center."

Although the Slavs adopted Byzantine, or Eastern Orthodox, Christianity, they were enemies of the Byzantine Empire, and for the next several centuries, the Byzantines, the Slavs, and later the Bulgars fought for control of Thrace and, eventually, of the Bulgarian kingdom. The Bulgars were yet another group of Asiatic invaders who arrived in the seventh century. They subdued the Slavic inhabitants, who greatly outnumbered them, and inaugurated the First Bulgarian Kingdom, which existed from 681 to 1018. Eventually, the entire region would be named after them. In 862, the Bulgarian czar Boris I adopted Christianity, speeding the assimilation of his

Alexander Nevsky Church
Photo courtesy of Embassy of Bulgaria, Washington, D.C.

tribe with the Slavs until there was little if any trace left of the Bulgars. Christianity became the official religion of the new Bulgarian state, although the majority of Slavs already were Christian; this new state would be a sworn enemy of Byzantium, so much so that the Bulgarian Church refused to subordinate itself to the patriarch of Constantinople, and succeeded in acquiring its own patriarch. Although Byzantium deeply influenced Bulgarian culture, rivalry with the Byzantine state, together with its independent church, fostered a distinctive Bulgarian identity. It was during the time of the First Bulgarian Kingdom, in the late ninth century, that a remarkable spiritual figure began his mission, which culminated in the establishment of a religious community in the foothills of the beautiful Rila Mountains, seventy-five miles from Sofia. At that time, Ivan Rilski ("of Rila") sought refuge in the Rila mountains. Because of his great personal integrity, he soon attracted a devoted following. There are several medieval biographies of the saint, which depict him as a man of humble origins who adopted the spiritual life of the hermit against the wishes of his family, and who possessed an extraordinary gift of healing. His renown reached the czar of Bulgaria, who desired an audience with him. But Ivan eschewed such attention and refused the czar's request, as well as the donations proffered him and his community. Ivan was eventually interred in the famous monastery of Rila, begun in 1147. By the fourteenth century, the monastery was a sprawling compound of some 300 rooms and numerous churches. Ivan in turn was canonized and attained fame throughout the Balkans.

The imposing monastery of Rila, the largest in Bulgaria, was the spiritual heart of Bulgaria, as well as a center of learning and study. Even the Turks granted Rila autonomy, and after it burned to the ground in 1833, it was beautifully restored. It fared far worse under Communist rule. In 1960, Rila was proclaimed a museum, which discouraged its mission as a monastery and a center for spiritual studies. In the 1990s, with a non-Communist government in power, Rila's spiritual role has become prominent once again. Religious services, open to the public, are held twice daily. Guided tours of the monastery also are available. During the tourist season, the throngs of visitors to Rila— some 300,000 in a typical year—disrupt its historic peacefulness. If one can overlook the din and congestion, there is much to gain from a visit to this historic shrine. The monastery library contains thousands of medieval manuscripts saved from the damaging fire, and valuable icons. In 1965, a history museum, with many priceless artifacts, also opened on the premises.

The Byzantines eventually put an end to the once-powerful Bulgarian kingdom in the eleventh century. Sofia continued to be invaded and ransacked by the Byzantines, as well as by the Pechenegs and Magyars. Another powerful Bulgarian kingdom was established in 1187 and would last 200 years. During this period, Bulgarian spirituality and religiosity were expressed in the feverish building of churches and monasteries; seventy-six of the monasteries have survived to this day.

Bulgaria's most famous medieval treasure is located in Boyana, a mere six miles from the center of Sofia. Situated on Mount Vitosha, this picturesque town is famous for the small Church of St. Panteleimon. Declared a museum by the Communist regime, the church is filled with priceless frescoes and icons and the unusual thirteenth-century portraits of two wealthy donors, Czar Kaloyan and his bride. Five miles from downtown Sofia lies the historic monastery at Dragalevsti, of which only the church has survived. Full of precious icons, it, too, was declared a museum in the Communist era. Nearby Samokov contains an ancient synagogue. The Jewish population, always a small one in Bulgaria, grew perceptibly after the expulsion of Jews from Spain in 1492. They concentrated in Sofia, where they encountered little anti-Semitism down to modern times (even after Bulgaria sided with Nazi Germany, the Bulgarian government refused to deport its Jewish minority).

In 1382, after determined resistance, the Ottoman Turks took Sofia and extinguished the Second Bulgarian Kingdom. For the next 500 years, Bulgaria, followed by the rest of the Balkan Peninsula, was under Turkish domination. Less than a century after its conquest, Sofia resembled a Turkish town, full of mosques, Turkish baths, and caravanseries. In theory, Christians could not be forced to convert to Islam. In practice, there were many penalties imposed on those who refused to convert. For instance, the most prominent churches in Sofia, St. Sophia and the Church of St. George, were turned into mosques. Some church congregations went into hiding or moved to the cellars of buildings to escape notice. At least two of these "basement" churches have survived to this day. Other churches fell into a state of neglect and disrepair. Moreover, all Christians and Jews were subject to a poll tax. Although Bulgarian history books are reluctant to bring the fact to light, many Bulgarians did indeed convert to Islam, automatically freeing themselves of all discriminatory penalties and taxes. As late as 1950, there were more than 1 million Bulgarian Moslems, or Pomoks, out of a population of no more than 5 million, evidence of how deeply Islam had taken root in the course of five centuries.

The Turks made Sofia into a regional capital, and constructed notable buildings, a few of which have survived the ravages of time. Sofia prospered during the period of Turkish expansion, which lasted to the mid–seventeenth century. Of the approximately 150 mosques in Sofia at the height of the Turkish occupation, only three have survived to this day; two have been turned into museums, and only one still serves the Moslem community. The most noteworthy of the three is the Banya-Bashi Mosque, designed and built in the sixteenth century by the famous Turkish architect Hadji Mimar Sinan, and characterized by a huge central dome and intricate interior designs. Close by is a Moslem cemetery, surrounded by a fifteen-foot-high Turkish-style wall. Besides Turkish architecture, Turkish customs, mores, food, and attire became so prevalent in Sofia that visitors from western Europe considered it a thoroughly Turkish city. Bulgaria, like the rest of its Balkan neighbors (with the exception of the Dalmatian coast), was isolated from major currents in western European thought and government for 500 years.

Resistance to Turkish domination gathered strength in the eighteenth and nineteenth centuries with the weakening of the Ottoman Empire and the financial and political disorders that attended it. With no army of its own, Bulgarians turned to the Russian government for help, and in 1878 the Russian army liberated the country from Turkish rule. Nominally, Bulgaria still remained a part of the Ottoman Empire until 1908, but to all intents and purposes, it was self-governing.

A constitutional convention in Veliko Tŭrnovo designated Sofia as the new capital. The new ruler, German prince Alexander von Battenberg, despised Sofia and was oblivious to the city's scenic location and its multicultural charms. Moreover, the Russian government in St. Petersburg connived to oust him from the throne and install someone more to its liking, but not before a monument was erected to him in Sofia to memorialize Bulgaria's first ruler since liberation from Turkey.

Sofia changed drastically in the period before World War I. While it never lost its Turkish flavor, by 1914, it had come in many ways to resemble a small European city; mosques were closed or reverted into churches, and new churches were built, including the grandiose, cathedral-like Alexander Nevsky Church that takes up a city block. The huge church, with its priceless Byzantine interior, was finished in 1913 and paid for by voluntary subscriptions. It was dedicated to the Russian czar Alexander II, who engineered the liberation of Bulgaria from Turkish rule in 1878, and to his favorite saint, Alexander Nevsky. A far less ostentatious, more moving tribute to Alexander II is his equestrian statue in front of the Narodno Sŭbranie, or national assembly, on Parliament Square.

In a lovely little secluded park stands yet another monument of Bulgarian gratitude, this one dedicated to the nearly 600 Russian doctors and medical personnel who died during the Russo-Turkish conflict that resulted in Bulgarian freedom. The tiny park is appropriately and simply named Doctor's Park.

A handsome, western-style royal palace for the Bulgarian ruler, which since 1948 has been the site of Sofia's art museum, came to occupy a prominent square, as did the new parliament building and the Academy of Sciences that lay next to it. The beautiful new National Theatre was designed and built by Austrian and German architects. Elegant shops opened their doors on Lege Street.

After World War I, primarily during the 1930s, major boulevards were created and Sofia University was constructed at its present location, designed by a French architect. Bulgaria gradually was reentering the mainstream of European culture. Unfortunately, politics, as they played themselves out in the nation's capital, were far from being democratic or even parliamentary, although, technically, Bulgaria was a constitutional monarchy. Bulgaria under Czar Boris III allied itself with Hitler, and Sofia suffered severely during World War II. On September 9, 1944, the Bulgarian Communist Party took over the government, aided and abetted by the Soviet Red Army that occupied the country. Free elections, which Stalin promised Roosevelt at the Yalta Conference the following winter, were suppressed. For the next forty-five years, totalitarianism held sway.

The Communist dictator and party chief, Georgi Dimitrov, died suddenly in 1948, possibly from poisoning. His body was duly embalmed and placed in a newly constructed mausoleum (the non-Communist government in 1992 removed the body and cremated it. This structure was followed by the new national library building and the immense Communist Party headquarters, as well as other public constructions. Sofia was gradually transformed from a graceful, unique Balkan city to one of singular drabness, with one residential city block after another bulldozed for the construction of apartment buildings. Although the fall of Communism put a halt to the erection of such eyesores, the Communist regime, to its credit, also had undertaken some positive steps to enhance the city, such as improving sanitation, renovating historic buildings (including St. Sophia and the Church of St. George), planting thousands of trees, constructing new parks, and fostering archaeological studies that revealed a wealth of information about Sofia's origins.

Communism, in short, was a blight less severe than the ravages of the barbarians and their wholesale destruction of the city over a period of hundreds of years. Fortunately, it was also short-lived. What lies ahead, one hopes, will be the resurrection of Sofia as the quaint Balkan city it once was.

Further Reading: All major guidebooks that focus on Eastern Europe in English cover Sofia, with varying degrees of thoroughness. Two excellent recent guidebooks on Sofia and Bulgaria are authored by British travel writer Philip Ward: *Sofia: Portrait of a City* (Cambridge: Oleander, 1993) and *Bulgaria: A Travel Guide* (Cambridge: Oleander, 1989; Gretna, Louisiana: Pelican, 1991), both highly readable and illustrated, but lacking in historical perspective. For those interested in the spiritual center of Bulgaria, the Rila Monastery, a good source is Ancho Anchev's short translated book, *The Rila Monastery* (Sofia: Sofia Press, 1989). A concise, excellent history of Bulgaria is Professor R. J. Crampton's *A Short History of Bulgaria* (Cambridge and New York: Cambridge University Press, 1987), which, of course, highlights major happenings in Bulgaria's chief city. A fascinating sidelight in Bulgarian history has been the history of the Jewish minority. The reasons for the near absence of anti-Semitism in Bulgaria down to World War II, when the government, alone among Eastern European states, refused to deport the country's Jews, are explored in a perceptive study by Professor Frederick Chary, *The Bulgarian Jews and the Final Solution, 1940–1944* (Pittsburgh: University of Pittsburgh Press, 1972).

—Sina Dubovoy

Sparta (Laconia, Greece)

Location: Historical Sparta was located in southwestern Greece, in a valley on the bank of the Eurotas River.

Description: Historical Sparta, also called Lacedaemon, was the capital of Laconia, once the most powerful city-state of ancient Greece. Modern Sparta, a small town of slightly more than 10,000 residents, is situated near the site of the ancient city.

Site Office: Archaeological Museum of Sparta
Dionysiou Dafnis Street
Sparta, Laconia
Greece
(731) 25-363

Ancient Sparta, on the bank of the Eurotas River valley, was at one time the most powerful city-state in all of ancient Greece. The citizens of Sparta governed a state called Lacedaemon, which included the region of Laconia. Sparta was settled by the Dorians in 1050 B.C. Before their arrival, Sparta was little more than a group of four villages in the Eurotas River plain.

Throughout the history of ancient Greece, Sparta's power fluctuated. Yet despite hundreds of years of fierce battle, Sparta was the only city-state that was never forced to overthrow a tyrant. The city-state was considered to have the ideal form of hoplite (military) government. In fact, the Athenian philosopher Plato based his *Republic* on an interpretation of Sparta's constitution.

Spartan government was based on a hoplite assembly and a council. Unlike other city-states, Sparta had two active kings who both claimed ancestral royalty. Due to the lack of true monarchy, most of the political power lay in the hands of five executive officers, called *ephoroi,* who were elected annually. These officers were responsible for keeping any form of radicalism in check, especially moves toward autocracy by the kings or generals.

Society was organized into a rigid hierarchy with a group of citizens, known as the Spartiatai, who had elite privileges and sole political rights. The primary role of the army was to maintain the superiority of this elite group.

Dorians, called *perioikoi,* dwelt near the Spartan village and were bound to military service. Other area dwellers, called Helots, acted as slaves to the Spartiatai. At the height of Spartan power, there were as many as 280,000 people living in the Spartan region.

In keeping with its founding principle of a strong military elite, the Spartan state forced its youth to undergo a rigorous communal upbringing. Education focused on physical training and discipline. Family life was overshadowed in importance by loyalty to the state. Young females were made to go through the same demanding exercise as males, their attention drawn away from marriage, which in turn was seen only as an institution in which to create more able-bodied children.

In 700 B.C., Sparta defeated Tegea in battle, and as part of the treaty, forced the nearby region of Messenia to join its allied forces, known as the Peloponnesian League. Soon after, Sparta followed suit with the cities of Arcadia. By 500 B.C., Sparta had brought nearly all of the cities in southern and central Greece into the league, achieving its status as a world power.

With its new allies in order, Sparta began to receive appeals for defense. When Persia began an invasion of Athens in 490 B.C., the Athenians came to Sparta for help. Sparta agreed, and in an alliance with Athens, known as the League of Greek States, was victorious.

After the defeat of Persia, the League of Greek States decided to maintain its alliance in order to seek revenge. Operations against Persia were now being conducted by the Spartan allies' navy and Sparta was to lead the force, despite the fact that naval operations were the weakest element of its defense. Sparta had few ships and little money to purchase supplies. It was ill-prepared to provide adequate manpower.

In addition, Sparta's focus on internal security took its attention away from leading the allied fleet. The government had looming suspicions that the generals would use their authority to overthrow the government and establish autocracy. This paranoia even led to the execution of a popular Spartan general.

Eventually, the allies became dissatisfied with Sparta's leadership. Its naval commander, Pausanias, was recalled to Sparta and executed. Soon after, the League of Greek States turned to Athens for naval leadership.

But the league never realized revenge against Persia, as Sparta was distracted by a revolt of the Messenian Helots. At the same time, trouble developed within the league. These troubles eventually led to the Peloponnesian War between Athens and Sparta, which lasted from 431 B.C. to 404 B.C., and included almost all the major city-states as allies of one side or the other. Most of Sparta's allies remained the city-states of the Peloponnesian League, thus the name of the war.

Although a series of events led to the outbreak of the war, the reason behind initial hostilities was Sparta's fear of Athens's expanding imperialism. As always, Sparta was very concerned about an internal overthrow of the mainland by its subjects, as such a rebellion would most likely mean the end of Spartiatai rule. As its imperialism expanded, Athens posed an increasing threat to Sparta's dominance.

Sparta maintained its dominance over the Greek mainland by ensuring that other city-states stayed relatively weak. Because its own army, composed of Spartiatai, was too

The countryside around Sparta
Photo courtesy of Greek National Tourist Organization

small to defend the entire mainland, Sparta had to balance its rivals against each other and used force (combined with other hoplite allies) only if a certain city-state became a threat. The city of Argos, in particular, caused Sparta to attack on several occasions.

Meanwhile, Athens pursued imperialism in several areas, including the Greek mainland. Although it could not allow its balance of power to be upset, Sparta was hesitant to enter into war with Athens. But two of its allies, Corinth and

Thebes, had their own hostilities toward Athens, hostilities that led to the outbreak of war.

In the 430s, Corinth and its former colony, Corcyra (Corfu), were in dispute. Corcyra turned to Athens, which was not in the Peloponnesian League, for support. In keeping with its interest in expansion, Athens agreed, as such an alliance would safeguard its sea passage to the west. However, the agreement broke a treaty between Athens and Sparta that allowed each side to make new alliances only with neutrals. Corinth, in return, demanded aid from its ally, Sparta. At the same time, Athens had captured one of Corinth's colonies north of the Aegean.

The Peloponnesian League came together for a meeting in which Corinth voiced its complaints against Athens and demanded Sparta's aïd. Thebes, known as Athens's traditional rival, supported Corinth in an appeal for a declaration of war, considered long overdue. Worried that Corinth would pull out of the league, Sparta had no choice but to consent.

While Sparta had a much larger hoplite army than Athens and therefore a great advantage on land, Athens, with its immense naval fleet, had a strong advantage at sea. In addition, the city of Athens was protected by sound fortifications and parallel defensive walls that stretched all the way to the port of Piraeus. Thus, even if Athens were besieged by land, it still had access to the sea. At the start of the war in 431 B.C., Athens was confident that if it remained behind its walls and its navy could control the seas, it could exhaust Sparta's forces into seeking peace.

During the first phase of the war, Sparta and its allies invaded Attica every year, destroying the city's olive groves, its main agricultural product. The people of Attica fled to Athens for safety, but overcrowding and deterioration within the walls soon led to an outbreak of the plague. As a result, Athens not only lost its leader, Pericles, but suffered a great loss of manpower.

Despite its weakened state, Athens managed to survive, and the two sides found themselves in an exhausted stalemate. Each battle won served only to negotiate a minor truce. This situation changed somewhat in 425 B.C., when Athens captured a group of 120 Spartiatai. This was considered a very notable achievement given the Spartans' endurance and determination to die rather than fall into enemy hands. The Spartiatai had tried to besiege the city of Athens, but were cut off from the mainland by an Athenian fleet, surrounded, and trapped on the island. The Spartiatai were forced to surrender.

However, the Spartans, led by General Brasidas, were inflicting great damage on Athens in the north. Athens's new leader, Cleon, who had succeeded Pericles, sought to attack Brasidas, and although Sparta won the battle, both leaders were killed. Sparta took this opportunity to negotiate a truce that included the release of the 120 captured Spartiatai. The Peace of Nikias was signed in 421 B.C., returning both sides to their prewar formation.

What the treaty failed to address, however, was the issue of Athenian imperialism. As a result, neither side honored the treaty. Not long after, Athens sought alliance with Argos, which was bound by a treaty of neutrality imposed on it by Sparta, as Argos was Sparta's main rival in the Peloponnese. When the treaty of neutrality expired in 418 B.C., Argos met Sparta in battle, hoping to reassert its power. Sparta's forces proved superior, allowing it to impose another treaty on Argos preventing it from engaging in war.

Meanwhile, in Athens, a young aristocrat politician named Alcibiades was gaining much influence with his un-

conventional ideas about pursuing further imperialistic ventures. When Athens received a plea for military help from Sicily, which was involved in a dispute with a neighboring city-state, Athens recognized an opportunity for expansion to the west. Encouraged by Alcibiades, Athens agreed to help plan a major expedition. However, the night before the Athenian navy set sail, a desecration of religious icons occurred in the city that many believed was caused by Alcibiades. He was allowed to sail with the expedition, but while at sea, was ordered back for a trial. On the return journey, Alcibiades jumped ship and defected to Sparta.

When the Athenians reached Sicily, they aimed their expedition primarily at Syracuse, the largest city on the island. Athenians believed that if the city could be captured, the entire island of Sicily would become an ally of Athens. But Alcibiades had advised the Spartans to send a commander to defend the city of Syracuse, and Sparta did support Sicily in its fight against the invading Athenians. In fact, the entire Athenian army was eventually overtaken by disease, forced to retreat, and massacred while crossing the hinterland of Sicily.

Sparta, in an effort to redirect its war effort against Athens, offered support to defecting Athenian allies and sought an alliance with Persia to build a navy strong enough to challenge the Athenian fleet.

Now in a position of extreme strategic advantage, Sparta made the moral proclamation that it was fighting the war to liberate Greek city-states from Athenian imperialism. It was not long before Sparta tried to persuade these subject states to rebel against Athens, in order to destroy the financial structure of the Athenian empire. However, if such a rebellion were to be effective, Sparta would need to support the states, both on land and at sea. Its economy still too weak to build the necessary fleet, Sparta appealed to Persia to assist in the financing. The Persian king, Darius II Ochus, was eager to put a check on the expansion of Athens's power, and Sparta was finally able to build a fleet capable of confronting the Athenians.

Shortly thereafter, Athens defeated the Spartan fleet in a major battle off the coast of Asia Minor. However, the people of Athens disapproved of many tactics of the generals in charge, particularly allowing hundreds of sailors to die when rescue could have been easily administered. The angry citizens demanded a trial, after which six generals were executed.

In 405 B.C., the Spartans decided to offer terms of a peace agreement, but were rejected by the Athenians. However, the Athenian forces were somewhat demoralized and allowed themselves to fall into a vulnerable position. In 404 B.C., in a surprise attack at Aegospotami, the Spartan fleet caught the Athenians beached and captured more than 170 of their ships. Only 10 of Athens's ships escaped. With the Aegean Sea completely under Spartan control, no ships carrying food were allowed to reach Athens. With nothing to feed its population, Athens was forced to surrender.

The Athenian empire was disbanded and the city-state was forced to become Sparta's ally. Athens turned over its remaining ships as the walls of its city were torn down in the name of Greek freedom. Despite this victory, the Spartans' concern was still in the suppression of Laconia and Messenia, leaving it ill-prepared to foster a harmonious relationship among the Greeks. Instead of strengthening the city-states, the Peloponnesian War left them weakened and vulnerable to outside forces.

In 404 B.C., immediately after the war, democracy was abolished in Sparta and an oligarchy, known as the Thirty Tyrants, was established. The tyrants' rule proved to be arbitrary and vindictive, leading to the formation of a democratic opposition. The Thebans, who were dissatisfied by the Spartan peace agreement, supported the opposition, which seized a stronghold in Attica and then marched on the Pireaus to collect more support. The Thirty Tyrants tried to dislodge the opposition but were, as always, distracted by paranoid fear of Sparta's own generals. Finally, the Spartans abandoned the new oligarchy and settled on a compromise that would once again establish a democracy.

Despite its wartime proclamation of Greek liberation, Sparta soon assumed the mantle of Athenian imperialism and proved to be even a more strict and steadfast ruler to subject states. Discontent among the Greeks started to fester. At the same time, Sparta was disputing the fate of the Ionian cities with Persia. In 394 B.C., Corinth and Thebes allied with Athens and declared war against Sparta, in what came to be known as the Corinthian War. Neither side had a particular advantage of strength, yet with equal determination, they fought on to exhaustion. It was Persia, in 387 B.C., that ended the war and implemented a treaty known as The King's Peace. Although it brought temporary peace to the opposing forces, the treaty did not settle the issue of hegemony among the city-states. It was not long before Sparta's rival city-states began to prepare for yet another struggle.

As Thebes had now become its most feared rival, Sparta set out to put a check on Theban power by intervening in Thebes's government and destroying its control over cities in the Boiotian League, a group of cities constructed around Theban leadership. In 382 B.C., Sparta installed an oligarchy in Thebes but was overthrown when, in 379 B.C., seven Thebans, disguised as women, sneaked into a pro-Spartan celebration, assassinating all but the commander, who agreed to evacuate the area.

Thebes battled against Sparta until 371 B.C., when Sparta tried to impose a general peace. Thebes agreed to the terms but insisted on signing the treaty on behalf of all the cities in the Boiotian League. When Sparta refused, the Theban army, accompanied by the Boiotian calvary, marched on to drive Sparta and its Peloponnesian calvary from the field in the battle of Leuctra. This Theban victory was the event that finally put an end to Sparta's dominance in the Peloponnese. Not surprisingly, many of the formerly allied cities, which Sparta had kept weak to assure its own control, rebelled. Supported by the Thebans, the former Peloponnese allies freed the Messenians and built a center of resistance to the Spartans in Megalopolis. However, fearful of Thebes's expanding power, Athens formed a friendly alliance with Sparta, leaving Thebes unable to sustain its supremacy.

For the next few decades, none of the city-states was able to establish any substantial dominance, especially Sparta, which was no longer considered a significant power. From 356 B.C., to 346 B.C., it acted as an ally to Phocis in the Sacred War. In 331 B.C., it attempted to gain control of southern Greece, but was defeated by Antipater in a fierce battle at Macedonia.

In 146 B.C., Sparta came under the control of Rome. Both lines of dynasties had died out, replacing kings with a board of six governing figures called *patronomi*. The structure of Spartan society became highly aristocratic and most of its cultural traditions continued for the next several centuries as the city-state slowly deteriorated.

After the fall of the Roman Empire, the area experienced periods of Byzantine, Frankish, and Turkish rule. In the 1820s it became part of a newly independent and unified Greece, and in 1835 the modern town of Sparta was founded near the site of the ancient city-state.

The ancient Spartans left few imposing monuments, not surprising given their austere way of life. Traces of the Spartan acropolis and theatre remain, as do shrines to various deities. One of the latter is the Sanctuary of Artemis Orthia, where Spartan boys endured ritual floggings in honor of the goddess. About five miles northwest of Sparta is a steep rock called Kaiadas, where the Spartans were said to have abandoned weak or deformed infants. In the center of the modern town is the Archaeological Museum of Sparta, which showcases artifacts from ancient Sparta and neighboring sites.

Further Reading: *Sparta* by A. H. Martin Jones (Cambridge, Massachusetts: Harvard University Press, and Oxford: Blackwell, 1967), weaves cultural elements with the historical tracking of Sparta as an ancient power. *A History of Greece* by J. B. Bury and Russell Meiggs (London: Macmillan, and New York: St. Martin's, 1975), provides a comprehensive history of ancient Greece along with a detailed chronology. *A Traveller's History of Greece* by Timothy Boatswain and Colin Nicolson (Moreton-in-Marsh, Gloucestershire: Windrush, and Brooklyn, New York: Interlink, 1990), details the history of Sparta in the context of Greek history in general. *Sparta* by Humfrey Mitchell (Cambridge and New York: Cambridge University Press, 1964), focuses on cultural aspects of ancient Sparta.

—Cynthia L. Langston

Split (Split, Croatia)

Location: In the region of Dalmatia and county of Split in Croatia, along the Dinaric *karst* (coast). Dalmatia is the coastal strip of the Balkan Peninsula, approximately 200 miles long and only 50 miles wide at its widest. Split is on the mainland, roughly 20 miles north of the islands of Brac and Hvar.

Description: The town began as a fortified settlement within the walls of a Roman palace. This palace was situated on a short peninsula running out from the Dalmatian coast. Over time, the city grew beyond the walls of the palace, expanding westward toward the sea. Today, the harbor of Split is the commercial center of town.

Site Office: TZ Split
Trg Republike 2
58000 Split, Split
Croatia
(58) 46-270

The town of Split, located on the eastern bank of the Adriatic Sea, along the coastline of the former Yugoslavia, is the capital of Dalmatia. The original name of the town was Aspalathos, which was the Greek word for a yellow plant that bloomed along the Dalmatian hillsides in spring. Ethnically, the town is more Italian than Slavic, although, of course, there has been a significant Slavic presence in Split since the eleventh century. Split's history needs to be considered in four time periods: the Roman era, the medieval era, the Renaissance and modern era (from the beginning of the fifteenth century to the end of the nineteenth century), and the post-World War I era.

Before it was a town, Split served as the site upon which the Roman emperor Caius Valerius Aurelius Diocletianus built his palace between the third and fourth centuries A.D. In the twentieth century, Diocletian's palace has been excavated, and the remains have enabled historians to piece together the early history of Split. These excavations were carried out jointly by the Town Planning Institute of Dalmatia and the University of Minnesota beginning in 1968 and continuing throughout the early 1970s; the results have been published in *Diocletian's Palace.* The palace was actually an immense compound, covering nearly 75,000 acres. The building plans for the palace followed the ground plan of a Roman military camp. Diocletian, who was originally from the town of Salona four miles to the north, was so enamored of the spot that he spent his retirement there after abdicating in A.D. 305. Upon Diocletian's death, the palace became the property of the Roman Empire and was a well maintained and fortified outpost of Rome.

Almost 300 years later, in the year 612, Diocletian's palace provided shelter for Roman refugees fleeing from the Avar and Slavic invasion of Salona, then the capital of Dalmatia, and other villages along the coast. The wealthier refugees erected houses within the compound, while others took up residence in the towers and substructures of the old palace. This settlement became the original city of Split.

In 626, when the Slavs withdrew somewhat, Split's creation and growth marked the beginning of Dalmatia's rejuvenation. The Roman Catholic Church played a large part in the town's early development. In 640 a priest came from Ravenna to be elected archbishop and consecrated the pagan temple in the palace in Christ's name, dedicating the cathedral to St. Mary. The church of Split then replaced the church of Salona in the Roman hierarchy, making Split the de facto capital of Dalmatia. Although the Catholic Church appears to have been the central organizing force in Split, some scholars maintain that Jews were "co-founders" of the town, that they were among those who took refuge in Diocletian's palace, and that a synagogue was among the earliest buildings. In the 200 to 300 years that intervened between the late Roman period of Split's "birth" and its medieval period, the town came under the control of the Byzantine Empire.

The most significant event in Split during the medieval period involved the laity's struggle to have the papacy recognize its specific linguistic needs. In 863 the ruler of Moravia asked Byzantine Emperor Michael III to send him missionaries who could instruct his subjects in Christianity in the Slavonic language. The emperor sent Methodious and Cyril on this mission, and the latter devised a phonetic alphabet that made it possible to write the Gospel in Slavonic. This alphabet became known as the glagolitic alphabet, and its use spread well beyond Moravia. (The better known "Cyrillic" alphabet was named after the same Cyril; ironically, however, he played no direct part in its development.) By the tenth and eleventh centuries, the glagolitic script had become the preferred linguistic medium in Dalmatia.

Concurrently, the Roman Church had begun a campaign to recapture some of the provinces that had been lost to the Byzantine Church. By the middle of the eleventh century, according to Giuseppe Praga's *History of Dalmatia,* the reforming papacy had "set its sights on all the Byzantine lands which had been part of the Western Patriarchate and where Latin civilization had remained intact." Split was one such territory, and in 1050 "the legate of [Pope] Leo IX, John, cardinal of Porto, came to Spalato [Split] and deposed the highly pro-Byzantine archbishop Dabrale, putting archbishop Giovanni in his place." This move was only partially successful, so ten years later a second legate (Marinardo)

A view of Split
Photo courtesy of Embassy of Croatia, Washington, D.C.

came to strengthen Rome's hold by prohibiting the use of glagolitic script in religious ceremonies.

Led by Gregory, bishop of Nin, the people of Dalmatia resisted this prohibition, thus forcing Rome's hand. Finally, at the Council of Split in 1079, the papacy relented, conceding that it would no longer oppose the Old Church Slavonic language and glagolitic script.

Between 800 and 1100 the town belonged alternately to the Venetians and the Croatians, with fewer than fifty years usually separating the reversals in the ruling party. At the beginning of the twelfth century, Split, along with other Dalmatian towns, became a free commune answerable to a king. This proved beneficial to the economy, and the urban territory of the town doubled, spreading west of the original palace site (the sea was to the south). Around the same time, a new force introduced itself into Split's politics: Hungary.

In the spring of 1097, Hungary's King Kálmán invaded Croatia, killed the Croatian regulus Peter, and assumed power. This aggressive behavior prompted Venice to intervene in May of that year in order to protect Venetian interests along the coast. The Venetian doge sent two ambassadors, Badoero da Spinale and Faledro Storlato, along with a com-

plement of ships and soldiers to Split in order to protect the Dalmatian coast, and to reassert his own sovereignty. Venice was justified in its concern; Kálmán's ambitions certainly included the eventual conquest of Dalmatia.

In 1107 Kálmán exploited the division within Dalmatia between the rich towns (Arbe, Zara, and Split) and the poor towns (Veglia, Ossero, Nona, and Trau). He allied himself with the latter and attacked the former. Split was besieged; the conflict was ultimately a standoff, but Hungary had entered Dalmatian politics for good. In the aftermath of the conflict, the Hungarian counts *(comes)* replaced the Venetian priors; Simon became the count of Split. The people of Split tolerated this change because they were promised that Hungary would never interfere with Split's ecumenical hierarchy.

However, in 1111 Crescenzio, archbishop of Split, died and the Hungarian lay authorities replaced him with a Hungarian cleric. Properly outraged, the people of Split, led by rector Adriano da Treviso, drove Simon out. A series of battles ensued between Venice and Hungary in Dalmatia. Doge Ordelafo Falier recaptured Split in 1116. After Falier died, the new doge, Domenico Michiel, who was reputed to be a very cautious leader, signed a five-year truce with the

Hungarians. As soon as the truce expired, and while the doge was crusading in the Holy Land, the Hungarians took Split, only to lose it again once the Venetian fleet returned.

The next major event in Split occurred more than 200 years later. In 1357, the Hungarian *comes* began to reassert themselves in the areas surrounding Split. Fearful that they would be deposed, the Venetian counts fled. But while the Hungarians were enjoying considerable success in other areas, Split resisted. Only after they received assurance that their rights, privileges, and property would remain intact did the people of Split, in March 1358, submit themselves to Hungarian rule. The period of Hungarian rule, however, was not long-lived.

In June 1390, Split was overwhelmed by Tvrtko I, the Bosnian King; when he died a year later, Split was free. According to Praga, "[The] Greater Council of Spalato [Split] voted that 'from Tvartko's death on there should be no further mention of a king or other authorities other than the governors and judges of the commune.'" Politically, Split became one of the autonomous Venetian communes. This political change ushered Split into its Renaissance period.

The first stage of this Venetian Renaissance was profitable; however, due to the ongoing war between the Europeans and the Turks, the second stage, economically speaking, was fairly stagnant. Nevertheless, throughout the Renaissance and early modern period, Dalmatia was a quiet port in the storm for Eastern Europe. As Fred Singleton writes in *A Short History of the Yugoslav Peoples,* "Whilst Croatia and Slovenia were being laid waste during the centuries-long struggle between the Ottomans and the Habsburgs, Dalmatia, although not entirely at peace, enjoyed relative prosperity and order. In the Venetian cities . . . there was a unique flowering of the arts, blending the spirit of the Italian Renaissance with the native culture of the Slavs." This period of relative calm lasted almost 400 years.

With the fall of the Venetian Empire to Napoléon in 1797, Split was once again thrown into a political maelstrom. News of Napoléon's conquest reached Dalmatia on May 16, 1797, and a power struggle ensued in and around the city. What followed was a marginal occupation by the French forces. Then, in 1808, Split was overrun by the forces of the Austrian Empire in the war between Napoléon and the Habsburgs. The Austrians took the town on August 2, but subsequently returned Dalmatia (and hence Split) to the French when the armistice at Schoenbrunn was signed two months later. By the time of the Congress of Vienna in 1815, however, the Austrians had significantly infiltrated Dalmatia and were, for the most part, the party in power.

After a rocky beginning, the remainder of the first half of the nineteenth century in Split was relatively calm. But the course of events in Eastern Europe in 1848 destroyed Split's quiescence by institutionalizing ethnic discord. In that year, the Austro-Hungarian Empire suffered a spate of bloody ethnic revolutions. Emperor Ferdinand I was forced to make concessions in the direction of autonomy for ethnic groups throughout his empire. This created problems for Split because of the conflict between its geographical location and its predominant ethnicity. Although the Balkan Peninsula along which Split lies was almost entirely Slavic, Split's ethnic heritage was predominantly Italian.

While the Slavic peoples were presenting their case to Ferdinand, Split was adamant about its Italian heritage. According to Giuseppe Praga,

> On March 29 Spalato [Split] requested to share the fate of Lombardy-Venetia instead of Croatia's. "Dalmatia can only belong with the other Italian lands," read the petition to the Emperor. "For over six centuries it has always handled its public affairs in the Italian language, and still does so today. Italian is taught in the schools and is spoken and written almost exclusively in all the towns and villages of the coast and islands." And to the city of Zagreb, which had sent Spalato an invitation in Croatian to unite, Spalato replied: "Dalmatia is Italian. Of the 12,000 inhabitants of our city only one was able to translate your honorable words."

Praga is probably oversimplifying matters somewhat. As the nineteenth century wore on, the battle in Dalmatia between the Venetian-leaning loyalists (called "autonomists") and the ethnic Croatians (known as "annexationists") grew considerably hotter. While Split was noted as being heavily pro-autonomist, the infiltration of the annexationists was relentless. By 1879, the annexationists had defeated the autonomists in the Dalmatian provincial election. In Split, however, the *Lega Nazionale* was running Italian schools as late as 1913. Thus while Split struggled to remain a bastion for the pro-autonomist movement, the rest of the province favored annexation. Matters finally came to a head during World War I.

When Italy announced its neutrality at the beginning of World War I, the people of Split were put into a difficult position. Although they had been resisting Habsburg rule, they now found themselves more threatened by the Slavic and Serbian influence in Russia and the Triple Entente than by Austria and Germany. Their fear of being "Slavicized" drove them into the arms of Germany and Austria. Later, when Italy declared war on Austria, many of the Italians living in Split had their freedoms severely restricted. After Austria and Germany were defeated, a dangerous power vacuum ensued in Split.

Croatia separated from the Austro-Hungarian Empire on October 29, 1918; two days later the Croatian National Council sent a message to the United States, England, France, Italy, and Serbia declaring that in the "territory of the Austro-Hungarian southern slavs" there was being formed "a nation of Slovenes, Croats, and Serbs" that would "unite with Serbia and Montenegro in a common state." This state, of course, was to be Yugoslavia. Hoping to capitalize on the power vacuum in Split, pro-annexationists lobbied to keep Italy out of the occupational forces there. This prompted Italy to send the warship *Puglia* to Split in order to protect the Italians living there.

The tide, however, had turned in favor of the annexationists, and Dalmatia was ceded to the new Yugoslavia, though this did not come about without incident. Roberto Ghiglianovich, a leader of the Dalmatian pro-autonomist movement, complained to American journalists and President Woodrow Wilson during a trip to the United States that Split was being overrun by "a provisional Croatian government in the service of the National Committee of Zagreb, which was using unheard of violence to suppress the demonstrations of the large and important Italian population of the city."

During World War II, Split was occupied by Axis troops; after the war, a Communist government came to power in Yugoslavia, and Split became a major industrial city. Under the leadership of Marshal Josip Tito, Yugoslavia remained more independent of the Soviet Union than did other Eastern European countries in this period. Ethnic tensions did not vanish, however, and in light of the recent disintegration of Yugoslavia (Croatia became independent in 1990), one can only suspect that there never was a truly unified front among the various ethnic groups. Just as Split was forced by Zagreb into the fold, it is probable that many other ethnicities were suppressed in the name of national ambition. Today, Split, along with the rest of the former Yugoslavia, is once again suffering the eruption of ethnic violence. The violence is being played out against a beautiful and historic setting. Diocletian's palace and the city's historic churches, many of which are great examples of pre-Romanesque art and architecture, have earned Split a place on the UNESCO list of World Heritage Sites. Split also is home to numerous museums.

Further Reading: *History of Dalmatia* by Giuseppe Praga, edited by Franco Luxardo and translated by Edward Steinberg (Pisa: Giardini Editori e Stampatori, 1993) offers a thorough account of Split's history; however, Praga's analysis is certainly slanted toward the Italian perspective. Anthony Kneževič's *The Croatian Nation: A Short History*, fourth edition, translated by Rudolph Hraščanec (Philadelphia: Croatian Catholic Union, Lodge "Croatia," 1990) provides a less detailed account, but one from the Slavic perspective. Fred Singleton's *A Short History of the Yugoslav Peoples* (Cambridge and New York: Cambridge University Press, 1985) considers Split and its history in the context of the Yugoslavian state. Finally, for an in-depth history of Diocletian's palace, the single most significant historical landmark in Split, see Jerko Marasovic's, Tomislav Marasovic's, Sheila McNally's, and John Wilkes's *Diocletian's Palace: Report on Joint Excavations*, four volumes (Omis: Tiskarsko poduzece "Franjo Kluz," 1972).

—Lawrence F. Goodman

Syracuse (Siracusa, Italy)

Location: Located partly on the island of Ortygia and partly on the mainland, Syracuse lies on Sicily's east coast, some eighty miles south of Messina.

Description: Syracuse was founded by Greek colonists from Corinth in the eighth century B.C. and soon became the dominant power on Sicily. By the fifth century B.C., it rivaled Athens in wealth and military might. Although the city lost its independence under Roman and Byzantine rule, it retained its preeminence among Sicilian cities until the Arab conquest in the ninth century A.D., when Palermo was chosen as the administrative center of the Arab government. Syracuse retained some of its importance as a center of commerce, owing to its magnificent natural harbors, but since the ninth century the city has remained a shadow of its former self.

Site Office: A.A.S.T. (Azienda Autonoma di Soggiorno e Turismo)
Via Maestranza, 33
CAP 96100 Syracuse, Siracusa
Italy
(931) 65201 or 66932

The earliest inhabitants in the area surrounding Syracuse belonged to the Stentinello culture, named after a site at the modern village of Stentinello near Syracuse. They arrived in Sicily some time between 4600 and 4200 B.C., but whether they settled on the island of Ortygia, with its natural harbors and freshwater springs, is not known. They were apparently undisturbed until about 1200 B.C., when they abandoned their villages in the coastal plains to resettle deeper into the interior. This movement may have been motivated by the incursions of the Sicels, who came to Sicily from Italy. Remains of Sicel settlements found in Syracuse do not predate the tenth century B.C., however. By the eighth century B.C., when Greek colonists began to settle the east coast of Sicily, the Sicels were the sole inhabitants in the region.

Syracuse was one of the first Greek colonies to be established on Sicily. Its date of foundation is traditionally given as 733 B.C. For their new home, a group of Corinthian settlers chose the island of Ortygia, separated from the mainland by a very narrow channel. The small harbor to the north and the large harbor to the south also played a role in the Corinthians' choice of location. But since the colonists depended on agriculture for a livelihood, they relied on the fertile lands of the Anapo valley on the mainland. A small temple to the goddess Athena, hidden within the structure of the Syracuse cathedral, is now the only reminder on Ortygia of this early Greek period. Before long, building also began on the mainland

in the section known as Achradina, where the agora, or open-air market, was situated. A much larger temple to Apollo was built on Ortygia around 575 B.C., and a temple to Olympian Zeus on the opposite side of the Great Harbor followed soon afterward. Around 550 B.C., the first causeway to connect the island to the mainland was constructed.

Although the riches of the land soon brought Syracuse considerable prosperity, the city owed its dominance over the island to its warlike disposition, which had been apparent from the start, and its military might, which took several centuries to develop. Other Greek colonies pursued a policy of peaceful coexistence with the native Sicels, but the Syracusans immediately set out to subdue the natives, for reasons unknown. By the seventh century B.C., the Syracusans had built three strongholds in the interior and on the coast at some remove from Ortygia to maintain their control of the Anapo valley. The Syracusans also enslaved Sicels they captured. The first skirmish with the other Greek cities came in the early fifth century B.C., when the tyrant Hippocrates of Gela attacked Syracuse. The battle's outcome was undecided, and upon mediation from Corinth, Gela agreed to cease its hostilities in exchange for Kamarina, a Syracusan settlement on the southeast coast. A few years later, in 485 B.C., the ruling oligarchs of Syracuse called on Gelon, who had succeeded Hippocrates as tyrant of Gela, to help them put down a rebellion of commoners and slaves. Gelon took the opportunity to make himself tyrant of Syracuse and immediately proceeded to build up the city, transfering half the population of Gela and all of Kamarina's inhabitants to Syracuse.

Continued warring between the Greek cities brought the first foreign military intervention to Sicily in 480. Carthage, which maintained trading posts on the western end of the island, sent a force under Hamilcar against Syracuse and Akragas (modern Agrigento) on behalf of a number of embattled smaller cities. The opponents met at Himera. Syracuse delivered a resounding defeat to the Carthaginian forces, so confirming its dominance over eastern Sicily, particularly in the coastal areas. Gelon was succeeded by his brother Hiero I, who ruled during a period of great cultural enrichment. Poets and philosophers, among them Pindar, Simonides, and Aeschylus, visited Hiero's court and helped build Syracuse's reputation in the Greek world. Hiero was succeeded in 467 by the third of the Deinomenid brothers, Thrasybulus, who managed to make himself so unpopular that he was expelled from the city in 466. Following Thrasybulus's expulsion, Syracuse enjoyed a period of democratization, but the class distinctions among Syracusans were not erased. Oligarchs, commoners, and slaves remained sharply divided, undermining the stability of the democratic government.

The Greek Theatre at Syracuse
Photo courtesy of A.T.C.T., Regione Siciliana

The island was essentially in Sicilian Greek hands and under Syracusan control by the mid-fifth century B.C. Carthaginians and Phoenicians held trading centers in the northwest but were not much of a force on the island. Parts of the interior, particularly toward the west, were still in native hands. A large-scale Sicel uprising concentrated in the northeastern part of the island temporarily threatened Greek dominion, but it was the Sicels' last gasp. In the wake of the rebellion, Syracuse took the opportunity to expand its territories in the interior. In 445, when it defeated Akragas, the city eliminated its only real Sicilian Greek rival.

For the next thirty-five years, Syracuse continued as the undisputed ruling force on Sicily, despite two Athenian efforts to dislodge it. The first Athenian invasion came in 427 at the invitation of Leontini, a settlement some twenty miles north along the coast, which was under Syracusan attack. The invaders did not fare well and agreed to a settlement of the conflict in 424 before any major battles took place. In defiance of the treaty, Syracuse razed Leontini in 422 without drawing a response from Athens. The second Athenian invasion—even more ill-advised than the first—came not long

afterward, in 415, at the behest of the western city of Segesta, which was embroiled in a border dispute with the neighboring city of Selinus. Selinus in turn had managed to secure Syracusan support, and for this reason the Athenian fleet laid siege to Syracuse, leaving Selinus in peace. The Athenian forces built an encampment off the southern harbor, not far from the temple of Olympian Zeus. Syracuse responded by hastily improving its fortifications and walling in several newer sections of the expanding city. After two years of intermittent combat, during which Sparta sent forces in support of Syracuse, the city defeated the Athenians, slaughtering many and enslaving the rest.

Peace did not prevail for long. A mere three years later, in 410, Segesta again brought on a foreign invasion, this time persuading the Carthaginians to intervene. After razing Selinus, however, Carthage withdrew. Hermocrates, a Syracusan general who had been dismissed from his post and banished from the city, gathered an army of sorts and began rebuilding the fortifications of Selinus, so precipitating the next Carthaginian invasion. In 406, Carthage returned, making short work of Hermocrates in Selinus before beginning an eastward march to Syracuse, destroying Akragas and Gela, which lay along the way. By 405 the Carthaginians had advanced to the walls of Syracuse, which by then had reverted to tyranny. Dionysius, a young army officer, had seized the moment to assert himself as tyrant, and his first action was to agree to a treaty with Carthage, which put an end to the war on condition that Syracuse cede a swath of Sicily along the southwestern coast to the invaders.

Under the rule of Dionysius, who became known as Dionysius the Elder, the island of Ortygia was turned into a virtual military stronghold, encircled by double walls with towers at regular intervals. The mainland sections of Syracuse were protected by walls, parts of which survive, along the Epipolae ridge in the interior and along the coast. Dionysius also built, at the westernmost point of the Epipolae defenses, the Euryalus fort, which was later reconstructed several times and also survives. With the city's fortifications improved, Dionysius set out to recapture southwestern Sicily from Carthage, waging four separate campaigns during which he recouped much of the lost territory. However, he never succeeded in removing the Carthaginians from the island. Dionysius's interests were limited to warfare, and under his rule Syracusan culture declined. Although famous poets and philosophers visited his court, they rarely found much reason to stay, unless forcibly detained. The history of Plato's sojourn at Syracuse is uncertain, but it seems that Dionysius imprisoned the famous philosopher and at one point intended to kill him.

Dionysius died in 367 B.C. and was succeeded by his son, Dionysius the Younger, who lacked his father's determination and so brought on a period of chaos in Syracuse and in Sicily at large. He made peace with Carthage, but on unfavorable terms, and he proved a despotic ruler. He was driven from power by his uncle Dion in 356, only to return two years later, upon Dion's death. He continued in his oppressive ways, and another invader, the Corinthian Timoleon, landed on Sicilian shores in 344 in answer to a call for help by desperate Syracusan citizens. In short order, Timoleon conquered all of Greek Sicily, removing petty tyrants and installing democracies in all Greek cities except Syracuse. Making himself tyrant of the city in 343, he began a seven-year reign of peace, rebuilding cities, reforming laws, bringing in new immigrants to revive agricultural production, and restoring prosperity to Greek Sicily. Timoleon retired in 336 B.C., leaving a democratic government behind in Syracuse.

Again democracy did not last. By 317 a new tyrant, the unusually ruthless Agathocles, managed to take power during a bloody uprising that left 4,000 dead. Agathocles' reign was one of incessant warring. Not only did he make war on the Carthaginians, trying to dislodge them from Sicily and at one point marching on Tunis, but he also fought the other Greek cities, particularly Akragas, which chafed at his control. By the time he was assassinated in 289 B.C., almost all of Sicily, with the exception of the Carthaginian province and the Phoenician settlements in the northwest, had returned to Syracusan control. The balance of power was fragile, however. The presence of Carthage remained a

threat, Akragas was eager to regain its independence, and disbanded mercenaries known as Mamertines had established themselves in Messina and were plundering the Greek cities. Soon after Agathocles' death, a Carthaginian fleet sailed into the Great Harbor. Whether it was a plea from Syracuse or from the Mamertines that brought the next foreign invasion is unclear, but in 278 Pyrrhus of Epirus—just having won his proverbially exhausting victories in Italy—threw his forces into the fray.

In a few years Pyrrhus drove the Carthaginians out of most of Sicily, established his own henchman, Hiero II, in Syracuse, and then withdrew from the island. When Pyrrhus was defeated by the Romans in Italy in 275, Hiero was left in control of the city. A much less warlike man than some of his predecessors, Hiero was not much interested in continuing the conflict with Carthage and signed a treaty with the Africans in 264. By then, however, Rome was showing an interest in Sicily, then the major grain producer in the Mediterranean, and in 263 Hiero changed his mind and forged an alliance with Rome that lasted until his death in 215 B.C. Under Hiero's leadership and no longer sapped by military demands on its resources, Syracuse experienced another period of great prosperity, even in the midst of the First Punic War between Carthage and Rome (264–241 B.C.), which laid waste to most of Sicily. By 241 most of Sicily had fallen to the Romans, but Syracuse maintained its independence.

Some of the most magnificent ancient Greek structures that have been preserved at Syracuse date to Hiero's time. He rebuilt the Neapolis section of mainland Syracuse, laying it out in a grander manner. He also ordered the reconstruction and enlargement of the Greek Theatre, which had originally been built during Timoleon's reign or even earlier. This ancient theater, the largest of its kind in Sicily, is cut out of a hillside at the northern end of Neapolis. The Romans made modifications to the "scene" building (from the Greek *skene*, a building forming the backdrop for dramatic performances), and the upper tiers of seats were quarried for building stone in the sixteenth century; still, the surviving ruins give an excellent impression of the theatre's former glory. A Nymphaeum is situated immediately above the theater. Hiero's court attracted such men as Theocritus and Archimedes, who invented a number of siege engines for Syracuse. The Second Punic War broke out in 218 B.C., but initially the renewed conflict did not affect Syracuse. Shortly after Hiero's death in 215, however, Syracuse once again transfered its allegiance to Carthage, with the result that the Romans laid siege to the city. Despite support from Carthage, the city fell in 212, apparently betrayed from within, and was plundered by the Roman soldiers. Archimedes, who had been active in the city's defense, was killed. Becoming part of the Roman province, Syracuse suffered a serious decline together with the rest of Sicily for the next 200 years, despite the city's having become the seat of the Roman governor. In the late first century B.C., Syracuse was made a Roman *colonia*, which meant that it acquired some privileges and became subject to Roman law. New colonists were brought in by

Rome to revive the ailing agricultural economy. Syracuse became something of a tourist attraction to the Romans, and Caligula in particular seems to have enjoyed visiting the city. Roman remains are not extensive, but a small Roman theatre and a much larger amphitheatre still bear witness to the Roman presence.

Christianity seems to have arrived in Syracuse some time in the third century A.D. (despite later legends that St. Paul, who is known to have stopped at Syracuse on his way to Rome, created a community of the faithful in the city in the first century A.D.). Catacombs rivaling the ones in Rome testify to a flourishing Christian community by the late third century. Syracuse became an important religious center by the sixth century, which was also the time that the Temple of Athena was converted into a cathedral. Following the breakup of the Roman Empire, Sicily came under the control of Constantinople and the Eastern Orthodox Church in 535, while Syracuse became the chief Byzantine outpost in the west after the empire lost its Italian territories. As a result, the surviving Greek element in Syracuse, as elsewhere on the island, once again became more prominent. The slow process of Latinization that had occurred under Roman domination was reversed, and the city reoriented itself to the Greek world. In fact, for a brief period in the mid-seventh century Syracuse was the capital of the Byzantine Empire under Emperor Constans II Pogonatus. When the emperor was assassinated in 668, a rebellion broke out, and though it was quickly put down the Byzantine capital was moved back to Constantinople.

The Byzantine period was not a time of prosperity for Syracuse. Its brief glory as capital of the Byzantine Empire had put a serious strain on its resources. The empire was in disarray, and there was a chronic shortage of funds for the administration of the Sicilian territories. Taxes, as a consequence, were very high. Syracuse also suffered significant losses in population as its citizens, not feeling safe in the city, began moving into the interior. All of the island was subject to Arab raiding in the seventh and eighth centuries.

Finally, a popular revolt against Byzantine overlordship in 827 weakened the existing power structure to such an extent that the Arabs mounted a full-scale invasion. An Arab invading army laid siege to Syracuse, but had to withdraw after several years. However, the Arabs took Palermo in 831 and by mid-century the western part of the island was under Arabic control. Palermo was made the seat of government, so becoming the most important city on Sicily. Syracuse held out against the invaders until 878, when the city fell and much of the remaining population was massacred. Arabs moved in in force, and Syracuse quickly became Islamic, its churches converted to mosques, the official language changed to Arabic. Commerce and agriculture flourished, however, and the city's fortunes took a slow turn for the better during the nearly two centuries of Arab domination.

By the early eleventh century, the unity of Arab control began to disintegrate as rival families fought for the Sicilian territories. Constantinople saw a chance to recover Sicily and sent a force under the general George Maniaces, who managed to occupy eastern Sicily during much of the 1030s. At Syracuse he moved to improve the military defenses, building the Castle of Maniaces on the Great Harbor, but he was unable to entrench himself firmly and was removed by the Arabs in a matter of years. What Byzantium had been unable to accomplish was again attempted some thirty years later by the Normans, who invaded in 1060 and took all of Sicily in short order. With the Norman presence a slow Latinization was once again put in motion, but for another hundred years or so Syracuse remained predominantly Islamic. A slow but inexorable decline in Sicily's fortunes set in, and Syracuse suffered with the rest of the island. Continued foreign control, by the Austrians, Angevins, and Spanish, encouraged the survival of a feudal system that inhibited agricultural and commercial development. Rivalries between Sicilian cities obstructed efforts at reform and made unified action against foreign dominion a dream.

The Sicilian cities, Syracuse included, frequently rebelled against their foreign rulers, but never in concert and most often by bringing in other foreign forces. Syracuse, for instance, rebelled against Holy Roman Emperor Frederick II (king of Sicily as Frederick I), who had inherited Sicily through his Norman mother, in the mid-thirteenth century. To liberate themselves from Frederick, the city supported the Angevin invasion under Charles of Anjou, who became King Charles I of Naples and Sicily. Predictably, the French turned out to be no more tolerable rulers, and the city soon turned against them, looking out for another conqueror to relieve its distress. Under the Spanish, however, Syracuse did not enjoy any greater prosperity; it was actually sold as a feudal possession to a Spanish baron. Although Sicily acquired, under the Spanish, a system of government that included a parliament, Syracuse refused to send representatives, probably because doing so allowed the city to claim some tax exemptions. Disaffection with the government and its law courts, which were largely administered by local barons primarily interested in their own gain, encouraged a generalized resistance to the rule of law among the Sicilian population. In addition, frequent attacks by pirates contributed to the fear under which many ordinary Sicilians lived from day to day. In the mid-sixteenth century the Spanish made an effort to improve fortifications to defend the people from piracy. In Syracuse, the ancient Greek and Roman structures were unfortunately quarried for building stone to be used in improvements at the Castle of Maniaces. A ruinous earthquake in 1693 further advanced the destruction of the city's ancient treasures.

By the early nineteenth century, Syracuse was little more than a provincial town with a glorious past long gone. However, some political reorganization finally diminished the stranglehold of feudalism on the island and brought a mild upturn to the city's economy. Some industrialization got underway, although well into the nineteenth century the Great Harbor remained quiet for most of the year, despite the fact that it was one of the best harbors in the western Mediterranean. The most decisive turning point was still a century away. In the wake of the Allied invasion of Sicily of 1943, Syracuse, which sustained little damage, became more prosperous. Light industry and food processing plants, together with fisheries and tourism provide a livelihood to some 125,000 Syracusans today.

Further Reading: The most easily available sources on Syracuse concern its ancient past. Margaret Guido's *Sicily: An Archaeological Guide* (New York and Washington: Praeger, and London: Faber, 1967) contains a helpful account of Syracuse's early history as well as detailed descriptions of the surviving ancient monuments. M. I. Finley and Dennis Mack Smith place Syracuse in the broader context of Sicilian history in the three-volume *A History of Sicily* (London: Chatto and Windus, and New York: Viking, 1968). Francis M. Guercio's *Sicily: The Garden of the Mediterranean* (London: Faber, 1968; Levittown, New York: Transatlantic Arts, 1969) supplies a good impression of modern Syracuse.

—Marijke Rijsberman

Taormina (Messina, Italy)

Location: Situated on Monte Tauro on the eastern coast of Sicily, Taormina lies a few miles northeast of Mount Etna and about thirty miles southwest of the Strait of Messina that divides Sicily from mainland Italy.

Description: Ancient Greek city that prospered until Roman times but began a long decline with the Byzantine Conquest of 535. In the nineteenth century, the town was discovered by British tourists, and now Taormina is a fashionable resort.

Site Office: A.A.S.T. (Azienda Autonoma Soggiorno e Turismo)
Piazza San Caterina (Palazzo Corvaja)
Taormina, Messina
Italy
(942) 23243

The history of Taormina has been determined by its cliff-top location on the slopes of Monte Tauro. For centuries its steep cliffs made the city a virtually impregnable stronghold and gave it great military importance. Its breathtaking views of the Sicilian coastline and of Mount Etna have endeared it to more recent generations of tourists. Taormina, or Tauromenion as it was first called, seems to have had a rather late start among Sicilian Greek settlements; at any rate its early history remains unclear. Ancient historians give inadequate and sometimes conflicting information, and historical conjecture leads to two possible accounts of its founding.

According to one account, Taormina was a stronghold built by the native Sicel population. Native grave sites attest to their early presence. Dionysius the Elder, tyrant of Syracuse, captured the stronghold in 392 B.C. and resettled it with his own mercenaries. According to information given by the historian Diodorus Siculus, however, the story really begins with Naxos, the earliest Greek settlement on Sicily, just a few miles south of Taormina. This account gives rise to the notion that Taormina was then uninhabited and was used by the Naxians as a place of last resort in times of crisis. Naxos was destroyed in 403 B.C. by Dionysius the Elder, and its inhabitants dispersed. According to Diodorus, Dionysius gave the land to the native Sicels, who chose to build their settlement not on the lava peninsula where Naxos had stood but on the cliffs at Taormina. This account then converges with the other one, in that it holds that, in 392 B.C., Dionysius thought better of his magnanimity and removed the Sicels, to replace them with his mercenaries. Possibly, Dionysius changed his mind because the Taormina Sicels had reached an understanding with Carthage.

Greater certainty exists about developments a few decades later. In 358 B.C. a certain Andromachus—father of the historian Timaeus, who wrote a history of his father's reign—formed a small army out of the surviving families from Naxos and managed to take Taormina with these forces. Andromachus made himself tyrant of the newly Hellenized city, although he has gone down in history as being averse to tyranny. His reputation for justice may at least in part be due to the fact that his first biographer was undoubtedly partial.

In the mid-fourth century B.C., Greek Sicily was in a state of chaos as a result of incessant internecine squabbling. When Corinth sent an army under Timoleon to intervene in the conflicts and restore order in Sicily, Andromachus was ready to receive the invaders, who landed on the beaches of Taormina in 344 B.C. The Greek army swept through eastern Sicily in short order. Establishing himself in Syracuse, Timoleon revitalized the cities and installed democracies everywhere but in Taormina. Andromachus's enduring loyalty to the Corinthian leader apparently earned him the freedom to govern his city as he saw fit. Another factor in Timoleon's decision not to intervene in Taormina may have been Andromachus's having voluntarily placed limits on his own power and instituted an administrative system managed by elected officials. Education, financial management, and the construction of public buildings, for instance, were in the hands of magistrates whose one-year tenure ended with the obligation to render an account of their work. This administrative system endured for some 300 years, up to imperial Roman times.

Taormina flourished in the last decades of Sicilian Greek autonomy, even though Andromachus's son and successor Timaeus was banished from the island by Agathocles, by then tyrant of Syracuse. Tyndarion seized the opportunity to establish himself as tyrant of Taormina and began a reign that exhibits some interesting parallels to the career of Andromachus. In the early third century B.C., Greek Sicily was once again in disarray, this time seriously embroiled in a long-standing conflict with Carthage. In 278 B.C., Pyrrhus of Epirus in Greece, who has become proverbial for the spectacular but exhausting victories he had won at that time in Italy, landed at the Taormina beach. Tyndarion immediately made himself a loyal ally to Pyrrhus, who was also enthusiastically welcomed everywhere else among Greek Sicilians. Pyrrhus, they hoped, would free them from the Carthaginians. Two years later, Pyrrhus had taken all of Sicily with the exception of Lilybaeum (now Marsala), where Carthage stubbornly held out. The foreign adventurer was not quite the liberator the Greeks had hoped for, however, being little inclined to allow the Greek cities any autonomy. With his local support on the wane, Pyrrhus decided to return to Italy in 276, leaving Sicily in a predicament not dissimilar to its situation before his arrival. Taormina at this time had lost its independence and belonged to Syracuse, forming the northern boundary of

A view of Taormina
Photo courtesy of The Chicago Public Library

the more powerful city's territories. Syracuse deployed its naval forces from Taormina in an attempt to subdue the Mamertines—disbanded mercenaries operating from Messina—who conducted raids far and wide in Greek territory and had a strongly destabilizing effect.

Taormina nevertheless continued in prosperity, as may be evident from the construction of a theatre on a site that affords the most spectacular views that Taormina has to offer. The theatre audience would have been able to see Sicily's east coast and the mountainous interior, Calabria's mountains across the straits, and Mount Etna. Little of the original Greek structure remains but fragments of the seating area, as it was reconstructed by the Romans in later centuries. Other Greek buildings have even more completely given way to later construction, so that Taormina's Greek past has become virtually invisible to the modern-day visitor.

Rarely free from foreign intervention, Sicily, most disastrously, became the major theatre in the Punic Wars of the mid- and late third century B.C. The Mamertines were most directly responsible for the Roman invasion of 264, since it was they who appealed to Rome for assistance when their position became untenable. They can hardly have foreseen the devastating twenty-three years of fighting between Rome and Carthage that followed. Together with Syracuse, which wisely made a treaty with the Romans, but unlike most other Sicilian cities, Taormina managed to stay out of the conflict and came away unharmed, still a dependency of autonomous Syracuse. At the end of the Second Punic War, which lasted from 218 to 201 B.C., all of Sicily became a province of Rome, and so Taormina became Roman also. However, by a treaty, Taormina acquired the privileged status of *civitas*, which meant that the city had considerable freedom of self-government and that it was exempt from the tithe levied elsewhere. As a consequence, the city did not share in the decline many other Greek communities suffered.

Great numbers of slaves of many different nationalities were of considerable importance to the overwhelmingly agrarian Sicilian economy, since they performed the lion's share of the agricultural labor. They were treated exceptionally badly during Roman times, and they revolted sometime in the mid-second century B.C. The starting date for what is known as the first Slave War is uncertain, perhaps as early as 139 or as late as 134. The revolt broke out in Enna in the eastern interior under the leadership of the slave Eunus and spread to Agrigento, where a certain Cleon headed the rebels. Eunus and Cleon between them managed to raise an army of 15,000 men, all rebel slaves. Although they were not well armed and certainly untrained, these troops managed to take Enna and Morgantina in the interior and Taormina on the coast. Taormina proved of great value to the revolt, because neither the local free Sicilians nor the Roman forces later sent to crush the insurrection were able to dislodge the rebels from the stronghold, despite the latter's lack of training and military discipline. However, in the end, the rebels in Taormina were betrayed from within in 132 B.C. The former slaves were tortured and then thrown off the cliffs under orders from the Roman consul Publius Rupilius, who went on to write a new legal code for the Roman province that was presumably designed to prevent slave revolts in the future. However, less than thirty years later, Sicilian slaves rebelled again, although this time Taormina remained uninvolved in the conflict.

After some decades of peace, another major conflict was fought on Sicilian soil, and this time Taormina suffered heavily. The assassination of Julius Caesar in 44 B.C. led to a civil war in the Roman Empire. One of the principal combatants, Sextus Pompeius Magnus (Pompey the Younger), first took northeastern Sicily, including Taormina, which became a naval base. Sextus soon established control over the rest of the island. Since Sicily provided almost all of Rome's wheat, Sextus's control over Sicily gave him the power to starve his adversaries. Octavian immediately set sail to drive Sextus from the island, and he succeeded in doing so by 36 B.C. Because Taormina had held out for Sextus until the very last, Octavian (who had in the meantime become Emperor Augustus) punished the city by deporting all its inhabitants. Estab-

lishing his own military base in the city and making it a *colonia*, he distributed the land to veterans from his own armies.

For the next few centuries Taormina was an overwhelmingly Roman community, although a tiny Greek element seems to have survived. Its inhabitants were subject to Roman law, held Roman citizenship, were eligible for careers in the Roman Empire, and practiced the syncretic Roman cult (building a Temple of Isis, for instance). Its economy apparently depended heavily on the export of wines grown in the vicinity. It was at this time that the Greek theatre was reconstructed, first to conform to Roman theatrical conventions and later to accommodate a circus where gladiatorial and other games were put on. A smaller imperial Roman theatre located in the city proper also survives, as well as some fragments from a reservoir that probably served public baths that have not been uncovered. Remains of a complex that included another public bath building were excavated in the 1960s.

Exactly when Christianity arrived in Taormina is not clear, but in the fourth century A.D. the Temple of Isis was turned into the Church of St. Pancratius, patron saint of the city for several centuries of Roman and Byzantine control, until the Arab conquest. Sicily had been subject to Arab raiding throughout the eighth century, and then in 827 a full-scale invasion took place. For some eighty years a war raged on the island that brought famines and plague epidemics to many cities. By the middle of the ninth century, the Arabs still controlled only the western part of the island. Syracuse held out until 878, but its fall that year meant the end of Byzantine domination in Sicily. Nevertheless, Taormina held the Arabs at bay until 902, when the city was burned and its inhabitants massacred. Taormina was initially resettled by Christians from the Val Demone in northern Sicily, but in 962 the Arabs returned and sacked the city once again. This time Arab settlers took over and called the city Mu'izzīyah or Moezzia. No sign of the Arab period now remains at Taormina with the exception of some Arab names and some typical architectural detail, but the dominant religion probably became Islam (although other religious practices were tolerated everywhere in Sicily) and churches may have been converted into mosques. Besides their language and religion, the Arabs also brought improved agricultural techniques and made improvements in irrigation. They also introduced the cultivation of lemons to Taormina. The flourishing export trade at this time depended on the harbor at Giardini, now a small fishing village at the foot of Monte Tauro.

Sicily prospered under the Arabs' rule until their power began to disintegrate in the eleventh century. The Normans, who were expanding their dominion in many parts of Europe in the eleventh century, seized the moment and invaded Sicily in 1060, landing near the beaches near Messina. Under their leader Roger (who became Count Roger I of Sicily), the Normans conquered the island in approximately thirty years, taking Taormina in 1078. Taormina was

Christianized again, even though tolerance was extended to Sicilian Moslems. A slow Latinization was soon underway: new immigrants were drawn from France and Italy; the new bishoprics all over the island were held by clergy from the Roman Catholic Church; and the new monasteries were peopled with monks who also hearkened to Rome. The Normans left few physical traces of their presence at Taormina, although Taormina apparently did well under their rule. The lack of Norman remains may perhaps be explained by the earthquake of 1169, which brought extensive damage to the east coast. The citizens reported having seen the walls of Etna's crater cave in, but Taormina suffered much less from the disaster than did other coastal towns such as Catania, which was almost completely destroyed. Despite its proximity to Mount Etna, Taormina was always spared the ravages of volcanic eruptions because its elevated location took it out of the path of lava flows.

By the mid-thirteenth century a struggle for the succession brought civil war to Sicily, pitting the capital Palermo against Messina, then the wealthiest port. Taormina was dragged into the struggle when Messina captured and almost destroyed the city in 1255. The war lasted fifteen years and was soon followed by another upheaval in 1282, the famous Sicilian Vespers, which transferred Sicily from French (Angevin) to Spanish (Aragonese) control. A revolt against the French broke out in Palermo and spread across the island, while Messina remained loyal to the French. Messinese troops were dispatched to Taormina to make sure it would not join the revolt. When, after a few days, the French also made their way to Taormina, the leader of the Messinese troops took offense at what he interpreted as a lack of trust on the part of the French. The Messinese attacked and imprisoned the French relief at Taormina and then themselves immediately joined the revolt. Both Taormina and Messina followed suit shortly afterward.

Under Spanish domination, in the fourteenth and fifteenth centuries, Taormina's medieval churches and palaces were built, attesting to a degree of continuing prosperity. The Palazzo Corvaja, which now serves as the city hall, the Palazzo Duca di Santo Stefano, and the Palazzo Ciampoli all survive unchanged. They are notable for their use of pumice, limestone, and lava in the walls, which is probably part of the Arab heritage and gives the buildings an aspect reminiscent of mosaics. A modest cathedral and a Dominican monastery were also built at this time. The Badia Vecchia, or Old Abbey, dates from the fourteenth century and bears some traces of the Norman Romanesque style. It was converted to a hotel in the nineteenth century.

The sixteenth century brought a decline to Taormina, as the city's interests were subordinated to those of the much more powerful Messina. The peace was frequently disturbed by the bold raids of Barbary pirates. New village settlements arose in the area surrounding the city. These offered more favorable taxes and rents; as a consequence Taormina lost half its population between 1583 and 1653, when censuses were taken. By 1653, only 3,000 people remained. However, the city did retain some military importance, and it was made the center of a military district in 1592.

By the eighteenth century, Taormina was little more than a picturesque village, and it was appreciated as such by the occasional visitor making the Grand Tour. Goethe in particular was inspired by its views and by the Greek flavor of its landscape. It was not until the nineteenth century that tourists, especially the British, really discovered the city, which, besides its panoramic views, also has an extremely mild climate and access to beautiful beaches. Since then, Taormina has been entirely given over to the rhythms of the harvest and the tourist season, with the exception of a brief interlude during World War II, when the Germans established their Sicilian headquarters in the city. As a consequence, the city was bombed several times by the American forces, following their invasion in 1943. Little serious damage was done in these air raids, and the tourists returned within a few years.

Further Reading: Since Taormina was never wealthy or powerful enough to inspire major studies devoted solely to its history, the best places to find information about it are the general histories of Sicily. The three-volume *History of Sicily* by Moses I. Finley and Denis Mack Smith (London: Chatto and Windus, and New York: Viking, 1968) is particularly balanced in its account, which begins with the Greek colonization. A condensed version of this history is available in the one-volume *History of Sicily* by Moses I. Finley, Denis Mack Smith, and Christopher Duggan (London: Chatto and Windus, 1986; New York: Elisabeth Sifton Books-Viking, 1987). This version also extends the history to the 1980s. Margaret Guido's *Sicily: An Archaeological Guide* (New York and Washington: Praeger, and London: Faber, 1967) gives the best information about the city's antique ruins and also contains a brief account of the city's early history. The most readable among books of more limited scope is Steven Runciman's *The Sicilian Vespers: A History of the Mediterranean World in the Later Thirteenth Century* (Cambridge and New York: Cambridge University Press, 1958).

—Marijke Rijsberman

Taranto (Taranto, Italy)

Location: A coast city on the instep of Italy's heel in the southeastern region of the country on the Gulf of Taranto; in the region of Apulia and province of Taranto.

Description: Once the greatest city of Magna Graecia and a center of Pythagorean philosophy, Taranto was famous in the ancient world for its production of purple dye, luxurious fabric, and wool, but suffered decline under the Roman Empire. Taranto served as a French naval base during the Napoleonic Wars and joined the Kingdom of Italy in 1860. The city is now home to a famous naval academy and a thriving steel industry.

Site Office: E.P.T. (Ente Provinciale per il Turismo)
Corso Umberto, 121
CAP 74100 Taranto, Taranto
Italy
(99) 432397

The Italian seacoast city of Taranto was once the jewel of Magna Graecia, or Greater Greece. Located on the instep of Italy's heel in the Apulia region, Taranto was known as Taras to the ancient Greeks, Tarentum to the Romans. The city was founded in about 708 B.C. by the Parthenians, believed to be the illegitimate children of Spartan women, born while their husbands were away at war with Messene from 743 to 723 B.C. and exiled so as not to cause embarrassment. The Parthenian founders met with great resistance from the native population of Japygians-Messapians and were never able to entirely subdue them.

Tarentum became an important center for trade and fishing, and provided the ancient world with some of its most luxurious commodities. A purple dye derived from the murex (a mussel) and a diaphanous fabric woven from the filaments of a shellfish called pinna marina were among the treasures crafted in Tarentum. The city was also famous for its wool, salt, wine, figs, and terra-cottas. During the fifth century B.C., Tarentum was the wealthiest city of Magna Graecia and established Thurii on the Italian toe, and Heraclea, near the Gulf of Taranto. Tarentum became a center of Pythagorean philosophy and was home to the fourth-century scholar, statesman, and mathematician, Archytas, as well as to Aristoxenus, the author of the first treatise on music. With the decline of Croton, the victorious rival of Sybaris, Tarentum became the region's most important city.

In 338 B.C., Tarentum was threatened with domination by Messapians and Lucanians and turned to its mother city, Sparta, for help. King Archidamus III of Sparta answered the call and lost his life in the struggle. In 332 B.C., Alexander of Epirus came to the city's rescue and moved an army into southern Italy, winning several victories over the Italian armies. Soon, however, the Tarentines began to fear Alexander's growing power and withheld their support.

In 328 B.C., Alexander of Epirus was defeated at Pandosia and was killed during the retreat. The Greek cities in southern Italy were amazed and frightened by the rise of Rome, which had long been an unknown city, inhabited merely by another of the native Italian tribes. Although some Greek cities (such as Neapolis) thought it best to form an alliance with Rome, Tarentum again looked abroad for assistance. Agathocles, master of Syracuse since 316 B.C., responded. However, the Tarentines had no desire to be taken away from their prosperous lifestyles to actively pursue the creation of a united front against the Romans. The aging Agathocles finally gave up his efforts and left Italy. Tarentum was once again left alone in the face of the Roman forces, which were now stronger than ever.

In 282 B.C., Thurii called for Roman help against the Italian tribes of Lucania and was promptly occupied by its would-be rescuers. With the Roman threat now in the heart of Magna Graecia, the Tarentines were finally prompted to take action. When ten Roman warships appeared in the Gulf of Taranto, the Tarentines sank five of them and executed the admiral. Encouraged by this victory, they sent an army to Thurii and expelled the small Roman garrison. At this time Rome was occupied with more pressing affairs farther north and decided to avoid a fight. Roman envoys were sent to Tarentum to request the return of Thurii and to sue for peace. The Tarentines ridiculed the Roman manner of speaking Greek, and, as the ambassadors were leaving the government center, a ruffian in the crowd deliberately urinated on one of their togas. The enraged Roman envoy proclaimed that the stain would be washed out in blood, later displaying the toga to the senate in Rome.

Rome declared war on Tarentum in 281 B.C. The frightened city formed an alliance with the Samnites, Lucanians, and Messapians, and once again looked abroad for a rescuer. A year later, Pyrrhus, king of Epirus, brought 25,000 veteran troops, trained in the phalanx technique, an innovative infantry formation that the Romans would now face for the first time. He also came equipped with twenty elephants to be used in battle. Unlike the city's previous champions, Pyrrhus organized the Tarentines and prepared them to help in their own defense. He closed the places of entertainment in the city and began to drill the citizens.

That same year, Pyrrhus marched out to Heraclea, where he met the Romans midway between Tarentum and Thurii. He selected a level battleground that was well suited to his use of cavalry, phalanxes, and elephants. The terrified Romans had never seen elephants and called them "Lucanian oxen." Unprepared to face the ingenious phalanx fighting

The Aragonese Castello at Taranto
Photo courtesy of A.A.T., Regione Puglia

formation and the giant beasts, the Romans were compelled to retreat. This victory emboldened some of Rome's newly conquered enemies, particularly the Samnites, who did not hesitate to join forces with Pyrrhus.

Despite his victory, Pyrrhus thought it best to make peace with Rome. Unfortunately, his attempt was to prove unsuccessful. Pyrrhus moved northward in 279 B.C. and fought the Romans again at Ausculum 100 miles northwest of Tarentum. He won the battle, but at the price of very heavy losses, especially among the troops he had brought with him from home. These men had been far more reliable and trustworthy than the Greeks of Magna Graecia and the native Italian allies, hence the term "Pyrric victory," which refers to any triumph that is costly enough to have the effect of a defeat.

Pyrrhus was in no position to pursue the Romans in their retreat. The next step for the severely weakened king was to form an alliance with Rome to fight the Carthaginians.

In 278 B.C., he left for Sicily and waged a successful campaign against the Carthaginians and the marauding Mamertine mercenaries, only to be called back by the Tarentines. Pyrrhus sailed back in 276 B.C., and in 275 he engaged in a battle with the Romans at Beneventum.

The Romans had learned their lesson, and this time they came well prepared. They shot arrows dipped in burning wax at the elephants, which turned tail and tore through Pyrrhus's lines. Because the battle took place on uneven ground, the phalanx technique was unsuccessful, and Pyrrhus suffered a terrible defeat. He washed his hands of Tarentum and left for Epirus and for wars in Greece. He was killed three years later in a Greek city, when a woman dropped a roof tile on his head. In the meantime, the Romans had occupied and plundered Tarentum in 272 B.C., destroying the city's capacity for warfare, but leaving intact the system of self-government.

Several decades later, during the Second Punic War, Tarentum fell into the hands of Rome's great nemesis, the Carthaginian general, Hannibal. While making an offering to the gods of the dead on the banks of Lake Avernus, the crater of an old volcano, Hannibal was approached by a deputation of five young noblemen from Tarentum, who entreated him to march on the city, assuring him a quick surrender. Attracted by this prospect, he set out south for Tarentum, burning and ravaging the land of Roman allies along the way.

By the time he reached the city, the situation had changed. Three days earlier, a Roman officer by the name of Marcus Livius had come to Tarentum and enrolled all the men of military age, posting them along every gate and wall. A shift in political influence had apparently taken place within the city, and the presence of Livius and his detachment assured that the pro-Roman party would be restored to power. With his eye on future possibilities, Hannibal made no attempt to take the city by storm or to ravage the territory.

A new opportunity to take Tarentum would soon arise. Hannibal spent most of the summer of 213 B.C. just outside the city. While the Tarentines had not, up to this point, dared to betray their Roman rulers, Roman cruelty toward their townsfolk caused them to change their minds. The Romans had taken hostages from Tarentum and Thurii in order to secure the cities' loyalty. A few of the hostages escaped but were recaptured. In retaliation, the Romans savagely whipped and then threw them to their deaths from the Tarpeian Rock. News of this cruelty infuriated the townsfolk, who hatched a plot against the Romans.

Two young men from Tarentum, Philemenus and Nicon, left the city one night with a dozen or so accomplices under the pretext of going out for a hunting expedition. Their scheme was successful, as they were arrested by Hannibal's sentries and taken to him. They explained their city's dilemma, and Hannibal promised his support in securing Tarentum's independence. He even gave the men some cattle to bring back into the city, in order to lend credibility to their cover story. This deception was repeated many times, and always resulted in a gift of booty to the Roman prefect and guard commander. The Romans began to look upon these nightly excursions as accepted practice, and eventually the gate was automatically opened to Philemenus at his whistle call.

In the meantime, Hannibal had feigned illness to account for his inactivity as he waited for the right moment to pounce on the city, which arrived when the garrison commander was otherwise occupied at a party. Hannibal selected a contingent of 10,000 infantry and cavalry, gave them four days' rations, and instructed them to march to Tarentum, but to conceal themselves in a deep river gully not far from the city. The move was masked by a detachment of Hannibal's Numidian cavalry, which raided the countryside around the marching column.

Hannibal approached the city's eastern wall and, as darkness fell and at an agreed-upon signal, Nicon silently disposed of the guard and opened one of the gates from within through which Hannibal's infantry began to file. The cavalry detachment remained outside the walls as a reserve. In the meantime, Philemenus went to a different gate, the one he normally used, whistled, and called out that he had caught a boar too heavy to carry. When the sentries came out, the Carthaginians poured into the city, falling on the hapless Romans. At dawn they realized what had happened and headed for their citadel, which from its position high upon a promontory controlled the entrance to the inner harbor.

Hannibal saw that the citadel was too formidable to be taken by assault. He built an earthwork to prevent the Romans from issuing in a sortie to attack Tarentum. He constructed a ditch and a double wall, then had Tarentine ships from the harbor dragged on wheels across the neck of the peninsula and used them to finish the blockade. The Romans finally lost Tarentum to Hannibal in either 213 or 212 B.C.

Hannibal had promised that the city would retain independence, and that he would require that only the Roman houses be given over to him. He gathered the Tarentines and told them to go home and write their names on their doors. The houses without names were then plundered by his army.

In 209 B.C. as Hannibal laid siege to Caulonia, the Roman Quintus Fabius Maximus marched to Tarentum and regained the city through the treachery of a lovesick garrison captain. The captain had a mistress whose brother was serving under Fabius, and he allowed the Romans to enter the city and to loot and murder at will. Enormous booty, including 30,000 captives, gold, silver, and works of art, was hauled off to Rome. Although severely punished, Tarentum later regained many of its privileges. The Romans established a colony in Tarentum in 123 B.C.. The city declined under the empire, in spite of Rome's attempts to repopulate Tarentum. Nonetheless, the city was able to keep alive its Greek language and customs for many years to come. Apollo and the underworld deities of Hades and Persephone enjoyed great popularity in Tarentum. Rome's special games held in honor of Pluto and Proserpina had been brought from Tarentum in 249 B.C., and were known as the *ludi Tarentini*.

Taranto was captured successively by Ostrogoths in A.D. 494, Byzantines in 540, Lombards in 675, and Arabs in 856. A monk named Bernard wrote that while on a pilgrimage to Jerusalem in 870, he saw thousands of Christian captives in Taranto being loaded onto galleys by the Saracens for shipment to Africa as slaves. Saracens destroyed the city in 927, but the Byzantine emperor Nicephorus II Phocas rebuilt it forty years later.

In 1063, Taranto was captured by Robert Guiscard of Normandy, whose army was not spared the sting of the region's scourge, the tarantula spider, named for the city and reputed to cause a peculiar frenzied madness called tarantism. The only "cure" for the disease was a dance known as the tarantella. Robert Guiscard's son, Bohemond I, became prince of Taranto. During the Middle Ages, the city served as a point of departure for many crusaders.

Taranto retained its importance under the Normans and subsequent rulers, including the Kingdom of Naples (controlled by the Angevins and then the Aragonese). By the fourteenth century, its territory had come to include much of Basilicata and Apulia. It was subjected to numerous attacks by the Turks in the sixteenth and seventeenth centuries. A popular uprising inspired by Tomaso Aniello's tax protest in Naples tore Taranto apart in 1647–48. In 1734 Taranto came under Bourbon rule and in 1799 it became part of Napoléon's Parthenopean Republic.

The French occupied the city in 1801, and the city proved to be one of Napoléon's strongest bases in his campaigns against the Russians and the English. After Napoléon's downfall, it was joined to the Kingdom of the Two Sicilies in 1815 and remained a subject thereof until the city joined the newly unified kingdom of Italy in the year 1860. With the opening of the Suez Canal, Taranto's harbor became one of Italy's most strategic points. Taranto became an important naval base, and during World War II, British aerial bombardments there caused heavy damage to the Italian fleet. On September 9, 1943, the Royal Navy entered the harbor unopposed and landed troops. Taranto is now home to a famous naval academy as well as to a thriving steel industry.

The Citta' Vecchia (old city) is located on an island between the Mare Grande (the bay) and the Mare Piccolo (a large lagoon). The Mare Piccolo is divided by a peninsula into two bays, one used as a naval harbor and the other for oyster cultivation. The industrial Borgo section of Taranto is located on the mainland to the northwest, and is connected with the Citta' Vecchia by the long Ponte di Porta Napoli.

Sadly, only a Doric column, a few traces of the old fortification wall, and a capital said to be from the temple of Poseidon remain of the Greek city. Although the ancient structures have disappeared, however, a wealth of literary sources describe the ancient city's famous statues. For example, Strabo mentions a colossal bronze statue of the god Helios, moved by levers and rollers, that was fifty-six feet high. Pliny describes a seated bronze image of Hera, "as high as a tower." Both works were ascribed to Lysippus of Sikyon. Dionysius of Syracuse bestowed on the city a great candelabrum that held a flame for each day of the year. This masterpiece was kept in the prytaneum, a great palace that housed fifty city judges, as well as the more distinguished embassies, the city keys, seals, and archives. Some scholars believe that the prytaneum stood on the spot on which the Cathedral was later built.

The Cathedral was dedicated to San Cataldo (St. Cathal of Munster), who lived in the city after returning from a pilgrimage to the Holy Land in the seventh century. Built in the eleventh century upon the remains of an older structure, the Cathedral was rebuilt several times, notably in 1596 and 1657. At the eastern end of the island stands the massive Castello, constructed by Ferdinand II of Aragon. Taranto's Museo Nazionale houses a large collection of prehistoric, Greek, and Roman artifacts from the immediate vicinity, including many examples of Greek jewelry from around 500 B.C.

Further Reading: *The Romans and Their World* by Peter Arnot (London: Macmillan, and New York: St. Martin's, 1970) and *The Roman Republic* by Isaac Asimov (Boston: Houghton Mifflin, 1966) are especially engaging studies of the subject. *The Punic Wars* by Nigel Bagnall (London: Hutchinson, 1990) and *Hannibal: Challenging Rome's Supremacy* by Sir Gavin De Beer (London: Thames and Hudson, and New York: Viking, 1969) offer fascinating accounts of the life of the Carthaginian Hannibal. *Magna Graecia* by Leonard Von Matt and Umberto Zanotti-Bianco, translated from the Italian by Herbert Hoffman (Genoa: Stringa, 1951; New York: Universe Books, 1962) focuses on the works of art produced in Magna Graecia and is copiously illustrated.

—Caterina Mercone Maxwell

Tarquinia (Viterbo, Italy)

Location: In the province of Viterbo, in the Lazio (historic Latium) region, west-central Italy, about forty-five miles northwest of Rome, four miles inland along the Marta River from the Tyrrhenian Sea.

Description: One of the principal cities of the Etruscan League, or Confederation, of semiautonomous city-states of historic Etruria; famous for its Etruscan necropolis, which contains the most notable painted tombs in Etruria; the richly decorated and remarkably preserved underground frescoes have helped historians document the Etruscan civilization.

Site Office: A.A.S.T. (Azienda Autonoma di Soggiorno e Turismo)
Piazza Cavour, 1
CAP 01016 Tarquinia, Viterbo
Italy
(766) 856384

Modern Tarquinia, known as Corneto until the 1920s, lies near the site of the ancient city named Tarquinii in Latin and Tarchuna or Tarchna in Etruscan; the city was once the principal, and possibly earliest, member of the twelve-city Etruscan Confederation against Rome. Tarquinii was the primary political and religious center of the Etruscans, and it was also the home of the Tarquin kings who ruled Rome before the creation of the Roman Republic. The Etruscans were the ancient peoples of Etruria (now largely in the Tuscany and historic Latium regions), located in west-central Italy. They were called Tyrrhenoi, or Tusci by the Greeks and Romans. The Etruscans occupied this area sometime before the eighth century B.C., dominating it from the seventh to third centuries, and slowly developed into a nation with distinct language, religion, customs, and art. Their urban civilization, which was culturally and commercially influenced by Greece, reached its peak in the sixth century B.C. before being overcome by Rome in the fourth century B.C., following a series of wars. Rome adopted many features of Etruscan art and culture.

Tarquinii was also an important Roman town, though it declined during the late Roman Empire. The city's ancient site was abandoned after invasions in the sixth and eighth centuries A.D., and the inhabitants moved to a lower-lying site, which became the fortressed town of Corneto; from 1922, it was called Tarquinia. Today, the rather small town retains a medieval character, with fortifications, castle remains, and a twelfth-century church. But it is mostly known for the Etruscan necropolis, which is situated just southwest of the ancient city. These subterranean burial grounds contain the most important and most remarkably preserved painted tombs in Etruria, primarily rock-cut chamber tombs dating from the sixth to the fourth centuries B.C.

In excavations conducted after 1900, the earliest archaeological remains found on the site of modern Tarquinia were ninth-century B.C. Villanovan (early Iron Age) pit tombs containing cremation urns. Considered by some to be precursors of the Etruscans, the Villanovans probably had sparse settlements in the area as far back as the Italian Early Bronze Age (c. 2000 B.C.). Excavations indicate that the Villanovans were a peaceful, agricultural people who lived in small hut villages, which were to become Etruscan towns or cities. But where did the Etruscans come from, and when? While it is fairly certain that they were of a prehistoric Mediterranean and Asiatic or Aegean stock, the subject has been debated since antiquity, and no definite answer has been found.

Scholars throughout history have divided their opinions among three camps, each of which has been modified over the years. Roman historian Herodotus theorized in the fifth century B.C. that Etruscans descended from peoples who migrated to Etruria from Lydia in Anatolia (Asia Minor, or modern Asian Turkey) as early as the thirteenth century B.C. and established themselves over the region's native Iron Age inhabitants. It is possible that ancestral Etruscans were among the migratory "sea peoples" who overran Asia Minor and coastal east Mediterranean lands shortly after 1200 B.C. and after the fall of Troy; these refugees then landed on the eastern coast of Italy's Umbria region around 800 B.C. But many centuries separate the sea peoples from the emergence of a distinct Etruscan culture in Italy. Dionysius of Halicarnassus, in the first century B.C., however, proposed that the Etruscans (who called themselves the Rasna, or Rasenna) were a native Italic people, aboriginal inhabitants of the area; or, as adopted by some of his modern followers, Neolithic survivors driven into Etruria by the arrival of Indo-European peoples. (Some, like Herodotus, say they arrived as a single group; others say they arrived in small groups over a period of centuries, migrating between the fourteenth and seventh centuries B.C.) A little-supported eighteenth-century theory has the Etruscans migrating from north of the Alps. Modern scholars maintain that further research in linguistics will be of greater value than that in archaeology in helping to resolve the origins of the Etruscans. It is known, for example, that the Etruscans' language is of non-Indo-European stock, though it was later influenced by Indo-European languages such as Greek. The Etruscan alphabet was derived from an early Greek writing system, originally learned from the Phoenicians, as early as the eighth century B.C.

In any event, the major Etruscan towns, including Tarquinii, had all been established by the mid–seventh century B.C. At that time, Etruria was roughly located between

Palazzo Vitelleschi
Photo courtesy of The Chicago Public Library

the Tiber and Arno Rivers, west and south of the Apennines, in west-central Italy. They subsequently expanded beyond their classical frontiers, in a series of conquests, to form a sphere of influence extending north to the plains of the Po River valley, and south to the Latium (Roman) and Campania regions. The Etruscans quickly subjugated the peninsula's native Latin tribes, who subsequently supported them as an aristocratic class of princes. The conquests were not undertaken in a coordinated collective manner, but independently by single cities and groups of individuals. While Greek and Roman historians speak of the twelve cities, or peoples, of Etruria, the list doubtlessly differed as cities rose and fell in power over the centuries.

The semi-autonomous city-states of Etruria were ruled in early times by kings who had both secular and religious authority. They were united by a common language and religion, but each had its own character and customs that extended over a broad territorial area. City representatives seem to have met yearly for a religious festival, where important matters would be discussed. Although the cities often allied for a common cause, they apparently never formed a close-knit confederation, even in crucial times. From the

outset of the fourth century B.C., as the Roman menace loomed, gatherings and alliances between threatened cities appear to have taken place only sporadically. Because they lacked a single leader, Rome was able to overthrow the cities one by one. As historian Luisa Banti writes, "The cities were much too individualistic, jealous and constantly bickering with each other to support the same political or religious structure for very long, especially if they had to submit to the command of an individual from some other city."

According to popular tradition, the principal Etruscan cities—including Tarquinii—were said to have been founded by the wise Tarchon, son of the Etruscan hero Tyrrhenus, whom legend names as the leader of the Etruscans when they left Lydia, Asia Minor. Another local legend involves Tages, a boy-priest who sprang from the earth near the city and taught the Etruscans religious rites and the divination sciences. Both stories, however, have been shown to be later inventions meant to glorify the city's antiquity and supremacy under the powerful Tarquin family, an Etruscan royal dynasty of Rome before the establishment of the Roman Republic.

Lucius Tarquinius Priscus, called Lucomo by Roman historians, was the fifth king of Rome, reigning from 616 to

578 B.C. According to historian Livy, his Greek father, a Corinthian nobleman, fled his native city during a political crisis and settled in Tarquinii in the mid–seventh century B.C. Lucomo was born there and married the Etruscan prophetess Tanaquil. The couple moved to Rome, where Lucomo assumed the throne as Lucius Tarquinius the Elder. He is credited with great building projects in Rome, including the draining of the Forum and starting the temple of Jupiter. After he was assassinated, Tanaquil helped install her son-in-law, Servius Tullius, in power. Tullius, in turn, was murdered by his son-in-law, Lucius Tarquinius Superbus, who then ruled as the seventh and last king of Rome, from 534 to 510. An absolute despot, Superbus enlarged Rome's prestige and leadership but initiated a reign of terror. Leading senators revolted, expelling the Tarquins and abolishing the monarchy. Etruscans fought to win back the throne for the Tarquins, but were defeated.

Certain aspects of the legends related to Tarquinii's antiquity and leadership position, as later espoused by the Tarquinian dynasty, have been corroborated by archaeological fact. Excavations have shown it to be the earliest city founded in Etruria, at least as far as Villanovan-period settlement is concerned. Even before the Etruscans arrived, the town, which they would call Tarchuna, was renowned for its wealth, decorative style, and metal-smithing skill, and had already been in contact with other Mediterranean peoples. Greek and Roman accounts gave Tarquinii primacy in religious and political affairs over other Etruscan cities. Until the seventh century B.C., as indicated by objects unearthed in cemeteries, it was also the richest city in Etruria. (Though it had a flourishing bronze industry, only one large bronze artifact, and not a very good one, has ever been found in Tarquinii.) In the first half of the seventh century, however, Tarquinii was surpassed in supremacy by Caere (the modern Cerveteri). Around 600 B.C., its importance was eclipsed even by Vulci. One hypothesis given for the city's decline in wealth is that its territory's rich ore-producing mines eventually passed to Caere.

The ancient site of Tarquinii, about four miles from the sea, was first uncovered in archaeological excavations conducted in the 1930s. It was situated upon the steep rocky slope of a low-lying hill called La Civita, or Plan di Civita, where the coastal plains met the uplands of the Marta River valley. Like other Etruscan cities, Tarquinii was built on a naturally fortified site that could be easily defended. The city's territory extended up the valley inland to the lake of Bolsena, and included a number of smaller towns as well as a harbor town where the Marta River met the sea. The port later became the Roman colony of Graviscae, the modern Porto Clementino. Tarquinii was also a busy port, and ruled the Etrurian coast in the sixth century B.C. Recent excavations in the port have shown evidence of Greek merchants who resided there as early as the late sixth century B.C.

The excavations conducted from 1934 to 1938 also confirmed that the Villanovan and Etruscan Tarquinii were on the same La Civita site, and not on the site of the medieval Cornet (where modern Tarquinia is), as had once been theo-

rized. Excavations in the old city area have also revealed the remains of imposing five-mile-long perimeter walls built in the fourth century B.C.; Etruscan and Roman street areas; buildings from the late fourth and third centuries; architectural terra-cottas from the sixth to first centuries; fragments of bucchero (wheel-made pots with shiny dark surfaces) and vases from the seventh century B.C. to Roman times; and numerous votive objects and terra-cottas. But the most noteworthy urban finds were the foundations of a great Etruscan temple now called the Ara della Regina (Altar of the Queen), dating from the fourth to third century B.C. It had a long, narrow floor plan with columns on three sides. The temple's pediment was graced with two magnificent winged terra-cotta horses in Hellenistic style (probably drawing the chariot of a god). Considered a masterpiece of Etruscan art, the frieze is exhibited at the Tarquinia National Museum. Although Villanovan huts have never been uncovered in the Tarquinii area, a small village containing pre-Etruscan tombs has been found on the site of modern Tarquinia.

But it is the Etruscan necropolis for which Tarquinia is world renowned, and it is the necropolis by which the world has largely come to know the Etruscans who lived—and died—there. The necropolis contains the most important painted tombs in Etruria, most dating from the sixth to fourth centuries B.C. Many ancient cemeteries succeed each other along the hills near La Civita; the earliest ones are Villanovan and date from the ninth to seventh centuries B.C. The main necropolis is located on Monterozzi Hill, a long ridge just southwest of the ancient city. Monterozzi contains the largest and most significant of the Tarquinian cemeteries. Begun in the second half of the eighth century B.C., tombs from the late Villanovan to the Roman periods have been found there. The earliest are pit or well tombs dug in the earth or rock, with ossuary urns, and often containing a variety of objects given to the dead. Later examples include trench and corridor tombs.

In the sixth century B.C., rock-cut burial chambers with entrance corridors began to dominate: these are the Etruscan cemeteries, the painted tombs of Tarquinia. Although only a fraction have been excavated, about 600 one-room underground tombs are scattered over the three-mile area of rough, rocky terrain. The tombs are decorated with rich, well-preserved frescoes depicting banquets, dancing and music-playing, athletic games, religious ceremonies, funerals, fishing and hunting scenes, war scenes, nature scenes, and the journey of souls to the Underworld. Most of the paintings have been attacked by weather, fungus, moisture, and looters to some degree, but they nevertheless vividly reveal a people living with fullness, who believed the rich feast of life on earth continued in the afterlife as well. By the fourth century B.C., the chamber tombs tended to be larger, and wooden coffins or sarcophagi for inhumation burials, or terra-cotta vases for cremation burials, were common. Tarquinii has more sarcophagi than any other Etruscan city; almost all of the painted sarcophagi and urns of Etruria have been found here.

Since the Etruscans lacked a written history, the painted tombs have helped historians document the development of Etruscan civilization from the sixth century B.C. One of the oldest is the Tomb of the Bulls, the only one inspired by a Greek myth. It depicts the ambush laid by the Greek hero Achilles for Troilus, a son of Priam, during the siege of Troy. Perhaps the most famous of all tombs is the Tomb of Hunting and Fishing, whose polychrome frescoes show animal scenes of dolphins and birds, and where man's presence is almost secondary. It dates to about 520. The most common wall-painting subjects of the sixth- and fifth-century tombs are athletic games and banquet scenes—ceremonies that evidently accompanied the burial of the wealthy. The Tomb of the Leopards (c. 450), the best preserved of all the tombs, and the Tomb of the Lionesses (c. 520), for example, show dancing, music-making, and banquet merriment. The famous Tomb of the Augurs is one of the finest tombs of the sixth century and the first depicting typically Etruscan games (wrestling, boxing) no longer under Greek influence. The Tomb of the Bacchants, from the end of the sixth century, is another notable piece.

The first tomb to show characteristically Greek athletic contests is the Tomb of the Olympic Games (520–500), with chariot races, foot races, jumping, and discus throwing. The most outstanding fifth-century tomb is the Tomb of the Triclinium (465–460), which contains the most beautiful and skillfully painted of all the dancing scenes. In the second half of the fourth century, the motifs of the painted tombs changed and the subject of the Underworld, with its male and female demons, began to appear. Though badly deteriorated, the Hellenistic Tomb of the Shields, depicting a man and woman feasting in the kingdom of the dead, is the masterpiece from this era. The polychrome frescoes from the Tomb of Orcus, also seriously damaged, are notable from this period as well. The first chamber shows banqueting couples reclining with demons, beasts, and snakes, while the wall paintings of the second chamber show an Etruscan interpretation of the Greek underworld, with its gods, demons, and condemned mortal heroes and culprits. The Hellenistic period (third to first century B.C.) was the last great era of Etruscan art before it merged with that of Rome. Tomb paintings are rare, since the Etruscans had by now been conquered by Rome; but the Tomb of the Cardinal and the Tomb of the Typhon, of which only copied drawings remain, originally displayed fine paintings of underworld figures. The practice of tomb painting seems to have died out by the beginning of the first century B.C.

In the spring of 1927, English writer D. H. Lawrence sojourned in Etrurian Italy and was instinctively attracted to the Etruscans' art and culture. He found in them the same pagan mysticism and "blood consciousness" that suffused his own literary work. The Etruscans were a people with a religion, he wrote, "that had not yet invented gods or goddesses, but live[d] by the mystery of the elemental powers in the Universe." He blamed the cold, overbearing Romans for taking all the vital lifeblood out of the Etruscans. Lawrence was especially captivated by the painted tombs of Tarquinia, and their depictions of the earthy, vigorous, sex-oriented life among the wealthy Etruscans. His Tarquinian travels and tomb explorations take up half of the posthumously published *Etruscan Places* (1932) and anticipated the great wave of archaeological interest in Etruria. Lawrence noted how fresh and alive the paintings were:

> Once it was all bright and dancing: the delight of
> the underworld; honouring the dead with wine,
> and flutes playing for a dance, and limbs whirl-
> ing and pressing. And it was deep and sincere
> honour rendered to the dead and to the myster-
> ies. It is contrary to our ideas; but the ancients
> had their own philosophy for it. As the pagan old
> writer says: 'For no part of us nor of our bodies
> shall be, which doth not feel religion.'

Before the Etruscans' arrival, Rome, in the Latium region, was just a strategic collection of small shepherd villages. Latin tribes resisting Etruscan influence began gravitating to Rome about 800 B.C., though the Etruscans had established a stronghold there by 600. In 510 and 509, Rome-centered Latin tribes revolted against the Etruscan kings. During the fifth and fourth centuries B.C., the city-states of Etruria faced the rising military power of Rome, giving rise to the "twelve-city confederation." After the largest Etruscan city, Veii (or Veio), fell to the newly founded Latin empire in 396 B.C. following an epic ten-year siege, Rome turned its sights on Tarquinii, which either alone or in alliance with other towns fought the Roman advance. We know of the city's long, bitter fourth-century struggle for political independence from the Romans primarily through the historian Livy. Tarquinii had fought the Romans as early as 509 B.C., when the Etruscans attempted to restore the last Tarquin king to power; this led to a series of feuds between Etruscan League cities and Rome. But the Roman-Tarquinian wars began in earnest in 394; others have been recorded for 388, 357, and 356 to 351. In this latter campaign, according to Livy, the Romans executed all the Tarquinian prisoners, and 354 Etruscan nobles sent to Rome were killed in the Forum as revenge for Roman prisoners executed in the 357 B.C. campaign.

In 351, with their territory devastated, Tarquinians sued for peace and were granted a truce for forty years. Following a short war waged by most Etruscan cities at the end of the peace period, in 311 B.C., Tarquinii finally fell into the Roman sphere. It became a *civitas foederata*, or federated state, but retained its oligarchic institutions.

Despite the wars, the fourth and third centuries were a period of great prosperity in Tarquinii; there was a profound Greek and, later, Roman influence in art and culture. By 280 B.C., all of the Etruscan city-states apparently were subject-allies of Rome, which had also annexed the Tarquinian Tyrrhenian coast. While most sources say that Rome established its first colony in Tarquinii's territory in 181 B.C.—the port of Graviscae—Livy relates that it already had been in Roman possession for a long time. In any event, the territory was divided into small Roman municipia as early as the second century B.C., definitely by the second half of the first century. In 90 B.C., Tarquinii observed Roman citizenship. The Etruscans were now subdued and assimilated: Latin was the accepted language, and Etruscan art was largely Roman in character.

Tarquinii later became a free municipium, and achieved some importance in the latter part of the first century A.D. Though it declined during the late Roman Empire, Tarquinii became the seat of a bishopric in the fifth century A.D. A number of factors contributed to the abandonment of the ancient city to a lower-lying site, about two miles to the northeast, sometime during the early Middle Ages. The coastal areas of historic Etruria, including Tarquinii, were decimated by malaria in the fourth century; and the city was sacked by Lombards (or the Langobardi) in the sixth century, and by the Saracens, a nomadic European Arab people, in the eighth century. It then became the medieval fortified town of Corneto.

By the eleventh century, the Tuscany region had fallen into disunity, evolving into a collection of independent city-states, each seeking to dominate its neighbors. During the late Middle Ages, Corneto played an important role in the complicated, seemingly endless power struggles between the papacy and the empire. Beginning in the twelfth century, the Holy Roman emperors and aristocracy—the Ghibellines—came in conflict with the Guelphs, creators of the powerful Papal States in central Italy and allies of the middle-class trade guilds. Corneto sided with the Guelphs and was part of the papal domain under Cardinal Giovanni Vitelleschi, whose family ruled the city in the fifteenth century. The Guelphs held the upper hand until the rise of the Medici grand dukes, who ruled the region from Florence later in the century.

Besides the necropolis and the painted tombs, other sites of historical interest in Tarquinia are the city's medieval fortifications; the fifteenth-century Palazzo Vitelleschi, which houses the archaeological museum; castle remains; and many Romanesque buildings. The Palazzo Vitelleschi, a huge Gothic-Renaissance palace, houses the Tarquinia National Museum, which contains the most remarkable collection of Etruscan treasures outside Rome. The museum includes antiquities, sarcophagi, and frescoes removed from excavations in the necropolis, as well as several reconstructed tombs. The masterpiece is the terra-cotta frieze of winged horses taken from the pediment of the Altar of the Queen, a Tarquinian temple where archaeological work is still under way. The Vitelleschi Palace was almost destroyed during World War II, but has since been completely restored. Inside the remnants of the castle of the Countess Matilda of Tuscany is the Church of Santa Maria di Castello. This Romanesque church, built between 1121 and 1208, stands near a tall tower built in the Middle Ages, and was part of the fortified citadel that once guarded the town. Tarquinia's high walls, numerous turrets, and narrow streets still lend the city a medieval character.

Further Reading: Etruscan and Italic scholar Luisa Banti's *The Etruscan Cities and Their Culture* (Berkeley and Los Angeles, California: University of California Press, 1973; London: Batsford, 1974) is widely regarded as one of the authoritative texts in the field, synthesizing literary, art, and archaeological sources to present an interpretation of Etruscan urban culture. A member of the Department of Greek and Roman Antiquities at the British Museum, Dr. Ellen MacNamara is an Etruscan and Italian Bronze Age expert whose books *The Etruscans* (London: British Museum Publications, 1990; Cambridge, Massachusetts: Harvard University Press, 1991) and *Everyday Life of the Etruscans* (London: Batsford, and New York: Putnam's, 1973) reconstruct the history and character of the people through artistic and archaeological evidence. English novelist D. H. Lawrence was also an accomplished travel writer. Of his three non-fiction books about Italy, *Etruscan Places* (London: Secker, and New York: Viking, 1932) is the most revealing as an important statement of Lawrence's philosophy; his extensive descriptions of Tarquinia and the nearby necropolis's painted tombs are particularly penetrating and lyrical.

—Jeff W. Huebner

Thermopylae (Phthiotis, Greece)

Location: On the south shore of the Gulf of Maliakós, about nine miles southeast of the town of Lamia in eastern Greece.

Description: The site of a historic battle between Sparta and Persia in 480 B.C. and several other military actions, most recently in World War II. In ancient times a narrow valley on the coast, Thermopylae is now a flat, rocky plain. The silt thrown up from area hot springs and soil deposited by the Spercheios River have created a six-mile long marsh between Thermopylae and the coast.

Contact: Greek National Tourist Organization
Plateia Laou
85100 Lamia, Phthiotis
Greece
(231) 30065

The valley of Thermopylae was a crucial pass for any invader wishing to travel from northern to southern Greece. Because of its location, the valley has been the site of several pivotal battles over the centuries. The most famous of these battles occurred in August of 480 B.C. when King Leonidas of Sparta, commanding 6,000 Greek soldiers against a much larger force of Persians, led a last-ditch defense while the bulk of his army made its retreat. Of the 300 Spartan soldiers who stayed behind, all—including Leonidas—were killed in battle against the Persians, earning them a lasting place in Greek and Western history for their bravery. Thermopylae was again the site of military battles in 352 B.C., 279 B.C., 191 B.C., and in 1941 during World War II.

Thermopylae is named after the hot springs found at the site, its name literally meaning "hot gates" in Greek. Situated between Mount Oeta and the Gulf of Maliakós, Thermopylae in ancient times was a narrow pass about nine miles long and southeast of the town of Lamia. At its center, the pass opened into a wide flat area where a defensive wall was first constructed by the Phoceans—Greeks who lived in the area. Because of the constant activity of the hot springs, which expel deposits of alluvial soil, and the silts deposited by the nearby Spercheios River, the original pass of Thermopylae is now a flat plain some six miles from the sea.

During the Persian Wars, in which the Greek city-states fought against the invading armies of the Persian empire to the east, the Greek colonies in Asia Minor were conquered while those in the far north were forced to become Persian allies. Eager for further conquest in the West, the Persians then turned their attention to the European mainland. In the mid-480s B.C., King Xerxes I of Persia spent four years preparing for an invasion of Greece. Once Greece was conquered, Xerxes hoped to expand into the rest of Mediterra-

nean Europe with the assistance of his ally, Carthage, in North Africa.

Xerxes' preparations for war were extensive, calling for two feats of engineering unparalleled for the time. The first of these feats was a canal he ordered dug through the Acte Peninsula just north of Mount Athos in northeastern Greece. The massive undertaking, involving thousands of conscripted workers, was meant to shorten the time that a Persian invasion fleet would spend in the northern Aegean Sea. The Persian king Darius, in an invasion attempt some ten years earlier, had lost his fleet to a sudden storm in the area and Xerxes was determined to avoid that mistake. In addition to building the canal, the Persian king also set up a number of supply camps along the northeastern Greek coast. These camps were stocked with cattle, salted meat, and other supplies the Persian army would require once they entered the area. The stores of salted meat were piled so high, the Greek historian Theopompus recorded, that from a distance they seemed to be a series of small hills.

The second feat of engineering was the construction of two floating wooden bridges across the Dardanelles dividing Asia Minor from the European mainland in what is modern-day Turkey. These bridges consisted of moored floats strung across the narrow straits. Harpalus, a Greek engineer credited with constructing the bridges, set them at an angle to the Black Sea to the north of the straits so as to lessen the impact of the strong current. Huge anchors were affixed to both ends of each bridge to insure that no sudden gale, a frequent occurrence in the area, could knock the bridges free from their moorings. Held together with cables of flax and papyrus, the bridges stood some 8 feet above the waterline and stretched some 1,400 yards across the straits. A flooring of wooden planks covered with dirt allowed the Persian supply wagons pulled by oxen to traverse the bridge easily.

In the spring of 480 B.C. Xerxes led his army across the Dardanelles and into northern Greece. At the same time, a Persian naval force followed along the coast. Although the Greeks had been well aware of the Persian preparations for war, they naively believed the elaborate construction efforts to be signs of the decadence of King Xerxes, who they assumed to be an Oriental tyrant in the classic mode. The bridges and canal, the Greeks argued, were evidence that Xerxes did not want to suffer even the slightest inconvenience during his invasion. This notion of Persian softness blinded the Greeks to the true dimensions of their enemy's strength until the Persian army was on the march. Accustomed to armies of a few hundred or few thousand soldiers, such as were normally mustered by Greek city-states, the Greeks were astonished by the Persian invasion force. Greek accounts of the time wildly claim that Xerxes led an army of

Monument to Spartan King Leonidas at Thermopylae
Photo courtesy of Consulate General of Greece, Chicago

some 5 million men into battle. Although contemporary historians estimate the true number at closer to 200,000 men (with some 75,000 pack animals carrying weapons and supplies), the Persian army was undoubtedly an enormous force for the time. With the advantages of bridges across the Dardanelles and well-stocked supply camps for the land forces and a canal through the Acte Peninsula for the naval forces, the Persians moved quickly, taking the Greeks by surprise.

A hasty meeting among the Greek cities in the south raised an army of about 6,000 men to meet the oncoming Persians. Led by King Leonidas, who claimed descent from the legendary hero Hercules, the Greek army occupied the pass at Thermopylae before the Persian army had advanced that far. Quickly rebuilding the stone wall across the pass constructed by the Phoceans years before, the Greeks arranged for their spearmen to take the frontmost rank upon the wall.

When Xerxes arrived to find Thermopylae defended by the Greeks, he encamped his army on the Malian plain just west of the pass. For four days he waited patiently, expecting that the fearful Greeks would eventually decide against a battle with the enormous army spread out across the plain before them and quietly withdraw. But Xerxes' hopes for a bloodless victory were dashed. The Greeks held firm.

On the fifth day Xerxes finally called for action. Relying heavily upon his archers, who were feared throughout the Middle East for the accuracy of their weapons, Xerxes launched an attack on the narrow pass and the stone wall behind which the Greeks were gathered. The ensuing battle was a decisive victory for the outnumbered Greeks. Their spearmen, able to hit the enemy from high above while the Persian archers could not reach the top of the wall with their arrows, drove the Persians back with heavy losses.

On the second day of the battle, Xerxes sent in his elite troops—the Immortals, as they were called—to clear the pass of the stubborn Greeks. But they too were driven back. The Greek historian Herodotus claims that Xerxes, watching the battle from a distance, jumped from his throne three times in alarm as he watched his best soldiers fall to Greek spears in the narrow pass.

On the third day, Xerxes agreed to a new strategy to defeat the Greeks. Following the instructions of a Greek traitor named Ephialtes, he sent a contingent of his Immortals across Mount Oeta by way of a twisting mountain path, called the Anopaea, so as to come at the Greeks from behind. Several thousand Immortals, led by their commander Hydarnes, marched single file during the night to the rear of the Greek encampment. Leonidas, knowing of this path and fearing that it might be used by the Persians, had posted some Phoceans to guard it, but they fled upon seeing the enemy approaching. Fortunately, some of the deserters ran to alert Leonidas of the ploy. (Legend claims that the Greeks were alerted that the Persians were walking along the mountain path by the sound of the dry leaves under their feet.)

However, Leonidas learned of the maneuver and he quickly ordered the bulk of his army to retreat towards the east and avoid the entrapment. He and some 300 Spartan and Thespian troops would stay at the wall to hold off the Persian troops there while another 1,100 troops, primarily Thebans, held the eastern end of the pass against the Immortals. Historians disagree as to Leonidas's strategy. The usual explanation is that Leonidas and his men knowingly sacrificed themselves so that most of their fellow soldiers could escape safely. But other historians argue that Leonidas may have planned a trap, pretending to send away the bulk of his army so that the Immortals who came around to the east end of Thermopylae could themselves be entrapped between his forces in the pass and the rest of the army farther to the east. They contend that Leonidas was betrayed by his own troops, who failed to join the battle out of cowardice and left him to be killed by the Persians.

What is beyond argument is the outcome of the final day's battle at Thermopylae. Surrounded by Persian troops in the narrow pass, Leonidas and his soldiers fought bravely. Abandoning their position at the wall, they charged the Persians who were assaulting them. Two brothers of Xerxes were killed in the fight. Leonidas himself was slain as well, a battle raging over his fallen body. The Spartans were finally forced back within the pass and onto a small hillock called Kolonis where they made their final stand. Surrounded and far outnumbered, the Greeks were massacred. Meanwhile the Thebans, seeing the battle lost and their own city of Thebes an inevitable casualty of the Persians now that the pass was taken, surrendered.

The battle at Thermopylae immediately spawned a number of legends. A Spartan soldier named Dieneces, when told that so many Persians were involved in the battle that the flight of their arrows blocked the sun, replied, "So much the better, we shall fight in the shade." One Spartan named Pantites had been sent as a messenger by Leonidas to Thessaly and missed the tragic battle. Upon arriving back home in Sparta he was scorned as a coward, despite his explanation, and hanged himself in shame. Following the battle, Xerxes ordered the head of Leonidas to be brought to him and displayed on a pole so that his entire army could see that the Greeks were mortal and could be killed. Forty years after the battle, Leonidas's body was taken back to Sparta for a hero's burial.

A stone lion was erected on the hillock of Kolonis as a memorial to the fallen. A plaque inscribed with the names of all 300 soldiers who died there was also put on display. Perhaps the most famous comment about the Battle of Thermopylae was written by the Greek poet Simonides and inscribed on that plaque: "Tell them in Lacedaemon, passer-by / That here, obedient to their laws we lie."

The loss of Thermopylae allowed the Persian army to descend into southern Greece, sacking Athens and burning the temples on the Acropolis. Just west of Athens, in the Battle of Salamis, the Greeks won a major victory against the Persian naval force. While the Persian land force was enormous in size and undefeated, they relied on their ships to keep them supplied. With many of those ships gone, the army could not be guaranteed provisions. Worse, Xerxes feared that news of a major Persian loss might stir the Greeks in Asia Minor to revolt, cutting off his supply lines entirely. He ordered his army to turn back to Thessaly and his navy to the Dardanelles to guard the bridges erected there.

Although the battle at Thermopylae between the armies of Xerxes and Leonidas has attained a lasting place in Western history as an outstanding example of bravery in battle, the site has seen a number of other major clashes over the centuries. In 352 B.C. Philip of Macedon used a conflict between the cities of Phocis and Thebes over control of the oracle at Delphi as an excuse to invade central Greece. Taking Methone, Philip moved south towards Thessaly. At Thermopylae he was prevented from going any farther south by an Athenian army. Philip turned his army north into neighboring Thrace instead.

In 279 B.C. the Gallic leader Brennus invaded Macedonia and conquered northern Greece. A Greek army at Thermopylae blocked the advance of the Gauls into Thessaly for three months before Brennus sent a diversionary force toward the city of Aetolia in western Greece. The Greeks were forced to withdraw from Thermopylae to defend Aetolia. Brennus then led his army as far as Delphi, where he was seriously wounded. When he retreated north, the Gauls replaced Brennus with a new leader, and he committed suicide.

In 193 B.C., Antiochus the Great, leader of Syria, wished to extend his empire into the Greek peninsula. The Greek cities of the north were already his allies and the Carthaginian general Hannibal lived in exile in Antiochus's court. In addition, Antiochus had recently married his daughter to Egyptian pharaoh Ptolemy V, gaining a strong ally to the south. Invited by the Aetolians to join them in an alliance with Macedon and Sparta, Antiochus sent a small army into Greece with the idea that a popular uprising would secure the region for Syria. The uprising did not materialize, however, and the allied cities were soon squabbling among themselves. Antiochus found himself the target of Roman attention. Rome

considered the small Greek city-states to be under Roman protection. Cato the Censor, a Roman tribune and later a leading statesman, was sent to drive Antiochus out of Greece. In 191 B.C., Antiochus erected a double wall with trenches across Thermopylae and built forts on the nearby mountainside in an effort to thwart the Romans. But when Cato took the forts, the Syrian army was virtually destroyed. Antiochus escaped with a mere 500 of his troops.

The most recent military engagement at Thermopylae occurred in April of 1941 during the Nazi invasion of Greece. A combined force of Australian and New Zealander soldiers held Thermopylae against an attack by German armored forces. Although they inflicted heavy casualties on the Nazi tanks, the Expeditionary Force was eventually outflanked and obliged to retreat.

A monument to Leonidas and the Greeks who died with him today marks the site of the pass of Thermopylae. Erected in 1955 by King Paul of Greece, the white marble monument is surmounted by a bronze figure of Leonidas. On the base are reliefs of scenes from the famous battle and epigrams about the soldiers who fell there. Opposite the statue is a tumulus, or small mound, which legend claims is where the Spartan dead were buried. The stone lion erected at the site in memory of the battle has long since vanished. The hot springs, located next to the tumulus, are still popular with travelers to the site.

Further Reading: *Thermopylae: The Battle for the West* by Ernle Bradford (New York: McGraw-Hill, 1980; as *Year of Thermopylae*, London: Macmillan, 1980) is a detailed account of the battle of Thermopylae and of the larger campaign by King Xerxes to conquer Greece. *Atlas of the Greek World* by Peter Levi (New York: Facts on File, and Oxford: Phaidon, 1980) provides descriptions of the major battles of the Persian wars as well as a number of maps showing the placement of troops in each battle. *A History of Greece: To the Death of Alexander the Great* by J. B. Bury and Russell Meiggs (New York: St. Martin's, 1975) covers the Battle of Thermopylae and the related sea battles between Persian and Greek forces.

—Thomas Wiloch

Thessaloníki (Thessalonike, Greece)

Location: Northern Greece, in the province of Macedonia and department of Thessalonike, at the head of the Gulf of Salonika.

Description: Founded in 315 B.C., Thessaloníki grew from a small town in ancient Greece to the second city of the Byzantine Empire. Conquered by the Turks in the fifteenth century, Thessaloníki remained part of the Ottoman Empire until it was returned to Greece in the early 1900s. Today's Thessaloníki is a mix of the modern and the ancient, and remains an architectural testament to all three powers (the Greeks, the Romans, and the Turks) that once ruled the Greek province of Macedonia.

Site Office: Greek National Tourist Organization
34 Mitropoleos Street
Thessaloníki, Thessalonike
Greece
(31) 271-888

Archaeological evidence found in a number of ancient mounds and settlements indicates the area around Thessaloníki (also known as Thessalonika or Salonika) has been inhabited since about 1000 B.C. The town itself dates to approximately 315 B.C., when Macedonia's King Cassander (one of Alexander the Great's generals, who succeeded the conqueror to the throne) incorporated several small villages in the area, including the town of Thermae, into a larger town and named his new city after his wife Thessaloníki, daughter of Philip II and half-sister of Alexander the Great. Philip's daughter was born during his successful expansion of his domain, and her name literally translates as "Victory in Thessaly." During Macedonian rule, Pella was the capital of the realm, but newly formed Thessaloníki grew quickly. A mere twenty-five miles from the capital, Thessaloníki soon became the kingdom's "second city," and was home to the land's navy.

After the Romans conquered the area in 146 B.C., the young city of Thessaloníki was declared capital of the province of Macedonia Prima. Thessaloníki already enjoyed a strategic location on the Thermaic Gulf, and its prosperity was further enhanced by the building of the Via Egnatia, the main artery leading across the empire from Rome to Constantinople. The proximity to important land and water routes, coupled with the existing trade route up the Axios Valley into the Balkans, ensured Thessaloníki's development and success.

During the early Roman period, the provincial capital of Thessaloníki saw a number of well-known guests. Exiled from Rome, Cicero spent some time in Macedonia at Thessaloníki in the year 58 B.C. Pompey, too, used Thessaloníki as a refuge in 49 B.C., taking cover in the Macedonian town while eluding Caesar. Thessaloníki's easy accessibility via the Roman road and the sea enabled the Apostle Paul to travel from Philippi to visit the town and spread the newly established Christian faith.

During Paul's first visit in A.D. 50, the apostle's teachings and actions enraged local Jews, who attacked the home where Paul had been staying. Despite the initial opposition, Christianity took hold in Thessaloníki, and the number of converts grew quickly; Paul's Thessaloníki epistles praise the group for their faith and determination. Paul returned to continue preaching to the local church in A.D. 56.

The relative peace enjoyed by Thessaloníki was destroyed in the third century when the town came under siege by the Goths. The Thessalonians successfully defended their city from the invaders in 253 and again in 269. The Roman Emperor Galerius, ruler of the eastern half of the Roman Empire, finally defeated the Goths and made Thessaloníki his capital in 300. Galerius did not allow his subjects the same religious freedom given by earlier Roman emperors and quickly became known for his zealous persecution of the Christians. In 306, the Greek Demetrius, both a Roman officer and a lay-pastor in the Thessalonian church, was arrested, jailed, and ultimately killed for his faith. The Christian community was furious and insisted that Demetrius be canonized. Today, St. Demetrius is Thessaloníki's patron saint, protecting the city from war, hunger, and sickness; the church of St. Demetrius was built over his tomb.

A later emperor, Theodosius the Great, made Thessaloníki his base for maneuvers against the continuing menace of the Goths. In poor health, Theodosius converted to Christianity and in 380 issued the Edict of Thessalonia, a proclamation condemning the worship of pagan gods and denouncing the Arian heresy at Constantinople. Despite Theodosius's open support of Christianity, he is responsible for one of the most heinous crimes committed against the residents of Thessaloníki.

The people of Thessaloníki were great fans of chariot racing. The popular favorite among the racers had been imprisoned for making improper advances toward an attendant of the Gothic commander. Enraged that their hero had been incarcerated, the Thessalonians lynched the town's military commander, Botheric, a Goth, along with several of his men. When Theodosius returned to the city and learned of the tragic events, he plotted a horrific revenge. The citizens of the town were all invited to the Hippodrome to watch the games held in the emperor's honor. Once the populace was inside the circus, Theodosius's Gothic troops slaughtered an estimated 7,000 citizens.

The city of Thessaloníki enjoyed continued prosperity under the Byzantine Empire. Favored by Justinian I, Thessaloníki became the empire's second city after Constan-

The White Tower, now a museum
Photo courtesy of Greek National Tourist Organization

tinople. But as with the rest of Greece during this era, peace there was frequently broken by periods of violence, often from outside. The wealth of Thessaloníki attracted the unwelcome attention not only of the Goths, but of the Avars and the Slavs as well. Thessaloníki defended itself five times against these invaders and remained a Byzantine stronghold in widely Slavic-held territory until the late seventh century. Despite Justinian II's efforts to drive out the Slavs, Thessaloníki did not again fall under direct Byzantine control until Staricius's successful offensive around 783. Despite the ongoing struggles, Thessaloníki prospered—not only economically, but spiritually as well. Ninth-century brothers and saints, Cyril and Methodius, two Thessalonians sent out as Apostles of the Slavs, are credited with converting the pagan Slavs to Christianity.

The city's wealth in a time of unrest continued to entice marauders, and Thessaloníki endeavored to maintain its self-defense. The citizens repelled invasions by the Bulgars and the Serbs, but in 904 Thessaloníki fell to the Saracens, led by the renegade Leo of Tripoli, himself a Greek. The Saracen conquerors rejoiced with a ten-day celebration of plunder and slaughter, taking an estimated 22,000 Thessalonians and selling them into slavery.

Thessaloníki survived the Saracens' pillaging only to be under siege nearly continuously for the next few centuries. Basil II used the town to mount his attack against the Bulgars, who had advanced on Thessaloníki in 1041. A century later, in 1185, Thessaloníki was once again invaded. The armed forces of the Norman William II of Sicily, under command of Tancred, captured the city and continued the tradition of looting the prosperous town. The devastation caused by the Norman siege was recorded by the renowned Homeric scholar and archbishop of Thessaloníki, Eustathius.

In 1204, as the period of the Fourth Crusade drew to an end, Boniface, margrave of Montferrat, was awarded territory as a consolation for having the imperial crown of Constantinople elude his grasp. Boniface established a bishopric in Thessaloníki and once again made the town a capital, this time of the Latin Kingdom of Thessaloníki.

Boniface's rule was short-lived: the town was taken by Theodore Angelus, the despot of Epirus, in 1222. Theodore promptly declared himself emperor of the land. While Thessaloníki was in the hands of Theodore, the rest of Macedonia was conquered by Ivan Asen II of Bulgaria. A few decades later in 1246, John Vatatzes declared both Thessaloníki and the rest of Macedonia to be once again under control of the imperial crown at Constantinople.

The fourteenth century was particularly chaotic for the empire's "second city," since Thessaloníki was seen as the first goal for anyone seeking to usurp the imperial title. The citizens of the town were again forced to defend their city against foreign aggression, this time from the Catalans in 1308, and later from the marauding Serbs and Turks.

Thessaloníki also faced problems from within. In 1342, religious insurrectionists called the Zealots took over Thessaloníki and ruled the area as a people's republic. Their rule was chaotic and brutal, marked by the murder of the town's nobles. Zealot rule was relatively short-lived; order was reinstated and control returned to the Byzantine Empire in 1349.

For the next few decades, control of Thessaloníki shifted back and forth between the Turks and the Byzantines. The town was taken by the Ottoman Turks in 1387 and again in 1394. Thessaloníki was restored to Byzantine control in 1403 when the Mongol Timur (Tamerlane) defeated the Turk Bayezid I at Ankara. In 1423, when Byzantine defeat at the hands of the Turks appeared inevitable, Andronicus Palaeologus, son of the Emperor Manuel II, handed control of Thessaloníki over to Venice, a move designed to protect the town from Turkish pillage.

But despite the protective moves of Andronicus Palaeologus, the Turks did come to Thessaloníki. In 1430, Thessaloníki fell prey to the worst siege in its long and chaotic history, led by Murad II of the Ottoman Empire. Ottoman takeover brought an end to the Greek way of life in Thessaloníki. Churches were destroyed or forcibly converted into mosques, a number of Greek inhabitants were cast out, the area was subjected to extraordinary taxes, and the men of the town were coerced into Turkish military service.

Later in the fifteenth century, the population of Thessaloníki expanded rapidly due to the arrival of thousands of Jews. Fleeing persecution, a group of Bavarian Jews arrived first around 1452, accompanied by a small group from Hungary. Later in the century, more than 20,000 Jews came to Thessaloníki, escaping the Edict of Alhambra, the 1492 declaration by Ferdinand and Isabella expelling all Jews from Spain and Portugal.

The heavy influx of Jews brought increased trade and prosperity to Thessaloníki; this prosperity helped the city recover from the destruction wrought by Murad II. The immigrants were by and large highly skilled craftsmen, specializing in the cutting of jewels and the manufacture of silk and woolen textiles. These crafts led quickly to trade throughout Europe. By the sixteenth century, the Jewish sector constituted nearly half of Thessaloníki's population. Despite their integration into Thessalonian life and commerce, the Jewish population continued to speak Ladino, a mixed language similar to Castilian, but written using Hebrew characters.

Along with its prosperity, however, Thessaloníki still had its share of crisis. Fires damaged the city in 1545 and 1617. The town suffered again in 1659 when religious discord broke out among the Jewish population. Ultimately, the Donmeh sect broke away and converted to Islam.

As the Ottoman hold on the Balkans and Macedonia grew weaker, the rule became harsher. During the eighteenth century, the citizens of Selanik—the Turkish name of Thessaloníki—protested the growing inhumanity of their oppressors, rising to revolt in 1720, 1753, 1758, and again in 1789.

The Greek struggle for independence continued into the next century. Throughout the period of Ottoman rule, a large number of Greeks had moved back into Macedonia and into Thessaloníki. The town itself over the course of time was allowed some measure of autonomy by the Turkish overlords. Reprisal, however, was fierce when these Greeks rose up again seeking their freedom. In 1821, when Emmanuel Pappas' rebellion on the Chalcidice Peninsula sparked open discord throughout the land, discontent swept through Thessaloníki and a number of citizens took part in supporting the Greek insurrection. The Turks moved swiftly to put down the Thessalonian supporters of the Greek War of Independence, killing at least 3,000 people in the city alone and jailing hundreds of others.

The seeds of discontent had been planted, and despite Turkish efforts, sentiments of rebellion against Ottoman rule continued to grow. In 1906, Thessaloníki became the headquarters of the Turkish Committee of Union and Progress, a group responsible for organizing the Macedonian revolt in 1908. One year later, in 1909, Ottoman Sultan Abdülhamid II was overthrown and exiled to Thessaloníki.

After the Balkan Wars in the early twentieth century, Thessaloníki, along with much of Macedonia, was reunited with Greece. At the time of the reunification, Thessaloníki's population was approximately 180,000, with nearly half of those citizens being Jews. Nearly 80,000 people were left homeless in the most devastating fire of Thessaloníki's long history, which broke out in 1917. Since no fire department had been established, the flames spread quickly, destroying ancient buildings, churches, and mosques, as well as nearly 10,000 homes. The housing situation was again strained several years later in 1923, when Thessaloníki received 115,000 refugees from Asia Minor.

Thessaloníki escaped extensive damage during World War I but unfortunately suffered greatly during World War II. The Germans advanced through Yugoslavia by way of the Vardar Valley, entering Thessaloníki three days later on April 9, 1941. During the occupation, nearly all of Thessaloníki's Jews were deported to Nazi death camps and killed. Once in the tens of thousands, Thessaloníki's Jewish population today is minimal.

Thessaloníki's reconstruction after the devastations of World War II was disrupted by the Greek Civil War, and again by a severe earthquake in 1978. The city has been rebuilt, however, and today modern buildings stand alongside archaeological and architectural treasures from Thessaloníki's long and varied history, from its early days as an ancient Greek village, to its zenith as second city of the Byzantine Empire, to the period when bazaars and mosques still testify to the widespread influence of the Ottoman Empire. Notable sites include the church of St. Demetrius, rebuilt many times since the fifth century, most recently between 1926 and 1948; numerous other historic churches; portions of city walls, dating from the Byzantine period; the triumphal Arch of Galerius, built in 297 to honor the Roman emperor; Roman baths, theatre, and marketplace; the Archaeological Museum, housing many ancient artifacts found in the area; the White Tower Museum, displaying Byzantine arts; and several museums dedicated to Macedonian history and culture.

Further Reading: *Macedonia: 4000 Years of Greek History and Civilization,* part of the Greek Lands in History series edited by M. B. Sakellariou, (Athens: Ekdotike Athenon S.A., 1993), provides extensive information on the political, religious, and artistic history of the region, including Thessaloníki. The oversized book is also beautifully illustrated.

—Monica Cable

Thíra (Cyclades, Greece)

Location: In the Aegean Sea; the most southerly of the larger Cyclades Islands; approximately seventy miles north of Crete and fifteen miles west of Anaphi.

Description: Also commonly known as Santoríni; site of Bronze Age settlements destroyed by a volcanic eruption that sank the center of the island in approximately 1500 B.C.; reinhabited two centuries later and ruled alternately by Greeks, Italians, and Turks; site of archaeological excavations revealing remarkably preserved buried settlements, most notably the Bronze Age city of Akrotiri.

Site Office: Thíra Archaeological Museum
Thíra, Cyclades
Greece
0286-22217

Thíra was a single, round island until a huge volcanic eruption in approximately 1500 B.C. fragmented the land mass into a ring of island remnants roughly 28 square miles in area surrounding a central bay-crater 1,300 feet deep. The largest of the island fragments is the crescent-shaped Thíra on the east; the much smaller Therasia, which separated from Thíra in an eruption of 236 B.C., is to the northwest; tiny Aspronisi (White Island) is to the southwest. The currently active volcanic cones Palai (or Hiera) Kameni and Nea Kameni (Old and New Burned Islands, respectively) appear as small islands in the middle of the basin. Palai Kameni first formed in 196 B.C. Nea Kameni was once two separate cones: Mikra, which appeared in A.D. 1573, and Nea, which was created in A.D. 1711–12; the two merged in an eruption in July 1925. Other small islands have appeared at various times since the initial eruption, only to disappear again after further volcanic or seismic activity. In 1570, just prior to the appearance of Mikra Kameni, the south coast of the main island, including the port of Eleusis, collapsed into the sea. Over half the buildings of the west coast of the main island were destroyed in a July 1956 earthquake measuring 7.8 on the Richter scale.

As Thíra's shape has continually changed over time, so has its name. In ancient times it was called Stronghyle (The Round One) because of its pre-eruption shape, as well as Kalliste (the Most Beautiful). The name Thíra comes from Theras, son of Autesion, who, along with his fellow Spartans, colonized the island. The Venetians, who ruled the Cyclades during the thirteenth, fourteenth, and fifteenth centuries, named the island Santoríni (also seen as Santorin and Sandoríni) after its patron saint, Irene of Salonika, who died there in exile in A.D. 304.

Carians probably inhabited Thíra in the third millennium B.C., followed by the Achaeans who settled on the island about 1900 B.C., later to be driven out by the Phoenicians. In the first half of the second millennium B.C. Thíra was a flourishing community. Excavations at the ancient town of Akrotiri show trade contact with Minoan Crete and possibly with places as far away as Libya, although the island maintained a distinct culture and artistic style of its own.

Some scholars view the Zeus and Europa rape myth as a reflection of these early contacts among Thíra, Crete, and mainland Greece and the eastern Mediterranean. According to legend, after Zeus abducted Europa from Tyre in Phoenicia and took her to the south coast of Crete, Europa's father, Agenor, sent his five sons out in search of her. They each took a different route. One son, Cadmus, along with his mother Telephassa, sailed to Rhodes and then to Thíra. There Cadmus built a temple to Poseidon and left his kinsman Membliarus to dwell on the island. According to Greek historian Herodotus, Membliarus and his descendants lived on Thíra for eight generations before Theras arrived from Sparta.

The flourishing civilization on Thíra ended abruptly with a volcanic eruption on the island in the middle of the second millennium B.C. Archaeological excavations at Akrotiri, a Bronze Age city at the southwest of what is today the main island remnant, reveal that there were two evacuations of the site prior to the eruption. The first one came before a major earthquake that damaged buildings in the settlement. Apparently, there were warning tremors that alerted the townspeople to flee. The earthquake was followed by a period of calm, possibly lasting several months, during which the townspeople returned to their homes. Excavated buildings reveal some makeshift repairs, and debris had been removed from streets and deposited into piles in the post-quake clean-up. Archaeologists have found stone tools resembling hammers on top of some of these piles; the large, thirty-pound stones have rope grooves in them, which would have enabled workmen to suspend them from buildings and use them as battering rams or wrecking balls to knock down damaged walls. Clay buckets of fresh mortar were also discovered.

Archaeologists have found virtually no precious materials at the site. Only two thin silver rings and a piece of gold leaf, apparently forgotten in some jeweler's shop, have surfaced. Neither were there any human or animal skeletons found, as there were at the suddenly ash-inundated Pompeii. There must have been some prior warning, possibly fumes or gases, before the eruption came on Thíra. The inhabitants of Akrotiri had time to take their valuables during this second, and final, evacuation of the city. The discovery of an abandoned gold-inlaid dagger at an excavation site one-half mile

A view of Thíra
Photo courtesy of Greek National Tourist Organization

from Akrotiri suggests that not all communities on the island were as fortunate.

The volcanic explosion itself was so violent that it exhausted the magma chamber below it under the sea, causing the area above the chamber to collapse and creating a caldera (collapsed area) one-third mile deep. The surrounding sea then flooded into the depression, creating the central bay. The island lost thirty-three square miles of land in the eruption, either blown up or sunk. Enormous tidal waves generated by the event obliterated sites along the north coast of Crete about seventy miles away. (The demise of Minoan Crete has often been attributed to the eruption on Thíra, but evidence today shows that Cretan civilization survived the eruption.)

Microscopic glass shards from the eruption have been found as far away as Egypt's northern coast, 500 miles to the southeast. Deep sea cores show a tephra dispersal in an easterly and southeasterly direction running past the island of Karpathos and including areas of the southwestern Turkish coast and the eastern half of Crete.

The volcano covered the island with a 1.5-inch layer of pellet-like pumice, which was made crusty, it is speculated, by post-eruption rains. Later, in a second stage, the volcano ejected much larger pumice stones, some exceeding 6 inches in diameter. This layer varied between 1.5 and 3.3 feet thick; it had a "snow" effect, falling more or less level, sometimes into houses through open windows or accumulating against closed doors.

In the final stage the volcano ejected a thick layer of volcanic ash called tephra or pozzuolana, deposited in some areas over sixteen feet thick. Along with the tephra came basaltic boulders, some of which did much damage to buildings. In some places fine tephra filled whole rooms in buildings, preserving everything inside for discovery centuries later. Organic materials such as wood furniture left impressions in the ash after they disintegrated, enabling archaeologists to make plaster-of-Paris casts. Some artifacts—wicker and rush baskets, for example—did not disintegrate at all but were preserved in a powdery condition in the ash.

The traditional method of archaeological dating, done by comparing styles of excavated pottery, dates the catastrophe at Thíra at about 1500 B.C. Carbon-14 tests from carbonized tree trunks at the bottom of the island's Firā quarry, however, date the eruption at about 1640 B.C., plus or minus thirty years. The latter date is supported by evidence from volcanic dust in Greenland icesheets, narrow growth rings in

California bristlecone pines, and oaks growing in Irish peat bogs, and by ancient Chinese climatic records. Controversy still exists today among adherents to the conventional dating system and those preferring these newer methods.

After the island was abandoned—presumably people either got away to other islands in boats or perished in the sea—at least two centuries passed before Thíra was inhabited again. Mycenaean shards from site surfaces indicate that Greek mainlanders had reestablished contact with Thíra by the end of the fourteenth century B.C.

In the eighth century B.C. Dorians from Crete settled Thíra. Their capital, at Mesa Vouna on Mt. Profitis Elias, survived until the early decades after Christ. In 630 B.C. Grinos, their king, founded a colony at Kyrene, largest of the Greek colonies in North Africa. Thíra and Melos were the only two Cycladic islands that sided with Sparta, and not Athens, during the Peloponnesian War (431–404 B.C.). Nevertheless, Thíra was later absorbed into the Athenian Empire.

Between approximately 300 and 145 B.C., the Ptolemies of Egypt made the ancient city of Thíra, founded on the southeast coast of the main island sometime before the ninth century B.C., into a naval base from which they could control the Aegean. The island cluster then came under Roman rule for some time. The Byzantines considered Thíra of strategic importance as well; fortifications they built there have since been destroyed by earthquakes. From 1207 to 1335 Thíra was held as a fief of Marco Sanudo, duke of Naxos; the medieval capital, Skoros, was located to the northwest of Firā, the modern capital. Thíra was under Italian rule until 1537, and Turks held the island until 1830, when it was reunited with Greece.

The ancient settlements of Thíra had been long forgotten by 1860, when workers quarrying tufa for the Suez Canal Company discovered buried walls on the south coast of Therasia. News of the discovery soon reached French vulcanologist Ferdinand-André Fouqué, who brought experts from the French School of Archaeology to the island. In 1869, the archaeologists discovered a prehistoric city in southern Therasia, buried by pumice from the great eruption; their findings proved that Thíra had been a center of Cycladic civilization. Baron Hiller von Gärtringen excavated the ruins of Thíra city from 1896 to 1903, an important excavation exploring the time of the Ptolemies during the Hellenistic period. He also uncovered artifacts from earlier periods, including archaic Greek rock inscriptions with some of the earliest known examples of Greek characters. The excavation of a neighboring necropolis carries on today.

The most important excavations, however, have taken place south of the village of Akrotiri, in the southwest part of the main island. Archaeologists of the French School began work at the site in the late 1860s. Working from a cut in the tufa caused by a ravine, they uncovered a frescoed room (which collapsed soon after excavation), but they lacked the technology to dig through the tons of volcanic debris covering the rest of the site. Work was not resumed until 100 years later, when Professor Spyridon Marinatos began excavating Akrotiri in

1967. After Marinatos was killed by the collapse of an excavated wall in 1974, Christos Doumas took over the operations. It soon became apparent that the job would take several generations of archaeologists, and substantial facilities were built to accommodate storage, repair, treatment, and examination of the finds. Test digs showed that there are ruins one-half mile away from the current excavations. Only one city block has been exhumed. It is estimated that it will take up to 300 years at the current rate of digging to complete the excavation.

The site was buried in a thick mantle of volcanic debris. Marinatos began to experiment with tunneling into the buildings, which were preserved to a height of more than one story. His idea was to provide entry to the houses through their doors, but to keep it all underground, presenting the excavation as a living museum with its modern vineyards still maintained aboveground. He had to abandon this idea due to the danger of the houses collapsing as their surrounding supports were tunneled out from around them. The tunnels also became a hazard when volcanic ash dried out and became powdery and weak. Tunneling, moreover, the least scientific method of exploration, disregards stratigraphy and sequence, and tunneling tools can easily destroy artifacts.

Because the buildings, some of them two or three stories high, were made of rubble and clay supported by now-disintegrated timber, Marinatos constructed a corrugated roof supported by steel pillars over the site to protect the remains from erosion and rains. Wooden wall reinforcements and door- and window-frames on the Bronze Age structures were reduced to nothing more than negative impressions in the surrounding pumice, and so Marinatos poured concrete into the empty impressions to bolster the fragile buildings. A large number of finds—pottery, stone artifacts, and some negatives of pieces of furniture into which liquid plaster of Paris was poured—have all been cleaned, mended, cataloged, and photographed.

The plan of Akrotiri, as much as has been revealed through excavation, resembles that of the present-day villages on Thíra, with their narrow, winding streets. The prehistoric site, however, shows separate mansions and building complexes. Ancient sewers, narrow stone-lined ditches covered with slabs, ran under the flagstones of the streets. Lavatories were connected to the system by means of clay pipes in the house walls. Few open spaces for village squares stand among the buildings. Presumably the city's population was large and tightly packed.

No palace ruins have so far appeared, indicating the possible absence of a single ruler or king on the island. All the buildings appear to be comfortable, private houses of large, possibly extended, families. There is no provision for the accommodation of animals within individual houses or in the town as a whole, indicating that Akrotiri was an urban community.

Most notable of all the finds are the well-preserved wall paintings discovered in various buildings on the site. These murals include pictures of boxing boys and women in Minoan dress gathering saffron. A three-wall painting known as the *Spring Fresco,* now in the National Archaeological Museum in Athens, shows soaring swallows and lilies blooming out of a rocky landscape. One room, in what is known as the West House, contains murals all pertaining to water: a sea battle, a fisherman holding up his catch, a river landscape, and a flotilla mural in which several ships are shown rowing from one harbor to another.

These wall paintings are, in fact, not true frescoes. The late Bronze Age painters at Akrotiri used the same techniques as those known to have been employed in Crete. In the upper-story walls, they applied a thin coat of lime plaster over normal wall surfaces of clay mixed with straw. Sometimes they used the natural white plaster as background color. They used ochre for the walls of their baths. A taut string or sharp instrument made guidelines of the images to be illustrated in the prepared wet plaster wall surface. The colors they employed, derived from minerals and thus well preserved, included natural plaster white, yellow, red, brown, blue, and black, sometimes in intermediate shades produced by mixing. Certain colors were used for different subjects; for example, blue was used to depict water, brown for the male body and for landscapes and petals, red to delineate bands as "frames" above and below murals, white for female bodies, and black for borders, details, and hair.

Archaeologists have also discovered thousands of intact pottery pieces. Large storage containers for food, liquids, and clothing; cooking, eating, and drinking vessels; and bathtubs, braziers, oil-lamps, flower pots, and beehives have all been found at Akrotiri. Stone tools and vessels, vases, grinders, pestles, hammers, anvils, millstones, and demolition hammerstones and anchors have been found as well.

Theories abound regarding possible connections between the volcanic eruption on Thíra and ancient stories of tidal waves, complete darkness, ash-falls, storms, and poisonous gases. Perhaps the most intensely investigated has been the myth of the sinking of Atlantis. Because Plato asserted that Atlantis was destroyed "in a single day and night," long enough for a tremendous volcanic eruption to bury a town or an island, many have been tempted to try to make the facts of Thíra's history fit the legends. In 1909 K. T. Frost suggested Knossos in Crete as the site of Plato's Atlantis, citing the harbor, the baths, the stadium, and the bull sacrifices as evidence linking the real and the mythical cities. Spyridon Marinatos suggested that perhaps when news of the island of Thíra "sinking into the sea" reached Egypt, the Egyptians confused Thíra with Crete, thus contributing to the Atlantis legend. Other scholars have searched Thíra's history for evidence, not of Atlantis, but of the Egyptian plagues or the parting of the Red Sea described in the Biblical Book of Exodus.

Drawn by such storied associations, as well as by the archaeological discoveries and the scenic beauty, over 1,000 people visit Thíra every day of the summer tourist season. In addition to the Bronze Age excavation sites such as Akrotiri, tourists can visit the Hellenistic ruins at Thíra; the ruined medieval capital of Skaros, with its castle of the Crispi; and the museums of Fira, the modern capital.

No ships cast anchor now in the bay created by the caldera. The water is too deep, over 200 fathoms (nearly 1,300 feet) in some places. Near Nea Kameni, anchors can still be lost or caught on irregularly shaped outcroppings of underwater pumice. Instead, ships moor to buoys secured to the sea-bed by long chains. Cliffs, dramatically colored red, white, and black with volcanic materials, rise 650 to 1,300 feet on all sides of the caldera. Because the islands themselves are made of porous volcanic tufa, rainwater will not stay on the surfaces unless caught in cisterns. The rain soaks down fast, allowing the growth only of vines, with their capillary roots that can reach to great depths to tap water. Tomatoes, grapes, pistachios and fava beans are grown, but the island is treeless. The island exports a potent, sweet wine as well as tufic dust used for water-resistant cement.

Further Reading: *Thera: Pompeii of the Ancient Aegean* by Christos G. Doumas (London and New York: Thames and Hudson, 1983), the current leader of the Akrotiri excavation, gives a first-rate survey of Thíra's geologic and cultural history, as well as specifics about archaeological finds and the Atlantis legend. *Unearthing Atlantis: An Archaeological Odyssey* by Charles Pellegrino (New York: Random House, 1991) speculates on the possibilities of Thíra's connection to Atlantis from a paleontological and volcanological perspective. *Lost Atlantis: New Light on an Old Legend* by J. V. Luce (London: Thames and Hudson, and New York: McGraw-Hill, 1969) was published only two years after Professor Marinatos began his dig and thus has limited information on the area of Thíra. However, it is full of geologic facts and photos of artifacts and the early stages of the excavation. *Atlantis: The Truth behind the Legend* by A. G. Galanopoulos and Edward Bacon (Indianapolis and New York: Bobbs-Merrill, and London: Nelson, 1969), written by a seismologist and an archaeologist, provides a well-illustrated account, full of geophysical and historic facts; it also deals with the eruption on Thíra and the Atlantis legend.

—L. R. Naslund

Tiryns (Argolis, Greece)

Location: Tiryns is located in the Argolis plain about four miles southeast of modern Argos, toward the seaport of Nauplia. It lies about twelve miles from Mycenae.

Description: The strongly-built citadel of Tiryns dominates the seaward end of the Argolid plain. It stands isolated on a small hill, about eighty feet high, that nonetheless dominates the surrounding territory. At the height of its power, Tiryns probably protected a coastal settlement, although it is now more than a mile from the sea. Its most distinctive feature is its Cyclopean architecture, so impressively massive that the classical writer Pausanias ranked the fortifications with the Pyramids of Egypt. The walls were also celebrated by the poet Homer in both the *Iliad* and the *Odyssey,* and by the later poet Pindar.

Site Office: Tiryns Archaeological Site
Tiryns, Argolis
Greece
(752) 22657

"Well-walled Tiryns" dominates the seaward portion of the plain of Argos. The site, occupied since Neolithic times, may have originally been a small rocky hill surrounded by marshland. Beneath the ruins of the Mycenaean-era citadel, archaeologists have uncovered the remains of a large circular building made of brick and dating to the Early Bronze Age. The structure may have been used as a granary. Tiryns was fortified for the first time in the fourteenth century B.C. The legend of the town's origin as related by the Greek poet Pausanias states that the walls were built by the Cyclops, giants with tremendous strength, whom the founder Proitos had called from their home in Lycia. However, its name, again according to Pausanias, was taken from a son of the hero Argos, who in turn was believed to be descended from the god Zeus. Tiryns is also closely associated with the legend of Perseus, who gained control over the city and the nearby fortress of Mycenae after slaying the sea monster that threatened Andromeda. One of their sons, Alkaios, became king of the Argolis area. He was followed by Amphitryon, husband of the beautiful Alkmene, who was seduced by the god Zeus during her husband's absence from the city. The product of this union was the hero and demigod Hercules, who was ordered to perform twelve tasks as penance for having killed his children during a mad rage.

One of Hercules' labors involved taming a pair of flesh-eating horses for Diomedes—perhaps a relative of the "Diomedes of the great war cry" mentioned by Homer. During the thirteenth century B.C., Tiryns was an outlier of the citadel of Mycenae. Diomedes, the man with whom Homer associates the city's troops, is represented as a faithful adherent of Agamemnon's. After the period of the Trojan War, however, Tiryns escaped for a time from its Mycenaean overlords. Even though it fell before 1200 B.C. in the disasters that overtook the other Mycenaean palace cultures of the Aegean, Tiryns was resettled and occupied. According to Greek tradition, the exiled descendants of Hercules placed themselves at the head of large Dorian bands and returned to conquer not only Tiryns and the Argolis, but the entire Peloponnese and eventually all of Greece itself.

Between the heroic age and the classical period in Greek history Tiryns was often under the control of the polis of Argos. According to the Greek historian Herodotus, King Pheidon of Argos, who reigned in the seventh century B.C., was the most powerful ruler in the Peloponnese. However, as the polis of Sparta increased in power, pressures on the Argives correspondingly increased, and the towns of the Argolis plain began to assert some independence. After Sparta established control over its home territory of Laconia, its citizens warred against and conquered Messenia, a polis toward the southwestern end of the peninsula. The Spartans then began expanding northwards, encroaching on Argive territory. In 547 and 494 B.C., the Spartan army won two decisive victories over the Argives. In the last of these two victories, known as the battle of Sepeia, the Spartans under King Cleomenes I exterminated all the male citizens of Argos who were of fighting age. When the Spartans called for soldiers to fight in the Persian Wars (490–479 B.C.), the Argives were unable to respond. However, Tiryns, which had not participated in the massacre of Sepeia, sent forces to serve with the Spartans against the Persians at the battle of Plataea. Both Tiryns and Mycenae maintained friendly relations with Sparta after the Persian Wars, and in 468 B.C. the Argives overran both towns and laid them waste in revenge.

Although the Argives seem to have been largely responsible for the destruction of Tiryns and its reduction to its present barren state, a few inhabitants still occupied the site as much as a thousand years later. A little Byzantine church and its connected cemetery on the acropolis, on top of the Mycenaean ruins, served a tiny village below the hill during the medieval period. Also, unlike many other Homeric sites, the location of Tiryns remained common knowledge through the centuries. Occupation of the Peloponnese Peninsula by the Moslem Ottoman Turks in the fifteenth century cut the site off from Western visitors at about the same time that interest in classical antiquities was rising. It was not until 1668 that a European, a Frenchman named Des Mouseaux, was allowed to visit the site and to publish the first modern

The east gallery at Tiryns

Gate to the upper acropolis
Photos courtesy of The Chicago Public Library

description of its walls. Many other travelers followed him, but the most significant of them all was the excavator of Troy, Heinrich Schliemann.

Some limited excavations were carried out at Tiryns by a German named Thiersch in 1831, but scientific investigation of the ruins only began when Schliemann and his assistant, the architect Wilhelm Dörpfeld, began digging at the site in 1884. Perhaps because of Dörpfeld's influence, much of Mycenaean Tiryns emerged undamaged, unmarred by Schliemann's enthusiastic, if by modern standards unscientific, excavation technique. However, Dörpfeld and Schliemann interpreted their discoveries conservatively, according to the commonly held theories of the time, which stated that the Mycenaean palaces were Phoenician in origin. It remained for later diggers, aware of the great extent of Mycenaean civilization, to change this view. Current excavations are under the aegis of the Archaeological Institute of Athens. They have uncovered Linear B texts and pottery with clear Aegean connections.

The famous walls of the city, measuring in places almost thirty-three feet across, make a circuit of nearly half a mile around the site. They consist of huge individual blocks of stone, neither trimmed to fit together nor held in place with mortar. Instead, spaces between the blocks are filled with smaller stones and clay. The entrance, on the eastern side of the acropolis, is approached by a ramp rather than a staircase—a ramp structured to leave the shieldless side of an attacker's body vulnerable to the defenders on the wall above. It was wide enough for a chariot, allowing citizens to drive into the lower level of the citadel.

The acropolis itself is subdivided into three levels: the lower, middle, and upper acropolis. Of the three levels, the middle and lower levels remain mostly unexcavated. However, two tunnels, which served as cisterns for water storage during times of siege, have been uncovered. The upper acropolis contains many of the riches that attracted Schliemann in the first place: the remains of the king's palace, the royal family's home, and the town's citadel, which were all completed around 1375 B.C. The middle acropolis contains a central courtyard immediately behind the palace, as well as remnants of buildings designed for craftsmen. The original wall surrounding the upper acropolis was expanded to enclose this area sometime between 1300 and 1250 B.C. The lower acropolis probably formed a place of refuge in times of war for people who lived on the plain.

A path from the east entrance leads to the left toward the main gate of the upper acropolis, while the remnants of the lower citadel lie to the right. The original wall separating the lower acropolis from the middle and upper acropolis gives the leftward path a tunnel-like feeling. The gate to the upper acropolis is very much like Mycenae's famous Lion Gate; the remains of the doorposts, the pivots for the doors, and a hole for a locking bar can still be seen. Beyond the inner gate is a wider area, leading to the courtyard of the upper acropolis. The outer wall at this point is pierced by six casements, leading into storerooms that now lack their eastern walls due

to the effects of erosion and time. The stones exhibit a curious polish, left by the fleece of sheep that local shepherds penned here over the centuries. The southern wall of the acropolis contains a series of intact casements.

The outer court marking the entrance to the palace leads to an interior room. Two columns directly ahead indicate the beginning of the "Great Court." A passage off to the right leads to the royal family's private chambers. After a slight jog to the left, the passage continues to a smaller yard known as the "Women's Court," and leads into the private apartments. From the "Great Court" a small propylon leads north into the "King's Court," surrounded by a colonnade and dominated by an altar where the king could offer sacrifices or libations—an altar, according to Homer's *Odyssey,* dedicated to Zeus Herkaios, the god of the fenced court. Farther north still is the "King's Megaron," containing a large central hearth and a place for the king's throne on the eastern wall. To the west of the propylon marking the entrance to the megaron is a passage leading to a bathroom, as well as storage areas and rooms for the palace staff.

North of the palace proper is an open yard, surrounded on three sides by the great inner wall (the eastern wall has collapsed since ancient times). The western wall of the yard holds a cistern and the remains of what might have been a tower. In addition, a doorway leads from the yard to a staircase that runs down from the acropolis. Some scholars believe that this might have been a special "servant's entrance," giving domestic workers access to their workplace without the necessity of opening the fortified gates. It is not, as some speculate, a secret passage; its presence is clearly revealed by the shape of the huge outer wall, which bulges out to the west around the staircase.

What actually caused the collapse of the palace culture of Tiryns remains unknown. Some modern scholars believe that the acropolis was destroyed by a terrible earthquake, which leveled Mycenae at the same time. The buildings of the upper acropolis were all demolished around 1200 B.C., and temporary shelters were constructed in the lower acropolis. However, the town was rebuilt and grew; by about 1150 B.C. it supported a population of around 15,000 citizens. Many fragments of Mycenaean culture—offering supporting details for both the *Iliad* and the *Odyssey*—survived to be discovered by modern excavators. The throne room with its hearth was still decorated with fragments of alabaster and blue glass. Bits of frescoes depicting battles and scenes of the hunt adorned the walls of the palace. Even the layout of the palace on its acropolis reflected the glory of Homer's Heroic Age. "This palace," write A. R. and Mary Burn in *The Living Past of Greece,* "is above all the place where a reader of the *Odyssey* feels on familiar ground."

Further Reading: The historic guide to the earliest excavations at Tiryns is by Heinrich Schliemann and Wilhelm Dörpfeld: *Tiryns* (n.p., 1886). Konstantinos P. Kontorlis's *Mycenaean Civilization: Mycenae, Tiryns, Pylos* (Athens: K. Kontorlis, 1974) is a good general overview of the Mycenaean period written for British first-form students. Raymond V. Schoder's *Ancient Greece from the Air* (London: Thames and Hudson, 1974; as *Wings over Hellas: Ancient Greece from the Air,* New York: Oxford University Press, 1974), and A. R. and Mary Burn's *The Living Past of Greece* (Boston: Little, Brown, 1980) both provide generalized guides to important sites in Greek history beyond the Mycenaean era. For specific archaeological information about Tiryns, see *Mycenae-Epidaurus-Tiryns-Nauplion* (Athens: Clio Editions, 1978), and E. Karpodini-Dimitraidi's *The Peloponnese: A Traveller's Guide to the Sites, Monuments, and History* (Athens: Ekdotike Athenon S.A., 1984).

—Kenneth R. Shepherd

Tivoli (Roma, Italy)

Location: Perched on a foothill of the Sabine Mountains on the Aniene River, seventeen miles east of Rome.

Description: Tivoli is a small but busy city, built on the eastern slope of one of the Sabine Mountains and bordered on three sides by the Aniene River. Hot springs, olive groves, the cascading waters, mountain views, and easy access to Rome combined to make Tivoli a favorite resort of Roman gentry. Historic and scenic remnants of buildings from at least 80 B.C. to the Renaissance dot the hillside. Tourism has been a major industry since the late nineteenth century.

Site Office: A.A.S.T. (Azienda Autonoma di Cura Soggiorno e Turismo)
Largo Garibaldi
CAP 00019 Tivoli, Roma
Italy
(774) 293522

Rome is the eternal city, and Tivoli is its eternal suburb. Unlike most suburbs, however, Tivoli came first. Estimated to be at least four centuries older than Rome, the ancient city then called Tibur was established and occupied well before Romulus, the legendary founder of Rome, built his straw hut on the edge of the Campagna in 753 B.C. Tools dating to the Iron Age have been found nearby.

The site Tivoli occupies is a natural fortress. Protected by the Sabine Mountains on one side and the Aniene River on three others, the independent city stood at the head of a large, flat plain, commanding the approach to Rome from the east. Tivoli was a natural target for Roman expansion.

In 390 B.C. Rome was sacked by Gauls. Ten years later Rome marched on Tivoli and took it. Over the next three centuries, the Romans mined the area surrounding the mountain city for its resources. The waters from the river were diverted to Rome by an aqueduct built nearby in 273 B.C. Stone from the next mountain was quarried and carried to Rome over the Via Tiburtina. Much of Rome under the Republic and later was built over hundreds of years with "lapis tibertinus," a stone that hardened after it was cut. The Colosseum, built between A.D. 70 and 80, was built from lapis tibertinus. Tivoli itself had Roman citizenship bestowed upon it in 90 B.C.

While roads, monumental buildings, and aqueducts were among the benefits that Roman engineering brought to its subjects, Tivoli itself became more than a military sentinal and provider of resources. Soon after gaining citizenship, Roman gentry began to build summer homes there, and Tivoli became Rome's playground.

Rome was wealthy, but it was also crowded and noisy. By the first century B.C., the population approached 1 million, and the city had grown chaotically. Money could buy a quiet place outside the city, and many of the city's elite began searching neighboring areas for a place to build country retreats. The first Roman leader to build a villa in Tivoli was Marius, who ruled Rome from 107 to 100 B.C. For subsequent generations, whose individual wealth grew along with the power of the state, Tivoli presented several favorable enticements.

Their desire for superior baths led many Romans just a few miles east of the city to a particularly opportune spot. Hot springs outside of Tivoli were fed by two lakes naturally inundated with sulphuretted hydrogen, thought to be helpful in easing afflictions of the urinary tract. But many of those who came to the Tivoli region for the waters stayed for the view.

Surveying Tivoli's surroundings, elite Romans saw olive groves, the Sabine Mountains, and powerful cascades of rushing water from the river snaking around the town at the base of the hill. Beyond the river was the wide, flat Campagna and, in the distance, Rome. While private homes in Rome were built with little slits on the ground floor, which allowed little light (and prevented entrance to thieves), wealthy men could build villas in Tivoli that admitted light and were open to a vista that not only impressed them by its beauty but also reminded them of their power.

After Marius built his villa, his rival, Sulla, who was briefly dictator of the Republic from 82 to 80 B.C., went one better: he had temples built there. These were the Sanctuary of Hercules Victor, a massive structure housing an oracle; and the Temple of Vesta, still acclaimed as one of the most beautiful examples of circular temple (called peripteroi) architecture that survives in the twentieth century. The temples owed much to Greek religion: Greek gods were considered idealized people, and needed houses; Roman religion, by contrast, was based on augury (reading the signs sent by the gods), and had little need for temples. Near the Temple of Vesta was the Temple of the Sibyl Albunae, a small rectangular building, erected sometime during the late Republic.

By the time Julius Caesar's nephew, Octavian, declared a new Republic and was himself declared Augustus, the first Roman emperor, in 27 B.C., Tivoli was the residence of the most powerful amd wealthy men in the world. Augustus himself was a frequent visitor; Tivoli was one of his favorite places.

By then the town also occupied a secure place in literature. The Roman poet Horace was a humble man whose fortunes had been unmade with the fall of the Republic (although not as badly as Cassius, a summer resident of Tivoli who had committed suicide after the Republican defeat at the battle of Phillipi in 42 B.C.). Horace had consciously set about to ingratiate himself with the wealthy and powerful men of

Ovato Fountain at the Villa d'Este, Tivoli
Photo courtesy of A.A.S.T., Tivoli

the new order. One of these became his patron. Maecenas, a minister under Augustus who owned a villa in Tivoli, bought a farm for Horace near the resort, which the poet celebrated in his verse.

Horace's poems, odes, epodes, and satires are still read 2,000 years after they were written, and still have the power to inform, enlighten, and amuse. Another monument from the same time, the buildings that comprise Hadrian's Villa, also stands. Unlike other villas, Hadrian's was designed by the emperor himself. Hadrian's influence on subsequent architecture has been enormous. The buildings, like the poems, enlighten the observer about a feature of Roman character—its love of the large—and also illuminate some of the darkness within the man who planned them and had them built.

Born in Rome and raised in Spain, Hadrian was the second provincial to become emperor. His father died when he was ten years old, and his father's cousin, Trajan, became his guardian. Trajan, who was childless, was the first man from the provinces to become emperor. The two shared many characteristics, but there is also some evidence of discord between them. Their personal styles were dissimilar, and Trajan did not name Hadrian as his successor until he was on his death bed. Even then, there is some possibility that it was Trajan's wife who formed the succession, but this cannot be proved.

There were strong similarities between the two, and an equally strong desire on Hadrian's part to outdo his predecessor. Both men were accomplished hunters. Trajan loved young men and boys; Hadrian's affair with the adolescent Greek, Antinous, was publicized throughout the world. In public policy, Hadrian continued and improved upon Trajan's policies; both are considered among Rome's greatest emperors. Trajan hired the best architects to build some of the great monuments of the empire; Hadrian designed or built some of the great monuments of the empire, from the Pantheon in Rome to Hadrian's Wall, which marked the northernmost point of the empire in Britain. Trajan had a sumptuous villa at Tivoli; Hadrian's villa, built just below the mountain town, grew to be the largest private dwelling known to the ancient world.

Hadrian started building his villa the year after he became emperor, in A.D. 118, and was still working on it in 134 (he died in 138). Influenced and enthralled by Greek art and architecture, Hadrian created new forms from ancient ideas. When his innovations were criticized by Trajan's architect, Apollodorus of Damascus ("go and build your pumpkins," Apollodorus is reported as having said to Hadrian), he reacted as many artists would to a critic, if they had the power. He had Apollodorus exiled and later put to death.

In size and scope, Hadrian's Villa has been compared to Versailles; as a restless, eclectic expression of its creator, it is similar to William Randolph Hearst's San Simeon. The massive grounds (1,000 x 500 yards) contain two palaces, one large and one small; two bathhouses, also large and small; a stadium; a temple (dedicated to Venus and Rome); a library; gardens; pools; guest quarters and slave quarters, among

other structures. Yet Hadrian spent a total of perhaps four years there, at most.

The first of the structures he built below the mountain, the Teatro Marittimo, was an ideal spot for solitude. Circular, with an indoor moat and a drawbridge controlled from an island in the center, the place was an essay in isolation. Later buildings, designed during his affair with Antinous, were open, happy places that have been described as "feminine." Hadrian traveled frequently throughout the empire, and built portions of the estate to reflect the lands he had visited. He also built a portion he called "Hades," which he felt he had visited in spirit. When Antinous died under mysterious circumstances, Hadrian built a shrine to him at the villa.

In A.D. 130, Antinous's body was found floating in the Canopus, an arm of the Nile at Alexandria. In his grief, Hadrian declared Antinous a god, had a new city, called Antinous (or Antinoopolis), built on the spot where his body was found, and commissioned sculptures of Antinous from Greece to Egypt. Coins bearing Antinous's likeness were found over a thousand years later as far north as England. At his villa, Hadrian built a pool modeled after the Canopus and lined it with statues of the boy. Somewhere between sixteen and thirty of these statues are in museums in Europe. The Antinous cult was attacked after Hadrian's death in 138, most severely by Christians. After Constantine the Great made Christianity the official religion of the empire in 313, the cult was largely wiped out.

Tivoli was probably at the height of its wealth during the reign of Hadrian. The emperors who followed him witnessed a slow decline of Roman power. While Tivoli remained a prosperous resort, by the middle of the fourth century Rome's empire had been halved. The political center of the empire moved east, to Constantinople in Byzantium. During the fifth century, the only bulwark against the complete eclipse of Rome was the institution partly responsible for its fall: the Roman Church.

Rome's military defenses were no match for the Goths, who swept into Italy in the fifth and sixth centuries. Tivoli was not spared. Sometime around 545, Totila, king of the Ostrogoths, took the city. Gothic armies had enlisted Italians in their campaigns, and those who remained loyal were rewarded; those who did not were severely punished. Tivoli was among the cities in the latter category. The Ostrogoth army plundered Tivoli, and Totila briefly made the city his base of operations. He was not the last foreign king to hole up there.

The Western Christian empire was consolidated under Charlemagne, who was crowned emperor of the Holy Roman Empire by Pope Leo III in 800. The spiritual and temporal power that had moved east to Byzantium returned to Rome, and Tivoli, too, rose with it. The Vatican inherited the mantle of the ancient empire. By agreement, all of the new kingdoms and empires ruled by consent of the church, and as a result the area controlled by the papacy took on a more earthly than heavenly cast. Spiritual leaders dictated to kings, most of whom controlled less wealth than the church. It was just a matter of time until a king would attempt to dictate church policy.

In the thirteenth century, after the death of Innocent III, Frederick II (who, like Totila, was from Germany) invaded Italy. He camped with his armies in Tivoli, and from there controlled the selection of the next pope. The ten cardinals responsible for choosing the new pope could not agree. With Frederick threatening to make the choice himself, the Roman governor, Senator Matteo Rosso Orsini, forced the cardinals to a decision in what became the first conclave.

Orsini's method was simple: each cardinal was pushed around by Orsini's soldiers, then the group of them was locked into a room in Rome, armed sentries surrounded them, toilets inside were not allowed to be emptied until a new pope was chosen, and food was to be as scarce and indigestible as possible. When one elderly cardinal took sick, he was placed inside a coffin, still alive, while the other cardinals sang hymns to the dead. When he died, his putrefying body remained in the room. After fifty-five days, the cardinals elected a pope who was acceptable to both Frederick in Tivoli and Orsini in Rome: Godfrey, who took the name Celestine IV. However, Celestine IV died soon after his election. The cardinals had scattered at the first chance, none of them wanting to chose between Frederick and Orsini again. Frederick remained in Tivoli, forced the issue, and one year later another pope was chosen—Innocent IV, who also died soon after his election.

The election of popes played a large role in the shape of Tivoli (as well as Europe), but the most far-reaching physical changes in Tivoli were brought about in the sixteenth century by a man who would have been pope but that he lost the election. When Cardinal Ippolito d'Este (Ippolito II) was not elected pope, he was given an appointment instead: he was made Governor of Tivoli.

As the son of Duchess Lucrezia Borgia, Ippolito came from a family that had, until his loss, controlled the papacy, and was still on the same plateau with kings and emperors. When d'Este found himself unable to occupy the world stage, he grew the world's finest garden in Tivoli. Cardinal d'Este converted a Benedictine convent into his residence and hired an artist, Pirro Ligorio, and an architect, G. Alberto Galvani, to renovate it. A district of Tivoli was torn down to make room for the villa's garden.

The gardens of the Villa d'Este were in part a return to the classicism of ancient Rome that inspired the Renaissance (one part of the garden is a depiction of the legendary founding of Rome), but with new twists. Using the waters of the Aniene River, d'Este created a rival to Hadrian's Villa up on the hill, with several terraced gardens sporting fountains, spouts, and sculpture, all in a widespread green landscape. A mechanical "water organ" was built: compressed air controlled by rushing water caused pipes to play and set figures moving as if in a dance.

Successive generations added to the Villa d'Este, and ownership eventually passed to Austria. In 1865, the composer Franz Liszt moved in to the top floor, working there periodically until his death in 1886. One of his piano compositions is called *Les Jeux d'Eau à la Villa d'Este*. While Liszt composed in a house owned by a sympathetic landlord, the rest of Italy was in the midst of the rebellion that transformed the Papal States (in which Tivoli was located) and the rest of the conglomeration of kingdoms, duchies, and republics on the Italian peninsula into one state.

Napoléon Bonaparte, who had conquered most of the peninsula by 1812, was the first to call the land "Italy." But once Napoléon was dead, and with democratic movements on the rise all over Europe, Italians rose up in numbers to demand a voice in their destiny. Led by Camillo Benso, Giuseppe Mazzini, and Giuseppe Garibaldi, the mass movement caused most of Italy's nobles—foreign and native born—to give way. The Papal States were the last holdout.

With the first stirrings of rebellion in 1815, the part of Italy controlled by the church had become a not so benevolent dictatorship. Commoners were put to death for small offenses, and those satellites that had remained within the papacy's realm were held more zealously than during times when challenges came from foreign powers. Tivoli benefitted in the form of public works undertaken by Pope Gregory XVI, who built flood tunnels there in 1831. This benefit came with a price, however: until 1816, Tivoli had for the most part been in control of its destiny. And while most of the rest of Italy welcomed reforms, the Papal States still resisted. They succumbed after a general vote held in 1870. The people accepted the pope as spiritual but not temporal ruler. In 1918, the Italian government took possession of the Villa d'Este.

Writers, artists, composers, and philosophers in the Victorian period "rediscovered" a new Arcadia in Italy: a place where life was lived as it was supposed to be lived. Henry James, in his book *Italian Hours*, called Tivoli "a pictorial felicity that was almost not of this world, but of a higher degree of distinction altogether."

Further Reading: While books written in English specifically about Tivoli are few and far between, many books treat its major personalities and landmarks in detail. *Beloved and God: The Story of Hadrian and Antinous* by Royston Lambert (New York: Viking, and London: Weidenfeld and Nicolson, 1984) is one of the few books on Hadrian to thoroughly explore his relationship with Antinous, the effects of Antinous' death on Hadrian, and the effects of his subsequent deification on the ancient world. *The Mute Stones Speak: The Story of Archaeology in Italy* by Paul Mackendrick (London: Metheun, and New York: St. Martin's, 1960) links the buildings in Tivoli to the builders, to Roman history, and to the present day. Of the many books published specifically about Renaissance gardens, the photographs of portions of the Villa d'Este reproduced in *Garden Architecture in Europe: 1450 - 1800* by Carl Friedrich Schroer (Cologne: Benedikt Taschen, 1990) are most exquisite; the accompanying text is illuminating. Among other descriptions of Tivoli, *Italian Hours* by Henry James (Boston: Houghton Mifflin, and London: Heinemann, 1909) is an enraptured remembrance of journeys undertaken by the author over a period from 1861 to 1900.

—Jeffrey Felshman

Todi (Perugia, Italy)

Location: In the southern portion of the central Italian region of Umbria, on a ridge above the Tiber valley, less than sixty miles north of Rome, twenty-five miles south of the city of Perugia, in the province of Perugia.

Description: Todi, now a town of about 17,000, covers a triangular site partially encircled by rings of Etruscan, Roman, and medieval walls; known as one of the best-preserved medieval towns in Italy; also famous for its Renaissance Church of Santa Maria della Consolazione, based on the plan of St. Peter's in Rome.

Site Office: A.P.T. (Azienda di Promozione Turistica del Tuderte)
Piazza Umberto I, 6
CAP 06059 Todi, Perugia
Italy
(75) 8943395

The earliest inhabitants of Todi were the Umbri, who, according to legend, established the city of Tuter after an eagle showed them the town's future location atop an isolated, triangular hill, 1,353 feet high, surrounded by rolling countryside. The Etruscans followed the Umbrians, calling the town Tutere. It became known as Tuder in Roman times. "Tud" is a term for community or town. The word "Tular," known from several ancient inscriptions, means "boundary." The town of Todi lies close to the boundary between Etruria and Umbria, along the eastern border of what was once the Etruscan kingdom, a well-organized league whose headquarters were at Volsinii (Orvieto) on the Tiber. The other Etruscan borders were the Tiber to the south, the Arno to the north, and on the west the coastal area from Pisa to Cerveteri. The Etruscans built their city at Todi at a lower altitude than did the Umbrians, locating it near today's Piazza del Popolo, Todi's main square. According to legend, the Etruscans attacked the Umbrians, slaughtering some and taking others as slaves.

Many Etruscan artifacts have been excavated in and around Todi in the last 200 years. The Etruscans' best artworks were those done in bronze. Of these, only a few larger pieces have survived, the *Mars of Todi* being one of them. This young warrior figure, wearing a tunic and corselet, stands well-balanced, lifting a spear in his left hand; he probably once held a libation bowl in his right. His head, which is now missing his helmet, is turned to the right, and he has a grave expression on his face. Possibly once a votive offering, today this statue, a work of the early fourth century B.C., is in the Vatican's Etruscan Museum. Other early artifacts are on display in Todi's Etruscan-Roman Museum in the Palazzo del Popolo.

The Romans followed the Etruscans; by 42 B.C. the town was known as Colonia Julia Fida Tuder. It prospered under Roman rule, its hill site keeping it safe from barbarian invaders.

In the Piazza Mercato Vecchio is still extant a series of arches topped with a Doric frieze. Known as "The Niches," the structure is believed to be the wall of a basilica of the Augustan era. Portions of the Roman amphitheatre remain, near the eastern Porta Romana, one of the three gates named for the cities of Rome, Perugia, and Orvieto through which one enters Todi. The walls connecting these gates form a triangle.

Three separate fortification systems were built at three different periods in history. The Porta Romana dates from the Middle Ages. Five hundred yards in, toward the center of the city, is the Gate of the Chain, built by the ancient Romans, near some still-visible Roman town walls. The foundations of the Porta Marzia, 120 yards closer still to the center of town, and part of Todi's oldest defense wall still somewhat discernible, were built by the Etruscans in about the fourth century B.C. Etruscan wall-building techniques consisted of superimposing large blocks of stone onto one another. The current Porta Marzia, while it is a medieval archway, is built of Roman materials, on top of its Etruscan foundations. Roman walls and those built during the Middle Ages were stronger than those of the Etruscans and thus more effective militarily.

Todi is known as one of the best-preserved medieval towns in Italy. It was one of the first free communes in the Middle Ages. There is no evidence that Todi was part of the Lombard duchy of Spoleto, and it may have been an independent entity all along. The eleventh-century church of San Niccolò is the oldest religious building extant in Todi and was built on the ruins of the Roman amphitheatre.

Lombard artists began the cathedral, with its pink and white marble Romano-Gothic facade and impressive staircase approach, on the site of an ancient temple of Apollo in the beginning of the twelfth century. It was redesigned and altered a number of times up until the seventeenth century. A Gothic arcade was added in the 1300s. It houses an unusual altarpiece showing the Madonna's head in high relief against the flat surface of the painting.

A powerful city during the twelfth and thirteenth centuries, Todi by the thirteenth century had control over Amelia and Terni, two towns about twenty miles to the south and southwest, and there was some skirmishing between Todi's soldiers and those of nearby Spoleto, Narni, and Orvieto.

In 1213 Lombard artists also began the Palazzo del Popolo, with its Gothic facade, triple windows, and delicate marble columns. It is Todi's oldest palace, and one of the

Church of Santa Maria della Consolazione
Photo courtesy of A.P.T., Todi

oldest town halls in Italy still used for that function along with the Palazzo del Capitano del Popolo next door. Built above a series of Roman cisterns, the Palazzo del Popolo is on the site of a Roman forum. At the center of Todi is the Piazza del Popolo, a square surrounded by the palaces, known as one of the most perfect medieval squares in Italy.

A great deal of building went on in the thirteenth century in Todi. Other thirteenth-century structures are the Palazzo Chiaravalle and the Church of Santa Maria in Camuccia, with Roman remains beneath and two Roman columns on either side of its doors.

In 1228 Jacopo dei Benedetti was born in Todi. He became a Franciscan poet and mystic who worked in Latin and the Italian vernacular. Before he joined the priesthood and became known as Jacopone da Todi, he had been a wealthy lawyer married to a noblewoman. One day, the platform on which his wife stood during a public festival collapsed. Benedetti tore aside her rich garments to examine her injuries only to find that she wore a hair shirt as penance underneath. She died, and Benedetti sought to join the religious life, but the Franciscans at first refused him entrance into the order due to his eccentricities. Instead he joined the Spirituals, a rather unworldly order, and lived at a monastery at Collazzone near Perugia. It is believed he wrote the Latin *Stabat mater dolorosa* on the sorrow of the Virgin Mary at the Cross and the *Stabat mater speciosa* as well as some early Christmas carols. The inscription in the crypt gives his death date at 1296, but the chronicles say he died in 1306. The Church of San Fortunato in the Piazza della Repubblica, begun by the Franciscans in 1292, was completed in 1460 and houses the tomb of Jacapone da Todi.

In 1241 Podesta Scarnabecco from Bologna built a fountain and arcades in Todi. The Palazzo dei Priori, or Palace of the Priors, was begun in 1283 and its construction continued into the fourteenth century. It was the seat of the city's medieval rulers and later that of the papal governors after Todi was absorbed into the Papal States in 1330, by which time the political effectiveness of the free commune had begun to decline.

During the fourteenth century Giovanni Gigliacci worked on the facade of the Palazzo dei Priori and in 1339 erected the bronze eagle, the city's heraldic emblem. The castle La Rocca (The Rock), now ruined, also dates to the fourteenth century. In 1432 Masolino, an artist from the nearby town of Panicale, south of Lake Trasimeno, created what is sometimes considered to be Todi's greatest work of art, the fresco *Madonna and Child,* also known as *Virgin of Todi,* inside the fourth chapel in the Church of San Fortunato. The fresco was restored in 1987.

In 1508 Cola di Mateuccio da Caprarola, who worked previously on papal buildings in Latium, began the Greek cross-style Church of Santa Maria della Consolazione outside the medieval walls, a half-mile downhill, and southwest of the town center. The origins of the Church of Santa Maria della Consolazione lie in the veneration of a Virgin and Child wall-painting, possibly done by the canonized monk

Giovanni di Rannuccio under commission of a member of the noble Atti family, at the monastery of Santa Margherita near Todi in the province of Perugia. The reputation of this particular Virgin and Child image as a source of miracles prompted pilgrimages and town processions to Santa Margherita. The church was built to house the image.

The church has commonly been considered to be a simplified version of Donato Bramante's plan for St. Peter's in Rome, the foundation of which had been laid two years earlier. Santa Maria della Consolazione is situated a little below the town on a terrace with a good view of the Umbrian landscape. It is a square with apses on all four sides; the apse containing the sacred image is semicircular, the rest are polygonal. The church is considered to be one of the finest centrally planned buildings of the sixteenth century. A 1574 visitation report mentions a model designed by Bramante, and says that the vaults of the church were to resemble those of the new St. Peter's. Whether the model, which was in Todi in 1574, was actually based on Bramante is a matter of conjecture. The plan is also close to a drawing by Leonardo da Vinci, and although the idea may have been one of Bramante's, it is known that the design's execution was done by the otherwise almost unknown Cola da Caprarola, with help from Baldassare Peruzzi, who had been one of Bramante's assistants at the Vatican in 1503.

A compagnia of lay noblemen, chaired by three rectors, was formed to preside over the building of the church. By October 7, 1508, Cola da Caprarola was under contract to work on the building. Excavations for the foundation were begun and by March 17, 1509, they had laid the first stone.

The community, particularly the nobility of Todi and its leader, Ludovico degli Atti, an astute politician, wealthy and well-learned, followed closely the construction of the church. Atti, who had established ties with Pope Julius II in 1507, with the bishop's approval guided the council in matters of the building and kept close watch over its construction.

Renaissance humanists considered the church's central plan to be symbolically and religiously meaningful. The layout of the building also enabled its creators to avoid having to build out on the treacherous cliffs of the church's site. The sacred image of Santa Maria della Consolazione was to be displayed beneath the dome, and entrances were arranged to facilitate the flow of visitors viewing the icon. Ritual processions were held each year on the third day after Pentecost from 1509, a tradition probably begun by Bishop Moscardi and Ludovico degli Atti. The latter, after his election of signore of Todi, made every effort to transform the holy image into a Maesta, a protector of the town and consoler of its people. The church became a common point of contact and place of accord for the entire town.

Plans for the building itself changed and evolved over the years. Between 1509 and 1512, work on its construction slowed down, possibly due to economic hardships or to uncertainties as to how to proceed, provoked by the presentation of new project plans having an altered layout from the structure already started.

Atti proposed, and the rest of the council immediately ratified on September 21, 1511, the authorization of the prior to oversee the distribution of stone and bricks from the town's old hilltop fort to be carried to the building site by townspeople and peasants. A new agreement was struck with Cola da Caprarola, providing new impetus to the work on the site. Cola da Caprarola subcontracted to master stonemason Pietro da Corsonio the completion of walls and pillars. A wooden model now in the municipal museum of Todi corresponds to this phase of the project's construction. Pope Julius II seems to have taken a personal interest in the building of the church. The litanies of the Virgin of Loreto, of special interest to Julius, appear in the windows of Santa Maria della Consolazione.

In 1531 the town council, looking for a suitable religious order to take over Santa Maria della Consolazione, chose the order of San Pietro dei Benedettini Cassinensi of Perugia. The following year, the stonemason and sculptor Filippo da Meli began to prepare the building for its rose windows. At this point the building was ready for its apse domes and the main dome, but funds were low. Pope Paul III, after visiting Todi to see the construction, allocated more money for the project. The town ordered a procession to be held on the first Sunday of every month in an effort to collect additional funds. Nevertheless, from 1545 to 1565 the work was slow, and frequently halted altogether. During the 1560s work began on the vaults of the apses and tribune.

The gilded cross at Santa Maria della Consolazione was raised in April 1606. The dome itself was raised in 1606 and 1607. The sacristy was begun in 1613. The sacred image of Santa Maria della Consolazione was transported from its old surroundings to the new church altar, and the church was inaugurated on April 20, 1617.

Since the Renaissance, Todi has remained largely unaltered and out of the historical spotlight. In the nineteenth century a number of haphazard excavations took place in and near Todi. The remarkable Etruscan and Roman finds made at that time are now in the Museo Archeologico in Florence or the Villa Giulia in Rome. The Etruscan bronze statue of Mars dating from the beginning of the fourth century B.C. was found in 1835 and is now in the Vatican.

Today Todi, while maintaining its medieval and Renaissance appearance, has begun to acquire the reputation of being a fashionable place to live. It has little industry, and agriculture from the surrounding countryside is important to its economy. It is the home of the Mostra Nazionale dell'Artigianato (National Exhibit of Crafts), a festival held each August and September. The town hosts the Mongolfieristico, a three-day hot-air-balloon show in mid-July. Todi is also the home of a cooking school featuring regional cuisine.

Further Reading: *Cento Città: A Guide to the Hundred Cities and Towns of Italy* by Paul Hofmann (New York: Henry Holt, 1988) gives a brief description of the principal characteristics of Todi and its places of interest. *The Renaissance from Brunelleschi to Michelangelo: The Representation of Architecture,* edited by Henry A. Millon and Vittorio Magnago Lampugnani (New York: Rizzoli, 1994; as *Italian Renaissance Architecture from Brunelleschi to Michelangelo,* London: Thames and Hudson, 1994) and *Architecture in Italy 1400–1600* by Ludwig H. Heydenreich and Wolfgang Lotz, translated by Mary Hottinger (Harmondsworth, Middlesex, and Baltimore, Maryland: Penguin, 1974) provide detailed discussions of the building and design of the Church of Santa Maria della Consolazione and other noteworthy examples of Italian Renaissance architecture.

—L. R. Naslund

Toledo (Toledo, Spain)

Location: South-central Spain, forty-five miles south-southwest of Madrid; surrounded on the east, south, and west sides by a precipitous gorge in the Tagus River.

Description: Capital of the province of Toledo; a town with Roman origins, considered a prime example of the influence of Christian, Jewish, and Moorish cultures on Spanish history.

Site Office: Oficina de Turismo
Puerta de Visagra, s/n
45003 Toledo, Toledo
Spain
(25) 22 08 43

Toledo is built on a granite outcrop 2,400 feet above sea level, surrounded on three sides by the river gorge and overlooking the bare Castilian plateau to the north. The limited possibilities for expansion and the consequent limitation of space within the old walled town have resulted in the retention of the old and twisting narrow streets. The houses are massive and somber, and the town has preserved an essentially Gothic and medieval atmosphere with the addition of characteristically Moorish architectural features.

Formerly the "Toletum" mentioned by the Roman historian Livy, Toledo may have been a Carthaginian trading station. It was certainly a Celtic settlement captured by the Romans in 193 B.C. and built up by them into a *colonia,* or town, on its admirable defensive site. Toledo avoided capture by the Vandals in the early fifth century only to fall in the sixth century to the Visigoths, who by A.D. 554 had made it their capital. The Visigoths immediately began the construction of numerous temples, the remains of which can still be seen today along with the Roman aqueduct, amphitheatre, and temple. Toledo became an important holy city during the Christianization of Spain by St. Eugene, an early bishop of Toledo rumored to have been a disciple of St. Paul.

The Visigothic relics include the primitive cathedral dedicated to the Virgin on April 12, 569 by King Reccared, who had been converted two years earlier; the fourth-century basilica built over the tomb of St. Leocadia and now known as Cristo de la Vega; and six small churches built in the period when the city was peacefully inhabited by its Hispanic Roman population, Goths, and Hispanic Jews. A series of church councils was held in the town, the most notable occurring in 396, 400, and 589, which repudiated Arianism, the doctrine that the second person in the Trinity is of a lower order than God the Father, as heretical.

In 712, after defeat at Guadelete, the Visigoths abandoned Toledo to the Moors. Moorish chroniclers are eloquent about the treasures that the army, led by Tarek, seized from Toledo. The Moors called the town Tolaitola. It was first a provincial capital governed by an emir under the caliphate of Córdoba. From 1035 to 1085 it was an independent state. The influence of the Arabs was so great that, even after the city was taken by Alfonso VI in 1085 with the help of El Cid, the Moors continued to worship in the Visigothic cathedral. The cathedral had been converted into a mosque and used by the Moors as their principal mosque from 712 until 1227.

The Moorish supremacy in Toledo was a time of unrest but also an era in which an immensely influential culture in Spain's history reached the peak of its importance. The Moorish rulers protected the largest Jewish colony in Spain, founded silk and woolen industries, and made the city an important center of both Arab and Jewish cultures. The fusion of Moorish and Arabic cultures, for which the term "Mozarabic" was coined, was instrumental in the importation to western Europe of the thought of Aristotle, already loaded with Neoplatonist glosses, and hence in the foundation of western scholasticism and the whole of post-medieval Christian philosophy and theology. In architecture, Mozarabic style came to denote a fusion of Christian folk art with Islamic style and design. It often featured horseshoe arches, ribbed cupolas, and turned modillons or scrolls underneath a cornice. In Toledo, as elsewhere in Spain, Mozarabic style was replaced by a series of architectural forms inspired by those of Roman antiquity, all of them now generically and indiscriminately known as Romanesque, the dominating style of the eleventh and twelfth centuries.

The Moors spanned the Tagus with two fortified bridges, the Panda de Alcantara to the northeast of the city, rebuilt in the thirteenth and seventeenth centuries, and the Panda de San Martin to the northwest, founded in 1212 and rebuilt in 1390. Modern load-bearing traffic bridges have been added on the north to facilitate crossing the river on the east and west sides of town. The Moors were also responsible for the inner walls of the city, said to have replaced a Visigothic construction. The outer wall was built by Alfonso VI in 1109.

Between the twelfth and fifteenth centuries, the Iberian Peninsula was gradually reclaimed from the Moors. The Islamic-inspired architecture and decorative work of the Christian reconquest period, principally in tile, wood, plaster, or brick, is known as Mudéjar style. Toledo was home to the largest Jewish community in Spain, some 12,000 in the twelfth century before the 1355 pogrom of Henry II of Trastámara, and the two most notable Mudéjar buildings in Toledo were among the ten synagogues of the old Jewish quarter. The first is the fourteenth-century El Tránsito synagogue built by Samuel Ha-Levi, treasurer under Peter the Cruel; it is now a church dedicated to the Virgin's dormition.

El Tránsito Synagogue, Toledo
Photo courtesy of Tourist Office of Spain, Chicago

The second is Santa Maria la Blanca, the principal synagogue of twelfth-century Toledo, given to the knights of Calatrava, who turned it into a church. The sixteenth-century modifications to the east end of Santa Maria have left intact the five astonishing, tiered aisles with twenty-four octagonal pillars supporting the horseshoe arches and superbly carved capitals. Mudéjar decoration of the interior of El Tránsito, especially the upper part of the walls and the east end, is of breathtaking intricacy and skill.

The town contains other magnificent examples of Mudéjar architecture, including the cathedral chapterhouse, with its particularly remarkable multicolored Mudéjar ceiling and stucco doorways. More important, no doubt, is the Church of Cristo de la Luz, a mosque erected in 1000 on the site of a Visigothic church and turned into a church in the twelfth century. Visigothic interior pillars support superimposed arches, and square bays support the church's nine domes. Alfonso VI passed by the church on his entry into Toledo. Also important for their Mudéjar decoration are the old palace, known as the Taller del Moro and used as a workyard to collect building materials for the cathedral, and the Church of Santiago del Arrabel, now restored, in which the pulpit catches the Mudéjar style as its point of transformation into Gothic. The church also has a sixteenth-century altarpiece.

At Toledo's highest point is the thirteenth-century Mudéjar Church of San Román, now a museum, with a fine tower similar to the fourteenth-century Mudéjar tower of the Church of Santo Tomé and aisles divided by arches reminiscent of Santa Maria. The thirteenth-century frescoes commemorate Christian subjects, and there have naturally been subsequent architectural and decorative additions, such as the sixteenth-century cupola and the eighteenth-century altarpiece. But the church, like the museum, recalls Toledo's greatest period when it was the capital from which radiated the fusion of the three great cultures that made up Spain. The museum's Visigothic collection is of more historical than aesthetic interest, but like the Mudéjar east end of San Vicente and the later doorway opposite San Román belonging to the monastery of San Clemente, it is a memorial to Toledo's great era. The doorway is in the style known as "Plateresque," another art historical category used loosely for the intricately wrought designs based on foliage and pilaster. Plateresque was Italian in origin but was heavily cultivated in Spain in the late fifteenth and early sixteenth centuries.

Dominating the city from its high point on the eastern edge of the old town is the Alcázar, a large square building with a tower at each corner and a fine arcaded patio erected on the site of a defensive Roman fort. On this site the seventh-century Visigothic king Wamba built a citadel subsequently used by the Moors. It was converted into a palace by St. Ferdinand III, king of Castile in 1217 and of León in 1230. El Cid, around whom so many legends revolve in Spanish literature, was its first governor. The Alcázar was enlarged several times in the sixteenth century, by Ferdinand and Isabella, Holy Roman Emperor Charles V, and King Philip II.

Over the centuries, the Alcázar was destroyed and rebuilt many times. Burned down in 1710 during the War of the Spanish Succession, it was restored in 1775. Destroyed by fire again, by the French, in 1810, it was again rebuilt, and in 1882 became a military academy. In 1887, a third fire was followed by a third restoration. Through all three fires, the facade by Juan de Herrera, the gateway by Alonso de Covarrúbias, and the great staircase by Francisco de Villalpando were preserved.

The fortress was again ruined during the Civil War in 1936. For eight weeks beginning on July 21 infantry cadets held out against republican forces, while their families, about 600 women and children, took shelter in the underground galleries. The commander, Colonel Moscardó, was ordered by telephone to surrender or see his son shot. His son was executed on August 23. The Alcázar was relieved on September 27.

The most important surviving monument to Toledo's history and to the era of its greatness is, however, the cathedral, whose own layered development reflects the cultural waves of the city's history. Construction, intended to follow French Gothic models, began under Ferdinand III in 1227, but the plans were modified in the course of building, which lasted until the last decade of the fifteenth century. The cathedral, just over 394 feet long, is immensely wide (177 feet), and is divided by eighty-four pillars into a nave and four aisles. A full set of side chapels was added in the fifteenth and sixteenth centuries. The 750 superb stained-glass windows, mostly of Flemish origin, also date from that period.

The choir stalls are among the most magnificent in Europe. Carved in wood, they are set in alabaster recesses divided by columns of red jasper and white marble, and recall in fifty-four detailed scenes the conquest of Granada. The sixteenth-century upper alabaster carvings by Alonso Berruguete and Felipe Bigarny portray Old Testament figures. There is a French fourteenth-century white marble Virgin. To the east of the chancel, a spectacular retable stretches up to the building's roof. An opening has been pierced in the roof to allow light to fall on the tabernacle, although in fact it illuminates the baroque Transparente by Narciso Tomé.

To the north of the building is the thirteenth-century doorway now known as the Puerto del Reloj. The main entrance is now the central doorway of the three pierced in the west front, known as the Puerta del Perdón. It is surrounded by statues and surmounted by a sculpted tympanum in which the Virgin presents the seventh-century bishop of Toledo, San Ildefonso, with an embroidered chasuble. The cathedral's fifteenth-century, 295-foot tower is accompanied by a seventeenth-century dome that replaced a second tower above the western facade. The fifteenth-century lion doorway into the transept on the south side of the church, designed by Master Hanequin of Brussels, is historically important for the evidence it provides about the importation of renaissance decoration via the Low Countries.

It is usual, convenient, and not entirely misleading to look on the joint reigns of Isabella, queen of Castile from 1474, and her husband, Ferdinand, king of Aragon from 1479, the "Catholic monarchs," as a watershed in the history of Spain. They united Castile and Aragon by their marriage, pushed the re-Christianization of Spain farther south by taking Granada in 1492, and achieved religious unity by expelling in 1492 all Jews and in 1502 all Moslems who would not become Catholics. In 1478 the Inquisition had been established in Spain not as an ecclesiastical tribunal, but as a council of state, with civil jurisdiction for the enforcement of religious conformity. Isabella died in 1504 and was succeeded in Castile by her eldest daughter, Joan I (Joan the Mad), whose husband reigned as Philip I until his own death in 1506. Ferdinand then ruled both kingdoms until his death in 1516, when Joan's son became Charles I of Spain and then, in 1519, Holy Roman Emperor Charles V. Charles inherited not only Spain, but also the Low Countries, Franche-Comté, Naples, Sicily, Austria, Silesia, and Moravia, making a Spanish hegemony in central and western Europe a very real likelihood. In 1556 the emperor abdicated in favor of his son, Philip II.

Toledo's decline dates from the sixteenth century. After losing much of its Jewish and Moslem population, it was at the center of the 1519 to 1521 revolt of the *comuneros,* partly in protest at the levies imposed by Charles I. Charles had been brought up in the Low Countries, could not speak Spanish, and regarded the lands he had inherited as a source of funds with which to secure election as emperor. However, the uprising resolved into private feudal hostilities and, in the end, into a conflict between the forces of popular rebellion and the nobles who repressed the movement at Villalar in 1521; that same year, the movement's two leaders, Juan Bravo from Segovia and Juan de Padilla from Toledo, were executed in Segovia. Commercial and political decline had already set in when Philip II transferred his court to Madrid in 1561.

The archbishops of Toledo once possessed immense power and wealth, but reform of ecclesiastical life, in particular the restraint of abuses of power and opulent styles of living, was instituted by the Catholic monarchs. Ferdinand entrusted the archbishopric of Toledo in 1494 to the austere and devout Francisco Jiménez de Cisneros, who had been imprisoned by one of his predecessors, resigned his benefices, joined the observant Franciscans in 1484, and became Isabella's confessor in 1492. In 1495 de Cisneros was made primate of Spain; although he had originally opposed the establishment of the Inquisition, he became inquisitor general and was made a cardinal in 1507. On Ferdinand's death, he also became regent of Castile, having acted as a genuinely reformatory prelate, and founded the University of Alcalá to further scholarly biblical studies. He also took Oran from Moslem hands in May 1509, in what he regarded as a divine crusade.

The painter El Greco (Doménikos Theotokópoulos) made Toledo the center of his activity. Among the rich treasures of Toledo's museums, his paintings, particularly the collection in the cathedral sacristy and *The Burial of the Count of Orgaz* of about 1586 in Santo Tomé, deserve special mention, if only because of their concentration in the city. A museum devoted to El Greco's work is situated in a house similar to the one in which he lived from 1585.

The town's collections of paintings, sculptures, and other artistic objects, some of which, by their nature, are not transportable, are exceedingly rich, and Toledo is today still important for its rich cultural heritage. The town also retains its reputation for the excellence of its forged steel, which once made it Spain's center of sword production.

Further Reading: There is little in English on Toledo as a historic town other than in the twentieth century, although guide books, gazetteers, and encyclopedia articles are often helpful. Otherwise, help must be sought from political histories of Castile or of Spain, and from histories of art and architecture. Among the works specific to Toledo is Geoffrey McNeill-Moss's *The Epic of the Alcazar: A History of the Siege of the Toledo Alcazar 1936* (London: Rich and Cowan, 1937; as *The Siege of Alcazar,* New York: Knopf, 1937).

—Clarissa Levi

Troy (Çanakkale, Turkey)

Location: The excavated ruins of ancient Troy, the modern Hissarlik, are in northwestern Turkey about four miles east of the Aegean Sea and a little closer to the Hellespont, or Dardanelles, to the north. Troy lies in the province of Çanakkale, about eighteen miles southeast of the coastal town of Çanakkale and a few miles west of the main E24 road that runs from Çanakkale generally south along the Aegean coast to İzmir. The village of Tevfikiye is less than one-half mile from the site on the access road from E24.

Description: The archaeological site consists of the ruins of nine ancient cities atop one another dating from about 3000 B.C. to about A.D. 400 when the town was largely abandoned and eventually disappeared. It was rediscovered in 1871 by the amateur archaeologist Heinrich Schliemann, who carried out somewhat inept excavations. The excavated structures are well marked and reveal the tumbled remains of a series of small fortified cities whose ruins created a mound the Turks named Hissarlik. The location was of strategic importance since it commanded clear views of the Dardanelles Straits and thus controlled traffic from the Aegean Sea to the Bosphorus and the Black Sea. Troy is most famous for its association with the Trojan War.

Contact: Çanakkale Museum
Zfer Mey., No. 4
Çanakkale, Çanakkale
Turkey
(286) 73252

Troy, known in ancient times as Ilium, endured for about three and one-half millennia, having been destroyed and rebuilt on several occasions. Despite this long span, Troy is known to history and myth for a single event that may have lasted about a decade. The Trojan War fought between the Achaeans (Greeks) and the inhabitants of Troy was immortalized by the great poet Homer in the *Iliad* and the *Odyssey*. In the *Iliad,* which tells of events during the last year of the conflict, Homer summarized the religious and historical memories of the Greeks in a powerful narrative that has endured since its composition around the last half of the eighth century B.C.

According to Homer and Greek legends, the origin of the Trojan War involved a scheme of the gods Zeus and Themis. Eris, goddess of discord, felt insulted and revenged herself by throwing before the goddesses Athena, Hera, and Aphrodite a golden apple labeled "to the fairest." All three claimed the apple and the choice was left to Paris, son of Priam, king of Troy. After Aphrodite offered Paris the most beautiful woman in the world in exchange for his vote, he awarded her the apple. This decision earned Troy the hatred of Athena and Hera and led to the onset of the war, for Paris was told that Helen, the wife of Menelaus, king of Sparta, was the most beautiful of all women. Paris kidnapped her, or perhaps she went willingly, then sailed off to Troy taking along her young son and a great treasure.

Menelaus appealed for help to Agamemnon, ruler of Mycenae, in accordance with the mutual-aid pact then existing between the Greek city-states. Agamemnon and many other Greek princes sent military forces to assist Sparta. An oracle prophesied that Troy would fall in ten years, and with this assurance the Greek armies set off to raid the coasts of Asia Minor on the way to attack Troy. The Greeks eventually landed on the beach before the city and undertook the long siege that was marked by famous battles between heroic warriors on both sides. The fierce duels of Achilles, Hector, Ajax, Odysseus or Ulysses, and others form a large part of the drama of the *Iliad*. The siege is said to have ended with the now-famous incident of the Trojan horse. The Greeks, pretending to give up and leave, left behind a large wooden horse, supposedly as an offering to Poseidon and Athena, in which a number of Greek warriors were hidden. The Trojans unwisely brought the horse into the city. The Greeks emerged at night, killed the guards, and opened the fortress gates. The full Greek army returned and Troy was sacked and burned with heavy casualties among the Trojans. After Troy fell, Odysseus left, and his long, eventful voyage home to Ithaca made up the subject matter of the *Odyssey*.

The dramatic history of Troy and the Trojan War became known to ancient and modern cultures solely through the epic poetry of Homer. But of Homer himself nothing is known except through the content of his writings. The date of composition of his works is uncertain, the source of his information is unclear, and even his very existence has been called into question. The ancient historian Josephus reports that in his day it was believed that Homer's poems were not left in written form. This is possible since it is known now that Greek writing was still crude and undeveloped in the eighth century B.C. Perhaps they were transmitted by memory and by oral recitations and were not composed into written form until several centuries after Homer lived, if in fact they even originated with a single person. Modern studies of bardic tradition show situations in other cultures in which epic poems have been orally composed and transmitted without the aid of writing. The text of the *Iliad* known to modern scholars derives from a tenth-century A.D. manuscript from Constantinople. In addition, papyrus fragments have been found in Egypt that contain parts of Homer's text. The remaining surviving manuscripts of Homer's works date from

Replica of the Trojan horse
Photo courtesy of Embassy of Turkey, Washington, D.C.

the fourteenth and fifteenth centuries A.D. No contemporary writings of Homer have ever been found.

Homer's works were not widely known until the first printed book of his writings appeared in Florence in 1488, followed by the great printed edition of Homer published in Venice in 1504. These publications began the modern critical discussion of the text that has continued up to the present day. Current opinion holds that Homer was probably a famous oral poet who flourished around the middle of the eighth century B.C. when the story of Troy was widely narrated in Greek courts. His works, then, may have been recorded in writing in his own time or, more likely, in the following century when the Greek language is known to have been better developed.

Homer's *Iliad* is the basis of the history of Troy during the Trojan War, which is believed to have taken place during the early or middle thirteenth century B.C. Modern archaeologists believe that after Troy's destruction by the Greeks or by earthquake, or by a combination of both, the surviving city was sacked again by the Sea Peoples around 1180 B.C. The site appears to have been abandoned after 1000 B.C., then reoccupied as the Greek colony of Ilion during the eighth century B.C. Ilion was a small market town built in a combination of archaic and classical styles. Ilion continued an obscure existence until Alexander the Great crossed the Hellespont with his army in 334 B.C. and conquered Anatolia together with the Troad, the area surrounding Ilion. Alexander carried a copy of the *Iliad* with him during his campaigns and was said to have slept with it under his pillow. He visited Troy, dedicated his armor to the Trojan Athena and took from her shrine arms and a shield that he believed to have been left from the Trojan War.

Alexander's successors built a new wall around Ilion about 300 B.C., but the city was overshadowed by a new city founded on the coast called Alexandria Troas. Ilion gradually decayed and by Roman times the town, now called Ilium, was derelict. Julius Caesar, who believed the Trojan Aeneas was his ancestor, visited the site in 48 B.C. and found the city in ruins. It was reconstituted as New Ilium. In the fourth century A.D., Constantine I considered building a major city at the site, and some construction was undertaken nearby. But by then the bay that had provided Troy with access to the sea was silted up, leaving the site with no harbor. Emperor Julian the Apostate visited Troy in 355 A.D. and found the citizens still making offerings at the shrines of Hector, Achilles, and Athena. But the Roman Empire itself was declining, and with it the economic life of New Ilium came to an end. City life ended by the sixth century A.D.; despite limited occupation thereafter, the site of Troy was deserted by the Middle Ages and knowledge of the exact location of the site was lost.

Yet interest in the story of the Trojan War continued and even grew from medieval times onward. Although the exact site of Troy was forgotten, the general location of the city was remembered. However, medieval and early modern travelers were often shown the ruins of Alexandria Troas as being the remains of Troy. A fifteenth-century traveler named Cyriac spent years exploring the eastern Mediterranean and

the Troad using Homer as a guide. This procedure was followed by many later explorers. The eighteenth-century traveler Robert Wood concluded correctly that the harbor in front of Troy had silted up over the centuries. But since little trace remained of the site of New Ilium, it was overlooked. The Frenchman Jean Baptiste Lechevalier carried out the first modern topographical survey of the Troad in the 1780s, but incorrectly concluded that Troy was at a town called Bunarbashi in the valley of the Scamander or Menderes River. It was left to an English resident of the area named Frank Calvert to conclude in the 1850s that Troy would be found at the mound called Hissarlik.

Calvert showed a representative of the British Museum traces of extensive ruins hidden under the soil at Hissarlik. He asked for a grant of £100 from the museum to do preliminary work at the site, but the museum turned him down. Calvert then bought the northern part of Hissarlik and in 1865 carried out trial excavations. He found remains of the Roman and Greek cities and exposed Bronze-Age levels enough to verify that the mound was deeply stratified. However, he lacked the money to continue excavations. Thus the glory of excavating Troy was left to a German amateur archaeologist, Heinrich Schliemann.

Schliemann was born in Germany in 1822 and eventually became a wealthy businessman from dealings in gunpowder, gold, and cotton in Russia and the United States. He retired from business at the age of forty-six and decided to make a reputation in some branch of science. Familiar with Homer and the story of the Trojan War, he visited Greece and Turkey in the summer of 1868. He met Frank Calvert, who inspired him to turn to archaeology and search for Homer's Troy. Calvert's earlier research led him to choose Hissarlik as the site for his exploration. Schliemann made a preliminary excavation in 1870 and undertook major digs from 1871 to 1873. Unfortunately, his method consisted of digging large trenches through the Hissarlik mound. Thus he demolished many structures that should have been carefully excavated and preserved. He identified four successive cities below the classical Ilium and concluded that the second from the bottom, which had been destroyed by fire, was Homeric Troy. Calvert disagreed with this conclusion, and the two quarreled. Later excavations revealed five more strata or cities with Homeric Troy most likely being the sixth or early seventh layer. Schliemann had also been misled by finding in the second layer a treasure of gold, silver, electrum, and bronze artifacts together with thousands of gold rings, bracelets, and earrings. In violation of his agreement with the Turkish government, Schliemann smuggled the treasure out of Turkey, and it was eventually deposited in a museum in Berlin. The entire treasure vanished when the Russians captured Berlin in 1945 and is believed to have ended up in Moscow.

Schliemann, however, continued to be privately troubled by Troy II as Homeric Troy. The size of this prehistoric city, about 100 by 80 yards, seemed too small to fit Homer's description, and the pottery found in the second layer seemed too primitive. He began to negotiate with the angry Turkish government for a permit to continue excavating, which he finally received in 1876. First he carried out important excavations at Greek Mycenae that brought him worldwide fame. Schliemann was assisted in Greece by an architect and archaeologist named Wilhelm Dorpfeld, who later participated in further excavations at Hissarlik. These began in 1878 and Dorpfeld joined him in 1882. Schliemann's last work at Troy was in 1889–90. At the very end of this campaign, buildings and artifacts were unearthed that linked later cities at Hissarlik with Mycenaean Greece, proving that Troy II could not have been Homeric Troy. Schliemann was preparing for the 1891 season of digging when he died suddenly near the end of 1890.

Dorpfeld carried on the excavations in 1893 and 1894, uncovering a large portion of the sixth city south of the mound together with two strata of the seventh city. The fortifications, substantial houses, and Mycenaean pottery of these levels led Dorpfeld to identify Troy VI and the lower level of Troy VII with Homeric Troy. An eighth city was identified as Hellenic Ilion and a ninth as Roman Ilium. An archaeologist from the United States, Carl Blegen, carried out further excavations at Hissarlik between 1932 and 1939. Blegen, able to analyze the strata in greater detail, concluded that Troy VIIa, the lower half of Troy VII, was the city destroyed in the Trojan War. Troy IX was believed to be the Hellenistic and Roman cities.

Major excavations are no longer being carried out at Troy, but minor excavations, restoration, and archaeological research have continued at the site using modern dating procedures and modern excavating methods. Controversy still exists, but it is generally agreed that Troy I, the lowest level, was a small fortress around 3000–2600 B.C. Troy II, dating from about 2600 to 2300 B.C., was a small city surrounded by walls. Troy III, IV, and V have been generally dated between 2300 and 1900 B.C. Troy VI, approximately 1900–1300 B.C., was a larger city influenced by Hittite architecture with large sloping walls. Troy VIIa, the lower level of Troy VII, existed from about 1300 to 1220 B.C. and is considered to be the Homeric Troy destroyed by Greeks around 1220 B.C. Troy VIIb was a reconstruction, carried out mostly by Greeks, and Troy VIII, dating from 900 to 350 B.C., was primarily a Greek city. Troy IX, dating from 350 B.C. to A.D. 400, was a Hellenistic and Roman settlement that had disappeared by the seventh century A.D.

In modern times Troy has evolved into one of the more popular tourist destinations in Turkey. The visitor come to view the excavated ruins first sees a huge replica of the wooden Trojan horse built as a tourist attraction. The ruins, partly as a result of Schliemann's destructive excavations, are a bewildering maze of walls, gates, and broken foundations. The identifiable structures are clearly labeled. Walls exist from several periods, in addition to the remains of an ancient council chamber, and the temple of Athena rebuilt by the Romans. The south gate faces the plain, and a building near it may have been a king's palace. Near a great chariot ramp is the place where Schliemann unearthed the spectacular treasure that later vanished in Berlin. There is also a partially restored shrine, and a Roman theatre and meeting place dating from the final stage of Troy's long history.

Further Reading: The best recent source for the history and archaeology of Troy and the Trojan War is Michael Wood's *In Search of the Trojan War* (London: British Broadcasting Corporation, and New York: Facts on File, 1986), a well-written and well-illustrated volume with an excellent bibliography. J. M. Cook's *The Troad: An Archaeological and Topographical Study* (Oxford: Clarendon, and New York: Oxford University Press, 1973) is a scholarly, well-illustrated archaeological and topographical study of Troy and its surroundings. It is required reading for anyone interested in Troy. Tom Brosnahan's *Turkey: A Travel Survival Kit* (Hawthorne, Victoria, and Berkeley, California: Lonely Planet, 1993) is a combination history and travel guidebook to Troy and the Troad as well as other parts of Turkey.

—Bernard A. Block

Turin (Torino, Italy)

Location: Northwestern Italy, 320 miles northwest of Rome; occupies the left bank of the Po River, where it merges with the Dora Riparia; capital of the Piedmont region and Torino province.

Description: Once the capital of the duchy of Savoy and an important participant in the events leading to the unification of Italy; center of the Italian automobile industry (Giovanni Agnelli opened the Fiat plant in Turin in 1899); has a military base, an international airport, a university founded in 1404, an art academy founded in 1652, a musical conservatory founded in 1867, the Egyptian Museum with a rich collection of Egyptian antiquities, and numerous libraries. Perhaps Turin's chief claim to fame is its possession of the Holy Shroud, believed by many to be the burial cloth of Jesus Christ, although evidence has indicated it was not; the shroud is housed in a chapel adjacent to the Duomo San Giovanni.

Site Office: A.P.T. (Azienda di Promozione Turistica)
Corso Farrucci, 122/128
CAP 10141 Turin, Torino
Italy
(11) 3352440 or 3859675

Turin, located in Italy's northwestern Piedmont region, has played a significant role in the country's history. It is frequently overlooked by travelers to Italy because it is perceived to be a dirty industrial city. This, however, is a misconception. There are numerous historically significant points of interest to visit, and much of the city retains its baroque ambiance.

Turin was originally founded by the Taurini, a group of people most often thought to be of Celtic descent (some claim that their ancestry was linked to the Ligurians), several hundred years before the birth of Christ. At that time it was known as Taurasia. This initial settlement was subsequently destroyed by Hannibal in 218 B.C. when he and his army of nearly 40,000 men crossed over the Alps and into the Piedmont region at the start of the Second Punic War. In 30 B.C. the city was seized by Augustus to serve as a colony for military veterans and it acquired the name Augusta Taurinorium. The initial squared street plan (orthogonal grid) of Turin was instituted during this period.

In the fifth century A.D., at the beginning of the Middle Ages, Turin was converted to Christianity and became the seat of a bishopric. A century later, in 568, the Lombards crossed the Alps to secure ruling power over most of northern Italy, marking, at the outset, a period of Italian disunity that would last for nearly thirteen centuries. This invasion, which was led by the Lombard king, Alboin, proceeded slowly and seemingly without opposition across the Po plain where dukes were installed in the major cities. In 572 Alboin was assassinated and his successor, Cleph, was murdered two years later. For the next ten years there occurred an interregnum in which the Lombards had no king at all and their collective power was administered by the independent dukes of the various cities and regions. They remained divided into dukedoms until sometime between 584 and 590 when they reunited under Agilulf against the threat of Frankish invasion. Arioald of Turin took the throne as king of the Lombard territories in 626 and was succeeded by Alakis of Trento in about 688. The Lombards continued to rule the territories and cities of northern Italy, an occupation that reached its height under Liutprand during the eighth century.

What happened to the Romans and Italians in the regions governed by the Lombards is not accurately known. A genuine "Dark Age" in the early history of Lombard-occupied Italy ensued. Archaeological evidence suggests that there was a fusion of cultures. Languages, laws, and people blended to such a degree that by the eighth century the cities of northern Italy were inhabited by a citizenry that was no longer easily distinguishable as Lombard or Roman.

In the second half of the eighth century, the Franks invaded Italy under their king, Pépin, and seized the Lombard kingdom. With their arrival came the foundation of the Papal States, a unique principality that would be confirmed by Charlemagne when he became king of the Franks in 774. Under this system the popes theoretically retained their state, but these states were, in most cases, actually controlled by the Carolingians. This led to struggle between pope and emperor and between church and state that would begin in the Middle Ages, and would continue in one form or another into modern times.

Turin begins its most significant individual history sometime in the eleventh century when, through marriage, the French dynasty of Savoy extended its territories to the Italian side of the Alps to include the counties of Susa and Turin. From this point on, the Savoys became increasingly interested in and oriented toward regions in Italy. Through a succession of Savoy counts, they gradually acquired more territory eastward in Piedmont. In 1391 the count of Savoy, Amadeus VIII, was granted the status of duke and the county of Savoy became a duchy. Throughout the following century, France and the duchy of Milan disputed over much of the Piedmont area held by the Savoys and from 1536 until 1559 the territory actually lost its independence to France.

Under the Treaty of Cateau-Cambrésis, signed in 1559, the duchy was returned to the duke, Emmanuel Philibert, who revived the prosperity of the state and moved

Dome of the Chapel of the Holy Shroud, rising above Turin
Photo courtesy of A.P.T., Turin

its capital from Chambery to Turin. Italian was made the official language spoken in public affairs and a permanent army was established. The transfer of the capital and the language conversion were instituted by Philibert in a plan designed to align his duchy politically and ideologically with the Italian states which had established them as the early crusaders for a united Italy.

The new capital at Turin necessitated urban development and military fortification. Emmanuel Philibert had regained his duchy through diplomatic rather than military means, and its location in the Piedmont region, sandwiched between France and the rest of Italy, placed an ever-present threat on its independence. From the very beginning of his residency, Philibert focused his energies and resources toward building Turin into an unbreachable fortress. The newest form of architect—the military engineer—was commissioned by the duke to design and oversee the building of a pentagonal citadel and other fortification projects. The initial chessboard pattern

(castrum) of Turin's layout, which had been started under the Romans, was continued and refined by a succession of such celebrated architects as Filippo Juvarra, Guarino Guarini, and Bernardo Vittone under the Savoy dynasty's directive.

The renowned Shroud of Turin, believed by many to be the burial cloth of Jesus Christ, was transferred to this city from France by Philibert in 1578 in an attempt to raise the spiritual level of its inhabitants. From a political point of view he believed that religion had coercive power and maintained that it reinforced the power of the prince because the pious were more orderly and obedient. The relocation of the shroud to Turin turned the city into a center of important pilgrimages yielding greater income for ducal and community treasuries. But suitable sheltering and exhibition of the shroud also placed new architectural demands on the city that were not resolved until the end of the seventeenth century when King Charles Emmanuel II of Savoy commissioned Guarino Guarini to design the Capella della Santa Sindone (Chapel of the Holy Shroud), which was annexed to Duomo San Giovanni (Cathedral of St. John) specifically for housing the shroud.

A period of warfare and neglect ensued during the fifty-year reign of Philibert's son, Duke Charles Emmanuel I, from 1580 to 1630. In 1588 the duke attempted to occupy French-ruled Saluzzo believing King Henry IV of France had not asserted his authority satisfactorily after the French Wars of Religion and was therefore vulnerable. However, Henry did not give up claims to Saluzzo easily. In 1600, after long negotiations, the French invaded Savoy and reacquired Saluzzo.

Victor Amadeus I held the reins of the Savoy duchy from 1630 until 1637, a period marked by continued conflict between Spain and France in which both countries occupied and used Piedmont as a convenient battleground and buffer zone. In addition, the population of Turin and neighboring villages was cut by more than half when an epidemic of bubonic plague swept through the area. In 1637, upon the sudden death of Victor Amadeus, the recovering duchy was passed to his young son, Charles Emmanuel II, who required the regency of his mother, Cristina, to hold dominion over the Savoy state during his youth. Cristina was the daughter of Henry IV of France and during her reign French influence in Turin was strong.

Savoy reached a turning point in the War of the Spanish Succession, which was fought from 1701 to 1714. Commanded by Prince Eugène, a distant cousin of Victor Amadeus, the Savoy army was able to prevent the loss of Turin to France in 1706. In 1713, under the Treaty of Utrecht, the Savoy dynasty secured Monferrato and the Lomellina, stretching its frontiers to the Ticino canton at the Swiss border. This treaty also established Sicily as a Savoy possession and named Victor Amadeus II king of Sicily, a title he held until 1720, when he relinquished Sicily to Austria in exchange for Sardinia. He then took the title of king of Sardinia.

In 1792, when war began between revolutionary France on one side, and the Austrian and Prussian monarchies on the other, the reigning duke of Piedmont was Victor Amadeus III. He had close connections with the powerful French Bourbon family in that his two daughters had married the two younger brothers of Louis XVI. When the Bastille was attacked by insurgents in July of 1789, the youngest brother, the comte d'Artois, who would one day be Charles X of France, brought his wife home to Turin. A French offensive against Piedmont followed in 1794, but did not succeed in conquering the territory until 1798, when Italy was invaded by French armies under the twenty-seven-year-old general, Napoléon Bonaparte.

Following a period of French control that lasted from 1798 until 1814, the monarchs of Italy were reestablished with absolute powers. The ruling authority of Victor Emmanuel I, son of Victor Amadeus III, was reinstated in Turin. This, however, did not bring about peace for the city. In 1821, members of the Carbonari, an expanding secret society hos-

tile to the restored regimes, played a part in the revolution in Piedmont which compelled Victor Emmanuel I to abdicate in favor of his brother, Charles Felix. The crisis was compounded when Charles Albert, prince of Carignan, was persuaded by revolutionaries to seize the throne in the absence of Charles Felix, and to grant a constitution. Within days of the uprising Charles Felix returned to Turin and Charles Albert quickly withdrew. The revolt continued for a few weeks on the outskirts of Turin, but was ultimately subdued by an invading Austrian army.

For the next forty years Turin contended with the dilemmas brought on by the Risorgimento, a nationalist movement in favor of the unification of Italy that was spearheaded by the Savoy dynasty. The term "risorgimento" was introduced by Vittorio Alfieri, an eighteenth-century dramatist who had predicted the arrival of a political and cultural "resurgence" in Italy after long years of rule by foreign autocrats.

One of the dominant figures in the initial struggle for Italian independence and unity was Giuseppe Mazzini, who was born in Genoa in 1805, and graduated with a law degree from the University of Genoa where he had developed a sharp awareness of the movement for Italian unification and political freedom. As a member of the Carbonari he participated in rebel activities when revolution broke out in central Italy in 1831. Mazzini was arrested by the Piedmontese police and imprisoned for several weeks. During this time he decided that the Carbonari were not clear enough in their aims and he began to develop his own plan for Italian nationalism. Mazzini was exiled upon his release from prison but continued to influence Italian insurgents through propaganda campaigns.

The Austrian chancellor, Prince Klemens Metternich, viewed Italian nationalism as a formidable threat and believed that liberal or nationalist sympathies would pose a threat to the survival of the Habsburg monarchy. When further revolutions broke out in the Papal States in 1831, Metternich arranged for Austrian military intervention in spite of French and British protests. The newly established French government, the government of the July Monarchy, which was itself the outgrowth of recent revolution, declared that a policy of non-intervention in Italy should be observed by the great powers. Austria ignored the warning and continued to intervene in Italy's political affairs.

A movement toward more conservative theory in Italian nationalism appeared in the 1840s. Vincenzo Gioberti, a liberal Piedmontese Catholic, suggested that the Italian states should retain their identities and their rulers, but should consolidate into a confederation presided over by the pope. Count Cesare Balbo (whose *Storia d'Italia* was published in 1856) also favored a confederation preserving the independence of the existing states, but envisioned Piedmont conferring military leadership to Italy and diplomatically ejecting Austria by according its Balkan territory. The third of these conservative theorists, Massimo Taparelli, who had lived in Rome, Florence, and Milan, as well as in his native Turin, was perhaps more experienced than his contemporaries. He theorized that the continuous pressure of public opinion in favor of Italian independence would decisively coerce the Austrians to leave Italy and he advocated passive resistance, a revolution "with our hands in our pockets."

The final phase of the Risorgimento was guided to a great extent by Count Camillo Benso di Cavour, a native of Turin who was trained at the city's Royal Military Academy and was very much self-educated in economic and political works. He came into contact with radical ideas while posted in Genoa and in the early part of his life he regarded himself a rebel against the autocratic regime of Turin that included his father. In 1835 he travelled to Paris and London where he was introduced to moderate parliamentary government that he found more compatible with his beliefs. An excursion on the Liverpool-Manchester railway (the first passenger line in the world) delighted him and he took an interest in various aspects of the Industrial Revolution.

Benso was elected to the parliament in Turin after the *Statuto*, the constitutional basis for the kingdom of Italy, was granted in 1848. In 1850 he was assigned the ministries of agriculture, commerce, and industry and two years later was asked by King Victor Emmanuel II of Sardinia and Piedmont (later, the first king of the unified Italy) to head the government when Massimo Taparelli resigned. When the Crimean War broke out in 1854, Benso was pressured into giving military assistance to France and Britain in the fight against Russia. The Russians surrendered early in the struggle. Benso's efforts in the affair were rewarded when he was invited to take part in the Congress of Paris that same year, where he met Napoléon III.

The struggles between Italy and Austria continued throughout these years. In April 1859, the Austrian government sent an ultimatum to Turin demanding a guarantee that within three days there would be a demobilization of the Piedmontese forces that had been assembled in the face of escalating tensions. Benso refused the request and major battles ensued between Austria and the allied forces of Piedmont and France. An abrupt decision by Napoléon III to call a truce was marked by the signing of the armistice of Villafranca, and this precipitated Cavour's resignation. The resignation was transient, however, and in January 1861, under the first parliament of the kingdom of Italy, Cavour was overwhelmingly elected the first premier. Turin was temporarily named the capital city of the kingdom from 1861 to 1865, but that status was later given to Florence and then Rome.

Conflict has plagued Turin from its inception. This was not to change even post-Risorgimento. After World War I, industrial wealth was centered in the Turin-Milan-Genoa triangle, and this area saw the rapid growth of socialist trade unions. Workers' councils at the Fiat automobile factories in Turin were encouraged by Antonio Gramsci, a leader in the communist wing of the Socialist Party, to adopt tactics based on the Russian Soviet model. In 1920 there were widespread strikes, followed by mass occupation of the factories by the workers. The workers ran the factories without their old bosses, and it appeared that a socialist revolution would succeed. Tough police action was taken by the government

of Francesco Nitti but the social chaos continued. When Nitti's government lost power in June, 1920, the successive ministry, under the aging Giovanni Giolitti, attempted the previously employed tactic of maintaining neutrality in industrial disputes. In Turin a modified proletarian society began to emerge and the majority of the city's population seemed to sympathize with the workers. Shopkeepers supplied goods for free or at reduced prices. Workers claimed that without the presence of management, they were manufacturing more cars than ever, but the cars piled up on the assembly lines because factory owners refused to market them. Eventually the socialist revolution of 1920 failed, because the workers had not seized political power, and the old management was restored at the factories. Giolitti had promised that workers' councils would remain in existence, but the promise was not kept by his successors.

In March 1943, at the height of World War II, the factories of Turin were again swept by large-scale strikes, and hundreds of arrests were made. In addition, Turin, like all of northern Italy in World War II, experienced intense Allied air raids and suffered heavy damages. During the economic depression that followed the war, Turin was flooded by the unemployed poor who migrated from southern Italy.

In 1970, fifty years after Gramsci conducted his factory councils experiments, the Red Brigade was born in a suburb of Turin known as Mirafiori. These Italian terrorists, organized into small but extremely disciplined cells stationed throughout Italy, proclaimed themselves vanguards of a coming revolution. They carried out numerous terrorist acts including the kidnapping and murder of political leader Aldo Moro in 1978.

Since 1980, thousands of Fiat workers have lost their jobs in the move toward the automated factory. They have been absorbed into the welfare state, taken jobs in the black-market labor force where no taxes are paid and no social regulations or minimum wages are honored, or have moved to other areas of Italy where they can find employment.

In 1988 the archbishop of Turin announced that scientific researchers using carbon dating determined that the Holy Shroud of Turin is actually an icon produced sometime between 1260 and 1390. Discussion and debate about its origins continue nevertheless. Visitors to Turin cannot view the shroud, which is housed in a silver casket locked inside an iron box and enclosed in a marble case inside the Chapel of the Holy Shroud. The keys are held by the Archbishop of Turin and the Chief of the Palatine Clergy. Visitors can view the magnificent chapel and cathedral as well as dozens of baroque and other architectural wonders that have survived from the period of the Savoy dynasty. These include the eighteenth-century basilica of Superga with the burial chapel of the house of Savoy; the fourteenth-century Palazzo Madama; the seventeenth-century Palazzo Reale; and the seventeenth-century Palazzo Carignano, which was the meeting place of the first Italian Parliament in 1861. The city's numerous museums include the Egyptian Museum, which houses the third most important collection of Egyptian antiquities in existence, including several copies of the funerary papyrus known as the Book of the Dead.

Further Reading: *Turin 1564–1680: Urban Design, Military Culture, and the Creation of the Absolutist Capital* by Martha D. Pollak (Chicago and London: University of Chicago Press, 1991)

is a comprehensive examination of the evolution of Turin during the years mentioned. It discusses personality and thinking of significant public figures, what was taking place in terms of evolving military theory, technological advances, and the architectural changes that were necessitated in Turin to maintain its fortification.

—Holly E. Bruns

Urbino (Pesaro e Urbino, Italy)

Location: Central Italy, about twenty-five miles inland from the Adriatic Sea and about twenty miles southeast of San Marino.

Description: Urbino stands on two summits, isolated in its hill territory. The town is neatly contained within imposing old walls and is dominated by the grand ducal palace built for Federico da Montelfeltro in the second half of the fifteenth century and home of a Renaissance court renowned throughout Europe as a center of art and learning. The palace now contains the fine collections of the Galleria Nazionale delle Marche. Numerous interesting religious buildings fill the town's limited space. The painter Raphael was born here; there is a museum in the house of his birth. The city has an old university, expanded in the 1960s with new buildings by architect Giancarlo De Carlo.

Site Office: A.P.T. (Azienda di Promozione Turistica)
Piazza Rinascimento, 1
CAP 61029 Urbino, Pesaro e Urbino
Italy
(722) 26 13

Pliny the Elder, a noted source for the origins of Urbino, wrote that the city was founded by Umbrians in the sixth century B.C. Etruscans, Celts, and in particular the Gallic Senones tribe all overran and occupied the area at various times. The first reference to Roman incursions into the territory come from the battle of Sentino in 295 B.C., in which the Romans fought alongside the local tribe of the Piceni to defeat the Senones. The Romans then took on the Piceni in 269–68 B.C., defeating them and establishing themselves more firmly on the site of Urbino. The settlement was named Urvinum Metaurense, after the Umbrian ruler Metaurus Suassus, possibly the founder of the original settlement.

While the settlement did not lie on a principal consular road, it did stand on a route of some importance; the single hill on which Roman Urbino developed rose like a natural fortress along the way. It was a rare example of a Roman chief town situated in mountainous territory. Town walls were erected during the third and second centuries B.C. The walls of Roman Urbino lay on the southernmost of the two summits upon which medieval Urbino grew; some buildings were probably constructed on the outside of these walls. The pattern of the walls was irregular owing to their having been built to conform to the natural terrain. As an administrative center, Urbino was provided in due course with public amenities such as a temple, a theatre, baths, and a basilica.

The town was given the status of a Roman municipium around 46 B.C., after Julius Caesar's consularship and the passing of the Lex Julia Municipalis. Pliny then gives the name of the town as Urbinum Hortense. It had many characteristics typical of a municipium, such as a municipal constitution, a people's assembly, a senate with magistrates, a particular jurisdiction, and a censor's register. Urbino's archaeological museum contains examples of Roman epigraphs that yield details about some of the leading town figures and their roles. Some indication of Urbino's strategic importance is given by the mention in Tacitus's *Annals* of Vitellius's vain attempt in A.D. 69 to take the town from Vespasian. The theatre was built in the first century A.D., making use of a natural depression on the east side of the town; the forum and other civic buildings had been erected on the western side.

The Goths took over the settlement at the end of the Western Roman Empire. With the imperial revival under the Byzantine Emperor Justinian I, the latter's famed general Belisarius laid siege to the town over a period of three years at the end of the 530s. The Goths, after running out of water, were finally forced to submit. Traveling with Belisarius was the historian Procopius of Caesarea, who recorded the siege and included many details on the plan of the city.

Throughout the period of Frankish and Carolingian rule in the eighth and ninth centuries, Urbino was dominated by the church and by the feudal families who oversaw the areas into which the region was divided. The bishop and the Capitol of Urbino were major land holders. Chapels were built in the Urbino territories, helping spread the church's teaching and power. In the 1069 *Pagina Confirmationes*, the bishop Mainardo clearly indicates the vastness of the church properties, many of them farmed out. Along the Metauro valley the Benedictines built a chain of monasteries. To help in administration, the bishops called increasingly upon members of important families in the region. Their involvement led to a shift in power, particularly in the tenth century in the time of the Berengars: papal problems weakened church control, allowing feudal lords to consolidate their power.

The creation of a liberated town commune caused feuding between town and rural lords for control of the county. These feuds lasted through the twelfth and thirteenth centuries, during which the commune tried to open up trade with other regions. Under the commune, more income and more people flowed into Urbino, which until the end of the eleventh century had hardly expanded outside the area of the Roman walls. A Benedictine convent and the church of San Sergio, first seat of the bishops, consecrated in 1021, were important exceptions. By 1069, the bishop's seat had become San Maria della Rocca within the city and on the site of the present cathedral's apse and transept. By this time a village had begun to develop on the second hill of Urbino.

Holy Roman Emperor Frederick II changed the course of Urbino's history with his consolidation of imperial power

The ducal palace at Urbino
Photo courtesy of A.P.T., Urbino

in Italy. Rewarding the loyalty of the counts Taddeo and Buonconte da Montefeltro, imperial documents of 1234 established the Montefeltro family as rulers of Urbino. The papacy fought back on behalf of the church's interests. Pope Innocent IV tried to remove Taddeo's privileges, leading to a bloody and highly destructive conflict between the Montefeltros, who supported the Ghibelline (imperial) cause, and papal legates along with their Guelph allies, in particular the Malatestas of Rimini. The town saw rule switch at least six times between 1283 and 1323.

Commerce continued in Urbino, and a new group of lords arose to run the lands farmed out by the bishops. Both public and private buildings were constructed, enhancing the importance of Urbino. Documents dating back to 1219, which formalized transformations that had occurred through the twelfth century, reveal the tight control the commune had imposed on commerce. A new wall would change the size of Urbino, including within it the buildings atop the second hill. This expansion also incorporated the area to the north, which had offered the easiest attack on the old town. New entrance gates were added, and a castle was later built on the top of this hill. A second wall of defenses was also built to extend a

little way beyond the limits the old Roman walls had traced. At the pass between the two hills, which had become a central crossroads, the square called the Pian di Mercato developed. It became the commercial center of Urbino. The administrative center was the Piazza Grande, or Maggiore. The bishop's seat lay on one side, and the communal offices were built on the other.

Religious orders settled across the town. The Dominicans, the first of the great new orders to establish themselves in Urbino at the start of the thirteenth century, benefited not only from their central location, but also from strong church support, particularly evident in a papal bull of 1297 granting them many privileges. The Augustinians settled in Urbino in 1225. In the eastern quarter of the new northern walled section was a large tract of land on which the Franciscans built in the mid-thirteenth century; in 1286 they took up the site of the former Benedictine monastery. The city was divided into quarters for administrative purposes, and three of these quarters housed a different religious order: the Franciscans in the north, the Augustinians in the south, and the Dominicans in the east; the bishop's and commune's area lay to the west. The various orders were supported by different leading families, the Montefeltri for example donating much to the Franciscans, several of their number being buried on their ground.

After the mid-thirteenth century and despite Ghibelline and Guelph feuding, Urbino's economy expanded, as did the physical town itself. The city walls were moved once more, this time in large part because of the destruction of war, having been demolished by the papal vicar Guillaume Durand after the siege of 1284–85 by papal troops. Corrado di Montefeltro won back Urbino and reestablished Ghibelline power in 1289. A new papal vicar ordered the submission of Urbino, but instead Corrado started building new defenses, only to be driven out before the end of the year. In 1292, Guido da Montefeltro won back Urbino, erecting some wooden defenses as an emergency measure and rebuilding the walls destroyed by Durand. These were damaged again in 1323 in the uprising against the count Federico. He was besieged in the fortress and killed on his surrender.

Learning began to flourish in Urbino in the first half of the fourteenth century. Fazio degli Uberti was an outstanding man of letters who wrote an allegorical epic entitled *Dittamondo* in imitation of Dante. Another important figure, Fra Bartolomeo dei Carusi of the Augustinian Order, dedicated his theological treatise for men of war, *De re bellica spirituale*, to Count Galasso. Less formal religious institutions, which had already begun to arise in the second half of the thirteenth century, continued to develop throughout the fourteenth: the first of these was Santa Maria della Misericordia di Pian di Mercato. They offered social assistance, Santa Maria in particular becoming a very important institution benefiting from large donations.

Church power grew after the mid-fourteenth century. San Bartolomeo was built in Urbino around this time. The papal legate Angelico Grimoard, very much in command of Urbino, erected a fortress on land formerly occupied by Montefeltro houses; this new fortress symbolically dominated the city from its height, standing on the site of the present fortress. Most of medieval Urbino would be replaced by subsequent building.

In 1375 Antonio da Montefeltro reasserted his family's power in the city, thwarting attempts by the pope to maintain authority over the territory. Montefeltro rule expanded in the 1380s, first over Gubbio and its territory, then over Cagli. The state of Urbino was thus formed, covering some 770 square miles. The area's principal economic activities involved the production of lumber, as well as agricultural produce and livestock. Gubbio, the area's principal town and a major commercial center, was on an important route for trade from the Adriatic Sea to western Italy. The state of Urbino formed a kind of buffer zone between the Marches and the Romagna regions, and as such maintained a relative degree of stability.

The church changed its position toward the Montefeltros; recognizing Antonio as apostolic vicar, a 1390 bull by Pope Boniface IX made clear Antonio's control of investitures across his lands. The papacy withdrew support from the Malestas, who maintained ambitions in the region, even giving Antonio the castles of Corinaldo and Mondolfo previously ruled for the church by Galeotto Malatesta.

Under Antonio, Urbino knew prosperity and reform. He succeeded in dissolving the powers of the feudal groups that remained. Formal new statutes were issued in the 1396 *Constitutiones appellationum*. Antonio was succeeded in 1404 by his son, Guidantonio, who was lord of Urbino until 1443. He too pushed civic reform. Markets and fairs brought an influx of goods and people to Urbino. Trades thrived. At the start of the century, wool makers and merchants from neighboring regions began doing business in Urbino, the wool trade forming a powerful society in 1427. Banking also developed to some extent; the first record of a lending bank in Urbino dates to 1407. Art too was being commissioned, for example in the fine frescoes by the Salimbeni brothers in the Oratorio di San Giovanni.

Military campaigns became a major source of revenue for the town, with the Montefeltros often serving others in their warring ambitions. Oddantonio distinguished himself so well that when he took over lordship of Urbino he was given the title of duke. He died within a year, in 1444.

Federigo da Montefeltro, half-brother of Oddantonio, succeeded him as count and ruled until his death in 1482. His achievements dominate the history of Urbino. Federigo's first exploits involved military campaigns outside Urbino's territory. He was captain to a string of popes, an ally of the Aragonese in Naples, served the Sforzas of Milan, became *gonfaloniere* of the church and captain general of the *Lega Italiana*. Federigo brought together a considerable number of loyal, highly trained citizens to participate in his warring campaigns, which yielded prosperous years and a cornucopia of culture for Urbino.

The fifteenth century was a period of great building programs for Urbino, financed by fighting. The state was defended by new rows of fortresses guarding its frontiers. The great architect was Francesco di Giorgio Martini, who developed coherent plans for Federigo. Federigo had a facility for developing extremely good relationships with those who served him, in his humanist development showing great respect for men of culture, whom he cultivated. Francesco di Giorgio's *Trattato di architettura civile e militare* speaks glowingly of Federigo, who treated the architect as a son, while commissioning from him 136 buildings. His best known work is Federigo's ducal palace, frequently cited as a marvel of the Italian Renaissance. It is not rigidly classical, however. Work on the main parts of the palace were begun from the mid-1460s by Luciano Laurana, with the eastern facade begun earlier by the Florentine Maso di Bartolomeo. The well-known, distinctive entrance with its twin towers is generally attributed to Laurana, who designed the magnificent Cortile d'Onore courtyard. Vespasiano da Bisticci, one of many writers on Urbino and author of a life of Federigo, recorded in detail the buildings erected during Federigo's rule. Among the non-military works were the reconstruction of the cathedral, the seat of the Jesuits, and the beginning of the San Chiara convent. To give some idea of the expenditure on the ducal palace, apparently some 200,000 scudi were spent on its construction; a house in Urbino at that time might have cost 50 to 100 scudi.

Federigo succeeded in fostering loyal patronage in his court. While abroad during his military campaigns, he made many contacts and later attracted artists from other territories to Urbino to adorn his palace with fine rooms and fine art. In the last twenty years of his life, the duke (a title Federigo was given in 1474), his money, and his love of learning attracted great men and great art to Urbino like a magnet. He knew many of the leading humanists, such as Alberti, Landino, and Ficini, and scientists like Paul von Middelburg. Among the long list of artists he commissioned was Piero della Francesca, whose portraits of Federigo and his wife Battista Sforza, from around 1465 (now in Florence's Uffizi Gallery), are renowned. So too are his *Flagellation of Christ* and *Madonna di Senigallia,* still in the museum of the ducal palace, while the depiction of the *Ideal City* is possibly by him or Laurana. Another famous portrait of Federigo with his son Guidobaldo is attributed to the Spaniard Pedro Berruguete. Giovanni Santi, father of Raphael, was from Urbino and painted for the duke. Federigo's library was a storehouse of humanist learning, containing one of the greatest manuscript collections in Europe.

Apart from his close relationships with popes and Italian potentates, Federigo maintained ties with the English court (he was made a knight of the garter by Edward IV), with Matthias Corvinus of Hungary, with Louis XI of France, and with the lords of Aragon. From his death, his legacy became a magnet for visitors to Urbino, and his name continues to be inseparable from the town. His rule was a time of extraordinary flowering for such a small town, which despite the spread of its reputation throughout Europe, did not expand physically beyond the limits it had established in the 1300s.

Guidobaldo I, handicapped son of Federigo, succeeded him and married Elisabetta Gonzaga. Art continued to flourish under his rule. The reconstruction of the cathedral begun at the end of Federigo's time was continued. Federigo's daughter Elisabetta, widowed early by the death of Roberto Malatesta, became a nun in the order of St. Clare and had its convent transformed into a building on a grand scale. The great painter Raphael was born in Urbino. His *Portrait of a Lady* is one of the great works held in the ducal palace. Raphael also painted a famous portrait of Baldesar Castiglione, whose book *The Courtier* evokes the court of Urbino at the end of the fifteenth century and is regarded as one of the great literary works of those times.

Guidobaldo's government was rudely interrupted by Cesare Borgia's taking of Urbino at the very end of the fifteenth century. During this occupation, in 1502, Leonardo da Vinci paid a visit to Urbino, making plans of the town. In 1503 Guidobaldo was restored to power after an uprising. In 1506 the Collegio dei Dottori was established, precursor of the university. Guidobaldo died without heir in 1508, and the Montefeltro line of rulers came to an end.

The nephew of Pope Julius II, Francesco Maria della Rovere, the first of a line of della Rovere lords of Urbino, was appointed as the new duke, ruling until 1538. A soldier of some standing, he like Federigo before him was made a knight of the garter, by Henry VII of England, an event marked by a painting of St. George (the patron saint of England) by Raphael. Francesco Maria's rule was interrupted by the incursion of the Medicis between 1517 and 1521. Urbino began to fade from the time the little state's court moved to Pesaro on the Adriatic coast in 1523. The ducal palace was given an extra story by Girlamo Genga for Guidobaldo II della Rovere. Certain notable artists continued to give Urbino importance in the sixteenth century. Federico Barocci is widely considered the second most important artist produced by Urbino after Raphael, but unlike those of Raphael, many of Barocci's works were painted and remain in Urbino.

Urbino itself suffered economic decline. It remained a center of letters, but the tone used to describe it was of idealized nostalgia, as reflected for example in the works by Bernardino Baldi or in an imaginary republic by Ludovico Agostini. The town fortifications were redone after mid-century. The 1590s saw famine and exodus.

At the end of the sixteenth century, Urbino contained numerous religious institutions, including five parish churches, six monasteries, eight friaries, ten confraternities, and four oratories, as well as the cathedral, whose dome was finally completed in 1604. That year Girolamo Campagna's statue honoring Duke Federigo was produced. The last of the della Rovere rulers was Federico Ubaldo, who died in 1626.

Urbino then became part of the papal states. In 1631, the dukedom was turned into a legation. The first legate, Antonio Barberini, brother of Pope Urban VIII, held a splendid festival on his arrival, but Urbino would suffer neglect under papal rule until Napoléon's invasion. A council of forty from four different classes put decisions for approval to the *gonfaloniere* and three *priori* chosen by the legate. The papal legates gravitated toward Pesaro. The rural population was miserably burdened with taxes, while the economy depended on grain, a commodity open to the corrupt speculations of wealthy landowners and the vagaries of weather and harvests. The growth of the school established by Guidobaldo I was encouraged by various popes until in 1671 Clement X gave the institution the status of a university. Various collections of the Urbino dukes were moved to the Vatican, including Federigo's magnificent library in 1657.

A notable inhabitant of Urbino, Gian Francesco Albani, was elected as Pope Clement XI in 1700, holding the seat until 1720. His power enabled him to revive Urbino's fortunes somewhat, improving conditions and encouraging architectural restoration. Two of Rome's major architects, Carlo Fontana and Alessandro Specchi, were called to the city. The nephew of Clement XI, the cardinal Annibale Albani, continued this work after the pope's death.

The mid-eighteenth century saw liberating economic reforms, and after famine in the 1760s the economy began to pick up. More restoration work was carried out, much of it by the architect Giuseppe Tosi. The earthquakes of 1781 and 1787 necessitated further work, notably the rebuilding of the cathedral by Giuseppe Valadier, following the facade design by Carlo Morigia.

By this time Napoléon Bonaparte had invaded Italy. The territories were divided in their support: the coastal areas quickly adopted anticlerical and anti-aristocratic initiatives, while the inland territories were more divided, some hostile to foreign and antireligious forces, others supporting them. In Urbino, two important figures, the *gonfaloniere* Fulvio Corboli and the archbishop Spiridione Bertoli, followed the French, but there was a popular uprising against the invaders in February 1797. With annexation to the Roman Republic, many church properties were confiscated. Liberalizing municipal and educational reforms were carried out. The road system of the area was rapidly developed, with the trade routes concentrating on exchange with the coast rather than looking inland. Bad harvests between 1814 and 1818 brought economic need, while the restoration of conservative church policy under Leo XII led to an alliance of the previously divided territories, which in 1831 united against Pope Gregory XVII. Cardinal Giuseppe Albani established stability, divided the area in two, and made the legate serve six months of the year in Pesaro, six in Urbino.

In the mid-nineteenth century, the major architectural projects came with the creation of the theatre by Vincenzo Ghinelli di Senigallia and the transformation of the former Dominican monastery into a new seminary by the architect and archbishop Alessandro Angeloni.

In 1860, the uniting force of the Piedmont troops entered Urbino, which became part of the Kingdom of Italy in 1861. At first the effects of centralization were not promising: part of the ducal palace was turned into a prison, the rest into prefectural offices. But the cityscape remained unaltered and began to benefit from protective measures. In 1869 the Raphael Academy was opened in the ducal palace, reestablishing it as a cultural center. Then in 1912 the Galleria Nazionale delle Marche was inaugurated as a fine museum within the palace.

In World War II much of the artistic heritage of the town was dangerously threatened when departing German forces laid mines around the town walls. What made the situation still more serious was that great art from other parts of Italy had been sent to Urbino, a small, remote location, for safekeeping. Only a few mines exploded, however, and the British forces defused the rest.

There is a tendency to view the whole of Urbino as a museum and reduce its history to Federigo's times. It remains a thriving cultural center, however, with a respected university that was expanded in 1966 by a leading Italian architect, Giancarlo De Carlo.

Further Reading: There are specialized studies in English of Renaissance Urbino, for example Cecil Clough's *The Duchy of Urbino in the Renaissance* (London: Variorum Reprints, 1981). For a fuller history of the town, the Italian work by Leonardo Benevelo and Paolo Boninsegna, *Urbino* (Rome: Editori Laterza, 1986), offers detail and is well illustrated, with numerous plans showing the city's development.

—Philippe Barbour

Valencia (Valencia, Spain)

Location: One and one-half miles west of the Mediterranean Sea on the Turia River, approximately 220 miles southeast of Madrid in eastern Spain.

Description: Capital of the province of Valencia; site of Roman and Moorish settlements; captured by the Spanish soldier-hero, El Cid, in the eleventh century. Situated on a fertile alluvial plain, Valencia is also known for rice cultivation and the introduction of *paella,* a popular Spanish dish made of rice and seafood.

Site Office: Municipal Tourist Office
Plaza del Pais Valenciano
Valencia, Valencia
Spain
(34) 351 04 17

Alluvium-rich, undulating farmland surrounds the city of Valencia. This is the "huerta," the agricultural phenomenon that drew various peoples to southern Spain. It is the beautiful backdrop to a succession of governmental regimes, religious institutions, and contradictory ideologies. Valencian history, marked by foreign invasions and assimilations, has produced a distinctive culture; sensitivity concerning regional identity reflects this. Valencians argue that their language, Valenciano, a dialect of Catalan, should be the official language. Recently, organizers of an International Film Festival held in Valencia used Castilian Spanish instead of Valenciano to subtitle foreign films. This choice was perceived, by those of particular Valencian pride, as a gesture of the state to challenge regional hegemony.

Seafaring Greeks established a colony in the region of Valencia, perhaps wrested from earlier Phoenicians, called Thuria; hence, the name of the river, Turia, that runs through the city today. The Romans obtained jurisdiction in 137 B.C. and established Valentia, or Valencia. They built a community for retired soldiers in the temperate land, embellishing the beginnings of an inherited irrigation system with their trademark ducts and channels.

During the dissolution of the Roman Empire, Visigoths, a western Germanic people, alighted upon the wealth of the region in A.D. 470. Arian Christians, the Visigoths denied the existence of the trinity and revelled in nature, presenting the Hispano-Romans with a source of theological agitation. In other respects, however, the Visigoths worked to maintain Roman civilization in matters of economy and government. With the arrival of the Moors (north African Berbers and Moslem Arabs), the eighth century was marked by social and political turbulence. Much of the cultural unity established by the Romans and Visigoths

was lost, though the Visigoth script was used into the twelfth century.

The splendors of Moorish Spain, the abundance of wealth and architectural beauty, were not realized until centuries after the invasion began in 711. Irrigation development, however, was furthered by the ninth century; most language associated with early hydraulic engineering in modern Spanish is of Arabic derivation. Crops such as artichokes and spinach were introduced by the eleventh century, but the most significant, early Arabic addition to the landscape (the Arabs being more influential and aristocratic than the Berbers, though in the minority) was the introduction of rice cultivation in the huerta, from which eventually came the famed gastronomical delight of Valencian *paella*.

The quest to recapture the country from the Moors gained momentum in the eleventh century, and in 1010 the Christians captured, and held for a short time, the caliphate at Córdoba. When the Arabic emirate al-Andalus disintegrated from the pressure of internal strife, Valencia became a *taifa,* one of a number of factional kingdoms. Arab occupation lasted there in various capacities from 725 to 1235 except for a brief yet momentous hiatus. In the summer of 1094 Rodrigo Díaz de Vivar, a Spanish soldier who came to be known as El Cid Campeador, entered the city of Valencia as its conqueror. "Cid" is derived from the Arabic *sidi,* meaning "leader" or "lord" and "Campeador" from the Latin for "winner of battles." Given the mix of cultures in this part of Spain, it is not surprising that Díaz's pseudonym combined the languages of the Arabs and the Christians.

A single manuscript copy exists of the *Poem of the Cid.* It was written around 1140; its author's identity is unknown. The praise it received from Christians and the anti-Islamic readings they gave it were matched by the angry Arab responses fueled by the economic and political significance of losing the great prize of Valencia. As for the historical Cid, Rodrigo Díaz was a mercenary soldier who served both Moslem and Christian kings. His campaigns were for the sake of victory, and for the additional prizes of booty, slaves, and dominion. Díaz died in Valencia in 1099, leaving control of the city to his wife, Doña Jimena. Valencia fell back into the hands of the Arabs but was finally reconquered by James I of Aragon in 1238, and the Kingdom of Valencia became a new state in the Christian Catalan-Aragonese Confederation, a community under the Crown of Aragon.

For the next five centuries Valencia enjoyed autonomy, guaranteed under the Act of the Union in 1319, and was governed by a parliamentary body called the Cortes. From its inception, independent Valencia had a potentially powerful economy, with its tremendous food production and silk industry. In 1483 La Lonja de la Seda, quite possibly Valencia's most outstanding piece of architecture, was erected to house

Statue of King James I at Valencia
Photo courtesy of Instituto Nacional de Promocion del Turismo

the Maritime Consulate and Commercial Exchange. Four centuries previously another edifice, the Gate of the Silk Exchange, had occupied the same spot; it was known in Arabic as Bab al-Qaysariya.

The kingdom eventually headed toward collapse, however, due to the wide gulf between the feudal lords and the common people. Rebellions characterized the sixteenth and seventeenth centuries. Also, the official expulsion of the Moriscos (Moors nominally converted to Christianity) in 1609, reduced the population by one quarter or more. The decline of agricultural and industrial productivity that resulted from this edict lasted for several centuries and further contributed to the top-heavy social structure. With a plethora of aristocrats, clerics, and lawyers, few laborers were available to cultivate the land. The Cortes, primarily a defender of elitist causes, was disunited, and the Kingdom of Valencia was gradually absorbed by Castilian legal institutions. In 1707 Philip V abolished Valencian autonomy altogether with the "New Plan," and Valencia added yet another process of assimilation to its history of political and cultural transformations.

The new order of the Valencian province coincided with a period of relative good fortune in the eighteenth century. The periphery of the city grew in population as agriculture became more commercial, and new industries flourished. Tenant farmers enjoyed a degree of monarchical protection from landowners for a while, and the Valencian region became one seat of the Spanish Enlightenment, especially in the fields of humanism and science.

Discord between center and periphery, a standard element of Spanish history, made its mark once again on Valencia in the nineteenth century. City merchants acquired capital in the country, land was expropriated, and rural wealth was withdrawn to the urban center. The investments of wealthy merchants into the cultivation of rural areas contributed greatly to the success of Valencia as an exporter of wine, citrus fruits, and rice. The tenants and laborers, however, were excluded from the benefits of this new capitalism.

Fervor for an autonomous Valencia developed throughout the region, although class divisions created somewhat different political camps. Much of the bourgeoisie obtained power through affiliation with the monarch while the laboring classes abandoned the mainstream Spanish liberalism for a determined republicanism. Valencia remained a center of republican and otherwise anti-monarchical organizations into the twentieth century. One of the leading republican voices of the late nineteenth and early twentieth centuries was Valencian author-activist Vicente Blasco Ibáñez, who was imprisoned several times and eventually exiled for his views.

The 1931 national elections revealed overwhelming support for a non-monarchical state. This ushered in the second Spanish Republic, characterized by progressive politics and modern idealism, but it eventually proved unstable. Valencia continued to be a hotbed of political parties embracing new ideologies and a battleground of divided interests between the landowners and laborers. The economy depreci-

ated as it tried to accommodate changes brought about by the republic. The Cortes expropriated private property and allowed for the formation of independently operated farming collectives. Economic experimentation, however, constituted poor planning; it gave rise to domestic conflicts and disrupted the income of the struggling republic.

During the Spanish Civil War of 1936–39, as nationalist forces put pressure on the capital of the republic in Madrid, the government seat was moved for a time to Valencia. By 1939 Alicante and Valencia were all that remained of the republic. They were simultaneously occupied on March 30, and the regime of General Francisco Franco began. Because of its strong affiliation with the republic, Valencia was bombed often by the nationalist forces during the Civil War. The city also suffered from postwar suspicion and the suppression of its regional identity and the causes it had supported. Irrigation works were preserved, however, and additional industries were soon incorporated. Despite periodic strikes, the region prospered under the new government.

There were two major events in Valencia in the late twentieth century. In marked contrast to the liberal politics usually associated with Valencia, it was from the city's Santo Domingo barracks that General Milans del Bosch launched his unsuccessful attempt at a military coup in 1981. The action lost the Spanish military much of its prestige. Then in 1982 Valencia suffered severe flooding; since then canals have been constructed on the west and south sides of the city.

The complexion and contours of city and coast have evolved significantly since a description of them was recorded in the time of the Cid. The city proper, once defined by walls, is now an unbound urban sprawl. The one and one-half miles of land that have emerged between the city and the port of El Grao at the sea are developed and populated. In the city center are such historic buildings as the Cathedral, which displays a mix of Romanesque, Gothic, and baroque architectural styles; the Basilica of La Virgen de los Despamparados, patroness of Valencia; the Church of San Esteban, said to be where the Cid's daughters were married; the Almudin, a medieval granary that now houses the Paleontological Museum; the birthplace of St. Vincent Ferrer, the Dominican who helped end the Great Schism; La Lonja; and numerous other historic churches, residences, and monuments, including an imposing equestrian statue of the Christian reconqueror James I. Valencia also is known for its arts and crafts and for its many festivals. Lladro porcelain is a Valencian product, and ceramics in the classical Spanish design, which draws its splendor in large part from Roman and Arabic influence, are exhibited each year at the Cevider Fair in Valencia. Most significant of festivals is the *fallas*. A week-long extravaganza in early to mid-March, with the culmination on March 19, the festival dates back hundreds of centuries. It is a Christian tribute consecrated to San Jose (St. Joseph), reminiscent of pagan saturnalias and derived specifically from the custom of carpenters to periodically clear their workshops of debris and to create bonfires of this waste. Celebrations include the parading and burning of enormous papier-mâché carnival figures on floats that comically and satirically represent events of the past year.

Further Reading: *The Kingdom of Valencia in the Seventeenth Century* by James Casey (Cambridge and New York: Cambridge University Press, 1979) is a study of Valencia that pays particular attention to how the city and the region evolved politically. *The Quest for El Cid* by Richard Fletcher (New York: Knopf, and London: Hutchinson, 1989) is an exceptional approach to history through myth, extremely comprehensive and inclusive of materials that preceded the events in the *Poem of the Cid*. Fletcher's manner is dynamic and personal, allowing the reader to experience his enthusiasm for Valencia in the eleventh century.

—Patricia Trimnell

Van (Van, Turkey)

Location: In province of Van in eastern Turkey on the country's largest lake, Lake Van.

Description: Ancient city, first mentioned in the ninth century B.C.; capital of Urartian kingdom from ninth to seventh centuries B.C., until supplanted by other small states; part of Armenian kingdom in Middle Ages, with many ruins from this period; conquered by Ottoman Turks in sixteenth century; center of Armenian mass murders 1915–16; subsequently deserted and replaced by new town a few miles away; today a center of Kurdish unrest.

Site Office: Tourist Information Office
Cumhuriyet Cad., No. 19
Van, Van
Turkey
(432) 2162018 or 2163675

The city of Van is among the oldest in the world. It was already a well established urban enclave at the time of its first recorded mention in the ninth century B.C. A tribe known as the Urartians had settled in the harsh terrain, cleared the land and irrigated it, turning a rugged wilderness into a blooming and fertile plain. The setting still is a gorgeous one—a scenic lake of volcanic origin, surrounded by high mountains. It is noteworthy that settlers were attracted to the lakeside even though there was no fishing in the lake because of its high saline content (although fish abound in nearby rivers); this factor also made the water undrinkable. Moreover, because of the lack of an outlet, the excess water of the lake has been known to overflow into treacherous floods. The climate is severe, with long, freezing winters, and the area is subject to frequent earthquakes. In modern times, a geological fault has been discovered running the length of eastern Turkey, accounting for the recorded destruction over time of more than 300 towns and villages. Nonetheless the hardy Urartians managed to establish a literate civilization on the shores of Lake Van and beyond, leaving behind many reminders of their existence.

The Urartian fortress that originally was Van is clearly visible, situated on a steep cliff overlooking the lake. It probably was built in the ninth century B.C. or even earlier, and continuously occupied even after the population expanded outside it. What remains of the fortress city are large slabs of stone bearing Urartian inscriptions and a long outdoor staircase leading to catacombs, also inscribed in Urartian. Little remains of Van's ancient walls, although the hulks of buildings erected hundreds of years after the disappearance of the Urartians are clearly visible, including a Christian chapel and an Islamic mosque.

Van became the capital of the kingdom of Urartu around 844 B.C., during the reign of Sharduri I. The kingdom was called "Ararat" in the Bible. In addition to developing the agricultural uses of their region, the Urartians also excelled at metalwork, masonry, and commerce. Much of their culture, including their cuneiform system of writing, was based on that of their neighbors, the Assyrians. Despite this sharing of culture, Urartu and Assyria were frequently at war. Van's height and fortifications enabled it to resist conquest by the more powerful Assyrians, however, until other peoples, including Cimmerians, Scythians, and Medeans, became allies of the Assyrians. The Urartian kingdom finally collapsed in the seventh century B.C. The region came under Medean, then Persian rule, and was resettled by people who came to be known as Armenians.

The Indo-European tribe that became the Armenian nation grew numerous in the area many centuries before the appearance of either Turks or Kurds. One of the Armenians' epic tales centers around their hero, Haik, who slew a giant warrior in a huge battle by the skill of his bow and arrow, right on the shores of Lake Van. From then on, it is believed, Haik's tribe took on the name of their patriarch and called themselves "Hai" and their land "Haiastan," although this is not the name by which foreigners such as Herodotus, Xenophon, and Strabo knew them then or now. For centuries Armenian society remained feudal, with aristocratic lords, a mass of peasants, and a weak king. They owed political allegiance to whatever neighboring power was strongest, the Roman Empire or the nearby Persian Empire, the Byzantines or their successors, the Turks.

The Armenians adopted Christianity en masse in the early fourth century, and developed a distinctively Armenian brand of the religion, distinguishing themselves from their Greek Orthodox neighbors. To this day, the city of Van bears reminders of the Christian presence. When the Armenians adopted an alphabet in the early fifth century, there took place a flowering of Armenian religious culture that was reflected in the many monasteries that developed around Van. Most of these are now in ruins.

The Armenians saw themselves as a bulwark of Christianity in the east. A combination of missionary zeal, the power of literacy, and the weakness of Van's neighbors led to the establishment of an Armenian state in the Middle Ages; it reached its height in the tenth century. It was known as the Kingdom of Vaspurakan, which included Van and the surrounding region, although the capital was situated approximately fifty miles west of the lake, at the city of Mush. Van and the nearby town of Bitlis became prosperous commercial centers. Most of present-day eastern Turkey was included in the Armenian state, which was a feudal kingdom of aristocratic lords, peasants, and a merchant class.

Shore of Lake Van
Photo courtesy of Embassy of Turkey, Washington, D.C.

In the heyday of their kingdom, the Armenians erected a great many buildings in and around the city and lake region, the remains of which can still be seen. A couple of miles from Lake Van's shore, on its own island within the lake, lies the astonishingly well preserved Church of the Holy Cross, whose tenth-century exterior is covered with sculpted biblical scenes. The church is surrounded by the ruined hulks of what were monastic buildings. This island, called Aktamar, became such a vital center of Armenian artistry and learning that the bishop of Aktamar was elevated to the rank of catholicos, reserved for only two other bishops in the Armenian church. He retained this high status until the late nineteenth century, when the numbers of congregants and monks declined dramatically. Also in the Van area lie the ruins of the famous monastery of St. Lazarus, and holy Mount Varak, which was studded with many monasteries, akin to an Armenian Mount Athos, all of them now in ruins. A particularly famous one was Surp Grigor, built in the tenth century, whose patron was the last Armenian king, Senekerim-Hohvanes. Even under the Turks, this monastery compound, with its noteworthy (and now ruined) churches of St. Sophia, St. John, and the better-preserved Church of the Mother of God, continued to be important.

Many of these houses of worship and cultural enclaves were damaged from time to time in their hostile surroundings, less because of the endemic earthquakes and floods than from the depredations of Armenia's aggressive neighbors, robber bands and Kurdish nomads. The Kurds, like the Armenians, were an Indo-European people who had wandered into the lake region from central Asia in the middle ages. They had adopted Islam, and perhaps were lured to the area by the wealth of the cities and isolated monastery compounds. They resisted all attempts to subordinate them to authority, until forced to under Ottoman rule, although this process took centuries. By the nineteenth century, Kurds constituted roughly one-third of the population of eastern Turkey, and many of them had become settled in cities and involved in trade and moneylending.

The fragile Armenian state also suffered from the invasions of the Seljuk Turks in the thirteenth century, followed by the Mongol raids under Timur (Tamerlane) in the fourteenth. The Mongol invasions resulted in the death and destruction of countless Armenians, Kurds, and Turks, making the area ripe for takeover by the Ottoman Turks in the sixteenth century. The Armenians in particular welcomed these "new" Turks, although they practiced a religion hostile to their own, because of the promise of order and stability.

Van became Turkish, and the region of Van an Ottoman province; until the eighteenth century, the area was peaceful and prosperous. Two impressive mosques were built in the surroundings of the lake, constructed in the sixteenth century of black and white stone that reflected the muted colors of the landscape. Since 1921, they have been abandoned, but are virtually intact. Right on the lake shore, a now ruined mosque of nine domes and four columns was built, attesting to the prosperity and relative stability of the times. Turks, Kurds, and Armenians, while not friends, lived together peaceably, though separately. Obstreperous Kurdish nomads were resettled along Turkey's long border with the Russian and Persian empires, where they served as border guards. As Moslems, Kurds were not discriminated against, could not be taxed, and were favored in Islamic courts, while Christians, besides being subjected to taxation, were not even allowed to serve as witnesses, and were forbidden to bear arms. With the economic and military weakening of the Ottoman Empire that set in during the eighteenth century and accelerated thereafter, the oppressions of both Turkish and Kurdish officialdom grew unbearable for the Armenians in eastern Turkey. They were by no means the only non-Moslem minority in the sprawling Ottoman Empire to suffer. Others, like the Greeks, had rebelled successfully and even achieved independence from Ottoman rule as early as 1829.

Like the Greeks also, Armenians had many ties with western Europe, many Armenians having emigrated from Anatolia to seek their freedom and fortune abroad. By the late nineteenth century, modern nationalism had spread even among the Turks. Blaming the empire's problems on its minorites, Sultan Abdülhamid II sanctioned the brutal massacres of Armenians in the 1890s. This did more than any form of oppression to awaken slumbering Armenian nationalism in the vilayet of Van, where most of the massacres occurred. Shortly after World War I broke out, the Armenians in Van rose up in revolt in the spring of 1915. Using this as an excuse, the Turkish government planned their systematic mass murder, although the Turks were not as thorough and efficient in the chaotic conditions of war as they might have been in peacetime. Hence, the Armenians in İstanbul, most of them wealthy, escaped the horrors that befell the entire region of Van. Armenians there were ordered out of their homes and men, women, and children were deported to distant areas where they were murdered. By the time the mass murders ended in 1916, in large part because of foreign pressure, more than 1 million Armenians had been slaughtered, their monasteries and churches demolished. Those who could, fled to the Russian-held portion of Armenia, which the Ottomans had ceded to Russia almost a century earlier and which would soon become a new Soviet republic.

In eastern Turkey, death and destruction had eliminated the majority of Armenians. The depredations of aggressive neighbors, such as the Soviet Union, wreaked further havoc in the province of Van, as Turkish and Red Army forces clashed in the aftermath of World War I. Depopulated, ancient Van was utterly laid waste. With the coming of peace and international recognition of the republic of Turkey in 1923, Van was rebuilt as a military base a few miles away from the original site of the city. Since then, because of the historic volatility and insecurity of the region, the new city has been sporadically opened and closed to tourists. The Kurds, who rebelled against the Turks in 1925, have never given up their desire for a nation-state of their own, which unofficially they dub Kurdistan, centered in the region of Van.

The new town of Van has grown into the largest city on the lake. It has evolved into a major military and commercial center, taking pride in its manufacture of bright carpets and oriental kilims. The town even boasts a small museum with artifacts, ironically, of Armenian provenance, as well as a store of relics from the Urartian period. It is a remote and unsettled area, but is rich in history, visible at every turn.

Further Reading: Van lies in the heart of the volatile Kurdish region in Turkey, and requires permission to visit. Nonetheless, many guidebooks in English describe the ancient lake region. One of the best is Joanne Ceruone's *A Kaleidoscope of Turkish Culture* (Chicago: J. C. Publications, 1991). Because Van will be forever associated with the Armenian massacres of 1915–16, accounts of the infamous slaughters and a history of Armenian settlement in the area are a must. Among the most objective and readable are Ronald Grigor Suny's *Looking toward Ararat: Armenia in Modern History* (Bloomington and Indianapolis: Indiana University Press, 1993) and Christopher J. Walker's *Armenia: The Survival of a Nation* (New York: St. Martin's, 1980; as *Armenia: A Modern History*, London: Croom Helm, 1980; second edition, New York: St. Martin's, and London: Routledge, 1990).

—Sina Dubovoy

Velia (Salerno, Italy)

Location: In southern Italy, about twenty-five miles southeast of Paestum, halfway between the Gulf of Salerno and Gulf of Policastro; in the region of Campania and province of Salerno.

Description: A former port city once surrounded by two harbors, now silted in. In ancient times a Greek colony, known for its philosophers. Ruins include walls that encircled the original colony, two sea gates, a colonaded marketplace (agora) and remains of an Ionic temple.

Contact: E.P.T. (Ente Provinciale per il Turismo)
Via Velia, 15
CAP 84100 Salerno, Salerno
Italy
(89) 224322 or 224539

Elea (later called Velia) was known in the ancient world more for the minds of its inhabitants than for its size or political influence. Founded later than most other Greek colonies in Italy, Elea became a final refuge for unpopular philosophers driven from Athens during the fifth and fourth centuries B.C. Here Xenophanes of Colophon, Parmenides, and Zeno formed what became known as the "Eleatic School" of pre-Socratic philosophy. Greek historian Diogenes Laertius wrote that Elea, though an inconsiderable city, produced great men.

Excavations at the site of Elea during the 1960s uncovered some of the most significant archaeological finds in Magna Graecia (the Greek settlements in Italy). An entire town plan was unearthed, as well as a town center (agora) and two sea gates. The original town, always somewhat small, had been abandoned by its inhabitants in the ninth century A.D., leaving the site undisturbed for later researchers. For many centuries the location of the ancient town was unknown.

The founding of the town around 535 B.C. was dramatic. According to the historian Herodotus, Elea's founders were Ionians from the city of Phocaea, the northernmost Greek city on the west coast of Asia Minor. The Phocaeans were sophisticated, politically independent sea merchants. In their special boats called "pentecoters" they made remarkably long sea voyages for their time, including visiting Spain and colonizing Massilia, which became Marseilles, in France.

Such independent people could not bear the thought of being enslaved by a conquering power. But they were the first under attack when Persian general Harpagus marched on Greece. The general built a mound of dirt against the city walls within which the Phocaeans had barricaded themselves. Finding their town surrounded, the Phocaeans asked for one day to deliberate on their surrender. Harpagus pulled his troops back for twenty-four hours, during which the Phocaeans launched their pentecoters with "their wives and children, their household goods, and even the images of their gods, with all the votive offerings" on board, Herodotus writes. The Persians marched into a deserted town.

At sea, the Phocaeans dropped a heavy piece of iron overboard and made oaths that they would never return to Phocaea to live in Persian slavery until the iron floated back to the surface. Half of them, however, became so homesick that they broke their oath and sailed back to Phocaea. The rest continued onward to find a new homeland.

The pentecoters finally docked in southern Corsica at a Phocaean settlement called Alalia. But their arrival alarmed the local populations of Etruscans and Carthaginians, who formed a naval alliance and attacked the Phocaean fleet. The Phocaeans won the battle, but many of their ships were lost or severely damaged.

Once again, the Phocaeans loaded as many goods as their ships could bear and set sail, this time for Rhegium (Reggio), on the very toe of the Italian boot. From Rhegium, they continued northward up the peninsula until they reached a promontory occupied by the Oenotrians. They conquered it and there founded the city of Elea.

The Oenotrians were no match for the Phocaeans, and neither were the fierce Lucanians from the surrounding hills, who never succeeded in overrunning the city. The Phocaeans' security was aided by the natural fortifications the hills provided, as well as by the site's location between two rivers, the Palistro and the Fiumarella Santa Barbara. In fact, the people of Elea did not seem to have much contact with the inland populations. They looked outward to the sea, concentrating on fishing and trade with other ports. The Greek language remained dominant. Greek customs, religious practices, and culture went unchanged, even later under Roman rule. Elea's coins circulated all over the Italian peninsula. Elean contacts with Massalia were so strong that the town became almost an outpost of the Gallic city.

Elea's claim as an intellectual center began in the late sixth century B.C. with the arrival of exiled philosopher Xenophanes of Colophon. An epic poet, early theologian, and wandering bard, Xenophanes had been banished from Athens for criticizing Hesiod and Homer and for attacking the gods. He pointed out that people always envisioned gods to be much like themselves; the gods in Homer's works, Xenophanes said, had human vices such as "theft, adultery, deceit, and other lawless acts." He enraged his contemporaries by suggesting that if animals such as oxen, lions, and horses had hands they would surely draw gods that had bodies resembling their own.

The Porta Rosa at Velia
Photo courtesy of E.P.T., Salerno

The foundation for the Eleatic school was based on Xenophanes' assertion of a single God, "neither in shape nor in thought like unto mortals." A version of the monotheism articulated by the Eleatic school later reappeared in the Western world in the form of Christianity. But Xenophanes and his later proteges also advanced theories that flew in the face of common sense. Among them was the theory that all sensory perception was an illusion. Because the all-powerful God was unchanging, the universe was unchanging, therefore matter did not move. Xenophanes also postulated that a new sun was formed each day from fiery particles and that the world contained an infinite number of suns and moons. Later philosophers acknowledged the Eleatic school as a precursor of early metaphysics, but academic quibbling over points such as the existence of motion tarnished the group's reputation and influence.

By the time Xenophanes reached Elea, he was an old man; according to various sources, he lived into his eighties or nineties. Known for reciting his poetry all over the Greek Empire, he was commissioned to compose a poem about the founding of Elea. This he did, but unfortunately the verses are not among the fewer than 200 surviving fragments of his work.

Elea's most famous mind was the philosopher Parmenides (born about 515 B.C.). A student of Xenophanes who built on his teacher's work while also contradicting it, Parmenides was widely respected. Sources mention that Parmenides, as an old man visiting Athens, met the young Socrates. In fact, he is the only person to have won an argument with Socrates (in Plato's dialogue *Parmenides*).

Parmenides was known for his views on being and non-being. He discussed these in a short poem called "Nature," fragments of which survive. This work was later the object of contempt by both Cicero and Plutarch, who considered the poem lacking in literary merits. Building on Xenophanes' theory of monotheism, Parmenides proposed that the truth behind the universe was "All is One." His philosophy was seen to merge two schools of thought: the

Ionian school of physical philosophers who believed that all existence came from primary matter, and Heracleitus's theory that all existence consisted of perpetual change.

Parmenides was also a great civil figure in Elea. The son of a wealthy family, he became governor of Elea and is said to have written the town's constitution, which lasted for centuries. He is credited with the rise in prominence of the city. Modern excavations unearthed a marble bust of Parmenides among the ruins of Elea, as well as evidence of a medical school founded in his name.

Parmenides' pupil, Zeno (born about 495 B.C.), furthered the Eleatic school. In a series of famous "Paradoxes," Zeno attempted to apply his teacher's principles even more broadly to the physical world with results that often contradicted common sense. His most famous contribution was the assertion that motion did not exist. In his paradox of Achilles and the hare, he claimed that in a race, Achilles (the faster runner) would always lose because the moving body must go half the distance before it goes the whole distance and there would be an infinite number of half-distances to reach, thus the hare (which started out ahead) would always be in the lead.

Under the moderate government of Parmenides' successors, the city of Elea thrived. It stayed aloof from most of the political and military struggles of the day, but in 387 B.C., Elea joined the Italiote League to oppose the aggressions of Dionysius the Elder, tyrant of Syracuse. The town also allied itself with Rome and sent ships to the First and Second Punic Wars.

Around 300 B.C. a serious flood buried the city. Residents rebuilt on the ruins, using the original town plans. Subsequent excavations of greatest interest date from this rebuilding period of between 300 B.C. and A.D. 62. The finds from this period display a distinctive checker-board style of brickwork in which large blocks alternate with squares of stonework. The bricks themselves are an archaeological curiosity. Made of heavy local clay, each is stamped with a mark denoting what appears to be a state tax.

Excavated sites include a gymnasium, baths, and a marketplace (agora) surrounded by a portico upheld by three pillars. Also unearthed was a small temple containing statuary and imperial portraits. Excavators were impressed by the engineering required to build a large aqueduct from this period that runs into the agora. In addition, unearthed roads reveal a paving pattern on inclined limestone that increases traction and maintains its levelness in spite of bends in the road.

The site's most impressive find, however, was the Porta Rosa, one of the greatest pieces of Greek architecture in Magna Graecia. The Porta Rosa is a grand archway cut into a fortification wall that served both as a viaduct and as a gate to the city. The arch is formed by perfectly cut dry-stone masonry without grout or mortar between the blocks.

After the rebuilding, the town remained politically independent for a time before becoming a Roman *municipium* in the first century B.C. The Romans called the town Velia. Brutus used it as a military base during the Civil War in 44 B.C. and built a villa nearby. Cicero is said to have visited him

there. The Roman poet Horace was advised to take the curative waters at Velia. Priestesses to Ceres, required to speak Greek, were culled from the ancient Greek settlement.

The town maintained its Greek character and language, but lost some of its influence as time went on. It did retain some of its intellectual prominence into the first century A.D. The grammarians Statius (father of the poet of the same name) and Palamedes were both born in Velia. The medical school, based on the experiments of Parmenides, flourished in the city.

Then in the late first century, natural disaster struck again. Experts surmise that the town was devastated by the severe earthquake in A.D. 62 that leveled many cities in the Campania region. Again, residents rebuilt, but Velia never quite rebounded. As the town continued to decline economically as trade routes changed, the two Velian harbors filled with silt. The nearby land was inadequate for agriculture, and the residents began to depend almost entirely on fishing. By the time the Roman Empire collapsed, Velia had become a small village.

In the fifth century A.D., the town was once again buried quite deeply in a landslide or flood. Residents rebuilt the town—possibly after some time had passed—following a plan totally unrelated to the ancient site. Few ruins from this period remain. By the ninth century, the area had become a malarial swampland. Arab pirates repeatedly stormed the town and the inhabitants moved inland to a town called Novi, which became known as Novi Velia. Another town, Castellammare della Bruca, developed near the site of ancient Velia during the twelfth century, but was abandoned during the seventeenth century.

In later years historians marveled at how malaria, borne by mosquitoes, could wipe out an advanced civilization. "How puzzled we were to explain why the brilliant life of Magna Graecia was snuffed out suddenly, like a candle, without any appreciably efficient cause," writes British antiquarian Norman Douglas in his *Old Calabria*. "How we listened to our preachers cackling about the inevitable consequences of Sybaritic luxury . . . and now a vulgar gnat is declared to be at the bottom of the whole mystery!"

During the nineteenth century, scholars speculated about the exact location of Velia. In 1818, a Danish scholar made the connection between the ruins of Castellammare and Velia. A few excavations took place thereafter, and in 1889 the first modern plan of the city's remains was produced. Excavations continued into the twentieth century, on a much more systematic basis. The agora and other sites were discovered in the 1950s. The Porta Rosa was uncovered during excavations made between 1962 and 1972 by Mario Napoli. The numerous statues that have been unearthed include a bust of Parmenides. The Velia ruins are accessible by road or rail; buses run to the site from Salerno.

Further Reading: *Guide to the Excavations at Velia* by Mario Napoli (Salerno: Di Mauro, 1973) provides a comprehensive, illustrated description of the site and its significance by one of the top archaeologists at Velia. For an excellent overview of early pre-Socratic Greek culture and its relationship to philosophy, see Victor Ehrenberg's *From Solon to Socrates: Greek History and Civilization During the Sixth and Fifth Centuries B.C.* (London: Methuen, and New York: Routledge, 1973). Norman Douglas's *Old Calabria* (London: Secker, and New York: Modern Library, 1915) is a quintessential post-Victorian travel diary documenting the author's walking tours of southern Italy. His opinionated descriptions are colored by expertise in natural sciences as well as a thorough knowledge of ancient history.

—Jean L. Lotus

Venice (Venezia, Italy): Rialto/Grand Canal

Location: The Rialto district is located at the topographical center of Venice, on the right bank of the Grand Canal. The Pescheria (fish market), Fabbriche Nuove and Vecchie, and Erberia (fruit and vegetable markets) follow the bend in the canal's S-curve. The Ponte di Rialto (Rialto Bridge) spans the canal at this point, linking the market area with the Fondaco dei Tedeschi, on the left bank.

Description: The Rialto was a driving force behind the transformation of Venice from a marshy lagoon settlement into a supreme maritime empire. During the Middle Ages, the high embankment ("Rivo-Alto") of the mud-flat islands that became Venice crystallized into the city-state's commercial nucleus, with its signature markets, bridge, and financial quarter. The ocean-going vessels of the medieval Venetian merchant have given way to tourist excursions and petrochemical refineries, but the Rialto continues to be the thriving heart of a city of incomparable beauty and historical significance.

Site Office: I.A.T. (Uffici di Informazioni e Aceoglienza Turistica)
San Marco
CAP 30124 Venice, Venezia
Italy
(41) 5226356

During Europe's "Dark Age," when Alaric's Visigoths and the more predatory Vandals swept down into northern Italy, residents of Padua, Aquileia, and other prosperous Roman towns bordering the Adriatic coast sought refuge in the comparatively secure, mud-flat islands of the Venetian Lagoon. Bringing what possessions and building materials they could carry, they settled at Cavarzere and Chiogga at the lagoon's southern tip; farther north, at Malamocco, Torcello, and Grado; and at the Rivo-Alto, or "high bank," of the cluster of islands in its center. A church dedicated to the Apostle James and built at the bend of the great S-shaped canal bisecting the Rialtine archipelago marked the legendary foundation of their *Città di Rivo Alto,* the future Venice, on the Feast of the Annunciation, March 25, 421, at noon.

The onslaught of Atilla's Huns, and the steady incursion of Lombards, begun in 568, triggered further mass migrations of refugees, who, like their predecessors, had guessed correctly that the marauders would not follow them into this uninviting spot. Each time the danger abated, many of the migrants returned to their native cities; others remained in the lagoons to carve out a living alongside the aboriginal inhabitants, from whom they learned that timber, salt distilled from the brackish waters, and bricks baked from local clay could be transported to their former residences in the Venetia-Histria province. The value of salt as a food preservative was almost inestimable. Writing in 523 to the military tribunes governing the disparate island communities, Cassiodorus, secretary of Theodoric, the Ostrogothic king of Italy, conveyed a bucolic picture of their lifestyle: "There lie your houses built like sea-birds' nests, half on sea and half on land . . . . Your inhabitants have fish in abundance. There is no distinction between rich and poor . . . . All your activity is devoted to the salt-works, whence comes your wealth . . . . From your gains you repair your boats which, like horses, you keep tied up at your house doors."

Commerce and sea power, the cornerstones of the Venetian Republic's future greatness, were well under way in the lagoon as early as the fifth century. In return for aiding Emperor Justinian's restoration of Ravenna to Byzantine rule in 539, the Venetians were granted military protection and trading privileges throughout the empire; fish, salt, Baltic amber, and slaves were the mainstays of this early trade. At that time, Torcello was the commercial nucleus, Grado the seat of ecclesiastical power, and the mainland town of Heraclea the administrative center. The sparsely populated Rialtine islands, which comprised Dorsoduro, Luprio, and the present-day Giudecca and Castello, and the other communities of the "Venetiae," crystallized into a more coherent political entity in the eighth century with their appointment of one Orso of Heraclea as their *dux* (*doge* in the Venetian dialect), in opposition to the Byzantine Emperor Leo III's mandate that all religious images be destroyed; the widely accepted version of this event holds that Paoluccio Anafesto assumed the title of doge in 697.

The events of 810 prompted the early Venetians to transfer their seat of government to the Rialto, a move formalized by the 814 Treaty of Ratisbone. Doge Obelario degli Antenori and his brothers had invited Pépin, son of Charlemagne, to garrison Venice. Anxious to preserve the favorable trading concessions granted by the Byzantines, the Venetians overthrew Obelario; their new leader, Angelo Participazio, led them to the Rialto, where a hefty bribe and the defenses they had prepared (stakes driven into the lagoon and a bombardment of arrows and even loaves of bread hurled in defiance of Pépin's threat to starve them out) drove the Frankish king away. The Rialtine settlements began to take on their present form: canals were dug, wooden piles were driven into the boggy earth to buttress the foundations of newly constructed stone buildings, and distinct communities emerged, parishes with their own churches, guilds, and town squares, interlaced with channels and bridges. Shortly afterward, the functions of political and legal administration were transferred to the newly built Ducal Palace and its adjoining

The Grand Canal
Photo courtesy of The Chicago Public Library

chapel, the Basilica di San Marco; the area around the Church of San Giacomo di Rialto became the commercial nucleus of the developing city-state.

Local markets sometimes stimulated the growth of medieval towns by attracting merchants and artisans. The Rialto commercial district and the Serenissima, the Most Serene Republic, nourished each other on a colossal scale. Poised between the Byzantine and Moslem East and the Latin and Frankish West, Venice maintained its independence from the Western Roman Empire while profiting from trade with both empires. Venetians set up merchant colonies on the mainland and utilized the ports of Sicily, Spain, Greece, Egypt, and Syria. Through astute political maneuvering and naval victories over Narentine pirates, Slavs, Saracens, Normans, and other aggressors, they gained more favorable trading concessions and privileges and consolidated their bases for commercial expansion. The Venetians' participation in the Fourth Crusade, prompted more by commercial interest than religious fervor, brought them sacred relics, a colonial empire in the Aegean, and a wider conduit to the Levant trade. Traversing the Po River Valley and the Alpine passes, Venetians sold to the northern Europeans products of their own manufacture, especially salt, glass, gold and silver ware, and

the coveted Chinese damask, Byzantine velvet, Cypriot cloth of gold, Asian cloves, Persian pearls, Russian furs, and other Eastern luxuries; they used the grain, woolens, wine, metals, and slaves they received to buy more luxuries for re-export from Venice.

Vendors' stalls, bankers' booths, slave traders' auction blocks, and barges hauling cargoes from ocean-going vessels all crowded the busy Rialto markets, which were established in 1097 near the Church of San Giacomo di Rialto, with its twelfth-century inscription exhorting merchants to honesty. Slave auctions at the Rialto were prohibited in 1366, but the slave trade continued to flourish, mainly through private sales of slaves imported from the Black Sea area for use as domestic servants and concubines. Tanners and silk weavers, jewelers and glassmakers, and a host of other craftspeople, trade associations, and guilds proliferated throughout the city, but the fruits of their labor were sold in the Rialto. Marino Sanudo, a patrician Renaissance chronicler of his city's history, exclaimed over the profusion of foodstuffs in the Pescheria, Beccaria, and Naranzeria, the fish, meat, and fruit markets: "In this land, where nothing grows, you will find an abundance of everything; for all manner of things from every corner and country of the earth ... are brought to this place;

and there are plenty to buy, since everyone has money." Venetians vigorously enforced their city's "staple rights" over the northern Adriatic; beginning in the thirteenth century, native and foreign merchants plying those waters had to unload their wares in the Rialto markets, then pay the requisite tolls and sell their merchandise there, an arrangement that channeled profits into Venice.

Marco Polo's pioneering sojourn in the Mongol Empire of Kublai Khan epitomizes the merchant's questing spirit and acquisitive values, but elements of community enterprise and individual initiative intertwined in the careers of medieval Venetian merchants. They were responsible for purchasing and outfitting their "round ships," which were constructed in the Arsenal and other government-sponsored shipyards. Twelfth-century merchants usually traveled with their wares or entrusted them to colleagues, but their choice of routes, cargoes, and sailing dates was subject to government regulation. Some consulted manuals, such as the fourteenth-century *Zibaldone da Canal,* which listed standards for converting weights, measures, and monies into their Venetian equivalents and instructed readers in the selection of quality merchandise; typical entries in the *Zibaldone* are, "If you carry gold ducats, you will have 18 and 1/4 bezants for 5 of them in Tunis," and, "Cinnamon ought to be of a reddish color, and it ought to be light and strong to the taste; and the good kind is a little sweet, and the common kind ought to be kept to [no more than] a third of the good." The book also instructs us that a pallid or reddish new moon produces rainy or foggy weather, that wearing the heart of a vulture, wrapped in the skin of a lion or wolf, wards off demons, and other vital astrological and medical lore.

Merchants congregated daily at the Rialto to cement their investment partnerships, contracted mainly through "sea loans," *colleganze,* and commission agents. With the sea loan, the merchant repaid the money borrowed to finance a voyage at interest levels as high as twenty percent, but investors risked loss through shipwreck, piracy, and other disasters. Under the *colleganza* system, which gained currency in the twelfth century, a merchant's investors pledged their funds in return for three-fourths of the profit gained from the venture. Commission agents, who became more common in the fourteenth century, received a percentage of the actual turnover they handled. Since many people eagerly pooled their resources to invest in international trading operations, merchants could make staggering fortunes. Many achieved political prominence, and their stately palaces line the Grand Canal. One such individual was Pietro Soranzo, who amassed an estate comprising 3,000 ducats worth of pepper, an equivalent quantity of nutmegs, cloves, tin, lead, and iron, gold from Russia totaling 1,478 ducats, raw silk valued at 3,810 ducats, and Russian skins worth 1,900 ducats, in addition to sugar, wax, honey, and pearls.

The Venetian aristocratic system, firmly in place by the thirteenth century, did not prevent members of the citizen class and immigrants from profiting by commerce; the republic reserved special warehouses called *fondaci* for foreign traders, but it cast a wary eye on their activities. German merchants transacted their business in the Fondaco dei Tedeschi, at the foot of the Rialto Bridge, from 1228 to 1812. Farther west, the Fondaco dei Turchi served as the Turkish emporium from 1621 to 1838.

Although the authorities imposed numerous legal restrictions on the conduct of both native-born and foreign merchants, they made exceptions when Venice would profit. For instance, Jews classified by the Venetians as German Jews were permitted to work only as pawnbrokers and dealers in second-hand clothing; Jews from other countries could engage in international trade, but all Jews were consigned to wear distinctive clothing and live in "ghettos," formerly the sites of iron foundries. In 1579 the Venetian Trade Board relaxed some of these prohibitions in anticipation that the Jewish merchants' commerce would reap substantial customs revenues.

During the sixteenth century, Venice tenaciously weathered political and economic crises precipitated by wars with other European powers and the Ottoman Empire, the diminution of its territory on the mainland, and the discovery by the Portuguese of new trade routes to India. Accompanying Venice's "Golden Age" of artistic and architectural splendor was a commercial revolution characterized by the *giro,* or transfer, system of banking. With the increase in permanent overseas trading bases, more Venetian merchants were conducting international business from their home city, by having payers verbally instruct a banker to rotate (*girare*) credit to their accounts; they then made purchases with the credit newly inscribed in the banker's ledger. Private banks were established on the Rialto as early as 1157 and consolidated by the first government-sponsored bank, the Banco della Piazza di Rialto, which opened in 1587; the second, the Banco del Giro, was instituted in 1619 to finance wars and other public debt. Bankers kept their silver *grossi* and gold *zecchini* at the nearby Palazzo Camerlenghi, the State Treasurer's office; Palazzo dei Diece Savi, on the opposite bank of the Grand Canal, was built in 1520 to house the tax collectors.

The Rialto market area underwent major renewal. Antonio Abbondi lo Scarpagnino reconstructed the porticoed market building, the Fabbriche Vecchie, after devastating fires engulfed the whole island of Rialto in 1514. Jacopo Sansovino designed the Fabbriche Nuove around 1555. In 1591 the Ponte di Rialto became the first stone bridge across the Grand Canal, succeeding a series of wooden bridges. The first one, constructed in 1264, perished in 1310 during an unsuccessful insurrection against the doge; its replacement collapsed in 1444 under the weight of crowds observing a wedding procession. The senate held a competition for a new stone bridge and heatedly debated the merits of the designs submitted, rejecting those of Michelangelo, Palladio, Sansovino, and other illustrious architects before finally settling on the largely functional design of a virtual unknown aptly named Antonio da Ponte (Antonio of the Bridge). The Ponte di Rialto remained the only footbridge spanning the Grand Canal until 1854, when the Accademia Bridge was built. The banking quarter widened out from the Rialto Bridge along the Merceria, Venice's busiest thoroughfare and the shortest route from the Rialto to San Marco.

After the fall of the Venetian Republic in 1797, the character of the Rialto district changed dramatically as tourism supplanted international trade. Shipping and industry later shifted to Port Marghera and Mestre on the mainland, and hotels and souvenir shops sprang up along the Grand Canal. More recently, rising tides are also affecting the Rialto district; one startling indication of this is a marker on the Ca' Loredan, near the Rialto Bridge, showing that the palace is "sinking at a rate of four inches every fifty years." Following the devastating floods of 1966, the Italian government, with UNESCO's backing, initiated campaigns to preserve the city's treasures.

Still, the Rialto district remains the vibrant core of Venice's commercial and mercantile activity, a magnet for natives and tourists alike. Throngs gather daily around the Rialto Bridge, the shops of the Merceria, and the Fabbriche Nuove, where the colorful Erberia (fruit and vegetable market) and modern Pescheria are concentrated. The Fondaco dei Tedeschi, Palazzo Camerlenghi, and other familiar landmarks still stand, and the financial quarter continues to occupy the Merceria.

Further Reading: Peter Lauritzen's *Venice: A Thousand Years of Culture and Civilization* (New York: Atheneum, and London: Weidenfeld and Nicolson, 1978) and John Julius Norwich's *A History of Venice* (New York: Knopf, and London: Lane, 1982) trace the development of the Rialto from a marshy lagoon settlement to the commercial heart of a maritime empire. *Venice: The Biography of a City* by Christopher Hibbert (New York and London: Norton, 1989) is a very readable history of the city with lively sections on the Rialto. Frederic C. Lane's authoritative study, *Venice: A Maritime Republic* (Baltimore, Maryland, and London: Johns Hopkins University Press, 1973), details the roles of commerce and sea power in the creation, ascendance, and decline of the Venetian Republic. *Medieval Trade in the Mediterranean World* by Robert S. Lopez and Irving W. Raymond (New York and London: Columbia University Press, 1955) and *Merchant Culture in Fourteenth-Century Venice: The Zibaldone da Canal,* translated by John E. Dotson (Albany: MRTS Press and State University of New York Press, 1994), afford a more personal glimpse into the world of the medieval Venetian merchant.

—Maria Chiara

Venice (Venezia, Italy): St. Mark's Square

Location: The square is located on the eastern side of Venice, at the end of the Grand Canal, making it the entrance to the very heart of Venice. Venice is located in the Adriatic Sea, just off the coast of the northern Italian mainland.

Description: The square of St. Mark's is actually comprised of the square proper (the piazza) and a smaller square adjacent to that (the piazzetta). The buildings that surround the piazza are the church of St. Mark, the Procuratie Vecchie and adjoining clock tower, the Procuratie Nuove, and the Fabbriche Nuove; the piazzetta is the wide corridor between the Doge's Palace (which is adjacent to the church of St. Mark's to the east) and the Library of St. Mark's.

Site Office: I.A.T. (Uffici di Informazioni e Accoglienza
Turistica)
San Marco
CAP 30124 Venice, Venezia
Italy
(41) 5226356

In many ways, the square of St. Mark's represents the Venetian Republic itself. To begin with, the mixture of eastern and western architecture of the buildings that comprise the square accurately reflects Venice's initial suzerainty to, and gradual secession from, the Byzantine Empire. Then, too, the layers of adornment that cover the facades of both the St. Mark's church and the Doge's Palace tell the story of the tremendous wealth Venice amassed throughout the Middle Ages. Finally, the overwhelming number of great works of art contained in these buildings attests to the republic's impact on the visual arts. Even the eventual fall of the republic to Napoléon's army in 1797 is represented by the palace Napoléon had built on the square, the Fabbriche Nuove. Indeed, the history of Venice's rise to power is almost inseparable from the growth of the square itself.

In 679, the twelve lagoon communities living in and around what we now call Venice declared themselves a republic, formally created a parliamentary body, and elected a *dux* to rule the republic; over the years, this Latin term was bastardized to "doge" in Italian and "duke" in English. In 774, under attack from Frankish forces, the republic moved its capital from Malamocco to Rialto—the site we now refer to as Venice. Agnello Parteciaco, the first doge to be elected on Rialto, ordered a palace to be built there. The Doge's Palace would grow and be rebuilt several times in the centuries that followed.

Agnello's son, Giustiniano, was elected Doge after his father and played a crucial role in the evolution of the Venetian Republic and of the square of St. Mark's. It was Giustiniano who, in 828, vowed to build a magnificent church for the remains of St. Mark when that saint's relics were secretly smuggled out of Alexandria. It was also Giustiniano who decided to change the patron saint of the republic from the Byzantine warrior saint, Theodore, to St. Mark. Symbolically, this change marked Venice's secession from Byzantine suzerainty. Giustiniano died, however, before the church was completed, and it fell to his brother, Doge Giovanni Parteciaco, to oversee the completion of the church.

In the early Middle Ages, a region could gain political clout by laying claim to the relics, and thus the patronage, of a particular saint. It was quite a coup, therefore, when Venice was able to obtain the relics of St. Mark. It seems that the prince of Alexandria wanted to build a large palace, and needed to tear down the chapel housing the remains of St. Mark in order to do so. A monk named Staurazio and a priest named Theodore, not wanting the sacred remains to be profaned, arranged with two merchants, Buono of Malamocco and Rufino of Torcello, to transport the remains to Venice. The body was placed in a basket and covered with pork to fool Egyptian customs. The relics arrived safely in Venice in 828 and were temporarily placed in a little wooden chapel until construction of the Doge's Chapel—which would eventually become the church of St. Mark's—was completed.

The Doge's Chapel was the first of three churches to be built on that site. It was finally completed in 836, St. Mark's relics were entombed, and the basilica was consecrated. Until recently, scholars have assumed that the first church looked significantly different from the present church; however, an archaeological survey done in the 1950s by Ferdinando Forlati, the procurator of the basilica, shows that the original church was similar in design to the current one. Like its successor, the first church was built on the pattern of a Greek cross; its architecture is supposed to have been Byzantine. Unfortunately, the first church was destroyed by fire.

In 976 Doge Pietro Candiano IV was attempting to convert the dogeship to a hereditary position and to increase the doge's power. The people rose up against Candiano, setting fire to the palace and the church. Pietro, together with his son, was captured in the atrium of the church while they were trying to escape; both were killed on the spot. St. Pietro Orseolo I paid for the reconstruction of both church and palace. Then, in 1063, Doge Contarini decided to "upgrade" the basilica. This "upgraded" version, the third church to be built, is the church which stands today in the square.

The design was a cruciform, as had been the previous church, made truly spectacular by five cupolas—one at each endpoint of the cross and one at the point of intersection. The combination of the cruciform and the cupolas was borrowed from the Church of the Holy Apostles in Constantinople. As

St. Mark's Square
Photo courtesy of The Chicago Public Library

many critics have pointed out, the semi-oriental character of Venice is reflected in St. Mark's, perhaps the most famous building in a city famed for its architecture. The structure was completed under Doge Domenico Selvo, and the Venetians set about facing the structure.

The elaborate facing is one of the most remarkable features of the Basilica. The entire architectural structure is covered by mosaics. In all, there are more than 8,000 square meters of mosaic covering St. Mark's. There are five mosaics on the facade alone. The central mosaic depicts the story of Christ as judge, while the four lateral ones portray the various deeds of St. Mark. This awesome facade commands the attention of all eyes in the square, making it, quite naturally, the focal point of the square.

Doge Domenico Selvo died, Doge Vitale Falier was elected, and at length, the facing was completed. In 1094 the new basilica was finally consecrated. Roughly a century later, the atrium was built. From this point forward, doges and wealthy merchants lavished incredible amounts of wealth on St. Mark's, reflecting the massive wealth that Venice was rapidly accumulating. Doge Falier commissioned the Pala d'Oro (Golden Altarpiece) in Constantinople in 1105 for the main altar. Although the bejeweled altar has been plundered somewhat over the years, it originally contained 1300 large pearls, 400 garnets, 90 amethysts, 300 sapphires, 300 emeralds, and 15 rubies. Throughout the basilica there are over 500 decorative columns (usually of rare marble) of normal size and more than 2,000 smaller ones.

The church also benefited greatly from the republic's military prowess. In 1201, Venice put its resources at the disposal of the Fourth Crusade. All the forces gathered first in St. Mark's before venturing forth, and when they did set sail for the east, it was under the command of ninety-three-year old Doge Enrico Dandolo. When Constantinople fell to the crusaders in 1204, Venice received the lion's share of the spoils. Dandolo decided to take with him from Constantinople a very large bronze sculpture of four Roman horses, referred to today as the famous quadriga of St. Mark's. Initially, the majestic quadriga was placed over the main portal. Today, however, a copy sits atop the portal, while the original is on display in the museum of St. Mark's. In 1344 Petrarch himself had been invited to the church as Doge Andrea Dandolo's guest to witness the surrender of Crete; referring to the quadriga, the great poet said in amazement, "they seem to neigh, and to paw the ground with their hooves."

Again in 1258, Venice's military success translated into a boon for St. Mark's. The Pillars of Acre and the Pietra del Bando, both trophies of Venice's successful war against its arch-rival Genoa, were brought to Venice from Acre in 1258 and placed next to the side entrance of St. Mark's. In short, as one scholar put it, "Everything that could serve to embellish San Marco was looted, from enamel plaques for the enlargement of the Pala d'oro and precious relics, reliquaries, chalices, icons, etc. for the Tesoro to reliefs for the walls of the church and the gilded horses which were placed,

after a brief sojourn at the Arsenal, on the facade of San Marco."

At the end of the fourteenth century, the Rood screen, made of rare marble, was erected, dividing the church from the chancel, and separating the baptistery from the chapel of St. Isidore; and at the end of the fifteenth century, Angelo Spavento added the vestry and the chapel of St. Theodore. In 1575, many of the mosaics were systematically replaced, resulting in a loss of two thirds of the original mosaics. The suspected author of this iconoclasm was no less an artist than Titian.

The actual tomb of St. Mark has been moved around quite a bit (and was even "lost" for roughly eighty years during the tenth and eleventh centuries), but today his coffin rests in the presbytery, in an altar behind panels of green parian marble. The altar is covered by a large ciborium—a marble canopy supported by four oriental alabaster columns; the columns probably date from the fifth century. From Doge Domenico Selvo's death in 1084, until Doge Andrea Dandolo's death in 1354, most doges were also buried in St. Mark's; after Dandolo, however, it was decreed that there could be no more burials of any kind in the basilica.

Over the centuries, the basilica has witnessed many historic events. In 1177, the reconciliation between the Holy Roman Emperor Frederick Barbarossa and Pope Alexander III took place in St. Mark's, mediated by Doge Sebastiano Ziani. In 1339, the peace between Venice and Verona was forged at the altar in St. Mark's. In 1377, Verttore Pisani was liberated from his imprisonment and led by the people to that same altar, where he was vested with the supreme command of Venetian forces in the war with Genoa.

The Doge's Palace is connected to the church via a door near the southern transept. Built on the site of the original Doge's Palace, the building that now bears that name is actually a structure incorporating three buildings that had previously been separate, but neighboring, structures: the Palatium Ducis (the doge's Byzantine mansion, built in the early eleventh century); the Palazzo Pubblico (the seat of the supreme parliamentary body called the Maggior Consiglio, or Grand Council); and the Palazzo di Giustizia (the law courts). Merging these three entities in one building reflects the Venetian attempts to streamline government in the fourteenth century.

The palace was built between 1340 and 1424; it was begun in the Gothic style, and finished in the Renaissance style. The first wave of construction concentrated on the Sala de Maggior Consiglio (the heart of the Venetian government, and the true seat of power), and was finished by 1365. In 1424, the newly elected Doge Francesco Foscari decided to rebuild the wing of the palace adjacent to the Grand Council because it was in need of repair. The palace facade is lined with slabs of pale rose and white marble, and is "broken up by two magnificent Gothic arcades of 107 columns—36 below and 71 above—and 14 large pointed windows."

The famous Porta della Carta (literally, "door of paper"), which provides an entrance into the palace, was finished in 1441. This entrance leads to a courtyard adorned by two large well-heads. Sometimes at the end of Carnival, the doge would stage bullfights in the courtyard. At one end of the courtyard stands the well known Scala dei Giganti (Giants' staircase), the staircase leading to the loggia. In 1556 Jacopo Sansovino executed the two huge sculptures of Mercury and Neptune that adorn the Giants' Staircase, giving it its name. It was this staircase the doge would descend after his election to show himself to the townspeople.

In the middle of the loggia is another staircase, the Scala d'Oro (the Golden Staircase), so named because of the extensive encrustation with jewels and gold leaf on and around the staircase. The first flight leads to the doge's apartments; the second flight leads to the Sala dell quatro porte (the Room of Four Doors), beyond which lies the very heart of the palace.

A devastating fire swept through the palace in 1577, and is considered "the most serious disaster which the palace suffered." In the wake of the fire, fifteen of Venice's leading architects were consulted as to how to rebuild. Some wanted to scrap the whole structure and start over, but the more prudent minds realized that the structure was still sound, and recommended simple reconstruction. The latter plan was adopted and swiftly executed .

Spatially, the church of St. Mark's faces the piazza, while the Doge's Palace, which is adjacent to the church and extends from the southern transept of the church to the water's edge, looks out over the piazzetta. Directly across from the palace, on the other side of the piazzetta, is the Library of St. Mark's. This building was designed by Jacopo Sansovino in 1536, but was not completed until 1588, after Sansovino had died. The Sansovino Library is considered one of the most perfect pieces of Renaissance architecture still in existence.

At the beginning of the piazzetta, almost at the water's edge and at the farthest point from the piazza, stand two imposing pieces of sculpture upon two very tall columns. One sculpture is of St. Theodore, the former patron saint of the city, and the other is of the winged lion that is symbolic of St. Mark. These columns form a gateway of sorts to Venice and are of symbolic significance. As scholar Harry Alt puts it:

"The two mighty emblems on the giant Greek columns of the Piazzetta which seem to guard the entrance to the seat of government and holy of holies, are symbolic of the origins of Venice; for the marble figure of St. Theodore, the patron saint of the Byzantine hosts, embodies the power under whose aegis the city grew up on the barren islands; the bronze chimera on the opposite pillar, as the lion of St. Mark, is emblematic of the heavenly mediator whose banner Venice bore over land and sea."

In 1617, the unraveling of the "Spanish Plot" created the bloodiest scene the piazzetta has ever witnessed. Spain, with the aid of Venetian co-conspirators, planned to land an army on the piazzetta during Ascension Day. Because of a Venetian ritual known as the "Wedding to the Sea" ceremony, the entire Venetian fleet would have been unprepared to fight. Spain planned to destroy the palace and the mint and take over the Great Council. However, a careless Venetian senator left a compromising note in a church, which eventually reached the Venetian authorities. By the next morning, a string of corpses was strung between the columns of St. Mark and St. Theodore; overall more than 300 traitors were executed.

The square proper, called the piazza, was provided for in 1172 when the farsighted Doge Sebastiano Ziani ordered several buildings demolished in order to clear the area in front of the basilica. Over the centuries, the piazza evolved. Although the tall clock tower known as the Campanile had been in place since the ninth century, its imposing marble top was not added until 1514. Then, in 1517, the Campanile was crowned with a bronze angel. The tower that exists today, however, is not the same one that was built in the ninth century; in 1902, the tower collapsed, injuring no one, and was promptly rebuilt.

Also in 1517, Pietro Lombardo's offices of the Procuratie Vecchie (the Procurator of St. Mark's), which stand at the corner of the piazza at a right angle to the basilica, were finished. The adjoining clock tower, which stands closest to St. Mark's and chimes the hour with the help of two bronze figures on top of the tower, was finished a few years earlier, in 1499. Finally, in 1584, Scamozzi and Longhena's huge Procuratie Nuove, which stands directly across from the Procuratie Vecchie, and which was the last building to be built in the square until Napoléon's corruption of it in 1797, was finished.

Of all the historic events that transpired in and around the square, perhaps none were as stunning or as final as those that took place in May and June of 1797. Feeling tremendous pressure from Napoléon, the Great Council met for the last time on May 12, 1797. The next day, placards hung in the piazza informing Venetians that the patricians had voluntarily renounced their hereditary rights, that the government had been disbanded, and that the republic had been dissolved; in June, there was a bonfire in the piazza into which the doge's clothes were symbolically thrown. Although the republic came to an end, its famed square remains as a testament to its former glory.

Further Reading: John H. Davis' *Venice* (New York: *Newsweek* Book Division, and London: Reader's Digest, 1973) is a very accessible source providing ample overview material and useful specific information about the buildings in the square. Also helpful, if oriented more for the tourist than the historian, is Harry Alt's *Venice: A Splendid City* (Munich: Wilhelm Andermann Verlag, 1955). Otto Demus' *The Church of San Marco in Venice* (Washington: Dumbarton Oaks, 1960) is a clear and very scholarly account of the church's history, architecture, and sculpture, while P. Angelo M. Caccin's *St. Mark's: The Basilica of Gold*, translated by Richard Creese-Parsons (Venice: Edizione Zanipolo, 1968) offers an equally complete account which is perhaps more readable. Also of interest is Maria Da Villa Urbani's *St. Mark's Basilica* (Milan: Kina Italia, n.d.), although this source is more useful for its pictorial representation than its actual text. Finally, for an in-depth account of the history of the Doge's Palace see Elena Bassi and Egle Renata Trincanato's *The Palace of the Doges in the History and Art of Venice* (Milan: Aldo Martello Editore, 1966.)

—Lawrence F. Goodman

Vergina (Hematheia, Greece)

Location: In northern Greece, adjacent to the modern town of Veroia; twelve miles inland from the northern tip of the Thermaic Gulf (Gulf of Salonika), and roughly forty miles southwest of Thessaloníki. In ancient times, Vergina was surrounded by the kingdoms of Illyria to the west, Thessaly to the south, and Thrace to the north and east.

Description: The city, originally named Aigai (alternately spelled "Aegae"), was the ancient capital of Macedon (Macedonia). It stands at the foot of the Vermion mountains to the south and west and overlooks the vast Haliakmon plains to the north.

Site Office: Vergina Archaeological Site
Vergina, Hematheia
Greece
(31) 830538

The ancient name for Vergina, "Aigai," translates as "land of goats," but the city was initially founded for its strategic value rather than its game. During the eighth and seventh centuries B.C., the area that would become the kingdom of Macedon was dominated by Illyrian tribes. Aigai was strategically positioned at the foot of the mountains, and thus sheltered from attack, while the location also provided ample opportunity to survey the Haliakmon plains. For these reasons, the Illyrian tribes established a base at Aigai. When in the early seventh century the indigenous tribes of Thracians and Paionians rose up against the Illyrians, the Illyrians pulled out.

In approximately 650 B.C., the descendants of the Temenid family that ruled Peloponnesian Argos staked their claim as rulers of the area. Led by Perdiccas I, the first of the Argead dynasty to rule Macedon, a capital was set up at the Illyrian site in Aigai, and Macedon was born. At some point after the Roman conquest—no one knows quite when—the name was changed to Vergina, after a certain queen of Beroea. This name change created quite a problem for modern historians. It was known from extant literature that Aigai was the site of the royal tombs of Macedon, of Philip II's assassination, and of the royal palace. However, there were no written records directly linking Aigai to Vergina, or to any other city for that matter. Other towns including Edessa and Verroia (Beroea) had also been suggested as possible sites for the location of Aigai.

The mystery was solved by two scholars, Professor N. G. L. Hammond and Professor Manolis Andronicos, whose separate research, taken together, confirmed that Vergina was none other than the ancient capital city of Aigai. Hammond had, on mostly topographical evidence, suggested

in 1972 that Vergina was Aigai. When Andronicos discovered in 1977 what appeared to be royal tombs at Vergina, confirmation of Hammond's theory promised to be close at hand. All the other tombs that had been discovered in and around Vergina had been thoroughly looted, making it impossible for scholars to determine whether or not the tombs contained royalty. The subterranean tombs Andronicos discovered, however, were sufficiently intact for him to confirm that they held royalty and that they dated from the middle of the fourth century B.C. Here at last were the Macedonian royal tombs; the supposition that Vergina and Aigai were one and the same was confirmed.

Andronicos and his team excavated an area that is now referred to as the "Great Mound" in search of tombs he hoped would be found underneath. The artificial mound is 361 feet in diameter and 49 feet high. It is made up of successive layers of waterproof clay, packed earth, and compacted rock, the lowest layer of which was ten feet thick. Underneath this large mound, Andronicos did indeed find three tombs, each covered with its own smaller tumulus. That these tombs held royal personages from the fourth century B.C. has been proven; pinpointing the identity of these personages, however, is a matter of some debate.

The first tomb, Tomb I, is a large cist-tomb decorated with three frescoes. It measures eleven and one-half feet long by seven feet wide by ten feet high and is covered with removable flat slabs of porous stone. The subject of the frescoes is Pluto's rape of Persephone and Demeter's consequent mourning. Two theories have been proposed about who was entombed here. Green, believing that the person in Tomb I must have been a woman (he claims, but does not cite, forensic evidence), argues that the only truly acceptable candidate is Phila, one of Philip II's wives. She would have been in her thirties when she died, so Green believes that the story of Persephone was chosen for the frescoes in order to illustrate the plight of a woman cut down in her prime. On the other hand, Hammond makes a more convincing case that the entombed individual was Amyntas III, Philip II's father.

Hammond reminds us that among Andronicos's findings was a shrine adjacent to Tomb I. This shrine measures thirty-one feet by twenty-six feet, and was probably, Hammond conjectures, used to worship a dead king as a god, there being some precedent for this kind of idolatry in fourth-century Macedonia. Since we know that the tomb dates from the middle of the fourth century, the only possibilities for royal occupants would be Philip II (who is almost certainly buried in Tomb II) and Amyntas III. Furthermore, it would make sense that Amyntas III would be buried in Tomb I, the oldest of the three tombs. Amyntas III was the first of his side of the Argead dynasty to be king, and he was a very well respected king. That he would have been "rewarded" at his

Golden wreath unearthed at Vergina
Photo courtesy of Greek National Tourist Organization

death by inaugurating a new section of the royal burial ground at Aigai is probable. With Amyntas III as the corpse, the interpretation of the frescoes changes as well. Hammond interprets the Persephone and Demeter frescoes to prophecy a life after death for King Amyntas.

The second tomb, Tomb II, is not a cist-tomb, but a barrel-roofed tomb complete with a facade and a door. Stylistically it is what is known as a "Macedonian tomb," made of porous stone and plaster. Tomb II consists of two chambers: an antechamber in which the bones of a middle-aged woman were contained within a funerary chest, and a main chamber in which a man's remains were wrapped in purple cloth and placed inside a gold funerary chest that was itself inside a marble sarcophagus. Each chamber is fourteen and one-half feet wide and has a vaulted roof. The facade, Doric in style, with a half-column on either side of the doorway, is inset to support an architrave, above which is a fresco portraying a hunting scene. About the identity of the remains in Tomb II there is more (yet still not total) agreement. While there is some dispute pertaining to the female, it is fairly well agreed that the male is none other than Philip II, the great king of Macedonia, and father of Alexander the Great. Some scholars suggest that the remains may be those of Philip III Arrhidaeus, Alexander the Great's half-brother, but the evidence rests on the side of Philip II.

The tomb appears to have been sealed before the plaster had a chance to dry, indicating that the burial was conducted in haste. This would certainly have been true following Philip II's assassination. Furthermore, in excavating the tomb, Andronicos found the remains of two men in the small tumulus covering the tomb. According to the extant written accounts, these would correspond to the bodies of the king's two assassins, reported to have been cremated on top of Philip II's remains. Other contents, such as a set of body armor and five miniature heads depicting members of Philip II's family, further support the claim that this is the tomb of Philip II. The only substantive debate involves the identity of the female corpse. It is almost certainly one of Philip's wives (of which there were seven), but scholars are split between those who consider it to be Meda, Philip's next-to-last wife, or Cleopatra-Eurydice, his last wife.

Tomb III, another barrel-roofed "Macedonian tomb," contains bones that have been shown to belong to a young boy, aged between twelve and fourteen. It is generally agreed that the boy in question is Alexander IV, son of Alexander the Great, who was executed (as was his mother, Roxane) by Cassander. Tomb III also contains a painting that is of some interest: on the main wall is a mural depicting a two-horse chariot race.

Taken en masse, all the evidence suggests the following chronology. In 369/368 B.C., following the death of his father, Amyntas III, Alexander II (Philip II's older brother) built Tomb I to receive the remains of his father, the first member from his branch of the Argead dynasty to become king. A separate area of the royal cemetery was chosen to honor Amyntas III. To further honor him, a shrine was built

in order to worship him, even in death. In 336, following the assassination of Philip II, Alexander the Great built Tomb II for *his* father, placing there many fine offerings. The king's chamber of Tomb II was shut before the inside was entirely finished because Alexander was in a great hurry to return order to the Greek states, and to implement the plans for the invasion of Asia. Finally, after Alexander the Great died in Asia, and his son was executed, the boy was placed in a nearby tomb. As Alexander IV was the last of Amyntas III's line of the Argead dynasty, it is possible that the Great Mound was placed over all three tombs to commemorate this line of Temenid kings.

Eleven years after Professor Andronicos's discovery of the royal tombs, he unearthed a fourth tomb, widely believed to be the tomb of Philip II's mother, Eurydice. The tomb of Eurydice was another two-chamber Macedonian tomb. The antechamber is eight feet deep and fourteen and three-quarters feet wide, while the main chamber is eighteen feet deep, eighteen feet wide, and nineteen feet in height. Just as it was on the facade of Tomb I, the rape of Persephone again provides the subject matter for a painting on the tomb's walls (the continuity, of course, further suggesting that the corpse in Tomb I was Amyntas III, Eurydice's husband). The tomb also contains an exceptionally rich throne in the main chamber.

All four tombs provided a wealth of artifacts and archaeological information about Macedon, and thus about northern Greece. Until Professor Andronicos's discoveries, the main sources of information about Macedonia were Athenian. As Athens and Macedonia were always antagonistic toward each other and at times outright enemies, information derived from the Athenian sources was certainly prejudiced. The discovery of the royal tombs helped to change this. As the scholar René Ginouvès writes, "since the discovery by Manolis Andronicos of the royal tombs at Vergina from 1977 onwards, a change of perspective has taken place in favour of Macedonia."

Vergina is also historically significant as the site of Philip II's assassination. He was killed in 336 B.C., and the story surrounding his assassination is remarkable for both its melodrama and its intrigue. Philip, who had been king since 359 B.C., had far surpassed any of his predecessors in his imperial ambitions for Macedon. Under his rule, the empire grew considerably, and in 336 the time was ripe for his most ambitious move yet: an invasion of Asia. A preliminary force was sent to Asia, but before Philip personally led the full Macedonian forces overseas, he planned a huge send-off in Aigai.

Although the gathering was occasioned by the marriage of his daughter, Cleopatra, to the Molossian king Alexander, Philip intended to use the gathering to build momentum and enthusiasm for the upcoming Asian campaign. Guests and dignitaries from across Greece came to Aigai to honor the king. So intent was Philip on using the wedding to catalyze military zeal that he sent a page to the oracle at Delphi, hoping to get a slogan from Apollo that he could use later as a rallying cry on the battlefield. Ominously, the following was returned: "The bull has been garlanded, the end is come, the sacrificer is at hand."

The end, at least for Philip, did indeed come. At dawn on the morning following the wedding ceremony, there was to be a ceremony in the stadium as a preface to an athletic competition. The stadium filled up, and the official procession entered led by Philip dressed in a white cloak. Suddenly a young man, one of the king's bodyguards, rushed forward, stabbed Philip with a Celtic sword, and fled toward the city gate where he had horses waiting for him. He may very well have escaped had he not tripped and fallen, at which point he was overtaken and killed on the spot.

Although there have always been people who believed that there was some conspiracy to kill Philip, evidence suggests that the assassin, a bodyguard named Pausanias, killed Philip out of personal vengeance: a year earlier Pausanias had been the king's lover, but was replaced in Philip's affections by another young boy (also named Pausanias). The first Pausanias began to taunt the second Pausanias rather severely. The second Pausanias became distraught and told Attalus, a senior and influential noble, that he could not bear the taunts, and that he would commit suicide rather than suffer them. Shortly thereafter, the second Pausanias stepped in front of a blow intended for the king during a battle with Pleurias (the Illyrian king). Attalus blamed the first Pausanias for the second's apparent suicide. He thus invited the surviving Pausanias over, plied him with liquor until he passed out, and then gave Pausanias to his muleteers to punish. Upon regaining consciousness, Pausanias was so upset that he went to Philip seeking vengeance on Attalus. But since Philip needed Attalus's political support, he sought to appease Pausanias with gifts rather than with action. For this, Pausanias swore he would kill the king, which he did one year later.

This story is reported by Aristotle (who was Alexander's tutor and certainly a reliable source). Justin's version of the story has Olympias, Philip's disgruntled wife, encouraging Pausanias in his regicidal plans, thereby becoming a co-conspirator; however, Justin's account is almost universally considered too incredible to be true. Two other co-conspirators, Heromenes and Arrhabaeus, were executed; their bodies are said to have been cremated upon Philip's tomb in order to purify his spirit. When Professor Andronicos published his findings of Tomb II at Aigai, he produced physical evidence of the remains of two men within the tumulus of what is believed to be Philip's tomb. By verifying key parts of the literary account, this evidence increases the veritability of the whole.

The third historically significant feature of Vergina is the royal palace that was built there, the Palace of Palatitsa. The ruins of this palace were first discovered by a French expedition in 1861. Because palaces had fallen out of favor in democratic southern Greece, the Macedonian palaces are something of a novelty in Greek culture. The building plan for Palatitsa is much like a house (albeit in grand style) with an internal courtyard. The main building is rectangular, measuring 343 feet by 290 feet. According to Ginouvès, this makes the palace, as a residence, of "a size without equal in Greek architecture." Within the palace lies a courtyard roughly 21,500 square feet. It was surrounded by a colonnade with sixteen columns to a side.

The palace had two stories, with a staircase at the west end providing access to the second level. The ruins suggest that a mix of architectural styles was used, and that the architecture leading from one room to the next was truly regal. One of the banquet rooms had a fairly large roof (roughly 3,230 square feet), pieces of which have been found, that apparently needed no supports (none have been found). This suggests that Macedonian architecture was fairly well advanced. In one of the rooms, which may have been a sanctuary, was found an inscription to Heracles (Hercules), who was believed to have been half human and half god, and who was also the mythical ancestor of the Temenid kings.

Further Reading: Unfortunately, Professor Andronicos's seminal report on the royal tombs at Vergina has not yet been translated into English, but for an overall account of Vergina's place in Macedonian history, with solid and readable accounts of the physical structures there, see *Macedonia: From Philip II to the Roman Conquest*, edited by René Ginouvès and translated by David Hardy (Princeton, New Jersey: Princeton University Press, 1994). Also helpful when investigating the royal tombs are two articles, N.G.L. Hammond's "The Evidence for the Identity of the Royal Tombs at Vergina," and Peter Green's "The Royal Tombs of Vergina: A Historical Analysis," which can both be found in *Philip II, Alexander the Great and the Macedonian Heritage*, edited by W. Lindsay Adams and Eugene N. Borza (Washington, D.C.: University Press of America, 1982). If, however, one wants to read accounts that tie the history more closely to the two great Macedonian leaders, Philip II and Alexander the Great, see N. G. L. Hammond's very clear and quite complete *Philip of Macedon* (Park Ridge, New Jersey: Noyes, 1980; London: Chatto and Windus, 1981; revised, Baltimore: Johns Hopkins University Press, 1994); or J.R. Ellis's somewhat more narrative account in *Philip II and Macedonian Imperialism* (Princeton, New Jersey: Princeton University Press, 1986; London: Thames and Hudson, 1987).

—Lawrence F. Goodman

Verona (Verona, Italy)

Location: On the banks of the Adige, at the base of the Lessini hills, Verona lies at an important junction in northern Italy, on the main route from Germany and Austria to central and southern Italy, and on the route across northern Italy from Venice to Milan and Turin. It is not far east of the southern end of Lake Garda.

Description: Verona's old town center is filled with beautiful buildings, many in the characteristic "peach-blossom" marble. It is famed for its well preserved amphitheatre and contains a number other Roman remains. San Zeno is among the most renowned Romanesque churches in Italy. Much of Verona's interest lies in its wealth of medieval buildings, both secular and religious, for instance those around the Piazza delle Erbe and Piazza dei Signori, dating to the times when Verona held its greatest power. Off the Piazza delle Erbe is the so-called Casa di Giulietta connected to the story of Romeo and Juliet. Verona is the second most important city of the Veneto region after Venice.

Site Office: A.P.T. (Azienda di Promozione Turistica)
Piazza Erbe, 38
37121 Verona, Verona
Italy
(45) 8000065 or 8006997

According to Pliny the Elder, Verona was first settled by the Euganei and the Reti. Others have claimed that the Gallic Cenomani under Duke Brenno were the original settlers. Around 100 B.C., the defeated Cimbri may have established colonies across what later became the Veronese district.

The settlement, whoever its founders may have been, received a Roman franchise in 89 B.C. The lyric poet Catullus lived in Verona in the first century B.C., as did the Roman architect Vitruvius, author of *De architectura*, the only writing on Roman architectural theory to have survived; Vitruvius had a profound influence on Renaissance architects.

Under the Roman Empire, Verona became a colony called Colonia Augusta. Verona thrived, as the few great remaining monuments attest, especially the amphitheatre from the first century. Its 44 tiers could contain an audience of over 20,000. A Roman theatre also survives, together with a couple of monumental gates, while the Piazza delle Erbe was the location of the Roman Forum. The amphitheatre lay outside the initial Roman town wall, but under Emperor Gallienus it was incorporated into the new walls. Constantine the Great won a battle against Maxentius's troops at Verona in 312.

The first Veronese Christian martyrs, the saints Fermo and Rustico, probably died sometime during the reign of Diocletian. Verona certainly had one of the few north Italian bishops prior to the fourth century. The best known early bishop was St. Zeno, whose status in Verona equaled that of a patron saint. He converted many by his example, and after he died around 380, legends grew of miracles he had performed, as well as a widely held belief, owing to his North African origins, that he was a negro. The Veronese bishops stood under the archbishop of Milan at first, but from the fifth century they looked to the Patriarchate of Aquileia.

The Ostrogoth king Theodoric defeated the barbarian Italian ruler Odoacer at Verona in 489. Ravenna became Theodoric's capital, but he lived in part in Verona. He may have had the church of San Stefano dismantled to put up new fortifications. His reign saw much building. Traces of his palace have been found on San Pietro Hill. Arcades and public baths were built and the aqueduct repaired.

The leader of the Lombard invaders, Alboin, arrived in 569 and was killed in Verona in 572. Paul the Deacon, an eighth-century historian, recounts how Alboin killed an enemy named Cunimund and took his daughter Rosamund as his wife, forcing her to drink from a cup made from her father's head. To avenge herself, Rosamund got one of Alboin's men, Peredeo, to kill Alboin in his bed. The story has been adapted in innumerable versions. Verona's brief period at the center of the Lombard kingdom came to an end once Pavia became the capital in the 620s. Verona itself became one of thirty-six duchies.

Charlemagne's Frankish expansion put an end to Lombard authority. The Franks took Verona in 774. While Pavia continued as the capital of the region, Charlemagne's eldest son Pépin favored life in Verona. A twelfth-century manuscript records that Pépin ordered the moving of St. Zeno's body from its humble chapel to a beatiful, specially commissioned church, possibly on the site of the present abbey. An eighth-century poem eulogizing Pépin's government of Verona describes a city encircled by high walls including forty-eight towers and containing many fine buildings, with nearly thirty churches situated around it. Carolingian times saw learning flourish; Charlemagne's grandson Lothair founded a series of schools, including one in Verona. Pacificus, a scholar of Hebrew and Greek, was a notable Veronese archdeacon of the ninth century. He collected manuscripts, which were donated to the Veronese cathedral and formed the start of the present Biblioteca Capitolare. Carolingian government was imposed by imperial officials, counts, and *scabini* (judges), elected from outside the town.

Berengar, duke of Friuli, became king of Italy in 888, making Verona his capital for a time, but the Franks drove him out briefly from 901 to 905. Berengar, after bribing the

View of Verona, with Roman amphitheatre at center
Photo courtesy of A.P.T., Verona

city guards to let him retake Verona, then ruled Italy until 922. At the end of his reign, he had lost his Italian kingdom to Rudolf II of Burgundy and had to retreat to Verona, where he was killed by the people in 924. His vassal Milo tracked down the leaders of this uprising. Six years later Milo was officially appointed count of Verona—an executive title that did not carry territorial claims, although he dominated the city's life until his death in 962.

Milo shifted his allegiance among the various pretenders to the Italian crown. When Arnold of Bavaria took the kingdom in 935, both Milo and the Veronese bishop Ratherius supported him. Ratherius was politically active, a reformer who fulminated against heresies. Milo switched support to Hugo of Provence, but Ratherius stuck to Arnold, leading to his imprisonment.

Around mid-century, the German Otto came to claim the Holy Roman Empire and made a settlement in the 952 Compromise of Augsburg with the new king of Italy since 949, Berengar II. The northeastern section of the Italian kingdom, the Mark of Verona and Aquileia, was transfered to the German kingdom, securing the route from Germany through to Italy via the Brenner pass and the Chiusa defile (the latter only ten miles north of Verona), the only way in these times by which a large army could enter.

The Mark was theoretically governed by three figures, a count, a marquis, and a duke, in ascending order of power. This situation only really lasted the ten years Milo continued to exercise power, the count controlling affairs largely independently, especially since, after Ratherius, the bishops of Verona, unlike those in other Lombard cities, were unable to command political power. In 967 Otto had convened an imperial diet at Verona, handing temporal control of the north Italian towns to the bishops, with the exception of Verona, perhaps an indication of Verona's strategic importance, perhaps also to avoid giving Ratherius, recently returned, too much control.

Another diet was held in Verona in 983, when Otto II made the gathered princes agree to elect his three-year-old son Otto king of Germany and Italy. After him, the Veronese supported Henry II and Conrad the Salic, and in the battle over investitures between emperor and pope they nearly always supported the emperor. The German court elected one German Veronese bishop, Cadolaus, as the schismatic Pope Honorius II in opposition to Alexander II in 1061, although his challenge only lasted three years.

Imperial control of official appointments in northern Italy had rapidly disappeared toward the end of the eleventh century, and the initially republican city communes began to emerge. The first documented evidence of Verona asserting its independent status is in an 1107 commercial and military treaty with Venice. Titled local officials, three consuls, first appear on a document of 1136. The consuls were the leading officials of the north Italian independent towns.

Trade flourished under the commune. Verona's production of high-quality woollen cloth became renowned throughout Italy. Each branch of trade and each handicraft

had a representative organization, the *ars* or art, headed by a *gastaldo*. The merchants' art quickly established extraordinary influence over the others, controlling them and in many respects representing them. It gathered such power that it became effectively an independent authority in the town. In the second half of the century, the republican nature of the communes was reduced in favor of more centralized government, with the top official often given the title of rector, a precursor of the *podestà*.

The involvement of the Holy Roman Empire in northern Italy was to be dramatically revived with the election of Frederick Barbarossa as Emperor Frederick I in 1152. Verona was exceptional in its almost unwavering support for the imperial, or Ghibelline, cause against that of the pope and the Guelphs. Verona's support of Barbarossa proved important.

When Frederick arrived for the first time in Italy in 1154, Verona was among his most evident supporters. Distrust arose, however, when, as Frederick was returning to Germany in 1155, rumors spread that a temporary bridge to carry his army over the Adige had been weakened deliberately by some Veronese citizens for a possible attack. Then at the strategic Chiusa pass, a Veronese rebel, Alberico, who controlled the pass, barred the way, demanding substantial payment. Frederick would not yield to such blackmail and with the help of two Veronese leaders took the fort. The incident seemed minor, but the next winter the Veronese bishop Tebaldo journeyed to the imperial court to placate Frederick, who accepted reconciliation with a large fine. Frederick came to Italy again in 1158, this time well received in Verona. At the Diet of Roncalia, he got the northern Italian cities to accept the imperial appointment of their top officials.

The tyranny of the imperial appointees, Frederick's support for Pope Victor IV during the schism while Bishop Tebaldo was a close friend of Pope Alexander III, and the brutal suppression of the Milanese after Frederick took Milan in 1162 led to the formation of regional alliances against Frederick. The *Societa Veronensis*, or Veronese League, was founded in 1164, allying Verona, Vicenza, Padua, and Venice.

Frederick amassed troops to lay waste to Veronese territory, but the allies' army proved just as strong, and Frederick withdrew. The great Lombard League of fifteen towns was created in 1167; cities such as Milan, Cremona, Brescia, Mantua, Ferrara, and Treviso combined with the Veronese alliance. The league fought extremely effectively to combat imperial power, culminating in the crushing of the imperial troops at Legnano in 1176. The cities regained powerful rights and appointments, confirmed by the Treaty of Constance in 1183. Frederick kept certain rights, for instance to invest consuls every five years and to exercise, via an envoy, appellate jurisdiction over serious cases, and he accepted a general pardon in exchange for unchallenged passage to Rome. Frederick's visit to Verona in 1184 solidified the peace, and he met with Pope Lucius III. The latter had been elected in 1181, but was forced from Rome and settled in Verona. There he convened a council on combatting heresy. He died in Verona in 1185 and was buried in the cathedral, which was nearing completion and was finished under his successor, Urban III, who consecrated the building in 1187 and remained in Verona for most of his brief pontificate.

The earliest documented reference to a Veronese *podestà* dates to 1169. From the middle to the end of the twelfth century, Verona's top official was either a consul or a *podestà*. Verona's growing wealth in the twelfth century is attested by many splendid buildings. The year 1172 saw the start of the tower on the municipal building, while the *podestà* Guglielmo da Ova erected a palace next to it—to be added to in later centuries—forming one of the finest city offices in Italy. The magnificent San Zeno Abbey was finished in 1178. That year a new city wall, begun in 1134 and doubling the protected area, was completed. The characteristic "peach-blossom" marble from which so much of medieval Verona was built was quarried at San Ambragio, south of Chiusa.

Verona's position, its climate and fertility, as well as firm civic control made it a desirable place to live. By now urban supremacy had been firmly established over the Veronese district, which by the end of the twelfth century included the whole of the eastern shore of Lake Garda. It was master of half of the navigable part of the Adige, an important trade route with Venice, while its possession of Peschiera assured control of much of the Mincio River.

Further trade treaties were signed with Venice in the second half of the twelfth century. An 1191 treaty was also made between Verona and Mantua, including what might be viewed as the first extradition agreement. Such treaties formalized friendships between city-states in times characterized by suspicion. The unity of the Lombard League had dissipated, replaced by frequently shifting alliances and minor struggles with neighboring city-states.

Internal strife was increasing as well. The feuding between the Montecchi and the counts of San Bonifacio (descended from Egelric, nephew of Milo) was already violent in 1206. All of Verona was divided. This feuding would be immortalized in Shakespeare's *Romeo and Juliet*. The parties called for outside help, the counts of San Bonifacio to the lords of Este in Paduan land, the Montecchi to the Ferrarese Salinguerra Torelli.

Ezzelino III da Romano, descendant of German nobles who had settled in the Brenta gorge, joined the Montecchi in 1207, helping expel the San Bonficacios and Azzo VI, marquis d'Este. Azzo won back Verona in September, briefly taking Ezzelino prisoner. Azzo and San Bonifacio ruled Verona from 1207 until 1212 when both died, but feuding between the rival parties continued. The marriage in 1221 of Ezzelino III to Count San Bonifacio's daughter sealed a shortlived peace. The Veronese people rebelled against their leaders, however, in 1225, and the Montecchi supported them. The *podestà* was expelled, Leo delle Carceri being elected in his place, with more far-reaching authority. A consultative council of eighty, the Quattuorviginta, was also formed. Leo delle Carceri did not hold his post for long: Ezzelino III ousted him in 1226 and became *podestà*.

At first Ezzelino III was not a Ghibelline, quite the contrary. When Frederick II called a meeting of envoys of the Italian cities, Ezzelino declined to send one, and later blocked Frederick's son Henry from traveling through the Chiusa pass. Shortly after his election to *podestà*, Ezzelino was in fact rejected by the two major figures of the Western medieval world, the Holy Roman emperor and the pope (the latter excommunicated him). Ezzelino had a supporter elected *podestà* in 1227, but another popular uprising ousted the latter and the extraordinary Veronese Communanza was formed, a kind of people's state within a state, in great part detached from the commune, even imposing some statutes of its own and giving some Veronese who did not have citizen status (i.e., those who were neither noble nor tax-paying) the chance to hold powerful posts.

The Veronese Communanza defended its position for some three years, with Ezzelino currying favor with it. In 1230 pressure on the commune from the San Bonifacios was severe, but Ezzelino rescued it and imprisoned Rizardo di San Bonifacio and several other leaders. On Ezzelino's recommendation, Salinguerra Torelli was made *podestà* and became a puppet of Ezzelino.

Ezzelino was all for starving his prisoners to death, but allies warned him against it. However, he kept the prisoners. This led to the intervention of the Lombard League, which called for San Bonifacio's release in 1231. Ezzelino agreed to release him in exchange for the strategic castle of San Bonifacio, but the league did not enforce the castle's handover and had its representative elected *podestà*. His lands under attack by the Paduans, Ezzelino was too weak to fight back; he resorted to calling on the Lombard League for assistance. Help was promised but deliberately delayed. The league feared Ezzelino's ambitions and wanted to curb his power.

This treatment of Ezzelino by the Lombard League caused his shift of allegiance. He asked the emperor for support in taking control of Verona, and the emperor, interested in Chiusa as well, agreed. Ezzelino was quickly successful. The emperor's support was of little consequence when the league intervened again, insisting that Ezzelino welcome the peace-making Dominican friar Giovanni da Schio in 1233. Fra Giovanni established peace, while Rizardo was given far-reaching political control.

Fra Giovanni had sixty heretics from well-established Veronese families burned, but Ezzelino's excommunication was annuled. Then on August 28, 1233, a massive gathering of reconciliation of the cities of the Trevisan Mark was held at Paquara, four miles south of Verona. Some say 400,000 people attended, many of them barefoot in imitation of Fra Giovanni, who declared peace between Ezzelino and the Estes. Fra Giovanni's peace was shortlived, his quickly growing powers envied and undermined. Ezzelino, supported by people from Bologna whom the friar had brought to garrison the area, reasserted his authority in Verona. From now until his death in 1259 he would acquire control and spread terror across the whole of the Trevisan Mark.

Verona would remain loyal to Ezzelino while he struck out on expeditions. By 1237 most towns of the Mark had submitted to him and the Ghibelline side. While the emperor ruled in name, Ezzelino exercised most of the practical power. Frederick appointed a vicar-general at the head of the Mark, but still Ezzelino held most of the strings, filling all the lower posts in the Mark. Wars between Ezzelino and the combined papal supporters continued for twenty years. Ezzelino asserted his control more and more: in 1244 he ousted the vicar-general and appointed his own man. The next year Frederick came to Verona to hold a general gathering of imperial leaders. A quarrel broke out between men of Verona and the duke of Austria's men. There were suggestions that it was part of an attempt by Frederick to oust Ezzelino, but the emperor soon left, never to return to Ezzelino's lands, although Ezzelino did support the emperor further with Veronese troops.

In the Trevisan Mark, Ezzelino gained a terrible reputation for cruelty, terrorizing townships, torturing, and imprisoning in atrocious conditions vast numbers, while his network of spies created further fear. Upon Frederick's death in 1250, Ezzelino was essentially seen as the leader of the Ghibellines in northern Italy. From 1255 the towns began to liberate themselves from him. In 1258 Pope Alexander IV ordered a crusade against Ezzelino, but no one would head it. Ezzelino allied himself to Uberto Pelavicino of Cremona. Together they won Brescia, but Ezzelino attempted to assert his authority at Uberto's expense. The latter turned against Ezzelino, who was defeated and mortally wounded by Uberto's allies at Cassano in 1259.

With celebrations at Ezzelino's death, the San Bonifacios reentered Verona. Leonardino della Scala, better known as Mastino, was chosen as *podestà* in the year Ezzelino died. He retained the title and was to start the della Scala (Scaligeri) dynasty of Verona. He controlled foreign policy and organized commercial initiatives. Verona continued to support the imperial cause when Conradin, Frederick II's grandson, sought to claim Sicily. He came to Verona, where his cause drew Mastino's support and assistance; it ultimately failed, however. Verona was ruled by Mastino's brothers while he was away; at this time rebels tried to eject the della Scalas. They did not succeed, but those who escaped persuaded the once-again exiled San Bonifacio to take on Verona. Verona then fought for its independence some four years. The city gradually regained control of its district, particularly with the help of Pinamonte Bonaccolsi, the new Mantuan leader.

The long struggle between emperor and pope was diminishing. Mastino sought Verona's reconciliation with the church. A heretical sect called the Patarenes had thrived in the Veronese district under Ezzelino and his loathing for the church. In 1276 the Patarenes' major center at Sermione was attacked and 166 members taken back to Verona as prisoners. However, Mastino would not burn them, as was required by canon law, so Verona was still not forgiven.

Mastino was assassinated, perhaps by Guelphs, in 1277. Remains in the city's Piazza dei Signori are said traditionally to be those of Mastino's palace. Certainly no trace exists of the frescoes Giotto is purported to have painted within it. During Mastino's time, the Veronese statutes had been brought up to date and tripled in number. While the *podestà's* power was theoretically increased, in practice the della Scala would monopolize it. Independent of this structure was the Council of the Gastaldi of the Arts, but again the della Scalas came to dominate this body. The city's control of the surrounding district grew tighter.

On Mastino's death, his brother Alberto was immediately elected to extraordinary power, the citizenry basically handing him the lordship of Verona. The imprisoned heretics were burned and the pope removed his interdict. Alberto was dedicated to the church, although his appointment to abbot of San Zeno of an illegitimate son, Giuseppe, distressed the monks and angered the poet Dante. Alberto's reign was marked by conflict with Padua and a Guelph-led rebellion in Verona. Alberto encouraged building: he ordered the construction of fortifications to enclose the Campo Marzo, of embankments to stop the Adige flooding, and the start of a marble palace for the Art of the Merchants, ruined later by fire. Alberto died in 1301. Bartolomeo della Scala, his eldest son, automatically became ruler of Verona. Bartolomeo ruled well for only a few years, dying in 1304. His brother Alboino succeeded him.

The new Holy Roman emperor, Henry VII, decided to revive imperial interest and control in Italy in the early fourteenth century. Few Italian leaders were enthusiastic, Alboino and his brother Can Franceso (Cangrande) being exceptions, lending useful military support. Alboino bought the new title of Imperial Vicar, an honor the della Scalas cherished. Cangrande was on his way to take a place of honor at Henry's imperial coronation at Rome when he had to rush back to Verona, where Alboino was dying.

Since Alboino's two sons were very young, Cangrande became ruler of Verona. With warrior ambitions, he devoted much of his rule to waging campaigns outside the Veronese district. Vicenza quickly came under his rule, followed by prolonged fighting with Padua. Eventually Henry officially pronounced against Padua, but the city reacted by attacking Verona. Federigo della Scala, a Scaligeri by marriage, mounted a firm defense, while for a time Cangrande would not leave Vicenza for fear of losing it. Henry's death in 1313, though depriving Cangrande of verbal authoritative support, left him much freer to pursue his own ambitions in wider territories.

Effectively, Cangrande came to be seen as head of the Ghibellines during this period. The issue of imperial succession was left undecided; of the two claimants, Frederick of Austria and Louis the Bavarian, Cangrande initially gave his support to Frederick in 1317. Cangrande gained the power to appoint *podestàs* in several cities at various times and was officially appointed captain and rector of the League of the Imperial Party in Lombardy. The pope excommunicated him. While Ghibelline factions in northern Italy hoped Cangrande would help fight for their causes, the della Scala ruler was set on his own ambitions, which led to a further assault on Padua. Almost a year of fighting ended in humiliation. The Veronese were surprised and dispersed, Cangrande was wounded in the leg, and a two-year truce was made.

When Louis captured Frederick in 1322, Cangrande changed his allegiance. While off campaigning, he fell ill, and Federigo della Scala and Cangrande's two nephews Mastino and Alberto attempted to take over authority in the city. Cangrande's mercenaries halted this attempt, but it was a loss to him to have to banish Federigo.

The pope and Robert of Naples sent ambassadors to Verona in 1326 to try to sway Cangrande to their side, but that year a great Ghibelline gathering was held at San Zenone in Mozzo in the Veronese district, where it was decided that to counter the Guelph supremacy, Louis should be called to Italy. He came. Cangrande was haughty with power and pride, and while paying homage to Louis in Trent at the start of 1327, also insisted that he should be given the vicariate of Padua. Louis declined even the offer of 200,000 florins. Cangrande left, saying he might switch allegiance, but was persuaded back and appointed vicar of seven cities: Verona, Vicenza, Feltre, Belluno, Monselice, Gassano and Conegliano. At Louis's crowning in Milan, the Veronese contingent was by far the most impressive and imposing, with Cangrande's power much in evidence.

Eventually, in 1328, Cangrande became lord of Padua, whose famine-stricken citizens elected him. The Ghibellines of north Italy rejoiced and an extravagant *Curia* was held at Verona. Cangrande still aimed to rule the entire Trevisan Mark, with Treviso submitting as he died.

Cangrande was a celebrated figure in medieval Italy, famed as a warrior, even if later he would be remembered more as a patron of Dante. He was a courageous, impetuous fighter, a quality much admired in those warring times. Dante's own connections with Verona have been difficult to establish. Various descriptions in his writings indicate he was familiar with it. Praising references to the "gran Lombardo" in the *Divine Comedy* are generally taken to refer to Cangrande. Certainly Cangrande's feats were sung in a great number of fourteenth-century poems.

Mastino II and Alberto (Cangrande's nephews) succeeded Cangrande, and all the towns under Cangrande pledged renewed allegiance. Mastino quickly established himself as the ambitious one, imposing his authority. While the cities that came under della Scala control maintained their customs, the della Scalas controlled them by selecting the top officials, and central taxes were imposed on them. Verona remained the headquarters of della Scala power, but as with Cangrande, most of Mastino's time was spent warring in other territories.

As the new leaders took office, two illegitimate sons of Cangrande, Ziliberto and Bartolomeo, tried to seize power but failed. A more significant problem was the threat the Venetians felt from della Scala conquests, particularly of Padua and Treviso, which meant the Veronese family held territory from the Alps to the lagoons around Venice. The new della Scala leaders placated Venice with a treaty in 1330, paying damages and upholding former treaties.

Mastino went on the offensive against Brescia, but it and numerous other northern Italian Guelph towns extraordinarily submitted to the imperial side, calling on John of

Bohemia's support. A joint Italian Ghibelline and Guelph league was constituted in response, the greatest force coming from Verona. John quickly withdrew. A plot to do away with the league's leaders was discovered and they disbanded the army. Mastino returned to Verona where there had been a serious fire, possibly started deliberately. Mastino tortured officials in an attempt to discover the truth.

Parma and Lucca submitted to Mastino, but Lucca, according to the league's agreement, ought to have been handed to Florence. Much of Italy now saw the Veronese family as a threat. Mastino and Alberto controlled half of the eastern Po Valley, eight major cities (Verona, Vicenza, Padua, Treviso, Feltre, Belluno, Parma, and Lucca), and land from the Alps to the Appenines. The historian Villani estimated their annual income at this time at around 700,000 florins, more than any king except that of France. Although without a port, Mastino now flouted commercial agreements with Venice, in particular its important control of salt. He built a castle to control disputed salt lakes around Padua.

In 1336 Venice and Florence, enraged over della Scala expansion, formed a pact to eliminate them. Mastino called on Lombard lords to help him, but they joined the enemy. A massive army was gathered against Verona, but it moved too slowly to take the town by surprise. The della Scala-controlled towns were easily converted to the league. Louis did not send help from Germany, and the Venetians refused to discuss a treaty. Instead, they advanced to Verona's walls and taunted the Veronese. Montecchio, key to the control of the Verona-Vicenza road, fell, and at this the beleaguered Mastino pawned his treasures (only Verona and Lucca of all the former della Scala towns had continued to contribute taxes) to build up another army. This led to a peace with Venice whereby the della Scala retained Verona, Vicenza, Parma, and Lucca, Venice took Treviso, and Florence received important Lucchese castles. Alberto, who had been captured in Padua, was reunited with Mastino.

Mastino, previously excommunicated, now took the radical step of changing his allegiance completely from the Holy Roman emperor to the Avignon pope Benedict XII. Mastino wanted Benedict to confirm his vicariate of Verona, Vicenza, and Parma; the latter agreed for money and forces, lifted the excommunication and officially founded a university at Verona in 1339. Mastino's power continued to decline rapidly, however. Parma fell, and he sold Lucca to Florence. His last ten years were taken up by general small-scale squabbles, except in 1343, when Italian towns united against the threat of the German soldier of fortune, Werner. Then the succession of Giovanni Visconti, the war-mongering cardinal/archbishop of Milan, created another focus for joint concern, and Mastino was made the leader of a new league to combat him. Mastino died, however, in 1351 before he had time to regain any of his dominions, leaving the della Scala and Verona much reduced in power. Still, he was buried in one of the most magnificent tombs in Europe.

Mastino's three sons were appointed new lords of Verona. Of them, Cangrande II was effectively the only ruler as his two brothers were very young. He changed his father's advisers and introduced economic changes, but he did not reduce taxes since they supported the extravagant life he and his court enjoyed. In 1354 he travelled to see his brother-in-law Louis, elector of Brandenburg. In his absence, Fregnano, Mastino's favorite bastard, claimed Cangrande had died. Fregnano himself assumed power, and the people rioted, destroying many public records. Cangrande swiftly arrived back and managed to regain control, ruthlessly executing the disloyal. To strengthen the ruler's position against rebels, he ordered the building of a mighty castle on the Adige's south bank, to become the Castelvecchio, and had a fortified bridge built across the river. He also saw that the church of San Dionigi was finished. He fluctuated in his foreign policy, but when Charles IV came to be crowned emperor, Cangrande escorted him to Mantua and his title of Imperial Vicar was upheld.

Cangrande's offspring consisted only of three illegitimate sons, but he wished them to succeed. His younger brother Cansignorio, however, perhaps prompted by rumors that he was going to be put to death, struck first and killed Cangrande. The Veronese decided to elect him their new ruler and even confirmed that the dynasty should be hereditary by law. The will of the citizens to govern themselves without a tyrant seemed to have evaporated.

Verona thrived under Cansignorio. He contributed to the beautification of the town, for example expanding his palace and commissioning two of the greatest artists in Veronese history, Altichiero da Zevio and Avanzo, to paint frescoes for him. He also supported church building. An intrigue to kill him in 1365 failed, and those implicated were publicly beheaded in the amphitheatre. There were bad harvests in several years, during which Cansignorio sold stores of corn cheaply (but still at a profit) to the poor. Earthquakes and plague caused further devastation. Cansignorio fell ill in 1375 and, wishing to secure his two illegitimate sons as the next rulers, had his brother Paolo Alboino killed and his sons Bartolomeo and Antonio declared lords of Verona and Vicenza before he died.

A period of peace and prosperity ensued, while the brothers gathered a brilliant court. Antonio, however, began to assume more and more power. When Bartolomeo was found mysteriously murdered in a Veronese street, Antonio pinned the blame on others, although he was widely considered to be the perpetrator. Injustice grew in Verona, especially following Antonio's sumptuous marriage to Samaritana da Polenta in 1382. Samaritana appointed favorites, spent outrageously, and had a crown stitched onto every one of her garments.

Spurred on by Venetian money, Antonio became embroiled with Francesco da Carrara of Padua. The campaign was disastrous for the Veronese; nearly the entire army was captured at the battle of the Brentelle in 1386. Antonio tried to amass another huge army, but it was ineffective and was easily beaten.

A change of power in Milan would put a quick end to the della Scalas. Gian Galeazzo Visconti became the Milanese leader and asked both Verona and Padua whether they wanted his support. Antonio, to his own cost, declined to enter into any treaty with Visconti. Milan joined forces with Padua, and in a short time Verona was crushed and taken. Antonio's citizens would not help him, and in October 1387 he escaped down the river to Venice, where he died. Della Scala rule in Verona came to a humiliating end.

From then on Verona lost its independent status, although Gian Galeazzo Visconti did not change local customs. Visconti went on to take Padua. The exiled Paduan Francesco II da Carrara met the traveling Guglielmo della Scala, Cangrande II's illegitimate male survivor in Munich. Francesco led an army to take Padua and Verona. The former he managed easily, but the Veronese rose against him and sent for Antonio's legitimate son Canfrancesco. However, the Visconti leader Ugulotto Biancarlo, learning of the Veronese uprising, returned surreptitiously and wrought a terrible massacre on the Veronese. The Veronese made no more attempt to free themselves, although Francesco came to the walls with Samaritana, who reputedly arrived clad in armor.

While a good system of administration was set up in Verona under Gian Galeazzo, he undermined Verona's university by favoring Padua's. The end of a separate Veronese army also signaled clearly the end of Veronese autonomy. Francesco continued to put pressure on Verona's new rulers, pestering the region once Gian Galeazzo's widow had become regent, to the extent that she tried to hand Verona to Venice. Guglielmo della Scala organized the taking of Verona with Francesco, and their success caused joyous celebrations. Guglielmo died a mere week later, many suspecting Francesco of treachery. But in fact the latter had Guglielmo's sons Brunoro and Antonio placed as lords of Verona. These two, after foolishly attempting to gain Venetian support to oust the Paduans and have Verona for themselves, were imprisoned. Francesco himself became lord of Verona, but for a very brief time, as Venice had decided to go to war against him. Very soon Verona fell to the Venetians.

After 1405, the Venetians exercised almost uninterrupted rule in Verona until Napoléon's arrival. The leading officials appointed by Venice controlled virtually every aspect of city life, although the Veronese were allowed to occupy lesser posts. During the war in which Venice fought alongside Florence and the pope against Milan and Mantua, from 1437 to 1441, a fleet was gathered in Verona that was taken up the Adige and, by an extraordinary achievement of engineering, transported overland across hills to take action on Lake Garda. At this time the Veronese painter Pisanello became a major proponent of the International Gothic style.

The one period when Verona briefly escaped Venice's clutches was during the upheaval of the League of Cambrai. Pope Julius II made the most of French royal expansionist hopes in Italy, beating the Venetians, who in 1509 abandoned Verona. The Veronese, realizing the French were approaching, appealed to the league, which had changed in nature; Julius abandoned solidarity with the French. The league explained that in its agreement, land east of the Mincio went to the emperor, so Maximilian briefly became ruler of Verona. In 1510, 1513, and 1516, Verona was besieged. The area suffered famine and other hardships, and its supposed protectors, the imperial troops, also wrought havoc. According to the Peace of Brussels, Venice was granted rule over Verona.

Venice erected massive new fortifications. The winged lion of St. Mark was reinstated on a great column in the Piazza delle Erbe. Commerce and culture flourished. Michele Sanmicheli designed grand military architecture and civic buildings. Verona produced, among others, the painter Paolo Caliari, better known as Veronese, and the sculptor Girolamo Campagna. In medicine, Girolamo Fracastoro was distinguished as a founder of modern pathology, while the mathematician Anton Mari Lorgno started the renowned *Società italiana delle scienze*. Verona was caught up in Venetian wars, but its most terrible suffering was occasioned by the plague of 1630–31, which reduced the population by almost two-thirds and severely weakened the city. The scholar and dramatist Scipione Maffei was perhaps Verona's leading light in the eighteenth century.

In 1796 the Napoleonic army took control of Verona. The Austrians captured it for a week, but then it reverted to the French. The Veronese revolted against their occupation at Easter 1797, known as the *Pasque Veronese*, taking hold of the city in a violent few days before being crushed by the French. After this, Venice formally gave up Verona to Napoléon, but Austria continued to fight for it and by century's end had gained control. The peace of Lunéville split the city in two, the French taking Verona right of the Adige, Austria taking the section to the left, named Veronetta.

With his creation of an Italian kingdom in 1805, Napoléon reasserted complete French rule, which lasted, unhappily for the Veronese, who suffered heavy taxes, conscription, and repression, until 1814. After the Congress of Vienna, Austrian rule replaced the French, lasting until 1866. The revolutionary year 1848 saw Charles Albert of Sardinia-Piedmont try unsuccessfully to free Verona. The Austrians responded by securing the defenses of the Quadrilateral, four heavily fortified corners to oversee their territory, one in Verona. Freedom arrived in 1866, after Austrian defeat at the battle of Sadowa. By the treaty of Vienna, Verona became part of the Italian kingdom.

Following terrible floods, in 1882 great embankments, known as the *muraglioni*, were erected along the Adige and gave Verona a new look. From 1898 the Veronese fairground grew, and since then Verona has hosted some of the largest exhibitions in Italy, such as Fieragricola and Vinitaly. In the twentieth century, Verona is perhaps most recognized internationally for its opera festival in the amphitheatre, founded in 1913. With World War II the city suffered badly from bombing, all its bridges being destroyed. After the war, the Shakespearean drama festival began. (In addition to *Romeo and Juliet*, Shakespeare's *The Two Gentlemen of Verona* is partly set in medieval Verona.) Today Verona remains an extremely important Italian trading crossroads, as well as a major tourist town.

Further Reading: A good work in English is *A History of Verona* by Alice Maud Allen (London: Methuen, 1910). Although dated, this is a substantial work of scholarship, concentrating on Verona through the Middle Ages and placing it in the wider context of the power struggles between Italy's city states. In Italian, *Verona, panorama storico* by Nerina Cremonese Alessio (Verona: Edizioni di "Vita Veronese," 1978) is a good introduction. Both these titles contain useful, detailed bibliographies. The Vita Veronese organization has published a large number of specialized pieces.

—Philippe Barbour

Vicenza (Vicenza, Italy)

Location: Approximately forty miles west-northwest of Venice, on the Bacchiglione River at the foot of the Berico hills.

Description: Vicenza is famed for its magnificent architecture, most notably that by the Renaissance master, Palladio. Among Palladio's most famous works in Vicenza are the Basilica, the Palazzo Chiericati, and the Olympic Theatre. Vicenza also contains fine Gothic buildings, as can be seen along the Corso Palladio, the main street. On Vicenza's outskirts stands the famed Palladian Villa Rotonda (properly known as the Villa Almerico). The nearby Villa Valmarana (or *dei Nani*) was built by the Muttonis, and is known for its frescoes by the Tiepolos. The Basilica di Monte Berico became a place of pilgrimage because of claims of apparitions of the Virgin Mary there in the fifteenth century. The present building dates from the end of the seventeenth century. From its hill there are fine views over Vicenza.

Site Office: A.P.T. (Azienda di Promozione Turistica)
Piazza Duomo, 5
CAP 36100 Vicenza, Vicenza
Italy
(444) 544122

Vicenza, or Vicentia as it was originally known, was settled by Ligurians sometime in the first century B.C. No clear records exist of this settlement prior to its coming under Roman rule. It received rights of Roman citizenship together with municipal regulations in 49 B.C., maintaining a town council structure throughout the Roman Empire. A very few remains have been found of a first-century palace under the present cathedral, and the forum was on the Piazza dei Signori, still the heart of Vicenza. Parts of Vicenza's oldest church, Sts. Felice e Fortunato, date to the fourth century; this church contains an early Christian martyr's shrine.

The Lombard king Alboin invaded northern Italy at the end of the 560s, and Vicenza became the seat of one of the Lombard duchies, subsequently becoming a Frankish county. From the tenth century it became a part of the Mark of Verona. With the Holy Roman Emperor appointing bishops to oversee much of his land, the Vicenza bishops wielded extensive power, causing numerous struggles with the citizens.

By the start of the twelfth century, Vicenza gained a great degree of autonomy as a commune, evidenced for example in a treaty of 1115 with Padua and by a record of a consulate in 1122. But extended feuding recommenced and only ended with the Peace of Fontaniva in 1140, when Vicenza and Verona signed a truce with Padua and Treviso.

The revival of the Holy Roman Empire's interest in northern Italy in the mid-twelfth century would cause the Italian city communes to form leagues against the German power. At the Diet of Ronacalia in 1158, Frederick I asserted his authority over the north Italian cities of the so-called Trevisan Mark, which included Vicenza, assuming the right to elect their leaders. Growing discontent with the authoritarian imperial rulers led Vicenza in 1164 to join the Veronese League, or *Societas Veronensis*, which included Padua and Venice, and which the emperor did not dare to fight. In 1167 this organization came together with other cities to form the greater Lombard League. Government of the cities was now in the hands of Rectors, each citizen swearing allegiance to him and to the common purpose of the League. Vicentine cavalry would play its part in this league's famous victory over the Holy Roman Emperor at Legnano in 1176. This success allowed the League cities to demand wide-ranging powers in the peace of the following year, which included building their own fortifications, choosing their own rulers, or consuls, and maintaining their long-established customs and tolls. The Treaty of Constance in 1183 confirmed these rights.

The leaders of Vicenza would for a time shift between assuming the old Roman title of consul and that of *podesta*, or mayor. As was the general pattern in the northern Italian cities, various parties fought bitter internal struggles for civic control, leading to a time when two podestas were elected at the same time to try to accommodate differences. Endless internal disputes were matched with petty skirmishes with neighboring cities. In 1213, Vicenza, Verona, and Padua signed an agreement putting a halt to long and harsh imprisonment of captives, and arranging for the return to their respective cities of people exiled due to internal city squabbles.

For centuries thereafter, control of Vicenza would fall under the various rule of powerful families from neighboring cities. The local fighting between the long-established counts of Vicenza and Vivario extended to surrounding districts, with all sides calling for outside assistance. By answering such calls, the da Romano family from the Brenta gorge would acquire extraordinary power across the Trevisan Mark. Ezzelino II da Romano became the leader of one of Vicenza's factions. Although temporarily ousted in 1207, he managed with the aid of Otto IV to become podesta in 1210, his authoritarian rule lasting some three years.

Power continued to shift until 1236, when Emperor Frederick II destroyed the city. For much of this time, those opposed to Ezzelino II and his son Ezzelino III prevailed under Azzo VII d'Este. In 1227 Ezzelino III was able to appoint his brother, Alberico, podesta of Vicenza. Ezzelino's cruel warring caused other cities of the Lombard League to act against him, but in 1231 peace was signed between the da

Loggia of the Capitano and Palladian Basilica at Vicenza
Photo courtesy of Italian Government Tourist Board-E.N.I.T.

Romano cities Verona and Vicenza and the others in the Trevisan Mark. In 1232, however, embittered by the way the neighboring cities had surreptitiously conspired against him to make him lose control of Verona, Ezzelino called on the emperor for support and took them on. In 1232 he had to help Alberico defend Vicenza.

Hostilities between the city-states continued, with Vicenza falling into the hands of Azzo d'Este. There was a brief respite in 1235 when peace was signed between the factions. Then in October 1235 Vicentine forces joined Paduans and Trevisans in besieging Rivalta in Veronese territory, but this action backfired on Vicenza as Ezzelino called on imperial aid. On November 1, the combined imperial and Ezzelino forces arrived at the walls of Vicenza, and Emperor Frederick II requested admission. This was not granted, so Frederick attacked Vicenza and took control. Azzo meanwhile had fled. The town was terribly pillaged, then set ablaze. Ezzelino, who normally forbade the plundering of captured towns, was linked with this devastating action. Imperial troops commanded by Gebhard of Arnstein were left to guard the peace, and Ezzelino was handed overall control of the city. He was free to choose the leaders of Vicenza's commune.

As his period of rule over Vicenza and neighboring regions continued, Ezzelino became increasingly tyrannical. Serious uprisings against him occurred in 1255. In Vicenza, where Ezzelino thought there was a plot to liberate the town from his rule, he treated the suspects with typical cruelty and then left a garrison of Veronese soldiers in the town. Ezzelino would die shortly after being wounded and captured at Cassino in autumn 1259. Vicenza rejoiced at the event and decided to make the day of Ezzelino's fall an annual celebration.

Vicenza regained its independence for only a short time, during which the commune, which Ezzelino had failed to eliminate, reinvigorated the government. A commercial agreement was signed with Verona that ensured safe passage through each other's territories. A set of new statutes was drawn up in 1264. But in 1266 Verona, now under a della Scala family leader, decided to go to war with Vicenza, which asked the Paduans to help in their defense. While Verona was unsuccessful, Padua ended up taking charge of Vicenza, appointing podesta, judges, and military men to the town.

The Paduans apparently suspected Alberto della Scala of conspiring to remove them from Vicenza in 1299, for that year they issued an official call for his death. Increasingly, Vicenza would be fought over by its stronger neighbors, Padua and Verona. The Vicentines themselves first rose against Paduan domination, following the election in 1308 of Henry of Luxembourg as the new Holy Roman Emperor, Henry VII, keen on imposing himself in Italy. Via the Scaligeri and the bishop of Verona, Tebaldo, they entered talks with the imperial side. Combined forces under the imperial representative Aymo, bishop of Geneva, Cangrande I della Scala, and Vanni Zeno advanced to Vicenza, where the Ghibelline faction let them into the city. For a brief period autonomous rule was handed back to the town, although its citizens swore allegiance to Aymo as the figure of imperial authority.

Cangrande I della Scala, who had won the favors of the emperor for the support he had given him in his campaigns, was appointed imperial vicar of Vicenza in 1312. Cangrande ruled Vicenza as a tyrant, and no one in the town dared challenge his authority. Eventually his might extended to Padua. Toward the end of Cangrande's rule, a fire swept through Vicenza, destroying perhaps one-quarter of its buildings.

Vicenza would remain firmly under della Scala rule until the family's demise in 1387. Important regional figures were appointed as podestas, for example Marsilio da Carrara, a Paduan whose support for the della Scalas led to his Vicentine position. Under Cangrande's successors Mastino and Alberto, Marsilio da Carrara became the third most important man in the Trevisan Mark with the title Alter Dominus Marchiae. The della Scalas saw their territories under attack, however: Venice and Florence together formed a pact against their expansionist plans. The suburbs of Vicenza were taken before peace was made in 1339. The della Scalas kept control of Verona and Vicenza, Parma and Lucca. Soon only Verona and Vicenza remained under their control.

While the della Scala family was continually involved in warring, commerce developed in Vicenza. The first mention of the Vicentine goldsmiths dates from 1339, and their craft was to become a particularly prosperous one in Vicenza, statutes for their guild existing from 1352. The poet and scholar Petrarch spent much time at Vicenza in the second half of the 1340s. Cangrande II della Scala, who succeeded Mastino and Alberto, supported church building, such as San Agostino outside Vicenza. Several churches from this medieval period in fact survive, such as San Corona and San Lorenzo.

But della Scala power was becoming increasingly insecure. Cangrande II was killed and succeeded by his brother Cansignorio, still an extremely wealthy lord. Cansignorio released corn cheaply and replaced houses in Vicenza with granaries. He also built a new bridge and a church and monastery for the Carmelite order. Cansignorio's only heirs were two illegitimate sons, Bartolomeo and Antonio, whose succession he tried to smooth by having them proclaimed lords of Verona and Vicenza before his death. Holy Roman Emperor Charles IV would give his blessing to the two illegitimate sons as vicars of Verona and Vicenza in 1376.

The rise of Gian Galeazzo Visconti in Milan would lead to the end of della Scala rule over Vicenza. Allied to Padua, he agreed that in the case of a successful campaign, he should receive Vicenza, Padua, and Verona. Antonio della Scala's power collapsed, and before fleeing to Venice he officially passed Verona and Vicenza to the emperor's representative. The Vicentines welcomed the troops from Milan, opening the city to them. Gian Galeazzo's small son Filippo Maria inherited rule of much of his father's territories, including Vicenza, in 1402, but his mother Caterina had to act as regent and deal with the Paduans wreaking havoc in the region.

Caterina resorted to Venice for an alliance against Padua in exchange for control of Vicenza and Verona. Guglielmo della Scala united with the Paduans to attack, but their attempt on Vicenza was thwarted by the Vicentines who, not wanting Paduan rule, requested and received protection from Venice. From then until 1797, Vicenza was governed by Venice, save for a brief period under Holy Roman Emperor Maximilian I, from 1509 to 1515, during which Vicenza suffered greatly.

Under the tight control of the Venetian government, Vicenza in the fifteenth and sixteenth centuries saw a flourishing of magnificent architecture that remains the town's principal claim to fame. The supposed apparition in the 1420s of the Virgin Mary on the Monte Berico above the town was marked by the building of a chapel there that became a place of pilgrimage. Bartolommeo Montagna, Vicenza's finest painter and the founder of the school of Vicenza, practiced at the end of the fifteenth and the beginning of the sixteenth centuries. Fine fourteenth- and fifteenth-century Gothic buildings were built in Vicenza, such as the Casa Navarotto, Casa Fontana, Sta. Maria dei Servi, Palazzo Regaù, Palazzo Brunello or Palazzo da Schio (known as the Ca' d'Oro). Many of the later structures show distinct Venetian Gothic influence. The Gothic-style Vicenza cathedral was constructed from the fourteenth to the sixteenth centuries. The early Renaissance style that began to emerge toward the end of the fifteenth century includes the fine cathedral tribune by Lorenzo da Bologna, as well as the Casa Pigafetta. The latter bears the family name of the adventurer Antonio Pigafetta, who joined Magellan on his 1519–22 expedition and wrote the story of the journey. One of the columns on the Piazza dei Signori is topped by a 1520 winged lion of St. Mark, symbol of Venice's dominance.

Vicenza's most splendid architectural legacy stems from the work of the sixteenth-century Andrea di Pietro della Gondola, better known as Palladio. Born in Padua in 1508, Palladio settled in Vicenza, first joining the stone-cutters and dressers' guild. He discovered the theories of ancient Roman architecture by reading Vitruvius, a one-time inhabitant of Verona. He then enjoyed the patronage of the humanist poet Gian Giorgio Trissino.

Inspired by classical models and earlier exponents of Renaissance architecture, Palladio developed a distinctive style combining elegance and grandeur, a style particularly marked by geometry and symmetry. Vicenza is full of his magnificent works, in particular the Basilica—his first building, which led to many commissions—the Palazzo Chiericati, the Villa Rotonda, and the Olympic Theatre, begun in 1580, the year of his death. Palladio also published many drawings together with his theory in the 1570 *Quattro Libri dell'architettura*. He is recognized as the earliest modern professional architect. After his death, his unfinished work was taken over by Vincenzo Scamozzi, who faithfully completed many projects and designed some fine buildings in his own right, such as the Palazzo del Commune. The permanent stage set of the Olympic Theatre, offering an extraordinary example of the mastery of perspective, was in fact by Scamozzi. Palladio's influence was particularly strong on eighteenth-century English and American architecture, for example Burlington's Chiswick House in London and Thomas Jefferson's home, Monticello. The finest flowering of art in Vicenza in the eighteenth century was in the Villa Valmarana (or *dei Nani*) built by the Muttoni, decorated inside with remarkable frescoes by the Tiepolos.

After Napoléon's invasion of Italy at the end of the eighteenth century, power shifted between French and Austrian forces for a time, until Napoléon's fall left Austria ruling Vicenza from 1814 until 1866. There was a famed uprising against the Austrians in 1848, which the Austrian overlords had great difficulty in containing, only reestablishing authority after the battle of Vicenza in June. Vicenza gained independence in 1866 and became part of the Kingdom of Italy.

During World War I, the hills to the north of Vicenza formed the Italian front line. Bombing in World War II severely damaged the city, but much has been well restored. The goldsmiths' trade has continued to mark the city's image, some 700 companies operating in that field, and the city has been promoted as the Città d'Oro (Golden City). The man who developed the silicon chip, Federico Faggin, was born in Vicenza, which established itself as a center of Italian electronics. Both these industries have made Vicenza a generally wealthy modern city.

Further Reading: In English, a vivid picture of medieval Vicenza can be gleaned from Alice Maud Allen's *A History of Verona* (London: Methuen, 1910). In Italian, there is the massive four-volume *Storia di Vicenza* by A. Broglio, R. L. Cracco, G. de Rosa, F. Barbieri, and P. Preto (Vicenza: Accademia Olimpica, 1987–1993), the first volume covering prehistory and Roman times, the second volume medieval times, the third volume Vicenza under Venice, the fourth volume on modern times.

—Philippe Barbour

Zadar (Zadar-Knien, Croatia)

Location: Adriatic coast of Croatia, on end of a low-lying peninsula that is separated by the Zadar Channel from the islands of Ugljan and Pašman.

Description: Also known in Italian as Zara; in ancient times as Iadera or Jadera. Picturesque historical town in Croatia, the former Yugoslavia; former capital of Dalmatia. Important port city with natural deepwater harbor between peninsula and mainland. Largest and most important town on the northern Dalmatian coast; industries include tourism (severely hampered because of wars in the mid-1990s) shipbuilding, and the production of liqueur, tobacco, and jute.

Site Office: TZO Zadar
I. Smiljanića b.b.
57000 Zadar, Zadar
Croatia (57) 25-948 or 25-040

Tied to the fate of the Illyrians, the Roman Empire, Byzantium, Venice, Austria-Hungary, Italy, Yugoslavia, and ultimately Croatia, this stronghold port on the Adriatic Sea has long been a pawn in the intricate game of Balkan history. As with other coastal cities in Dalmatia and Istria, Zadar's strategic location was of great interest to any power that wished at least to establish a presence on the Adriatic, if not actually to control the area. The Venetians, conveniently based just across the Adriatic, turned out to be the most adept at maintaining control of Zadar (known in Italian as Zara) during the crucial period of the transition of power in the Balkans from the Byzantines to a host of diverse powers that included the Serbs, Bulgarians, Hungarians, Venetians, and Ottomans.

Yet considering the tumultuous tug-of-war that frequently brought major adversaries to the town's very gates throughout the Middle Ages and into the Renaissance, the securely walled Zadar was able to maintain its Roman character. Centuries of incursions by Slavs, Turks, Hungarians, and Byzantines into the Balkans had little impact on the town's physical environment; even today it appears as a typical Italian town.

The fertile, temperate coastal zone near Zadar is only about eighteen miles wide; the sharply rising mountains isolate Zadar from the more continental Balkan interior. This geography, which follows the Adriatic coastline, made it natural for the coastal cities and islands to develop Mediterranean—specifically Roman and Italian—rather than Slavic cultures.

Zadar was founded before the ninth century B.C. by the Illyrian tribe known as Liburnians, who called the place Jader. When Greek traders began exploring the Adriatic coast around the sixth century B.C., they found a fertile and pros-

perous district that was home to a remarkable shipbuilding industry. The town was recorded as Idassa in the fourth century B.C., when Greeks from the colony of Pharos on the island of Hvar apparently defended the site against invaders.

In general, the Greeks had little effect on the Illyrian culture. Only with the arrival of the Romans did Zadar begin to develop the civic character that would define the town for the next five centuries and beyond. One such distinction was the autonomy that the Romans allowed many of their towns, a prominent trait of Zadar even as the medieval powers in the region later traded the town back and forth.

The Romans granted Zadar the status of *municipium* in 69 B.C., and in 58 B.C. the empire began sending colonists to the site, mostly military veterans. Jader was the largest in area, if not in population, among the numerous Roman cities in Dalmatia settled in this fashion. The Romans developed a rectangular plan for the peninsular city, in which two main avenues intersected, dividing Jader into four nearly equal quadrants. A monumental forum was located in the western quarter. From landside, entry to the stone-walled colony was through a triumphal gate.

In about A.D. 300 the Roman Catholic Church became the dominant force in Zadar's political life, a role that institution enjoyed throughout the rest of the millennium. As civic leader, the bishop was responsible for the general well-being of the population.

Considering the Italian orientation of the coast, it was only natural when the Ostrogoth Theodoric conquered Italy in 493 that Dalmatia—including Zadar—would join the new kingdom. The takeover by Goths of the erstwhile heart of the Western Roman Empire rankled the Byzantine emperor Justinian I (ruled 527–565), who hoped to recreate the once-glorious empire by rejoining East and West. He did succeed in recovering Dalmatia by 537.

In the second decade of the seventh century, the relatively undisturbed Dalmatian coast suffered the onslaughts of both Slavs and Avars (sources of the day suggest that the two tribes colluded in these southern thrusts), and much of the population fled to the coastal islands. Only a few towns—Dubrovnik, Kotor, Split, Trogir, and Zadar being the most prominent—were spared the plundering that ravaged the surrounding countryside. The prominent coastal city of Salona fell in 614, leaving Zadar as the largest city in the northern Adriatic. The fortified Zadar was a suitable selection as capital of the Byzantine archonate of Dalmatia.

Throughout the late seventh century, the Byzantines established the *theme* system to control military activities within the empire. Originally the word *theme* referred to an army corps, but in time it also came to include the province housing the military personnel. A general known as a *strategos* commanded each military province.

Church of St. Donato at Zadar
Photo courtesy of Art and Culture Council of America, Toronto

After 803 the Franks figured prominently in the governance of northern Dalmatia, which began paying tribute to them that year. In 805, Paul, *dux Iaderae*, and Bishop Donatus appeared at the court of Charlemagne as representatives of Dalmatia. In 810 Charlemagne promised to relinquish his claim to Byzantium's Adriatic provinces, including the towns on the Dalmatian coast, in exchange for acknowledgment from the "legitimate" Roman emperor that Charlemagne was emperor of the West. This was accomplished through the Treaty of Aachen in 810, and a Byzantine *strategos* soon was based in Zadar once again.

The Byzantine emperor Basil I (the Macedonian, ruled 867–886) created the *theme* of Dalmatia in the 870s, which comprised—in addition to the capital at Zadar—the empire's remaining properties along the coast: the towns of Dubrovnik, Kotor, Split, and Trogir, and the islands of Cres, Krk, Lošinj, and Rab. The *strategos* in Zadar had access to these other points, mainly the islands to the north, only by sea. Unlike the case in other *themes* throughout the empire, where the *strategos* had control of local administration, it appears that the Dalmatian *strategos* operated from Zadar only in a military capacity. From the base in Zadar, he could protect Constantinople's trade route to Venice, defend Byzantine ships against Arab pirates, control Slav populations along the coast, and provide a base for the empire's naval campaigns to recapture southern Italy from the Arabs.

From the 950s, the head of Zadar's town government—the prior—also served as head of the Byzantine administration in Dalmatia. The fact that the prior acquired also the title of *strategos* reflected the increasingly civil nature of the post as other problems forced the Byzantines to recall their Adriatic fleet for duty elsewhere; in effect the Byzantines no longer administered Dalmatia directly. In 1067 the prior also became known as *katepan* (the Croatian equivalent of *strategos*). In religious matters, however, Zadar's bishopric was subordinate to the town of Split's leadership. After about 1100, the citizenry and local nobility began to establish a political base, too; the Great Council, comprising a rector, three judges, and an assembly of nobles, began to determine policy in Zadar.

It was under Croatian auspices in the late ninth century and throughout the tenth century that important architectural projects such as the Church of St. Donato and Cathedral of St. Anastasius were begun. The architectural heritage of Zadar maintains a blending of styles from both East and West, with an especially strong preference for Italian styles.

The Church of the Holy Trinity (later St. Donato), commissioned at the close of the eighth century or in the early ninth century, occupied the site of the city's Roman forum. Modeled on the Church of San Vitale in Ravenna (c. 547), the central-plan church with three apses also reflected the ground plan of the Royal Chapel built at Aachen for Charlemagne. The church, perhaps the finest example of medieval Croatian architecture, was largely built from materials taken from the Roman forum.

Also in the ninth century was begun the Cathedral of St. Stosije (later St. Anastasia). According to legend, the cathedral was built to house the relics of the virgin saint Anastasia, which the emperor Nicephorus I had presented to Bishop Donatus in Constantinople. Construction continued at the cathedral until 1324. The cathedral, facing a public square, was constructed atop the walls of an earlier church, which had itself been constructed upon the foundation walls of an ancient cistern. The Cathedral of St. Stosije is significant as the largest and best example of Croatian Romanesque design in Dalmatia. Prominent features include three ornately carved portals in the gabled front facade, two rose windows, and a five-story, square bell tower at the rear; the bottom two stories were begun in the fifteenth century, the top three stories not completed until 1892.

In plan, the cathedral is a basilica with nave and two side aisles, each terminating in an apse. A Gothic sacristy was added in the fourteenth century, and a seminary was built adjacent to the cathedral in the eighteenth century. Features of note in the interior include an outstanding marble sarcophagus, ninth-century marble seats, a thirteenth-century Romanesque fresco and a ninth-century high altar, located in the central apse; the fourteenth-century sacristy; and a Gothic carved wooden choir stall installed in the fifteenth century. Below the main altar is a three-part crypt from the twelfth century.

Throughout the Middle Ages, Zadar was the focus of frequent territorial disputes among Venice, Hungary, and Croatia. Even though in 998 Zadar swore allegiance to the Venetian doge Pietro Orseolo, the town remained a Byzantine protectorate until 1001, when Byzantium transferred Zadar to Venice.

Venetian overlordship of the Dalmatian towns and islands chiefly maintained "staple rights," which meant that a vassal town could sell goods only at home (for domestic needs) or in Venice; further, foreign merchants could not trade directly with the vassal town but had to purchase that town's goods in Venice. The Dalmatians actually managed to trade in a limited manner on their own—with each other, with the Italian towns of Ancona and Apulia, and with the inland Slavic areas. Particularly dismaying to Zadar, however, was the Venetians' proscription of trading rights with Italian ports on the northern Adriatic. Considering Zadar's proximity to these ports, the Venetian monopoly was especially humiliating. Naturally, Venice's strategy fostered opposition within Zadar, a dangerous situation in light of the fact that Venice was in a nearly continual state of contest with other powers for control of the Dalmatian coast. Zadar favored the Hungarians in particular, and in an effort to switch allegiances, the town revolted against Venice many times during eight centuries of nearly continual rule by the ambitious republic. Each time, the Venetians brutally reined in the unruly town, and promptly tightened controls.

Various kings and princes made a point of endowing Zadar's churches and monasteries, but after 1200 more and more money was directed toward temporal endeavors such as sumptuous palaces and ornate town halls; these were built in the transitional style between the Romanesque and the Gothic.

Zadar proclaimed allegiance to King Kálmán (ruled 1095–1116) of Hungary, who had added Croatia to his realm in 1097 and many of the imperial cities in Dalmatia in 1102. Although after 1102 Zadar belonged formally to the Hungarian-Croatian kingdom, the royal administration repeatedly was obliged to acknowledge the town's autonomy within the realm. As well, Hungary regularly had to fight for its new possession against the encroachments of Venice and Byzantium. Frequently Hungary had to relinquish control of Dalmatia, especially the key port of Zadar, to another power.

According to Andreas Dandolo, in 1112 the doge Ordelafo Falieri asked Emperor Alexius I Comnenus (ruled 1081–1118) to transfer to Venice supremacy over Zadar. Evidently the response was negative, and in 1116 the Venetians attacked Zadar and defeated the Hungarians who were stationed there to defend the town.

Venice, eager to regain lost ground, moved in on Zadar after Kálmán's death in 1116. Even as most of the coast reverted once again to Hungary in 1133, the tenacious Venetians managed to control the town for most of the century. Fueling the overall power shift in the Adriatic area was the decade-long war between Venice and Byzantium that began in 1171; this was the beginning of the end for the two powers. The Venetians regained control of Zadar, but by 1181 the Hungarians were back in Dalmatia, and they battled successfully for the town in February of that year. Venice tried to recoup its loss in 1193, but to no avail.

From the 1190s the duke of Croatia—usually a son or brother of the Hungarian king—had a residence in Zadar, whenever Hungary controlled the town. Hungary was indeed in control when, in an audacious bid to recapture the Dalmatian port, the Venetians under the feisty doge Enrico Dandolo (ruled 1192–1205) went so far as to recruit the elite members of the Fourth Crusade to reconquer Zadar in 1202; they besieged the city and sacked it, an act against fellow Roman Catholics that outraged Pope Innocent III, who condemned the expedition. Adding to the insult was the fact that the Hungarian king himself had agreed to join that very crusade to the Holy Land. In any event, on November 24, 1202, Zadar was forced to submit to the crusaders—and once again to Venice.

The thirteenth century had begun tumultuously for Zadar and continued in the same manner. Domald, prince of Split, took Zadar from Venice in 1209. Undaunted, Venice soon regained the town. Zadar tried to free itself from Venetian rule by a revolt in 1242–43, at which time the town submitted to Hungary. The attempt failed, and Venice went so far as to install troops in Zadar in 1247.

Meanwhile, local possibilities loomed for the neighboring Croatian state. For Paul Subić, ban of Croatia and Dalmatia, Zadar presented an irksome Venetian intrusion into territory that he thought should be Croatian. Emboldened by a papal interdict against Venice, early in 1310 his forces advanced to the walls of Zadar. The Venetian garrison held the town until in 1311 another major uprising against Venice raged inside the city. The rebels took control of the town, but the Venetians, boasting a large contingent of foreign mercenaries, besieged Zadar from the sea. The Croatian troops defended the city, and the town appealed to the Hungarian king Charles Robert (ruled 1301–42), offering him its submission. The king assented, granting the citizens' request that they be allowed to elect their own prince; they chose Paul Subić's son, Mladen II. Neither the Venetians nor the combined forces of Zadar and the Croatians could drive the other power out, and a stalemate was reached. Venice soon made peace with the pope, however, allowing the republic to concentrate greater attention on Zadar. Mladen held the town through 1312, but in September 1313, with no relief in sight from Hungary, Zadar finally surrendered to Venice.

The nature of the power plays in the Balkans among the established empires and emerging states was complex. By the mid-fourteenth century, Serbia under Stephen Dušan was in the throes of becoming an empire, too. Dušan sought Venetian support against Byzantium, and Venice needed his assistance in its continual struggle with Hungary over Dalmatia. In the late fourteenth century, Hungary concluded peace with Serbia and decided that Dalmatia should once more be a priority. Inconveniently for Hungary, Venice had slowly but surely assured the vassalage of most of the fortified towns along the Dalmatian coast: Zadar in 1202; Dubrovnik, through three treaties, in 1232, 1235, and 1252; the islands of Hvar and Brač in 1278; Šibenik and Trogir in 1322; Split in 1327; Nin in 1329. While most of these properties enjoyed a large degree of autonomy in local affairs, they did have to pay tribute to Venice, provide military assistance for Venice's frequent campaigns, and trade within the strict commercial limits dictated by Venice.

Hungary nevertheless continued to agitate for supremacy in Dalmatia, and barely three decades later did reenter the fray. With the assistance of yet another ascendant Balkan neighbor, Bosnia, the Hungarian king Louis I marched into Dalmatia in 1345. Zadar's citizens eagerly joined in the anti-Venetian fervor by launching their seventh revolt. The republic under Andrea Dandolo (doge from 1343 to 1354) responded with another of its seaside sieges, finally routing the Hungarian forces in July 1346. When the Venetians regained control of Zadar soon after that, they rendered the town defenseless by destroying the frequently advantageous sea walls, by confiscating all weapons from the citizenry, and by limiting the food on hand to a four-month supply. No less humiliating was Venice's interference with Zadar's longstanding tradition of internal civil autonomy by sending Venetians to handle key administrative roles. The Hungarians, by then occupied elsewhere, had to abandon hopes for sovereignty in Zadar and along the coast, and in 1348 signed an eight-year peace pact with Venice.

Not even a decade later, the Hungarian king was back on the coast. In 1357 Louis massed his troops ostensibly for an attack on Serbia, but turned them against Venice instead, much to the latter's surprise. Though unprepared for the sort of counteroffensive that would maintain its coastal properties, Venice still had its garrison in Zadar, which, after much skirmishing within the town, did manage to hold on to the prized port. In fact, keeping Zadar was Venice's only victory, and a hard-won success at that. Hungary had essentially routed the rival republic from Dalmatia, however, and was not content to let Venice keep even one town; Louis forced Venice to submit to the Peace of Zadar in February 1358, whereby the Venetians finally surrendered Zadar.

The Hungarians held tenaciously to Zadar. Even when the Bosnians took over the coast between Zadar and Dubrovnik under Tvrtko I (reigned 1353–91), who was co-operating with the Croatians against the Hungarians, Zadar and Dubrovnik themselves remained under Hungarian control. Presumably the independent-minded citizens of Zadar were just as happy not to have to submit to an immediately adjacent neighbor such as Bosnia, which could have been more intrusive than Hungary farther to the north.

Zadar oscillated between subjection to Hungary or Venice until 1409, when Hungary formally ceded control of the city to Venice; on this occasion Venice purchased Zadar from Ladislas, king of Naples from 1386 to 1414, who also claimed the Hungarian throne. Venice made Zadar center of the republic's high military command for Dalmatia and Albania, retaining its own *katepan* to head the garrison there; a separate official called the *knez* was responsible for all civil matters.

Subjected once again to the iron hand of Venice, Zadar was oppressed from within by the Venetians, and from without by a new power on the horizon, the Turks, who occasionally advanced to the strategic port's very walls. In time, Zadar became the most heavily fortified town on the Adriatic (owning at least one cannon by 1351), although Venice's desire to make the town impregnable resulted mainly in sporadic bouts of construction when the Turks were nearly at the gate. During the 1463–79 war with Turkey, Venice sent additional infantry to Zadar as Turkish pressure increased, but often defensive manpower in the town was lacking, making adequate fortifications even more crucial. Venice and the Ottoman Empire frequently maintained good relations, too, at which times the Turks did not pursue Venetian properties in Dalmatia; even when the two powers were at war, which also was often enough, the Turks were likely to concentrate on targets closer to Constantinople. Zadar otherwise stagnated, as the resources of the church and the citizenry dwindled, and as elsewhere in Europe the plague and famines took their toll.

Zadar continued to be involved in the Turkish-Venetian conflicts during the sixteenth and seventeenth centuries, and its fortunes were generally bound up with sustaining and defending against various attacks. Not until the eighteenth century did Zadar enjoy renewed prosperity as war debts were paid off, and new baroque buildings, such as the splendid Church of St. Simeon, were completed. This church, a reconstruction of a much older Church of St. Stephen, changed its name when the sarcophagus of St. Simeon was moved here in 1632; an earlier silver shrine of St. Simeon was given to Zadar in 1308 by Queen Elisabeth of Hungary. The Church of St. Simeon also contains many fine seventeenth-century paintings.

In 1797, after the Treaty of Campo Formio, Zadar passed to the Austrians, who held the city until World War I, except for a period of French rule under Napoléon from 1805 to 1813. In 1813 Austria regained the town by advancing from landside as the British blockaded the port, and from 1815 to 1918 Zadar was capital of the Austrian crownland of Dalmatia. After World War I, Italy sought to wrest control of the city, and the difficult Treaty of Rapallo in 1920 transferred Zadar to Italy, making it the lone Italian holding on the Dalmatian coast. Zadar was designated a free port in 1923, but by this time Split had superseded its northern rival as the chief town in Dalmatia.

Zadar was a stronghold for the Axis powers during World War II, and as such was subjected to heavy air raids by the Allies, which damaged the port and destroyed 75 percent of the city's buildings. Many important churches, including the Church of St. Donato, survived those assaults, however. Yugoslav partisans liberated Zadar in October 1944, and the Italians ceded Zadar to Yugoslavia in a treaty signed on February 10, 1947.

With excellent connections to both Yugoslavia and Italy, Zadar in the postwar decades thrived as the cultural and economic centerpiece of Dalmatia. Advanced educational institutions, shipyards, industrial concerns, and tourism have made Zadar one of the richest and most important cities in the region. Most postwar industrial expansion and construction occurred on the mainland, leaving the old town on the peninsula as a tourist attraction.

Despite its relative isolation on the Dalmatian coast, Zadar—staunchly Roman Catholic—has fallen victim to the hostilities between the Orthodox Serbs and Catholic Croats. As with many other historic and resort sites in the former Yugoslavia, the tourist economy suffered greatly in the early 1990s, and the Serbian offensive against Croatia's cultural monuments damaged a number of historic buildings in Zadar.

Further Reading: *History of Yugoslavia* by Vladimir Dedijer et al. (New York: McGraw-Hill, 1974; Maidenhead, Berkshire: McGraw-Hill, 1975) provides an informative, detailed summary of the turbulent history of the nations and states that have determined Zadar's regional milieu. *Croatia: Land, People, Culture,* edited by Francis H. Eterovich and Christopher Spalatin, volume one (Toronto: University of Toronto Press, 1964) more specifically addresses the national heritage of which Zadar is now a part. *The Early Medieval Balkans: A Critical Survey from the Sixth to the Late Twelfth Century* (Ann Arbor: University of Michigan Press, 1983) and *The Late Medieval Balkans: A Critical Survey from the Late Twelfth Century to the Ottoman Conquest* (Ann Arbor: University of Michigan Press, 1987), both by John V. Fine, provide a comprehensive, clear analysis of the complex politics among the various powers—including the Venetians, Byzantines, and Hungarians—

that struggled for control of the Adriatic coast in the pre-Ottoman Balkans. Specialized studies include *The Illyrians* by John Wilkes (Oxford and Cambridge, Massachusetts: Blackwell, 1992), which examines one of the earliest cultures to inhabit the Adriatic coast, and *Dalmatia* by J. J. Wilkes (Cambridge, Massachusetts: Harvard University Press, and London: Routledge and Kegan Paul, 1969), one in a series of wonderful, scholarly volumes called *History of the Provinces of the Roman Empire*. M. E. Mallett and J. R. Hale's *The Military Organization of a Renaissance State: Venice c. 1400 to 1617* (Cambridge and New York: Cambridge University Press, 1984) offers an extensive discussion of the strategic necessities of a seafaring republic, and how important Zadar and the Adriatic coast were to Venice's survival. Two highly readable personal narratives are *Black Lamb and Grey Falcon: A Journey through Yugoslavia* by Rebecca West (London: Macmillan, and New York: Viking, 1941), an incisive commentary on conditions in pre-World War II Yugoslavia by the grande dame of Balkan travel, and *A Paper House: The Ending of Yugoslavia* by Mark Thompson (New York: Pantheon, and London: Vintage, 1992), a first-person journalistic account including Thompson's visit to Zadar in the midst of hostilities there in the early 1990s.

—Randall J. Van Vynckt

INDEX

Listings are arranged in alphabetical order. Entries in bold type have historical essays on the page numbers appearing in bold type. Page numbers in italic indicate illustrations.

NOTES ON CONTRIBUTORS

BARBOUR, Philippe. Freelance writer and editor. Commissioning editor, Gale Research International, 1992–94; editor, St. James Press, 1991; editor, Wine Buyers Guides, 1988–90. Co-author of *Wine Buyers Guide: Saint Emilion,* 1991. Editor of *The European Union Handbook,* to be published 1995.

BASTIN, Richard. Lecturer in Spanish history, International Institute of Seville. Author of *Introduction to the History of Spain,* to be published 1995.

BLOCK, Bernard A. Freelance writer. Reference and documents librarian, Ohio State University, Columbus, 1969–92.

BOWEN, Jessica M. Marketing administrative assistant, Anixter, Skokie, Illinois.

BRICE, Elizabeth. Special collections librarian, Miami University, Oxford, Ohio.

BROADRUP, Elizabeth E. Freelance writer and picture researcher.

BRUNS, Holly E. Graduate student in fiction writing, Columbia College, Chicago.

CABLE, Monica. Office manager, Above the Clouds, Worcester, Massachusetts. Recipient of Watson Foundation fellowship for independent research in southeast Asia, 1991–92.

CHENOWETH, Dellzell. Freelance writer and editor. Editor-in-chief of *Hawaii Review,* 1983–84.

CHIARA, Maria. Anthropologist and freelance researcher/writer.

CLASSE, Olive. Freelance writer and translator. Lecturer, then senior lecturer in French, University of Glasgow, 1965–90. Translator of *Mission to Marseilles* by Leo Malek, 1991. Editor of *Encyclopedia of Literary Translation,* to be published 1996.

COLLIER, Christopher P. Freelance writer.

DUBOVOY, Sina. Independent scholar and freelance writer specializing in history and biography.

DZIRLO, Amira. Freelance writer and architect.

FELSHMAN, Jeffrey. Freelance writer. Recipient of 1993 Peter Lisagor Award for exemplary journalism, from the Society of Professional Journalists, Chicago Headline Club, for "The Fall and Rise of Anita Brick," Chicago *Reader,* December 3, 1993.

FLINK, John A. Freelance writer.

GOODMAN, Lawrence F. English instructor, Illinois Institute of Technology, Chicago.

HALL, Sarah M. Freelance writer and editor. Research editor, Gale Research International, 1991–94. Co-editor of *Reference Guide to World Literature,* 1995.

HANAFEE, Mark D. Researcher, The Catholic Charities of the Archdiocese of Chicago.

HARMS, William. Senior writer, news office, University of Chicago.

HEENAN, Patrick. Research student, University of London. Editor of *1992,* 1989.

HOLLISTER, Pam. Freelance writer. Executive speechwriter and editorial supervisor, corporate communications department, Centerior Energy Corporation, Cleveland, 1980–93.

HUEBNER, Jeff W. Freelance writer.

JAROS, Tony. Copy editor, *Vegetarian Times.*

KLAWINSKI, Rion. Freelance writer.

KLOBUCHAR, Lisa. Editor, TitleWorks. Associate editor, World Book Publishing, 1986–94.

LAMONTAGNE, Manon. Assistant director, National Film Board of Canada.

LAMONTAGNE, Monique. Research student, University of London. Co-editor of *The Voice of the People: Reminiscences of Early Settlers, 1866-1895,* 1984.

LANGSTON, Cynthia L. Strategic planner, TBWA Advertising, New York.

LEDGER, Gregory J. Freelance writer; contributing writer, *Windy City Times.*

LEVI, Clarissa. Freelance writer.

LEVI, Claudia. Freelance writer.

LOIZOU, Nicolette. Student, University of London.

LOTUS, Jean L. Freelance writer and editor.

McNULTY, Mary F. Freelance writer and editor. Editor, American Association of Law Libraries newsletter, 1988–93.

MAXWELL, Caterina Mercone. Freelance writer, editor, and proofreader. Co-author of *A Survey of Family Literacy in the United States,* 1995.

MILLER, Julie A. Freelance writer.

MIN, Hyunkee. Operations supervisor, Chubb Group of Insurance Companies, Chicago.

MINSKY, Laurence. Senior copywriter, Frankel and Company, Chicago. Co-author, with Emily Calvo, of *How to Succeed in Advertising When All You Have is Talent,* 1994.

MOONEY, Paul. English professor, Oakton Community College, Des Plaines, Illinois.

NASLUND, L. R. Freelance writer.

PHILLIPS, Michael D. Instructor in humanities, Brigham Young University, Provo, Utah. Editor of *Snapshots of Belize: An Anthology of Short Fiction,* 1995, and *Six Belizean Plays,* 1995.

RIJSBERMAN, Marijke. Freelance writer and editor; English instructor, Chicago State University.

RING, Trudy. Commissioning editor, Fitzroy Dearborn Publishers.

SALKIN, Robert M. Commissioning editor, Fitzroy Dearborn Publishers.

SAWYERS, June Skinner. Associate editor, Loyola University Press. Co-author of *The Chicago Arts Guide,* 1994; author of *Chicago Portraits: Biographies of 250 Famous Chicagoans,* 1991.

SCHELLINGER, Paul E. Freelance writer and editor. Editor of *St. James Guide to Biography,* 1991; co-editor of *Twentieth-Century Science-Fiction Writers,* third edition, 1991. Editor of *Encyclopedia of the Novel,* to be published 1997.

SHEPHERD, Kenneth R. Freelance writer and editor. Adjunct instructor in history, Henry Ford Community College, Dearborn, Michigan. Associate editor, Gale Research, 1987–94.

SHTULMAN, Jill I. President, JSA Creative Services.

SULLIVAN, James. Freelance writer.

SY-QUIA, Hilary Collier. Doctoral candidate in German literature, University of California at Berkeley. Translator of and author of introduction to Schiller's *Don Carlos and Maria Stuart,* to be published 1996.

TEGGE, Jeffrey M. Geography editor, New Standard Encyclopedia.

TRIMNELL, Patricia. Freelance writer.

VAN VYNCKT, Randall J. Editor of *International Dictionary of Architects and Architecture,* 1993.

VASUDEVAN, Aruna. Freelance writer. Editor, St. James Press, London, 1991-94. Editor of *Twentieth-Century Romance and Historical Writers,* 1994, and *International Dictionary of Films and Filmmakers, Volume 1: Films,* to be published 1995.

WHARTON, Patricia. Freelance writer.

WILOCH, Thomas. Freelance writer.

WOOD, Beth F. Freelance writer, researcher, and editor. Lecturer, University of Central London, 1984–90.

XANTHEAS, Peter C. Membership coordinator and researcher, Preservation Wayne, Detroit, Michigan.